Asthma

Basic Mechanisms and Clinical Management

Asthma

Basic Mechanisms and Clinical Management

THIRD EDITION

Edited by

PETER J. BARNES
National Heart and Lung Institute, London, UK

IAN W. RODGER
*Merck Frosst Centre Canada Inc.
Point Claire-Dorval, Quebec, Canada*

NEIL C. THOMSON
*Department of Respiratory Medicine,
West Glasgow Hospitals University Trust*

ACADEMIC PRESS
San Diego London Boston
New York Sydney Tokyo Toronto

This book is printed on acid-free paper.

Copyright © First published 1988
Second edition 1992
Third edition 1998 by ACADEMIC PRESS

All Rights Reserved
No part of this publication may be reproduced or transmitted in any form or by any means electronic or mechanical, including photocopy, recording, or any information storage and retrieval system, without permission in writing from the publisher.

Academic Press
525 B Street, Suite 1900, San Diego, California 92101-4495, USA
http://www.apnet.com

Academic Press Limited
24–28 Oval Road, London NW1 7DX, UK
http://www.hbuk.co.uk/ap/

ISBN 0–12–079027–0

A catalogue record for this book is available from the British Library

Typset by Paston Press Ltd, Loddon, Norfolk
Printed in Great Britain by The University Printing House, Cambridge
98 99 00 01 02 03 CU 8 7 6 5 4 3 2 1

Preface

Asthma continues to be a major health problem throughout the world and affects over 5% of the population in industrialized countries. There is epidemiological evidence for an increasing prevalence of asthma throughout the world, particularly in children, and evidence for increasing morbidity and mortality in many countries. The reasons for this universal epidemic are not yet certain, and there is a need for better understanding of the fundamental pathophysiological and molecular mechanisms of asthma. We also need to understand how current treatment for asthma works and how best to apply it to control this important disease. The first edition of this book appeared in 1988 and provided a comprehensive overview of asthma mechanisms and therapy. A second edition, which encompassed the many changes in understanding and therapy of asthma, appeared in 1992. Both editions were widely acclaimed and became 'best-sellers' in the field. Because knowledge about asthma has continued to advance very rapidly, with new concepts, the development of new therapeutic agents and changes in asthma management, we thought that it was timely to produce a third edition. This new edition incorporates the many changes in our understanding of asthma that have occurred in recent years; this has involved radical revision of most chapters and the addition of several new chapters. We have selected internationally recognized experts who are at the forefront of research as contributors and we are very grateful to all of them for their outstanding and up-to-date contributions and for producing their manuscripts on time. This book brings together all the recent information on basic mechanisms of asthma and also covers clinical aspects and therapy in depth. This integrated approach should appeal to both basic researchers and to clinicians who need to keep abreast of recent developments in this rapidly expanding field.

We would like to thank Susan Lord of Academic Press for all her help in putting together this third edition. We are very grateful to our secretaries for all their help in chasing up manuscripts and to our wives for their patience and understanding. We hope that the third edition will be even more successful than the first two!

Peter J. Barnes, London
Neil C. Thomson, Glasgow
Ian W. Rodger, Montreal

Foreword to the second edition

Scientific, epidemiological and clinical information on asthma is expanding rapidly. In consequence, it has become necessary to revise *Asthma: Basic Mechanisms and Clinical Management* only four years after the first edition.

Asthma remains one of the most common and treatable conditions in medicine. It affects some 5% of the adult population in the Western world and around 10% of children. The condition is always distressing and occasionally lethal. It is therefore obligatory for every patient to receive the best possible treatment currently available.

One of the greatest challenges in modern medicine is the fact that, while the understanding of pathogenesis and proven improvements in therapy has developed rapidly, there is considerable evidence that morbidity from asthma is increasing and no reduction in mortality is occurring, in spite of attempts to identify and tackle factors which should prevent death. Thus, there is still much to learn. No doubt new information will emerge, but there may also be a need to re-evaluate some of our preconceived ideas and re-examine some of our previous treatment policies. In the meantime this radically revised and expanded second edition will help in this important re-evaluation process.

Peter Barnes, Ian Rodger and Neil Thomson must be congratulated in drawing together a group of specialist authors of international standing. Let us hope that their efforts to inform and educate both scientists and clinicians will be rewarded by a better quality of life for all asthmatic patients.

January 1992 *Dame Margaret Turner-Warwick, DBE, DM, PhD, PRCP*
President of the Royal College of Physicians, London NW1, UK

Contributors

Muntasir M. Abdelaziz Academic Department of Respiratory Medicine, St Bartholomew's and the Royal London School of Medicine and Dentistry, The London Chest Hospital, London E2 9JX, UK

Ian M. Adcock Department of Thoracic Medicine, National Heart and Lung Institute, Imperial College of Science, Technology and Medicine, London SW3 6LY, UK

Gary P. Anderson Department of Pharmacology, University of Melbourne, Parkville, 3052 VIC, Australia

Sandra D. Anderson Department of Respiratory Medicine, Royal Prince Alfred Hospital, Camperdown NSW, Australia 2050

Tony R. Bai Respiratory Health Network of Centres of Excellence, University of British Columbia Pulmonary Research Laboratory, St Paul's Hospital, Vancouver, Canada

Peter J. Barnes Department of Thoracic Medicine, National Heart and Lung Institute, Imperial College of Science, Technology and Medicine, London SW3 6LY, UK

Eugene R. Bleecker Division of Pulmonary and Critical Care Medicine, University of Maryland School of Medicine, Baltimore MD 21201, USA

William W. Busse University of Wisconsin Hospital, Madison WI 53792–3244, USA

Anoop J. Chauhan University Medicine, Southampton General Hospital, Southampton SO16 6YD, UK

K. Fan Chung National Heart and Lung Institute, Imperial College of Science, Technology and Medicine, London SW3 6LY, UK

D.W. Cockcroft Department of Medicine, Royal University Hospital, Saskatoon, Saskatchewan, S7N 0W8 Canada

C.J. Corrigan Department of Medicine, Imperial College School of Medicine, Charing Cross Campus, London W6 8RF, UK

Contributors

Graham K. Crompton Respiratory Unit, Western General Hospital, Edinburgh EH4 2DU, UK

Adnan Custovic North West Lung Centre, Wythenshawe Hospital, Manchester M23 9LT, UK

E.E. Daniel Department of Medicine, Division of Respirology, Faculty of Health Sciences, McMaster University, 1200 Ma Street West, Room HSC-3U 30, Hamilton, Ontario, Canada L8N 325

Robert J. Davies Academic Department of Respiratory Medicine, St Bartholomew's and the Royal London School of Medicine and Dentistry, The London Chest Hospital, London E2 9JX, UK

Jagdish L. Devalia Academic Department of Respiratory Medicine, St Bartholomew's and the Royal London School of Medicine and Dentistry, The London Chest Hospital, London E2 9JX, UK

Elliot C. Dick University of Wisconsin Hospital, Madison WI 53792–3244, USA

Jeffrey M. Drazen Division of Pulmonary and Critical Care Medicine, Brigham and Women's Hospital, Boston MA 02115, USA

James E. Gern University of Wisconsin Hospital, Madison WI 53792–3244, USA

David J. Godden Aberdeen Royal Hospitals NHS Trust, Aberdeen Royal Infirmary, Aberdeen AB25 2ZN, UK

Nicholas J. Gross Departments of Medicine and Molecular Biochemistry, Stritch School of Medicine, Loyola University of Chicago, USA

Ian P. Hall Division of Therapeutics, University Hospital, Nottingham NG7 2UH, UK

Catherine M. Hawrylowicz Department of Allergy and Respiratory Medicine, 5th Floor, Thomas Guy House, Guy's Hospital, United Medical and Dental Schools, London SE1 9RT, UK

Stephen T. Holgate University Medicine, Southampton General Hospital, Southampton SO16 6YD, UK

L.J. Janssen Department of Medicine, Division of Respirology, Faculty of Health Sciences, McMaster University, 1200 Ma Street West, Room HSC-3U 30, Hamilton, Ontario, Canada L8N 325

Peter K. Jeffery Lung Pathology Unit, Histopathology, Royal Brompton: National Heart and Lung Institute, Imperial College of Science, Technology and Medicine, London, UK

Rudolf A. Jörres Krankenhaus Grosshansdorf, Zentrum für Pneumologie und Thoraxchirurgie, D-22927 Grosshansdorf, Germany

A. M. Khawaja Department of Thoracic Medicine, National Heart and Lung Institute, Imperial College of Science, Technology and Medicine, London SW3 6LY, UK

Thriumala M. Krishna University Medicine, Southampton General Hospital, Southampton SO16 6YD, UK

Tak H. Lee Department of Allergy and Respiratory Medicine, Guy's Hospital, United Medical and Dental Schools, London SE1 9RT, UK

Robert F. Lemanske Jr University of Wisconsin Hospital, Madison WI 53792–3244, USA

Y.-C. Liu Department of Thoracic Medicine, Chang Gung Memorial Hospital, Taipei, Taiwan, Republic of China

Contributors

Helgo Magnussen Zentrum für Pneumologie und Thoraxchirurgie, LVA Freie und Hausestadt Hamburg, Wöhrendanum 80, D-22927 Grosshansdorf, Germany

Boaz Meijer Beatrixoord, 9751 ND Haren, The Netherlands

A.J. Newman Taylor Department of Occupational and Environmental Medicine, National Heart and Lung Institute, 1b Mannesa Road, London SW3 6LR, UK

Robert Newton Department of Thoracic Medicine, National Heart and Lung Institute, Imperial College of Science, Technology and Medicine, London SW3 6LY, UK

Paul M. O'Byrne Asthma Research Group, Department of Medicine, McMaster University, Hamilton, Ontario, Canada

P.D. Paré Respiratory Health Network of Centres of Excellence, University of British Columbia Pulmonary Research Laboratory, St Paul's Hospital, Vancouver, Canada

Martyn R. Partridge Whipps Cross Hospital, London E11 1NR; UK National Asthma Campaign, UK

F.L. Pearce Department of Chemistry, University College London, London WC1H 0AJ, UK

Søren Pedersen Department of Paediatrics, University of Odense, Kolding Sygehus, DK-6000 Kolding, Denmark

Carl G.A. Persson Department of Clinical Pharmacology, University Hospital, Lund, Sweden

Dirkje S. Postma Department of Pulmonology, University Hospital, 9713 GZ Groningen, The Netherlands

Neil B. Pride Department of Thoracic Medicine, National Heart and Lung Institute, Imperial College of Science, Technology and Medicine, London SW3 6LY, UK

Dr David Proud Johns Hopkins Asthma and Allergy Center, Baltimore MD 21224, USA

Clive R. Roberts Respiratory Health Network of Centres of Excellence, University of British Columbia Pulmonary Research Laboratory, St Paul's Hospital, Vancouver, Canada

Douglas S. Robinson Allergy and Clinical Immunology, National Heart and Lung Institute, Imperial College of Science, Technology and Medicine, London SW3 6LY

I.W. Rodger Pharmacology Department, Merck Frosst Canada Inc., Point Claire-Dorval, Quebec, Canada

D.F. Rogers Department of Thoracic Medicine, National Heart and Lung Institute, Imperial College of Science, Technology and Medicine, London SW3 6LY, UK

Malcolm R. Sears Department of Medicine, McMaster University, Firestone Regional Chest and Allergy Unit, St Joseph's Hospital, Hamilton, Ontario, Canada

Sean D. Sullivan Departments of Pharmacy and Health Services, University of Washington, Seattle, Washington 98195, USA

Andrzej Szczeklik Department of Medicine, Jagiellonian University, School of Medicine, 31–066 Krakow, Poland

Anne E. Tattersfield Division of Respiratory Medicine, City Hospital, Nottingham NG5 1PB, UK

Neil C. Thomson Department of Respiratory Medicine, West Glasgow Hospitals University NHS Trust, 1053 Gt Western Road, Glasgow G12 0YN, UK

Professor Per Venge, MD Department of Clinical Chemistry, University Hospital, S-751 85 Uppsala, Sweden

Dr A. Wardlaw Respiratory Medicine, University of Leicester Medical School, Glenfield Hospital, Leicester LE3 9QP, UK

Kevin B. Weiss Center for Health Services Research, Rush Primary Care Institute; Rush Presbyterian St Luke's Medical Center, Chicago, Illinois 60612, USA

Nicholas J. Withers Department of Respiratory Medicine, Bristol Royal Infirmary, Bristol BS2 8HW, UK

Ashley Woodcock North West Lung Centre, Wythenshawe Hospital, Manchester M23 9LT, UK

Ann J. Woolcock Institute of Respiratory Medicine, Level 8, Building 82, Royal Prince Alfred Hospital, Camperdown NSW 205, Australia

Contents

Preface v
Foreword to the second edition vii
Contributors ix

1 Epidemiology   1
Malcolm R. Sears

    Introduction. 1
    Diagnosis of asthma 1
    Asthma in children 2
    Factors influencing development of childhood asthma 5
    Childhood asthma and AHR. 11
    Prognosis of childhood asthma 11
    Asthma in adults 12
    AHR in adults 14
    Risk factors for adult asthma 15
    Evidence for increasing severity of asthma 16
    Mortality from asthma 17
    Risk factors for asthma mortality 19
    References 21

2 Genetics   35
Boaz Meijer, Eugene R. Bleecker and Dirkje S. Postma

    Introduction. 35
    Methods of finding disease genes in asthma 35
    Problems in genetic studies of asthma 37

Modes of inheritance of atopy, BHR and asthma	37
Genetic studies of atopy and asthma	38
Genetic studies of BHR and asthma	41
Conclusion	43
References	44

3 Airway Pathology in Asthma — 47
Peter K. Jeffery

Introduction	47
Sputum and bronchoalveolar lavage	48
Appearances at post-mortem	48
Loss of surface epithelium	49
Thickening of the epithelial 'basement membrane'	51
Increased numbers of mucus-secreting cells	51
Enlargement of bronchial smooth muscle mass	53
Bronchial vasculature, congestion and oedema	56
Recruitment of inflammatory cells	57
Airway wall nerves	58
Conclusions	59
Acknowledgements	60
References	60

4 Physiology — 65
Neil B. Pride

Introduction	65
Factors restricting and amplifying induced airway narrowing *in vivo*	65
Sites of airway narrowing in asthma	69
Effects of posture and sleep on airway function	76
Response to increasing severity of airway narrowing	77
Conclusions	83
References	83

5 Airway Smooth Muscle Cells: Structure and Function — 89
L.J. Janssen, E.E. Daniel and I.W. Rodger

Introduction	89
Morphology	89
Ion channels	96
Phosphatidylinositides and calcium	100
Physiological mechanisms	101
Conclusion	105
References	106

6 Mast Cells and Basophils — 113
F.L. Pearce

Introduction	113
Mast cell heterogeneity	114
Mast cells, basophils and adhesion molecules	115
Distribution and morphology of human lung mast cells	116
BAL in extrinsic asthma	117

Contents

Immunologically induced mediator release	118
BAL mast cells steroid therapy in asthma	120
Antiasthmatic drugs and the inhibition of histamine release from pulmonary mast cells	120
Role of basophils in allergy and asthma	121
Role of mast cells in early asthmatic reactions	122
Role of mast cells in late asthmatic reactions and cytokine production	122
Acknowledgements	123
References	123

7 Monocytes, Macrophages and Dendritic Cells — 127
Catherine M. Hawrylowicz and Tak H. Lee

Introduction	127
Role of mononuclear phagocytes in inflammatory mechanisms of asthmatic disease	128
Changes in phenotype and function of monocytes and macrophages in asthma	128
Effects of glucocorticoids on monocyte/macrophage function	134
Conclusion	135
References	137

8 Eosinophils — 141
Per Venge

Introduction	141
Biochemistry and function	142
Receptors and degranulation	147
Mechanisms of eosinophil accumulation	148
The eosinophil in asthma	149
Monitoring of asthma by eosinophil markers	150
Pharmacological control of the eosinophil	150
Conclusions	151
References	151

9 Lymphocytes — 159
Gary P. Anderson

Introduction	159
Lymphocytes are strongly implicated in the pathogenesis of asthma	161
Overview of lymphocyte responses to inhaled foreign antigens	163
Perspective: defects in termination of immune responses may cause chronic disease	176
Acknowledgement	177
References	177

10 Epithelial Cells — 187
Jagdish L. Devalia, Muntasir M. Abdelaziz and Robert J. Davies

Introduction	187
Morphology of the airway epithelium	187
Airway epithelium and hyperresponsiveness	190
Epithelial cell-derived mediators	190
Adhesion molecules and the airway epithelium	194
Immunoregulation	195

Culture of human airway epithelial cells *in vitro* 196
Summary 197
References 199

11   Pathophysiology of Airway Mucus Secretion in Asthma   205
Y.-C. Liu, A.M. Khawaja and D.F. Rogers

Introduction 205
Airway mucus 205
Airway mucus-secreting cells 207
Mucus abnormalities in asthma 208
Consequences of airway mucus hypersecretion and hyperviscosity 213
Inducers of airway mucus secretion 216
Pharmacological treatment of mucus hypersecretion in asthma 218
Conclusions 221
Acknowledgements 222
References 222

12   Tracheobronchial Circulation   229
David J. Godden

Introduction 229
Organization and control of the tracheobronchial circulation 229
Role of the tracheobronchial circulation in asthma 231
Conclusion 234
References 234

13   Adhesion Molecules   239
A. Wardlaw

Introduction 239
Leucocyte adhesion receptors 240
Expression of adhesion receptors in allergic disease 242
Role of adhesion receptors in leucocyte migration in allergic disease 244
In vivo studies of adhesion receptor antagonists in models of allergic
   inflammation 246
Summary 246
References 248

14   Microvascular–Epithelial Exudation of Plasma   253
Carl G.A. Persson

Introduction 253
Exudation pathways 254
Acute challenge-induced microvascular–epithelial exudation 257
Mucosal exudation of plasma in disease 259
Roles of exuded plasma 261
On the inward perviousness in asthma 262
Airway epithelial restitution in a plasma-derived gel 263
References 265

Contents

15 Prostaglandins and Thromboxane — 269
Paul M. O'Byrne

Introduction — 269
Arachidonic acid metabolism — 270
Cyclooxygenase products — 271
Prostaglandin and thromboxane receptors — 271
Role of COX products in asthma — 272
Stimulatory prostaglandins and thromboxane — 273
Inhibitory prostaglandins — 275
Conclusions — 277
References — 278

16 Cysteinyl Leukotrienes — 281
Jeffrey M. Drazen

Introduction — 281
Formation and metabolism of the leukotrienes — 281
Leukotrienes in asthma — 283
Conclusions — 289
References — 290

17 Kinins — 297
David Proud

Introduction — 297
Structure, formation and metabolism — 298
Receptors and general pharmacological properties — 300
Kinin formation in airway inflammation — 301
Effects of kinins on airways — 301
Mechanisms of action — 302
Summary — 303
Acknowledgements — 304
References — 304

18 Chemokines — 309
K. Fan Chung

Introduction — 309
Discovery and structure — 310
Cell sources — 312
Regulation — 312
Chemokines as chemoattractants and cell activators — 313
Chemokine receptors — 316
Expression and release of chemokines in asthma — 317
Conclusion — 318
References — 319

19 Lymphokines — 329
Douglas S. Robinson

Introduction — 329

Type 1 and type 2 T–cells	329
Actions of type 2 cytokines relevant to asthma	331
Evidence of T–cell cytokine production in asthma	333
T–cell cytokines in non-atopic asthma	335
Factors determining type 1 or type 2 T-cell development	336
Alteration of established type 1 or type 2 cytokine profile	336
Potential for intervention	337
Conclusion	337
References	337

20 Other Mediators of Asthma — 343
K. Fan Chung and Peter J. Barnes

Introduction	343
Histamine	343
Platelet-activating factor	346
Oxygen radicals	348
Complement	351
Serotonin	353
Eosinophil proteins	354
Endothelin	354
References	357

21 Nitric Oxide — 369
Peter J. Barnes

Introduction	369
Generation of NO	369
Effects of NO on airway function	371
Exhaled NO	375
Therapeutic implications	382
References	383

22 Neural Control of Airway Function in Asthma — 389
Peter J. Barnes

Introduction	389
Afferent nerves	392
Parasympathetic nerves	393
Adrenergic control	399
References	403

23 Humoral Control of Airway Tone — 409
Neil C. Thomson

Introduction	409
Vasoactive peptides	409
Hormones	413
Circulating inflammatory mediators	415
Oxygen and carbon dioxide	415
Conclusions	417
References	417

Contents

24 NANC Nerves and Neuropeptides — 423
Peter J. Barnes

Introduction	423
Non-adrenergic non-cholinergic (NANC) nerves	425
VIP and related peptides	428
Tachykinins	434
CGRP	436
Neurogenic inflammation	438
Other neuropeptides	444
Role of neuropeptides in asthma	446
References	447

25 Transcription Factors — 459
Robert Newton, Peter J. Barnes and Ian M. Adcock

Basal and regulated transcription	459
NF-κB, the Rel family of proteins and IκB proteins	460
AP-1 and related transcription factors	464
CCAAT/enhancer-binding proteins and NF-IL-6	464
JAK–STAT pathway	465
Glucocorticoid receptors	466
Cross-talk between transcription factors and their transduction pathways	466
Transcription factors in asthma	469
References	471

26 Airway Remodelling — 475
Tony R. Bai, Clive R. Roberts and P.D. Paré

Introduction	475
Structural changes in the airway walls in asthma	475
Extracellular matrix	478
Smooth muscle	481
Summary	483
Acknowledgements	483
References	484

27 Pathophysiology of Asthma — 487
Peter J. Barnes

Introduction	487
Asthma as an inflammatory disease	488
Inflammatory cells	489
Structural cells	491
Inflammatory mediators	493
Effects of inflammation	495
Transcription factors	499
Anti-inflammatory mechanisms in asthma	499
Genetic influences	500
Unanswered questions	501
References	503

28 Allergens — 507
D.W. Cockcroft

Introduction	507
Atopy	508
Inhaled allergens	508
Allergic bronchopulmonary aspergillosis	518
Ingested/injected allergens	520
Acknowledgements	520
References	520

29 Occupational Asthma — 529
A.J. Newman Taylor

Introduction: initiators and provokers of asthma	529
Irritant-induced asthma	530
Causes of hypersensitivity-induced occupational asthma	531
Importance of hypersensitivity-induced occupational asthma	532
Occupational asthma and hypersensitivity	532
Determinants of hypersensitivity-induced occupational asthma	534
Diagnosis of hypersensitivity-induced occupational asthma	536
Investigation of hypersensitivity-induced occupational asthma	537
Outcome of hypersensitivity-induced occupational asthma	541
Management of occupational asthma	542
References	543

30 Infections — 547
William W. Busse, Elliot C. Dick, Robert F. Lemanske Jr and James E. Gern

Introduction	547
Epidemiology of respiratory infections and wheezing	547
Mechanisms of virus-induced airway hyperresponsiveness	556
Summary	564
Acknowledgements	564
References	564

31 Asthma Provoked by Exercise, Hyperventilation and the Inhalation of Non-isotonic Aerosols — 569
Sandra D. Anderson

Introduction	569
Respiratory water loss and conditioning of inspired air	570
Role of the bronchial circulation	572
Generation and deposition of non-isotonic aerosols in the respiratory tract	574
Comparison between challenge with exercise and hyperventilation and challenge with non-isotonic aerosols	574
Mechanisms by which a change in osmolarity and airway drying induce airway narrowing	576
Effect of pharmacological agents	578
References	581

Contents

32 Atmospheric Pollutants — 589
Rudolph A. Jörres and Helgo Magnussen

- Ozone — 589
- Nitrogen dioxide — 591
- Sulphur dioxide — 592
- Summary — 593
- References — 593

33 Drug-induced Asthma — 597
Peter J. Barnes and Neil C. Thomson

- β-Blockers — 597
- Additives — 600
- ACE inhibitors — 601
- Local anaesthetics — 602
- Other drugs — 602
- References — 603

34 Aspirin-induced Asthma — 607
Andrzej Szczeklik

- History and definition — 607
- Pathogenesis — 608
- Clinical presentation — 611
- Diagnosis — 612
- Differential diagnosis — 613
- Prevention and treatment — 613
- References — 614

35 Allergen Avoidance — 617
Adnan Custovic and Ashley Woodcock

- Introduction — 617
- Indoor allergens as a cause of asthma — 618
- Primary sensitization — 621
- Allergen exposure and asthma severity — 622
- Are threshold values useful? — 623
- Allergen avoidance: reported effectiveness — 625
- Allergen avoidance — 633
- Allergen avoidance in asthma prevention — 640
- Conclusions — 640
- References — 641

36 β-Adrenoceptor Agonists — 651
I.P. Hall and A.E. Tattersfield

- Introduction — 651
- Molecular pharmacology — 651
- Clinical pharmacology — 658
- Differences between β-agonists — 661
- Efficacy and safety of inhaled β-agonists — 662

	Oral β-agonists	669
	References	670

37 Anticholinergic Bronchodilators — 677
Nicholas J. Gross

Introduction	677
Rationale for use of anticholinergic bronchodilators	677
Pharmacology	679
Clinical efficacy	681
Side-effects	685
Clinical recommendations	686
References	686

38 Theophylline — 689
Peter J. Barnes

Introduction	689
Historical background	689
Chemistry	690
Molecular mechanisms of action	690
Effects	693
Pharmacokinetics	696
Routes of administration	697
Clinical use	698
Side-effects	701
Future of theophylline	701
References	702

39 Cromones — 707
Anoop J. Chauhan, Nicholas J. Withers, Thriumala M. Krishna and Stephen T. Holgate

Introduction	707
Pharmacokinetics	708
Effects on inflammatory cells and nerves	708
Evidence for anti-inflammatory actions	711
Mechanisms of action	711
Clinical studies in asthma	712
References	716

40 Glucocorticosteroids — 725
Peter J. Barnes

Introduction	725
Molecular mechanisms	725
Effects on cell function	732
Effects on asthmatic inflammation	735
Clinical efficacy of inhaled steroids	736
Pharmacokinetics	738
Side-effects of inhaled steroids	739
Clinical use of inhaled steroids	745

Contents

	Systemic steroids	746
	Glucocorticoid resistance in asthma	748
	References	755

41 Mediator Antagonists — 767
Peter J. Barnes and K. Fan Chung

Introduction	767
Antihistamines	767
Leukotriene antagonists	768
Prostaglandin inhibitors	772
Platelet-activating factor (PAF) antagonists	773
Phospholipase inhibitors	774
Bradykinin antagonists	775
Antioxidants	775
Adenosine antagonists	776
NO synthase inhibitors	776
Endothelin antagonists	777
Basic protein inhibitors and heparin	777
Inflammatory enzyme inhibitors	777
Combined inhibitors	778
References	778

42 Immunomodulators — 783
C.J. Corrigan

Clinical need	783
Immunosuppressive therapy in asthma	784
Newer immunosuppressive agents	790
Conclusions	791
References	791

43 Future Therapies for Asthma — 795
Peter J. Barnes, Ian W. Rodger and Neil C. Thomson

Introduction	795
New bronchodilators	796
Mediator antagonists	801
Cytokines and cytokine inhibitors	805
Anti-inflammatory drugs	807
Gene therapy	813
Conclusions	814
References	814

44 Management of Severe Asthma — 821
Graham K. Crompton

Introduction	821
General assessment and management	821
Specific treatment for severe acute asthma	823
Assisted ventilation	827
Management of catastrophic asthma	828
Other measures	830

Summary 831
References 831

45 Management of Asthma in Adults   835
Ann J. Woolcock

Introduction 835
Classification of asthma for purposes of management 836
Aims of management 837
Asthma management plan 837
Treatment of the patient with severe persistent asthma 852
Likely future changes to management 853
References 854

46 Asthma in Children   859
Søren Pedersen

Introduction 859
Anatomical and physiological factors 860
Wheezing illness 860
Natural history 861
Risk factors and prevention 863
Growth 864
Pharmacokinetics and pharmacodynamics 864
Assessment of the clinical condition 865
Irreversible airway obstruction 868
Treatment 869
Special age groups 883
Severe acute asthma/status asthmaticus 884
Immunotherapy 888
General measures 889
References 890

47 Pharmacoeconomics of Asthma Treatments   903
Sean D. Sullivan and Kevin B. Weiss

Introduction 903
Principles and applications of pharmacoeconomics 904
Cost–benefit analysis 905
Cost-effectiveness analysis 906
Asthma outcomes for pharmacoeconomic evaluation 907
Pharmacoeconomics of asthma pharmacotherapy 907
Conclusions 914
References 914

48 Education and Self-management   917
Martyn R. Partridge

Introduction 917
Health professional education and guidelines 917
Compliance 918
How do we improve communication? 919
Self-management 922

Contents

    Conclusion 924
    References 924

Index 927

Plates appear between pages 326 and 327.

1

Epidemiology

MALCOLM R. SEARS

INTRODUCTION

Asthma remains one of the commonest disorders encountered in clinical medicine in both children and adults. The prevalence and pathogenesis of asthma have been extensively studied, but its fundamental causes and the interplay of factors inducing and inciting airway inflammation are still unclear. International variations in morbidity and mortality have become subjects of intense investigation and debate. Recent studies of the epidemiology of asthma have helped to increase our understanding of the heterogeneity of the condition and, to some extent, its causes (including genetic factors) and prognosis.[1,2]

DIAGNOSIS OF ASTHMA

Asthma defies precise definition.[3–5] The most recent consensus document[6] provides the following operational definition:

> Asthma is a chronic inflammatory disorder of the airways in which many cells play a role, in particular mast cells, eosinophils, and T lymphocytes. In susceptible individuals this inflammation causes recurrent episodes of wheezing, breathlessness, chest tightness, and cough particularly at night and/or in the early morning. These symptoms are usually associated with widespread but variable airflow limitation that is at least partly reversible either spontaneously or with treatment. The inflammation also causes an associated increase in airway responsiveness to a variety of stimuli.

ASTHMA: BASIC MECHANISMS AND CLINICAL MANAGEMENT (3rd Edn)
ISBN 0-12-079027-9

Copyright © 1998 Academic Press Limited
All rights of reproduction in any form reserved

Asthma or 'variable airflow obstruction' is recognized from a pattern of one or more characteristic symptoms, including wheeze, cough, chest tightness and dyspnoea, and is best confirmed by evidence of variable or reversible airflow obstruction accompanying symptoms.[7] In children, asthma causing nocturnal or post-exercise cough may be misdiagnosed as bronchitis or infection.[8,9]

While a carefully taken history is very helpful in diagnosing asthma, there are several causes of similar respiratory symptoms and a history is less than fully repeatable over time. Symptom histories checked after 3 months resulted in changes in diagnostic categories in 7–13% of children, although the prevalence rate remained stable as directions of change were virtually balanced.[10] Nevertheless a physician diagnosis of asthma remains the 'gold standard' with which questionnaires and objective measurements are compared.[11]

Increased responsiveness of the airways to non-allergic stimuli usually accompanies asthma symptoms[12] but is not synonymous with asthma. Airway hyperresponsiveness (AHR) is absent in some subjects with other clear evidence of asthma, but may be variably present in some children and adults without significant respiratory symptoms.[13–17] Asthma is characterized by inflammation,[18] although evolving methods for the non-invasive detection of airway inflammation[19] have not yet been used in community studies. Evidence of variable or reversible airflow obstruction, such as a 20% increase in forced expiratory volume in 1 s (FEV_1) or peak expiratory flow (PEF) occurring spontaneously or with treatment,[20] is helpful if present. However, asthmatic patients who smoke, or work in highly polluted atmospheres, may develop less reversible disease, while some patients develop irreversible disease despite being lifetime non-smokers.[21]

Peak flow variability has been compared with methacholine challenge in community populations.[22] While both measures related to each other, to respiratory symptoms and to atopy, methacholine responsiveness was more closely related to asthma than was peak flow. Among Danish children, methacholine challenge was a more sensitive marker of diagnosed asthma than peak flow or exercise challenge.[23] Of symptomatic subjects with any of these three challenge tests positive, methacholine identified 75% and combined methacholine test and peak flow monitoring identified 89%.

Toelle et al.[24] have proposed a working definition of current asthma for epidemiological purposes, namely recent symptoms (within the last 12 months) together with an appropriate degree of airway responsiveness. While this excludes some with characteristic symptoms without AHR at the time of testing, these form a minority of asthmatic patients where asthma is relatively mild, and the working definition does provide a sound basis for comparative studies across regions and across times.

ASTHMA IN CHILDREN

In a series of cross-sectional studies, between 1.6 and 26.5% of children in the UK,[9,14,25–28] the USA,[29–31] Canada,[32,33] Australia,[15,34–39] New Zealand,[40–42] Scandinavia,[43–45] Switzerland,[39] Chile[39] and Costa Rica[46] report recent respiratory symptoms suggesting asthma (Table 1.1). When studied cumulatively, up to 48.5% of young children may wheeze at some time to age 6 years.[47] The variations in questions asked, age groups studied and local factors all affect comparability of responses. A uniform approach

1 Epidemiology

Table 1.1 Selected studies of prevalence of asthma and wheezing in children.

Location	Age (mean or range)	Prevalence	Year published	Reference
Tyneside, UK	7	9.3% current, 11.1% cumulative	1983	14
London, UK	9	11.1% current, 18.2% cumulative	1983	25
National, UK	7, 11, 16	Current 8.3%, 4.7%, 3.5%, respectively Cumulative 18.3%, 21.9%, 24.7%	1987	26
Southampton, UK	7, 11	Asthma 9.5%, wheeze 12.1% current	1989	9
Southampton, UK	5–18	24.3% male, 15.6% female	1995	27
UK	5–17	15% wheeze last year, 13.1% asthma	1994	28
Boston, USA	5–9	9.2% persistent wheeze	1980	29
Baltimore, USA	Grades 1, 6	7.2% current, 10.5% cumulative	1982	31
National, USA	6–11	7.6% asthma and wheeze with colds	1988	30
Tucson, USA	0–6	48.5% cumulative to 6, 28.7% at 6	1995	47
Hamilton, Canada	7–10	4.4% asthma, 21.2% wheezing most days	1986	32
South-west Ontario, Canada	9	20.5% wheezing (3/year or more)	1996	33
Sydney, Australia	12.6* 8.9*	7.2% male, 4.2% female 8.6% male, 4.4% female	1980	34
Queensland, Australia	8.1* 12.1*	20.5% wheeze 18.1% wheeze	1983	35
South Australia	8	27.4% cumulative (asthma + wheeze)	1986	36
New South Wales	8–10	12.4% asthma, 24.3% wheezing	1987	15
Eastern Australia	?	17.1% asthma, 19.5% wheezing	1992	37
Australia	Adolescent	16.5% asthma	1992	38
Melbourne	7, 12, 15	23.1%, 20.9%, 18.6% wheeze last 12 months	1993	39
Lower Hutt, NZ	11–13	13.5% asthma	1983	40
Dunedin, NZ	9	18.1% wheezing (3/year or more)	1987	41
Auckland, NZ	8–10	European 13.5% (symptoms + AHR) Maori 10.8%	1988	42
Norway	7–15	1.6% current, 3.7% cumulative, 9% wheeze	1984	43
Sweden	4, 7, 10, 14	5.1% asthma	1989	44
Sweden	7–16	4.0% asthma	1988	45
Switzerland	7, 12, 15	7.4%, 6.09%, 4.5% wheeze last 12 months	1993	39
Chile	7, 12, 15	26.5%, 21.1%, 17.7% wheeze last 12 months	1993	39
Costa Rica	5–17 (?)	23.4% asthma	1994	46

AHR, airway hyperresponsiveness.

to enable detection of true differences in asthma prevalence between countries and between regions has recently been implemented. Early reports from the International Study of Asthma and Allergies in Childhood (ISAAC), in which each centre or country used a standard protocol in 6–7 and 13–14-year-old children,[48,49] not only suggest there are real differences between regions but also point to the likely underdiagnosis of asthma in many areas considered to have low prevalence rates. Some of the reported studies using the ISAAC protocol among 13–14 year olds are summarized in Table 1.2. The ISAAC questionnaire item regarding wheezing or whistling in the chest in the last 12 months was found to have good validity when compared with an assessment of a paediatric pulmonologist making a clinical diagnosis of asthma aided by measurements of airway responsiveness to hypertonic saline.[54]

The prevalence of wheezing in childhood has increased over the last decade;[55] however, differences in methodology, definitions and use of the label 'asthma' have exaggerated the true increase in prevalence.[56,57] In Australia in 1969, 19.1% of 7-year-old children had experienced recurrent episodes of wheezing;[58] in New Zealand in 1973, 23% of 9-year-old children had a history of one or more attacks of wheezing.[59] In these and other early studies, parents of half to two-thirds of the children reporting wheeze denied their child had 'asthma'; hence a change in use of that label in subsequent years could greatly alter the prevalence of reported asthma with little change in the burden of disease. In the 1994–95 Canadian ISAAC study, over 80% of those with wheezing four or more times a year had been given the label 'asthma'.[60]

Burney et al.[55] found an increase in the prevalence of asthma for English children of both sexes between 1973 and 1986; however, the increase in prevalence of wheeze was much less and significant only in girls. These authors later surveyed 36 regions in the UK in 1982 and 1992; they found a three-fold increase in asthma attacks in 5–11-year-old children, with a 30–60% increase in occasional wheeze and a 30–40% increase in persistent wheeze,[61] suggesting a real increase in the prevalence of asthma. In the USA, reported prevalence among 6–11-year-old children increased significantly from 4.8 to 7.6% between the first (1971–74) and second (1976–80) National Health and Nutrition Examination Surveys.[30] In each of these surveys, 'asthma' included recurrent wheezing not associated with colds as well as physician-diagnosed asthma. The increase in

Table 1.2 Prevalence of childhood asthma among 13–14 year olds using the ISAAC questionnaire.

Location	Asthma at any time (%)	Wheeze at any time (%)	Wheeze in last 12 months (%)	Reference
Hamilton, Canada	19	44	30	50
Saskatoon, Canada	12	36	23	50
Adelaide, Australia	22	40	29	48
Sydney, Australia	26	45	30	48
Sussex, UK	15	48	29	48
Bochum, Germany	4	33	20	48
Wellington, NZ	18	44	28	48
Bag of Plenty, NZ	22	30	–	51
Singapore	21	29	10	52
Hong Kong	–	20	12	53

1 Epidemiology

prevalence therefore cannot be explained by a greater use of the label 'asthma' and is probably real.

In Aberdeen, Scotland, the prevalence of wheezing among children rose from 10% in 1964 to 20% in 1989, shortness of breath increased from 6 to 10%, and diagnosed asthma increased from 4 to 10%.[62] There was a similar increase in eczema, from 6 to 12%, and in hayfever, from 3 to 12%, indicating a generalized increase in atopy. The labelling of wheezing as 'asthma' increased from 27 to 50% between 1966 and 1989. Hence increased 'asthma' is related to both a true increase in symptoms and greater use of the diagnostic label. In Finland, the increase in asthma detected at medical examination in army conscripts rose from 0.08% in 1961 to 0.29% in 1966 and then 20-fold to 1.79% in 1989.[63] It seems unlikely that a change of this proportion relates only to diagnostic fashion, as this would mean 95% of cases of asthma were undiagnosed before 1966.

Bauman and colleagues[64] compared the trends in prevalence of current and cumulative wheezing with that of 'asthma' among Australian children between 1968 and 1992; they found a less pronounced increase in wheeze than for reported asthma, suggesting substantial diagnostic transfer and greater acceptance of the label 'asthma'.[64] In Hong Kong, the prevalence of asthma at any time and of current wheeze increased 71% and 240% over a few years, also suggesting a much greater recognition of wheezing in the population or a very substantial new burden of illness for unexplained reasons.[65]

Substantial increases in hospital admissions for asthma in children in many countries suggest that the prevalence of severe asthma has increased.[66-68] In the USA, the rate of hospitalization for children under 15 years with asthma increased at least 145% between 1970 and 1984.[69] In England, admissions increased 167% in 5–14 year olds over a period of 8 years.[70] The increase in hospital admissions for asthma has occurred despite falling admission rates for other respiratory conditions and a considerable increase in the use of antiasthmatic medications;[71] this is due in part to a doubling of the readmission rate.[71-72] In one New Zealand paediatric unit, admissions rose dramatically from 21 in 1965 to 186 in 1975 and 609 in 1985, while the severity on admission (based on wheezing, pulse rate and accessory muscle use) also increased considerably in the latter 10 years.[73] These facts all suggest that asthma has increased in severity, although there is some evidence of increased self-referral to hospital especially for nebulizer treatments for children.[74] In an Australian paediatric hospital, the ratio of attendances at the emergency room for asthma to all attendances increased steadily over 10 years, as did the ratio of admissions for asthma to all other admissions, whereas these trends were not seen for non-asthma respiratory conditions.[75] However, in some countries in recent years asthma admissions have begun to decline, a trend attributed to improved treatment especially with inhaled corticosteroids.[76]

FACTORS INFLUENCING DEVELOPMENT OF CHILDHOOD ASTHMA

Factors relevant to the development of asthma may be inherent (sex, genetic predisposition, lung size) or external (diet, exposure to allergens, air pollution including environmental tobacco smoke exposure, infection, socioeconomic status and region of residence).[77,78]

Sex

The male predominance of diagnosed asthma in children[31,79] decreases with increasing age.[9,80] Its relationship to the prevalence of atopy has been variably reported, some studies finding no male predominance[81] while others clearly found atopy was related to male sex.[82,83] In Boston children, the male:female ratio for 'asthma' was 1.8:1, but for 'recurrent wheeze most days or nights' was 1:1, suggesting a sex-related diagnostic bias in use of the label 'asthma'.[29] In Switzerland, wheezing girls were less often diagnosed as asthmatic across all grades of severity,[84] but in Canada this was true only in those with very mild wheezing.[85]

Genetic and ethnic factors

The most powerful influence determining development of childhood asthma is the parental (especially maternal) history of asthma and atopy.[86] Twin studies have shown serum IgE levels in adults and children are genetically influenced,[87] while other twin studies suggest a genetic effect on AHR, even when there is not clinical concordance for asthma.[88] However the slope of the dose–response curve to methacholine is not different between monozygotic and dizygotic twins, suggesting that environmental factors are more important than genetic factors in determining individual airway responsiveness.[89]

Differences in prevalence between races may be due to both genetic and environmental factors. Among children of different races living in the same environment, the prevalence of respiratory symptoms is similar. There were no significant differences in the proportions with a history of wheezing illness among London children of European (15%), African (18%) or Indian (17%) descent, although the European children reported a higher prevalence of croup, whooping cough and bronchitis.[90] In a New Zealand study, the prevalence of current respiratory symptoms accompanied by AHR to histamine was slightly higher in European (13.5%) compared with Maori children (10.8%).[42] In the USA, the prevalence of asthma is consistently some 50% higher among black children compared with white children,[30,31] suggesting ethnic factors influence the expression of asthma, although differences in lifestyle may modify exposure to environmental factors despite residence in the same community. Wide variations in the prevalence of asthma, rhinitis and eczema in Chinese children in China and Hong Kong suggest that environmental factors are important in determining the prevalence of allergic disorders in different geographical locations given a similar genetic background.[91]

Region of residence

Movement from a rural to an urban environment appears to increase substantially the likelihood of developing childhood asthma. In South Africa, 3.17% of Xhosa children resident in Cape Town, but only 0.14% of Xhosa children remaining in rural Transkei, demonstrated exercise-induced asthma.[92] Over 25% of children who moved to New Zealand from the Tokelauan Islands developed asthma compared with 11% of those remaining in the islands.[93] These increases suggest that environmental factors have provoked expression of asthma symptoms in susceptible individuals in the new location. The prevalences of rhinitis and eczema in Tokelauan Islanders living in New Zealand were 28.3% and 8.5% respectively compared with 13.7% and 0.1% in their peers living in the islands.[93] Hence allergens in the new environment were more potent sensitizing agents, and may be significant factors in inducing asthma as well as rhinitis and eczema. Among Swedish conscripts aged 17–20 years, the prevalence of asthma was significantly higher in city dwellers (3.3%) compared with rural residents (2.5%), again suggesting an influence of urbanization.[94] Asians moving from Hong Kong to Australia develop higher prevalences of atopy and asthma.[95]

Atopy

The development of asthma in children is related to atopy, whether documented by positive skin tests[96] or elevated serum IgE levels.[97] Furthermore, the persistence of wheezing into later childhood is very strongly associated with atopy. Hayfever and eczema in a 7-year-old child with wheeze increased four-fold the risk of persistence of asthma to age 20.[98] Atopic children had an increased risk of developing AHR, especially those sensitive to more than one allergen group;[99,100] the severity of the AHR increased with the degree of atopy. House-dust mite sensitivity is one of the most important risk factors for development of asthma in children and adults.[101–103] Striking increases in asthma morbidity and mortality in New Guinea were attributed to the introduction of house-dust mites to areas where they were previously not resident.[103]

The likelihood of diagnosed asthma is increased not only by the presence of atopy but also by the intensity of atopy, as judged by the size of the weals to relevant allergens.[83] There is a strong relationship between serum IgE, clinical asthma and AHR; in children with serum IgE < 32 IU/ml asthma was not reported, while the prevalences of diagnosed asthma and of a methacholine $PC_{20}FEV_1 < 8$ mg/ml increased as serum IgE increased.[104] Furthermore, a clear relationship exists between serum IgE and AHR in children never admitting to asthma or wheezing.[104]

Multivariate modelling used to analyse data from 4366 Australian children in four regions and 878 adults showed a high prevalence of hyperresponsiveness at 7–9 years (16–18%), decreasing to 7–8% at 11–14 years and then increasing to 12–14% in adults. Atopy was the most important risk factor at all ages. Parental asthma, early respiratory illness and being born in Australia had significant effects, while eating fish more than once a week had a protective effect.[105]

The differing prevalence of atopy between males and females may explain much of the gender difference in childhood asthma.[83] Boys aged 13 years had a greater prevalence of positive skin tests to all allergens tested compared with girls; when this was included in risk analysis, gender difference in asthma prevalence was insignificant. House-dust mite and cat dander sensitivity were major predictors not only for diagnosed asthma[83] but also for AHR to methacholine.[106] Furthermore, the effect of atopy greatly overshadowed the effect of baseline airway calibre in predicting methacholine responsiveness. Non-atopic children showed AHR only when baseline lung function was substantially impaired, whereas among atopic children AHR was seen at all levels of lung function.[107]

Even in an environment at altitude free of house-dust mites, such as Los Alamos, New Mexico, asthma is strongly related to allergen with increased serum IgE and sensitization to other allergens, particularly those from cats[108] and dogs.[109] In the UK, children exposed to furry pets had twice the risk of having symptoms of asthma.[110]

The home environment contains other risk factors for asthma apart from mite and pet allergens. Dampness in the home increased the risks of asthma associated with passive smoking (odds ratio 1.3), and increased the risks associated with exposure to cats or dogs in combination with passive smoking (odds ratio 8.0), suggesting an interaction among these factors.[111] For Kenyan children aged 9–11 years, dampness in the child's sleeping area, and the presence of rugs or carpet in the bedroom, increased the risk of asthma (odds ratios >2.5).[112]

Passive smoking

Parental, especially maternal, smoking is associated with reduction of pulmonary function,[113] lower respiratory tract illness[114] and with wheezing illness and asthma in children. Cogswell et al.[115] followed a cohort of high-risk children born to parents at least one of whom was atopic. By age 5 years 62% of parents who smoked had children who had experienced episodes of wheeze compared with 37% of parents who did not smoke. Gortmaker et al.[116] found the prevalence of parent-reported asthma in children aged 0–17 years increased from 5.0 to 7.7% if the mother smoked, and the prevalence of 'functionally impairing' (i.e. severe) asthma doubled from 1.1 to 2.2%. In Boston children aged 5–9 years, persistent wheeze occurred in 1.85% of children from households where neither parent smoked, 6.85% of children from households with one parent currently smoking and 11.8% of children from households with two parents currently smoking.[29] The trend was independent of parental wheezing and was associated with a reduction in lung function in these children.

Italian 9-year-old males had more than two-fold increase in AHR if one or other parent smoked.[117] European children exposed to maternal smoking in their first year of life showed increased airway responsiveness to exercise (odds ratio 2.82, 95% confidence interval (CI) 1.25–6.34), especially among those with asthma (odds ratio 20.6, 95% CI 2.9–168.9).[118] Exposure to environmental tobacco smoke (ETS) measured by urinary cotinine levels was associated with exacerbations of asthma in children and with decreased lung function.[119] Among children presenting to a Montreal emergency room, maternal heavy smoking was an independent risk factor (odds ratio 2.77, 95% CI 1.35–5.66).[120] Other home environmental factors that increased the risk for troublesome childhood asthma were the use of a humidifier in the child's room and electric heating in the home.

The weight of evidence suggests that home exposure to ETS is second only to atopy as a major risk factor for childhood asthma,[121–125] recently confirmed by a meta-analysis.[126] Mechanisms suggested include increased propensity for lower respiratory tract infection,[127] and an increased risk of developing atopy.[128] Newborn infants without parental atopy had increased cord blood IgE and a four-fold higher risk of developing atopic disease by age 18 months if the mother smoked.[128] The relationship between passive exposure to cigarette smoke and increased allergen sensitization in children remains somewhat uncertain.[129–131] Passive smoking worsens peak flow variability in children with mild to moderately increased airway responsiveness.[132] The Boston study identified

1   Epidemiology

maternal asthma and exposure *in utero* to maternal smoking as the most important risk factors for the development and persistence of asthma after age 2 years.[133]

Early respiratory infection

Lower respiratory tract infections in early childhood may have a lasting effect on respiratory symptoms and pulmonary function in childhood.[114,134–143] Children with bronchiolitis due to respiratory syncytial virus (RSV) had a greater exercise lability than control children with the same degree of atopy.[144] Wheezing and AHR were more common in Tyneside children after RSV bronchiolitis, although the prevalence of recognized and treated asthma, and of atopy, did not differ from that of a control group.[145] Similar findings have been reported after 'croup'.[146] Children with lower respiratory tract infections in infancy[147,148] subsequently had an increased prevalence of cough, wheeze, colds going to the chest, use of medication, absences from school and medical consultations compared with controls, increased airway responsiveness to exercise and reduced pulmonary function, but no difference in the prevalence of atopy. However, it is possible that these children had pre-existing AHR which predisposed them to more severe and symptomatic lower respiratory tract infections.

Low lung function

Martinez *et al*.[149] prospectively studied 124 newborn infants and found that the risk of a wheezing illness was 3.7 times higher among infants whose airway conductance was in the lowest third compared with those in the highest two-thirds. Girls whose lung volumes at end-expiration were in the lowest third had a 16-fold risk of a wheezing illness in the first year of life. Follow-up studies showed impaired airflow and lower lung volumes in those with repeated respiratory illness, further suggesting that diminished initial airway function may be a predisposing factor to recurrent wheezing illness.[150] Children with transient wheezing that had resolved by age 6 were more likely to have had initially low lung function, whereas those with persistent asthma were more likely to be atopic and low lung function was not a significant factor.

Diet

Among Australian children, eating fish more than once per week had a protective effect on development of asthma.[105] After adjusting for sex, ethnicity, country of birth, atopy, respiratory infection in the first 2 years of life and a parental history of asthma or smoking, children who ate fresh oily fish with > 2% fat had a reduced risk for current asthma (odds ratio 0.26, 95% CI 0.09–0.72), while no other food groups or non-oily fish or processed fish gave any protection.[151] The adoption of a 'Western' diet with less fish (and higher salt intake) has been hypothesized to account for the development of asthma and atopy in South-East Asian immigrants in Australia.[95] Demissie *et al*.[152] reported greater methacholine responsiveness among children with a higher salt intake, but no relationship with asthma or exercise responsiveness. Studies in the UK have shown inconsistent relationships between adult asthma, AHR and salt intake,[153,154] and the influence of sodium in the development of childhood asthma remains unclear.

Family size

The prevalence of atopy has been found in some studies[155] but not others[86] to decline with increasing number of siblings. If this is confirmed, decreasing family size may contribute to increased atopy in Western countries. Possibly local factors make family size an important risk factor in some areas but not in others.

Socioeconomic status

Anderson et al.[26] found a striking similarity in the prevalence of asthma in different social classes in Britain. Likewise, Mitchell et al.[156] found no relationship between socioeconomic status and asthma diagnosis, AHR or the combination. In the longitudinal study of New Zealand children, socioeconomic status was not a significant predictor for any characteristic of asthma.[157] However, while the prevalence of asthma may not be clearly associated with socioeconomic status, clinical manifestations, treatment and subsequent effects, e.g. school absenteeism and hospitalization, may differ among socioeconomic groups.[25,31,158]

Air pollution

While there is little doubt that episodes of air pollution increase morbidity from asthma, the evidence that outdoor air pollutants play a role in the development of asthma is less impressive. Among several key studies in recent years are those of von Mutius et al.,[159] who compared respiratory illness in the former East and West Germany. Children in the less polluted city of Munich had higher rates of wheezing (20% vs. 17%), diagnosed asthma (9.3% vs. 7.3%) and hayfever (8.6% vs. 2.4%) compared with children in Leipzig, who had more bronchitis (30.9% vs. 15.9% in Munich). Atopy was more common in West Germany (36.7% vs. 18.2%) ($P < 0.0001$). These findings suggested that the 'Western' lifestyle predisposed to atopy more than air pollution predisposed to asthma.[160] Airway responsiveness to cold air was higher in West Germany (8.3% vs. 5.5%) ($P < 0.0001$). However, among 9–11-year-old children in Leipzig, upper respiratory symptoms were related to SO_2 (odds ratio 1.72), NO_2 (odds ratio 1.53) and particulates (odds ratio 1.62) after controlling for passive smoke exposure, number of siblings, temperature and humidity.[161] Exposure to SO_2 and NO_2 as well as increased ozone decreased lung function, and areas with increased ozone may have a higher prevalence of asthma.[162]

The effects of air pollution are likely very small compared with the indoor environments. The ISAAC study found no consistent evidence for an effect of industrial air pollution in an area of New Zealand exposed to emissions from paper mills and sulphur fumes.[51]

CHILDHOOD ASTHMA AND AHR

Children with frequent symptoms of asthma almost invariably show AHR to histamine or methacholine,[12–15,163] and frequent variation in peak flow measurements. Siersted et al.[23] examined Danish children using three methods for detection of AHR; while agreement between diagnosed asthma and methacholine AHR was the strongest, there were additional cases of AHR detected by serial peak flow monitoring for 14 days. A substantial number of children show AHR to methacholine or histamine but have no current or past symptoms suggestive of asthma. This finding was consistent among studies in New Zealand,[13,42] Australia,[15] Canada,[33] the USA[164] and the UK[14] and was similar in degree whether the challenge was performed using histamine,[14,15,42] methacholine,[13,33] or hyperventilation with cold air.[164] On the other hand, some children with a history of current symptoms suggesting asthma (wheezing, cough, chest tightness on exercise) do not have demonstrable non-specific AHR.[13–15,33]

Among Australian children aged 8–11 years[15,99,100] 17.9% had hyperresponsive airways, 6.7% of the cohort had hyperresponsiveness without current or past respiratory symptoms, and 5.6% had been diagnosed with asthma yet did not show AHR. AHR was more prevalent than diagnosed asthma but less prevalent than wheezing or a report of any respiratory symptom. The association between symptoms and AHR was nevertheless highly significant ($P < 0.001$). Of children with a marked increase in airway responsiveness, all but one had diagnosed asthma. Factors associated with the development of AHR included a history of early respiratory illness, a history of asthma in either parent, and atopy.[99,100] If all three factors were present, the risk of moderate or severe AHR increased six-fold.

In New Zealand children, skin sensitivity to the house-dust mite was associated with a five-fold risk of developing AHR by age 13, while grass pollen sensitivity alone was not a significant risk factor either for hyperresponsiveness or for asthma symptoms.[102] Among Southampton children, AHR decreased with age from 29.1% at 7 years to 16.5% at 11 years, while prevalence of atopy increased from 26 to 31.6% over the same period.[165]

PROGNOSIS OF CHILDHOOD ASTHMA

Several factors may play a role in determining the prognosis of childhood asthma.[166] Martinez et al.[47] followed a birth cohort of infants to age 6 years and reported that 19.9% developed wheezing before age 3 that did not persist beyond 3 years, 13.7% developed wheezing before age 3 that did persist to age 6 years, while 15% began to wheeze between age 3 and 6 years. Factors predicting early (before age 3) but not later (in remission by age 6) wheeze were low lung function, maternal smoking but not maternal asthma, and absence of elevated IgE or skin test reactivity. Factors associated with a more persistent wheezing illness were maternal asthma and atopy, an elevated IgE, normal lung function at age 1 year and decreased flow rates at age 6 years. They concluded that the majority of infantile wheezing is transient, associated with decreased lung function at birth, but that atopy in the family history and in the child are important predictors of persistence.

The long-term outlook of childhood asthma has been reviewed recently.[167] Children in Australia and the UK have been followed from age 7 into their early thirties. There was considerable variability in the outcome, but overall some 25% of wheezing children reported wheeze as adults, some of whom had experienced significant periods of remission before again developing wheezing as an adult. Cigarette smoking by teenagers was associated with a poorer outcome.

Among predictors of persistent wheezing from childhood to adulthood are low lung function in childhood and persistent AHR.[167–170] Those with persistent asthma had lower lung function as adults, even after using a bronchodilator. In a 10-year follow-up of 85 children in a Copenhagen clinic, outcome in children with non-atopic asthma was predicted by the initial frequency of symptoms, initial FEV_1, active smoking and age at onset of symptoms, whereas the initial FEV_1 was the strongest predictor of outcome in those with atopic asthma.[171]

In a UK study of 235 subjects classified at age 10–15 as having asthma, wheeze with infections or no symptoms, predictors for adult wheezing at age 34–40 included original group, atopy and current smoking. Methacholine responsiveness was independently associated with original group, atopy and female sex. The difference in outcome for children with asthma versus those with wheeze only in the presence of infection could not be explained by atopy alone.[172]

In the national British cohort study, the cumulative prevalence of wheezing was 18% at age 7, 24% at age 16 and 43% by age 33. Of those with asthma or wheeze before age 7, 50% had experienced attacks in the previous year at age 7, 18% at age 11, 10% at age 16, 10% at age 23 and 27% at age 33.[173] Relapse at age 33 after prolonged remission of childhood wheezing was more common among current smokers and atopic subjects. The incidence of new asthma between ages 17 and 33 was associated strongly with active cigarette smoking, a history of hayfever and (less strongly) with female sex.

ASTHMA IN ADULTS

There are fewer studies of the prevalence and characteristics of asthma in adults. The difficulties of obtaining a random sample in adults are greater than in childhood, and asthma may be confused with symptoms due to airway obstruction caused by smoking-related diseases.[174,175] Dodge and Burrows[176] found very different prevalence figures for physician-diagnosed asthma and wheezing, even in younger age groups: in 20–24 year olds, 5% reported asthma, 12% had attacks of shortness of breath with wheeze, 25% had wheezed even without colds and 42% had experienced wheezing with colds. As chronic bronchitis and emphysema are rare below age 35, the majority of the symptomatic subjects are likely to have had asthma. In older subjects, however, many diagnosed as having asthma also carried the diagnosis of chronic bronchitis or emphysema. Differentiation may be difficult without detailed investigation and a prolonged test of reversibility. Asthma may be significantly underdiagnosed in adults, as well as overdiagnosed in those suffering from chronic bronchitis and emphysema.[177] Dyspnoea and poor lung function due to asthma may be incorrectly ascribed to chronic bronchitis or cardiac failure in the elderly.

1 Epidemiology

Studies of the prevalence of 'asthma' in adulthood conducted in the last two decades are summarized in Table 1.3. Problems of differing methodologies make comparisons between or even within countries difficult to interpret. More recent studies conducted through the European Community and elsewhere have used a rigorously tested and standardized questionnaire and ancillary diagnostic aids (European Community Respiratory Health Study, ECRHS).[188,189] Early results confirm substantial variations in asthma prevalence rates between countries.[190-199] Data selected from the preliminary report of results from 48 centres in 22 countries[200] are shown in Table 1.4. There is a degree of similarity in prevalence rates of wheezing in the last 12 months, most countries reporting between 20 and 27%, with more variability in nocturnal dyspnoea, attacks of asthma and current medications for asthma, the proportion receiving treatment varying from 70% (India) to under 3% (Estonia). India reported a very low prevalence rate for wheezing (4.1%) whereas the prevalence of treatment was close to that of other centres, suggesting that there may be a substantial prevalence of unrecognized or unreported wheezing in India. The very low treatment rate in Estonia likely reflects the economic difficulties of that country.

Further analyses of the ECRHS data, using information from detailed questionnaires, lung function and methacholine challenge, skin tests and smoking histories will provide

Table 1.3 Studies of prevalence of asthma and wheezing in adults.

Location	Age (range)	Prevalence	Year published	Reference
Busselton, Australia	18–88	5.9% current (symptoms + AHR)	1987	178
Busselton, Australia	18–55	9.0% (1981), 16.3% (1990) diagnosed asthma 26.4% (1981), 36.7% (1990) wheeze ever	1992	183
Victoria, Australia	18–93	22% current wheeze, 12.1% asthma	1992	182
Eastern Australia	18–50	19.1% wheeze, 7.1% diagnosed asthma	1996	185
Goroka, Papua New Guinea	Adult	2.5% wheeze and breathlessness	1985	186
Tucson, USA	>20	3.0–7.9% asthma, depending on age, sex >30% wheezing most ages	1980	176
Lebanon, USA	>7	5.6% asthma, 21.3% wheeze	1984	179
National, USA (NHANES)	25–74	2.6% asthma	1989	175
Saskatoon, Canada	20–29	2.7% current, 9.3% cumulative	1983	180
Finland	18–64	1.84% asthma	1988	181
Finland (army)	19	1.8% asthma	1990	62
Sweden (army)	17–20	2.8% asthma	1989	94
Singapore	20–74	4.7% men, 4.3% women	1994	184
South-west France	Over 65	6.1% (cumulative), 2.5% (current) asthma	1996	187

AHR, airway hyperresponsiveness.

Table 1.4 Self-reported prevalence rates for wheezing and asthma among adults aged 20–44 years, ECRHS data.

Location	Wheeze in last 12 months (%)	Woken by dyspnoea in last 12 months (%)	Attack of asthma in last 12 months (%)	Current medicines for asthma (%)
England (Cambridge)	25.2 (23.5–26.9)	8.4 (7.3–9.5)	5.7 (4.8–6.6)	6.8 (5.8–7.8)
USA (Portland)	25.7 (24.0–27.3)	7.7 (6.7–8.7)	5.8 (4.9–6.7)	4.8 (4.0–5.6)
France (Montpelier)	14.4 (13.2–15.5)	4.1 (3.5–4.8)	3.6 (3.0–4.2)	3.5 (2.9–4.1)
Netherlands (Groningen)	21.1 (19.7–22.6)	7.6 (6.7–8.5)	3.0 (2.4–3.6)	3.6 (2.9–4.2)
Germany (Hamburg)	21.1 (19.7–22.5)	5.0 (4.3–5.8)	3.0 (2.4–3.6)	3.4 (2.8–4.1)
Spain (Barcelona)	19.2 (17.7–20.7)	4.6 (3.8–5.4)	2.1 (1.5–2.6)	2.2 (1.7–2.8)
India (Bombay)	4.1 (3.1–5.2)	6.8 (5.5–8.2)	2.6 (1.7–3.5)	2.8 (1.8–3.8)
Australia (Melbourne)	28.8 (27.2–30.5)	11.4 (10.2–12.5)	9.7 (8.7–10.8)	9.3 (8.3–10.4)
New Zealand (Auckland)	25.2 (23.7–26.8)	9.9 (8.8–11.0)	6.8 (5.8–7.7)	8.5 (7.5–9.6)
Sweden (Uppsala)	19.2 (17.8–20.7)	4.9 (4.2–5.7)	3.3 (2.7–4.0)	5.0 (4.2–5.7)
Estonia (Tartu)	26.8 (25.0–28.6)	8.1 (7.0–9.3)	1.8 (1.3–2.4)	0.6 (0.3–0.9)

Reproduced from ref. 200, with permission.

additional insight into reasons for similarities and differences between different regions, and risk factors explaining high and low prevalence rates, such as differences in atopy and in smoking. The strength of the ECRHS study is the use of standardized methods for questionnaires and objective measurements, although different response rates[191,200] and problems in translation of some terms may make comparisons still somewhat imperfect. Preliminary reports show a relationship between IgE and methacholine responsiveness (greater responsiveness in those with higher IgE) and differences in these relationships between smokers and non-smokers.[198]

The incidence of asthma (new cases of the disease) in adult life is seldom measured. In the follow-up of 14-year-old Melbourne children, 15 of 82 'control' children had developed asthma symptoms by age 21, an incidence of 18.3% in 7 years or 2.6% per annum,[201] a much higher incidence than reported in the USA. The incidence in Connecticut was 1.4% in 6 years or 0.2% per annum,[179] similar to the incidence of 1% in 4 years or 0.25% per annum found in the Tecumseh study.[202] The National Health and Nutrition Survey follow-up study detected an incidence of 0.21% per annum; incidence was significantly higher in females than males.[175]

AHR IN ADULTS

Measurements of airway responsiveness in adults are no more specific or sensitive than they are in children for confirming the diagnosis of asthma.[16] The difference in methacholine responsiveness between Italian adults with obvious histories of asthma and those with no history was not well defined, although the difference in mean airway responsiveness between the two groups was highly significant.[203] Among adults in England, AHR was associated with age, smoking (the effect of which increased with age) and atopy (the effect of which decreased with age).[204] AHR was least obvious in 35–44 year olds and increased in older and younger subjects. In Norwegian adults, independent predictors for AHR included male sex, younger age, smoking, level of airway calibre (FEV_1) and rural residence.[205] While clear correlations exist between the presence of non-specific AHR and chronic respiratory symptoms in adults,[206,207] the presence of hyperresponsiveness is not specific for asthma.

RISK FACTORS FOR ADULT ASTHMA

Curiously, in some developing countries the prevalence of asthma in adults is much higher than the prevalence in children; in such countries the age of onset of asthma is often after age 20. Among the South Fore people of Papua New Guinea, the point prevalence of asthma in children in the 1970s was nil and the adult prevalence only 0.28%.[208] A decade later, the prevalence was 0.6% in children but a striking 7.3% in adults.[209] Severe asthma, provoked by previously benign factors (exercise, stress, infection), now occurs in these people, who also have a high level of allergy to the house-dust mite[210] and are exposed to a four-fold greater mite density in blanket dust compared with that found in a similar village which has not experienced the dramatic increase in asthma prevalence.[211]

The number of recognized occupational causes of asthma is ever increasing.[212–214] The prevalence of occupational asthma depends not only on the nature of the inciting agent but also on the circumstances of exposure. In some industries, over 30% of workers develop asthma, e.g. animal handlers and workers with proteolytic enzymes, whereas lower rates occur with other agents, e.g. 5% with isocyanates and 4% with western red cedar. Occupational asthma may account for a significant part of the total burden of asthma in a population: some 15% of all cases of asthma in Japan are attributed to occupational causes.[215] 'Occupational' asthma may affect persons remote from the work site; a series of epidemics of asthma in Barcelona, resulting in considerable morbidity, was traced to exposure to soybean dust released during unloading of ships in the harbour.[216,217]

Dietary factors as a reason for geographic variations in adult asthma have been studied over recent years. Burney[154] suggested a relationship with dietary sodium, based on sales of table salt and asthma mortality. Subsequent clinical trials reported increased airway responsiveness to methacholine, physiological deterioration and increased asthma morbidity following increased dietary sodium intake.[153] Differences in peak flow rates between sodium- and placebo-treated groups were 5.6% (95% CI 2.2–9.8) and 7.8% (95% CI 3.9–12.9) for morning and evening PEF respectively. However, other authors found no relationship between hyperresponsiveness to methacholine and 24-h urinary sodium[218,219] and no relationship between sodium excretion and self-reported wheeze, hayfever, eczema or asthma.[218] In a prospective study of over 77 000 women, vitamin E appeared protective against asthma, but only when part of the diet and not when given as a supplement, suggesting other factors may be responsible for this finding.[220]

Epidemics of severe asthma and increased hospital admissions have occurred during or following weather changes. Outbreaks of asthma were associated with thunderstorms in Australia[221] and in England.[222] Asthma admissions in Birmingham correlated with daily spore counts measured 60 km away in Derby.[223] Some fungi depend on weather conditions for spore release (e.g. *Didymella*); hence fungal aeroallergens may cause epidemics of clinical asthma, similar to those seen in countries that experience ragweed seasons.[224] The effects of weather changes on asthma are probably largely related to the indirect effects on local allergens, e.g. fungal spores, pollens or house-dust mite populations, rather than to direct cold or irritant effects on the airways. However, Bermuda has little air pollution and pollen counts are low all year, yet asthma attendances increased when there was lower humidity, cooler air temperature and winds from the north-east ocean with no appreciable aeroallergens, suggesting a direct effect of climatic factors on the expression of asthma.[225]

EVIDENCE FOR INCREASING SEVERITY OF ASTHMA

Data relating to hospital admissions for asthma[65-73] and to the use of antiasthma drugs[226-228] suggest that either the prevalence or severity of asthma, or both, have increased. The increase in admissions is not due to earlier presentation of patients with milder asthma[229] nor is there evidence for change in prevalence suffficient to account for the increase in hospitalization. An increase in the prevalence of severe asthma is therefore more likely.

The marked increase in use of bronchodilator drugs and inhaled corticosteroids in New Zealand, Australia and the UK between 1975 and 1981[226] could reflect an increased severity of asthma or improved treatment of asthma. A community survey of antiasthma drug use in New Zealand in 1984 found 80% of prescriptions for salbutamol were for patients with asthma rather than chronic bronchitis and/or emphysema.[227] The same study found that inhaled corticosteroid therapy was still underused; only 42% of subjects with daily symptoms of asthma were prescribed beclomethasone. Hence the increase in sales of antiasthma drugs was not necessarily the result of improved treatment of asthma. Given that the total prevalence of asthma is not greatly changed, a considerable part of the increase in drug use must relate to an increase in the prevalence of more severe disease. In five Nordic countries, asthma drug sales have approximately doubled over 10 years. Even more importantly, the number of drugs used per patient has also increased significantly.[228]

The changing epidemiology of asthma may be related in part to an increase in severity of asthma, possibly due to environmental pollutants, house-dust mite exposure or occupational exposures. There is also evidence to relate this change in severity to use of pharmacological agents. Short-acting β-agonists are highly effective in relieving symptoms of asthma, but in the long term appear to maintain or increase the activity of the disease they treat, resulting in increased morbidity and increased mortality.[230,231] In a randomized, controlled, year-long crossover study of 64 subjects, regular use of an inhaled β-agonist was associated with deterioration in control of asthma,[230] reduced lung function, increased AHR and a shorter time to first exacerbations.[231]

Although this study of regular fenoterol was the first designed to look specifically for this long-term adverse effect of β-agonists, other prior studies involving shorter periods of regular bronchodilator use contain evidence suggesting that other β-agonists share this adverse effect, perhaps to a lesser degree.[232-235] Several studies have suggested that

1 Epidemiology

regular β-agonist in standard doses can increase airway responsiveness to methacholine[234,236–238] and exercise[239] and both the early[240] and late[241] allergic asthmatic response. Even among subjects with very mild asthma not using inhaled corticosteroids, regular use of salbutamol was associated with trends towards increased symptoms and worsened lung function including methacholine airway responsiveness.[242] Debate continues as to the magnitude of effect, and clinical significance, of these findings.

MORTALITY FROM ASTHMA

Mortality rates in asthma are generally low. However, on two occasions in the last three decades substantial increases in reported asthma mortality have occurred.[243–249] Between 1964 and 1966, asthma mortality in England and Wales, Australia and New Zealand rose markedly, especially in young people. This increase was variously attributed to a direct toxic effect of high-dose sympathomimetic bronchodilator drugs,[247] delay in obtaining more effective treatment due to over-reliance on symptomatic relief from bronchodilator therapy,[250] increased exposure to aeroallergens,[251] or diagnostic transfer.[244] The second 'epidemic' in New Zealand from 1977 took mortality in young people to a peak of over 4.0 per 100 000 in 5–34 year olds in 1979.[252]

Increases in asthma mortality in other countries during the 1980s have been gradual and more difficult to quantify,[253–255] but appear to be real (Table 1.5). A significant increase in mortality occurred in young people in the UK between 1974 and 1985.[256] In Canada, mortality in 5–34 year olds more than doubled from 0.2 per 100 000 in 1974 to 0.5 per 100 000 in 1984.[257] In the USA, asthma mortality rates also doubled from 0.15 to 0.36 per 100 000 in this age group and appear still to be increasing.[255] The increase there, as in New Zealand, has been more obvious in non-Caucasians than in Caucasians and in younger rather than older persons.[258,259] In the last 5 years, there appears to have been a plateau, with a decline in asthma mortality rates in young people in many countries, although mortality is still increasing in the USA and Japan.

The accuracy of certification of death due to asthma has been studied in England[260] and New Zealand,[261] with similar results. More false negatives and fewer false positives were identified in England compared with New Zealand, although both countries had a net overestimate of asthma mortality of 13% when age groups were matched,[262] indicating that the higher New Zealand mortality rate was not due to reduced accuracy of certification. In a regional study within New Zealand, a very low rate of false-negative reporting of asthma deaths was found in young people, again verifying 5–34 year olds as a suitable group for longitudinal study of trends.[246] In the USA, a significant degree of under-reporting of asthma deaths has been identified, making it likely that the national statistics underestimate true mortality rates.[263]

The introduction in 1979 of the ninth revision of the WHO International Classification of Diseases (ICD) confounded analysis of trends in mortality. Under the ICD9 rules, deaths certified as due to asthma but with mention of bronchitis, previously coded under ICD8 to bronchitis, were now coded to asthma.[264] In older age groups, this caused an apparent increase in asthma mortality of 35% or more.[265] However, in 5–34 year olds the effect of the introduction of ICD9 was negligible, and trends through 1979 in this group can be accepted virtually without adjustment.[253] Furthermore, although the 1979 change

Table 1.5 Annual mortality rates from asthma in persons aged 5–34 years in 13 countries, 1975–94 (rate per 100 000 population).

	1975	1976	1977	1978	1979	1980	1981	1982	1983	1984	1985	1986	1987	1988	1989	1990	1991	1992	1993	1994
Australia	0.78	0.81	0.98	0.93	0.87	0.94	1.10	1.03	0.95	1.28	1.37	1.50	1.40	1.16	1.25	1.13	0.94	0.54	0.86	0.78
Canada	0.31	0.36	0.33	0.28	0.31	0.44	0.46	0.45	0.50	0.50	0.42	0.51	0.49	0.45	0.40	0.37	0.34	0.33	0.25	0.21
England/Wales	0.59	0.52	0.61	0.68	0.68	0.68	0.91	0.82	0.90	0.81	0.88	0.83	0.99	0.90	0.79	0.74	0.72	0.52	0.51	0.51
France	0.26	0.26	0.20	0.28	0.19	0.27	0.29	0.31	0.26	0.37	0.53	0.50	0.50	0.45	0.47	0.37	0.55	0.45	0.40	0.38
Italy	0.03	0.06	0.06	0.02	0.05	0.12	0.16	0.15	0.11	0.11	0.15	0.21	0.15	0.21	0.18	0.16	0.16	0.19	0.23	0.14
Japan	0.44	0.46	0.42	0.33	0.33	0.36	0.40	0.39	0.47	0.39	0.53	0.61	0.62	0.63	0.62	0.69	0.71	0.61	0.73	0.69
Netherlands	0.26	0.09	0.26	0.18	0.21	0.31	0.23	0.36	0.35	0.20	0.26	0.18	0.21	0.25	0.11	0.11	0.20	0.20	0.14	0.11
New Zealand	1.42	2.02	3.64	2.96	4.12	3.46	3.27	3.14	2.28	2.57	1.85	2.87	1.94	2.55	1.65	0.80	0.74	0.87	0.50	0.74
South Africa	0.75	0.79	0.71	0.99	1.04	0.89	1.05	1.19	1.09	0.99	0.98	0.74	0.99	0.88	0.69	0.55	0.63	0.78	0.86	
Spain	0.11	0.14	0.13	0.12	0.09	0.16	0.12	0.12	0.15	0.18	0.19	0.19	0.28	0.23	0.30	0.21	0.17	0.20	0.23	
Sweden	0.45	0.37	0.28	0.45	0.71	0.57	0.69	0.44	0.56	0.39	0.48	0.60	0.54	0.42	0.27	0.63	0.39	0.20	0.12	0.12
USA	0.22	0.19	0.17	0.21	0.22	0.26	0.29	0.35	0.36	0.32	0.38	0.39	0.42	0.42	0.43	0.43	0.49	0.42	0.47	
West Germany*	0.49	0.66	0.63	0.70	0.78	0.85	0.83	0.86	0.84	0.80	0.86	0.73	0.73	0.56	0.59	0.55	0.56	0.45	0.34	0.41

* For consistency, all figures to 1994 are for West Germany.

in ICD produced a significant step-increase in reported asthma mortality in older age groups, this does not explain the continued upward trend in mortality rates in following years in many countries. An increase in total prevalence, an increase in prevalence of severe disease, a reduction in efficacy of treatment or an adverse effect of treatment must be proposed as the cause, or causes, of the increase in asthma mortality seen in many countries. Changes in diagnostic fashion, accuracy of certification or coding rules cannot explain the significant changes in mortality, especially in young people.

RISK FACTORS FOR ASTHMA MORTALITY

Identified risk factors for death from asthma include young and old age, ethnicity, psychosocial disturbances, a previous history of severe or life-threatening attacks, previous hospital admissions and emergency room visits, and discontinuity of physician care.[266–268] A recent comparison of asthma deaths and near-fatal asthma in South Australia again highlighted these risk factors, but found that near-fatal cases were younger, more likely to be male, less likely to have other medical conditions and less likely to experience delay in receiving medical care compared with fatal cases.[269] Asthma mortality rates for Maori and Pacific Island Polynesians in New Zealand in 1981–83 were respectively 5.6 and 2.8 times higher than the rate for Europeans.[252] Similar higher mortality rates are evident among black compared with white children in the USA.[258,259] Differing attitudes to medical care, and cultural and economic barriers to good-quality medical care, may explain part of this difference. In New Zealand, asthma mortality in low socioeconomic areas was twice than in high socioeconomic areas.[270]

The dominant pattern in English,[271] American,[272] Australian,[273] Canadian,[276] and New Zealand,[252,267,274,275] descriptive studies of asthma mortality was of a fatal outcome associated with inadequate assessment and inappropriate treatment of severe asthma, with over-reliance on high doses of bronchodilator therapy and insufficient use of corticosteroid therapy. However, these studies did not determine why those at risk had developed such severe asthma.

Half of the deaths in children under 15 years old identified in the New Zealand national study were associated with disturbed home and family relationships, resulting in less attention being given to the child's asthma.[277] Similar findings have been reported in a case-control study of childhood asthma deaths in the USA.[278] Medical management risk factors included the prescription of several different classes of drugs for asthma and inadequacy of prescribed drug therapy.[267]

Ulrik and Fredericksen[279] followed a cohort of Copenhagen asthmatic patients and examined risk factors for mortality. Significantly increased risks were related to cigarette smoking, age, presence of blood eosinophilia (hence atopic status), degree of impairment of lung function and degree of reversibility to β-agonist. Somewhat contrary to expectations, the higher relative risks for mortality were associated with moderate rather than severe airflow obstruction and greater rather than lesser degrees of reversibility. Hence the atopic highly reversible asthmatic patient is at higher risk than the asthmatic patient with more fixed airflow obstruction.

Three case-control studies in New Zealand associated prescription of a specific β_2-agonist, fenoterol, with a greater risk of mortality.[280–282] These studies were somewhat difficult to interpret because of the use of different information sources for drug exposure,[280] possible control selection bias[280,281] and difficulties in selecting appropriate markers for 'severe' asthma,[280–282] although each succeeding study used an improved design and more relevant controls to reduce these concerns. To clarify risks associated with drug use, a large nested case-control study was conducted using computerized prescription records in Saskatchewan, Canada.[283] Cases of fatal or near-fatal asthma ($n = 129$) were each compared with two to eight controls ($n = 655$) also receiving medication but not having a fatal or near-fatal attack. Use of fenoterol was associated with a significantly higher risk of death (odds ratio 5.4 per canister per month) than salbutamol (odds ratio 2.4 per canister per month); however, when adjustment was made for the two-fold higher dose in the fenoterol canister, the odds ratios were not different (2.3 vs. 2.4). Use of more than two metered-dose inhaler canisters per month increased the risk of mortality over 10-fold.[284,285] Whether these associations are explained by use of higher doses or more potent agents in more severe asthma (confounding by indication) or reflect the direct deleterious effect of β-agonist on asthma severity cannot be answered directly from these studies, although randomized clinical trials have pointed towards the latter explanation. Some authors have reported channelling of more potent β-agonists to those with more severe asthma,[286–288] but others have not found evidence of this.[289]

Given that β-agonist treatments may increase the severity of asthma,[230,231,240,241] it seems plausible that the epidemics of asthma mortality that occurred in the 1960s and in New Zealand in the 1980s reflect an overall increase in severity of asthma due to chronic use of high-dose inhaled bronchodilator.[249] This increased the proportion of asthmatic patients who were at risk of death from asthma when circumstances were less than ideal during an acute attack. The association between a specific potent β-agonist and mortality probably has two components: the selective prescription of a higher dose β_2-agonist for more severe asthma,[286–288] and the subsequent effect of that β_2-agonist in further increasing the severity of asthma.[230,231]

Use of anti-inflammatory medication was associated with a reduced risk of asthma mortality in the Saskatchewan case-control study.[290] Subjects using more than one canister per month of low-dose inhaled corticosteroid had an odds ratio for death of 0.1 (95% CI 0.02–0.6). As it is likely that inhaled corticosteroids were given to those with more severe asthma, this argues against the association between β-agonist use and mortality being simply due to confounding by severity, and for a causal association.

That overuse of potent β-agonists might be a factor in both increased morbidity and mortality from asthma is strongly supported by trends in New Zealand over the 5 years since fenoterol was suggested to be a risk factor for asthma mortality. Effective withdrawal of fenoterol was followed within the year by an abrupt fall not only in asthma mortality but also in morbidity as reflected by hospital admissions for asthma (Fig. 1.1). Mortality has continued to decline to levels below those of 30 years ago (before the 'first' epidemic associated with use of high-dose isoprenaline), while hospital admissions likewise have remained much lower than in the 1980s.[291]

The gradual reduction in mortality in many countries during the early 1990s is likely due to several factors, including increased recognition of problems associated with severe asthma,[266–269] increased use of inhaled corticosteroids,[290] reduced reliance on high doses of potent β-agonists[230,249,291] and increased education of professionals and patients alike.[292] Given that there is no evidence of a reduction in asthma prevalence, rather the converse, and that there are many factors which may increase asthma severity, there is no

1 Epidemiology

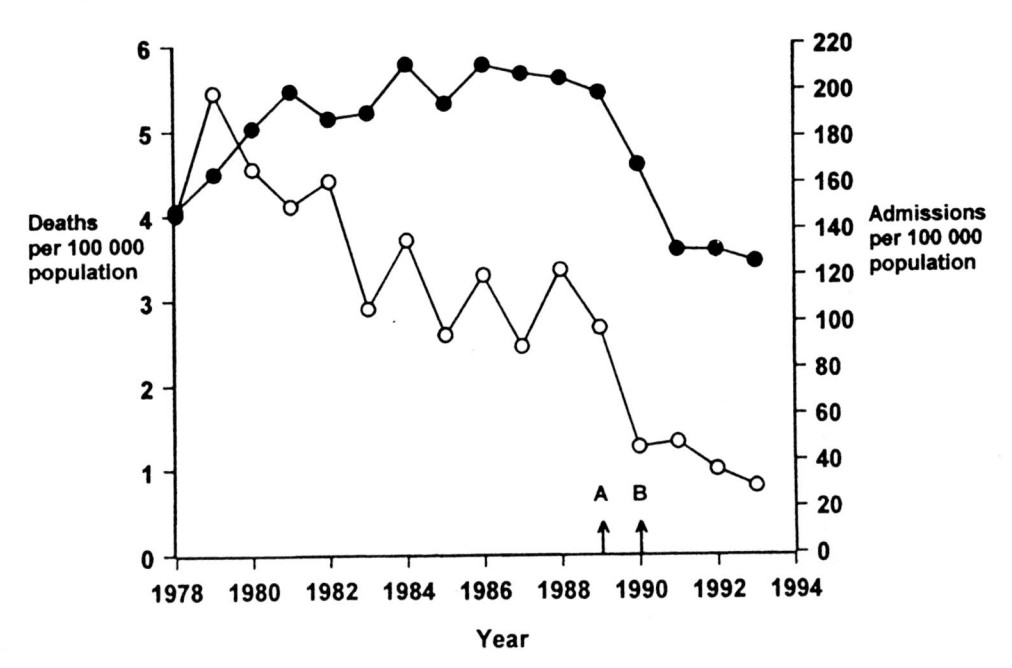

Fig. 1.1 Abrupt decrease in asthma-related hospital admissions and mortality in New Zealanders aged 15–44 years following withdrawal of fenoterol in 1990. ○, Deaths per 100 000 population; ●, admissions per 100 000 population. A, case-control study published; B, fenoterol regulated. Reproduced from ref. 291, with permission.

room for complacency despite these downward mortality trends. It is especially worrying that there is as yet no decline in the upward trends in mortality rates in the USA and Japan, where use of β-agonists remains markedly greater than use of inhaled corticosteroids.

REFERENCES

1. Burney P: Why study the epidemiology of asthma? *Thorax* (1988) **43**: 425–428.
2. Burney P: Epidemiology of asthma. *Allergy* (1993) **48**: 17–21.
3. Scadding JG: Definition and clinical categories of asthma. In Clark TJH, Godfrey S (eds) *Asthma*. London, Chapman & Hall, 1983, p. 5.
4. Godfrey S: What is asthma? *Arch Dis Child* (1985) **60**: 997–1000.
5. International Consensus Report on Diagnosis and Management of Asthma. National Heart, Lung, and Blood Institute, National Institutes of Health, publication no. 92–3091, 1992, p. 1.
6. Global Strategy for Asthma Management and Prevention. NHLBI/WHO Workshop Report March 1993. National Heart, Lung and Blood Institute, National Institutes of Health, publication no. 95–3659, 1995, p. 6.
7. Venables KM, Burge PS, Davison AG, Newman-Taylor AJ: Peak flow rate records in surveys: reproducibility of observers' reports. *Thorax* (1984) **39**: 828–832.
8. Konig P: Hidden asthma in childhood. *Am J Dis Child* (1981) **135**: 1053–1055.
9. Clifford RD, Radford M, Howell JB, Holgate ST: Prevalence of respiratory symptoms among 7- and 11-year-old schoolchildren and association with asthma. *Arch Dis Child* (1989) **64**: 1118–1125.

10. Peat JK, Salome CM, Toelle BG, Bauman A, Woolcock AJ: Reliability of a respiratory history questionnaire and effect of mode of administration on classification of asthma in children. *Chest* (1992) **102**: 153–157.
11. Jenkins MA, Clarke JR, Carlin JB, et al.: Validation of questionnaire and bronchial hyperresponsiveness against respiratory physician assessment in the diagnosis of asthma. *Int J Epidemiol* (1996) **25**: 609–616.
12. Hargreave FE, Ryan G, Thomson NC, et al.: Bronchial responsiveness to histamine or methacholine in asthma: measurement and clinical significance. *J Allergy Clin Immunol* (1981) **68**: 347–355.
13. Sears MR, Jones DT, Holdaway MD, et al.: The prevalence of bronchial reactivity to inhaled methacholine in New Zealand children. *Thorax* (1986) **41**: 283–289.
14. Lee DA, Winslow NR, Speight ANP, Hey EN: Prevalence and spectrum of asthma in childhood. *Br Med J* (1983) **286**: 1256–1258.
15. Salome CM, Peat JK, Britton WJ, Woolcock AJ: Bronchial hyperresponsiveness in two populations of Australian schoolchildren. I. Relation to respiratory symptoms and diagnosed asthma. *Clin Allergy* (1987) **17**: 271–281.
16. Enarson DA, Vedal S, Schulzer M, Dybuncio A, Chan-Yeung M: Asthma, asthmalike symptoms, chronic bronchitis, and the degree of bronchial hyperresponsiveness in epidemiological surveys. *Am Rev Respir Dis* (1987) **136**: 613–617.
17. Pattemore PK, Asher MI, Harrison AC, et al.: The interrelationship among bronchial hyperresponsiveness, the diagnosis of asthma, and asthma symptoms. *Am Rev Respir Dis* (1990) **142**: 549–554.
18. Barnes PJ: Pathophysiology of asthma. *Br J Clin Pharmacol* (1996) **42**: 3–10.
19. Pizzichini E, Pizzichini MMM, Efthimiadis A, et al.: Indices of airway inflammation in induced sputum: reproducibility and validity of cell and fluid-phase measurements. *Am J Respir Crit Care Med* (1996) **154**: 308–317.
20. Editorial: Airflow limitation—reversible or irreversible? *Lancet* (1988) **i**: 26–27.
21. Brown PJ, Greville HW, Finucane KE: Asthma and irreversible airflow obstruction. *Thorax* (1984) **39**: 131–136.
22. Higgins BG, Britton JR, Chinn S, Cooper S, Burney PGJ, Tattersfield AE: Comparison of bronchial reactivity and peak expiratory flow variability measurements for epidemiologic studies. *Am Rev Respir Dis* (1992) **145**: 588–593.
23. Siersted HC, Mostgaard G, Hyldebrandt N, Hansen HS, Boldsen J, Oxhoj H: Interrelationships between diagnosed asthma, asthma-like symptoms, and abnormal airway behaviour in adolescence: the Odense Schoolchild Study. *Thorax* (1996) **51**: 503–509.
24. Toelle BG, Peat JK, Salome CM, Mellis CM, Woolcock AJ: Toward a definition of asthma for epidemiology. *Am Rev Respir Dis* (1992) **146**: 633–637.
25. Anderson HR, Bailey PA, Cooper JS, Palmer JC, West S: Morbidity and school absence caused by asthma and wheezing illness. *Arch Dis Child* (1983) **58**: 777–784.
26. Anderson HR, Bland JM, Peckham CS: Risk factors for asthma up to 16 years of age: evidence from a national study. *Chest* (1987) **91**: 127S-130S.
27. Begishvili B, Doull IJM, Freezer NJ, et al.: Relation between serum IgE, bronchial hyperresponsiveness and asthma in randomly selected children. *Eur Respir J* (1995) **8** (Suppl 19): 314s.
28. Strachan DP, Anderson HR, Limb ES, O'Neill A, Wells N: A national survey of asthma prevalence, severity, and treatment in Great Britain. *Arch Dis Child* (1994) **70**: 174–178.
29. Weiss ST, Tager IB, Speizer FE, Rosner B: Persistent wheeze. Its relation to respiratory illness, cigarette smoking, and level of pulmonary function in a population sample of children. *Am Rev Respir Dis* (1980) **122**: 697–707.
30. Gergen PJ, Mullally DI, Evans R: National survey of prevalence of asthma among children in the United States, 1976 to 1980. *Pediatrics* (1988) **81**: 1–7.
31. Mak H, Johnston P, Abbey H, Talamo RC: Prevalence of asthma and health service utilization of asthmatic children in an inner city. *J Allergy Clin Immunol* (1982) **70**: 367–372.
32. Kerigan AT, Goldsmith CH, Pengelly LD: A three-year cohort study of the role of environmental factors in respiratory health of children in Hamilton, Ontario. *Am Rev Respir Dis* (1986) **133**: 987–993.

33. FitzGerald JM, Fester DE, Morris MM, Schulzer M, Hargreave FE, Sears MR: Relation of airway responsiveness to methacholine to parent and child reporting of symptoms suggesting asthma. *Can Respir J* (1996) **3**: 115–123.
34. Peat JK, Woolcock AJ, Leeder SR, Blackburn CRB: Asthma and bronchitis in Sydney schoolchildren. I. Prevalence during a six-year study. *Am J Epidemiol* (1980) **111**: 721–727.
35. Mitchell C, Miles J: Lower respiratory tract symptoms in Queensland schoolchildren. The questionnaire: its reliability and validity. *Aust NZ J Med* (1983) **13**: 264–269.
36. Crockett AJ, Ruffin RE, Schembri DA, Alpers JH: The prevalence rate of respiratory symptoms in schoolchildren from two South Australian rural communities. *Aust NZ J Med* (1986) **16**: 653–657.
37. Bauman A, Mitchell CA, Henry RL, et al.: Asthma morbidity in Australia: an epidemiological study. *Med J Aust* (1992) **156**: 827–831.
38. Forero R, Bauman A, Young L, Larkin P: Asthma prevalence and management in Australian adolescents: results from three community surveys. *J Adolescent Health* (1992) **13**: 707–712.
39. Robertson CF, Bishop J, Sennhauser FH, Mallol J: International comparison of asthma prevalence in children: Australia, Switzerland, Chile. *Pediatr Pulmonol* (1993) **16**: 219–226.
40. Mitchell EA: Increasing prevalence of asthma in children. *NZ Med J* (1983) **96**: 463–464.
41. Jones DT, Sears MR, Holdaway MD, et al.: Asthma in New Zealand children. *Br J Dis Chest* (1987) **81**: 332–340.
42. Asher MI, Pattemore PK, Harrison AC, et al.: International comparison of the prevalence of asthma symptoms and bronchial hyperresponsiveness. *Am Rev Respir Dis* (1988) **138**: 524–529.
43. Skarpaas IJK, Gulsvik A: Prevalence of bronchial asthma and respiratory symptoms in schoolchildren in Oslo. *Allergy* (1984) **40**: 295–299.
44. Holmgren D, Aberg N, Lindberg U, Engstrom I: Childhood asthma in a rural community. *Allergy* (1989) **44**: 256–259.
45. Braback L, Kalvesten L, Sundstrom G: Prevalence of bronchial asthma among schoolchildren in a Swedish district. *Acta Paediatr Scand* (1988) **77**: 821–825.
46. Soto-Quiros M, Bustamante M, Gutierrez I, Hanson LA, Strannegard IL, Karlberg J: The prevalence of childhood asthma in Costa Rica. *Clin Exp Allergy* (1994) **24**: 1130–1136.
47. Martinez FD, Wright AL, Taussig LM, Holberg CJ, Halonen M, Morgan WJ: Asthma and wheezing in the first six years of life. *N Engl J Med* (1995) **332**: 133–138.
48. Pearce N, Weiland S, Keil U, et al.: Self-reported prevalence of asthma symptoms in children in Australia, England, Germany and New Zealand: an international comparison using the ISAAC protocol. *Eur Respir J* (1993) **6**: 1455–1461.
49. Asher MI, Keil U, Anderson HR, et al.: International study of asthma and allergies in childhood (ISAAC): rationale and methods. *Eur Respir J* (1995) **8**: 483–491.
50. Pizzichini MMM, Taylor B, Habbick B, Senthilselvan A, Rennie D, Sears MR: Prevalence of childhood asthma in two age groups in two regions of Canada: the ISAAC study. *Am J Respir Crit Care Med* (1997) **151**: A71.
51. Moyes CD, Waldon J, Ramadas D, Crane J, Pearce N: Respiratory symptoms and environmental factors in schoolchildren in the Bay of Plenty. *NZ Med J* (1995) **108**: 358–361.
52. Goh DYT, Chew FT, Quek SC, Lee BW: Prevalence and severity of asthma, rhinitis, and eczema in Singapore schoolchildren. *Arch Dis Child* (1996) **74**: 131–135.
53. Lai CKW, Douglass C, Ho SS, Chan J, Lau J, Wong G: Asthma epidemiology in the Far East. *Clin Exp Allergy* (1996) **26**: 5–12.
54. Clarke JR, Jenkins MA, Robertson CF, et al.: Validation of asthma questionnaires with respiratory physician assessment. *Am J Respir Crit Care Med* (1995) **151**: A568.
55. Burney PGJ, Chinn S, Rona RJ: Has the prevalence of asthma changed? Evidence from the national study of health and growth 1973–86. *Br Med J* (1990) **300**: 1306–1310.
56. Anderson HR: Is the prevalence of asthma changing? *Arch Dis Child* (1989) **64**: 172–175.
57. Hill R, Williams J, Tattersfield A, Britton J: Change in use of asthma as a diagnostic label for wheezing illness in schoolchildren. *Br Med J* (1989) **299**: 898.
58. Williams H, McNicol KN: Prevalence, natural history and relationship of wheezy bronchitis and asthma in children: an epidemiological study. *Br Med J* (1969) **4**: 321–325.

59. Anyon CP, Kiddle GB: The prevalence of wheezy children in Lower Hutt. *NZ Med J* (1974) **79**: 822–823.
60. Pizzichini MMM, Faulkner T, Tedesco R, Faulman G, Sears MR: Prevalence, severity and diagnosis of asthma in 6–7 year old children. *Eur Respir J* (1995) **8** (Suppl 19): 283s.
61. Rona RJ, Chinn S, Burney PGJ: Trends in the prevalence of asthma in Scottish and English primary school children 1982–92. *Thorax* (1995) **50**: 992–993.
62. Ninan TK, Russell G: Respiratory symptoms and atopy in Aberdeen schoolchildren: evidence from two surveys 25 years apart. *Br Med J* (1992) **304**: 873–875.
63. Haahtela T, Lindholm H, Bjorksten F, Koskenvuo K, Laitinen LA: Prevalence of asthma in Finnish young men. *Br Med J* (1990) **301**: 266–301.
64. Bauman A: Has the prevalence of asthma symptoms increased in Australian children? *J Paediatr Child Health* (1993) **29**: 424–428.
65. Leung R, Ho SS, Chan J, Wong G, Lau J, Lai CKW: Increasing trend of asthma and allergic disease in Hong Kong schoolchildren. *J Allergy Clin Immunol* (1996) **97**: 375.
66. Mitchell EA: International trends in hospital admissions rates for asthma. *Arch Dis Child* (1985) **60**: 376–378.
67. Anderson HR: Increase in hospital admissions for childhood asthma: trends in referral, severity, and readmissions from 1970 to 1985 in a health region of the United Kingdom. *Thorax* (1989) **44**: 614–619.
68. Richards W: Hospitalization of children with status asthmaticus: a review. *Pediatrics* (1989) **84**: 111–118.
69. Halfon N, Newacheck PW: Trends in the hospitalization for acute childhood asthma, 1970–84. *Am J Public Health* (1986) **76**: 1308–1311.
70. Anderson HR, Bailey P, West S: Trends in the hospital care of acute childhood asthma 1970–8: a regional study. *Br Med J* (1980) **281**: 1191–1194.
71. Mullally DI, Howard WA, Hubbard TJ, Grauman JS, Cohen SG: Increased hospitalizations for asthma among children in the Washington, DC area during 1961–1981. *Ann Allergy* (1984) **53**: 15–19.
72. Mitchell EA, Cutler DR: Paediatric admissions to Auckland Hospital for asthma from 1970 to 1980. *NZ Med J* (1984) **97**: 67–70.
73. Dawson KP: The severity of asthma in children admitted to hospital: a 20-year review. *NZ Med J* (1987) **100**: 520–521.
74. Storr J, Barrell E, Lenney W: Rising asthma admissions and self referral. *Arch Dis Child* (1988) **63**: 774–779.
75. Kun HY, Oates RK, Mellis CM: Hospital admissions and attendances for asthma: a true increase? *Med J Aust* (1993) **159**: 312–313.
76. Wennergen G, Kristjansson S, Strannegard I-L: Decrease in hospitalization for treatment of childhood asthma with increased use of antiinflammatory treatment, despite an increase in the prevalence of asthma. *J Allergy Clin Immunol* (1996) **97**: 742–748.
77. Burney PGJ: Epidemiology. *Br Med Bull* (1992) **48**: 10–22.
78. Newman-Taylor A: Environmental determinants of asthma. *Lancet* (1995) **345**: 296–299.
79. Milne GA: The incidence of asthma in Lower Hutt. *NZ Med J* (1969) **70**: 27–29.
80. Dawson B, Horobin G, Illsley R, Mitchell R: A survey of childhood asthma in Aberdeen. *Lancet* (1969) **i**: 827–830.
81. Verity CM, VanHeule B, Carswell F, Hughes AO: Bronchial lability and skin reactivity in siblings of asthmatic children. *Arch Dis Child* (1984) **59**: 871–876.
82. Astarita C, Harris RI, de Fusco R, et al.: An epidemiological study of atopy in children. *Clin Allergy* (1988) **18**: 341–350.
83. Sears MR, Burrows B, Flannery EM, Herbison GP, Holdaway MD: Atopy in childhood. I. Gender and allergen related risks for development of hay fever and asthma. *Clin Exp Allergy* (1993) **23**: 941–948.
84. Sennhauser FH, Kuhni CE: Prevalence of respiratory symptoms in Swiss children: is bronchial asthma really more prevalent in boys? *Pediatr Pulmonol* (1995) **19**: 161–166.
85. Pizzichini MMM, Faulkner T, Tedesco R, Faulman G, Sears MR: Is asthma underdiagnosed in girls? *Am J Respir Crit Care Med* (1996) **153**: A429.

1 Epidemiology

86. Sears MR, Holdaway MD, Flannery EM, Herbison GP, Silva PA: Parental and neonatal risk factors for atopy, airway hyper-responsiveness and asthma. *Arch Dis Child* (1996) **75**: 392–398.
87. Bazaral M, Orgel HA, Hamburger RN: Genetics of IgE and allergy: serum IgE levels in twins. *J Allergy Clin Immunol* (1974) **54**: 288–304.
88. Godfrev S, Konig P: Exercise-induced bronchial lability in atopic children and their families. *Ann Allergy* (1974) **33**: 199–205.
89. Zamel N, Leroux M, Vanderdoelen JL: Airway response to inhaled methacholine in healthy nonsmoking twins. *J Appl Physiol: Respir Physiol* (1984) **56**: 936–939.
90. Johnston IDA, Bland JM, Anderson HR: Ethnic variation in respiratory morbidity and lung function in childhood. *Thorax* (1987) **42**: 542–548.
91. Lai CKW, Chen YZ, Zhong NS, et al.: Comparison of prevalence of asthma symptoms, allergic rhinitis and eczema in Chinese children in China and Hong Kong using the International Study of Asthma and Allergies in Childhood (ISAAC) protocol. *Am J Respir Crit Care Med* (1996) **153**: A856.
92. Van Niekerk CH, Weinberg EG, Shore SC, Heese HdeV, van Schalkwyk DJ: Prevalence of asthma: a comparative study of urban and rural Xhosa children. *Clin Allergy* (1979) **9**: 319–324.
93. Waite DA, Eyles EF, Tonkin SL, O'Donnell TV: Asthma prevalence in Tokelauan children in two environments. *Clin Allergy* (1980) **10**: 71–75.
94. Aberg N: Asthma and allergic rhinitis in Swedish conscripts. *Clin Exp Allergy* (1989) **19**: 59–63.
95. Leung R: Asthma, allergy and atopy in South-East Asian immigrants in Australia. *Aust NZ J Med* (1994) **24**: 255–257.
96. Zimmerman B, Feanny S, Reisman J, et al.: Allergy in asthma. I. The dose relationship of allergy to severity of childhood asthma. *J Allergy Clin Immunol* (1988) **81**: 63–70.
97. Stempel DA, Clyde WA, Henderson FW, Collier AM: Serum IgE levels and the clinical expression of respiratory illness. *J Pediatr* (1980) **97**: 185–190.
98. Giles GG, Gibson HB, Lickiss N, Shaw K: Respiratory symptoms in Tasmanian adolescents: a follow up of the 1961 birth cohort. *Aust NZ J Med* (1984) **14**: 631–637.
99. Peat JK, Britton WJ, Salome CM, Woolcock AJ: Bronchial hyperresponsiveness in two populations of Australian schoolchildren. II. Relative importance of associated factors. *Clin Allergy* (1987) **17**: 283–290.
100. Peat JK, Britton WJ, Salome CM, Woolcock AJ: Bronchial hyperresponsiveness in two populations of Australian schoolchildren. III. Effect of exposure to environmental allergens. *Clin Allergy* (1987) **17**: 291–300.
101. Platts-Mills TAE, de Weck AL (Chairmen): Dust mite allergens and asthma—a worldwide problem. *J Allergy Clin Immunol* (1989) **83**: 416–427.
102. Sears MR, Herbison GP, Holdaway MD, et al.: The relative risks of skin sensitivity to grass pollen, house dust mite and cat dander in the development of childhood asthma. *Clin Exp Allergy* (1989) **19**: 419–424.
103. Sporik R, Chapman MD, Platts-Mills TAE: House dust mite exposure as a cause of asthma. *Clin Exp Allergy* (1992) **22**: 897–906.
104. Sears MR, Burrows B, Flannery EM, Herbison GP, Hewitt CJ, Holdaway MD: Relation between airway responsiveness and serum IgE in children with asthma and in apparently normal children. *N Engl J Med* (1991) **325**: 1067–1071.
105. Peat JK, Salome CM, Woolcock AJ: Factors associated with bronchial hyperresponsiveness in Australian adults and children. *Eur Respir J* (1992) **5**: 921–929.
106. Sears MR, Burrows B, Herbison GP, Holdaway MD, Flannery EM: Atopy in childhood. II. Relationship to airway responsiveness, hay fever and asthma. *Clin Exp Allergy* (1993) **23**: 949–956.
107. Sears MR, Burrows B, Herbison GP, Flannery EM, Holdaway MD: Atopy in childhood. III. Relationship with pulmonary function and airway responsiveness. *Clin Exp Allergy* (1993) **23**: 957–963.
108. Sporik R, Ingram JM, Price W, Sussman JH, Honsinger RW, Platts-Mills TAE: Associations of asthma with serum IgE and skin test reactivity to allergens among children living at high altitude. Tickling the dragon's breath. *Am J Respir Crit Care Med* (1995) **151**: 1388–1392.

109. Ingram JM, Sporik R, Honsinger R, Chapman MD, Platts-Mills TAE: Quantitative assessment of exposure to dog (*Can f* 1) and cat (*Fel d* 1) allergens: relation to sensitization and asthma among children living in Los Alamos, New Mexico. *J Allergy Clin Immunol* (1995) **96**: 449–456.
110. Strachan DP, Carey IM: Home environment and severe asthma in adolescence: a population based case-control study. *Br Med J* (1995) **311**: 1053–1056.
111. Lindfors A, Wickman M, Hedlin G, Pershagen G, Rietz H, Nordvall SL: Indoor environmental risk factors in young asthmatics: a case-control study. *Arch Dis Child* (1995) **73**: 408–412.
112. Mohamed N, Mg'ang'a L, Odhiambo J, Nyamwaya J, Menzies R: Home environment and asthma in Kenyan schoolchildren: a case-control study. *Thorax* (1995) **50**: 74–78.
113. Tager IB, Weiss ST, Munoz A, Rosner B, Speizer FE: Longitudinal study of the effects of maternal smoking on pulmonary function in children. *N Engl J Med* (1983) **309**: 699–703.
114. Leeder SR, Corkhill R, Irwig LM, Holland WW, Colley JRT: Influence of family factors on the incidence of lower respiratory illness during the first year of life. *Br J Prev Soc Med* (1976) **30**: 203–212.
115. Cogswell JJ, Mitchell EB, Alexander J: Parental smoking, breast feeding, and respiratory infection in development of allergic diseases. *Arch Dis Child* (1987) **62**: 338–344.
116. Gortmaker SL, Walker DK, Jacobs FH, Ruch-Ross H: Parental smoking and the risk of childhood asthma. *Am J Public Health* (1982) **72**: 574–579.
117. Martinez FD, Antognoni G, Macri F, *et al.*: Parental smoking enhances bronchial responsiveness in nine-year-old children. *Am Rev Respir Dis* (1988) **138**: 518–523.
118. Frischer T, Huehr J, Meinert R, *et al.*: Maternal smoking in early childhood: a risk factor for bronchial responsiveness to exercise in primary school children. *J Pediatr* (1992) **121**: 17–22.
119. Chilmonczyk BA, Salmun LM, Megathlin KN, *et al.*: Association between exposure to environmental tobacco smoke and exacerbations of asthma in children. *N Engl J Med* (1993) **328**: 1665–1669.
120. Infante-Rivard C: Childhood asthma and indoor environmental risk factors. *Am J Epidemiol* (1993) **137**: 834–844.
121. Couriel JM: Passive smoking and the health of children. *Thorax* (1994) **49**: 731–734.
122. Weiss ST, Sparrow D, O'Connor GT: The interrelationship among allergy, airways responsiveness, and asthma. *J Asthma* (1993) **30**: 329–349.
123. Morgan WJ, Martinez FD: Risk factors for developing wheezing and asthma in childhood. *Pediatr Clin North Am* (1992) **39**: 1185–1203.
124. Peat JK: The rising trend in allergic illness: which environmental factors are important? *Clin Exp Allergy* (1994) **24**: 797–800.
125. Stoddard JJ, Miller T: Impact of parental smoking on the prevalence of wheezing respiratory illness in children. *Am J Epidemiol* (1995) **141**: 96–102.
126. Lipsett M, Hurley S: Exposure to environmental tobacco smoke and induction of childhood asthma: a meta-analysis. *Am J Respir Crit Care Med* (1996) **153**: A431.
127. Mannino D, Siegel M, Husten C, Rose D, Etzel R: Environmental tobacco smoke exposure and respiratory diseases in children. *Chest* (1994) **106**: 115S.
128. Magnusson CGM: Maternal smoking influences cord serum IgE and IgD levels and increases the risk for subsequent infant allergy. *J Allergy Clin Immunol* (1986) **78**: 898–904.
129. Murray AB, Morrison BJ: Passive smoking by asthmatics: its greater effect on boys than on girls and on older than younger children. *Pediatrics* (1989) **84**: 451–459.
130. Ownby DR, McCullough J: Passive exposure to cigarette smoke does not increase allergic sensitization in children. *J Allergy Clin Immunol* (1988) **82**: 634–637.
131. Ronchetti R, Macri F, Ciofetta G, *et al.*: Increased serum IgE and increased prevalence of eosinophilia in 9-year-old children of smoking parents. *J Allergy Clin Immunol* (1990) **86**: 400–407.
132. Mejer GG, Postma DS, Van Der Heide S, *et al.*: Exogenous stimuli and circadian peak expiratory flow variation in allergic asthmatic children. *Am J Respir Crit Care Med* (1996) **153**: 237–242.
133. Hanrahan JP, Speizer FE, Tager IB: A prospective study of risk factors for asthma in

urban children >2 followed from early in pregnancy. *Am J Respir Crit Care Med* (1996) **153**: A254.
134. Samet JM, Tager IB, Speizer FE: The relationship between respiratory illness in childhood and chronic air-flow obstruction in adulthood. *Am Rev Respir Dis* (1983) **127**: 508–523.
135. Douglas JWB, Waller RE: Air pollution and respiratory infection in children. *Br J Prev Soc Med* (1966) **20**: 1–8.
136. Colley JRT, Douglas JWB, Reid DD: Respiratory disease in young adults: influence of early childhood lower respiratory tract illness, social class, air pollution, and smoking. *Br Med J* (1973) **3**: 195–198.
137. Kiernan KE, Colley JRT, Douglas JWB, Reid DD: Chronic cough in young adults in relation to smoking habits, childhood environment and chest illness. *Respiration* (1976) **33**: 236–244.
138. Leeder SR, Corkhill RT, Wysocki MJ, Holland WW: Influence of personal and family factors on ventilatory function of children. *Br J Prev Soc Med* (1976) **30**: 219–224.
139. Holland WW, Kasap HS, Colley JRT, Cormack W: Respiratory symptoms and ventilatory function: a family study. *Br J Prev Soc Med* (1969) **23**: 77–84.
140. Colley JR, Holland WW, Leeder SR, Corkhill RT: Respiratory function of infants in relation to subsequent respiratory disease: an epidemiological study. *Bull Eur Physiopathol Respir* (1976) **12**: 651–657.
141. Leeder SR, Corkhill RT, Irwig LM, Holland WW, Colley JRT: Influence of family factors on asthma and wheezing during the first five years of life. *Br J Prev Soc Med* (1976) **30**: 213–218.
142. Voter KZ, Henry MM, Stewart PW, Henderson FW: Lower respiratory illness in early childhood and lung function and bronchial reactivity in adolescent males. *Am Rev Respir Dis* (1988) **137**: 302–307.
143. Gold DR, Tager IB, Weiss ST, Tosteson TD, Speizer FE: Acute lower respiratory illness in childhood as a predictor of lung function and chronic respiratory symptoms. *Am Rev Respir Dis* (1989) **140**: 877–884.
144. Sims DG, Downham MAPS, Gardner PS, Webb JKG, Weightman D: Study of 8-year-old children with a history of respiratory syncytial virus bronchiolitis in infancy. *Br Med J* (1978) **1**: 11–14.
145. Pullan CR, Hey EN: Wheezing, asthma, and pulmonary dysfunction 10 years after infection with respiratory syncytial virus in infancy. *Br Med J* (1982) **284**: 1665–1669.
146. Gurwitz D, Corey M, Levison H: Pulmonary function and bronchial reactivity in children after croup. *Am Rev Respir Dis* (1980) **122**: 95–99.
147. Mok JYQ, Simpson H: Symptoms, atopy, and bronchial reactivity after lower respiratory infection in infancy. *Arch Dis Child* (1984) **59**: 299–305.
148. Mok JYQ, Simpson H: Outcome for acute bronchitis, bronchiolitis, and pneumonia in infancy. *Arch Dis Child* (1984) **59**: 306–309.
149. Martinez FD, Morgan WJ, Wright AL, *et al*.: Diminished lung function as a predisposing factor for wheezing respiratory illness in infants. *N Engl J Med* (1988) **319**: 1112–1117.
150. Martinez FD, Morgan WJ, Wright AL, *et al*.: Initial airway function is a risk factor for recurrent wheezing respiratory illness during the first three years of life. *Am Rev Respir Dis* (1991) **143**: 312–316.
151. Hodge L, Salome CM, Peat JK, Haby MM, Xuan W, Woolcock AJ: Consumption of oily fish and childhood asthma. *Med J Aust* (1996) **164**: 137–140.
152. Demissie K, Ernst P, Gray Donald K, Joseph L: Usual dietary salt intake and asthma in children: a case-control study. *Thorax* (1996) **51**: 59–63.
153. Carey OJ, Locke C, Cookson JB: Effect of alterations of dietary sodium on the severity of asthma in men. *Thorax* (1993) **48**: 714–718.
154. Burney P: A diet rich in sodium may potentiate asthma. Epidemiologic evidence for a new hypothesis. *Chest* (1987) **91**: 143S-148S.
155. von Mutius E, Martinez FD, Fritzsch C, Nicolai T, Reitmeir P, Thiemann H-H: Skin test reactivity and number of siblings. *Br Med J* (1994) **308**: 692–695.
156. Mitchell EA, Stewart AW, Pattemore PK, Asher MI, Harrison AC, Rea HH: Socioeconomic status in childhood asthma. *Int J Epidemiol* (1989) **18**: 888–890.
157. Silva PA, Sears MR, Jones DT, *et al*.: Some family social background, developmental, and

behavioural characteristics of nine year old children with asthma. *NZ Med J* (1987) **100**: 318–320.
158. Wissow LS, Gittelsohn AM, Szklo M, Starfield B, Mussman M: Poverty, race, and hospitalization for childhood asthma. *Am J Public Health* (1988) **78**: 777–782.
159. von Mutius E, Fritzsch C, Welland SK, Roll G, Magnusson H: Prevalence of asthma and allergic disorders among children in united Germany: a descriptive comparison. *Br Med J* (1992) **305**: 1395–1399.
160. von Mutius E, Martinez FD, Fritzsch C, Nicolai T, Roell G, Thiemann H-H: Prevalence of asthma and atopy in two areas of West and East Germany. *Am J Respir Crit Care Med* (1994) **149**: 358–364.
161. von Mutius E, Sherrill DL, Fritzsch C, Martinez FD, Lebowitz MD: Air pollution and upper respiratory symptoms in children from East Germany. *Eur Respir J* (1995) **8**: 723–728.
162. Schmitzberger R, Rhomberg K, Buchele H, *et al*.: Effects of air pollution on the respiratory tract of children. *Pediatr Pulmonol* (1993) **15**: 68–74.
163. Hopp RJ, Bewtra AK, Nair NM, Watt GD, Townley RG: Methacholine inhalation challenge studies in a selected pediatric population. *Am Rev Respir Dis* (1986) **134**: 994–998.
164. Weiss ST, Tager IB, Weiss JW, Munoz A, Speizer FE, Ingram RH: Airways responsiveness in a population sample of adults and children. *Am Rev Respir Dis* (1984) **129**: 898–902.
165. Clifford RD, Radford M, Howell JB, Holgate ST: Prevalence of atopy and range of bronchial response to methacholine in 7- and 11-year-old schoolchildren. *Arch Dis Child* (1989) **64**: 1126–1132.
166. von Mutius E: Progression of allergy and asthma through childhood to adolescence. *Thorax* (1996) **51** (Suppl 1): S3-S6.
167. Strachan D, Gerritsen J: Long-term outcome of early childhood wheezing: population data. *Eur Respir J* (1996) **9**: 42–47.
168. Carey VJ, Weiss ST, Tager IB, Leeder SR, Speizer FE: Airways responsiveness, wheeze onset, and recurrent asthma episodes in young adolescents. The East Boston Childhood Respiratory Disease Cohort. *Am J Respir Crit Care Med* (1996) **153**: 356–361.
169. Roorda RJ, Gerritsen J, Van Aalderen WMC, *et al*.: Risk factors for the persistence of respiratory symptoms in childhood asthma. *Am Rev Respir Dis* (1993) **148**: 1490–1495.
170. Burrows B, Sears MR, Flannery EM, Herbison P, Holdaway MD, Silva PA: Relation of the course of bronchial responsiveness from age 9 to age 15 to allergy. *Am J Respir Crit Care Med* (1995) **152**: 1302–1308.
171. Ulrik CS, Backer V, Dirksen A, Pedersen M, Koch C: Extrinsic and intrinsic asthma from childhood to adult age: a 10 year follow-up. *Respir Med* (1995) **89**: 547–554.
172. Ross S, Godden DJ, Abdalla M, *et al*.: Outcome of wheeze in childhood: the influence of atopy. *Eur Respir J* (1995) **8**: 2081–2087.
173. Strachan DP, Butland BK, Anderson HR: Incidence and prognosis of asthma and wheezing illness from early childhood to age 33 in a national British cohort. *Br Med J* (1996) **312**: 1195–1199.
174. Fletcher CM, Pride NB: Definitions of emphysema, chronic bronchitis, asthma, and airflow obstruction: 25 years on from the Ciba symposium. *Thorax* (1984) **39**: 81–85.
175. McWhorter WP, Polis MA, Kaslow RA: Occurrence, predictors, and consequences of adult asthma in NHANESI and follow-up survey. *Am Rev Respir Dis* (1989) **139**: 721–724.
176. Dodge RR, Burrows B: The prevalence and incidence of asthma and asthma-like symptoms in a general population sample. *Am Rev Respir Dis* (1980) **122**: 567–575.
177. Stellman JL, Spicer JE, Cayton RM: Morbidity from chronic asthma. *Thorax* (1982) **37**: 218–221.
178. Woolcock AJ, Peat JK, Salome CM, *et al*.: Prevalence of bronchial hyperresponsiveness and asthma in a rural adult population. *Thorax* (1987) **42**: 361–368.
179. Schachter EN, Doyle CA, Beck GJ: A prospective study of asthma in a rural community. *Chest* (1984) **85**: 623–630.
180. Cockcroft DW, Berscheid BA, Murdock KY: Unimodal distribution of bronchial responsiveness to inhaled histamine in a random human population. *Chest* (1983) **83**: 751–754.
181. Vesterinen F, Kaprio J, Koskenvuo M: Prospective study of asthma in relation to smoking habits among 14 729 adults. *Thorax* (1988) **43**: 534–539.

182. Abramson M, Kutin J, Bowes G: The prevalence of asthma in Victorian adults. *Aust NZ J Med* (1992) **22**: 358–363.
183. Peat JK, Haby M, Spijker J, Berry G, Woolcock AJ: Prevalence of asthma in adults in Busselton, Western Australia. *Br Med J* (1992) **305**: 1326–1329.
184. Ng TP, Hui KP, Tan WC: Prevalence of asthma and risk factors among Chinese, Malay, and Indian adults in Singapore. *Thorax* (1994) **49**: 347–351.
185. Comino EJ, Mitchell CA, Bauman A, Henry RL, Robertson CF: Asthma management in eastern Australia, 1990 and 1993. *Med J Aust* (1996) **164**: 403–406.
186. Dowse GK, Smith D, Turner KJ, Alpers MP: Prevalence and features of asthma in a sample survey of urban Goroka, Papua New Guinea. *Clin Allergy* (1985) **15**: 429–438.
187. Nejjari C, Tessier JF, Letenneur L, Dartigues JF, Barberger-Gateau P, Salamon R: Prevalence of self-reported asthma symptoms in a French elderly sample. *Respir Med* (1996) **90**: 401–408.
188. Burney PGJ, Luczynska C, Chinn S, Jarvis D: The European Community Respiratory Health Survey. *Eur Respir J* (1994) **7**: 954–960.
189. Burney P: Asthma prevalence and risk factors. In Vuylsteek K, Hallen M (eds) *Epidemiology*. Amsterdam, IOS Press, 1994, pp. 241–251.
190. Boezen HM, Schouten JP, Postma DS, Rijcken B: Distribution of peak expiratory flow variability by age, gender and smoking habits in a random population sample aged 20–70 years. *Eur Respir J* (1994) **7**: 1814–1820.
191. de Marco R, Verlato G, Zanolin E, Bugiani M, Drane JW: Nonresponse bias in EC Respiratory Health Survey in Italy. *Eur Respir J* (1994) **7**: 2139–2145.
192. Bjornsson E, Plaschke P, Norrman E, *et al.*: Symptoms related to asthma and chronic bronchitis in three areas of Sweden. *Eur Respir J* (1994) **7**: 2146–2153.
193. Boezen HM, Schouten JP, Postma DS, Rijcken B: Relation between respiratory symptoms, pulmonary function and peak flow variability in adults. *Thorax* (1995) **50**: 121–126.
194. Crane J, Lewis S, Slater T, *et al.*: The self reported prevalence of asthma symptoms amongst adult New Zealanders. *NZ Med J* (1994) **107**: 417–421.
195. Jarvis D, Luczynska C, Chinn S, Burney P: The association of age, gender and smoking with total IgE and specific IgE. *Clin Exp Allergy* (1995) **25**: 1083–1091.
196. Neukirch F, Pin I, Knani J, *et al.*: Prevalence of asthma and asthma-like symptoms in three French cities. *Respir Med* (1995) **89**: 685–692.
197. Droste JHJ, Kerkhof M, de Monchy JGR, *et al.*: Association of skin test reactivity, specific IgE, total IgE, and eosinophils with nasal symptoms in a community-based population study. *J Allergy Clin Immunol* (1996) **97**: 922–932.
198. Sunyer J, Munoz A: Concentrations of methacholine for bronchial responsiveness according to symptoms, smoking, and immunoglobulin E in a population-based study in Spain. *Am J Respir Crit Care Med* (1996) **153**: 1273–1279.
199. Abramson M, Kutin J, Czarny D, Walters EH: The prevalence of asthma and respiratory symptoms among young adults: is it increasing in Australia? *J Asthma* (1996) **33**: 189–196.
200. Burney P, Chinn S, Luczynska C, Jarvis D, Lai E: Variations in the prevalence of respiratory symptoms, self-reported asthma attacks, and use of asthma medication in the European Community Respiratory Health Survey (ECRHS). *Eur Respir J* (1996) **9**: 687–695.
201. Martin AJ, McLennan LA, Landau LI, Phelan PD: The natural history of childhood asthma to adult life. *Br Med J* (1980) **1**: 1397–1400.
202. Broder I, Higgins MW, Mathews KP, Keller JB: Epidemiology of asthma and allergic rhinitis in a total community, Tecumseh, Michigan. IV. Natural history. *J Allergy Clin Immunol* (1974) **54**: 100–110.
203. Cerveri I, Bruschi C, Zoia MC, *et al.*: Distribution of bronchial nonspecific reactivity in the general population. *Chest* (1988) **93**: 26–30.
204. Burney PGJ, Britton JR, Chinn S, *et al.*: Descriptive epidemiology of bronchial reactivity in an adult population: results from a community study. *Thorax* (1987) **38**: 38–44.
205. Bakke PS, Baste V, Gulsvik A: Bronchial responsiveness in a Norwegian community. *Am Rev Respir Dis* (1991) **143**: 317–322.
206. Ricjken B, Schouten JP, Weiss ST, Speizer FE, van der Lende R: The relationship of nonspecific bronchial responsiveness to respiratory symptoms in a random population sample. *Am Rev Respir Dis* (1987) **136**: 62–68.

207. Sparrow D, O'Connor G, Colton T, Barry CL, Weiss ST: The relationship of nonspecific bronchial responsiveness to the occurrence of respiratory symptoms and decreased levels of pulmonary function. The normative aging study. *Am Rev Respir Dis* (1987) **135**: 1255–1260.
208. Anderson HR: The epidemiological and allergic features of asthma in the New Guinea Highlands. *Clin Allergy* (1974) **4**: 171–183.
209. Woolcock AJ, Dowse GK, Temple K, et al.: The prevalence of asthma in the South Fore people of Papua New Guinea. A method for field studies of bronchial reactivity. *Eur J Respir Dis* (1983) **64**: 571–581.
210. Dowse GK, Turner KJ, Stewart GA, Alpers MP, Woolcock AJ: The association between *Dermatophagoides* mites and the increasing prevalence of asthma in village communities within the Papua New Guinea highlands. *J Allergy Clin Immunol* (1985) **75**: 75–83.
211. Turner KJ: Changing prevalence of asthma in developing countries. In Michel FB, Bousquet J, Godard P (eds) *Highlights in Asthmology*. New York, Springer-Verlag, 1987, pp 37–43.
212. Fabbri LM, Maestrelli P, Saetta M, Mapp CM: Mechanisms of occupational asthma. *Clin Exp Allergy* (1994) **24**: 628–635.
213. Chan-Yeung M, Malo J-L: Occupational asthma. *N Engl J Med* (1995) **333**: 107–112.
214. Newman-Taylor AJ: Respiratory irritants encountered at work. *Thorax* (1996) **51**: 541–545.
215. Kobayashi S: Different aspects of occupational asthma in Japan. In Fraser CA (ed.) *Occupational Asthma*. New York, Van Nostrand Reinhold, 1980, pp 229–244.
216. Anto JM, Sunyer J, Rodrigues-Roisin R, et al.: Community outbreaks of asthma associated with inhalation of soybean dust. *N Engl J Med* (1989) **320**: 1271–1273.
217. Sunyer J, Anto JM, Rodrigo MJ, et al.: Case-control study of serum immunoglobulin-E antibodies reactive with soybean in epidemic asthma. *Lancet* (1989) **i**: 179–182.
218. Britton J, Pavord I, Richards K, et al.: Dietary sodium intake and the risk of airway hyperreactivity in a random adult population. *Thorax* (1994) **49**: 875–880.
219. Devereaux G, Bech JR, Bromly C, et al.: Effect of dietary sodium on airways responsiveness and its importance in the epidemiology of asthma: an evaluation in three areas of northern England. *Thorax* (1995) **50**: 941–947.
220. Troisi RJ, Willett WC, Weiss ST, Trichopoulos D, Rosner B, Speizer FE: A prospective study of diet and adult-onset asthma. *Am J Respir Crit Care Med* (1995) **151**: 1401–1408.
221. Egan P: Weather or not. *Med J Aust* (1985) **142**: 330.
222. Packe GE, Archer PStJ, Ayres JG: Asthma and the weather. *Lancet* (1983) **i**: 281.
223. Brown HM, Jackson F: Asthma and the weather. *Lancet* (1983) **ii**: 630.
224. Salvaggio JE, Kundur VG: New Orleans epidemic asthma: relationship between outbreaks and influx of ragweed pollen. *J Allergy* (1968) **41**: 90–91.
225. Carey MJ, Cordon I: Asthma and climatic conditions: experience from Bermuda, an isolated island community. *Br Med J* (1986) **293**: 843–844.
226. Keating G, Mitchell EA, Jackson R, Beaglehole R, Rea H: Trends in sales of drugs for asthma in New Zealand, Australia and the United Kingdom, 1975–81. *Br Med J* (1984) **289**: 348–351.
227. Sinclair BL, Clark DWJ, Sears MR: Use of anti-asthma drugs in New Zealand. *Thorax* (1987) **42**: 670–675.
228. Klaukka T, Peura S, Martikainen J: Why has the utilization of antiasthmatics increased in Finland? *J Clin Epidemiol* (1991) **44**: 859–863.
229. Rea H, Sears M, Mitchell E, Garrett J, Mulder J, Anderson R: Is asthma becoming more severe? *Thorax* (1987) **42**: 736.
230. Sears MR, Taylor DR, Print CG, et al.: Regular inhaled beta-agonist treatment in bronchial asthma. *Lancet* (1990) **336**: 1391–1396.
231. Taylor DR, Sears MR, Herbison GP, et al.: Regular inhaled β-agonist in asthma: effect on exacerbations and lung function. *Thorax* (1993) **48**: 134–138.
232. Beswick KBJ, Pover GM, Sampson S: Long-term regularly inhaled salbutamol. *Curr Med Res Opin* (1986) **10**: 228–234.
233. Van Arsdel PP, Schraffin RM, Rosenblatt J, Sprenkle AC, Altman LC: Evaluation of oral fenoterol in chronic asthmatic patients. *Chest* (1978) **73**: 6 (Suppl) 997–998.
234. van Schaych CP, Graafsma SJ, Visch MB, Dompeling E, van Weel C, van Herwaarden CLA: Increased bronchial hyperresponsiveness after inhaling salbutamol during 1 year is not caused by subsensitization to salbutamol. *J Allergy Clin Immunol* (1990) **86**: 793–800.

1 Epidemiology

235. Harvey JE, Tattersfield AE: Airway response to salbutamol: effect of regular salbutamol inhalations in normal, atopic and asthmatic subjects. *Thorax* (1982) **37**: 280–287.
236. Kraan J, Koeter GH, van der Mark ThW, Sluiter HJ, de Vries K: Changes in bronchial hyperreactivity induced by 4 weeks of treatment with antiasthmatic drugs in patients with allergic asthma: a comparison between budesonide and terbutaline. *J Allergy Clin Immunol* (1985) **76**: 628–636.
237. Wahedna I, Wong CS, Wisniewski AFZ, Pavord ID, Tattersfield AE: Asthma control during and after cessation of regular beta$_2$ agonist treatment. *Am Rev Respir Dis* (1993) **148**: 707–712.
238. Taylor DR, Sears MR: Bronchodilators and bronchial hyperresponsiveness. *Thorax* (1994) **49**: 190.
239. Inman MD, O'Byrne PM: The effect of regular inhaled albuterol on exercise-induced bronchoconstriction. *Am J Respir Crit Care Med* (1996) **153**: 65–69.
240. Cockcroft DW, McPaarland CP, Britto SA, Swystun VA, Rutherford BC: Regular inhaled salbutamol and airway responsiveness to allergen. *Lancet* (1993) **342**: 833–837.
241. Cockcroft DW, O'Byrne PM, Swystun VA, Bhagat R: Regular use of inhaled albuterol and the allergen-induced late asthmatic response. *J Allergy Clin Immunol* (1995) **96**: 44–49.
242. Drazen JM, Israel E, Boushey HA, *et al.*: Comparison of regularly scheduled with as-needed use of albuterol in mild asthma. *N Engl J Med* (1996) **335**: 841–847.
243. Speizer FE, Doll, R, Heaf P: Observations on recent increase in mortality from asthma. *Br Med J* (1968) **1**: 335–339.
244. Esdaile JM, Feinstein AR, Horwitz RI: A reappraisal of the United Kingdom epidemic of fatal asthma. *Arch Intern Med* (1987) **147**: 543–549.
245. Gandevia B: Pressurised sympathomimetic aerosols and their lack of relationship to asthma mortality in Australia. *Med J Aust* (1973) **i**: 273–277.
246. Jackson RT, Beaglehole R, Rea HH, Sutherland DC: Mortality from asthma: a new epidemic in New Zealand. *Br Med J* (1982) **285**: 771–774.
247. Stolley PD, Schinnar R: Association between asthma mortality and isoproterenol aerosols: a review. *Prev Med* (1978) **7**: 519–538.
248. Pearce N, Crane J, Burgess C, Jackson R, Beasley R: Beta agonists and asthma mortality: deja vu. *Clin Exp Allergy* (1991) **21**: 401–410.
249. Sears MR, Taylor DR: The β_2-agonist controversy. Observations, explanations and relationship to asthma epidemiology. *Drug Safety* (1994) **11**: 259–283.
250. Beaupre A: Death in asthma. *Eur J Respir Dis* (1987) **70**: 259–260.
251. Jenkins PF, Mullins J, Davies BH, Williams DA: The possible role of aero-allergens in the epidemic of asthma deaths. *Clin Allergy* (1980) **11**: 611–620.
252. Sears MR, Rea HH, Beaglehole R, *et al.*: Asthma mortality in New Zealand: a two-year national study. *NZ Med J* (1985) **98**: 271–275.
253. Jackson R, Sears MR, Beaglehole R, Rea HH: International trends in asthma mortality 1970 to 1985. *Chest* (1988) **94**: 914–918.
254. Buist AS: Is asthma mortality increasing? *Chest* (1988) **93**: 449–450.
255. Robin ED: Death from bronchial asthma. *Chest* (1988) **93**: 614–618.
256. Burney PGJ: Asthma mortality in England and Wales: evidence for a further increase, 1974–1984. *Lancet* (1986) **ii**: 323–326.
257. Mao Y, Semenciw R, Morrison H, MacWilliam L, Davies J, Wigle D: Increased rates of illness and death from asthma in Canada. *Can Med Assoc J* (1987) **137**: 620–624.
258. Sly RM: Increases in deaths from asthma. *Ann Allergy* (1984) **53**: 20–25.
259. Sly RM: Mortality from asthma in children 1979–1984. *Ann Allergy* (1988) **60**: 433–442.
260. A subcommittee of the BTA Research Committee: accuracy of death certificates in bronchial asthma. *Thorax* (1984) **39**: 505–509.
261. Sears MR, Rea HH, de Boer G, *et al.*: Accuracy of certification of deaths due to asthma. A national study. *Am J Epidemiol* (1986) **124**: 1004–1011.
262. Sears MR, Rea HH, Rothwell RPG, *et al.*: Asthma mortality: comparison between New Zealand and England. *Br Med J* (1986) **293**: 1342–1345.
263. Hunt LW, Silverstein MD, Reed CE, O'Connell EJ, O'Fallon WM, Yunginger JW: Accuracy of the death certificate in a population-based study of asthmatic patients. *JAMA* (1993) **269**: 1947–1952.

264. World Health Organisation: Manual of the international statistical classification of diseases, injuries and cases of death: based on the recommendations of the ninth revision conference, 1975, Vol 1. Geneva, World Health Organisation, 1977.
265. Stewart CJ, Nunn AJ: Are asthma mortality rates changing? *Br J Dis Chest* (1985) **79**: 229–234.
266. British Thoracic Association: Death from asthma in two regions of England. *Br Med J* (1982) **285**: 1251–1255.
267. Rea HH, Scragg R, Jackson R, Beaglehole R, Fenwick J, Sutherland DC: A case-control study of deaths from asthma. *Thorax* (1986) **41**: 833–839.
268. Buist AS, Sears MR, Reid LM, Boushet HA, Spector SL, Sheffer AL: Asthma mortality: trends and determinants. *Am Rev Respir Dis* (1987) **136**: 1037–1039.
269. Campbell DA, McLennan G, Coates JR, et al.: A comparison of asthma deaths and near-fatal asthma attacks in South Australia. *Eur Respir J* (1994) **7**: 490–497.
270. Jackson GP: Asthma mortality by neighbourhood of domicile. *NZ Med J* (1988) **101**: 593–595.
271. Johnson AJ, Nunn AJ, Somner AR, Stableforth DE, Stewart CJ: Circumstances of death from asthma. *Br Med J* (1984) **288**: 1870–1872.
272. Barger LW, Vollmer WM, Felt RW, Buist AS: Further investigation into the recent increase in asthma death rates: a review of 41 asthma deaths in Oregon in 1982. *Ann Allergy* (1988) **60**: 31–39.
273. Robertson CF, Rubinfield AR, Bowes G: Deaths from asthma in Victoria: a 12-month survey. *Med J Aust* (1990) **152**: 511–517.
274. Rea HH, Sears MR, Beaglehole R, et al.: Lessons from the national asthma mortality study: circumstances surrounding death. *NZ Med J* (1987) **100**: 10–13.
275. Rothwell RPG, Rea HH, Sears MR, et al.: Lessons from the national asthma mortality study: deaths in hospitals. *NZ Med J* (1987) **100**: 199–202.
276. Tough SC, Green FHY, Paul JE, Wigle DT, Butt JC: Sudden death from asthma in 108 children and young adults. *J Asthma* (1996) **33**: 179–188.
277. Sears MR, Rea HH, Fenwick J, et al.: Deaths from asthma in New Zealand. *Arch Dis Child* (1986) **61**: 6–10.
278. Strunk RC, Mrazek DA, Fuhrmann GSW, LaBrecque JF: Physiological and psychological characteristics associated with deaths due to asthma in childhood: a case-controlled study. *JAMA* (1985) 254: 1193–1198.
279. Ulrik CS, Frederiksen J: Mortality and markers of risk of asthma death among 1,075 outpatients with asthma. *Chest* (1995) **108**: 10–15.
280. Crane J, Pearce N, Flatt A, et al.: Prescribed fenoterol and death from asthma in New Zealand, 1981–83: case-control study. *Lancet* (1989) **i**: 918–922.
281. Pearce N, Grainger J, Atkinson M, et al.: Case-control study of prescribed fenoterol and death from asthma in New Zealand, 1977–81. *Thorax* (1990) **45**: 170–175.
282. Grainger J, Woodman K, Pearce N, et al.: Prescribed fenoterol and death from asthma in New Zealand, 1981–87: a further case-control study. *Thorax* (1991) **46**: 105–111.
283. Spitzer WO, Suissa S, Ernst P, et al.: The use of β-agonists and the risk of death and near death from asthma. *N Engl J Med* (1992) **326**: 501–506.
284. Ernst P, Habbick B, Suissa S, et al.: Is the association between inhaled beta-agonist use and life-threatening asthma because of confounding by severity? *Am Rev Respir Dis* (1993) **148**: 75–79.
285. Suissa S, Blais L, Ernst P: Patterns of increasing β-agonist use and the risk of fatal or near-fatal asthma. *Eur Respir J* (1994) **7**: 1602–1609.
286. Petri H, Urquhart J, Herings R, Bakker A: Characteristics of patients prescribed three different inhalational β_2-agonists: an example of the channeling phenomenon. *Post Marketing Surveillance* (1991) **5**: 57–66.
287. Wilson JD: Selective prescribing of fenoterol and salbutamol in New Zealand general practice. *Post Marketing Surveillance* (1991) **5**: 105–118.
288. Garrett JE, Turner P: The severity of asthma in relation to beta agonist prescribing. *NZ Med J* (1991) **104**: 39–40.
289. Beasley R, Burgess C, Pearce N, Woodman K, Crane J: Confounding by severity does not

explain the association between fenoterol and asthma death. *Clin Exp Allergy* (1994) **24**: 660–668.
290. Ernst P, Spitzer W, Suissa S, *et al.*: Risk of fatal and near fatal asthma in relation to inhaled corticosteroid use. *JAMA* (1992) **268**: 3462–3464.
291. Sears MR: Epidemiological trends in asthma. *Can Respir J* (1996) **3**: 261–268.
292. Garrett J, Kolbe J, Richards G, Whitlock T, Rea H: Major reduction in asthma morbidity and continued reduction in asthma mortality in New Zealand: what lessons have been learned? *Thorax* (1995) **50**: 303–311.

2

Genetics

BOAZ MEIJER, EUGENE R. BLEECKER AND DIRKJE S. POSTMA

INTRODUCTION

It has been well established that asthma has an important hereditary component. In 1936 Bray presented an overview of several studies reporting an increased risk of developing asthma when the family history was positive for allergy.[1] Furthermore, twin studies have shown a greater concordance for asthma in monozygotic compared with dizygotic twins.[2-4] In the last decade, there have been numerous studies confirming the hereditary component in asthma. However, there is still controversy concerning the mode of inheritance and the loci on the genome that might be linked with asthma and closely associated phenotypes.

METHODS OF FINDING DISEASE GENES IN ASTHMA

The human genome is composed of approximately 3×10^9 base pairs and is estimated to contain between 60 000 and 80 000 genes.[5] There are two different approaches to finding genes involved in asthma: linkage studies and association studies.

Linkage studies test the linkage of a genetic marker with a trait. Polymorphic genetic markers used in linkage studies usually are not disease genes but any identifiable site where the DNA sequence is highly variable between individuals. The methods used in these studies are maximum likelihood approaches and allele-sharing methods. Maximum likelihood approaches require the construction of a transmission model to explain the inheritance of a disease in pedigrees, which is usually achieved by segregation analysis.

Segregation analysis compares the observed numbers of affected individuals with the expected number of affected individuals, using various modes of inheritance (e.g. recessive and dominant). The most parsimonious model identified by the analysis is used for the maximum likelihood approach. Genetic linkage is assessed by cotransmission of markers on known locations of the chromosome with the disease. The statistical significance of the linkage is expressed by the LOD score (base 10 *logarithm of the odds* ratio). The odds ratio is the likelihood of linkage of the marker with the disease gene under the transmission model divided by the likelihood of no linkage. Traditionally, a LOD score of $\geqslant 3$ is taken as evidence for linkage and a LOD score of $\leqslant -2$ is taken as evidence for exclusion of a disease gene from the area around the marker. An advantage of this approach is that not only the LOD score is calculated but also the distance between the marker and the disease gene, expressed as recombination fraction (θ).

In allele-sharing methods, one tries to prove that the transmission of a marker in a family is not random by showing that affected relatives inherit the same allele of a marker more often than expected by chance. The most frequently used form of allele-sharing methods is performed on affected pairs of siblings within a family (affected sib-pair analysis). If affected sib-pairs share the same allele of a marker more often than expected by chance, this marker is thought to be linked to the phenotype. The advantage of this kind of analysis is that the mode of inheritance does not have to be defined, as is the case for linkage studies in pedigrees.

Association studies test the association of a specific allele in a gene with a certain phenotype. In this approach, affected and unaffected individuals from a population are studied. A specific allele at a gene of interest is regarded to be associated with the phenotype if it occurs at a significantly higher frequency among affected individuals compared with unaffected individuals. This method is usually used to evaluate the role of genes that are thought to be important in the pathophysiological mechanisms in asthma, so-called candidate genes. Some examples of candidate genes for asthma are illustrated in Table 2.1.

Table 2.1 Some candidate genes in asthma.

Candidate gene	Chromosomal location
Interleukin-4	5q
β_2-Adrenergic receptor	5q
Lymphocyte specific glucocorticoid receptor 1	5q
HLA class II	6p
Tumour necrosis factor-β	6p
High-affinity IgE receptor β-subunit	11q
Tumour necrosis factor receptor 1	12q
Interferon-γ	12q
Mast-cell chymase	14q
T-cell receptor α/δ complex	14q

q, long arm of the chromosome.
p, short arm of the chromosome.

PROBLEMS IN GENETIC STUDIES OF ASTHMA

A major obstacle in studies on the genetics of asthma appears to be the definition of asthma. Asthma is a clinical diagnosis and no single clinical parameter always delineates asthma from other pulmonary diseases or a healthy state. However, certain well-defined phenotypes exist that are strongly associated with asthma. They provide a means for dissecting the genetic components of asthma. The two phenotypes that are especially closely linked to asthma, i.e. bronchial hyperresponsiveness (BHR) and atopy, have been applied in genetic studies. An issue that must be considered when studying related phenotypes is whether these characteristics are causally related to asthma or whether they represent a less important association.

Problems may further arise when individuals who inherit a predisposing allele do not manifest the disease (incomplete penetrance) or when others who do not inherit the predisposing allele develop the disease (phenocopy). These problems are very relevant for asthma, where expression of the disease is known to be influenced by age, gender and environmental factors such as allergen exposure, (viral) infections, cigarette smoke, air pollutants and possibly diet.

Mutations in different genes may result in identical phenotypes (genetic heterogeneity), for example when the gene products are required for a common biochemical pathway. Genetic heterogeneity seriously impedes genetic mapping, because a genetic marker may transmit with a trait in some families but not in others. It may be one of the explanations why, in complex genetic diseases such as asthma, evidence of strong linkage to a specific chromosomal region in one study is difficult to replicate in another population. This may be especially the case if one studies isolated populations where a linkage is found that is not as relevant in general populations.

The expression of some traits may require the presence of mutations in multiple genes simultaneously (polygenic inheritance). This phenomenon is known to occur in human diseases, such as some forms of Hirschsprung's disease,[6] and there is evidence that this may also be the case with total serum IgE levels.[7,8] Polygenic inheritance may be even more complicated by gene–gene interactions, where only a combination of certain mutations in multiple genes may interact with environmental influences to produce a trait.

MODES OF INHERITANCE OF ATOPY, BHR AND ASTHMA

Atopy is characterized by elevated levels of total serum IgE, detection of specific IgE against common aeroallergens and/or positive skin tests to common aeroallergens. Several studies have shown that genetic predisposition has a major role in the development of atopy. A twin study in 1984 reported an intra-pair correlation coefficient for total serum IgE of 0.82 in monozygotic twins compared with 0.52 in dizygotic twins, yielding a heritability of 61%.[4] Most studies using segregation analysis of total serum IgE levels have agreed that a major locus contributes to this trait, but different modes of inheritance have been suggested. Meyers et al.[9,10] reported evidence for recessive inheritance of high total serum IgE levels. Dizier et al.[11] found recessive inheritance of high total serum IgE after accounting for specific responses to common aeroallergens, suggesting the involvement of a major gene in basal IgE production. In a study of a Dutch population, a recessive model best fitted the inheritance of high total serum IgE levels;[12] since there was significant residual familial correlation in the analysis, a two-locus segregation analysis was performed. The two-locus segregation analysis suggested that a portion of the residual variance in total serum IgE levels is due to a second locus.[13] When the heterozygote is phenotypically distinct from the two homozygotes, the trait is said to be codominant; this mode of inheritance is difficult to distinguish from recessive inheritance. Evidence for codominant inheritance of total serum IgE levels has been reported in Amish families from Pennsylvania[14] and in a Hispanic and Caucasian population in Arizona.[15] Hasstedt et al.[7] found evidence for polygenic inheritance in a population of Mormons, a finding that was confirmed in a Caucasian population from Arizona.[8] Blumenthal et al.[16] studied three large family pedigrees and found a different mode of inheritance in each family, suggesting genetic heterogeneity.

BHR is found in virtually all patients with asthma. The mechanisms responsible for BHR in asthma include individual susceptibility and the effects of environmental exposures to pro-inflammatory stimuli.[17,18] It is now generally accepted that BHR is linked to an ongoing inflammatory process in the airway wall. Therefore, one can focus on the genes that predispose an individual to develop BHR as a logical method to investigate the hereditary factors that cause asthma. BHR is usually expressed as the provocation concentration on a cumulative dose–response curve of methacholine or histamine that causes a fall in forced expiratory volume in 1 s of 20% ($PC_{20}FEV_1$). In a twin study, Hopp et al.[4] reported that BHR to methacholine is under genetic control, with a heritability of 66%. These same investigators demonstrated in a study of non-asthmatic parents of asthmatic individuals that BHR to methacholine may be inherited from one generation to the other independently of asthma.[19] Segregation analysis indicated that although a familial component exists in the transmission of BHR to methacholine, this is not due to a single autosomal locus.[19] In contrast, Longo et al.[20] found evidence for dominant inheritance of BHR to carbachol. A further study in which a segregation analysis of BHR to histamine was performed also showed a dominant model to be the best-fitting model.[21] Although the data suggest a genetic predisposition to BHR, the genetic regulation of BHR, the interaction with environmental stimuli and the number of genes involved are not clear.

Only a few segregation analyses of asthma have been published. Holberg et al.[22] performed a segregation analysis on physician-diagnosed asthma in a Hispanic and Caucasian population in Arizona. The study population comprised 906 nuclear families recruited through a child, regardless of the child having an asthma diagnosis. When the children were approximately 6 years of age, the prevalence of physician-diagnosed asthma was assessed for each member in the nuclear family using a questionnaire. Their findings were not conclusive, i.e. there was evidence for an oligenic mode of inheritance with possibly a recessive component and an additional polygenic component.

GENETIC STUDIES OF ATOPY AND ASTHMA

The first report of linkage of atopy to a chromosome was reported by Cookson et al.[23] who defined atopy as a positive skin test (2 mm greater than the negative control), and/or elevated total serum IgE and/or elevated specific serum IgE.[23] Linkage analysis, using a dominant mode of inheritance, showed linkage of atopy underlying asthma to a marker on chromosome 11q. The study provoked much controversy over its broad definition of atopy, the failure to replicate this linkage in other populations[24–27] and even exclusion of this region in one population.[28] A subsequent study by Cookson et al. indicated that the transmission of atopy with the locus on chromosome 11q was only detectable through the maternal line; the reason for the maternal inheritance was not elucidated.[29] A study by Sandford et al. suggested that the β-subunit of the high-affinity IgE receptor (FcεRI-β) was the responsible candidate gene on chromosome 11q.[30] The high-affinity IgE receptor, which is expressed on mast cells and basophils, represents a good candidate gene for asthma, since it controls IgE-mediated mast cell degranulation.[31] Shirakawa et al. found association of atopy with a common polymorphism (Leu181) in the FcεRI-β gene, with evidence of maternal inheritance.[32] In a study of 1004 persons from 230 families in Australia, Hill et al. found a greater risk for atopy in individuals with the Leu181/Leu183 polymorphism in the FcεRI-β gene when this polymorphism was inherited through the maternal line.[33] A study from Japan found evidence for association of atopy with a restriction fragment polymorphism in the FcεRI-β gene.[34] Daniels et al.[34] performed a genome-wide search of trait loci underlying asthma in a general population from Australia.[35] They found potential linkage of total serum IgE with a marker in the FcεRI-β gene and a marker on chromosome 16. Both linkages were replicated in an additional group of families, recruited from clinics in the UK. In this study, markers from chromosome 13 showed evidence of linkage to the atopy phenotype, which was also confirmed in the British families. Despite the evidence for the involvement of the FcεRI-β gene in the development of atopy, many studies have not been able to support this finding. A study in pairs of siblings with asthma and atopy in The Netherlands revealed increased sharing of alleles of a marker on chromosome 11q, but not of a marker in the FcεRI-β gene.[36] Doull et al. showed an association of a polymorphism on chromosome 11q with atopy, but this locus was too distant from the FcεRI-β gene to explain the association.[37] In a study of 95 nuclear families, no evidence for linkage of atopy underlying atopic dermatitis to the FcεRI-β gene was found.[38] Other linkage studies from Australia[39] and England,[40] and association studies from Italy,[41] England[42] and Japan,[43] also failed to demonstrate the importance of the FcεRI-β gene in atopy.

Another region on the genome of major interest has been an area on chromosome 5q. This region contains several genes that may be important in the development and progression of inflammation in atopy and asthma, e.g. interleukin (IL)-3, IL-4, IL-5, IL-9, IL-13, granulocyte–macrophage colony-stimulating factor and fibroblast growth factor-1.[44] Linkage of five markers in this region of chromosome 5q with total serum IgE has been reported to be present in 11 extended Amish families;[45] none of these markers showed linkage with specific IgE concentrations. Meyers et al.[12] found similar results in a study of families from The Netherlands that were ascertained through a parent with asthma. Evidence for linkage with markers on chromosome 5q was obtained by both sib-pair analysis and LOD score approaches, with evidence for recessive inheritance of high serum IgE levels. As the two studies were performed in two distinct populations, i.e. an inbred population and a population that was ascertained through a proband with asthma, they may identify a different locus in each population. Further analysis by Xu et al.[13] of the total serum IgE levels in the Dutch population showed that a two-locus model fitted the inheritance of high serum IgE levels better than a one-locus model. Using this two-locus model also increased the LOD score of the most informative marker on chromosome 5q. These findings indicate that at least two different genes are required for the expression of this trait, either on chromosome 5 or another chromosome. The possible involvement of genes on chromosome 5q in asthma was recently supported by a study in a random population in England that found association of polymorphism in the IL-9 gene with total IgE levels.[37] However, studies from Minnesota[46] and Australia[35] have been unable to find linkage of total serum IgE with this region on chromosome 5q. The Minnesota study comprised a limited number of large families and might therefore be unable to find linkage to markers that are important in a general population.

Recently, two studies have attempted to implicate other genomic regions in the development of atopic disorders and the regulation of total serum IgE levels. A Japanese study attempted to associate atopic disorders with polymorphisms in a region on chromosome 14q, which contains genes for cellular proteases.[47] These authors selected patients who had only one atopic disorder at the time of investigation: asthma, allergic rhinitis or atopic dermatitis. No association of asthma and allergic rhinitis was found with polymorphisms in this region. However, eczema was associated with a polymorphism in the mast-cell chymase gene, a serine protease secreted by mast cells in the skin. A study conducted in an Afro-Caribbean population from Barbados reported linkage of total serum IgE levels with markers on chromosome 12q; this linkage was replicated in a Caucasian population from the Pennsylvania Amish.[48] In the Afro-Caribbean population, markers on chromosome 12q were also linked with asthma, which was defined as a positive history of asthma using a questionnaire *and* a confirmation of asthma by an interview with a physician.

Genetic regulation of specific IgE responses may be different from the regulation of total serum IgE levels. Specific IgE levels might be governed by the variation in HLA or T-cell receptor (TCR) molecules, since these proteins play a pivotal role in the recognition and handling of foreign antigens. Genetic studies have focused on HLA class II types (HLA-DP, HLA-DQ and HLA-DR). Association of HLA-DR5 and specific IgE against short ragweed was reported in a study where the HLA type was determined serologically.[49] In contrast, Young *et al.*[50] who determined HLA type using a polymerase chain reaction (PCR), found little evidence to support an important role for HLA class II genotypes in specific IgE responses. There was a possible association of certain HLA-DR genotypes in individuals with specific IgE response to the domestic cat and in subjects responsive to the mould *Alternaria alternata*. However, there was no association of HLA class II genotypes and specific IgE response to the house-dust mite *Dermatophagoides pteronyssinus* I and II, domestic dog and timothy grass. Similarly, Li *et al.*[51] found no association of PCR-determined HLA class II genotypes and skin test reactivity to six common aeroallergens. Two studies did associate certain HLA-DQ genotypes with toluene diisocyanate-induced asthma, the most common form of occupational asthma.[52,53]

Moffatt *et al.*[54] investigated whether specific serum IgE was linked to the TCR α and β complex genes on chromosomes 14q and 7 respectively. They studied a set of affected sib-pairs from the UK and a set of sib-pairs from Australia. No linkage of IgE serotypes and a marker in the TCR-β complex was found. However, a marker in the TCR-α complex was found to be linked to IgE response to *D. pteronyssinus* I and II and domestic cat in the British subjects. In the Australian subjects, the marker in the TCR-α complex was found

2 Genetics

to be linked to IgE response to *D. pteronyssinus* I, domestic cat and grass; furthermore, linkage to high total serum IgE was found.

GENETIC STUDIES OF BHR AND ASTHMA

Genetic studies that have attempted to link BHR to a specific locus in the genome are summarized in Table 2.2. The first linkage study of BHR failed to find linkage with a locus on chromosome 11q.[27] Another study excluded linkage of this phenotype with loci on chromosomes 6p and 11q.[28] In the same population, Postma *et al.*[55] found evidence for linkage of BHR to markers on chromosome 5q, near the region of the IL-4 cytokine gene cluster, with the best linkage found at the locus that regulates total serum IgE levels.[12,45] This study involved sib-pair analysis in the offspring of probands in 84 Dutch families. The probands were evaluated for obstructive airways disease between 1962 and 1970 and were identified as having asthma in that period by the presence of BHR to histamine and symptoms compatible with asthma. Van Herwerden *et al.*[56] studied 123 affected sib-pairs, recruited from a general population in Australia. They found evidence of linkage of BHR with a marker in the FcεRI-β gene. Linkage was also observed in siblings sharing BHR when those with atopy were excluded. In a genome-wide search, Daniels *et al.*[35] studied 172 sib-pairs from 80 families ascertained for atopy in Australia. They screened the families with 269 markers, covering all the chromosomes. Regions on chromosomes 4 and 7 were linked to bronchial responsiveness, whereas no linkage was found with a marker in the FcεRI-β gene nor with markers in the cytokine gene cluster on chromosome 5q.

Genetic studies that have attempted to associate BHR with certain polymorphisms are summarized in Table 2.3. Hall *et al.*[57] studied 65 patients with mild to moderate asthma.

Table 2.2 Linkage studies of bronchial hyperresponsiveness.

Year	Reference	Stimulus	Bronchial hyperresponsiveness	Result
1992	27	Methacholine	$PD_{20}FEV_1 \leq 8\,\mu mol$	No linkage with a marker on chromosome 11q
1992	28	Histamine	$PC_{20}FEV_1 \leq 32\,mg/ml$	Exclusion of linkage with markers on chromosomes 6p and 11q
1995	55	Histamine	$PC_{20}FEV_1 \leq 32\,mg/ml$ and as a quantitative trait	Linkage with markers on chromosome 5q
1995	56	Methacholine	$PD_{20}FEV_1 \leq 2\,mg$	Linkage with the FcεRI-β gene on chromosome 11q
1996	35	Methacholine	Slope on dose–response curve as a quantitative trait	Linkage with markers on chromosomes 4 and 7

FcεRI-β, β-subunit of the high-affinity IgE receptor; $PC(D)_{20}FEV_1$, provocative concentration (dose) causing a 20% fall in forced expiratory volume in 1 s.

Table 2.3 Association studies of bronchial hyperresponsiveness.

Year	Reference	Stimulus	Bronchial hyperresponsiveness	Result
1995	55	Methacholine	$PD_{20}FEV_1$ as a quantitative trait	Association with a polymorphism in the β_2-adrenergic receptor gene
1995	50	Histamine	$PC_{20}FEV_1 < 1$ mg/ml	No association with HLA class II genotypes
1995	56	Methacholine	$PD_{35}G_{rs}$ as a quantitative trait	No association with a polymorphism in the β_2-adrenergic receptor gene
1995	32	Methacholine	$PD_{20}FEV_1 \leq 10\,\mu\text{mol}$	Association with a polymorphism in the FcεRI-β gene
1996	36	Histamine	$PD_{20}FEV_1 \leq 2.5\,\mu\text{mol}$	Association with a polymorphism on chromosome 11q (not in the FcεRI-β gene)

FcεRI-β, β-subunit of the high-affinity IgE receptor; G_{rs}, respiratory conductance; $PC(D)_{20}FEV_1$, provocative concentration (dose) causing a 20% fall in forced expiratory volume in 1 s.

Patients who were homozygous for the Glu27 polymorphism in the β_2-adrenergic receptor gene had a four-fold higher geometric mean $PD_{20}FEV_1$ for methacholine than the individuals who were homozygous for the Gln27 polymorphism (the wild type); heterozygous individuals had an intermediate value. In a Japanese population, no association of BHR with a polymorphism in the β_2-adrenergic receptor gene was found.[58] Li et al.[51] studied the association of HLA-DQ and HLA-DR genotypes and asthma in a southern Chinese population of patients with asthma; no association was found between any of these genotypes and BHR. Hill et al.[33] found an association of BHR with the Leu181/Leu183 polymorphism in the FcεRI-β gene in an Australian population. A study of 131 families, both with and without family members with asthma, from a general population in England found that BHR was associated with a specific allele in a locus on chromosome 11q; this locus was too distant from the FcεRI-β gene to account for the association.

Reversibility of airflow limitation has been long established as an important phenotype of asthma. Most patients have some reversibility, whereas it may diminish after corticosteroid therapy (the so-called ceiling effect); with normal lung function, reversibility is difficult to assess.[59] Inhalation of a β_2-adrenergic agonist is generally used to assess the level of reversibility. The central role of β_2-adrenergic receptors in brochodilation makes the β_2-adrenergic receptor gene an important candidate gene for asthma; this is further strengthened by the reported linkage of atopy and BHR to the region on chromosome 5q that contains the β_2-adrenergic receptor gene. Reihsaus et al.[12,45,55] investigated whether mutations in the gene could be detected and whether any of these mutations were associated with asthma. Nine different mutations were discovered, four of which caused changes in the encoded amino acids (Fig. 2.1). None of the mutations was

2 Genetics

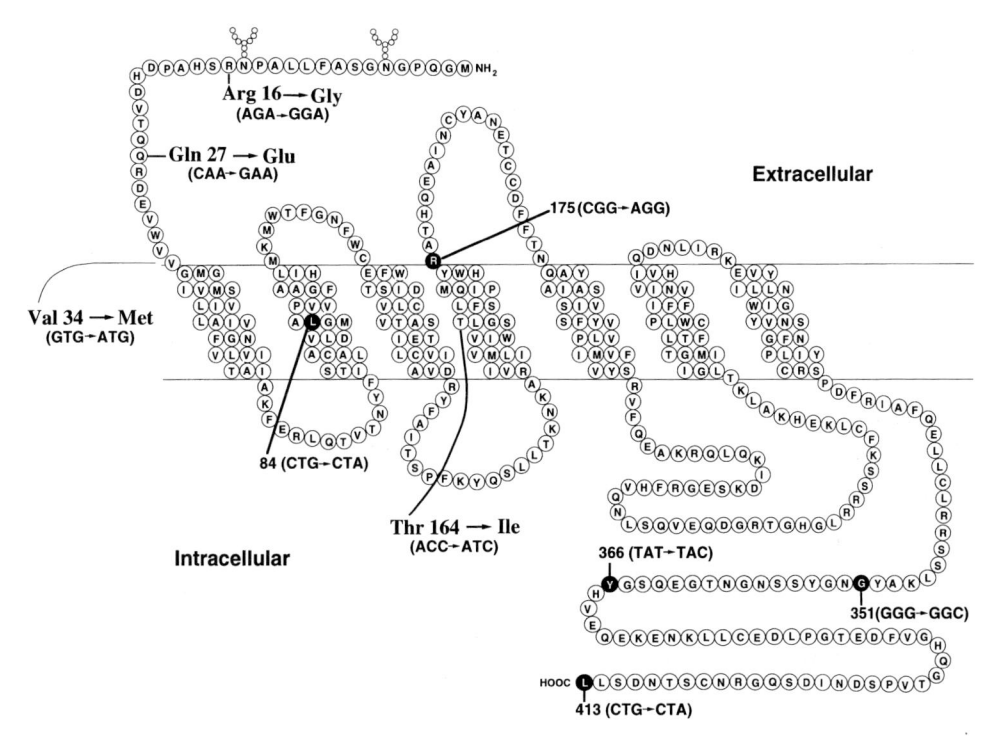

Fig. 2.1 The human β_2-adrenergic receptor and the position of the nine mutations. The mutations that do not alter the amino acid sequence are indicated by circles, whereas the mutations that cause changes in the encoded amino acids are labelled. Reproduced with permission from Dr S.B. Liggett.

found more often in the asthma patients than in the healthy control subjects. In the asthma group, one mutation (substitution of Gly by Arg at position 16) identified a subset of patients with more severe asthma, characterized by higher use of corticosteroid and requiring immunization therapy more often.[60] The same mutation was found to be associated with nocturnal asthma in a study of 45 asthmatic patients.[61] A group from Japan investigated two polymorphisms in the β_2-adrenergic receptor gene in four three-generation families recruited through a proband with asthma.[58] A polymorphism in the β_2-adrenergic receptor gene was reported that was associated with lower airways responses to salbutamol and a higher incidence of asthma. The β_2-adrenergic receptor displays an unexpectedly high degree of genetically based structural variance in the population. Since these receptors differ in certain properties, they may play a role in the expression of certain asthmatic phenotypes or determine how patients respond to therapy. This has to be further explored.

CONCLUSION

Important accomplishments have been achieved during the last few years in understanding the genetic basis of asthma. However, no gene has yet been clearly implicated in the development or progression of asthma. Future research will need the united effort of clinicians, genetic epidemiologists and molecular biologists in order to achieve the ultimate goal of finding genes involved in asthma. New developments, such as the completion of the Human Genome Project, in which all the genes on the human genome will be sequenced and mapped, may facilitate the search for genes. This may eventually lead to new diagnostic methods and new strategies for therapeutic interventions in asthma.

REFERENCES

1. Bray GW: The hereditary factor in asthma and other allergies. *Br Med J* (1930) 384–387.
2. Edfors-Lubs ML: Allergy in 7000 twin pairs. *Acta Allergol* (1971) **26**: 249–285.
3. Duffy DL, Martin NG, Battistutta D, Hopper JL, Mathews JD: Genetics of asthma and hay fever in Australian twins. *Am Rev Respir Dis* (1990) **142**: 1351–1358.
4. Hopp RJ, Bewtra AK, Watt GD, Nair NM, Townley RG: Genetic analysis of allergic disease in twins. *J Allergy Clin Immunol* (1984) **73**: 265–270.
5. Antequera F, Bird A: Predicting the total number of human genes. *Nature Genetics* (1994) **8**: 114.
6. Puffenberger EG, Kauffman ER, Bolk S, et al.: Identity-by-descent and association mapping of a recessive gene for Hirschsprung disease on human chromosome 13q22. *Hum Mol Genet* (1994) **3**: 1217–1225.
7. Hasstedt SJ, Meyers DA, Marsh DG: Inheritance of immunoglobulin E: genetic model fitting. *Am J Med Genet* (1983) **14**: 61–66.
8. Borecki IB, McGue M, Gerrard JW, Lebowitz MD, Rao DC: Familial resemblance for immunoglobulin levels. *Hum Genet* (1994) **94**: 179–185.
9. Meyers DA, Beaty TH, Freidhoff LR, Marsh DG: Inheritance of total serum IgE (basal levels) in man. *Am J Hum Genet* (1987) **41**: 51–62.
10. Meyers DA, Beaty TH, Colyer CR, Marsh DG: Genetics of total serum IgE levels: a regressive model approach to segregation analysis. *Genet Epidemiol* (1991) **8**: 351–359.
11. Dizier MH, Hill M, James A, et al.: Detection of a recessive major gene for high IgE levels acting independently of specific response to allergens. *Genet Epidemiol* (1995) **12**: 93–105.
12. Meyers DA, Postma DS, Panhuysen CI, et al.: Evidence for a locus regulating total serum IgE levels mapping to chromosome 5. *Genomics* (1994) **23**: 464–470.
13. Xu J, Levitt RC, Panhuysen CIM, Postma DS, et al.: Evidence for two unlinked loci regulating total serum IgE levels. *Am J Hum Genet* (1995) **57**: 425–430.
14. Meyers DA, Bias WB, Marsh DG: A genetic study of total IgE levels in the Amish. *Hum Hered* (1982) **32**: 15–23.
15. Martinez FD, Holberg CJ, Halonen M, Morgan WJ, Wright AL, Taussig LM: Evidence for Mendelian inheritance of serum IgE levels in Hispanic and non-Hispanic white families. *Am J Hum Genet* (1994) **55**: 555–565.
16. Blumenthal MN, Namboodiri K, Mendell N, Gleich G, Elston RC, Yunis E: Genetic transmission of serum IgE Levels. *Am J Med Genet* (1981) **10**: 219–228.
17. O'Connor GT, Sparrow D, Weiss ST: The role of allergy and nonspecific airway hyperresponsiveness in the pathogenesis of chronic obstructive pulmonary disease. *Am Rev Respir Dis* (1989) **140**: 225–252.
18. Holgate ST, Beasley R, Twentyman OP: The pathogenesis and significance of bronchial hyperresponsiveness in airways disease. *Clin Sci* (1987) **73**: 561–572.
19. Townley RG, Bewtra A, Wilson AF, et al.: Segregation analysis of bronchial response to methacholine inhalation challenge in families with and without asthma. *J Allergy Clin Immunol* (1986) **77**: 101–107.
20. Longo G, Strinati R, Poli F, Fumi F: Genetic factors in nonspecific bronchial hyperreactivity. An epidemiologic study. *Am J Dis Child* (1987) **141**: 331–334.

21. Panhuysen CIM, Xu J, Postma DS, Bleecker ER, Meyers DA: Evidence for a major locus for bronchial hyperresponsiveness independent of the locus regulating total serum IgE levels. *Am J Hum Genet* (1996) **59**: A231.
22. Holberg CJ, Elston RC, Halonen M, *et al.*: Segregation analysis of physician-diagnosed asthma in Hispanic and non-Hispanic white families. A recessive component? *Am J Respir Crit Care Med* (1996) **154**: 144–150.
23. Cookson WOCM, Sharp PA, Faux JA, Hopkin JM: Linkage between immunoglobulin E responses underlying asthma and rhinitis and chromosome 11q. *Lancet* (1989) **i**: 1292–1295.
24. Lympany P, Welsh KI, Cochrane GM, Kemeny DM, Lee TH: Genetic analysis of the linkage between chromosome 11q and atopy. *Clin Exp Allergy* (1992) **22**: 1085–1092.
25. Rich SS, Roitman-Johnson B, Greenberg B, Roberts S, Blumenthal MN: Genetic analysis of atopy in three large kindreds: no evidence of linkage to D11S97. *Clin Exp Allergy* (1992) **22**: 1070–1076.
26. Hizawa N, Yamaguchi E, Ohe M, *et al.*: Lack of linkage between atopy and locus 11q13. *Clin Exp Allergy* (1992) **22**: 1065–1069.
27. Lympany P, Welsh K, MacCochrane G, Kemeny DM, Lee TH: Genetic analysis using DNA polymorphism of the linkage between chromosome 11q13 and atopy and bronchial hyperresponsiveness to methacholine. *J Allergy Clin Immunol* (1992) **89**: 619–628.
28. Amelung PJ, Panhuysen CI, Postma DS, *et al.*: Atopy and bronchial hyperresponsiveness: exclusion of linkage to markers on chromosomes 11q and 6p. *Clin Exp Allergy* (1992) **22**: 1077–1084.
29. Cookson WOCM, Young RP, Sandford AJ, *et al.*: Maternal inheritance of atopic IgE responsiveness on chromosome 11q. *Lancet* (1992) **340**: 381–384.
30. Sandford AJ, Shirakawa T, Moffatt MF, *et al.*: Localisation of atopy and beta subunit of high-affinity IgE receptor (Fc epsilon RI) on chromosome 11q. *Lancet* (1993) **341**: 332–334.
31. Galli SJ: New concepts about the mast cell. *N Engl J Med* (1993) **328**: 257–265.
32. Shirakawa T, Li A, Dubowitz M, *et al.*: Association between atopy and variants of the beta subunit of the high-affinity immunoglobulin E receptor. *Nature Genetics* (1994) **7**: 125–129.
33. Hill MR, James AL, Faux JA, *et al.*: Fc epsilon RI-beta polymorphism and risk of atopy in a general population sample. *Br Med J* (1995) **311**: 776–779.
34. Shirakawa T, Mao XQ, Sasaki S, Kawai M, Morimoto K, Hopkin JM: Association between Fc epsilon RI beta and atopic disorder in a Japanese population. *Lancet* (1996) **347**: 394–395.
35. Daniels SE, Bhattacharrya S, James A, *et al.*: A genome-wide search for quantitative trait loci underlying asthma. *Nature* (1996) **383**: 247–250.
36. Collee JM, ten Kate LP, de Vries HG, *et al.*: Allele sharing on chromosome 11q13 in sibs with asthma and atopy. *Lancet* (1993) **342**: 936.
37. Doull IJ, Lawrence S, Watson M, *et al.*: Allelic association of gene markers on chromosomes 5q and 11q with atopy and bronchial hyperresponsiveness. *Am J Respir Crit Care Med* (1996) **153**: 1280–1284.
38. Coleman R, Trembath RC, Harper JI: Chromosome 11q13 and atopy underlying atopic eczema. *Lancet* (1993) **341**: 1121–1122.
39. Brereton HM, Ruffin RE, Thompson PJ, Turner DR: Familial atopy in Australian pedigrees: adventitious linkage to chromosome 8 is not confirmed nor is there evidence of linkage to the high affinity IgE receptor. *Clin Exp Allergy* (1994) **24**: 868–877.
40. Watson M, Lawrence S, Collins A, *et al.*: Exclusion from proximal 11q of a common gene with megaphenic effect on atopy. *Ann Hum Genet* (1995) **59**: 403–411.
41. Martinati LC, Trabetti E, Casartelli A, Boner AL, Pignatti PF: Affected sib-pair and mutation analyses of the high affinity IgE receptor beta chain locus in Italian families with atopic asthmatic children. *Am J Respir Crit Care Med* (1996) **153**: 1682–1685.
42. Hall IP, Wheatley A, Dewar J, Wilkinson J, Morrison J: Fc epsilon RI-beta polymorphisms unlikely to contribute substantially to genetic risk of allergic disease. *Br Med J* (1996) **312**: 311.
43. Fukao T, Kaneko N, Teramoto T, Tashita H, Kondo N: Association between FcεRIβ and atopic disorder in Japanese population? *Lancet* (1996) **348**: 407.
44. Chandrasekharappa SC, Rebelsky MS, Firak TA, Le Beau MM, Westbrook CA: A long-range

restriction map of the interleukin-4 and interleukin-5 linkage group on chromosome 5. *Genomics* (1990) **6**: 94–99.
45. Marsh DG, Neely JD, Breazeale DR, et al.: Linkage analysis of IL4 and other chromosome 5q31.1 markers and total serum immunoglobulin E concentrations. *Science* (1994) **264**: 1152–1156.
46. Blumenthal MN, Wang Z, Weber JL, Rich SS: Absence of linkage between 5q markers and serum IgE levels in four large atopic families. *Clin Exp Allergy* (1996) **26**: 1–5.
47. Mao XQ, Shirakawa T, Yoshikawa T, et al.: Association between genetic variants of mast-cell chymase and eczema. *Lancet* (1996) **348**: 581–583.
48. Barnes KC, Neely JD, Duffy DL, et al.: Linkage of asthma and total serum IgE concentration to markers on chromosome 12q: evidence from Afro-Caribbean and Caucasian populations. *Genomics* (1996) **37**: 41–50.
49. Marsh DG, Freidhoff LR, Ehrlich Kautzky E, Bias WB, Roebber M: Immune responsiveness to *Ambrosia artemisiifolia* (short ragweed) pollen allergen *Amb a* VI (Ra6) is associated with HLA-DR5 in allergic humans. *Immunogenetics* (1987) **26**: 230–236.
50. Young RP, Dekker JW, Wordsworth BP, et al.: HLA-DR and HLA-DP genotypes and immunoglobulin E responses to common major allergens. *Clin Exp Allergy* (1994) **24**: 431–439.
51. Li PK, Lai CK, Poon AS, Ho AS, Chan CH, Lai KN: Lack of association between HLA-DQ and -DR genotypes and asthma in southern Chinese patients. *Clin Exp Allergy* (1995) **25**: 323–331.
52. Bignon JS, Aron J, Ju LY, et al.: HLA class II alleles in isocyanate-induced asthma. *Am J Respir Crit Care Med* (1994) **149**: 71–57.
53 Balboni A, Baricordi OR, Fabbri LM, Gandini E, Ciaccia A, Mapp CE: Association between toluene diisocyanate-induced asthma and DQB1 markers: a possible role for aspartic acid at postion 57. *Eur Respir J* (1996) **9**: 207–210.
54. Moffatt MF, Hill MP, Cornelis F, et al.: Genetic linkage of T-cell receptor alpha/delta complex to specific IgE responses. *Lancet* (1994) **343**: 1597–1600.
55. Postma DS, Bleecker ER, Amelung PJ, et al.: Genetic susceptibility to asthma: bronchial hyperresponsiveness coinherited with a major gene for atopy. *N Engl J Med* (1995) **333**: 894–900.
56. van Herwerden L, Harrap SB, Wong ZY, et al.: Linkage of high-affinity IgE receptor gene with bronchial hyperreactivity, even in absence of atopy. *Lancet* (1995) **346**: 1262–1265.
57. Hall IP, Wheatley A, Wilding P, Liggett SB: Association of Glu 27 beta-2 adrenoceptor polymorphism with lower airway reactivity in asthmatic subjects. *Lancet* (1995) **345**: 1213–1214.
58. Ohe M, Munakata M, Hizawa N, et al.: Beta-2 adrenergic receptor gene restriction fragment length polymorphism and bronchial asthma. *Thorax* (1995) **50**: 353–359.
59. Kerstjens HAM, Brand PLP, Hughes MD, et al.: A comparison of bronchodilator therapy with or without inhaled corticosteroid therapy for obstructive airways disease. Dutch Chronic Non-Specific Lung Disease Study Group. *N Engl J Med* (1992) **327**: 1413–1419.
60. Reihsaus E, Innis M, MacIntyre N, Liggett SB: Mutations in the gene encoding for the beta-2 adrenergic receptor in normal and asthmatic subjects. *Am J Respir Cell Mol Biol* (1993) **8**: 334–339.
61. Turki J, Pak J, Green SA, Martin RJ, Liggett SB: Genetic polymorphisms of the beta-2 adrenergic receptor in nocturnal and nonnocturnal asthma. Evidence that Gly16 correlates with the nocturnal phenotype. *J Clin Invest* (1995) **95**: 1635–1641.

3

Airway Pathology in Asthma

PETER K. JEFFERY

INTRODUCTION

This chapter reports and illustrates the salient structural and inflammatory changes of asthma. The definition of asthma will be that of the American Thoracic Society.[1] For an overview of the morphology of the normal airway the reader is referred elsewhere.[2]

Occlusion of the airway lumen by tenacious secretions, tissue eosinophilia, loss of airway surface epithelium, thickening of the reticular basement membrane (also referred to as the lamina reticularis) and enlargement of bronchial smooth muscle and submucosal gland mass are pathological features that have been consistently reported in cases of fatal asthma since the end of the last century.[3-11] Alterations to the airway mucosa occur early in the disease process as evidenced by examination of bronchial biopsies obtained from subjects with newly diagnosed mild asthma; these include loss of surface epithelium, tissue eosinophilia and thickening of the reticular basement membrane.[11-17] Tissue eosinophilia and eosinophil activation and degranulation are also key features of the late-phase reaction (LPR), which may occur in the hours following allergen challenge.[18-21] Infiltration of the mucosa by inflammatory cells including eosinophils is a feature of the airway mucosa in all forms of asthma, whether mild or severe, allergic (also called extrinsic or atopic) or non-allergic (intrinsic or non-atopic), or that due to occupation (e.g. exposure to isocyanates).[16,22-25]

To date the pathologist recognizes only one form of asthma yet clinically the condition is clearly heterogeneous.

SPUTUM AND BRONCHOALVEOLAR LAVAGE

Examination of sputa and airway fluids obtained at bronchoalveolar lavage (BAL) have provided clues to the underlying abnormalities in each condition. The critical examination of spontaneously produced or saline-induced sputum will probably become a much used and relatively non-invasive method for determining the extent of inflammation in the asthmatic airway.[26-29] Corkscrew-shaped twists of condensed mucus (Curschmann's spirals),[30] clusters of surface airway epithelial cells (called Creola bodies),[31] and the presence of Charcot–Leyden crystals (composed of eosinophil cell and granule membrane lysophospholipase)[32] together with eosinophils and metachromatic cells are characteristic features of sputa obtained from asthmatic but not bronchitic patients.[33] However, sputum eosinophilia has also been reported in the absence of the airway hyperresponsiveness (AHR) characteristic of asthma.[34]

APPEARANCES AT POST-MORTEM

Post-mortem examination of cases of fatal asthma has shown that the lungs are hyperinflated and remain so on opening the pleural cavities due to the widespread presence of markedly tenacious plugs in intrapulmonary bronchi (see Plate 3a). On intrabronchial inflation with fixative, even a 1.5-m head of fluid often fails to move these airway plugs.[8,35] Histologically, the airway plugs in asthma are a mixture of inflammatory exudate and mucus in which lie desquamated surface epithelial cells, lymphocytes and eosinophils. The arrangement of the cellular elements of the plug often takes the form of several concentric lamella, suggesting that several episodes of inflammation have led to their formation rather than a single (terminal) event (see Plate 3b). The non-mucinous, proteinaceous contribution is the result of increased vascular permeability and includes a fibrinous component. Interaction of constituents of serum and mucin is likely to lead to increased viscosity of the airway plug.[36] Interestingly, there are reports of sudden asthma death in which intraluminal plugs may be absent[37] but these are rare.[9,38] The combination of tissue, blood and BAL/sputum eosinophilia is strongly associated with asthma but there may also be marked heterogeneity in the numbers of tissue eosinophils identified in fatal asthma.[39] This may be due, in part, to eosinophil degranulation, which makes cell identification difficult, or to the reported variation in the numbers and relative proportions of neutrophils and eosinophils with progressive duration of the terminal episode.[17,40] Unlike chronic obstructive pulmonary disease (COPD), there is little evidence of destructive emphysema in fatal asthma and right ventricular hypertrophy is uncommon when the diagnosis of asthma is uncomplicated by COPD. Areas of atelectasis and petechial haemorrhages may be present due to bronchial obstruction, reabsorption collapse and repeated forced inspiratory efforts.

LOSS OF SURFACE EPITHELIUM

Histologically, shedding and damage of airway surface epithelium is prominent in asthma, both in fatal asthma (see Plate 1a,b) and in biopsy specimens of patients with mild disease.[13-15] Loss of epithelium is followed by areas of mitotic activity[41] and epithelial regeneration, which first appears in the form of simple or stratified squamous epithelium[8] prior to its differentiation and maturation to form new ciliated and mucous (goblet) cells. In symptomatic asthma, there may be platelet aggregation together with fibrillary material, thought to be fibrin, at sites of damage (Fig. 3.1). Such deposits of fibrillary material are also seen during the LPR following allergen challenge (P.K. Jeffery, unpublished results). The greater the loss of surface epithelium in biopsy specimens, the greater appears to be the degree of AHR.[14] It is recognized that there is an inevitable artefactual loss of surface epithelium during the taking and processing of these small (2 mm diameter) biopsy pieces, even in normal healthy subjects, which makes interpretation of the extent of epithelial sloughing controversial.[42] The suggested fragility of the epithelium in asthma *in vivo* is supported by the frequent reports of Creola bodies in sputa[31] and the reported association between the numbers of bronchial epithelial cells recovered by BAL and the degree of AHR in asthmatic patients with mild disease.[15]

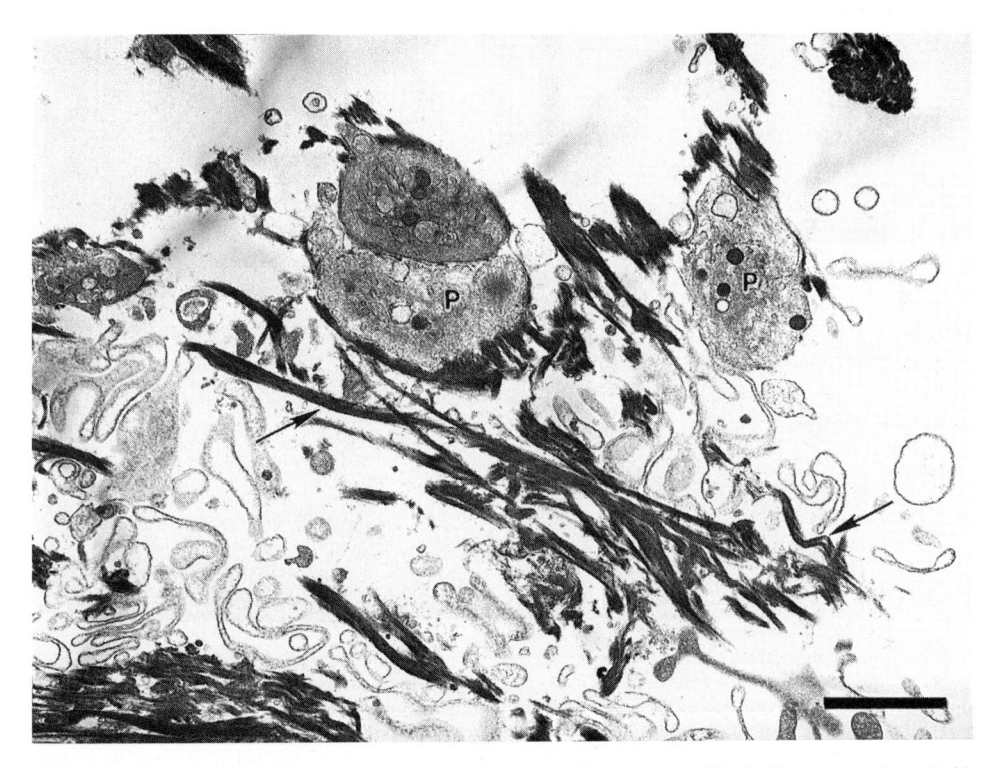

Fig. 3.1 In symptomatic subjects, aggregations of platelets (P) together with fibrillary material, probably fibrin (arrows), is formed where the surface epithelium is lost. Scale bar = 2.0 μm.

The fragility of the surface may involve disruption of tight junctions,[43,44] which act as a selective epithelial barrier to the passage of ions, molecules and water between cells: this disruption may enhance stimulation of intraepithelial nerves (Fig. 3.2) leading to axonal reflexes, stimulation of secretion by mucous glands, vasodilatation and oedema through the release of sensory neuropeptides (i.e. neurogenic inflammation).[45,46] There is also experimental evidence that the sensitivity of bronchial smooth muscle to substances placed in the airway lumen correlates strongly with the integrity of the surface epithelium.[47] Loss or damage of surface epithelium would thus lead to a reduction in the concentration of factors normally relaxant to bronchial smooth muscle with resultant increased sensitivity and 'reactivity'.[48,49]

When superficial cells are lost in asthma the preferential plane of cleavage appears to be between superficial and basal cells,[50] leaving basal cells still attached to their basement membrane. Epithelial cells may also act as effector cells by their synthesis and release of cytokines such as interleukin (IL)-6, IL-8 and granulocyte–macrophage colony-stimulating factor (GM-CSF)[51] (see Chapter 10). Disruption of the epithelium and attempts at repair may increase production of these pro-inflammatory cytokines by those cells which remain.

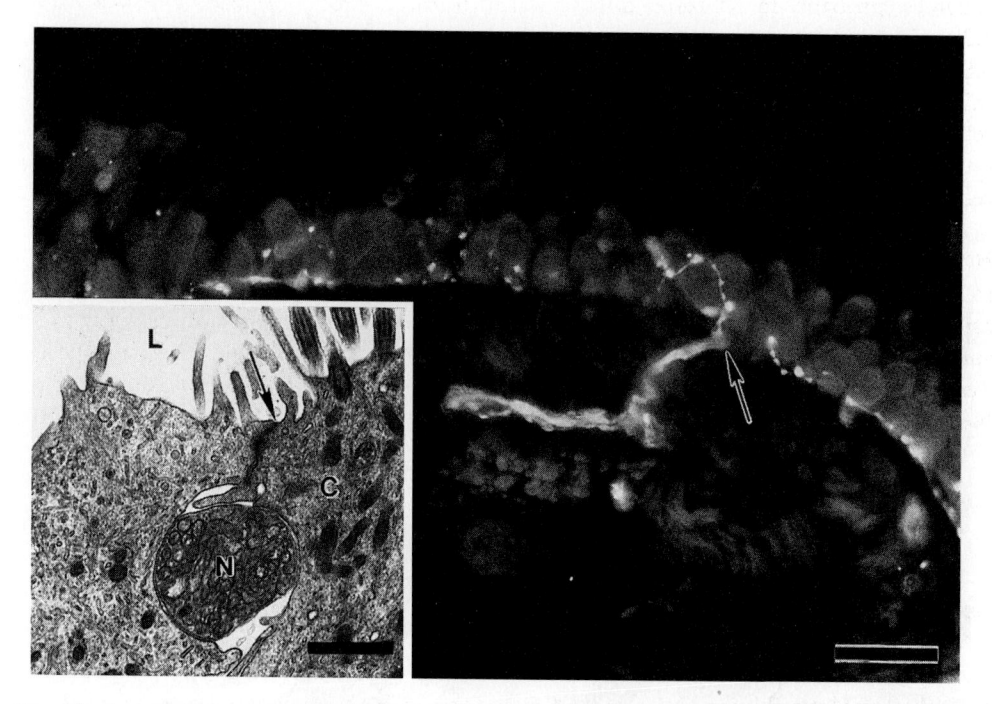

Fig. 3.2 Immunofluorescence of the localization of PGP 9.5, an antibody directed against nerve terminals. A large subepithelial nerve bundle reduces in thickness to a point where it enters (arrow) the surface epithelium of a human intrapulmonary bronchus. Scale bar = 50 μm. The inset (bottom left) shows the ultrastructure of an intraepithelial nerve (N) lying immediately beneath the tight junction (arrow) of one ciliated (C) and a non-ciliated epithelial cell. The nerve has lost its Schwann cell covering and basement membrane to become enclosed by the epithelial cells. Disruption of the tight junction would allow entry of allergen from the airway lumen (L). Scale bar = 1.0 μm.

3 Airway Pathology in Asthma

THICKENING OF THE EPITHELIAL 'BASEMENT MEMBRANE'

Observed by light microscopy, thickening of the reticular basement membrane (i.e. lamina reticularis) has long been recognized as a consistent change in all forms of asthma[8,14,52-56] (see Plate 1b). Whilst there may also be focal and variable thickening in COPD and other inflammatory chronic diseases of the lung such as bronchiectasis and tuberculosis,[56] the lesion, when homogeneous and hyaline in appearance, is highly characteristic and present in both fatal and mild asthma and in patients with a long history of asthma but who have not died of their asthma (Fig. 3.3a). The thickening of the reticular layer, which is immunopositive for collagen types III and V together with fibronectin but not laminin, has been referred to as 'subepithelial fibrosis'.[53] However, its thickening is distinct from the fibrosis associated with scar formation as, ultrastructurally, it does not resemble the underlying interstitial collagen of a scar. The reticular layer is composed of thinner fibres of reticulin linked to a matrix rich in sugars together with entrapped exogenous molecules such as tenascin, heparin sulphate and serum-derived components (Fig. 3.3b). Swelling of this layer may also contribute to its thickening and increased rigidity. Interestingly, the thickened layer does not behave as a barrier to the transmigration of inflammatory cells which, by the release of enzymes (probably metalloproteases), can pass through it with apparent ease (see Fig. 3.2b). In contrast, the 'true' epithelial basement membrane (i.e. the basal lamina), which consists mainly of type IV collagen, glycosaminoglycans and laminin, is not thickened (Fig. 3.4). It has been predicted that thickening and increased rigidity of the reticular basement membrane may reduce its capacity to fold during bronchial smooth muscle contraction and that reduced folding would result in increased airway responsiveness.[57]

Adjacent subepithelial fibroblasts may contribute to the thickening of the reticular layer (Fig. 3.5a). An association between the numbers of myofibroblasts underlying the reticular layer and thickening of the reticular layer has also been demonstrated in asthma.[58] Gizycki and colleagues[59] have also observed that myofibroblasts appear in substantial numbers during the LPR following allergen challenge (Fig. 3.5b) and these may contribute, via secretion of additional reticulin, to the thickening of the reticular basement membrane.

INCREASED NUMBERS OF MUCUS-SECRETING CELLS

Bronchial goblet cell hyperplasia and submucosal gland enlargement have been reported as histological hallmarks and are the correlate of hypersecretion of mucus in chronic bronchitis.[60] There is also significant submucosal gland enlargement seen in fatal asthma[35] and this may contribute to the excessive production of mucus and the plugging of airways usually associated with a fatal attack (see Chapter 11).[61] Dilatation of gland ducts, referred to as bronchial gland ectasia, is also described.[62]

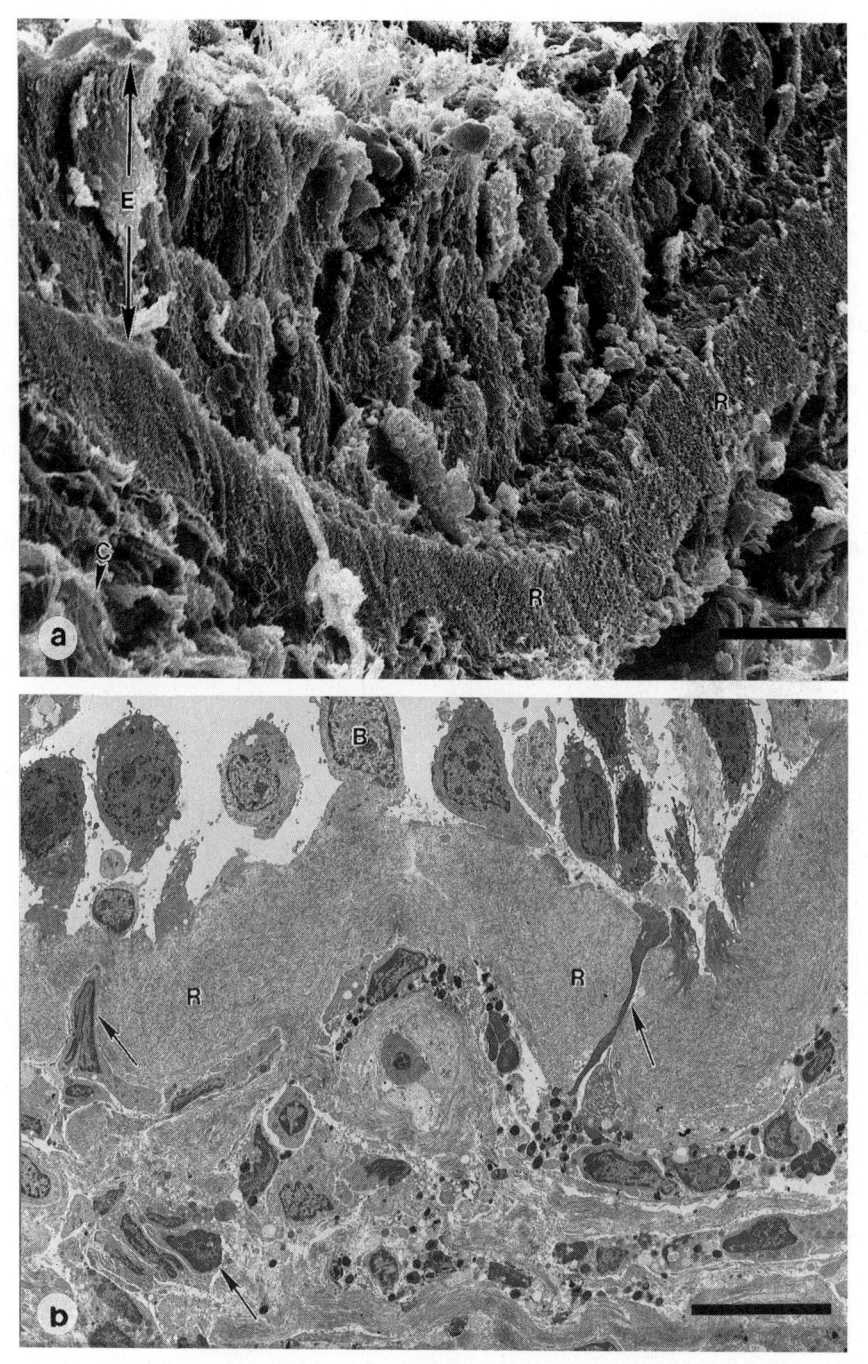

Fig. 3.3 (a) Scanning electron micrograph of the airway of a patient with a 25-year history of asthma but who died postoperatively of another cause. The fracture plane shows disrupted surface epithelium (E) beneath which there is a dense layer of reticulin (R) with strands of interstitial collagen (C) below. Scale bar = 10 μm. (b) Transmission electron micrograph of the subepithelial zone in an atopic asthmatic patient during a late-phase response. There is a recruitment of inflammatory cells (arrows), which appear to be able to migrate across the thickened reticular lamina (R). B, remaining basal cells of the surface epithelium. Scale bar = 10 μm.

3 Airway Pathology in Asthma

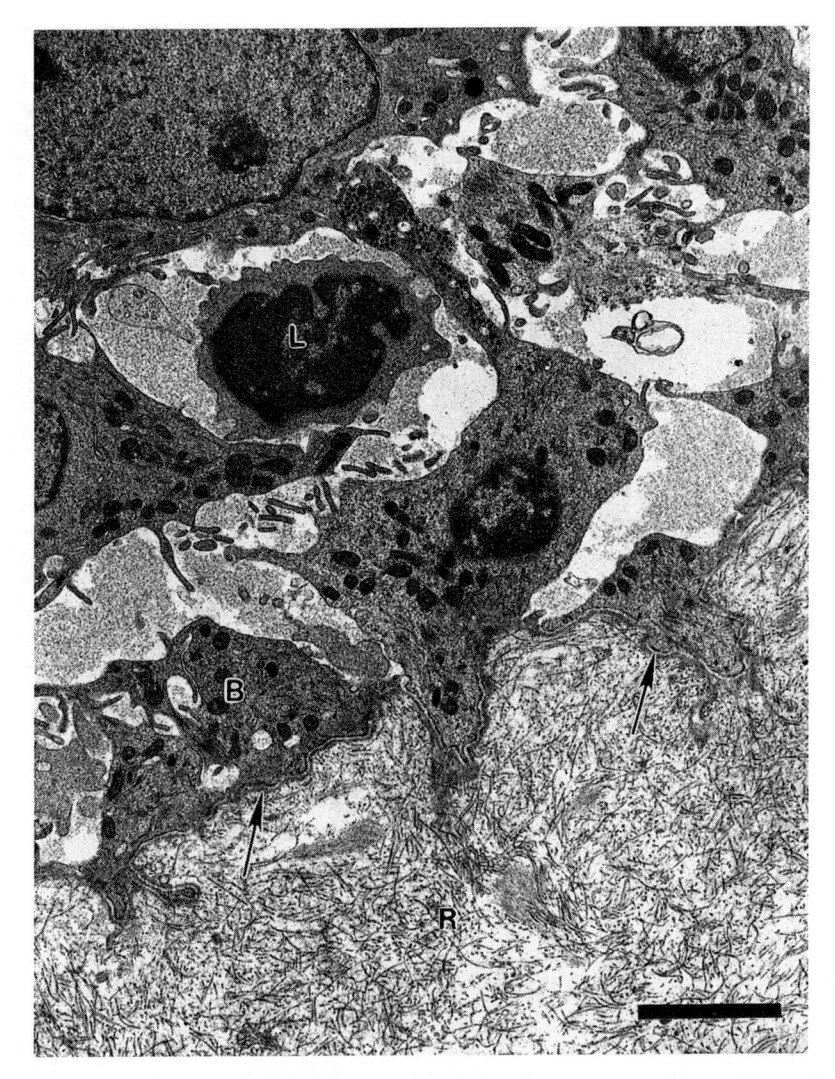

Fig. 3.4 Transmission electron micrograph of the basal zone of the surface epithelium of human bronchus, illustrating the basal lamina (also referred to as the 'true' basement membrane) to which the epithelial basal cells (B) attach (arrows). It is the lamina reticularis (R) (i.e. the reticular basement membrane) comprised of reticular fibres that is thickened in asthma. L, intraepithelial lymphocyte. Scale bar = 2.5 μm.

ENLARGEMENT OF BRONCHIAL SMOOTH MUSCLE MASS

The percentage of bronchial wall occupied by bronchial smooth muscle shows a marked increase in fatal asthma[35] (Fig. 3.6). Importantly, the increase in muscle mass is reported to be in the larger rather than the smaller intrapulmonary bronchi of lungs obtained following a fatal attack compared with those of asthmatic subjects dying of other causes:[63] it is likely a major contributor to the thickening of the airway wall and hence to the increased resistance to airflow.[64–67] Using a morphometric technique Dunnill

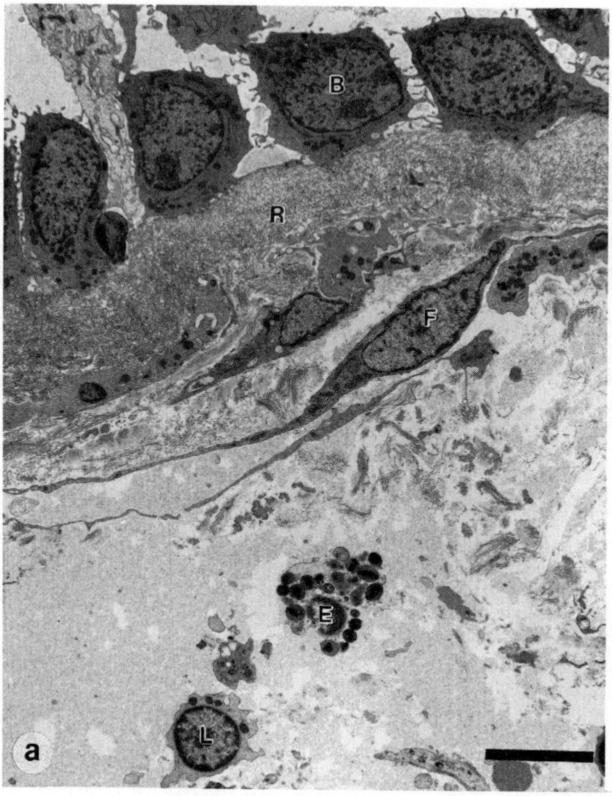
B
R
F
E
L
a

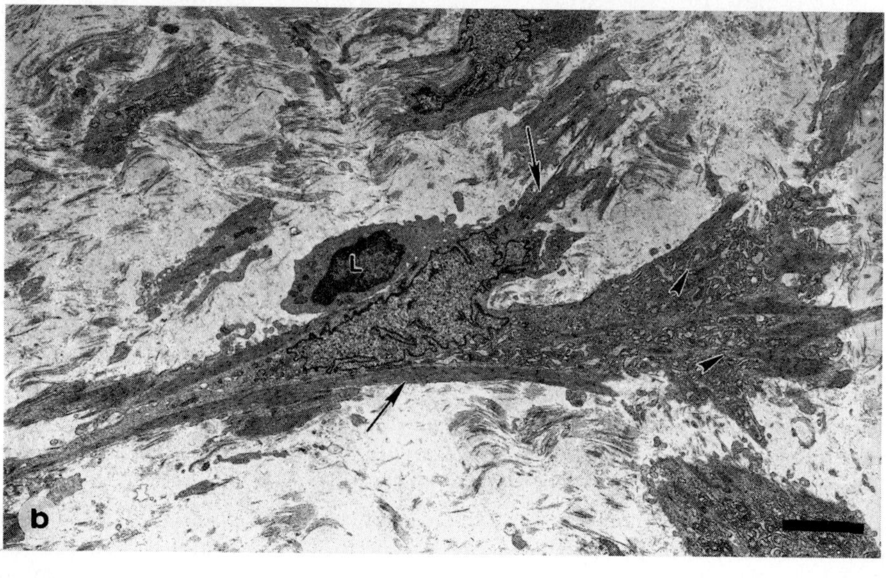
L
b

3 Airway Pathology in Asthma

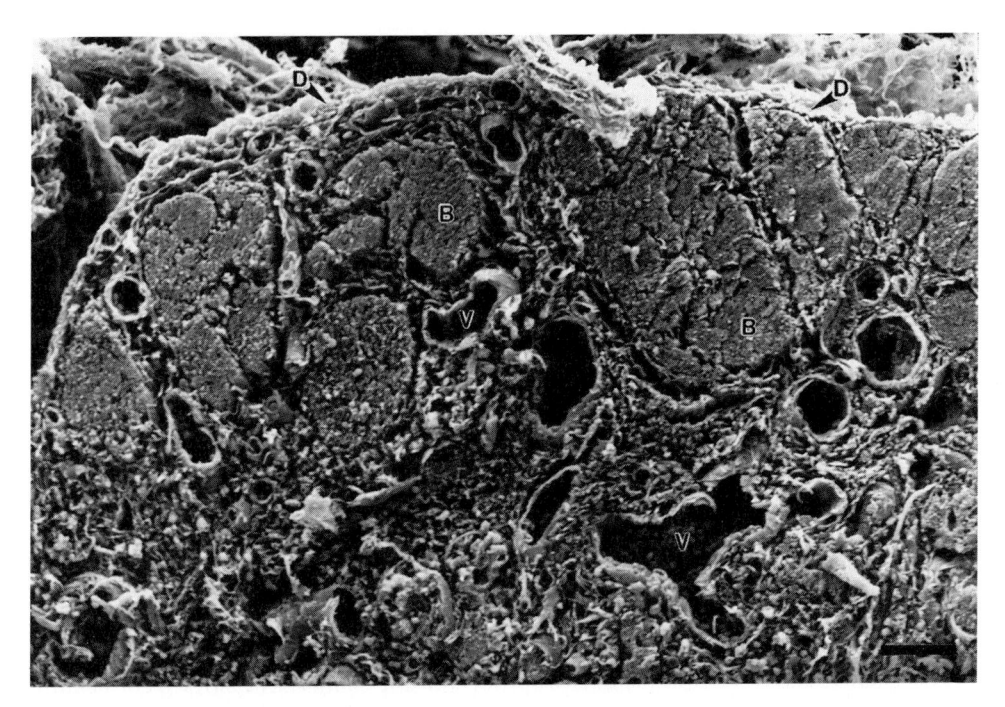

Fig. 3.6 Scanning electron micrograph of part of the bronchial wall in a case of fatal asthma. The mucosa is thickened by marked dilatation of bronchial vessels (V) and an increase in the mass of bronchial smooth muscle (B). The surface epithelium has been completely lost leaving a denuded area (D). Scale bar = 50 μm.

showed that approximately 12% of the wall in segmental bronchi obtained from cases of fatal asthma was composed of muscle compared with a figure of approximately 5% in the normal.[35] Hogg and colleagues[68] have confirmed this contribution by bronchial muscle to airway wall thickening in airways larger than 2 mm diameter and demonstrated a three- to four-fold increase over normal in the area of the wall occupied by bronchial smooth muscle. In asthma the increase in muscle mass does not appear to extend to airways of less than 2 mm in diameter.[69] In the absence of wheeze, values for muscle mass in segmental bronchi in chronic bronchitis and emphysema fall largely within the normal range but intermediate levels are present in so-called wheezy bronchitis, where there appears to be an asthmatic component.[70]

Whether the increase in muscle mass is due to muscle fibre hyperplasia[71] or hypertrophy is at present unclear. Two patterns of distribution of increased muscle mass have been described, one in which the increase occurs throughout the airways and

Fig. 3.5 (a) Mild atopic asthma: the release of eosinophil (E) granule and membrane products may damage the mucosa and its epithelium. In addition, factors such as transforming growth factor β may stimulate nearby fibroblasts (F) to produce increased reticulin (R). B, epithelial basal cell; L, subepithelial lymphocyte. Scale bar = 5.0 μm. (b) Transmission electron micrograph of the bronchial mucosa during the late-phase response. There is a substantial increase of the numbers of myofibroblasts: each is characterized by a highly crenated nucleus, dilated rough endoplasmic reticulum (arrowheads) and bundles of myofibrils (arrows). L, adjacent lymphocyte. Scale bar = 5.0 μm.

another in which the increase is restricted to the largest airways:[72,73] it is suggested that in the former there is muscle fibre hyperplasia and hypertrophy, in the latter hypertrophy alone. Recent observations of the LPR to allergen have demonstrated the increased presence of cell forms that share ultrastructural features of fibroblast, myofibroblast and bronchial smooth muscle. These cells may represent the precursors of the additional blocks of bronchial smooth muscle seen in fatal asthma. These observations of airway wall remodelling in asthma[74] show similarity to those seen in vascular disease.[75] Myofibroblast differentiation and their role in bronchial smooth muscle mass enlargement may become a novel target for antiasthma treatment in the future.[74,76]

BRONCHIAL VASCULATURE, CONGESTION AND OEDEMA

The increase in thickness of the bronchial wall in asthma is unlikely accounted for by the increase in bronchial smooth muscle and mucous gland mass alone. Dilatation of the mucosal bronchial vasculature, congestion of its vessels, new vessel growth and wall oedema are also features of fatal asthma (see Plate 3 and Fig. 3.6). Subepithelial oedema has been suggested to be responsible for lifting and sloughing of the surface epithelium.[8] The onset of vasodilatation, congestion and mucosal oedema in response to a variety of mediators of inflammation,[77] and perhaps that which occurs in response to exercise, can be rapid and, equally, should be relatively rapidly reversed by appropriate treatment.

James and colleagues[64] have shown that airway wall thickening (due to one or more of the above changes) need only be relatively minor to have dramatic consequences on airflow limitation. The increased wall thickness in airways greater than 2 mm diameter was described in fatal asthma by Huber and Koessler[69] in 1922. The relevance of the thickening to reduced airflow and AHR was discussed by Freedman[78] and Benson,[79] who suggested that for a given degree of smooth muscle shortening the effect on luminal narrowing and hence resistance to airflow (to the fourth power) would be considerably greater if the airway wall were thickened. The concept and link of airway geometry to AHR has been supported by the results of computer modelling: the model predicts that when the airway wall is thickened there will be only moderate effects on airflow resistance in the absence of muscle contraction but, in contrast, there will be profound effects when bronchial smooth muscle shortens, even normally.[65–67] The extent to which muscle shortening can occur is determined, in part, by the force opposing it: in the airways the opposing force is largely generated by the elastic recoil of the attached alveolar walls which surround each airway. Airway wall oedema has been suggested to uncouple muscle from its surrounding lung tissue, thereby allowing a greater force, and maximal extent, of contraction.[80] In a similar way, destruction of alveolar attachments to bronchioli in smoker's COPD[81] may lead to reduction in forced expiratory volume in 1 s (FEV_1) and the development of apparent AHR. Loss of lung elastic recoil in asthma would be predicted to have a similar effect but a study of the elastic fibre content of asthmatic lung and bronchi has shown that there is at least no reduction of elastic fibre content as determined by morphometry of tissue sections.[82] The association of structure and function is an interesting and important area requiring much further study.

3 Airway Pathology in Asthma

RECRUITMENT OF INFLAMMATORY CELLS

In fatal asthma there is a marked inflammatory cell infiltrate throughout the airway wall and also in the occluding plug (see Plates 3b and 1b): lymphocytes are abundant,[8,17,83] eosinophils are characteristic and neutrophils are usually absent. The inflammation may spread to surrounding alveolar septae and affect adjacent arteries.[83] The association of tissue eosinophilia and asthma is a strong one but the extent of tissue eosinophilia varies with each case and the duration of the terminal episode.[17,39,40]

Asthma is now recognized as an inflammatory condition of the airways in which there is tissue eosinophilia and a predominance of T-lymphocytes of the CD4 (helper) subset (see Chapters 9 and 19). The activation of the T-helper cells results in the release of cytokines, particularly IL-4, IL-5 and IL-10 which characterize an 'allergic' profile of inflammation. Release of these cytokines leads to the recruitment of eosinophils (not neutrophils) from bronchial vessels, their activation and the release of a range of highly charged molecules that damage mucosal tissue (Fig. 3.7). However, in some cases of asthma death eosinophils may not be identified at all.[17,39] Extensive eosinophil degranulation and 'disintegration' may make cell identification difficult. Neutrophils appear to predominate in acute attacks of asthma associated with sudden death,[40] and whilst there are increased numbers of T-cells in fatal asthma this is not unique to asthma as it occurs to a similar extent in cystic fibrosis.[17]

Studies of biopsies obtained by flexible fibreoptic bronchoscopy or at open lung biopsy in asthma demonstrate the very early involvement of inflammatory cells,[12] including the presence of and interaction between, (T) lymphocytes, eosinophils and plasma cells (Fig. 3.8).[13–15,84] Our own studies of bronchial biopsies in asthma have shown that the increase in leucocytes, including lymphocytes and eosinophils, occurs in relatively mild atopic, occupational and intrinsic asthma and that it is associated with an increase in 'activation' markers for both lymphocytes ($CD25^+$ cells) and eosinophils ($EG2^+$ cells).[14,16,23,24,84] In symptomatic atopic asthmatic patients, irregularly shaped lymphomononuclear cells have been identified. EG2 is a marker for the cleaved ('secreted') form of eosinophil cationic protein, which can be found both within eosinophils and diffusely in the wall, often in association with the reticular layer beneath the epithelium. Eosinophil-derived products such as major basic protein,[85] together with toxic oxygen radicals and proteases, probably all contribute to epithelial fragility: release of granules and of cytokines such as IL-4 and transforming growth factor β (TGF-β) may also stimulate nearby fibroblasts to produce additional reticulin and thicken the reticular basement membrane (see Fig. 3.5a). Studies of BAL show increased numbers of eosinophils and T-helper cells with evidence of mast cell and eosinophil degranulation.[86–88]

Macrophages may also increase in number, particularly in the more severe intrinsic form of asthma.[24] Mast cells initiate the immediate response to allergen exposure. Mast cells may also be an important source of IL-4 and other pro-inflammatory cytokines whose secretion may act as a trigger to the induction of subsequent persistent production of IL-4 and IL-5 by lymphocytes.[89,90] Little is known of the role of basophils in asthma, albeit there is evidence for increased recruitment of basophils and their precursors to sites of allergic reaction in atopic patients[91] (for a more detailed description of individual cell types see Chapters 4–11).

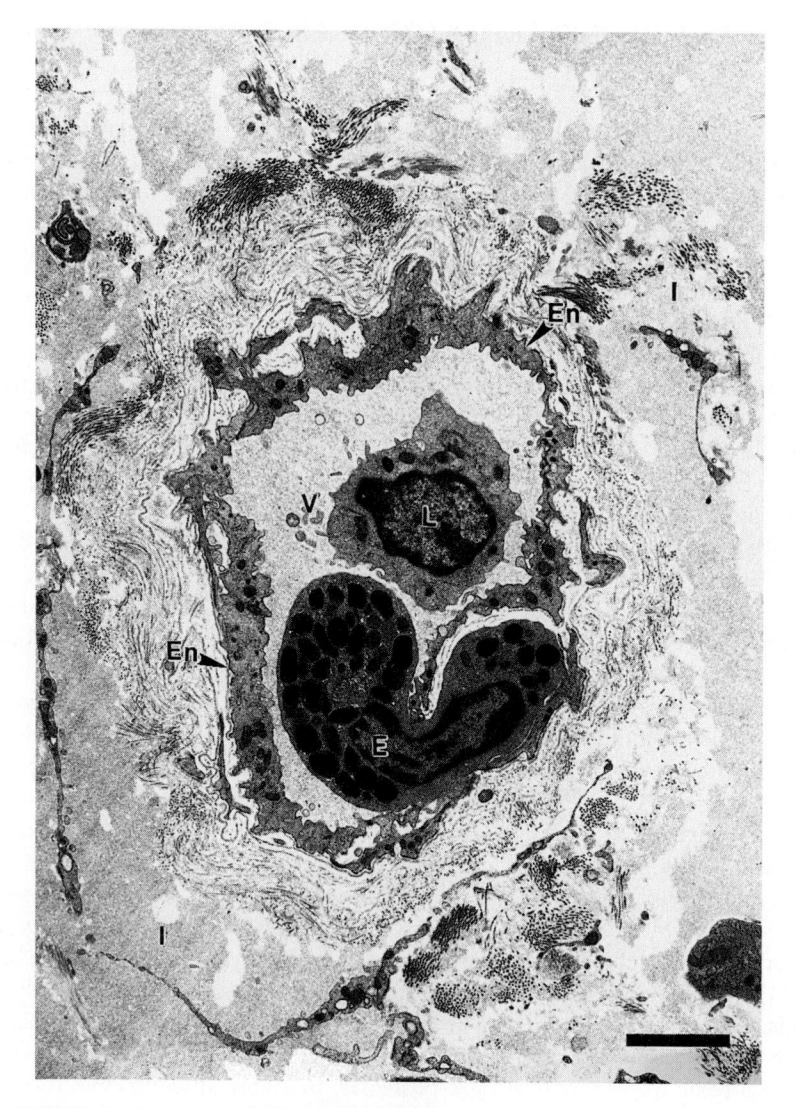

Fig. 3.7 TEM illustrating eosinophil (E) diapedesis in response to a chemokine gradient in the interstitium (I) of a subject with mild atopic asthma. The eosinophil is leaving the vessel (V) following its adhesion to the endothelium (En) by cell-surface very late activation (VLA)-4 and its endothelial ligand vascular cell adhesion molecule (VCAM)-1. L, adjacent lymphomononuclear cell. Scale bar = 5.0 μm.

AIRWAY WALL NERVES

The topic of airway wall innervation and its relationship with asthma is a large one.[45,46] There are data showing that in *fatal* asthma there is an absence of (relaxant) vasoactive intestinal polypeptide (VIP)-containing nerve fibres and an increase in the numbers of substance P-containing fibres (stimulatory to bronchial smooth muscle), contrasting

3 Airway Pathology in Asthma

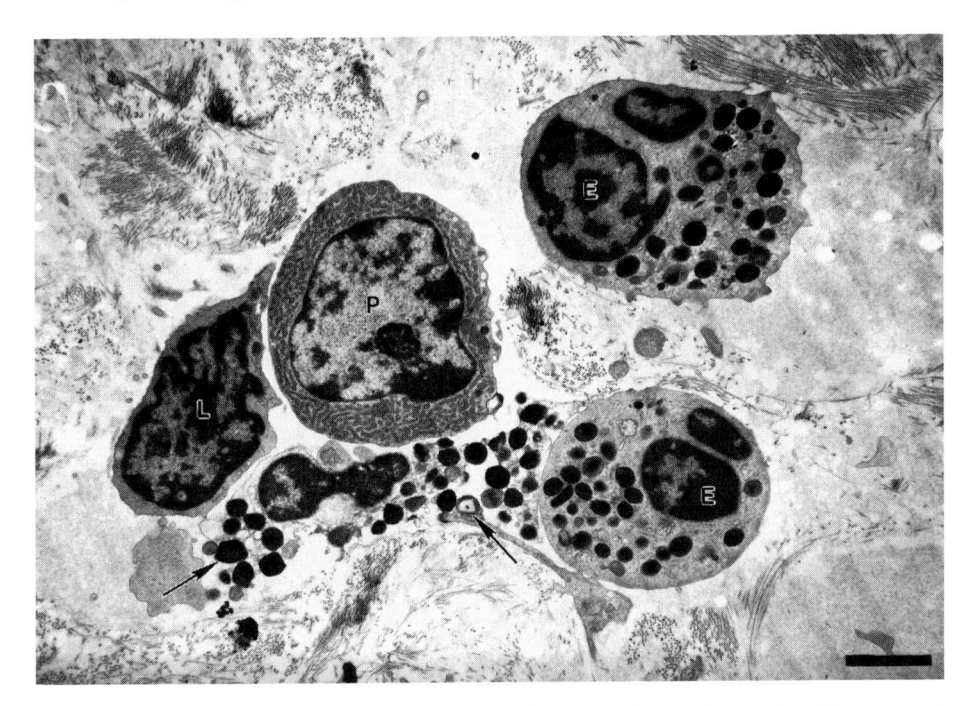

Fig. 3.8 Transmission electron micrograph of a subepithelial zone in a patient with mild atopic asthma demonstrating the close association of lymphocyte (L), eosinophils (E) and a plasma cell (P). One eosinophil appears to be disintegrating and has released its granules, which lie free in the interstitium (arrows). Scale bar = 2.5 μm.

markedly with the innervation of the control lungs taken at resection from chronic smokers.[92,93] However, the reduction has not been confirmed in examination of bronchial biopsies in mild asthma.[94] Whilst Sharma and colleagues[95,96] have described a reduction of airway VIP and β-adrenergic receptors in cystic fibrosis, the densities of both VIP and β-adrenergic receptors are reported to be similar in asthma to those of grossly normal tissue of lungs resected for carcinoma.

CONCLUSIONS

Airway mucosal inflammation is part of the very early pathology of asthma. Structural changes, which occur even in mild stable asthma, include thickening of the reticular basement membrane and fragility of the surface epithelium. Exposure to allergen in many allergic asthmatic subjects induces an allergic reaction in which the immediate response is one of mast cell degranulation; there may be a late phase response also, comprising eosinophil degranulation and the induction of a phenotypic change in fibroblasts, which become contractile and may be the origin of increased amounts of bronchial smooth muscle in chronic severe asthma. The enlargement of bronchial smooth muscle mass is a characteristic change in medium and large intrapulmonary airways obtained from cases of fatal asthma. Together with vasodilatation and congestion of the bronchial vasculature and consequent oedema, these pathological changes lead to a thickening of the airway wall that reduces the airway lumen and markedly increases the resistance to airflow, especially when bronchial smooth muscle contracts. The additional secretion of mucus further impedes airflow and the admixture of an inflammatory exudate leads to a highly tenacious airway plug and to the severe life-threatening attack, which in some cases is fatal.

ACKNOWLEDGEMENTS

I thank Leone Oscar for invaluable secretarial assistance and Andy Rogers for his willing help with the illustrations. I am grateful to the National Asthma Research Campaign and Cystic Fibrosis Trust (UK) for their support and I acknowledge the numerous clinical and research colleagues who work with me and with whom I have the pleasure to collaborate.

REFERENCES

1. Summary and recommendations of a workshop on the investigative use of fibreoptic bronchoscopy and bronchoalveolar lavage in asthmatics. *Am Rev Respir Dis* (1985) **132**: 180–182.
2. Jeffery PK: Structural, immunologic, and neural elements of the normal human airway wall. In Busse WW, Holgate ST (eds) *Asthma and Rhinitis*. Oxford, Blackwell Scientific Publications, 1995, pp 80–106.
3. von Leyden E: Ueber Bronchialasthma. *Dtsch Militararztl Z* (1886) **15**: 515–538.
4. Fraenkel A: Zur Pathologischen Anatomie des Bronchialasthmas. *Dtsch Med Wochenschr* (1900) **16**: 269–272.
5. Huber HL, Koesssler KK: The pathology of bronchial asthma. *Arch Intern Med* (1992) **30**: 689–760.
6. Kountz WB, Alexander HL: Death from bronchial asthma. *Arch Pathol* (1928) **5**: 1003–1019.
7. Houston JC, de Nevasquez S, Trounce JR: A clinical and pathological study of fatal cases of status asthmaticus. *Thorax* (1953) **8**: 207–213.
8. Dunnill MS: The pathology of asthma, with special reference to changes in the bronchial mucosa. *J Clin Pathol* (1960) **13**: 27–33.
9. Cardell BS, Pearson RSB: Deaths in asthmatics. *Thorax* (1959) **14**: 341–352.
10. Hogg JC: The pathology of asthma. *Clin Chest Med* (1984) **5**: 567–571.
11. Jeffery PK: Pathology of Asthma. *Br Med Bull* (1992) **48**: 23–39.
12. Laitinen LA, Laitinen A, Haahtela T: Airway mucosal inflammation even in patients with newly diagnosed asthma. *Am Rev Respir Dis* (1993) **147**: 697–704.
13. Laitinen LA, Heino M, Laitinen A, Kava T, Haahtela T: Damage of the airway epithelium and bronchial reactivity in patients with asthma. *Am Rev Respir Dis* (1985) **131**: 599–606.
14. Jeffery PK, Wardlaw A, Nelson FC, Collins JV, Kay AB: Bronchial biopsies in asthma: an ultrastructural quantification study and correlation with hyperreactivity. *Am Rev Respir Dis* (1989) **140**: 1745–1753.
15. Beasley R, Roche W, Roberts JA, Holgate ST: Cellular events in the bronchi in mild asthma and after bronchial provocation. *Am Rev Respir Dis* (1989) **139**: 806–817.

3 Airway Pathology in Asthma

16. Azzawi M, Bradley B, Jeffery PK, et al.: Identification of activated T lymphocytes and eosinophils in bronchial biopsies in stable atopic asthma. *Am Rev Respir Dis* (1990) **142**: 1407–1413.
17. Azzawi M, Johnston PW, Majumdar S, Kay AB, Jeffery PK: T lymphocytes and activated eosinophils in asthma and cystic fibrosis. *Am Rev Respir Dis* (1992) **145**: 1477–1482.
18. De Monchy JG, Kauffman HF, Venge P, et al.: Bronchoalveolar eosinophils during allergen-induced late asthmatic reactions. *Am Rev Respir Dis* (1985) **131**: 373–376.
19. Aalbers R, Smith M, Timens W: Immunohistology in bronchial asthma. *Respir Med* (1993) **87**: 13–21.
20. Bentley AM, Qui Meng, Robinson DS, Hamid Q, Kay AB, Durham SR: Increase in activated T lymphocytes, eosinophils and cytokine messenger RNA expression for IL-5 and GM-CSF in bronchial biopsies after allergen inhalation challenge in atopic asthmatics. *Am J Respir Cell Mol Biol* (1993) **8**: 35–42.
21. Adelroth E, Rogers AV, O'Byrne PM, Jeffery PK: Airway eosinophil infiltration, degranulation and loss of granular crystalline core electron-density is a feature of allergen-induced late phase reaction (abstract). *Am J Respir Crit Care Med* (1994) **149**: A529.
22. Bradley BL, Azzawi M, Jacobson M, et al.: Eosinophils, T-lymphocytes, mast cells, neutrophils and macrophages in bronchial biopsies from atopic asthmatics: comparison with atopic non-asthma and relationship to bronchial hyperresponsiveness. *J Allergy Clin Immunol* (1991) **88**: 661–674.
23. Bentley AM, Maestrelli P, Saetta M, et al.: Activated T lymphocytes and eosinophils in the bronchial mucosa in isocyanate-induced asthma. *J Allergy Clin Immunol* (1992) **89**: 821–829.
24. Bentley AM, Menz G, Storz Chr, et al.: Identification of T-lymphocytes, macrophages and activated eosinophils in the bronchial mucosa in intrinsic asthma: relationship to symptoms and bronchial responsiveness. *Am Rev Respir Dis* (1992) **146**: 500–506.
25. Di Stefano A, Saetta M, Maestrelli P, et al.: Mast cells in the airway mucosa and rapid development of occupational asthma induced by toluene diisocyanate. *Am Rev Respir Dis* (1993) **147**: 1005–1009.
26. Hargreave FE, Popov T, Kidney J, Dolovich J: Sputum measurements to assess airway inflammation in asthma. *Allergy* (1993) **48**: 81–83.
27. Hargreave FE, Wong BJO, Popov T, Dolovich J: Noninvasive methods to examine the anti-inflammatory effects of drugs. In T. Hansel, J. Morley (eds) *New Drugs in Allergy and Asthma*. Basel, Birkhauser Verlag, 1993, pp 291–295.
28. Hargreave FE: The investigation of airway inflammation in asthma: sputum examination. *Clin Exp Allergy* (1997) (Suppl.) **1**: 36–40.
29. Pavord ID, Pizzichini MMM, Pizzichini E, Hargreave FE: The use of induced sputum to investigate airway inflammation. *Thorax* (1997) **52**: 498–501.
30. Curschmann H: Uber Bronchiolitis exsudatira und ihr Verhaltuis zum Asthma nervosum. *Dtsch Arch Klin Med* (1883) **32**: 1–34.
31. Naylor B: The shedding of the mucosa of the bronchial tree in asthma. *Thorax* (1962) **17**: 69–72.
32. Weller PF, Bach DS, Austen KF: Biochemical characterization of human eosinophil Charcot–Leyden crystal protein (lysophospholipase). *J Biol Chem* (1984) **259**: 15 100–15 105.
33. Gibson PG, Girgis-Gabardo A, Morris MM, et al.: Cellular characteristics of sputum from patients with asthma and chronic bronchitis. *Thorax* (1989) **44**: 693–699.
34. Gibson PG, Dolivich J, Denburg J, Ramsdale EH, Hargreave FE: Chronic cough: eosinophilic bronchitis without asthma. *Lancet* (1989) **i**: 1346–1348.
35. Dunnill MS, Massarella GR, Anderson JA: A comparison of the quantitative anatomy of the bronchi in normal subjects, in status asthmaticus, in chronic bronchitis, and in emphysema. *Thorax* (1969) **24**: 176–179.
36. List SJ, Findlay BP, Forstner GG, Forstner JF: Enhancement of the viscosity of mucin by serum albumin. *Biochem J* (1978) **175**: 565–571.
37. Reid LM: The presence or absence of bronchial mucus in fatal asthma. *J Allergy Clin Immunol* (1987) **80**: 415–416.
38. Messer JW, Peters GA, Bennett WA: Causes of death and pathologic findings in 304 cases of bronchial asthma. *Dis Chest* (1960) **38**: 616–624.

39. Gleich GJ, Motojima S, Frigas E, Kephart GM, Fujisawa T, Kravis LP: The eosinophilic leucocyte and the pathology of fatal bronchial asthma: evidence for pathologic heterogeneity. *J Allergy Clin Immunol* (1980) **80**: 412–415.
40. Sur S, Crotty TB, Kephart GM, *et al*.: Sudden onset of fatal asthma: a distinct entity with few eosinophils and relatively more neutrophils in the airway submucosa? *Am Rev Respir Dis* (1993) **148**: 713–719.
41. Ayers M, Jeffery PK: Proliferation and differentiation in adult mammalian airway epithelium: a review. *Eur Respir J* (1988) **1**: 58–80.
42. Soderberg M, Hellstrom S, Sandstrom T, Lungren R, Bergh A: Structural characterization of bronchial mucosal biopsies from healthy volunteers: a light and electron microscopical study. *Eur Respir J* (1990) **3**: 261–266.
43. Elia C, Bucca C, Rolla G, Scappaticci E, Cantino D: A freeze-fracture study of tight junctions in human bronchial epithelium in normal, bronchitic and asthmatic subjects. *J Submicrosc Cytol Pathol* (1988) **20**: 509–517.
44. Godfrey RWA, Severs NJ, Jeffery PK: Freeze-fracture morphology and quantification of human bronchial epithelial tight junctions. *Am J Respir Cell Molec Biol* (1992) **6**: 453–458.
45. Barnes PJ: State of art: neural control of human airways in health and disease. *Am Rev Respir Dis* (1986) **134**: 1289–1314.
46. Jeffery PK: Innervation of the airway mucosa: structure, function and changes in airway disease. In Goldie R (ed.) *Immunopharmacology of Epithelial Barriers. The Handbook of Immunopharmacology* (series ed. C. Page). London, Academic Press, 1994, pp 85–118.
47. Sparrow MP, Mitchell HW: The epithelium acts as a barrier modulating the extent of bronchial narrowing produced by substances perfused through the lumen. *Br J Pharmacol* (1991) **103**: 1160–1164.
48. Hogg JC, Eggleston PA: Is asthma an epithelial disease? *Am Rev Respir Dis* (1984) **129**: 207–208.
49. VanHoutte PM: Epithelium-derived relaxing factor(s) and bronchial reactivity. *J Allergy Clin Immunol* (1989) **83**: 855–861.
50. Montefort S, Roberts JA, Beasley R, Holgate ST, Roche WR: The site of disruption of the bronchial epithelium in asthmatic and non-asthmatic subjects. *Thorax* (1992) **47**: 499–503.
51. Bellini A, Yoshimura H, Vittori E, Marini M, Mattoli S: Bronchial epithelial cells of patients with asthma release chemoattractant factors for T-lymphocytes. *J Allergy Clin Immunol* (1993) **92**: 412–424.
52. Callerame MD, Condemi MD, Bohrod MD, Vaughan JH: Immunologic reactions of bronchial tissues in asthma. *N Engl J Med* (1971) **284**: 459–464.
53. Roche WR, Beasley R, Williams JH, Holgate ST: Subepithelial fibrosis in the bronchi of asthmatics. *Lancet* (1989) **i**: 520–523.
54. Nowak J: Anatomopathologic changes in the bronchial walls in chronic inflammation, with special reference to the basement membrane, in the course of bronchial asthma. *Acta Med Pol* (1969) **2**: 151–172.
55. Sobonya RE: Quantitative structural alterations in long-standing allergic asthma. *Am Rev Respir Dis* (1984) **130**: 289–292.
56. Crepea SB, Harman JW: The pathology of bronchial asthma. I. The significance of membrane changes in asthmatic and non-allergic pulmonary disease. *J Allergy* (1955) **26**: 453–460.
57. Lambert RK: Role of bronchial basement membrane in airway collapse. *J Appl Physiol* (1991) **71**(2): 666–673.
58. Brewster CEP, Howarth PH, Djukanovic R, Wilson J, Holgate ST, Roche WR: Myofibroblasts and subepithelial fibrosis in bronchial asthma. *Am J Respir Cell Mol Biol* (1990) **3**: 507–511.
59. Gizycki MJ, Adelroth E, Rogers AV, O'Byrne PM, Jeffery PK: Myofibroblast involvement in the allergen-induced late response in mild atopic asthma. *Am J Respir Cell Mol Biol* (1997) **16**: 664–673.
60. Reid L: Pathology of chronic bronchitis. *Lancet* (1954) **i**: 275–279.
61. Wanner A: Airway mucus and the mucociliary system. In Middleton E, Reed CE, Ellis EF, Adkinson NF, Uunginer JW (eds) *Allergy: Principles and Practice*. St Louis, Mosby, 1988, pp 541–548.

3 Airway Pathology in Asthma

62. Cluroe A, Holloway L, Thomson K, Purdie G, Beasley R: Bronchial gland duct ectasia in fatal bronchial asthma: association with interstitial emphysema. *J Clin Pathol* (1989) **42**: 1026–1031.
63. Carroll N, Elliot A, Morton A, James A: The structure of large and small airways in nonfatal and fatal asthma. *Am Rev Respir Dis* (1993) **147**: 405–410.
64. James AL, Pare PD, Hogg JC: The mechanics of airway narrowing in asthma. *Am Rev Respir Dis* (1989) **139**: 242–246.
65. Moreno RH, Hogg JC, Pare PD: Mechanisms of airway narrowing. *Am Rev Respir Dis* (1986) **133**: 1171–1180.
66. Wiggs BR, Moreno R, Hogg JC, Hilliam C, Pare PD: A model of the mechanics of airway narrowing. *J Appl Physiol* (1990) **69**: 849–860.
67. Wiggs BR, Bosken C, Pare PD, James A, Hogg JC: A model of airway narrowing in asthma and in chronic obstructive pulmonary disease. *Am Rev Respir Dis* (1992) **145**: 1215–1218.
68. Hogg J: The pathology of asthma. In Austen KF, Lichtenstein L, Kay AB, Holgate ST (eds) *Asthma, Vol. IV, Physiology, Immunopharmacology and Treatment*. Oxford, Blackwell Scientific Publications, 1993, pp 17–25.
69. Huber HL, Koessler K: The pathology of bronchial asthma. *Arch Intern Med* (1922) **30**: 689–760.
70. Thurlbeck WM: Chronic airflow obstruction. Correlation of structure and function. In Petty TL (ed) *Chronic Obstructive Pulmonary Disease*. Dekker, 1985, pp 129–203.
71. Heard BE, Hossain S: Hyperplasia of bronchial muscle in asthma. *J Pathol* (1983) **110**: 319–331.
72. Ebina M, Takahashi T, Chiba T, Motomiya M: Cellular hypertrophy and hyperplasia of airway smooth muscles underlying bronchial asthma: a 3-D morphometric study. *Am Rev Respir Dis* (1993) **148**: 720–726.
73. Ebina M, Yaegashi H, Chiba R, Takahashi T, Motomiya M, Tanemura M: Hyperreactive site in the airway tree of asthmatic patients revealed by thickening of bronchial muscles. *Am Rev Respir Dis* (1990) **141**: 1327–1332.
74. Bousquet J, Chanez P, Lacoste JY, et al.: Asthma: a disease remodeling the airways. *Allergy* (1992) **47**: 3–11.
75. Jeffery PK: Structural changes in asthma. In Page C, Black J (eds) *Airways and Vascular Remodelling in Asthma and Cardiovascular Disease*. London, Academic Press, 1994, pp 3–19.
76. Stewart AG, Tomlinson PR, Wilson J: Airway wall remodelling in asthma: a novel target for the development of anti-asthmatic drugs. *Trends Pharmacol Sci* (1993) **14**: 275–279.
77. Widdicombe J: New perspectives on basic mechanisms in lung disease: 4. Why are the airways so vascular? *Thorax* (1993) **48**: 290–295.
78. Freedman BJ: The functional geometry of the bronchi. *Bull Eur Physiopathol Respir* (1972) **8**: 545–551.
79. Benson MK: Bronchial hyperreactivity. *Br J Dis Chest* (1975) **69**: 227–239.
80. Ding DJ, Martin JG, Macklem PT: Effects of lung volume on maximal methacholine-induced broncho-constriction in normal humans. *J Appl Physiol* (1987) **62**: 1324.
81. Saetta M, Ghezzo H, Wong Dong Kim, et al.: Loss of alveolar attachments in smokers. A morphometric correlate of lung function impairment. *Am Rev Respir Dis* (1985) **132**: 894–900.
82. Godfrey RWA, Lorimer S, Majumdar S, Adelroth E, Johansson S-A, Jeffery PK: Airway and lung parenchyma content of elastic fibre is not reduced in asthma (abstract). *Am Rev Respir Dis* (1992) **145**: A463.
83. Saetta M, Di Stefano A, Rosina C, Thiene G, Fabbri LM: Quantitative structural analysis of peripheral airways and arteries in sudden fatal asthma. *Am Rev Respir Dis* (1991) **143**: 138–143.
84. Jeffery PK, Godfrey RWA, Adelroth E, Nelson F, Rogers A, Johansson S-A: Effects of treatment on airway inflammation and thickening of reticular collagen in asthma: a quantitative light and electron microscopic study. *Am Rev Respir Dis* (1992) **145**: 890–899.
85. Filley WV, Holley KE, Kephart GM, Gleich GJ: Identification by immunofluorescence of eosinophil granule major basic protein in lung tissue of patients with bronchial asthma. *Lancet* (1982) **i**: 11–16.
86. Gerblich AA, Campbell AE, Schuyler MR: Changes in T-lymphocyte subpopulations after antigenic bronchial provocation in asthmatics. *N Engl J Med* (1984) **310**: 1349–1352.
87. Wardlaw AJ, Dunnett S, Gleich GJ, Collins JV, Kay AB: Eosinophils and mast cells in

bronchoalveolar lavage in mild asthma: relationship to bronchial hyperreactivity. *Am Rev Respir Dis* (1988) **137**: 62–69.
88. Adelroth E, Rosenhall L, Johansson S-A, Linden M, Venge P: Inflammatory cells and eosinophilic activity in asthmatics investigated by bronchoalveolar lavage: the effects of anti-asthmatic treatment with budesonide or terbutaline. *Am Rev Respir Dis* (1990) **142**: 91–99.
89. Bradding P, Feather IH, Howarth PH, *et al.*: Interleukin 4 is localized to and released by human mast cells. *J Exp Med* (1992) **176**: 1381–1386.
90. Bradding P, Feather IH, Wilson S, *et al.*: Immunolocalization of cytokines in the nasal mucosa of normal and perennial rhinitic subjects. *J Immunol* (1993) **151**: 3853–3865.
91. Denburg JA, Telizyn S, Belda A, Dolovich J, Bienenstock J: Increased numbers of circulating basophil progenitors in atopic patients. *J Allergy Clin Immunol* (1985) **76**: 466–472.
92. Ollerenshaw SL, Woolcock AJ: Quantification and location of vasoactive intestinal peptide immunoreactive nerves in bronchial biopsies from subjects with mild asthma. *Am Rev Respir Dis* (1993) **147**: A285.
93. Ollerenshaw SL, Jarvis D, Sullivan CE, Woolcock AJ: Substance P immunoreactive nerves in airways from asthmatics and non-asthmatics. *Eur Respir J* (1991) **4**: 673–682.
94. Haworth PH, Djukanovic R, Wilson JW, Holgate ST, Springall DR, Polak JM: Neuropeptide-containing nerves in endobronchial biopsies from asthmatic and non-asthmatic subjects. *Am J Cell Molec Biol* (1995) **13**: 288–296.
95. Sharma R, Jeffery PK: Airway β-adrenoceptor number in cystic fibrosis and asthma. *Clin Sci* (1990) **78**: 409–417.
96. Sharma RK, Jeffery PK: Airway VIP receptor number is reduced in cystic fibrosis but not asthma. *Am Rev Respir Dis* (1990) **141**: A726.

4

Physiology

NEIL B. PRIDE

INTRODUCTION

Although the primary pathophysiology of asthma is in the sublaryngeal airways, airway obstruction inevitably impairs pulmonary gas exchange. Extrapulmonary factors, such as increases in ventilation and cardiac output, the performance of the respiratory muscles and the perception of obstruction, play a critical part in sustaining gas exchange during severe asthma. In this chapter some recent developments in understanding of these changes are discussed but no attempt is made to give a comprehensive description of applied physiology in asthma.

FACTORS RESTRICTING AND AMPLIFYING INDUCED AIRWAY NARROWING *IN VIVO*

Airway smooth muscle

In population studies there is a very wide range of airway responsiveness to inhaled histamine or methacholine; although only a minority of normal subjects produce significant narrowing in response to the largest doses of methacholine, the distribution in the population is probably unimodal,[1] with a group of hyperresponsive individuals superimposed on a broadly normal distribution. If this model is correct, the hyper-responsiveness of identified asthma subjects merges imperceptibly into the normal range.

The conventional view of the enhanced constrictor responsiveness of asthmatic airways has been that airway smooth muscle (ASM) is unduly 'twitchy' and primed to contract due to some undefined combination of enhanced mediator or neural stimulation or enhanced ASM response due to increased contractility or mass. These large differences in responsiveness *in vivo* cannot be explained by the differences in the contractility of human ASM *in vitro* as conventionally measured; human ASM excised from normal lungs invariably contracts in response to bronchoconstrictor drugs such as histamine and it has been difficult to demonstrate large interindividual differences in this response. There is relatively little information on the *in vitro* mechanical properties of ASM in asthmatic subjects[2] and how these differ from those of normal subjects (see Chapter 5), although the effects of contraction will be amplified by the increase in ASM mass. While most studies support an increase in ASM in asthma,[2] a recent study of axially sectioned large bronchi has failed to find this.[3] As discussed below, contraction of ASM may immediately reduce airway wall compliance. In normal subjects, induced ASM contraction is readily removed by a deep inflation or by β-adrenergic agonists; indeed it is relatively difficult to induce airway narrowing with inhaled methacholine when tidal volume is increased, as during exercise[4] or voluntary hyperventilation. Recently it has been found that when deep inflations are avoided for several minutes, considerable airway narrowing can be induced by methacholine aerosols in normal as well as in asthmatic subjects.[5] Furthermore, after such narrowing develops it is not removed by a single deep inflation, but may require a sequence of large breaths. Removal of the induced airway narrowing takes longer in asthmatic than in normal subjects. A suggested mechanism is that the normal tidal excursions of breathing (with occasional deep breaths) keeps ASM in a relatively high-compliance state but that its state stiffens with lack of stretch.[2,6] These findings indicate the need for a detailed reappraisal of static, dynamic and time-dependent behaviour of ASM.

Mechanisms restricting airway narrowing

In 1984 Woolcock *et al.*[7] suggested that dose–response curves to inhaled histamine in normal subjects differed from those in asthmatic subjects not only in position (much larger doses of histamine being required to induce airway narrowing in normal than in asthmatic subjects) but also in shape. Whereas progressive airway narrowing could be induced in subjects with asthma, only limited reduction of forced expiratory volume in 1 s (FEV_1) could be induced in most normal subjects before a near-plateau of bronchial narrowing developed.[7,8] Convincing evidence of the development of a true plateau of airway narrowing in most normal humans was subsequently obtained (Fig. 4.1).[9,10] These findings led to the alternative hypothesis that the basic abnormality in asthma might be loss of normal mechanisms restricting airway narrowing rather than amplification of mechanisms of narrowing.

Restraints on luminal narrowing depend on the mechanical properties of the total airway wall and of surrounding lung tissue. In central intrathoracic but extrapulmonary airways, cartilage restricts the extent of luminal narrowing. In the trachea, the attachments of the muscle to the cartilage rings, at least in experimental animals, are such that smooth muscle contraction results in the formation of complete cartilage rings encircling the lumen.[11,12] In the central conducting airways there are separate plates of

4 Physiology

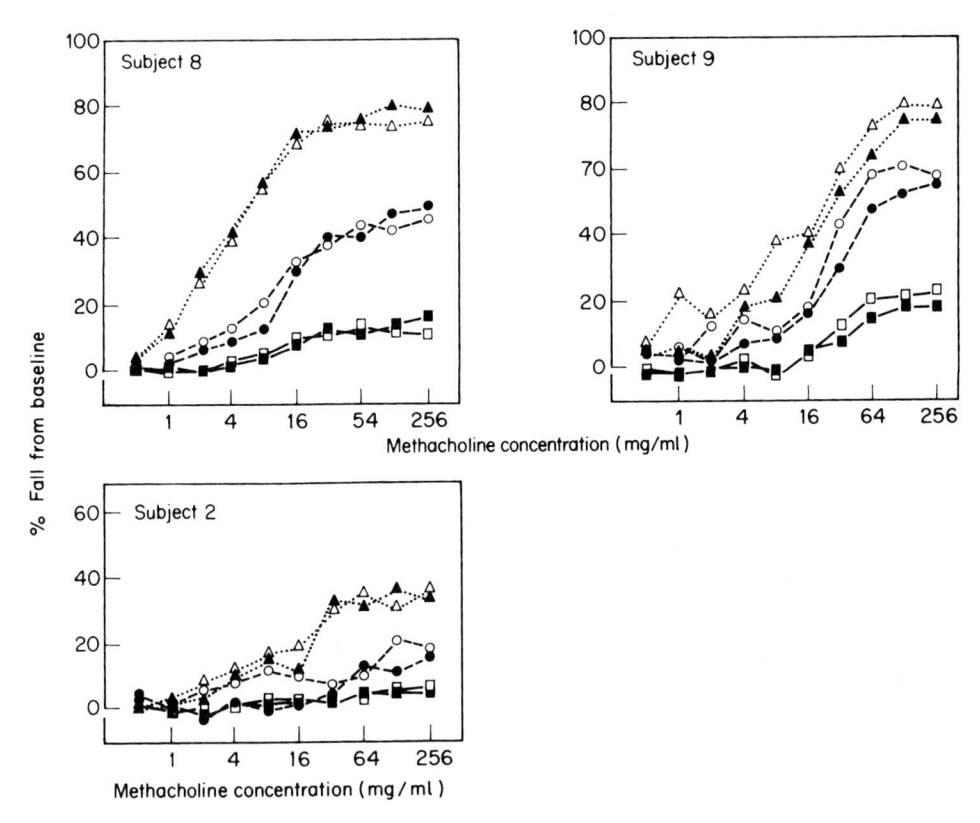

Fig. 4.1 Dose–response curves showing plateau of response developing to inhaled methacholine in three normal subjects on two separate days. Open symbols, day 1; closed symbols, day 2. Response assessed by forced expiratory volume in 1 s (■ □), maximum expiratory flow at 40% vital capacity on complete (● ○) and partial (▲ △) forced exhalations. Reproduced from ref. 9, with permission.

cartilage but these are also brought closer together to 'fortify' the airway walls by ASM contraction. Contraction of ASM itself may directly decrease wall compliance.[13] Conversely, occasional paradoxical decreases in maximum expiratory flow after bronchodilators in normal subjects have been attributed to enhanced collapsibility of central airways due to loss of the stabilization provided by the normal tonic contraction of ASM.[14] A further factor preventing complete closure of the lumen is folding of the mucosa, preserving some lumen between the folds.[15]

For the intrapulmonary airways, the most important factor stabilizing the airway wall against the effects of ASM contraction is the attachment of alveolar walls to the external perimeter of the airway wall.[16] By these attachments the forces distending the alveoli are transmitted to the external airway wall, promoting airway distension as the alveoli expand. Theory and most experimental work suggest that these extra-airway forces would have their greatest stabilizing effect at large lung volumes, so that a given amount of activation of ASM would lead to greater shortening at small rather than large lung volume. In excised lobes of dogs, methacholine can induce complete airway closure in collapsed lung[17] but not when the lungs are inflated. In humans *in vivo*, the magnitude of maximum bronchoconstriction to inhaled methacholine is greater at small rather than large lung volume and is quite sensitive to small changes in volume above and below functional residual capacity (FRC).[18] However, the magnitude of airway narrowing produced by submaximal doses of methacholine appears to be unchanged by moderate changes in lung volume, suggesting that ASM initially may contract freely and that the restraints applied by surrounding lung chiefly act to prevent extreme narrowing or closure.[18,19] Inducing emphysema by intratracheal elastase enhances airway response to methacholine in rats, and this change has been related to loss of lung elastic recoil.[20] In human asthma there is also some loss of lung elastic recoil,[21] which would reduce the parenchymal forces distending the airway. Although alveolar attachments to the airway perimeter are probably intact, coupling between alveolar and airway inflation may also be reduced due to enlargement of the outer perimeter of the airway wall or surrounding inflammatory changes.[16]

Modification of the luminal effects of ASM contraction in asthma

Enhanced reductions in luminal calibre can occur with normal shortening of ASM if there is thickening of the airway wall internal to the contracting muscle.[22] Workers in Vancouver have modelled the effects of wall thickness[23] and developed methods[24] to estimate this in collapsed human lungs at post-mortem. They found on average a doubling of the thickness of airway wall in lungs from 18 subjects with fatal asthma.[24] These changes involved all sizes of airways. Calculations suggest that these changes would not be sufficient to increase basal airway resistance but would greatly enhance the effects of constrictor challenge.[24] Subsequent studies of lungs from subjects with asthma who died of non-respiratory causes showed less dramatic airway wall thickening compared with fatal asthma;[25] in particular there was less increase in ASM. It has also been hypothesized that rapid expansion of the blood volume in and around the airway wall, acutely increasing its thickness, might be responsible for the airway narrowing that occurs after exercise or isocapnic hyperventilation in asthmatic subjects.[26] Other amplifying factors would include increased airway wall compliance or reduced effects of extra-airway distension.[16] In fact, although there is little direct evidence, effective airway compliance probably is reduced. Some information can be obtained from studying the increase in anatomical dead space (which indicates the total volume of the larger intrathoracic and extrathoracic airways) or in total airway conductance as lung volume is increased. Increases in both dead space[27] and airway conductance[28,29] are reduced in asthmatic subjects compared with normal subjects. These changes might be due to a true reduction in airway wall compliance, perhaps reflecting wall thickening and/or stiffening (Table 4.1), but increases in the airway lumen with lung inflation also might be reduced by total or partial occlusion of some parallel airways by muco-inflammatory plugs, or by a reduction in extra-airway distending forces. The direct effect of loss of lung elastic recoil can be examined by measuring change in dead space or conductance[28,29] with increase in lung recoil pressure rather than lung volume; however, even at a standard lung recoil pressure, enlargement of the outer perimeter of the airway wall or peri-airway inflammation might reduce coupling between alveolar and airway expansion.[16] The number and depth of mucosal folds that develop with ASM contraction are also likely to be influenced by changes in wall structure, including the thickening of the collagen layer below the basement membrane.

4 Physiology

Table 4.1 Factors reducing effective compliance of individual airways.

Initial luminal area
Encroachment by wall thickening, secretions or increased mucosal folding

Wall properties
Thickening due to structural or transient (e.g. increased blood volume, inflammatory oedema) changes
Stiffening of wall (e.g. state of airway smooth muscle, increased collagen below basement membrane, cartilage 'fortification')

Extra-airway distending forces
Reduction in lung recoil pressure
Impaired coupling between alveolar attachments and airway perimeter

Implications for the development and resolution of airway hyperresponsivensss in asthma

Currently, airway hyperresponsiveness is often regarded as a consequence of airway inflammation, which results in an enhanced mediator or neural stimulus to ASM contraction from a standard provocation. This view was attractive because of the difficulty in identifying abnormalities of ASM contractility in asthma. Nevertheless increased ASM mass[30] and increased wall thickness,[23] even without encroaching significantly on the airway wall lumen under basal conditions, can be expected to enhance the luminal effects of a given amount of ASM contraction and may reduce the restraint to narrowing imposed by alveolar attachments to the external airway wall.[23,24] Airway wall compliance probably is reduced, narrowing the range over which luminal size varies with changes in lung inflation. These changes in the airway wall indicate that there is a strong structural component to airway hyperresponsiveness in asthma; unless these changes regress with treatment of inflammation, hyperresponsiveness may persist even when basal airways resistance is normal. In addition, as discussed below, there may be increased peripheral lung resistance even when overall airway function is normal. These findings attenuate the distinction often made between the importance of geometric factors in determining airway hyperresponsiveness in chronic obstructive pulmonary disease (COPD) and its relative unimportance in the genesis of the hyperresponsiveness of asthma. Several studies have now shown a relation between baseline FEV_1 and the intensity of airway responsiveness in asthma which extends into the normal range of FEV_1;[31,32] nevertheless for a given level of FEV_1, responsiveness is more intense in asthma than in COPD.

SITES OF AIRWAY NARROWING IN ASTHMA

Intrathoracic airway narrowing

The airway narrowing in episodic asthma is due to varying combinations of smooth muscle contraction, mucosal swelling and luminal secretions. Pathological studies suggest that inflammation of the airway mucosa extends throughout the tracheobronchial tree, indicating the potential for narrowing to develop in all sizes of airways. However, studies of regional ventilation and of gas exchange imply that airway narrowing varies greatly between parallel airways. Heterogeneity of individual airway constrictor responses after inhaling histamine has been shown in animals using high-resolution computed tomography. Presumably there can be non-uniformity on a longitudinal basis also. Direct visualization of airways, either at bronchoscopy or by external imaging, inevitably is restricted to the larger conducting airways. Narrowing of the large airways has been observed on occasions when bronchoscopy or bronchography have precipitated an asthmatic attack. Attempts to determine the serial site of narrowing have been made by measuring pressures with various types of intrabronchial catheter. Indirect methods (changes during helium breathing or after a deep inflation) have also been used to deduce this information. Little useful information can be obtained from standard tests of lung function while breathing air.

Response to breathing a helium–oxygen mixture

In the 1970s there was considerable interest in trying to localize the serial site of airflow limitation in asthma by measuring the increase in maximum expiratory flow when the density of the expired gas was reduced by breathing an 80% helium–20% oxygen mixture.[33] In some (but not all) asthmatic subjects, maximum expiratory flow does not show the normal increase when breathing helium–oxygen; this suggests that the major site of flow limitation is no longer in the central airways as in normal subjects, but has moved to more peripheral airways where flow is presumed to be laminar and independent of density.[34] This change is usually attributed to increased frictional pressure losses in narrowed peripheral airways. Some asthmatic subjects consistently lose or consistently retain density dependence of maximum flow with repeated attacks, but this is not invariably the case; in general, loss of density dependence becomes more common as expiratory airflow limitation increases in severity[35,36] and is particularly observed in asthmatic subjects who smoke.[37] Reduced density dependence of maximum expiratory flow should not be interpreted as indicating *only* peripheral airways are involved even if they are the site of flow limitation.

Subsequently, interest in the use of helium response on maximum expiratory flow waned. The method analysed the changes in the airways between the alveoli and the sites of expiratory flow limitation ('choke-points') but experimental studies in dogs showed that relatively small changes in geometry and position of choke-points could profoundly affect the helium response.[38] There is considerable variation in size of the baseline helium response in the normal population and in disease; changes in helium response in an individual before and after an acute intervention may be more reliable. Alternatively, the changes in pulmonary resistance (measured with an oesophageal balloon catheter) when switching from breathing air to a helium–oxygen mixture can be studied.[39] A large reduction in resistance suggests an important narrowing of the central conducting airways.

Effects of a deep inflation on airway function

More recently attempts have been made to localize the site of dominant airway narrowing by another indirect technique, the effects of a deep inflation (DI) on airway function. For

4 Physiology

30 years it has been recognized that a DI may transiently affect airway dimensions when tidal breathing is resumed. The first studies demonstrated airway widening after DI in normal subjects when narrowing had been induced by inhaled histamine or methacholine. Later Gayrard et al.[40] pointed out that DI could lead to subsequent narrowing in subjects with acute asthma. Initially, changes after DI were attributed to a change in bronchial muscle activity (direct or reflex) produced by stretch. An alternative hypothesis is that the relation between parenchymal and airway hysteresis explains the variable changes in airway function after DI.[41,42] When airway hysteresis exceeds that of parenchyma, DI results in temporary bronchial widening when tidal breathing is resumed. When parenchymal hysteresis exceeds that of the airways, DI results in bronchial narrowing. Equal degrees of hysteresis result in no effect of DI on resting airway calibre, as found in most normal subjects. A major interest of this idea is that it might localize the site of disease within the lung. Contraction of bronchial muscle in conducting airways would be expected to increase airway hysteresis without affecting parenchymal hysteresis. Increased tone in the extreme periphery of the lung (respiratory bronchioles and alveolar ducts) would be expected to increase parenchymal hysteresis with only a small increase in airway hysteresis.

The effects of DI can be examined by measuring the effects on airway resistance, or more simply by comparing maximal expiratory flow at 30–40% vital capacity above residual volume (RV) from forced expirations begun from just above FRC (partial curve, P) with isovolumic flow during manoeuvres started from total lung capacity (TLC) (maximal curve, M) (Fig. 4.2). From these two expirations the results are expressed as M:P ratios. In normal subjects under basal conditions M:P ratios on average are a little greater than 1.0.[43] In spontaneous episodes of asthma, M:P ratios are less than 1.0 (airway function worse after DI) and tend to fall as FEV_1 (per cent predicted) falls.[43–45] This suggests an important obstruction of the most peripheral airways in spontaneous asthma. In contrast, when acute obstruction is induced by challenge with inhaled short-acting drugs, M:P ratios rise, sometimes to very high values, indicating that DI removes obstruction. This is characteristic of a conducting airway response. Similar rises in M:P

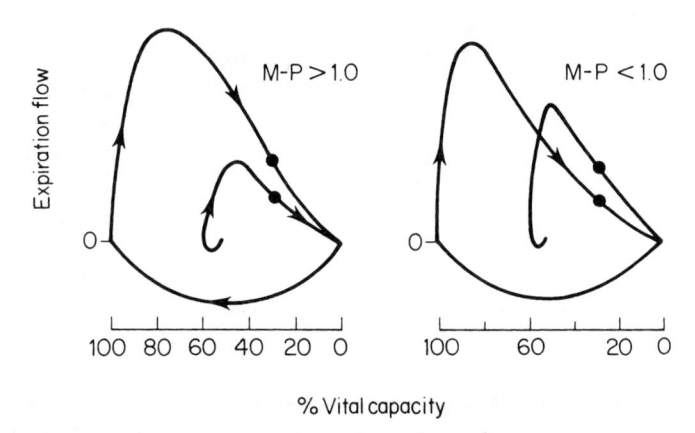

Fig. 4.2 Use of forced expirations started from about 60% vital capacity (partial curve) and full inflation (maximum curve) to derive isovolumic maximum expiratory flow at about 30% vital capacity on maximum (M) and partial (P) curves. Results expressed as M:P ratios in Fig. 4.3. Reproduced from ref. 42, with permission.

ratios have been found when airway narrowing is induced by other acute, short-lived challenges such as exercise or hyperventilation; with antigen challenge, however, rises in M:P ratios are less pronounced for the impairment of FEV_1 in the early phase of the reaction and tend to be lower (although still more than 1.0) in the late phase.[42,46] Thus changes after allergen challenge are intermediate between those found with a simple pharmacological challenge and spontaneous asthma (Fig. 4.3). The results have been interpreted as providing evidence for the presence of inflammatory changes in the peripheral airways in spontaneous asthma and to a lesser extent in the late response to allergen. The tendency to airway narrowing after a DI has been shown to relate to elevated concentrations of eosinophils and histamine in bronchoalveolar lavage fluid in asthmatic subjects.[47] A similar response to DI to that found in spontaneous asthma is also found in smokers with airflow obstruction[48] in whom the dominant role of peripheral airway changes is well established.

Although different patterns of changes in the M:P ratio must indicate differences in mechanical behaviour of the airways in response to DI, their reliability as surrogate markers of an inflammatory response remains to be established. As mentioned above, a DI-induced contraction of ASM by a reflex or a direct 'myogenic' response could reduce the M:P ratio without being associated with inflammation. Reduced airway compliance in asthma could reduce the change with DI. A more direct approach to monitoring airway inflammation is to measure expired nitric oxide.[49]

These studies of DI have usually been made after a 45–60 s period of normal tidal breathing without DI. When ASM contraction is induced experimentally its persistence depends on the timing and frequency of DI. When DI is avoided for several minutes after

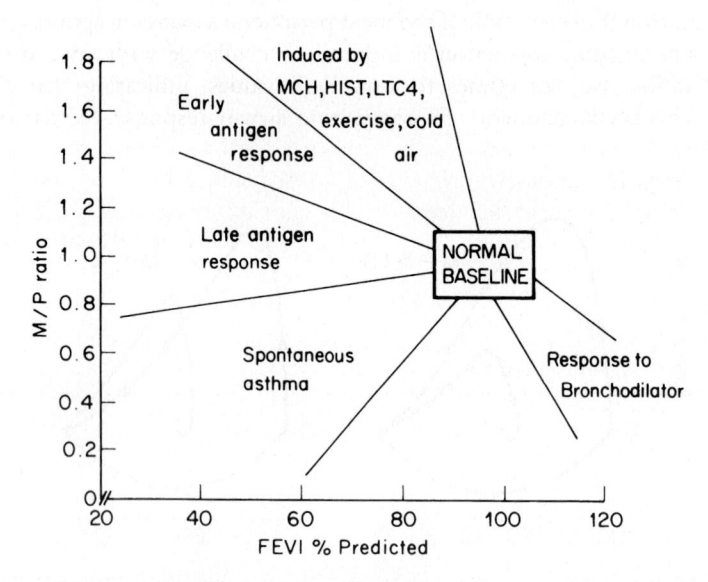

Fig. 4.3 Scheme of changes in isovolumic maximum expiratory flow on maximum and partial flow–volume curves (M:P ratios). In the normal subject this ratio is close to 1.0. With acute bronchoconstrictor challenges the ratio rises as forced expiratory volume in 1 s (FEV_1) falls, but in spontaneous asthma M:P ratios are <1.0 and fall as FEV_1 falls. Changes with antigen challenge are intermediate. MCH, methacholine; HIST, histamine; LTC_4, leukotriene C_4. Reproduced from ref. 42, with permission.

inhaling an aerosol of methacholine, a much greater airway narrowing is induced in both normal and asthmatic subjects.[5] This narrowing is removed by a series of DIs more readily in normal compared with asthmatic subjects, possibly because reduced effective airway compliance in asthma reduces the stretch with DI. A further possibility is that the poor response to DI, and to tidal inflations, may itself allow a change in the state of ASM so that it stiffens, amplifying the reduction in airway wall compliance.[6]

Direct intrabronchial measurements

There are obvious limits to deducing the serial distribution of resistance from measurements of gas flow at the mouth and progress will probably depend on methods making more direct measurements of bronchial dimensions or the distribution of resistance. Unfortunately, the acoustic reflection technique, which obtains a distance–area function of the airway, at present can only provide reliable data from the mouth to just beyond the carina,[50,51] while hopes that tantalum dust could be used to obtain inhalation bronchograms *in vivo* have not been fulfilled. High-resolution computed tomography at present only shows airways in cross-section.[52] While endobronchial pressure measurements are the obvious approach to define the distribution of airflow resistance, their invasive nature restricts their use and there are important physiological limitations. Endobronchial catheters may stimulate mucosal receptors and induce reflex bronchoconstriction and mucus production and can only be used in central airways without compromising airflow through the catheterized airway. Two different wedged-catheter techniques have been developed that measure pressure at a more peripheral site; in one a 3-mm catheter is wedged in a right lower lobe bronchus and lateral pressure measured proximally.[53] With this technique total lung resistance in normal subjects appears about 50% higher than without the presence of a catheter. In remission of asthma, inspiratory (but not expiratory) peripheral airway resistance was slightly increased; in middle-aged patients with persistent airflow obstruction due to chronic asthma, there were increases in both central and peripheral resistance (Table 4.2). Allowing for the contribution of the extrathoracic airway to total pulmonary resistance, peripheral resistance accounted for about one-third of intrathoracic resistance in normal subjects and for about 50% in the subjects with chronic, persistent asthma.[53] In another study of asymptomatic young subjects with asthma and an average FEV_1 of 89% predicted, peripheral airway resistance

Table 4.2 Peripheral and total pulmonary resistance in normal subjects and chronic persistent asthma (data from ref. 53).

	M/F	Mean age (years)	FEV_1 (% predicted)	R_L ($cmH_2O \cdot litre^{-1}$ s)	R_P ($cmH_2O \cdot litre^{-1}$ s)	R_P/R_L (%)
Normal	5/0	56 (± 6)	88 (± 2)	3.1 (± 0.1)	0.7 (± 0.3)	23
Chronic persistent asthma	7/3	57 (± 3)	54 (± 3)	9.3 (± 1.9)	4.1 (± 1.0)	44

Values are mean ($\pm$ SE); M/F, male/female; FEV_1, forced expiratory volume in 1 s; R_L, total pulmonary resistance, measured during tidal breathing; R_P, peripheral resistance of airways <3 mm diameter and lung tissue.
1 $cmH_2O \approx 98$ kPa.

appeared significantly increased.[54] Another technique measures pressure–flow relations in the occluded lung beyond a bronchoscope wedged in a segmental bronchus. In asthmatic subjects in remission with normal total airways resistance and FEV_1, a considerable increase in peripheral lung resistance has been found;[55] this technique measures the combined resistance of peripheral airways and collateral channels. These results directly confirm earlier suggestions from pathological and physiological studies that there are residual changes in the peripheral airways even in remission of asthma, when overall airway function is normal.

In summary, functional abnormality of the peripheral airways is usually present in spontaneous asthma, even when in remission; the extent to which larger conducting airways are involved may vary between individuals and between attacks in an individual, but usually they appear to be narrowed in chronic asthma with persistent airway narrowing. Measurements before and after a DI suggest that easily reversible airway narrowing, such as that induced by histamine, methacholine or exercise, may chiefly involve central conducting airways, while the less reversible changes of spontaneous asthma and the late reaction to allergens may be in more peripheral airways.

Involvement of the extrathoracic airway

Trachea, glottis and pharynx

Narrowing of the extrathoracic airway may add to the intrathoracic airway obstruction in asthma and sometimes extrathoracic airway narrowing is mistaken for asthma.[56] Structural narrowing of the larynx or trachea usually leads to persistent symptoms and distinctive changes in maximum expiratory and inspiratory flow–volume curves[57] that indicate the need for airway endoscopy. In contrast isolated functional obstruction of the upper airway is characteristically episodic and may mimic acute asthma. Two distinct patterns have been described.[58,59] Most commonly, inspiratory stridor is associated with reduction in maximum inspiratory flow throughout the vital capacity and normal maximum expiratory flow; sometimes, however, there is reduction in both maximum inspiratory and expiratory flow. Endoscopic examinations suggest that the larynx is the site of obstruction.[60] Typically patients are young women, who may be admitted repeatedly to hospital with attacks of noisy acute breathlessness. The second type of functional wheezing is mainly expiratory and is produced by forceful breathing close to RV; this is associated with completely normal maximum flow–volume curves; the wheeze probably arises from excessive tidal narrowing of central intrathoracic airways during expiration. Other aspects of lung function, such as blood gases, single-breath nitrogen test and FRC are normal.

Narrowing of the glottis[61] and extrathoracic airway also accompanies some attacks of asthma, but it is uncertain how often it plays an important role. Even in normal subjects, inhaled or intravenous histamine results in some narrowing of the glottis; this narrowing is more pronounced in asthmatic subjects. Detailed studies of resistance during air and helium–oxygen breathing suggest involvement of the glottis and/or extrathoracic trachea in a significant proportion of asthmatic attacks.[39] Expiratory narrowing of the supraglottic airway has also been demonstrated.[62]

Table 4.3 Effects of change in posture on total respiratory resistance while breathing via the mouth or via the nose in normal and asthmatic subjects.

	Total respiratory resistance		Nasal airway resistance (cmH$_2$O·litre^{-1} s)
	Mouth breathing (cmH$_2$O·litre^{-1} s)	Nose breathing (cmH$_2$O·litre^{-1} s)	
Normal subjects ($n = 10$)			
Sitting	2.2 (0.1)	4.2 (0.4)	2.1
Supine	3.0 (0.1)	5.7 (0.4)	2.7
Asthma alone ($n = 8$)			
Sitting	4.6 (0.8)	7.1 (0.6)	2.5
Supine	5.3 (0.9)	8.8 (0.6)	3.4
Asthma with nasal symptoms ($n = 10$)			
Sitting	3.7 (0.2)	7.6 (1.4)	3.9
Supine	5.4 (0.3)	11.2 (1.9)	5.8

Values are mean (SE). Resistance was measured by forced oscillation at 6 Hz at the airway opening. 1 cmH$_2$O ≈ 98 Pa.

Nose

Nasal obstruction due to rhinitis and/or polyps is common in asthmatic subjects. In normal subjects during tidal breathing the nose provides a resistance of about 2 cmH$_2$O·litre^{-1} s (0.2 kPa·litre^{-1} s), approximately 50% of the total airflow resistance. Nasal resistance may be considerably higher in the presence of rhinitis, particularly when lying down (Table 4.3). Presumably if nasal resistance is greatly raised, tidal breathing is divided between the oral and nasal routes, with reduction of the protective and humidifying effects of the nose.

Reflexes between the nose and intrapulmonary airways have been investigated extensively in experimental animals but their importance in humans is uncertain. The better established interaction is between acute nasal stimulation and transient narrowing of intrapulmonary airways, but the reverse relationship—nasal congestion following the provocation of intrapulmonary airway narrowing—has been described.[63] Whether these interactions can lead to sustained airway narrowing or nasal congestion and whether they are exaggerated by chronic nasal or bronchial disease is unknown.

Involvement of lung parenchyma

Conventionally, asthma is regarded as a disease of the airways without significant changes in the lung tissue or air spaces, although one recent biopsy study has found alveolar tissue inflammation.[64] Adults who have had asthma in childhood may develop slightly enlarged TLC and small unexplained reductions in lung recoil pressure during growth.[65] However, most recent studies using body plethysmography have failed to find an acute increase in TLC in spontaneous or induced attacks of asthma;[66] older studies probably overestimated TLC in the presence of airway obstruction due to lack of equilibration of mouth with alveolar pressure during panting.[67] Emphysema is believed not to develop in non-smokers who develop persistent airflow obstruction due to chronic asthma; this is supported by the consistent preservation of a normal or even increased value of carbon monoxide transfer coefficient (CO transfer per litre of lung volume).[68] Nevertheless, the modest loss of lung recoil pressure[21] and, as discussed above, expansion of the airway outer diameter and peri-airway inflammation may reduce the distension of the smaller airways produced by alveolar attachments to the perimeter,[16] even although the alveolar attachments are themselves normal.

A more controversial area is the involvement of lung tissue in the constrictor response. In animals, changes in lung tissue resistance have accounted for a large part of the increase in pulmonary resistance after bronchoconstrictor aerosols, sometimes exceeding the contribution of narrowing of the conducting airways.[69,70] The origins of the increased lung tissue resistance are uncertain; possibilities are contractile elements in the lung periphery, such as smooth muscle in alveolar ducts or small pulmonary vessels. Alternatively, contracted conducting airways may act as struts within the lung parenchyma, directly impeding its expansion.[71] But recent work suggests that homogeneity of airway narrowing may contribute to apparent changes in lung tissue.[72] A detailed study of the effects of inhaled methacholine in humans with asthma failed to show any increase in lung tissue resistance when panting at 1 Hz but a considerable increase in resistance of the conducting airways;[73] however, the effects of tissue resistance are much greater at tidal breathing frequencies and a further study using inhaled histamine in humans has found some peripheral lung effects during very slow breathing manoeuvres.[74] More studies are required to establish whether animal models of asthma involve very different sites of increase in resistance from those found in humans.

EFFECTS OF POSTURE AND SLEEP ON AIRWAY FUNCTION

In normal subjects, airways resistance during tidal breathing increases in the supine posture by about 50% compared with seated values (see Table 4.3). This increase is appropriate for the reduction in end-expired lung volume (FRC).[75]

In asthmatic subjects in remission, increases in resistance on lying down are a little greater than in normal subjects but there is a normal fall in FRC and resistance reverts to previous levels rapidly on sitting up. However, when asthma is active, adopting the supine posture leads to absolute and proportionately larger increases in total respiratory resistance than in normal subjects despite a smaller than normal reduction in FRC, implying a true rise in resistance at isovolume. This could be due to an increase in extrathoracic or chest wall resistance or a true decrease in isovolumic intrapulmonary airway dimensions; there are accompanying changes in maximum expiratory flow–volume curves that suggest a true decrease in airway dimensions. Some studies have also claimed that when the supine posture is maintained for several hours airway function progressively deteriorates in asthmatic but not normal subjects;[76,77] however, a controlled study in which groups of asthmatic subjects remained in either the sitting or supine posture showed similar deterioration in airway function in both groups.[78] In our own laboratory, we have found that airway function in some subjects with asthma does not revert to initial sitting values in the first 10 min after lying supine; these subjects

4 Physiology

appear to be those with symptomatic asthma at the time of the study. The supine posture undoubtedly exacerbates the degree of airway narrowing and breathing discomfort; patients invariably seek relief by sitting up, although this leads to less dramatic improvement than in the orthopnoea of left ventricular failure.

During sleep there are in addition progressive increases in both supralaryngeal and infralaryngeal airway resistance through the night[79] and FRC falls, particularly during rapid eye movement sleep.[80,81]

RESPONSE TO INCREASING SEVERITY OF AIRWAY NARROWING

As airways obstruction becomes more severe, there is increasing difficulty in sustaining ventilation and gas exchange. Minute ventilation has to be increased because of increased oxygen consumption and increase in physiological dead space. Additional factors maintaining oxygenation are local pulmonary vasoconstriction in poorly ventilated areas of lung and an increase in cardiac output. Expiratory flow limitation means that adequate expiratory flow can only be attained at larger lung volumes (Fig. 4.4). Hyperinflation places a heavy burden on the inspiratory muscles. Thus while there is usually no intrinsic

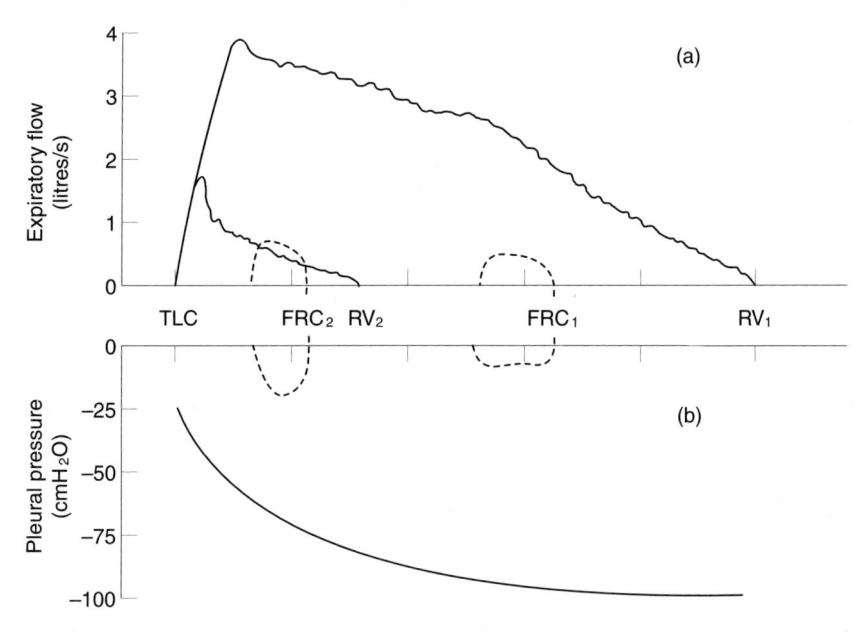

Fig. 4.4 Hyperinflation and expiratory flow limitation in exacerbation of asthma. (a) Tidal (dashed lines) and maximum (continuous lines) expiratory flow–volume curves during remission and exacerbation. Residual volume in exacerbation (RV_2) is larger than functional residual capacity (FRC_1) during remission. Total lung capacity (TLC) is unchanged. Tidal expiratory flow reaches maximum values during exacerbation. (b) Inspiratory pleural pressure during static maximum inspiratory efforts (lower continuous line) and during tidal breathing in remission and exacerbation (dashed lines). The ability of inspiratory muscles to generate negative pleural pressures declines rapidly as TLC is approached.

abnormality of the gas-exchanging part of the lung, pulmonary circulation, control of ventilation or respiratory muscles, these systems play a vital role in sustaining oxygenation during severe asthma.

Ventilatory pattern and tidal intrathoracic pressures

Typically in severe asthma there is a small tidal volume and rapid frequency of respiration, reflecting the decrease in vital capacity and inspiratory capacity. Total ventilation is increased. The limited studies of the intrathoracic pressures generated during severe asthmatic attacks emphasize the very negative inspiratory pressures with only small positive pressures on expiration.[82] Optimum expiratory pressures to generate maximum expiratory flow are lower in attacks of asthma than during remission. At this optimum pressure there will be little dynamic narrowing of central airways and therefore expiratory wheezing may not be pronounced. When excessive expiratory pressures are generated, there is no increase in ventilation but work done by the respiratory muscles will be increased and wheezing more pronounced.

Pulmonary gas exchange and oxygenation

The basic abnormalities of pulmonary gas exchange in asthma were established 25 years ago. There is grossly uneven ventilation, which can be visualized between lung regions on regional ventilation scans but is also found within small regions of lung. This leads to inequality of ventilation–perfusion ($\dot{V}/\dot{Q}$) ratios and moderate reduction in arterial P_{O_2} (Pa_{O_2}), which is usually above 50 mmHg (6.7 kPa) even when FEV_1 is <25% of predicted values.[83] Increase in total minute ventilation usually results in arterial P_{CO_2} (Pa_{CO_2}) levels being normal or below normal until airways obstruction becomes very severe[83] and attenuates, but does not eliminate, the fall in Pa_{O_2}. Small increases in inspired oxygen concentration restore Pa_{O_2} to normal values. More recent analyses of the precise $\dot{V}/\dot{Q}$ abnormality using the multiple inert gas elimination technique (MIGET)[84] emphasize the importance of the absence of anatomical shunt, compensatory changes in pulmonary blood flow and of the ventilatory (and possibly also the circulatory) response in maintaining oxygenation.

In asymptomatic asthmatic subjects, the distribution of $\dot{V}/\dot{Q}$ ratios is often as narrow as in normal subjects, although occasional patients show increased dispersion of ventilation or perfusion (Fig. 4.5). No anatomical shunt ($\dot{V}/\dot{Q}$ ratio = 0) has been demonstrated.[84] Recent studies have shown less abnormality in $\dot{V}/\dot{Q}$ dispersion in asymptomatic patients than found in the initial report.[85]

In chronic symptomatic asthma, the distribution of $\dot{V}/\dot{Q}$ ratios becomes broader, but still without very low ratios or anatomical shunt despite the expected presence of many severely obstructed airways.[86,87] These results suggest considerable compensatory reductions in perfusion to poorly ventilated areas, diminishing any fall in Pa_{O_2}; patients with similar airway obstruction due to COPD have more pronounced chronic hypoxaemia. In acute attacks requiring emergency admission, there is a much wider dispersion of $\dot{V}/\dot{Q}$ ratios, which is often bimodal with alveolar units with low $\dot{V}/\dot{Q}$ ratios but again without significant increase in shunt.[88] In none of these studies was there an increase in

4 Physiology

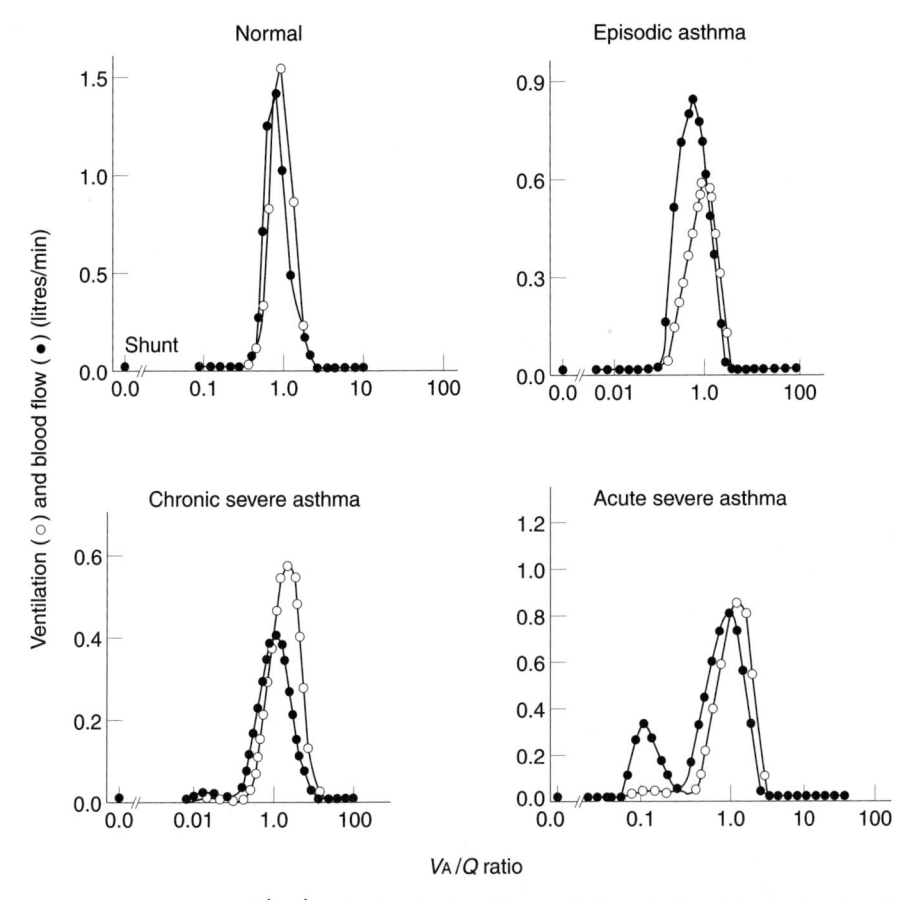

Fig. 4.5 Different patterns of $\dot{V}_A/\dot{Q}$ ratio distributions (○, ventilation; ●, blood flow) plotted against a $\dot{V}_A/\dot{Q}$ ratio on a log scale. Healthy young individuals have narrow unimodal distributions centred around 1.0; in contrast, patients with episodic or chronic asthma have broader unimodal distributions and an increase in units with high $\dot{V}_A/\dot{Q}$ ratios; those with acute severe asthma in addition develop a bimodal pattern of blood-flow distribution and shunt is absent. Note different scaling of axes. Reproduced from ref. 84, with permission.

alveolar units with continuing ventilation but zero blood flow ($\dot{V}/\dot{Q} = \infty$). However, the increased dispersion of $\dot{V}/\dot{Q}$ ratios includes an increase in alveolar units with higher $\dot{V}/\dot{Q}$ ratios, so that physiological dead space, as conventionally measured from the difference between arterial and mixed expired $P\text{CO}_2$, is increased and the ventilatory efficiency for excreting carbon dioxide is decreased. Even in the most severe asthma, requiring mechanical ventilation, no significant anatomical shunt has been found. Pathological studies in fatal cases of asthma show multiple occlusions of airways by muco-inflammatory plugs, sometimes extending into major segmental or even lobar airways. Presumably main airways are occluded by plugs in less severe attacks, yet obvious atelectasis on the chest radiograph is unusual. The most attractive explanation of the absence of shunt is that low $\dot{V}/\dot{Q}$ units receive some collateral ventilation whose efficiency may depend on the accompanying hyperinflation. An important consequence is that small increases in inspired oxygen almost always correct hypoxaemia.

When breathing air there appears to be an efficient compensatory reduction in blood flow to poorly ventilated areas. Breathing increased concentrations of inspired oxygen broadens the dispersion of blood flow and breathing 100% oxygen, fortunately never required in clinical practice, may result in the development of atelectasis and anatomical shunt.[86,88] These changes may follow removal of hypoxic pulmonary vasoconstriction in poorly ventilated areas. Bronchodilators, particularly when given intravenously, may also remove pulmonary vasoconstriction by a direct effect on pulmonary vascular smooth muscle and so increase abnormality of $\dot{V}/\dot{Q}$ ratios, and sometimes reduce Pa_{O_2}.[84]

Control of breathing and perception of obstruction

A vigorous drive to breathing is essential to overcome the mechanical and gas exchange inefficiency and prevent severe hypoxaemia and carbon dioxide retention. Some of the increase in ventilation may be due to the effects of hypoxaemia, although hypocapnia often persists when inspired oxygen is increased, suggesting that stimulation of mechanoreceptors in the lung is also important.[89] There is a wide variation in the perception of airways obstruction;[90] patients who perceive the severity of the obstruction poorly and/or show relatively low increase in ventilation in response to obstruction (or accompanying changes in blood gases) are at particular risk of repeated near-fatal or fatal attacks of asthma.[91] Survivors of near-fatal attacks, studied later when in remission, have shown reduced chemosensitivity to hypoxia and blunted perception of added loads to breathing.[92] Rather than having acquired blunted responses following the development of asthma, such patients may lie in the lower part of the wide normal range of hypoxic chemosensitivity. Although Pa_{CO_2} remains in or slightly below the normal range[83] until a late stage in most patients, a significant minority develop hypercapnia, and tend to present with a similar Pa_{CO_2} in each episode.[93] These patients may have a blunted ventilatory response to carbon dioxide.[94] An intriguing recent finding is that some asthmatic subjects have difficulty in fully activating the diaphragm[95] and that this difficulty is related to a depressed mood,[96] providing a possible physiological explanation for the known association between depression and increased risk of death from asthma.

Hyperinflation and respiratory muscle function

FRC consistently rises during an attack of asthma (Fig. 4.6). In the normal subject, FRC is the volume at which the inward recoil of the lungs is equal to the opposing outward recoil of the chest cage. During a passive expiration, expiratory flow ceases at this relaxation volume because alveolar pressure equals atmospheric pressure. In an asthmatic attack the end-expired volume is larger than the relaxed volume of the chest. Several different mechanisms contribute to this increase in end-expired volume. First, there is usually a large increase in RV, presumably due to enhanced airway closure or near closure; RV may sometimes be larger than the control FRC. Additional dynamic mechanisms maintain end-expired volume above that dictated by airway closure. Thus if there is expiratory flow limitation during tidal expiration, expiration may be then terminated by the initiation of the following inspiration rather than by the cessation of passive expiratory flow. However, even in the absence of expiratory flow limitation, a large end-expiratory

4 Physiology

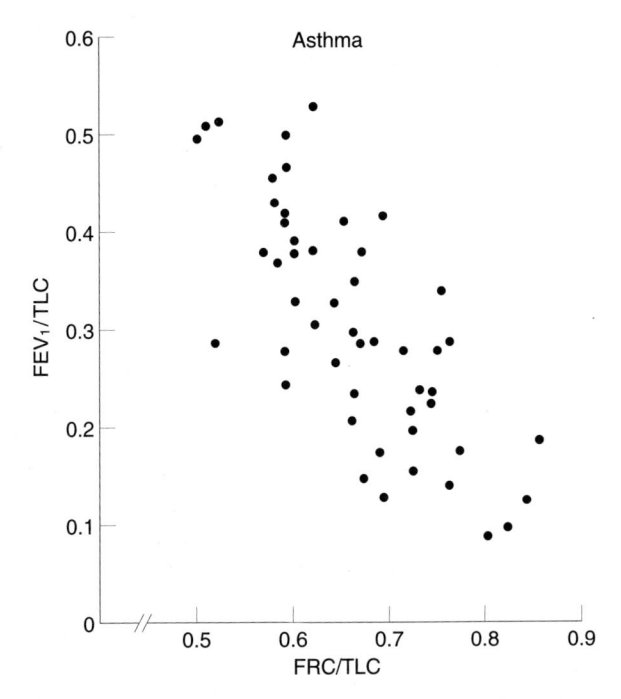

Fig. 4.6 Relation between expiratory airflow obstruction (expressed as forced expiratory volume in 1 s, FEV_1/total lung capacity, TLC) and hyperinflation (expressed as functional residual capacity, FRC/TLC) in subjects with asthma. Lung volumes measured with body plethysmography.

volume may be maintained by persistent inspiratory muscle activity throughout expiration.[97-99] When end-expired volume is above the relaxation volume, the passive effect of respiratory system recoil generates an intrinsic positive end-expiratory pressure (intrinsic PEEP) in the alveoli, which has to be overcome before inspiratory flow begins.[100] Although hyperinflation is relatively well maintained when awake in the supine posture, this depends on an appropriate adjustment of respiratory muscle activity. Hyperinflation reduces during sleep,[80,81] particularly during rapid eye movement sleep,[80] and this must contribute to the worsening of pulmonary function that occurs at night.

Table 4.4 Hyperinflation.

Advantages	Disadvantages
Widens airways, reducing resistive work and improving tidal flow capacity	Increases elastic work of breathing
Improves collateral ventilation, preventing shunt and improving PaO_2 response to increase in inspired O_2 concentration	Shortens initial length of inspiratory muscles, decreasing their capacity to generate force
	Imposes an inspiratory threshold load (intrinsic PEEP) that has to be overcome before inspiratory flow commences

PEEP, positive end-expiratory pressure.

As airway size enlarges and collateral ventilation is enhanced with increase in lung volume, breathing at a larger lung volume partially overcomes the effects of airway narrowing, but this compensation is achieved at the cost of the inspiratory muscles, which have to develop increased tidal pressures to overcome the airway narrowing, maintain increased total ventilation and cope with hyperinflation (Table 4.4). A simple means of measuring hyperinflation would greatly amplify the assessment of respiratory mechanics during the asthmatic crisis. Studies with experimental bronchoconstriction suggest that the degree of hyperinflation may be adjusted to minimize the total work of breathing.[101] When moderate bronchoconstriction is induced by histamine, spontaneous hyperinflation reduces total work of breathing, a slight increase in total inspiratory work (the increase in elastic work outweighing the decrease in resistive work) above that at the original FRC being offset by a larger reduction in expiratory flow-resistive work. However, with severe spontaneous asthma hyperinflation is less likely to reduce airways resistance[28,29] sufficiently to prevent a considerable increase in total inspiratory work, which may account for much of the discomfort of the asthmatic attack[102] and lead to eventual fatigue.

There is surprisingly little systematic information on respiratory muscle performance in asthma, even though inability to sustain ventilation is the usual reason for initiating mechanical ventilation in severe asthma. Respiratory muscle strength in asthma is influenced by non-specific factors such as age, gender and nutrition and, more specifically, by hyperinflation and systemic drugs used in treatment (Table 4.5). Most studies have been of younger asthmatic subjects with moderate airway obstruction and hyperinflation;[103] in such circumstances, strength and endurance of inspiratory and expiratory muscles is normal or enhanced.[103–105] However, respiratory muscle strength declines with age without any corresponding decline in the elastic and resistive load of severe asthma. Sustained treatment with oral corticosteroids may sometimes lead to loss of skeletal muscle mass in patients with airflow obstruction (the effect on limb muscles probably being greater than on the respiratory muscles) but this effect seems less consistent and probably less common in patients with asthma compared with COPD.[106–109] The inspiratory pressure produced by maximal inspiratory efforts declines progressively at volumes greater than about 75% TLC (see Fig. 4.4). Fortunately FRC/TLC is >0.75 only in very severe obstruction; nevertheless as airway obstruction

Table 4.5 Factors affecting respiratory muscle strength in asthma.

Non-specific
Age:declines with increasing age
Gender:women less strong than men
Muscle mass:influenced by nutrition and habitual activity

Specific
Hyperinflation and mechanical load
  Reduced length of inspiratory muscles possibly attenuated by adaptation or training effect
Systemic drugs used in treatment
  Corticosteroids:chronic treatment may cause loss of mass; acute high doses may rarely cause acute myopathy
  β-Adrenergic agonists:hypokalaemia weakens; long-term high dose may increase muscle mass
  Theophyllines:possible increase in strength and protection against fatigue

increases respiratory muscle performance must become a limiting factor eventually, even when strength and endurance in the tidal breathing range is normal during remission. Despite much research, no techniques are established for simply detecting central or peripheral fatigue or the need for pressure support[100] or assisted ventilation in the presence of severe airway obstruction.

CONCLUSIONS

Routine physiological monitoring of the severity of asthma is currently based on simple tests of airway function and arterial oxygen saturation or blood gases. There are as yet no simple reliable tests to anticipate impending central or respiratory muscle fatigue. The degree of hyperinflation is rarely followed. At a more basic level, little is known about collateral ventilation, why narrowing varies so greatly between parallel airways, the production and removal of airway secretions or the control of airway mucosal blood flow. Understanding of the pathophysiology of the asthma attack remains very incomplete.

REFERENCES

1. Cockcroft DW, Berscheid BA, Murdock KY: Unimodal distribution of bronchial responsiveness in a random population. *Chest* (1983) **83**: 751–754.
2. Solway J, Fredberg JJ: Perhaps airway smooth muscle dysfunction contributes to asthmatic bronchial hyper-responsiveness after all. *Am J Respir Cell Mol Biol* (1997) **171**: 144–146.
3. Thomson RJ, Bramley AM, Schellenberg RR: Airway muscle stereology: implications for increased shortening in asthma. *Am J Respir Crit Care Med* (1996) **154**: 749–757.
4. Freedman S: Exercise as a bronchodilator. *Clin Sci* (1992) **83**: 383–389.
5. Skloot G, Permutt S, Togias A: Airway hyperresponsiveness in asthma: a problem of limited smooth muscle relaxation with inspiration. *J Clin Invest* (1995) **96**: 2393–2403.
6. Fredberg JJ, Jones KA, Nathan M, Raboudi S, Prakash Y, Shore SA, Butler JP, Sieck GC: Friction in airway smooth muscle:mechanism, latch, and implications in asthma. *J Appl Physiol* (1996) **81**: 2703–2712.
7. Woolcock AJ, Salome CM, Yan K: The shape of the dose–response curve to histamine in asthmatic and normal subjects. *Am Rev Respir Dis* (1984) **130**: 71–75.
8. Michoud MC, Lelorier J, Amyot R: Factors modulating the interindividual variability of airway responsiveness to histamine. The influence of H_1 and H_2 receptors. *Bull Eur Physiopathol Respir* (1981) **17**: 807–821.
9. Sterk PJ, Daniel EE, Zamel N, Hargreave FE: Limited bronchoconstriction to methacholine using partial flow–volume curves in non-asthmatic subjects. *Am Rev Respir Dis* (1985) **132**: 272–277.
10. James A, Lougheed D, Pearce-Pinto G, Ryan G, Musk B: Maximal airway narrowing in a general population. *Am Rev Respir Dis* (1992) **146**: 895–899.
11. Olsen CR, Stevens AE, Pride NB, Staub NC: Structural basis for decreased compressibility of constricted trachea and bronchi. *J Appl Physiol* (1967) **23**: 35–39.
12. James AL, Paré PD, Moreno RH, Hogg JC: Quantitative measurement of smooth muscle shortening in isolated pig trachea. *J Appl Physiol* (1987) **63**: 1360–1365.
13. Olsen CR, Stevens AE, McIlroy MB: Rigidity of trachea and bronchi during muscular constriction. *J Appl Physiol* (1967) **23**: 27–34.

14. Bouhuys A, Hunt VR, Kim BM, Zapletal A: Maximum expiratory flow rates in induced bronchoconstriction in man. *J Clin Invest* (1969) **48**: 1159–1168.
15. Lambert RK, Codd SL, Alley MR, Pack RJ: Physical determinants of bronchial mucosal folding. *J Appl Physiol* (1994) **77**: 1206–1216.
16. Macklem PT: Theoretical analysis of the effect of airway smooth muscle load on airway narrowing. *Am J Respir Crit Care Med* (1996) **153**: 83–89.
17. Murtagh PS, Proctor DF, Permutt S, Kelly B, Evering S: Bronchial closure with mecholyl in excised dog lobes. *J Appl Physiol* (1971) **31**: 409–415.
18. Ding DJ, Martin JG, Macklem PT: The effects of lung volume on maximal methacholine-induced bronchoconstriction in normal humans. *J Appl Physiol* (1987) **62**: 1324–1330.
19. Wang YT, Coe CI, Pride NB: The effect of reducing airway dimensions by altering posture on histamine responsiveness. *Thorax* (1990) **45**: 530–535.
20. Bellofiore S, Eidelman DH, Macklem PT, Martin JG: Effects of elastase-induced emphysema on airway responsiveness to methacholine in rats. *J Appl Physiol* (1989) **66**: 606–612.
21. Pride NB, Macklem PT: Lung mechanics in disease. In Macklem P, Mead J (eds) *Handbook of Physiology: Respiratory System, vol III, Mechanics of Breathing*. Bethesda, American Physiological Society, 1986, pp 678–679.
22. Freedman BJ: The functional geometry of the bronchi. The relationship between changes in external diameter and calibre, and a consideration of the passive role played by the mucosa in bronchoconstriction. *Bull Eur Physiopathol Respir* (1972) **8**: 545–551.
23. Moreno RH, Hogg JC, Paré PD: Mechanics of airway narrowing. *Am Rev Respir Dis* (1986) **133**: 1171–1180.
24. James AL, Paré PD, Hogg JC: The mechanics of airway narrowing in asthma. *Am Rev Respir Dis* (1989) **139**: 242–246.
25. Carroll N, Elliot J, Morton A, James A: The structure of large and small airways in nonfatal and fatal asthma. *Am Rev Respir Dis* (1993) **147**: 405–410.
26. McFadden ER Jr: Hypothesis: exercise-induced asthma as a vascular phenomenon. *Lancet* (1990) **i**: 880–883.
27. Wilson JW, Li X, Pain MCF: The lack of distensibility of asthmatic airways. *Am Rev Respir Dis* (1993) **148**: 806–809.
28. Butler J, Caro CG, Alcala R, DuBois AB: Physiological factors affecting airway resistance in normal subjects and in patients with obstructive respiratory disease. *J Clin Invest* (1960) **39**: 584–591.
29. Colebatch HJH, Finucane KE, Smith MM: Pulmonary conductance and elastic recoil relationship in asthma and emphysema. *J Appl Physiol* (1973) **34**: 143–153.
30. Lambert RK, Wiggs BR, Kuwano K, Hogg JC, Paré PD: Functional significance of increased airway smooth muscle in asthma and COPD. *J Appl Physiol* (1993) **74**: 2771–2781.
31. Ramsdale EH, Roberts RS, Morris MM, Hargreave FE: Differences in responsiveness to hyperventilation and methacholine in asthma and chronic bronchitis. *Thorax* (1985) **40**: 422–426.
32. Dirksen A, Madsen F, Engel T, Frølund L, Heinig JH, Mosbech H: Airway calibre as a confounder in interpreting bronchial responsiveness in asthma. *Thorax* (1992) **47**: 702–706.
33. Despas PJ, Leroux M, Macklem PT: Site of airway obstruction in asthma as determined by measuring maximal expiratory flow breathing air and helium–oxygen mixture. *J Clin Invest* (1972) **51**: 3235–3243.
34. Ingram RH Jr, McFadden ER Jr: Localization and mechanisms of airway responses. *N Engl J Med* (1977) **297**: 596–600.
35. Fairshter RD, Wilson AF: Relationship between the site of airflow limitation and localization of the bronchodilator response in asthma. *Am Rev Respir Dis* (1980) **122**: 27–32.
36. Partridge MR, Saunders KB: The site of airflow limitation in asthma: the effect of time, acute exacerbations of disease and clinical features. *Br J Dis Chest* (1981) **75**: 263–272.
37. Antic R, Macklem PT: The influence of clinical factors on site of airway obstruction in asthma. *Am Rev RespirDis* (1976) **114**: 851–859.
38. Jadue C, Greville H, Coalson JJ, Mink SN: Forced expiration and HeO$_2$ response in canine peripheral airway obstruction. *J Appl Physiol* (1985) **58**: 1788–1801.

39. Lisboa C, Jardim J, Angus E, Macklem PT: Is extra-thoracic airway obstruction important in asthma? *Am Rev Respir Dis* (1980) **122**: 115–121.
40. Gayrard P, Orehek J, Grimaud C, Charpin C: Bronchoconstrictor effects of a deep inspiration in patients with asthma. *Am Rev Respir Dis* (1975) **111**: 433–439.
41. Ingram RH Jr: Site and mechanism of obstruction and hyperresponsiveness in asthma. *Am Rev Respir Dis* (1987) **136**: S62–564.
42. Ingram RH Jr: Physiological assessment of inflammation in the peripheral lung of asthmatic patients. *Lung* (1990) **168**: 237–248.
43. Berry RB, Fairshter RD: Partial and maximal flow–volume curves in normal and asthmatic subjects before and after inhaltion of metaproterenol. *Chest* (1985) **88**: 697–702.
44. Lim TK, Pride NB, Ingram RH Jr: Effects of volume history during spontaneous and acutely induced air-flow obstruction in asthma. *Am Rev Respir Dis* (1987) **135**: 591–596.
45. Lim TK, Ang SM, Rossing TH, Ingenito EP, Ingram RH Jr: The effects of deep inhalation on maximal expiratory flow during intensive treatment of spontaneous asthmatic episodes. *Am Rev Respir Dis* (1989) **140**: 340–343.
46. Pellegrino R, Violante B, Crimi E, Brusasco V: Effects of deep inhalation during early and late asthmatic reactions to allergen. *Am Rev Respir Dis* (1990) **143**: 822–825.
47. Pliss IB, Ingenito EP, Ingram RH Jr: Responsiveness, inflammation, and effects of deep breaths on obstruction in mild asthma. *J Appl Physiol* (1989) **66**: 2298–2304.
48. Pride NB, Ingram RH Jr, Lim TK: Interaction between parenchyma and airways in chronic obstructive pulmonary disease and in asthma. *Am Rev Respir Dis* (1991) **143**: 1446–1449.
49. Barnes PJ, Kharitonov SA: Exhaled nitric oxide: a new lung function test. *Thorax* (1996) **51**: 233–237.
50. Molfino NA, Slutsky AS, Hoffstein V, et al.: Changes in cross-sectional airway areas induced by methacholine, histamine and LTC_4 in asthmatic subjects. *Am Rev Respir Dis* (1992) **146**: 577–580.
51. Molfino NA, Slutsky AS, Julià-Serdà G, et al.: Assessment of airway tone in asthma. *Am Rev Respir Dis* (1993) **148**: 1238–1243.
52. Okazawa MO, Müller N, McNamara AE, Child S, Verburgt L, Paré PD: Human airway narrowing measured using high resolution computed tomography. *Am J Respir Crit Care Med* (1996) **154**: 1557–1562.
53. Yanai M, Sekizawa K, Ohrui T, Sasaki H, Takishima T: Site of airway obstruction in pulmonary disease: direct measurement of intrabronchial pressure. *J Appl Physiol* (1992) **72**: 1016–1023.
54. Ohrui Y, Sekizawa K, Yanai M, et al.: Partitioning of pulmonary responses to inhaled methacholine in subjects with asymptomatic asthma. *Am Rev Respir Dis* (1992) **146**: 1501–1505.
55. Wagner EM, Liu MC, Weinmann GG, Permutt S, Bleecker ER: Peripheral lung resistance in normal and asthmatic subjects. *Am Rev Respir Dis* (1990) **141**: 584–588.
56. Bucca C, Rolla G, Brussino L, De Rose V, Bugiani M: Are asthma-like symptoms due to bronchial or extrathoracic airway dysfunction? *Lancet* (1995) **346**: 791–795.
57. Miller RD, Hyatt RE: Evaluation of obstructive lesions of the trachea and larynx by flow–volume loops. *Am Rev Respir Dis* (1973) **108**: 475–481.
58. Rodenstein DO, Francis C, Stanescu DC: Emotional laryngeal wheezing: a new syndrome. *Am Rev Respir Dis* (1983) **127**: 354–356.
59. Goldman J, Muers M: Vocal cord dysfunction and wheezing. *Thorax* (1991) **46**: 401–404.
60. Newman KB, Mason UG III, Schmaling KB: Clinical features of vocal cord dysfunction. *Am J Respir Crit Care Med* (1995) **152**: 1382–1386.
61. Collett PW, Brancatisano T, Engel LA: Changes in the glottic aperture during bronchial asthma. *Am Rev Respir Dis* (1983) **128**: 719–723.
62. Collett PW, Brancatisano AP, Engel LA: Upper airway dimensions and movements in bronchial asthma. *Am Rev Respir Dis* (1986) **133**: 1143–1149.
63. Yap JCH, Pride NB: Effect of induced bronchoconstriction on nasal airflow resistance in patients with asthma. *Clin Sci* (1994) **86**: 55–58.
64. Kraft M, Djukanovic R, Wilson S, Holgate ST, Martin RJ: Alveolar tissue inflammation in asthma. *Am J Respir Crit Care Med* (1996) **154**: 1505–1510.

65. Greaves IA, Colebatch HJH: Large lungs after childhood asthma: a consequence of enlarged airspaces. *Aust NZ J Med* (1985) **15**: 427–434.
66. Kirby JG, Juniper EF, Hargreave FE, Zamel N: Total lung capacity does not change during methacholine-stimulated airway narrowing. *J Appl Physiol* (1986) **61**: 2144–2147.
67. Stanescu DC, Rodenstein D, Cauberghs M, Van de Woestijne KP: Failure of body plethysmography in bronchial asthma. *J Appl Physiol: Environ Exercise Physiol* (1982) **52**: 939–948.
68. Keens TG, Mansell A, Krastins IRB, *et al*.: Evaluation of the single-breath diffusing capacity in asthma and cystic fibrosis. *Chest* (1979) **76**: 41–44.
69. Ludwig MS, Romero PV, Bates JHT: A comparison of the dose–response behaviour of canine airways and parenchyma. *J Appl Physiol* (1989) **67**: 1220–1225.
70. Ludwig MS: Role of lung parenchyma. In *Asthma* vol 2, Barnes PJ, Grunstein MM, Leff AR, Woolcock AJ (eds). Philadelphia, Lippincott-Raven, 1997, pp. 1319–1334.
71. Mitzner W, Blosser S, Yager D, Wagner E: Effect of bronchial smooth muscle contraction on lung compliance. *J Appl Physiol* (1992) **72**: 1900–1907.
72. Lutchen KR, Hantos Z, Petak F, Adamicza A, Sukli B: Airway inhomogeneities contribute to apparent lung tissue mechanics during constriction. *J Appl Physiol* (1996) **80**: 1841–1849.
73. Kariya ST, Thompson LM, Ingenito EP, Ingram RH: Effects of lung volume, volume history, and methacholine on lung tissue viscance. *J Appl Physiol* (1989) **66**: 977–982.
74. Pellegrino R, Violante B, Crimi E, Brusasco V: Effects of aerosol methacholine and histamine on airways and lung parenchyma in healthy humans. *J Appl Physiol* (1993) **74**: 2681–2686.
75. Linderholm H: Lung mechanics in sitting and horizontal postures studied by body plethysmographic methods. *Am J Physiol* (1963) **204**: 85–91.
76. Jonsson E, Mossberg B: Impairment of ventilatory function by supine posture in asthma. *Eur J Respir Dis* (1984) **65**: 495–503.
77. Larsson K, Bevegård S, Mossberg B: Posture-induced airflow limitation in asthma: relationship to plasma catecholamines and an inhaled anticholinergic agent. *Eur Respir J* (1988) **1**: 458–463.
78. Whyte KF, Douglas NJ: Posture and nocturnal asthma. *Thorax* (1989) **44**: 579–581.
79. Ballard RD, Saathoff MC, Patel DK, Kelly PL, Martin RJ: Effect of sleep on nocturnal bronchoconstriction and ventilatory patterns in asthmatics. *J Appl Physiol* (1989) **67**: 243–249.
80. Ballard RD, Irvin CG, Martin RJ, Pak J, Pandey R, White DP: Influence of sleep on lung volume in asthmatic patients and normal subjects. *J Appl Physiol* (1990) **68**: 2034–2041.
81. Ballard RD, Clover CW, White DP: Influence of non-REM sleep on inspiratory muscle activity and lung volume in asthmatic patients. *Am Rev Respir Dis* (1993) **147**: 880–886.
82. Hedstrand U: Ventilation, gas exchange, mechanisms of breathing and respiratory work in acute bronchial asthma. *Acta Soc Med Ups* (1971) **76**: 248–270.
83. McFadden ER Jr, Lyons HA: Arterial blood gas tension in asthma. *N Engl J Med* (1968) **278**: 1027–1032.
84. Rodriguez-Roisin R, Roca J: Contributions of multiple inert gas elimination technique to pulmonary medicine. 3. Bronchial asthma. *Thorax* (1994) **49**: 1027–1033.
85. Wagner PD, Dantzker DR, Iacovoni VE, Tomlin WC, West JB: Ventilation–perfusion inequality in asymptomatic asthma. *Am Rev Respir Dis* (1978) **118**: 511–524.
86. Corte P, Young I: Ventilation–perfusion relationships in symptomatic asthma. Response to oxygen and clemastine. *Chest* (1985) **88**: 167–175.
87. Ballester E, Roca J, Ramis LI, Wagner PD, Rodriguez-Roisin R: Pulmonary gas exchange in chronic asthma. Response to 100% oxygen and salbutamol. *Am Rev Respir Dis* (1990) **141**: 558–562.
88. Rodriguez-Roisin R, Ballester E, Roca J, Torres A, Wagner PD: Mechanisms of hypoxemia in patients with status asthmaticus requiring mechanical ventilation. *Am Rev Resp Dis* (1989) **139**: 732–739.
89. Hudgel DW, Capehart M, Hirsch JE: Ventilation response and drive during hypoxia in adult patients with asthma. *Chest* (1979) **76**: 294–299.
90. Rubinfeld AR, Pain MCF: Perception of asthma. *Lancet* (1976) **i**: 882–884.

91. Rea HH, Scragg R, Jackson R, Beaglehole R, Fenwick J, Sutherland DC: A case-control study of deaths from asthma. *Thorax* (1986) **41**: 833–839.
92. Kikuchi Y, Okabe S, Tamura G, *et al.*: Chemosensitivity and perception of dyspnea in patients with a history of near-fatal asthma. *N Engl J Med* (1994) **330**: 1329–1334.
93. Mountain RD, Sahn SA: Clinical features and outcome in patients with acute asthma presenting with hypercapnia. *Am Rev Respir Dis* (1988) **138**: 535–539.
94. Rebuck AS, Read J: Patterns of ventilatory response to carbon dioxide during recovery from severe asthma. *Clin Sci* (1971) **41**: 13–21.
95. Allen GM, McKenzie DK, Gandevia SC, Bass S: Reduced voluntary drive to breathe in asthmatic subjects. *Respir Physiol* (1993) **93**: 29–40.
96. Allen GM, Hickie I, Gandevia SC, McKenzie DK: Impaired voluntary drive to breathe: a possible link between depression and unexplained ventilatory failure in asthmatic patients. *Thorax* (1994) **49**: 881–884.
97. Martin J, Powell E, Shore S, Emrich J, Engel LA: The role of respiratory muscles in the hyperinflation of bronchial asthma. *Am Rev Respir Dis* (1980) **121**: 441–447.
98. Muller N, Bryan AC, Zamel N: Tonic inspiratory activity as a cause of hyperinflation in histamine-induced asthma. *J Appl Physiol: Respir Environ Exercise Physiol* (1980) **4**: 869–874.
99. Muller N, Bryan AC, Zamel N: Tonic inspiratory muscle activity as a cause of hyperinflation in asthma. *J Appl Physiol: Respir Environ Exercise Physiol* (1981) **50**: 279–282.
100. Lougheed, MD, Webb KA, O'Donnell DE: Breathlessness during induced lung hyperinflation in asthma: the role of the inspiratory threshold load. *Am J Respir Crit Care Med* (1995) **152**: 911–920.
101. Wheatley JR, West S, Cala SJ, Engel LA: The effect of hyperinflation on respiratory muscle work in acute induced asthma. *Eur Respir J* (1990) **3**: 625–632.
102. Lougheed MD, Lam M, Forkert L, Webb KA, O'Donnell DE: Breathlessness during acute bronchoconstriction in asthma: pathophysiologic mechanisms. *Am Rev Respir Dis* (1993) **148**: 1452–1459.
103. De Bruin PF, Ueki J, Watson A, Pride NB: Size and strength of the respiratory and quadriceps muscles in patients with chronic asthma. *Eur Respir J* (1997) **10**: 59–64.
104. McKenzie DK, Gandevia SC: Strength and endurance of inspiratory, expiratory and limb muscles in asthma. *Am Rev Respir Dis* (1986) **134**: 999–1004.
105. Gorman RB, McKenzie DK, Gandevia SC, Plassman BL: Inspiratory muscle strength and endurance during hyperinflation and histamine induced bronchoconstriction. *Thorax* (1992) **47**: 922–927.
106. Melzer E, Souhrada JF: Decrease of respiratory muscle strength and static lung volumes in obese asthmatics. *Am Rev Respir Dis* (1980) **121**: 17–22.
107. Picado C, Fiz JA, Montserrat JM, *et al.*: Respiratory and skeletal muscle function in steroid-dependent bronchial asthma. *Am Rev Respir Dis* (1990) **141**: 14–21.
108. Decramer M, Lacquet LM, Fagard R, Rogiers P: Corticosteroids contribute to muscle weakness in chronic airflow obstruction. *Am J Respir Crit Care Med* (1994) **150**: 11–16.
109. Perez T, Becquart L-A, Stach B, Wallaert B, Tonnel A-B: Inspiratory muscle strength and endurance in steroid-dependent asthma. *Am J Respir Crit Care Med* (1996) **153**: 610–615.

5

Airway Smooth Muscle Cells: Structure and Function

L.J. JANSSEN, E.E. DANIEL AND I.W. RODGER

INTRODUCTION

It is well recognized that contraction of airway smooth muscle (ASM) is the principal component underlying the bronchoconstriction that characterizes the acute phase of an asthmatic attack. It is also well established that the airway structure(s) of asthmatic subjects is abnormal. Such abnormalities are, in all likelihood, consequent upon an exaggerated inflammatory process that remodels the airway architecture, promotes interstitial oedema and induces hypersecretion of mucus into the airway lumen. Associated with these changes is an increase in ASM mass that may also function abnormally as evidenced by its hyperresponsiveness to a wide range of provoking stimuli.

The objective of this chapter is to provide an overview of the structure and function of ASM cells. Given the space constraints, this chapter is not an exhaustive review of the literature. Rather it is an attempt to provide basic information on cellular morphology and physiology and to direct the more interested reader to authoritative key references on particular aspects of the topic.

MORPHOLOGY

The focus in this section is directed at those structures that are expected to control ASM function at a cellular level. For details on the innervation of ASM the interested reader is directed to the review by Gabella.[1]

ASM cells

ASM cells possess most features in common with other smooth muscle cells[2,3] and only some unusual features will be noted here. Cell division is uncommon in normal ASM of adults. As in other smooth muscles, near the nucleus are mitochondria at the nuclear poles, a Golgi apparatus and some rough endoplasmic reticulum (ER). Peripherally, the usual contractile filaments are present, namely 4–7-nm actin filaments, thicker 12–20-nm myosin filaments and 8–12-nm intermediate filaments. The contractile filaments are located in the periphery of all cells and make contact intermittently with the cell membrane at electron-dense regions. The cell membrane, or plasmalemma, contains numerous caveolae located in multiple longitudinal rows in regions without attachments to contractile filaments. In electron micrograph sections of profiles of smooth muscle, caveolae have been found to increase the plasmalemma length by up to 30% and to increase the calculated cell surface area by up to 70% (also see ref. 1). ASM cells have many invaginations of the cell membrane and irregular cell profiles compared with other smooth muscle cells. Near the plasmalemma are profiles of smooth ER, often in close relationship to it and caveolae; smooth ER is also sometimes found distant from the plasmalemma. The two populations of ER, near the plasmalemma and distant from it, may have distinct functions. There also are mitochondria near the plasmalemma, which may represent a population distinct from those near the nuclear poles. Inferences based on location are not yet supported by any direct evidence. Smooth muscle cells are surrounded by a basal lamina, and between them are found collagen fibres and elastin. Evidence of the production of these connective tissue components within smooth muscle cells has been reported in similar cells in culture.[2]

Cell–cell junctions

Among smooth muscle cells, several types of junction are recognized:[4–7] gap junctions (or nexuses), close oppositions (sometimes called adherens junctions), and intermediate contacts or desmosome-like structures. Although gap junctions and close appositions are common in adult canine airway from trachea to sixth-order bronchi, the desmosome-like intermediate contacts are uncommon compared with other smooth muscles; they seem to be more common in trachea than in small bronchi. Since these junctions are presumed[2,6,8–10] to provide a mechanical linkage between cells, their unusual near-absence from bronchi implies that the muscle cells may move relative to one another during respiration. The only other junction with some stability to strain is the gap junction;[8,9] there are many of these throughout the airway down to the sixth-order bronchi. However, there is reason to believe that gap junctions in airway may turn over very rapidly.[11,12] They may need to break apart and reform rapidly as muscle cells move relative to one another during bronchoconstriction and bronchodilation.

There is no information about the function of close appositions in smooth muscle cells. In contrast, there is good evidence[13–17] that the gap junctions alone are sufficient to provide excellent cell–cell coupling, both metabolic and electrical. Whether they are necessary for such coupling is unclear because several smooth muscles that lack visible gap junctions have good electrical coupling.[13,16,18] There is good theoretical evidence

5 Airway Smooth Muscle Cells: Structure and Function

that field coupling across narrow extracellular gaps can occur,[19–21] and some experimental evidence that small gap junctions invisible to electron microscopy may provide coupling.[5] Agents that inhibit gap junction conductance, such as heptanol and octanol, appear to have effects on other (K^+ or Ca^{2+}) conductances as well;[22] thus, functional changes in the presence of these alcohols cannot be conclusively associated with loss of gap junction conductances.[23] In ASM, clearly evident gap junctions are common and probably do provide the excellent cell–cell coupling that is observed.

The proteins comprising gap junctions have been cloned from a number of tissues and exist in smooth muscle as well as in other cell types.[24–29] In smooth muscle, including ASM, the protein commonly expressed is connexin 43.[30–42] Connexins constitute a family

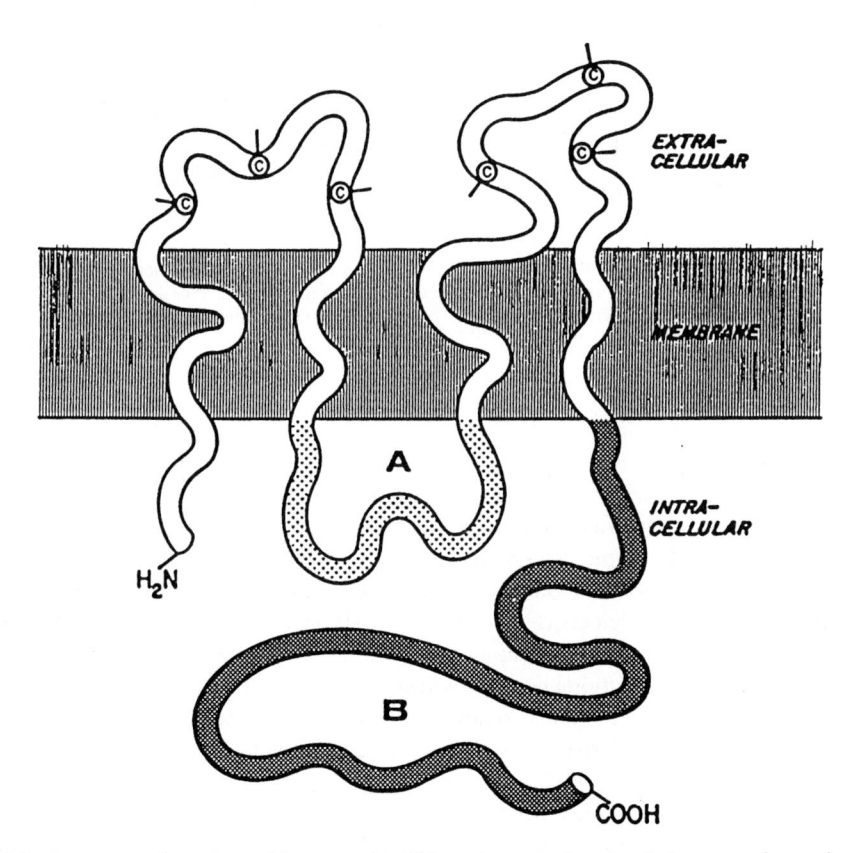

Fig. 5.1 Structure and topology of the connexins (Cx) relative to the junctional plasma membrane. A model of the topology of the subunit gap junction proteins has been developed based on hydropathy plots and tested by proteolysis and immunocytochemical studies. That model predicts that the connexins have four transmembrane spans and have both their amino and carboxyl termini located on the intracellular face of the functional membrane. Unshaded portions represent the regions of connexins that are relatively more conserved among all members of the family: the four transmembrane and two extracellular domains. The two extracellular domains each contain three invariant cysteines (represented by circled C). In contrast, the cytoplasmic loop in the middle of the connexins (A) and the cytoplasmic carboxyl-terminal tail (B) are entirely different among the connexins, both in sequence and in length. These regions, especially segment B, contain multiple sites for phosphorylation by kinases. The lengths of domains A and B for Cx43 are 55 and 154 and for Cx32 are 36 and 74 (rat connexins). Reproduced from ref. 29, with permission.

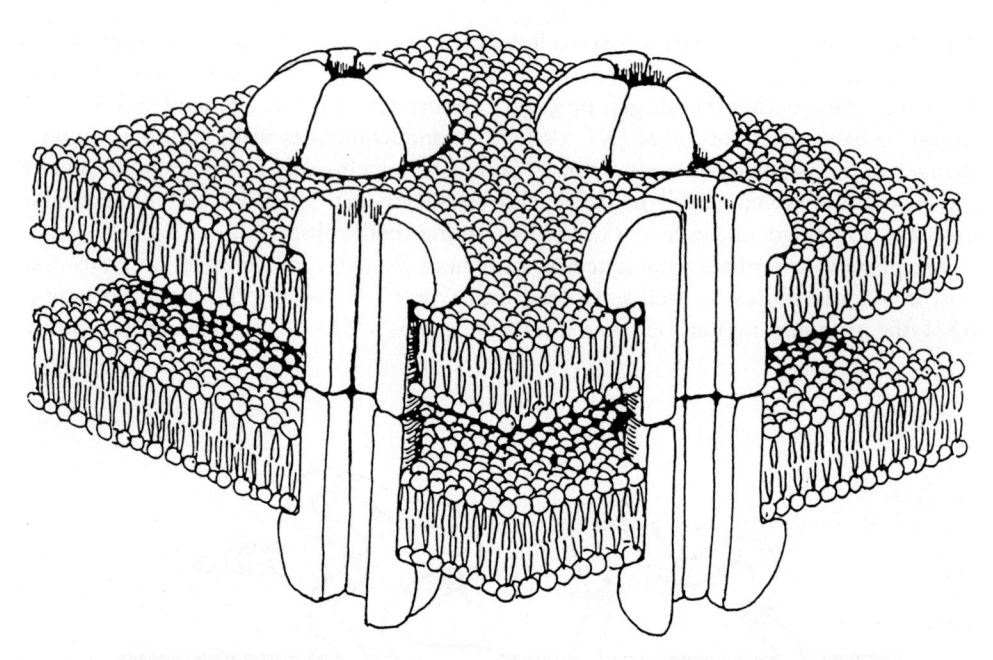

Fig. 5.2 Small part of a gap junction showing two plasma membranes closely apposed (distance 2–4 nm). Junctional channels traversing both membranes are formed by connexons in each membrane. Molecules up to 1 kDa and iron can pass through open channels. Each connexon is made up of six subunits. Reproduced from ref. 43, with permission.

of related proteins[24,43] with the general structure shown in Fig. 5.1. Size aligned molecules (Fig. 5.2) compose a channel and, when inserted in a plasmalemmal membrane, constitute a hemi-gap junction. When aligned with another hemi-gap junction the two form a channel, a connexon, connecting two cells. Arrays of these (up to 100 or more), which allow flow of ions (electrical current) as well as intracellular molecules (up to 1 kDa) between cells, constitute a gap junction.

The coupling by gap junctions can be modified by changes in gap junction number, unitary conductance or their open probability.[43] Coupling between ASM cells has not been studied directly, but in other smooth muscle cells it has been shown to be inhibited by cyclic AMP (cAMP) elevation.[44,45] Cell–cell coupling is increased by protein kinase C (PKC) in some preparations[46] but decreased in others[47–52] (but see ref. 53). In other cell types, tyrosine kinase activation may inhibit cell–cell coupling.[53–55] Modulations by other protein kinases are probably effected through phosphorylation of serines or threonines located mostly on the C-terminal cytoplasmic tail of the connexin molecule. In ASM, the expression of connexin 43 can be modulated. Prostaglandin E_2 (PGE_2) can increase the density of gap junctions[12] and the expression of its mRNA (Z. Li, personal communication). Other products of arachidonate metabolism may also play a role in the control of gap junction expression in ASM.[56] Whether prostanoids or, in particular, leukotrienes (see below) play a role in airway hyperresponsiveness by affecting gap junction expression is an intriguing possibility that remains to be determined.

5 Airway Smooth Muscle Cells: Structure and Function

Species differences in ASM

The structure of human, bovine and guinea-pig airway has been studied in most detail. Human[57] and bovine[58] trachea are nearly identical to canine trachea in terms of the structural arrangements that have been described. In particular, the presence of similar cell-organelle contractile filaments, cell-surface features and cell–cell junctions has been described. Studies of fresh human bronchial smooth muscle (E. E. Daniel, I. Berezin and G. Cox, unpublished data) found normal gap junctions in abundance (see fig. 8 in ref. 59). It will be of particular interest (and of possible relevance) to our understanding of airway responsiveness in disease to determine whether the cell's ability to contract in response to depolarizing current inputs and to chemical stimulants is affected by agents that uncouple gap junctions selectively. To date, no such selective agents exist.

Guinea-pig trachea also contains a supply of gap junctions, apparently somewhat smaller and less dense than in other species.[7,60] Gabella[1] has reported data suggesting that there are moderately high densities of gap junctions (about 10 per 100 smooth muscle cell profiles) in mouse, rat and rabbit tracheal smooth muscle. Cell–cell coupling or its role in airway responsiveness has not been evaluated directly in these species because there is no selective inhibitor of gap junction conductances. Tables 5.1 and 5.2 summarize these and other data.

Innervation

In most species,[1] nerves penetrate ASM mainly from the adventitial side.[61] Cholinergic excitatory and any non-adrenergic inhibitory efferent nerves derive from the ganglia of the plexus of the tracheal wall, which receive preganglionic innervation via the vagus. Adrenergic nerves derive from the sympathetic chain and in some species send branches

Table 5.1 Density of gap junction and innervation in airway smooth muscle.

Animal	No. of gap junctions*	No. of axons*	No. of varicose axons*
Canine†			
Bronchi, 1st	2.7	2.15	1.3
Bronchi, 2nd	3.1	16.8	8.7
Bronchi, 3rd	3.7	15.6	8.25
Bronchi, 4th	2.3	27.3	15.1
Human			
Trachea	2.7	3.0	0.8
Bronchi, 3rd–4th	2.2	35.7	17.9
Guinea-pig			
Trachea, proximal	1.2	–	19.2
Trachea, distal	2.1	–	12.7
Bovine trachea	8.0	1.1	–

* Values for canine are 10^3 μM; all other values are per 100 smooth muscle cells.
† From E.E. Daniel, unpublished results; data normalized to length of plasma membrane.

Table 5.2 Structures of trachea.

Animal*	No. of gap junctions (per 100 SMC)	No. of axons (per 100 SMC)	No. of varicose axons† (per 100 SMC)
Mouse	10.5	42	18
Rat	8.1	59	24
Rabbit	10.1	16	7
Sheep	1.8	4	1

* Based on study of two animals except sheep (one animal).
† Distances between varicose endings and muscle vary widely (2–50 mm to <2.5 µm).
SMC, smooth muscle cells.

to the ganglia, which contain many fewer nerve cells than in the enteric nervous system (hundreds instead of millions) before passing to the airway muscle. All species studied contain cholinergic nerves as their primary excitatory motor nerve supply. All species studied except the dog and rat also have non-adrenergic inhibitory nerves supplying ASM.[62] Sensory nerves have cell bodies in the jugular and nodose ganglia and run in the vagus and recurrent laryngeal nerves to the airway; initially most are myelinated but lose their myelin before entering the airway.[1]

Trachea in most large animal species have a sparse innervation, and first-order bronchi have only slightly more dense innervation.[61,63] Nerves do not penetrate into the bundles and are rare at the adventitial surfaces. Canine trachea produce acetylcholine-mediated excitatory junction potentials (EJPs) only if stimulated with high field strength in the sucrose gap.[14,63,64] These results reflect both the low density of innervation and the good coupling between muscle cells in canine trachea, similar to bovine trachea.[58,65] The varicose nerve profiles present contain predominantly small (30–40 nm diameter) clear (agranular) vesicles.[11,63] The cholinergic EJPs or contractions induced by nerve stimulation are abolished by atropine or other muscarinic antagonists. A few varicose nerve profiles with small granular vesicles, presumably adrenergic, are also found.[11,63] Functional evidence of adrenergic nerves can be found.[66–69] Rarely, nerve varicosities containing predominantly large granular vesicles are found. These may be peptidergic. Although no non-adrenergic inhibitory motor fibres have been found in dogs, there is evidence that vasoactive intestinal polypeptide (VIP) and nitric oxide (NO) may participate in mediating inhibition in guinea-pig and human airways.[62] VIP is found in large granular vesicles. However much evidence suggests that NO is the chief non-adrenergic mediator; it is produced by a cytosolic enzyme often located at the same nerves as VIP.[62] Sensory nerves, not identifiable on the basis of their ultrastructure, are also present and would be expected to contain substance P and calcitonin gene-related peptide (CGRP) which, if released, could affect airway function by causing bronchconstriction and oedema. However, the density of substance P- and CGRP-containing nerves varies between species.

The density of the innervation of canine bronchi is much higher, up to 10-fold higher at the fourth- to fifth-order bronchi[63] (E. E. Daniel, unpublished observations). Moreover, these nerve profiles are found within, as well as at the periphery of, muscle bundles, sometimes close to profiles of muscle cells. There are not only varicose profiles with small

5 Airway Smooth Muscle Cells: Structure and Function

clear (agranular) vesicles (presumably cholinergic) but also a substantial minority (up to 60%) with small granular vesicles (presumably adrenergic). There is evidence to support the functional importance of both nerve types in affecting bronchial muscle function by postsynaptic and presynaptic actions.[70,71] The close proximity of bare putative cholinergic and putative adrenergic varicosities within small nerve bundles in bronchi provides a structural basis for prejunctional controls; in canine airways these include (a) M_1 muscarinic autoreceptors inhibiting acetylcholine output from cholinergic nerves (although M_2 receptors seem to provide this function in guinea-pig and human airways) and (b) β_1- and β_2-adrenergic heteroreceptors inhibiting acetylcholine release.[70–75] No evidence of prejunctional control of mediator release from adrenergic nerves has been obtained, but technical impediments preclude definite conclusions.

Very few nerve profiles with mostly large granular vesicles or many mitochondria have been found in tracheal smooth muscle of dog, human and cow.[11,57,63,76] Paucity of large granular vesicles is consistent with the absence of functional evidence of peptidergic nerves in this tissue; paucity of nerve profiles with many mitochondria probably reflects failure of this as a criterion to identify sensory nerves.

In guinea-pig tracheal smooth muscle, however, the innervation is much denser, located within (sometimes very close to muscle cells), as well as at the periphery of, muscle bundles and contains some varicose endings with small granular (putative adrenergic) or large granular (putative peptidergic) vesicles as well as a majority of varicose endings with small agranular (putative cholinergic) vesicles.[7,60,77] Gabella[1] has reported that trachea of mouse and rat have a high density of innervation (36–74 axons and 15–27 varicose nerve profiles per 100 smooth muscle cell profiles). This makes them comparable to (or somewhat higher than) the guinea-pig in terms of nerve density. He also reports that the rabbit trachea has an intermediate density of innervation (about 16 axons and 6 varicosities per 100 muscle cell profiles), whereas the sheep has a very low density of innervation (like human, bovine and canine trachea). These numbers suggest that there may be an inverse correlation between tracheal nerve density and animal size (i.e. low in large animals and high in small ones). The functional significance of this is obscure, since Gabella[1] also reports that all of these species except the sheep have a moderately high density of gap junctions. The notion that low nerve density must be compensated by spread of responses through a high density of gap junctions does not apply.

In human bronchi, as in canine, the density of innervation is much greater than in trachea, and more nerve profiles appear to be adrenergic.[57,78] These profiles with mainly small clear synaptic vesicles that appear to be cholinergic may sometimes have some large granular vesicles, but these are relatively rare. Structural evidence allowing identification of a significant peptidergic or sensory innervation directly to human or canine bronchial muscle is still lacking. The dense innervation of guinea-pig tracheal smooth muscle belies statements that it is an irrelevant model for studies of the neural control of airway. The ultrastructure of nerves in bronchi of guinea-pig has not been reported, but functional studies suggest that varicosities of peptidergic nerves with large granular vesicles will be present.

Tables 5.1 and 5.2 summarize the present knowledge of major features of airway muscle judged to be relevant to its myogenic and neurogenic control at a cellular level. Clearly, there needs to be more evidence relevant to (a) structures related to mechanical coupling between muscle cells, (b) electrical and metabolic coupling between muscle cells of human bronchi, (c) the precise locus of presynaptic and postsynaptic receptors to neural mediators, (d) the nature of peptidergic and sensory innervation at an ultrastructural level, (e) the occurrence of synthesis and degradation of collagen and elastin by smooth muscle cells, and (f) the structural bases for series elastic resistances in these muscles.

ION CHANNELS

The recent development of patch clamp electrophysiological techniques has allowed the direct examination of individual ion channels in single cells. In the following section, the properties of the various channels in ASM will be outlined; their roles in physiological responses is discussed on p. 101.

K⁺ channels

The most common type of K^+ channel is both voltage and Ca^{2+}-dependent and has a conductance of 100–270 pS.[79-86] These Ca^{2+}-dependent K^+ channels [K(Ca; maxi-K)] are blocked non-selectively by tetraethylammonium (TEA) or Ba^{2+},[82-85,87,88] or selectively by charybdotoxin and iberiotoxin.[81,86,89-91] There is a complex regulation of K(Ca) activity in ASM. First, the G-protein G_s can interact with the K(Ca) channels directly[92] or it can stimulate cAMP-dependent protein kinase (PKA) leading to phosphorylation of the channel,[85,86,92,93] in both cases resulting in increased channel activity. Second, many spasmogens evoke a transient activation of K(Ca) due to release of internal Ca^{2+} and consequent shift in the K(Ca) voltage-activation curve towards more negative potentials;[79,85,94] this activation is followed by a prolonged suppression, which may involve decreased sensitivity of the channels to Ca^{2+}, a shift in the kinetic scheme towards shorter duration open times and/or pH-related changes in channel activity.[81,82,95-97]

ASM cells also exhibit delayed rectifier K^+ current [K(DR)]. Activation of K(DR) is Ca^{2+} independent, much slower than that of other voltage-dependent currents such as K(Ca) or Ca^{2+} currents, and occurs only at potentials more positive than ≈ -40 mV.[84,87,88,94,98,99] Its inactivation is also voltage dependent (half maximal at resting membrane potential) and time dependent (time constant of 0.1–1 s.[87,94,98] K(DR) channels have a relatively small conductance of 13 pS,[87] are insensitive to charybdotoxin, apamin, glibenclamide or low concentrations of TEA, but can be blocked by low millimolar concentrations of 4-aminopyridine (to which the other types of K^+ channel are insensitive).[87,88,94,98]

While there is a plethora of indirect data implicating the presence of ATP-dependent K^+ channels [K(ATP)] in ASM, particularly the inhibitory effects of cromakalim,[89,90,91,100-102] there is little direct data available. In rabbit tracheal smooth muscle,[103] a K(Ca) channel with conductance of 155 pS was activated following exposure to the mitochondrial uncoupler 2,4-dinitrophenol (leading to decreased levels of ATP) and was reversibly inhibited by ATP. The authors ruled out a mechanism involving chelation of Ca^{2+} by ATP and concluded that ATP decreased the Ca^{2+} sensitivity of the K(Ca) channels, thereby shifting the Ca^{2+}–P_0 relationship to the right. However,

5 Airway Smooth Muscle Cells: Structure and Function

decreased K(Ca) activity may also be explained by ATP-induced acidification of the cytosol (leading to a shift in the Ca^{2+}–P_0 curve to the right) or stimulation of Ca^{2+}-pump activities (leading to decreased $[Ca^{2+}]_i$ and a shift in the voltage-activation curve to the right).

Ca^{2+}-dependent Cl^- channels

ASM cells also exhibit a Ca^{2+} dependent Cl^- current [Cl(Ca)].[95–97,104–110] While these channels have not yet been studied at the single channel level, their properties can be inferred from whole-cell studies. Their unitary conductance must be less than 20 pS, since unitary events greater than 2 pA cannot be resolved even with a driving force on Cl^- of 100 mV.[105] In addition, the kinetics of their activation and inactivation are slow compared to those of K(Ca) channels. For example, following a spontaneous burst of Ca^{2+} from the sarcoplasmic reticulum, K(Ca) currents activate and inactivate within 50–100 ms, while Cl(Ca) currents do so over several hundred milliseconds;[95,105] a comparable difference is noted in the rates of activation and inactivation of these currents when $[Ca^{2+}]_i$ is elevated following agonist stimulation[95,106] or voltage-dependent Ca^{2+} influx.[88,94,98,108] These observations are consistent with Cl(Ca) exhibiting considerably slower kinetics than the K(Ca) current, or that the Cl^- channels require higher levels of $[Ca^{2+}]$ to be activated than is the case for the K^+ channels. Finally, there is evidence that activation and inactivation of these channels is in part voltage dependent.[105,108]

Non-selective cation channels

When K^+ and Cl^- conductances are blocked, agonists evoke a current that is reduced in magnitude by removal of extracellular Na^+;[95,96,107,110] the reversal potential of this current is ≈ 0 mV under 'physiological' conditions, but is displaced from 0 mV when the transmembrane gradient of various cations is altered. These observations suggest the presence of non-selective cation channels. Little is known about these currents, in part due to the lack of specific antagonists. It seems that Ca^{2+} does not trigger this current directly (as is the case for the accompanying Cl^- current), since it is not activated by caffeine.[96,110] In addition, this current may be physiologically quite important, since it seems to be able to conduct Ca^{2+}.[95,96,110]

Voltage-activated Na^+ channels

In cultured human bronchial smooth muscle cells, membrane depolarization evokes a large current (100–1000 pA) with properties of voltage-dependent Na^+ channels: (a) fast kinetics (time constant of 2.5 ms); (b) peak activation at -10 to 0 mV; (c) inactivation that is voltage dependent (decreasing by $\approx 40\%$ at a holding potential of -60 mV); (d) sensitivity to tetrodotoxin but not to replacement of Ca^{2+} with Cd^{2+} or to 10 mM TEA. It should be noted, however, that these currents could not be recorded consistently in this study and that no other study of ASM (including freshly dissociated human bronchial smooth muscle) has described such currents.[97,111]

Voltage-dependent Ca²⁺ channels

Voltage-dependent Ca^{2+} currents in ASM cells of the human,[111] dog,[88,112] cow,[113] pig,[114] horse[115] and guinea-pig[98] all exhibit typical 'L-type' characteristics. These characteristics are (a) threshold and maximal activation at ≈ -40 and $\approx +20$ mV respectively, (b) an inactivation that is voltage dependent (half maximal at ≈ 20 mV) and time dependent (current half-inactivated with ≈ 50 ms), and (c) a defined sensitivity to 1,4-dihydropyridine (DHP) Ca-channel antagonists.

The biochemical structure of the L-type calcium channel has been widely studied and DHP agonists and antagonists have been used as probes to isolate the protein constituents of the channel. This calcium channel is a multimeric protein complex composed of five different polypeptide subunits each with different molecular masses (Fig. 5.3); these subunits are referred to as α_1, α_2, β, γ and δ. The α_1-subunit (175 kDa), which is encoded by three genes, apparently forms the ion-selective pore and contains essential phosphorylation sites and binding sites for some calcium antagonists, for example DHPs and phenylalkylamines (Fig. 5.3). The protein has four repeating motifs each containing six putative transmembrane-spanning regions (termed S1–S6); the S4 region constitutes the putative voltage sensor. The α_1-subunit closely resembles that found in sodium channels and is approximately 55% homologous with the sodium channel in the transmembrane-spanning domain. However, the cytosolic regions of the

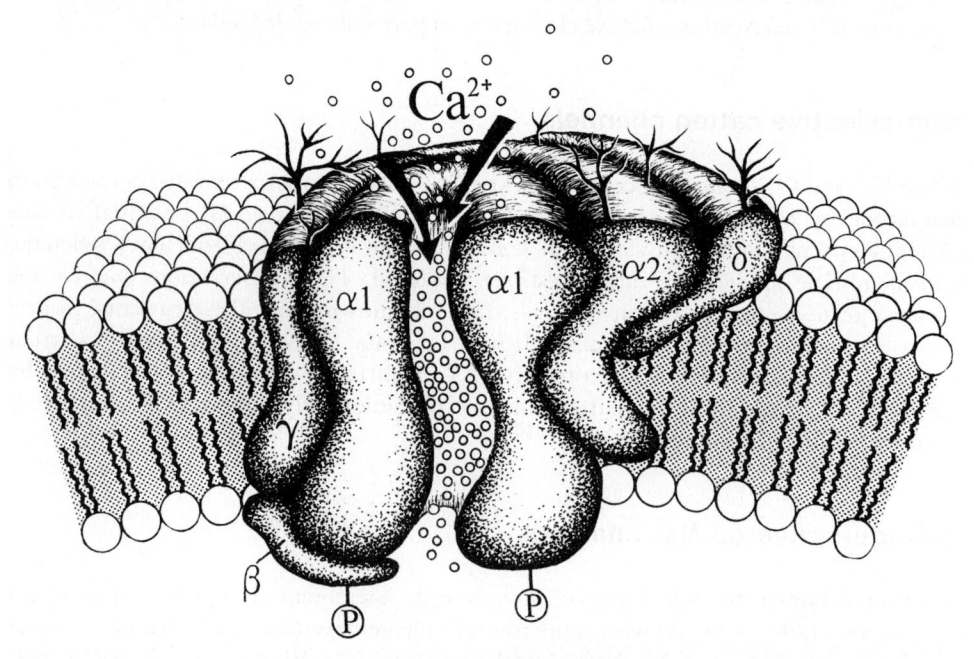

Fig. 5.3 Structural organization of the L-type calcium iron channel. The calcuim channel is a pentameric protein complex. Each of the five different polypeptide subunits (α_1, α_2, β, γ, δ) has a different molecular mass. The α_1-subunit is depicted as the ion channel or pore. It contains the dihydropyridine and phenylalkylamine calcium antagonist-binding sites, essential regulatory (cAMP-dependent) phosphorylation sites as well as the voltage sensor apparatus. Like the α_1-subunit, the β-subunit has similar phosphorylation sites for cAMP-dependent protein kinase. The α_2, γ and δ subunits are involved in modulating channel conductance. Reproduced from ref. 116 with permission.

5 Airway Smooth Muscle Cells: Structure and Function

α_1-subunit are significantly different from those in the sodium channel. The α_1-subunit has been cloned from a variety of tissue types, including lung, and northern blot analysis has revealed specific messenger RNA (mRNA) transcript sizes that are tissue specific. This indicates that, in all likelihood, distinct isoforms of the L-type calcium channel exist, most probably arising through the process of alternative splicing. A comparison of the deduced amino acids from the nucleotide sequences is also consistent with the existence of different isoforms. For more detailed information see ref 116.

Electrophysiological studies performed at the single-channel level indicate the presence of two types of Ca^{2+} channel with unitary conductances of 21–26 pS and 10 pS respectively.[117,118] Recently, a second type of whole-cell Ca^{2+} current has been identified in canine bronchial smooth muscle. This current exhibits 'T-type' properties, including threshold and maximal activation at ≈ -60 mV and ≈ -20 mV respectively, and inactivation that occurs much more rapidly (half maximal within 10 ms) and at more negative potentials ($V_{0.5} \approx -70$ mV) than is the case for L-type current[119] (Fig. 5.4).

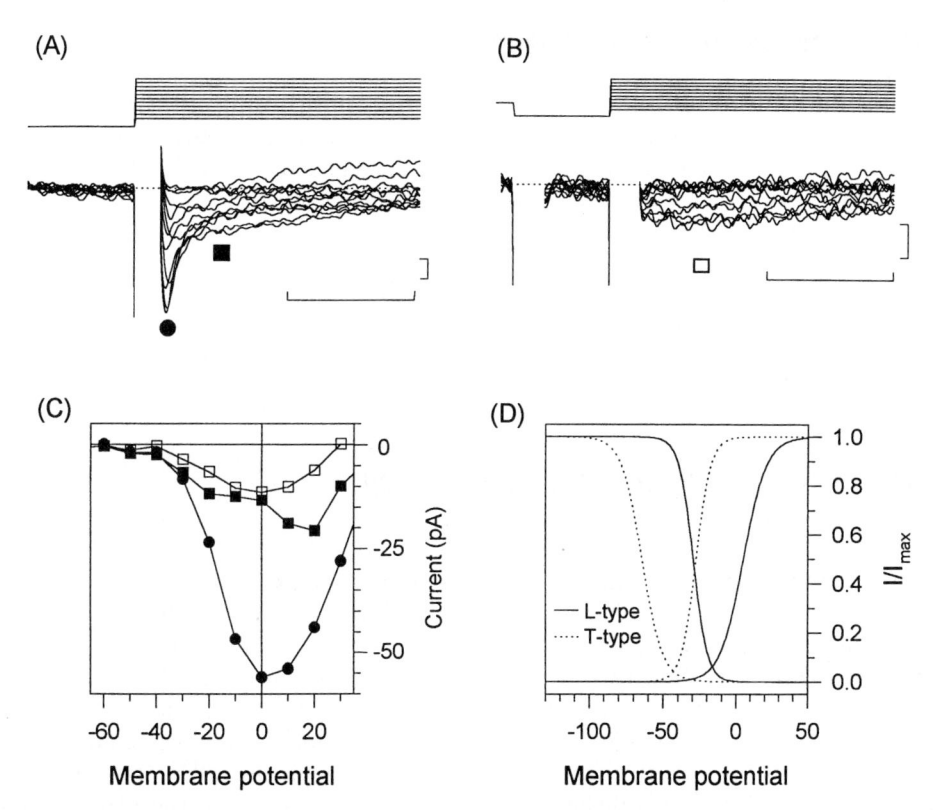

Fig. 5.4 Voltage-dependent Ca^{2+} currents in canine bronchial smooth muscle. (A) Depolarizing pulses from a holding potential of -80 mV evoke a rapidly inactivating current (●; T-type) and a second slowly inactivating current (■; L-type). (B) When this voltage protocol is repeated from a holding potential of -40 mV (with brief step to -80 mV to ensure capacitative transients were the same as in A), the T-type current is lost, leaving only L-type current (□). (C) Current–voltage relationships for currents shown in (A) and (B). (D) Boltzmann fits for activation and inactivation of T-type current (dotted line) and of L-type current (solid line); 'window current' exists at -60 to -30 mV for T-type current, and at -35 to 0 mV for L-type current. Reproduced from ref. 119, with permission.

'Window current' is predicted over the physiologically relevant range of membrane potentials (i.e. -60 to -20 mV; Fig. 5.4). Compared with L-type currents, these are somewhat less sensitive to DHPs (reduced by about 50% by 1 μM nifedipine) or replacement of Ca^{2+} with Ba^{2+}, and slightly more sensitive to 40 μM Ni^{2+}.

There is complex regulation of L-type Ca^{2+} currents by agonists. For example, some describe suppression by cholinergic agonists, due to Ca^{2+}-induced inactivation (following release of internally sequestered Ca^{2+})[120] and/or to phosphorylation by PKC,[114] while others find augmentation of these currents (i.e. channel activity at a given voltage is greater in the presence of agonist).[118,121] Similarly, β-agonists, acting through PKA, suppress Ca^{2+} currents in human and equine ASM,[122] but augment those in bovine tracheal smooth muscle by a mechanism that does not involve cAMP or PKA.[113] T-type currents, on the other hand, do not exhibit Ca^{2+}-induced inactivation and are not suppressed during cholinergic stimulation.[119]

Voltage-independent Ca^{2+} channels

Recently, Ca^{2+}-fluorescence studies using cultured human ASM cells have provided evidence for a Ca^{2+}-permeable pathway that is activated by bradykinin and histamine.[123,124] A wide range of polyvalent cations, including Ba^{2+}, Mg^{2+}, Mn^{2+}, Ni^{2+}, Co^{2+}, Cd^{2+} and La^{3+}, seem to be impermeable, suggesting that the pathway is not a non-selective cation channel. Furthermore, the Ca^{2+} entry is distinct from voltage-dependent Ca^{2+} channels, since it is DHP insensitive and enhanced by hyperpolarization (apparently via increased inward driving force on Ca^{2+}). Finally, ouabain does not influence Ca^{2+} entry, suggesting that the Na^+/K^+ pump (possibly coupled to the Na^+/Ca^{2+} exchanger) does not contribute.

PHOSPHATIDYLINOSITIDES AND CALCIUM

Many agonists stimulate the activity of a Ca^{2+}-dependent inositol-specific phospholipase C (PLC), leading to inositol phosphate metabolism; inositol 1,4,5-trisphosphate (IP$_3$) in turn opens Ca^{2+}-permeable channels on the sarcoplasmic reticulum, leading to a rapid elevation of $[Ca^{2+}]_i$ to a level 200–1000 nM above baseline, after which $[Ca^{2+}]_i$ falls to a plateau level 50–400 nM above baseline as long as the agonist is applied. Responses of this type are evoked by cholinergic agonists,[110,125–131] substance P or neurokinin A,[109,132] histamine,[123–125,129–131] leukotrienes C$_4$ and D$_4$,[132,133] bradykinin,[124,132,134,135] PGF$_{2\alpha}$,[129] serotonin[125,136–138] and endothelin.[139] In most ASM preparations, the plateau phase involves influx of Ca^{2+} through voltage-dependent Ca^{2+} channels.[126,131,134–136] In cultured human and rat ASM, however, the plateau phase is insensitive to blockade of voltage-dependent Ca^{2+} channels;[123,124,132,138,139] Ca^{2+} influx is nonetheless important, since the plateau phase is reduced by removal of external Ca^{2+} and augmented by hyperpolarization (which increases the driving force on Ca^{2+}). This voltage-independent influx may involve non-selective cation channels, as is the case in equine ASM.[110]

5 Airway Smooth Muscle Cells: Structure and Function

PHYSIOLOGICAL MECHANISMS

In this section we have attempted to integrate the information on how the morphological structures, ion channels and signalling pathways described above are orchestrated to provide the physiological functioning of ASM. While reference is made to data obtained from studies using intact tissues, emphasis is placed on studies using single cells.

Resting membrane potential

Resting membrane potential (V_R) in ASM ranges from -70 to -30 mV, which does not correspond to the equilibrium potential (E_{ion}) of any single ion species: E_K is ≈ -80 mV, E_{Cl} is ≈ -30 to 0 mV,[140] and those for Na^+ and Ca^{2+} are in the very positive range. This would suggest that V_R is determined by K^+ and at least one other ion. Consistent with this, blockade of K^+ channels leads to membrane depolarization and contraction.[65] The specific type of K^+ channel(s) involved is not entirely clear. Maxi-K channels are unlikely to be active at V_R and resting levels of $[Ca^{2+}]_i$, and their blockade generally has no effect on V_R or resting tone,[86,92,102] though there are exceptions;[89] in addition, small transunit outward currents (STOCs) are not recorded until V_m exceeds -30 to -40 mV in smooth muscle cells of human,[97] canine,[95] guinea-pig[95,96,105] and swine[80] airways. While small conductance K(Ca) channels are active at V_R[111] and may account for TEA-evoked depolarization, apamin has no effect on V_R.[88] K(ATP) is also not involved in setting V_R, since V_R and tone are unaffected by glibenclamide.[100,101]

Replacement of external Cl^- with the impermeant anion isethionate (which shifts E_{Cl} in the positive direction) elicits depolarization and increases electrotonic potentials in canine tracheal smooth muscle;[141,142] in addition, spontaneous Cl^- currents are recorded at V_R.[95,80] On the other hand, a persistent Ca^{2+} influx ('window current') is predicted at potentials of -60 to -20 mV for T-type Ca^{2+} currents and at -30 to 0 mV for L-type Ca^{2+} currents (see p. 101); consistent with this prediction, removing external Ca^{2+} causes $[Ca^{2+}]_i$ to drop.[128,139] In contrast, removal of external Na^+ has little or no effect on tone or membrane currents at rest.[143] Thus, V_R in ASM is likely set primarily by a mixture of K^+, Cl^- and Ca^{2+} currents.

$[Ca^{2+}]_i$ homeostasis and refilling of the sarcoplasmic reticulum

Resting $[Ca^{2+}]_i$ ranges from 50 to 250 nM in ASM of the cow,[125,126,135] sheep,[133] human,[123,124,129,132,139] dog,[128,134,136,143,144] guinea-pig[131] and rat.[138] In general, $[Ca^{2+}]_i$ is maintained at low levels by: (a) extrusion of Ca^{2+} from the cytoplasm via a Ca^{2+}-ATPase and Na^+/Ca^{2+} exchanger on the plasmalemma; (b) sequestration of Ca^{2+} into intracellular organelles such as the endoplasmic or sarcoplasmic reticulum and mitochondria; and (c) buffering of $[Ca^{2+}]_i$ by various cytosolic Ca^{2+}-binding proteins. The relative contribution of these Ca^{2+} homeostatic pathways varies from tissue to tissue. The internal Ca^{2+}-ATPase maintains the filling state of the sarcoplasmic reticulum and compensates for a constant 'leak' of Ca^{2+} from the sarcoplasmic reticulum; the latter may account for spontaneous Cl^- and K^+ currents (STICs and STOCs, respectively).[80,95,98,105] As a result, inhibition of the sarcoplasmic reticulum Ca^{2+}-ATPase (e.g. by cyclopiazonic acid or thapsigargin) leads to an elevation of $[Ca^{2+}]_i$ and to functional depletion of the sarcoplasmic reticulum within 15 min.[104,131,143] Ca^{2+} homeostasis must also include extrusion of Ca^{2+} from the cell in order to compensate for the constant influx of the Ca^{2+} into the cell (i.e. 'window' current; see p. 99). Na^+/Ca^{2+} exchange makes little or no contribution to Ca^{2+} extrusion, since Na^+ removal has no effect on $[Ca^{2+}]$ nor on recovery from cholinergic stimulation.[143] The plasmalemmal Ca^{2+}-ATPase, on the other hand, has been proposed to play a key role in this respect,[143] but this has not been tested directly due to the lack of specific inhibitors.

Refilling of the sarcoplasmic reticulum following agonist stimulation can involve voltage-dependent Ca^{2+} channels, since depolarizing pulses in cells depleted of internal Ca^{2+} allow partial recovery from depletion of sarcoplasmic reticulum, and this recovery is DHP sensitive.[104] The DHP sensitivity of refilling led to the suggestion that L-type Ca^{2+} channels were involved; however, L-type currents are suppressed during cholinergic stimulation and require potentials well beyond the physiologically relevant range (≈ -70 to -30 mV) for activation. T-type currents, on the other hand, are also DHP sensitive but unaffected during cholinergic stimulation, and exhibit 'window' current at ≈ -60 to -20 mV (see p. 99). Uptake of Ca^{2+} into the sarcoplasmic reticulum of ASM is also stimulated by cAMP and β-adrenergic stimulation.[144,145]

Spontaneous mechanical and electrical activity

In the absence of stimulation by neurotransmitters or other pharmacological agents, ASM of the dog,[87] cat[146] and horse[147,148] is mechanically and electrically quiescent. Bovine tracheal smooth muscle is also generally quiescent, though one study reported occasional spontaneous mechanical and electrical activity.[149] Human[150–152] and guinea-pig[149,153–156] ASM, on the other hand, exhibit considerable spontaneous mechanical and electrical activities, the latter of which are referred to as 'slow waves' (frequency ≈ 1 Hz; amplitude 10–25 mV). These activities are unaffected by suppression of neurotransmission[146,151,155,157,158] or by antagonists of cholinergic,[151,155,158] α- or β-adrenergic[155,156,159] or histaminergic receptors,[151] suggesting that this activity is not due to spontaneous release of mediators from nerves. Spontaneous myogenic activity seems to be dependent on constitutive metabolism of arachidonic acid (AA), although the enzymatic pathway involved is apparently species dependent. For example, in guinea-pig ASM, it is greatly reduced by cyclooxygenase inhibitors but is relatively unaffected by inhibitors of lipoxygenase[150,152] or thromboxane synthase.[155] Spontaneous activity in human ASM, on the other hand, is not reduced by inhibition of cyclooxygenase but is abolished by inhibition of lipoxygenase.[150–152] In equine ASM, which is normally devoid of mechanical activity, indomethacin evokes contractions that are antagonized by FPL 55712,[147,148] suggesting that basal production of excitatory AA metabolites balances that of inhibitory AA metabolites and that inhibition of cyclooxygenase leads to an excess production of the former, resulting in contraction.

In quiescent tissues, spasmogens can evoke changes in mechanical and electrical activity that resemble the spontaneous activities described above. For example, slow waves are evoked in canine ASM by cholinergic agonists,[160] thromboxanes,[87] leukotrienes[68] or K^+-channel blockers.[160] Similarly, bovine tracheal smooth muscle displays phasic contractions and slow waves in response to histamine.[149] Thus, even tissues normally quiescent exhibit slow wave activity identical to that recorded from spontaneously active tissues if appropriately provoked. This suggests that a myogenic oscillatory mechanism is resident in all ASM tissues and is invoked by excitatory stimulation by acetylcholine and histamine (canine and bovine ASM), leukotrienes (human ASM) or prostanoids (guinea-pig and equine ASM), or by blockade of K^+ channels (see below).

The depolarizing phase of slow waves involves opening of voltage-dependent Ca^{2+} channels, since blockers of these channels eliminate slow wave activity and leave the cell in a relatively hyperpolarized state (i.e. ≈ -45 mV).[151,153,155,157,160] L-type Ca^{2+} channels have been proposed to mediate slow waves since both are DHP sensitive, even though the threshold potentials for these channels and for slow waves are substantially different (≈ -30 and ≈ -45 mV, respectively); in addition, L-type currents are suppressed during agonist stimulation while slow waves are not. T-type Ca^{2+} currents, however, may contribute to slow wave activity, since (a) they are also DHP sensitive but unaffected by cholinergic agonists; (b) slow waves sweep the membrane continuously between -45 and -30 mV, which overlap the threshold and peak potentials for T-type 'window current'; and (c) recovery from inactivation of T-type currents occurs within 1 s, consistent with a role in oscillations that have a frequency of ≈ 1 Hz (recovery of L-type currents can take up to 30 s).[119]

However, the ionic conductance changes underlying the repolarizing phase of the slow waves are controversial. In gastrointestinal smooth muscle, this is attributed to opening of K(Ca) channels subsequent to influx of Ca^{2+} through voltage-dependent channels.[161] In ASM, however, slow waves persist in the presence of K^+-channel blockers such as TEA.[160] The repolarizing phase may instead be mediated by Ca^{2+}-dependent Cl^- currents, which are also triggered by voltage-dependent Ca^{2+} influx.[108] Alternatively, Na^+/K^+ ATPase activity may play a role, since slow waves are reduced upon cooling or exposure to ouabain.[149,153,156,158]

Agonist-mediated excitation

In general, spasmogens activate PLC leading to the generation of IP_3 that triggers release of internal Ca^{2+} (see p. 100; Fig. 5.5); the latter, in turn, leads to activation of the $Ca^{2+}/$calmodulin-dependent myosin light chain kinase and contraction.[162,163] Coincident generation of diacyglycerol by PLC[139] enhances PKC activity, leading to a plethora of effects including contraction.[162,163] In addition, certain spasmogens inhibit adenylate cyclase activity, thereby removing an inhibitory influence on the contractile apparatus (i.e. 'disinhibition'); it is of interest that the cholinergic EC_{50} for suppression of adenylate cyclase is 10- to 100-fold lower than that for stimulation of inositol phosphate metabolism and contraction;[125,127,128,130] thus, acetylcholine acts by first suppressing relaxations if they are present before stimulating contraction.

The IP_3-mediated elevation of $[Ca^{2+}]$ also leads to transient activation of Ca^{2+}-dependent Cl^- channels (see p. 97) accompanied by non-selective cation currents (see p. 97), with subsequent depolarization (Fig. 5.5). This is followed by a prolonged suppression of K^+ conductances, apparently including K(Ca) and K(DR):[95–97,106,109]

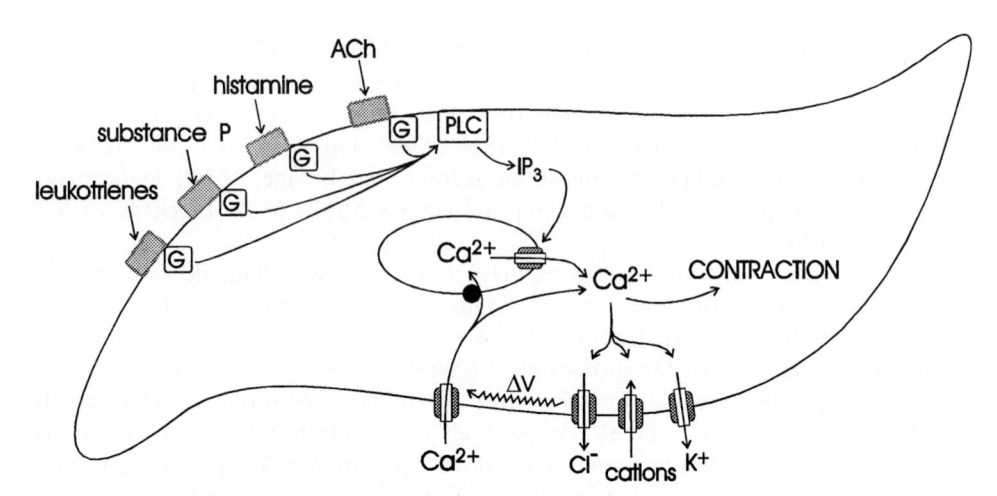

Fig. 5.5 Mechanisms underlying agonist-induced excitation. Excitatory agonists stimulate G-protein and phospholipase C (PLC) activities, resulting in generation of inositol 1,4,5-trisphosphate (IP$_3$) and release of internally sequestered Ca^{2+}. Elevation of [Ca^{2+}]$_i$ in turn leads to changes in activity of Cl$^-$, non-selective cation and K$^+$ channels (see p. 103 for details) resulting in membrane depolarization (ΔV) and influx of Ca^{2+}, thereby increasing/prolonging the elevation of [Ca^{2+}]$_i$ and contributing to refilling of the internal Ca^{2+} pool. The net result is contraction via activation of myosin light chain kinase. ACh, acetylcholine.

since V_R is determined primarily by K$^+$,[87,94,99] this suppression will result in prolonged depolarization. Depolarization in turn results in opening of voltage-dependent Ca^{2+} channels, which contribute to the plateau phase of agonist-evoked Ca^{2+} transients (see p. 100), refilling of the agonist-depleted Ca^{2+} pool (see p. 101) and to contraction. It is important to note, however, that voltage-dependent Ca^{2+} influx is not strictly necessary for contraction: release of internal Ca^{2+} alone is sufficient to mediate contractions, as attested to by their persistence during exposure to Ca^{2+}-channel blockers, removal of external Ca^{2+} or voltage clamp at negative membrane potentials.[95,104]

Agonist-mediated inhibition

Many relaxant agonists, acting through adenylate cyclase activation, can cause down-regulation of myosin light chain kinase activity and relaxation or inhibition of contraction.[162,163] In addition, they can act through voltage-dependent mechanisms. For example, relaxations evoked by β-agonists are antagonized by charybdotoxin or iberiotoxin or by elevation of E_K using high KCl;[86,89–92,102] activation of PKA by β-agonists leads to phosphorylation of K(Ca) channels, resulting in increased opening and hyperpolarization[93] (Fig. 5.6). Cyclic GMP may also activate the channels, since NO and related NO-donating agents cause bronchodilation, hyperpolarization, abolition of slow waves, and increased ^{86}Rb$^+$ efflux by a mechanism sensitive to TEA, procaine, charybdotoxin and iberiotoxin (Fig. 5.6).

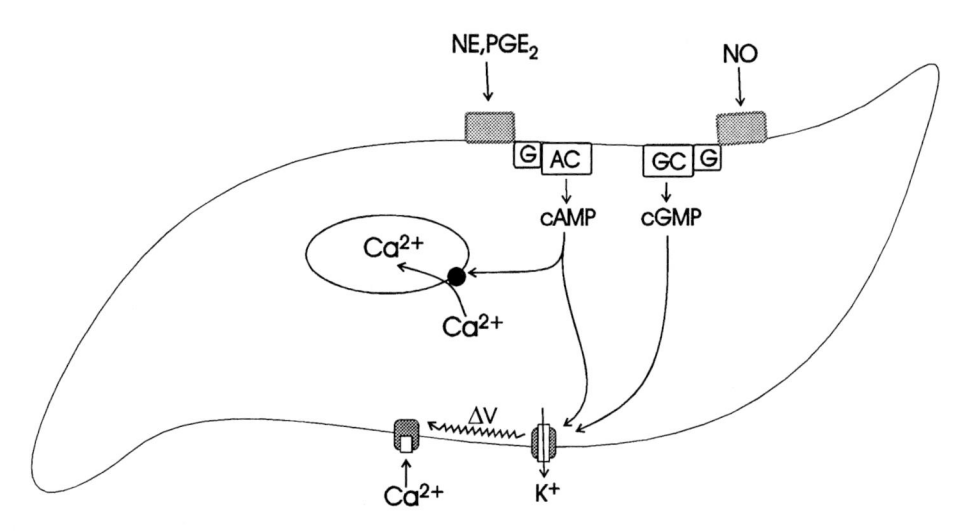

5 Airway Smooth Muscle Cells: Structure and Function

Fig. 5.6 Mechanisms underlying agonist-induced relaxation. Norepinephrine (NE) acting through adenylate cyclase (AC), stimulates uptake of cytosolic Ca^{2+} into the sarcoplasmic reticulum and activates Ca^{2+}-dependent K^+ channels; the latter results in membrane hyperpolarization (ΔV) and closure of voltage-dependent Ca^{2+} channels. The net result is reduced activity of myosin light chain kinase and relaxation. Nitric oxide (NO), acting through guanylate cyclase (GC), also relaxes airway smooth muscle, in part via activation of K^+ channels. PGE_2, prostaglandin E_2.

CONCLUSION

In ASM, responsiveness is likely to be determined by the interplay of myogenic and neural controls, modulated by inflammatory and other stimuli. To date, however, the functional range and the extent of modulation of myogenic and neural control systems has not been clarified. The structures likely to control airway muscle at a cellular level are gap junctions, other cell–cell junctions, nerve endings in muscle and sometimes nearby immune cells. There is wide variation between species in nerve density in trachea, with small animals having a notably higher density than humans. There is also considerable variation in nerve density between different components of the airway musculature; in humans and dogs, the bronchi are much more densely innervated and the bronchi contain a significant component of their innervation from nerves that appear to be adrenergic. The canine ASM also exhibits some structural features that suggest the importance of internal Ca^{2+} stores (much ER both near to and distant from the cell membrane). There is a suggestion that mechanical stress is not transmitted between bronchial cells by way of intermediate contacts, which are sparse, especially in bronchi. This may allow more movement of cells relative to one another and also may account for the highly irregular muscle cell profiles. As a result, mechanical stress during breathing and with bronchoconstriction/bronchodilation may be transmitted mainly by gap junctions, explaining their rapid turnover. Much has been learned regarding the wide variety of ion channels and signalling pathways present in ASM, as well as how these are orchestrated to produce physiological (and in certain cases pathological) functioning. Notwithstanding, further research is necessary to elucidate precisely how these events might underpin airway hyperresponsiveness and asthma.

REFERENCES

1. Gabella G: Innervation of airway smooth muscle: fine structure. *Annu Rev Physiol* (1987) **49**: 583–594.
2. Gabella G: Structure of smooth muscles. In Bülbring E, Brading AN, Jones AW, Tomita T (eds) *Smooth Muscle: an Assessment of Current Knowledge*. Austin, University of Texas Press, 1981, pp 1–52.
3. Somlyo AP, Somlyo AV: Ultrastructure of smooth muscle. In Danial EE, Paton DM (eds) *Methods in Pharmacology*, vol 3. New York, Plenum Press, 1975; pp 3–45.
4. Henderson RM: Cell-to-cell contacts. In Daniel EE, Paton DW (eds) *Methods in Pharmacology*, vol 3. New York, Plenum Press, 1975; pp 47–77.
5. Gabella G, Blindell D: Gap junction of the muscles of the small and large intestine. *Cell Tissue Res* (1981) **219**: 469–488.
6. Henderson RM, Duchon G, Daniel EE: Cell contacts in duodenal smooth muscle layers. *Am J Physiol* (1971) **221**: 564–574.
7. Jones TR, Kannan MS, Daniel EE: Ultrastructural study of guinea pig trachea smooth muscle and its innervation. *Can J Physiol* (1980) **58**: 974–983.
8. Daniel EE, Duchon G, Henderson RM: The ultrastructural bases for co-ordination of intestinal motility. *Dig Dis* (1972) **17**: 289–298.
9. Daniel EE, Robinson K, Duchon G, Henderson RM: The possible role of close contacts (nexuses) in the propagation of control electrical activity in the stomach and small intestine. *Dig Dis* (1971) **16**: 611–622.
10. Daniel EE, Taylor GS, Daniel VP, Holman ME: Can non-adrenergic inhibitory varicosities be identified structurally? *Can J Physiol Pharmacol* (1977) **55**: 243–250.
11. Kannan MS, Daniel EE: Formation of gap junctions by treatment *in vitro* with potassium conductance blockers. *J Cell Biol* (1978) **78**: 338–348.
12. Agrawal R, Daniel EE: Control of gap junction formation in canine trachea by arachidonic acid metabolites. *Am J Physiol* (1986) **250**: C495–C505.
13. Daniel EE, Daniel VP, Duchon G, *et al.*: Is the nexus necessary for cell-to-cell coupling in smooth muscle? *J Membr Biol* (1976) **28**: 207–239.
14. Daniel EE, Davis C, Sharma V: Effects of endogenous and exogenous prostaglandin in neurotransmission in canine trachea. *Can J Physiol Pharmacol* (1987) **65**: 1433–1441.
15. Garfield RE, Kanan MS, Daniel EE: Gap junction formation in myometrium: control by estrogens, progesterone and prostaglandins. *Am J Physiol* (1980) **238**: C81–C89.
16. Sims SM, Daniel EE, Garfield RE: Improved electrical coupling in uterine smooth muscle is associated with increased numbers of gap junctions at parturition. *J Gen Physiol* (1982) **80**: 353–375.
17. Garfield RE: Cell-to-cell communication in smooth muscle. In Grover AK, Daniel EE (eds) *Calcium and Contractility Smooth Muscle*. Clifton, NJ, Humana Press, 1985, pp 143–173.
18. Daniel EE: Ultrastructure of airway smooth muscle. In Armour CL, Black JL (eds) *Mechanisms in Asthma: Pharmacology, Physiology and Management*. New York, Alan R Liss, 1988, pp 167–176.
19. Sperelakis N, Mann JE: Evaluation of electric field changes in the cleft between excitable cells. *J Theor Biol* (1977) **64**: 71–96.
20. Sperelakis N, Marshall R, Mann JE: Propagation down a chain of excitable cells by electric field interactions in the junctional clefts: effects of variation in extracellular resistances including a 'sucrose-gap' simulation. *IEEE Trans Biomed Eng* (1983) **30**: 658–664.
21. Sperelakis N, Tarr M: Weak electrotonic interaction between neighboring contiguous visceral smooth muscle cells. *Am J Physiol* (1965) **66**: 119–134.

22. Parez-Armendariz ENM, Spray PC, Bennett MVL: δ-Octanol reduces calcium, potassium and gap junctional currents in mouse pancreatic B-cells. *Biophys J* (1989) **55**: 218a.
23. Serio R, Barajas-Lopez C, Daniel EE, Berezin I, Huizinga JD: Pacemaker activity in the colon: role of interstitial cells of Cajal and smooth muscle cells. *Am J Physiol* (1991) **260**: G636–G645.
24. Fishman GI, Eddy RL, Shows TB, Rosenthal L, Leinwand LA: The human connexin gene family of gap junction proteins: distinct chromosomal locations but similar structures. *Genomics* (1991) **10**: 250–256.
25. Sullivan R, Ruangvoravat C, Joo D, et al.: Structure, sequence and expression of the mouse Cx43 gene encoding connexin 43. *Gene* (1993) **130**: 191–199.
26. Willecke K, Jungbluth S, Dahl E, Hennemann H, Heynkes R, Grzeschik K-H: Six genes of the human connexin gene family coding for gap junctional proteins are assigned to four different human chromosomes. *Eur J Cell Biol* (1990) **53**: 275–280.
27. Winterhager E, Stutenkemper R, Traub O, Beyer E, Willecke K: Expression of different connexin genes in rat uterus during decidualization and at term. *Eur J Cell Biol* (1991) **55**: 133–142.
28. Kanter HL, Saffitz JE, Beyer EC: Molecular cloning of two human cardiac gap junction proteins, connexin40 and connexin45. *J Mol Cell Cardiol* (1994) **26**: 861–868.
29. Beyer EC, Paul DL, Goodenough DA: Connexin family of gap junction proteins. *J Membr Biol* (1990) **116**: 187–194.
30. Campos de Carvalho AC, Roy C, Moreno AP, et al.: Gap junctions formed of connexin43 are found between smooth muscle cells of human corpus cavernosum. *J Urol* (1993) **149**: 1568–1575.
31. Chow I, Lye SJ: Expression of the gap junction protein connexin-43 is increased in the human myometrium toward term and with the onset of labor. *Am J Obstet Gynecol* (1994) **170**: 788–795.
32. Garfield RE, Thilander G, Blennerhassett MG, Sakai N: Are gap junctions necessary for cell-to-cell coupling of smooth muscle: An update. *Can J Physiol Pharmacol* (1992) **70**: 481–490.
33. Hendrix EM, Mao SJT, Everson W, Larsen WJ: Myometrial connexin 43 trafficking and gap junction assembly at term and in preterm labor. *Mol Reprod Dev* (1992) **33**: 27–38.
34. Li Z, Zhou Z, Daniel EE: Expression of gap junction connexin 43 and connexin 43 mRNA in different regional tissues of intestine in dog. *Am J Physiol* (1993) **265**: G911–G916.
35. McNutt CM, Nicholson BJ, Lye SJ: ACTH-induced preterm labour in the ewe is associated with increased mRNA and protein levels of myometrial gap junction protein, connexin-43. *J Endocrinol* (1994) **141**: 195–202.
36. Mikkelsen HB, Huizinga JD, Thuneberg L, Rumessen JJ: Immunohistochemical localization of gap junction protein (connexin43)in the muscularis externa of murine, canine, and human intestine. *Cell Tissue Res* (1993) **274**: 249–256.
37. Nnamani C, Godwin A, Ducsay CA, Longo LD, Fletcher WH: Regulation of cell–cell communication mediated by connexin 43 in rabbit myometrial cells. *Biol Reprod* (1994) **50**: 377–389.
38. Petrocelli T, Lye SJ: Regulation of transcripts encoding the myometrial gap junction protein, connexin-43, by estrogen and progesterone. *Endocrinology* (1993) **133**: 284–290.
39. Rennick RE, Connat J-L, Burnstock G, Rothery S, Severs NJ, Green CR: Expression of connexin43 gap junctions between cultured vascular smooth muscle cells is dependent upon phenotype. *Cell/Tissue Res* (1993) **271**: 323–332.
40. Sakai N, Tabb T, Garfield RE: Studies of connexin 43 and cell-to-cell coupling in cultured human uterine smooth muscle. *Am J Obstet Gynecol* (1992) **167**: 1267–1277.
41. Thilander G, King GJ, Garfield RE: Connexin43 and gap junction content in the porcine myometrium during the estrous cycle. *Theiogenology* (1993) **40**: 323–332.
42. Li Z, Zhou Z, Daniel EE, O'Byrne PM: Expression of gap junction CX43 and mRNA in canine trachea smooth muscle, bronchus and lungs. *Am Rev Respir Dis* (1993) **147**: A53.
43. Jongsma HJ, Gross D: The cardiac connection. *News Physiol Sci* (1991) **6**: 34–40.
44. Cole WC, Garfield RE: Evidence for physiological regulation of myometrial gap junction permeability. *Am J Physiol* (1986) **251**: C207–C239.
45. Dookwah HD, Barhoumi R, Narasimhan TR, Burghardt RC: Gap junctions in myometrial

cell cultures: evidence for modulation by cyclic adenosine 3':5'-monophosphate. *Biol Reprod* (1992) **47**: 397–407.
46. Moreno AP, Campos de Carvalho AC, Christ G, Melman A, Spray DC: Gap junctions between human corpus cavernosum smooth muscle cells: gating properties and unitary conductance. *Am J Physiol* (1993) **264**: C80–C92.
47. Kenne K, Fransson-Steen R, Honkasalo S, Wärngård L: Two inhibitors of gap junctional intercellular communication, TPA and endosulfan: different effects on phosphorylation of connexin 43 in the rat liver epithelial cell line, IAR 20. *Carcinogenesis* (1994) **15**: 1161–1166.
48. Oh SY, Grupen CG, Murray AW: Phorbol ester induces phosphorylation and down-regulation of connexin 43 in WB cells. *Biochim Biophys Acta Mol Cell Res* (1991) **1094**: 243–245.
49. Asamoto M, Oyamada M, El Aoumari A, Gros D, Yamasaki H: Molecular mechanisms of TPA-mediated inhibition of gap-junctional intercellular communication: evidence for action on the assembly or function but not the expression of connexin 43 in rat liver epithelial cells. *Mol Carcinog* (1991) **4**: 322–327.
50. Fitzgerald DJ, Murray AW: Inhibition of intercellular communication by tumour-promoting phorbol esters. *Cancer Res* (1980) **40**: 2925–2937.
51. Takeda A, Hashimoto E, Yamamura H, Shimazo T: Phosphorylation of liver gap junction protein by protein kinase C. *FEBS Lett* (1987) **210**: 169–172.
52. Gainer HC, Murray AM: Diacylglycerol inhibits gap junction communication in cultured epidermal cells: evidence for a role of protein kinase C. *Biochem Biophys Res Commun* (1985) **126**: 1109–1113.
53. Kanemitsu MY, Lau AF: Epidermal growth factor stimulates the disruption of gap junctional communication and connexin43 phosphorylation independent of 12-O-tetradecanoylphorbol 13-acetate-sensitive protein kinase C: the possible involvement of mitogen-activated protein kinase. *Mol Biol Cell* (1993) **4**: 837–848.
54. Pelletier DB, Boynton AL: Dissociation of PDGF receptor tyrosine kinase activity from PDGF-mediated inhibition of gap junctional communication. *J Cell Physiol* (1994) **158**: 427–434.
55. Kurata WE, Lau AF: p130$^{gag\text{-}fps}$ disrupts gap junctional communication and induces phosphorylation of connexin43 in a manner similar to that of pp60$^{v\text{-}src}$. *Oncogene* (1994) **9**: 329–335.
56. Agrawal R, Daniel EE: Control of gap junction formation in canine trachea by archidonic acid metabolites. *Am J Physiol* (1986) **250**: C495–C505.
57. Daniel EE, Kannan M, Davis C, Posey-Daniel V: Ultrastructural studies on the neuromuscular control of human tracheal and bronchial muscle. *Respir Physiol* (1986) **63**: 109–128.
58. Cameron AR, Bullock CG, Kirkpatrick CT: The ultrastructure of bovine tracheal muscle. *J Ultrastruct Res* (1982) **81**: 290–305.
59. Janssen L, Daniel EE: Myogenic control of airway smooth muscle and cell-to-cell coupling. In Raeburn D, Giembycz MA (eds) *Airway Smooth Muscle: Development and Regulation of Contractility*. Basel, Birkhauser Verlag, 1994, pp 101–135.
60. Hoyes AD, Barber P: Innervation of the trachealis muscle in the guinea pig: a quantitative ultrastructural study. *J Anat* (1980) **130**: 789–900.
61. Kannan MS, Daniel EE: Structural and functional study of control of canine tracheal muscle. *Am J Physiol* (1980) **238**: C27–C33.
62. Ellis JL, Undem BJ: Pharmacology of non-adrenergic, non-cholinergic nerves in airway smooth muscle. *Pulmon Pharmacol* (1995) **7**: 205–223.
63. Daniel EE, Serio R, Jury J, Pashley M, O'Byrne P: Effects of inflammatory mediators on neuromuscular transmission in canine trachea *in vitro*. In Armour CL, Black JL (eds) *Mechanisms in Asthma: Pharmacology, Physiology and Management*. New York, Alan R Liss, 1988, pp 167–176.
64. Serio R, Daniel EE: Thromboxane effects on canine trachealis neuromuscular function. *J Appl Physiol* (1988) **64**: 1979–1988.
65. Kirkpatrick CT: Excitation and contraction in bovine tracheal smooth muscle. *J Physiol* (1975) **244**: 263–281.
66. Ito Y, Tajima K: Actions of indomethacin and prostaglandins on neuroeffector transmission in the dog trachea. *J Physiol* (1981) **319**: 379–392.

5 Airway Smooth Muscle Cells: Structure and Function

67. Janssen LJ, Daniel EE: Pre and postjunctional effects of a thromboxane-mimetic in canine bronchi. *Am J Physiol* (1991) **261**: L271–L282.
68. Abela A, Daniel EE: Neural and myogenic effects of leukotrienes C4, D4 and E4 on canine bronchial smooth muscle. *Am J Physiol* (1994) **266**: L414–L425.
69. Abela A, Daniel EE: The neural and myogenic effects of cyclooxygenase products on canine bronchial smooth muscle. *Am J Physiol* (1995) **268**: L47–L55.
70. Janssen LJ, Daniel EE: Characterization of the prejunctional β-adrenoceptors in canine bronchial smooth muscle. *J Pharmacol Exp Ther* (1990) **254**: 741–749.
71. Janssen LJ, Daniel EE: Pre- and post-junctional muscarinic receptors in canine bronchi. *Am J Physiol* (1990) **259**: L304–L314.
72. Watson N, Barnes PJ, Maclagan J: Actions of methoctramine, a muscarinic M_2 receptor antagonist, on muscarinic and nicotinic cholinoceptors in guinea-pig airways *in vivo* and *in vitro*, *Br J Pharmacol* (1992) **105**: 107–112.
73. Killingsworth CR, Robinson NE: The role of muscarinic M_1 and M_2 receptors in airway constriction in the cat. *Eur J Pharmacol* (1992) **210**: 231–238.
74. Minette P, Barnes PJ: Prejunctional muscarinic receptors on cholinergic nerves in human and guinea pig airways. *J Appl Physiol* (1988) **64**: 2532–2537.
75. Blaber LC, Fryer AD: Neuronal muscarinic receptors attenuate vagally-induced contraction of feline bronchial smooth muscle. *Br J Pharmacol* (1985) **86**: 723–728.
76. Cameron AR, Kirkpatrick CT: A study of excitatory neuromuscular transmission in the bovine trachea. *J Physiol* (1977) **270**: 733–745.
77. Mansour S, Danniel EE: Structural changes in tracheal nerves and muscles associated with *in vivo* sensitization of guinea pigs. *Respir Physiol* (1988) **72**: 282–294.
78. Pack RJ, Richardson PS: The adrenergic innervation of the human bronchus: a light and electron microscopic study. *J Anat* (1984) **138**: 493–502.
79. Wade GR, Sims SM: Muscarinic stimulation of tracheal smooth muscle cells activates large-conductance Ca^{2+}-dependent K^+ channel. *Am J Physiol* (1993) **265**: C658–C665.
80. Saunders H-MH, Farley JM: Spontaneous transient outward currents and Ca^{++} activated K^+ channels in swine tracheal smooth muscle cells. *J Pharmacol Exp Ther* (1991) **257**: 1114–1120.
81. Kume H, Kotlikoff MI: Muscarinic inhibition of single K_{Ca} channels in smooth muscle cells by a pertussis-sensitive G protein. *Am J Physiol* (1991) **261**: C1204–C1209.
82. Kume H, Takagi K, Satake T, *et al.*: Effects of intracellular pH on calcium-activated potassium channels in rabbit tracheal smooth muscle. *J Physiol* (1990) **424**: 445–457.
83. McCann JD, Welsh MJ: Calcium-activated potassium channels in canine airway smooth muscle. *J Physiol* (1986) **372**: 113–127.
84. Green KA, Foster RW, Small RC: A patch-clamp study of K^+-channel activity in bovine isolated tracheal smooth muscle cells. *Br J Pharmacol* (1991) **102**: 871–878.
85. Muraki K, Imaizumi Y, Watanabe M: Ca-dependent K channels in smooth muscle cells permeabilized by β-escin recorded using the cell-attached patch-clamp technique. *Pflügers Arch* (1992) **420**: 461–469.
86. Savaria D, Lanoue C, Cadieux A, Rousseau E: Large conducting potassium channel reconstituted from airways smooth muscle. *Am J Physiol* (1992) **262**: L327–L336.
87. Boyle JP, Tomasik M, Kotlikoff MI: Delayed rectifier potassium channels in canine and porcine airways smooth muscle cells. *J Physiol* (1992) **447**: 329–350.
88. Muraki K, Imaizumi Y, Kojima T, *et al.*: Effects of tetraethylammonium and 4-aminopyridine on outward currents and excitability in canine tracheal smooth muscle cells. *Br J Pharmacol* (1990) **100**: 507–515.
89. Murray MA, Berry JL, Cook SJ, *et al.*: Guinea-pig isolated trachealis: the effects of charybdotoxin on mechanical activity, membrane potential changes, and the activity of plasmalemmal K^+ channels. *Br J Pharmacol* (1991) **103**: 1814–1818.
90. Jones TR, Charette L, Garcia ML, Kaczorowski GJ: Selective inhibition of relaxation of guinea-pig trachea by charybdotoxin, a potent Ca^{++}-activated K^+ channel inhibitor. *J Pharmacol Exp Ther* (1990) **255** 697–706.
91. Jones TR, Charette L, Garcia ML, Kaczorowski GJ: Interaction of iberiotoxin with β-adrenoceptor agonist and sodium nitroprusside on guinea pig trachea. *J Appl Physiol* (1993) **74**: 1879–1884.

92. Kume H, Graziano MP, Kotlikoff MI: Stimulatory and inhibitory regulation of calcium-activated potassium channels by guanine nucleotide-binding proteins. *Proc Natl Acad Sci USA* (1992) **89**: 11051–11055.
93. Kume H, Takai A, Tokuno H, Tomita T: Regulation of Ca^{2+}-dependent K^+ channel activity in tracheal myocytes by phosphorylation. *Nature* (1989) **341**: 152–154.
94. Kotlikoff MI: Potassium currents in canine airway smooth muscle cells. *Am J Physiol* (1990) **259**: L384–L395.
95. Janssen LJ, Sims SM: Acetylcholine activates non-selective cation and chloride conductances in canine and guinea-pig tracheal myocytes. *J Physiol* (1992) **453**: 197–218.
96. Janssen LJ, Sims SM: Histamine activates Cl^- and K^+ currents in guinea-pig tracheal myocytes: convergence with cholinergic signalling pathway. *J Physiol* (1993) **465**: 661–677.
97. Janssen LJ: Acetylcholine and caffeine activate Cl^- and suppress K^+ conductances in human bronchial smooth muscle. *Am J Physiol* (1996) **270**: L772–L781.
98. Hisada T, Kurachi Y, Sugimoto T: Properties of membrane currents in isolated smooth muscle cells from guinea-pig trachea. *Pflügers Arch* (1990) **416**: 151–161.
99. Fleischmann BK, Washabau RJ, Kotlikoff MI: Control of resting membrane potential by delayed rectifier potassium currents in ferret airway smooth muscle cells. *J Physiol* (1993) **469**: 625–638.
100. Murray MA, Boyle JP, Small RC: Cromakalim-induced relaxation of guinea-pig isolated trachealis: antagonism by glibenclamide and by phentolamine. *Br J Pharmacol* (1989) **98**: 865–874.
101. Black JL, Armour CL, Johnson PRA, *et al*.: The action of a potassium channel activator, BRL 38227 (lemakalim), on human airway smooth muscle. *Am Rev Respir Dis* (1990) **142**: 1384–1389.
102. Miura M, Belvisi MG, Stretton CD, *et al*.: Role of potassium channels in bronchodilator responses in human airways. *Am Rev Respir Dis* (1992) **146**: 132–136.
103. Groschner K, Silberberg SD, Gelband CH, vanBreeman C: Ca^{2+}-activated K^+ channels in airway smooth muscle are inhibited by cytoplasmic adenosine triphosphate. *Pflügers Arch* (1991) **417**: 517–522.
104. Janssen LJ, Sims SM: Emptying and refilling of Ca^{2+} store in canine tracheal myocytes as indicated by membrane currents and contractions. *Am J Physiol* (1993) **265**: C877–C886.
105. Janssen LJ, Sims SM: Spontaneous transient inward currents and rhythmicity in canine and guinea-pig tracheal myocytes. *Pflügers Arch* (1994) **427**: 473–480.
106. Janssen LJ, Sims SM: Substance P activates Cl^- and K^+ conductances in guinea-pig tracheal smooth muscle cells. *Can J Physiol Pharmacol* (1994) **72**: 705–710.
107. Fleischmann BK, Kotlikoff MI: Methacholine-induced currents and $[Ca^{++}]_i$ in freshly dissociated airway smooth muscle cells. *Biophys J* (1995) **68**: A449.
108. Janssen LJ, Sims SM: Ca^{2+}-dependent Cl^- currents in canine tracheal smooth muscle cells. *Am J Physiol* (1995) **269**: C163–C169.
109. Nakajima T, Hazama H, Hamada E, *et al*.: Ionic basis of neurokinin-A-induced depolarization in single smooth muscle cells isolated from guinea-pig trachea. *Pflügers Arch* (1995) **430**: 552–562.
110. Fleischmann BK, Wang Y-X, Kotlikoff MI: Muscarinic activation and calcium permeation of nonselective cation currents in airway myocytes. *Am J Physiol* (1997) **272**: C341–C349.
111. Marthan R, Martin C, Amédée T, Mironneau J: Calcium channel currents in isolated smooth muscle cells from human bronchus. *J Appl Physiol* (1989) **66**: 1706–1714.
112. Kotlikoff MI: Calcium currents in isolated canine airway smooth muscle cells. *Am J Physiol* (1988) **254**: C793–C801.
113. Welling A, Felbel J, Peper K, Hofmann F: Hormonal regulation of calcium current in freshly isolated airway smooth muscle cells. *Am J Physiol* (1992) **262**: L351–L359.
114. Yamakage M, Hirshman CA, Croxton TL: Cholinergic regulation of voltage-dependent Ca^{2+} channels in porcine tracheal smooth muscle cells. *Am J Physiol* (1995) **269**: L776–L782.
115. Fleischmann BK, Wang YX, Pring M, Kotlikoff MI: Voltage-dependent calcium currents and cytosolic calcium in equine airway myocytes. *J Physiol* (1996) **492**: 347–358.
116. Rodger IW: Voltage-dependent and receptor-operated calcium channels. In Raeburn D,

5 Airway Smooth Muscle Cells: Structure and Function

Giembycz MA (eds) *Airway Smooth Muscle: Peptide Receptors, Ion Channels and Signal Transduction.* Birkhauser Verlag, 1995, pp 155–168.
117. Worley JF, Kotlikoff MI: Dihydropyridine-sensitive single calcium channels in airway smooth muscle cells. *Am J Physiol* (1990) **259**: L468–L480.
118. Tomasik M, Boyle JP, Worley FJ III, Kotlikoff MI: Contractile agonists activate voltage-dependent calcium channels in airway smooth muscle cells. *Am J Physiol* (1992) **263**: C106–C113.
119. Janssen LJ: T-type and L-type voltage-dependent Ca^{2+} currents in canine bronchial smooth muscle: characterization and physiological roles. *Am J Physiol* (1997) **272**: C1757–C1765.
120. Wade GR, Barbera J, Sims SM: Cholinergic inhibition of Ca^{2+} current in guinea-pig gastric and tracheal smooth muscle cells. *J Physiol* (1996) **491**: 307–319.
121. Kamishima T, Nelson MT, Patlak JB: Carbachol modulates voltage sensitivity of calcium channels in bronchial smooth muscle of rats. *Am J Physiol* (1992) **263**: C69–C77.
122. Tomasik M, Kotlikoff ML: A signal transduction pathway for inhibition of voltage dependent calcium channel activity in smooth muscle cells. *Biophys J* (1995) **68**: A206.
123. Murray RK, Fleischmann BK, Kotlikoff MI: Receptor-activated calcium influx in human airway smooth muscle cell: use of Ca imaging and perforated patch-clamp techniques. *Am J Physiol* (1993) **264**: C485–C490.
124. Murray RK, Kotlikoff MI: Receptor-activated calcium influx in human airway smooth muscle cells. *J Physiol* (1991) **435**: 123–144.
125. Felbel J, Trockur B, Ecker T, Landgraf W, Hofmann F: Regulation of cytosolic calcium by cAMP and cGMP in freshly isolated smooth muscle cells from bovine trachea. *J Biol Chem* (1988) **263**: 16764–16771.
126. Kajita J, Yamaguchi H: Calcium mobilization by muscarinic cholinergic stimulation in bovine single airway smooth muscle. *Am J Physiol* (1993) **264**: L496–L503.
127. Yang CM, Chou S-P, Sung T-C: Muscarinic receptor subtypes coupled to generation of different second messengers in isolated tracheal smooth muscle cells. *Br J Pharmacol* (1991) **104**: 613–618.
128. Yang CM, Chou S-P, Wang Y-Y, *et al.*: Muscarinic regulation of cytosolic free calcium in canine tracheal smooth muscle cells: Ca^{2+} requirement for phospholipase C activation. *Br J Pharmacol* (1993) **110**: 1239–1247
129. Marmy N, Mottas J, Durand J: Signal transduction in smooth muscle cells from human airways. *Respir Physiol* (1993) **91**: 295–306.
130. Widdop S, Daykin, K, Hall IP: Expression of muscarinic M2 receptors in cultured human airway smooth muscle cells. *Am J Respir Cell Mol Biol* (1993) **9**: 541–546.
131. Sims SM, Jiao Y, Zheng ZG: Intracellular calcium stores in isolated tracheal smooth muscle cells. *Am J Physiol* (1996) **271**: L300–L309.
132. Panettieri RA, Murray RK, DePalo LR, *et al.*: A human airway smooth muscle cell line that retains physiological responsiveness. *Am J Physiol* (1989) **256**: C329–C335.
133. Mong S, Miller J, Wu H-L, Crooke ST: Leukotriene D$_4$ receptor-mediated hydrolysis of phosphoinositide and mobilization of calcium in sheep tracheal smooth muscle cells. *J Pharmacol Exp Ther* (1988) **244**: 508–515.
134. Yang CM, Hsia H-C, Hsieh J-T, *et al.*: Bradykinin-stimulated calcium mobilization in cultured canine tracheal smooth muscle cells. *Cell Calcium* (1994) **16**: 59–70.
135. Marsh KA, Hill SJ: Des-Arg^9-bradykinin-induced increases in intracellular calcium ion concentration in single bovine tracheal smooth muscle cells. *Br J Pharmacol* (1994) **112**: 934–938.
136. Yang CM, Hsieh J-T, Yo Y-L, *et al.*: 5-Hydroxytryptamine-stimulated calcium mobilization in cultured canine tracheal smooth muscle cells. *Cell Calcium* (1994) **16**: 194–204.
137. Yang CM, Yo Y-L, Hsieh, J-T, Ong R. 5-Hydroxytryptamine receptor-mediated phosphoinositide hydrolysis in canine cultured tracheal smooth muscle cells. *Br J Pharmacol* (1994) **111**: 777–786.
138. Tolloczko B, Jia YL, Martin JG: Serotonin-evoked calcium transients in airway smooth muscle cells. *Am J Physiol* (1995) **269**: L234–L240.
139. Mattoli S, Soloperto M, Mezzetti M, Fasoli A: Mechanisms of calcium mobilization and

phosphoinositide hydrolysis in human bronchial smooth muscle cells by endothelin 1. *Am J Respir Cell Mol Biol* (1991) **5**: 424–430.
140. Aickin CC: Chloride transport across the sarcolemma of vertebrate smooth and skeletal muscle. In Alvarez Leefmans FJ, Russell JM (eds) *Chloride Channels and Carriers in Nerve, Muscle and Glial Cells*. New York, Plenum Press, 1990, pp 209–249.
141. Daniel EE, Jury J, Bourreau J-P, Jager L: Chloride and depolarization by acetylcholine in canine airway smooth muscle. *Can J Physiol Pharmacol* (accepted).
142. Daniel EE, Bourreau J-P, Abela A, Jury J: The internal calcium store in airway muscle: emptying, refilling, and chloride. *Biochem Pharmacol* (1992) **43**: 29–37.
143. Janssen LJ, Walters DK, Wattie J: Regulation of $[Ca^{2+}]_i$ in canine airway smooth muscle by Ca^{2+}-ATPase and Na^+/Ca^{2+} exchange mechanisms. *Am J Physiol* (1997) **273**: L322–L330.
144. Madison JM, Yamaguchi H: Muscarinic inhibition of adenylyl cyclase regulates intracellular calcium in single airway smooth muscle cells. *Am J Physiol* (1996) **270**: L208–L214.
145. Twort CHC, vanBreeman C: Human airway smooth muscle in culture: control of the intracellular calcium store. *Am Rev Respir Dis* (1988) **137**: 10A.
146. Ito Y, Itoh T: The roles of stored calcium in contractions of cat tracheal smooth muscle produced by electrical stimulation, acetylcholine, and high K^+. *Br J Pharmacol* (1984) **83**: 667–676.
147. Gill KK, Kroeger EA: Effects of indomethacin on neural and myogenic components in equine airway smooth muscle. *J Pharmacol Exp Ther* (1990) **252**: 358–364.
148. Tesarowski DP, Kroeger EA: Effects of indomethacin and leukotrienes on equine airway smooth muscle tone. *Am Rev Respir Dis* (1987) **135**: A91.
149. Souhrada M, Souhrada JF, Cherniack RM: Evidence for a sodium electrogenic pump in airway smooth muscle. *J Appl Physiol* (1981) **51**: 346–352.
150. Honda K, Tomita T: Electrical activity in isolated human tracheal muscle. *Jpn J Physiol* (1987) **37**: 333–336.
151. Davis C, Kannan MS, Jones TR, Daniel EE: Control of human airway smooth muscle: *in vitro* studies. *J Appl Physiol* (1982) **53**: 1080–1087.
152. Ito M, Baba K, Takagi K, *et al.*: Some properties of calcium-induced contraction in the isolated human and guinea-pig tracheal smooth muscle. *Respir Physiol* (1985) **59**: 143–153.
153. Small RC: Electrical slow waves and tone of guinea-pig isolated trachealis muscle: effects of drugs and temperature changes. *Br J Pharmacol* (1982) **77**: 45–54.
154. Allen SL, Foster RW, Small RC, Towart R: The effects of the dihydropyridine Bay K 8644 in guinea-pig isolated trachealis. *Br J Pharmacol* (1985) **86**: 171–180.
155. Mansour S, Daniel EE: Maintenance of tone, role of arachidonate metabolites and effects of sensitization in guinea-pig trachea. *Can J Physiol Pharmacol* (1986) **64**: 1096–1103.
156. McCaig DJ: Electrophysiology of neuroeffector transmission in the isolated, innervated trachea of the guinea-pig. *Br J Pharmacol* (1986) **89**: 793–801.
157. Foster RW, Okpalugo BI, Small RC: Antagonism of Ca^{2+} and other actions of verapamil in guinea-pig isolated trachealis. *Br J Pharmacol* (1984) **81**: 499–507.
158. McCray PB, Joseph T: Spontaneous contractility of human fetal airway smooth muscle. *Am J Respir Cell Mol Biol* (1993) **8**: 573–580.
159. Allen SL, Beech DJ, Foster RW, *et al.*: Electrophysiological and other aspects of the relaxant action of isoprenaline in guinea-pig isolated trachealis. *Br J Pharmacol* (1985) **86**: 843–854.
160. Janssen LJ, Daniel EE: Depolarizing agents induce oscillations in canine bronchial smooth membrane potential: possible mechanisms. *J Pharmacol Exp Ther* (1991) **259**: 110–117.
161. Himpens B, Somlyo AP: Free-calcium and force transients during depolarization and pharmacological coupling in guinea-pig smooth muscle. *J Physiol* (1988) **395**: 507–530.
162. Coburn RF, Baron CB: Coupling mechanisms in airway smooth muscle. *Am J Physiol* (1990) **258**: L119–L133.
163. Gerthoffer WT: Regulation of the contractile element of airway smooth muscle. *Am J Physiol* (1991) **261**: L15–L28.

6

Mast Cells and Basophils

F.L. PEARCE

INTRODUCTION

Human bronchial asthma is characterized by a widespread and variable intrathoracic airflow obstruction and an enhanced responsiveness of the airways to non-specific stimulation. Manifestation of the asthmatic response may be conveniently divided into three stages: a rapid spasmogenic phase, a late sustained phase and a subacute, chronic inflammatory phase.[1,2] The immediate response to inhaled allergen has traditionally been associated with the activation of pulmonary mast cells and the release of histamine and spasmogenic products of arachidonic acid metabolism, including prostaglandins and leukotrienes.[1,2] Other studies have indicated that the IgE-dependent activation of alveolar macrophages[3] and platelets[4] may also be involved. The release of chemotactic factors leads to the recruitment of further inflammatory cells including neutrophils, eosinophils and monocytes.[1,2] Activation of all three cell types may be involved in late-phase responses and in the associated induction of non-specific bronchial hyperreactivity. The recruitment of eosinophils appears to be critical for the development of many of the features of chronic asthma, including desquamation of the surface respiratory epithelium.[5] T-lymphocytes of the Th2 phenotype are probably also essentially involved in the disorder.[6] Nitric oxide derived from epithelial cells (and possibly also from macrophages, Th1 cells and mast cells) could be involved in amplifying and perpetuating the Th2 cell-mediated inflammatory response.[7]

There is currently considerable interest in the cellular basis of acquired airway hyperresponsiveness and the critical role of inflammation in the pathogenesis of asthma. The inflammatory changes in the chronic asthmatic patient include occlusion of the airways with viscid mucus, infiltration of the luminal secretions and mucosa with inflammatory calls, thickening of the bronchial basement membrane, hypertrophy of the airway smooth muscle and breakdown of the bronchial epithelium[8,9] The latter effect may expose afferent nerve endings and generate local axon reflexes to amplify and propagate the inflammatory response.[10]

From the above brief discussion, it is clear that no single cell type can be responsible for all the manifestations of human bronchial asthma and that a diversity of inflammatory cells, mediators and neuronal mechanisms are likely to be involved. This chapter considers some of the properties of human pulmonary mast cells and discusses their possible role in asthma. In so doing, it should be appreciated that mast cells from different locations may exhibit marked variations in their morphological, histochemical and functional properties.[11–14] That is, they are biochemically heterogeneous.

MAST CELL HETEROGENEITY

The concept of mast cell heterogeneity has now become firmly established, largely as a result of the development of methods for the enzymic dispersion of free mast cells from diverse target tissues, including the lung, of experimental animals and humans.[11,12] These preparations complement murine serosal mast cells, human basophil leucocytes and tissue culture-derived mast cells, which have been widely used in the study of mediator release.

The best example of mast cell heterogeneity derives from the pioneering work of Enerbäck and his colleagues[15] on the distribution of this cell in the gastrointestinal tract of the rat. Two distinct subpopulations may be identified. The mast cells in the lower layers of the intestinal wall resemble those found in other connective tissues (connective tissue mast cell(s), CTMC), whereas the cells in the mucosa (mucosal mast cell(s), MMC) show very different properties: they are smaller in size and more variable in shape than the CTMC, contain a unique proteolytic enzyme (rat mast cell protease (RMCP) II rather than RMCP I), have a lower content of histamine and 5-hydroxytryptamine, and possess fewer granules. These granules contain the less highly sulphated glycosaminoglycan chondroitin sulphate di-B instead of heparin. These properties require that special conditions of fixation and staining be used to reveal this cell type. Most importantly, the granules may become resistant to metachromatic staining after routine processing in some common formalin-based fixatives. The cells may be distinguished by sequential staining with combinations of dyes such as alcian blue and safranin. The mature rat CTMC stains with safranin, whereas the MMC stains with alcian blue, consistent with the lower degree of sulphation of its proteoglycan matrix. Again, the fluorescent dye berberin stains only the CTMC. Finally, the two mast cell types differ grossly in their responses to histamine liberators and to antiallergic drugs.[11,12]

It must be emphasized, however, that the above histochemical criteria for distinguishing between subpopulations of mast cells have been developed exclusively for the rat. The extent to which these findings may be extrapolated to other species, and especially to humans, is by no means clear. It would currently appear that there are at least two types of histochemically distinct mast cell in both the intestine and lung of humans.[16–18] However, the distinction between the cell types is more subtle and less striking than in the rodent.[18] Moreover, the subpopulations are no longer confined to particular anatomical areas of the target organ. Under these conditions, the terms CTMC and MMC, which are anatomical descriptions, may be incorrect and misleading.[11]

The observed histochemical differences may again reflect variations in the proteoglycan content of the cells. Thus, chondroitin sulphate E has been associated with the intestinal cell, both chondroitin sulphate E and heparin with the lung cell, and heparin with the skin cell.[19,20]

A more distinct separation of human mast cells into two subtypes may be made on the basis of their neutral protease composition and the associated ultrastructure of their secretory granules. The predominant mast cell (MC) present in the mucosa of the bowel and the interalveolar septa of the lung contains tryptase with little, if any, chymase and has thus been designated MC_T. In contrast, the more abundant type in the submucosa of the intestine and in the skin contains both proteases and has been designated MC_{TC}.[19,21] Human MC_T have often been considered to resemble murine MMC, and MC_{TC} to resemble CTMC. The MC_T have varying numbers of irregularly shaped granules with discrete scrolls or particulate or beaded material, while the MC_{TC} have more regularly shaped, electron-dense granules with characteristic grating or lattice substructures.[21]

Evidence for the functional heterogeneity of human mast cells is also much less compelling than in the rodent. It would appear, however, that human skin mast cells differ from the lung and intestinal cells in producing prostaglandin D_2 (PGD_2) but little or no leukotriene C_4 (LTC_4) and in in being particularly responsive to basic histamine liberators such as morphine, compound 48/80, anaphylatoxins and substance P.[22,23] However, this dichotomy is not complete and mast cells from human heart[24] and bladder[25] also respond to these polyamines.

MAST CELLS, BASOPHILS AND ADHESION MOLECULES

Murine MMC and CTMC do not represent a state of terminal differentiation but instead comprise elastic populations that may undergo phenotypic alteration depending upon the tissue microenvironment in which the cells are located. Mast cells of either phenotype can, upon injection into appropriate sites in mast cell-deficient mice, reconstitute both the MMC and CTMC populations.[13] The two phenotypes derive from precursors in the bone marrow that enter the circulation and further differentiate within specific tissues in response to endogenous cytokines. Stem cell factor (SCF) appears to be particularly important for the development of CTMC and interleukin (IL)-3 for the MMC.[13,26] The tissue-specific localization of mast cells may involve the expression of defined homing or adhesion receptors similar to those involved in the trafficking of T-lymphocytes.[26] In similar fashion, basophils can emigrate from the intravascular compartment into specific tissues, and an influx of basophils accompanies the late-phase response that occurs after antigen challenge of the skin and airways.[27] Human basophils express a typical pattern of cell adhesion molecules on their surface, which presumably facilitates their penetration into the tissue space.[27]

DISTRIBUTION AND MORPHOLOGY OF HUMAN LUNG MAST CELLS

Mast cells are widely distributed throughout the human respiratory tract and are found in large numbers in the walls of the alveoli and airways. Most of the mast cells in the conducting airways are located below the bronchial epithelium but appreciable numbers of cells are found intercalated between the epithelial cells and adjacent to the surface of the lumen. These latter cells would come into immediate contact with inhaled antigens and might be expected to be of major importance in modulating the initial phases of the allergic response. More deeply situated mast cells and other cell types may then become progressively involved in the chronic disease, as damage to the mucosal surface allows an increased penetration of inhaled antigen. Mast cells may be recovered from the mucosal, luminal surface of the airways by bronchoalveolar lavage (BAL) and from the parenchyma of the tissue by enzymic dispersion.

Mast cells comprise $0.32 \pm 0.05\%$ (mean $\pm$ SEM, $n = 20$) of the total nucleated population recovered by BAL.[28] The majority of the cells obtained are alveolar macrophages ($86.0 \pm 2.5\%$) with appreciable numbers of lymphocytes ($8.0 \pm 1.5\%$), neutrophils ($4.0 \pm 1.0\%$) and eosinophils ($2.0 \pm 0.5\%$). In five experiments,[28] suspensions of cells obtained by enzymic dissociation of whole lung were again shown to contain large numbers of macrophages ($77.6 \pm 7.3\%$), significant numbers of lymphocytes ($5.7 \pm 2.3\%$) and neutrophils ($10.2 \pm 6.7\%$), and an increased proportion of mast cells ($3.9 \pm 1.2\%$) relative to the BAL fluid. In a further series of experiments, the histamine content of the parenchymal mast cell (2.6 ± 0.1 pg/cell, $n = 12$) was found to be significantly greater than that of the corresponding BAL cell (1.2 ± 0.3 pg/cell, $n = 20$, $P < 0.01$).[28] Both populations stain with alcian blue dye, but do not counterstain with safranin, a property that seems to be characteristic of most human mast cells.[18]

The ultrastructure of the human parenchymal lung cell has been studied in some detail[21,29,30] and a representative example is shown in Fig. 6.1. The cytoplasm of the cell contains large numbers of secretory granules that can exhibit a variety of different ultrastructural patterns. The most common granule type contains cylindrical scrolls,[29] which may be characteristic of the MC_T phenotype,[21] while other mast cells contain granules whose matrices appear as highly ordered crystals or electron-dense particles. In addition, the cells also contain cytoplasmic lipid bodies. On activation, the granules become swollen and amorphous and their membranes fuse to produce chains that enlarge to form tortuous cytoplasmic channels. The latter eventually open to the exterior through multiple points on the cell surface, thereby permitting the release of histamine. These pores progressively widen, ultimately allowing the entry of extracellular markers. The opening of the degranulation channels is accompanied by both increasingly prominent filaments in the intervening cytoplasm and the convolution of the plasma membrane into multiple folds and projections.

The BAL cell broadly resembles that from the lung parenchyma. However, there are generally fewer granules, some of which are characteristically 'basket-shaped' or partially disrupted, and numerous lipid bodies and cytoplasmic folds and projections. Overall, the cell appears to be in a partially activated state.

6 Mast Cells and Basophils

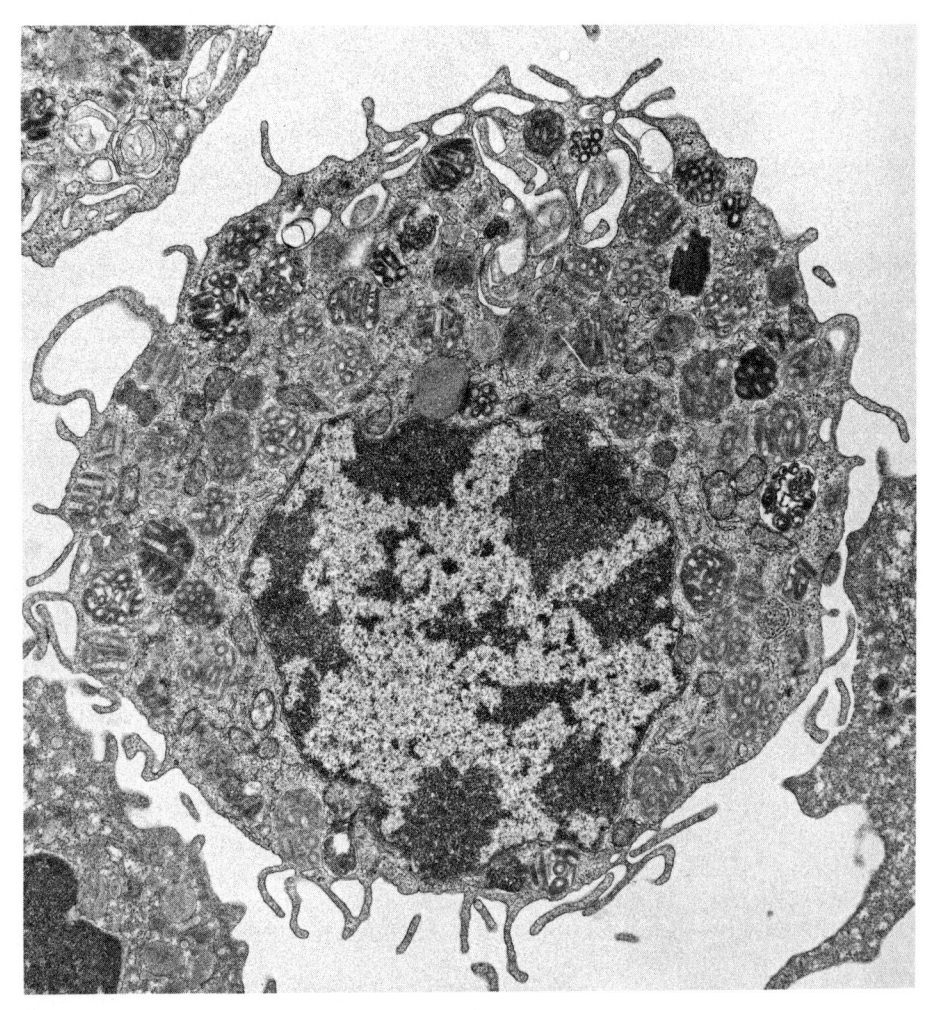

Fig. 6.1 Low-power electron micrograph of a parenchymal mast cell obtained by enzymic dissociation of human lung. Reproduced from ref. 29, with permission.

BAL IN EXTRINSIC ASTHMA

The BAL fluid of 10 extrinsic asthmatic subjects contained a significantly ($P < 0.01$) higher proportion of both eosinophils ($8.0 \pm 2.7\%$) and mast cells ($1.41 \pm 0.27\%$) compared with that of normal controls.[28,31] The histamine content of the lavage increased in parallel with the number of mast cells. The forced expiratory volume in 1 s (FEV_1) and forced vital capacity (FVC) were measured in these subjects, together with the concentration of histamine required to produce a 20% reduction in FEV_1 (PC_{20} histamine). Strikingly, there was a highly significant correlation between the percentage of mast cells in the lavage and the severity of the disease as indicated by measured indices of both airway obstruction (FEV_1 expressed as a percentage of the predicted, and the FEV_1/FVC ratio) and of hyperresponsiveness (PC_{20} histamine).[31]

IMMUNOLOGICALLY INDUCED MEDIATOR RELEASE

Mast cells obtained by BAL of normal subjects and by enzymic dissociation of whole lung released histamine in a dose-dependent fashion on challenge with anti-human IgE (Table 6.1). The rate of release was rather more rapid for the BAL than for the parenchymal cell, requiring 2 and 5 min for completion respectively. The spontaneous release of histamine was also greater for the BAL cell than for the tissue cell.

In addition to histamine, anti-IgE induced a dose-dependent release of immunoreactive PGD_2 and LTC_4 from both mast cell populations (Table 6.1). The antiserum was more effective in inducing PGD_2 production than histamine release and maximal amounts of the prostanoid were generated at lower dilutions of antibody. As in the case of histamine, the spontaneous release of PGD_2 was higher for the BAL cells than for the dispersed lung cells. The spontaneous generation of LTC_4 was rather variable, particularly for the BAL cells, and higher concentrations of anti-IgE were required to evoke the *de novo* production of eicosanoid. For both cell populations, PGD_2 was the predominant eicosanoid produced and exceeded the amount of LTC_4 formed by about one order of magnitude. As might be expected, the rates of release of the newly generated mediators were slower than that of histamine for both BAL and parenchymal lung cells and required 10–15 min for completion.

The spontaneous release of histamine from the BAL mast cells of extrinsic asthmatic patients was $18.7 \pm 3.7\%$ ($n = 10$) compared with $7.4 \pm 0.8\%$ ($n = 30$, $P < 0.01$) for control subjects.[31] The BAL cells of asthmatic individuals thus appear to be inherently unstable. Most interestingly, mast cells obtained by BAL of asthmatic individuals showed an enhanced reactivity towards anti-IgE and exhibited a greater release of histamine at all

Table 6.1 Immunologically induced release of mediators from bronchoalveolar lavage (BAL) and dispersed lung (DL) cells.

Anti-IgE (dilution)	Histamine (% release) BAL	Histamine (% release) DL	LTC_4 (ng/10^6 mast cells) BAL	LTC_4 (ng/10^6 mast cells) DL	PGD_2 (ng/10^6 mast cells) BAL	PGD_2 (ng/10^6 mast cells) DL
100	39.0 ± 6.2	49.5 ± 6.5	18.2 ± 3.5	16.1 ± 5.0	242 ± 50	145 ± 49
300	34.8 ± 5.7	41.0 ± 9.0	10.5 ± 3.5	11.8 ± 3.2	251 ± 60	139 ± 39
1 000	31.9 ± 4.9	31.7 ± 7.3	7.0 ± 2.5	8.8 ± 2.9	219 ± 47	142 ± 30
10 000	17.1 ± 10.2	10.0 ± 5.0	2.2 ± 1.2	2.9 ± 1.8	127 ± 34	63 ± 20
100 000	0.0 ± 2.0	0.5 ± 0.5	–	0.6 ± 0.3	61 ± 33	36 ± 20

All values are means $\pm$ SEM for 10 (BAL) or 8 (DL) experiments and are corrected for the spontaneous releases in the absence of inducer. Spontaneous releases for the BAL and DL cells, respectively, were: histamine 11.5 ± 2.0 and 4.4 ± 1.4; LTC_4 11.0 ± 7.3 and 1.8 ± 0.6; PGD_2 55 ± 10 and 20 ± 14.
Reproduced from ref. 32, with permission.

6 Mast Cells and Basophils

effective dilutions of the antiserum (Fig. 6.2a). This effect was strikingly localized: it was confined to the BAL cells and was not apparent in the basophil leucocytes, which behaved identically to the controls (Fig. 6.2b). Specific antigen also led to histamine release from BAL cells and basophils of asthmatic subjects but not from those of controls (Fig. 6.2c).

Changes in mast cell reactivity in the course of allergic inflammation appears to be a general phenomenon and we have observed an increased immunological responsiveness of the cells recovered from areas of active disease not only in asthma but also in Crohn's disease and interstitial cystitis.[33] In some cases, these changes were paralleled by an increase in the number of mast cells. This phenomenon was studied in more detail in the case of interstitial cystitis.[25] Bladder mast cells from inflamed tissue showed increased reactivity to IgE-directed ligands and calcium ionophores, suggesting that they were globally upregulated, and decreased reactivity to compound 48/80 and substance P,

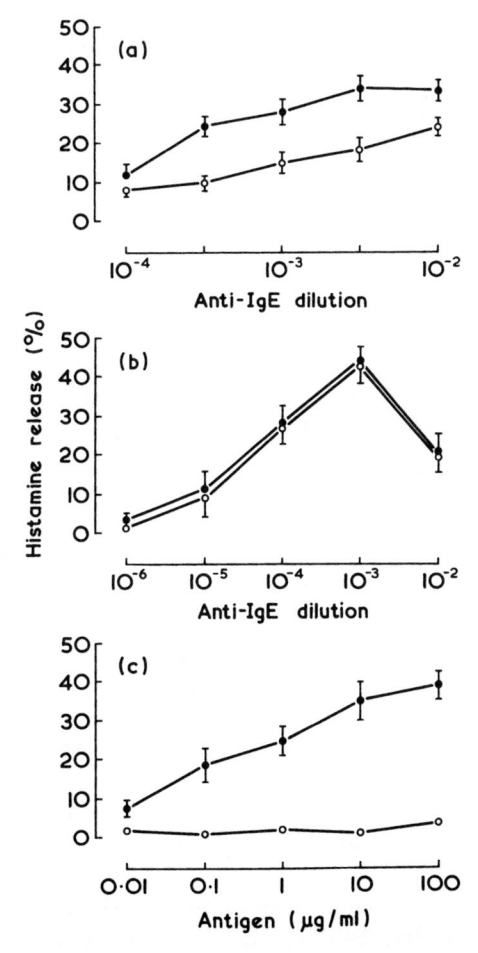

Fig. 6.2 Histamine release from bronchoalveolar lavage (BAL) mast cells and basophil leucocytes from control (○) and asthmatic (●) subjects. (a) Release from BAL cells stimulated with anti-human IgE; (b) release from basophils stimulated with anti-human IgE; and (c) release from BAL cells stimulated with specific antigen to *Dermatophagoides pteronyssinus*. Values are means ± SEM. Reproduced from the data in ref. 31 and from unpublished results.

indicating some phenotypic switch. The origin of this effect remains obscure but it may reflect the influence of cytokines released in the course of allergic inflammatory disease. In any event, it clearly further complicates the issue of mast cell heterogeneity.

BAL MAST CELLS AND STEROID THERAPY IN ASTHMA

Inhaled corticosteroids are one of the most effective groups of drugs available for the management of human bronchial asthma. Their action is undoubtedly complex and their efficacy probably arises from a number of anti-inflammatory effects. Corticosteroids reportedly inhibit histamine release from human basophils but not from lung parenchymal cells.[34] We therefore thought it to be of interest to study the effect of steroid treatment on BAL mast cells.

In a preliminary investigation,[35] seven asthmatic subjects were given a 2-week course of oral prednisolone (30 mg/day). This treatment led to a small improvement in lung function, with the FEV_1 rising from 70 ± 22 litres to 81 ± 23 litres, and a sharp drop in both the percentage of mast cells ($0.18 \pm 0.04\%$ to $0.11 \pm 0.03\%$) in, and the histamine content (6.6 ± 2.9 ng/10^6 cells to 3.3 ± 0.7 ng/10^6 cells) of, the lavage fluid. The spontaneous release of histamine from the isolated mast cells was also dramatically reduced ($17.1 \pm 2.7\%$ to $8.5 \pm 3.6\%$). These data suggest that one effect of steroids in asthma therapy may be to reduce the number, and suppress the spontaneous reactivity, of the luminal mast cells.

ANTIASTHMATIC DRUGS AND THE INHIBITION OF HISTAMINE RELEASE FROM PULMONARY MAST CELLS

Disodium cromoglycate has an established place in the treatment and prophylaxis of bronchial asthma. However, despite intensive research, the mode of action of the chromone is still uncertain. Its activity was originally attributed simply to the inhibition of mediator release from mast cells but it is now clear that the compound can attenuate the activity of a range of inflammatory cells, including neutrophils, eosinophils and monocytes.[36] The drug can also inhibit reflex bronchoconstriction and influence C-fibre activity in the lungs and bronchi.[37] In addition, its effects on histaminocytes are very site specific and vary strikingly from one mast cell subtype to another.[11,12] For these reasons, we examined the effect of sodium cromoglycate, together with its congener nedocromil sodium, on histamine release from BAL cells and lung parenchymal cells. For comparison, the effects of the methylxanthine theophylline and the β-adrenoceptor agonist salbutamol were also tested.[28,38]

Immunologically induced histamine releases from both cell types were comparably inhibited by the latter two drugs, the IC_{30} values (the concentrations required to produce 30% inhibition of release) for BAL and parenchymal cells being respectively 500 μM and 300 μM for theophylline and 20 nM and 100 nM for salbutamol. However, both sodium cromoglycate and nedocromil sodium were strikingly more active against the BAL cell compared with the dispersed lung cell, and nedocromil sodium was about one order of

6 Mast Cells and Basophils

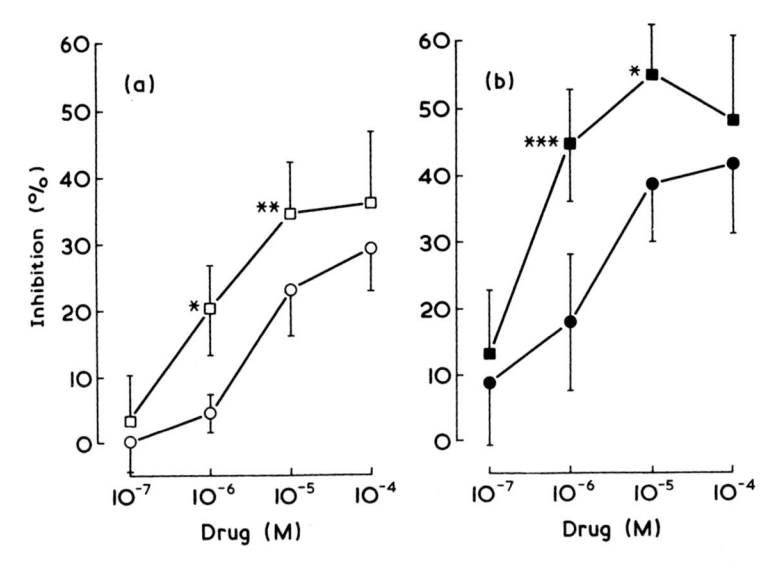

Fig. 6.3 Effect of sodium cromoglycate (○, ●) and nedocromil sodium (□, ■) on immunologically induced histamine release from (a) dispersed human lung mast cells and (b) human bronchoalveolar lavage mast cells. The drugs were added to the dispersed lung cells simultaneously with the secretory stimulus (open symbols) or preincubated with the lavage cells for 10 min before challenge (closed symbols). These conditions were shown to be optimal for activity in each case. Values are means ± SEM for eight (a) or five (b) experiments. Asterisks denote values that were significantly different: *$P < 0.5$, **$P < 0.01$, ***$P < 0.001$. Reproduced from ref. 38, with permission.

magnitude more effective than sodium cromoglycate against both cell types (Fig. 6.3). The IC_{30} values for BAL and parenchymal cells were, respectively, 0.5 μM and 5 μM for nedocromil sodium and 7 μM and 420 μM for sodium cromoglycate.[38]

The characteristics of the inhibition produced by sodium cromoglycate and nedocromil sodium also varied according to the mast cell. A marked tachyphylaxis was observed with the parenchymal cell whereas the activity against the BAL cell increased with preincubation.[38] The latter observation is, of course, more in keeping with the clinical use of the drug, where it is ideally administered prophylactically before antigen exposure. Given the superficial location of the BAL cell within the airways, and its greater exposure to drugs given by inhalation, these findings have particular clinical significance.

ROLE OF BASOPHILS IN ALLERGY AND ASTHMA

The role of the basophil in the mediation of allergic inflammation has not been clearly elucidated. The basophil leucocyte has long been considered to be the circulating equivalent of the tissue mast cell but recent studies have emphasized the differences between these histaminocytes.[39] However, late-phase reactions in the lung,[40] skin[41] and nose[42] are associated with an influx into the target organ of inflammatory leucocytes including basophils. As discussed, this egress is facilitated by the expression of specific adhesion molecules on the surface of the basophil.[27] A role for basophils rather than mast cells in the mediation of late-phase reactions in the nose has been suggested by the finding of the basophil markers histamine and LTC_4, but not PGD_2 which is mast cell derived, in nasal washings in late antigen-induced rhinitis.[43] Broadly similar results have been reported in the skin.[44] Clearly, the role of the basophil in other allergic conditions requires clarification.

ROLE OF MAST CELLS IN EARLY ASTHMATIC REACTIONS

Current evidence suggests that the immediate bronchoconstrictor response to inhaled allergens is largely mediated by mast cell products. Allergen challenge of extrinsic asthmatic subjects leads to secretion of histamine, together with the other mast cell-associated mediators tryptase and PGD_2, into the BAL fluid.[45-47] Pulmonary mast cells are thus clearly activated in the course of the asthmatic response.

Elucidation of the exact role of histamine as a bronchoconstrictor mediator in asthma was originally rendered difficult by the central sedative action and questionable specificity of conventional antihistaminic drugs. The development of newer, non-sedative and more potent and selective histamine H_1-receptor antagonists such as cetirizine, astemizole and terfenadine has, however, rendered such a study possible. These findings have recently been reviewed.[48] In particular, administration of these agents prior to allergen challenge dramatically attenuates the immediate phase of bronchoconstriction. The available data indicate that about half of this response is due to liberated histamine and the remainder to leukotrienes, thromboxanes and prostaglandins.

ROLE OF MAST CELLS IN LATE ASTHMATIC REACTIONS AND CYTOKINE PRODUCTION

Late-phase asthmatic reactions have been intimately associated with the development of bronchial hyperreactivity and airway inflammation. As such, they may be more relevant to the situation in chronic clinical asthma. The role of mast cells in such reactions has been the subject of considerable debate. Release of histamine into the systemic circulation in late-phase responses is controversial but has been reported by some authors.[2] However, as discussed above, allergen-induced late-phase reactions in the nose are accompanied by the release of histamine but not PGD_2, suggesting that the amine originates from basophils recruited into the nasal mucosa.[43,44]

The efficacy of sodium cromoglycate, which blocks immediate and late-phase responses to inhaled allergens and the development of airway hyperreactivity, has been widely used to implicate the mast cell in the progression of bronchial asthma. However, it is now clear that the activity of the chromone is not confined to the mast cell and that the drug may inhibit a range of inflammatory cells.[36] Moreover, a diversity of cromoglycate-like drugs have been developed, many of which are more potent than the chromone itself in preventing histamine release from lung mast cells.[49] However, with the exception of nedocromil sodium, none has proved to be clinically useful. Moreover, as shown above, β-adrenoceptor agonists such as salbutamol are much more potent than sodium cromoglycate in preventing histamine release from human pulmonary mast cells. However, these agents do not block late-phase responses or the development of bronchial hyperreactivity.

The above data would appear to militate against a role for the mast cell in chronic clinical asthma. However, interest in this field has been resurrected by several papers demonstrating the production of various cytokines by mast cells.[50-60] In total, these studies have shown that immunological activation of tissue culture-derived murine and human mast cells leads to increased levels of mRNA and/or secretion of a large range of cytokines including tumour necrosis factor (TNF)-α, granulocyte–macrophage colony-stimulating factor (GM-CSF), interferon (IFN)-γ, IL-1, IL-3, IL-4, IL-5, IL-6, IL-13, and four members of the macrophage inflammatory protein (MIP) gene family, namely T-cell activator (TCA)-3, JE, MIP-1α and MIP-1β.[50-60] Identification of these molecules raises the possibility of a wide range of potential roles for mast cells in pathological responses. Release of cytokines could recruit, prime and activate neutrophils, macrophages, basophils and eosinophils, increase immunoglobulin secretion and regulate the proliferation and phenotype of other mast cells. It should be recognized, however, that most of the available data are derived from murine mast cell lines and the situation in the human is less clear cut. The presence of preformed IL-4, IL-5, IL-6 and TNF-α has been demonstrated by immunohistochemical techniques in human mast cells from the bronchial and nasal mucosa, and from the skin.[56-58] These cells also expressed mRNA for certain cytokines following immunological or other activation. However, the cells were heterogeneous with respect to their cytokine content and, in both the bronchial and nasal mucosa, the IL-4-positive mast cells were predominantly of the MC_{TC} phenotype whereas IL-5 and IL-6 were almost exclusively confined to the MC_T phenotype.

The ability of human histaminocytes to secrete cytokines is more controversial. It is clear that human basophils release substantial amounts of IL-4 in response to IgE receptor stimulation.[61,62] Similar findings have also been reported for human lung mast cells.[56] However, studies in our laboratories[63] have been unable to confirm the latter findings and we failed to observe any immunological release of IL-6 or GM-CSF and only a very moderate (<10 pg/10^6 mast cells) release of IL-4 and TNF-α. The origins of these discrepancies are at present unknown but they suggest that caution should be exercised in assigning a role to mast cell-derived cytokines in human allergic disease.

ACKNOWLEDGEMENTS

Work from the author's laboratory was supported by grants from the National Asthma Campaign, Fisons plc, the Medical Research Council and the Wellcome Trust.

REFERENCES

1. Holgate ST, Kay AB: Mast cells, mediators and asthma. *Clin Allergy* (1985) **15**: 221–234.
2. Kay AB: Mediators and inflammatory cells in asthma. In Kay AB (ed) *Asthma: Clinical Pharmacology and Therapeutic Progress*. Oxford, Blackwell Scientific Publications, 1986, pp 1–10.

3. Joseph M, Tonnel AB, Torpier G, Capron A, Arnoux B, Benveniste J: Involvement of immunoglobulin E in the secretory processes of alveolar macrophages from asthmatic patients. *J Clin Invest* (1983) **71**: 221–230.
4. Morley J, Sanjar S, Page CP: The platelet in asthma. *Lancet* (1984) **ii**: 1142–1144.
5. Makino S, Fukuda T: Eosinophils and allergy in asthma. *Allergy Proc* (1995) **16**: 13–21.
6. Austen KF: Interactions of cells, cytokines, and mediators in bronchial asthma. *Allergy Proc* (1994) **15**: 183–187.
7. Barnes PJ, Liew FY: Nitric oxide and asthmatic inflammation. *Immunol Today* (1995) **16**: 128–130.
8. Barnes PJ, Fan Chung K, Page CP: Inflammatory mediators and asthma. *Pharmacol Rev* (1988) **40**: 49–84.
9. Kay AB: Inflammatory cells in acute and chronic asthma. *Am Rev Respir Dis* (1987) **135**: S63–S66.
10. Barnes PJ: Airway neuropeptides. In Barnes PJ, Roger IW, Thomson NC (eds) *Asthma: Basic Mechanisms and Clinical Management*. London, Academic Press, 1988, pp 395–413.
11. Pearce FL: On the heterogeneity of mast cells. *Pharmacology* (1986) **32**: 61–71.
12. Barrett KE, Pearce FL: Heterogeneity of mast cells. *Hbook Exp Pharmacol* (1991) **97**: 93–117.
13. Kitamura Y: Heterogeneity of mast cells and phenotypic change between subpopulations. *Annu Rev Immunul* (1989) **7**: 59–76.
14. Bienenstock J: An update on mast cell heterogeneity. *J Allergy Clin Immunol* (1988) **81**: 763–769.
15. Enerbäck L: The gut mucosal mast cell. *Monogr Allergy* (1981) **17**: 222–232.
16. Strobel S, Miller HRP, Ferguson A: Human intestinal mucosal mast cells: evaluation of fixation and staining techniques. *J Clin Pathol* (1981) **34**: 851–858.
17. Befus D, Goodacre R, Dyke N, Bienenstock J: Mast cell heterogeneity in man. I. Histological studies in the intestine. *Int Arch Allergy Appl Immunol* (1985) **76**: 232–236.
18. Greenwood B: The histology of mast cells. In Engström I, Lindholm N (eds) *Current Views on Bronchial Asthma*. Stockholm, Fisons Sweden AB, 1985, pp 143–149.
19. Schwartz LB: Mediators of human mast cells and human mast cell subsets. *Ann Allergy* (1987) **58**: 226–235.
20. Thompson HL, Schulman ES, Metcalfe DD: Identification of chondroitin sulphate E in human lung mast cells. *J Immunol* (1988) **140**: 2708–2713.
21. Craig SS, Schwartz LB: Human MC_{TC} type mast cell granule: the uncommon occurrence of discrete scrolls associated with focal absence of chymase. *Lab Invest* (1990) **63**: 581–585.
22. Lawrence ID, Warner JA, Cohan VL, Hubbard WC, Kagey-Sobotka A, Lichtenstein LM: Purification and characterization of human skin mast cells. Evidence for human mast cell heterogeneity. *J Immunol* (1987) **139**: 3062–3069.
23. Benyon RC, Lowman MA, Church MK: Human skin mast cells: their dispersion, purification, and secretory characterization. *J Immunol* (1987) **138**: 861–867.
24. Patella V, de Crescenzo G, Ciccarelli A, Marino I, Adt M, Marone G: Human heart mast cells: a definitive case of mast cell heterogeneity. *Int Arch Allergy Immunol* (1995) **106**: 386–393.
25. Frenz AM, Christmas TJ, Pearce FL: Does the mast cell have an intrinsic role in the pathogenesis of interstitial cystitis? *Agents Actions* (1994) **41**: C14–C15.
26. Smith TJ, Weis JG: Mucosal T cells and mast cells share common adhesion receptors. *Immunology Today* (1996) **17**: 60–63.
27. Bochner BS, Sterbinsky SA, Knol EF, et al.: Function and expression of adhesion molecules on human basophils. *J Allergy Clin Immunol* (1994) **94**: 1157–1162.
28. Hammond MD, Brostoff J, Geraint-James D, Johnson NMcI: Some studies on human pulmonary mast cells obtained by bronchoalveolar lavage and by enzymic dissociation of whole lung tissue. *Int Arch Allergy Appl Immunol* (1987) **82**: 507–512.
29. Caulfield JP, Lewis RA, Hein A, Austen KF: Secretion in dissociated human pulmonary mast cells. Evidence for solubilization of granule contents before discharge. *J Cell Biol* (1980) **85**: 299–311.
30. Dvorak AM, Schulman ES, Peters SP, et al.: Immunoglobulin E-mediated degranulation of isolated human lung mast cells. *Lab Invest* (1985) **53**: 45–56.
31. Flint KC, Leung KBP, Hudspith BN, Brostoff J, Pearce FL, Johnson NMcI: Bronchoalveolar

mast cells in extrinsic asthma: a mechanism for the initiation of antigen specific bronchoconstriction. *Br Med J* (1985) **291**: 923–926.
32. Leung KPB, Flint KC, Hudspith BN, *et al.*: Some further properties of human pulmonary mast cells recovered by bronchoalveolar lavage and enzymic dispersion of lung tissue. *Agents Actions* (1987) **20**: 213–215.
33. Pearce FL, Frenz AM, Shah PM: Changes in mast cell reactivity in the course of allergic inflammation. *Inflamm Res* (1996) **45** (Suppl): S31–S32.
34. Fox CC, Kagey-Sobotka A, Schleimer RP, Peters SP, MacGlashan DW, Lichtenstein LM: Mediator release from human basophils and mast cells from lung and intestinal mucosa. *Int Arch Allergy Appl Immunol* (1985) **77**: 130–136.
35. Millar AB, Hudspith BN, Lau A, Pearce FL, Johnson NMcI: A mechanism for the role of steroids in the treatment of asthma? *Thorax* (1989) **44**: 359P.
36. Kay AB, Walsh GM, Moqbel R, *et al.*: Disodium cromoglycate inhibits activation of human inflammatory cells *in vitro*. *J Allergy Clin Immunol* (1987) **80**: 1–8.
37. Richards IM, Dixon M, Jackson DM, Vendy K: Alternative modes of action of sodium cromoglycate. *Agents Actions* (1986) **18**: 294–300.
38. Leung KPB, Flint K, Brostoff J, *et al.*: Effects of sodium cromoglycate and nedocromil sodium on histamine secretion from human lung mast cells. *Thorax* (1988) **43**: 756–761.
39. Henderson WR: Basophils. *Immunol Allergy Clin North Am* (1990) **10**: 273–282.
40. Pepys J, Hargreave FE, Chan M, McCarthy DS: Inhibitory effects of disodium cromoglycate on allergen-inhalation tests. *Lancet* (1968) **ii**: 134–137.
41. Solley GO, Gleich GJ, Jordon RE, Schroeter AL: The late phase of the immediate wheal and flare reaction. Its dependence on IgE antibodies. *J Clin Invest* (1976) **58**: 408–420.
42. Bascom R, Wachs M, Naclerio RM, Pipkorn V, Galli SJ, Lichtenstein LM: Basophil influx occurs after nasal antigen challenge: effects of topical corticosteroid pretreatment. *J Allergy Clin Immunol* (1988) **81**: 580–589.
43. Naclerio RM, Proud D, Togias AG, *et al.*: Inflammatory mediators in late antigen-induced rhinitis. *N Engl J Med* (1985) **313**: 65–70.
44. Massey WA, Lichtenstein LM: Role of basophils in human allergic disease. *Int Arch Allergy Immunol* (1992) **99**: 184–188.
45. Holgate ST, Benyon RC, Howarth PH, *et al.*: Relationship between mediator release from human lung mast cells *in vitro* and *in vivo*. *Int Arch Allergy Appl Immunol* (1985) **77**: 47–56.
46. Murray JJ, Tonnel AB, Brash AR, Roberts LJ,. Gosset EP, Workman R: Release of prostaglandin D$_2$ into human airways during antigen challenge. *N Engl J Med* (1986) **315**: 800–804.
47. Wenzel SE, Fowler AA, Schwartz LB: Activation of pulmonary mast cells by bronchoalveolar allergen challenge: *in vivo* release of histamine and tryptase in atopic subjects with and without asthma. *Am Rev Respir Dis* (1988) **137**: 1002–1008.
48. Wood-Baker R, Church MK: Histamine and asthma. *Immunol Allergy Clin North Am* (1990) **10**: 329–336.
49. Church MK: Cromoglycate-like-antiallergic drugs: a review. *Drugs Today* (1978) **14**: 281–341.
50. Wodnar-Filipowicz A, Heusser CH, Morani C: Production of the haemopoietic growth factors GM-CSF and interleukin-3 by mast cells in response to IgE receptor-mediated activation. *Nature* (1989) **339**: 150–152.
51. Plaut M, Pierce JH, Watson CJ, Hanley-Hyde J, Nordan RP, Paul WE: Mast cell lines produce lymphokines in response to cross-linkage of Fc$_e$RI or to calcium ionophores. *Nature* (1989) **339**: 64–67.
52. Young JD-E, Liu C-C, Butler G, Cohn ZA, Galli SJ: Identification, purification and characterization of a mast cell-associated cytolytic factor related to tumour necrosis factor. *Proc Natl Acad Sci USA* (1987) **84**: 9175–9179.
53. Burd PR, Rogers HW, Gordon JR, *et al.*: Interleukin 3-dependent and independent mast cells stimulated with IgE and antigen express multiple cytokines. *J Exp Med* (1989) **170**: 245–257.
54. Gordon JR, Galli SJ: Mast cells as a source of both preformed and immunologically inducible TNF-α/catechin. *Nature* (1990) **346**: 274–276.
55. Burd PR, Thompson WC, Max EE, Mills FC: Activated mast cells produce interleukin 13. *J Exp Med* (1995) **181**: 1373–1380.

56. Bradding P, Feather IH, Howarth PH, *et al.*: Interleukin 4 is localised to and released by human mast cells. *J Exp Med* (1992) **176**: 1381–1386.
57. Bradding P, Feather IH, Wilson S, *et al.*: Immunolocalization of cytokines in the nasal mucosa of normal and perennial rhinitic subjects. *J Immunol* (1993) **151**: 3853–3865.
58. Bradding P, Okayama Y, Howarth PH, Church MK, Holgate ST: Heterogeneity of human mast cells based on cytokine content. *J Immunol* (1995) **155**: 297–307.
59. Jaffe JS, Schulman ES: Activation of human lung mast cells induces early expression of IL-5 mRNA but not IL-3, IL-4 and GM-CSF. *J Immunol* (1993) **150**: 147.
60. Ikayama Y, Petit-Frère C, Kassel O, *et al.*: Expression of IL-4 and IL-5 mRNA in human mast cells via Fc$_\varepsilon$RI in the presence of SCF. *Allergy Clin Immunol News* (1994) **37**: 131.
61. Schroeder JT, MacGlashan DW, Kagey-Sobotka A, White JM, Lichtenstein LM: Cytokine generation by human basophils. *J Allergy Clin Immunol* (1994) **94**: 1189–1195.
62. MacGlashan DM, White JM, Huang, S-K, Ono SJ, Schroeder JT, Lichtenstein LM: Secretion of interleukin-4 from human basophils: the relationship between IL-4 mRNA and protein in resting and stimulated basophils. *J Immunol* (1994) **152**: 3006–3016.
63. Gibbs BF, Arm JP, Gibson K, Lee TH, Pearce FL: Human lung mast cells release small amounts of interleukin-4 and tumour necrosis factor-α in response to stimulation by anti-IgE and stem cell factor. *Eur J Pharmacol* (1997) **327**: 73–78.

7

Monocytes, Macrophages and Dendritic Cells

CATHERINE M. HAWRYLOWICZ AND TAK H. LEE

INTRODUCTION

Cells of the monocyte/macrophage lineage, or the mononuclear phagocyte system, are of primary importance in the innate defence systems of the body to fight infection, in tumour surveillance, clearance of non-inflammatory debris, tissue maintenance and repair, regulation of haemopoiesis, as well as in the recruitment and activation of other inflammatory cells. These functions are mediated through phagocytosis of particulate antigens and microorganisms, synthesis and release of numerous soluble mediators, as well as cytotoxic activity. These cells also play an important role in the acquired immune response through their capacity to process and present antigen and regulate T-lymphocyte responses. Whilst these processes are critical for the survival of the host, imbalances in their regulation can lead to acute injury or chronic inflammation of the lung.

Several distinct populations of macrophages coexist in the lung: alveolar, interstitial and intravascular macrophages,[1–3] as well as dendritic cells (DCs). The first are the most intensively studied in humans since they are readily available in bronchoalveolar lavage (BAL) and are morphologically very similar to tissue macrophages. Alveolar macrophages are unique as they are located within the interphase between air and lung tissue and represent the first line of defence against the continuous exposure to inhaled constituents of the air. They are highly phagocytic and have strong microbicidal activity, yet in normal healthy individuals express poor antigen-presenting cell activity, features likely to be critical for the well-being of the host. In asthma, distinct changes in the phenotype and function of these cells, as well as in peripheral blood monocytes, is observed. Samples available for the study of human cells in allergic and asthmatic disease include:

(1) peripheral blood, which provides a readily available source of monocytes;
(2) BAL (80–95% macrophages), which is generally only available from mild to moderate asthmatic individuals.[4] Nevertheless significant alterations in function are observed that are presumed to be more pronounced in severe disease;
(3) endobronchial biopsy samples, which are very small, are useful for immunohistochemical, but provide limited opportunity for functional analyses.

ROLE OF MONONUCLEAR PHAGOCYTES IN INFLAMMATORY MECHANISMS OF ASTHMATIC DISEASE

Chronic inflammation of the bronchial wall is a characteristic feature of asthma and there is an association between the presence of airway inflammatory cells and airway hyperresponsiveness in patients with stable disease and following exacerbation of asthma.[4-14] In both atopic and non-atopic patients, influxes into the lung of eosinophils and CD4$^+$ T-lymphocytes, the majority of which bear the CD45RO primed or memory phenotype, are particularly pronounced, although increases in neutrophils and mast cells also occur. Macrophages represent a major cell population within the lung and although the majority of studies using BAL suggest their numbers are not markedly increased in asthmatic compared with control subjects, bronchial biopsy samples demonstrate increased numbers in the submucosa. However, an increased number of cells with a monocytic/immature macrophage phenotype expressing elevated levels of MHC class II antigens is observed.[3,11] There is growing consensus that failure to control T-cell immune functions underlies the disease process in physiologically hyperresponsive individuals and that the macrophage population plays a central role in the regulation of local T-cells. The mononuclear phagocyte system is likely to contribute to the disease process in two ways: firstly, through the release of soluble mediators in the local microenvironment that can induce and exacerbate the inflammatory condition; secondly, monocytes, macrophages and DCs regulate the activation of T-lymphocytes in a process where granulocytes will play little if any role.

CHANGES IN PHENOTYPE AND FUNCTION OF MONOCYTES AND MACROPHAGES IN ASTHMA

Non-specific functions

Following activation, macrophages have the potential to produce a broad array of pro-inflammatory mediators.[2,3,15] Whilst eosinophils and mast cells probably represent the major source of mediators that regulate inflammatory asthmatic responses, the precise magnitude of the contribution by macrophages relative to granulocytes is unclear and difficult to assess in humans. It is likely to vary in atopic versus non-atopic asthmatic individuals and to increase with disease severity.

Pathways of mononuclear phagocyte activation

Macrophages are activated by several different mechanisms and cell surface receptors, including bacterial products and particulate antigens, cytokines, complement components, antibodies, immune complexes, following cognate interactions with T-lymphocytes and adherence.[2,3,16,17] Macrophage activation has been most extensively studied in infectious systems, where Th1-derived cytokines such as interferon (IFN)-γ are important.[16,17] In contrast, in asthma a Th2 response tends to predominate.[13,14] Interleukin (IL)-4, a major Th2 product, is associated with increased IgE production but reduced cytokine production by mononuclear phagocytes.[18] However, recent studies suggest that in asthma these cells may become less prone to IL-4 inhibition.[19]

The enhanced activation of macrophages observed in asthma shows some correlation with higher expression of several receptors and their ligands. For example, increased levels of the low-affinity receptor for IgE, FcϵRII (and IgE), are seen in atopic versus control donors. Approximately 5–15% of blood monocytes and alveolar macrophages normally express FcϵRII, and this increases to around 20% in mildly atopic individuals and up to 80% of peripheral blood monocytes in severely atopic individuals.[20–23] Furthermore, IL-4 and other cytokines (IFN-α, IFN-γ, granulocyte–macrophage colony-stimulating factor (GM-CSF), macrophage colony-stimulating factor (M-CSF)) as well as IgE itself enhance expression of FcϵRII on human peripheral blood monocytes and alveolar macrophages.[21,24] GM-CSF, which is increased in asthmatic lung,[25] primes macrophages to respond to IgE.[26] Spiegelberg[20] suggests that enhanced FcϵRII expression is associated with increased responsiveness of peripheral blood monocytes to IgE immune complexes in experiments using IgE-coated red blood cells and measuring cytotoxic function. Additional observations include IgE immune complex-mediated release of IL-1β by peripheral blood monocytes;[27] increased spontaneous and IgE-induced release of tumour necrosis factor (TNF) and IL-6 by alveolar macrophages of allergic asthmatic subjects;[28] and IgE stimulation of leukotriene (LT)B$_4$, LTC$_4$, prostaglandin (PG)E$_2$ and superoxide anion in human monocytes.[29,30] Finally FcϵRI, the high-affinity receptor for IgE, has been reported on activated monocytes, and may enhance presentation of allergen to T-lymphocytes.[31]

In atopic patients peripheral blood monocytes exhibit increased expression of the complement receptors CR1 and CR3, which are important for the binding of C3b and C3bi opsonized particles respectively.[32,33] Elevated levels of expression of MHC class II antigens, which are essential for antigen-specific interactions with CD4 T-lymphocytes, is also seen.[11,33,34] Increased activation of monocytes and macrophages through complement receptors, following T-cell interactions, by cytokines and immune complexes are therefore all potential pathways for increased mononuclear phagocyte activation at inflammatory sites.

Macrophage products that modulate inflammation

Numerous monocyte- and macrophage-derived products have the potential to recruit and activate inflammatory cells in asthma and include cytokines, complement components,

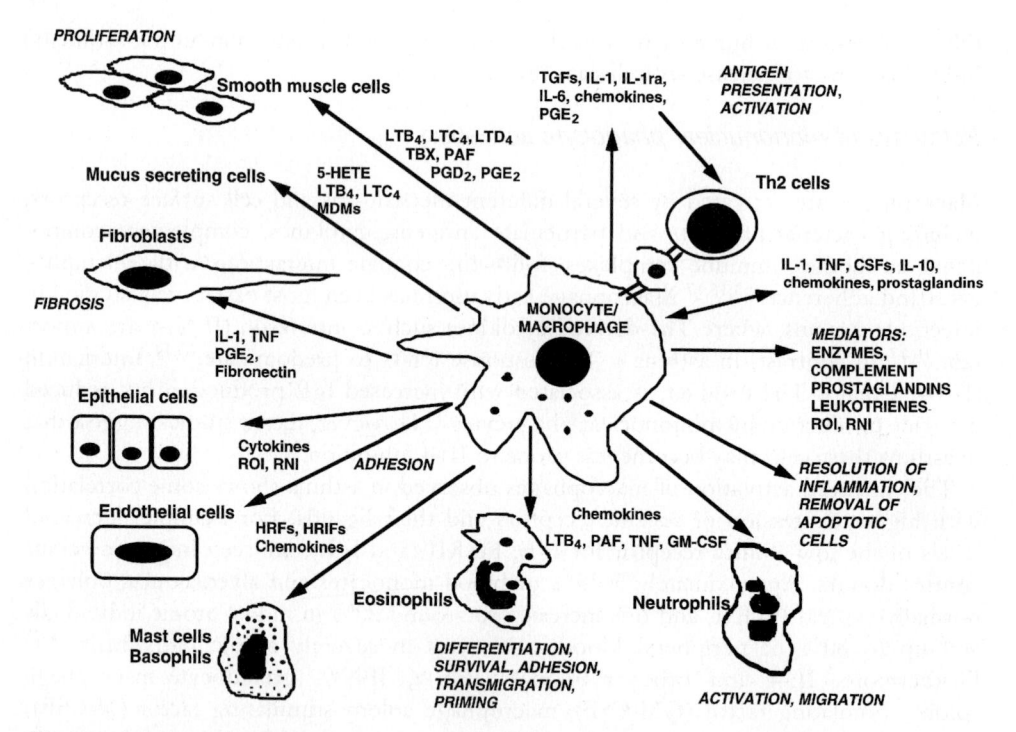

Fig. 7.1 Soluble mediators synthesized by mononuclear phagocytes and their potential role in regulating events in asthmatic inflammation. CSFs, colony-stimulating factors; GM-CSF, granulocyte–macrophage colony-stimulating factor; 5-HETE, 5-hydroxyeicosatetraenoic acid; HRFs, histamine-releasing factors; HRIFs, histamine-releasing inhibition factors; IL, interleukins; IL-1ra, IL-1 receptor antagonist; LTB$_4$, leukotriene B$_4$; LTC$_4$, leukotriene C$_4$; LTD$_4$, leukotriene D$_4$; MDMS, macrophage-derived mucus secretagogue; PAF, platelet-activating factor; PGD$_2$, prostaglandin D$_2$; PGE$_2$, prostaglandin E$_2$; ROI, reactive oxygen intermediates; RNI, reactive nitrogen intermediates; TBX, thromboxanes; TGFs, transforming growth factors; TNF, tumour necrosis factor.

various enzymes, oxygen and nitrogen metabolites as well as products of the arachidonic acid cascade.[2,3,15–17] Whilst this capacity is generally not unique to macrophages, these cells represent a major source of several of these mediators (Fig. 7.1).

Cytokines. Monocytes/macrophages synthesize an enormous array of cytokines. Selective upregulation of several pro-inflammatory cytokines by peripheral blood monocytes and/or alveolar macrophages is observed in asthma, including IL-1, IL-6, IL-8, TNF-α and GM-CSF.[12,14,25,28,35–38] *Ex vivo* studies demonstrate enhanced basal production of GM-CSF but not IL-1β, IL-8 or TNF-α in monocytes, but not macrophages, from asthmatic subjects. Following lipopolysaccharide (LPS) stimulation asthmatic monocytes showed increased GM-CSF and IL-1β, whilst macrophages demonstrated enhanced TNF-α, GM-CSF and IL-8 secretion.[38] In addition, mRNA for platelet-derived growth factor (PDGF) is also increased.[39]

An important function of both IL-1 and TNF in asthma is likely to be the early induction of chemokines and of adhesion molecules such as intercellular adhesion molecule (ICAM-1) and E-selectin on vascular endothelium that promote the recruitment

7 Monocytes, Macrophages and Dendritic Cells

and adhesion of granulocytes and lymphocytes.[40] These cytokines, together with fibronectin and PGE$_2$, are also important regulators of fibroblast proliferation. TNF and IL-1 activate mononuclear phagocytes themselves for cytokine release, adhesion molecule expression and many additional functions, as well as regulating the function of lymphocytes. TNF induces many neutrophil functions including degranulation and release of oxygen metabolites, which contribute to inflammatory processes (reviewed in refs 41 and 42). Neutralization of, or deficiency in, TNF, IL-1 or IL-4 in animal models of airway inflammation support their role in the early recruitment of eosinophils and neutrophils.[43,44]

A second family of cytokines, the chemokines, are produced by mononuclear phagocytes and control specific cellular recruitment to inflammatory sites. They are divided into two groups: the C-X-C family primarily acts on neutrophils, the C-C family on mononuclear phagocytes, eosinophils and lymphocytes. A strong association between C-C family and chronic inflammatory responses is proposed.[45] Increased expression of a newly described chemoattractant, IL-16,[46] the C-X-C chemokine IL-8 and the C-C chemokines monocyte chemotactic protein (MCP-1), macrophage inflammatory protein (MIP-1α) and regulated upon activation in normal T cells expressed and secreted (RANTES) are variously reported in mouse and/or human lung homogenates, BAL or sputum during airway inflammation.[43,44,47] LTB$_4$, platelet-activating factor (PAF), C5a, PDGF, TNF and IL-1, which are all produced by mononuclear phagocytes, also act in a less specific manner as chemoattractants of likely importance in asthma.[42,43,48]

Macrophages represent a major source of GM-CSF within the asthmatic lung where expression is greatly increased.[25,38] A major function of GM-CSF in asthma is in promoting eosinophil survival and activation.[13,49] GM-CSF also regulates neutrophil function as well as macrophage maturation and activation and is therefore likely to indirectly influence T-cell function (discussed below).

Many additional macrophage-derived cytokines are modulated during inflammation, including those with the capacity to inhibit immune/inflammatory processes. For example, production of IL-1 receptor antagonist (IL-1ra) is increased,[50] although transforming growth factor (TGF)-β1, which is involved in wound healing and suppression of lymphocyte activation, is unchanged.[51] In contrast, spontaneous and induced production of IL-10, a potent anti-inflammatory cytokine that inhibits cytokine production by both Th1 and Th2 subsets as well as mononuclear phagocytes, is reduced in blood monocytes and alveolar macrophages from asthmatic individuals.[52] As always, the relative levels of the different cytokine activities within the local microenvironment will determine the functional consequences of this expression.

Arachidonic acid pathway metabolites. Human mononuclear phagocytes synthesize and release metabolites of both the cyclooxygenase pathway (thromboxanes and prostaglandins) and the lipoxygenase pathway (leukotrienes and hydroxyeicosatetraenoic acids (HETEs); reviewed in ref. 48). The levels produced by mononuclear phagocytes vary and again the relative contribution of macrophage-derived products to the asthmatic inflammatory process is very difficult to dissect. Stimuli for production include LPS, silica, calcium ionophores and immune complexes. IFN-γ does not directly induce eicosanoid synthesis, but primes for greater release following stimulation with agonists such as LPS. Thromboxane A$_2$ and LTB$_4$ are produced in particularly large amounts by alveolar macrophages. Levels of LTB$_4$ exceed those of blood monocytes and neutrophils, and alveolar macrophages from asthmatic subjects release increased levels of LTB_4 *in vitro*.[48,53–55] Leukotrienes modulate smooth muscle cell function; constriction is regulated by LTC_4, LTD_4 and LTE_4.[48] These are produced at low levels by human alveolar macrophages but in greater quantities by blood monocytes (which increase in number in asthmatic lungs).[29,48,56] The relative amounts of mediators such as thromboxane, PGE_2, PGD_2 and PAF are also likely to determine smooth muscle tone and function in airways.[48] PGE_2 is released in large quantities[57] and causes bronchoconstriction as does PAF.[48] PGD_2 is reportedly increased in BAL following allergen provocation.[48,58,59] It is produced in large amounts by mast cells, although release by human alveolar macrophages has also been documented.[48,57] Macrophage production of 5-HETE and LTB_4 may further contribute to the inflammatory process by enhancement of mucous secretion.[48] Eosinophil function is also regulated by several lipid mediators produced by macrophages. In addition to GM-CSF, LTB_4 and PAF act as eosinophil chemotaxins and induce mediator release.[48] Human alveolar macrophages also spontaneously release products that induce histamine release from basophils and these products are found in BAL of both normal and asthmatic individuals. Similarly, blood monocytes and BAL produce histamine release inhibition factors.[60,61]

Production of reactive oxygen metabolites and reactive nitrogen intermediates. Superoxide anions (O_2^-), hydrogen peroxide (H_2O_2) and hydroxyl radicals (OH˙) represent major products of activated macrophages and neutrophils. Expression of reactive oxygen metabolites is increased in alveolar macrophages and monocytes from asthmatic versus control subjects.[62,63] Their function has been most extensively studied in relation to antimicrobial defence mechanisms,[2,3,15] but they are known to cause tissue damage in a number of disease conditions and disruption of epithelial cells in asthmatic airways is likely to be mediated, at least in part, by reactive oxygen metabolites and lysosomal enzymes relased by activated macrophages. Relevant stimuli for production by macrophages include immune complexes and cytokines (either indirectly by priming or directly) such as IFN-γ, PDGF, GM-CSF and TNF as well as LTB_4.[2]

Important reactive nitrogen intermediates include nitric oxide, nitrites and nitrates, with nitric oxide predominating. Expression of reactive nitrogen intermediates is activation dependent, but regulated independently of reactive oxygen metabolites. In animal models of infection IFN-γ increases, whilst the Th2 cytokines IL-4 and IL-10 inhibit, production of reactive nitrogen intermediates. Nevertheless, increased expression is observed in airway inflammatory disease.[64] Reactive nitrogen intermediates have potent antimicrobial activity, are immunosuppressive for lymphocytes and may also contribute to tissue damage. A role in immunosuppression of T-lymphocytes in rodent lung has been proposed by Holt and colleagues and is discussed below.

Antigen processing and presentation

CD4$^+$ T-lymphocytes are detected in increased numbers in asthmatic airways and play a central role in the disease process through the secretion of cytokines such as GM-CSF, IL-3, IL-4, IL-5 and IL-10. These cytokines in turn control immunoglobulin production (particularly IgE), eosinophil and mast cell function.[13,14] CD4$^+$ T-cells are activated following interaction with specialized antigen-presenting cells (APCs), the major types

7 Monocytes, Macrophages and Dendritic Cells

being DCs, monocytes/macrophages and B cells. APCs must express MHC class II antigens at their cell surface, take up and process exogenous antigens, and re-express antigenic peptides at their cell surface bound to the peptide-binding groove of the MHC class II molecule.[65] They also produce cytokines that are important for T-cell function and phenotype development (e.g. IL-1, IL-10, IL-12). In addition 'professional' APCs are required to express costimulatory and adhesion molecules such as B7-1, B7-2 and ICAM-1, required for optimal activation of T-cells.[66] Failure to express costimulatory function may lead to the induction of non-responsiveness or anergy in the T-cell population or may only permit activation of recently stimulated T-cells, which have less stringent activation requirements. Several other cells, so called 'non-professional' APCs, e.g. eosinophils, appear less effective in T-cell activation. This is probably due to the lack of constitutive expression of MHC class II molecules and the very low or lack of costimulatory molecule expression.

Alveolar macrophages generally function poorly as APCs.[3,67,68] These cells are heterogeneous and several laboratories have shown that fractionation based on size, density and cell surface antigen expression defines populations with distinct abilities for T-cell activation.[2,3,34,69,70] Small, low-density monocyte-like cells show good APC activity. In contrast mature macrophages, which are larger and of lower density, are poor APCs. Evidence for immunosuppression of local T-cells by alveolar macrophages and other resident APCs is suggested by *in vivo* depletion experiments in rodents.[3] When compared with monocytes, alveolar macrophages show reduced capacity to 'cluster' with T-cells, adhesion molecule expression and MHC class II antigen expression. Studies on synthesis of cytokines that enhance T-cell activation are difficult to interpret functionally since macrophages also produce mediators such as IL-1ra, IL-10, PGE$_2$ and TGF-β that are inhibitory. Costimulatory molecule expression by human alveolar macrophages is low and poorly induced with IFN-γ.[71] Monocytes constitutively express B7-2, but require activation for B7-1 expression, and the importance of B7-1 versus B7-2 for responsiveness to different antigens remains unclear.[66] In asthma, increased numbers of immature/monocytic-like cells are observed and these are likely to contribute, at least in part, to the increased T-cell activation observed. In addition, changes in the resident alveolar macrophage immunomodulatory phenotype have been reported (see below).

DCs are present in very low numbers in the lung (<1% in BAL), but on isolation show excellent APC function and are believed to represent the major APC population in this organ.[3,70,72] DCs and monocytes/macrophages originate from a common bone marrow precursor and blood monocytes *in vitro* can differentiate into macrophages or DCs depending on culture conditions: GM-CSF or M-CSF favours maturation to a macrophage phenotype, whilst GM-CSF and high concentrations of IL-4, and other conditions, favours a dendritic-like phenotype. Stimulation with TNF or by CD40 cross-linking lead to the generation of a 'mature' DC phenotype.[73] Since GM-CSF production is markedly increased in the asthmatic lung and IL-4 is a cytokine associated with the development of atopy, it is possible, but unproven, that maturation of newly recruited monocytes into DCs in the asthmatic lung may also contribute to the enhanced T-cell activation status observed. DCs show heterogeneity depending on both location in the lung and maturational status: immature DCs show good phagocytic and antigen-processing function but poor T-cell stimulation, whilst in mature cells that have migrated to tissues during inflammatory responses this is reversed.

Alveolar macrophages normally suppress DC activity and T-lymphocyte function in the lungs and this has been postulated to occur through several mechanisms.[3,72,74] These include via immunosuppressive mediators such as PGE_2, TGF-β, reactive oxygen metabolites and reactive nitrogen intermediates. Studies by Holt and colleagues favour a role for reactive nitrogen intermediates, but additional mechanisms appear likely.[3] Although the majority of these studies are in rodents, similar mechanisms are thought to occur in humans. Suppression by alveolar macrophages appears reduced in inflammation, which may be due to increased production of GM-CSF. This influences DC maturation and may modulate alveolar macrophage APC activity itself. The interactions between resident and newly recruited APCs and the influences of the asthmatic inflammatory microenvironment are extremely complex and have been studied most extensively in rodents.

EFFECTS OF GLUCOCORTICOIDS ON MONOCYTE/MACROPHAGE FUNCTION

Glucocorticoids represent one of the major therapies for the treatment of inflammatory disease and are extremely effective in the management of allergic and asthmatic disease in the vast majority of patients.[75] Although their mechanism of action is not fully defined, potent inhibitory effects on pro-inflammatory cytokine production, cell migration and lymphocyte activation are well documented and likely to play an important role. However while it is clear that certain functions are downregulated, many others are upregulated by these agents.[76] For example, studies in mice have documented increased production of macrophage inflammatory protein 1[77] and IL-10[78] by glucocorticoids, both *in vitro* and *in vivo*. Studies by Daynes and colleagues[79] also demonstrate that *in vivo* administration of glucocorticoids in mice leads to skewing of the T-helper response to a Th2 phenotype. Glucocorticoids can also enhance cytokine receptor expression *in vitro*, including the IL-6, GM-CSF, IL-1 type II, IL-2Rα and IFN-γ receptors, certain Fc receptors and some complement proteins (see ref. 80 and references therein).

Glucocorticoids upregulate some cellular functions associated with antigen presentation to T-lymphocytes. Dexamethasone, or hydrocortisone, act in synergy with GM-CSF or IL-3 to upregulate MHC class I and II mRNA and cell surface antigen expression on human blood monocytes.[80,81] This effect is as potent as the classically described IFN-γ pathway but appears independent of IFN-γ as assessed by the addition of neutralizing antibodies in cultures of highly purified human monocytes (Fig. 7.2) and failure to detect IFN-γ in culture supernatant. However, in contrast to IFN-γ, this pathway either fails to induce or inhibits costimulatory molecule expression and may differentially regulate antigen-processing events. These cells are therefore likely to have very different abilities to stimulate T-cell activation. Thus it seems probable that the overall effect of glucocorticoids will involve complex modulation rather than gross inhibition of immune/inflammatory function, an important consideration with respect to the design of new glucocorticoid derivatives and improvement to current treatment regimens.

A. HLA-DR

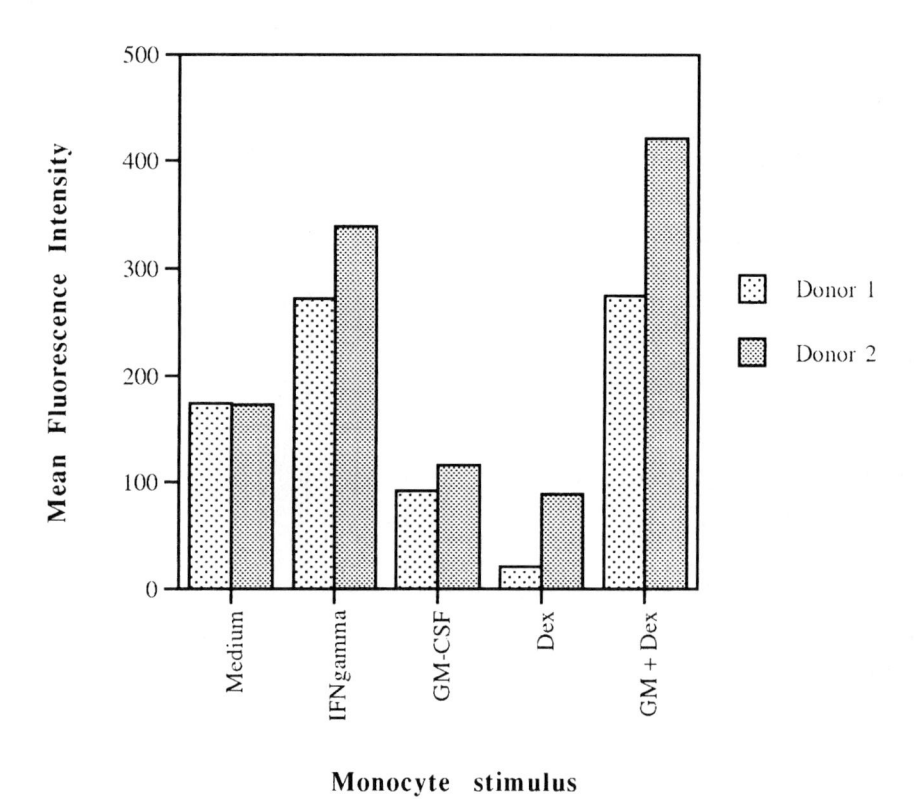

Fig. 7.2 Monocytes purified from human blood (>95%) were cultured for 3 days with combinations of 50 units/ml granulocyte–macrophage colony-stimulating factor (GM-CSF), 10^{-7} M dexamethasone (Dex) and 200 ng/ml interferon (IFN)-γ in the presence or absence of control or anti-IFN-γ (μg/ml) antibodies. Expression of HLA-DR or HLA-DQ antigen expression was assayed by immunostaining and analysis by flow cytometry. Similar results were obtained for HLA-DP (data not shown). (A) GM-CSF and dexamethasone induce comparable levels of MHC class II expression to IFN-γ on monocytes purified from human peripheral blood. Data from two representative donors is shown. (B) See page 136.

In contrast to the efficacy of glucocorticoids in most individuals, a minority of patients present with a steroid-resistant phenotype.[82] These individuals have a chronic severe disease with no improvement of airflow obstruction even with high doses of glucocorticoids. The defect does not appear to be due to abnormal absorption or clearance of glucocorticoids, and steroid-dependent metabolic functions appear normal. Instead, defective responses of monocytes and T-cells have been reported (reviewed in ref. 83). Recent studies using peripheral blood cells suggest that the defect is not due to large differences in receptor number, receptor–ligand binding or affinity, but rather that there is reduced binding to DNA and abnormal binding to transcription factors such as AP-1 by the glucocorticoid receptor.[84]

B. HLA-DQ

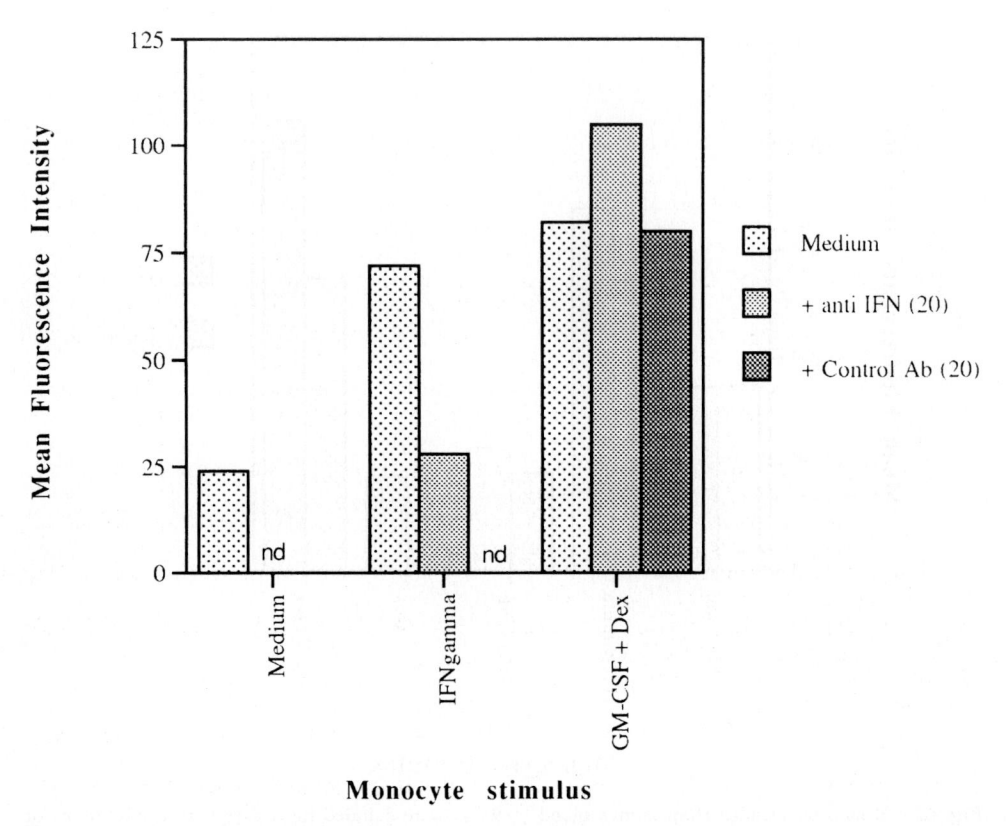

Fig. 7.2 (B) Induction of MHC class II antigens of human monocytes by GM-CSF and dexamethasone is IFN-γ independent.

CONCLUSION

We have attempted to provide a broad overview of the potential role of cells of the mononuclear phagocyte system in asthmatic disease. These cells secrete a large number of pro-inflammatory mediators likely to contribute to the disease process. Whilst this function is not unique to these cells, their capacity to synthesize large quantities of these products suggests that they are likely to contribute to asthmatic inflammation. A second function, the capacity to present antigen to T-lymphocytes, represents a primary action of monocytes, macrophages and DCs in asthma. Studies reveal that alveolar macrophages show a highly specialized APC phenotype and changes in this capacity play a central role in the disease process. Finally, studies with glucocorticoids both *in vitro* and *in vivo* demonstrate complex actions of these mediators on monocytes and macrophages and highlight the likely importance of the mononuclear phagocyte system in determining lung homeostasis and induction of the disease process.

7 Monocytes, Macrophages and Dendritic Cells

REFERENCES

1. Brain JD: Lung macrophages: how many kinds are there? What do they do? *Am Rev Respir Dis* (1988) **137**: 507–509.
2. Lohmann-Matthes M-L, Steinmuller C, Franke-Ullman G: Pulmonary macrophages. *Eur Respir J* (1994) **7**: 1678–1689.
3. Bilyk N, Upham JW, Holt PG: Pulmonary macrophages. In Kradin RL, Robinson BWS (eds) *Immunopathology of Lung Disease*. Boston, Butterworth-Heinemann, 1996, pp 57–72.
4. McLennan G, Walsh RL, Robinson BWS: Bronchoalveolar lavage. In Kradin RL, Robinson BWS (eds) *Immunopathology of Lung Disease*. Boston, Butterworth-Heinemann, 1996, pp 529–540.
5. Dunnill MS: The pathology of asthma with special reference to the bronchial submucosa. *J Clin Pathol* (1960) **13**: 27–34.
6. Metzger WJ, Zavala D, Richerson HB, et al.: Local allergen challenge and BAL of allergic asthmatic lungs. *Am Rev Respir Dis* (1987) **135**: 433–440.
7. Jeffery PK, Wardlaw AJ, Nelson FC, et al.: Bronchial biopsies in asthma: an ultrastructural, quantitative study and correlation with hyperreactivity. *Am Rev Respir Dis* (1989) **140**: 1745–1753.
8. Holgate ST, Djukanovic R, Wilson J, et al.: Inflammatory processes and bronchial hyperresponsiveness. *Clin Exp Allergy* (1991) **21** (Suppl 1): 30–36.
9. Laitinen LA, Laitinen A, Haahtela T: Airway mucosal inflammation even in patients with newly diagnosed asthma. *Am Rev Respir Dis* (1993) **147**: 697–704.
10. Bentley AM, Durham SR, Kay AB: Comparison of the immunopathology of extrinsic, intrinsic and occupational asthma. *J Allergy Clin Immunol* **4**: 222–232.
11. Poston RN, Chanez P, Lacoste JY, et al.: Immunohistochemical characterization of the cellular infiltration in asthmatic bronchi. *Am Rev Respir Dis* (1992) **145**: 918–921.
12. Mattoli S, Mattoso VL, Soloperto M, et al.: Cellular and biochemical events of BAL in symptomatic non-allergic asthma. *J Allergy Clin Immunol* (1991) **87**: 794–802.
13. Corrigan CJ, Kay AB: T cells and eosinophils in the pathogenesis of asthma. *Immunol Today* (1992) **13**: 501–506.
14. Bochner BS, Undem BJ, Lichtenstein LM: Immunological aspects of allergic asthma. *Annu Rev Immunol* (1994) **12**: 295–335.
15. Nathan CF: Secretory products of macrophages. *J Clin Invest* (1987) **79**: 319–326.
16. Bancroft GJ, Kelly JP, Kaye PM, et al.: Pathways of macrophage activation and innate immunity. *Immunol Lett* (1994) **43**: 67–70.
17. Adams DO, Hamilton TA: The cell biology of macrophage activation. *Annu Rev Immunol* (1987) **2**: 283–318.
18. Zurawski G, de Vries JE: IL-13, an IL-4-like cytokine that acts on monocytes, B cells, but not T cells. *Immunol Today* (1994) **15**: 19–26.
19. Chanez P, Vignola AM, Paul-Eugene N et al.: Modulation by IL-4 of cytokine release from mononuclear phagocytes in asthma. *J Allergy Clin Immunol* (1994) **94**: 997–1005.
20. Spiegelberg HL: Structure and function of Fc receptors for IgE on lymphocytes, monocytes and macrophages. *Adv Immunol* (1984) **135**: 61–88.
21. Williams J, Johnson S, Mascali JJ, et al.: Regulation of CD23 on mononuclear phagocytes in normal and asthmatic subjects. *J Immunol* (1992) **149**: 2823–2829.
22. Melewicz FM, Zeiger RS, Mellon MH, et al.: Increased peripheral blood monocytes with Fc receptors for IgE in patients with severe allergic disorders. *J Immunol* (1981) **126**: 1592–1595.
23. Melewicz FM, Kline LE, Cohen AB, et al.: Characterization of Fc receptors for IgE on human alveolar macrophages. *Clin Exp Immunol* (1982) **49**: 364–370.
24. Kawabe T, Takami M, Hosoda M, et al.: Regulation of FcR2/CD23 gene expression by cytokines and specific ligands (IgE and anti-FcER2 monoclonal antibody). *J Immunol* (1988) **141**: 1376–1382.
25. Sousa AR, Poston RN, Lane SJ, et al.: Detection of GM-CSF in asthmatic bronchial epithelium and decrease by inhaled corticosteroids. *Am Rev Respir Dis* (1993) **147**: 1557–1561.

26. Matz J, Williams J, Rosenwasser LJ, et al.: GM-CSF stimulates macrophages to respond to IgE via the low affinity IgE receptor (CD23). *J Allergy Clin Immunol* (1994) **93**: 650–657.
27. Borish L, Mascali JJ, Rossenwasser LJ: IgE dependent cytokine production by human monocytes. *J Immunol* (1991) **146**: 63–67.
28. Gosset P, Tsicopoulos A, Wallaert B, et al.: TNFα and IL-6 production by human mononuclear phagocytes from allergic asthmatics after IgE-dependent stimulation. *Am Rev Respir Dis* (1992) **146**: 768–774.
29. Ferreri NR, Howland WC, Spiegelberg H: Release of leukotrienes C4 and B4 and PGE2 from human monocytes stimulated with aggregated IgG, IgA and IgE. *J Immunol* (1986) **136**: 4188–4193.
30. Demoly P, Vachier I, Pene J, et al.: IgE produces monocytes superoxide anion release: correlation with CD23 expression. *J Allergy Clin Immunol* (1994) **93**: 108–116.
31. Maurer D, Ebner C, Reiningen B, et al.: 1995. The high affinity IgE receptor (FcεRI) mediates IgE-dependent allergen presentation.
32. Kay AB, Diaz P, Carmichael J, et al.: Corticosteroid resistant chronic asthma and monocyte complement receptors. *Clin Exp Immunol* (1981) **44**: 576–580.
33. Wilkinson JR, Lane SJ, Lee TH: Effects of corticosteroids on cytokine generation and expression of activation antigens by monocytes in bronchial asthma. *Int Arch Allergy Appl Immunol* (1991) **94**: 220–221.
34. Poulter LW, Janossy G, Power C, et al.: Immunological/physiological relationships in asthma: potential regulation by lung macrophages. *Immunol Today* (1994) **15**: 258–261.
35. Broide DH, Lotz M, Cuomo AJ, et al.: Cytokines in symptomatic asthma airways. *J Allergy Clin Immunol* (1992) **89**: 958–967.
36. Gosset P, Tsicopoulos A, Wallaert B, et al.: Increased secretion of TNFα and IL-6 by alveolar macrophages consecutive to LAR. *J Allergy Clin Immunol* (1991) **88**: 561–571.
37. Borish L, Mascali JJ, Dishuck J, et al.: Detection of alveolar macrophage-derived IL-1β in asthma. Inhibition with corticosteroids. *J Immunol* (1992) **149**: 3078–3082.
38. Hallsworth MP, Soh CP, Lane SJ, et al.: Selective enhancement of GM-CSF, TNF-alpha, IL-1-beta and IL-8 production by monocytes and macrophages of asthmatic subjects. *Eur Respir J* (1994) **7**: 1096–1102.
39. Taylor IK, Sorooshian M, Wangoo A, et al.: PDGFβ mRNA in human alveolar macrophages *in vivo* in asthma. *Eur Respir J* (1994) **7**: 1966–1972.
40. Lassalle P, Gosset P, Delneste Y, et al.: Modulation of adhesion molecule expression on endothelial cells during the late asthmatic reaction: role of macrophage-derived TNFα. *Clin Exp Immunol* (1993) **94**: 105–110.
41. Vassalli P: The pathophysiology of TNF. *Annu Rev Immunol* (1992) **10**: 411–420.
42. Dayer JM: IL-1, TNF and their specific inhibitors. *Eur Cytokine Netw* (1994) **5**: 563–571.
43. Lukacs NW, Strieter RM, Chensue SW, et al.: Activation and regulation of chemokines in allergic airway inflammation. *J Leukoc Biol* (1996) **59**: 13–17.
44. Drazen JM, Arm JP, Austen KF: Sorting out the cytokines of asthma. *J Exp Med* (1996) **183**: 1–5.
45. Oppenheim JJ, Zachariae COC, Mukaida N, et al.: Properties of the novel proinflammatory 'intercrine' cytokine family. *Annu Rev Immunol* (1991) **9**: 817–848.
46. Cruikshank WW, Long A, Tarpy RE, et al.: Early identification of IL-16 and MIP-1α in BAL of antigen-challenged asthmatics. *Am J Respir Cell Mol Biol* (1995) **13**: 738–747.
47. Kurashima K, Mukaida N, Fujimura M, et al.: Increase of chemokine levels in sputum precedes exacerbation of acute asthma attacks. *J Leukoc Biol* (1996) **59**: 313–316.
48. Nasser SM, Lee TH: Lipid mediators. In Kay AB (ed) *Allergy and Allergic Diseases*, vol. 1. Oxford, Blackwell Science, 1997, pp 380–417.
49. Clark SC, Kamen R: The human haematopoietic colony-stimulating factors. *Science* (1987) **236**: 1229–1237.
50. Sousa AR, Lane SJ, Nakhosteen JA, et al.: The expression of IL-1β and IL-1ra on asthmatic bronchial epithelium. *Am J Respir Crit Care Med* (1996) **154**: 1061–1066.
51. Aubert JD, Dalal BI, Bai TR, et al.: TGFβ1 gene expression in human airways. *Thorax* (1994) **49**: 225–232.

7 Monocytes, Macrophages and Dendritic Cells

52. Borish L, Aarons A, Rumbyrt J, et al.: Interleukin-10 regulation in normal subjects and patients with asthma. *J Allergy Clin Immunol* (1996) **97**: 1288–1296.
53. Bigby T, Holtzman MJ: Enhanced 5-lipoxygenase activity in lung macrophages compared to monocytes from normal subjects. *J Immunol* (1987) **138**: 1546–1550.
54. Balter MS, Toews GB, Peters-Golden M: Different patterns of arachidonate metabolism in autologous human blood monocytes and alveolar macrophages. *J Immunol* (1989) **142**: 602–608.
55. Damon M, Chavis C, Daures JP: Increased generation of the arachidonic acid metabolites LTB$_4$ and 5-HETE by human alveolar macrophages in patients with asthma. *Eur Respir J* (1989) **2**: 202–209.
56. Williams JD, Czop JK, Austen KF: Release of leukotrienes by human monocytes on stimulation of their phagocytic receptor for particulate activators. *J Immunol* (1984) **132**: 3034–3040.
57. MacDermot J, Kelsey CR, Waddell KA, et al.: Synthesis of LTB4 and prostanoids by human alveolar macrophages: analysis by gas chromatography/mass spectrometry. *Prostaglandins* (1984) **27**: 163–179.
58. Murray JJ, Tonnel AB, Brash AR, et al.: Release of prostaglandin D2 into human airways during acute antigen challenge. *N Engl J Med* (1986) **315**: 800–804.
59. Wenzel SE, Westcott JY, Smith HR, et al.: Spectrum of prostanoid release after bronchoalveolar allergen challenge in atopic asthmatics and control groups. *Am Rev Respir Dis* (1989) **139**: 450–457.
60. Alam R, Welter J, Forsythe P, et al.: Detection of histamine release inhibitory factors and histamine releasing factor-like activities in BAL. *Am Rev Respir Dis* (1990) **141**: 666–671.
61. McClean DSP, Forsythe PA, Grant JA, et al.: Synthesis of histamine release inhibitory factors and histamine releasing factor-like activities in BAL cells. *J Allergy Clin Immunol* (1989) **83**: 236A.
62. Vachier I, Le Douchen C, Loubatiere J, et al.: Imaging reactive oxygen species in asthma. *J Biolumin Chemilumin* (1994) **9**: 171–175.
63. Calhoun WJ, Reed HE, Moest DR, et al.: Increased superoxide by alveolar macrophages and air-space cells after bronchoprovocation. *Am Rev Respir Dis* (1992) **145**: 317–325.
64. Barnes PJ: Nitric oxide and airway disease. *Ann Med* (1995) **27**: 389–393.
65. Cresswell P: Assembly, transport, and function of MHC class II molecules. *Annu Rev Immunol* (1994) **12**: 259–293.
66. June CH, Bluestone JA, Nadler LM, et al.: The B7 and CD28 receptor families. *Immunol Today* (1994) **15**: 321–331.
67. Lipscomb MF, Lyons GCR, Nunez EJ, et al.: Human alveolar macrophages: HLA-DR-positive macrophages that are poor stimulators of a primary MLR. *J Immunol* (1986) **136**: 497–504.
68. Gant V, Cluzel M, Shakoor P, et al.: Alveolar macrophage accessory cell function in bronchial asthma. *Am Rev Respir Dis* (1992) **146**: 900–904.
69. Spiteri MA, Poulter LW: Characterization of immune inducer and suppressor macrophages from the normal human lung. *Clin Exp Immunol* (1991) **83**: 157–162.
70. Van Haarst JM, Hoogsteden HC, de Wit HJ, et al.: Dendritic cells and their precursors isolated from human BAL: immunocytologic and functional properties. *Am J Respir Cell Mol Biol* (1994) **11**: 344–350.
71. Chelen CJ, Fang Y, Freeman GJ, et al.: Human alveolar macrophages present antigen ineffectively due to defective expression of B7 costimulatory cell surface molecule. *J Clin Invest* (1995) **95**: 1415–1421.
72. Schneeberger EE, Gong JL: Dendritic cells in the lung. In Kradin RL, Robinson BWS (eds) *Immunopathology of Lung Disease*. Boston, Butterworth-Heinemann, 1996, pp 73–89.
73. Peters JH, Gieseler R, Thiele B, et al.: Dendritic cells: from ontogenetic orphans to myelomonocytic descendents. *Immunol Today* (1996) **17**: 273–278.
74. Holt PG, Oliver J, Bilyk N, et al.: Downregulation of the antigen presenting cell function(s) of pulmonary DC *in vivo* by resident alveolar macrophages. *J Exp Med* (1993) **177**: 397–407.
75. Barnes PJ: Inhaled glucocorticoids for asthma. *N Engl J Med* (1995) **332**: 868–873.

76. Mason D: Genetic variation in the stress response. *Immunol Today* (1991) **12**: 57–60.
77. Calandra T, Bernhagen J, Metz CN, *et al*.: MIF as a glucocorticoid-induced modulator of cytokine production. *Nature* (1995) **377**: 68–71.
78. Sigola L, Guida L, Hawrylowicz CM *et al*.: Glucocorticoids differentially regulate macrophage production of TNF versus IL-10 in response to *Mycobacterium bovis* (ms in preparation).
79. Daynes RA, Araneo BA: Contrasting effects of glucocorticoids on the capacity of T cells to produce the growth factors interleukin 2 and interleukin 4. *Res Immunol* (1991) **142**: 40–45.
80. Hawrylowicz CM, Guida L, Paleolog E: Dexamethasone upregulates GM-CSF receptor expression on human monocytes. *Immunology* **83**: (1994) 274–280.
81. Sadeghi R, Feldmann M, Hawrylowicz CM: Upregulation of HLA class II but not ICAM-1 by GM-CSF or IL-3 in synergy with dexamethasone. *Eur Cytokine Netw* (1991) **3**: 373–378.
82. Cypcar D, Busse WW: Steroid-resistant asthma. *J Allergy Clin Immunol* (1993) **92**: 362–372.
83. Lane SJ, Lee TH: Corticosteroid resistant asthma. In Chanez P (ed) *From Genetics to Quality of Life*. Proc XVth World Congress of Asthmology, Montpellier, April 1996. Oxford, Blackwell Scientific Publications, pp 52–57.
84. Adcock IM, Lane SJ, Brown CR, *et al*.: Abnormal glucocorticoid receptor–activator protein-1 interaction in steroid-resistant asthma. *J Exp Med* (1995) **182**: 1951–1958.

8

Eosinophils

PER VENGE

INTRODUCTION

The eosinophil granulocyte is a myeloid cell produced in the bone marrow. The mature cell has a bilobed nucleus and an abundance of various-sized granules; hence the classification of the cell as a granulocyte. Some of the granules contain a typical crystal structure when examined at the electron microscopic level. It has been estimated that between 10^{10} and 10^{11} eosinophils are produced each day. Some of the cells may never leave the bone marrow, but the majority is released to the circulation where the half-life is 13–18 h.[1–3] Most eosinophils enter the tissues and stay there for another several days to weeks. It is believed that for each cell present in the circulation there are 100–300 in the tissues. Thus, the eosinophil should probably be regarded as a tissue cell. Accumulation of eosinophils normally occurs in the skin and the intestine, and to a minor extent also in the airways. At these locations the eosinophils most likely take part in the primary defence against intruding microbes. Production of eosinophils in the bone marrow is regulated by a number of growth factors, such as interleukin (IL)-3, IL-5 and granulocyte–macrophage colony-stimulating factor (GM-CSF).[4,5] Of these growth factors, IL-3 and GM-CSF are important for the expansion of the stem cell population, whereas IL-5 seems to be the growth factor particularly responsible for the actual commitment of the myeloid cells to an eosinophilic direction. It is noteworthy that administration of GM-CSF to humans regularly produces a profound eosinophilia.[6] In addition to these growth factors, granulocyte colony-stimulating factor (G-CSF) seems to affect eosinophil behaviour, since administration of this growth factor produces a two- to four-fold increase in blood eosinophil numbers and changes in the state of activation of the cells.[7]

The eosinophil was discovered and named by the German biologist Paul Ehrlich in 1879 due to the affinity of the cell to the acid dye eosin. This discovery created an immense interest in the eosinophil for some decades. It was found that in many diseases the number of eosinophils in the blood was greatly increased. These diseases included parasitic infestations, asthma and other allergic disorders, cancer, etc. The early interpretation of these data was that eosinophils might play a role in parasite defence, a notion that is very much supported by fairly recent findings. However, the interpretation of the presence of eosinophils in the lung in asthma and the raised numbers in blood in allergic diseases was rather that the eosinophil played a protective role against the harmful factors operative in these diseases. In fact this view of the eosinophil was still common in major medical textbooks during the late 1980s. Even if the role of the eosinophil is somewhat enigmatic, the current view is that it is a pro-inflammatory cell with a substantial tissue destructive potency. It was also noted early on that the number of eosinophils was reduced in blood in diseases such as acute infections and during acute myocardial infarctions. Together with a much later observation, that the number was reduced during treatment with high doses of corticosteroids, this led to the conclusion that during such conditions eosinophils were redistributed in the body due to endogenous production of cortisol. This is probably only partly true, since we know today that eosinophils are quickly attracted to sites of inflammation and actively participate in the process. Thus, the reason for the transient eosinopenia during acute inflammation is quick drainage of blood eosinophils to the site of inflammation and an insufficient but increased bone marrow production to compensate for the loss. Typically, when the inflammatory process has subsided, the eosinopenia is followed by a transient eosinophilia, due to the continued increase of bone marrow production but a normalization of the turnover of eosinophils. After the initial enthusiasm with the discovery of the eosinophil during the last century, interest faded away and very little attention was devoted to the eosinophil over the next 60–70 years. However, today several hundred research groups worldwide are studying various aspects of the eosinophil, mostly in relation to its role in asthma and allergy. According to Medline, the number of reports published on eosinophils in 1970 was 23, while during 1996 it was 542.

BIOCHEMISTRY AND FUNCTION

The eosinophil is primarily a secretory cell. The biological activities exerted by the eosinophil are related to the biological activities of the products produced by and released from the cell.[1,3,8] Basically one can distinguish between those mediators produced in the bone marrow during eosinophil development, stored in the secretory granules and released from the eosinophil after exposure to various secretagogues and those mediators produced and released as a direct consequence of activation of the mature cell in the peripheral blood or in tissues. The preformed mediators are stored in the eosin-stained granules of the eosinophil and may be separated into several subpopulations. One major subpopulation is peroxidase positive and another peroxidase negative, i.e. they either store or lack the presence of eosinophil peroxidase (EPO). The peroxidase-positive granules also store the other three major proteins of the eosinophil: eosinophil cationic protein (ECP), eosinophil protein X/eosinophil-derived neurotoxin (EPX/EDN) and major basic protein (MBP). In contrast, the peroxidase-negative granules seem to store only ECP and EPX/EDN. The peroxidase-positive granules are the heavier granules, probably due to the presence of the typical crystal structure in the granules, which is formed by MBP. One characteristic of the four proteins is their cationic charge and high isoelectric point, between pH 10 and 11. These extreme isoelectric points make them heavily and positively charged in the near-neutral environment in tissues and also make them very sticky and apt to bind to negatively charged molecules such as those found on cell membranes. Two of the proteins, ECP and EPX/EDN, are closely related based on extensive homologies found in their amino acid sequences. As described below, these two proteins also share a number of biological activities. The close homology is also demonstrated by the monoclonal antibody EG2, which recognizes a common epitope on the two proteins.[9]

Among the non-preformed mediators are reactive oxygen species and various lipid mediators. Indeed the eosinophil is a very capable producer of such mediators and fully comparable, if not superior, to the neutrophil in the production of reactive oxygen species such as O_2^-, H_2O_2 and $OH^{\cdot}$. As to the production of lipid mediators such as prostaglandins (PGE_2), leukotrienes (LTC_4) and platelet-activating factor (PAF), the eosinophil may actually be as capable as mast cells and basophils.

In addition to the mediators described above, the human eosinophil contains in its granules several other enzymatic activities of unclear biological relevance. These include a gelatinase and an arylsulphatase B, and histaminase and phospholipase activities.[10] The presence of these activities was taken as an indication of a regulatory role of eosinophils in allergy, since these enzymes have the potential to inactivate many of the putative mediators of the allergic reaction, such as histamine, PAF and leukotrienes. Whether these mechanisms are operative *in vivo* remains enigmatic. Another group of proteins produced and stored by the human eosinophil is the group of molecules collectively called cytokines. Thus, recent findings have shown that eosinophils produce mRNA for cytokines such as IL-2–IL-6, IL-8, transforming growth factor (TGF)-α, TGF-β, GM-CSF and also express and produce the respective cytokines and store them in their granules for subsequent release.[11–16] This capacity of the eosinophil emphasizes its substantial potential to interfere in many central processes of inflammation. It should also be emphasized that most of these cytokines act as growth factors for eosinophils and activators of the mature eosinophil. Thus, by autocrine and paracrine mechanisms the eosinophils may greatly affect their own degree of activation and behaviour in the body.

Other molecules produced by the eosinophil, that may be of considerable interest are the Charcot–Leyden crystal (CLC) protein[17,18] and the α-subunit of the high-affinity IgE receptor (T. Bjerke, personal communication). The CLC protein, which presumably is a plasma membrane protein, is shed from eosinophils and forms typical extracellular needle-like crystals in tissues with heavy eosinophil infiltration. The CLC protein is a lysophospholipase and has interesting biological activities, which may be important in allergic inflammation. The function of the storage and secretion of the α-subunit of this high-affinity IgE receptor is not yet clear, but may be related to the regulation of IgE synthesis.

Molecular characteristics and biological activities of the four major granule proteins

Eosinophil cationic protein

ECP is a one-chain, zinc-containing protein with a molecular mass varying from 16 to 22 kDa.[19,20] This heterogeneity is partly due to differences in glycosylation of the molecule. At least six distinct variants have been recognized in preparations of ECP originating from pools of cells from several hundred blood donors (P. Venge, unpublished results). ECP has also been named 'ribonuclease 3' (RNase 3) due to homologies with both human RNases and RNases of other vertebrate origins.[21] ECP displays a 70% amino acid sequence homology with EPX/EDN. Besides being an RNase, ECP is a potent cytotoxic molecule with the capacity to kill mammalian as well as non-mammalian cells such as parasites. The cytotoxic activity of ECP is attributed to one particular variant, which is uniquely identified by a monoclonal antibody produced against ECP, Mab 652. The mechanism of cytotoxicity is due to the capacity of ECP to make pores in cell membranes[22] (Fig. 8.1). These pores are about 5 nm wide and allow the passage of water and smaller molecules, resulting in osmotic lysis of the affected cell. The non-cytotoxic properties of ECP include the alteration of glycosaminoglycan production and storage by human fibroblasts[23] and the induction of receptors for insulin-like growth

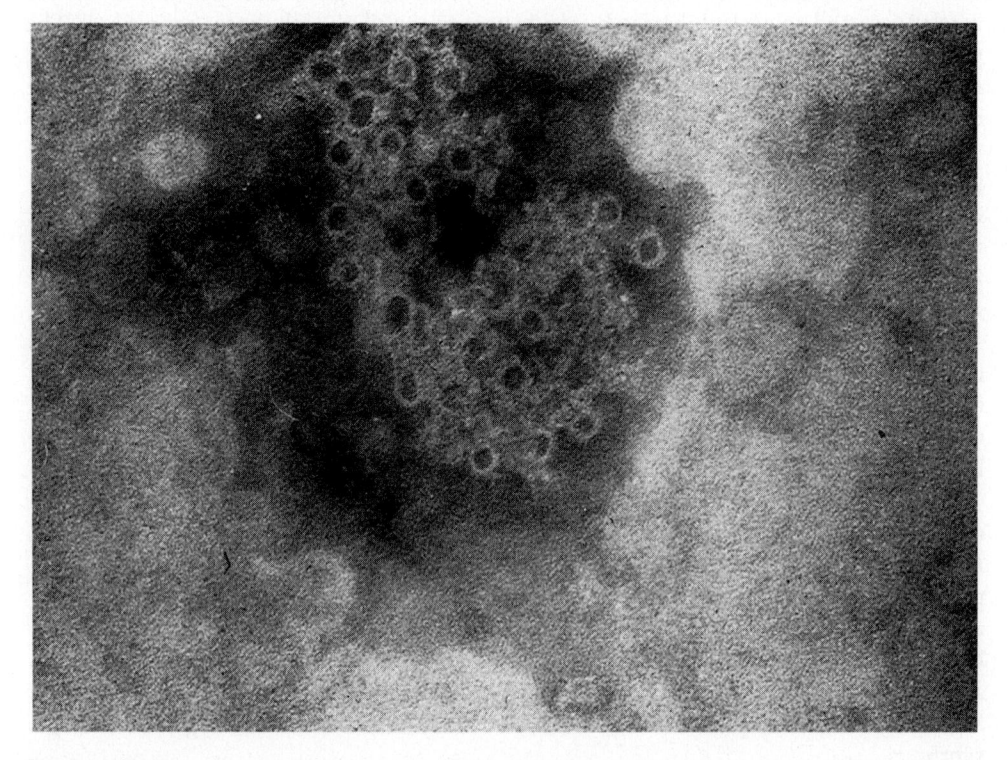

Fig. 8.1 Electron microscopic demonstration of the effect of eosinophil cationic protein (ECP) on cellular membranes. Addition of ECP to artificial lipid membranes produces pores with a diameter of about 5 nm (J. Ding-Young and P. Venge, unpublished results).

factor (IGF)-1 on bronchial epithelial cells.[24] These findings point to a role for ECP in tissue repair processes and may explain the presence of eosinophils in fibrotic lesions. ECP also stimulates airway mucus secretion,[25] a finding that may be of importance for our understanding of the role of eosinophils in diseases such as asthma, since asthma is characterized by, among other things, hypersecretion of the airways. Furthermore, ECP increases microvascular transport of macromolecules.[26] Another finding of potential interest is the capacity of ECP to inhibit T-lymphocyte proliferation and to inhibit immunoglobulin synthesis by B-cells,[27-29] since such findings emphasize the potential of the eosinophil as an immunomodulator. An early finding was the observation that ECP shortens the coagulation time *in vitro* by a mechanism related to the activation of coagulation factor XII.[30,31] Such studies were performed in an attempt to explain the findings in patients with the hypereosinophilic syndrome, in whom thromboembolic phenomena are common. It is also noteworthy that factor XII is activated in blood after allergen challenge of allergic asthmatic subjects along with the activation of eosinophils. Another study showed that ECP enhanced fibrinolysis.[32] The mechanism was preactivation of plasminogen, resulting in the enhancement of plasminogen activator activation of plasminogen. However, the biological meaning of this finding remains uncertain. Recent findings indicate that ECP may activate human mast cells and basophils and induce the secretion of both histamine and tryptase, emphasizing the close link between the activation of eosinophils and these cells in humans.[33] The mechanisms involved in extracellular protection against the cytotoxic actions of ECP are not clear. *In vitro* ECP binds to heparin in a one-to-one molecular fashion.[34] ECP specifically binds to α_2-macroglobulin, but only after brief exposure of α_2-macroglobulin to various proteases such as cathepsin G and thrombin.[35] In very recent studies we have also shown that ECP is taken up by neutrophils and stored in these cells (P. Venge, unpublished results).

Eosinophil protein X and eosinophil-derived neurotoxin

EPX and EDN are two names for the same protein.[36-38] Since there has been no agreement as to the nomenclature, the protein is provisionally called EPX/EDN. EPX/EDN is a one-chain basic protein, but which is less basic than ECP. The molecular mass is 18 kDa and, as opposed to ECP, only one variant has been described. As indicated above, EPX/EDN and ECP share a large degree of amino acid sequence homology. EPX/EDN is a far more potent RNase than ECP,[39] but much less cytotoxic when tested against cancer cells *in vitro*. Also, the neurotoxic activity of EPX/EDN is far less than that of ECP, even though EDN is named because of this activity. Thus, when injected into the brains of experimental animals EPX/EDN produces damage to the tissues reminiscent of the so-called Gordon phenomenon, i.e. destruction of Purkinje cells of the cerebellum and the development of ataxia. However, the concentrations needed to achieve these effects were about 100 times higher than those needed for ECP.[40] Such data also indicate that there is no relationship between the cytotoxic activities and the RNase activities of the two molecules. In analogy with ECP, EPX/EDN also inhibits T-lymphocyte proliferation in a non-cytotoxic fashion and at concentrations similar to those of ECP. The effect of EPX/EDN on parasites, such as the larvae of *Schistosoma mansoni*, is distinct from that of ECP, since EPX/EDN does not kill the parasite. In contrast, EPX/EDN seems to reversibly paralyse the movements of the parasite, a phenomenon that could be an important defence mechanism and facilitate eradication of the parasite.[41]

Eosinophil peroxidase

EPO is a two-chain protein with a total molecular mass of 67 kDa.[42] The light chain has a molecular mass of 15 kDa and the heavy chain a molecular mass of 52 kDa. EPO is a very basic and sticky protein and its functions are largely attributed to its peroxidase activity, although a number of peroxidase-independent activities have been suggested as well. Together with halides or SCN^- and H_2O_2, EPO constitutes a potent cytotoxic mechanism, which among other things kills parasites.[43] However, together these molecules may also cause increased microvascular permeability in a non-cytotoxic fashion.[44] The peroxidase activity of EPO is involved in the inactivation of some lipid mediators, notably the leukotrienes. The presumably peroxidase-independent biological activities of EPO include damage to nasal sinus mucosa. EPO was also shown to induce mast cell release of histamine and cause platelet aggregation.[44] The latter finding may be relevant for the signs and symptoms of patients with hypereosinophilia and their propensity for thrombotic manifestations. An interesting finding, which may have relevance for the role of eosinophils in asthma, is the demonstration that EPO functions as an allosteric antagonist of the muscarinic M_2 receptor.[1] EPO, together with MBP, may act as autocrine inducers of eosinophil degranulation. Thus, secretion of EDN was potently induced by EPO, as was the production and secretion of the cytokine IL-8 from eosinophils.[45] This cytokine is a strong chemoattractant for neutrophils, but also for blood eosinophils obtained from allergic subjects. Neutrophils take up EPO by a specific and probably receptor-dependent mechanism. This uptake could be an important regulatory mechanism, which actively neutralizes the toxic effects of EPO. However, EPO also increases the adhesiveness of neutrophils. Thus, there seem to be important links between eosinophils and neutrophils that may be mediated partly by EPO.

Major basic protein

The name 'major basic protein' derives from the fact that in guinea-pig eosinophils this is the dominant protein, making up about 50% of the content in the granules.[46,47] MBP forms the typical crystalloid structures in human eosinophil granules. MBP is a single-chain protein with a molecular mass of 13.9 kDa. It is a very basic protein (isoelectric point about pH 11) and exists in the cell partly as a preprotein. MBP is not entirely unique to the eosinophil, since it is also found in basophils and placenta cells.[1] The major biological functions of MBP are related to the cytotoxic activities of the molecule and involve the killing and damage of a variety of parasites and mammalian cells such as pneumocytes, nasal mucosal cells, etc. In analogy to the other major eosinophil granule proteins, MBP also has some non-cytotoxic biological activities on various cells. These include the degranulation of mast cells and basophils, platelet aggregation, induction of neutrophil superoxide anion production, enhancement of the expression of the neutrophil receptors CR3 and p150,95, contraction of airway smooth muscle and inhibition of airway mucus production and secretion, and induction of the secretion of EPX/EDN and IL-8 from eosinophils.[1] In the context of asthma, MBP has some interesting effects on the respiratory epithelium, which may partly explain the development of the hyperresponsiveness of the airways of asthmatic subjects. Similar to ECP, MBP also causes increased microvascular transport of macromolecules and, also similar to EPO, MBP acts as an allosteric antagonist of the muscarinic M_2 receptor. MBP, and to some extent also the

8 Eosinophils

other eosinophil granule proteins, impairs thrombomodulin function, which may be one further mechanism underlying the thromboembolism seen in patients with hypereosinophilia.[48]

RECEPTORS AND DEGRANULATION

A biological response from the eosinophil may be initiated through a large number of different receptors. Table 8.1 lists some of the best-known inducers of eosinophil responses. Molecules can be identified that induce growth and differentiation, adhesion, migration, secretion, phagocytosis, apoptosis and anti-apoptosis, and priming. As seen from Table 8.1 some molecules affect several of these responses, particularly the eosinophil-specific cytokines, IL-5 and eotaxin.[49–51]

The receptors most commonly involved in the secretory response are probably IgE, IgA and C3b receptors, but lipid mediators such as platelet-activating factor (PAF) and cytokines such as GM-CSF may also induce eosinophil degranulation.[52–54] Secretion of granule proteins is brought about by several mechanisms: one is by granule fusion with the plasma membrane and another is by vesicular transport. These mechanisms are important in our understanding of the selective release of granule proteins, which may occur in response to certain stimuli. Granule fusion with the plasma membrane and reversed exocytosis occurs at the parts of the plasma membrane where the receptor–ligand interaction takes place.[55] This receptor–ligand-directed granule fusion is obligatory for granule protein secretion to occur and also explains why secretagogues such as the phorbol ester PMA, which activates eosinophils by intracellular mechanisms, is unable to cause any substantial eosinophil degranulation.

Priming is the phenomenon where some molecule makes the eosinophil more prone to react to a secondary stimulus. Priming is probably a very important regulatory process of eosinophil activity *in vivo*. In asthmatic individuals, circulating eosinophils are primed for

Table 8.1 Some inducers of eosinophil responses.

Inducers	Eosinophil response
Interleukin (IL)-2	Inhibition of chemotaxis
IL-3	Growth and differentiation, priming, anti-apoptosis
IL-5	Growth and differentiation, priming, migration, respiratory burst, anti-apoptosis
IL-8	Chemotaxis (not normal cells)
Granulocyte–macrophage colony-stimulating factor (GM-CSF)	Growth and differentiation, priming, degranulation, anti-apoptosis
RANTES	Chemotaxis, degranulation, respiratory burst
Eotaxin	Chemotaxis
Monocyte chemotactic protein (MCP)-3	Chemotaxis
Platelet-activating factor (PAF)	Chemotaxis, degranulation
Leukotriene (LT)B$_4$	Chemotaxis
Immune complexes	Degranulation, respiratory burst

an enhanced response to adhesion molecules, chemotactic signals and secretagogues.[56–58] This enhanced propensity of the eosinophil is a prerequisite for the attraction of the cell to the allergic inflammatory process in the asthmatic lung and for the enhanced secretion of granule proteins. It is likely that IL-5, possibly together with GM-CSF, is the most important priming molecule for eosinophils in asthma. From a mechanistic point of view, one important cellular consequence of the exposure of the eosinophil to priming molecules is an increased cellular uptake of glucose (P Venge, unpublished results) and upregulation of membrane receptors. Taken together these activities increase the readiness of the eosinophil to respond to a stimulus, partly because of the increased availability of necessary receptors and partly because of a more favourable energy situation.

MECHANISMS OF EOSINOPHIL ACCUMULATION

Accumulation of eosinophils at sites of allergic inflammation involves several steps (Fig. 8.2). One is the above-mentioned priming of the eosinophil for enhanced responsiveness to the molecules necessary. The next step is adhesion to the endothelial cells followed by transmigration through the vascular wall and subsequently migration through tissue structures. Eosinophil adhesion involves several distinct steps including specific molecular interactions at the surface of eosinophils and endothelial cells. Most important for selective eosinophil adhesion and transmigration is the interaction between the adhesion molecule very late antigen (VLA)-4 exposed on the eosinophil plasma membrane and vascular cellular adhesion molecule (VCAM)-1 on the endothelial surface.[59–61]

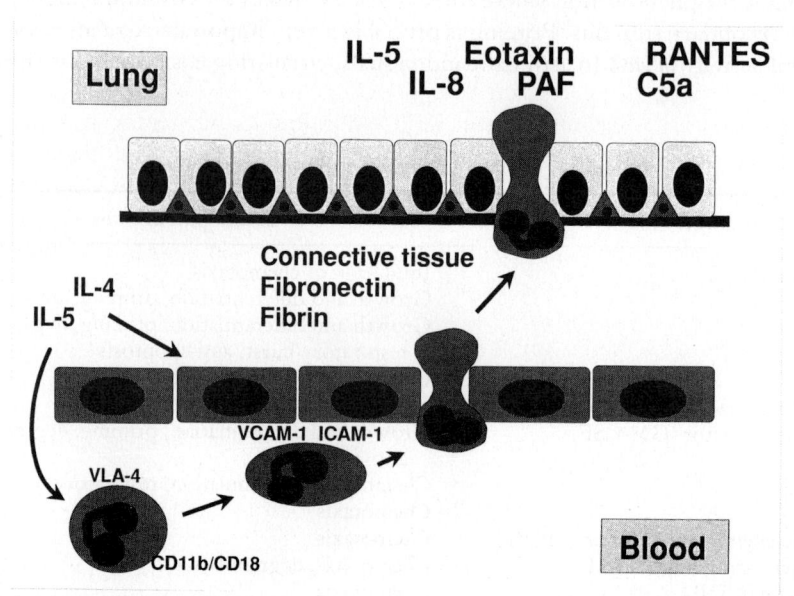

Fig. 8.2 A schematic illustration of the events leading to the accumulation of eosinophils in the lung in asthma.

Certain chemotactic molecules produced in the lung by T-lymphocytes, mast cells and some other cells govern the actual transmigration and further migration in tissues. In the allergic asthmatic lung, IL-5 in conjunction with the chemokines RANTES[62] and IL-8 have been shown to be responsible for the majority of the chemotactic activity (M. Lampinen and P. Venge, unpublished results). However, other molecules such as LTB_4, PAF and eotaxin may also be important.[63–67] An interesting finding is that IL-2 downregulates the response of eosinophils to most chemotactic molecules (M. Lampinen and P. Venge, unpublished results).

Alternative mechanisms of local accumulation of eosinophils have been proposed recently and involve the local proliferation of eosinophils and prolonged survival in the tissue.[68] The former hypothesis is based on the demonstration of increased numbers of circulating progenitors of eosinophils in patients with allergy[69] and the latter hypothesis on the finding that molecules such as IL-5 and GM-CSF are anti-apoptotic to eosinophils and thereby may increase their survival quite substantially. Increased production of both these cytokines has been demonstrated in the lungs of asthmatic subjects.[70,71]

THE EOSINOPHIL IN ASTHMA

As indicated above, the presence of eosinophils in the lungs of asthmatic patients has been known for more than a century, but the actual role of the eosinophil in asthma has been, and still is, unclear. One important early observation was the finding of a correlation between severity of asthma and blood eosinophil numbers, suggesting a causal role.[72] However, it was not until the mid-1980s, when the use of bronchoalveolar lavage and bronchial biopsies became common research tools in asthma, that a possible causal relationship was clearly suggested.[73,74] These studies include results after allergen provocation, but also with chronic severe asthma. In the first such study, a group of allergic asthmatic subjects were lavaged during the late asthmatic reaction. It was found that there was a clear relation between the development of the late asthmatic reaction and the number of eosinophils in the lavage fluid. Thus, eosinophil increase was only seen in those patients who developed a late asthmatic reaction and not in those who experienced an early asthmatic reaction only. Furthermore, the lavage of the same patients after allergen challenge, but several hours before the late reaction, showed no increase in eosinophil numbers, clearly indicating the active accumulation of eosinophils as part of the late asthmatic reaction. In studies on chronic asthma, the number of eosinophils in the lavage fluid and the concentration of ECP were significantly related to the severity of asthma. In this study, the infiltration of increased numbers of activated eosinophils was also seen adjacent to injured epithelium, suggesting a role for the eosinophil. In addition to these two pioneering studies, numerous studies with various designs have been performed, many of which have also used sputum as a source of eosinophils and eosinophil products.[75–104] All these studies have clearly shown the association of eosinophils and eosinophil activation in both allergic and non-allergic asthma. However, since no unique means exists today to specifically counteract the accumulation and activity of eosinophils in humans, the conclusion of a causal relationship between eosinophil accumulation and activity in the lung on the one hand and the development of asthma on the other is still based on circumstantial evidence. It should be emphasized that the suggestions from animal studies of such a causal relationship have to be interpreted with caution, since it is questionable whether any of these animal studies are relevant models of human asthma.

MONITORING OF ASTHMA BY EOSINOPHIL MARKERS

With the knowledge of an important role of the eosinophil in asthma it has become of interest to monitor the accumulation and activity of this cell in asthma. One such means is the estimation of the number of eosinophils in blood or sputum. Blood eosinophil count has been, and still is, an important clinical tool in asthma care. High blood counts suggest an ongoing asthmatic inflammation. A more direct measure of eosinophil involvement in asthma is sputum measurement.[84,98,105–107] Several studies have shown good relationships between severity of asthma and eosinophil numbers. The drawback of this method are difficulties in standardization of the sputum sampling procedure and the reproducibility of counting the eosinophils in sputum. Several ongoing studies are trying to solve these problems and suggest appropriate protocols. Another approach for eosinophil monitoring is to measure some of the products secreted from the eosinophil during activation.[108–110] Such an approach is very attractive, since the mere enumeration of cells does not provide any information on the extent of activation of the eosinophil population and since it is reasonable to believe that it is the extent of activation of the cells that may be related to asthma development. Sensitive immunoassays have been developed for all four major secretory granule proteins of the eosinophil, i.e. ECP, EPO, EPX/EDN and MBP. Of these proteins, the measurement of ECP is the method that has been most widely used, both for measurements in serum and for measurements in sputum and bronchoalveolar lavage fluid. Serum measurements have shown that levels of ECP are related to the degree of activation of the blood pool of eosinophils and that this activation in turn seems to be related to the process ongoing in the lung. It has been shown by numerous studies that serum measurements are related to the severity of asthma as estimated by various means, such as lung function, the development of late asthmatic reaction after allergen challenge or exercise-induced asthma. ECP has also been used successfully in the follow-up of asthmatic patients during anti-inflammatory treatment with inhaled corticosteroids and as an objective indicator of allergen exposure.[111,112] Measurements of such activation markers of eosinophils in asthma therefore seem to be interesting and useful clinical tools, which provide the clinician with unique information about asthmatic inflammation. Together with conventional means to judge the severity of the asthmatic disease and the efficacy of treatment, such information should guide clinicians to better care of their patients.

PHARMACOLOGICAL CONTROL OF THE EOSINOPHIL

Corticosteroids reduce the number of circulating eosinophils and serum ECP levels. Corticosteroids also reduce the accumulation of eosinophils in the lungs of asthmatic patients. *In vitro*, however, the effects of corticosteroids on eosinophil degranulation and migration are marginal or non-existent.[113] A direct inhibitory effect of corticosteroids on eosinophils, therefore, does not seem very likely. The inhibitory effect is probably indirect through inhibition of the production of molecules that activate eosinophils, such as T-lymphocyte production of IL-5 and other cytokines.[113] It has also been shown that corticosteroids inhibit upregulation of adhesion molecules on endothelial cells and the production of chemotactic molecules in the lung. Short-acting β_2-agonists were shown to reduce ECP levels and the number of blood eosinophils. The long-acting β_2-agonist salmeterol prevented the expected increase in serum ECP levels after allergen challenge in those patients who had a late asthmatic reaction.[114–116] Even though these effects were fairly marginal and transient, the data suggest that β_2-agonists may inhibit degranulation of eosinophils through a cyclic AMP-dependent pathway. In a 1-year study we found a significant reduction in serum ECP levels after treatment with oral theophylline.[111] Together with unpublished data that show a transient but profound reduction of serum ECP levels after intravenous injection of theophylline, this suggests an effect on eosinophil degranulation by this type of drug. Other drugs that may affect eosinophil activity are the cromoglycates and the antihistaminic drug ceterizine, although their effects *in vivo* seem to be weak. Finally, it should be mentioned that allergen hyposensitization has been shown to downregulate several aspects of eosinophil activities such as adhesion, migration and secretion.[58,117,118] These effects are most likely mediated through the reduced production of the Th2 cytokines.[58,117,118]

CONCLUSIONS

The eosinophil granulocyte is a bone marrow-derived cell that spends most of its lifetime in tissues. It is an inflammatory cell that has the capacity to secrete and produce a number of molecules with many different potent activities, of which the cytotoxic activities are the most conspicuous. Thus, by virtue of the secretion of the cytotoxic proteins ECP, EPO and MBP and the production of oxygen radicals, the eosinophil is capable of destroying most tissues including epithelial cells in the lung. It is most likely that the eosinophil is one of the cells of major importance for the development of asthma. The accumulation and activation of eosinophils in the lung are governed by the upregulation of adhesion molecules on lung endothelial cells and the production of various cytokines and chemotactic molecules by T-lymphocytes and other cells. Of these cytokines, IL-5 seems to play a central role, since it regulates most aspects of eosinophil behaviour, such as growth, apoptosis, adhesion, migration and secretion. The potent inhibitory effect of corticosteroids and allergen hyposensitization on these eosinophil activities is not only a direct effect on the eosinophil, but also indirect and probably mediated through inhibition of the production of such eosinophil-activating cytokines.

REFERENCES

1. Gleich GJ, Adolphson CR, Leiferman KM: The biology of the eosinophilic leukocyte. *Annu Rev Med* (1993) **44**: 85–101.

2. Spry CJF: *Eosinophils. A Comprehensive Review and Guide to the Scientific and Medical Literature.* Oxford, Oxford University Press, 1988.
3. Venge P, Bergstrand H, Håkansson L: Neutrophils and eosinophils. In Kelley WN, Harris ED, Ruddy S, Sledge CB (eds) *Textbook of Rheumatology*, 5th edn. Philadelphia, WB Saunders, 1996, pp 146–160.
4. Sanderson CJ: Interleukin-5: an eosinophil growth and activation factor. *Dev Biol Stand* (1988) **69**: 23–29.
5. Clutterbuck EJ, Hirst EMA, Sanderson CJ: Human interleukin-5 (IL-5) regulates the production of eosinophils in human bone marrow cultures: comparison and interaction with IL-1, IL-3, IL-6 and GMCSF. *Blood* (1989) **73**: 1504–1512.
6. Höglund M, Simonsson B, Smedmyr B, Öberg G, Venge P: The effect of rGM-CSF on neutrophil and eosinophil regeneration after ABMT as monitored by circulating levels of granule proteins. *Br J Haematol* (1994) **86**: 709–716.
7. Karawajczyk M, Höglund M, Ericsson J, Venge P: Administration of G-CSF to healthy subjects: the effects on eosinophil counts and mobilization of eosinophil granule proteins. *Br J Haematol* (1997) **96**: 259–265.
8. Venge P: Human eosinophil granule proteins: structure function and release. In *Immunopharmacology of Eosinophils*. London, Academic Press, 1993, pp 43–55.
9. Tai P-C, Spry CJ, Petterson C, Venge P, Olsson I: Monoclonal antibodies distinguish between storage and secreted forms of eosinophil cationic protein. *Nature* (1984) **309**: 182–184.
10. Weller PF: Eosinophils: structure and functions. *Curr Opin Immunol* (1994) **6**: 85–90.
11. Levi Schaffer F, Barkans J, Newman T, *et al.*: Identification of interleukin-2 in human peripheral blood eosinophils. *Immunology* (1996) **87**: 155–161.
12. Levi Schaffer F, Lacy P, Severs NJ, *et al.*: Association of granulocyte–macrophage colony-stimulating factor with the crystalloid granules of human eosinophils. *Blood* (1995) **85**: 2579–2586.
13. Moqbel R, Ying S, Barkans J, *et al.*: Identification of messenger RNA for IL-4 in human eosinophils with granule localization and release of the translated product. *J Immunol* (1995) **155**: 4939–4947.
14. Lim KG, Wan HC, Resnick M, *et al.*: Human eosinophils release the lymphocyte and eosinophil active cytokines, RANTES and lymphocyte chemoattractant factor. *Int Arch Allergy Immunol* (1995) **107**: 342.
15. Costa JJ, Matossian K, Resnick MB, *et al.*: Human eosinophils can express the cytokines tumor necrosis factor-a and macrophage inflammatory protein-1a. *J Clin Invest* (1993) **91**: 2673–2684.
16. Wong DTW, Donoff RB, Yang J, *et al.*: Sequential expression of transforming growth factors a and b_1 by eosinophils during cutaneous wound healing in the hamster. *Am J Pathol* (1993) **143**: 130–142.
17. Leonidas DD, Elbert BL, Zhou Z, Leffler H, Ackerman SJ, Acharya KR: Crystal structure of human Charcot–Leyden crystal protein, an eosinophil lysophospholipase, identifies it as a new member of the carbohydrate-binding family of galectins. *Structure* (1995) **3**: 1379–1393.
18. Gomolin HI, Yamaguchi Y, Paulpillai AV, Dvorak LA, Ackerman SJ, Tenen DG: Human eosinophil Charcot–Leyden crystal protein: cloning and characterization of a lysophospholipase gene promoter. *Blood* (1993) **82**: 1868–1874.
19. Olsson I, Venge P: Cationic proteins of human granulocytes. II. Separation of the cationic proteins of the granules of leukemic myeloid cells. *Blood* (1974) **44**: 235–246.
20. Olsson I, Venge P: Cationic proteins of human granulocytes. I. Isolation of the cationic proteins from the granules of leukaemic myeloid cells. *Scand J Haematol* (1972) **9**: 204–214.
21. Rosenberg HF, Dyer KD, Tiffany HL, Gonzalez M: Rapid evolution of a unique family of primate ribonuclease genes. *Nature Genetics* (1995) **10**: 219–223.
22. Ding-E Young J, Peterson CGB, Venge P, Cohn ZA: Mechanism of membrane damage mediated by human eosinophil cationic protein. *Nature* (1986) **321**: 613–616.
23. Hernäs J, Särnstrand B, Lindroth P, Peterson CGP, Venge P, Malmström A: Eosinophil cationic protein alters proteoglycan metabolism in human lung fibroblast cultures. *Eur J Cell Biol* (1992) **59**: 352–363.

24. Chihara J, Urayama O, Tsuda A, Kakazu T, Higashimoto I, Yamada H: Eosinophil cationic protein induces insulin-like growth factor I receptor expression on bronchial epithelial cells. *Int Arch Allergy Immunol* (1996) **111**: 43–45.
25. Lundgren JD, Davey RT Jr, Lundgren B, et al.: Eosinophil cationic protein stimulates and major basic protein inhibits airway mucus secretion. *J Allergy Clin Immunol* (1991) **87**: 689–698.
26. Minnicozzi M, Duran WN, Gleich GJ, Egan RW: Eosinophil granule proteins increase microvascular macromolecular transport in the hamster cheek pouch. *J Immunol* (1994) **153**: 2664–2670.
27. Kimata H, Yoshida A, Ishioka C, Jiang Y, Mikawa H: Inhibition of ongoing immunoglobulin production by eosinophil cationic protein. *Clin Immunol Immunopathol* (1992) **64**: 84–88.
28. Kimata H, Yoshida A, Ishioka C, Jiang Y, Mikawa H: Eosinophil cationic protein inhibits immunoglobulin production and proliferation *in vitro* in human plasma cells. *Cell Immunol* (1992) **141**: 422–432.
29. Peterson CG, Skoog V, Venge P: Human eosinophil cationic proteins (ECP and EPX) and their suppressive effects on lymphocyte proliferation. *Immunobiology* (1986) **171**: 1–13.
30. Venge P, Dahl R, Hällgren R: Enhancement of factor XII dependent reactions by eosinophil cationic protein. *Thromb Res* (1979) **14**: 641–649.
31. Dahl R, Venge P: Activation of blood coagulation during inhalation challenge tests. *Allergy* (1981) **36**: 129–133.
32. Dahl R, Venge P: Enhancement of urokinase-induced plasminogen activation by the cationic protein of human granulocytes. *Thromb Res* (1979) **14**: 599–608.
33. Zheutlin LM, Ackerman SJ, Gleich GJ, Thomas LL: Stimulation of basophil and rat mast cell histamine release by eosinophil granule-derived cationic proteins. *J Immunol* (1984) **133**: 2180–2185.
34. Fredens K, Dahl R, Venge P: *In vitro* studies of the interaction between heparin and eosinophil cationic protein. *Allergy* (1991) **46**: 27–29.
35. Peterson CG, Venge P: Interaction and complex formation between the eosinophil cationic protein (ECP) and alpha-2-macroglobulin. *Biochem J* (1987) **245**: 781–787.
36. Slifman NR, Peterson CG, Gleich GJ, Dunette SL, Venge P: Human eosinophil-derived neurotoxin and eosinophil protein X are likely the same protein. *J Immunol* (1989) **143**: 2317–2322.
37. Peterson CG, Venge P: Purification and characterization of a new cationic protein—eosinophil protein-X (EPX)—from granules of human eosinophils. *Immunology* (1983) **50**: 19–26.
38. Durack DT, Ackerman SJ, Loegering DA, Gleich GJ: Purification of human eosinophil-derived neurotoxin. *Proc Natl Acad Sci USA* (1981) **78**: 5165–5169.
39. Gullberg U, Widegren B, Arnason U, Egesten A, Olsson I: The cytotoxic eosinophil cationic protein (ECP) has ribunoclease activity. *Biochem Biophys Res Commun* (1986) **139**: 1239–1242.
40. Fredens K, Dahl R, Venge P: The Gordon phenomenon induced by the eosinophil cationic protein and eosinophil protein-X. *J Allergy Clin Immunol* (1982) **70**: 361–366.
41. McLaren DJ, Peterson CG, Venge P: *Schistosoma mansoni*: further studies of the interaction between schistosomula and granulocyte-derived cationic proteins *in vitro*. *Parasitology* (1984) **88**: 491–503.
42. Carlson MGC, Peterson CGB, Venge P: Human eosinophil peroxidase: purification and characterization. *J Immunol* (1985) **134**: 1875–1879.
43. Slungaard A, Mahoney JR Jr: Thiocyanate is the major substrate for eosinophil peroxidase in physiologic fluids. Implications for cytotoxicity. *J Biol Chem* (1991) **266**: 4903–4910.
44. Yoshikawa S, Kayes SG, Parker JC: Eosinophils increase lung microvascular permeability via the peroxidase–hydrogen peroxide–halide system: bronchoconstriction and vasoconstriction unaffected by eosinophil peroxidase inhibition. *Am Rev Respir Dis* (1993) **147**: 914–920.
45. Kita H, Abu Ghazaleh RI, Sur S, Gleich GJ: Eosinophil major basic protein induces degranulation and IL-8 production by human eosinophils. *J Immunol* (1995) **154**: 4749–4758.
46. Gleich GJ, Loegering DA, Mann KG, Maldonado JE: Comparative properties of the Charcot–Leyden crystal protein and the major basic protein from human eosinophils. *J Clin Invest* (1976) **57**: 633–640.

47. Gleich GJ, Loegering DA, Maldonado JE: Identification of a major basic protein in guinea pig eosinophil granules. *J Exp Med* (1973) **137**: 1459–1471.
48. Slungaard A, Vercellotti GM, Tran T, Gleich GJ, Key NS: Eosinophil cationic granule proteins impair thrombomodulin function. A potential mechanism for thromboembolism in hypereosinophilic heart disease. *J Clin Invest* (1993) **91**: 1721–1730.
49. Sanderson CJ: Interleukin-5, eosinophils, and disease. *Blood* (1992) **79**: 3101–3109.
50. Mould AW, Matthaei KI, Young IG, Foster PS: Relationship between interleukin-5 and eotaxin in regulating blood and tissue eosinophilia in mice. *J Clin Invest* (1997) **99**: 1064–1071.
51. Tenscher K, Metzner B, Schöpf E, Norgauer J, Czech W: Recombinant human eotaxin induces oxygen radical production, Ca^{2+}-mobilization, actin reorganization, and CD11b upregulation in human eosinophils via a pertussis toxin-sensitive heterotrimeric guanine nucleotide-binding protein. *Blood* (1996) **88**: 3195–3199.
52. Wardlaw AJ, Moqbel R, Kay AB: Eosinophils: biology and role in disease. *Adv Immunol* (1995) **60**: 151–266.
53. Capron M: Eosinophils: receptors and mediators in hypersensitivity. *Clin Exp Allergy* (1989) **19** (Suppl 1): 3–8.
54. Carlson M, Peterson C, Venge P: The influence of IL-3, IL-5 and GM-CSF on normal human eosinophil and neutrophil C3b-induced degranulation. *Allergy* (1993) **48**: 437–442.
55. Lindau M, Hartmann J, Scepek S: Three distinct fusion processes during eosinophil degranulation. *Ann NY Acad Sci* (1994) **710**: 232–247.
56. Håkansson L, Carlson M, Stålenheim G, Venge P: Migratory responses of eosinophil and neutrophil granulocytes from patients with asthma. *J Allergy Clin Immunol* (1990) **85**: 743–750.
57. Håkansson L, Heinrich C, Rak S, Venge P: Priming of eosinophil adhesion in patients with birch pollen allergy during pollen season: effect of immunotherapy. *J Allergy Clin Immunol* (1997) **99**: 551–562.
58. Carlson M, Håkansson L, Kämpe M, Stålenheim G, Peterson C, Venge P: Degranulation of eosinophils from pollen-atopic patients with asthma is increased during pollen season. *J Allergy Clin Immunol* (1994) **89**: 131–139.
59. Gosset P, Tillie Leblond I, Janin A, *et al.*: Expression of E-selectin, ICAM-1 and VCAM-1 on bronchial biopsies from allergic and non-allergic asthmatic patients. *Int Arch Allergy Immunol* (1995) **106**: 69–77.
60. Håkansson L, Björnsson E, Janson C, Schmekel B: Increased adhesion to vascular cell adhesion molecule-1 and intercellular adhesion molecule-1 of eosinophils from patients with asthma. *J Allergy Clin Immunol* (1995) **96**: 941–950.
61. Smith CH, Barker JNWN, Lee TH: Adhesion molecules in allergic inflammation. *Am Rev Respir Dis* (1993) **148** (Suppl): S75–S78.
62. Venge J, Lampinen M, Rak S, *et al.*: Identification of IL-5 and RANTES as the major eosinophil chemoattractants in the asthmatic lung. *J Allergy Clin Immunol* (1996) **97**: 1110–1115.
63. Alam R, York J, Boyars M, *et al.*: Increased MCP-1, RANTES, and MIP-1a in bronchoalveolar lavage fluid of allergic asthmatic patients. *Am J Respir Crit Care Med* (1996) **153**: 1398–1404.
64. Ponath PD, Qin SX, Ringler DJ, *et al.*: Cloning of the human eosinophil chemoattractant, eotaxin. Expression, receptor binding, and functional properties suggest a mechanism for the selective recruitment of eosinophils. *J Clin Invest* (1996) **97**: 604–612.
65. Schweizer RC, Van Kessel-Welmers BAC, Warringa RAJ, *et al.*: Mechanisms involved in eosinophil migration. Platelet-activating factor-induced chemotaxis and interleukin-5-induced chemokinesis are mediated by different signals. *J Leukoc Biol* (1996) **59**: 347–356.
66. Spada CS, Nieves AL, Krauss AH-P, Woodward DF: Comparison of leukotriene B_4 and D_4 effects on human eosinophil and neutrophil motility *in vitro*. *J Leukoc Biol* (1994) **55**: 183–191.
67. Resnick MB, Weller PF: Mechanisms of eosinophil recruitment. *Am J Respir Cell Mol Biol* (1993) **8**: 349–355.
68. Stern M, Meagher L, Savill J, Haslett C: Apoptosis in human eosinophils: programmed cell

death in the eosinophil leads to phagocytosis by macrophages and is modulated by IL-5. *J Immunol* (1992) **148**: 3543–3549.
69. Denburg JA: Basophils, mast cells and eosinophils and their precursors in allergic rhinitis. *Clin Exp Allergy* (1991) **21** (Suppl. 1): 253–258.
70. Humbert M, Durham SR, Ying S, *et al.*: IL-4 and IL-5 mRNA and protein in bronchial biopsies from patients with atopic and nonatopic asthma: evidence against 'intrinsic' asthma being a distinct immunopathologic entity. *Am J Respir Crit Care Med* (1996) **154**: 1497–1504.
71. Konno S, Gonokami Y, Kurokawa M, *et al.*: Cytokine concentrations in sputum of asthmatic patients. *Int Arch Allergy Immunol* **109**: 73–78.
72. Horn BR, Robin ED, Theodore J, Van Kessel A: Total eosinophil counts in the management of bronchial asthma. *N Engl J Med* (1975) **292**: 1152–1155.
73. Bousquet J, Chanez P, Lacoste JY, *et al.*: Eosinophilic inflammation in asthma. *N Engl J Med* (1990) **323**: 1033–1039.
74. De Monchy JG, Kauffman HF, Venge P, *et al.*: Bronchoalveolar eosinophilia during allergen-induced late asthmatic reactions. *Am Rev Respir Dis* (1985) **131**: 373–376.
75. Hoshino M, Nakamura Y: Relationship between activated eosinophils of the bronchial mucosa and serum eosinophil cationic protein in atopic asthma. *Int Arch Allergy Immunol* (1997) **112**: 59–64.
76. Kaminuma O, Mori A, Ogawa K, *et al.*: Successful transfer of late phase eosinophil infiltration in the lung by infusion of helper T cell clones. *Am J Respir Cell Mol Biol* (1997) **16**: 448–454.
77. Louis R, Van Tulder L, Poncelet M, Corhay JL, Mendez P, Radermecker M: Correlation between bronchoalveolar lavage (BAL) fluid cell lysate histamine content and BAL fluid eosinophil count in atopic and nonatopic asthmatics. *Int Arch Allergy Immunol* (1997) **112**: 309–312.
78. Shi HZ, Qin SM, Huang GW, *et al.*: Infiltration of eosinophils into the asthmatic airways caused by interleukin 5. *Am J Respir Cell Mol Biol* (1997) **16**: 220–224.
79. Simon HU, Yousefi S, Schranz C, Schapowal A, Bachert C, Blaser K: Direct demonstration of delayed eosinophil apoptosis as a mechanism causing tissue eosinophilia. *J Immunol* (1997) **158**: 3902–3908.
80. Endo S, Suzui H, Tomita S, *et al.*: The study of the relationship between serum eosinophil cationic protein and bronchial responsiveness in the patients with bronchial asthma. *Jpn J Allergol* (1996) **45**: 11–16.
81. Konno S, Gonokami Y, Kurokawa M, *et al.*: Cytokine concentrations in sputum of asthmatic patients. *Int Arch Allergy Immunol* (1996) **109**: 73–78.
82. Lantero S, Sacco O, Scala C, Morelli MC, Rossi GA: Eosinophil locomotion and the release of IL-3 and IL-5 by allergen-stimulated mononuclear cells are effectively downregulated *in vitro* by budesonide. *Clin Exp Allergy* (1996) **26**: 656–664.
83. Martin LB, Kita H, Leiferman KM, Gleich GJ: Eosinophils in allergy: role in disease, degranulation, and cytokines. *Int Arch Allergy Immunol* (1996) **109**: 207–215.
84. Pizzichini MMM, Popov TA, Efthimiadis A, *et al.*: Spontaneous and induced sputum to measure indices of airway inflammation in asthma. *Am J Respir Crit Care Med* (1996) **154**: 866–869.
85. Synek M, Beasley R, Frew AJ, *et al.*: Cellular infiltration of the airways in asthma of varying severity. *Am J Respir Crit Care Med* (1996) **154**: 224–230.
86. Teran LM, Carroll MP, Frew AJ, *et al.*: Leukocyte recruitment after local endobronchial allergen challenge in asthma: relationship to procedure and to airway interleukin-8 release. *Am J Respir Crit Care Med* (1996) **154**: 469–476.
87. Gleich GJ, Jacoby DB, Fryer AD: Eosinophil-associated inflammation in bronchial asthma: a connection to the nervous system. *Int Arch Allergy Immunol* (1995) **107**: 205–207.
88. Jarjour NN, Busse WW: Cytokines in bronchoalveolar lavage fluid of patients with nocturnal asthma. *Am J Respir Crit Care Med* (1995) **152**: 1474–1477.
89. Oosterhoff Y, Kauffman HF, Rutgers B, Zijlstra FJ, Koëter GH, Postma DS: Inflammatory cell number and mediators in bronchoalveolar lavage fluid and peripheral blood in subjects with asthma with increased nocturnal airways narrowing. *J Allergy Clin Immunol* (1995) **96**: 219–229.
90. Robinson D, Assoufi B, Durham S, Kay A: Eosinophil cationic protein (ECP) and eosinophil

protein X (EPX) concentrations in serum and bronchial lavage fluid in asthma: effect of prednisolone treatment. *Clin Exp Allergy* (1995) **25**: 1118–1127.
91. Bousquet J, Chanez P, Vignola AM, Lacoste J-Y, Michel FB: Eosinophil inflammation in asthma. *Am J Respir Crit Care Med* (1994) **150**: S33–S38.
92. Kroegel C, Liu MC, Hubbard WC, Lichtenstein LM, Bochner BS: Blood and bronchoalveolar eosinophils in allergic subjects after segmental antigen challenge: surface phenotype, density heterogeneity, and prostanoid production. *J Allergy Clin Immunol* (1994) **93**: 725–734.
93. Maestrelli P, Calcagni PG, Saetta M, *et al.*: Sputum eosinophilia after asthmatic responses induced by isocyanates in sensitized subjects. *Clin Exp Allergy* (1994) **24**: 29–34.
94. Calhoun WJ, Jarjour NN, Gleich GJ, Stevens CA, Busse WW: Increased airway inflammation with segmental versus aerosol antigen challenge. *Am Rev Respir Dis* (1993) **147**: 1465–1471.
95. Lacoste J-Y, Bousquet J, Chanez P, *et al.*: Eosinophilic and neutrophilic inflammation in asthma, chronic bronchitis, and chronic obstructive pulmonary disease. *J Allergy Clin Immunol* (1993) **92**: 537–548.
96. Virchow JC Jr, Kroegel C, Hage U, Kortsik C, Matthys H, Werner P: Comparison of sputum-ECP levels in bronchial asthma and chronic bronchitis. *Allergy* (1993) **48**: 112–118.
97. Sedgwick JB, Calhoun WJ, Vrtis RF, Bates ME, McAllister PK, Busse WW: Comparison of airway and blood eosinophil function after *in vivo* antigen challenge. *J Immunol* (1992) **149**: 3710–3718.
98. Virchow JC Jr, Holscher U, Virchow CS: Sputum ECP levels correlate with parameters of airflow obstruction. *Am Rev Respir Dis* (1992) **146**: 604–606.
99. Broide DH, Gleich GJ, Cuomo AJ, *et al.*: Evidence of ongoing mast cell and eosinophil degranulation in symptomatic asthma airway. *J Allergy Clin Immunol* (1991) **88**: 637–648.
100. Frigas E, Motojima S, Gleich GJ: The eosinophilic injury to the mucosa of the airways in the pathogenesis of bronchial asthma. *Eur Respir J* (1991) **13**: (Suppl): 123s–135s.
101. Rak S, Björnson A, Håkanson L, Sörenson S, Venge P: The effect of immunotherapy on eosinophil accumulation and production of eosinophil chemotactic activity in the lung of subjects with asthma during natural pollen exposure. *J Allergy Clin Immunol* (1991) **88**: 878–888.
102. Laitinen LA, Laitinen A, Haahtela T: Airway mucosal inflammation even in patients with newly diagnosed asthma. *Am Rev Respir Dis* (1993) **147**: 697–704.
103. Laitinen LA, Laitinen A, Heino M, Haahtela T: Eosinophilic airway inflammation during exacerbation of asthma and its treatment with inhaled corticosteroid. *Am Rev Respir Dis* (1991) **143**: 423–427.
104. Ädelroth E, Rosenhall L, Johansson SA, Linden M, Venge P: Inflammatory cells and eosinophilic activity in asthmatics investigated by bronchoalveolar lavage. The effects of antiasthmatic treatment with budesonide or terbutaline. *Am Rev Respir Dis* (1990) **142**: 91–99.
105. Pizzichini E, Pizzichini MMM, Efthimiadis A, Dolovich J, Hargreave FE: Measuring airway inflammation in asthma: eosinophils and eosinophilic cationic protein in induced sputum compared with peripheral blood. *J Allergy Clin Immunol* (1997) **99**: 539–544.
106. Pizzichini E, Pizzichini MMM, Efthimiadis A, *et al.*: Indices of airway inflammation in induced sputum: reproducibility and validity of cell and fluid-phase measurements. *Am J Respir Crit Care Med* (1996) **154**: 308–317.
107. Turner MO, Hussack P, Sears MR, Dolovich J, Hargreave FE: Exacerbations of asthma without sputum eosinophilia. *Thorax* (1995) **50**: 1057–1061.
108. Venge P: Monitoring of asthma inflammation by serum measurements of eosinophil cationic protein (ECP). A new clinical approach to asthma management. *Respir Med* (1995) **89**: 1–2.
109. Venge P: Eosinophil activity in bronchial asthma. *Allergy Proc* (1994) **15**: 139–141.
110. Hällgren R, Venge P: Clinical impact of the monitoring of allergic inflammation. In Matsson P, Ahlstedt S, Venge P: Matsson P, Ahlstedt S, Venge P, Thorell J (eds) *The Eosinophil in Inflammation*. London, Academic Press, 1991, pp 119–140.
111. Pedersen B, Dahl R, Karlström R, Peterson CGB, Venge P: Eosinophil and neutrophil activity in asthma in a one-year trial with inhaled budesonide: the impact of smoking. *Am J Respir Crit Care Med* (1996) **153**: 1519–1529.
112. Boner AL, Peroni DG, Piacentini GL, Venge P: Influence of allergen avoidance at high

altitude on serum markers of eosinophil activation in children with allergic asthma. *Clin Exp Allergy* (1993) **23**: 1021–1026.
113. Kita H, Abu-Ghazaleh R, Sanderson CJ, Gleich GJ: Effect of steroids on immunoglobulin-induced eosinophil degranulation. *J Allergy Clin Immunol* (1991) **87**: 70–77.
114. Pedersen B, Dahl R, Håkansson L, Venge P: The influence of inhaled salmeterol on bronchial inflammation. A bronchoalveolar lavage study in patients with bronchial asthma. Unpublished data.
115. Pedersen B, Dahl R, Larsen BB, Venge P: The effect of salmeterol on the early- and late-phase reaction to bronchial allergen and postchallenge variation in bronchial reactivity, blood eosinophils, serum eosinophil cationic protein, and serum eosinophil protein X. *Allergy* (1993) **48**: 377–382.
116. Dahl R, Venge P: Blood eosinophil leucocyte and eosinophil cationic protein. *In vivo* study of the influence of beta-2-adrenergic drugs and steroid medication. *Scand J Respir Dis* (1978) **59**: 319–332.
117. Rak S: Effects of immunotherapy on the inflammation in pollen asthma. *Allergy* (1993) **48**: 125–128.
118. Rak S, Löwhagen O, Venge P: The effect of immunotherapy on bronchial hyperresponsiveness and eosinophil cationic protein in pollen-allergic patients. *J Allergy Clin Immunol* (1988) **82**: 470–480.

9

Lymphocytes

GARY P. ANDERSON

INTRODUCTION

The major physiological role of lung lymphocytes is to coordinate and execute protective adaptive immune responses that are at once highly specific for an infectious pathogen or inhaled antigen yet also self-limiting and non-injurious to the person. In contrast, the airways of asthmatic individuals are intensively infiltrated by lymphocytes that cause disease. This chapter discusses the biology of how T-lymphocytes become activated to be armed effector cells able to direct cellular inflammation of the airways and inappropriate antibody secretion from B-cells. New research on the molecular pathways of lymphocyte activation is highlighted to indicate where future therapeutic interventions can be made.

Surface markers identify major lymphocyte subpopulations

Lymphocytes are classified by the molecules they express on their surface and, increasingly in the case of T-cells, are subclassified by the cytokines they secrete. In turn, these surface molecules and cytokines largely mediate their highly specialized functions. The major divisions of lymphocytes are shown in Fig. 9.1 and their relative proportions in lung compared with blood are shown in Table 9.1.

T-cells express the surface marker CD3 (designated $CD3^+$ T-cells), a component of the T-cell receptor (TCR) complex used to respond to foreign antigen. The two most common subpopulations of $CD3^+$ T-cells have either the CD4 or CD8 surface markers ($CD4^+$ or $CD8^+$ T-cells, respectively) which cluster with CD3 in the TCR and serve as coreceptors for antigenic peptides presented by major histocompatibility complex (MHC)

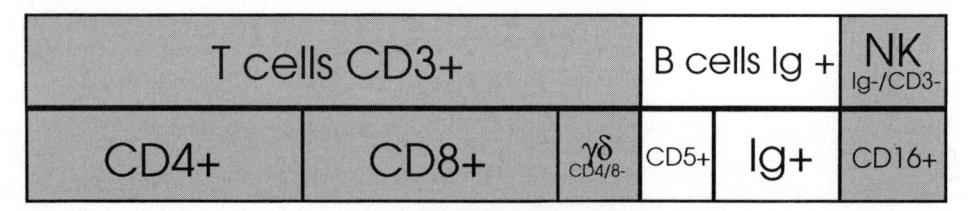

T cells CD3+			B cells Ig +	NK Ig-/CD3-
CD4+	CD8+	γδ CD4/8-	CD5+	Ig+	CD16+

Fig. 9.1 Principle divisions of lymphocytes. The approximate proportions of each subgroup of lymphocytes defined by surface markers is represented diagrammatically. Note that discrete functional subsets defined by cytokine production profiles are not shown.

molecules on antigen-presenting cells (APCs). Although some exceptions have recently been discovered, CD4+ cells almost always recognize foreign antigens presented by MHC class II molecules (MHC II-restricted responses) on APCs, whereas CD8+ cells recognize MHC I-restricted antigens.[1a] MHC I-restricted antigens are self-antigens (to which T-cells are tolerant or non-responsive) and antigens from intracellular pathogens, notably viruses, mycobacteria and intracellular parasites, which elicit adaptive immune responses. The major function of CD4 cells is to direct immune responses by secreting regulatory cytokines. The acquisition of CD4 cell functional phenotype (e.g. Th1 and Th2 pattern responses is discussed in detail below). CD8 cells also produce some regulatory cytokines but can develop into cytotoxic T-cells (CTLs).

The peptide recognition region of the TCR on the vast majority of CD4 and CD8 cells is encoded by three α and β TCR chain genes (VDJ genes) and these cells are therefore designated α/β TCR T-lymphocytes. However a third T-cell population that lacks CD4 and CD8 (i.e. CD3+, CD4−, CD8−) and uses separate genes to encode alternate chains of γ and δ TCR has been identified. These $\gamma\delta$ T-cells respond to antigen presented by minor MHC molecules, notably CD1,[1b] and have a predilection for localizing to epithelia. Further minor but potentially important functional subgroups within CD4+, CD8+ and $\gamma\delta$ cells also exist.

The single known effector function of B-lymphocytes (B-cells) is to produce antibody. Surface immunoglobulin is the B-cell antigen receptor and is also a B-cell surface marker.

Table 9.1 Number and percentage of lymphocytes in the lung and blood of normal humans (modifed from Saltini et al., 1991).

Population	Lung (%)	Venous blood (%)
Epithelial lining fluid (x $10^3/\mu l$)*	3.9 ± 0.7	NA
CD3+ T-cells	81 ± 2	70 ± 5.0
$\alpha\beta$ TCR	95 ± 1	99 ± 2.0
$\gamma\delta$ TCR†	4 ± 1	5 ± 1.0
CD4+	48 ± 2	52 ± 3.0
CD8+	38 ± 3	35 ± 2.0
B cells	2 ± 1	9 ± 2.0
NK cells	2 ± 1	12 ± 1.0

* Substantial numbers of lymphocytes are also present as a sequestered pool in the pulmonary microcirculation, in peribronchial nodes, lymphoid aggregates and in the lung interstitium
† Underestimates total $\gamma\delta$ cells which are enriched in the epithelium

9 Lymphocytes

A subpopulation of B-cells bearing the surface marker CD5 arises early in development and has a distinct immunoglobulin repertoire. The role of CD5 cells in the lung is still unclear, although they are enriched in the pleural cavity and share some functional properties with $\gamma\delta$ T-cells. Although primed B cells are highly efficient APC they seem unimportant for T cell reactivation.[1c]

Natural killer (NK) cells are neither T-cells nor B-cells. These cells lack a specific antigen recognition receptor analogous to T-cell TCR or B-cell immunoglobulin. However they express the marker CD16 (which designates the third class of IgG receptor (FcγRIII) able to bind IgG1 and IgG3). NK cells are highly granular due to storage of the lytic enzymes perforin and granzymes and kill a limited range of IgG-bound targets, e.g. virus-infected epithelial cells, in an antibody-dependent cytotoxicity (ADCC) reaction.

In addition to TCR, CD4, CD5, CD8 and CD16, a very large number of additional molecules, the majority of which have been sufficiently characterized to assign a systematic cluster of differentiation (CD) designation, are variably expressed (Table 9.2). These molecules include receptors allowing responses to cytokines, adhesion molecules directing migration and homing, and activation markers whose expression reflects the functional state of the lymphocytes. It is also important to note that marker expression is labile and markers such as CD8 or CD4 may be lost due to downregulation[2] or, in severe inflammation, by proteolytic cleavage.

Lymphocytes are strongly implicated in the pathogenesis of asthma

The evidence that T-cells cause asthma is very strong but not unequivocal. Airways of asthmatic patients who have died from their disease are intensely infiltrated with lymphocytes, the majority of which are T-cells.[3-6] T-cells are present in mucosal biopsies or lavages of asthmatic patients and show evidence of both activation and active cytokine secretion.[7-33] Furthermore, drugs that control asthma, such as glucocorticosteroids, and more powerful immunosuppressive agents, such as cyclosporin A, FK506 and rapamycin, improve asthma symptoms, at least to some extent, and concurrently inhibit T-cell number and activity.[34-38]

Several mechanisms have been identified whereby aberrant T-cell activity causes asthma.

(i) Naive CD4$^+$ T-cells acquire a functional phenotype, termed a Th2 pattern phenotype, where they are able to secrete the key cytokines IL-4, IL-13 and IL-5. IL-4 (and IL-13)[39] provide essential signals to B-lymphocytes instructing gene rearrangement to produce immunoglobulin E. IL-5 is critical to sustain eosinophilic inflammation.[40,41] IL-5 also promotes IgA production which contributes to eosinophil degranulation. Cytokines from CD4$^+$ cells influence not only eosinophilic inflammation but also arming of macrophages and monocytes, chemokine (eotaxin, MIP-Iα, RANTES) secretion,[42] and changes in tissue structure and function.[43,44] Further evidence has implicated CD4 cells in the induction of bronchial hyperreactivity.[45-48]

(ii) CD8$^+$ T-cells precipitate worsening of asthma by mediating cellular defence against acute viral infection and possibly chronic coinfection by pathogens (such as *Myco-*

Table 9.2 Markers found exclusively or predominantly on T-Cells.[1d]

CD Number	Common Name(s)	Main Function	Ligand
CD1	–	Antigen presentation	-
CD2	–	T Cell activation	-
CD4	–	Helper/inducer function	MHCII
CD8	–		MHCI
CD10	Neutral endopeptidase	Peptidase	-
CD16	FcγR III	Phagocytosis	Fcγ
CD24	HSA	Proliferation	-
CD27	–	Costimulation	-
CD28	–	Costimulation	-
CD30	–	Death (apoptosis)	CD153
CD45	–	Signalling	CD22
CD69	Very early antigen-1	Signalling	–
CD90	Thy1	Recirculation, activation	-
CD95	FAS/APO-1	Death (apoptosis)	FAS-L
CDw137	4–1BB	Costimulation	4–1BBL
CDw150	SLAM	Activation	-
CD152	CTLA-4	Negative costimulation	CD80/86
CD153	CD30-L	Costimulation	CD30
CD154	CD40	Costimulation	CD40L,gp39

Cytokine receptors

CD Number	Common Name(s)	Main Function	Ligand
CD25	IL-2Rα	Interleukin 2 receptor	IL-2

Adhesion molecules expressed on, but not restricted to T cells

CD Number	Common Name(s)	Main Function	Ligand
CD2	–	APC and targets	LFA-3
CD11a/CD18	LFA-1	APC, endothelium	ICAM-1,-2,-3
CD34	–	Endothelium	-
CD44	Pgp-1/HERMES	Cell-cell-matrix adhesion	Hyluronate
CD49a/CD29	VLA-1	Cell-matrix adhesion	Laminin, collagen

L, ligand; FcγR, receptor for IgG; HSA, heat shock antigen; APC, antigen presenting cell.

plasma pulmonis and *Candida albicans*). CD8+ cells cause cytotoxic responses that kill infected epithelial cells and cause more severe mononuclear cell-mediated inflammation by secreting IFNγ. There is some evidence that CD8+ cells, like CD4 cells, may also acquire a 'TH2-like' cytokine pattern.[2,49] A subpopulation of CD8 cells which can suppress CD4 responses may also fail to adequately check CD4 activation.[50]

(iii) γδ T-cells have a less certain role in asthma but secrete growth factors that alter the cellular composition of the epithelium[43] and respond directly to some viral or bacterial pathogens. Although γδ cells can adopt a TH2 like pattern of cytokines[51] it is not known if they release sufficient IL-5 to drive eosinophilic inflammation or sufficient IL-4 to drive IgE responses in human asthma.[52,53]

9 Lymphocytes

The most obvious role of B-cells in asthma is the generation of IgE antibodies. IgE levels correlate broadly with asthma risk and triggering IgE-dependent activation mechanisms recapitulates at least some of the cellular and physiological abnormalities of asthma. The importance of B-cell production of IgG isotypes and IgA may have been underestimated due to the predominance of current IgE-driven allergic models of disease.[54,55] IgG isoforms can participate in macrophage/mononuclear cell activation, which may be particularly important in more chronic or severe forms of asthma where high levels of interferon (IFN)-γ have been found. IgG and IgG3 trigger NK-mediated cytolysis, which may precipitate asthma crisis in viral infection. IgA remains one of the most effective known stimuli for eosinophil activation. Secretory IgA dimers (sIgA) are vectorially secreted across epithelia and may therefore have a key role in selective degranulation of eosinophils near the epithelial basement membrane.

OVERVIEW OF LYMPHOCYTE RESPONSES TO INHALED FOREIGN ANTIGENS

A general model of a normal T-cell response, indicating potential defects that may cause asthma, is shown in Fig. 9.2.

Inhaled antigen is presented to naive T-cells on MHC molecules by the major airway APC, the dendritic cell, after migration to the specialized microenvironment of a lymph node draining the airway mucosa where the probability of T-cell encounters with antigen is increased by simple proximity.[56,57] The lymph node also focuses soluble factors such as cytokines and chemokines draining from the airways. APCs and T-cells bind via adhesion molecules allowing the TCR complex to scan the APC surface for an antigenic peptide complementary in shape and surface properties to its own complementarity-determining region (CDR).[58,59] If no antigen of the correct specificity is formed the APC and T-cell dissociate. The TCR dimer has no intrinsic signalling ability but antigen recognition leads to rapid T-cell signalling because the TCR forms a complex with CD3 dimers, the CD4 (or CD8) coreceptor, dimers of the protein ζ (all of which have cytoplasmic docking domains for signalling tyrosine kinases) and also with the phosphatase CD45. CD4 also serves as a receptor for interleukin (IL)-16.[60] T-cell activation is checked at this stage by the availability of costimulatory molecules on both T-cell and APC.[61] In the absence of correct costimulatory signals, the T-cell response aborts and the T-cell may die or become permanently non-responsive (anergic). Costimulation leads to rapid IL-2 and IL-2 receptor (CD25) expression, allowing T-cell expansion to occur (clonal expansion because all the progeny have the same antigen specificity as the parent).[62-67] During this stage of primary activation soluble signals, especially cytokines such as IL-4, or conversely IFN-γ and IL-12, and possibly some physical costimulatory signals have a strong influence on the acquisition of a Th2 functional phenotype. T-cells alter their expression of surface adhesion molecules, shedding the lymph node homing molecule L-selectin (CD62L), upregulating CD2 and LFA-1 to facilitate interaction with targets that engage with B-cells and gain very late antigen (VLA)-4 to selectively transmigrate across inflamed endothelium. Cytokines, together with T-cell physical signals, instruct B-cells to make IgE, IgA and IgG isotype antibodies. On re-exposure to antigen, memory cells give rise rapidly to clonally expanded pools of effector T-cells.[68] Cytokines and chemokines,

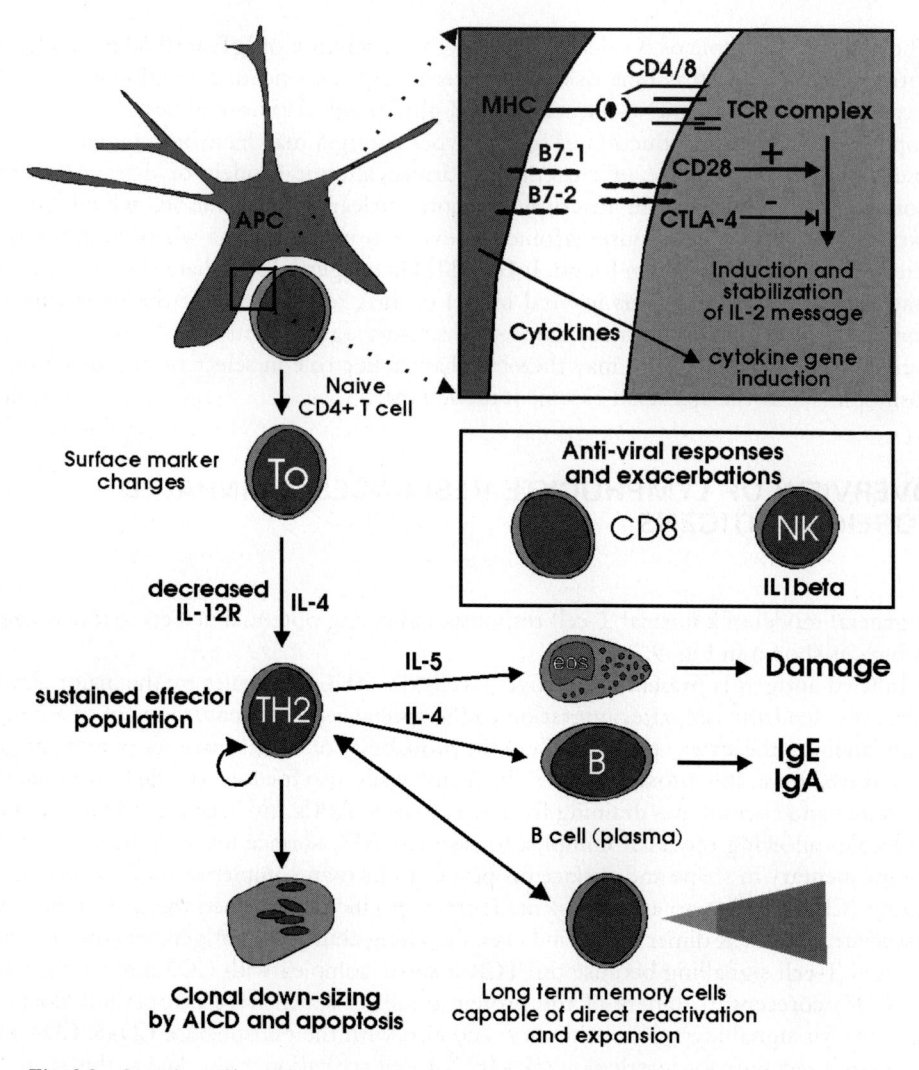

Fig. 9.2 Overview of lymphocyte activation in airway disease. The diagram D detailed in the text.

notably IL-1, can also directly reactivate quiescent effector cells, whereas cytokine release from effector T-cells causes bystander activation of neighbouring lymphocytes. The upregulation of MHCII on epithelium, mast cells, eosinophils, basophils and other cells allows *in situ* activation of effector T-cells in the airways.[69] During the termination of immune responses, proliferation and activation are checked by negative costimulatory signals and death of the expanded T-cell population, sparing only a small population of long-lived memory cells. A separate population of memory B-cells also persisits.

Understanding how this normally tightly regulated series of activation and termination mechanisms goes awry is central to understanding asthma and designing new therapies. As discussed in more detail below there is evidence that defects in the both the induction and termination of lymphocyte responses are abnormal in asthma.

Primary activation of T-cells requires tightly matched TCR CDR3 region complementary to the processed antigenic peptide presented by APCs on MHC molecules

As dendritic cells migrate from the mucosa to regional lymph nodes loaded with captured antigen that they process into short peptides for expression on their surface bound to MHCII, they alter their surface properties by increasing expression of adhesion molecules for counter-receptors on T-cells (LFA-3/CD58, which binds to T-cell CD2 and intercellular adhesion molecule (ICAM) 1, 2 and 3, which bind to T-cell LFA-1). Although dendritic cells are extremely effective in triggering primary T-cell activation, quantitative affinity and stochastic studies have revealed that recognition between the TCR and MHC–peptide is surprisingly transient and weak. However, the very few processed peptides specific for a given CDR on a reactive T-cell are able to trigger a full response because as few as 1000 peptide–MHC complexes per cell can serially trigger signal transduction events in more than 100 000 TCR complexes.[58] To achieve the necessary diversity to recognize the myriad peptides derived from antigens in the environment, highly variable regions of the TCR, the CDRs, are randomly encoded by rearrangement of VDJ gene elements focusing on the CDR3 region formed by regions of the α and β TCR chains. CDR1 and CDR2 are less varied in structure and recognize regions of the MHC molecule around the peptide. The TCR VDJ genes used to generate diverse CDR structures can now be very accurately measured, broadly using monoclonal antibodies and very specifically using PCR methods. Application of these methods has revealed that the very large numbers of T-cells in the airways of human asthmatic subjects or in the lungs of experimental animals represent expanded populations (clones) of a limited number of T-cells with distinct TCR gene usage.[73–75,63] In humans, house dust mite-specific CD4$^+$ T-cells predominantly use $V_\alpha 8$ and $V_\beta 3$ genes to encode TCR. In mice, $V_\beta 8$ genes predominate in responses to inhaled ovalbumin and confer the capacity to mount asthma-like inflammation and acquire bronchial hyperreactivity which can be adoptively transferred with $V_\beta 8$ CD4$^+$ T-cells alone. Animal studies further suggest that $V_\beta 2$ T-cells may negatively regulate $V_\beta 8$-mediated responses.[75]

Altered peptides subvert T cell activation

The necessity for very high specificity in the TCR CDR3/MHC-peptide recognition event in order for successful T-cell activation has been exploited therapeutically by designing altered peptides that provide an incomplete activation signal to the TCR complex resulting in anergy (permanent non-responsiveness) to the antigenic determinant. This strategy uses an immunodominant peptide (i.e one giving a strong T-cell response) from a complex antigen such as house dust mite Der p1 (which has multiple antigenic epitopes that give rise to different peptides when processed by APCs) and systemically alters its structure to produce a recognized but imperfect fit (reduced affinity) for the complementary CDR3 region of TCR. This strategy has been used in experimental systems to anergize TH cells at the point before they polarize to Th1 or Th2 pattern reponses[70] and to increase IFN-γ production in human CD4$^+$ T-cells after they have adopted a Th2 cytokine pattern.[71,72] The concept is attractive because analysis of TCR usage by sensitive PCR methods suggests that the large numbers of antigen-specific T-cells are actually a clonally expanded population of a relatively small number of antigen specific progenitors.[62]

This strategy, however, has some important limitations in asthma therapy in that it is confined to a single major allergen and even then it is most effective only when the modified immunodominant peptide causes 'spillover' modulation of the minor epitopes (i.e. the other antigenic peptides from a complex antigen protein) and when the altered peptide retains sufficiently high affinity for the MHC-binding pocket to competitively displace its target tightly bound native peptide prototype. However, altered peptides may still prove useful adjuvants to influence cytokine production in early disease or as preventative therapies.

Signal transduction from the TCR complex requires multiple intermediates

The TCR $\alpha\beta$ (or $\gamma\delta$) dimer is unable to signal T-cell activation itself but peptide recognition triggers prompt T-cell activation because the TCR forms a complex with six other known molecules (Fig. 9.3). Several of these signalling intermediates are the targets of broad spectrum immunosuppressors. Each TCR dimer aggregates with two CD3 dimers (composed of $\varepsilon\delta$ chains and $\varepsilon\gamma$ chains), a dimer of the protein ζ (or a heterodimer of ζ and the protein η in mice or ζ and the IgE receptor γ-chain in humans), the CD4 or CD8 MHC coreceptor and the phosphatase CD45. CD3 and ζ have intracellular domains containing immunoreceptor tyrosine-activation motifs (ITAMs), which allow interaction with cytoplasmic tyrosine kinases. These kinases include:

(1) ZAP-70, which is unassociated at rest but binds avidly to ζ ITAMs;
(2) Lck, physically associated with the CD4 or CD8 coreceptor;
(3) Fyn, physically associated with ζ;
(4) Syk.

In addition src kinase probably contributes to each ITAM phosphorylation. It is currently thought that the aggregation by antigen of the TCR complex allows Lck and Fyn to doubly phosphorylate ζ ITAMs, greatly increasing affinity for ZAP-70 and Syk, which utilize specific src-homology 2 (SH2) domains to bind to the phosphorylated ITAM residues. Subsequent autophosphorylation of ZAP-70 provides specific SH2 docking sites for effector molecules.[76] These include Lck, Fyn, FAK-related PTK, ras-GTP, abl, vav, cbl, Lnk and SLP76. Lnk and SLP76 in turn activate PLCγ1 linked to phospholipase Cγ. Diacylglycerol and inositol trisphosphate engendered by PLCγ activate protein kinase C and increase intracellular calcium that activates, in due course, translocation of transcription factors and kinases coordinating IL-2 gene induction, IL-2 receptor (CD25) upregulation and cyclin kinases supporting entry into cell cycle, division and proliferation.

The integrity of TCR complex signalling is critical to T-cell activation. Several classes of broadly active immunosuppressor agent, for example cyclosporin A and FK506/fujimycin, prevent IL-2 gene induction by interfering with early TCR complex signals. IL-2 gene induction requires binding of a dimer comprising the nuclear factor NF-AT$_n$ and the cytoplasmic factor NF-ATc, which is constitutively inactive and must be activated by the phosphatase calcineurin before it can translocate to the nucleus.

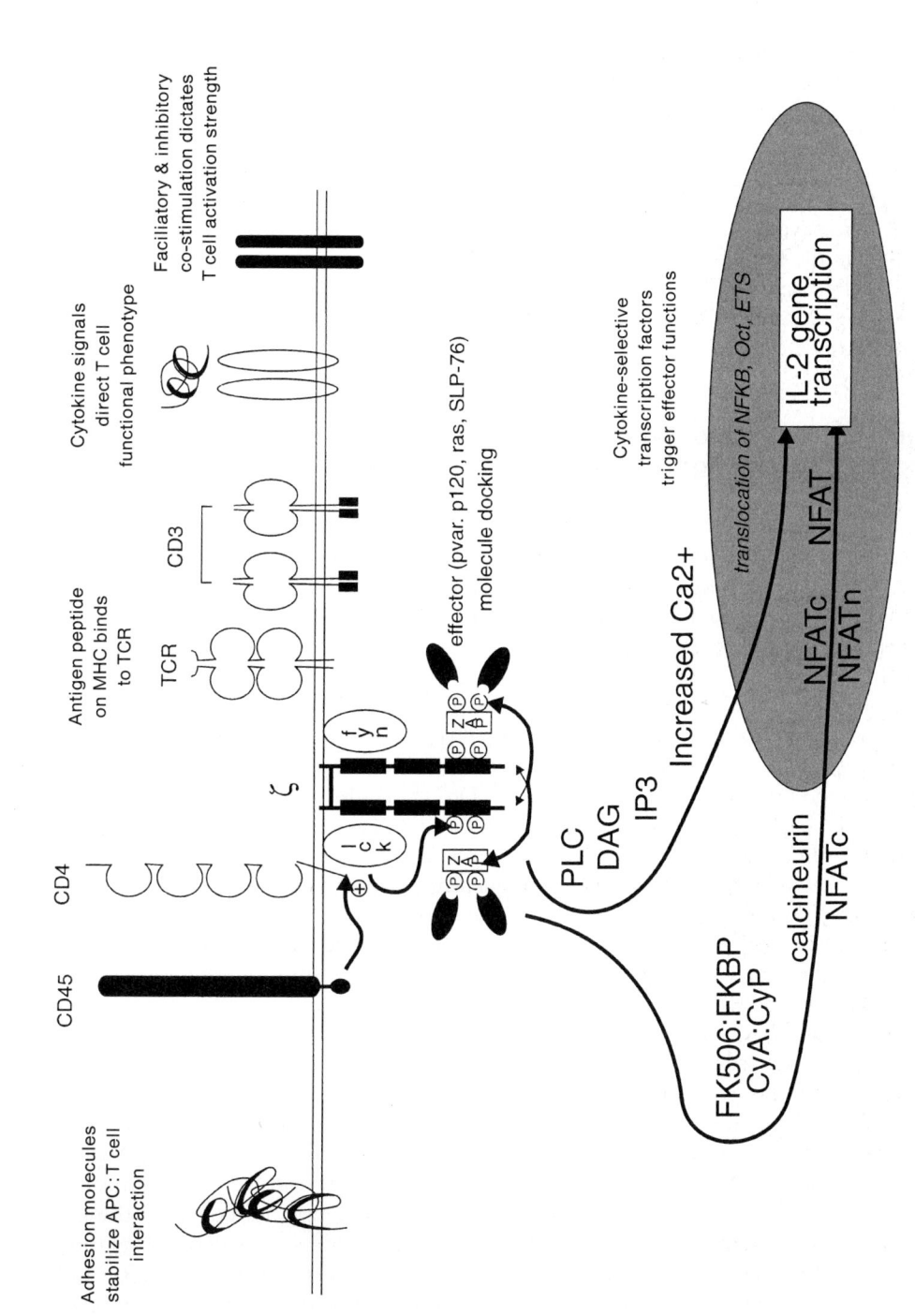

Fig. 9.3 Signalling from the TCR complex, costimulation and site of action of immunosuppressive agents.

Cyclosporin A and FK506 inactivate calcineurin by binding to it as complexes with cyclophilin or FK-binding protein 12, respectively. A third broad immunosuppressor rapamycin, does not prevent IL-2 induction but instead blocks IL-2 signalling by inhibiting IL-2 receptor-specific kinases. Several other broadly acting immunosuppressors are targeted at early TCR signalling events, including meclofenamic acid.

Analysis of the biochemistry of altered TCR peptides that induce T-cell anergy has demonstrated that the ζ-chain can be actively phosphorylated without inducing ZAP-70 phosphorylation despite its physical recruitment to the TCR complex.[77,78] This, and related data, have prompted the search for specific ITAM, SH2, ZAP-70, Lck and Fyn inhibitors in the hope that more selective immunosuppressor agents could be designed. Conceptually, these would offer no advantage in terms of specificity over cyclosporin A, FK506 or rapamycin. Cyclosporin A has shown only moderate benefit in asthma in trials performed to date.

'Memory' cells have differential responsiveness to chemoattractant signals and IL-1

After primary sensitization has occurred, lymphocytes alter their functional and surface properties. Downregulation of CD62L (L-selectin) alters the trafficking pattern such that lymphocytes no longer migrate through regional nodes but rather accumulate in inflamed tissue. Activated T-cells classically alter the isoform of CD45 from CD45RA to CD45RO; increase expression of CD25, an IL-2 receptor component; may express MHCII molecules (HLA-DR);[79,8,17,30,16] and upregulate VLA-4[80] and VLA-4/CD49d, a counter-receptor for vascular cell adhesion molecule (VCAM)-1 expressed in inflamed endothelium in airway mucosa.[81] In addition, T-cells upregulate expression of CD44, a receptor for tissue matrix components including fibronectin, ICAM-1 (CD54), LFA-1 (CD11a); and, the early activation marker CD69.[28] Traffic to the airway mucosa is promoted by expression of $\alpha E\beta 7$ integrins.[82] Similar changes are observed in animal models of asthma.[83,28,81]

T-cells with these altered markers are frequently called 'memory' cells, although it is unlikely they represent the true long-term memory cells that maintain antigen recall for decades after antigen exposure. Rather 'memory' cells are more usefully thought of as recently activated effector cells. These effector cells also differ in their responsiveness to chemotactic factors.

Recently activated lymphocytes migrate in response to growth factors, such as hepatocyte growth factor,[84] and differentially respond to chemokines. T-cell subsets differentially express five human chemokine receptors (CCR 1–5), the activation of which is likely to be an important coordination mechanism of airway inflammation.[84,85] 'Memory' T-cells (defined by CD45RO expression) respond to the chemokines MCP-1 via CCR-1 and RANTES and CXCR-2 which recognize IL-8.[86,87,60,87a] Similarly, 'memory' T-cells recognize the novel ligands IP-10 and mig (which are strongly upregulated by IFN-γ) via preferential CXCR-3 receptor expression, whereas migration of antigen-experienced B-cells is promoted via the novel chemokine receptors BLR-1 and BLR-2 whose ligand(s) are unknown.[88] Thus the recruitment and traffic of antigen-experienced lymphocytes (and probably prompt antigen-independent recruitment of NK cells) is tightly regulated on at least three levels: lymphocyte surface adhesion receptor expression, endothelial counter-receptor expression and responsiveness to soluble recruitment molecules especially chemokines.

It seems likely that chemokines will also be discovered to have an important role in modulating effector function of lymphocytes within the asthmatic airway. A very similar role has already been discovered for IL-1. IL-1α and IL-1β are promptly produced by mononuclear/macrophage lineages and many other cell types in response to airway insult. Th2, but not Th1, cells express IL-1 receptors and 'memory' Th2 cells proliferate strongly in response to IL-1, which, may even be produced by Th2 cells as an autocrine growth factor.[89–92] This relationship between Th2 cells and IL-1 provides a rationale for the unexpected effectiveness of IL-1 receptor antagonists in animal models of asthma. However the situation in human asthma may be much more complex as IL-4 has recently been discovered to upregulate soluble IL-1 receptor type II, which sequesters IL-1 reducing its effects.[93–95]

It should also be noted that 'memory' cells differ in their susceptibility to pharmacological inhibition, being substantially less sensitive to suppression of IL-5 synthesis by glucocorticosteroids (Brinkmann, personal communication). This may be due in part to the ability of IL-4 to downregulate glucocorticosteroid receptor responses.[95a] Furthermore IL-4-independent B-cell IgE production may be sustained by IL-13 in established disease.[96]

HLA polymorphism and asthma

Like the TCR, the structure of MHCI and MHCII peptide binding domains is extremely diverse to ensure the recognition of myriad self and non-self antigens. To achieve this diversity MHC molecules are polygenic, i.e. encoded by multiple genes. Human MHCI molecules are encoded by HLA-A, HLA-B and HLA-C. MHCII molecules are encoded by HLA-DR, HLA-DP and HLA-DQ genes. These genes are also polymorphic, i.e. encoded by a very large number of alleles. Since the structure of the dominant peptides derived from major allergens is (theoretically) constant and some peptides are bound by certain by HLA molecules in preference to others, there has been great interest in studying the genetic HLA haplotype of individuals in relation to disease susceptibility or severity. MHCI haplotypes are clearly associated with increased risk of autoimmune diseases, since MHCI molecules present the self-antigen presumed to drive these diseases. Results in human asthma have been less satisfactory. Studies of reactivity to house dustmite in a British general population failed to identify any relationship between the MHCII alleles HLA-DRBI, DQAI, DQBI or DPBI and response to skin prick with extracts of *Dermatophagoides pteronyssinus* or *D. farinae*.[97] This lack of correlation may be due to the large number of immunogenic epitopes that could be derived from crude *Dermatophagoides* extracts. Studies on purified major allergenic proteins have been more informative. Heterodimers of HLA-DRαβ1*1501 and related HLA-DRαβ1*1502 regulate T-cell responses to *Amb a* V, an allergenic protein released by ragweed pollen.[98] Similarly the majority of patients responding to the house dust mite *Der p* II protein have a positive association with HLA-DQ7 and negative association with HLA-DQ2 alleles. However, analysis of the actual peptide within the *Der p* II protein that triggers T-cell responses revealed that approximately 60% of patients respond to multiple peptides derived from the same allergen.[99] The multiplicity of potentially stimulatory peptides derived from one allergenic protein (in turn only one component of the mixtures of allergenic proteins released from the allergens that provoke asthma) is a serious limitation on current strategies aimed at manipulating peptide structure to alter lung mucosal immunity.

Studies in this field may prove useful in distinguishing allergic patients at risk of developing asthma: HLA-DR2 [HLA-B7, SC31, DR2] is strongly associated with asthma induced by ragweed pollen (measured as response to *Amb a* V protein), whereas the extended haplotype HLA-D8, SC01, DR3 is confined to rhinitic patients.[100] Since multiple genes are found within the MHC complex where HLA genes are located, it cannot be excluded that these HLA haplotypes simply mark the influence of products from unidentified genes within the same cluster and linked to the respective HLA alleles.

Costimulation is essential for T-cell activation

T-cells that fail to receive a second costimulation signal during MHC–TCR complex triggering do not contribute to immune responses because they are either permanently rendered unresponsive (made tolerant or anergic to the stimulus) or they die by apoptosis (programmed cell death).[101–104] Costimulation is therefore critical for the initiation of immune responses to inhaled allergen.[65] Several molecular costimulation pathways have been identified, each using molecules of stable invariant structure for interactions between T-cells and APCs. Manipulation of costimulation is therefore conceptually highly attractive because whereas the MHC–peptide–TCR complex is almost endlessly varied, the relatively more stereotyped costimulation interactions provide excellent therapeutic targets.

The best-characterized costimulatory receptor on the T-cell is CD28[105] which binds to the related molecules B7-1 and B7-2 expressed on the surface of APCs.[106,107] B7-1 and B7-2 are designated CD80 and CD86, respectively, and are highly expressed on lung dendritic cells and also on B-cells.[108] B7 family molecules can also bind to a second receptor called CTLA-4, expressed on T-cells. B7-1 and CD28 are constitutively expressed on the APC and T-cell respectively. CTLA-4 is closely related to CD28 but instead of providing a CD28-like positive effect CTLA-4 inhibits T-cell activation.[109] B7-2 upregulation quickly follows APC–peptide–TCR complex engagement. CTLA-4 in contrast is not constitutively expressed on T-cells. The negative signal provided by CTLA-4, which prevents T-cell IL-2 production and entry into the cell cycle, occurs only after initial T-cell activation and serves to dampen responses. Recombinant versions of CTLA-4 fused to an immunoglobulin carrier (CTLA-4-Ig) have been constructed. CTLA-4-Ig binds B7 molecules approximately 20 times more avidly than CD28 and therefore prevents B7–CD28 costimulation.

Treatment of A/J mice, which are genetically predisposed to develop airway hyperresponsiveness, with CTLA-4-Ig, either before primary antigen sensitization or immediately before allergen challenge in immunized animals, prevents the induction of bronchial hyperresponsiveness and eosinophilic inflammation as well as diminishing IL-4 production and IgE levels.[110,65] Studies using monoclonal antibodies to inhibit B7-1 or B7-2 suggested initially that B7-2 was more important in this suppression of eosinophilia and reactivity;[111] this is consistent with *in vitro* studies suggesting that B7-2 is a physical signal directing acquisition of a Th2 functional phenotype.[112,113] However, CTLA-4-Ig does not prevent Th2 cell induction in mice *in vivo*.[114] This issue remains unresolved: a mutated CTLA-4-Ig that selectively blocks B7-1/CD28, but spares B7-2, interactions prevents allergen-induced eosinophilic inflammation but not IgE production, a Th2-driven process.[115]

The suppression of inflammation and airway hyperreactivity by B7/CD28 blockade after primary sensitization to allergen is important because it indicates that this approach would be useful in established disease where T-cells have already become sensitized to inhaled allergen. Conceptually, suppression of a costimulation signal should lead to long-lasting anergy. It is unclear at present if this occurs *in vivo*. If it is demonstrated that anergy does not occur *in vivo*, it may mean that the armed effector cells do not require costimulation for the maintenance of reactivity to antigen once they have undergone primary sensitization. In this case, costimulation in established disease would serve mainly to expand and reactivate 'memory' T-cells.

In addition to B7-1/B7-2/CD28 interaction, several other costimulatory molecules have been identified of which 4-1BB, a member of the tumour necrosis factor (TNF)/nerve growth factor (NGF) receptor superfamily, is the most important.[116] 4-1BB is activated by 4-1BB ligand, a member, of the TNF/NGF family, and provides a positive costimulation that can substitute, at least partially, for CD28/B7 interactions. The role of 4-1BB in asthma is unknown.

TCR complex component expression is variable and controlled by transcription factors

It is of great significance that the expression of TCR components is coordinately and tightly regulated by transcription factors,[117] notably the T-cell-lineage-specific factor GATA-3. The expression of the TCR β-chain is governed by three transcription enhancing sites in its promoter region,[118] each of which can recognize GATA-3. GATA-3 is a transcription factor containing a 'zinc-finger' DNA-binding motif in its tertiary structure. GATA-3 is unusually confined in its expression to T-cell lineages.[119] Although dominant negative mutant studies (where an inactive copy of GATA-3 was overexpressed to swamp normal GATA-3 and thereby block its activity) suggest that some functional redundancy may occur between GATA-3 and the related GATA-1 and GATA-2 molecules,[120] GATA-3 may influence other T-cell responses or responsiveness. These include regulation of CD8 α-chain gene expression,[121,122] expression of IFN-γ in activated T-cells[123] and TCR δ-chain expression.[124] Although no specific defects or polymorphisms have been reported to date, changes to GATA-3 are prime candidates for risk and severity determinants of asthma. The high specificity of these factors for regulating T-cell activation suggests that defects in their biology may be linked to susceptibility or progression of human asthma.

Cytokines direct acquisition of functional Th1 and Th2 phenotypes

During primary activation, T-cells receive additional signals from soluble cytokines that have a strong influence on their subsequent cytokine production and effector function. In mice, CD4$^+$ T-cells derived from a common precursor undergo commitment to two functional subsets, designated Th1 and Th2, which are not readily distinguished by surface markers but which produce polarized and mutually antagonistic cytokine panels. Th2 cells are associated most closely with allergic asthma because they produce IL-4, IL-5 and IL-13, whereas Th1 cells produce IFN-γ and lymphotoxin. Although it is convenient to discuss Th2 cells in human asthma it is essential to recognize that individual CD4 cells rarely coexpress IL-4 and IL-5[11] and human CD4 cells coexpressing IL-5 and IFN-γ have unequivocally been identified.[96] 'Th2 pattern response' is used here to represent the aggregate cytokine production in which cytokine profile is biased towards that resembling murine Th2 cells (high IL-4, IL-5, IL-10, IL-3; low IFN-γ, IL-2, lymphotoxin).

IL-4 has a central role in the induction of strong Th2 pattern responses since neutralizing anti-IL-4 antibodies or inactivation of the IL-4 gene profoundly suppresses Th2 mucosal immunity, IgE production and eosinophilic lung inflammation.[125,126] Since IL-4 is itself a Th2 cytokine, identifying the initial source of IL-4 that drives these responses has been a major research focus.

Although initially identified as a major IL-4 producer,[127,128] the NK1.1 cell (a rare mouse T-cell subset with highly restricted V_α and V_β usage that responds to non-classical MHCI molecules including CD1 and thymus leukaemia antigen)[129,130] is unlikely to be important. Mice made deficient in this cell type by gene manipulation or antibodies develop normal Th2 responses and an asthma-like inflammation of the airways when sensitized and challenged.[131,132,133] Instead, it is now believed that CD4 cells themselves produce autocrine IL-4 during primary responses[134] which is augmented by IL-6.[135] The involvement of IL-6, an early response acute cytokine of the LIF/IL-11 family, is potentially important in asthma. IL-6 is produced by perturbed airway epithelial cells[136,137] and, conceivably, would readily drain with chemokines to regional nodes[138] where primary CD4 cell activation occurs. Most recently, B-cells have been found to have a role in instructing primary IL-4 induction in a CD1-restricted subset of CD4$^+$ T-cells in amounts that would be sufficient to allow bystander spillover.[139] However, B-cells are unlikely to be the primary trigger for T-cell IL-4 production *in vivo* as dendritic cells more than adequately perform this function in the absence of B-cells.[140] The role of B-cells in providing T-cell costimulation in chronic asthma is unexplored.

Th2 commitment in human asthma is probably a progressive and summative process, drawing on cumulatively increasing contributions from some committed T-cells and IL-4 for non-T-cell sources. After primary disease induction, IL-4 is almost certainly released from non-lymphocyte sources, notably mast cells, basophils and eosinophils, particularly after they have been primed by airway inflammation.[141–143] *In vitro* studies suggest that only a few antigen-experienced (CD62L$^{lo\ (L-selectin)}$, CD4$^+$, CD3$^+$) CD4$^+$ cells can drive commitment of naive (CD62L$^+$) CD4$^+$ cells. Furthermore, eosinophils, mast cells and basophils can also be induced to express MHCII, allowing direct T-cell activation, and FcεRI, allowing IgE-mediated responses, at least *in vitro*.[144,145]

Extinction of IL-12 signalling is a prerequisite for Th2 commitment and is regulated by genes in the 5q31.1 allergy cluster

Cytokines from Th1 cells, notably IFN-γ and IL-12, prevent or reverse Th2 responses.[146] Since IFN-γ, whose production is induced by IL-12, is produced by Th0 cells and IL-12 is classically produced by APCs during primary T-cell activation,[147,148] a highly efficient escape mechanism must exist to allow THp → Th2 phenotype commitment to occur. Analysis of transgenic mice where the TCR is held constant by genetic manipulation suggested that Th2 responses occurred due to deficient IL-12 signalling, revealed as failure to phosphorylate the transcription factors Jak-2, Stat-1, Stat-3 and Stat-4 that could not be attributed to downregulation of these proteins or the T-cell kinase tyk2.[149] Subsequently, selective loss of the β2-subunit of the IL-12 receptor[149] was identified as the basis of IL-12 non-responsiveness. These studies also revealed that stabilization of IL-12 receptorβ2-subunit expression underlies the ability of IFN-γ to redirect Th cell commitment[149] and extended these findings to humans.[150] These findings are of great diagnostic and therapeutic significance for human asthma. Studies with recombinant IFN-γ, IL-12 or with genetically altered mice lacking these cytokines or their receptors predict that these cytokines have a strongly protective influence preventing Th2-mediated airway inflammation.[151,151a] Genetic analysis in mice has linked a candidate region syntenic (matched) to human 5q31.1, which contains a locus coding for interferon response factor (IRF)-1, to control of the persistence of IL-12 receptor β2-subunit during Th2 commitment,[152] suggesting the possibility of designing susceptibility screens for humans. At least one further Th2 commitment susceptibility locus is predicted[152a] and a major IL-5-regulating locus is predicted.[153] Manipulation of Th phenotype with, for example, IFN-γ, IL-12 or adjuvants that induce this cytokine coupling may be more rationally executed and monitored by measuring CD4 IL-12 receptor β2-subunit levels.

Cytokine and cognate (physical) signals from Th cells direct B-cell antibody production

B-cell antibody response to proteins in inhaled allergen or derived from airway pathogens such as virus only occur when both the naive mature B-cell and a preprimed (i.e. antigen-experienced) CD4$^+$ Th cell recognize the same antigen, although the B-cell and T-cell need not recognize the same antigenic epitope of that antigen.

The vast diversity of B-cell immunoglobulin specificities for antigen recognition develops during B-cell ontogeny in the bone marrow. Here B-cells mature and randomly rearrange immunoglobulin *VDJ* genes to generate antibodies that have monospecificity for each B-cell yet are also extremely diverse across the B-cell population. Immature B-cells express surface IgM (and IgD), which serves as the antigen receptor, and traffic to lung draining lymph nodes as well as other secondary lymphoid organs (spleen, mucosal-associated lymphoid organs). B-cells are highly efficent APCs and antibody responses are initiated when B-cells capture antigen with surface immunoglobulin and then present processed peptides on MHCII to T-cells of the same antigen specificity. Th cells then upregulate the coactivation molecule CD40L, a member of the TNF family, which triggers B-cell CD40 and drives the B-cell from a quiescent state into cell cycle and proliferation.[154] Secondary cytokine production from Th cells supports this proliferative response. Later the same cytokines cause IgM$^+$ mature B-cells to switch immunoglobulin isotype from IgM to either IgE, IgG isotypes or IgA. Th2 cells usually provide B-cell help for antibody production: IL-4 from these cells directs IgE class switch and IL-5 directs IgA production, However Th1-derived IFN-γ can also influence antibody production by suppressing Th2-driven isotypes and inducing IgG isotypes. In addition,

Th2-like CD8 cells (Tc2 cells; which lose cytotoxicity during commitment) have been shown, at least *in vitro*, to be able to provide B-cell help.[2]

Isotype switch and a process called affinity maturation, which selects B-cells producing the highest affinity antibodies generated by somatic hypermutation of the immunoglobulin genes, occur in specialized structures within secondary lymphoid organs called germinal centres. It has recently been discovered that germinal centres form in inflamed lung tissue outside nodes in a mouse model of asthma. Such local germinal centres may give rise to local antibody-secreting cells in the airways. Mature isotype-switched cells then leave the secondary lymphoid organ and either differentiate into plasma cells, which make large amounts of antibody, or, long-term memory B-cells able to rapidly give rise to large numbers of antibody-secreting plasma cells on subsequent antigen exposure.

IL-4 induction and signalling utilize molecular pathways different to those most cytokines

As described above, T-cell activation requires MHC–peptide–TCR signalling leading to increase in Ca^{2+} concentration and PLCγ1 activation. However, functional phenotype acquisition requires soluble signals, most importantly IL-4. The central role of IL-4 in Th2 phenotype acquisition,[155–157] MHCII upregulation and B-cell IgE switch has led to intense research on its signal transduction mechanisms. These differ markedly from other cytokines. IL-4 signals via a receptor complex composed of (at least) two IL-4 receptor α-chains and the IL-2 receptor γ-chain. Additionally, the IL-13 α-chain, which is homologous to the IL-4 receptor α-chain, may be involved in some cases. Receptor chain complexing leads to phosphorylation by JAK1 kinases,[158] facilitating interaction with the transcription factor STAT-6.[159,95,160] Phosphorylated STAT-6 dimerizes and translocates to the nucleus, where it binds to the promoter regions of IL-4-sensitives genes.[161–164] Concurrently IL-4 signals via phosphorylation of insulin-related substrate (IRS)-2, which is coupled to growth and proliferative responses. IL-4 induces its own expression. Translocation of NF-AT$_p$ from the cytosol to the cell nucleus is promoted by increased intracellular Ca^{2+} concentration and NF-ATp cooperates with AP-1 dimers, probably via an NFATp/AP-1 composite binding site in IL-4-sensitive promoters.[165,166] Tissue-specific expression of IL-4 is regulated by the protooncogene c-maf,[167] which is preferentially expressed in Th2 but not Th1 cells and acts synergistically with NF-AT$_p$ to induce IL-4 production. NF-AT$_p$ in turn is potentiated by a novel protein called NIP45 (NF-AT interacting protein).[168–170]

The importance of STAT-6 in the regulation of IL-4 responses and Th2 induction has been reinforced by gene ablation experiments in mice, which produces animals with markedly deficient Th2 immune responses. Paradoxically, similar results were not found with c-maf.

Although IL-4 appears to have a central role in allergy and probably asthma, several caveats must be noted. Firstly there is strong evidence to suggest that while IL-4 may govern allergic inflammation induction, established disease may proceed in an IL-4-independent manner.[96] Th2 responses that follow chronic stimulation are difficult or impossible to reverse with IL-12/IFN-γ and persist in an IL-4-independent manner. In the case of B-cell IgE production, there is evidence that maintenance of antibody production may be driven by IL-13, a close homologue of IL-4. However, T-cells lack IL-13 receptors and persistence of strong Th2 pattern responses is likely to be IL-4 independent,[171] which questions the value of blocking this cytokine in chronic disease. Furthermore there is at least *in vitro* evidence that primary Th2 pattern responses can be induced in a strictly IL-4-independent manner in some cases. Human CD4⁺ CD45RO⁻ T-cells can be driven into proliferation and Th2 commitment by triggering CD28 in the presence of IL-2 in a TCR- and IL-4-independent manner.[96] It is also now clear that IL-4 deficient mice, which fail to mount IgE responses or acute eosinophilic lung inflammation when challenged with antigen[126] can be coerced to develop eosinophilic lung inflammation when chronically challenged with antigen.[172] These results parallel similar findings in B-cell-deficient mice unable to make any antibodies[173] or IgE-deficient mice. The lung has an intrinsic capacity to respond to some inhaled insults by triggering IL-5 production from non-T-cell sources and by upregulating chemokines and other innate immunity defences to coordinate eosinophil infiltration. It is unlikely that the true role of IL-4 in human asthma will be unravelled until new IL-4 inhibitors currently in early development are tested in clinical trials.

T-cell responses in viral exacerbation and coinfection

The cytokine pattern observed in human asthma, whether measured as immunoreactive protein in biopsies or lavages or as mRNA message, does not match the classical Th2 pattern in that IFN-γ is consistently elevated and concomitantly expressed with IL-4 and IL-5.[174] In contrast, the majority of allergen-specific T-cell clones from human asthmatic lungs express predominantly Th2-like cytokines. This difference underscores the inescapable fact that asthma cannot be a purely atopic disease. Indeed, epidemiologically, asthma prevalance continues to rise whereas allergy prevalence is constant. The discrepancy begs explanation. A simple answer is provided by the observations that some humans T-cells coexpress IL-5 and IFN-γ or IL-4 and IFN-γ. A more sophisticated unproven explanation, predicted by recent studies in mice, suggests that defects in IFN-γ receptor signalling lead to increased IFN protein levels due to defective feedback inhibition.[151a]

IFN-γ expression, however, is also a very early response to viral infection and an essential component of cell-mediated immune defences to numerous pathogens. It is highly plausible therefore that the coexistence of Th2 cytokines with IFN-γ (and IL-12) reflects concurrent immune responses to inhaled antigen and one or more pathogens present as a coinfection of the lung.[175] It is suggested here that the severity and progression of asthma might therefore be better modelled as a problem of polysusceptibility to multiple antigens derived from both inhaled antigens and pathogens.

While the cosusceptibility concept presented here is speculative and must be assessed experimentally, there is more than ample evidence to suggest that asthmatic individuals are predisposed to viral infections that may precipitate exacerbations and, not infrequently, catastrophic crises. This increased susceptibility has been related to upregulated ICAM-1 expression in inflamed epithelium, since ICAM-1 is a rhinovirus receptor. We have previously proposed that Th2 immune responses compromise IFN-mediated host defences, leading to swifter, more severe and more frequent viral infections.[176] This concept is consistent with studies in mice showing that concurrent helminth infection, which induces a strong Th2 response, decreases viral clearance, CD8⁺ T-cell cytotoxicity and protective Th1 immune responses.[177]

Lymphocytes are sources of growth factors

As outlined in this chapter, lymphocytes are central agents in the induction of antibody production and the ability to mount cell-mediated host defence. It should be noted that very recent data suggests that lymphocytes may have another important function in chronic asthma: regulation of tissue bulk and tissue phenotype via the secretion of growth factors. It is of considerable interest that Panettieri et al. recently demonstrated that T-cells stimulate cell cycle progression in airway smooth muscle via a cognate interaction involving CD44 and hyaluronic acid.[44,178] More recently, epithelial homing gut $\gamma\delta$ T-cells have been found to produce hepatocyte growth factor-like molecules able to contribute to epithelial integrity. It will therefore be of great interest to determine the role of T-cells as growth factor sources in chronic asthma.

PERSPECTIVE: DEFECTS IN TERMINATION OF IMMUNE RESPONSES MAY CAUSE CHRONIC DISEASE

Under normal circumstances, T-cell responses to antigen initially amplify then contract spontaneously and terminate. This intrinsic negative regulation of immune responses is thought to protect the person from illnesses resulting from unchecked T-cell activation. In contrast, in asthma T-cell responses smoulder over decades. The simplest explanation for this is that chronic antigen exposure in the environment drives continuous T-cell activation. However animal studies suggest that T-cell responses downregulate during chronic antigen exposure. Surprisingly, nothing is known of the termination mechanisms that *end* airway immune responses and how their defects might lead to chronic airway disease.

Several mechanisms that terminate T-cell immune responses have been identified. Negative costimulation through CTLA-4, as discussed above, is one of the most important mechanisms. It is now also known that, at least in mouse model systems, chronic T-cell stimulation leads to downregulation of immune responses accompanied by apoptotic cell death of T-cells. This death by apoptosis follows two types of induction: (a) so-called 'activation-induced cell death' (AICD) and (b) active apoptosis induced by ligation of the FAS death receptor.[179,180] Strikingly, nothing is yet known about the role of lymphocyte ACID- and FAS-mediated apoptosis as determinants of chronic asthma. The author has recently proposed that defects in T-cell apoptosis are central pathogenic faults allowing chronic airway inflammation to develop.[180] (Fig. 9.4).

Much is now understood about how lymphocytes acquire distinct functional phenotypes, how they become activated and how they contribute to antibody production and cellular inflammation. Almost nothing is known about the molecular basis of disease severity and progression. Intuitively, the answers that are needed are not those to the questions of what determines atopy/allergy. Atopy is a risk factor for asthma not a cause of asthma. As proposed here, understanding two aspects of T-cell biology; i.e. defects that lead to more persistent immune responses and the problem of poly-susceptibility to *allergen and pathogen*, are at the forefront of advanced research into lymphocyte biology in asthma

9 Lymphocytes

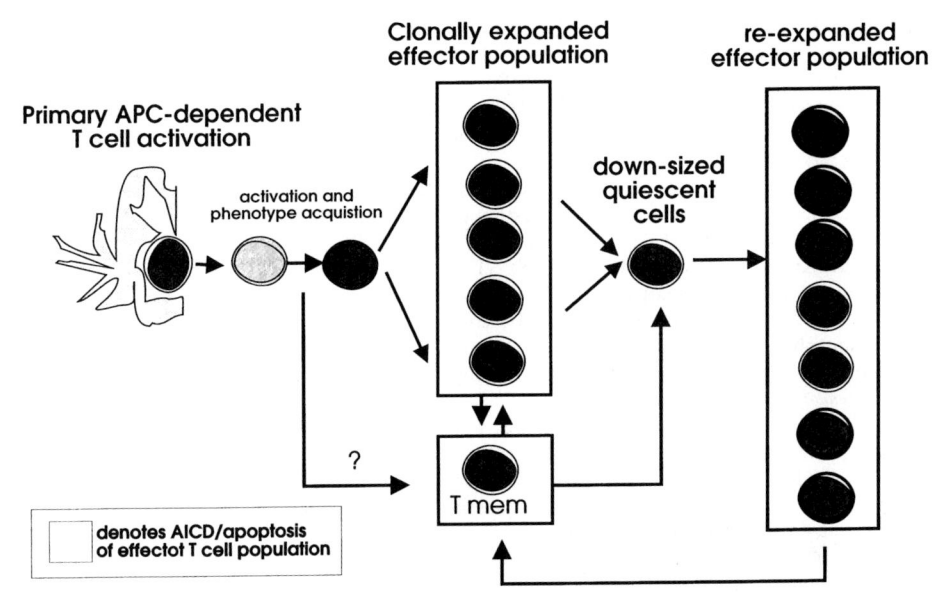

Fig. 9.4 Posible role of apoptosis in chronic-airway inflammation.

ACKNOWLEDGEMENT

The author is most grateful to Ms V. Tresidder for patient and expert assistance in the preparation of this manuscript.

REFERENCES

1a. Zingernagel RM, Doherty PC: Major transplant antigens viruses and specificity of surveillance T-cells. *Contemp Topic Immunol* (1977) **7**: 179.

1b. Strominger, JL: The $\gamma\delta$ T-cell receptor and class 1b MHC-related proteins: emigmatic molecules of immune recognition. *Cell* (1989) **57**: 895–898.

1c. Korsgren M, Erjefalt JS, Korsgren O, Sundler F, Persson CG: Allergic eosinophil-rich inflammation develops in lungs and airways of B-cell-deficient mice. *J Exp Med* (1997) **185**: 885–892.

1d. Saltini C, Richeldi L, Holroyd KJ, du Bois RM, Crystal RG: Lymphocytes. In *The Lung. Scientific Foundations* (eds) R.G. Crystal & J.B. West. New York: Raven Press, 1991, pp 459–482.

2. Erard F, Wild MT, Garcia-Sanz-JA, Le-Gros G: Switch of CD8 T-cells to noncytolytic CD8-CD4$^-$ cells that make Th2 cytokines and help B-cells. *Science* (1993) **260**: 1802–1805.

3. Azzawi M, Bradley B, Jeffery PK, *et al.*: Identification of activated T-lymphocytes and eosinophils in bronchial biopsies in stable atopic asthma. *Am Rev Respir Dis* (1990) **142**: 1410–1413.

4. Kirby JG, Hargreave FE, Gleich GJ, O'Byrne PM: Bronchoalveolar cell profiles of asthmatic and nonasthmatic subjects. *Am Rev Respir Dis* (1987) **136**: 379–383.

5. Jeffery PK, Wardlaw AJ, Nelson FC, Collins JV, Kay AB: Bronchial biopsies in asthma: an ultrastructural, quantitative study and correlation with hyperreactivity. *Am Rev Respir Dis* (1989) **140**: 1745–1753.

6. Hamid Q, Barkans J, Robinson DS, Durham SR, Kay AB: Co-expression of CD25 and CD3 in atopic allergy and asthma. *Immunology* (1992) **75**: 659–663.
7. Robinson D, Hamid Q, Bentley A, Ying S, Kay AB, Durham SR: Activation of CD4$^+$ T-cells, increased Th2-type cytokine mRNA expression, and eosinophil recruitment in bronchoalveolar lavage after allergen inhalation challenge in patients with atopic asthma. *J Allergy Clin Immunol* (1993) **92**: 313–324.
8. Robinson DS, Bentley AM, Hartnell A, Kay AB, Durham S: Activated memory T helper cells in bronchoalveolar lavage from atopic asthmatics. Relationship to asthma symptoms, lung function and bronchial responsiveness. *Thorax* (1993) **48**: 26–32.
9. Beasley R, Roche W, Roberts JA, Holgate ST: Cellular events in the bronchi in mild asthma and after bronchial provocation. *Am Rev Respir Dis* (1989) **139**: 806–817.
10. Bradley BL, Azzawi M, Assoufi B, *et al.*: Eosinophils, T-lymphocytes, mast cells, neutrophils and macrophages in bronchial biopsies from atopic asthmatics: comparison with atopic nonasthma and normal controls and relationship to bronchial hyperresponsiveness. *J Allergy Clin Immunol* (1991) **88**: 661–674.
11. Jung T, Schauer U, Rieger C: Interleukin-4 and interleukin-5 are rarely co-expressed by human T-cells. *Eur J Immunol* (1995) **25**: 2413–2416.
12. Walker C, Kaego MK, Braun MD, Blaser K: Activated T-cells and eosinophils in bronchoalveolar lavages from subjects with asthma correlated with disease severity. *J Allergy Clin Immunol* (1991) **88**: 935–942.
13. Virchow JC, Walker C, Hafner D, *et al.*: T-cells and cytokines in bronchoalveolar lavage fluid after segmental allergen provocation in atopic asthma. *Am J Respir Crit Care Med* (1995) **151**: 960–968.
14. Robinson DS, Hamid Q, Ying S, *et al.*: Evidence for a predominant 'Th2-type' bronchoalveolar lavage T-lymphocyte population in atopic asthma. *N Engl J Med* (1992) **326**: 298–304.
15. Robinson DS, Ying S, Bentley AM, *et al.*: Relationships among numbers of bronchoalveolar lavage cells expression messenger ribonucleic acid for cytokines, asthma symptoms, and airway methacholine responsiveness in atopic asthma. *J Allergy Clin Immunol* (1993) **92**: 397–403.
16. Bentley AM, Menz G, Storz C, *et al.*: Identification of T-lymphocytes, macrophages, and activated eosinophils in the bronchial mucosa in intrinsic asthma. Relationship to symptoms and bronchial responsiveness. *Am Rev Respir Dis* (1992) **146**: 500–506.
17. Bentley AM, Meng Q, Robinson DS, Hamid Q, Kay AB, Durham SR: Increases in activated T-lymphocytes, eosinophils and cytokine messenger RNA for IL-5 and GM-CSF in bronchial biopsies after allergen inhalation challenge in atopic asthmatics. *Am J Respir Cell Mol Biol* (1993) **8**: 35–42.
18. Till SJ, Li B, Durham S, *et al.*: Secretion of the eosinophil-active cytokines IL-5, GM-CSF and IL-3 by bronchoalveolar lavage CD4$^-$ and CD8$^-$ T-cell lines in atopic asthmatics and atopic and nonatopic controls. *Eur J Immunol* (1995) **25**: 2727–2731.
19. Aalbers R, Kauffman HF, Vrugt B, Koeter GH, de-Monchy JG: Allergen-induced recruitment of inflammatory cells in lavage 3 and 24 h after challenge in allergic asthmatic lungs. *Chest* (1993) **103**: 1178–1184.
20. Bellini A, Vittori E, Marini M, Ackerman V, Mattoli S: Intraepithelial dendritic cells and selective activation of Th2-like lymphocytes in patients with atopic asthma. *Chest* (1993) **103**: 997–1005.
21. Corrigan CJ, Hamid Q, North J, *et al.*: Peripheral blood CD4, but not CD8 T-lymphocytes in patients with exacerbation of asthma transcribe and translate messenger RNA encoding cytokines which prolong eosinophil survival in the context of a Th2-type pattern: effect of glucocorticoid therapy. *Am J Respir Cell Mol Biol* (1995) **12**: 567–578.
22. Djukanovic R, Feather I, Gratziou C, *et al.*: Effect of natural allergen exposure during the grass pollen season on airways inflammatory cells and asthma symptoms. *Thorax* (1996) **51**: 575–581.
23. Doi S, Murayama N, Inoue T, *et al.*: CD4 T-lymphocyte activation is associated with peak expiratory flow variability in childhood asthma. *J Allergy Clin Immunol* (1996) **97**: 955–962.
24. Gemou-Engesaeth V, Kay AB, Bush A, Corrigan CJ: Activated peripheral blood CD4 and

9 Lymphocytes

CD8 T-lymphocytes in child asthma: correlation with eosinophilia and disease severity. *Pediatr Allergy Immunol* (1994) **5**: 170–177.

25. Gratziou C, Carroll M, Montefort S, Teran L, Howarth PH, Holgate ST: Inflammatory and T-cell profile of asthmatic airways 6 hours after local allergen provocation. *Am J Respir Crit Care Med* (1996) **153**: 515–520.
26. Gratziou C, Carroll M, Walls A, Howarth PH, Holgate ST: Early changes in T-lymphocytes recovered by bronchoalveolar lavage after local allergen challenge of asthmatic airways. *Am Rev Respir Dis* (1992) **145**: 1259–1264.
27. Kay AB, Ying S, Durham SR: Phenotype of cells positive for interleukin-4 and interleukin-5 mRNA in allergic tissue reactions. *Int Arch Allergy Immunol* (1995) **107**: 208–210.
28. Kennedy JD, Hatfield CA, Fidler SF, *et al.*: Phenotypic characterization of T-lymphocytes emigrating into lung tissue and the airway lumen after antigen inhalation in sensitized mice. *Am J Respir Cell Mol Biol* (1995) **12**: 613–623.
29. Mori A, Suko M, Tsuruoka N, *et al.*: Allergen-specific human T-cell clones produce interleukin-5 upon stimulation with the Th1 cytokine interleukin-2. *Int Arch Allergy Immunol* (1995): 220–222.
30. Oehling AG Jr, Walker C, Virchow JC, Blaser K: Correlation between blood eosinophils, T-helper cell activity markers and pulmonary function in patients with allergic and intrinsic asthma. *J Invest Allergol Clin Immunol* (1992) **2**: 295–299.
31. Oosterhoff Y, Hoogsteden HC, Rutgers B, Kauffman HF, Postma DS: Lymphocyte and macrophage activation in bronchoalveolar lavage fluid in nocturnal asthma. *Am J Respir Crit Care Med* (1995) **151**: 75–81.
32. Richmond I, Booth H, Ward C, Walters EH: Intrasubject variability in airway inflammation in biopsies in mild to moderate stable asthma. *Am J Respir Crit Care Med* (1996) **153**: 899–903.
33. Synek M, Beasley R, Frew AJ, *et al.*: Cellular infiltration of the airways in asthma of varying severity. *Am J Respir Crit Care Med* (1996) **154**: 224–230.
34. Robinson DS, Hamid Q, Ying S, *et al.*: Prednisolone treatment in asthma is associated with modulation of bronchoalveolar lavage cell interleukin-4, interleukin-5 and interferon-γ cytokine gene expression. *Am Rev Respir Dis* (1993) **148**: 402–406.
35. Fukuda T, Asakawa J, Motojima S, Makino S: Cyclosporin A reduces T-lymphocyte activity and improves airway hyperresponsiveness in corticosteroid-dependent chronic severe asthma. *Ann Allergy Asthma Immunol* (1995) **75**: 65–69.
36. Mori A, Suko M, Nishizaki Y, *et al.*: IL-5 production by $CD4^+$ T-cells of asthmatic patients is suppressed by glucocorticoids and the immunosuppressants FK506 and cyclosporin A. *Int Immunol* (1995) **7**: 449–457.
37. Alexander AG, Barnes NC, Kay AB, Corrigan CJ: Clinical response to cyclosporin in chronic severe asthma is associated with reduction in serum soluble interleukin-2 receptor concentrations. *Eur Respir J* (1995) **8**: 574–578.
38. Wilson JW, Djukanovic R, Howarth PH, Holgate ST: Inhaled beclomethasone dipropionate downregulates airway lymphocyte activation in atopic asthma. *Am J Respir Crit Care Med* (1994) **149**: 86–90.
39. Del Prete GF, De Carli M, D'Elios MM, *et al.*: Allergen exposure induces the activation of allergen-specific Th2 cells in the airway mucosa of patients with allergic respiratory disorders. *Eur J Immunol* (1993) **23**: 1445–1449.
40. Krouwels FH, Hol BE, Bruinier B, Lutter R, Jansen HM, Out TA: Cytokine production by T-cell clones from bronchoalveolar lavage fluid of patients with asthma and healthy subjects. *Eur Respir J* Suppl. (1996) **22**: 95S–103S.
41. Krug N, Madden J, Redington AE, *et al.*: T-cell cytokine profile evaluated at the single cell level in BAL and blood in allergic asthma. *Am J Respir Cell Mol Biol* (1996) **14**: 319–326.
42. MacLean JA, Ownbey R, Luster AD: T-cell-dependent regulation of eotaxin in antigen-induced pulmonary eosinophila. *J Exp Med* (1996) **184**: 1461–1469.
43. Boismenu R, Havran WL: Modulation of epithelial cell growth by intraepithelial gamma delta T-cells. *Science* (1994) **266**: 1253–1255.
44. Lazaar AL, Albelda SM, Pilewski JM, Brennan B, Pure E, Panettieri RA Jr: T-lymphocytes adhere to airway smooth muscle cells via integrins and CD44 and induce smooth muscle cell DNA synthesis. *J Exp Med* (1994) **180**: 807–816.

45. Bloemen PG, Buckley TL, van den Tweel MC, *et al*.: LFA-1, and not Mac-1, is crucial for the development of hyperreactivity in a murine model of nonallergic asthma. *Am J Respir Crit Care Med* (1996) **153**: 521–529.
46. Corry DB, Folkesson HG, Warnock ML: Interleukin 4, but not interleukin 5 or eosinophils, is required in a murine model of acute airway hyperreactivity. *J Exp Med* (1996) **183**: 109–117.
47. Lukacs NW, Strieter RM, Chensue SW, Kunkel SL: Interleukin-4-dependent pulmonary eosinophil infiltration in a murine model of asthma. *Am J Respir Cell Mol Biol* (1994) **10**: 526–532.
48. Montefort S, Gratziou C, Goulding D, *et al*.: Bronchial biopsy evidence for leukocyte infiltration and upregulation of leukocyte–endothelial cell adhesion molecules 6 hours after local allergen challenge of sensitized asthmatic airways. *J Clin Invest* (1994) **93**: 1411–1421.
49. Maestrelli P, Del Prete GF, De Carli M, *et al*.: CD8 T-cell clones producing interleukin-5 and interferon-gamma in bronchial mucosa of patients with asthma induced by toluene diisocyanate. *Scand J Work Environ Health* (1994) **20**: 376–381.
50. Laberge S, Wu L, Olivenstein R, Xu LJ, Renzi PM, Martin JG: Depletion of $CD8^+$ T-cells enhances pulmonary inflammation but not airway responsiveness after antigen challenge in rats. *J Allergy Clin Immunol* (1996) **98**: 617–627.
51. Ferrick DA, Schrenzel MD, Mulvania T, Hsieh B, Ferlin WG, Lepper H: Differential production of interferon-gamma and interleukin-4 in response to Th1- and Th2- stimulating pathogens by gamma delta T-cells *in vivo*. *Nature* (1995) **373**: 255–257.
52. Spinozzi F, Agea E, Bistoni O, *et al*.: Increased allergen-specific, steroid-sensitive gamma delta T-cells in bronchoalveolar lavage fluid from patients with asthma. *Ann Intern Med* (1996) **124**: 223–227.
53. Spinozzi F, Agea E, Bistoni O, *et al*.: Local expansion of allergen-specific $CD30^+$ Th2-type gamma delta T-cells in bronchial asthma. *Mol Med* (1995) **1**: 821–826.
54. Haczku A, Chung KF, Sun J, Barnes PJ, Kay AB, Moqbel R: Airway hyperresponsiveness, elevation of serum-specific IgE and activation of T-cells following allergen exposure in sensitized Brown-Norway rats. *Immunology* (1995) **85**: 598–603.
55. Huang XZ, Wu JF, Cass D, *et al*.: Inactivation of the integrin beta 6 subunit gene reveals a role of epithelial integrins in regulating inflammation in the lung and skin. *J Cell Biol* (1996) **133**: 921–928.
56. Lambert LE, Berling JS, Kudlacz EM: Characterization of the antigen-presenting cell and T-cell requirements for induction of pulmonary eosinophilia in a murine model of asthma. *Clin Immunol Immunopathol* (1996) **81**: 307–311.
57. Cella M, Scheidegger D, Palmer-Lehmann K, Lane P, Lanzavecchia A, Alber G: Ligation of CD40 on dendritic cells triggers production of high levels of interleukin-12 and enhances T-cell stimulatory capacity: T–T help via APC activation. *J Exp Med* (1996) **184**: 747–752.
58. Valitutti S, Muller S, Cella M, Padovan E, Lanzavecchia A: Serial triggering of many T-cell receptors by a few peptide-MHC complexes. *Nature* (1995) **375**: 148–151.
59. Viola A, Lanzavecchia A: T-cell activation determined by T-cell receptor number and tunable thresholds. *Science* (1996) **273**: 104–106.
60. Cruikshank WW, Long A, Tarpy RE, *et al*.: Early identification of interleukin-16 (lymphocyte chemoattractant factor) and macrophage inflammatory protein 1 alpha (MIP1 alpha) in bronchoalveolar lavage fluid of antigen-challenged asthmatics. *Am J Respir Cell Mol Biol* (1995) **13**: 738–747.
61. Iwamoto I, Nakao A: Induction of Th2 cell tolerance to a soluble antigen by blockade of the LFA-1-dependent pathway prevents allergic inflammation. *Immunol Res* (1995) **14**: 263–270.
62. Burastero SE, Crimi E, Balbo A, *et al*.: Oligoclonality of lung T-lymphocytes following exposure to allergen in asthma. *J Immunol* (1995) **155**: 5836–5846.
63. Gelder CM, Morrison JF, Chung KF, Barnes PJ, Adcock IM: T-cell receptor repertoire in peripheral blood and bronchial biopsies from normal and asthmatic subjects. *Biochem Soc Trans* (1996) **24**: 316S.
64. Krinzman SJ, De Sanctis GT, Cernadas M, *et al*.: T-cell activation in a murine model of asthma. *Am J Physiol* (1996) **271**: L476–L483.
65. Krinzman SJ, De Sanctis GT, Cernadas M, *et al*.: Inhibition of T-cell costimulation abrogates airway hyperresponsiveness in a murine model. *J Clin Invest* (1996) **98**: 2693–2699.

9 Lymphocytes

66. Lai CK, Chan CH, Leung JC, Lai KN: Serum concentration of soluble interleukin 2 receptors in asthma. Correlation with disease activity. *Chest* (1993) **103**: 782–786.
67. Park CS, Lee SM, Chung SW, Uh S, Kim HT, Kim YH: Interleukin-2 and soluble interleukin-2 receptor in bronchoalveolar lavage fluid from patients with bronchial asthma. *Chest* (1994) **106**: 400–406.
68. Laberge, Cruikshank WW, Kornfeld H, Center DM: Histamine-induced secretion of lymphocyte chemoattractant factor from CD8+ T-cells is independent of transcription and translation. Evidence for constitutive protein synthesis and storage. *J Immunol* (1995) **155**: 2902–2910.
69. Vignola AM, Chanez P, Campbell AM, *et al*.: Quantification and localization of HLA-DR and intercellular adhesion molecule-1 (ICAM-1) molecules on bronchial epithelial cells of asthmatics using confocal microscopy. *Clin Exp Immunol* (1994) **96**: 104–109.
70. Tsitoura DC, Holter W, Cerwenka A, Gelder CM, Lamb JR: Induction of anergy in human T helper 0 cells by stimulation with altered T-cell antigen receptor ligands. *J Immunol* (1996) **156**: 2801–2808.
71. Tsitoura DC, Verhoef A, Gelder CM, O'Hehir RE, Lamb JR: Altered T-cell ligands derived from a major house dust mite allergen enhance IFN-gamma but not IL-4 production by human CD4+ T-cells. *J Immunol* (1996) **157**: 2160–2165.
72. Akdis CA, Akdis M, Blesken T, *et al*.: Epitope-specific T-cell tolerance to phospholipase A2 in bee venom immunotherapy and recovery by IL-2 and IL-15 *in vitro*. *J Clin Invest* (1994) **98**: 1676–1683.
73. Molfino NA, Doherty PJ, Suurmann IL, *et al*.: Analysis of the T-cell receptor Vgamma region gene repertoire in bronchoalveolar lavage (BAL) and peripheral blood of atopic asthmatics and healthy subjects. *Clin Exp Immunol* (1996) **104**: 144–153.
74. Wedderburn LR, O'Hehir RE, Hewitt CR, Lamb JR, Owen MJ: *In vivo* clonal dominance and limited T-cell receptor usage in human CD4+ T-cell recognition of house dust mite allergens. *Proc Natl Acad Sci USA* (1993) **90**: 8214–8218.
75. Renz H, Bradley K, Gefland EW, Production of interleukin-4 and interferon-gamma by TCR-V beta-expressing T-cell subsets in allergen-sensitized mice. *Am J Respir Cell Mol Biol*: (1996) **14**: 36–43.
76. Neumeister EN, Zhu Y, Richard S, *et al*.: Binding of ZAP-70 to phosphorylated T-cell receptor zeta and eta enhances its autophosphorylation and generates specific binding sites for SH2 domain-containing proteins. *Mol Cell Biol* (1995) **15**: 3171–3178.
77. Madrenas J, Wange RL, Wang JL, Isakov N, Samelson SE, Germain RN: Zeta-phosphorylation without ZAP70 activation induced by TCR antagonists or partial agonists. *Science* (1995) **267**: 515–518.
78. Sloan-Lancaster J, Shaw AS, Ruthbard JB, Allen PM: Partial T-cell signalling: altered phospho-zeta and lack of ZAP-70 recruitment in APL-induced T-cell anergy. *Cell* (1994) **79**: 913–922.
79. Walker C, Bode E, Boer L, Hansel TT, Blaser K, Virchow JC Jr: Allergic and nonallergic asthmatics have distinct patterns of T-cell activation and cytokine production in peripheral blood and bronchoalveolar lavage. *Am Rev Respir Dis* (1992) **146**: 109–115.
80. Werfel S, Massey W, Lichtenstein LM, Bochner BS. Preferential recruitment of activated, memory T-lymphocytes into skin chamber fluids during human cutaneous late-phase allergic reactions. *J Allergy Clin Immunol* (1995) **96**: 57–65.
81. Rabb HA, Olivenstein R, Issekutz TB, Renzi PM, Martin JG: The role of the leukocyte adhesion molecules VLA-4, LFA-1, and Mac-1 in allergic airway responses in the rat. *Am J Respir Crit Care Med* (1994) **149**: 1186–1191.
82. Rihs S, Walker C, Virchow JC Jr, *et al*.: Differential expression of alpha E beta 7 integrins on bronchoalveolar lavage T-lymphocyte subsets: regulation by alpha 4 beta 1-integrin cross-linking and TGF-beta. *Am J Respir Cell Mol Biol* (19xx) **15**: 600–610.
83. Krinzman SJ, De-Sanctis GT, Cernadas M, *et al*.: Inhibition of T cell costimulation abrogates airway hyperresponsiveness in a murine model. *J Clin Invest* (1996) **98**: 2693–2699.
84. Adams DH, Harvath L, Bottaro DP, *et al*.: Hepatocyte growth factor and macrophage inflammatory protein 1 beta: structurally distinct cytokines that induce rapid cytoskeletal

changes and subset-preferential migration in T-cells. *Proc Natl Acad Sci USA* (1994) **91**: 7144–7148.
85. Alam R, York J, Boyars M, et al.: Increased MCP-1, RANTES, and MIP-1alpha in bronchoalveolar lavage fluid of allergic asthmatic patients. *Am J Respir Crit Care Med* (1996) **153**: 1398–1404.
86. Qin S, LaRosa G, Campbell JJ, et al.: Expression of monocyte chemoattractant protein-1 and interleukin-8 receptors on subsets of T-cells, correlation with transendothelial chemotactic potential. *Eur J Immunol* (1996) **26**: 640–647.
87. Caw MW, Roth SJ, Luther E, Rose SS, Springer TA: Monocyte chemoattractant protein 1 acts as a T-lymphocyte chemoattractant. *Proc Natl Acad Sci USA* (1994) **91**: 3652–3656.
87a. Schall TJ, Bacon K, Toy KJ, Goeddel DV: Selective attraction of monocytes and T-lymphocytes of the memory phenotype by cytokine RANTES. *Nature* (1990) **347**: 669–671.
88. Forster R, Emrich T, Kremmer E, Lipp M: Expression of the G-protein-coupled receptor BLR1 defines mature, recirculating B-cells and a subset of T-helper memory cells. *Blood* (1994) **84**: 830–840.
89. Weaver CT, Hawrylowics CM, Unanve ER: T helper subsets require the expression of distinct costimulatory signals by antigen-presenting cells. *Proc Natl Acad Sci USA* (1988) **85**: 8181–8185.
90. Kline JN, Fisher PA, Monick MM, Hunninghake GW: Regulation of interleukin-1 receptor antagonist by Th1 and Th2 cytokines. *Am J Physiol* (1995) **269**: L92–L98.
91. Salari R, Smithers N, Page K, Bolton E, Champion BR: Interleukin 1 responsiveness and receptor expression by murine Th1 and Th2 clones. *Cytokines* (1990) **2**: 129–141.
92. Zubiaga AM, Munoz E, Huber BT: Production of IL-1α by activated Th type 2 cells. Its role as an autocrine growth factor. *J Immunol* (1991) **146**: 3849–3856.
93. Colotta F, Muzio FRM, Beutini R, et al.: Interleukin-1 type II receptor: a decoy target for IL-1 that is regulated by IL-4. *Science* (1993) **261**: 472–475.
94. Colotta F, Orlando S, Fadlon EJ, et al.: Chemoattractants induce rapid release of the interleukin 1 type II decoy receptor in human polymorphonuclear cells. *J Exp Med* (1995) **181**: 2181–2186.
95. Colotta F, Saccani S, Giri JG, et al.: Regulated expression and release of the IL-1 decoy receptor in human mononuclear phagocytes. *J Immunol* (1996) **156**: 2534–2541.
95a. Kam JC, Szefler SJ, Surs W, Sher ER, Leung DY: Combination IL-2 and IL-4 reduces glucocorticoid receptor-binding affinity and T-cell response to glucocorticoids. *J Immunol* (1993) **151**: 3460–3466.
96. Brinkmann V, Kinzel B, Kristofic C: TCR-independent activation of human CD4$^+$ 45RO$^-$ T-cells by anti-CD28 plus IL-2: induction of clonal expansion and priming for a Th2 phenotype. *J Immunol* (1996) **156**: 4100–4106.
97. Holloway JW, Doull I, Begishvili B, Beasley R, Holgate ST, Howell WM: Lack of evidence of a significant association between HLA-DR, DQ and DP genotypes and atopy in families with HDM allergy. *Clin Exp Allergy* (1996) **26**: 1142–1149.
98. Huang SK, Yi M, Palmer E, Marsh DG: A dominant T-cell receptor beta-chain in response to a short ragweed allergen, *Amb a* 5. *J Immunol* (1995) **154**: 6157–6162.
99. O'Brien RM, Thomas WR, Nicholson I, Lamb JR, Tait BD: An immunogenetic analysis of the T-cell recognition of the major house dust mite allergen *Der p* 2: identification of high- and low-responder HLA-DQ alleles and localization of T-cell epitopes. *Immunology* (1995) **86**: 176–182.
100. Blumenthal M, Marcus-Bagley D, Awdeh Z, Johnson B, Yunis EJ, Alper: HLA-DR2, [HLA-B7, SC31, DR2], and [HLA-B8, SC01, DR3] haplotypes distinguish subjects with asthma from those with rhinitis only in ragweed pollen allergy. *J Immunol* (1992) **148**: 411–416.
101. Schwartz RH: A cell culture model for T-lymphocyte clonal anergy: *Science* (1990) **248**: 1349–1356.
102. Noel PJ, Boise LH, Green JM, Thompson CB: CD28 costimulation prevents cell death during primary T cell activation. *J Immunol* (1996) **157**: 636–642
103. Harding FA, McArthur JG, Gross JA, Raulet DH, Allison JP: CD28 mediated signalling costimulates murine T-cells and prevents induction of anergy in T-cell clones. *Nature* (1992) **356**: 607–609.

9 Lymphocytes

104. Krummel MF, Allison JP: CTLA-4 engagement inhibits IL-2 accumulation and cell cycle progression upon activation of resting T-cells. *J Exp Med* (1996) **183**: 2533–2540.
105. Jenkins MK, Taylor PS, Norton SD, Urdahl KB: CD28 delivers a costimulatory signal involved in antigen specific IL-2 production by human T-cells. *J Immunol* (1991) **147**: 2461–2466.
106. Linsley PS, Brady W, Grosmaire L, Aruffo A, Damle NK, Ledbetter JA: Binding of B-cell activation antigen B7 to CD28 costimulates T-cell proliferation and interleukin-2 mRNA accumulation. *J Exp Med* (1991) **173**: 721–730.
107. Freeman GJ, Girbben JG, Boussiotis VA, *et al.*: Cloning of B7-2: a CTLA-4 counter receptor that costimulates human T-cell proliferation. *Science* (1993) **262**: 909–911.
108. Masten BJ, Yates JL, Pollard Koga AM, Lipscomb MF: Characterization of accessory molecules in murine lung dendritic cell function: roles of CD80, CD86, CD54 and CD40L. *Am J Respir* (1997) **16**: 335–342.
109. Walunas TL, Bakker CY, Bluestone JA: CTLA-4 ligation blocks CD28-dependent T-cell activation. *J Exp Med* (1996) 2541–2550.
110. Keane-Myers A, Gause WC, Linsley PS, Chen SJ, Wills-Karp M: B7-CD28/CTLA-4 costimulatory pathways are required for the development of T helper cell 2-mediated allergic airway responses to inhaled antigens. *J Immunol* (1997) **158**: 2042–2049.
111. Tsuyuki S, Bertrand C, Erard F, *et al.*: Activation of the Fas receptor on lung eosinophils leads to apoptosis and the resolution of eosinophilic inflammation of the airways. *J Clin Invest* (1995) **96**: 2924–2931.
112. Kucheroo VK, DAS MP, Brown JA, *et al.*: B7-1 and B7-2 costimulatory molecules activate differentially the Th1/Th2 developmental pathways: application to autoimmune disease therapy. *Cell* (1995) **80**: 707–718.
113. Rooney JW, Hodge MR, McCaffrey PG, Rao A, Glimcher LH: A common factor regulates both Th1- and Th2-specific cytokine gene expression. *EMBO J* (1994) **13**: 625–633.
114. Harris N, Campbell C, Le-Gros G, Ronchese F: Blockade of CD28/B7 co-stimulation by mCTLA4-Hygamma1 inhibits antigen-induced lung eosinophilia but not Th2 cell development or recruitment in the lung. *Eur J Immunol* (1997) **27**: 166–161.
115. Harris N, Peach R, Naemura J, Linsley PS, Le-Gros G, Ronchese F: CD80 costimulation is essential for the induction of airway eosinophilia. *J Exp Med* (1997) **185**: 177–182.
116. DeBenedette MA, Chahinian A, Mak TW, Watts TH: Costimulation of CD28-0 T-lymphocytes by 4-1BB ligand. *J Immunol* (1997) **158**: 551–559.
117. Leiden JM: Transcriptional regulation of T-cell receptor genes. *Annu Rev Immunol* (1993) **11**: 539–570.
118. Henderson AJ, McDougall S, Leiden J, Calame KL: GATA elements are necessary for the activity and tissue specificity of the T-cell receptor beta-chain transcriptional enhancer. *Mol Cell Biol* (1994) **14**: 4286–4294.
119. Ho IC, Vorhees P, Marin N, *et al.*: Human GATA-3: a lineage-restricted transcription factor that regulates the expression of the T-cell receptor alpha gene. *EMBO J* (1991) **10**: 1187–1192.
120. Smith VM, Lee PP, Szychowski S, Winoto A: GATA-3 dominant negative mutant. Functional redundancy of the T-cell receptor alpha and beta enhancers. *J Biol Chem* (1995) **270**: 1515–1520.
121. Hambor JE, Mennone J, Coon ME, Hanke JH, Kavathas P: Identification and characterization of an Alu-containing, T-cell-specific enhancer located in the last intron of the human CD8 alpha gene. *Mol Cell Biol* (1993) **13**: 7056–7070.
122. Landry DB, Engel JD, Sen R: Functional GATA-3 binding sites within murine CD8 alpha upstream regulatory sequences. *J Exp Med* (1993) **178**: 941–949.
123. Penix L, Weaver WM, Pang Y, Young HA, Wilson CB: Two essential regulatory elements in the human interferon gamma promoter confer activation specific expression in T-cells. *J Exp Med* (1993) **178**: 1483–1496.
124. Ko LJ, Yamamoto M, Leonard MW, George KM, Ting P, Engel JD: Murine and human T-lymphocyte GATA-3 factors mediate transcription through a *cis*-regulatory element within the human T-cell receptor delta gene enhancer. *Mol Cell Biol* (1991) **11**: 2778–2784.
125. Kopf M, Le-Gros G, Bachmann M, Lamers MC, Bluethmann H, Kohler G: Disruption of the murine IL-4 gene blocks Th2 cytokine responses. *Nature* (1993) **362**: 245–8.

126. Coyle AJ, Le Gros G, Bertrand C, et al.: Interleukin-4 is required for the induction of lung Th2 mucosal immunity. *Am J Respir Cell Mol Biol* (1995) **13**: 54–59.
127. Von der Weid T, Beebe AM, Roopenian DC, Coffman RL: Early production of IL-4 and induction of Th2 responses in the lymph node originate from an MHC class I-independent CD4+NK1.1− T-cell population. *J Immunol* (1996) **157**: 4421–4427.
128. Yoshimoto T, Paul WE: CD4pos, NK1.1pos T-cells promptly produce interleukin 4 in response to *in vivo* challenge with anti-CD3. *J Exp Med* (1994) **179**: 1285–1295.
129. Joyce S, Negishi I, Boesteanu A, et al.: Expansion of natural (NK1+) T-cells that express alpha beta T-cell receptors in transporters associated with antigen presentation-1 null and thymus leukemia antigen positive mice. *J Exp Med* (1996) **184**: 1579–1584.
130. Bendelac A, Lantz O, Quimby ME, Yewdell JW, Bennink JR, Brutkiewicz: CD1 recognition by mouse NK1+ T-lymphocytes. *Science* (1995) **263**: 863–865.
131. Brown Dr, Fowell DJ, Corry DB, et al.: Beta 2-microglobulin-dependent NK1.1+ T-cells are not essential for T helper cell 2 immune responses. *J Exp Med* (1996) **184**: 1295–1304.
132. Zhang Y, Rogers KH, Lewis DB: Beta 2-microglobulin-dependent T-cells are dispensable for allergen-induced T helper 2 responses. *J Exp Med* (1996) **184**: 1507–1512.
133. Guery JC, Galbiati F, Smiroldo S, Adorini L: Selective development of T helper (Th)2 cells induced by continuous administration of low dose soluble proteins to normal and beta (2)-microglobulin-deficient BALB/c mice. *J Exp Med* (1996) **183**: 485–497.
134. Kamogawa W, Minasi LA, Carding SR, Bottomly K, Flavell RA: The relationship of IL-4- and IFN gamma-producing T-cells studied by lineage ablation of IL-4-producing cells. *Cell* (1993) **75**: 985–995.
135. Rincon MJ, Anguita, J, Nakamura T, Fikrig E & Flavell RA: IL-6 directs the differentiation of IL-4-producing CD4+ T-cells. *J Exp Med* (1997) **185**: 461–469.
136. Cromwell O, Hamid Q, Corrigan CJ, et al.: Expression and generation of interleukin-8, IL-6 and granulocyte–macrophage colony-stimulating factor by bronchial epithelial cells and enhancement by IL-1 beta and tumour necrosis factor-alpha. *Immunology* (1992) **77**: 330–337.
137. Adler KB, Fischer BM, Wright DT, Cohn LA, Becker S: Interactions between respiratory epithelial cells and cytokines: relationships to lung inflammation. *Ann NY Acad Sci* (1994) **725**: 128–145.
138. Gretz, JE, Kaldijan EP, Anderson AO, Shaw, S: Sophisticated strategies for information encounter in the lymph node. The reticular network as a conduit of soluble information and a highway for cell traffic. *J Immunol* (1996) **157**: 495–499.
139. Smiley ST, Kaplan MH, Grusby MJ: Immunoglbulin E production in the absence of interleukin-4-secreting CD1-dependent cells. *Science* (1997) **275**: 977–979.
140. Ronchese F, Hausmann B, Le-Gros G, et al.: Interferon-gamma- and interleukin-4-producing T-cells can be primed on dendritic cells in vivo and do not require the presence of B-cells. *Eur J Immunol* (1994) **24**: 1148–1154.
141. Paul WE, Seder RA, Plaut M: Lymphokine and cytokine production by Fc epsilon RI+ cells. *Adv Immunol* (1993) **53**: 1–29.
142. Moqbel R, Ying S, Barkans J, et al.: Identification of messenger RNA for IL-4 in human eosinophils with granule localization and release of the translated product. *J Immunol* (1995) **155**: 4939–4947.
143. Sabin EA, Kopf MA, Pearce EJ: *Schistosoma mansoni* egg-induced early IL-4 production is dependent upon IL-5 and eosinophils. *J Exp Med* (1996) **184**: 1871–1878.
144. Del Pozo V, De Andre B, Martin E, et al.: Eosinophil as antigen-presenting cell: activation of T-cell clones and T-cell hybridoma by eosinophils after antigen processing. *Eur J Immunol* (1992) **22**: 1919–1925.
145. Frandji P, Oskeritzian C, Cacaraci F, et al.: Antigen-dependent stimulation by bone marrow-derived mast cells of MHC class II-restricted T-cell hybridoma. *J Immunol* (1993) **151**: 6318–6328.
146. Kips JC, Brusselle GJ, Joos GF, et al.: Interleukin-12 inhibits antigen-induced airway hyperresponsiveness in mice. *Am J Respir Crit Care Med* (1996) **153**: 535–539.
147. Manetti R, Parronchi P, Giudizi MG, et al.: Natural killer cell stimulatory factor (interleukin 12 [IL-12]) induces T helper type 1 (Th1)-specific immune responses and inhibits the development of IL-4-producing Th cells. *J Exp Med* (1993) **177**: 1199–1204.

9 Lymphocytes

148. Seder RA, Gazzinelli R, Sher A, Paul WE: Interleukin 12 acts directly on CD4+ T-cells to enhance priming for interferon gamma production and diminishes interleukin 4 inhibition of such priming. *Proc Natl Acad Sci USA* (1993) **90**: 10188–10192.
149. Szabo SJ, Jacobson NG, Dighe AS, Gubler U, Murphy KM: Developmental commitment to the Th2 lineage by extinction of IL-12 signaling. *Immunity* (1995) **2**: 665–675.
150. Rogge L, Barberis-Maino L, Biffi M, *et al.*: Selective expression of an interleukin-12 receptor component by human T helper 1 cells. *J Exp Med* (1997) **185**: 825–831.
150a. Szabo SJ, Dighe AS, Gubler U, Murphy KM: Regulation of the interleukin (IL)-12R beta 2 subunit expression in developing T helper 1 (Th1) and Th2 cells. *J Exp Med* (1997) **185**: 817–824.
151. Magram J, Connaughton SE, Warrier RR, *et al.*: 12-deficient mice are defective in IFN gamma production and type 1 cytokine responses. *Immunity* (1996) **4**: 471–481.
151a. Coyle-AJ; Tsuyuki-S; Bertrand-C; Huang-S; Aguet-M; Alkan-SS; Anderson-GP. Mice lacking the IFN-gamma receptor have impaired ability to resolve a lung eosinophilic inflammatory response associated with a prolonged capacity of T-cells to exhibit a Th2 cytokine profile.J Immunol. 1996; **156**: 2680–5.
152. Gorham JD, Guler ML, Steen RG, *et al.*: Genetic mapping of a murine locus controlling development of T helper 1/T helper 2 type responses. *Proc Natl Acad Sci USA* (1996) **93**: 12467–12472.
152a. Conboy IM, DeKruyff RH, Tate KM, *et al.*: Novel genetic regulation of T helper 1 (Th1)/Th2 cytokine production and encephalitogenicity in inbred mouse strains. *J Exp Med* (1997) **185**: 439–451.
153. Rodrigues V Jr, Abel L, Piper K, Dessein AJ: Segregation analysis indicates a major gene in the control of interleukin-5 production in humans infected with *Schistosoma mansoni*. *Am J Hum Genet* (1996) **59**: 453–461.
154. Lane P, Brocker T, Hubele S, Padovan E, Lanzavecchia A, McConnell F: Soluble CD40 ligand can replace the normal T-cell-derived CD40 ligand signal to B-cells in T-cell-dependent activation. *J Exp Med* (1993) **177**: 1209–1213.
155. Berton MT, Linehan LA: IL-4 activates a latent DNA-binding factor that binds a shared IFN-gamma and IL-4 response element present in the germ-line gamma 1 Ig promoter. *J Immunol* (1995) **154**: 4513–4525.
156. Seder RA, Paul WE: Acquisition of lymphokine-producing phenotype by CD4+ T cells. *Annu Rev Immunol* (1994) **12**: 635-673.
157. Kretsovali A, Papamatheakis J: A novel IL-4 responsive element of the E alpha MHC class II promoter that binds to an inducible factor. *Nucleic Acids Res* (1995) **23**: 2919–2928.
158. Wang HY, Zamorano J, Yoerkie JL, *et al.*: The IL-4-induced tyrosine phosphorylation of the insulin receptor substrate is dependent on JAK1 expression in human fibrosarcoma cells. *J Immunol* (1997) **158**: 1037–1040.
159. Mikita T, Campbell D, Wu P, Williamson K, Schindler U: Requirements for interleukin-4-induced gene expression and functional characterization of Stat6. *Mol Cell Biol* (1996) **16**: 5811–5820.
160. Lai SY, Molden J, Liu KD, Puck LM, *et al.*: Interleukin-4-specific signal transduction events are driven by homotypic interactions of the interleukin-4 receptor alpha subunit. *EMBO J* (1996) **15**: 4506–4514.
161. Hou J, Schindler U, Henzel WJ, Ho TC, Brasseur M, McKnight L: An interleukin-4-induced transcription factor: IL-4 Stat. *Science* (1994) **265**: 1701–1706.
162. Lai SY, Molden J, Liu KD, Puck JM, White MD, Goldsmith MA: Interleukin-4-specific signal transduction events are driven by homotypic interactions of the interleukin-4 receptor alpha subunit. *EMBO J* (1996) **15**: 4506–4514.
163. Brunn GJ, Falls EL, Nilson AE, Abraham T: Protein-tyrosine kinase-dependent activation of STAT transcription factors in interleukin-2- or interleukin-4-stimulated T-lymphocytes. *J Biol Chem* (1995) **270**: 11 628–11 635.
164. Kotanides H, Moczygemba M, White MF, Reich NC: Characterization of the interleukin-4 nuclear activated factor/STAT and its activation independent of the insulin receptor substrate proteins. *J Biol Chem* (1995) **270**: 19 481–19 486.

165. Rooney JW, Hoey T, Glimcher LH: Coordinate and cooperative roles for NF-AT and AP-1 in the regulation of the murine IL-4 gene. *Immunity* (1995) **2**: 473–483.
166. Hodge MR, Chun HJ, Rengarajan J, *et al*.: NF-AT_Driven interleukin-4 transcription potentiated by NIP45. *Science* (1996) **274**: 1903–1905.
167. Ho IC, Hodge MR, Rooney JW, Glimcher LH: The proto-oncogene c-maf is responsible for tissue-specific expression of interleukin-4. *Cell* (1996) **85**: 973–983.
168. Zhao J, Freeman GJ, Gray GS, Nadler LM, Glimcher LH: A cell type-specific enhancer in the human B7.1 gene regulated by NF-kappaB. *J Exp Med* (1996) **183**: 777–789.
169. Hodge MR, Chun HJ, Rengarajan J, Alt A, Lieberson R, Glimcher LH: NF-AT-driven interleukin-4 transcription potentiated by NIP45. *Science* (1996a) **274**: 1903–1905.
170. Hodge MR, Ranger AM, Charles de la Brousse F, Hoey T, Grusby MJ, Glimcher LH: Hyperproliferation and dysregulation of IL-4 expression in NF-ATp-deficient mice. *Immunity* (1996b) **4**: 397–405.
171. Hu-Li J, Huang H, Ryan J, Paul WE: In differentiated CD4$^+$ T-cells, interleukin 4 production is cytokine–autonomous, whereas interferon γ production is cytokine-dependent. *Proc Natl Acad Sci USA* (1997) **94**: 3189–3194.
172. Hogan SP, Mould A, Kikutani H, Ramsay AJ, Foster PS: Aeroallergen-induced eosinophilic inflammation, lung damage, and airways hyperreactivity in mice can occur independently of IL-4 and allergen-specific immunoglobulins. *J Clin Invest* (1997) **99**: 1329–1339.
173. Korsgren M, Erjefalt JS, Korsgren O, *et al*.: Allergic eosinophil-rich inflammation develops in lungs and airways of B cell-deficient mice. *J Exp Med* (1997) **185**: 885–892.
174. King CL, Stupi RJ, Craighead N, June CH, Thyphronitis G: CD28 activation promotes Th2 subset differentiation by human CD4$^+$ cells. *Eur J Immunol* (1995) **25**: 587–595.
175. Gern JE, Vrtis R, Kelly EA, Dick EC, Busse WW: Rhinovirus produces nonspecific activation of lymphocytes through a monocyte-dependent mechanism. *J Immunol* (1996) **157**: 1605–1612.
176. Anderson GP, Coyle AJ: Th2 and 'Th2-like' cells in allergy and asthma: pharmacological perspectives. *Trends Pharmacol Sci* (1994) **15**: 324–332.
177. Actor JK, Shirai M, Kullberg MC, Buller RMC, Sher A, Berzofisky JA: Helminth infection results in decreased virus specific CD8$^+$ cytotoxic T-cells and Th1 cytokine responses as well as delayed virus clearance. *Proc Natl Acad Sci USA* (1993) **90**: 948–952.
178. Panettieri RA Jr, Lazaar AL, Pure E, Albelda SM: Activation of cAMP-dependent pathways in human airway smooth muscle cells inhibits TNF-alpha-induced ICAM-1 and VCAM-1 expression and T-lymphocyte adhesion. *J Immunol* (1995) **154**: 2358–2365.
179. Simon HU, Yousefi S, Dommann-Scherrer CC, *et al*.: Expansion of cytokine-producing CD4$^-$CD8$^-$ T-cells associated with abnormal Fas expression and hypereosinophilia. *J Exp Med* (1996) **183**: 1071–1082.
180. Anderson GP: Resolution of chronic inflammation by therapeutic induction of apoptosis. *Trends Pharmacol Sci* (1996) **17**: 438–442.

10

Epithelial Cells

JAGDISH L. DEVALIA, MUNTASIR M. ABDELAZIZ AND
ROBERT J. DAVIES

INTRODUCTION

The last few decades have greatly increased our understanding of the role of epithelial cells in airway inflammation and the pathogenesis of bronchial asthma. Although these cells have traditionally been seen to play an important role in preventing the entry of noxious inhaled substances into the body and clearing particulates out of the airways, recent studies have demonstrated that airway epithelial cells can synthesize and release several biologically active mediators, which can modulate the function of other inflammatory cells implicated in the pathogenesis of bronchial asthma.

MORPHOLOGY OF THE AIRWAY EPITHELIUM

Although a variety of recognizable epithelial cell types are identified in the airways, ciliated and goblet cells make up the bulk of the tracheobronchial epithelium[1] (Fig. 10.1). Other cell types include serous, basal, Clara, brush, and neuroendocrine cells.[2-4] In addition migratory cells, such as lymphocytes, and neural elements may migrate into or pierce the epithelial basement membrane, respectively.[3,4] Ciliated cells, the predominant cell type in the airway epithelium, are characterized by electron-lucent cytoplasm and are responsible for propelling the tracheobronchial secretion toward the pharynx. More recently, it has been suggested that these cells may also play an important role in epithelial electrolyte transport processes,[5] thus influencing the composition of the periciliary fluid and subsequently the beat of the cilia. Goblet cells and serous cells, unlike ciliated

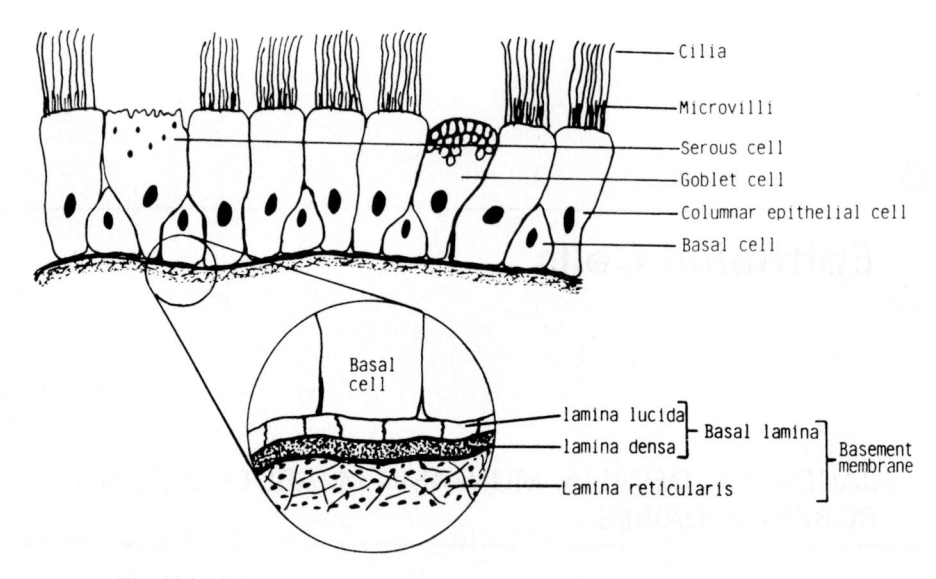

Fig. 10.1 Schematic view of the major epithelial cell types in airway epithelium.

epithelial cells, have electron-dense granules containing, respectively, acidic and neutral mucin,[1] which provide the mucous blanket with specific viscoelastic properties required for effective ciliary beating and mucociliary clearance.[4,5] In addition, serous cells are thought to be involved in the formation of periciliary fluid.[4,5] The presence of the basal cells contributes to the pseudostratified appearance of the epithelium in the large bronchi and trachea and are thought to play a role in the attachment of the superficial cells to the airway basement membrane.[6] Although mucous and serous cells have been shown to be capable of division, basal cells are regarded as the stem cell for the other epithelial cells.[4] The number of basal cells gradually decreases in the lower airways until there are none in the terminal and respiratory bronchioles, where Clara cells are thought to take over as the progenitor cells for the bronchiolar cells.[7] These cells are also believed to be involved in the production of the bronchiolar surfactant.[8] In contrast, brush cells, which derive their name from the presence of microvilli on the luminal surface, are identified in the airway at all levels from the nose to the bronchioli in several species.[2] However, these cells have not been observed in humans and their function remains unknown.[2,4] In addition, the airway epithelium contains a small number of neuroendocrine cells which, although basal in position, possess cytoplasmic projections extending to the luminal surface.[9] Clusters of such cells, known as neuroepithelial bodies, are sometimes associated with intraepithelial nerves. Although the function of these cells is not well understood, some studies have shown that they are rich in bioactive amines such as somatostatin, serotonin, endothelin and calcitonin.[9,10]

The integrity of the epithelium is maintained by intercellular junctional complexes formed between the adjacent epithelial cells and composed of tight junctions, intermediate junctions and desmosomes.[11-14] The tight junction is an area where the external leaflets of the membranes of the adjacent cells are closely apposed and form a continuous belt circumscribing each cell just below the luminal surface.[12,13] The permeability of the epithelial barrier is thought to be dependent on the integrity of this continuous

10 Epithelial Cells

structure.[12–14] Just below the tight junction is the intermediate junction, which forms another continuous structure encircling each cell. Its function is to strengthen the cell attachment, as well as anchor the terminal web of actin filaments to the plasma membrane. Its electron-dense intercellular space contains calcium-dependent adhesion molecules, E-cadherins.[12] Subadjacent to the intermediate junction is the desmosome, where cytoplasmic filaments overlap and penetrate the membrane on either side and bridge the gap between adjacent cells. Desmosomes are thought to enhance cohesion between cells and are found particularly in areas where tissues are subject to disruption. Unlike tight and intermediate junctions, desmosomes form a discontinuous structure between the cells. Hemi-desmosomes (half-desmosomes) consist of peptides called integrins and anchor the basal cells to the basement membrane[14] (Fig. 10.2).

The gap junctions or nexus, on the other hand, represent low-resistance intercellular pathways between cells, permitting the passage of ions and small molecular substances.

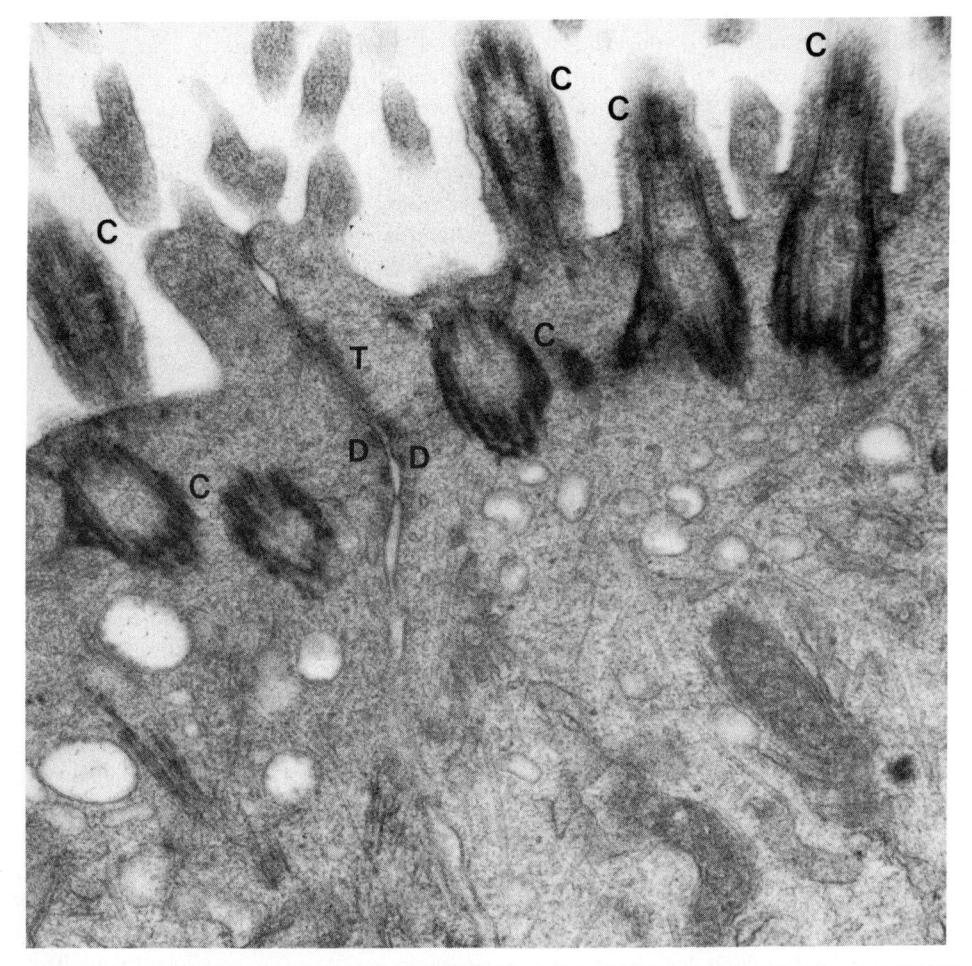

Fig. 10.2 Transmission electron micrograph of human bronchial epithelium showing cilia (C), microvilli (M), tight junction (T) and desmosomes (D) (magnification × 33 600).

They are found particularly during the developmental period and repair following injury.[15,16]

AIRWAY EPITHELIUM AND HYPERRESPONSIVENESS

Bronchial asthma is a syndrome characterized by variable airflow obstruction, airway inflammation and bronchial hyperresponsiveness.[17] Although a cardinal feature of bronchial asthma, increased airway responsiveness is also associated with several other disease states, where damage to the airway mucosa is well established. Consequently, it has been suggested that bronchial hyperresponsiveness may be related to the disruption of airway epithelium and airway inflammation.[14,18] It is not clear, however, whether the loss of epithelial integrity contributes to the increased bronchial responsiveness or whether it is itself a result of the airway inflammation and the consequential hyperreactivity. Nevertheless, there are several possible mechanisms that may explain how epithelial abnormalities could lead to increased bronchial reactivity, including (a) increased permeability to allergen;[14,19,20] (b) changes in osmolarity of the bronchial surface lining fluid;[19] (c) exposure of sensory nerve fibres to irritants and potentiation of local axon reflexes;[14,21] (d) increased production of inflammatory mediators[22-27] and reduction of putative 'protective' (both anti-inflammatory and relaxing) mediators;[28-32] and (e) modulation of the immune system.[33] Of these mechanisms, the ones involving synthesis and release of agents with either inflammatory or protective properties and immunoregulation by epithelial cells have been the most widely studied and are discussed in greater detail below.

EPITHELIAL CELL-DERIVED MEDIATORS

Nitric oxide

Nitric oxide (NO) is a highly reactive compound with multiple roles in immune effector mechanisms, neurotransmission and intracellular and intercellular communication.[34,35] It is produced in the lung by many cell types, including epithelial cells, from L-arginine by the action of the enzyme nitric oxide synthase (NOS).[34-37] NOS is present as three isoforms, two of which are expressed constitutively and one of which is inducible. The constitutive isoforms are found in endothelial cells (ecNOS) and neurones (ncNOS) and are activated by calcium influx in response to physiological stimuli.[34,35,37] In contrast, the inducible isoform (iNOS), which is expressed in many cell types including epithelial cells, is calcium independent and transcribed in response to endotoxin and cytokines such as interferon (IFN)-γ interleukin (IL)-1β and tumour necrosis factor (TNF)-α.[34,37-39] Endogenous NO is produced at picomolar amounts by cNOS and is responsible for the maintenance of physiological homeostasis.[34,35,40] In contrast, production of NO by iNOS occurs at nanomolar concentrations and may lead to generation of peroxynitrites and hydroxyl radicals, which damage the tissues on reaction with superoxides.[40,41] It has been suggested that bronchial epithelial cells of asthmatic patients may express increased levels of iNOS and consequently are likely to play a role in the increased epithelial damage and shedding observed in these individuals.[40,42]

Endothelin

The endothelin (ET) family comprises three 21 amino acid peptides, ET-1, ET-2 and ET-3, which are thought to be derived from larger precursors, namely proendothelin-1, proendothelin-2 and proendothelin-3, respectively.[40,43,44] The ETs are potent vasoconstrictors that cause a slow developing but prolonged bronchocontraction.[43,44] In addition, ETs increase mucus secretion from both bronchial and nasal epithelial cells.[45,46] The biological effects of ETs are believed to be mediated via at least two receptors, ET_A and ET_B. ET_A is selective for ET-1 and ET-2, while ET_B is non-selective with equipotent binding of ET-1, ET-2 and ET-3.[47] Although ET-1 was first isolated in the medium of porcine aortic endothelial cells,[43] it is now evident that the three isoforms are widely distributed in various animal and human tissues. Animal studies have shown that canine and porcine airway epithelial cells can synthesize ET-1 and ET-3.[48] Similarly, studies by Marciniak and co-workers[49] have shown that human airway epithelial cells can synthesize proendothelin-1 and proendothelin-2. These authors have also demonstrated that immunoreactivity for all three proendothelins can be detected in human airway submucosal glands.[49]

Although the exact role of ETs in bronchial asthma remains to be established, increased expression and level of ET-1 have recently been reported in bronchial biopsies and bronchoalveolar lavage fluids collected from asthmatic patients compared with those from non-asthmatic subjects.[43,50,51] It has also been suggested that ETs cause bronchoconstriction through the release of secondary mediators such as platelet activating factors and prostaglandins;[52] therefore an exaggerated effect is likely to be seen in asthmatic airways, which are infiltrated with inflammatory cells capable of generating these secondary mediators.[52]

Lipid mediators

The airway epithelial phospholipase oxygenation system is a potential source of inflammatory mediators that may have autocrine and paracrine functions. Animal and human studies have shown that epithelial cells contain abundant stores of fatty acid substrates, produced via phospholipase activity, and express high levels of cyclooxygenase and lipoxygenase activities.[22,23,53,54] It has been demonstrated that prostaglandin $(PG)E_2$ is the main product of the cyclooxygenase system in animal and human airway epithelium[54] and serves to inhibit bronchoconstriction directly and indirectly via inhibition of cholinergic neurotransmission and mast-cell mediator release.[55] Flavahan and co-workers[28,29] originally reported that mechanical removal of epithelium from canine bronchial rings rendered these hyperresponsive to histamine, 5-hydroxytryptamine and acetylcholine; they suggested that this bronchoconstrictive response was a consequence of the removal of a naturally occurring relaxing factor released by the epithelium. Although the nature of this epithelium-derived relaxing factor(s) still remains to be elucidated fully,[14] animal studies have suggested that this may involve a cyclooxygenase metabolite of arachidonic acid, possibly PGE_2.[28,55–57] Although airway epithelial cells are also capable of generating small amounts of $PGF_{2\alpha}$ and PGD_2, which lead to increased smooth muscle contraction and bronchial hyperresponsiveness,[54] it is thought that the effects produced by these compounds would be negligible in view of PGE_2-induced brochodilation.[56,57]

In addition to cyclooxygenase pathway products, animal and human studies have demonstrated that airway epithelial cells can also use the lipoxygenase pathway. Animal studies have shown that tracheal epithelial cells from dog and sheep do not generate detectable 15-lipoxygenase products but do have an active 5-lipoxygenase pathway, leading to production of leukotriene (LT)B_4.[22,23] In contrast, studies with cultured human tracheal and bronchial epithelial cells have demonstrated that these cells generate 15-lipoxygenase metabolites, such as 12- and 15-hydroxyeicosatetraenoic acids (12-HETE and 15-HETE), which have weak neutrophil chemotactic activity.[23,25,26] Studies from our laboratory have demonstrated that human bronchial epithelial cells also generate PGE_2 and LTC_4 in addition to 15-HETE and 12-HETE, and that exposure to NO_2 at concentrations as low as 0.4 p.p.m., occasionally found at the kerbside in heavy summer traffic, significantly enhances the synthesis of LTC_4.[58] Although LTC_4, 15-HETE and 12-HETE are predominantly neutrophil chemoattractants, LTC_4 additionally leads to increased vascular permeability, excessive mucus production and smooth muscle contraction in the airways and consequently may play a role in the early-phase reaction in bronchial asthma.[59]

Cytokines

Production of pro-inflammatory cytokines by the airway epithelium has been of particular interest in allergic conditions such as asthma and allergic rhinitis, since these cytokines influence the activity of inflammatory cells such as eosinophils, T-lymphocytes and mast cells, the infiltration of which in the airways is a characteristic feature of these disorders.[17,60,61] We and others have demonstrated that airway epithelial cells generate a wide variety of cytokines that, either directly or in conjunction with one another, influence the growth, differentiation, activation, migration and survival of other inflammatory cells.[27,62–66] The airway epithelial cell-derived cytokines can be divided into four groups according to their functions: (a) chemotactic factors, which influence the chemotaxis of other inflammatory cells; (b) colony-stimulating factors, which promote the differentiation and survival of the recruited inflammatory cells; (c) pro-inflammatory multifunctional cytokines, which initiate and amplify inflammatory events via their influence on a variety of target cells; and (d) growth factors that regulate the growth and differentiation of airway epithelial cells as well as immune and inflammatory processes.

Chemotactic factors

Chemotactic factors generated by human airway epithelial cells include lymphocyte chemoattractant factor, granulocyte–macrophage colony-stimulating factor (GM-CSF) and members of the chemokine superfamily.[27,64–66] The latter is divided into α (or C-X-C) and β (or C-C) subfamilies, depending on the presence or absence of one amino acid

10 Epithelial Cells

between the first two of four conserved cysteines.[67,68] The two subfamilies exhibit differential target selectivity, with neutrophils as the main target of C-X-C chemokines, and eosinophils, basophils, monocytes and T-lymphocytes as the main targets for C-C chemokines. Studies of airway epithelial cells and cell lines have demonstrated that these cells are capable of expressing and releasing large amounts of IL-8, a member of the C-X-C subfamily,[63,64] which on a molar basis is one of the most potent chemoattractant for neutrophils.[69,70] It also stimulates neutrophil activation and adherence to endothelial cells, induces T-cell chemotaxis and inhibits IL-4-induced IgE production by B-cells.[66,69,70] Recently we have demonstrated that, in addition to being a potent neutrophil chemoattractant, IL-8 may also be chemoattractant for eosinophils in the presence of other mediators, such as GM-CSF.[71] This is in accordance with the finding of others who have shown that IL-8 possesses chemotactic activity for eosinophils obtained from atopic subjects or eosinophils primed with IL-3 and GM-CSF.[72–75] More recently, we and others have shown that airway epithelial cells also release other chemotactic factors including RANTES (regulated on activation, normal T-cell expressed and secreted) and monocyte chemotactic protein-1 (MCP-1), both of which belong to the C-C chemokine subfamily.[65,66] Whilst RANTES is a potent eosinophil chemotaxin, MCP-1 induces monocyte and basophil activation and chemotaxis.[66,76,77] Studies from our laboratory have demonstrated that conditioned medium from cultured bronchial epithelial cells has potent eosinophil and neutrophil chemoattractant properties, which can be attenuated by neutralizing antibodies against IL-8, GM-CSF and/or RANTES.[71] These studies suggest that IL-8, GM-CSF and RANTES released by human bronchial epithelial cells are likely to be involved in the chemotaxis of these inflammatory cell types.[71]

Colony-stimulating factors

Studies of GM-CSF have demonstrated that in addition to being chemoattractant for eosinophils and neutrophils, this cytokine also potentiates the differentiation and survival of these cells.[66,72,78,79] Cox and coworkers[80,81] have demonstrated that survival of eosinophils and neutrophils is prolonged by incubation with conditioned medium from airway epithelial cell cultures and that this increase in survival can be blocked by antibodies against GM-CSF, in the case of eosinophils, and by antibodies against GM-CSF and granulocyte colony-stimulating factor (G-CSF), in the case of neutrophils. More recently, we have demonstrated that there is significant correlation between epithelial expression of GM-CSF and the number of activated eosinophils in the airway epithelium of asthmatic individuals.[82]

Multifunctional cytokines

The multifunctional cytokines synthesized and released by airway epithelial cells include IL-1β, IL-6, IL-11 and TNF-α, which have pleiotropic pro-inflammatory effects on a variety of target cells.[63,64,66,71] These cytokines are involved in activation of B-lymphocytes and monocytes and induce acute-phase protein synthesis.[66,78] In addition TNF-α augments the capacity of other cells to produce other inflammatory mediators such as IL-6, IL-8 and RANTES.[65,66,78] Furthermore, TNF-α and IL-1β interact with endothelial cells and upregulate the expression of intercellular adhesion molecule-1 (ICAM-1), vascular cell adhesion molecule 1 (VCAM-1) and E-selectin, which are involved in the adherence and transendothelial migration of inflammatory cells.[83-85] We have recently demonstrated that conditioned medium from human bronchial epithelial cell cultures significantly increases the adherence of eosinophils and neutrophils to endothelial cell cultures. Additionally, we have demonstrated that conditioned medium-induced adherence of eosinophils and neutrophils to the endothelial cells is significantly attenuated by neutralizing antibodies to IL-1β, TNF-α, ICAM-1, VCAM-1 and E-selectin.[71] These studies suggest that the expression of cell adhesion molecules involved in the adherence of eosinophils and neutrophils to endothelial cells is likely to be modulated by epithelial cell-derived mediators. However, our finding that the conditioned medium-induced adherence of eosinophils and neutrophils to human endothelial cells *in vitro* could not be reduced to basal levels by the combination of anti-IL-1β and anti-TNF-α, despite these antibodies having an additive effect, suggests that other factors are also likely to be involved in this process. Indeed, several studies showed that IL-6 can also influence the expression of cell adhesion molecules.[86,87]

Growth factors

Animal and human studies have shown that bronchial epithelial cells can produce transforming growth factor β (TGF-β), which is important for cell growth and differentiation.[66,78,88] In allergic conditions, TGF-β may be associated with the fibrosis observed in longstanding asthma and the subendocardial fibrosis associated with the hypereosinophilic syndrome.[78] Contrary to having inflammatory effects, TGF-β also has anti-inflammatory effects in that it inhibits the activity of B-lymphocytes, T-helper (CD4) and cytotoxic (CD8) lymphocytes.[66,78,88] It may also lessen allergic inflammation by suppressing IL-4-induced IgE production by B-cells and inhibiting mast cell proliferation.[78]

ADHESION MOLECULES AND THE AIRWAY EPITHELIUM

More recently, interest has focused on the role of adhesion molecules in the pathogenesis of bronchial asthma and allergic conditions. Despite the commonly held view that cell adhesion molecules are widely involved in the pathogenesis of allergic respiratory diseases,[89-93] comparatively few studies have examined their expression on airway epithelial cells. Whilst ICAM-1 has been shown to be expressed on bronchial epithelial cells, it has not been possible to demonstrate the expression of VCAM-1 and E-selectin on these cells.[94,95] Studies of ICAM-1 have demonstrated that although this molecule is expressed at low levels on both bronchial epithelium and vascular endothelium under normal conditions, it is markedly upregulated by inflammatory stimuli such as endotoxin, IL-1β, TNF-α and IFN-γ.[93-96] In addition, studies in animals have demonstrated that ICAM-1 expression is upregulated in inflamed airway epithelium *in vivo* and thus may play an important role in airway eosinophilia and hyperresponsiveness, since antibodies against ICAM-1 attenuated both the eosinophilia and hyperresponsiveness in these animals.[93] Furthermore, studies from our laboratory have suggested that the expression of ICAM-1 may be upregulated on bronchial tissue of patients with bronchial asthma.[97] Monterfort and colleagues[98] have also demonstrated that the expression of ICAM-1 and

10 Epithelial Cells

VCAM-1 is significantly increased on nasal mucosa in patients with perennial allergic rhinitis compared with non-rhinitic individuals. More recently, studies by Koizumi and coworkers[99,100] have demonstrated increased serum levels of soluble ICAM-1, VCAM-1 and E-selectin in asthmatic patients compared with non-asthmatic subjects and suggested that the levels of these adhesion molecules are increased even further during exacerbations of asthma.

The observation that eosinophils are present in large numbers at sites of inflammation in airway tissue in allergic disease suggests that mechanisms exist which favour their recruitment in these conditions. Studies have shown that cell adhesion molecules, including E-selectin, VCAM-1 and ICAM-1 expressed on endothelial cells and leucocyte function-associated antigen 1 (LFA-1), macrophage antigen-1 (MAC-1), very late antigen 4 (VLA-4) and L-selectin expressed on eosinophils, play an important role in this process.[89–93] We have studied the adhesion of eosinophils to IL-1β- and TNF-α-treated human endothelial and human bronchial epithelial cells and demonstrated that whilst eosinophil adhesion to endothelial cells is mediated via ICAM-1, VCAM-1 or E-selectin, eosinophil adhesion to bronchial epithelial cells is mediated mainly via ICAM-1.[101] However, several studies have shown that the first step in eosinophil recruitment involves reversible margination or rolling of eosinophils along the endothelial surface, a process mediated via carbohydrate–selectin interactions.[91,92,95] This is followed by eosinophil activation, possibly as a consequence of cytokines secreted by other cells in the vicinity, leading to firm adhesion and transendothelial migration of the eosinophils.[92,95] The latter process involves the interaction between ICAM-1/VCAM-1 present on endothelial cells and LFA-1/VLA-4 present on eosinophils[91,92,95,96] (Fig. 10.3).

IMMUNOREGULATION

Cytokines have been shown to influence the expression of MHC class II antigens on the surface of epithelial cells, thereby conferring upon these cells the potentially important role of antigen processing and presentation to T-lymphocytes. Studies of cultured human retinal pigment epithelial cells[102] and fetal pancreatic duct epithelial cells[103] have demonstrated that these cells are capable of expressing HLA-DR antigens following exposure to IFN-γ. Immunohistochemical studies of human fibrotic lung[104] and human bronchial tissue, obtained from patients with peripheral lung cancer,[105] have also shown that type II alveolar epithelial cells and ciliated bronchial epithelial cells express HLA-DR antigens and the genes encoding these antigens.

The putative mechanisms and consequences of the interaction between antigen-presenting cells (APCs) and T-lymphocytes have been reviewed recently.[33,106] It has been suggested that preferential binding of specific allergenic peptides to HLA class II antigens may lead to recognition by, and activation and proliferation of, specific T-cell clones (either Th1 or Th2 clones), which predispose the individual to the development of certain diseases.[33,78,106] Th1 lymphocytes produce predominantly IL-2 and/or IFN-γ and are thought to be involved in delayed-type hypersensitivity reactions and in the synthesis of IgM and some IgG subclasses. Th2 lymphocytes, on the other hand, have been shown to synthesize IL-3, IL-4, IL-5, IL-10 and IL-13,[78,106,107] and are thought to be important in allergic-type inflammatory reactions and defence against parasites. IL-3 and IL-5 are

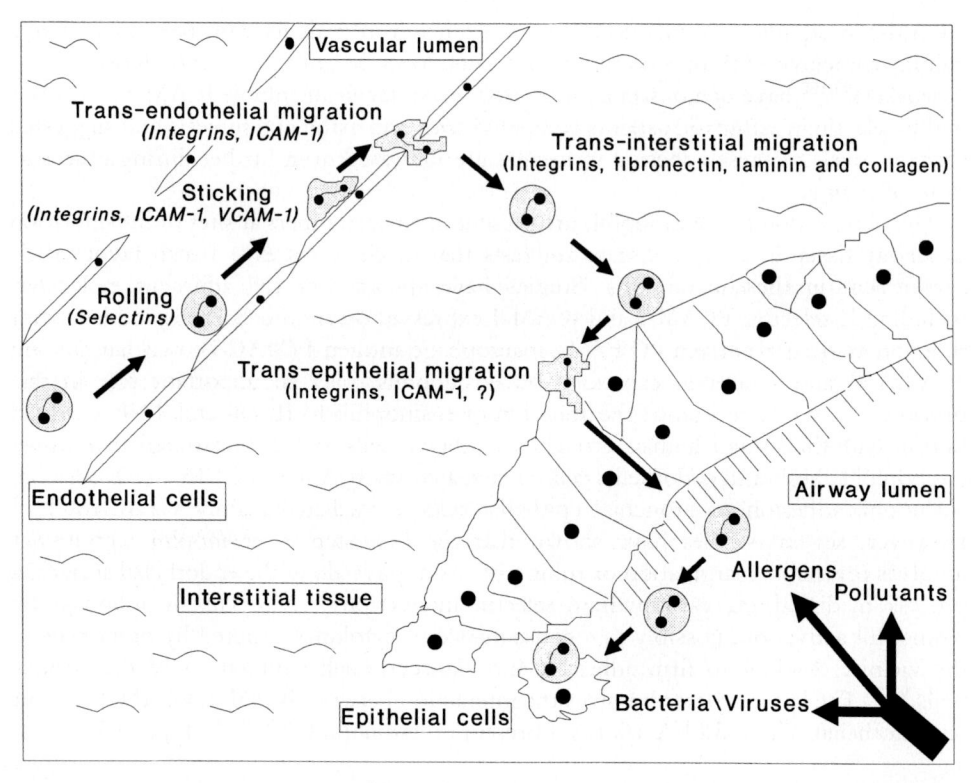

Fig. 10.3 Cell adhesion molecules thought to be involved in transendothelial and transepithelial migration of eosinophils in bronchial asthma. Modified from ref. 90.

potent eosinophil activators and stimulate eosinophil growth, differentiation and chemotaxis.[78,106] These effects are complemented by IL-4 and IL-13, which stimulate switching of B-cells to produce IgE and, through their selective induction of endothelial VCAM-1, may contribute in the selective recruitment of eosinophils.[78,106,107] Additionally IL-4 and IL-13, together with IL-10, act to suppress cell-mediated immunity by influencing the growth and function of Th1 lymphocytes.[78,107]

CULTURE OF HUMAN AIRWAY EPITHELIAL CELLS *IN VITRO*

In spite of increasing evidence for an important physicochemical role of the airway epithelium *in vivo*, it has not proved easy to assign a specific pathogenic role to the airway epithelium, due to the presence of other cell types and underlying tissues. Human airway epithelial cell cultures offer an ideal *in vitro* model system for the study of the epithelium in the aetiology of airway diseases and the underlying mechanism(s) of inflammation and hyperresponsiveness. Although nasal and bronchial epithelial cells have been cultured *in vitro* by several groups,[107-111] a major difficulty experienced by many of the workers in the field has been to get these cells to grow to confluency consistently, such that large

10 Epithelial Cells

numbers can be subsequently harvested for further study. Also, the epithelial cells have often been isolated from enzyme-dispersed tissue, which itself is known to bring about detrimental morphological and biochemical changes. In the case of earlier studies investigating bronchial cells, the bronchial tissue had been obtained 8–12 hours after death, again questioning the physiological status of the cells. Indeed, Farber and Young[112] and others have demonstrated that anoxic conditions lead to accelerated degradation of membrane phospholipids and consequently cause irreversible cell injury. We have addressed these difficulties and have demonstrated that it is possible to produce confluent cultures of fully differentiated ciliated nasal and bronchial epithelial cells from surgical tissues[113] (Fig. 10.4). More recently, we have cultured nasal epithelial cells from nasal biopsy specimens of well-characterized groups of atopic rhinitic, atopic non-rhinitic (patients with atopic eczema) and non-atopic non-rhinitic subjects, and investigated the cytokine profiles generated by these cells.[114] Our studies have demonstrated that the cells from atopic rhinitic and atopic non-rhinitic subjects generate significantly larger amounts of GM-CSF, TNF-α, RANTES and IL-8 compared with epithelial cells from non-atopic non-rhinitic individuals. Similarly, preliminary studies of epithelial cells cultured from bronchial biopsies of well-characterized groups of asthmatic and non-asthmatic subjects suggest that the cells of asthmatic subjects may also release greater quantities of inflammatory cytokines compared with the cells of non-asthmatic subjects.

SUMMARY

Taken together these studies indicate that human epithelial cells have the capacity to generate a range of inflammatory compounds and to regulate the expression of others, such as cytokines and cell adhesion molecules. In view of the importance of the airway epithelium as the first line of defence against airborne dusts, vapours, gases and fumes, and its capacity to play a role in the maintenance of a physicochemical homeostasis, it is not difficult to envisage how perturbation of this barrier, and particularly the epithelial cells that predominate within this barrier, may bring about adverse changes in and around the surrounding tissues and possibly help to explain the pathogenesis of asthma.

It is tempting to hypothesize that in bronchial asthma dysfunction of the bronchial epithelium, either resulting either from acute exposure to irritants such as air pollutants, cigarette smoke, viruses and bacteria or as a consequence of genetic predisposition to a specific allergen, itself results in the initiation, maintenance and potentiation of inflammation at the site(s) of exposure. This may be expressed either in the generation of pro-inflammatory mediators, which interact with mediators derived from other inflammatory cells such as mast cells and Th2 lymphocytes and act as potent eosinophil and neutrophil chemoattractants and activators, or in the upregulation of cell adhesion molecules involved in the inter-tissue trafficking of these and other 'inflammatory' cell types (Fig. 10.5). Alternatively, depletion of any naturally occurring anti-inflammatory mediators and smooth muscle relaxing agents, which help to maintain the integrity of the bronchus and the surrounding tissues, may ensue.

In view of the advances that have been made in the understanding of the putative mechanisms that may be of importance in the aetiology of asthma, it should be possible to

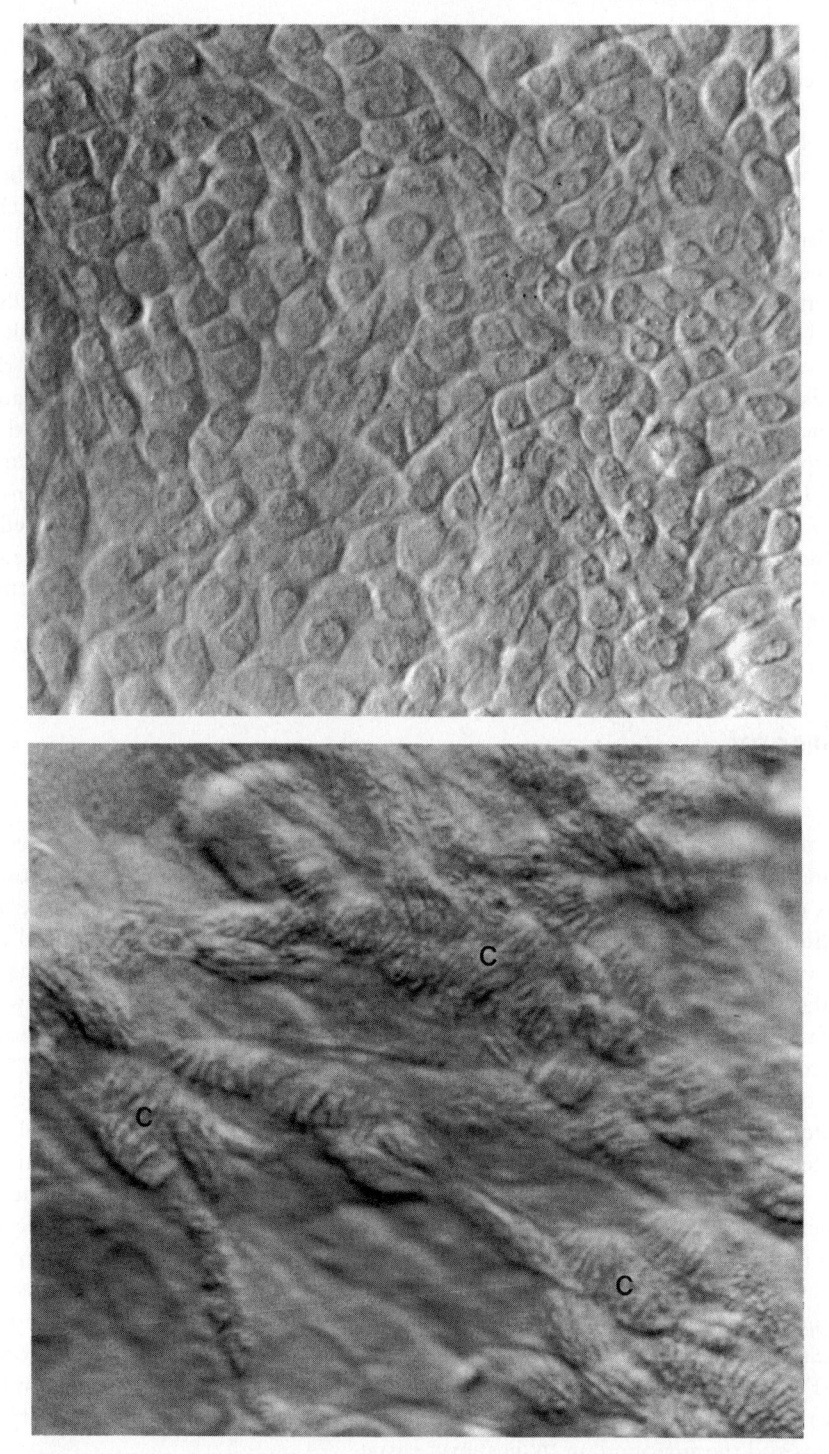

Fig. 10.4 Typical culture of human bronchial epithelial cells, viewed by light microscopy incorporating Hoffman Modulation Contrast optics at (a) low power (magnification ×300) and (b) high power, showing ciliated cells (c) (magnification ×600).

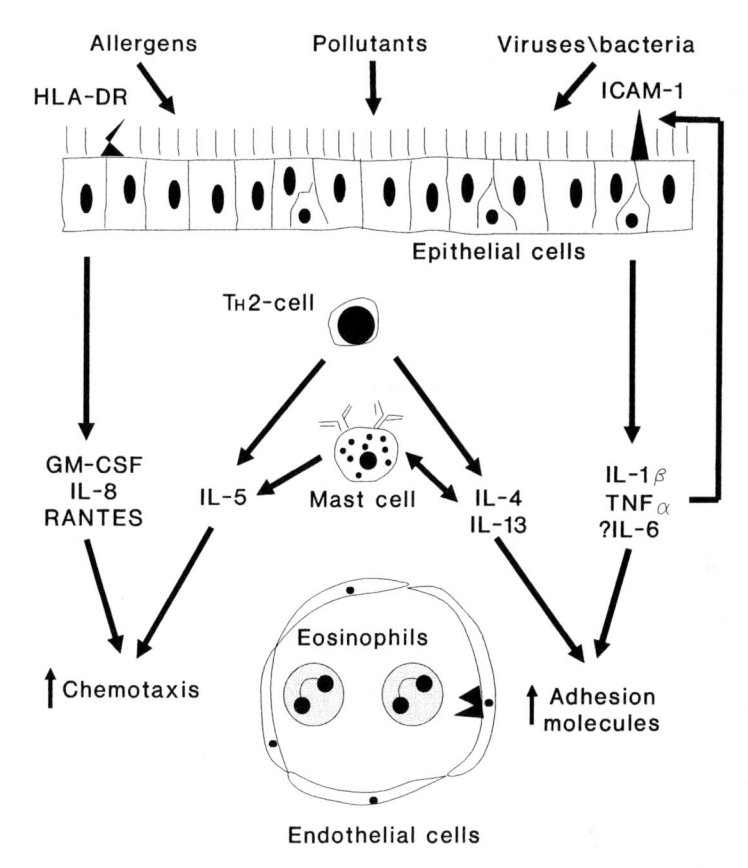

Fig. 10.5 Putative role of airways epithelium in the aetiology of asthma.

formulate specific therapies to counteract not only the bronchoconstriction so characteristic of asthma but also the events leading to bronchoconstriction. It is probable that this new generation of therapeutic agents will address the question of the specific inflammatory cell types involved and ways of downregulating their growth and activation. These novel agents may take the form of specific monoclonal antibodies directed against specific cytokines and cell adhesion molecules.[115,116]

REFERENCES

1. Breeze RG, Wheeldon EB: The cells of the pulmonary airways. *Am Rev Respir Dis* (1977) **116**: 705–776.
2. St George J, Hyde DM, Plopper CG: Epithelial cells of the conducting airways. In Farmer SG, Hay DWP (eds) *The Airway Epithelium: Physiology, Pathophysiology and Pharmacology*. New York, Marcel Dekker, 1991, pp 3–39.
3. Jeffery PK: Morphologic features of airway surface epithelial cells and glands. *Am Rev Respir Dis* (1983) **128**: S14–S20.
4. Jeffery PK: Structural, immunologic, and neural elements of the normal human airway wall.

In Busse WW, Holgate ST (eds) *Asthma and Rhinitis*. Boston, Blackwell Scientific Publications, 1995, pp 80–106.
5. Johnson CW, Larivee P, Shelhamer JH: Epithelial cells: regulation of mucus secretion. In Busse WW, Holgate ST (eds) *Asthma and Rhinitis*. Boston, Blackwell Scientific Publications, 1995, pp 584–598.
6. Evans MJ, Plopper CG: The role of basal cells in adhesion of columnar epithelium to airway basement membrane. *Am Rev Respir Dis* (1988) **138**: 481–483.
7. Evans MJ, Cabral-Anderson LJ, Freeman G: The role of the Clara cell in renewal of bronchiolar epithelium. *Lab Invest* (1978) **38**: 648–655.
8. Niden AH: Bronchiolar and large alveolar cell in pulmonary phospholipid metabolism. *Science* (1980) **158**: 1323–1324.
9. Becker LK: The coming of age of a bronchial epithelial cell. *Am Rev Respir Dis* (1993) **148**: 1166–1168.
10. Giaid A, Polak JM, Gaitonade V, *et al*.: Distribution of endothelin-like immunoreactivity and mRNA in the developing and adult human lung. *Am J Respir Cell Mol Biol* (1991) **4**: 50–58.
11. Elia C, Bucca C, Rolla G, Scappaticci E, Cantino D: A freeze-fracture study of human bronchial epithelium in normal, bronchitic and asthmatic subjects. *J Submicrosc Cytol Pathol* (1988) **20**: 509–517.
12. Farquhar MG, Palade GE: Junctional complexes in various epithelia. *J Cell Biol* (1963) **17**: 375–412.
13. Carson JL, Collier AM, Boucher RC: Ultrastructure of the respiratory epithelium in the human nose. In Mygind N, Pipkorn U (eds) *Allergic and Vasomotor Rhinitis. Pathological aspects*. Munksgaard, Copenhagen, 1987, pp 11–27.
14. Hulsmann AR, De Jongste JC: Modulation of airway responsiveness by airway epithelium in humans: putative mechanisms. *Clin Exp Allergy* (1996) **26**: 1236–1242.
15. Gordon RE, Lane BP, Marin M: Regeneration of rat tracheal epithelium: changes in gap junctions during specific phases of each cell cycle. *Exp Lung Res* (1982) **3**: 47–56.
16. McNutt NS, Weinstein RS: The ultrastructure of the nexus. A correlated thin section and freeze cleave study. *J Cell Biol* (1970) **47**: 666–688.
17. Djukanovic R, Roche WR, Wilson JW, *et al*.: Mucosal inflammation in asthma. *Am Rev Respir Dis* (1990) **142**: 434–457.
18. Burrows B. Bronchial hyperresponsiveness, atopy, smoking, and chronic obstructive pulmonary disease. *Am Rev Respir Dis* (1989) **140**: 1515–1517.
19. Lozewicz S, Wells C, Gomez E, *et al*.: Morphological integrity of the bronchial epithelium in mild asthmatics. *Thorax* (1990) **45**: 12–15.
20. Hogg JC, Eggleston PA: Is asthma an epithelial disease? *Am Rev Respir Dis* (1984) **129**: 207–208.
21. Barnes PJ: Asthma is an axon reflex. *Lancet* (1986) **i**: 242–245.
22. Holtzman MJ, Aizwa H, Nadel JA, Goetzl EJ: Selective generation of leukotriene B4 by tracheal epithelial cells from dogs. *Biochem Biophys Res Commun* (1983) **114**: 1071–1076.
23. Hunter JA, Finkbeiner WE, Nadel JA, Goetzl EJ, Holtzman MJ: Predominant generation of 15-lipoxygenase metabolites of arachidonic acid by epithelial cells from human trachea. *Proc Natl Acad Sci USA* (1985) **82**: 4633–4637.
24. Lazarus SC: Role of inflammation and inflammatory mediators in airways disease. *Am J Med* (1986) **81**: 2S–7S.
25. Churchill L, Chilton FH, Resau JH, Bascom R, Hubbard WC, Proud D: Metabolism of endogenous arachidonic acid by cultured human tracheal epithelial cells. *Am Rev Respir Dis* (1989) **140**: 449–459.
26. Denberg JA, Dolovich J, Harnish D: Basophil, mast cell and eosinophil growth and differentiation factors in human allergic disease. *Clin Exp Allergy* (1989) **19**: 249–254.
27. Mattoli S, Miante S, Calabro F, Mazzetti M, Allegra L: Human bronchial epithelial cells exposed to isocyanates potentiate the activation and proliferation of T cells induced by antigen receptor triggering through the release of IL-1 and IL-6. In Johanssen SGO (ed) *Pharmacia Allergy Research Foundation, Award Book 1990*. Uppsala, AW Grafiska, 1990, pp 25–35.

10 Epithelial Cells

28. Flavahan NA, Aarhus LL, Rimete TJ, Vanhoutte PM: Respiratory epithelium inhibits bronchial smooth muscle. *J Appl Physiol* (1985) **58**: 834–838.
29. Flavahan NA, Vanhoutte PM: The respiratory epithelium releases a smooth muscle relaxing factor. *Chest* (1985) **87**: S189–S190.
30. Barnes PJ, Cuss FM, Palmer JB: The effect of airway epithelium on smooth muscle contractibility in bovine trachea. *Br J Pharmacol* (1985) **86**: 685–691.
31. Vanhoute PM: Epithelium-derived relaxing factor(s) and bronchial reactivity. *Am Rev Respir Dis* (1988) **138**: 24S–30S.
32. Wilkens JH, Wilkens H, Forstermann U, Frolich JC: The effect of bronchial epithelium on bronchial contractility. *Pneumologie* (1990) **44**: 373–374.
33. Ricci M, Rossi O: Dysregulation of IgE responses and airway allergic inflammation in atopic individuals. *Clin Exp Allergy* (1990) **20**: 601–609.
34. Lyons CR: The role of nitric oxide in inflammation. *Adv Immunol* (1996) **60**: 323–370.
35. Barnes PJ, Belvisi MG: Nitric oxide and lung disease. *Thorax* (1993) **48**: 1034–1043.
36. Kobzik L, Bredt DS, Lowenstein CJ, *et al.*: Nitric oxide synthase in human and rat lung: immunocytochemical and histochemical localization. *Am J Respir Cell Mol Biol* (1993) **9**: 371–377.
37. Barnes PJ, Belvisi MG: Exhaled nitric oxide: a new lung function test. *Thorax* (1996) **51**: 233–237.
38. Robbins RA, Barnes PJ, Springall DR, *et al.*: Expression of inducible nitric oxide in human lung epithelial cells. *Biochem Biophys Res Commun* (1994) **203**: 209–218.
39. Nathan C, Xie Q: Regulation of biosynthesis of nitric oxide. *J Biol Chem* (1994) **269**: 13725–13728.
40. Howarth PH, Redington AE, Sringall DR, *et al.*: Epithelially derived endothelin and nitric oxide in asthma. *Int Arch Allergy Immunol* (1995) **107**: 228–230.
41. Hogg N, Darley-Usmar VM, Wilson MT, Moncada S: Production of hydroxyl radicals from the simultaneous generation of superoxide and nitric oxide. *Biochem J* (1992) **281**: 419–424.
42. Hamid Q, Springall DR, Riveros-Moreno V, *et al.*: Induction of nitric oxide synthase in asthma. *Lancet* (1993) **342**: 1510–1513.
43. Yanagisawa M, Kurihara H, Kimura S, *et al.*: A novel potent vasoconstrictor peptide produced by vascular endothelial cells. *Nature* (1988) **332**: 411–415.
44. Inoue A, Yanagisawa M, Kimura S, *et al.*: The human endothelin family: three structurally and pharmacologically distinct isopeptides predicted by three separate genes. *Proc Natl Acad Sci USA* (1989) **86**: 2863–2867.
45. Johnson CW, Logun C, Wu T, Shelhamer JH: Endothelin-1 induced secretion of mucin type glycoprotein from human airways is enhanced by inhibitors of neutral endopeptidase and is independent of eicosanoid generation. *Am Rev Respir Dis* (1992) **145**: A362.
46. Mullol J, Chowdhury BA, White MV, *et al.*: Endothelin in human nasal mucosa. *Am J Respir Cell Mol Biol* (1993) **8**: 393–402.
47. Zamora MA, Dempsey EC, Walchak S, Stelzner TJ: BQ123 an ET_A receptor antagonist, inhibits endothelin-1-mediated proliferation of human pulmonary artery smooth muscle cells. *Am J Respir Cell Mol Biol* (1993) **9**: 429–433.
48. Vittori E, Marini M, Fasoli A, De Franchis R, Mattoli S: Increased expression of endothelin in bronchial epithelial cells of asthmatic patients and effect of corticosteroids. *Am Rev Respir Dis* (1992) **146**: 1320–1325.
49. Marciniak SJ, Plumpton C, Barker PJ, Huskisson NS, Davenport AP: Localization of immunoreactive endothelin and proendothelin in the human lung. *Pulmon Pharmacol* (1992) **5**: 175–182.
50. Springall DR, Howarth PH, Counihan H, Djukanovic R, Holgate ST, Polak JM: Endothelin immunoreactivity of airway epithelium in asthmatic patients. *Lancet* (1991) **337**: 697–701.
51. Redington AE, Springall DR, Ghatei MA, *et al.*: Endothelin in bronchoalveolar lavage fluid and its relationship to airflow obstruction in asthma. *Am J Respir Crit Care Med* (1995) **151**: 1034–1039.
52. Wu T, Mullol J, Rieves RD, *et al.*: Endothelin-1 stimulates eicosanoid production in cultured human nasal mucosa. *Am J Respir Cell Mol Biol* (1992) **6**: 168–174.
53. Eling TE, Danilowicz RM, Henke DC, Sivarajah K, Yankaskas JR, Boucher RC: Arachi-

donic acid metabolism by canine tracheal epithelial cells. Product formation and relationship to chloride secretion. *J Biol Chem* (1986) **261**: 12841–12849.
54. Holtzman MJ: Arachidonic acid metabolism in airway epithelial cells. *Annu Rev Physiol* (1992) **54**: 303–329.
55. Hay DWP, Farmer SC, Raeburn D, Robinson VA, Flemming WW, Feaden JS: Airway epithelium modulates the reactivity of guinea pig respiratory smooth muscle. *Eur J Pharmacol* (1986) **129**: 11–18.
56. Barnett K, Jacopy DB, Nadel JA, Lazarus SC: The effects of epithelial cell supernatant on contractions of isolated canine tracheal smooth muscle. *Am Rev Respir Dis* (1988) **138**: 780–783.
57. Butler GB, Adler KB, Evans JN, Morgan DW, Szarek JL: Modulation of rabbit airway smooth muscle responsiveness by respiratory epithelium: involvement of an inhibitory metabolite of arachidonic acid. *Am Rev Respir Dis* (1987) **135**: 1099–1104.
58. Devalia JL, Sapsford RJ, Cundell DR, Rusznak C, Campbell AM, Davies RJ: Human bronchial epithelial cell dysfunction following *in vitro* exposure to nitrogen dioxide. *Eur Respir J* (1993) **6**: 1308–1316.
59. Piacentini GL, Kaliner MA: The potential roles of leukotrienes in bronchial asthma. *Am Rev Respir Dis* (1991) **143**: 96S–99S.
60. Corrigan CJ, Kay AB: The role of inflammatory cells in the pathogenesis of asthma and chronic obstructive pulmonary disease. *Am Rev Respir Dis* (1991) **143**: 1165–1168.
61. Venge P, Dahl R: Are blood eosinophil number and activity important for the development of the late asthmatic reaction after allergen challenge? *Eur Respir J* (1989) **2**: 430S–434S.
62. Denburg JA, Jordana M, Gibson P, Hargreave F, Gauldie J, Dolovich J: Cellular and molecular basis of allergic airways inflammation. In Johanssen SGO (ed) *Pharmacia Allergy Research Foundation, Award Book 1990*. Uppsala, AW Grafiska, 1990, pp 15–22.
63. Cromwell O, Hamid Q, Corrigan CJ, *et al.*: Expression and generation of interleukin-8, IL-6, granulocyte–macrophage-colony stimulating factor by bronchial epithelial cells and enhancement by interleukin-1β and tumour necrosis factor-α. *Immunology* (1992) **77**: 330–337.
64. Devalia JL, Campbell AM, Sapsford RJ, *et al.*: The effect of nitrogen dioxide on synthesis of inflammatory cytokines expressed by human bronchial epithelial cells *in vitro*. *Am J Respir Cell Mol Biol* (1993) **9**: 271–278.
65. Wang JH, Devalia JL, Xia C, Sapsford RJ, Davies RJ: Expression of RANTES in human bronchial epithelial cells *in vitro* and *in vivo* and the effect of corticosteroids. *Am J Respir Cell Mol Biol* (1996) **14**: 27–35.
66. Levine SJ: Bronchial epithelial cell–cytokine interactions in airway inflammation. *J Invest Med* (1995) **43**: 241–249.
67. Schall TJ: Biology of the RANTES/SIS cytokine family. *Cytokine* (1991) **3**: 165–183.
68. Miller MD, Krangel MS: Biology and biochemistry of the chemokines: a family of chemotactic and inflammatory cytokines. *Crit Rev Immunol* (1992) **12**: 17–46.
69. Baggiolini M, Walz A, Kunkel SL: Neutrophil activating peptide-1/interleukin-8, a novel cytokine that activate neutrophils. *J Clin Invest* (1989) **84**: 1045–1049.
70. Baggiolini M: Neutrophil activation and the role of interleukin-8 and related cytokines. *Int Arch Allergy Immunol* (1992) **99**: 196–199.
71. Abdelaziz MM, Devalia JL, Khair OA, Calderon M, Sapsford RJ, Davies RJ: The effect of conditioned medium from cultured human bronchial epithelial cells on eosinophil and neutrophil chemotaxis and adherence, *in vitro*. *Am J Respir Cell Mol Biol* (1995) **13**: 728–737.
72. Resnick MB, Weller PF: Mechanisms of eosinophil recruitment. *Am J Respir Cell Mol Biol* (1993) **8**: 349–355.
73. Shute J: Interleukin-8 is a potent eosinophil chemo-attractant. *Clin Exp Allergy* (1994) **24**: 203–206.
74. Warringa RA, Koenderman L, Kok PTM, Kreukniet J, Bruijnzeel PL: Modulation and induction of eosinophil chemotaxis by granulocyte–macrophage colony-stimulating factor and interleukin-3. *Blood* (1991) **77**: 2694–2700.
75. Warringa RA, Mengelers HJ, Kuijper PH, Raaijmakers JA, Bruijnzeel PL, Koenderman L: *In vivo* priming of platelet-activating factor-induced eosinophil chemotaxis in allergic asthmatic individuals. *Blood* (1992) **79**: 1836–1841.

10 Epithelial Cells

76. Alam R, Stafford S, Forsythe P, *et al.*: RANTES is a chemotactic and activating factor for human eosinophils. *J Immunol* (1993) **150**: 3442–3448.
77. Kameyoshi Y, Dorschner A, Mallet AI, Christophers E, Schroder JM: Cytokine RANTES released by thrombin-stimulated platelets is a potent attractant for human eosinophils. *J Exp Med* (1992) **176**: 587–592.
78. Borish L, Rosenwasser LJ: Update on cytokines. *J Allergy Clin Immunol* (1996) **97**: 719–734.
79. Owen WF Jr, Rothenberg ME, Silberstein DS, *et al.*: Regulation of human eosinophil viability, density, and function by granulocyte/macrophage colony-stimulating factor in the presence of 3T3 fibroblasts. *J Exp Med* (1987) **142**: 2424–2429.
80. Cox G, Ohtoshi T, Vancheri C, *et al.*: Promotion of eosinophil survival by human bronchial epithelial cells and its modulation by steroids. *Am J Respir Cell Mol Biol* (1991) **4**: 525–531.
81. Cox G, Gauldie J, Jordana M: Bronchial epithelial cell-derived cytokines (G-CSF and GM-CSF) promote the survival of peripheral blood neutrophils *in vitro*. *Am J Respir Cell Mol Biol* (1992) **7**: 507–513.
82. Trigg CJ, Manolitsas ND, Wang JH, *et al.*: Placebo-controlled immunopathological study of four months inhaled corticosteroids in asthma. *Am J Respir Crit Care Med* (1994) **150**: 17–22.
83. Wegner CD, Gundel RH, Rothlein R, Letts G: Expression and probable roles of cell adhesion molecules in lung inflammation. *Chest* (1992) **101**: 34S–39S.
84. Dobrina A, Menegazzi R, Carlos TM, *et al.*: Mechanisms of eosinophils adherence to cultured vascular endothelial cells. *J Clin Invest* (1991) **88**: 20–26.
85. Bochner BS, Luscinskas FW, Gimbrone MA, *et al.*: Adhesion of human basophils, eosinophils, and neutrophils to interleukin 1-activated human vascular endothelial cells: contribution of endothelial cell adhesion molecules. *J Exp Med* (1991) **173**: 1553–1556.
86. Hutchins D, Steel CM: Regulation of ICAM-1 (CD54) expression in human breast cancer cell lines by interleukin 6 and fibroblast-derived factors. *Int J Cancer* (1994) **58**: 80–84.
87. Ohteki T, Okamoto S, Nakamura M, Nemoto E, Kumagai K: Elevated production of interleukin-6 by hepatic MNC correlates with ICAM-1 expression on the hepatic sinusoidal endothelial cells in autoimmune MRL/1pr mice. *Immunol Lett* (1993) **36**: 145–152.
88. Robinson DS, Durham SR, Kay B: Cytokines in asthma. *Thorax* (1993) **58**: 845–853.
89. Macky CR, Imhof BA: Cell adhesion in immune system. *Immunol Today* (1993) **14**: 99–104.
90. Leff AR, Hamann KJ, Wegner CD: Inflammation and cell–cell interactions in airway hyperresponsivness. *Am J Physiol* (1991) **260**: L189–L206.
91. Albelda SM, Buck CA: Integrins and other cell adhesion molecules. *FASEB J* (1990) **4**: 2868–2880.
92. Calderon E, Lockey RF: A possible role of adhesion molecules in asthma. *J Allergy Clin Immunol* (1992) **90**: 852–865.
93. Wegner CD, Gundel RH, Reilly P, Haynes N, Letts G, Rothlein R: Intercellular adhesion molecule-1 (ICAM-1) in the pathogenesis of asthma. *Science* (1990) **247**: 456–459.
94. Tosi MF, Stark JM, Smith W, Hamedani A, Gruenert DC, Infeld MD: Induction of ICAM-1 expression on human airway epithelial cells by inflammatory cytokines: effects on neutrophil–epithelial cell adhesion. *Am J Respir Cell Mol Biol* (1992) **7**: 214–221.
95. Wegner CD, Gundel RH, Rothlein R, Letts G: Expression and probable roles of cell adhesion molecules in lung inflammation. *Chest* (1992) **101**: 34S–39S.
96. Dobrina A, Menegazzi R, Carlos TM, *et al.*: Mechanisms of eosinophils adherence to cultured vascular endothelial cells. *J Clin Invest* (1991) **88**: 20–26.
97. Manolitsas N, Trigg CJ, McAulay AE, *et al.*: Expression of intercellular adhesion molecule-1 and the β1-integrins in asthma. *Eur Respir J* (1994) **7**: 1439–1444.
98. Monterfort S, Feather IH, Wilson SJ, *et al.*: The expression of leukocyte–endothelial adhesion molecules is increased in perennial allergic rhinitis. *Am J Respir Cell Mol Biol* (1992) **7**: 393–398.
99. Kobayasshi T, Hashimoto S, Imai K, *et al.*: Elevation of serum soluble intercellular adhesion molecule-1 (sICAM-1) and sE-selectin levels in bronchial asthma. *Clin Exp Immunol* (1994) **96**: 110–115.
100. Koizumi A, Hashimoto S, Kobayasshi T, Imai K, Yachi A, Horie T: Elevation of serum soluble vascular cell adhesion molecule-1 (sVCAM-1) level in bronchial asthma. *Clin Exp Immunol* (1995) **101**: 468–473.

101. Davies RJ, Abdelaziz MM, Khair OA, Devalia JL: Cell adhesion molecules involved in eosinophil adhesion to human bronchial epithelial cells, in vitro (abstract) *J Allergy Clin Immunol* (1995) **95**: 219P.
102. Liversidge JM, Sewell HF, Forrester JV: Human retinal pigment epithelial cells differentially express MHC class II (HLA, DP, DR and DQ) antigens in response to in vitro simulation with lymphokine or purified IFN-γ. *Clin Exp Immunol* (1988) **73**: 489–494.
103. Motojima K, Matsuo S, Mullen Y: DR antigen expression on vascular endothelium and duct epithelium in fresh or cultured human fetal pancreata in the presence of gamma-interferon. *Transplantation* (1989) **48**: 1022–1025.
104. Komatsu T, Yamamoto M, Shimokata K, Nagura H: Phenotypic characterisation of alveolar capillary endothelial cells, alveolar epithelial cells and alveolar macrophages in patients with pulmonary fibrosis, with special reference to MHC class II antigens. *Virchows Arch (A)* (1989) **415**: 79–90.
105. Rossi GA, Sacco O, Lapertosa G, Corte G, Ravazzoni C, Allegra L: Human ciliated bronchial epithelial cells express HLA DR antigens and HLA DR genes (abstract). *Am Rev Respir Dis* (1988) **137**: 5.
106. Aebischer I, Stadler BM: T_H1–T_H2 cells in allergic responses: at the limits of a concept. *Adv Immunol* (1996) **61**: 341–403.
107. Defrance T, Carayon P, Billian G, et al.: Interleukin-13 is a B-cell stimulating factor. *J Exp Med* (1994) **179**: 135–143.
108. Lechner JF, Haugen A, McClendon IA, Pettis EW: Clonal growth of normal adult human bronchial epithelial cells in serum-free medium. *In Vitro* (1982) **18**: 633–642.
109. Wiesel JM, Gamiel H, Vlodavsky I, Gay I, Ben-Bassat H: Cell attachment, growth characteristics and surface morphology of human upper-respiratory tract epithelium cultured on extracellular matrix. *Eur J Clin Invest* (1983) **13**: 57–63.
110. Wu R, Yankaskas J, Cheng E, Knowles MR, Boucher R: Growth and differentiation of human nasal epithelial cells in culture. *Am Rev Respir Dis* (1985) **132**: 311–320.
111. Ayars GH, Altman LC, McManus MM, et al.: Injurious effect of the eosinophil peroxidase–hydrogen peroxidase–halide system and major basic protein on human nasal epithelium in vitro. *Am Rev Respir Dis* (1989) **140**: 125–131.
112. Farber JL, Young EE: Accelerated phospholipid degradation in anoxic rat hepatocytes. *Arch Biochem Biophys* (1981) **211**: 312–320.
113. Devalia JL, Sapsford RJ, Wells C, Richman P, Davies RJ: Culture and comparison of human bronchial and nasal epithelial cells in vitro. *Respir Med* (1990) **84**: 303–312.
114. Calderón MA, Devalia JL, Prior AJ, Sapsford RJ, Davies RJ: A comparison of cytokine release from epithelial cells cultured from nasal biopsy specimens of atopic patients with and without rhinitis and non-atopic subjects without rhinitis. *J Allergy Clin Immunol* (1997) **99**: 65–76.
115. Wein M, Bochner BS: Adhesion molecule antagonists: future therapies for allergic diseases? *Eur Respir J* (1993) **6**: 1239–1242.
116. Symon FA, Wardlaw AJ: Selectins and their counter receptors: a bitter sweet attraction. *Thorax* (1996) **51**: 1155–1159.

11

Pathophysiology of Airway Mucus Secretion in Asthma

Y.-C. LIU, A.M. KHAWAJA and D.F. ROGERS

INTRODUCTION

Mucus, morbidity and mortality are inextricably linked in the pathophysiology of asthma. Death in asthma ultimately results from sudden occlusion of conducting airways, leading to asphyxia (status asthmaticus). Part of this occlusion is due to mucus because at postmortem the lungs of these patients have 'mucus' plugs blocking the airway lumen.[1] The presence of the plugs indicates a chronic defect in either the mucus secretory system or mucociliary clearance, or both. Mucus hypersecretion also contributes to the chronic symptoms associated with the everyday morbidity of asthma, namely wheeze and the often embarrassing cough and sputum production. It should be noted, however, that airway mucus abnormalities are only one factor in the pathophysiology of asthma. Nevertheless, problems with mucus may predominate in some individuals more than others, or may be more important in a particular phase of an individual's condition. The aim of this chapter is to review the pathophysiology of the secretory system in the context of asthmatic airways.

AIRWAY MUCUS

The airway lumen is lined with a thin film of viscoelastic liquid that under normal conditions protects the respiratory system. The liquid is often referred to as 'mucus'. The primary role of airway mucus is to form a protective barrier between the external environment and the body's sterile internal environment. Physical and chemical aspects

Table 11.1 Constituents of airway mucus.

Water
Electrolytes
Mucous glycoproteins (mucins)
Lipids
Antimicrobial enzymes (lysozyme, lactoferricin,[124] transferrin, peroxidases)
Immunoglobulins (secretory IgA, IgG, IgM)
Antiproteases (α_1-proteinase inhibitor, SLPI,[125] α_1-antichymotrypsin, α_2-macroglobulin, TIMP)
Enzymatic antioxidants (superoxide dismutase,[126] glutathione peroxidase,[127] catalase)
Non-enzymatic antioxidants (reduced glutathione, ascorbic acid, uric acid, α-tocopherol)
Albumin
Proline-rich proteins
DNA
Cell-derived proteinases (e.g. neutrophil proteases: antibacterial)
Cell-derived inflammatory mediators
Plasma/interstitial-derived inflammatory mediators (e.g. bradykinin)
Endogenous anti-inflammatory molecules (e.g. lipocortin)

SLPI, secretory leukoprotease inhibitor; TIMP, tissue inhibitor of metalloproteinases.

of human airway surface liquid[2] and the functions of proteins and lipids in airway secretions[3] have been considered recently, and a brief overview only is given here.

Airway mucus is a 1–2% aqueous solution of glycoconjugates (predominantly mucous glycoproteins, termed mucins), proteoglycans, electrolytes, enzymes, antienzymes, antioxidants, antibacterial agents, other plasma-derived proteins, lipids and various cellular mediators (Table 11.1). Airway mucus is often considered to exist as a bilayer, comprising an upper 'gel' layer and a lower aqueous 'sol' layer. The liquid is in intimate contact with the surface epithelial cells, the most abundant of which are the ciliated cells.[4] This arrangement allows mucus to function as an escalator for removal of inhaled particles from the lung. The gel layer is dynamic, being continuously wafted towards the throat by the underlying beating cilia. The sol layer provides the cilia with a medium within which they can effectively beat. The efficiency of mucociliary clearance is directly related to the elasticity and inversely proportional to the viscosity of the gel layer, and is dependent upon the depth of the aqueous layer.[5] Therefore, airway hypersecretory pathophysiology is a result of ciliary motility dysfunction or of abnormal mucus, either its viscoelastic property or simply its quantity.

Mucins

The viscoelastic properties of airway mucus are attributed largely to high molecular weight glycoproteins known as mucins,[6] which are secreted by specialized cells in the epithelium and submucosa.[7] Mucins consist of a peptide backbone, termed apomucin, to which multiple oligosaccharide side-chains are bound. Apomucins are peptides of 100–400 kDa that account for 10–20% of the total glycoprotein weight. They are expressed in goblet cells and glandular mucous cells, encoded from several genes,[8,9] and are synthesized in the rough endoplasmic reticulum of these cells. Primary structure analysis shows that apomucin molecules are composed of numerous amino acid tandem repeats, with the number of repeats and number of amino acids per repeat varying with each mucin gene.[10] Currently, nine mucin genes are recognized, namely *MUC 1–4*, *MUC 5AC*, *MUC 5B*, *MUC 6–8*.[11] Mucin gene quantitation has been hampered by the variability in numbers of tandemly repeating nucleotides per mRNA molecule and by the polydispersity of mRNA transcripts. However, using new techniques, localization of mucin genes, e.g. *MUC 2* and *MUC 5*, in airway tissue is becoming possible.[12,13] In contrast, determination of the biophysical properties of the mucins encoded by these different genes is as yet unknown and is going to be an extremely difficult task.

Abundant serine and threonine residues within the apomucin act as glycosylation sites at which carbohydrate side-chains attach.[14] Carbohydrates account for 70–80% of the total mass of the mucin molecule. Five sugars, namely fucose, galactose, N-acetylglucosamine, N-acetylgalactosamine and N-acetylneuraminic acid, have been identified in the mucin molecules and are covalently attached, starting with N-acetylgalactosamine, to the apomucin backbone by post-translational modification in the Golgi. The side-chains vary from 2–20 sugars in length, with an average of eight. The complete mucin glycoprotein, which is highly sulphated, is accumulated into secretory granules awaiting appropriate stimuli.

AIRWAY MUCUS-SECRETING CELLS

The principal mucus-secreting cells of the airways are the surface epithelial goblet cells and the mucous cells of the submucosal glands. These cells are reservoirs of mucin, releasing large amounts of mucus into the lumen on stimuli, but secreting little if unprovoked. The goblet and mucous cells are considered in more detail below. Other secretory cells in the airways are the surface epithelial serous cells,[15] Clara cells and possibly ciliated cells, and the serous cells of the submucosal glands. Serous cells, although containing secretory granules, appear to be more involved with maintenance of airway sterility, via secretion of antimicrobial enzymes, than with mucus production.[16] The ciliated cell does not contain secretory granules, but may be involved with production of the 'glycocalyx', the highly Alcian blue-positive material that appears to be in intimate association with the surface of the ciliated cells.[17] The precise role of the Clara cell is not known,[18] but may be surfactant production.[19] In diseases associated with airway mucus hypersecretion, including asthma, goblet cell hyperplasia and submucosal gland hypertrophy are associated with reduced numbers of serous and ciliated cells, and possibly Clara cells.

Goblet cells

These are the principal mucus-secreting cells of the surface epithelium and as such contribute to 'first-line' defence of the airways.[20] Initially named according to their appearance,[21] the goblet shape may be an artefact of chemical fixation.[22] Their distribution within the respiratory tract is species dependent.[23] In humans, goblet cells are abundant in larger airways with a density estimated at 6000–7000/mm^2.[24] Their main secretory product is mucin, stored in electron-lucent secretory granules, which is released by an array of secretagogues.[20] In normal human airways, the ratio of submucosal glands to goblet cells has been calculated as 40:1,[25] indicating that they might not play a significant role in hypersecretion. However, it is known that goblet cells undergo hyperplasia and metaplasia in diseased conditions,[26] and that they predominate in distal airways where excess 'abnormal' mucus may be much harder to remove. It is in the pathophysiology of chronic airway hypersecretory disease that the importance of goblet cells becomes apparent.

Submucosal glands

Submucosal glands are a complex network of tubules and ducts opening on to the airway lumen.[27] Composed of secretory tubules lined with mucous and, more distally, serous cells, a collecting duct and a ciliated neck, they are predominant in large cartilaginous airways. The mucous cells are responsible for most of the mucus secretion in human large airways. They line the 'mucous tubules' in the submucosal gland and contain an abundance of electron-lucent, acid-staining secretory mucin granules. Morphologically, mucous cells are similar to goblet cells.

MUCUS ABNORMALITIES IN ASTHMA

Mucus abnormalities are generally considered to be a feature of asthma. However, it is not known precisely whether the abnormalities lie in overproduction of mucus, an intrinsic biochemical abnormality in asthmatic mucus, interactions between mucus and other airway components, or a combination of any or all of these factors. These possibilities are addressed below.

Is there mucus hypersecretion in asthma?

One fundamental indication of airway hypersecretion in asthma is found at post-mortem in the lungs of patients who have died in status asthmaticus: many generations of airway from bronchi to bronchioles are blocked or partially occluded with gelatinous plugs[1,28,29] (Fig. 11.1). It should be noted, however, that although a characteristic feature of asthma, mucus plugging is not found in all patients dying of asthma.[30] The 'mucus' plugs are erroneously named because mucus is only one component, together with plasma proteins, DNA, cells and proteoglycans.[1,31] However, mucins are the major gel-forming component.[32,33] Using morphometric techniques, the amount of mucus in both central and peripheral airways has been found to be increased in chronic asthmatic patients and in patients with severe fatal asthma compared with control subjects[26] (Fig. 11.2).

Mucus plug formation may not be an abrupt terminal process in end-stage asthma because incomplete plugs are found in the airways of asthmatic subjects who have died from causes other than their asthma.[34] These plugs can be coughed up and expectorated,

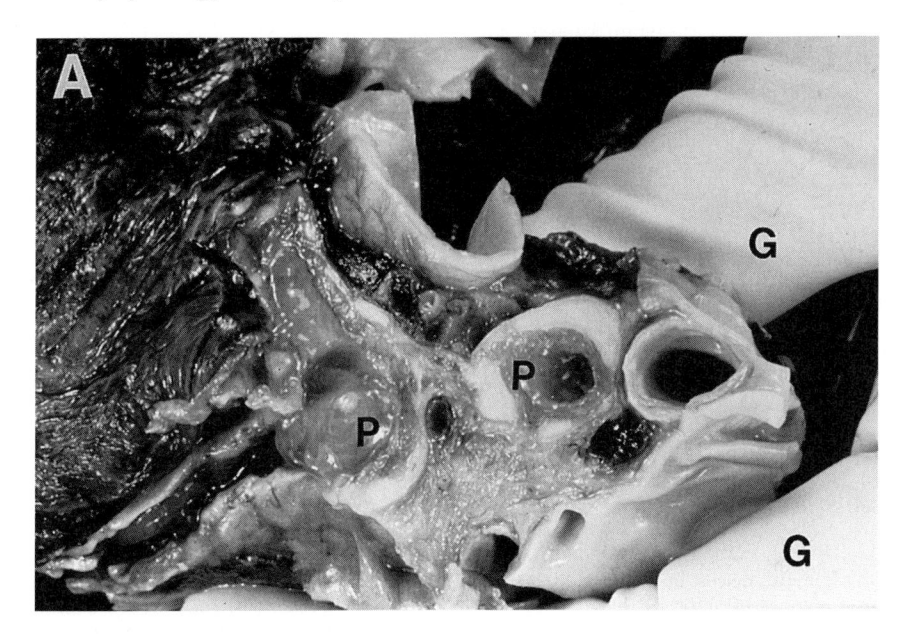

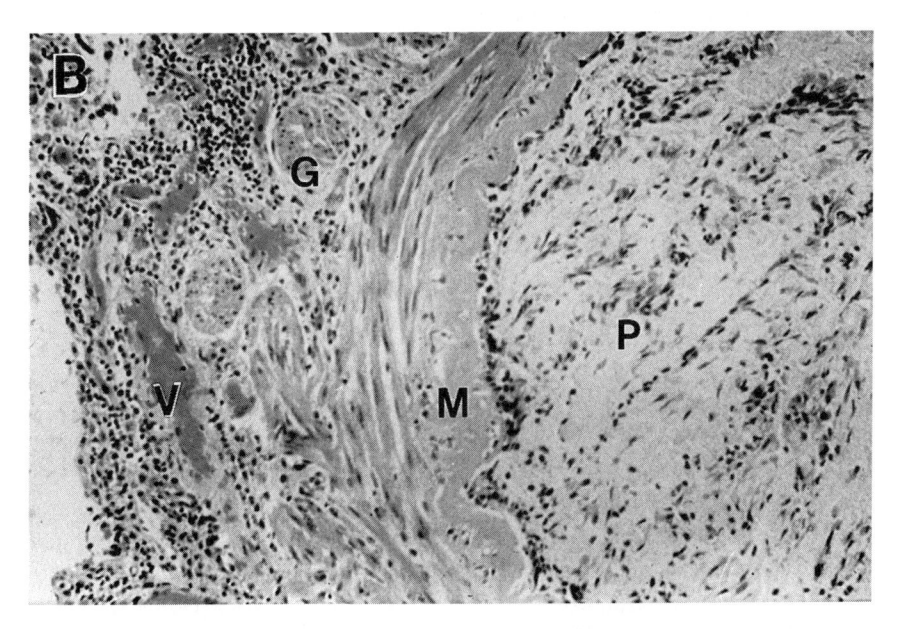

Fig. 11.1 Mucus plugging in asthma. (A) Gross pathology of the lung from an asthmatic patient cut through to show gelatinous plugs (P) blocking or partially occluding the large airways. Photo courtesy of Dr Catherine Corbishley, St George's Healthcare, London, whose gloved hands (G) are holding the specimen. (B) Histology of a small airway of an asthmatic patient showing occlusion of the lumen with a mucus plug (P) containing numerous inflammatory cells. M, basement membrane (thickened); G, submucosal gland; V, blood vessel (with evidence of vasodilatation).

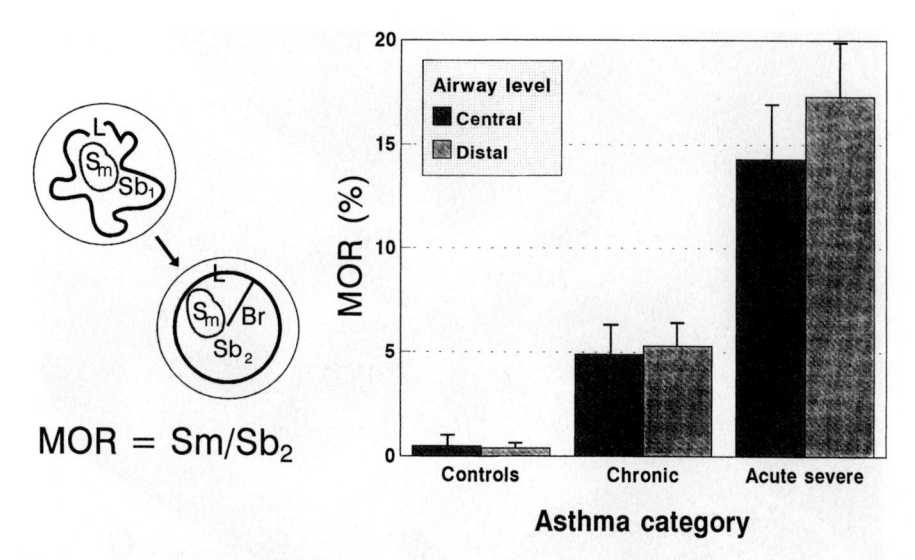

Fig. 11.2 Luminal mucus in asthma. Amount of luminal mucus was quantified morphometrically in histological sections of autopsied lungs of patients without lung disease (controls, $n = 4$), patients with chronic asthma ($n = 5$) or patients who had died of a sudden attack ($n = 3$). The size of stainable mucin (S_m) in the airway lumen was expressed, after digital conversion of the length (L) of the basement membrane to a circle, as a ratio of the size of the bronchus (S_{b2}): S_m/S_{b2} represents the mucus-occupying ratio (MOR). Luminal mucus increases with severity of asthma. Redrawn using data in ref. 26.

and histologically they resemble the plugs found in the airways of patients dying of asthma.[35] The latter observations indicate that plug formation is a chronic process that in some individuals progresses to the stage where it couples with some other aspect of airway dysfunction, e.g. bronchospasm, to catastrophic effect.

Another indication of mucus hypersecretion in asthma is that many, but not all, asthmatic individuals demonstrate a degree of sputum production, whilst most produce more sputum during acute attacks or during recovery from an attack.[36] The increased sputum production is associated with an increase in mucus secretion. Firstly, there is a three-fold increase in dry weight of asthmatic sputa compared with controls, due in part to enhanced mucus secretion as evidenced by increased markers for mucin (fucose and sulphate).[37] Secondly, elevations of mucin-like glycoprotein have been detected in induced sputum in stable asthmatic subjects compared with healthy control subjects.[38] It should be noted, however, that although the mucus content of asthmatic sputum is increased, a number of other molecules are also increased, including DNA, lactoferrin, eosinophil cationic protein and plasma proteins such as albumin and fibrinogen.[37–39] Thus, mucus hypersecretion is only one process associated with sputum production in asthma.

The increase in mucus secretion reflects an increased amount of airway secretory apparatus. Using morphometric techniques, it has been found that goblet cell hyperplasia and metaplasia are found throughout the lower airways of patients dying of acute severe asthma but not of chronic asthma[26] (Fig. 11.3). In contrast to the marked difference in magnitude of goblet cell increase between chronic asthma and acute severe asthma, the degree of gland hypertrophy was similar between the two patient groups[26] (Fig. 11.3). Although requiring confirmation in a larger population of patients, these observations

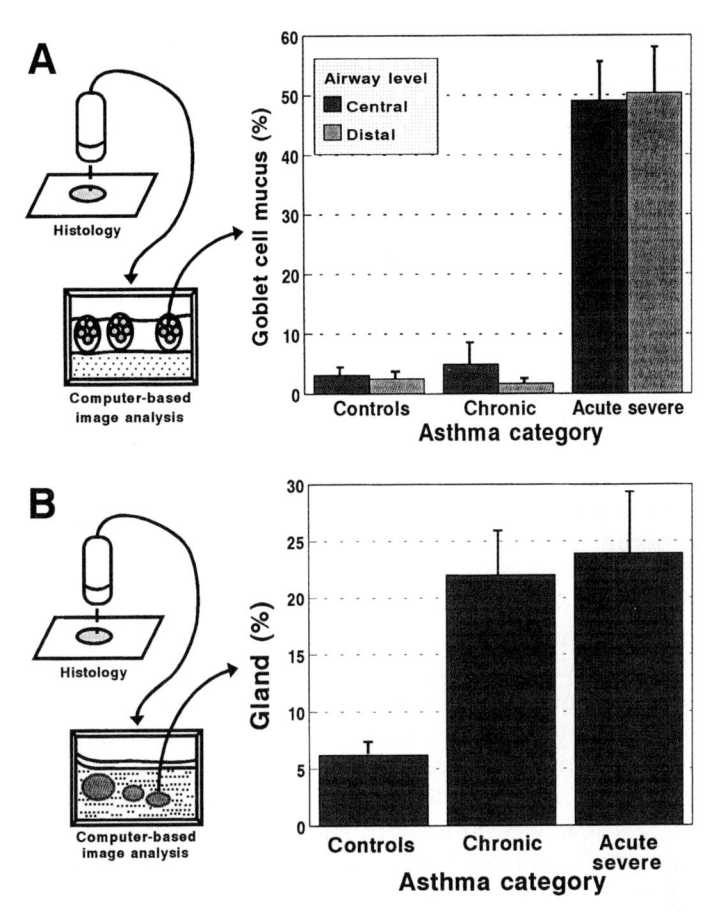

Fig. 11.3 Increased amounts of airways mucus-secreting tissue in asthma. Goblet cell hyperplasia (A) and submucosal gland hypertrophy (B) were quantified morphometrically in histological sections of autopsied lungs of patients without lung disease (controls, $n = 4$), patients with chronic asthma ($n = 5$) or patients who had died of a sudden attack ($n = 3$). Goblet cell hyperplasia was expressed as the proportion of stainable mucin to total epithelial layer. Submucosal gland hypertrophy was expressed as the proportion of gland to bronchial wall. Redrawn using data in ref. 26.

indicate that in patients dying of asthma airway submucosal gland hypertrophy is a nonspecific feature whereas disproportionate goblet cell hyperplasia is associated with, and possibly contributes to, fatal acute severe attacks.

Is there an abnormality of mucin in asthma?

Little is known about whether or not there is an intrinsic biochemical alteration of mucin in asthma, although the tenacious gelatinous nature of mucus plugs has often been taken to indicate that there is. Unlike airway plugs of patients with chronic bronchitis, those of asthmatic subjects require more than a 1.5-m head of water pressure to dislodge them from the airways.[40] Analysis of sputum viscosity in several bronchial diseases, including chronic bronchitis, bronchiectasis and asthma, showed that although there was considerable variability in viscosity in all diseases, the viscosity of asthmatic sputum was greater than that of the other diseases studied.[37,41] However, differences in sputum rheology may not reflect differences in mucus rheology. In addition, these early studies used a cone and plate viscometer. The high shear rates employed in analysis are likely to have induced changes in the rheology of the sputum, making it difficult to draw conclusions about differences in the original rheological properties between the different diseases. Nevertheless, a later study, using a coaxial cylinder rheometer, found similar data in 'bronchorrhoea' sputum from chronic asthmatic individuals.[42]

More recently, a detailed analysis of mucus plugs collected and pooled from a patient dying in status asthmaticus revealed a number of notable differences compared with mucus collected from the airways of non-asthmatic subjects.[33] The plugs were markedly resistant to dispersion in guanidinium chloride, which indicates a lack of covalent crosslinking maintaining the integrity of the mucus gel. Mucins were in high concentration (25% of the non-dialysable material), the majority (85%) were of extreme size (30 000–40 000 kDa compared with an average of 15 000 kDa in 'normals', cystic fibrosis and chronic bronchitis) and they appeared by electron microscopy as more highly complex networks than mucins from other respiratory diseases. The major mucin species was distinctly less acidic than mucins previously described from either normal or diseased airways. Otherwise, the architecture and general composition of asthmatic mucins were similar to other respiratory mucins. The authors concluded that the asthmatic mucus gel was stabilized by non-covalent interactions between extremely large mucins assembled from 'normal'-sized subunits, which suggests an abnormality in the mucus, possibly due to a defect in the processing steps of the assembly process required to make normal airway mucins. The authors further concluded that the combination of characteristics of the mucins could explain the solidity of the mucus plugs extracted from the patient's lungs. They acknowledge that studies on more patients are required to test this hypothesis. Nevertheless, this study represents the first reliable demonstration of an intrinsic abnormality in mucus in asthma.

It is clearly necessary to know whether the abnormality in the mucus is due to the presence, or absence, of a specific mucin species in the asthmatic secretions. Mucin products of *MUC 5AC* and two unknown mucin species, probably different glycoforms of the *MUC 5B* gene,[43] were detected in mucus retrieved from the lungs of a patient dying of asthma.[44] However, these mucins were also found in 'normal' secretions, collected via tracheal intubation, and in sputum from a patient with chronic bronchitis. Work will be continuing in this important area long into the future. Although MUC 5A was not specific to asthma, of possible importance was the finding that compared with its abundance in asthmatic and normal secretions it was only a minor component of secretions from the chronic bronchitic patient. It will be of great interest in the future to confirm or refute this result and to determine whether the level of MUC 5A, and other mucins, is linked to a pathophysiological process, such as mucus plug formation in asthma.

Interactions between mucus and other airway components

Airway mucus in asthma is composed of numerous constituents other than mucins (see above). A number of these constituents have been shown in experimental systems to

11 Pathophysiology of Airway Mucus Secretion in Asthma

interact with mucin in a way detrimental to airway homeostasis. These are considered below.

Increased plasma exudation from the bronchial microvasculature into the airway interstitium and lumen is a characteristic feature of asthma.[45] It should be noted, however, that other airway hypersecretory conditions also exhibit some degree of airways plasma exudation at some stage of the disease.[37] Plasma induces mucin secretion in human airways *in vitro*,[46] which would increase the amount of mucus in the airway lumen. Albumin, one of the plasma proteins found in asthmatic sputum,[39] synergistically increases mucus viscosity *in vitro*. There is no evidence that this happens *in vivo* in the asthmatic patient. The concentration of albumin may be insufficient or the interaction may be nullified by interactions with other macromolecules. Similarly, DNA is found in asthmatic sputum.[38,47] DNA (5 mg/ml) increases mucus viscosity *in vitro*,[48] although there is no evidence that this happens *in vivo* in the asthmatic patient. In fact, there is evidence that it does not happen. Firstly, the DNA content of asthmatic sputum has been found to be comparatively low, <1 mg/ml,[47] which is insufficient to markedly increase mucus viscosity.[48] Secondly, cystic fibrosis sputa usually have a higher DNA content, but without mucus plug formation. Thus, plasma exudation may lead to an increase in the amount of airway mucus, by increasing mucin secretion both directly and indirectly, whereas it is uncertain whether albumin or DNA significantly affect mucus viscosity.

Another possible interaction is between newly secreted mucus and inflammatory cell products. A recent post-mortem morphometric study has found that mucus release from goblet cells is 'restricted' in both central and peripheral airways from patients dying of their asthma compared with patients with chronic bronchitis (or control subjects without respiratory disease).[49] Mucus in the airway lumen appears to be 'anchored' in place by maintaining continuity with the goblet cells. The favoured hypothesis of the authors to explain this observation is that, in chronic bronchitis, neutrophil proteinases from neutrophils, the predominant inflammatory cell in the airways of patients with chronic bronchitis, cleave cell surface-attached mucins from the goblet cells. Eosinophils, rather than neutrophils, are the predominant inflammatory cell in asthmatic airways and certainly predominated in the airways of the patients in the above study.[49] There is no evidence for eosinophil products being able to cleave mucins. Thus, it would appear from this study that 'continuity' between goblet cells and their secreted mucus is specific to asthma and that the lack of neutrophils in the airways of patients dying of asthma contributed to the mucus plugging of their airways. It is possible that full release of goblet cell mucin may require a final cleavage mechanism because retention of secretory product is a feature of airway goblet cell exocytosis.[50] It should be noted that neutrophilic inflammation is a feature of the sputum from a number of asthmatic patients during an acute exacerbation.[47] It would be fascinating, albeit technically demanding, to determine the degree of goblet cell–mucus continuity in this group of patients.

CONSEQUENCES OF AIRWAY MUCUS HYPERSECRETION AND HYPERVISCOSITY

The discussion of the previous section has established, in broad terms, that in the airways in asthma there is overproduction of a hyperviscous mucus. Depending upon how much is present, increased mucus in the airway lumen may not noticeably affect airflow. Larger quantities may still not affect airflow, as in many patients with chronic bronchitis, but may induce cough that contributes markedly to morbidity of both asthmatic and bronchitic patients. In the asthmatic patient, there are two main potential consequences of abnormal airway mucus, namely airway obstruction and change in airway reactivity.

Airway obstruction

Obstruction of the airways with mucus and partially formed or complete mucus plugs is a feature of asthma (see above). The development of airway obstruction by mucus will be a combination of mucus abnormalities and ciliary function, leading to reduced mucociliary clearance and eventual mucostasis. Although problems associated with the variability of depth of aerosol deposition hampered the interpretation of early studies, use of an improved technique has demonstrated that mucociliary clearance in severe asthmatic patients is indeed slower than that in patients with milder disease.[51] It should be noted, however, that a number of factors other than excess mucus can reduce clearance, including epithelial shedding (with consequent loss of cilia) and generation of mediators that slow mucociliary clearance directly, e.g. leukotriene D_4.[52]

Mucus in small airways is considered to be more intractable for clearance because acidic mucus glycoproteins interfere with ciliary function, and in small airways ciliary function is normally reduced by a decreased number and length of cilia. There may also be loss of ciliary epithelium due to eosinophil major basic protein[53] and to goblet cell metaplasia replacing ciliated cells.[54] Owing to their smaller diameter, a high mucus occupancy compromises the patency of small airways more easily than that of large airways.

Obstruction of the airways leads to ventilation–perfusion mismatch.[55] Specifically, mucus obstruction in the asthmatic lung is patchy,[1,34] which diverts ventilation from some alveolar regions to others to produce mismatch. There follows arterial hypoxaemia and stimulation of chemoreceptors leading to hyperventilation and dyspnoea. In addition, luminal mucus contributes to increased airways resistance and consequently the work of breathing.

Change in airway reactivity

Any amount of reduction of airway luminal cross-sectional area will lead to a markedly exaggerated increase in airflow resistance in response to contraction of airways smooth muscle (i.e. bronchoconstriction).[56] This exaggerated bronchoconstriction is termed 'airway hyperresponsiveness' and is a characteristic of asthma.[57] Reductions in airway cross-sectional area can be due to a number of factors including airway wall thickening, increased surface tension at the air–liquid interface, reductions in the external support of the airway wall, and increased luminal mucus. In cats, introduction of an 'intraluminal space-occupying substance' (small glass beads) to mimic intraluminal mucus led to marked airway hyperresponsiveness to acetylcholine: the provocative concentration of acetylcholine causing a standard fall in pulmonary resistance was decreased four-fold in cats with beads.[58] Small increases in luminal liquid can lead to marked airflow limitation[59] (Fig. 11.4). In addition, intraluminal liquid not only narrows airways simply by filling the

11 Pathophysiology of Airway Mucus Secretion in Asthma

A r=1 → Bronchoconstriction → **B** r=0.5

$$R \alpha \frac{1}{1^4} = 1 \text{ unit} \qquad R \alpha \frac{1}{(0.5)^4} = 16$$

C t=0.1, r=0.9 → Bronchoconstriction → **D** t=0.26, r=0.24

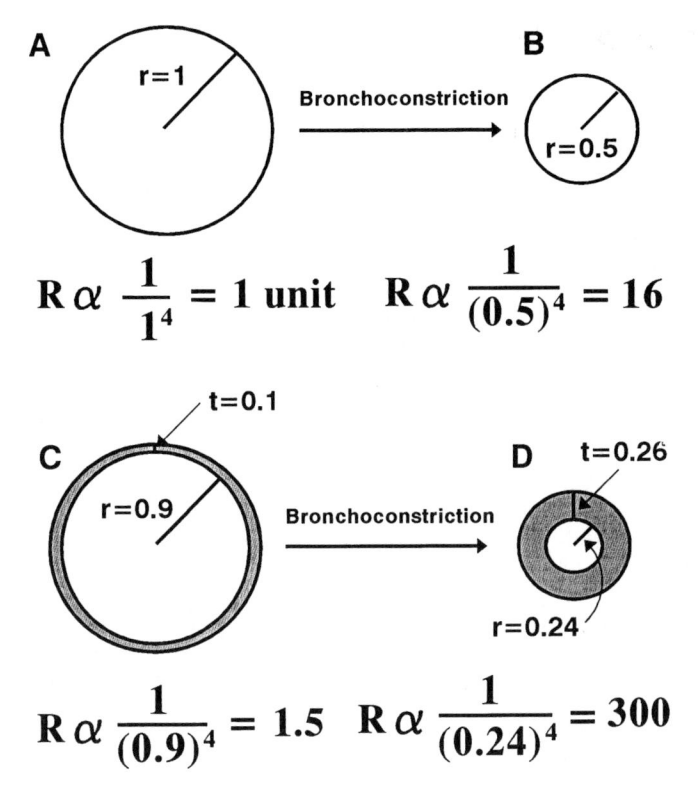

$$R \alpha \frac{1}{(0.9)^4} = 1.5 \qquad R \alpha \frac{1}{(0.24)^4} = 300$$

Fig. 11.4 Effect of luminal mucus on airflow resistance in asthma: theoretical explanation. (A) From Poiseuille's law, resistance to flow (R) is proportional to the reciprocal of the radius (r) raised to the fourth power. (B) Without luminal mucus, bronchoconstriction to reduce the airway radius by half increases airflow resistance 16-fold. (C) A small increase in thickness (t) of the mucus, which reduces the radius of the airway by only one-tenth, has a negligible effect on airflow in the unconstricted airway (cf. A). (D) With bronchoconstriction, luminal mucus, which is essentially non-compressible, markedly amplifies the airflow resistance of the airway.

interstices produced in the airway wall by bronchoconstriction, but also produces an inward force because of surface tension, further compromising airway patency.[59]

The degree of airway responsiveness is influenced by the distribution of an aerosol of spasmogen in the airways and by factors limiting the accessibility of bronchoconstrictors to the smooth muscle. Excessive luminal mucus can have variable effects on airway hyperresponsiveness to inhaled spasmogens.[60–63] Responsiveness can be either reduced or increased depending upon the pattern of distribution of mucus in the airways, which is usually not homogeneous. The increased depth of mucus would be expected to form a protective layer and reduce responsiveness. However, with mucus accumulation limited to particular regions of the lungs, resulting in uneven flow distribution, inhaled bronchoconstrictors will be directed to less resistant or normal airways and this amplifies their responsiveness. This effect obviously has implications for airway responsiveness testing of asthmatic subjects in the clinical laboratory and, depending upon the distribution of airway mucus in any one patient, could lead to overestimates or underestimates of reactivity.

INDUCERS OF AIRWAY MUCUS SECRETION

Numerous inflammatory mediators are generated in asthmatic airways[64] and a variety of neural mechanisms may contribute to asthma pathophysiology.[65] Discussions of the generation of particular mediators and of neural mechanisms in asthma can be found in other chapters of this book and will not be repeated below. Of the mediators that have been studied, all those generated in asthma affect mucus secretion in experimental conditions (Table 11.2) and have been reviewed in detail.[20,27,66] In the discussion below, only the most recent or relevant references will be given. One problem with the majority of studies on the effect of mediators or neural mechanisms on airway secretion is that only the acute secretory response to a single exogenous administration of drug, or nerve stimulation, in naive tissue (i.e. not previously exposed to the drug) is determined.[67] Thus, although being extremely useful in understanding the basic physiology of airway secretion, acute studies may have little relevance to asthma where mediators are likely to be generated over extended periods.

Table 11.2 Neural and humoral control of airway mucus secretion.

Stimulation	Magnitude of effect
Cholinergic nerves	++
Adrenergic nerves	0/+
Sensory–efferent nerves	0/++
Cholinoceptor agonists	++
α-Adrenoceptor agonists	+
β-Adrenoceptor agonists	+
Phosphodiesterase isotype IV inhibitors	+
Vasoactive intestinal polypeptide	+
Peptide histidine isoleucine/methionine	?
Neuropeptide tyrosine	?
Substance P	++
Neurokinin A	+
Cacitonin gene-related peptide	0/+
Nitric oxide	−ve/+
Histamine	0/+
5-Hydroxytryptamine	0
Prostaglandins	0/+
Leukotrienes	0/+
Platelet-activating factor	0/+
Bradykinin	+
Endothelin	0/+
Monocyte/macrophage-derived mucus secretagogue	+
Proteinases	+++
Purine nucleotides	+
Reactive oxygen species	0/+
Tumour necrosis factor α	++

+++, Highly potent; ++, marked effect; +, lesser effect; 0, minimal effect; ?, effect alone not published; −ve/+, decrease or increase secretion (dependent upon preparation).

Humoral inducers

The most potent inducers of airway mucus secretion, from both goblet cells and submucosal gland cells, are inflammatory cell proteinases, most notably mast cell chymase, neutrophil elastase and cathepsin G.[68,69] They are at least 10-fold more potent than isoprenaline or histamine. Proteinases can also strip surface proteoglycans from epithelial cells, which may alter the viscosity of airway mucus and affect airway patency.

Macrophages/monocytes are other inflammatory cells that produce a number of products which stimulate airway mucus secretion. Most notable of these are two macrophage/monocyte-derived mucus secretagogues, one 2 kDa (MMS-2), the other 68 kDa (MMS-68). MMS-68, in particular, induces mucus secretion in a variety of *in vitro* experimental systems with a fair degree of potency (60–90% increases in secretion at a concentration of 10 μg/plate),[70] although it is less potent than the proteinases. In cigarette smokers, patients with chronic bronchitis and patients with steroid-dependent asthma, MMS-68 can be detected at higher concentrations than in control subjects,[71] but not mild asthmatic subjects.[72] These observations indicate that MMS-68 is not specific for asthma but may contribute to mucus hypersecretion in severe airway conditions.

The pleiotropic cytokines tumour necrosis factor (TNF)-α, interleukin (IL)-1β and IL-6 dose dependently increase mucus secretion in human tracheal explants.[73–75] Mucus secretion induced by both TNF-α and IL-6 was biphasic with an initial peak at 4–8 h followed by a second peak at 48–72 h. TNF-α-induced mucus secretion is possibly mediated via the production of inducible nitric oxide synthase and IL-6 transactivated by the transcription factor nuclear factor κB (NF-κB).[76] TNF-α and IL-6 both induced concomitant *MUC 2* gene expression, with expression by TNF-α dependent upon activation of protein kinase C and tyrosine kinase.[75] Although the increase in mucin gene expression indicates that these cytokines contribute to the pathophysiology of the hypersecretory state, it should be noted that, unlike MUC 5AC mucin, MUC 2 mucin was not found in any great amount in mucus plugs or sputum from asthmatic subjects.[33,77] Also very interesting is the observation that, in contrast to the stimulatory effects of the pro-inflammatory cytokines above, the 'anti-inflammatory' cytokine IL-10 inhibited mucus release elicited by TNF-α stimulation.[78]

The bioactive lipid mediators, leukotrienes, prostaglandins, platelet-activating factor and hydroxyeicosatetraenoic acids (HETEs), are not especially potent as inducers of airway mucus secretion and have variable effects on airway mucus secretion. Similarly, neither bradykinin, endothelin, adenosine nor reactive oxygen species, such as hydrogen peroxide, have particularly marked effects on secretion. The actions of nitric oxide are variable, being stimulatory in goblet cells in culture[79] and inhibitory in an *in vitro* preparation where the principal source of mucus is the submucosal glands.[80] Further formal comparative studies are required to clarify these differences.

There are a variety of other cellular or plasma-derived secretagogues released during immediate hypersensitivity reactions or late-onset reactions such as prostaglandin-generating factor of anaphylaxis, eosinophil cationic protien, and complement components, which may elicit mucus secretion from human airways.[81]

Neural mechanisms

Three neural pathways mediate airway mucus secretion in the airways of experimental animals, namely sympathetic (adrenergic), parasympathetic (cholinergic) and non-adrenergic, non-cholinergic (NANC).[82] At a simple level, the latter neural mechanism can be visualized as two divisions.[83] The first comprises cholinergic and adrenergic nerves in which neuropeptides are colocalized with the classical neurotransmitter, for example vasoactive intestinal polypeptide (VIP) colocalized with acetylcholine in cholinergic nerves. The second division comprises a discrete population of C-fibres that, together with their sensory function, also subserve a motor function and so may be termed 'sensory–efferent' nerves. The neurotransmitters of these nerves are termed 'sensory neuropeptides' and include calcitonin gene-related peptide and the tachykinins substance P and neurokinin A. In contrast to experimental animals, in human airways *in vitro* it is only cholinergic neural control of mucus secretion that can be readily demonstrated.[84] Adrenergic control has not been demonstrated[84] and NANC neural control can only be demonstrated under special circumstances.[85] Although not formally studied, there is indirect evidence that cholinergic nerves contribute to mucus hypersecretion in asthma.[86]

Acetylcholine released from cholinergic nerves acts on muscarinic receptors, of which three types can be demonstrated pharmacologically, namely M_1, M_2 and M_3.[86] Classically, the M_1 receptor is localized to cholinergic ganglia and facilitates neurotransmission, the M_2 receptor is localized to the nerve endings where it acts as an autoinhibitory feedback receptor to limit acetylcholine release, and the M_3 receptor mediates the end-organ response, e.g. mucus secretion. In ferret trachea *in vitro*, the M_3 receptor mediated cholinergic mucus secretion and nerve activity was regulated by the M_2 receptor.[87] No evidence was found for involvement of the M_1 receptor. In asthma, there appears to be a defect in M_2 regulation of cholinergic bronchoconstriction.[86] Evidence from the ferret indicates that this defect may extend to cholinergic mucus secretion and this may contribute to airway mucus hypersecretion in asthma. Additional regulation of cholinergic mucus secretion may also be defective in asthma. In ferret trachea *in vitro*, endogenous VIP regulates the magnitude of cholinergic mucus secretion.[88] VIP-like immunoreactivity is markedly reduced in the lungs of patients with asthma.[89] Thus, if VIP regulates cholinergic mucus secretion in humans, loss of VIP in asthmatic subjects may also contribute to mucus hypersecretion.

There is indirect evidence that sensory–efferent nerves may contribute to asthma pathogenesis by setting up local (axon) reflexes,[90] although this has been disputed.[91] In human airways, stimulation of these nerves induces modest increases in mucus secretion.[85] The effect is most likely to be mediated via tachykinin NK_1 receptors on the secretory cells.[92,93] Tachykinin activity, particularly that of substance P, is elevated in asthma.[94] This may lead to enhanced airway mucus secretion.

PHARMACOLOGICAL TREATMENT OF MUCUS HYPERSECRETION IN ASTHMA

The pathophysiological mechanisms leading to development of mucus hypersecretion in asthma are unknown. However, the sequence of events must be linked to inflammation in

11 Pathophysiology of Airway Mucus Secretion in Asthma

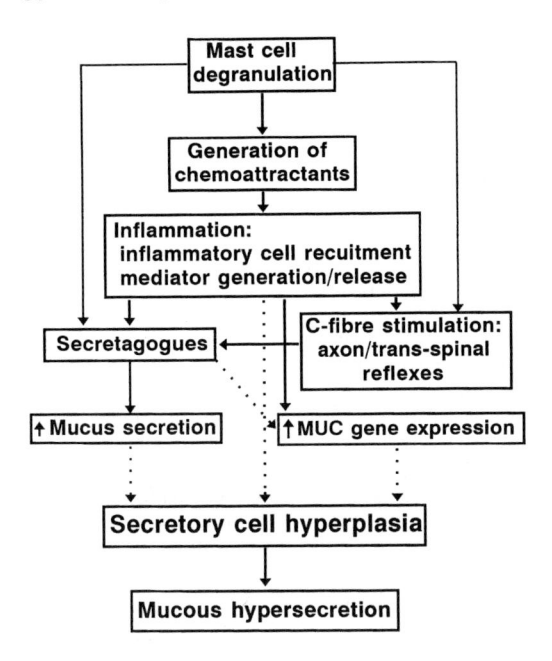

Fig. 11.5 Pathogenesis of airway mucus hypersecretion in asthma. Solid lines indicate experimental evidence for a link between boxes (e.g. the inflammatory cytokine tumour necrosis factor α is a secretagogue and increases mucin gene expression). Broken lines indicate hypothetical scenario without experimental evidence.

the airways and the generation of mediators of secretion, both humoral and from neural sources, which leads to acute increases in mucus secretion (Fig. 11.5). The impetus for transference from acute hypersecretion to the chronic hypersecretory state is unknown, but could be linked to increased mucin gene expression. Consequently, with the basic pathophysiological mechanisms ill defined, treatment of hypersecretion in asthma, as of asthma itself, is based less on empirical data than may be desirable.

Chapters 35–43 of this book are dedicated to discussions of the treatment of asthma. There are numerous treatments that may be of benefit in alleviating mucus hypersecretion in asthma and these are summarized in Table 11.3. In the following discussion, only treatments that have been shown to have direct effects on airway mucus secretion are considered.

Glucocorticosteroids, given orally or by inhalation, are currently the single most effective treatment of patients with asthma, presumably because they are so effective at reducing airway inflammation.[95] In cultured human airways, corticosteroids inhibit spontaneous mucus secretion and also inhibit secretion stimulated by histamine or 5-HETE, although the latter effect appeared to be due more to inhibition of baseline rather than stimulated secretion.[96] In addition to an acute inhibitory effect on mucus secretion, corticosteroids also inhibit goblet cell hyperplasia. In rats whose airways have been transformed into a 'bronchitic' state by subacute exposure to cigarette smoke, corticosteroids markedly inhibit goblet cell increase in both central and peripheral airways.[97] Corticosteroids also inhibit and reverse goblet cell hyperplasia in asthmatic patients[98] and reduce 'bronchorrhoea' sputum in hospitalized severe asthmatic patients.[99] Intriguingly,

Table 11.3 Therapeutic prospects for mucus hypersecretion in asthma.

Avoidance of sensitizing and trigger factors
    Pollutants, allergens (including occupational exposure)
Facilitation of removal of excess airway mucus
    Mucolytics and expectorants
    Bronchodilators
Reduction of airway inflammation
    Glucocorticosteroids
    Antibiotics (in particular erythromycin)
Receptor antagonists
    Anticholinergics (preferably selective muscarinic M_3-receptor antagonists)
    Inflammatory mediator antagonists (e.g. histamine, PAF, leukotrienes, cytokines)
    Tachykinin receptor antagonists
    Antagonists with actions at multiple receptors (e.g. the dual tachykinin NK_1/NK_2 receptor antagonist, FK224)
Inhibition of nerve activity
    Prejunctional inhibition of neurotransmitter release (e.g. including agonists at μ-opioid and $GABA_B$ receptors, and by K^+-channel openers, in particular large conductance Ca^{2+}-activated K^+ channels)
    Muscarinic M_2-receptor agonists (to activate the inhibitory autoreceptor on cholinergic nerves)
    Other inhibitors of neurotransmission or neural activity (e.g. nedocromil sodium, sodium cromoglycate, frusemide)
Antioxidants (e.g. N-acetylcysteine, spin-trap compounds[128])
Neutral endopeptidase replacement (by inhalation)
Suppression of mucin gene activation (e.g. antisense therapy)

GABA, γ-aminobutyric acid; PAF, platelet-activating factor.

dexamethasone has recently been shown to suppress *MUC 2* and *MUC 5AC* mucin gene expression in a human pulmonary mucoepidermoid carcinoma cell line.[100]

The antibiotic erythromycin inhibits baseline and histamine- or methacholine-induced glycoconjugate release from human airways *in vitro* and has an additive effect when incubated together with dexamethasone.[101] In the same study, a range of other antibiotics did not affect mucus secretion, indicating that erythromycin has, in addition to its antibiotic properties, unique antisecretory activity. Two isolated case reports of asthmatic patients, one a small boy,[102] the other an elderly man,[103] showed a beneficial effect of erythromycin administration on mucus hypersecretion. The late onset of the mucus-suppressing effect, starting 4 h after treatment, and the reduction of human airway hyperreactivity[104] indicate that the antisecretory action may depend partly on an anti-inflammatory activity. It would obviously be of interest to determine the effect of erythromycin on asthmatic mucus hypersecretion in a formal, controlled clinical trial, and to correlate the effect with clinical improvement.

Mucolytic agents should reduce the viscosity of sputum and enhance the efficiency of ciliary movement without causing secretion. Some mucolytic agents do reduce sputum viscosity, by breaking disulphide bonds (e.g. N-acetylcysteine),[105] by dissolving acidic glycoconjugates (e.g. bromhexine, ambroxol)[106,107] or by digesting sputa via proteolytic activity (e.g. iodide, iodinated glycerol).[108] Most studies in asthma have revealed that mucolytic agents stimulate mucociliary movements, help sputum clearance or improve lung function in these patients.[109–112] In status asthmaticus, bronchoscopy combined with a mucolytic agent may be an effective therapy for intractable mucous plugs.[113–115]

The cromones nedocromil sodium and sodium cromoglycate are antiasthma drugs that inhibit bronchospasm due to indirect challenge but do not inhibit mucus secretion from secretory cells.[116] However, the cromones, and also frusemide, have recently been shown to reduce either the single-channel conductance or the open probability state of a voltage- and Ca^{2+}-dependent chloride (Cl^-) channel in airway epithelium from a number of mammalian sources.[117] Although this Cl^- channel is most probably involved in epithelial cell volume regulation, it would be interesting to determine whether it also contributes to regulation of the hydration of airway mucus.

Anticholinergic agents suppress sputum production in various forms of airway diseases,[118] although atropine also inhibits ciliary activity. In perennial rhinitis, nasally inhaled ipratropium bromide is beneficial in relieving nasal discharge[119] and may have a similar action in acute asthmatic exacerabations. Currently, anticholinergic agents are non-selective for all muscarinic receptor subtypes: they not only inhibit end-organ M_3-mediated responses, e.g. mucus secretion, but also reduce the activity of the prejunctional autoinhibitory M_2 receptor.[86] Consequently, M_3-selective anticholinergics, lacking inhibitory effects at the M_2 receptor, may prove a better therapeutic option than the currently available non-selective anticholinergics. These drugs are in development, but so far have not reached the clinic.

Tachykininergic tone may be raised in asthma (see above). There are now a number of trials of tachykinin antagonists, including NK_1, NK_2 and dual NK_1/NK_2 antagonists, in asthma and there is disagreement between them as to the therapeutic benefit of these drugs.[94] There have been no formal studies of the effect of these compounds on mucus production in asthma. However, the dual NK_1/NK_2 antagonist FK224 has been found to reduce cough and sputum production in patients with chronic bronchitis[120] and inhibits bradykinin-induced cough and bronchoconstriction in asthmatic patients.[121] Controlled clinical trials are required to determine the effect of these drugs on mucus hypersecretion in asthma and other bronchial diseases.

Antisense oligodeoxynucleotide therapies are starting to enter a variety of basic research studies and clinical trials.[122] Treatment of rat tracheal cultures with an 18-base antisense oligomer effectively inhibited mucin mRNA expression.[123] This is obviously a new and exciting area for treatment of airways mucus hypersecretion.

CONCLUSIONS

Airway mucus is certainly abnormal in asthma. There is too much of it and it is viscid and difficult to clear from the airways. These abnormalities were thoroughly discussed in the second edition of this book. Since then, there have been a number of important advances in investigation of airway mucus in asthma. The first is that, at least in one patient dying of asthma, some mucin species in the mucus plug are intrinsically abnormal, with their biophysical profile consistent with high viscosity. Second, goblet cells retain continuity with their secreted mucin in asthmatic airways but not in normal or chronic bronchitic airways. Third, the identification of an abnormally low level of MUC 5AC in chronic bronchitic patients compared with normal and asthmatic subjects has indicated that there may be different levels of different mucins in different hypersecretory diseases. Fourth, MMS-68, although not specific for asthma, may be a marker for severe hypersecretory disease and may merit therapeutic targetting. Fifth, improved molecular biological techniques are beginning to unravel the complexities of mucin genes. This has enabled tissue localization and has shown that certain cytokines can upregulate mucin gene expression. Thus, the tools are becoming available to determine differences in gene expression between normal states and different diseases, and to study mucin gene regulation. Sixth, the technique of induced sputum may allow biochemical and molecular biological investigation of mucus from healthy individuals for comparison with disease conditions. Finally, advances in the pharmaceutical industry have allowed the development of a wide variety of selective molecules, including tachykinin antagonists, leukotriene antagonists and K^+ channel activators, which opens the options for treatment of asthma and of airways mucus hypersecretion. These important advances have been made over the last 5 years. Because of the rate of technological advancement and the recruitment of scientists from other disciplines, it is highly likely that the next few years will see the greatest advances in the understanding of mucus and mucus hypersecretion in disease for the last decade or more. We are already looking forward to the fourth edition of this book to see these developments.

ACKNOWLEDGEMENTS

The authors wish to thank the National Asthma Campaign (UK), Ciba-Geigy Pharmaceuticals (Basel, Switzerland) and Pfizer Central Research (Sandwich, Kent) for support. Y.-C. Liu is in receipt of a Biomedicine Science Scholarship from Chang Gung Memorial Hospital, Taipei, Taiwan.

REFERENCES

1. Dunnill MS: The pathology of asthma with special reference to changes in the bronchial mucosa. *J Clin Pathol* (1960) **13**: 27–33.
2. Widdicombe JH, Widdicombe JG: Regulation of human airway surface liquid. *Respir Physiol* (1994) **99**: 3–12.
3. Jacquot J, Hayem A, Galabert C: Functions of proteins and lipids in airway secretions. *Eur Respir J* (1992) **5**: 343–358.
4. Rhodin JAG: Ultrastructure and function of the human tracheal mucosa. *Am Rev Respir Dis* (1966) **93**: 1–15.
5. Puchelle E, Zahm JM, Duvivier C: Spinnability of bronchial mucus. Relationship with viscoelastic and mucus transport properties. *Biorheology* (1983) **20**: 239–249.
6. King M: Rheological requirements for optimal clearance of secretions: ciliary transport versus cough. *Eur J Respir Dis* (1980) **61** (Suppl 110): 39–45.
7. Hovenberg HW, Carlstedt I, Davies JR: Mucus glycoproteins in bovine trachea: identification of the major mucin populations in respiratory secretions and investigation of their tissue origins. *Biochem J* (1997) **321**: 117–123.
8. Gum JR: Mucin genes and the proteins they encode: structure, diversity, and regulation. *Am J Respir Cell Mol Biol* (1992) **7**: 557–564.
9. Perini JM, Vandamme-Cubadda N, Aubert JP, *et al.*: Multiple apomucin translation products from human respiratory mucosa mRNA. *Eur J Biochem* (1991) **196**: 321–328.

11 Pathophysiology of Airway Mucus Secretion in Asthma

10. Rose MC: Mucins: structure, function and role in pulmonary diseases. *Am J Physiol* (1992: **263**: L413–L429.
11. Rose MC, Gendler SJ: Airway mucin genes and gene products. In Rogers DF, Lethem MI (eds) *Airway Mucus: Basic Mechanisms and Clinical Perspectives*. Basel, Birkhäuser Verlag, 1997, pp 41–66.
12. Dohrman A, Tsuda T, Escudier E, et al.: Distribution of lysozyme and mucin (MUC2 and MUC3) mRNA in human bronchus. *Exp Lung Res* (1994) **20**: 376–380.
13. Voynow JA, Rose MC: Quantitation of mucin mRNA in respiratory and intestinal epithelial cells. *Am J Respir Cell Mol Biol* (1994) **11**: 742–750.
14. Lamblin G, Lhermitte M, Klein A, et al.: The carbohydrate diversity of human respiratory mucin: a protection of the underlying mucosa? *Am Rev Respir Dis* (1991) **144**: S19–S24.
15. Rogers AV, Dewar B, Corrin B, Jeffery PK: Identification of serous-like cells in the surface epithelium of human bronchioles. *Eur Respir J* (1993) **6**: 498–504.
16. Basbaum CB, Jany B, Finkbeiner WE: The serous cell. *Annu Rev Physiol* (1990) **52**: 97–113.
17. Spicer SS, Chakrin LW, Wardell SR: Histochemistry of mucosubstances in the canine and human respiratory tract. *Lab Invest* (1971) **25**: 483–490.
18. Widdicombe JG, Pack RJ: The Clara cell. *Eur J Respir Dis* (1982) **63**: 202–220.
19. Auten RL, Watkins RH, Shapiro DL, Horowitz S: Surfactant apoprotein A (SP-A) is synthesized in airway cells. *Am J Respir Cell Mol Biol* (1990) **3**: 491–496.
20. Rogers DF: Airway goblet cells: responsive and adaptable front-line defenders. *Eur Respir J* (1994) **7**: 1690–1706.
21. Schulze FE: Epithel und drusenzellen. *Arch Mikrosk Anat* (1867) **3**: 139–197.
22. Verdugo P: Goblet cell secretion and mucogenesis. *Annu Rev Physiol* (1990) **52**: 157–176.
23. Jeffery PK: Morphology of airway surface epithelial cells and glands. *Am Rev Respir Dis* (1983) **128**: S14–S20.
24. Ellefsen P, Tos M: Goblet cells in the human trachea: quantitative studies of a pathological biopsy material. *Arch Otolaryngol* (1972) **95**: 547–555.
25. Reid L: Pathology of chronic bronchitis. *Lancet* (1954) **i**: 275–278.
26. Aikawa T, Shimura S, Sasaki H, Ebina M, Takishima T: Marked goblet cell hyperplasia with accumulation in the airways of patients who died of severe acute asthma attack. *Chest* (1992) **101**: 916–921.
27. Fung DCK, Rogers DF: Airway submucosal glands: physiology and pharmacology. In Rogers DF, Lethem MI (eds) *Airway Mucus: Basic Mechanisms and Clinical Perspectives*. Basel, Birkhäuser Verlag, 1997, pp 179–210.
28. Houston JC, De Navasquez S, Trounce JR: A clinical and pathological study of fatal cases of status asthmaticus. *Thorax* (1953) **8**: 207–213.
29. Saetta M, Stefano AD, Rosine C, Thiene G, Fabbri LM: Quantitative structural analysis of peripheral airways and arteries in sudden fatal asthma. *Am Rev Respir Dis* (1991) **143**: 138–143.
30. Keal EE, Reid L: Pathological alterations in mucus in asthma within and without the cell. In Stein M (ed) *New Directions in Asthma*. Park Ridge, American College of Chest Physicians, 1975, pp 223–239.
31. Bhaskar KR, O'Sullivan DDF, Coles SJ, Kozakevich H, Vawter GP, Reid LM: Characterisation of airway mucus from a fatal case of status asthmaticus. *Pediatr Pulmonol* (1988) **5**: 176–182.
32. Feldhoff PA, Bhavanandan VP, Davidson EA: Purification, properties and analysis of human asthmatic bronchial mucin. Biochemistry (1979) **18**: 2430–2436.
33. Sheehan JK, Richardson PS, Fung DCK, Howard M, Thornton DJ: Analysis of respiratory mucus glycoprotein in asthma: a detailed study from a patient who died in status asthmaticus. *Am J Respir Cell Mol Biol* (1995) **13**: 748–756.
34. Dunnill MS: The morphology of the airways in bronchial asthma. In Stein M (ed) *New Directions in Asthma*. Park Ridge, American College of Physicians, 1975, pp 213–221.
35. Sanerkin NG, Evans DMD: The sputum in bronchial asthma: pathognomic patterns. *J Pathol Bacteriol* (1965) **89**: 535–541.
36. Openshaw PJM, Turner-Warwick M: Observation on sputum production in patients with variable airflow obstruction: implications for the diagnosis of asthma and chronic bronchitis. *Respir Med* (1989) **83**: 25–31.

37. Lopez-Vidriero MT, Reid L: Chemical makers of mucus and serum glycoproteins and their relation to viscosity in mucoid and purulent sputum from various hypersectory diseases. *Am Rev Respir Dis* (1978) **117**: 465–477.
38. Fahy JV, Steiger DJ, Liu J, Basbaum CB, Finkbeiner WE, Boushey HA: Markers of mucus secretion and DNA levels in induced sputum from asthmatic and from healthy subjects. *Am Rev Respir Dis* (1993) **147**: 1132–1137.
39. Fahy JV, Liu J, Wong H, Boushey HA: Cellular and biochemical analysis of induced sputum from asthmatic and from healthy subjects. *Am Rev Respir Dis* (1993) **147**: 1126–1131.
40. Dunnill MS, Massarella GR, Anderson JA: A comparision of the quantitative anatomy of the bronchi in normal subjects, in status asthmaticus, in chronic bronchitis and in emphysema. *Thorax* (1969) **24**: 176–179.
41. Charman J, Reid L: Sputum viscosity in chronic bronchitis, bronchiectasis, asthma and cystic fibrosis. *Biorheology* (1972) **9**: 185–189.
42. Shimura S, Sasaki T, Sasaki H, Takishima T, Umeya K: Viscoelastic properties of bronchorrhoea sputum in bronchial asthmatics. *Biorheology* (1988) **25**: 173–179.
43. Thornton DJ, Davies JR, Carlstedt I, Sheehan JK: Structure and biochemistry of human respiratory mucins. In Rogers DF, Lethem MI (eds) *Airway Mucus: Basic Mechanisms and Clinical Perspectives*. Basel, Birkhäuser Verlag, 1997, pp 19–39.
44. Thornton DJ, Carlstedt I, Howard M, Devine P, Price MR, Sheehan JK: Respiratory mucins: identification of core proteins and glycoforms. *Biochem J* (1996) **316**: 967–975.
45. Rogers DF, Evans TW: Plasma exudation in asthma. *Br Med Bulletin* (1992) **48**: 120–134.
46. Williams IP, Rich B, Richardson PS: Action of serum on the output of secretory glycoproteins from human bronchi *in vitro*. *Thorax* (1983) **38**: 682–685.
47. Fahy JV, Kim KW, Liu J, Boushey HA: Prominent neutrophilic inflammation in sputum from subjects with asthma exacerbation. *J Allergy Clin Immunol* (1995) **95**: 843–852.
48. Picot R, Das I, Reid L: Pus, deoxyribonucleic acid and sputum viscosity. *Thorax* (1978) **33**: 235–242.
49. Shimura S, Andoh Y, Haraguchi M, Shirato K: Continuity of airway goblet cells and intraluminal mucus in the airways of patients with bronchial asthma. *Eur Respir J* (1996) **9**: 1396–1401.
50. Newman TM, Robichaud A, Rogers DF: Microanatomy of secretory granule release from guinea pig tracheal goblet cells. *Am J Respir Cell Mol Biol* (1996) **15**: 529–539.
51. O'Riordan TG, Zwang J, Smaldone GC: Mucociliary clearance in adult asthma. *Am Rev Respir Dis* (1992) **146**: 594–603.
52. Russi W, Abraham WM, Chapman G, Stephenson J, Codias E, Wanner A: Effects of leukotriene D$_4$ on mucociliary and respiratory function in allergic and non-allergic sheep. *J Appl Physiol* (1985) **59**: 1416–1422.
53. Frigas E, Loegering DA, Gleich GJ: Cytotoxic effects of the guinea pig eosinophil major basic protein on tracheal epithelium. *Lab Invest* (1980) **42**: 35–43.
54. Ayers MM, Jeffery PK: Proliferation and differentiation in mammalian airway epithelium. *Eur Respir J* (1988) **1**: 58–80.
55. Wagner PD, Hedenstierna G, Rodriguez-Roisin R: Gas exchange, expiratory flow obstruction and the clinical spectrum of asthma. *Eur Respir J* (1996) **9**: 1278–1282.
56. Wiggs B, Moreno R, James A, Hogg JC, Paré PD: A model of the mechanics of airway narrowing in asthma. In Kaliner MA, Barnes PJ, Persson CGA (eds) *Asthma: its Pathology and Treatment*. New York, Marcel Dekker, 1991, pp 73–101.
57. Hargreave FE, Ryan G, Thomson NC, *et al*.: Bronchial responsiveness to histamine or methacholine in asthma: measurement and clinical significance. *J Allergy Clin Immunol* (1981) **68**: 347–355.
58. Suzuki T, Inoue H, Lin J-T, Takishima T: Intraluminal space occupying substance induces airway hyperresponsiveness. *Am Rev Respir Dis* (1992) **145** (Suppl): A50.
59. Yager D, Shore S, Drazen JM: Airway luminal liquid. Sources and role as an amplifier of bronchoconstriction. *Am Rev Respir Dis* (1991) **143**: S52–S54.
60. King M, Kelly S, Cosio S: Alteration of airway reactivity by mucus. *Respir Physiol* (1985) **62**: 47–59.

11 Pathophysiology of Airway Mucus Secretion in Asthma

61. Kim CS, Eldridge MA: Aerosol deposition in the airway model with excessive mucus secretions. *J Appl Physiol* (1985) **59**: 1766–1772.
62. Kim CS, Eldridge MA, Wanner A: Airway responsiveness to inhaled and intravenous carbachol in sheep: effect of airway mucus. *J Appl Physiol* (1988) **65**: 2744–2751.
63. Kim CS, Abraham WM, Garcia L, Sackner MA: Enhanced aerosol deposition in the lung with mild airways obstruction. *Am Rev Respir Dis* (1989) **139**: 422–426.
64. Barnes PJ, Chung KF, Page CP: Inflammatory mediators and asthma. *Pharmacol Rev* (1988) **40**: 49–84.
65. Barnes PJ: Overview of neural mechanisms in asthma. *Pulmon Pharmacol* (1995) **8**: 151–159.
66. Marin MG: Update: pharmacology of airway secretion. *Pharmacol Rev* (1994) **46**: 35–65.
67. Rogers DF: *In vivo* preclinical test models for studying airway mucus secretion. *Pulmon Pharmacol* (1997) **10**: 121–128.
68. Nadel JA: Role of mast cell and neutrophil proteases in airway secretion. *Am Rev Respir Dis* (1991) **144**: S48–S51.
69. Lundgren JD, Rieves RD, Mullol J, Logun C, Shelhamer JH: The effect of neutrophil proteinase enzymes on the release of mucus from feline and human airway cultures. *Respir Med* (1994) **88**: 511–518.
70. Gollub EG, Goswami SK, Sperber K, Marom Z: Isolation and characterisation of a macrophage-derived high molecular weight protein involved in the regulation of mucus-like glycoconjugate secretion. *J Allergy Clin Immunol* (1992) **89**: 696–702.
71. Sperber K, Gollub E, Goswami S, Kalb TH, Mayer L, Marom Z: *In vivo* detection of a novel macrophage-derived protein involved in the regulation of mucus-like glycoconjugate secretion. *Am Rev Respir Dis* (1992) **146**: 1589–1597.
72. Sperber K, Chanez P, Bousquet J, Goswami S, Marom Z: Detection of a novel marcophage-derived mucus secretagogue (MMS-68) in bronchoalveolar lavage fluid of patients with asthma. *J Allergy Clin Immunol* (1995) **95**: 868–876.
73. Levine SJ, Logun C, Larivee P, Shelhamer HJ: IL-1β induces secretion of respiratory mucous glycoprotein from human airways *in vitro*. *Am J Respir Dis* (1993) **147**: A437.
74. Levine SJ, Larivee P, Logun C, Shelhamer HJ: IL-6 induces respiratory mucous glycoprotein secretion and MUC-2 gene expression by human airways epithelial cells. *Am J Respir Crit Care Med* (1994) **149**: A27.
75. Levine SJ, Larivée P, Logun C, Angus CW, Ognibene FP, Shelhamer JH: Tumor necrosis factor-α induces mucin hypersecretion and MUC-2 gene expression by human airway epithelial cells. *Am J Respir Cell Mol Biol* (1995) **12**: 196–204.
76. Fischer BM, Krunkosky TM, Wright DT, Dolan-OKeefe M, Adler KB: Tumor necrosis factor-alpha (TNF-α) stimulates mucin secretion and gene expression in airway epithelium *in vitro*. *Chest* (1995) **107**: 133S–135S.
77. Hovenberg HW, Davies JR, Herrmann A, Linden CJ, Carlstedt I: MUC-5AC, but not MUC-2, is a prominent mucin in respiratory secretions. *Glycoconjugate J* (1996) **13**: 839–847.
78. Levine SJ, Larivee P, Logun C, Shelhamer HJ: IL-10 inhibits TNF-α mediated respiratory mucous glycoprotein secretion by human airways epithelial cells. *Am J Respir Crit Care Med* (1994) **149**: A986.
79. Adler KB, Fischer BM, Li H, Choe NH, Wright DT: Hypersecretion of mucin in response to inflammatory mediators by guinea pig tracheal epithelial cells *in vitro* is blocked by inhibition of nitric oxide synthase. *Am J Respir Cell Mol Biol* (1995) **13**: 526–530.
80. Ramnarine SI, Khawaja AM, Barnes PJ, Rogers DF: Nitric oxide inhibition of basal and neurogenic mucus secretion in ferret trachea *in vitro*. *Br J Pharmacol* (1996) **118**: 998–1002.
81. Lundgren JD, Shelhamer JH: Pathogenesis of airway mucus hypersecretion. *J Allergy Clin Immunol* (1990) **85**: 399–417.
82. Rogers DF: Neural control of airway secretions. In Barnes PJ (ed) *Autonomic Control of the Respiratory System*. Amsterdam B.V., Harwood Academic Publishers, 1997, pp 201–227.
83. Ramnarine SI, Rogers DF: Non-adrenergic, non-cholinergic neural control of mucus secretion in the airways. *Pulmon Pharmacol* (1994) **7**: 19–33.
84. Baker B, Peatfield AC, Richardson PS: Nervous control of mucin secretion into human bronchi. *J Physiol* (1985) **365**: 297–305.

85. Rogers DF, Barnes PJ: Opioid inhibition of neurally mediated mucus secretion in human bronchi. *Lancet* (1989) **i**: 930–932.
86. Barnes PJ: Muscarinic receptor subtypes in the airways. *Life Sci* (1993) **52**: 521–527.
87. Ramnarine SI, Haddad E-B, Khawaja AM, Mak JCW, Rogers DF: On muscarinic control of neurogenic mucus secretion in ferret trachea. *J Physiol* (1996) **494**: 577–586.
88. Liu YC, Khawaja AM, Rogers DF: Modulation of neurogenic mucus secretion in ferret trachea *in vitro* by endogenous vasoactive intestinal peptide (abstract). *Respir Med* (1997) **97**: A46.
89. Ollerenshaw S, Jarvis D, Woolcock A, Sullivan C, Scheibner T: Absence of immunoreactive vasoactive intestinal peptide in tissue from the lungs of patients with asthma. *N Engl J Med* (1989) **73**: 2505–2510.
90. Barnes PJ: Asthma as an axon reflex. *Lancet* (1986) **i**: 242–245.
91. Karlsson J-A: A role for capsaicin sensitive, tachykinin containing nerves in chronic coughing and sneezing but not in asthma: a hypothesis. *Thorax* (1993) **48**: 396–400.
92. Rogers DF, Aursudkij B, Barnes PJ: Effect of tachykinins on mucus secretion in human bronchi *in vitro*. *Eur J Pharmacol* (1989) **174**: 283–286.
93. Ramnarine SI, Hirayama Y, Barnes PJ, Rogers DF: 'Sensory–efferent' neural control of mucus secretion: characterization using tachykinin receptor antagonists in ferret trachea *in vitro*. *Br J Pharmacol* (1994) **113**: 1183–1190.
94. Khawaja AM, Rogers DF: Tachykinins: receptor to effector. *Int J Biochem Cell Biol* (1996) **28**: 721–738.
95. Barnes PJ: Mechanisms of action of glucocorticosteroids in asthma. *Am J Respir Crit Care Med* (1996) **154**: S21–S27.
96. Marom Z, Shelhamer J, Alling D, Kaliner M: The effects of corticosteroids on mucus glycoprotein secretion from human airways *in vitro*. *Am Rev Respir Dis* (1984) **129**: 62–65.
97. Rogers DF, Jeffery PK: Inhibition of cigarette smoke-induced airway secretory cell hyperplasia by indomethacin, dexamethasone, prednisolone, or hydrocortisone in the rat. *Exp Lung Res* (1986) **10**: 285–298.
98. Laitinen LA, Laitinen A, Haahtela T: A comparative study of the effects of an inhaled corticosteroid, budesonide, and a β_2-agonist, terbutaline, on airway inflammation in newly diagnosed asthma: a randomised, double-blind, parrallel-group controlled trial. *J Allergy Clin Immunol* (1992) **90**: 32–42.
99. Shimura S, Sasaki T, Sasaki H, Takishima T: Chemical properties of bronchorrhea sputum in bronchial asthma. *Chest* (1988) **94**: 1211–1215.
100. Kai H, Yoshitake K, Hisatsune A, *et al.*: Dexamethasone suppresses mucus production and MUC-2 and MUC-5AC gene expression by NCI-H292 cells. *Am J Physiol* (1996) **271**: L484–L488.
101. Goswami SK, Shmuel K, Marom Z: Erythromycin inhibits respiratory glycoconjugate secretion from human airways *in vitro*. *Am Rev Respir Dis* (1990) **141**: 72–76.
102. Suez D, Szefler S: Excessive accumulation of mucus in children with asthma: a potential role for erythromycin? A case discussion. *J Allergy Clin Immunol* (1986) **77**: 330–334.
103. Marom ZM, Goswami SK: Respiratory mucus hypersecretion (bronchorrhea): a case discussion. Possible mechanism(s) and treatment. *J Allergy Clin Immunol* (1991) **87**:1050–1055.
104. Miyatake H, Taki F, Taniguchi H, Suzuki K, Takagi K, Satake T: Erythromycin reduces the severity of bronchial hyperresponsiveness in asthma. *Chest* (1991) **99**: 670–673.
105. Martin J, Powell E, Shore S, Emrich J, Engel LA: The role of respiratory muscles in the hyperventilation of bronchial asthma. *Am Rev Respir Dis* (1980) **121**: 441–447.
106. Martin GP, Loveday BE, Marriot C: The effect of bromhexine hydrochloride on the viscoelastic properties of mucus from the mini-pig. *Eur Respir J* (1990) **3**: 392–396.
107. Dorow P: Mucolytics: when dispensible, when necessary? *Lung* (1990) **168**: 622S–626S.
108. Morgan EJ, Petty T: Summary of the national mucolytic study. *Chest* (1990) **97**: S24–S27.
109. Tekeres M, Horvath A, Bardosi L, Kenyeres P: Clinical studies on the mucolytic effect of mesna. *Clin Ther* (1981) **4**: 56–60.
110. Millman M, Millman FM, Goldstein IM, Mercandetti AJ: Use of acetylcysteine in bronchial asthma—another look. *Ann Allergy* (1985) **54**: 294–296.

111. Spicak V, Kacirek S, Pohunek P: Treatment of bronchial obstruction in asthmatic children. *Bull Eur Physiopathol Respir* (1987) **23**: 107S–109S.
112. Volkl KP, Schneider B: Therapy of respiratory tract diseases with *N*-acetylcysteine. An open therapeutical observation study of 2,513 patients. *Fortschr Med* (1992) **110**: 346–350.
113. Millman M, Goodman AH, Goldstein IM, Millman FM, Van Campen SS: Status asthmaticus: use of acetylcysteine during bronchoscopy and lavage to remove mucous plugs. *Ann Allergy* (1983) **50**: 85–93.
114. Karnik AM, Medhat M, Farah S: Therapeutic use of bronchoalveolar lavage in a very difficult asthmatic: a case report. *J Asthma* (1989) **26**: 181–184.
115. Henke CA, Hertz M, Gustafson P: Combined bronchoscopy and mucolytic therapy for patients with severe refractory status asthmaticus on mechanical ventilation: a case report and review of the literature. *Crit Care Med* (1994) **22**: 1880–1884.
116. Parnham MJ: Sodium cromoglycate and nedocromil sodium in the therapy of asthma, a critical comparison. *Pulmon Pharmcol* (1996) **9**: 95–105.
117. Alton EWFW, Kingsleigh-Smith D, Munkonge FM, *et al*.: Asthma prophylaxis agents alter the function of airway epithelial chloride channel. *Am J Respir Cell Mol Biol* (1996) **14**: 380–387.
118. Lopez-Vidriero MT, Costello J, Clark TJH, Das I, Keal EE, Reid L: Effect of atropine on sputum production. *Thorax* (1975) **30**: 543–547.
119. Bourm P, Mygind N, Schultz Larsen F: Intranasal ipratropium: a new treatment for perennial rhinitis. *Clin Otolaryngol* (1979) **4**: 407–411.
120. Ichinose M, Katsumata U, Kikuchi, R, *et al*.: Effect of tachykinin antagonist on chronic bronchitis patients. *Am Rev Respir Dis* (1993) **147** (suppl): A318.
121. Ichinose M, Nakajima N, Takahashi T, *et al*.: Protection against bradykinin-induced bronchoconstriction in asthmatic patients by neurokinin receptor antagonist. *Lancet* (1993) **340**: 1248–1251.
122. Bayever E, Iversen P, Smith L, Spinola J, Zon G: Systemic human antisense therapy begins. *Antisense Res Dev* (1992) **2**: 109–110.
123. Bhattacharyya SN, Ashbaugh P, Kaufman B, Manna B: Retinoic acid modulation of mucin mRNA in rat tracheal explants: response to actinomycin D, cylcoheximide, signal transduction effectors and antisense oligodeoxynucleotide. *Inflammation* (1994) **18**: 565–574.
124. Odell EW, Sarra R, Foxworthy M, Chapple DS, Evans RW: Antibacterial activity of peptides homologous to a loop region in human lactoferrin. *FEBS Lett* 1996; **382**: 175–178.
125. Bingle L, Tetley T: Secretory leukoprotease inhibitor: partnering α_1-proteinase inhibitor to combat pulmonary inflammation. *Thorax* (1996) **51**: 1273–1274.
126. Oury TD, Day BJ, Crapo JD: Extracellular superoxide dismutase in vessels and airways of humans and baboons. *Free Radic Biol Med* (1996) **20**: 957–965.
127. Avissar N, Finklestein JN, Horowitz S, *et al*.: Extracellular glutathione peroxidase in human lung epithelial lining fluid and in lung cells. *Am J Physiol* (1996) **14**: L173–L182.
128. Thomas CE, Ohlweiler, Carr AA, *et al*.: Characterisation of the radical trapping activity of a novel series of cyclic nitrone spin traps. *J Biol Chem* (1996) **271**: 3097–3104.

12

Tracheobronchial Circulation

DAVID J. GODDEN

INTRODUCTION

Despite recognition of the tracheobronchial circulation for over two millennia, study of its role in disease processes is relatively recent. This reflects its complex and variable anatomy, resulting in difficulties in physiological studies that led the late John Butler to refer to 'that pesky circulation!'.[1] However, the recognition that asthma is an inflammatory process has been a stimulus to further study of the vascular component of that process. This chapter describes the organization and control of the tracheobronchial circulation, the role of the circulation in drug and mediator distribution and conditioning of inspired air, and the mechanical contribution of the vasculature to airway narrowing.

ORGANIZATION AND CONTROL OF THE TRACHEOBRONCHIAL CIRCULATION

The blood supply to the airway wall is derived principally from the thyroid and bronchial arteries. In humans, the inferior thyroid arteries supply the upper portion of the trachea[2] and anastomose with the bronchial arteries which supply the major and minor bronchial walls, merging with pulmonary microvessels at the level of the terminal bronchioles. The origin of the bronchial arteries is variable,[3] the right bronchial arteries most commonly arising from the upper intercostal vessels and the left from the ventral surface of the aorta.[3,4] Within the airway wall, the vessels are arranged in an outer plexus of arterioles with branches passing through the muscular layer to provide a rich submucosal plexus.[2,5]

A similar arrangement is seen in dogs[6] and sheep.[7] Physiological studies suggest that airway smooth muscle perfusion is mainly derived from the systemic circulation,[8] but the pulmonary circulation also provides a contribution to airway blood flow, mainly to the adventitia rather than the mucosa and increasing towards more distal airways.[9,10] The subepithelial capillaries connect to a deeper plexus of mucosal capacitance vessels, described in some species as sinuses.[11] The mucosal capillaries are 7–10 μm in diameter[6,12] and are fenestrated in some species such as rats and guinea-pigs, but not in normal humans except at specific sites.[11] More generalized fenestrations have been described in capillaries of human asthmatic subjects.[11] Venous drainage from the trachea is by tracheal veins to the right heart, while from the distal central airways bronchial veins drain to the azygos, hemiazygos or intercostal veins[13] and thence to the right atrium. More peripherally, the bronchial circulation has extensive anastomoses with the pulmonary circulation at both precapillary and postcapillary levels.[7,14,15] Much of the venous blood from the bronchial circulation drains by that route to the pulmonary veins.[7,16] Within the postcapillary venules, the vascular endothelium is fenestrated, providing potential for solute and water exchange.[17]

Tracheobronchial vascular resistance is under neurohumoral control.[11,18] The cholinergic system is probably not important in determining resting vascular tone,[19] although intravascular injection of cholinergic agonists dilates both bronchial[19,20] and tracheal vessels.[21,22] Aerosol administration of methacholine also increases bronchial,[20] though not tracheal, blood flow.[23] Sympathetic innervation of the airway blood vessels varies amongst species and is sparse in humans. Resting vascular tone is also relatively independent of the adrenergic system,[23,24] although α-adrenoceptor agonists vasoconstrict tracheal[22,25] and bronchial vessels[19,25] while α-adrenoceptor antagonists increase mucosal blood flow,[23] as do β-adrenoceptor agonists.[26–29] A further important neural influence derives from the peptidergic system. Studies by Martling et al.[30,31] suggest that antidromic stimulation of sensory nerves increases tracheal blood flow; these findings are supported by Laitinen et al.[32] A range of neuropeptides act on the tracheal vasculature: neurokinin A, vasoactive intestinal polypeptide, calcitonin gene-related peptide and substance P mediate vasodilation, while bombesin and neuropeptide mediate vasoconstriction.[33] The role of these peptides in 'neurogenic inflammation' has been widely studied, although the relative importance of these mechanisms in human airway responses remains uncertain. Other inflammatory mediators, predominantly studied in animal models, also affect vascular resistance. Histamine has complex site- and species-specific actions including both vasoconstriction and vasodilation.[27,34,35] Bradykinin vasodilates tracheal and bronchial vessels in dogs and pigs.[27,28,36] The prostaglandins PGE_1 $PGF_{2\alpha}$ and PGD_2 all vasodilate tracheal vessels,[27] while leukotrienes C_4 and D_4 decrease bronchial blood flow. Allergen inhalation in sheep and pig models of allergic bronchial responsiveness results in bronchial vasodilation, which shows both early[37,38] and later components.[39,40] The mechanism of this response is complex, involving neuropeptides, leukotrienes and prostaglandins. Administration of exogenous nitric oxide (NO) vasodilates the bronchial circulation,[41] while studies using NO inhibitors indicate that endogenous NO mediates bronchial vascular resistance: N-monomethyl-L-arginite (LNNA) reduces baseline bronchial blood flow and the increase that occurs in response to acetylcholine;[42] L-N-nitroarginite methel ester (L-NAME) attenuates the increase that follows a nebulized β-agonist.[29]

12 Tracheobronchial Circulation

ROLE OF THE TRACHEOBRONCHIAL CIRCULATION IN ASTHMA

There are a number of ways in which the tracheobronchial circulation may be involved in the pathogenesis of asthma: vascular effects may influence drug and mediator distribution and heat and water exchange with inspired gas; changes in vascular volume may affect airway calibre directly; hydrostatic forces may result in airway oedema formation; changes in airway vascular permeability may also result in airway oedema as well as affecting the nature and composition of airway lining fluid (Fig. 12.1).

Drug and mediator distribution

The role of the airway vasculature in drug and mediator distribution is complex, since mucosal perfusion, thickness and permeability are closely linked. Enhanced perfusion of the mucosa is associated with an increase in mucosal thickness,[22] which in turn increases the diffusion barrier for topically applied agents. This may explain the reduction in clearance of ^{99m}Tc-DTPA (diethylenetriaminepentaacetic acid) from the lumen of the trachea when tracheal blood flow is increased.[43] These effects may be induced by instillation of hyperosmolar fluid into the tracheal lumen.[44] Conversely, reduced blood flow may result in slower clearance of drugs and mediators. Reducing bronchial perfusion prolongs the bonchoconstrictor effects of methacholine in sheep[45] and histamine in dog peripheral airways[46] and the effects of antigen challenge on smooth muscle.[47] Similar

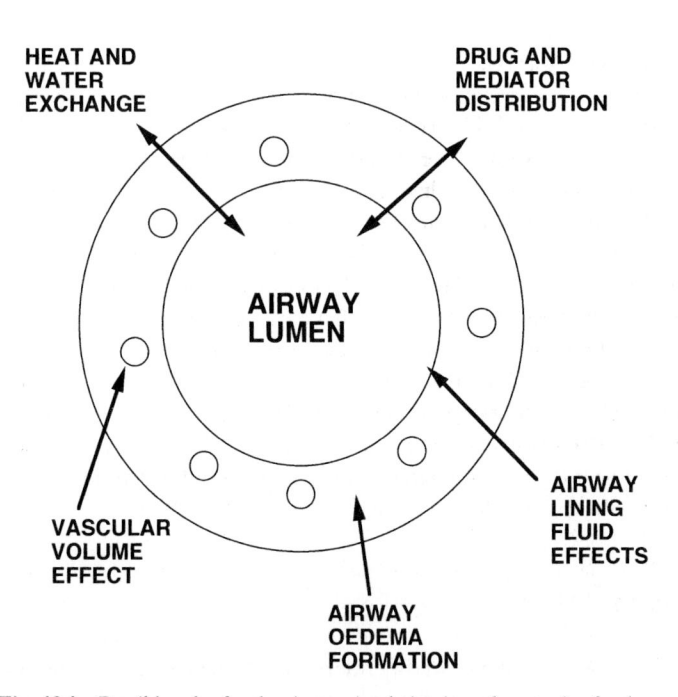

Fig. 12.1 Possible roles for the airway circulation in pathogenesis of asthma.

considerations probably apply for locally generated mediators. Permeability of the tracheal mucosa increases when the epithelium is injured.[48] The process of plasma exudation (see chapter 14) has a directional element, protein flux towards the airway lumen greatly exceeding flux from the lumen.[49] Since plasma exudation is an element of the asthmatic response, this may have implications for drug and mediator clearance from the airway lumen. In future, manipulation of airway perfusion may provide a method to enhance the bioactivity of drugs in the treatment of asthma.

Heat and water exchange

Heat and water fluxes are fundamental to the pathogenesis of exercise-induced asthma, which is related to the degree of hyperventilation during exercise.[50] The relative contributions of the pulmonary and bronchial vasculature to heat and water exchange in the airways remains a matter for debate,[50,51] as is the relative importance of heat versus water loss as the stimulus to exercise-induced asthma.[52,53] However, it is known that blood flow in the airway mucosa of the dog is increased up to seven-fold by hyperventilation of dry air[54] and by topical cooling.[55] This increase is a local effect, occurring only in mucosa directly exposed to dry air,[54] and is unaffected by α- or β-adrenergic blockade but attenuated by administration of topical lidocaine,[56] suggesting that it may be due to a sensory neuronal axon reflex. Supporters of the 'osmolarity theory' argue that hyperventilation of cold or dry gas increases osmolarity of airway lining fluid, leading to a release of mediators that then trigger vasodilation.[53] This is compatible with observations that nasal challenge with cold dry air releases local inflammatory mediators.[57] If this is the mechanism, it could, in theory, be simulated by topical application of hyperosmolar fluids. Where this has been done in animal models using fluids or aerosols the response has been varied, blood flow increasing in some models[44,58] but not in others.[59] In contrast, supporters of the 'temperature theory' argue that rapid rewarming of airways after a period of hyperventilation in asthmatic subjects is due to reactive hyperaemia in the airway wall, which causes bronchoconstriction.[60] Direct evidence for this in humans is lacking, although inhalation of a topical vasoconstrictor reduces the magnitude of airway rewarming that follows hyperventilation.[52]

Whether the response to hyperventilation is a result of temperature change or water loss, it has been assumed that increases in blood flow will contribute directly to airway narrowing. However, some animal studies have cast doubt on this assumption. In a dog model of peripheral airway response to ventilation with dry air, occluding the bronchial circulation to the ventilated segment worsened the degree of dry air-induced mucosal injury but had no effect on the degree of airflow obstruction that occurred.[61] The authors have suggested that the major role for the airway circulation in the response to dry air may be to protect against mucosal damage by extrusion of fluid and proteins and that it is not an important direct contributor to airflow obstruction.[62,63]

Airway calibre

The potential contribution of the airway microvasculature to airway narrowing has three components: the space-occupying effect of engorged airway vessels (the vascular volume effect), the development of airway wall oedema and the extravasation of fluid into the airway lumen.

The concept that bronchial mucosal swelling may contribute to the pathogenesis of an asthma attack has long been recognized,[64] and an increase in number and size of airway wall vessels is a feature of fatal asthma.[65,66] Studies in sheep demonstrated an association between vascular congestion and increased lung resistance during both histamine[67] and antigen challenge[38] but the relative contribution to airway narrowing of increased vascular volume compared with oedema and extravasation was not defined. Vasodilation is associated with a thickening of the tracheal mucosa, as measured by a hydraulic microactive probe method in dogs.[22,68] Although this is of small magnitude (10–15%), it could be highly significant if a similar absolute degree of thickening occurred in more distal airways and changes in airflow associated with vasomotion have been demonstrated in peripheral airways.[69] Human data are relatively sparse and necessarily indirect. A rapid infusion of saline can induce a degree of airway narrowing comparable to that induced by hyperventilation in humans and, if given after hyperventilation, can amplify the response.[70] Lockhart and colleagues demonstrated that administration of methoxamine, a vasoconstrictor α-agonist, can prevent exercise-induced asthma[71] and airway hyperresponsiveness to methacholine in subjects with heart failure.[72] These observations would be consistent with a direct contribution of vascular congestion to airflow obstruction. Further support comes from their observation that inflation of antishock trousers, a manoeuvre designed to increase intrathoracic vascular congestion, can induce mild bronchial hyperresponsiveness to methacholine in non-asthmatic subjects.[73] However, more recent work in animal models casts doubt on the importance of the vascular volume effect. In a sheep model in which the bronchial artery was perfused at control and at high levels, high-level perfusion increased vascular volume but did not affect baseline lung resistance, peripheral lung resistance or responsiveness to methacholine.[74] In dogs, the increase in peripheral lung resistance during volume loading appeared unrelated to vascular engorgement;[75] in a study using high resolution computed tomography (HRCT) imaging of airways, infusion of a large volume load induced only a modest change in airway calibre.[76] However, while the absolute magnitude of the vascular volume effect is small, it remains possible that it becomes important as an amplifier of smooth muscle contraction.

In contrast to the small or negligible direct influence of vascular volume, associated airway oedema may be an important contributor to airway narrowing.[75] Airway oedema may occur due to hydrostatic forces, such as those seen in left ventricular failure,[77,78] or due to increased vascular permeability secondary to inflammatory mediator effects. Oedema may occur in the peribronchial connective tissue of the adventitia, between the smooth muscle layer and the epithelium, or within the lumen and evidence for such effects has been obtained in volume loading studies in dogs.[79] As well as directly influencing airway calibre by a space-occupying effect, oedema in the adventitia may uncouple the airway from the lung parenchyma, leading to unopposed airway collapse.[80] In the airway wall, it may amplify airway narrowing occurring due to smooth muscle contraction, while in the airway lumen, in addition to occupying space, it may increase surface tension of airway lining fluid, promoting further airway collapse.[81] These amplification effects have been recently reviewed.[82]

The majority of inflammatory mediators involved in the asthmatic process result in increased permeability of the tracheobronchial vasculature. This results in plasma leakage into the airway lumen. This process, in addition to its mechanical effects described above, has important pharmacological consequences, and a large literature has now accumulated on the factors controlling vascular permeability (described in detail in Chapter 14).

CONCLUSION

The airways have a rich blood supply whose anatomy shows interindividual and interspecies variation. Both the systemic and pulmonary circulations supply the airway walls, the contribution of the latter increasing towards the more distal airways. A complex neurohumoral regulatory system controls vascular resistance and many inflammatory mediators have important effects on airway perfusion. Vascular factors therefore are an important component of the inflammatory response in the airways that forms the basis of clinical asthma. Mediator distribution and heat and water exchange with inspired gas may be influenced by circulatory factors. Airway calibre may be affected directly by changes in vascular volume and more importantly by airway oedema formation, which may be a consequence of hydrostatic or permeability changes in the vasculature. Secondary changes in the volume and composition of airway lining fluid may occur. These factors may serve to amplify the effect of airway smooth muscle shortening, which is the principal cause of bronchoconstriction in asthma. The potential to influence drug delivery to the airways by manipulating circulatory factors is as yet largely untapped but may be a fruitful area for future research.

REFERENCES

1. Butler J: Preface. In Butler J (ed) *The Bronchial Circulation*. New York, Marcel Dekker, 1992, p. vii.
2. Miura T, Grillo HC: The contribution of the inferior thyroid artery to the blood supply of the human trachea. *Surg Gynecol Obstet* (1966) **123**: 99–102.
3. Cauldwell EW, Siekert RG, Lininger RE, Anson BJ: The bonchial arteries: an anatomic study of 150 human cadavers. *Surg Gynecol Obstet* (1948) **86**: 395–412.
4. Pump KK: Distribution of bronchial arteries in the human lung. *Chest* (1972) **62**: 447–451.
5. Cudkowicz L, Armstrong JB: Observations on the normal anatomy of the bronchial arteries. *Thorax* (1951) **6**: 343–358.
6. Laitinen A, Laitinen LA, Moss R, Widdicombe JG: Organisation and structure of the tracheal and bronchial blood vessels in the dog. *J Anat* (1989) **166**: 133–140.
7. Charan NB, Turk GM, Dhand R: Gross and subgross anatomy of bronchial circulation in sheep. *J Appl Physiol: Respir Environ Exercise Physiol* (1984) **57**: 658–664.
8. Wagner EM, Mitzner WA: Contribution of pulmonary versus systemic perfusion of airway smooth muscle. *J Appl Physiol* (1995) **78**: 403–409.
9. Baile EM, Minshall D, Dodek PM, Pare PD: Blood flow to the trachea and bronchi: the pulmonary contribution. *J Appl Physiol* (1994) **76**: 2063–2069.
10. Barman SA, Ardell JL, Taylor AE: Effect of phorbol myristate acetate-induced lung injury on airway blood flow. *Respir Physiol* (1995) 99: 249–257.
11. Widdicombe J: Physiologic control. Anatomy and physiology of the airway circulation. *Am Rev Respir Dis* (1992) **146**: S3–S7.

12. MacDonald DM: The ultrastructure and permeability of the tracheobronchial blood vessels in health and disease. *Eur Respir J (Suppl)* (1990) **12**: 572s–585s.
13. Miller WS. *The Lung*. Springfield, IL Charles C Thomas, 1947, pp 69–83.
14. Tobin CE: The bronchial arteries and their connections with other vessels in the human lung. *Surg Gynecol Obstet* (1952) **95**: 741–750.
15. Wagenvoort CA, Wagenvoort N: Arterial anastomoses bronchopulmonary arteries and pulmobronchial arteries in perinatal lungs. *Lab Invest* (1967) **16**: 13–24.
16. Lockhart A, Marthan R, Charan N, Pare P: Airway circulation in health and disease. *Eur Respir J* (1996) **9**: 1105–1110.
17. Pietra GG, Magno M: Pharmacologic factors influencing permeability of the bronchial microcirculation. *Fed Proc* (1978) **37**: 2466–2470.
18. Godden DJ: Reflex and nervous control of the tracheobronchial circulation. *Eur Respir J (Suppl)* (1990) **12**: 602s–607s.
19. deLetona JML, de la Mata RC, Aviado DM: Local and reflex effects of bronchial artery injection of drugs. *J Pharmacol Exp Ther* (1961) **133**: 295–303.
20. Lakshminaryan S, Jindal SK, Kirk W, Butler J: Increases in bronchial blood flow following bronchoconstriction with methacholine and prostaglandin F_{2alpha} in dogs. *Chest* (1985) **87**: 183S–194S.
21. Himori N, Taira N: A method for recording smooth muscle and vascular responses of the blood perfused trachea *in situ*. *Br Pharmacol* (1976) **56**: 293–299.
22. Laitinen LA, Robinson NP, Laitinen A, Widdicombe JG: Relationship between tracheal mucosal thickness and vascular resistance in dogs. *J Appl Physiol* (1986) **61**: 2186–2193.
23. Barker JA, Chediak AD, Baier HJ, Wanner A: Tracheal mucosal blood flow responses to autonomic agonists. *J Appl Physiol* (1988) **65**: 829–834.
24. Baile EM, Osborne S, Pare PD: Effect of autonomic blockade on tracheobronchial blood flow. *J Appl Physiol* (1986) **65**: 520–525.
25. Onorato DJ, Demirozu MC, Breitenbucher A, Atkins ND, Chediak AD, Wanner A: Airway mucosal blood flow in humans. Response to adrenergic agonists. *Am J Respir Crit Care Med* (1994) **149**: 1132–1137.
26. Charan NB, Turk GM, Ripley R: Measurement of bronchial arterial blood flow and bronchovascular resistance in sheep. *J Appl Physiol* (1985) **59**: 305–308.
27. Laitinen LA, Laitinen MA, Widdicombe JG: Dose-related effects of pharmacological mediators on tracheal vascular resistance in dogs. *Br J Pharmacol* (1987) **92**: 703–709.
28. Corfield DR, Hanafi Z, Webber SE, Widdicombe JG: Changes in tracheal mucosal thickness and blood flow in sheep. *J Appl Physiol* (1991) **71**: 1282–1288.
29. Carvalho P, Johnson SR, Charan NB: Role of nitric oxide in beta-agonist induced bronchial arterial vasodilatation (abstract). *Am J Respir Crit Care Med* (1996) **153**: A814.
30. Martling C-R, Angaard A, Lundberg JM: Non-cholinergic vasodilation in the tracheobronchial tree of the cat induced by vagal nerve stimulation. *Acta Physiol Scand* (1985) **125**: 343–346.
31. Martling C-R, Gazelius B, Lundberg JM: Nervous control of tracheal blood flow in the cat measured by the laser Doppler technique. *Acta Physiol Scand* (1987) **130**: 409–417.
32. Laitinen LA, Laitinen A, Widdicombe JG: Parasympathetic control of tracheal vascular resistance in the dog. *J Physiol* (1987) **385** 135–146.
33. McCormack DG, Salonen RO, Widdicombe JG, Barnes PJ: Sensory neuropeptides are potent vasodilators of canine bronchial arteries *in vitro*. *Am Rev Respir Dis* (1988) **137**: 139.
34. Webber SE, Salonen RO, Corfield DR, Widdicombe JG: Effects of non-neural mediators and allergen on tracheobronchial blood flow. *Eur Respir J (Suppl)* (1990) **12**: 638s–643s.
35. Parsons GH, Villablanca AC, Brock, JM, *et al*.: Bronchial vasodilation by histamine in sheep: characterization of receptor subtype. *J Appl Physiol* (1992) **72**: 2090–2098.
36. Matran R, Alving K, Martling CR, Lacroix JS, Lundberg JM: Effects of neuropeptides and capsaicin on tracheobronchial blood flow of the pig. *Acta Physiol Scand* (1989) **135**: 335–342.
37. Alving K, Matran R, Lacroix JS, Lundberg JM: Allergen challenge induces vasodilation in pig bronchial circulation via a capsaicin-sensitive mechanism. *Acta Physiol Scand* (1988) **134**: 571–572.
38. Long WM, Yerger LD, Martinez H, *et al*.: Modification of bronchial blood flow during allergic airway responses. *J Appl Physiol* (1988) **65**: 272–282.

39. Long WM, Yerger LD, Abraham WM, Lobel C: Late-phase bronchial vascular responses in allergic sheep. *J Appl Physiol* (1990) **69**: 584–590.
40. Alving K, Matran R, Fornhem C, Lundberg JM: Late phase bronchial and vascular responses to allergen in actively-sensitized pigs. *Acta Physiol Scand* (1991) **143**: 137–138.
41. Alving K, Fornhem C, Weitzberg E, Lundberg JM: Nitric oxide mediates cigarette smoke-induced vasodilatory responses in the lung. *Acta Physiol Scand* (1992) **146**: 407–408.
42. Sasaki F, Pare P, Ernest D, *et al*.: Endogenous nitric oxide influences acetylcholine-induced bronchovascular dilation in sheep. *J Appl Physiol* (1995) **78**: 539–545.
43. Hanafi Z, Corfield DR, Webber SE, Widdicombe JG: Tracheal blood flow and luminal clearance of ^{99m}Tc-DTPA in sheep. *J Appl Physiol* (1992) **73**: 1273–1281.
44. Wells UM, Hanafi Z, Widdicombe JG: Osmolality alters tracheal blood flow and tracer uptake in anesthetized sheep. *J Appl Physiol* (1994) **77**: 2400–2407.
45. Wagner EM, Mitzner WA: Bronchial circulatory reversal of methacholine-induced airway constriction. *J Appl Physiol* (1990) **69**: 1220–1224.
46. Kelly L, Kolbe J, Mitzner W, Spannhake EW, Bromberger-Barnea B, Menkes H: Bronchial blood flow affects recovery from constriction in dog lung periphery. *J Appl Physiol* (1986) **60**: 1954–1959.
47. Csete ME, Chediak AD, Abraham WM, Wanner A: Airway blood flow modifies allergic airway smooth muscle contraction. *Am Rev Respir Dis* (1991) **144**: 59–63.
48. Wells UM, Woods AJ, Hanafi Z, Widdicombe JG: Tracheal epithelial damage alters tracer fluxes and effects of tracheal osmolality in sheep *in vivo*. *J Appl Physiol* (1995) **78**: 1921–1930.
49. Erjefalt I, Persson CG: Allergen bradykinin, and capsaicin increase outward but not inward macromolecular permeability of guinea-pig tracheobronical mucosa. *Clin Exp Allergy* (1991) **21**: 217–224.
50. Solway J: Airway heat and water fluxes and the tracheobronchial circulation. *Eur Respir J (Suppl)* (1990) **12**: 608s–617s.
51. McFadden ER Jr: Heat and water exchange in human airways. *Am Rev Respir Dis* (1992) **146**: S8–S10.
52. Gilbert IA, McFadden ER Jr: Airway cooling and rewarming. The second reaction sequence in exercise-induced asthma. *J Clin Invest* (1992) **90**: 699–704.
53. Anderson SD, Daviskas E: The airway microvasculature and exercise induced asthma. *Thorax* (1992) **47**: 748–752.
54. Baile EM, Guillemi S, Pare PD: Tracheobronchial and upper airway blood flow in dogs during thermally induced panting. *J Appl Physiol* (1987) **63**: 2240–2246.
55. Salonen RO, Webber SE, Deffebach ME, Widdicombe JG: Tracheal vascular and smooth muscle responses to air temperature and humidity in dogs. *J Appl Physiol* (1991) **71**: 50–59.
56. Baile EM, Godden DJ, Pare PD: Mechanism for increase in tracheobronchial blood flow induced by hyperventilation of dry air in dogs. *J Appl Physiol* (1990) **68**: 105–112.
57. Togias AG, Naclerio RM, Proud D, *et al*.: Nasal challenge with cold, dry air results in release of inflammatory mediators. Possible mast cell involvement. *J Clin Invest* (1985) **76**: 1375–1381.
58. Deffebach ME, Salonen RO, Webber SE, Widdicombe JG: Cold and hyperosmolar fluids in canine trachea: vascular and smooth muscle tone and albumin flux. *J Appl Physiol* (1989) **66**: 1309–1315.
59. Godden DJ, Baile EM, Okazawa M, Pare PD: Hypertonic aerosol inhalation does not alter central airway blood flow in dogs. *J Appl Physiol* (1988) **65**: 1990–1994.
60. Gilbert IA, Fouke JM, McFadden ER Jr: Heat and water flux in the intrathoracic airways and exercise-induced asthma. *J Appl Physiol* (1987) **63**: 1681–1691.
61. Freed AN, Amori C, Schofield BH: The effect of bronchial blood flow on hyperpnea-induced airway obstruction and injury. *J Clin Invest* (1995) **96**: 1221–1229.
62. Freed AN, Omori C, Schofield BH, Mitzner W: Dry air-induced mucosal cell injury and bronchovascular leakage in canine peripheral airways. *Am J Respir Cell Mol Biol* (1994) **11**: 724–732.
63. Freed AN: Models and mechanisms of exercise-induced asthma. *Eur Respir J* (1995) **8**: 1770–1785.
64. de Burgh Daly I: Interference of intrinsic pulmonary mechansisms as a potential cause of asthma. *Edinburgh Med J* (1935) **43**: 139–142.

65. Dunnill MS: The pathology of asthma, with special references to changes in the bronchial mucosa. *J Clin Pathol* (1960) **13**: 27–33.
66. Kuwano K, Bosken CH, Pare PD, Bai TR, Wiggs BR, Hoggs JC: Small airways dimensions in asthma and in chronic obstructive pulmonary disease. *Am Rev Respir Dis* (1993) **148**: 1220–1225.
67. Long WM, Sprung CL, el Fawal H, et al.: Effects of histamine on bronchial artery blood flow and bronchomotor tone. *J Appl Physiol* (1985) **59**: 254–261.
68. Laitinen LA, Laitinen A, Widdicombe J: Effects of inflammatory and other mediators on airway vascular beds. *Am Rev Respir Dis* (1987) **135**: S67–S70.
69. Csete ME, Abraham WM, Wanner A: Vasomotion influences airflow in peripheral airways. *Am Rev Respir Dis* (1990) **141**: 1409–1413.
70. Gilbert IA, Winslow CJ, Lenner KA, Nelson JA, McFadden ER Jr: Vascular volume expansion and thermally induced asthma. *Eur Respir J* (1993) **6**: 189–197.
71. Dinh Xuan AT, Chaussain M, Regnard J, Lockhart A: Pretreatment with an inhaled alpha 1-adrenergic agonist, methoxamine, reduces exercise-induced asthma. *Eur Respir J* (1989) **2**: 409–414.
72. Cabanes LR, Weber SN, Matran R, et al.: Bronchial hyperresponsiveness to methacholine in patients with impaired left ventricular function. *N Engl J Med* (1989) **320**: 1317–1322.
73. Regnard J, Baudrillard P, Salah B, Dinh Xuan AT, Cabanes L, Lockhart A: Inflation of antishock trousers increases bronchial response to methacholine in healthy subjects. *J Appl Physiol* (1990) **68**: 1528–1533.
74. Blosser S, Mitzner W, Wagner EM: Effects of increased bronchial blood flow on airway morphometry, resistance, and reactivity. *J Appl Physiol* (1994) **76**: 1624–1629.
75. Tang GJ, Freed AN: The role of submucosal oedema in increased peripheral airway resistance by intravenous volume loading in dogs. *Eur Respir J* (1994) **7**: 311–317.
76. Brown RH, Zerhouni EA, Mitzner W: Visualization of airway obstruction *in vivo* during pulmonary vascular engorgement and edema. *J Appl Physiol* (1995) **78**: 1070–1078.
77. Baier H, Onorato D, Barker J, Wanner A: Tracheal mucosal edema in hydrostatic pulmonary edema. *J Appl Physiol* (1994) **77**: 352–356.
78. Snashall PD, Chung KF: Airway obstruction and bronchial hyperresponsiveness in left ventricular failure and mitral stenosis. *Am Rev Respir Dis* (1991) **144**: 945–956.
79. Brown RH, Zerhouni EA, Mitzner W: Airway edema potentiates airway reactivity. *J Appl Physiol* (1995) **79**: 1242–1248.
80. Moreno RH, Hogg JC, Pare PD: Mechanics of airway narrowing. *Am Rev Respir Dis* (1986) **133**: 1171–1180.
81. Yager D, Butler JP, Bastacky J, Israel E, Smith G, Drazen JM: Amplification of airway constriction due to liquid filling of airway interstices. *J Appl Physiol* (1989) **66**: 2873–2884.
82. Yager D, Kamm RD, Drazen JM: Airway wall liquid. Sources and role as an amplifier of bronchoconstriction. *Chest* (1995) **107**: 105S–110S.

13

Adhesion Molecules

A. WARDLAW

INTRODUCTION

Adhesion receptors are involved in many leucocyte functions, including haematopoiesis, migration, activation, mediator generation and apoptosis. They comprise several gene families of membrane glycoproteins that are also involved in biological processes as diverse as wound healing, thrombogenesis, atherogenesis, embryogenesis and maintenance of tissue architecture. The literature on adhesion molecules is extensive and there are a number of comprehensive and up-to-date reviews that cover the biology of these molecules. In this chapter I will very briefly outline the adhesion receptor families involved in leucocyte function before concentrating on studies specifically relevant to asthma. Although adhesion receptors are involved in many areas of leucocyte biology, most work has concentrated on the role of these receptors in directing leucocyte migration through vascular endothelium. This is also the focus of this review. Current ideas about leucocyte emigration are based on the concept of a staged process, with leucocytes first becoming loosely tethered to the venular endothelium under flow conditions, followed by cellular activation, which in turns leads to firmer adhesion and transmigration. Each step is required, allowing both a diversity of signals to control migration and multiple targets for therapeutic intervention.[1] This model applies to migration through the systemic circulation and is therefore assumed to be relevant to the bronchial circulation. In contrast there is good evidence that migration through the alveolar bed is fundamentally different because of structural differences in the pulmonary capillary circulation. As asthma is primarily an airway disease this caveat is only important when interpreting data from animal challenge studies, where it is possible that some migration is occurring through the pulmonary circulation.

LEUCOCYTE ADHESION RECEPTORS

The major gene families of adhesion receptors involved in leucocyte adhesion are the *selectins* and their counter-receptors, members of the *integrin* family and members of the immunoglobulin family. The members of these families relevant to leucocyte adhesion are summarized in Fig. 13.1 and Table 13.1.

Selectins and their counter-receptors

There are three selectins: E-selectin expressed on endothelium, P-selectin expressed by platelets and endothelium and L-selectin expressed on most leucocytes.[2] All three selectins have a common structure, with an N-terminal lectin domain, an epidermal growth factor (EGF)-like domain and a variable number of consensus repeats related to complement-binding proteins (Fig. 13.1). E-selectin expression is induced on human umbilical vein endothelial cells (HUVEC) *in vitro* by stimulation with cytokines including interleukin (IL)-1 and tumour necrosis factor (TNF)-α with optimal expression at 4 h. *In vivo* expression is weak or absent on uninflamed tissue but induced during inflammatory processes. E-selectin is particularly well expressed in the skin where it is thought to act as an addressin for skin-homing lymphocytes through its ability to bind the carbohydrate antigen cutaneous lymphocyte antigen (CLA) expressed on a minority of blood T-cells but a majority of T-cells in inflamed skin.[3] Like the other selectins, E-selectin binds sialylated fucosylated sugar moieties such as sialyl Lewis X. Expression of these carbohydrate structures is controlled by cell-specific fucosyltransferases such as FucT V11.[4] The backbone structure that presents these carbohydrates to E-selectin have not been fully defined, but include on mouse neutrophils a receptor very closely related to a

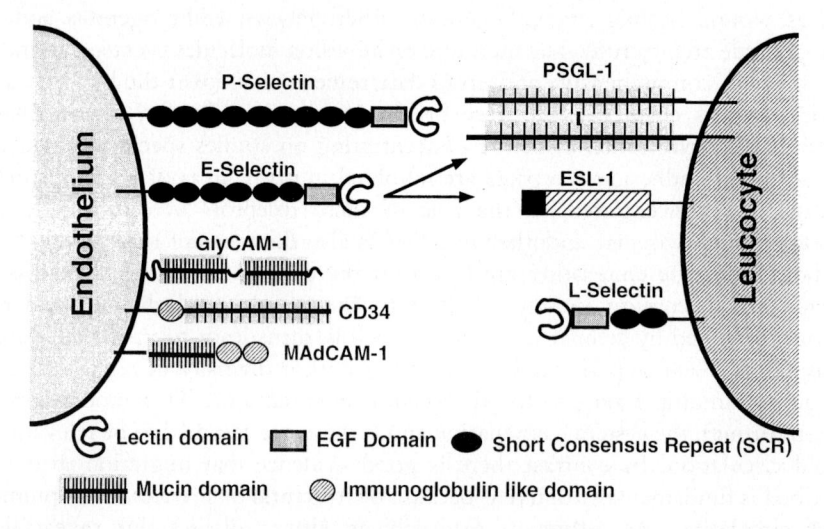

Fig. 13.1 Schematic representation of the structure of the selectins and their ligands. EGF, epidermal growth factor; ESL-1, E-selectin ligand-1; GlyCAM-1, glycosylated cell adhesion molecule 1; MAdCAM-1, mucosal addressin cell adhesion molecule 1; PSGL-1, P-selectin glycoprotein ligand 1.

13 Adhesion Molecules

Table 13.1 Leucocyte adhesion receptors and their counter-receptors involved in migration.

Leucocyte receptor	Endothelial receptor	Matrix protein
Integrin		
VLA-4 ($\alpha 4\beta 1$)	VCAM-1	Fibronectin
VLA-5 ($\alpha 5\beta 1$)		Fibronectin
VLA-6 ($\alpha 6\beta 1$)		Laminin
$\alpha 4\beta 7$	MAdCAM-1/VCAM-1	Fibronectin
LFA-1	ICAM-1, ICAM-2	
Mac-1	ICAM-1	Fibrinogen
p150,95	?	
CD11d/CD18	ICAM-3	
Immunoglobulin-like		
PECAM	PECAM/$\alpha v\beta 3$	
ICAM-3 (binds LFA-1)		
Selectins		
L-selectin	GlyCAM-1, CD34, mouse MAdCAM-1	
Carbohydrate		
PSGL-1	P-selectin/E-selectin	
E-selectin ligand	E-selectin	

GlyCAM-1; glycosylated cell adhesion molecule 1; ICAM-1, intercellular cell adhesion molecule 1; LFA-1, lymphocyte function associated receptor; MAdCAM-1, mucosal addressin cell adhesion molecule 1; PECAM, platelet–endothelial cell adhesion molecule; PSGL-1, P-selectin glycoprotein ligand 1; VCAM-1, vascular cell adhesion molecule 1.

chicken fibroblast growth factor receptor (E-selectin ligand (ESL)-1) and in humans P-selectin glycoprotein ligand (PSGL)-1, which E-selectin binds with a lower affinity than P-selectin. P-selectin is stored in intracellular granules and expression can be rapidly upregulated on HUVEC by several mediators such as histamine and thrombin. More long-term expression can be induced by IL-4 and IL-3.[5] P-selectin binds PSGL-1, which is expressed on most leucocytes, although expression and function are often dissociated, especially on T-cells.[6] L-selectin is constitutively expressed but shed on cellular activation as a result of the actions of a membrane-bound metalloproteinase. L-selectin is the peripheral lymph node homing receptor and several receptors for it have been identified on lymph node high endothelial venules (HEV). These include glycosylated cell adhesion molecule (GlyCAM)-1, CD34 and, in the mouse, mucosal addressin cell adhesion molecule (MAdCAM)-1. Like PSGL-1 these all contain mucin-like regions rich in O-linked sugars such as sialyl Lewis X. The ligand for L-selectin on inflamed venular endothelium has not been identified. Selectins mediate capture of leucocytes under flow conditions. Selectin–carbohydrate bonds mediate a rolling type of interaction. There is considerable overlap in selectin function and single gene deletion mice are relatively healthy. However, type II leucocyte adhesion deficiency (LAD), in which all three selectins are dysfunctional, and a combined E- and P-selectin 'knock out' mouse both show profound immunodeficiency.[7]

Integrins and their receptors

Integrins are a large superfamily of heterodimeric glycoproteins involved in a wide range of biological functions, including maintenance of tissue homeostasis through binding to matrix proteins.[8] Only a limited number of integrins have been shown to be involved in leucocyte migration. The $\beta 2$ (CD18) leucocyte integrins comprise four members, CD11a–d/CD18. CD11a/CD18 (LFA-1, lymphocyte function associated receptor) is expressed on all leucocytes and is involved in a range of functions, including transmigration through endothelium and T-cell activation. It has three receptors: intercellular adhesion molecule (ICAM)-1 and ICAM-2 expressed on endothelium and ICAM-3 expressed on most leucocytes. CD11b/CD18 (Mac-1) is expressed on myelocytes and CD11c/CD18 (p150,95) is well expressed on tissue macrophages. CD11d/CD18 has been little studied as yet. Mac-1 binds ICAM-1 and has a diverse number of other ligands; it also mediates ICAM-1-independent granulocyte binding to endothelium through an as yet undefined receptor. Impaired expression of the CD18 integrins as in type I LAD leads to profound immunodeficiency, largely as a result of impaired neutrophil migration. ICAM-1 expression is induced on a large number of cell types, including epithelium, endothelium and haematopoietic cells, by cytokine stimulation. $\alpha 4\beta 1$ (VLA-4) is expressed on all leucocytes except neutrophils and binds vascular cell adhesion molecule (VCAM)-1, whose expression on endothelium is selectively upregulated by IL-4 as well as IL-13.[9] VLA-4 also binds fibronectin through a C-terminal non-RGD domain. Both VLA-4 and VCAM-1 gene deletion mice die during embryonic development because of effects on haematopoiesis and cardiac development. $\alpha 4\beta 7$ also binds VCAM-1 and fibronectin as well as MAdCAM-1, a receptor largely expressed by gut endothelium, consistent with the role of $\alpha 4\beta 7$ as a gut lymphocyte homing receptor.[10] $\alpha E\beta 7$ is expressed on a subset of T-lymphocytes and binds E-cadherin, so mediating localization of intraepithelial lymphocytes.[11] $\alpha v\beta 3$ is expressed on endothelium and macrophages and mediates phagocytosis of apoptotic granulocytes.[12] It also binds the widely expressed immunoglobulin-like receptor platelet–endothelial cell adhesion molecule (PECAM), which is involved in leucocyte penetration of endothelial basement membrane.[13] The CD18 integrins are unable to capture leucocytes under flow conditions, only mediating binding after the cell has become attached to the endothelium and activated. In contrast, the $\alpha 4\beta 1$ and $\alpha 4\beta 7$ integrins can bind ligand under flow conditions, at least in the case of lymphocytes.[14]

EXPRESSION OF ADHESION RECEPTORS IN ALLERGIC DISEASE

Endothelial adhesion receptors

Adhesion molecule function is regulated in a number of ways, including increased expression as with E- and P-selectin, ICAM-1 and VCAM-1, shedding as with L-selectin and conformational changes in the binding affinity of the receptor as seen with many integrins. A number of groups have studied expression of E-selectin, ICAM-1 and VCAM-1 in asthma and other allergic inflammatory conditions. P-selectin expression has been less widely studied, partly because of the difficulty in distinguishing between intracellular and luminal staining. In general terms, studies using allergen challenge have been generally consistent with observations in cytokine-stimulated HUVEC. In the skin, low background expression of ICAM-1 is seen with absent expression of E-selectin and VCAM-1. After allergen challenge increased endothelial expression of all three receptors has been reported.[15,16] In the airway, Montefort et al.[17] found increased expression of ICAM-1 and E-selectin 6 h after local allergen challenge with no increase in VCAM-1 expression. Bentley et al.[18] reported a trend towards increased VCAM-1 expression (significance was lost through one outlier) with a good correlation between VCAM-1 expression and eosinophil infiltration 24 h after aerosol allergen challenge.

In clinical asthma, findings have been more variable, probably reflecting the inherent problems in accurately quantifying small changes in expression using immunohistochemistry. Montefort et al.[19] were unable to detect changes in adhesion receptor expression in atopic asthma. In a study of atopic and non-atopic asthma, Bentley et al. could only detect a modest increase in ICAM-1 and E-selectin expression in their non-atopic asthmatic subjects with relatively high background expression.[18] In contrast Gosset et al.[20] found low background expression in normal subjects and could detect increases in adhesion molecule expression in atopic but not non-atopic asthmatic subjects; Ohkawara et al.[21] agreed with these findings in six atopic asthmatic subjects. Fukuda et al.[22] detected no increase in ICAM-1 or E-selectin staining over controls; however, the E-selectin antibody they used cross-reacted with P-selectin. This group did find an increase in VCAM-1 expression that correlated with eosinophil counts but only in those subjects with detectable IL-4 in the bronchoalveolar lavage (BAL) fluid.[22] In nasal endothelium generally weak expression of VCAM-1 has been observed, although increased over normal controls, in patients with perennial rhinitis and nasal polyps.[23]

ICAM-1 expression on epithelial cells is consistently increased on bronchial epithelium of patients with asthma.[18,24,25] Ciprandi et al.[26,27] demonstrated induction of ICAM-1 within 30 min of allergen challenge on both nasal and conjunctival epithelium. This is considerably faster than the rate at which expression is induced in HUVEC. The role of ICAM-1 as a receptor for the major group of rhinoviruses means that the epithelium in asthmatic individuals may be more vulnerable to viral infection. Expression of CD44, a receptor for the matrix protein hyaluronate, is increased on the bronchial epithelium in asthma, although it is also found on normal epithelium.[28]

Soluble adhesion molecules

Several adhesion molecules can be detected in soluble form circulating in the plasma. Montefort et al.[29] found that concentrations of E-selectin, ICAM-1 and VCAM-1 were not elevated in stable asthma, but there was a significant increase compared with normal controls in concentrations of soluble (s)E-selectin and sICAM-1 in patients with acute severe asthma. However, concentrations of these molecules did not correlate with disease severity and were therefore not thought useful in clinical management. In another study of 45 atopic and non-atopic asthmatic subjects, serum concentrations of sICAM-1, sE-selectin and sVCAM-1 were increased during 'asthma attacks' when compared with stable periods.[30,31] Modest increases in concentrations of sICAM-1 and sE-selectin have also been detected in BAL fluid after segmental allergen challenge.[32,33] Zangrilli et al.[34] measured sVCAM-1 concentrations in BAL fluid 24 h after segmental allergen challenge in 27 ragweed-allergic asthmatic and 18 atopic non-asthmatic subjects. A marked increase in sVCAM-1 concentrations was observed in BAL fluid, which correlated with increased numbers of eosinophils and concentrations of IL-4 and IL-5. Most of the increase occurred in the late responders.[34] As yet there have been no clear-cut correlations between disease severity and concentration of soluble adhesion receptors so the significance of these findings in terms of disease pathogenesis or usefulness in monitoring inflammatory activity is still uncertain.

ROLE OF ADHESION RECEPTORS IN LEUCOCYTE MIGRATION IN ALLERGIC DISEASE

There is an extensive literature on the role of adhesion receptors in directing leucocyte traffic, much of which has concentrated on neutrophils. In the following section I summarize the findings on those leucocytes thought to play a prominent role in the inflammatory process in asthma, in particular eosinophils, mast cells and basophils, and T-cells (Table 13.2).

Eosinophils

This area has recently been reviewed and the literature is only briefly summarized here.[35] Eosinophils can bind both E-selectin and P-selectin, with eosinophils preferentially binding P-selectin and neutrophils showing enhanced binding to E-selectin.[36,37] Eosinophils express an isoform of PSGL-1 with greater avidity for P-selectin than neutrophil PSGL-1.[38] L-selectin on eosinophils is shed on activation both *in vitro* and on tissue eosinophils *in vivo*.[39] Anti-L-selectin monoclonal antibodies inhibited eosinophil rolling *in vivo* and adhesion to HUVEC *in vitro*.[40,41] Eosinophils express functional LFA-1, Mac-1, α4β1 and α4β7. Antibodies against the first three of these receptors have been shown to block adhesion and transmigration through HUVEC.[42–44] IL-4 or IL-13 stimulation of HUVEC biased migration towards a VLA/VCAM-1 pathway, whereas stimulation with IL-1 and TNF-α migration was predominantly CD18 integrin dependent and involved both an ICAM-1-dependent and ICAM-1-independent component. Using our frozen-section model of adhesion to nasal polyp endothelium, we have found that eosinophil binding was energy dependent and inhibited by antibodies against CD18 but not α4 integrins (A.J. Wardlaw, personal observation).

The β1 integrins α4β1 and α6β1 have been shown to play a role in binding to fibronectin and laminin respectively. These matrix proteins also supported eosinophil survival through triggering release of granulocyte–macrophage colony-stimulating factor, an event that is potently inhibited by glucocorticoids.[45] With other aspects of eosinophil activation more variable effects with fibronectin have been observed, with some authors reporting enhancement and others inhibition of eosinophil degranulation.[46,47] Eosinophils appear to bind less readily to epithelial cells than endothelial cells despite good epithelial expression of ICAM-1. Increased expression requires vigorous activation with both cytokines and phorbol myristate acetate and is CD18 dependent.[48]

13 Adhesion Molecules

Table 13.2 Role of adhesion receptors in asthma.

Selectins	
P-selectin	Expression upregulated by IL-4
	Constitutively expressed on luminal surface of nasal polyp endothelium
	Eosinophils bind more avidly than neutrophils
E-selectin	Endothelial expression upregulated after allergen challenge *in vivo*
	Lymphocyte addressin in the skin
	Neutrophils bind more avidly than eosinophils *in vitro*
L-selectin	No special role in asthma defined
Integrins	
$\beta 2$ integrins	Major family of receptors involved in endothelial transmigration of all leucocytes
$\beta 1$ integrins	Important in regulating T-cell, monocyte, eosinophil and mast cell/basophil interactions with matrix proteins
$\alpha 4\beta 1/\alpha 4\beta 7$	Alternate transmigration pathway available to all leucocytes other than neutrophils
	Monoclonal antibody against $\alpha 4$ inhibits eosinophil and lymphocyte tissue accumulation in allergen challenge animal models *in vivo*
$\alpha v\beta 3$	Important in macrophage recognition of senescent eosinophils and neutrophils
$\alpha \varepsilon \beta 7$	Localization of intraepithelial lymphocytes in bronchial epithelium
Cell adhesion molecules	
ICAM-1	Widespread expression in inflamed tissue (including epithelium) in allergic diseases with a range of functions defined *in vitro*
	No clearly defined specific function in allergic disease
	Anti-ICAM monoclonal antibody inhibited eosinophil transmigration and development of BHR in primate model of asthma
ICAM-2, ICAM-3	No defined role in asthma
VCAM-1	Endothelial expression upregulated *in vitro* by IL-4
	Ligand for VLA-4 and therefore involved in VLA-4-mediated transmigration
PECAM	Important in mediating transmigration through endothelial basement membrane. No specific role in asthma yet defined
MAdCAM-1	No specific role in asthma so far

BHR, bronchial hyperresponsiveness; ICAM, intercellular adhesion molecule; MAdCAM-1, mucosal addressin cell adhesion molecule 1; PECAM, platelet–endothelial cell adhesion molecule; VCAM-1, vascular cell adhesion molecule 1.

Mast cells and basophils

Possibly because they are seen as tissue-dwelling cells mast cell adhesion interactions have been relatively little studied. Human skin and lung mast cells have been shown to express VLA-3, VLA-4 and VLA-5, through which they spontaneously adhered to laminin and fibronectin. They did not express VLA-1, VLA-2 or VLA-6 and did not adhere to collagen types I or IV.[49,50] Basophils express L-selectin, which is shed on activation, and have a more neutrophil- than eosinophil-like pattern of sialyl Lewis X expression. The expression of PSGL-1 on basophils has not been reported. Antibodies against E-selectin inhibit adhesion of basophils to cytokine-activated HUVEC. Like eosinophils, basophils express VLA-4 and can bind to VCAM-1. Cross-linking of basophil $\beta 1$ integrin receptors from asthmatic but not non-asthmatic donors resulted in histamine release.[51]

T-lymphocytes

Lymphocytes recirculate from the blood into the lymphoid organs and back to blood in search of antigen, a process called lymphocyte homing. Lymphocyte homing is controlled by adhesion receptors on lymphocytes (homing receptors) and their counter-receptors (addressins) on vascular endothelium,[52] with L-selectin defined as a peripheral lymph node homing receptor binding principally to an HEV-specific glycoform of CD34, $\alpha 4\beta 7$ as the mucosal homing receptor binding to MAdCAM-1 and E-selectin/CLA as the skin addressin/homing receptor. Support for this model has been provided by L-selectin gene deletion mice, which have atrophic peripheral lymph nodes, and $\beta 7$ gene deletion mice, which show markedly impaired recruitment of lymphocytes into Peyer's patches.[53,54] A lung lymphocyte homing receptor has been proposed but remains unidentified. Indeed although the lung is a highly lymphocytic organ, the specific receptors that control T-cell migration are still poorly understood. There are differences between lung and peripheral blood T-cells in their adhesion receptor phenotype but the relevance of these observations to lymphocyte traffic into the lung remains unclear. Thus expression on lung versus blood CD3 cells of $\alpha\varepsilon\beta 7$, $\alpha 4\beta 1$, $\alpha 1\beta 1$ and ICAM-1 is increased whereas L-selectin is less well expressed.[55,56] Lymphocyte homing may be important in directing allergen-sensitized T-cells to the airway or skin. For example, caesin-reactive T-cells from patients with milk-induced eczema had higher expression of the CLA antigen than *Candida albicans*-reactive T-cells from the same patients or caesin-reactive T-cells from non-atopic controls.[57] When house dust mite-sensitive patients with asthma and atopic dermatitis were compared the house dust mite-responsive T-cells from the eczema patients, but not from the asthma group, were in the CLA-positive T-cell subset.[58]

IN VIVO STUDIES OF ADHESION RECEPTOR ANTAGONISTS IN MODELS OF ALLERGIC INFLAMMATION

In vivo models of allergen challenge in a variety of animal species have been used extensively to investigate whether inflammatory markers, particularly eosinophil counts, and measures of bronchial hyperresponsiveness can be modulated by adhesion receptor antagonists (usually monoclonal antibodies). These studies are detailed in Table 13.3. Inhibitory effects have been demonstrated using antibodies against a number of receptors, including VLA-4, Mac-1, LFA-1, ICAM-1, VCAM-1 and E-selectin, although differences in degree and pattern of inhibition have been observed depending on the species used and the exact conditions. Most support has been gained for an important role of VLA-4 and VCAM-1 in both eosinophil and T-lymphocyte migration into the airways after allergen challenge and this has given impetus to the development of VLA-4 antagonists in clinical trials.

SUMMARY

Considerable progress has been made in our understanding of the molecular mechanisms involved in leucocyte adhesion interactions. Migration through endothelium is a staged

13 Adhesion Molecules

Table 13.3 Use of anti-adhesion receptor monoclonal antibodies in animal models of allergen challenge.

Receptor	Reference	Species	Findings
E-selectin	59	Cynomolgus monkey	In this neutrophil-dependent model of late-phase response to *Ascaris* challenge, both cell migration and bronchoconstriction were inhibited
ICAM-1	60	Cynomolgus monkey	In this multiple-allergen (*Ascaris*) challenge model of asthma, both airway eosinophilia and development of BHR were inhibited
Mac-1	61	Cynomolgus monkey	In the same multiple-antigen challenge model, anti-Mac-1 inhibited development of BHR and ECP concentrations in BAL but not eosinophil counts
$\alpha 4$	62	Sheep	Late response to *Ascaris* challenge inhibited when antibody given both intravenously and by inhalation but no effect on BAL eosinophils
$\alpha 4$, Mac-1, LFA-1	63	Rat	Antibodies against all three receptors inhibited early and late response to ovalbumin challenge without any effect on cell counts in BAL at 8 h
$\alpha 4$	64	Rat	Antibody against VLA-4 inhibited eosinophil and T-lymphocyte infiltration at 24 h
$\alpha 4$	65	Guinea-pig	Inhibition into the skin of eosinophil infiltration induced by chemoattractants and PCA (passive cutaneous anaphylaxis) reaction
$\alpha 4$	66	Guinea-pig	Inhibition of BHR, EPO release and eosinophil infiltration into the airway of ovalbumin-challenged animals
ICAM-1/LFA-1, $\alpha 4$/VCAM-1	67	Mouse	Eosinophil and T-cell infiltration inhibited by anti-$\alpha 4$ and anti-VCAM-1 but not by anti-ICAM-1 or anti-LFA-1

BAL, bronchoalveolar lavage; BHR, bronchial hyperresponsiveness; ECP, eosinophil cationic protein; EPO, eosinophil peroxidase; ICAM-1, intercellular adhesion molecule 1; LFA-1, lymphocyte function associated receptor; VCAM-1, vascular cell adhesion molecule 1; VLA-4, very late antigen 4.

process, with each stage offering a level of control over the cell specificity and degree of migration. Although the structure and function of the receptors involved in leucocyte migration have been well characterized, the contribution each makes to the pattern of leucocyte accumulation in asthma has still not been completely defined, albeit a number of interesting observations have been made. There is good evidence for an important role for VLA-4/VCAM in mediating eosinophil transmigration into the lung in asthma. P-selectin may also be important. VLA-4 is attractive as a therapeutic target for eosinophils because of its lack of expression on neutrophils. The receptors controlling migration of other cells, in particular T-cells and monocytes, have been less well studied. Results using monoclonal antibodies in a number of animal models suggest that adhesion receptor blockade may be an effective anti-inflammatory strategy in asthma. The development of drugs that can be used to test this hypothesis in the clinic are awaited with considerable interest.

REFERENCES

1. Springer TA: Traffic signals for lymphocyte re-circulation and leukocyte emigration: the multi-step paradigm. *Cell* (1994) **76**: 310.
2. Rosen SD: Cell surface lectins in the immune system. *Semin Immunol* (1993) **5**: 237–247.
3. Picker LJ, Kishimoto TK, Smith CW, Warnock RA, Butcher EC: ELAM-1 is an adhesion molecule for skin-homing T cells. *Nature* (1991) **349**: 796.
4. Maly P, Thall AD, Petryniak B, et al.: The α(1,3) fucosyltransferase Fuc-TV11 controls leukocyte trafficking through an essential role in L-, E- and P-selectin ligand biosynthesis. *Cell* (1996) **86**: 643–653.
5. Yao L, Pan J, Setiadi H, Patel KD, McEver RP: Interleukin 4 or oncostatin induces a prolonged increase in P-selectin mRNA and protein in human endothelial cells. *J Exp Med* (1996) **184**: 81–92.
6. Sako D, Comess KM, Barone KM, Camphausen RT, Cumming DA, Shaw GD: A sulfated peptide segment at the amino terminus of PSGL-1 is critical for P-selectin binding. *Cell* (1995) **83**: 323–331.
7. Frenette PS, Mayadas TN, Rayburn H, Hynes RO, Wagner DD: Susceptibility to infection and altered hematopoiesis in mice deficient in both P and E-selectins. *Cell* (1996) **84**: 563–574.
8. Hynes RO: Integrins: versatility, modulation and signalling in cell adhesion. *Cell* (1992) **69**: 11–25.
9. Elices MJ, Osbourn L, Takada Y, et al.: VCAM-1 on activated endothelium interacts with the leukocyte integrin VLA-4 at a site distinct from the VLA-4/fibronectin binding site. *Cell* (1990) **60**: 577–584.
10. Erle DJ, Briskin MJ, Butcher ED, Garcia-Pardo A, Lazarovits AI, Tidswell M: Expression and function of the MAdCAM-1 receptor integrin α4/β7 on human leukocytes. *J Immunol* (1994) **153**: 517–528.
11. Cepek KL, Shaw SK, Parker CM, et al.: Adhesion between epithelial cells and T lymphocytes mediated by E-cadherin and the αεβ7 integrin. *Nature* (1994) **372**: 190–193.
12. Savill J, Dransfield I, Hogg N, Haslett C: Vitronectin receptor-mediated phagocytosis of cells undergoing apoptosis. *Nature* (1990) **343**: 170–173.
13. Liao F, Huynh HK, Eiroa A, Greene T, Polizzi E, Muller WA: Migration of monocytes across endothelium and passage through extracellular matrix involve separate molecular domains of PECAM-1. *J Exp Med* (1995) **182**: 1337–1343.
14. Berlin C, Bargatze RF, Campbell JJ, et al.: α4 Integrin mediates lymphocyte attachment and rolling under physiologic flow. *Cell* (1995) **80**: 413–422.
15. Kyan-Aung U, Haskard DO, Poston RN, Thornhill MH, Lee TH: Endothelial leukocyte adhesion molecule-1 and intercellular adhesion molecule-1 mediated the adhesion of eosinophils to endothelial cells *in vitro* and are expressed by endothelium in allergic cutaneous inflammation *in vivo*. *J Immunol* (1991) **146**: 521–528.
16. Leung YM, Pober JS, Cotran RS: Expression of endothelial–leukocyte adhesion molecule-1 in elicited late phase allergic reactions. *J Clin Invest* (1991) **87**: 1805–1809.
17. Montefort S, Gratziou C, Goulding D, et al.: Upregulation of leukocyte–endothelial cell adhesion molecules 6 hours after local allergen challenge of sensitised asthmatic airways. *J Clin Invest* (1993) **93**: 1411–1421.
18. Bentley AM, Durham SR, Robinson DS, et al.: Expression of endothelial and leukocyte adhesion molecules, intercellular adhesion molecule-1, E-selectin and vascular cell adhesion molecule-1 in the bronchial mucosa in steady state and allergen induced asthma. *J Allergy Clin Immunol* (1993) **92**: 857–868.
19. Montefort S, Roche WR, Howarth PH, et al.: Intercellular adhesion molecule-1 (ICAM-1) and endothelial leucocyte adhesion molecule-1 (ELAM-1) expression in the bronchial mucosa of normals and asthmatic subjects. *Eur Respir J* (1992) **5**: 815–823.
20. Gosset P, Tillie-Leblond I, Janin A, et al.: Expression of E-selectin, ICAM-1 and VCAM-1 on bronchial biopsies from allergic and non-allergic asthmatic patients. *Int Arch Allergy Immunol* (1995) **106**: 69–77.
21. Ohkawara Y, Yamauchi K, Maruyama N, et al.: *In situ* expression of the cell adhesion molecules

in bronchial tissues from asthmatics with air flow limitation: *in vivo* evidence of VCAM-1/VLA-4 interaction in selective eosinophil infiltration. *Am J Respir Cell Mol Biol* (1995) **12**: 4–12.
22. Fukuda T, Fukushima Y, Numao T, *et al*.: Role of interleukin-4 and vascular cell adhesion molecule-1 in selective eosinophil migration into the airways in allergic asthma. *Am J Respir Cell Mol Biol* (1996) **14**: 84–94.
23. Montefort S, Feather IH, Wilson SJ, *et al*.: The expression of leukocyte endothelial adhesion molecules is increased in perennial allergic rhinitis. *Am J Respir Cell Mol Biol* (1992) **7**: 393–398.
24. Vignola AM, Campbell AM, Chanez P, *et al*: HLA-DR and ICAM-1 expression on bronchial epithelial cells in asthma and chronic bronchitis. *Am Rev Respir Dis* (1993) **147**: 529–534.
25. Manolitsas ND, Trigg CJ, McAulay AE, *et al*.: The expression of intercellular adhesion molecule-1 and the β1-integrins in asthma. *Eur Respir J* (1994) **7**: 1439–1444.
26. Ciprandi G, Pronzato C, Ricca V, Passalacqua G, Bagnasco M, Canonica GW: Allergen specific challenge induces intercellular adhesion molecule-1 (ICAM-1/CD54) expression on nasal epithelial cells in allergic subjects. Relationship with early and late inflammatory phenomena. *Am J Resp Crit Care Med* (1994) **150**: 1653–1659.
27. Ciprandi G, Buscaglia S, Pesce GP, Villaggio B, Bagnesco M, Canonica GW: Allergic subjects express intracellular adhesion molecule 1 (ICAM-1 or CD54) on epithelial cells of conjunctiva after allergen challenge. *J Allergy Clin Immunol* (1993) **91**: 783–792.
28. Lackie PM, Baker JE, Gunthert U, Holgate ST: Expression of CD44 isoforms is increased in the airway epithelium of asthmatic subjects. *Am J Respir Cell Mol Biol* (1997) **16**: 14–22.
29. Montefort S, Lai CKW, Kapahi P, *et al*.: Circulating adhesion molecules in asthma. *Am J Respir Crit Care Med* (1994) **149**: 1149–1152.
30. Kobayashi T, Hashimoto S, Imai K, *et al*.: Elevation of serum soluble intercellular adhesion molecule-1 (sICAM-1) and sE-selectin levels in bronchial asthma. *Clin Exp Immunol* (1994) **96**: 110–115.
31. Koizumi A, Hashimoto S, Kobayashi T, Imai K, Yachi A, Horie T: Elevation of serum soluble vascular cell adhesion molecule-1 (sVCAM-1) levels in bronchial asthma. *Clin Exp Immunol* (1995) **101**: 468–473.
32. Georas SN, Liu MC, Newman W, Beall LD, Stealey BA, Bochner BS: Altered adhesion molecule expression and endothelial cell activation accompany the recruitment of human granulocytes to the lung after segmental antigen challenge. *Am J Respir Cell Mol Biol* (1992) **7**: 261–269.
33. Takahashi N, Liu MC, Proud D, Yu X-Y, Hasegawa S, Spannhake EW: Soluble intercellular adhesion molecule-1 in bronchoalveolar lavage fluid of allergic subjects following segmental antigen challenge. *Am J Respir Crit Care Med* (1994) **150**: 704–709.
34. Zangrilli JG, Shaver JR, Cirelli RA, *et al*.: sVCAM-1 levels after segmental allergen challenge correlates with eosinophil influx, IL-4 and IL-5 production and the late phase response. *Am J Respir Crit Care Med* (1995) **151**: 1346–1353.
35. Wardlaw AJ, Walsh GM, Symon FA: Mechanisms of eosinophil and basophil migration. *Allergy* (1994) **49**: 797–807.
36. Symon FA, Walsh GM, Watson SR, Wardlaw AJ: Eosinophil adhesion to nasal polyp endothelium is P-selectin dependent. *J Exp Med* (1994) **180**: 371–376.
37. Bochner BS, Sterbinsky SA, Bickel CA, Werfel S, Wein M, Newman W: Differences between human eosinophils and neutrophils in the function and expression of sialic acid containing counterligands for E-selectin. *J Immunol* (1994) **152**: 774–778.
38. Symon FA, Lawrence MB, Williamson M, Walsh GM, Watson SR, Wardlaw AJ: Characterisation of the eosinophil P-selectin ligand. *J Immunol* (1996) **157**: 1711–1719.
39. Georas SN, Liu MC, Newman W, Beall LD, Stealey BA, Bochner BS: Altered adhesion molecule expression and endothelial cell activation accompany the recruitment of human granulocytes to the lung after segmental antigen challenge. *Am J Respir Cell Mol Biol* (1992) **7**: 261–269.
40. Sriramarao P, von Adrian UH, Butcher EC, Bourdon MA, Broide DH: L-selectin and very late antigen-4 integrin promote eosinophil rolling at physiological shear rate *in vivo*. *J Immunol* (1994) **153**: 4238–4246.
41. Knol EF, Kansas GS, Tedder TF, Schleimer RP, Bochner BS: Human eosinophils use L-selectin to bind to endothelial cells under non static conditions. *J Allergy Clin Immunol* (1993) **91**: 334.

42. Ebisawa M, Bochner BS, Georas SN, Schleimer RP: Eosinophil transendothelial migration induced by cytokines. Role of the endothelial and eosinophil adhesion molecules in IL-1b induced transendothelial migration. *J Immunol* (1992) **149**: 4021–4028.
43. Schleimer RP, Sterbinsky SA, Kaiser J, et al.: IL-4 induces adherence of human eosinophils and basophils, but not neutrophils to endothelium. Association with expression of VCAM-1. *J Immunol* (1992) **148**: 1086–1092.
44. Walsh GM, Hartnell A, Mermod JJ, Kay AB, Wardlaw AJ: Human eosinophil, but not neutrophil adherence to IL-1 stimulated HUVEC is α4β1 (VLA-4) dependent. *J Immunol* (1991) **146**: 3419–3423.
45. Walsh GM, Wardlaw AJ: Dexamethasone inhibits prolonged survival and autocrine granulocyte macrophage colony stimulating factor production by human eosinophils cultured on laminin or tissue fibronectin. *J Allergy Clin Immunol* (1997) **100**: 208–215.
46. Neeley SP, Hamann KJ, Dowling T, McAllister KT, White SR, Leff AR: Augmentation of stimulated eosinophil degranulation by VLA-4 (CD49d)-mediated adhesion to fibronectin. *Am J Respir Cell Mol Biol* (1994) **11**: 206–213.
47. Kita H, Horie S, Gleich GJ: Extracellular matrix proteins attenuate activation and degranulation of stimulated eosinophils. *J Immunol* (1996) **156**: 1174–1181.
48. Godding V, Stark JM, Sedgwick JB, Busse WW: Adhesion of activated eosinophils to respiratory epithelial cell is enhanced by tumor necrosis factor-a and IL-1b. *Am J Respir Cell Mol Biol* (1995) **13**: 555–562.
49. Columbo M, Bochner BS, Marone G: Human skin mast cells express functional β1 integrins that mediate adhesion to extracellular matrix proteins. *J Immunol* (1995) **154**: 6058–6064.
50. Sperr WR, Agis H, Czerwenka K, et al.: Differential expression of cell surface integrins on human mast cells and human basophils. *Ann Hematol* (1992) **65**: 6–10.
51. Lavens SE, Goldring K, Thomas LH, Warner JA: Effects of integrin clustering on human lung mast cells and basophils. *Am J Respir Cell Mol Biol* (1996) **14**: 95–103.
52. Butcher EC, Picker LJ: Lymphocyte homing and homeostasis. *Science* (1996) **272**: 60–66.
53. Ley K, Tedder TF: Leukocyte interactions with vascular endothelium. New insights into selectin mediated attachment and rolling. *J Immunol* (1995) **155**: 525–528.
54. Wagner N, Lohler J, Kunkel EJ, et al.: Critical role for β7 integrins in formation of the gut associated lymphoid tissue. *Nature* (1996) **382**: 366–370.
55. Picker LJ, Martin RJ, Trumble AE, et al.: Control of lymphocyte re-circulation in man: differential expression of homing associated adhesion molecules by memory/effector T cells in pulonary versus cutaneous effector sites. *Eur J Immunol* (1994) **24**: 1269–1277.
56. Kennedy JD, Hatfield CA, Fidler SF, et al.: Phenotypic characterisation of T lymphocytes emigrating into lung tissue and the airway lumen after antigen inhalation in sensitised mice. *Am J Respir Cell Mol Biol* (1995) **12**: 613–623.
57. Abernathy-Carver KJ, Sampson HA, Picker LJ, Leung DYM: Milk-induced eczema is associated with the expansion of T cells expressing cutaneous lymphocyte antigen. *J Clin Invest* (1995) **95**: 913–918.
58. Babi LFS, Picker LJ, Soler MTP: Circulating allergen reactive T cells from patients with atopic dermatitis and allergic contact dermatitis express the skin-selective homing receptor the cutaneous lymphocyte associated antigen (CLA). *J Exp Med* (1995) **181**: 747–753.
59. Gundel RH, Wegner CD, Torcellini CA, et al.: ICAM-1 mediates antigen-induced acute airway inflammation and late phase obstruction in monkeys. *J Clin Invest* (1991) **88**: 1407–1411.
60. Wegner CD, Grundel RH, Reilly P, Haynes N, Letts GL, Rothlein R: ICAM-1 in the pathogenesis of asthma. *Science* (1990) **247**: 416–418.
61. Wegner CD, Gundel RH, Churchill L, Letts LG: Adhesion glycoproteins as regulators of airway inflammation: emphasis on the role of ICAM-1. In Holgate ST, Austen KF, Lichtenstein LF, Kay AB (eds) *Asthma: Physiology, Pharmacology and Treatment*. London, Academic Press, 1993, pp 227–242.
62. Abraham WM, Sielczak MW, Ahmed A, et al.: α4 Integrins mediate antigen-induced late bronchial responses and prolonged airway hyperresponsiveness in sheep. *J Clin Invest* (1994) **93**: 776–787.
63. Rabb HA, Olivenstein R, Issekutz TB, Renzl PM, Martin JG: The role of the leucocyte

adhesion molecules VLA-4, LFA-1 and Mac-1 in allergic airway in rat. *Am J Respir Crit Care Med* (1994) **149**: 1186–1191.
64. Richards IM, Kolbasa KP, Hatfield CA, *et al.*: Role of VLA-4 in the antigen induced accumulation of eosinophils and lymphocytes in the lungs and airway of sensitized brown Norway rats. *Am J Respir Cell Mol Biol* (1996) **15**: 172–183.
65. Weg VB, Williams TJ, Lobb PR, Nourshargh S: A monoclonal antibody recognizing the very late activation antigen-4 inhibits eosinophil accumulation *in vivo*. *J Exp Med* (1993) **177**: 561–566.
66. Pretolani MC, Ruffie C, de Silva L, Joseph D, Lobb R, Vargaftig B: Antibody to very late activation antigen 4 prevents antigen-induced bronchial hyperreactivity and cellular infiltration in the guinea pig airways. *J Exp Med* (1994) **180**: 795–805.
67. Nakajima H, Sano H, Nishimura T, Yoshida S, Iwanoto I: Role of vascular cell adhesion molecule 1/very late antigen 4 and intercellular adhesion molecule 1 interactions in antigen-induced eosinophil and T cell recruitment into the tissue. *J Exp Med* (1994) **179**: 1145–1154.

14

Microvascular–Epithelial Exudation of Plasma

CARL G.A. PERSSON

INTRODUCTION

Just beneath the airway epithelium there is a profuse microcirculation (Fig. 14.1). In humans, this microvascular–epithelial arrangement is seen in the nose and in almost the entire tracheobronchial tree.[1] In select experimental animals, such as the guinea-pig, the nose and trachea[2] are similar to humans, whereas more peripherally rodents may largely resort to a peribronchial microvascular bed. The epithelium and the subepithelial systemic microcirculation are two important end-organs exhibiting multiple interactions in asthma.[3] The focus of the present chapter is on airways plasma exudation in disease, defence and repair. As a corollary the need for exploratory *in vivo* approaches in studies of airway inflammation is underscored.[4] My objectives are as follows.

(1) Explain mechanisms and pathways involved in mucosal exudation of non-sieved plasma, and distinguish between epithelial barrier functions involved in paracellular exudation and absorption, respectively. For example, this point illustrates the potential for exuded plasma to contribute much of the active molecular milieu of mucosal tissue and surface *in vivo* not only in the severely disrupted mucosa but also in conditions with an uncompromised epithelial barrier function.
(2) Link plasma exudation and its derived adhesive and leucocyte-activating proteins, cytokines, peptide mediators, etc. with airway mucosal defence, repair and disease. Additionally, this point reinforces the importance of inhibiting (by anti-inflammatory treatment) the exudation that occurs as a sign and promoter of disease activity, whilst leaving unimpeded the exudation responses that need to be mounted in respiratory defence and repair.

ASTHMA: BASIC MECHANISMS AND CLINICAL MANAGEMENT (3rd Edn)
ISBN 0-12-079027-9

Copyright © 1998 Academic Press Limited
All rights of reproduction in any form reserved

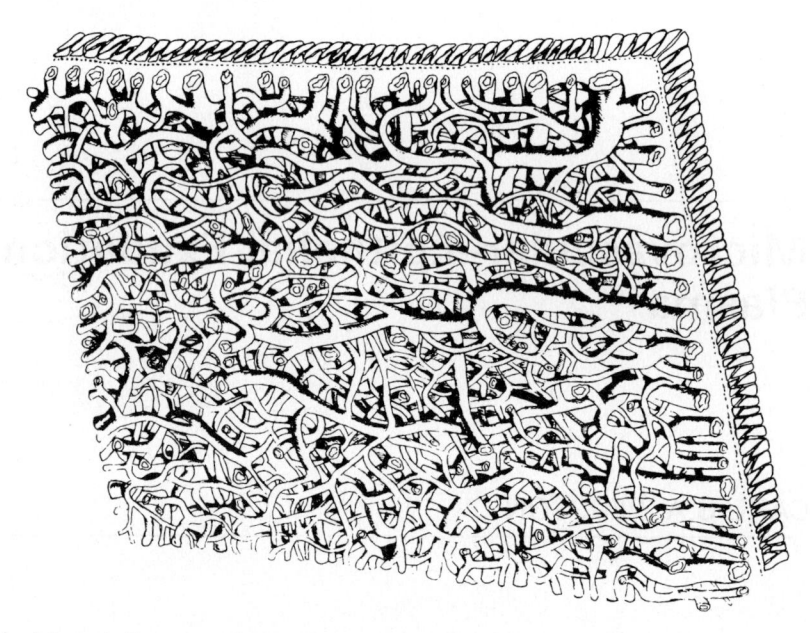

Fig. 14.1 Schematic illustration of a profuse subepithelial plexus of microvessels of nasal and tracheobronchial airways.

(3) Describe epithelial repair occurring *in vivo* (after shedding) and deduce potential roles of shedding–restitution processes in causing exudative and other asthma-like changes in the airways. Inferences to be noted here are that prompt and speedy processes of epithelial restitution may explain a maintained mucosal tightness at shedding, and that 'cytoprotection' may become an important aspect of the treatment of inflammatory airway diseases.

EXUDATION PATHWAYS

The acute plasma exudation response to airway mucosal challenges involves a series of events that all occur within minutes after challenge.[3,5,6] By cellular release or other mechanisms, the inflammatory challenge results in increased mucosal tissue levels of vasoactive agents. Vascular permeability-increasing agents such as histamine act directly on the venular wall endothelium.[7,8] It appears that cell–cell contact is interrupted at distinct points in the walls of the postcapillary venules.[8] The mechanism of this interendothelial gap formation has been widely accepted as a contractile event, producing 1-μm clefts between the cells. However, recent wholemount three-dimensional morphology reveals that the gaps are rather small round holes that appear as if there is transiently reduced adhesion along tiny stretches of the endothelial cell–cell contact[5,9] (Fig. 14.2). Another possibility is the presence of vesiculo-vacuolar 0.1-μm channels across thin parts of endothelial cells.[10] Through these gaps/holes in the venular wall non-sieved plasma is moved by the hydrostatic pressure gradient into extravascular sites, locally abolishing the

14 Microvascular–Epithelial Exudation of Plasma

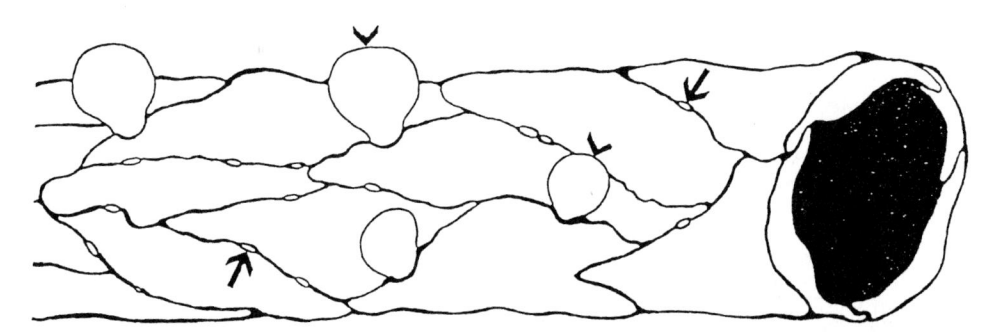

Fig. 14.2 A postcapillary venule actively involved in extravasation of plasma and leucocytes (see refs 5 and 9). The morphological correlates of increased vascular permeability are distinct sites of endothelial cell separation (arrows). At separate interendothelial sites, leucocytes are adhering and moving across the venular wall (arrowheads). Drawing by Jonas Erjefält.

colloid osmotic pressure gradient between the microvessels and the tissue.[11] The venular endothelial cell is also a target for agents that produce antiexudative effects by direct vascular antipermeability mechanisms.[12] In animal airways, acute steroid treatment has been demonstrated to produce such antipermeability effects,[13] but studies in human airways have not confirmed this action.[14] In asthma the steroids are effective antiexudative agents,[15] apparently as a reflection of their anti-inflammatory efficacy.[14,15] β-Agonists and theophylline have a capacity to acutely reduce inflammatory challenge-induced plasma exudation responses in the airways.[12,16–18] Whether this effect occurs to any important degree in the treatment of chronic asthma with these latter two classes of drugs remains unknown.

During the first 10–20 s after a proper topical airway challenge the lamina propria is flooded with plasma exudate (Fig. 14.3). Apparently unhindered, the exudate then passes through the epithelial basement membrane and further up between epithelial cells that normally are separated at the base.[19] At the apical pole the epithelial cells are tightly connected. However, not even the tight junctions of an intact epithelial lining appear to be significant obstacles to the further flux of bulk exudate into the airway lumen (reviewed in refs 3, 20 and 21). For example, within a few minutes after histamine challenge, albumin and α_2-macroglobulin appear together on the surface of the intact human nasal airways in the same concentration ratio that these extremely dissimilar (in size) proteins have in the circulating blood (C. Svensson et al., unpublished observations). The luminal entry of plasma has been described as a self-sustaining process that occurs as long as sufficient amounts of plasma impinge upon the basolateral aspects of epithelial cells[20] (see Fig. 14.3). Incidentally, this mechanism appears to exclude epithelial barrier function as an important direct target for exudative and antiexudative effects. Although the process is non-sieved, the bulk plasma that is moved to the mucosal surface cannot be identical to circulating plasma. Promptly after extravasation several protein systems of the blood plasma would be activated, generating a great variety of peptides and oligoproteins (Table 14.1). Preventing extravasation effectively inhibits the formation of plasma-derived active molecules in the airways.

Tissue flooding with extravasated plasma moving towards the lumen is not merely a passive lavage of the lamina propria. The extravasated plasma may also carry molecules

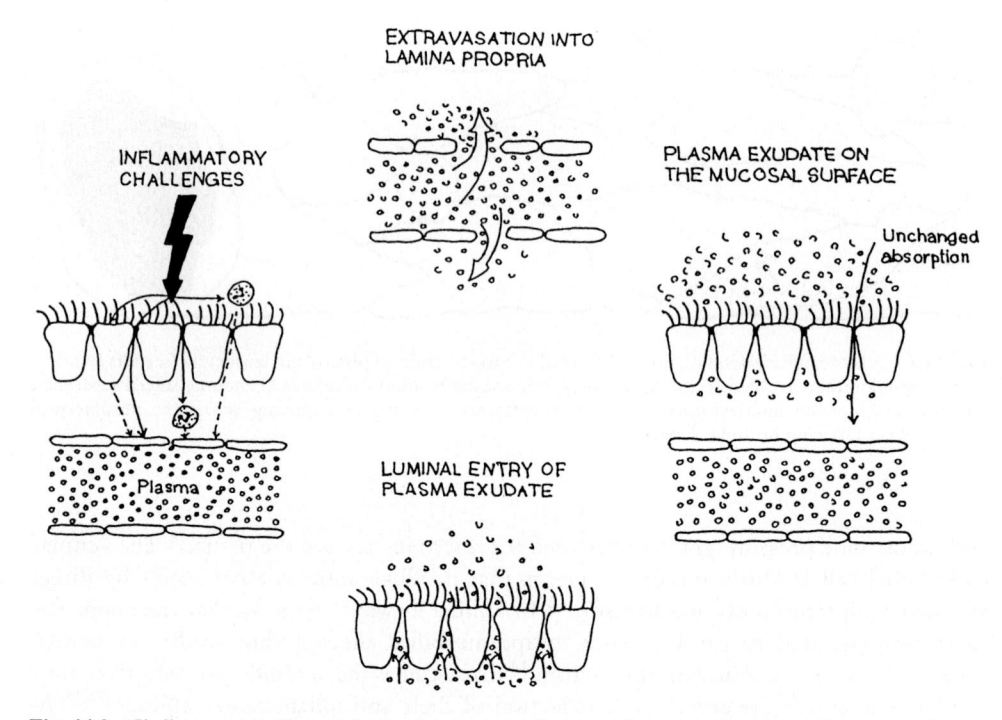

Fig. 14.3 Challenge with allergen, leukotriene-type mediators and several other pro-inflammatory factors produce dose-dependent extravasation, distribution into the lamina propria and luminal entry of 'bulk' plasma. This process may occur without causing significant oedema, without disrupting the epithelial lining and without increasing the absorption ability of the airway mucosa. Thus, in defence and inflammation all plasma protein systems, irrespective of molecular size, appear in the airway mucosa and on the surface of an intact epithelium.

Table 14.1 The contents of extravasated plasma.

Proteins
Adhesive molecules (fibrinogen, fibronectin, etc.)
Proteases and antiproteases
Cytokine-modulating proteins
Immunoglobulins

Cytokines
Growth factors (platelet-derived growth factor, insulin-like growth factor, transforming growth factor β, etc.)
Interleukins
Several cytokines bound, carried and targeted by α_2-macroglobulin and other plasma proteins

Peptides
Complement fragments
Bradykinins
Fibrinolysis peptides

residing in the airway tissue to the epithelium and to the airway surface.[3] Simple histamine challenges that produce graded exudations of bulk plasma may, through this action, also increase airway surface levels of subepithelial cytokines including interleukin (IL)-6.[22] Cytokines or other important molecules residing in the subepithelium in airway diseases may be bound by α_2-macroglobulin[23] or other plasma proteins and thus be distributed and targeted by the plasma exudation process. A lamina propria lavage effected by disease- or challenge-evoked plasma exudation will thus have to be taken into account in studies of the mechanisms of asthma. In short, lavage fluid and biopsy data on cytokine molecules cannot be properly interpreted if nothing is known about the plasma exudative condition at, or prior to, sampling.

Erjefält et al.,[19] employing colloidal gold (5 nm in diameter) as plasma tracer, have observed that the plasma exudate moves between all epithelial cells in the challenged area and all around each cell (see Fig. 14.3). Hence, the burden on each unit length of cell junction would be minute even at pronounced rates of exudation of bulk plasma. This finding tallies with previous observations demonstrating a non-injurious nature of the mucosal exudation process (reviewed in refs 3 and 31). Based on such data, generated in animal and human airways, luminal entry of plasma was recently suggested as a major first-line mucosal defence process.[21] A mechanism has been discovered that potentially explains how the extravasated bulk plasma may pass through epithelial junctions:[24,25] in intact airway tube preparations (mounted in organ baths that allow separate regulation of mucosal and serosal bathing fluids) a slightly increased hydrostatic pressure load (<5 cm H_2O) on the basolateral aspects of the epithelial lining cells is sufficient for moving macromolecular solutes to the mucosal surface. Indeed, this process is reversible and repeatable[25] similar to the *in vivo* exudation evoked by intermittent challenge with histamine-type mediators.[26] The epithelial junctions evidently yield and close so that luminal entry of macromolecules occurs without being associated with, or followed by, increased mucosal absorption of polar solutes (see Fig. 14.3) (reviewed in ref. 20).

In summary, the direct physiological and pharmacological regulation of mucosal exudation of plasma takes place at the level of the endothelial cells of the microvascular wall. This is the site where inflammatory factors produce the extravasation holes that govern the whole airways exudation process. Mediators and drugs may not affect the epithelial passage of plasma into the lumen, which seems logical because the exudate will get into the lumen anyway (by sustaining its own passage). Furthermore, drug-induced tightening effects on the epithelium would not be desirable because it could increase the tendency for oedema formation in the airway mucosa (see Figs 14.3 and 14.4).

ACUTE CHALLENGE-INDUCED MICROVASCULAR–EPITHELIAL EXUDATION

The human nasal mucosa lends itself to airway specific challenge and lavage studies *in situ*.[6] In several respects, nasal conditions may be far better controlled than those attainable in human bronchi *in vivo*. To take advantage of this a pool technique has been developed.[27] A compressible nasal pool device makes it possible to fill the entire ipsilateral nasal cavity with fluid and solutes. A relatively well-defined and large airway

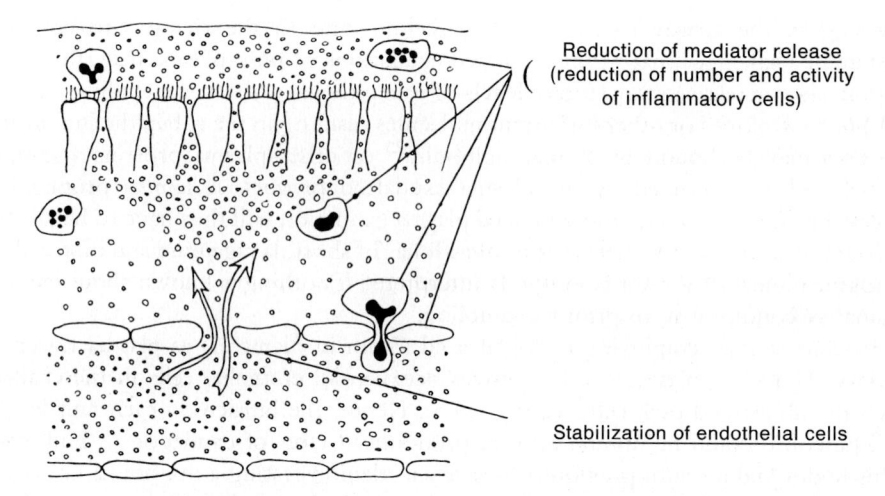

Fig. 14.4 When glucocorticoids reduce plasma exudation in human airways this probably reflects inhibition of the inflammatory process (above) rather than a direct vascular antipermeability effect (below).

mucosal surface area can thus be exposed to predetermined concentrations of agents and tracers. After a selected mucosal exposure time the pool fluid may be recovered, almost quantitatively, by the device. Thus the exposed mucosal surface is selectively and gently lavaged by the nasal pool fluid, providing the opportunity to sample mucosal parameters exclusively from the area of interest. This lavage procedure can be repeated without undue mucosal effects. Using the nasal pool device, Greiff et al.,[27] have demonstrated graded exudative effects of different mucosal surface concentrations of inflammatory agents. The mechanisms involved in the human airway mucosa in health and disease can be examined in the nose and it is possible that many of these nasal findings may be valid also in the tracheobronchial airways.[6]

A contractile and secretory neurotransmitter such as acetylcholine and its analogues (methacholine, carbachol, etc.) do not have exudative effects in the airways of animals and humans.[28,29] In human nasal airways, in health and allergic disease, irritants such as nicotine and capsaicin, which evoke strong neurogenic (tachykininergic) responses, appear to have no effects on plasma exudation.[30–32] These findings have recently been extended to human bronchi:[33] inhalation of histamine increased several-fold the luminal entry of α_2-macroglobulin, whereas significant cough-inducing capsaicin inhalations were completely devoid of this effect (measured as sputum concentrations in subjects who also received hypertonic saline inhalations for sputum induction). Such human data are in sharp contrast to findings in guinea-pig airways, where agents such as capsaicin produce pronounced exudation of plasma into the airway lumen.[34] Neurogenic airway inflammation (exudation) may thus be of relevance to rodents but not to humans. Another important inference is that plasma exudation in human airways is more specific to inflammation compared with rodent airways, because simple neural reflex mechanisms may not produce this response. Histamine-type mediators (histamine, bradykinin, leukotriene D_4, etc.) produce graded exudative responses over a wide range of concentrations in both guinea-pig and human airways.[6] Eosinophil granule proteins increase vascular permeability in the hamster cheek pouch.[35] Select cytokines, proteases, fibrinolysis peptides, etc. may also induce plasma exudation and so will any agent that has a capacity to release vasoactive agents *in vivo*.

Using nasal and endobronchial allergen challenge experiments it has been demonstrated that the human airway mucosa responds with luminal entry of bulk plasma (including α_2-macroglobulin and fibrinogen) within a few minutes of the challenge.[29,36,37] Allergen challenge in subjects with allergic airway disease may produce both immediate and late-phase plasma exudation responses.[15,36–38] Similarly, in sensitized guinea-pigs, allergen challenge of the tracheal mucosa produces dual plasma exudation responses.[39] The immediate exudation phase is over in about 1 h. A late airway exudation phase follows that peaks about 5 h after challenge and then fades.[26] Single-dose topical treatment with clinically effective antiasthmatic steroids prevents the late-phase exudation response.[40]

In contrast with allergens, the occupational small molecular weight chemical toluene diisocyanate (TDI) produces a strong, sustained plasma exudation response in airways that have not previously been exposed to TDI and thus have not been sensitized to this reactive agent.[28,41] Within a wide dose range (3 nl to 30 μl), TDI produces dose-dependent plasma exudation into tracheal airways of previously unexposed guinea-pigs. (These doses are comparable to the accepted exposure level for humans, which corresponds to a daily body burden of about 15 μl TDI.) The acute TDI-induced plasma exudation response in non-sensitized guinea-pigs peaks 5 h after challenge and continues for an additional 15 h.[41] This acute mucosal exudation response is insensitive to steroid treatment. Guinea-pigs that receive repeated challenges with 3 nl of TDI on the large tracheobronchial airways develop an increased inflammatory responsiveness to TDI.[42] Thus, challenge with exceedingly low doses of TDI (0.3 nl) in sensitized animals is associated with pronounced eosinophilia and a marked, sustained exudative response. This latter TDI-induced plasma exudation response, in contrast to that observed in non-sensitized animals, is inhibited by glucocorticoid pretreatment.[42] In patients with asthma due to TDI exposure, this occupational agent produces a late-phase response that includes a plasma exudation process.[43] TDI challenge-induced late-phase plasma exudation in patients is inhibited by pretreatment with either glucocorticoids or theophylline, whereas chromones are ineffective.[44]

MUCOSAL EXUDATION OF PLASMA IN DISEASE

Plasma exudation in inflammatory airway diseases was first demonstrated by the determination of plasma proteins in sputum samples obtained in asthma and chronic bronchitis.[15] Also, it was observed that steroid treatment significantly reduces the sputum level of different plasma parameters.[15] Interestingly, the inhibition of exudation seems to occur without the concomitant reduction of sputum levels of secretory parameters. Indeed, the latter may increase.[15] The relatively poor antisecretory effect of glucocorticosteroids is further evidence for significant and qualitative differences between airway secretory and exudative processes[45,46] and supports the notion that the (plasma) exudative response may reflect airway inflammation better than other physiological end-organ responses in the airways.

Albumin is usually the only plasma protein that has been analysed in the numerous studies of bronchoalveolar lavage (BAL) fluids obtained from asthmatic lungs. However, for several reasons BAL levels of albumin alone may not always be a useful indicator of the plasma exudation process.

(1) It has now been demonstrated in studies of the acute response to allergen challenge that albumin in BAL fluids may be unchanged, whereas large plasma proteins, such as fibrinogen and α_2-macroglobulin, are significantly increased.[35,36] Such a result could even be expected: the inflammatory stimulus-induced luminal entry of plasma is almost a bulk flux of proteins with little size restriction;[3,6] in contrast to albumin the much larger plasma proteins are normally present in low concentrations and therefore may more clearly reflect an exudative response.

(2) An additional confounding aspect is that BAL fluid contains material, particularly albumin, that has accumulated on the surface of the airways for variable and unknown periods of time.

(3) Furthermore, BAL samples both airway and alveolar surface material, with albumin being the most common protein in both locales. This non-specific sampling may pose a general problem because asthma is more an airway than a pulmonary disease.

(4) The observation, in *in vitro* experiments that albumin may be secreted by airway epithelia[47] may be of little importance. Despite its attractiveness, this mechanism remains to be demonstrated under proper *in vivo* conditions. Indeed, the most potent stimulants of *in vitro* albumin secretion, such as muscarinic cholinomimetics and β_2-adrenoceptor agonists, appear to lack the ability to evoke albumin secretion in the airways *in vivo*, even when given at extremely high doses.[18,28,29]

By analysing several proteins in BAL fluids, Van de Graaf et al.[48] have demonstrated that maintenance treatment with an inhaled steroid significantly reduces plasma exudation in chronic asthma. Svensson et al.,[49] measured a large plasma protein (fibrinogen) and a plasma-derived mediator (bradykinin) in nasal lavage fluids and also demonstrated inhibition of plasma exudation and plasma-derived peptides by topical steroid treatment in seasonal allergic rhinitis during several weeks of natural pollen exposure. Airway glucocorticoids may have potent anti-inflammatory effects but little action on airway mucosal defence processes.[50] Thus a complex treatment regimen involving topical and systemic steroid treatment did not seem to affect airways plasma exudation parameters in patients with rhinovirus inoculation-induced common cold.[51] A lack of antiexudative effects of steroids in both acute common cold and acute exposure to histamine-type mediators or inflammatory chemicals[14,42] can be interpreted as an inability of steroids to prevent plasma exudation when expressed as a defence mechanism during acute events. Similarly, observations in animal airways suggest that steroids may not reduce the accumulation and activation of protective and repair-promoting plasma exudates and leucocytes at sites of airway epithelial damage and shedding, nor do these drugs reduce the prompt restitution of epithelial cell cover that occurs *in vivo* after shedding.[52] The pharmacology of airway steroid drugs may thus involve a balance between potent effects on exudative inflammation (Fig. 14.4) on the one hand and limited interference with exudative defence and repair on the other. Such experimental observations may explain in part why long-term treatment with these anti-inflammatory drugs in asthma and rhinitis may be associated with unchanged or even reduced frequency of airway infections.[13,50]

14 Microvascular–Epithelial Exudation of Plasma

ROLES OF EXUDED PLASMA

The recent observations on mucosal exudation mechanisms in animal and human airways may call for a revision of several generally acknowledged roles of plasma exudation in airway diseases. Because of the swift luminal entry of bulk plasma, increased microvascular permeability in the airways may no longer, without further qualification, be equated with airway oedema. Also, the presence of plasma proteins in the airway lumen may no longer be interpreted as a convincing sign that the epithelium has been damaged. More specifically, just because plasma is exuded into the airway lumen tells us nothing at all about the inward perviousness of the airway mucosa to inhaled molecules. Further, it may be incorrect to assume the occurrence of tracheobronchial plasma exudation merely from measurements of albumin in BAL fluids; and it may be a mistake to conclude that a protein which increased in BAL fluids must come from a non-humoral source just because the level of albumin did not exhibit a significant increase.

Even if extravasation of plasma into the airway tissue does not always produce mucosal oedema, there are several other sequelae to be considered. Extravasated plasma may deposit its proteins as well as its fibrinous and adhesive macromolecules (Fig. 14.5; see Table 14.1) in the lamina propria, in the basement membrane, between the epithelial lining cells and on the mucosal surface. Plasma may thus be an important source of adhesive protein components for the mucosal extracellular matrix. For example, plasma-derived fibrin and fibronectin in the epithelium and on the mucosal surface may govern the traffic and activity of neutrophils and eosinophils in airway inflammation.[53–55] By

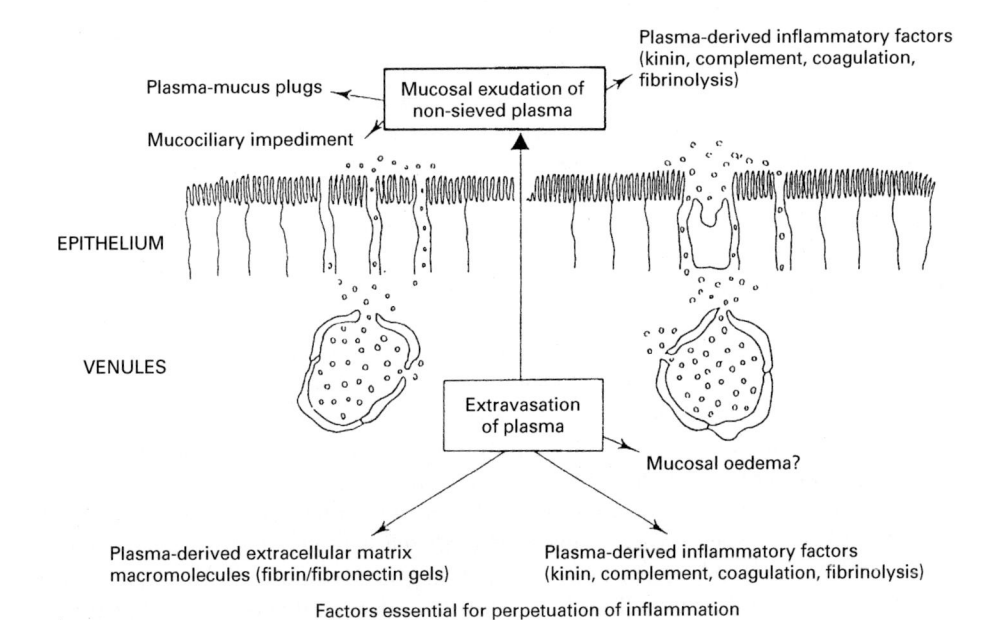

Fig. 14.5 Some of the potential pathophysiological roles of plasma exudation in nasal and tracheobronchial airways. An additional role concerns the usefulness of measuring plasma exudation parameters on the mucosal surface in order to monitor the intensity of the subepithelial inflammatory process.

continuously supplying these proteins, together with plasma-derived growth factors, complement fragments, kinins, fibrinolysis molecules and numerous other peptides (see Table 14.1), the extravasation process in the airways is likely a crucial component of airway inflammation.

By its physical properties and interactions plasma exudates may further impede the patency of the airway passages in several ways[56] (Fig. 14.5). Against a background of mucosal thickening and stagnated exudate–mucous material in the lumen (exuded plasma may accumulate in the airways particularly during the night and early morning hours),[57] an attack of asthma may cause extremely severe obstruction of the bronchi, even during moderate bronchial smooth muscle contraction.[58] The role of exuded plasma may be particularly marked in inflammatory airway diseases, because these conditions may be associated with the development of an exudative hyperresponsiveness of the subepithelial microcirculation.[59]

ON THE INWARD PERVIOUSNESS IN ASTHMA

It is the tight junctions at the apical pole of epithelial lining cells that constitute the main barrier to mucosal penetration of inhaled molecules. A disruption, and more so a shedding, of epithelial cells (as in asthma) would therefore be expected to produce marked increases in the mucosal absorption of a large variety of solutes, including polar molecules that would normally pass through narrow paracellular epithelial routes. However, Elwood et al.,[60] in a pioneering study on tracheobronchial (and alveolar?) absorption permeability in asthma, recorded a mean permeability index that was somewhat lower in hyperreactive asthmatic subjects compared with healthy individuals. A reduced permeability in asthma has also been reported by Halpin et al.,[61] who examined absorption of inhaled Tc-DTPA (diethylenetriaminepentaacetic acid) and who made corrections for the concomitant elimination of the tracer by mucociliary transport. Controlled nasal absorption studies by Greiff et al.,[62] employing the nasal pool device, have now demonstrated than an abnormally tight airway barrier may develop during several weeks of exudative, eosinophilic mucosal inflammation.[20] For individuals suffering from airway diseases, maintenance or improvement of the absorption barrier may be necessary to avoid uncontrolled aggravation of the disease: inhaled noxious molecules would penetrate less readily into highly reactive airway tissue components and the biologically active molecules exuded on to the mucosal surface would not be readily reabsorbed to activate the mucosal target cells that abound in disease.

The presence of an unimpeded airway absorption barrier in asthma evokes misgivings about the hypothesis of airway denudation in this disease. Such doubts also emanate from the demonstration that denudation can be produced artefactually by careful cryosectioning of such allergic airway tissue samples that in wholemount specimens exhibit no denudation.[63] It is possible that exuded bulk plasma may clot and form some kind of surface layer in damaged airways (Fig. 14.6). However, this would not constitute a proper barrier to solutes. Instead, the explanation for the mucosal tightness in desquamative and exudative airway diseases may be found in the epithelial restitution process that follow from shedding of epithelial cells.[20,63–65]

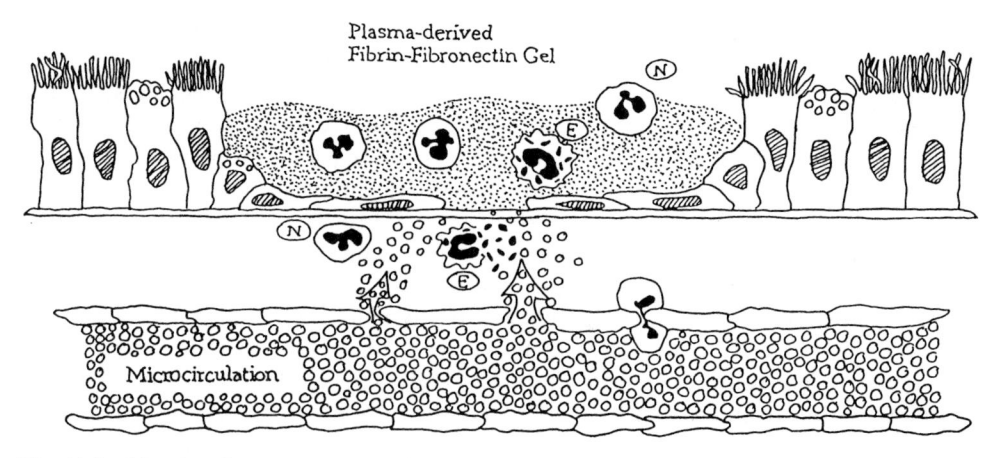

Fig. 14.6 After denudation, epithelial restitution occurs speedily under the provisional cover of a plasma-derived and leucocyte-rich gel. N, neutrophils; E, eosinophils. Drawing by Johan Erjefält.

AIRWAY EPITHELIAL RESTITUTION IN A PLASMA-DERIVED GEL

Early experimental studies have shown that traumatic removal of the tracheal epithelium, with severe mucosal damage and bleeding (not 'shedding-like'), is followed by a delayed and slow process whereby remaining epithelial cells in the margin of the damage eventually flatten and move medially to cover the wounded area.[64] An *in vivo* method similar to the denudation that would occur after shedding has now been developed. The new technique involves the oral insertion of a specially designed steel probe into the guinea-pig trachea, with the probe being gently stroked along the mucosal surface in a non-cartilaginous area of the airway. Without surgery and without causing damage to the basement membrane, or indeed, without causing any bleeding, an 800-μm denuded zone with distinct margins is thus reproducibly created along the mucosal surface above the trachealis muscle.[65] Immediately after epithelial removal using this technique both ciliated and secretory cells from the intact, remaining epithelium dedifferentiate (cilia being 'internalized' and secretory granules being released), flatten out and migrate over the basement membrane. Importantly, migration starts instantaneously and the migration rate is very fast (2–3 μm/min[65] (see Fig. 14.6).

An immediate response to the denudation *in vivo* is plasma exudation.[54] This result is expected because the epithelium appears to have a nitric oxide generation mechanism that tonically suppresses the permeability of the subepithelial microcirculation.[66] The exuded plasma creates a gel that attracts leucocytes. The gel is a dynamic structure that is supplied continuously by exuded plasma.[54] It covers the basement membrane until a new, flat and tight epithelium has been established.[65] Hence, migration of restitution cells occurs in close association with plasma-derived adhesive proteins such as fibronectin and fibrin,[54] as well as other plasma- and leucocyte-derived factors,[9] including such growth factors that may promote repair. The gel provides both a provisional cover and a proper supramembranal milieu for the *in vivo* restitution process (see Fig. 14.6).

If only columnar cells are being shed, the remaining cobbled surface of basal cells immediately change into flattened basal cells that establish cell–cell contact.[67] Indeed, the flattened repair cells that occur promptly after epithelial shedding, whether shedding has produced denudation or loss of only columnar cells, provide reduced junctional lengths per unit mucosal surface area. It is thus suggested that the prompt appearance of poorly differentiated, large, flat cells together with the ensuing epithelial metaplasia may explain observations of reduced airway absorption in disease.[20] In summary, abnormal degrees of epithelial shedding and plasma exudation may occur in airways that nevertheless exhibit maintained or even improved barrier functions.

In vivo animal studies indicate that shedding–repair processes alone produce a series of physiological and cellular responses in the airways. In health, these are functionally important and lead to repair and homeostasis.[68] However, in disease extensive epithelial shedding–restitution processes may cause part of the pathophysiology, cellular pathology and structural changes (Fig. 14.7) that are now regarded as characteristic of asthmatic bronchi.[55] Shedding–restitution may not only evoke a secretory response and a sustained plasma exudation.[65] Epithelial damage and repair will also cause traffic and activation of eosinophils[9] with significantly increased numbers of free eosinophil granules in the mucosa (secondary to eosinophil cytolysis, which appears to be the *in vivo* paradigm of activation of these cells[69] in asthma). There is accumulation and activation of neutrophils.[19] Apart from the epithelial disruption and the occurrence of epithelial restitution cells, there are several changes that may be regarded as remodelling effects.[55] These

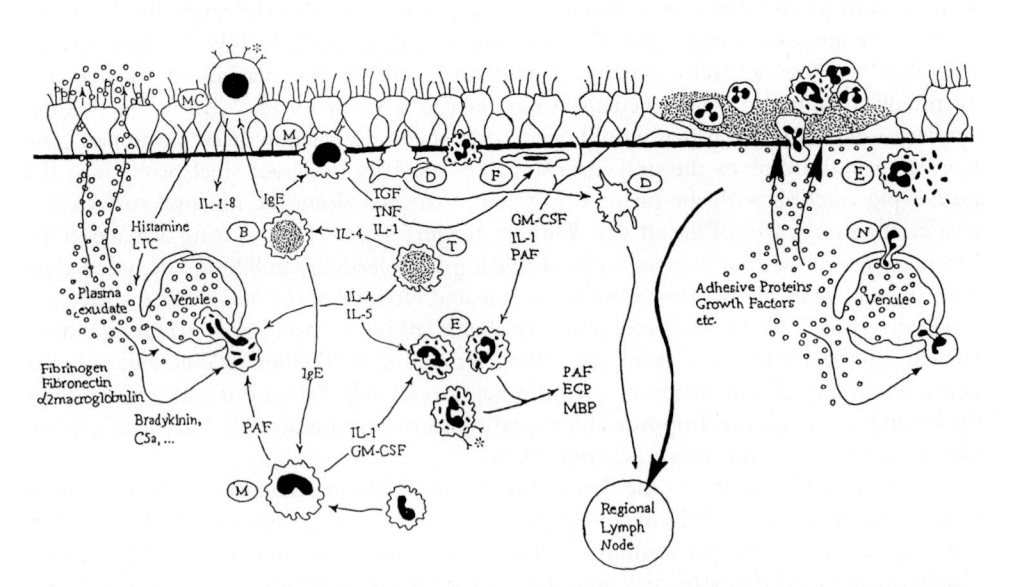

Fig. 14.7 An airway mucosa where exudative inflammation is proceedings either with the epithelium intact (left) or with shedding–restitution as a prominent feature (right). In both kinds of inflammatory conditions, plasma-derived adhesive proteins and other plasma-derived effector molecules contribute significantly to the molecular milieu of the lamina propria, the epithelium and the mucosal surface. B, B-cell; C5a, complement 5a; D, dendritic cell; E, eosinophil; ECP, eosinophil cationic protein; F, fibroblast; GM-CSF, granulocyte–macrophage colony-stimulating factor; IL, interleukin; LTC, leukotriene C; M, macrophage; MBP, major basic protein; MC, mast cell; N, neutrophil; PAF, platelet-activating factor; T, T-cell; TGF, transforming growth factor; TNF, tumour necrosis factor.

include epithelial metaplasia, thickened reticular basement membrane, enlargement of regional lymph nodes, proliferation of fibroblasts/smooth muscle cells and the laying down of plasma-derived adhesive proteins as extracellular matrix. Inferentially, it appears increasingly important to protect the airway epithelium from damage so that shedding–restitution-evoked processes can be reduced to a minimum in asthma (see Fig. 14.7).

REFERENCES

1. Jezierski PV: Zur Pathologie des Asthma Bronchiale. *Deutsch Arch Klin Med* (1906) **85**: 342–347.
2. Sobin SS, Frasher WG, Tremer HM, Madley GG: The microcirculation of the tracheal mucosa. *Angiology* (1963) **14**: 165–170.
3. Persson CGA: Airway epithelium and microcirculation. *Eur Respir Rev* (1994) **4**: 23, 352–362.
4. Persson GCA: *In vivo* veritas. *Thorax* (1996) **51**: 441–443.
5. McDonald M: Endothelial gaps and permeability of venules in rat tracheas exposed to inflammatory stimuli. *Am J Physiol* (1994) **266**: L61–L83.
6. Persson CGA, Svensson C, Greiff L, *et al.*: Editorial. The use of the nose to study the inflammatory response of the respiratory tract. *Thorax* (1992) **47**: 993–1000.
7. Cohnheim J: *Vorlesungen ueber Allgemeine Pathologie I*. Berlin, Hirschwald, 1882, pp 232–367.
8. Hulström D, Svensjö E: Intravital and electron microscopic study of bradykinin-induced vascular permeability changes using FITC-dextran as a tracer. *J Pathol* (1979) **129**: 125–133.
9. Erjefält JS, Sundler F, Persson CGA: Eosinophils, neutrophils and venular gaps in the airway mucosa at epithelial removal–restitution. *Am J Respir Crit Care Med* (1996) **153**: 1666–1674.
10. Feng D, Nagy JA, Hipp J, Dvorak HF, Dvorak AM: Vesiculo-vacuolar organelles and the regulation of venule permeability to macromolecules by vascular permeability factor, histamine, and serotonin. *J Exp Med* (1996) **183**: 1981–1986.
11. Grega GJ, Persson CGA, Svensjö E: Endothelial cell reactions to inflammatory mediators assessed *in vivo* by fluid and solute flux analysis. In Ryan US (ed) *Endothelial Cells*. Boca Raton, FL, CRC, 1988, pp 103–122.
12. Persson CGA, Svensjö E: Vascular responses and their suppression: drugs interfering with venular permeability. In Bonta IL, Bray MA, Parnham MJ (eds) *Handbook of Inflammation*, vol 5. Amsterdam, Elsevier, 1985, pp 61–82.
13. Brattsand R, Selroos O: Current drugs for respiratory diseases. Glucocorticoids. In Page C (ed) *Drugs and the Lung* (1994) 101–220.
14. Greiff L, Andersson M, Svensson C, Alkner U, Persson CGA: Glucocorticoids may not inhibit plasma exudation by direct vascular antipermeability effects in human airways. *Eur Respir J* (1994) **7**: 1120–1124.
15. Persson CGA: Plasma exudation and asthma. *Lung* (1988) **166**: 1–23.
16. Persson CGA, Erjefält I, Andersson P: Leakage of macromolecules from guinea-pig tracheobronchial microcirculation. Effects of allergen, leukotrienes, tachykinins and anti-asthma drugs. *Acta Physiol Scand* (1986) **127**: 95–105.
17. Erjefält I, Persson CGA: Pharmacological control of plasma exudation into tracheobronchial airways. *Am Rev Respir Dis* (1991) **143**: 1008–1014.
18. Svensson C, Greiff L, Andersson M, Alkner U, Grönneberg R, Persson CGA: Antiallergic actions of high topical doses of terbutaline in human nasal airways. *Allergy* (1995) **50**: 884–890.
19. Erjefält JS, Erjefält I, Sundler F, Persson CGA: Epithelial pathway for luminal entry of bulk plasma. *Clin Exp Allergy* (1995) **25**: 187–195.
20. Persson CGA, Andersson M, Greiff L, *et al.*: Airway permeability. *Clin Exp Allergy* (1995) **23**: 807–814.
21. Persson CGA, Erjefält I, Alkner U, *et al.*: Plasma exudation as a first line respiratory mucosal defence. *Clin Exp Allergy* (1991) **21**: 17–24.
22. Persson CGA, Alkner U, Andersson M, Greiff L, Linden M, Svensson C: Histamine challenge-

induced 'lamina propria lavage' and mucosal output of IL-6 in human airways. *Eur Respir J* (1995) **8** (Suppl 19): 125s.
23. James K: Interactions between cytokines and α₂-macroglobulin. *Immunol Today* (1990) **11**: 163–166.
24. Persson CGA, Erjefält I, Gustafsson B, Luts A: Subepithelial hydrostatic pressure may regulate plasma exudation across the mucosa. *Int Arch Allergy Appl Immunol* (1990) **92**: 148–153.
25. Gustafsson BG, Persson CGA: Asymmetrical effects of increases in hydrostatic pressure on macromolecular movement across the airway mucosa. A study in guinea-pig tracheal tube preparation. *Clin Exp Allergy* (1991) **21**: 121–126.
26. Svensson C, Baumgarten CR, Pipkorn U, Alkner U, Persson CGA: Reversibility and reproducibility of histamine-induced plasma leakage in the human nasal airways. *Thorax* (1989) **44**: 13–18.
27. Greiff L, Alkner U, Pipkorn U, Persson CGA: The 'nasal pool' device applies controlled concentrations of solutes on human nasal airway mucosa and samples its surface exudations/secretions. *Clin Exp Allergy* (1990) **20**: 253–259.
28. Erjefält I, Persson CGA: Inflammatory passage of plasma macromolecules into airway tissue and lumen. *Pulmon Pharmacol* (1989) **2**: 93–102.
29. Prescott T, Kaliner MA: Vascular mechanisms in rhinitis. In Busse WW, Holgate ST (eds) *Asthma and Rhinitis*. Oxford, Blackwell Science, 1995, pp 777–790.
30. Greiff L, Svensson C, Andersson M, Persson CGA: Effects of topical capsaicin in seasonal allergic rhinitis. *Thorax* (1995) **50**: 225–229.
31. Bascom R, Kagey-Sobotka A, Proud D: Effect of intranasal capsaicin on symptoms and mediator release. *J Pharmacol Exp Ther* (1991) **259**: 1323–1327.
32. Greiff L, Erjefält I, Wollmer P, *et al.*: Effects of nicotine on the human nasal airway mucosa. *Thorax* (1993) **48**: 651–655.
33. Haldursdottir H, Greiff L, Wollmer P, *et al.*: Inhaled histamine, but not inhaled capsaicin, evokes acute exudation of α₂-macroglobulin into human bronchi. *Am J Respir Crit Care Med* (1996) **153**: A290.
34. Persson CGA, Erjefält I: Inflammatory leakage of macromolecules from the vascular compartment into the tracheal lumen. *Acta Physiol Scand* (1986) **126**: 615–616.
35. Minnicozzi M, Durán WN, Gleich GJ, Egan RW: Eosinophil granule proteins increase microvascular macromolecular transport in the hamster cheek pouch. *J Immunol* (1994) **153**: 2664–2669.
36. Salomonsson P, Grönneberg R, Gilljam H, *et al.*: Bronchial exudation of bulk plasma at allergen challenge in allergic asthma. *Am Rev Respir Dis* (1992) **146**: 1535–1542.
37. Svensson C, Grönneberg R, Andersson M, *et al.*: Allergen challenge-induced entry of alpha-2-macroglobulin and tryptase into human nasal and bronchial airways. *J Allergy Clin Immunol* (1995) **96**: 239–246.
38. Pipkorn U, Proud D, Schleimer RP, *et al.*: Effects of short term systemic glucocorticoid treatment on human nasal mediator release after antigen challenge. *J Clin Invest* (1987) **80**: 957–961.
39. Erjefält I, Greiff L, Alkner U, Persson CGA: Allergen-induced biphasic plasma exudation responses in guinea-pig large airways. *Am Rev Respir Dis* (1993) **148**: 695–701.
40. Persson CGA: Airway microcirculation, epithelium and glucocorticoids. In Schleimer RP, Busse W, O'Byrne P (eds) *Topical Glucocorticoids in Asthma: Mechanisms and Clinical Actions*. New York, Dekker, 1996, pp 167–201.
41. Persson CGA, Gustafsson B, Luts A, Sundler F, Erjefält I: Toluene diisocyanate produces an increase in airway tone that outlasts the inflammatory exudation phase. *Clin Exp Allergy* (1991) **21**: 715–724.
42. Erjefält I, Persson CGA: Increased sensitivity to toluene diisocyanate (TDI) in airways previously exposed to low doses of TDI. *Clin Exp Allergy* (1992) **22**: 854–862.
43. Fabbri LM, Mapp C: Bronchial hyperresponsiveness, airway inflammation and occupational asthma induced by toluene diisocyanate. *Clin Exp Allergy* (1991) **21**: 42–47.
44. Boschetto P, Fabbri LM, Zocca E, *et al.*: Prednisone inhibits late asthmatic reactions and airway inflammation induced by toluene diisocyanate in sensitised subjects. *J Allergy Clin Immunol* (1987) **80**: 261–267.

45. Persson CGA: Airway mucosal exudation of plasma. In Takishima T, Shimura S (eds) *Airway Secretion. Physiological Bases for the Control of Mucous Hypersecretion*. New York, Dekker, 1994, pp 451–468.
46. Persson CGA: Permeability changes in obstructive airway diseases. In Sluiter HJ, Van der Lende R (eds) *Bronchitis IV*. Assen, Van Gorcum, 1989, pp 236–248.
47. Webber SE, Widdicombe JG: The transport of albumin across the ferret *in vitro* whole trachea. *J Physiol* (1989) **408**: 457–472.
48. Van de Graaf EA, Out TA, Roos CM, Jansen HM: Respiratory membrane permeability and bronchial hyperreactivity in patients with stable asthma. Effects of therapy with inhaled steroids. *Am Rev Respir Dis* (1991) **143**: 362–368.
49. Svensson C, Klementsson H, Andersson M, Pipkorn U, Alkner U, Persson CGA: Glucocorticoid-reduced attenuation of mucosal exudation of bradykinins and fibrinogen in seasonal allergic rhinitis. *Allergy* (1994) **49**: 177–183.
50. Persson CGA, Pipkorn U: Glucocorticoids. In Waksman BH (ed) *Fifty Years' Progress in Allergy*. Basel, Karger, 1990, pp 264–277.
51. Farr BM, Gwaltney JM Jr, Hendley JO: A randomized controlled trial of glucocorticoid prophylaxis against experimental rhinovirus infection. *J Infect Dis* (1990) **162**: 1173–1177.
52. Erjefält JS, Erjefält I, Sundler F, Persson CGA: Effects of topical budesonide on epithelial restitution *in vivo* in guinea-pig trachea. *Thorax* (1995) **50**: 785–792.
53. Tang L, Eaton JW: Fibrin(ogen) mediates acute inflammatory responses to biomaterials. *J Exp Med* (1993) **178**: 2147–2156.
54. Erjefält JS, Erjefält I, Sundler F, Persson CGA: Microcirculation-derived factors in airway epithelial repair *in vivo*. *Microvasc Res* (1994) **48**: 161–178.
55. Persson CGA, Erjefält JS, Erjefält I, Korsgren M, Nilsson M, Sundler F: Epithelial shedding–restitution as a causative process in airway inflammation. *J Clin Exp Allergy* (1996) **26**: 746–755.
56. Persson CGA: Role of plasma exudation in asthmatic airways. *Lancet* (1986) **ii**: 1126–1129.
57. Svensson C, Andersson M, Greiff L, Persson CGA: Day–night differences in mucosal plasma proteins in common cold. *Acta Otolaryngol* (1996) **166**: 85–90.
58. Hutt G, Wick H: Bronchial-lumen and Atemwiderstand. *Z Aerosol Forsch Ther* (1956) **5**: 131–140.
59. Svensson C, Andersson M. Greiff L, Alkner U, Persson CGA: Exudative hyperresponsiveness of the airway microcirculation in seasonal allergic rhinitis. *Clin Exp Allergy* (1995) **25**: 942–950.
60. Elwood RK, Kennedy S, Belzberg A, Hogg JC, Paré PD: Respiratory mucosal permeability in asthma. *Am Rev Respir Dis* (1983) **129**: 523–527.
61. Halpin DMG, Currie D, Jones B, Leigh TR, Evans TW: Permeability of bronchial mucosa to ^{113m}In-DTPA in asthma and the effects of salmeterol. *Eur Respir J* (1993) **6**: 512s.
62. Greiff L, Wollmer P, Svensson C, Andersson M, Persson CGA: Effect of seasonal allergic rhinitis on airway mucosal absorption of chromium-51-labelled EDTA. *Thorax* (1993) **48**: 648–650.
63. Erjefält JS, Korsgren M, Nilsson MC, Sundler F, Persson CGP: Prompt epithelial damage- and restitution-processes in allergen challenged guinea-pig trachea. *J Clin Exp Allergy* (1997 **27**: 1458–1470.
64. Wilhelm DL: Regeneration of tracheal epithelium. *J Pathol Bacteriol* (1953) **65**: 543–550.
65. Erjefält JS, Erjefält I, Sundler F, Persson CGA: *In vivo* restitution of airway epithelium. *Cell Tissue Res* (1995) **281**: 305–316.
66. Erjefält JS, Erjefält I, Sundler F, Persson CGA: Mucosal nitric oxide may tonically suppress airways plasma exudation. *Am J Respir Crit Care Med* (1994) **150**: 227–232.
67. Erjefält JS, Greiff L, Sundler F, Persson CGA: Basal cells promptly flatten out at detachment of the columnar epithelium in human and guinea-pig airways. *Eur Respir J* (1995) **8** (Suppl 19): 207s.
68. Persson CGA, Erjefält JS: Airway epithelial restitution following shedding and denudation. In Crystal RG, West JB, Weibel ER, Barnes PJ (eds) *The Lung: Scientific Foundations*, 2nd edn. New York, Raven, 1996, pp 2611–2627.
69. Persson CGA, Erjefält JS : Eosinophilcytolysis and free granule: an *in vivo* paradigm for cell activation and drug development. *Trends Pharmacol Sci* (1997) **18**: 117–123.

15

Prostaglandins and Thromboxane

PAUL M. O'BYRNE

INTRODUCTION

Asthma is a disease described by the presence of characteristic symptoms, the physiological abnormalities of variable airflow obstruction and airway hyperresponsiveness, and by persisting airway inflammation, which is identified by the presence of activated inflammatory cells,[1,2] the most obvious being eosinophils, mast cells and lymphocytes, and structural changes likely caused by the persisting cellular inflammation.[3]

The severity of asthma, as measured by symptoms and treatment requirements, and the physiological abnormalities, particularly airway hyperresponsiveness, correlate with the numbers of inflammatory cells in the airways.[4,5] The mechanisms by which airway inflammation causes asthma are still largely unknown. It is likely, however, that mediators released from effector cells in the airways cause the influx of inflammatory cells and subsequent activation of these cells. Further release of mediators from activated cells causes the bronchoconstriction, airway hyperresponsiveness and other manifestations of inflammation, such as airway oedema and excess airway secretions.

It has been difficult to convincingly implicate any mediator in the pathogenesis of asthma. The investigation of the potential role of a mediator has depended on three types of evidence. Generally, the first type of evidence is obtained when the mediator is synthesized and delivered by inhalation to asthmatic subjects to determine whether the inhaled mediator mimics a component of the asthmatic response. Secondly, attempts are made to measure the mediator (or its metabolite) in a biological fluid following induction of an asthmatic response. The best evidence is obtained when selective mediator receptor antagonists or synthetase inhibitors are developed, which can be studied to determine whether they inhibit some component of an asthmatic response. Many mediators have been proposed as playing a role in the pathogenesis of asthma; however, problems occur that have (until recently) precluded convincing evidence being obtained for any mediator. These were difficulties in measuring the mediator or its metabolite at its site of action in the airways; the lack of potent, specific mediator antagonists; and the absence of an animal model of asthma.

The purpose of this chapter is to examine the evidence that one group of mediators, prostaglandins and thromboxane, are involved in the pathogenesis of asthmatic responses. In addition, evidence is considered which suggests that some members of this group of mediators provide a protective function in asthmatic airways.

ARACHIDONIC ACID METABOLISM

Arachidonic acid (5,8,11,14-eicosatetraenoic acid) is an essential polyunsaturated fatty acid, normally formed from dietary linoleic acid. Arachidonic acid is stored bound to phospholipids in several different membranes and organelles in the cell. In the resting cell, the levels of free arachidonic acid are insignificant. The vast majority of arachidonic acid is found esterified in glycerolipids. When cells are activated, arachidonic acid is liberated and used as substrate for cellular synthesis of different oxygenated metabolites. The release of arachidonic acid from cell membrane phospholipids is through the action of a family of phospholipases, such as phospholipase A_2 or phospholipase C, and can result in the production of a wide variety of mediators, which may be relevant in the pathogenesis of asthma (Fig. 15.1). These lipid mediators have traditionally been considered in two

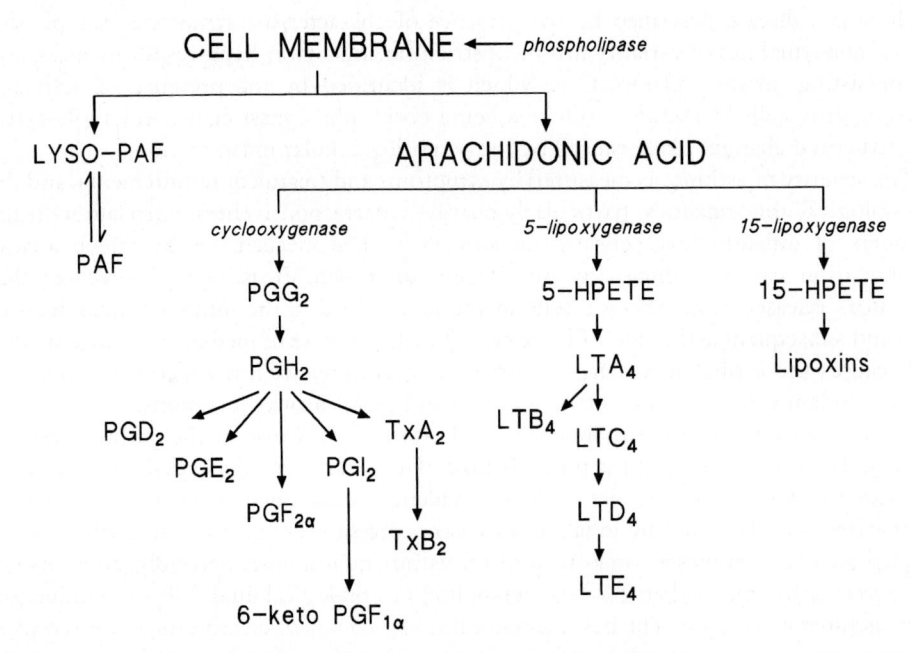

Fig. 15.1 The spectrum of eicosanoids produced as a result of arachidonic acid metabolism. HPETE, hydroperoxyeicosatetraenoic acid; PAF, platelet-activating factor.

classes: mediators that result from the action of the enzyme cyclooxygenase on arachidonic acid, i.e. prostaglandins (PG) or thromboxane (Tx); and mediators that result from the action of the enzyme 5-lipoxygenase on arachidonic acid, i.e. the leukotrienes (LT). More recently, however, other products have been identified that result from the activity of different enzymes, such as 12- and 15-lipoxygenase. In fact, the main arachidonic acid metabolite in the human lung is the 15-lipoxygenase product 15-hydroxyeicosatetraenoic acid (15-HETE).[6] Furthermore, biologically active eicosanoids may be found via complex pathways involving interactions between different lipoxygenases. Lipoxins are such lipoxygenase interaction products, formed by the sequential actions of 15- and 5-lipoxygenases or 12- and 15-lipoxygenases on arachidonic acid.[7] Another group of enzymatic reactions that lead to formation of eicosanoids is the monooxygenase pathway, i.e. cytochrome P450-catalysed reactions, which can insert oxygen into many different positions of arachidonic acid.[8]

CYCLOOXYGENASE PRODUCTS

The oxidative metabolism of arachidonic acid by the enzyme cyclooxygenase (COX) produces the cyclic endoperoxides PGG_2 and PGH_2. The subsequent action of prostaglandin isomerases produces either PGD_2 or PGE_2, reductive cleavage produces $PGF_{2\alpha}$, while the action of one of two terminal synthetases on the endoperoxide produces PGI_2 and TxA_2. COX appears to be present in most cells; however, the COX metabolites released from a particular cell are quite specific (e.g. TxA_2 from platelets and PGI_2 from endothelial cells). This suggests that terminal synthetases are cell specific. More recently, it has been identified that hydroperoxides generated during lung injury may give rise to a series of eicosanoids, the isoprostanes,[9] which are compounds structurally very similar to conventional prostaglandins. The only difference is that the x side-chain of the molecules connects to the prostaglandin cyclopentane ring in *trans* rather than *cis* configuration. Isoprostanes appear to exert their biological activities through the same receptors as conventional prostanoids.

The level of free arachidonic acid has, until recently, been considered as the rate-limiting factor in the synthesis of prostaglandins. However, cytokines and growth factors may induce the formation of prostaglandin synthesis in cells and tissues that are devoid of COX activity in the basal state. This is due to induction of an isoenzyme, COX-2, which differs structurally and functionally from the constitutive enzyme (COX-1).[10] Moreover, COX-2 appears to be expressed in cells participating in inflammatory reactions.

PROSTAGLANDIN AND THROMBOXANE RECEPTORS

The classification of prostanoid receptors has been difficult because the selectivity of the agonists is very poor. However, each of the prostanoids has a receptor where it is more potent in comparison with the others. These receptors are termed P receptors, with a prefix to indicate which prostanoid is the most potent agonist.[11] This gives DP, EP, FP, IP and TP receptors for PGD_2, PGE_2, $PGF_{2\alpha}$, PGI_2 and TxA_2, respectively. Among the

EP receptors, there are at least four subtypes (EP$_1$, EP$_2$, EP$_3$ and EP$_4$). Most of the prostanoid receptors have been cloned and found to belong to the family of G protein-coupled receptors with seven transmembrane domains. The bronchoconstrictive actions of TxA$_2$, PGD$_2$ and PGF$_{2\alpha}$ are mediated by the TP receptor. The DP receptor is preferentially activated by PGD$_2$ and mediates vasodilation. EP receptors mediate the relaxation of airway smooth muscle and inhibition of inflammatory cells in response to PGE$_2$; EP$_2$ receptors are probably most important in this context. Several of the receptors, namely DP, EP$_2$, EP$_4$ and IP receptors, couple to adenylyl cyclases via G$_S$ proteins, resulting in increased intracellular levels of cyclic AMP (cAMP). It is known that EP$_1$, EP$_3$, FP and TP receptors use phospholipase C pathways and G$_q$ protein transductions to induce inositol trisphosphate generation and increased levels of intracellular calcium.

ROLE OF COX PRODUCTS IN ASTHMA

All of the COX products of arachidonic acid metabolism have been synthesized and, with the exception of thromboxane, are readily available for study. Thromboxane has an exceedingly short half-life (about 30 s) and studies have been limited to a few, very limited, experimental preparations, none of them in the airways. Fortunately, several stable thromboxane mimetics have been synthesized. These are endoperoxides that activate the thromboxane receptor and mimic the biological actions of thromboxane. In addition, while a wide variety of COX inhibitors exist and have been extensively studied, with the exception of thromboxane no selective synthetase inhibitors or receptor antagonists are available for the other prostaglandins.

The prostaglandins are most easily considered in two classes in order to evaluate their possible role in asthma. These are the stimulatory prostaglandins, such as PGD$_2$, PGF$_{2\alpha}$ and TxA$_2$, which are potent bronchoconstrictors (Fig. 15.2), and the inhibitory

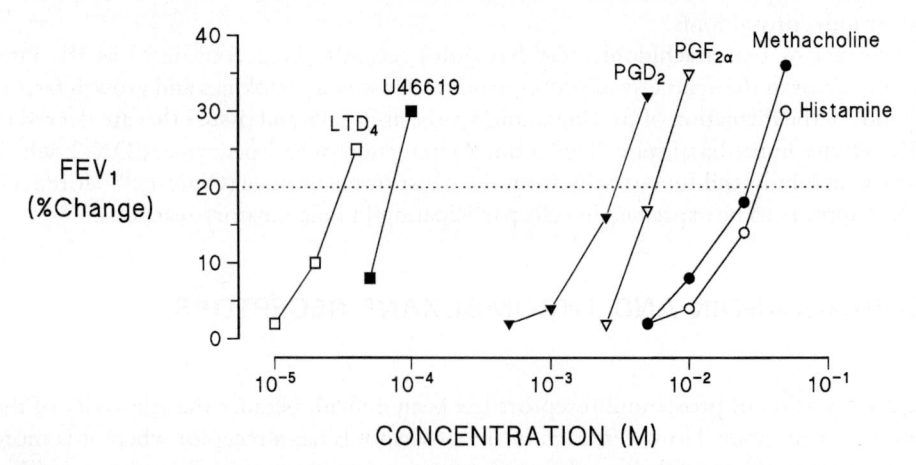

Fig. 15.2 Concentration–response curves to histamine, methacholine and a variety of eicosanoids in one mild, stable asthmatic subject. Leukotriene D$_4$ (LTD$_4$) and the thromboxane mimetic U46619 are the most potent bronchoconstrictors studied to date. FEV$_1$, forced expiratory volume in 1 s.

prostaglandins, such as PGE_2, which can reduce bronchoconstrictor responses and attenuate the release of acetylcholine from airway nerves. Prostaglandins have a variety of effects on airway function in asthma. Evidence has been obtained in both animal models of airway hyperresponsiveness and in human subjects with asthma that COX metabolites are involved in causing bronchoconstriction and also airway hyperresponsiveness after inhalation of stimuli, such as allergens. There is, however, little convincing evidence that COX metabolites are important in causing the ongoing, persisting airway hyperresponsiveness that is characteristic of asthma.

COX products have been implicated in the pathogenesis of allergen-induced early asthmatic as well as late asthmatic responses. This has been done by pretreating subjects with several different COX inhibitors. For example, Joubert et al.[12] reported that pretreatment with indomethacin inhibited the late response, without having a major effect on the early response. In another study, however, pretreatment with indomethacin (100 mg/day) did not influence either the early or late asthmatic responses[13] and therefore could not confirm the original observations. However, indomethacin did significantly inhibit the development of allergen-induced airway hyperresponsiveness, which suggests that a COX product is involved in the pathogenesis of this response. The most likely candidates are the stimulatory prostaglandins PGD_2, $PGF_{2\alpha}$ or TxA_2.

STIMULATORY PROSTAGLANDINS AND THROMBOXANE

PGD₂

PGD_2 is known to be released from stimulated dispersed human lung cells *in vitro*[14] and from the airways of allergic human subjects that have been stimulated by allergen.[15] PGD_2 is a bronchoconstrictor of human airways,[16] and is more potent than $PGF_{2\alpha}$ when inhaled by human subjects. PGD_2 causes bronchoconstriction, in part directly through stimulation of the TP_1 receptors[17] as well as indirectly through presynaptically stimulating acetylcholine release from airway cholinergic nerves.[18] It is not known whether the cholinergic component occurs through a cholinergic reflex or through a direct presynaptic effect causing the release of acetylcholine. Subthreshold contractile concentrations of PGD_2 have been demonstrated to increase airway responsiveness to inhaled histamine and methacholine in asthmatic subjects.[19] Thus, PGD_2 released in human airways after allergen inhalation has the potential to both cause acute bronchoconstriction and increase airway hyperresponsiveness to other constrictor mediators. However, specific receptor antagonists for PGD_2 or inhibitors of its production are not available to allow a precise evaluation of the importance of this COX metabolite in causing asthmatic responses.

PCF₂ₐ

$PGF_{2\alpha}$ also has the potential for being important in causing bronchoconstriction and airway hyperresponsiveness after inhaled allergen in human subjects. This is because it is released from human lungs, is a potent bronchoconstrictor in asthmatic airways[20] and inhaled subthreshold constrictor concentrations can increase airway responsiveness in dogs[21] and human subjects.[22] As with PGD_2, there are no selective $PGF_{2\alpha}$-receptor antagonists available that would allow identification of the importance of these metabolites in causing these responses. Indeed, because all contractile prostaglandins act via a single TP_1 receptor,[11] differentiation of the relative importance of the contractile prostaglandins in causing asthmatic responses may prove to be extremely difficult.

TxA₂

TxA_2 is a potent constrictor of smooth muscle. It was originally described as being released from platelets,[23] but is now known to be released from other cells, including macrophages and neutrophils.[24] As its biological half-life is very short, implicating TxA_2 in disease processes has depended on measurement of its more stable metabolite TxB_2 in biological fluids; on the use of the stable endoperoxides U44069 or U46619, which mimic most of the biological effects of TxA_2 and have been used as TxA_2 analogues; and on the use of inhibitors of TxA_2 synthesis and antagonists of the TxA_2 receptor. Using these techniques, TxA_2 has been implicated in the pathogenesis of airway hyperresponsiveness in dogs[25] and primates;[26] the late cutaneous response to intradermal allergen in humans;[27] the immediate response to inhaled allergen in dogs;[28] the late asthmatic response after inhaled allergen in humans;[29] and airway hyperresponsiveness in asthmatic subjects.[30] More information is available for a possible role of TxA_2 in asthma than for any of the other prostanoids. This is because of the availability of several potent and selective thromboxane synthetase inhibitors, as well as antagonists of the TP_1 receptor.

TxA_2 is a very potent bronchoconstrictor in human subjects. It is more potent than either PGD_2 or $PGF_{2\alpha}$, and is, next to the cysteinyl leukotrienes, the most potent bronchoconstrictor yet studied in humans. For example, inhaled U46619 has been delivered to asthmatic airways and shown to be 178 times more potent, on a molar basis, than inhaled methacholine.[31] Also, inhaled U46619 causes very transient (<1 h) airway hyperresponsiveness.[31] Lastly, as with TxA_2 in canine airways,[32] it causes bronchoconstriction in asthmatic airways, in part by stimulating acetylcholine release by presynaptic stimulation of cholinergic nerves[33] (Fig. 15.3).

The effects of TxA_2-synthetase inhibitors and TP_1-receptor antagonists on airway responsiveness have been studied by several investigators. In one study, the TxA_2-synthetase inhibitor OKY 046, administered orally, reduced acetylcholine airway hyperresponsiveness in stable asthmatic subjects (although these studies were uncontrolled), while a lipoxygenase inhibitor had no effect in these subjects.[30] However, several other studies that have used both TxA_2-synthetase inhibitors[34] or TP_1 antagonists[35] have not confirmed this observation. Also, a TP_1 antagonist demonstrated no activity in protecting against exercise-induced bronchoconstriction.[36] The effect of pretreatment with the thromboxane synthetase inhibitor CGS 13080 on airway responses after allergen challenge has also been reported.[37] CGS 13080 slightly but significantly inhibited the magnitude of the early but not the late responses after inhaled allergen. In addition, there was no effect on airway hyperresponsiveness to inhaled histamine measured 24 h after allergen. These studies, taken together, suggest that thromboxane may be released following allergen challenge and be partly responsible for the early asthmatic response, but is not important in causing the late response or airway hyperresponsiveness following allergen inhalation.

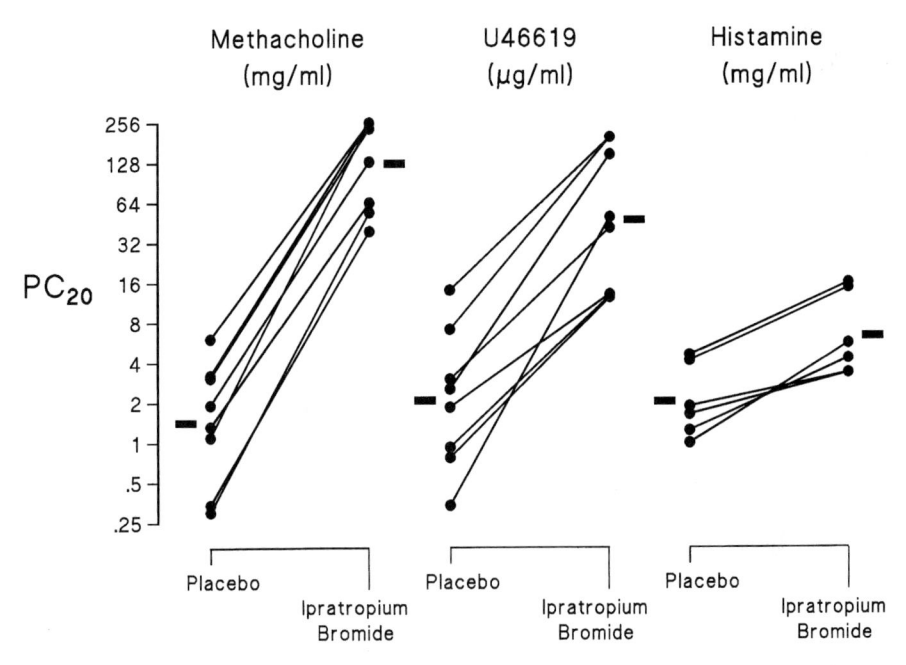

Fig. 15.3 The effect of pretreatment with the cholinergic antagonist ipratropium bromide on airway responses to inhaled methacholine, U46619 and histamine in mild, stable asthmatic subjects. The response to the agonist is expressed as the provocative concentration causing a 20% fall in the forced expiratory volume in 1 s (PC_{20}). Ipratropium bromide increased the PC_{20} for all agonists, but significantly more for the cholinergic agonist methacholine and for U46619 than for histamine. This suggests that part of the bronchoconstriction caused by U46619 is occurring through cholinergic mechanisms. Reproduced from ref. 33, with permission.

INHIBITORY PROSTAGLANDINS

The differentiation of the prostaglandins into stimulatory and inhibitory classes is somewhat inappropriate. For example, both PGE_2 and $PGF_{2\alpha}$ can have different effects on the airways depending on the time after inhalation at which the response is measured.[38,39] However, the main actions of PGE_2 and PGI_2 on airway function are to relax airway smooth muscle and to antagonize the contractile responses of other bronchoconstrictor agonists. In addition, PGE_2 is extremely potent at inhibiting the release of acetylcholine from airway cholinergic nerves.[40] This effect is thought to occur through stimulation of presynaptic receptors.

Histamine tachyphylaxis

The evidence that inhibitory prostaglandins play a role in modulating the contractile responses of agonists such as histamine and acetylcholine in asthmatic subjects comes from studies that have demonstrated that tachyphylaxis occurs following repeated challenges with inhaled histamine, when challenges are separated by up to 6 h.[41] Histamine tachyphylaxis is prevented by pretreatment with indomethacin,[41] which suggests that tachyphylaxis occurs through release of inhibitory prostaglandins in the airways. Also, pretreatment of asthmatic subjects with oral PGE_1, in doses that do not cause bronchodilation, reduces airway responsiveness to both histamine and methacholine.[42] In addition, histamine tachyphylaxis in asthmatic subjects is blocked by pretreatment with the H_2-receptor antagonist cimetidine in asthmatic subjects,[43] which suggests that H_2-receptor stimulation is involved with the development of histamine tachyphylaxis. Contraction of asthmatic airways by histamine also reduces airway responsiveness to acetylcholine[44] and exercise.[45] This lack of specificity suggests that either receptor downregulation or an alteration of the contractile properties of airway smooth muscle is occurring.

Exercise refractoriness

In most patients with exercise-induced bronchoconstriction, the episode of bronchoconstriction is followed by a period during which repeated exercise causes less bronchoconstriction (Fig. 15.4). This has been called the refractory period after exercise.[46] Exercise refractoriness is also attenuated by indomethacin pretreatment,[47] suggesting an important role for inhibitory prostaglandins in this response (Fig. 15.4). Also, the studies on histamine tachyphylaxis suggested that histamine released following exercise not only causes exercise-induced bronchoconstriction but also provides partial protection against subsequent exercise-induced bronchoconstriction, through PGE_2 released by stimulation of histamine H_2 receptors. However, this hypothesis is incorrect. This is because the marked attenuation of exercise-induced bronchoconstriction by pretreatment with

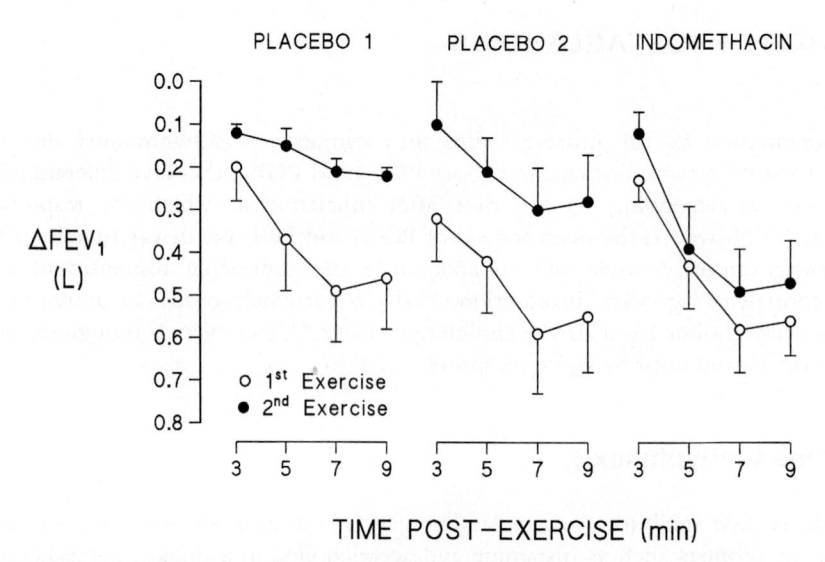

Fig. 15.4 The effect of pretreatment with indomethacin or placebo on exercise-induced bronchoconstriction and exercise refractoriness in asthmatic subjects. Indomethacin did not influence the magnitude of exercise-induced bronchoconstriction, but did attenuate exercise refractoriness. FEV_1, forced expiratory volume in 1 s. Reproduced from ref. 48, with permission.

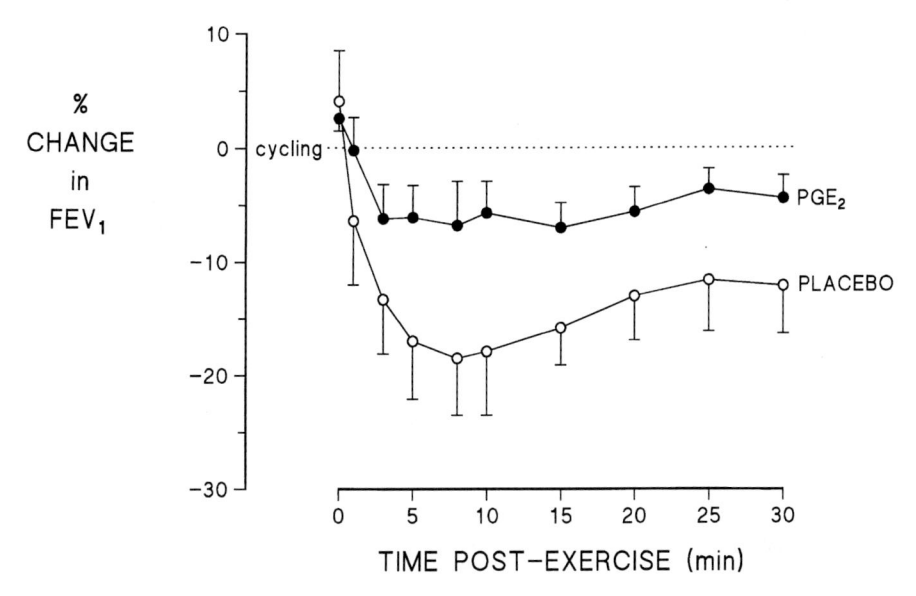

Fig. 15.5 The effect of pretreatment with inhaled PGE$_2$ or placebo on exercise-induced bronchoconstriction in asthmatic subjects. PGE$_2$ attenuated the magnitude of exercise-induced bronchoconstriction. FEV$_1$, forced expiratory volume in 1 s. Reproduced from ref. 51, with permission.

LTD$_4$-receptor antagonists[46] indicates that LTD$_4$, rather than histamine, is the main mediator responsible for exercise-induced bronchoconstriction. Also, exercise refractoriness is not prevented by pretreatment with the H$_2$-receptor antagonists cimetidine or ranitidine, which effectively prevent histamine tachyphylaxis.[47] This raised the possibility that exercise refractoriness is caused by leukotriene-stimulated inhibitory prostaglandin release. This was supported by a study which demonstrated that there is interdependence between the COX and lipoxygenase pathways of arachidonate metabolism in causing exercise-induced bronchoconstriction and refractoriness in asthmatic subjects.[48] The study demonstrated that exercise refractoriness and LTD$_4$ tachyphylaxis exists in the same subjects and that the magnitude of the protection afforded by exercise correlates with that afforded by LTD$_4$; that cross-refractoriness exists between exercise and LTD$_4$; and that all of these effects are attenuated by COX inhibition. In addition, pretreatment with inhaled PGE$_2$ has been demonstrated to markedly attenuate both exercise[49] (Fig. 15.5) and allergen-induced bronchoconstriction.[50–52] Thus it appears that the mechanism of exercise-induced bronchoconstriction and refractoriness is cysteinyl leukotriene release causing bronchoconstriction and leukotriene-induced PGE$_2$ release causing exercise refractoriness.

CONCLUSIONS

Despite more than 30 years of research on the release, metabolism and clinical relevance of prostaglandins and thromboxane in lung disease, no definitive role has been identified for these mediators in the pathogenesis of persisting asthma. However, it is likely that PGD_2 and TxA_2 are involved in causing acute bronchoconstriction after stimuli such as inhaled allergen in asthmatic patients. Also, there is evidence indicating that inhibitory prostaglandins can be released by asthmatic airways, which reduces bronchoconstrictor responses to stimuli such as exercise, leukotrienes and histamine. However, it is unlikely that prostaglandins are directly involved in causing the influx and maturation of the effector inflammatory cells, or involved in causing the ongoing airway hyperresponsiveness in asthma.

REFERENCES

1. Djukanovic R, Roche WR, Wilson JW, et al.: State of the art: mucosal inflammation in asthma. *Am Rev Respir Dis* (1990); **142**: 434–457.
2. Woolley KL, Adelroth E, Woolley MJ, Ellis R, Jordana M, O'Byrne PM: Granulocyte–macrophage colony-stimulating factor, eosinophils and eosinophil cationic protein in mild asthmatics and non-asthmatics. *Eur Respir J* (1994), **7**: 1576–1584.
3. Dunnill MS, Massarell GR, Anderson JA: A comparison of the quantitive anatomy of the bronchi in normal subjects, in status asthmaticus, in chronic bronchitis and in emphysema. *Thorax* (1969) **24**: 176–179.
4. Kirby JG, Hargreave FE, Gleich GJ, O'Byrne PM: Bronchoalveolar cell profiles of asthmatic and nonasthmatic subjects. *Am Rev Respir Dis* (1987) **136**: 379–383.
5. Bousquet J, Chanez P, Lacoste JY, et al.: Eosinophilic inflammation in asthma. *N Engl J Med* (1990) **323**: 1033–1039.
6. Hamberg M, Hedqvist P, Rådegran K: Identification of 16-hydroxy-5,8,11,13-eicosatetraenoic acid (15-HETE) as the major metabolite of arachidonic acid in human lung. *Acta Physiol Scand* (1980) **110**: 219–221.
7. Dahlen S-E, Serhan CN: Lipoxins: bioactive lipoxygenase interactive products. In Crooke ST, Wong A (eds) *Lipoxygenases and Their Products*. San Diego, Academic Press, 1992, pp 235–275.
8. Fitzpatrick FA, Murphy RA: Cytochrome P-450 metabolism of arachidonic acid: formation and biological actions of 'epoxygenase'-derived eicosanoids. *Pharmacol Rev* (1989) **40**: 229–241.
9. Morrow JD, Awad JA, Boss HJ, Blair IA, Jackson Roberts J: Non-cycloxygenase-derived prostanoids (F_2-isoprostanes) are formed *in situ* on phospholipids. *Proc Natl Acad Sci USA* (1992) **89**: 10 721–10 725.
10. Masferrer JL, Seibert K, Zweifel B, Needleman P: Endogenous glucocorticosteroids regulate an inducible cyclooxygenase enzyme. *Proc Natl Acad Sci USA* (1992) **89**: 3917–3921.
11. Coleman RA, Smith WL, Narumiya S: VIII International Union of Pharmacology classification of prostanoid receptors: properties, distribution, and structure of the receptors and their subtypes. *Pharmacol Rev* (1994) **46**: 205–229.
12. Joubert JR, Shephard E, Mouton W, Van Zyk L, Viljoen I: Non-steroid anti-inflammatory drugs in asthma: dangerous or useful therapy? *Allergy* (1985) **40**: 202–207.
13. Kirby JG, Hargreave FE, Cockcroft DW, O'Byrne PM: The effect of indomethacin on allergen-induced asthmatic responses. *J Appl Physiol* (1989) **66**: 578–583.
14. Yen SS, Mathe AA, Dugan JJ: Release of prostaglandins from healthy and sensitized guinea-pig lung and trachea by histamine. *Prostaglandins* (1976) **11**: 227–239.
15. Murray JJ, Tonnel AB, Brash AR, et al.: Release of prostaglandin D_2 into human airways during acute antigen challenge. *N Engl J Med* (1986) **315**: 800–804.
16. Hardy CC, Robinson C, Tattersfield AE, Holgate ST: The bronchoconstrictor effect of inhaled prostaglandin D_2 in normal and asthmatic men. *N Engl J Med* (1984) **311**: 209–213.
17. Johnston SL, Freezer NJ, Ritter W, O'Toole S, Howarth PH: Prostaglandin D_2-induced bronchoconstriction is mediated only in part by the thromboxane prostanoid receptor. *Eur Respir J* (1995) **8**: 411–415.

18. Beasley R, Varley J, Robinson C, Holgate ST: Cholinergic-mediated bronchoconstriction induced by prostaglandin D$_2$, its initial metabolite 9$_a$,11$_B$-PGF$_{2a}$, and PGE$_{2a}$ in asthma. *Am Rev Respir Dis* (1987) **136**: 1140–1144.
19. Fuller RW, Dixon CMS, Dollery CT, Barnes PJ: Prostaglandin D$_2$ potentiates airway responsiveness to histamine and methacholine. *Am Rev Respir Dis* (1986) **133**: 252–254.
20. Thomson NC, Roberts R, Bandouvakis J, Newball H, Hargreave FE: Comparison of bronchial responses to prostaglandin F2a and methacholine. *J Allergy Clin Immunol* (1981) **68**: 392–398.
21. O'Byrne PM, Aizawa H, Bethel RA, Chung KF, Nadel JA, Holtzman MJ: Prostaglandin F2a increases airway responsiveness of pulmonary airways in dogs. *Prostaglandins* (1984) **28**: 537–543.
22. Walters EH, Parrish RW, Bevan C, Smith AP: Induction of bronchial hypersensitivity: evidence for a role for prostaglandins. *Thorax* (1981) **36**: 571–574.
23. Hamberg M, Svensson J, Samuelsson B: Thromboxanes: a new group of biologically active compounds derived from prostaglandin endoperoxides. *Proc Natl Acad Sci USA* (1975) **72**: 2994–2998.
24. Higgs GA, Moncada S, Salmon JA, Seager K: The source of thromboxane and prostaglandins in experimental inflammation. *Br J Pharmacol* (1983) **79**: 863–868.
25. Aizawa H, Chung KF, Leikauf GD, et al.: Significance of thromboxane generation in ozone-induced airway hyperresponsiveness in dogs. *J Appl Physiol* (1985) **59**: 1918–1923.
26. McFarlane CS, Ford-Hutchinson AW, Letts LG: Inhibition of thromboxane (TxA$_2$)-induced airway hyperresponsiveness to aerosolized acetylcholine by the selective TxA$_2$ antagonist L655,240 in the conscious primate. *Am Rev Respir Dis* (1985) **137**: 100A.
27. Dorsch WD, Ring J, Melzer H: A selective inhibitor of thromboxane biosynthesis enhances immediate and inhibits late cutaneous allergic reactions in man. *J Allergy Clin Immunol* (1983) **72**: 168–174.
28. Kleeberger SR, Kolbe J, Adkinson NF Jr, Peters SP, Spannhake EW: Thromboxane contributes to the immediate antigenic response of canine peripheral airways. *J Apply Physiol* (1987) **62**: 1589–1595.
29. Shephard EG, Malan L, Macfarlane CM, Mouton W, Joubert JR: Lung function and plasma levels of thromboxane B$_2$, 6-ketoprostaglandin F$_{1a}$ and B-thromboglobulin in antigen-induced asthma before and after indomethacin pretreatment. *Br J Clin Pharmacol* (1985) **19**: 459–470.
30. Fujimura M, Sasaki F, Nakatsumi Y, et al.: Effects of a thromboxane synthetase inhibitor (OKY-046) and a lipoxygenase inhibitor (AA-861) on bronchial responsiveness to acetylcholine in asthmatic subjects. *Thorax* (1986) **41**: 955–959.
31. Jones GL, Saroea G, Watson RL, O'Byrne PM: The effect of an inhaled thromboxane mimetic (U46619) on airway function in human subjects. *Am Rev Respir Dis* (1992) **145**: 1270–1275.
32. Tamaoki J, Sekizawa K, Osborne ML, Ueki IF, Graf PD, Nadel JA: Platelet aggregation increases cholinergic neurotransmission in canine airway. *J Appl Physiol* (1987) **62**: 2246–2251.
33. Saroea HG, Inman M, O'Byrne PM: U46619-induced bronchoconstriction in asthmatic subjects is mediated by acetylcholine release. *Am J Respir Crit Care Med* (1995) **151**: 321–324.
34. Gardiner PV, Young CL, Holmes K, Hendrick DJ, Walters EH: Lack of short-term effect of the thromboxane synthetase inhibitor UK-38,485 on airway reactivity to methacholine in asthmatic subjects. *Eur Respir J* (1993) **6**: 1027–1030.
35. Stenton SC, Young CA, Harris A, Palmer JB, Hendrick DJ, Walters EH: The effect of GR32191 (a thromboxane receptor antagonist) on airway responsiveness in asthma. *Pulmon Pharmacol* (1992) **5**: 199–202.
36. Magnussen H, Boerger S, Templin K, Baunack AR: Effects of a thromboxane-receptor antagonist, BAY u 3405, on prostaglandin D2- and exercise-induced bronchoconstriction. *J Allergy Clin Immunol* (1992) **89**: 1119–1126.
37. Manning PJ, Stevens WH, Cockcroft DW, O'Byrne PM: The role of thromboxane in allergen-induced asthmatic responses. *Am Rev Respir Dis* (1990) **141**: A395.
38. Walters EH, Parrish RW, Bevan C, Parrish RW, Smith BH, Smith AP: Time-dependent effect of prostaglandin E$_2$ inhalation on airway responses to bronchoconstrictor agents in normal subjects. *Thorax* (1982) **37**: 438–442.

39. Fish JE, Newball HH, Norman PS, Peterman VI: Novel effects of PGF$_{2a}$ on airway function in asthmatic subjects. *J Appl Physiol* (1983) **54**: 105–112.
40. Walters EH, O'Byrne PM, Fabbri LM, Graf PD, Holtzman MJ, Nadel JA: Control of neurotransmission by prostaglandins in canine trachealis smooth muscle. *J Appl Physiol* (1984) **57**: 129–134.
41. Manning PJ, Jones GL, O'Byrne PM: Tachyphylaxis to inhaled histamine in asthmatic subjects. *J Appl Physiol* (1987) **63**: 1572–1577.
42. Manning PJ, Lane CG, O'Byrne PM: The effect of oral prostaglandin E$_1$ on airway responsiveness in asthmatic subjects. *Pulmon Pharmacol* (1989) **2**: 121–124.
43. Jackson PA, Manning PJ, O'Byrne PM: A new role for histamine H$_2$-receptors in asthmatic airways. *Am Rev Respir Dis* (1988) **138**: 784–788.
44. Manning PJ, O'Byrne PM: Histamine bronchoconstriction reduces airway responsiveness in asthmatic subjects. *Am Rev Respir Dis* (1988) **137**: 1323–1325.
45. Hamilec CM, Manning PJ, O'Byrne PM: Exercise refractoriness post histamine bronchoconstriction in asthmatic subjects. *Am Rev Respir Dis* (1988) **138**: 794–798.
46. Manning PJ, Watson RM, Margolskee DJ, Williams V, Schartz JI, O'Byrne PM: Inhibition of exercise-induced bronchoconstriction by MK-571, a potent leukotriene D$_4$ receptor antagonist. *N Engl J Med* (1990) **323**: 1736–1739.
47. Edmunds AT, Tooley M, Godfrey S: The refractory period after exercise-induced asthma: its duration and relation to the severity of exercise. *Am Rev Respir Dis* (1978) **117**: 247–254.
48. O'Byrne PM, Jones GL: The effect of indomethacin on exercise-induced bronchoconstriction and refractoriness after exercise. *Am Rev Respir Dis* (1986) **134**: 69–72.
49. Manning PJ, Watson RL, O'Byrne PM: The effect of H$_2$-receptor antagonists on exercise-induced refractoriness in asthma. *J Allergy Clin Immunol* (1992) **88**: 125–126.
50. Manning PJ, Watson RW, O'Byrne PM: Exercise-induced refractoriness in asthmatic subjects involves leukotriene and prostaglandin interdependent mechanisms. *Am Rev Respir Dis* (1993) **148**: 950–954.
51. Melillo E, Woolley KL, Manning PJ, Watson RM, O'Byrne PM: Effect of inhaled PGE$_2$ on exercise-induced bronchoconstriction in asthmatic subjects. *Am J Respir Crit Care Med* (1994) **149**: 1138–1141.
52. Pavord ID, Wong C, Williams J, Tattersfield AE: Effect of inhaled prostaglandin E$_2$ on allergen-induced asthma. *Am Rev Respir Dis* (1993) **148**: 87–90.

16

Cysteinyl Leukotrienes

JEFFREY M. DRAZEN

INTRODUCTION

The leukotrienes (LTB$_4$, LTC$_4$, LTD$_4$ and LTE$_4$) and lipoxins (LxA and LxB) are molecules derived by lipoxygenation of arachidonic acid. Although each of these molecules is potentially important in the pathogenesis of asthma, the evidence currently available suggests that of these molecules, LTC$_4$, LTD$_4$ and LTE$_4$ play a significant role in initiating and maintaining an asthmatic response; this chapter focuses primarily on these molecules.

FORMATION AND METABOLISM OF THE LEUKOTRIENES

Arachidonic acid is a normal component of many cell membrane phospholipids; it is commonly found esterified to such phospholipids in the *sn*2 position. In the presence of appropriately activated phospholipase A$_2$, (PLA$_2$), arachidonic acid is cleaved from the cell membrane (Fig. 16.1). There are at least two distinct forms of PLA$_2$ with this capacity.[1] Cytosolic PLA$_2$ (cPLA$_2$) is the enzyme activated when arachidonic acid is cleaved in the intracellular microenvironment. This form of the enzyme is catalytically active at calcium levels consistent with the intracellular microenvironment and cleaves arachidonic acid from phospholipids in the perinuclear membrane. In contrast, secretory PLA$_2$ (sPLA$_2$) operates in the extracellular microenvironment to cleave arachidonic acid[2–4] from the cell's exterior plasma membrane; precisely how the arachidonic acid so cleaved enters the cell to serve as a substrate for various enzymes is not clear.

ASTHMA: BASIC MECHANISMS AND CLINICAL MANAGEMENT (3rd Edn)
ISBN 0-12-079027-9

Copyright © 1998 Academic Press Limited
All rights of reproduction in any form reserved

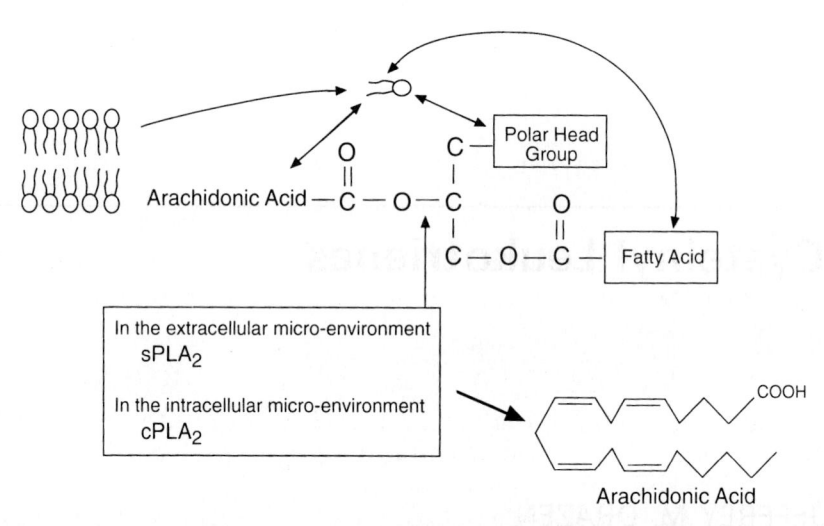

Fig. 16.1 The cleavage of arachidonic acid from membrane phospholipids. In the intracellular microenvironment, the perinuclear membrane is the likely substrate for cytosolic phospholipase A_2 (cPLA$_2$). In the extracellular microenvironment the cell's plasma membrane is the likely substrate for secretory phospholipase A_2 (sPLA$_2$). Regardless of the specific enzyme involved arachidonic acid is cleaved from the *sn*2 position on the phospholipid of interest.

Arachidonic acid, released as a result of PLA$_2$ action, enters into a series of reactions at the perinuclear membrane. The first of these requires a specific 5-lipoxygenase-activating protein (FLAP),[5–7] which allows arachidonate to serve as a substrate for the enzyme 5-lipoxygenase[8–12] (Fig. 16.2). 5-Lipoxygenase sequentially catalyses the addition of oxygen to arachidonic acid to form 5-hydroperoxyeicosatetraenoic acid (5-HPETE) and leukotriene A$_4$ (LTA$_4$) respectively; LTA$_4$ is a major branch-point in the formation of the leukotrienes.[13,14] A variety of cells, most notably neutrophilic polymorphonuclear leucocytes (PMNs), express a specific epoxide hydrolase that catalyses the formation of leukotriene B$_4$ (LTB$_4$) from LTA$_4$.[15–19] It is also possible for multiple lipoxygenases to act sequentially on arachidonic acid to form the lipoxins.[14,20] In distinction to the neutrophil, other cells, including eosinophils, mast cells and alveolar macrophages, not only have the capacity to form LTA$_4$ from arachidonic acid but also possess a unique and specific glutathionyl-*S*-transferase, LTC$_4$ synthase, which catalyses the conjugation of glutathione to LTA$_4$ at carbon 6 to form leukotriene C$_4$ (LTC$_4$). LTC$_4$ synthase has a high degree of homology with FLAP and is also an integral perinuclear membrane protein.[21–24] Once formed, LTC$_4$ exits the cell via a specific transmembrane transporter. In the extracellular microenvironment LTC$_4$ serves as a substrate for γ-glutamyl transpeptidase, which cleaves the glutamic acid moiety from its peptide chain to form leukotriene D$_4$ (LTD$_4$); LTD$_4$ is further processed by the removal of the glycine moiety from its peptide chain to form leukotriene E$_4$ (LTE$_4$).[25,26] LTC$_4$, LTD$_4$ and LTE$_4$ make up the material formerly known as slow-reacting substance of anaphylaxis of SRS-A and are collectively known as the cysteinyl leukotrienes.

Once formed, and in the presence of appropriately activated PMNs, the cysteinyl leukotrienes are degraded to their respective sulphoxides and 6-*trans* diastereoisomers of LTB$_4$.[27] In the absence of such cells the major degradation and excretion products of the

16 Cysteinyl Leukotrienes

cysteinyl leukotrienes are native LTE$_4$, N-acetyl LTE$_4$, or the products resulting from ω-oxidation and β-elimination of LTE$_4$.[28,29]

LEUKOTRIENES IN ASTHMA

Since the structural identification of the leukotrienes, a number of lines of evidence have accrued indicating that the cysteinyl leukotrienes may be involved in the asthmatic response; these are reviewed below.

The pathology of chronic mild asthma and production of cysteinyl leukotrienes by the cells found in the asthmatic lesion

The cysteinyl leukotrienes are synthesized and exported into the microenvironment by constitutive and infiltrating cells including mast cells and eosinophils;[30] these two cell types are known to be critical cells in the asthmatic lesion.[31–34] Because of the proximity of these cells to the airway microvasculature and smooth muscle, it seems likely that cysteinyl leukotrienes produced by these cells act at leukotriene receptors on airway smooth muscle and the bronchial vasculature, mediating airway obstruction and microvascular leak. In addition to the synthesis of leukotrienes by eosinophils and mast cells, it is also likely that the bronchial vascular endothelium will be exposed to cells such as PMNs capable of donating LTA$_4$. When LTA$_4$ is provided for effector cells, such as vascular endothelial cells or platelets containing LTC$_4$ synthase,[35–37] the cysteinyl leukotrienes can be produced by transcellular metabolism. Thus both resident cells as well as infiltrating cells found in the asthmatic lesion have the capacity to produce cysteinyl leukotrienes.

Biological effects of the cysteinyl leukotrienes relevant to the asthmatic response

The leukotrienes are known to have profound biochemical and physiological effects, even in picomolar concentrations, including induction of airway obstruction, tissue oedema, and expression of bronchial mucus from submucosal glands;[30,38,39] these pathobiological effects make them important candidate mediators of asthmatic responses. Prominent among the effects of the cysteinyl leukotrienes is their ability to mediate airway narrowing in normal individuals and persons with asthma. Indeed, when aerosols of leukotrienes are inhaled by normal subjects, airway obstruction, as manifested by a decrease in flow rates during a forced exhalation, occurs.[40–52] When flow rates at 30% of vital capacity measured from partial flow–volume curves ($\dot{V}_{30}P$) are used as the index of airway obstruction, it has been established that LTC$_4$ and LTD$_4$ have bronchoconstrictor effects that are prolonged compared with those induced by histamine or methacholine. When a 30% decrease in the $\dot{V}_{30}P$ is induced, the duration of bronchoconstrictor effect of histamine or methacholine is about 3–5 min. In contrast, the duration of bronchoconstriction resulting from the cysteinyl leukotrienes when an

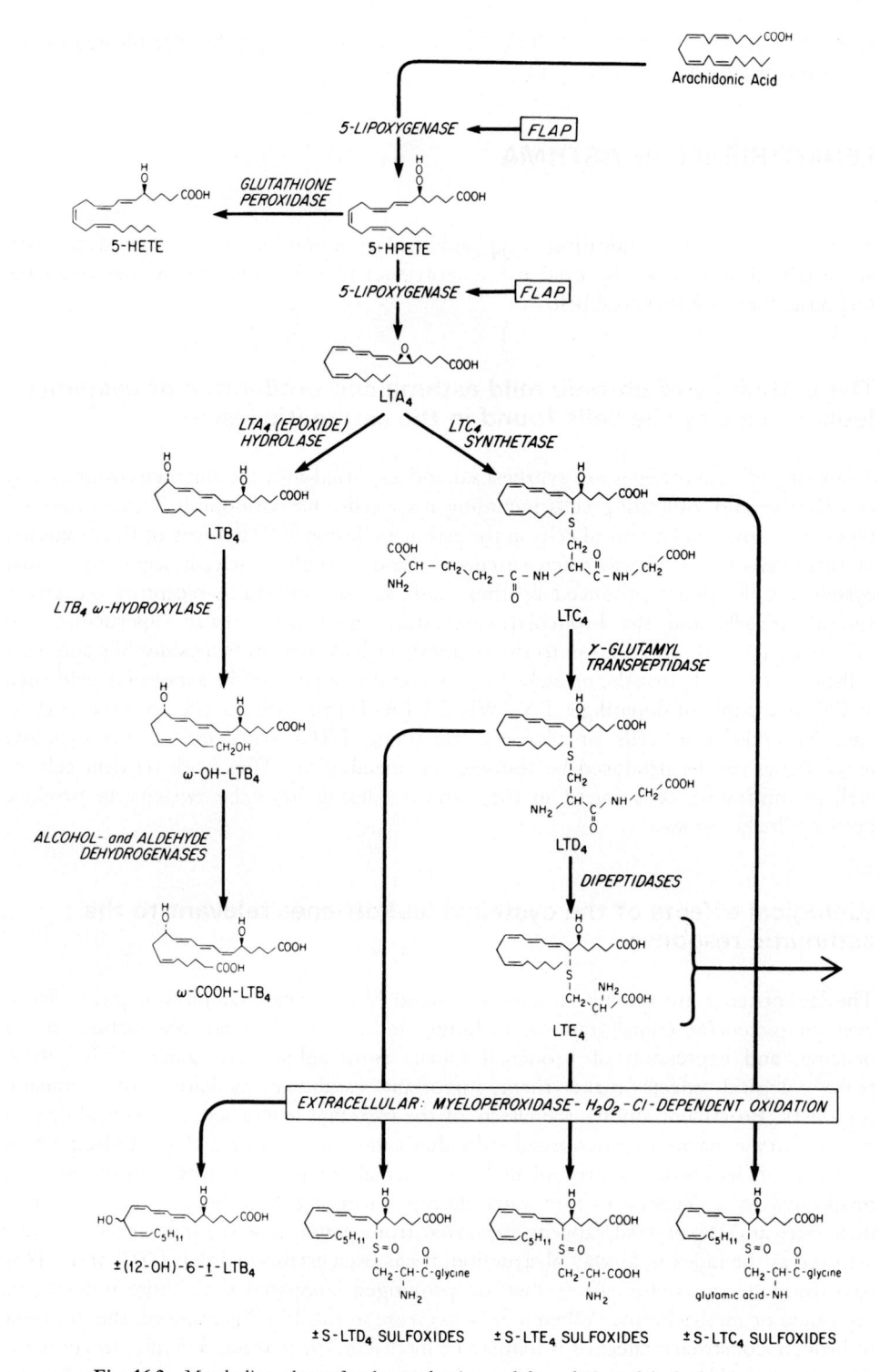

Fig. 16.2 Metabolic pathway for the production and degradation of the leukotrienes.

16 Cysteinyl Leukotrienes

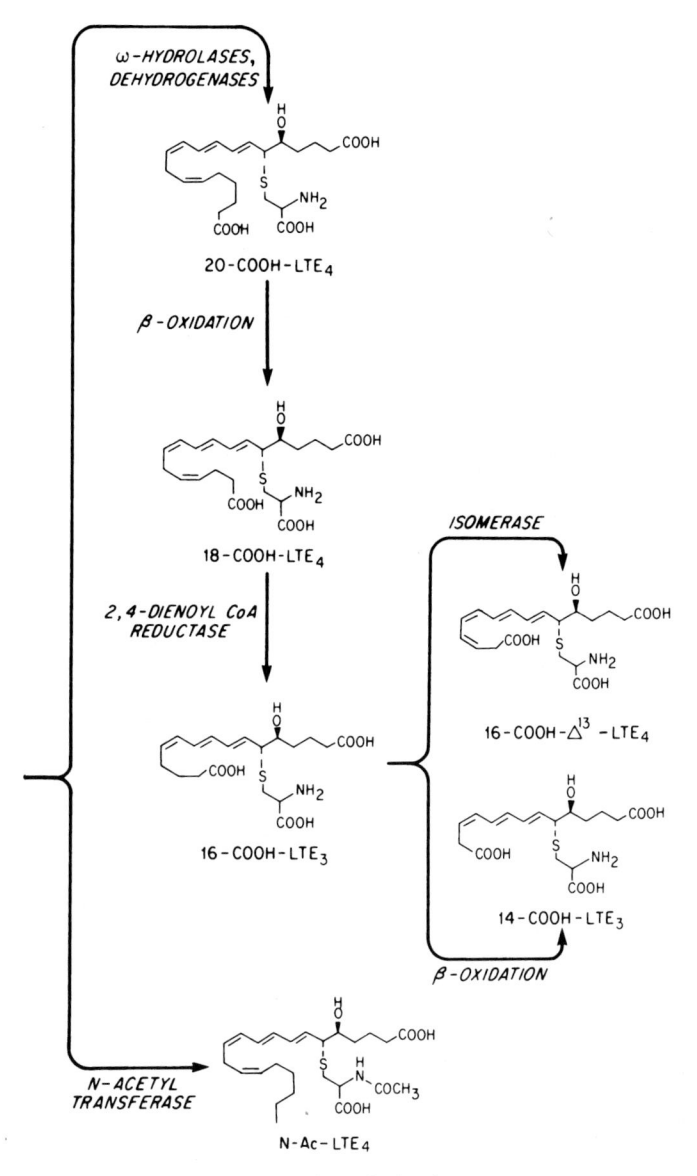

Fig. 16.2 *Continued.*

equivalent peak magnitude of effect is achieved is about 25–30 min. More important than differences in the duration of effect is the relative potency of the cysteinyl leukotrienes compared with agonists such as histamine and methacholine. The range of nebulizer concentrations required to achieve a 30–40% decrease in the $V_{30}P$ due to inhalation of LTD_4 in normal subjects varies about 100-fold from approximately 3 μM to 300 μM. These concentrations are about 3000-fold less than the nebulizer concentrations of histamine required to achieve an equivalent degree of airway narrowing. In addition, among normal subjects there is a relationship between responsiveness to the leukotrienes and responsiveness to reference agonists such as histamine or methacholine.[45,46] Subjects that are more responsive to histamine are those that are more responsive to the leukotrienes. When the forced expiratory volume in 1 s (FEV_1) is used as the outcome indicator rather than the $\dot{V}_{30}P$, approximately five times greater leukotriene nebulizer concentrations are required to achieve a 15–20% decrease in this airway response index. Although these data are consistent with the hypothesis that the airways which narrow in order to reduce the FEV_1 are less sensitive to the cysteinyl leukotrienes than the airways that narrow to decrease the $\dot{V}_{30}P$, this hypothesis has never been established by direct experiment.

In subjects with asthma, LTC_4 and LTD_4 are potent bronchoconstrictor agonists when administered by aerosol as indicated by induced decrements in the FEV_1, the $\dot{V}_{30}P$ or specific conductance.[45,46,48,50,52–54] When the $\dot{V}_{30}P$ or $\dot{V}_{40}P$ is used as the outcome indicator, nebulizer concentrations of LTD_4 of about 0.3–30 μM are required to decrease airflow rates by approximately 30%. These nebulizer concentrations are approximately one-tenth of those required by normal subjects in order to achieve the same decrement in airflow rates. Since normal subjects are approximately 100-fold less sensitive to histamine or methacholine than asthmatic subjects, while they are only 10-fold less responsive to the cysteinyl leukotrienes, this indicates that the relative degree of hyperresponsiveness to the cysteinyl leukotrienes is less than that observed when histamine or methacholine is used as the contractile agonist.

LTE_4 is also a potent bronchoactive agonist. The potency of LTE_4 relative to histamine differs between normal and asthmatic subjects. In normal subjects LTE_4 is about 30 times more potent than histamine[50] while in subjects with asthma LTE_4 is about 300 times more potent than histamine,[55] regardless of whether flow rates low in the vital capacity from partial flow–volume curves or specific conductance[47] are used as the outcome indicators. There is evidence indicating that patients with aspirin-induced asthma are hyperresponsive to LTE_4 when compared with other asthmatic subjects,[56,57] indicating a potentially unique role for this cysteinyl leukotriene in the pathogenesis of this uncommon form of asthma.

Leukotriene recovery in asthma

Cysteinyl leukotrienes have been recovered after experimental challenges which elicit clinical symptoms similar to those that occur in spontaneously occurring asthmatic conditions. Leukotrienes have also been recovered in the nasal lavage fluid after intranasal challenge with either antigen or cold air.[58,59] Leukotrienes are recovered in significantly greater amounts in the bronchoalveolar lavage (BAL) fluid from subjects with symptomatic asthma compared with subjects with asymptomatic asthma or normal subjects,[60–65] suggesting that the leukotrienes are produced locally in the lung of patients with asthma. Although leukotrienes can be recovered from BAL fluid of patients with active asthma or after airway challenge, because of the invasive nature of the procedure required to obtain the fluid it is unlikely that the extensive clinical use of BAL leukotriene levels as an index of leukotriene production will occur.

In this regard, indices of leukotriene production that rely on measurements made on blood or urine samples offer potential utility in the assessment of which asthmatic responses represent states in which the leukotrienes are among the effector molecules mediating bronchoconstriction. A number of investigators have detected cysteinyl leukotrienes in the plasma during asthma attacks,[66-68] although the methods used to assure the authenticity of the materials identified have been suboptimal and the findings have not been widely reproduced, probably for technical reasons.[69] In contrast, it has been shown that accurate and quantitative measurements of LTE_4 can be made in urine samples.[70-72] In normal human subjects, after intravenous administration of radiolabelled LTC_4, 12–48% of the counts are recovered in the urine with 4–13% as intact LTE_4.[73,74] Not only can exogenously administered leukotrienes be recovered in the urine but it is now common to measure leukotriene recovery from the urine as an index of endogenous production of leukotrienes.[70,71,75] For example, it has been shown that there is an increase in the recovery of authentic LTE_4 in the urine of asthmatic subjects in the early phase after antigen challenge.[76] After antigen challenge, the magnitude of the induced fall in the FEV_1 and the amount of LTE_4 in the urine are closely correlated.[75,77,78] Taylor et al.[71] used solid-phase extraction followed by reversed-phase high performance liquid chromatography and radioimmunoassay to measure LTE_4 in the urine. To allow quantitative determination of the amounts of LTE_4 in the urine, they used recovery of radiolabelled LTE_4 added as an internal standard. In normal subjects they recovered 23.8 ng of LTE_4 per mmol of creatinine, while in 20 asthmatic subjects during acute spontaneous attacks they recovered slightly over three times as much LTE_4 (78.3 ng/mmol creatinine). There was no relationship observed between the severity of the attack as measured by the FEV_1 and the amount of LTE_4 recovered in the urine. In six of the eight subjects in whom urinary LTE_4 measurements were available both before and after treatment for an acute asthmatic exacerbation (all received prednisolone), there was a decrease in the LTE_4 excretion rate. Drazen and coworkers[79] demonstrated that over two-thirds of subjects presenting for emergency treatment of asthma had elevated urinary LTE_4 compared with a reference group of normal subjects. Interestingly, among the individuals presenting for emergency asthma treatment those whose lung function responded initially to inhaled β-agonist treatment were those most likely to have elevated urinary LTE_4 levels.

Asano et al.[80] examined urinary LTE_4 excretion rates in eight patients with mild chronic stable asthma ($FEV_1 \sim 70\%$ predicted, inhaled β-agonists as the only asthma treatment) followed on a metabolic ward for 4 days. These authors were able to show that there was no diurnal variation in urinary LTE_4 excretion. Furthermore, they demonstrated that, on average, patients with asthma have significantly higher urinary LTE_4 levels than normal subjects. However, among the asthmatic patients studied there were individuals who were persistent hyperexcreters of urinary LTE_4 and others with urinary LTE_4 levels that were persistently within the normal range. Among the explanations for this finding is the possibility that among patients with asthma whose clinical phenotype is similar, there are individuals whose asthma is associated with leukotriene production and others for whom this is not the case. Of course, the difference among individuals could also reflect differences in renal excretion or production rather than differences in pulmonary production. These data clearly indicate that during spontaneous, induced or chronic stable asthma, at least in some individuals, there is enhanced urinary LTE_4 excretion and, by inference, increased cysteinyl leukotriene production.

Leukotriene receptor blockade and synthesis inhibition

Two classes of pharmacological agents have been available for probing the role of cysteinyl leukotrienes in asthma: leukotriene receptor antagonists and leukotriene synthesis inhibitors. These agents have been used in clinical trials in human asthma. With regard to leukotriene receptor antagonists, it is well established that there are at least two distinct receptors, the cysteinyl leukotriene receptors type 1 and type 2 ($CysLT_1$ and $CysLT_2$), for cysteinyl leukotrienes in contractile tissue.[81] Although over a dozen chemically distinct antagonists at the $CysLT_1$ receptor have been recognized, only a few have been studied in intact human subjects (Table 16.1). With regard to leukotriene synthesis inhibitors, agents have been developed that are direct inhibitors of 5-lipoxygenase (zileuton)[82] as well as agents that inhibit the interaction between arachidonic acid and FLAP (MK-0591, Bay x1005).[83,84]

Inhibition of induced asthma

A number of LTD_4 receptor antagonists or synthesis inhibitors have been tested for their effects on the bronchospasm that accompanies experimental asthma induced by antigen, cold air, exercise or aspirin in humans.[78,85–98] Although there is variation among agents and protocols, all of the agents tested inhibit the bronchospasm that accompanies exercise or cold air-induced asthma by 30–70%. These agents are also effective inhibitors of allergen-induced asthma; a 30–70% inhibition of the airway obstruction associated with the early-allergen response is achieved. For reasons that are not clear, leukotriene receptor antagonists have been more effective than 5-lipoxygenase inhibitors in reducing the severity of allergen-induced asthma. Also, in allergen-induced asthma with regard to 5-lipoxygenase inhibitors, FLAP antagonists have been somewhat more effective than direct inhibitors of 5-lipoxygenase in inhibiting allergen-induced asthma. Among the various forms of induced asthma, both leukotriene receptor antagonists and synthesis inhibitors have been very effective in preventing the physiological changes that accompany aspirin-induced asthma.[87,92,93,99–102] Indeed, current data indicate that virtually all the physiological effects of aspirin-induced asthma derive from the action of the cysteinyl leukotrienes. Taken together these data indicate that inhibition of the synthesis or action of the leukotrienes is associated with an amelioration of the physiological changes that occur in laboratory induced-asthma.

Table 16.1 Effects of leukotriene receptor antagonists on the bronchoconstrictor response to inhaled LTD_4 in intact humans.

Agent	Dose and route	Effects on the LTD_4 response	Ref.
L-649,923	1000 mg p.o.	3.8-fold decrease in LTD_4 responsiveness	109
LY-171,883	400 mg p.o.	4.5-fold decrease in LTD_4 responsiveness	110
L-648,051	12 mg by inhalation	Decrease in LTD_4 response in duration and magnitude by $\approx 50\%$	111
Zafirlukast	40 mg p.o.	117-fold decrease in LTD_4 responsiveness	112
Verlukast	28 mg i.v.	44-fold decrease in LTD_4 responsiveness	113

16 Cysteinyl Leukotrienes

Leukotriene inhibition in chronic stable asthma

The data reviewed above indicate that the leukotrienes mediate a portion of the physiological effects observed in induced asthma. If chronic stable asthma represents repeated recurrence of naturally occurring induced asthma and episodes, then one would postulate that chronic stable asthma could be improved by treatment with such agents. Such improvement is indeed noted. Three agents, LY171883, zileuton and zafirlukast, have been used in trials of 4–6 weeks' duration in patients with mild–moderate chronic stable asthma.[103–105] The general design of each of the trials was similar. Patients whose asthma could be somewhat controlled by use of inhaled β-agonists alone, who had FEV_1 values between 40 and 80% of predicted and who had moderate asthma symptoms, as judged from daily symptom diaries, were recruited and enrolled. There was a 1–3 week 'run-in' period when all patients had their asthma control monitored while on single-blind oral placebo. During this period baseline lung function, β-agonist use and symptom data were gathered. Patients then entered a randomized treatment period in which they received randomized treatment with active agent (four times a day for zileuton and twice a day for LY171883 and zafirlukast) or placebo. Patients returned on a weekly basis to their clinical centres where their lung function and asthma symptom data were recorded.

The patients recruited into these trials were similar; they had moderate asthma symptoms and FEV_1 values of approximately 60% of predicted. With chronic active treatment, there was an approximately 15% improvement over baseline (5–7% with LY171883) in the FEV_1 at the completion of the randomized treatment period; in all trials this treatment was significantly more effective than placebo in improving the FEV_1. In these trials there was a significant salutary treatment effect observed when asthma symptoms, β-agonist use or morning peak flow was used as the outcome indicator. The data from all three trials are consistent with continuously improving asthma control during active treatment. Further elaboration of this subject is contained in Chapter 4.

Placebo-controlled trials examining the effects of 3–6 months of active treatment in chronic stable asthma have also been completed and published in which zileuton was the active treatment. The enrolment criteria and overall design used were similar to the 4-week trial reviewed above, except that the duration of the randomized treatment was either 13 weeks[106] or 26 weeks.[107] In these trials, zileuton was shown to be an effective asthma treatment when compared with placebo. Active treatment was associated with about a 15% improvement in FEV_1 that was sustained over the trial duration as well as with decreased β-agonist use and improved asthma symptoms. In the 13-week trial, zileuton treatment significantly improved asthma-specific quality of life,[108] while placebo treatment did not. Most importantly, patients receiving active treatment required significantly fewer, more than 2.5-fold fewer, courses of 'rescue' steroid treatment than did patients receiving placebo treatment (Fig. 16.3). These data indicate that chronic inhibition of 5-lipoxygenase is associated with a prolonged salutary effect on asthma control.

CONCLUSIONS

These data support the following conclusions as to the role of cysteinyl leukotrienes in bronchial asthma: (a) they are produced by constitutive cells (mast cells/macrophages)

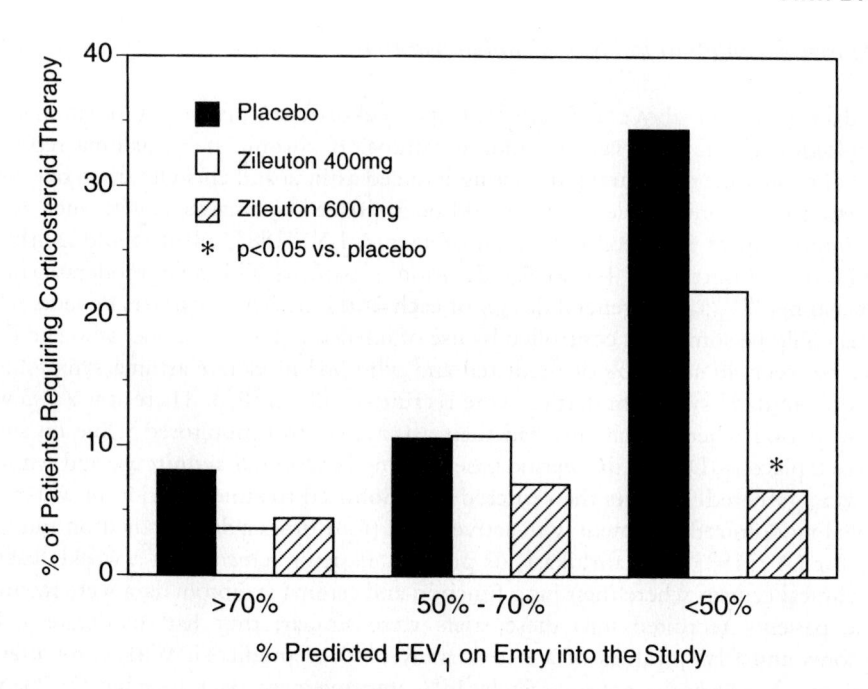

Fig. 16.3 'Steroid-sparing' effects of chronic 5-lipoxygenase inhibition in patients with mild-to-moderate chronic stable asthma. FEV_1, forced expiratory volume in 1 s. Reproduced from ref. 106, with permission.

and infiltrating cells (eosinophils) implicated in the asthma response; (b) they are potent bronchoconstrictor agonists; (c) laboratory-induced and spontaneous asthma is associated with an enhanced recovery of leukotrienes in the urine of subjects with asthma; (d) asthma control is enhanced, when compared with placebo, by agents capable of interfering with leukotriene action or synthesis. Although these data indicate that the leukotrienes play a pivotal role in the asthmatic response, how to most effectively use them in asthma treatment is not established and will require further clinical investigation.

REFERENCES

1. Murakami M, Kudo I, Inoue K: Molecular nature of phospholipases A2 involved in prostaglandin I_2 synthesis in human umbilical vein endothelial cells. Possible participation of cytosolic and extracellular type II phospholipases A2. *J Biol Chem* (1993) **268**: 839–844.
2. Gonzalez-Buritica H, Khamashita MA, Hughes GR: Synovial fluid phospholipase A2s and inflammation. *Ann Rheum Dis* (1989) **48**: 267–269.
3. White SR, Strek ME, Kulp GVP, *et al.*: Regulation of human eosinophil degranulation and activation by endogenous phospholipase-A_2. *J Clin Invest* (1993) **91**: 2118–2125.
4. Murakami M, Kudo I, Suwa Y, Inoue K: Release of 14-kDa group-II phospholipase A2 from activated mast cells and its possible involvement in the regulation of the degranulation process. *Eur J Biochem* (1992) **209**: 257–265.
5. Dixon RA, Diehl RE, Opas E, *et al.*: Requirement of a 5-lipoxygenase-activating protein for leukotriene synthesis. *Nature* (1990) **343**: 282–284.

16 Cysteinyl Leukotrienes

6. Miller DK, Gillard JW, Vickers PJ, et al.: Identification and isolation of a membrane protein necessary for leukotriene production. *Nature* (1990) **343**: 278–281.
7. Reid GK, Kargman S, Vickers PJ, et al.: Correlation between expression of 5-lipoxygenase-activating protein, 5-lipoxygenase, and cellular leukotriene synthesis. *J Biol Chem* (1990) **265**: 19 818–19 823.
8. Dixon RA, Jones, RE, Diehl RE, Bennett CD, Kargman S, Rouzer CA: Cloning of the cDNA for human 5-lipoxygenase. *Proc Natl Acad Sci USA* (1988) **85**: 416–420.
9. Keppler A, Orning L, Bernstrom K, Hammarstrom S: Endogenous leukotriene D4 formation during anaphylactic shock in the guinea pig. *Proc Natl Acad Sci USA* (1987) **84**: 5903–5907.
10. Matsumoto T, Funk CD, Radmark O, Hoog JO, Jornvall H, Samuelsson B: Molecular cloning and amino acid sequence of human 5-lipoxygenase. *Proc Natl Acad Sci USA* (1988) **85**: 26–30.
11. Rouzer CA, Rands E, Kargman S, Jones RE, Register RB, Dixon RA: Characterization of cloned human leukocyte 5-lipoxygenase expressed in mammalian cells. *J Biol Chem* (1988) **263**: 10 135–10 140.
12. Funk CD, Hoshiko S, Matsumoto T, Radmark O, Samuelsson B: Characterization of the human 5-lipoxygenase gene. *Proc Natl Acad Sci USA* (1989) **86**: 2587–2591.
13. Samuelsson B: Leukotrienes: mediators of immediate hypersensitivity reactions and inflammation. *Science* (1983) **220**: 568–575.
14. Samuelsson B, Dahlen SE, Lindgren JA, Rouzer CA, Serhan CN: Leukotrienes and lipoxins: structures, biosynthesis, and biological effects. *Science* (1987) **237**: 1171–1176.
15. Medina JF, Radmark O, Funk CD, Haeggstrom JZ: Molecular cloning and expression of mouse leukotriene-A4 hydrolase cDNA. *Biochem Biophys Res Commun* (1991) **176**: 1516–1524.
16. Wetterholm A, Haeggstrom JZ: Leukotriene-A4 hydrolase: an anion activated peptidase. *Biochim Biophys Acta* (1992) **1123**: 275–281.
17. Munafo DA, Shindo K, Baker JR, Bigby TD: Leukotriene A(4) hydrolase in human bronchoalveolar lavage fluid. *J Clin Invest* (1994) **93**: 1042–1050.
18. Orning L, Gierse JK, Fitzpatrick FA: The bifunctional enzyme leukotriene A_4 hydrolase is an arginine aminopeptidase of high efficiency and specificity. *J Biol Chem* (1994) **269**: 11 269–11 273.
19. Mancini JA, Evans JF: Cloning and characterization of the human leukotriene A_4 hydrolase gene. *Eur J Biochem* (1995) **231**: 65–71.
20. Serhar CN, Haeggstrom JZ, Leslie CC: Lipid mediator networks in cell signaling: update and impact of cytokinines. *FASEB J* (1996) **10**: 1147–1158.
21. Lam BK, Penrose JF, Freeman GJ, Austen KF: Expression cloning of a cDNA for human leukotriene C_4 synthase, an integral membrane protein conjugating reduced glutathione to leukotriene A_4. *Proc Natl Acad Sci USA* (1994) **91**: 7663–7667.
22. Welsch DJ, Creely DP, Hauser SD, Mathis KJ, Krivi GG, Isakson PC: Molecular cloning and expression of human leukotriene-C_4 synthase. *Proc Natl Acad Sci USA* (1994) **91**: 9745–9749.
23. Penrose JF, Spector J, Lam BK, et al.: Purification of human lung leukotriene C_4 synthase and preparation of a polyclonal antibody. *Am J Respir Crit Care Med* (1995) **152**: 283–289.
24. Penrose JF, Spector J, Baldasaro M, et al.: Molecular cloning of the gene for human leukotriene C_4 synthase: organization, nucleotide sequence, and chromosomal localization to sq35. *J Biol Chem* (1996) **271**: 11 356–11 361.
25. Lewis RA, Drazen JM, Austen KF, Clark DA, Corey EJ: Identification of the C(6)-S-conjugate of leukotriene A with cysteine as a naturally occurring slow reacting substance of anaphylaxis (SRS-A). Importance of the 11-*cis*-geometry for biological activity. *Biochem Biophys Res Commun* (1980) **96**: 271–277.
26. Parker CW, Falkenhein SF, Huber MM: Sequential conversion of the glutathionyl side chain of slow reacting substance (SRS) to cysteinyl-glycine and cysteine in rat basophilic leukemia cells stimulated with A-23187. *Prostaglandins* (1979) **18**: 863–886.
27. Lee CS, Lewis RA, Corey EJ, et al.: Oxidative inactivation of leukotriene C4 by stimulated human polymorphonuclear leukocytes. *Proc Natl Acad Sci USA* (1982) **79**: 4166–4170.
28. Sala A, Voelkel N, Maclouf J, Murphy RC: Leukotriene E4 elimination and metabolism in normal human subjects. *J Biol Chem* (1990) **265**: 21 771–21 778.

29. Stene DO, Murphy RC: Metabolism of leukotriene E4 in isolated rat hepatocytes. Identification of beta-oxidation products of sulfidopeptide leukotrienes. *J Biol Chem* (1988) **263**: 2773–2778.
30. Lewis RA, Austen KF, Soberman RJ: Leukotrienes and other products of the 5-lipoxygenase pathway. Biochemistry and relation to pathobiology in human diseases. *N Engl J Med* (1990) **323**: 645–655.
31. Bousquet J, Chanez P, Lacoste JY, *et al.*: Eosinophilic inflammation in asthma. *N Engl J Med* (1990) **323**: 1033–1039.
32. Galli SJ: New concepts about the mast cell. *N Engl J Med* (1993) **328**: 257–265.
33. Laitinen A, Laitinen LA: Cellular infiltrates in asthma and in chronic obstructive pulmonary disease. *Am Rev Respir Dis* (1991) **143**: 1159–1160.
34. Laitinen LA, Laitinen A, Haahtela T: Airway mucosal inflammation even in patients with newly diagnosed asthma. *Am Rev Respir Dis* (1993) **147**: 697–704.
35. Feinmark SJ, Cannon PJ: Endothelial cell leukotriene C4 synthesis results from intercellular transfer of leukotriene A4 synthesized by polymorphonuclear leukocytes. *J Biol Chem* (1986) **261**: 16466–16472.
36. Feinmark SJ, Cannon PJ: Vascular smooth muscle cell leukotriene C4 synthesis: requirement for transcellular leukotriene A4 metabolism. *Biochim Biophys Acta* (1987) **922**: 125–135.
37. Petersgolden M, Feyssa A: Transcellular eicosanoid synthesis in cocultures of alveolar epithelial cells and macrophages. *Am J Physiol* (1993) **264**: L438–L447.
38. Piper PJ: Formation and actions of leukotrienes. *Physiol Rev* (1984) **64**: 744–761.
39. Henderson WR: The role of leukotrienes in inflammation. *Ann Intern Med* (1994) **121**: 684–697.
40. Holroyde MC, Altounyan RE, Cole M, Dixon M, Elliott EV: Bronchoconstriction produced in man by leukotrienes C and D. *Lancet* (1981) **ii**: 17–18.
41. Weiss JW, Drazen JM, Coles N, *et al.*: Bronchoconstrictor effects of leukotriene C in humans. *Science* (1982) **216**: 196–198.
42. Bisgaard H, Groth S, Dirksen H: Leukotriene D4 induces bronchoconstriction in man. *Allergy* (1983) **38**: 441–443.
43. Weiss JW, Drazen JM, McFadden ER Jr, *et al.*: Airway constriction in normal humans produced by inhalation of leukotriene D. Potency, time course, and effect of aspirin therapy. *JAMA* (1983) **249**: 2814–2817.
44. Barnes NC, Piper PJ, Costello JF: Comparative effects of inhaled leukotriene C4, leukotriene D4, and histamine in normal human subjects. *Thorax* (1984) **39**: 500–504.
45. Smith LJ, Greenberger PA, Patterson R, Krell RD, Bernstein PR: The effect of inhaled leukotriene D4 in humans. *Am Rev Respir Dis* (1985) **131**: 368–372.
46. Adelroth E, Morris MM, Hargreave FE, O'Byrne PM: Airway responsiveness to leukotrienes C4 and D4 and to methacholine in patients with asthma and normal controls. *N Engl J Med* (1986) **315**: 480–484.
47. Drazen JM: Inhalation challenge with sulfidopeptide leukotrienes in human subjects. *Chest* (1986) **89**: 414–419.
48. Roberts JA, Giembycz MA, Raeburn D, Rodger IW, Thomson NC: *In vitro* and *in vivo* effect of verapamil on human airway responsiveness to leukotrienes D4. *Thorax* (1986) **41**: 12–16.
49. Bisgaard H, Groth S: Bronchial effects of leukotriene D4 inhalation in normal human lung. *Clin Sci* (1987) **72**: 585–592.
50. Davidson AB, Lee TH, Scanlon PD, *et al.*: Bronchoconstrictor effects of leukotriene E4 in normal and asthmatic subjects. *Am Rev Respir Dis* (1987) **135**: 333–337.
51. Smith LJ, Kern R, Patterson R, Krell RD, Bernstein PR: Mechanism of leukotriene D4-induced bronchoconstriction in normal subject. *J Allergy Clin Immunol* (1987) **80**: 340–347.
52. Arm JP, Lee TH: Sulphidopeptide leukotrienes in asthma. *Clin Sci* (1993) **84**: 501–510.
53. Griffin M, Weiss JW, Leitch AG, *et al.*: Effects of leukotriene D on the airways in asthma. *N Engl J Med* (1983) **308**: 436–439.
54. Pichurko BM, Ingram RH Jr, Sperling RI, *et al.*: Localization of the site of the bronchoconstrictor effects of leukotriene C4 compared with that of histamine in asthmatic subjects. *Am Rev Respir Dis* (1989) **140**: 334–339.

55. Arm JP, Spur BW, Lee TH: The effects of inhaled leukotriene E4 on the airway responsiveness to histamine in subjects with asthma and normal subjects. *J Allergy Clin Immunol* (1988) **82**: 654–660.
56. Arm JP, O'Hickey SP, Hawksworth RJ, et al.: Asthmatic airways have a disproportionate hyperresponsiveness to LTE4, as compared with normal airways, but not to LTC4, LTD4, methacholine, and histamine. *Am Rev Respir Dis* (1990) **142**: 1112–1118.
57. Arm JP, O'Hickey SP, Spur BW, Lee TH: Airway responsiveness to histamine and leukotriene E4 in subjects with aspirin-induced asthma. *Am Rev Respir Dis* (1980) **140**: 148–153.
58. Togias AG, Naclerio RM, Peters SP, et al.: Local generation of sulfidopeptide leukotrienes upon nasal provocation with cold, dry air. *Am Rev Respir Dis* (1986) **133**: 1133–1137.
59. Silber G, Proud D, Warner J, et al.: In vivo release of inflammatory mediators by hyperosmolar solutions. *Am Rev Respir Dis* (1988) **137**: 606–612.
60. Lam S, Chan H, LeRiche JC, Chan-Yeung M, Salari H: Release of leukotrienes in patients with bronchial asthma. *J Allergy Clin Immunol* (1988) **81**: 711–717.
61. Diaz P, Gonzalez MC, Galleguillos FR, et al.: Leukocytes and mediators in bronchoalveolar lavage during allergen-induced late-phase asthmatic reactions. *Am Rev Respir Dis* (1989) **139**: 1383–1389.
62. Zehr BB, Casale TB, Wood D, Floerchinger C, Richerson HB, Hunninghake GW: Use of segmental airway lavage to obtain relevant mediators from the lungs of asthmatic and control subjects. *Chest* (1989) **95**: 1059–1063.
63. Sladek K, Dworski R, Soja J, et al.: Eicosanoids in bronchoalveolar lavage fluid of aspirin-intolerant patients with asthma after aspirin challenge. *Am J Respir Crit Care Med* (1994) **149**: 940–946.
64. Wenzel SE, Trudeau JB, Kaminsky DA, Cohn J, Martin RJ, Westcott JY: Effect of 5-lipoxygenase inhibition on bronchoconstriction and airway inflammation in nocturnal asthma. *Am J Respir Crit Care Med* (1995) **152**: 897–905.
65. Kane GC, Pollice M, Kim CJ, et al.: A controlled trial of the effect of the 5-lipoxygenase inhibitor, zileuton, on lung inflammation produced by segmental antigen challenge in human beings. *J Allerg Clin Immunol* (1996) **97**: 646–654.
66. Okubo T, Takahashi H, Sumitomo M, Shindoh K, Suzuki S: Plasma levels of leukotrienes C4 and D4 during wheezing attack in asthmatic patients. *Int Arch Allergy Appl Immunol* (1987) **84**: 149–155.
67. Shindo K, Fukumura I, Miyakawa K: Plasma levels of leukotriene E(4) during clinical course of bronchial asthma and the effect of oral prednisolone. *Chest* (1994) **105**: 1038–1041.
68. Sampson AP, Castling DP, Green CP, Price JF: Persistent increase in plasma and urinary leukotrienes after acute asthma. *Arch Dis Child* (1995) **73**: 221–225.
69. Heavey DJ, Soberman RJ, Lewis RA, Spur B, Austen KF: Critical considerations in the development of an assay for sulfidopeptide leukotrienes in plasma. *Prostaglandins* (1987) **33**: 693–708.
70. Tagari P, Ethier D, Carry M, et al.: Measurement of urinary leukotrienes by reversed-phase liquid chromatography and radioimmunoassay. *Clin Chem* (1989) **35**: 388–391.
71. Taylor GW, Taylor I, Black P, et al.: Urinary leukotriene E4 after antigen challenge and in acute asthma and allergic rhinitis. *Lancet* (1989) **i**: 584–588.
72. Westcott JY, Johnston K, Batt RA, Wenzel SE, Voelkel NF: Measurement of peptidoleukotrienes in biological fluids. *J Appl Physiol* (1990) **68**: 2640–2648.
73. Orning L, Kaijser L, Hammarstrom S: In vivo metabolism of leukotriene C4 in man: urinary excretion of leukotriene E4. *Biochem Biophys Res Commun* (1985) **130**: 214–220.
74. Maltby NH, Taylor GW, Ritter JM, Moore K, Fuller RW, Dollery CT: Leukotriene C4 elimination and metabolism in man. *J Allergy Clin Immunol* (1990) **85**: 3–9.
75. Sladek K, Dworski R, Fitzgerald GA, et al.: Allergen-stimulated release of thromboxane A2 and leukotriene E4 in humans. Effect of indomethacin. *Am Rev Respir Dis* (1990) **141**: 1441–1445.
76. Tagari P, Rasmussen JB, Delorme D, et al.: Comparison of urinary leukotriene E4 and 16-carboxytetranordihydro leukotriene E4 excretion in allergic asthmatics after inhaled antigen. *Eicosanoids* (1990) **3**: 75–80.

77. Manning PJ, Rokach J, Malo JL, et al.: Urinary leukotriene E4 levels during early and late asthmatic responses. *J Allergy Clin Immunol* (1990) **86**: 211–220.
78. Manning PJ, Watson RM, Margolskee DJ, Williams VC, Schwartz JI, O'Byrne PM: Inhibition of exercise-induced bronchoconstriction by MK-571, a potent leukotriene D4-receptor antagonist. *N Engl J Med* (1990) **323**: 1736–1739.
79. Drazen JM, Obrien J, Sparrow D, et al.: Recovery of leukotriene-E4 from the urine of patients with airway obstruction. *Am Rev Respir Dis* (1992) **146**: 104–108.
80. Asano K, Lilly CM, Odonnell WJ, et al.: Diurnal variation of urinary leukotriene E(4) and histamine excretion rates in normal subjects and patients with mild-to-moderate asthma. *J Allergy Clin Immunol* (1995) **96**: 643–651.
81. Coleman RA, Eglen RM, Jones RL, et al.: Prostanoid and leukotriene receptors: a progress report from the IUPHAR working parties on classification and nomenclature. *Prostaglandins and Related Compounds* (1995) **23**: 283–285.
82. Carter GW, Young PR, Albert DH, et al.: 5-Lipoxygenase inhibitory activity of zileuton. *J Pharmacol Exp Ther* (1991) **256**: 929–937.
83. Depre M, Friedman B, Vanhecken A, et al.: Pharmacokinetics and pharmacodynamics of multiple oral doses of MK-0591, a 5-lipoxygenase-activating protein inhibitor. *Clin Pharmacol Ther* (1994) **56**: 22–30.
84. Gardiner PJ, Cuthbert NJ, Francis HP, et al.: Inhibition of antigen-induced contraction of guinea-pig airways by a leukotriene synthesis inhibitor, BAY x1005. *Eur J Pharmacol* (1994) **258**: 95–102.
85. Fuller RW, Black PN, Dollery CT: Effect of the oral leukotriene D4 antagonist LY171883 on inhaled and intradermal challenge with antigen and leukotriene D4 in atopic subjects. *J Allergy Clin Immunol* (1989) **83**: 939–944.
86. Israel E, Juniper EF, Callaghan JT, et al.: Effect of a leukotriene antagonist, LY171883, on cold air-induced bronchoconstriction in asthmatics. *Am Rev Respir Dis* (1989) **140**: 1348–1353.
87. Christie L, Lee TH: The effects of SKF104353 on aspirin induced asthma. *Am Rev Respir Dis* (1991) **144**: 957–958.
88. Dahlen SE, Dahlen B, Eliasson E, et al.: Inhibition of allergic bronchoconstriction in asthmatics by the leukotriene-antagonist ICI-204,219. *Adv Prostaglandin Thromboxane Leukotriene Res* (1991) **21A**: 461–464.
89. Taylor IK, O'Shaughnessy KM, Fuller RW, Dollery CT: Effect of cysteinyl-leukotriene receptor antagonist ICI 204,219 on allergen-induced bronchoconstriction and airway hyperreactivity in atopic subjects. *Lancet* (1991) **337**: 690–694.
90. Findlay SR, Barden JM, Easley CB, Glass M: Effect of the oral leukotriene antagonist, ICI 204,219, on antigen-induced bronchoconstriction in subjects with asthma. *J Allergy Clin Immunol* (1992) **89**: 1040–1045.
91. Finnerty JP, Wood-Baker R, Thomson H, Holgate ST: Role of leukotrienes in exercise-induced asthma: inhibitory effect of ICI 204219, a potent LTD4 receptor antagonist. *Am Rev Respir Dis* (1992) **145**: 746–749.
92. Dahlen B, Kumlin M, Margolskee DJ, et al.: The leukotriene-receptor antagonist MK-0679 blocks airway obstruction induced by inhaled lysine-aspirin in aspirin-sensitive asthmatics. *Eur Respir J* (1993) **6**: 1018–1026.
93. Dahlen B, Margolskee DJ, Zetterstrom O, Dahlen SE: Effect of the leukotriene receptor antagonist MK-0679 on baseline pulmonary function in aspirin sensitive asthmatic subjects. *Thorax* (1993) **48**: 1205–1210.
94. Friedman BS, Bel EH, Buntinx A, et al.: Oral leukotriene inhibitor (MK-886) blocks allergen-induced airway responses. *Am Rev Respir Dis* (1993) **147**: 839–844.
95. Makker HK, Lau LC, Thomson HW, Binks SM, Holgate ST: The protective effect of inhaled leukotriene-D(4) receptor antagonist ICI-204,219 against exercise-induced asthma. *Am Rev Respir Dis* (1993) **147**: 1413–1418.
96. Oshaughnessy KM, Taylor IK, Oconnor B, Oconnell F, Thomson H, Dollery CT: Potent leukotriene-D(4) receptor antagonist ICI-204,219 given by the inhaled route inhibits the early but not the late phase of allergen-induced bronchoconstriction. *Am Rev Respir Dis* (1993) **147**: 1431–1435.

97. Dahlen B, Zetterstrom O, Bjorck T, Dahlen SE: The leukotriene-antagonist ICI-204,219 inhibits the early airway reaction to cumulative bronchial challenge with allergen in atopic asthmatics. *Eur Respir J* (1994) **7**: 324–331.
98. Diamant Z, Timmers MC, Vanderveen H, et al.: The effect of MK-0591, a novel 5-lipoxygenase activating protein inhibitor, on leukotriene biosynthesis and allergen-induced airway responses in asthmatic subjects *in vivo*. *J Allergy Clin Immunol* (1995) **95**: 42–51.
99. Christie PE, Smith CM, Lee TH: The potent and selective sulfidopeptide leukotriene antagonist, SK&F 104353, inhibits aspirin-induced asthma. *Am Rev Respir Dis* (1991) **144**: 957–958.
100. Israel E, Fischer AR, Rosenberg MA, et al.: The pivotal role of 5-lipoxygenase products in the reaction of aspirin-sensitive asthmatics to aspirin. *Am Rev Respir Dis* (1993) **148**: 1447–1451.
101. Nasser SMS, Lee TH: Aspirin-induced early and late asthmatic responses. *Clin Exp Allergy* (1995) **25**: 1–3.
102. Nasser SM, Bell GS, Foster S, et al.: Effect of the 5-lipoxygenase inhibitor ZD2138 on aspirin-induced asthma. *Thorax* (1994) **49**: 749–756.
103. Cloud ML, Enas GC, Kemp J, et al.: A specific LTD4/LTE4-receptor antagonist improves pulmonary function in patients with mild, chronic asthma. *Am Rev Respir Dis* (1989) **140**: 1336–1339.
104. Israel E, Rubin P, Kemp JP, et al.: The effect of inhibition of 5-lipoxygenase by zileuton in mild to moderate asthma. *Ann Intern Med* (1993) **119**: 1059–1066.
105. Spector SL, Smith LJ, Glass M, et al.: Effects of 6 weeks of therapy with oral doses of ICI 204,219, a leukotriene D_4 receptor antagonist, in subjects with bronchial asthma. *Am J Respir Crit Care Med* (1994) **150**: 618–623.
106. Israel E, Cohn J, Dube L, Drazen JM: Effect of treatment with zileuton, a 5-lipoxygenase inhibitor, in patients with asthma: a randomized controlled trial. *JAMA* (1996) **275**: 931–936.
107. Liu M, Dube LM, Lancaster J: Acute and chronic effects of a 5-lipoxygenase inhibitor in asthma: a 6-month randomized multicenter trial. *J Allergy Clin Immunol* (1996) **98**: 859–871.
108. Juniper EF, Guyatt GH, Epstein RS, Ferric PJ, Jaeschke R, Hiller TK: Evaluation of impairment of health related quality of life in asthma: development of a questionnaire for use in clinical trials. *Thorax* (1992) **47**: 76–83.
109. Barnes N, Piper PJ, Costello J: The effect of an oral leukotriene antagonist L-649,923 on histamine and leukotriene D4-induced bronchoconstriction in normal man. *J Allergy Clin Immunol* (1987) **79**: 816–821.
110. Phillips GD, Rafferty P, Robinson C, Holgate ST: Dose-related antagonism of leukotriene D4-induced bronchoconstriction by p.o. administration of LY-171883 in nonasthmatic subjects. *J Pharmacol Exp Ther* (1988) **246**: 732–738.
111. Evans JM, Barnes NC, Zakrzewski JT, et al.: L-648,051, a novel cysteinyl-leukotriene antagonist is active by the inhaled route in man. *Br J Clin Pharmacol* (1989) **28**: 125–135.
112. Smith LJ, Geller S, Ebright L, Glass M, Thyrum PT: Inhibition of leukotriene D4-induced bronchoconstriction in normal subjects by the oral LTD4 receptor antagonist ICI 204,219. *Am Rev Respir Dis* (1990) **141**: 988–992.
113. Kips JC, Joos GF, Delepeleire I, Margolskee DJ, Buntinx A, Pauwels RA, Vanderstraeten ME: MK-571, a potent antagonist of leukotriene D4-induced bronchoconstriction in the human. *Am Rev Respir Dis* (1991) **144**: 617–621.

17

Kinins

DAVID PROUD

INTRODUCTION

Kinins are potent vasoactive peptides that are generated during inflammatory events *in vivo*. Although the history of the kallikrein–kinin system can be traced back to the observation of Abelous and Bardier in 1909 that intravenous injection into the dog of an alcohol-insoluble fraction of human urine caused a pronounced, but reversible, fall in systolic blood pressure,[1] it was not until almost two decades later that Frey[2] established that the substance responsible for this effect was non-dialysable and thermolabile. Further studies, in collaboration with Werle and Kraut, showed that a similar activity was present in blood and in the pancreas.[3,4] On the incorrect assumption that the active substance, in each case, was identical and was derived from the pancreas, it was named kallikrein (from the Greek *kallikreas*, meaning pancreas). Werle subsequently demonstrated that kallikrein enzymatically released a substance from plasma that was capable of contracting smooth muscle.[5] This generated material was later named 'kallidin'. Working independently, Rocha e Silva and colleagues coined the name 'bradykinin' for a similar smooth muscle spasmogen liberated from plasma by the action of either trypsin or the venom of the snake *Bothrops jararaca*.[6]

In the years since these pioneering studies, major progress has been made in delineating the pharmacological properties of kinins and in understanding the biochemical pathways by which these peptides are formed and metabolized in humans. Moreover, evidence has accumulated to suggest that kinins may be important mediators during inflammatory diseases of the airways, such as asthma. The present chapter reviews our current knowledge of the potential role of kinins in airway inflammation.

STRUCTURE, FORMATION AND METABOLISM

Kinins are generated from α_2-globulin precursor proteins called kininogens. The two precursors found in humans, high molecular weight (HMW) kininogen and low molecular weight (LMW) kininogen, are derived from a single gene as a consequence of alternative RNA splicing.[7] Kininogens are synthesized in the liver. HMW kininogen represents approximately one-third of the kininogen in blood, while LMW kininogen constitutes the remaining two-thirds. Both kininogens gain access to the interstitium and the lymph and extravascular LMW kininogen has also been detected in the distal nephron.[8]

Three kinins have been reported to exist in humans: bradykinin, lysylbradykinin (kallidin) and methionyllysylbradykinin. All three peptides contain the C-terminal nonapeptide sequence of bradykinin (Fig. 17.1). It is now accepted, however, that methionyllysylbradykinin is a laboratory artefact that is found only under conditions of acidification when it is produced, by the action of pepsin, from kininogen that has already undergone limited hydrolysis.[9] Thus, bradykinin and lysylbradykinin are the physiologically relevant kinins in humans.

Enzymes that release kinins from kininogens are generally referred to as kininogenases. Although plasmin, trypsin and mast cell tryptase[10] are capable of generating bradykinin *in vitro*, these enzymes are likely to play little role in kinin generation *in vivo*. The historical name, kallikrein, is still used to refer to the most physiologically important kininogenases from blood (plasma kallikrein) and from the major exocrine organs (tissue, or glandular, kallikrein). This shared nomenclature is unfortunate, since the plasma and tissue enzymes are derived from different genes and are biochemically and immunologically distinct from each other. Plasma kallikrein is synthesized in the liver and exists in the blood as a single-

Aminopeptidase M
↓
LYS-ARG-PRO-PRO-GLY-PHE-SER-PRO-PHE-ARG
(lysylbradykinin/kallidin)

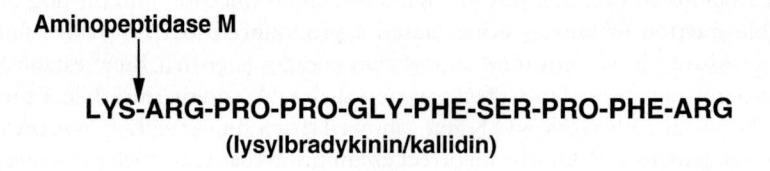

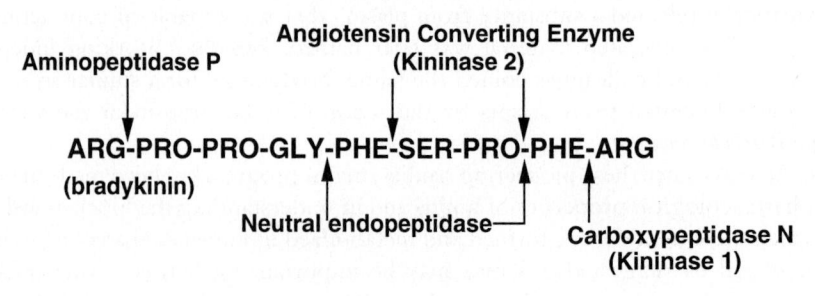

Aminopeptidase P   **Angiotensin Converting Enzyme (Kininase 2)**
↓   ↓
ARG-PRO-PRO-GLY-PHE-SER-PRO-PHE-ARG
(bradykinin)
              ↑   ↑
          Neutral endopeptidase   **Carboxypeptidase N (Kininase 1)**

Fig. 17.1 Structure of bradykinin and lysylbradykinin and the sites of hydrolysis of these peptides by some of the major kininases.

chain γ-globulin zymogen, prekallikrein, that circulates in a complex with HMW kininogen.[11] Activation of prekallikrein to kallikrein *in vivo* occurs principally as a result of the factor XII-dependent process, referred to as contact activation.[12] When kallikrein is generated by the interaction of prekallikrein, HMW kininogen and factor XII with certain negatively charged surfaces, this enzyme acts on its preferred substrate, HMW kininogen, to release bradykinin.[13] The kallikrein is then rapidly inactivated by the plasma protease inhibitors, C1-inactivator and α_2-macroglobulin. While plasma kallikrein generates bradykinin, tissue kallikreins are unique in that they hydrolyse two dissimilar bonds within either HMW or LMW kininogen to release lysylbradykinin. Although originally believed to be present only in the major exocrine organs, tissue kallikreins are acidic glycoproteins that are now known to enjoy a widespread distribution in exocrine and endocrine tissues.[12] Tissue kallikrein is present in both the upper and lower airways in humans[14,15] and has been localized to the serous cells of submucosal glands.[16] There are no effective, naturally occurring inhibitors of tissue kallikreins in humans. The ability of tissue kallikrein to generate kinins from both HMW and LMW kininogens provides it with more available substrate than plasma kallikrein. This, together with its resistance to inhibition, suggests that tissue kallikreins may be of particular importance in kinin generation in inflammatory events.

Once bradykinin and lysylbradykinin are generated *in vivo*, metabolic destruction is a major mechanism for regulating their actions. Virtually all tissues and biological fluids contain enzymes (kininases) that are capable of degrading kinins.[17] Hydrolysis of any of the peptide bonds within the bradykinin moiety leads to a loss of biological activity. Although peptidases derived from all of the major classes of proteolytic enzymes can degrade kinins, those kininases that are believed to be the most important regulators of these peptides during airway inflammation in humans are shown in Fig. 17.1. These peptidases are either derived from plasma and enter the airway mucosa by transudation during inflammatory events or are present on the surface of epithelial or endothelial cells. Hydrolysis of kinins by plasma peptidases has been reasonably well delineated but the full profile of peptidases on cells in the airways remains to be determined. An aminopeptidase M-like enzyme has been detected in airway secretions during allergic inflammation.[18] This enzyme does not result in the loss of biological activity of kinins but converts lysylbradykinin to bradykinin by removal of the N-terminal lysine residue. Aminopeptidase M in the airway originates, in part, from plasma but a similar activity is also present on the surface of respiratory epithelial cells.[19] The major plasma enzymes that would contribute to kinin degradation are carboxypeptidase N (kininase 1) and angiotensin-converting enzyme (ACE; kininase 2). Carboxypeptidase N-like activity has been shown to enter airway secretions from plasma during allergic inflammation[18] and degrades kinins by removal of the C-terminal arginine residue to produce the B_1 kinin receptor agonists, des(Arg^9)bradykinin and des(Arg^{10})lysylbradykinin (see below). Lower levels of ACE also enter the nasal mucosa from plasma,[18] although the major source of this enzyme is the endothelial cell surface. ACE degrades kinins by sequential removal of C-terminal dipeptides and will also degrade des(Arg) kinins by removal of the C-terminal tripeptide. Finally, a lot of recent attention has focused on the potential role of neutral endopeptidase in peptide hydrolysis during airway inflammation. This peptidase is present on the surface of the respiratory epithelial cell and hydrolyses kinins at the same site as ACE to release the C-terminal dipeptide.[19,20]

RECEPTORS AND GENERAL PHARMACOLOGICAL PROPERTIES

Bradykinin and lysylbradykinin display essentially the same pharmacological properties. Some minor differences in the potencies of these two peptides are seen in intact tissue preparations but this probably reflects minor variations in their rates of metabolism. Kinins exert their actions via two subtypes of receptors, originally defined in animal tissues by Regoli and Barabe in 1980.[21] In this initial definition, the B_1 kinin receptor was characterized by the fact that the carboxypeptide metabolites of kinins, des(Arg^9)-bradykinin and des(Arg^{10})lysylbradykinin, were more potent than the parent peptides and their actions could be antagonized by Leu^8-des(Arg^9)bradykinin. On the B_2 receptor, bradykinin and lysylbradykinin were equally active, but the carboxypeptidase metabolites were inactive. Another interesting difference between the two receptor subtypes is that, while the B_2 receptor is constitutively expressed on many cell types, B_1-receptor expression requires induction. In animals, *de novo* expression of B_1 receptors *in vivo* occurs upon exposure to noxious stimuli, including bacterial lipopolysaccharide and ultraviolet light, or to pro-inflammatory cytokines.[22,23] The ability of injurious and pro-inflammatory stimuli to induce B_1 receptors in animal models raises the possibility that this receptor may play a role in the actions of kinins in chronic inflammatory conditions. More recently, the development of selective antagonists confirmed the existence of these two receptor subtypes[24-26] and both human kinin receptors have now been cloned.[27,28] Expression of the human B_1 receptor has revealed some interesting properties. In contrast to the receptor from several animal species, des(Arg^9)bradykinin is an ineffective ligand at the human receptor, and it is clear that des(Arg^{10})lysylbradykinin is the natural ligand.[28,29] This has important implications for studies to evaluate the role of B_1 receptors in the actions of kinins in the human airways (see below). The other interesting feature of the human B_1 receptor is that it is not subject to ligand-induced desensitization.[29] Thus, once the receptor is induced by pro-inflammatory stimuli, functional responses can continue unabated as long as ligand is present, providing further support for a potential role of this receptor in chronic inflammatory conditions.

In general, the pharmacological properties of kinins would suggest that they are ideal mediators of inflammation.[30] Bradykinin and lysylbradykinin contract most types of intestinal smooth muscle and either contract or have no effect on isolated airway smooth muscle, depending on the species from which the tissue is derived (see below). Kinins may contract or relax isolated vascular smooth muscle depending on the vessel being examined. *In vivo*, however, the net effect of kinins are to cause peripheral vasodilation and hypotension. In addition to being vasodilators, bradykinin and lysylbradykinin are approximately 100-fold more potent than histamine in increasing the permeability of postcapillary venules and, as a direct consequence of this, they are potent inducers of oedema. Kinins are also potent stimulators of epithelial ion transport and can stimulate sensory nerves to cause pain and hyperalgesia. Moreover, they are clearly capable of stimulating the release of biologically active lipids, such as prostaglandins and platelet-activating factor, from a variety of cell types.[30]

KININ FORMATION IN AIRWAY INFLAMMATION

The first direct evidence that kinins could be generated during airway inflammation in humans was provided using a model of nasal provocation with allergen in which inflammatory mediators could be measured in secretions recovered by lavage. Insufflation of an appropriate allergen into the nasal cavity of allergic subjects resulted in the immediate manifestation of sneezing, nasal congestion and rhinorrhoea. Concomitant with this symptomatic response, strikingly increased levels of kinins could be measured in recovered lavages.[31] Increased kinin generation correlated with symptoms and with increases in other inflammatory mediators, including histamine. Kinin generation was not observed following allergen challenge of non-allergic subjects nor when allergic subjects were challenged with a non-relevant allergen. High performance liquid chromatography (HPLC) analysis showed that both bradykinin and lysylbradykinin were produced. Following these initial observations, kinin generation has now been documented in a variety of inflammatory conditions of the upper airways, including the late-phase allergic response[32] and both experimental and naturally occurring viral infections.[33,34] In the lower airways, kinin levels have been shown to increase in bronchoalveolar lavage fluids from asthmatic subjects who either had active symptoms of asthma or were responding to aerosolized allergen challenge when compared with lavage samples from normal subjects.[15] More recently, endobronchial challenge has been used to demonstrate increased kinin generation in asthmatic subjects after allergen challenge compared with saline challenge of the same subjects.[35,36]

EFFECTS OF KININS ON AIRWAYS

The effect of bradykinin on isolated airway smooth muscle varies depending on the species being examined. Kinins constrict isolated guinea-pig tracheal rings via a prostaglandin-dependent pathway but have no effect on airways from rabbits, rat or dogs.[37,38] At best, bradykinin is a weak constrictor of isolated large human airways,[38–41] although Simonsson and colleagues[40] reported an enhanced sensitivity of bronchial strips taken from patients with chronic airflow obstruction. Isolated human peripheral airways are contracted by bradykinin via a cyclooxygenase-dependent mechanism, but the maximal contraction observed is only 30–40% of that induced by cholinergic agonists.[41,42]

The effects of bradykinin on human airway function *in vivo* depend upon the route of administration. When given intravenously, bradykinin causes a modest, transient fall in airway function that has been suggested to be due to alveolar duct constriction.[39] When given by inhalation, however, bradykinin is a potent bronchoconstrictor in asthmatic, but not normal, subjects.[43–46] The aerosolized peptide also induces retrosternal discomfort and cough in all subjects.[45] When administered to the peripheral airways, bradykinin increases peripheral airway resistance in asthmatic, but not normal, subjects (Fig. 17.2). Interestingly, however, reactivity in the peripheral airways does not correlate with whole lung reactivity to bradykinin.[47]

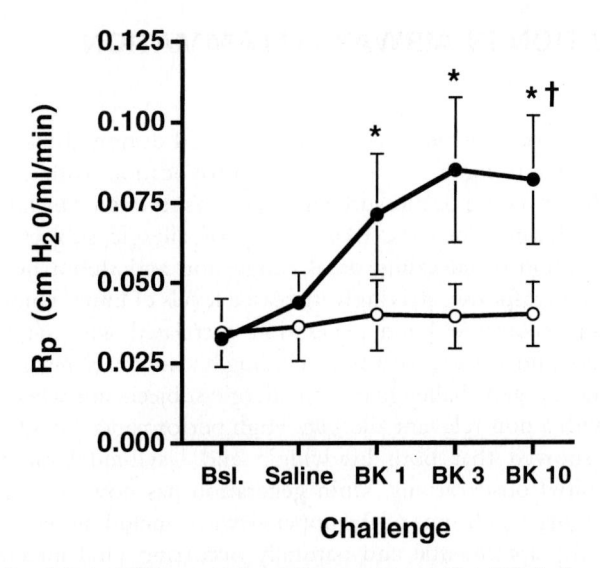

Fig. 17.2 Effect of bradykinin (BK) on peripheral airway resistance (R_p). BK challenges refer to concentrations in mg/ml. Data are expressed as mean ± SEM. ●, asthmatic subjects; ○, normal subjects. *, Statistically significant difference between R_p and saline control R_p, $P < 0.05$; †, mean R_p does not include a single asthmatic subject who reached the maximum allowable peripheral airway pressure after the 3 mg/ml dose of BK; Bsl, baseline. Reproduced from ref 47, with permission.

In addition to affecting bronchial tone, increased kinin generation during asthmatic reactions could lead to: (a) an increase in ciliary beat frequency via a prostaglandin-dependent pathway,[48] and (b) an increase in the volume of airway secretions. Lysylbradykinin has been shown to increase the production of respiratory mucus macromolecules in canine tracheal explants,[49] while bradykinin has been shown to increase canine tracheal gland secretion *in vivo* by a reflex mechanism after stimulation of bronchial C-fibres.[50] Moreover, since bradykinin can stimulate chloride secretion across the respiratory epithelium,[51,52] transepithelial water transport could also contribute to increased airway secretions. Kinins could also increase the volume of secretions as a result of increased vascular permeability and transudation of plasma.[47]

MECHANISMS OF ACTION

It has been implied that the bronchoconstrictor action of kinins in asthmatic subjects is mediated via B_2 kinin receptors, because bradykinin and lysylbradykinin are potent bronchoconstrictors, while the putative B_1-receptor agonist, des(Arg^9)bradykinin, is inactive.[53] The recent observation that des(Arg^9)bradykinin is an ineffective agonist at the human B_1 receptor[28,29] cast doubt upon this assertion, but a double-blind study comparing the reactivity of asthmatic subjects to bradykinin and the appropriate human B_1-receptor agonist ligand, des(Arg^{10})lysylbradykinin, confirmed that the B_1-receptor agonist is not a bronchoconstrictor, even in subjects with good sensitivity to bradykinin (D. Proud, unpublished observations). Bradykinin-induced bronchoconstriction is not inhibited by administration of cyclooxygenase inhibitors[45,46] or an antihistamine.[46] This latter finding is consistent with observations that bradykinin is not a secretagogue for human mast cells.[54] Interestingly, bronchoconstriction induced by inhalation of bradykinin is exacerbated by prior administration of an inhibitor of nitric oxide (NO) synthase, indicating that the effects of bradykinin are inhibited by the formation of NO in the airways.[55] Although the exact mechanism by which bradykinin induces bronchoconstriction remains to be elucidated, several pieces of evidence imply that neural reflexes are involved. Moreover, the capacity of bradykinin to induce these neural reflexes seems to be restricted to inflamed airways. This latter concept is supported by the fact that bradykinin does not cause bronchconstriction in normal subjects and by the observation that airway responsiveness to bradykinin correlates with the degree of eosinophilic inflammation in the airways.[56] Furthermore, exacerbation of airway inflammation in asthmatic subjects by allergen provocation increases airway reactivity to bradykinin to a much greater degree than that to methacholine.[57] The lack of a pronounced direct effect of bradykinin on isolated smooth muscle suggests the involvement of indirect pathways and the ability of bradykinin to evoke cough indicates that sensory nerves can be stimulated. Following bradykinin-evoked bronchoconstriction there is a decreased sensitivity to subsequent challenges with bradykinin.[45] Although the frequency of this tachyphylactic response is controversial,[58] it is consistent with neuronal desensitization. The ability of sodium cromoglycate and nedocromil sodium[59] to inhibit bradykinin-induced bronchoconstriction also has been interpreted as supporting a role for sensory nerve stimulation, while the inhibitory effect of ipratroprium bromide[45] in a limited number of subjects indicates a potential role of cholinergic reflexes in bradykinin-induced bronchoconstriction. This concept is further supported by studies in the human upper airways, where subjects with active allergic inflammation show hyperreactivity to bradykinin. This hyperreactivity is due to the stimulation of neural reflexes, including central cholinergic reflexes, that do not occur in normal, non-inflamed airways.[60] These data, together with the observation that bradykinin causes bronchoconstriction in dogs by vagal reflex following stimulation of sensory C-fibres,[61] suggest that the bronchoconstrictor actions of bradykinin in humans may also involve sensory stimulation, induction of parasympathetic reflexes and, possibly, the concomitant release of neuropeptides.

SUMMARY

In the last decade, there has been a renewed interest in the role of kinins in human airway inflammation. The correlation between kinin generation and symptoms of inflammation, together with the demonstration that administration of kinins to the airway mucosa can induce relevant symptoms, provides strong circumstantial support for a role of kinins in the pathogenesis of diseases such as asthma. To definitively establish the contribution of kinins to asthma, however, it will clearly be necessary to be able to use specific pharmacological interventions to block their actions and demonstrate a concomitant effect on symptoms. The development of the first competitive kinin antagonists[25] appeared to provide an exciting opportunity to perform such interventive studies. Initial observations using one such compound, NPC 567, in the sheep were encouraging, in that a bradykinin antagonist inhibited both antigen-induced hyperreactivity and the late-phase response.[62,63] Administration of the same compound to the upper airways of humans, however, failed to block the effects of a challenge with bradykinin.[64] This early compound suffered from a relatively low affinity for the kinin receptor and was, in addition, readily susceptible to hydrolysis by peptidases. These problems appeared to have been overcome with the development of a potent second-generation kinin antagonist, Hoe 140.[26] However, this compound still suffered from a relatively short half-life in the airways due to mucociliary clearance.[65] Despite this limitation, and some problems with study design, Hoe 140 did cause significant improvement in pulmonary function in moderately severe asthmatic subjects.[66] Given these encouraging results, and the fact that non-peptide orally active compounds are now in development, it seems likely that the next few years will provide exciting opportunities to delineate the role of kinins in the pathogenesis of asthma and other inflammatory diseases of the human airways.

ACKNOWLEDGEMENTS

Dr Proud acknowledges support from grant number HL 32272 from the National Institutes of Health.

REFERENCES

1. Abelous JE, Bardier E: Les substance hypotensives de l'urine humaine normale. *C R Soc Biol* (1909) **66**: 511–512.
2. Frey EK: Zusammenhange zwischen herzarbeit und nierentatigheit. *Arch Klin Chir* (1926) **142**: 663–669.
3. Frey EK, Kraut H: Ein neues kreislaufhormon und seine wirkung. *Arch Exp Path Pharm* (1928) **133**: 1–56.
4. Kraut H, Frey EK, Werle E: Der nachweis eines kreislaufhormons in der pankreasdrüse. *Hoppe-Seyler's Z Physiol Chem* (1930) **189**: 97–106.
5. Werle E, Götze W, Keppler A: Über die wirkung des kallikreins auf den isolierten darm und über eine neue darmkontrahierende substanz. *Biochem Z* (1937) **281**: 217–233.
6. Roche e Silva M, Beraldo WT, Rosenfeld G: Bradykinin, a hypotensive and smooth muscle stimulating factor released from plasma globulin by snake venoms and by trypsin. *Am J Physiol* (1949) **156**: 261–273.
7. Kitamura N, Kitagawa H, Fukushima D, Takagaki Y, Miyata T, Nakanishi S: Structural organization of the human kininogen gene and a model for its evolution. *J Biol Chem* (1985) **260**: 8610–8617.
8. Proud D, Perkins M, Pierce JV, *et al.*: Characterization and localization of human renal kininogen. *J Biol Chem* (1981) **256**: 16 034–16 039.
9. Guimaraes JA, Pierce JV, Hial V, Pisano JJ: Methionyl-lysyl-bradykinin: the kinin released by pepsin from human kininogens. *Adv Exp Med Biol* (1976) **70**: 265–269.
10. Proud D, Siekierski ES, Bailey GS: Identification of human lung mast cell kininogenase as tryptase and relevance of tryptase kininogenase activity. *Biochem Pharmacol* (1988) **37**: 1473–1480.
11. Mandle RJ Jr, Colman RW, Kaplan AP: Identification of prekallikrein and HMW-kininogen as a circulating complex in human plasma. *Proc Natl Acad Sci USA* (1976) **73**: 4179–4183.
12. Proud D, Kaplan AP: Kinin formation: mechanisms and role in inflammatory disorders. *Annu Rev Immunol* (1988) **6**: 49–83.

13. Pierce JV, Guimaraes JA: Further characterization of highly purified human plasma kininogens. In Pisano JJ, Austen KF (eds) *Chemistry and Biology of the Kallikrein–Kinin System in Health and Disease*. Washington, DC, DHEW Publ. No. (NIH)76-791, 1976, pp 121–127.
14. Baumgarten CR, Nichols RC, Naclerio RM, Proud D: Concentrations of glandular kallikrein in human nasal secretions increase during experimentally-induced allergic rhinitis. *J Immunol* (1986) **137**: 1323–1328.
15. Christiansen SC, Proud D, Cochrane CG: Detection of tissue kallikrein in the bronchoalveolar lavage fluids of asthmatic subjects. *J Clin Invest* (1987) **79**: 188–197.
16. Proud D, Vio CP: Localization of immunoreactive tissue kallikrein in the human trachea. *Am J Respir Cell Mol Biol* (1993) **8**: 16–19.
17. Erdos EG: Kininases. In Erdos EG (ed) *Bradykinin, Kallidin and Kallikrein. Handbook of Experimental Pharmacology*, vol 25 (Suppl). New York, Springer-Verlag, 1979, pp 427–487.
18. Proud D, Baumgarten CR, Naclerio RM, Ward PE: Kinin metabolism in nasal secretions during experimentally-induced allergic rhinitis. *J Immunol* (1987) **138**: 428–434.
19. Proud D, Subauste MC, Ward PE: Glucocorticoids do not alter peptidase expression on a human bronchial epithelial cell line. *Am J Respir Cell Mol Biol* (1994) **11**: 57–65.
20. Erdos EG, Skidgel RA: Neutral endopeptidase 24.11 (enkephalinase) and related regulators of peptide hormones. *FASEB J* (1989) **3**: 145–151.
21. Regoli D, Barabe J: Pharmacology of bradykinin and related peptides. *Pharmacol Rev* (1980) **32**: 1–46.
22. Davis AJ, Perkins MN: The involvement of bradykinin B_1 and B_2 receptor mechanisms in cytokine-induced mechanical hyperalgesia in the rat. *Br J Pharmacol* (1994) **113**: 63–68.
23. Perkins MN, Kelly D: Interleukin-1 beta-induced desArg9bradykinin-mediated thermal hyperalgesia in the rat. *Neuropharmacology* (1994) **33**: 657–660.
24. Marceau F: Kinin B_1 receptors: a review. *Immunopharmacology* (1995) **30**: 1–26.
25. Vavrek RJ, Stewart JM: Competitive antagonists of bradykinin. *Peptides* (1985) **6**: 161–164.
26. Wirth K, Hock FJ, Albus U, et al.: Hoe 140 a new potent and long acting bradykinin-antagonist: *in vivo* studies. *Br J Pharmacol* (1991) **102**: 774–777.
27. Hess JF, Borkowski JA, Young GS, Strader CD, Ransom RW: Cloning and pharmacological characterization of a human bradykinin (BK-2) receptor. *Biochem Biophys Res Commun* (1992) **184**: 260–268.
28. Menke JG, Borkowski JA, Bierilo KK, et al.: Expression cloning of a human B_1 bradykinin receptor. *J Biol Chem* (1994) **269**: 21 583–21 586.
29. Austin CE, Faussner A, Robinson HE, et al.: Stable expression of the human kinin B_1 receptor in CHO cells: characterization of ligand binding and effector pathways. *J Biol Chem* (1997) **272**: 11 420–11 425.
30. Bathon JM, Proud D: Bradykinin antagonists. *Annu Rev Pharmacol Toxicol* (1991) **31**: 129–162.
31. Proud D, Togias A, Naclerio RM, Crush SA, Norman PS, Lichtenstein LM: Kinins are generated *in vivo* following nasal airway challenge of allergic individuals with allergen. *J Clin Invest* (1983) **72**: 1678–1685.
32. Naclerio RM, Proud D, Togias AG, et al.: Inflammatory mediators in late antigen-induced rhinitis. *N Engl J Med* (1985) **313**: 65–70.
33. Naclerio RM, Proud D, Lichtenstein LM, et al.: Kinins are generated during experimental rhinovirus colds. *J Infect Dis* (1987) **157**: 133–142.
34. Proud D, Naclerio RM, Gwaltney JM Jr, Hendley JO: Kinins are generated in nasal secretions during natural rhinovirus colds. *J Infect Dis* (1990) **161**: 120–123.
35. Liu M, Hubbard WC, Proud D, et al.: Immediate and late inflammatory responses to ragweed antigen challenge of the peripheral airways in allergic asthmatics: cellular, mediator, and permeability changes. *Am Rev Respir Dis* (1991) **144**: 51–58.
36. Christiansen SC, Proud D, Sarnoff RB, Juergens U, Cochrane CG, Zuraw BL: Elevation of tissue kallikrein and kinin in the airways of asthmatic subjects after endobronchial allergen challenge. *Am Rev Respir Dis* (1992) **145**: 900–905.
37. Collier HOJ: The action and antagonism of kinins on bronchioles. *Ann NY Acad Sci* (1963) **104**: 290–298.
38. Bhoola KD, Collier HOJ, Schachter M, Shorley PG: Actions of some peptides on bronchial muscle. *Br J Pharmacol* (1962) **19**: 190–197.

39. Newball HH, Keiser HR, Webster ME, Pisano JJ: Effects of bradykinin on human airways. In Pisano JJ, Austen KF (eds) *Chemistry and Biology of the Kallikrein–Kinin System in Health and Disease*, Washington, DC, DHEW Publ. No. (NIH)76–791, 1976, pp 505–511.
40. Simonsson BG, Skoogh B-E, Bergh NP, Andersson R, Svedmyr N: *In vivo* and *in vitro* effects of bradykinin on bronchial motor tone in normal subjects and patients with airway obstruction. *Respiration* (1973) **30**: 378–388.
41. Molimard M, Martin CAE, Naline E, Hirsch A, Advenier C: Contractile effects of bradykinin on the isolated human small bronchus. *Am J Respir Crit Care Med* (1994) **149**: 123–127.
42. Hulsmann AR, Raatgep HR, Saxena PR, Kerrebijn KF, De Jongste JC: Bradykinin-induced contraction of human peripheral airways mediated by both bradykinin β_2 and thromboxane prostanoid receptors. *Am J Respir Crit Care Med* (1994) **150**: 1012–1018.
43. Herxheimer H, Stresemann E: The effects of bradykinin aerosol in guinea pigs and man. *J Physiol* (1961) **158**: 38–39.
44. Varonier HS, Panzani R: The effect of inhalation of bradykinin on healthy and atopic (asthmatic) children. *Int Arch Allergy* (1968) **34**: 293–296.
45. Fuller RW, Dixon CMS, Cuss FMC, Barnes PJ: Bradykinin-induced bronchoconstriction in humans. Mode of action. *Am Rev Respir Dis* (1987) **135**: 176–180.
46. Polosa R, Phillips GD, Lai CKW, Holgate ST: Contribution of histamine and prostanoids to bronchoconstriction provoked by inhaled bradykinin in atopic asthma. *Allergy* (1990) **45**: 174–182.
47. Berman AR, Liu MC, Wagner EM, Proud D: Dissociation of bradykinin-induced plasma exudation and reactivity in peripheral airways. *Am J Respir Crit Care Med* (1996) **154**: 418–423.
48. Tamaoki J, Kobayashi K, Saki N, Chiyotani A, Kanemura T, Takizawa T: Effect of bradykinin on airway ciliary motility and its modulation by neutral endopeptidase. *Am Rev Respir Dis* (1989) **140**: 430–435.
49. Baker AP, Hillegass LM, Holden DA, Smith WJ: Effect of kallidin, substance P, and other basic polypeptides on the production of respiratory macromolecules. *Am Rev Respir Dis* (1977) **115**: 811–817.
50. Davis B, Roberts AM, Coleridge HM, Coleridge JCG: Reflex tracheal gland secretion evoked by stimulation of bronchial C-fibers in dogs. *J Appl Physiol* (1982) **53**: 985–991.
51. Leikauf GD, Ueki IF, Nadel JA, Widdicombe JH: Bradykinin stimulates Cl secretion and prostaglandin E_2 release by canine tracheal epithelium. *Am J Physiol* (1985) **248**: F48–F55.
52. Widdicombe JH, Coleman DL, Finkbeiner WE, Tuet IK: Electrical properties of monolayers cultured from cells of human tracheal mucosa. *J Appl Physiol* (1985) **58**: 1729–1735.
53. Polosa R, Holgate ST: Comparative airway response to inhaled bradykinin, kallidin, and [des-Arg9]bradykinin in normal and asthmatic subjects. *Am Rev Respir Dis* (1990) **142**: 1367–1371.
54. Lawrence ID, Warner JA, Cohan VL, *et al.*; Induction of histamine release from human skin mast cells by bradykinin analogs. *Biochem. Pharmacol.* (1989) **38**: 227–233.
55. Ricciardolo FLM, Geppetti P, Mistretta A, *et al.*: Randomized double-blind placebo controlled study of the effect of inhibition of nitric oxide synthesis in bradykinin-induced asthma. *Lancet* (1996) **348**: 374–377.
56. Roisman GL, Lacronique JG, Desmazes-Dufeu N, Carré C, Le Cae A, Dusser DJ: Airway responsiveness to bradykinin is related to eosinophilic inflammation in asthma. *Am J Respir Crit Care Med* (1996) **153**: 381–390.
57. Berman AR, Togias AG, Skloot G, Proud D: Allergen challenge of atopic asthmatics enhances bronchial reactivity to bradykinin more than that to methacholine. *J Appl Physiol* (1995) **78**: 1844–1852.
58. Rajakulasingam K, Church MK, Howarth PH, Holgate ST: Factors determining bradykinin bronchial responsiveness and refractoriness in asthma. *J Allergy Clin Immunol* (1993) **92**: 140–142.
59. Dixon CMS, Barnes PJ: Bradykinin-induced bronchoconstriction: inhibition by nedocromil sodium and sodium cromoglycate. *Br J Clin Pharmacol* (1989) **27**: 831–836.
60. Riccio MM, Proud D: Evidence that enhanced nasal reactivity to bradykinin in patients with symptomatic allergy is mediated by neural reflexes. *J Allergy Clin Immunol* (1996) **97**: 1252–1263.

61. Kaufman MP, Coleridge HM, Coleridge JCG, Baker DG: Bradykinin stimulates afferent vagal C-fibers in intrapulmonary airways of dogs. *J Appl Physiol* (1980) **48**: 511–517.
62. Soler M, Sielczak M, Abraham WM: A bradykinin-antagonist blocks antigen-induced airway hyperresponsiveness and inflammation in sheep. *Pulmon Pharmacol* (1990) **3**: 9–15.
63. Abraham WM, Burch RM, Farmer SG, Sielczak MW, Ahmed A, Cortes A: A bradykinin antagonist modifies allergen-induced mediator release and late bronchial responses in sheep. *Am Rev Respir Dis* (1991) **143**: 787–796.
64. Pongracic JA, Naclerio RM, Reynolds CJ, Proud D: A competitive kinin receptor antagonist, [DArg^0, Hyp^3, DPhe^7]-bradykinin, does not affect the response to nasal provocation with bradykinin. *Br J Clin Pharmacol* (1991) **31**: 287–294.
65. Proud D, Bathon JM, Togias AG, Naclerio RM: Inhibition of the response to nasal provocation with bradykinin by Hoe 140: efficacy and duration of action. *Can J Physiol Pharmacol* 1995) **73**: 820–826.
66. Akbary AM, Wirth KJ, Schölkens BA: Efficacy and tolerability of icatibant (Hoe 140) in patients with moderately severe chronic bronchial asthma. *Immunopharmacology* (1996) **33**: 238–242.

ns/>
18

Chemokines

K. FAN CHUNG

INTRODUCTION

Chemotactic cytokines or chemokines, an entirely new class of leucocyte chemoattractants (*chemo*attractant cyto*kine*), consist of a superfamily of small secreted factors (8–10 kDa) with little similarity in structure and function to traditional immune cytokines such as tumour necrosis factors (TNF), interferons (IFN) and the interleukins (IL). Up to 20 human chemokines have been identified so far by cloning or biochemical purification and amino acid sequencing. Chemokines have sequences that have been conserved, indicating a common ancestral gene, and share four conserved cysteine residues that form disulphide bonds in the tertiary structures. This superfamily of chemokines has been classified into two main branches according to the position of the first two cysteines in the conserved motif (Table 18.1). The C-X-C branch (where X is an amino acid), also known as α chemokines, is characterized by the separation of the first two cysteines in the primary structure by an amino acid, while in the C-C branch (or β chemokines) the two cysteines are directly adjacent.

The chronic inflammatory response in asthma is characterized by a submucosal infiltration of eosinophils, T-lymphocytes, activated monocytes/macrophages and mast cells.[1-3] Activated T-cells have been identified as CD4$^+$ T-cells that express the Th2 profile of cytokines, i.e. IL-3, IL-4 and IL-5.[4] Some of these T-cells also express CD45RO, supporting their role as memory T-cells.[5] Following allergen challenge, there is a rise in the number of EG2$^+$ cells, indicating activated eosinophils, and CD4$^+$ T-cells[6,7] and recruitment of neutrophils.[8] Although eosinophils are most prominent in the airways of patients dying of severe asthma,[9] a predominant neutrophil infiltration of the airway submucosa has been observed in some cases of asthma deaths of sudden onset.[10] In

Table 18.1 Human chemokine superfamily.

C-X-C chemokines
Epithelial cell-derived neutrophil-activating protein (ENA-78)
Interleukin-8 (IL-8) or neutrophil-activating protein 1 (NAP-1)
Stromal cell-derived factor 1α (SDF-1α) and SDF-1β
Granulocyte chemotactic protein 2 (GCP-2)
Melanocyte growth stimulatory activity (MGSA/GRO-α, β or γ)
Platelet factor 4 (PF-4)
IP-10
Monokine induced by interferon-γ (mig)
Platelet basic protein
  β-Thromboglobulin
    Connective tissue protein III (CTAP-III)
    Neutrophil-activating protein 2 (NAP-2)
Macrophage inflammatory protein 2 (MIP-2)
Cytokine-induced neutrophil chemoattractant (CINC) (rat)

C-C chemokines
Regulated on activation, normal T-cell expressed, and secreted (RANTES)
C10
HC-14
I-309/ T-cell activation gene 3 (TCA3)
Macrophage inflammatory protein 1α (MIP-1α) and MIP-1β
Monocyte chemoattractant protein 1 (MCP-1), MCP-2, MCP-3 and MCP-4
Eotaxin

C chemokine
Lymphotactin

view of the pathophysiological significance of inflammatory cells in asthma, chemokines may play an important role as chemoattractants in this disease.

DISCOVERY AND STRUCTURE

C-X-C chemokines

Platelet factor 4 (PF-4) stored in platelet α-granules was the first member of the C-X-C chemokines to be described in 1955 but IL-8 (or NAP-1) has been the most intensively studied chemokine, with its major actions as a neutrophil chemoattractant and activator. The sequence of an IL-8 cDNA clone was first described in 1987 and expressed in activated human lymphocytes,[11] followed later by the partial amino acid sequences of identical neutrophil chemotactic factors,[12,13] which were identical to the deduced sequence of the IL-8 cDNA clone. The gene encoding IL-8 was cloned and sequenced in 1989.[14] Other C-X-C chemokines were discovered in rapid succession, including neutrophil-activating protein 2 (NAP-2),[15] growth-related oncogene (GRO)-α, GRO-β and GRO-γ,[16,17] epithelial cell-derived neutrophil-activating protein (ENA-78)[18] and

18 Chemokines

granulocyte chemotactic protein 2 (GCP-2).[19] A secreted protein produced by lipopolysaccharide-stimulated murine macrophages, called macrophage inflammatory protein 2 (MIP-2), was found to be a chemoattractant for human neutrophils and to be closely related to GRO.[20] Another cDNA clone, KC, identified a transcript induced in murine fibroblasts and its predicted amino acid sequence was similar to murine MIP-2.[21] Another C-X-C chemokine, cytokine-induced neutrophil chemoattractant (CINC), was first purified from an epithelial cell line of normal rat kidney stimulated by IL-1β or TNF-α, its closest reported human homologue being GRO/MGSA (melanocyte growth stimulating activity).[22] Mig (monokine induced by IFN-γ) was discovered by differential screening of a cDNA library prepared from IFN-γ-activated macrophages[23] and has specific chemoattractant activity for T-cells.[24]

C-C chemokines

The first human C-C chemokine gene, called LD78, was discovered by differential hybridization cloning of human tonsillar lymphocytes.[25] cDNA isoforms of a closely related chemokine, Act-2, were also described[26] and two similar proteins, MIP-1α and MIP-1β, were purified from culture media of endotoxin-stimulated mouse macrophages.[27] Because the close identity of the amino acid sequence between the murine and human proteins (75% identity) suggested that these two molecules were homologues, the terms human MIP-1α and MIP-1β have replaced LD78 and Act-2 respectively. In murine fibroblasts, platelet-derived growth factor induced two genes, one of which proved to be a murine homologue of the GRO gene (KC), the other being designated JE.[28] The human homologue of JE was found to encode a monocyte chemoattractant and activating factor, which led to the identification of monocyte chemoattractant protein 1 (MCP-1). MCP-1 is the best-characterized C-C chemokine, having been purified and cloned from different sources.[26,29,30] HC-14, discovered from IFN-γ-stimulated monocytes and now called MCP-2, has been isolated from osteosarcoma cell cultures;[31] these cell lines also yielded MCP-3, which has been cloned and expressed.[32,33] MCP-4 cDNA has recently been identified in a library constructed from human fetal RNA; MCP-4 has functional similarities to MCP-3 and eotaxin, with chemoattractant activities for monocytes, T-cells and eosinophils.[34]

Other C-C chemokines, I-309, RANTES and HC-14, were purified and cloned as products of activated T-cells.[35-37] Subtractive hybridization was used to find genes expressed uniquely in T-cells and this led to the discovery of RANTES (*r*egulated on *a*ctivation, *n*ormal *T*-cell *e*xpressed, and *s*ecreted) cDNA encoding a polypeptide of 91 amino acids with an 8-kDa secreted protein. RANTES gene is expressed in IL-2-dependent T-cell lines. In peripheral blood mononuclear cells, low but detectable levels of RANTES transcripts can be measured in unstimulated cells; an increase in mRNA is seen 5–7 days after antigen exposure or stimulation with phytohaemagglutinin.[36] RANTES has also been shown to be expressed and released from human eosinophils.[38] Eotaxin was first purified from bronchoalveolar lavage (BAL) fluid of allergen-challenged guinea-pigs.[39,40] Human eotaxin was first cloned as a human homologue of guinea-pig eotaxin.[41] Eotaxin is a highly specific chemoattractant for eosinophils.

C chemokine

The discovery of lymphotactin, which is a specific chemoattractant for lymphocytes,[42] has led to the description of another branch, the C branch. Lymphotactin lacks the first and third cysteines in the four-cysteine motif, but shares much similarity in amino acid sequence to the C-C chemokines.

CELL SOURCES

In general, monocytes and tissue macrophages are a rich source of C-X-C and C-C chemokines. Monocytes respond to a large variety of pro-inflammatory agents, including IL-1α, IL-1β, TNF-α, granulocyte–macrophage colony-stimulating factor (GM-CSF), IL-3, lipopolysaccharide and immune complexes, by releasing IL-8. IL-8 has also been induced following adherence of monocytes to plastic and by changes in ambient oxygen.[43,44] GRO-α, GRO-β and GRO-γ are expressed and secreted by monocytes and macrophages.[16,45,46] MCP-1 and MCP-2 are major stimulated products of monocytes. Both MIP-1α and MIP-1β genes can be coordinately expressed after stimulation of T-cells (e.g. with anti-CD3), B-cells or monocytes and macrophages (e.g. with lipopolysaccharide).[25,27,35,47–50] The MIP-1α gene is rapidly induced in human monocytes following adherence to endothelial cells and to other substrates.[51]

Lymphocytes are sources of some C-C chemokines, particularly RANTES,[35,36] I-309,[35] MIP-1α[47,52] and MIP-1β,[47,53] but are less prominent than mononuclear phagocytes as C-X-C chemokine producers. Neutrophils produce IL-8 in response to IL-1β, TNF-α, adherence,[54] GM-CSF,[55] GRO-α and GRO-β on adherence to fibronectin[16] and, in addition, the C-X-C chemokine MIP-1α.[56] Eosinophils release IL-8 after stimulation with calcium ionophore A23187[57] and after stimulation with RANTES or platelet-activating factor (PAF) after priming with GM-CSF.[58]

Epithelial cells stimulated with IL-1 or TNF-α produce IL-8,[59,60] GRO-α, GRO-β and GRO-γ,[16,18] ENA-78,[18] MCP-1[61,62] and RANTES,[63] but not MIP-1α. IL-8 expression by epithelial cells is increased by respiratory syncytial virus infections[64] and on exposure to neutrophil elastase.[65] MCP-1 and RANTES immunoreactivity has been reported in human airway epithelium.[63,66] RANTES is produced by human vascular endothelial cells and airway smooth muscle cells.[67,68] In addition, human airway smooth muscle cells are also an important source of IL-8 and, to a lesser extent, of MIP-1α and eotaxin.

REGULATION

The transcriptional control of the IL-8 gene remains the most studied. Several transcriptional regulatory elements can bind to the region preceding the first exon, including NF-κB, NF-IL-6, AP-1, glucocorticoid element and an octamer-binding motif.[14] NF-IL-6 and NF-κB-like factors may act as *cis*-acting elements in IL-8 mRNA expression.[69] IL-8 mRNA expression after stimulation with IL-1 or TNF-α is rapid and results at least partly from transcriptional activation.[60,70–72] A secondary phase of IL-8 mRNA expression

18 Chemokines

following an early rapid increase induced by IL-1 has been observed with cultured human airway epithelial cells. The stability of IL-8 mRNA may be influenced by RNA instability elements, AUUUA, found in the 3′-untranslated region.[73,74] IL-8 expression can be inhibited in blood monocytes[75] and in airway epithelial cells[76] by glucocorticoids; IFN-γ, IL-4 and IL-10 can inhibit IL-8 production in blood monocytes.[75,77,78]

The RANTES gene has several transcriptional consensus elements for DNA binding in its immediate upstream region, including NF-κB, NF-IL-6, AP-1 and AP-3.[79] Many of these potential regulatory sites were originally described in promoters expressed specifically in T-cells and myeloid or erythroid cells, whereas other elements were first described as consensus sites for factors responsive to specific second messenger stimulation. This large number of potential regulatory sites raises the possibility of a wide range of transcriptional control for RANTES expression in different tissues.

MIP-1α, but not RANTES, mRNA expression and protein release can be induced from blood monocytes and alveolar macrophages by IL-1β and lipopolysaccharide.[49,80] IL-1β- and lipopolysaccharide-induced expression and protein release of MIP-1α were inhibited by glucocorticoids through inhibition of transcription.[49] No glucocorticoid response element (GRE) sites have been found upstream of the transcription initiation site of the human MIP-1α gene[81] and the effect of glucocorticoids could be exerted by interaction at other regulatory sites or with other transcription factors such as AP-1. Part of the inhibition of MIP-1α mRNA resulted from a small increase in mRNA breakdown, probably related to repeating nucleotide motifs in the 3′-untranslated region of the MIP-1α mRNA.[47] In cultured human airway epithelial cells, RANTES but not MIP-1α mRNA and protein can be induced synergistically by the mixture of TNF-α, IL-1β and IFN-γ.[63] Although there does not appear to be a GRE consensus element in the upstream region of the RANTES gene, glucocorticoids also potently inhibit the induced expression and release of RANTES.[63] Both the Th2-derived cytokines, IL-10 and IL-13, inhibit MIP-1α release from stimulated monocytes and alveolar macrophages.[82,83]

CHEMOKINES AS CHEMOATTRACTANTS AND CELL ACTIVATORS

Migration of leucocytes from the vascular compartment into tissues occurs through the sequence of adhesion to the endothelial cell via the expression of integrins, diapedesis and migration in response to a chemoattractant gradient. Chemokines may play a major role in activating migrating leucocytes and endothelial cells to increase their adhesiveness and in establishing a chemotactic gradient. Interaction between chemokines and negatively charged proteoglycans may provide a solid phase for maintenance of a persistent chemotactic gradient following a brief burst of chemokine release.[84] The activity of IL-8 as a neutrophil chemoattractant has been shown to be potentiated by its binding to heparan sulphate or heparin, although the IL-8-activating activity is reduced.[85] MIP-1β, when immobilized by binding to proteoglycans, binds to endothelium to trigger the adhesion of T-cells, particularly CD8$^+$ T-cells, to vascular cell adhesion molecule (VCAM-1).[86] MIP-1β has been localized to lymph node endothelium and could act as a tethered ligand on endothelial cells; thus it could provide the required signals for activation of lymphocyte integrins for adhesion to endothelium and migration.

While eosinophil chemoattractant chemokines may have an important local role in eosinophil recruitment from blood microvessels, they may also cooperate with other cytokines such as IL-3, IL-5 and GM-CSF, which promote maturation, activation and prolonged survival of the eosinophil.[87-89] Thus eosinophils may be primed by these cytokines for an enhanced chemotactic response to chemokines. In addition, cytokines such as IL-5 may act remotely as a hormone to stimulate the release into the circulation of a rapidly mobile pool of bone marrow eosinophils.[90]

Neutrophils

IL-8, which has been the most studied of the C-X-C chemokines, induces shape change, a transient rise in intracellular free calcium concentrations ($[Ca^{2+}]_i$), exocytosis with release of enzymes and proteins from intracellular storage organelles and respiratory burst through activation of NADPH oxidase,[91] and as such behaves like a classical chemoattractant. IL-8 also upregulates the expression of two integrins (CD11b/CD18 and CD11c/CD18) during exocytosis of specific granules.[92,93] IL-8 activates neutrophil 5-lipoxygenase, with the formation of leukotriene B_4 and 5-hydroxyeicosatetraenoic acid,[94] and also induces the production of PAF.[95]

Eosinophils

IL-8 induces $[Ca^{2+}]_i$ elevation, shape change and release of eosinophil peroxidase from eosinophils of patients with hypereosinophilic syndrome,[96] and can induce eosinophil chemotaxis of primed eosinophils.[97] However, eosinophils are more responsive to C-C rather than C-X-C chemokines. RANTES is a powerful eosinophil chemoattractant, being as effective as C5a and two to three times more potent than MIP-1α.[98,99] RANTES upregulates the expression of CD11b/CD18 on eosinophils.[100] RANTES and MIP-1α induce exocytosis of eosinophil cationic protein from cytochalasin B-treated cells, although RANTES is relatively weak in this effect.[98] When injected into the skin of dogs, RANTES induced an infiltration of eosinophils and monocytes.[101] RANTES, but not MIP-1α, also elicited a respiratory burst from eosinophils.[98] MCP-3 and MCP-4 are as effective as chemoattractants for eosinophils as RANTES and eotaxin,[34,41,102] while eotaxin is specific for eosinophils. Eotaxin is more effective at inducing eosinophil infiltration than RANTES when injected into the skin of a rhesus monkey.[41]

T-lymphocytes

IL-8 has a small chemotactic activity for either $CD4^+$ or $CD8^+$ T-lymphocytes,[103] while RANTES is a chemoattractant for memory T-cells *in vitro*.[104] Human MIP-1α and MIP-1β are also chemoattractants for distinct subpopulations of lymphocytes, with MIP-1α acting on $CD8^+$ and MIP-1β on $CD4^+$ T-lymphocytes.[105] RANTES attracts both phenotypes and acts on resting and activated T-lymphocytes, while MIP-1α and MIP-1β are effective on anti-CD3 stimulated cells only.[106] On the other hand, MIP-1β, but not

MIP-1α, has been reported to be chemotactic for resting T-cells and enhances the adherence of CD8+ but not CD4+ cells to VCAM-1.[86]

MCP-1 induces T-cell migration.[107] Natural killer cells migrate vigorously in response to RANTES, MIP-1α and MCP-1.[108] Human recombinant IP-10 is a chemoattractant for human monocytes and promotes T-cell adhesion to endothelial cells,[109] while mig has effects on activated T-cells only.[24] The C chemokine lymphotactin also shares T-lymphocyte chemoattractant activity.[42] The selective chemoattractant activities for different subsets of lymphocytes suggest that specific members of the chemokine family may be involved in different immune and inflammatory responses.

Basophils

IL-8 induces the release of histamine[110,111] and sulphidopeptide leukotrienes[111] from human blood basophils, with enhanced release with IL-3, IL-5 or GM-CSF pretreatment.[112] C-C chemokines are more powerful stimulants of basophils. MCP-1 is as potent as C5a in stimulating exocytosis in human basophils,[113–115] with release of high levels of histamine. In the presence of IL-3, IL-5 or GM-CSF, there is enhanced release of histamine and production of leukotriene C$_4$.[113,115] RANTES and MIP-1α are less effective releasers of histamine from basophils. MIP-1β is inactive on basophils.[116] RANTES is the most effective basophil chemoattractant,[114,116,117] while MCP-1 is more effective as an inducer of histamine and leukotriene release.[116]

Monocytes

C-X-C chemokines are generally not active on monocytes, with IL-8 only being able to induce a small release of $[Ca^{2+}]_i$ and a respiratory burst.[118] By contrast, the C-C chemokines MCP-1, RANTES, I-309, HC14 (or MCP-2 and MCP-3 attract monocytes *in vitro*;[31,104,119–123] MCP-1, MCP-2 and MCP-3 induce a selective infiltration of monocytes in animal skin.[31,124] All C-C chemokines stimulate $[Ca^{2+}]_i$ release.[116,123,125] MCP-1 also induces a respiratory burst, expression of β2 integrins (CD11b/CD18 and CD11c/CD18) and the production of IL-1 and IL-6.[121,124,126] Growth of tumour cell lines cultured in the presence of human blood lymphocytes is inhibited by the addition of MCP-1.[30]

Other effects

High concentrations of RANTES have been reported to increase T-cell proliferation, IL-2 receptor expression, and IL-2 and IL-5 production.[127] RANTES and MIP-1α at low concentrations (0.1 nM) stimulate T-cells to express matrix metalloproteinases, which are enzymes that allow cells to migrate through the basement membrane.[128] RANTES and MIP-1α also directly stimulate surface IgE- and IgG4-positive B-cells for enhanced production of IgE and IgG4 production specifically.[129] Thus, these C-C chemokines may potentially regulate immunoglobulin synthesis.

CHEMOKINE RECEPTORS

The chemokine receptors form a family of structurally and functionally related proteins, being members of the superfamily of heptahelical, rhodopsin-like, G protein-coupled receptors. Responses of basophils, eosinophils and monocytes to C-C chemokines are prevented by pretreatment of these cells with *Bordetella pertussis* toxin,[116,122] which specifically inhibits GTP-binding proteins, indicating coupling to G proteins. Activation of heterotrimeric G proteins leads to dissociation of α-subunits from $\beta\gamma$-subunits and to activation of phospholipase C. Hydrolysis of phosphatidylinositol 4,5-bisphosphate produces the second messengers inositol 1,4,5-trisphosphate and diacylglycerol, trigerring cellular responses such as chemotaxis, degranulation and respiratory burst. Multiple and distinct signalling pathways exist for chemokine receptors, depending on cell type, receptor and the ligand involved.[127,130] C-X-C chemokine effects on neutrophils also induces G-protein activation.[131]

C-X-C chemokine receptors

Two receptors for IL-8, IL-8A and IL-8B, have been described, localized predominantly on neutrophils. Whereas IL-8A is specific for IL-8, IL-8B can also bind C-X-C chemokines such as NAP-2 and GRO-α. A cDNA encoding an MCP-1 receptor has been isolated from a human cell line and its 3-kb RNA found in monocytes but not neutrophils or lymphocytes.[132]

C-C chemokine receptors

Many C-C chemokine receptors have been identified by using orphan receptor cloning strategies and at least five distinct receptors have been described. The CC-CKR1 receptor, originally isolated from U937 cell lines, is activated by MIP-1α, RANTES and MCP-3.[133,134] CC-CKR2 receptors exist in two alternatively spliced forms and are highly expressed in blood monocytes;[135] these receptors bind MCP-1 and MCP-3, but not MCP-2. The CC-CKR3 receptor has been cloned from peripheral blood monocytes and is found in eosinophils and monocytes; it is the receptor for eotaxin.[136,137] CC-CKR3 receptors are highly expressed on eosinophils (40 000–400 000 receptors per cell)[137,138] compared with CC-CKR1 and CC-CKR2 receptors, which are expressed on monocytes and T-cells at <3000 receptors per cell.[139] The CC-CKR4 receptor, which is highly expressed in T-cells and IL-5-primed basophils, is activated by MIP-1α, RANTES and MCP-1.[140,141] The fifth chemokine receptor, CC-CKR5, responds to MIP-1α, MIP-1β and RANTES; its distribution is not currently known.[136,142,143] CC-CKR5 is one of the cofactors required for entry of HIV-1 into target cells in addition to surface CD4[144] and this may explain why RANTES, MIP-1α and MIP-1β inhibit HIV-1 replication.[145]

Promiscuous receptors

The erythrocyte chemokine receptor binds both C-X-C and C-C classes of chemokines[146,147] and has been identified as the Duffy blood group antigen.[148] This may be a mechanism by which chemokines are removed.

Another class of chemokine receptors include virally encoded receptors, such as one encoded by a cytomegalovirus open reading frame HCMV US28[133] and one from the herpes saimiri virus HSV ECRF3,[149] which are probably shared C-C and C-X-C receptors respectively. It is possible that these receptors have been transduced by viruses during evolution and may have an antiviral role.

EXPRESSION AND RELEASE OF CHEMOKINES IN ASTHMA

An early report has shown enhanced coexpression of IL-8 and GM-CSF in bronchial epithelial cells of patients with asthma,[150] which is of particular interest because GM-CSF, IL-3 and IL-5 can increase the responses of basophils and eosinophils to chemokines.[97,113,115] In addition, IL-8 appears to possess chemotactic activity for primed eosinophils.[97] Human IL-8 is able to induce accumulation of guinea-pig peritoneal eosinophils in guinea-pig skin,[151] and a human anti-IL-8 antibody inhibited IL-1-induced eosinophil accumulation in rat skin.[152] Enhanced release of IL-8 has been demonstrated from alveolar macrophages obtained from mild asthmatic subjects compared with those from normal subjects.[153] High levels of IL-8 are not specific for asthma because these have been reported in sputum samples obtained from patients with chronic bronchitis and bronchiectasis.[154]

Chemokines in BAL fluid can be detected, although at low levels even after the fluid has been concentrated. Elevated levels of MCP-1, RANTES, MIP-1α and IL-8 in BAL fluid of mild allergic asthmatic subjects has been measured.[58,155] Elevated levels of IL-8, MCP-1 and MIP-1α have been shown in patients with interstitial lung disease and pulmonary sarcoidosis.[156,157] Using a semi-quantitative reverse-transcription polymerase chain reaction, RANTES but not MIP-1α mRNA expression has been shown to be increased in bronchial biopsies of patients with mild asthma[158] (Fig. 18.1). No differences in MIP-1α mRNA expression were observed in alveolar macrophages and in bronchial biopsies obtained from nomal and asthmatic subjects. Although RANTES expression by immunohistochemistry can be demonstrated in the epithelium of the airway mucosa,[63,159] there do not appear to be differences between normal and asthmatic subjects[158] (see Plate 2). RANTES expression was also observed in airway smooth muscle cells and in submucosal T-cells.[158] The constitutive expression of RANTES in airway smooth muscle raises the question as to its role under normal circumstances. The C-C chemokine MCP-1 has been shown to be overexpressed in asthmatic epithelium.[66] Immunohistochemistry of human nasal polyps with anti-eotaxin monoclonal antibodies showed expression in eosinophils, lymphocytes, macrophages and respiratory epithelium.[41]

The role of C-C chemokines during allergic inflammation is supported by the observations that certain C-C chemokines can be expressed and released following exposure of sensitized individuals to allergen. In the skin of allergic individuals, allergen challenge induces expression of RANTES and MCP-3 mRNA, associated with eosinophil

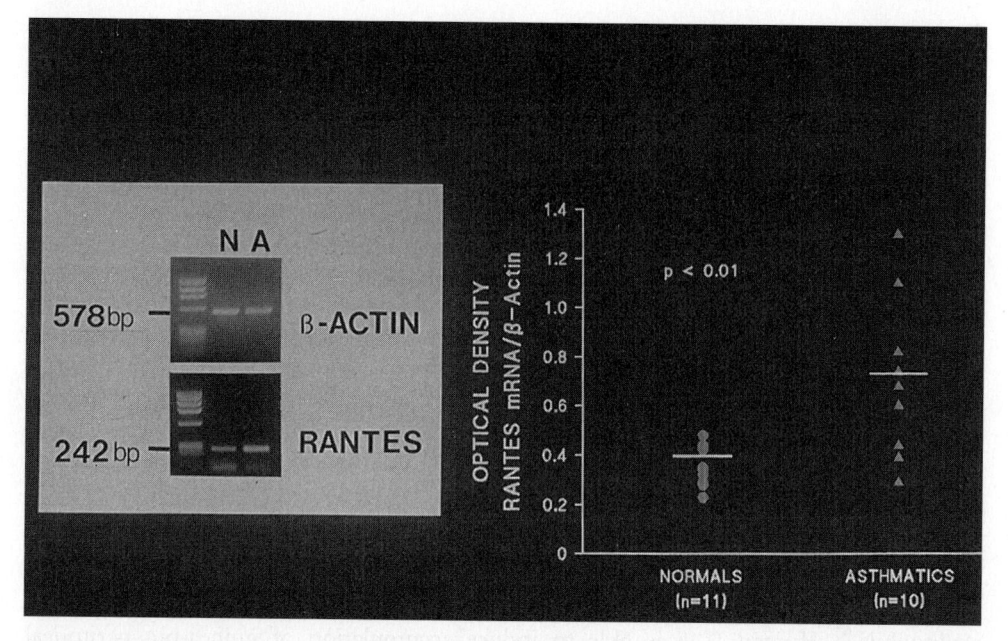

Fig. 18.1 RANTES and β-actin mRNA expression as assesed by reverse-transcription polymerase chain reaction in mucosal biopsies obtained from mild asthmatic and normal volunteers. Left, an example of RANTES and β-actin cDNA on a 2% agarose ethidium bromide-stained gel from an asthmatic (A) and a normal (N) subject. Right, the individual measurements expressed as a ratio of RANTES to β-actin in terms of abundance as measured by laser densitometry. Horizontal bars indicate mean values. There was a significant increased expression of RANTES in the asthmatic biopsies.

and lymphocyte infiltration.[160] Although in one study MIP-1α was not detectable in BAL fluid of asthmatic subjects, there was an increase in MIP-1α after segmental allergen challenge, contributing to lymphocyte chemoattractant activity.[161] Following segmental allergen challenge, increased levels of IL-5 and RANTES were detected in BAL fluid, but the increase in IL-5 levels was more marked.[162] Chemotactic activity for eosinophils of BAL fluid obtained from atopic asthmatic subjects with birch pollen allergy during the season was accounted for by IL-5 and RANTES.[163] These studies indicate that IL-5 and RANTES may act in concert as eosinophil chemoattractants and activators. Eotaxin mRNA is constitutively expressed in guinea-pig lung, but increases up to six-fold following allergen challenge of sensitized guinea-pigs.[139,164]

CONCLUSION

The chemokines form a diverse group of potent chemoattractants and cell activators, with effects on a wide range of cells. The central role of chemokines appears to be related to the trafficking of leucocytes but they may also have other roles. It is likely that chemokines would be involved in immunological and inflammatory processes as part of the normal or pathological response. Although there are many chemokines with overlapping and similar functions, the receptors and transduction mechanisms mediating their effects appear to be more limited. Inhibition of specific chemokine receptors may be a direct approach to blocking chemokine effects. One key question is whether inhibition of a single chemokine or of a single chemokine receptor can suppress eosinophil infiltration and prevent the ensuing pathophysiological consequences. The idea that eotaxin, which is highly specific for eosinophils and works to a large extent only on the CC-CKR3 receptor, could play a pivotal role is highly attractive and raises the possibility that blocking the CC-CKR3 receptor with pharmacological tools may be effective. Such tools will be important in dissecting the role of chemokines that are chemoattractants for eosinophils, monocytes and lymphocytes in the pathophysiology of allergic inflammation and asthma.

REFERENCES

1. Bousquet J, Chanez P, Lacoste JY, *et al.*: Eosinophilic inflammation in asthma. *N Engl J Med* (1990) **323**: 1033–1039.
2. Bentley AM, Menz G, Storz C, *et al.*: Identification of T-lymphocytes, macrophages and activated eosinophils in the bronchial mucosa of intrinsic asthma: relationship to symptoms and bronchial hyperresponsiveness. *Am Rev Respir Dis* (1992) **146**: 500–506.
3. Poston R, Chanez P, Lacoste JY, Litchfield P, Lee TH, Bousquet J: Immunohistochemical characterization of the cellular infiltration of asthmatic bronchi. *Am Rev Respir Dis* (1992) **145**: 918–921.
4. Hamid Q, Azzawi M, Ying S, *et al.*: Expression of mRNA for interleukins in mucosal bronchial biopsies from asthma. *J Clin Invest* (1991) **87**: 1541–1546.
5. Robinson DS, Bentley AM, Hartnell A, Kay AB, Durham SR: Activated memory T helper cells in bronchoalveolar lavage fluid from patients with atopic asthma: relation to asthma symptoms, lung function, and bronchial responsiveness. *Thorax* (1993) **48**: 26–32.
6. Bentley AM, Meng Q, Robinson DS, Hamid Q, Kay AB, Durham SR: Increases in activated T lymphocytes, eosinophils and cytokine mRNA expression for interleukin-5 and granulocyte/macrophage colony-stimulating factor in bronchial biopsies after allergen inhalation challenge in atopic asthmatics. *Am J Respir Cell Mol Biol* (1993) **8**: 35–42.
7. Broide DH, Firestein GS: Endobronchial allergen challenge: demonstration of cellular source of granulocyte macrophage colony-stimulating factor by *in situ* hybridization. *J Clin Invest* (1991) **88**: 1048–1053.
8. Diaz P, Gonzalez MC, Galleguillos FR, *et al.*: Leucocytes and mediators in bronchoalveolar lavage during allergen-induced late-phase asthmatic reactions. *Am Rev Respir Dis* (1989) **139**: 1383–1389.
9. Dunnill MS: The pathology of asthma with special reference to changes in the bronchial mucosa. *J Clin Pathol* (1960) **13**: 27–33.
10. Sur S, Crotty TB, Kephart GM, *et al.*: Sudden-onset fatal asthma: a distinct entity with few eosinophils and relatively more neutrophils in the airway submucosa? *Am Rev Respir Dis* (1993) **148**: 713–719.
11. Schmid J, Weissmann C: Induction of mRNA for a serine protease and a beta-thromboglobulin-like protein in mitogen-stimulated human leukocytes. *J Immunol* (1987) **139**: 250–256.
12. Schroder JM, Mrowietz U, Morita E, Christophers E: Purification and partial biochemical characterisation of a human monocyte-derived, neutrophil-activating peptide that lacks interleukin 1 activity. *J Immunol* (1987) **139**: 3474–3483.
13. Yoshimura T, Matsushima K, Tanaka S, Robinson EA, Appella E, Leonard EJ: Purification of a human monocyte-derived neutrophil chemotactic factor that has peptide sequence similarity to other host defense cytokines. *Proc Natl Acad Sci USA* (1987) **84**: 9233–9237.
14. Mukaida N, Shiroo M, Matsushima K: Genomic structure of the human monocyte-derived neutrophil chemotactic factor IL-8. *J Immunol* (1989) **143**: 1366–1371.

15. Walz A, Baggiolini M: Generation of the neutrophil-activating peptide NAP-2 from platelet basic protein or connective tissue-activating peptide III through monocyte proteases. *J Exp Med* (1990) **171**: 449–454.
16. Haskill S, Peace A, Morris J, et al.: Identification of three related human GRO genes encoding cytokine functions. *Proc Natl Acad Sci USA* (1990) **87**: 7732–7736.
17. Geiser T, Dewald B, Ehrengruber MU, Clark-Lewis I, Baggiolini M: The interleukin-8-related chemotactic cytokines GRO alpha, GRO beta, and GRO gamma activate human neutrophil and basophil leukocytes. *J Biol Chem* (1993) **268**: 15 419–15 424.
18. Walz A, Burgener R, Car B, Baggiolini M, Kunkel SL, Strieter RM: Structure and neutrophil-activating properties of a novel inflammatory peptide (ENA-78) with homology to interleukin 8. *J Exp Med* (1991) **174**: 1355–1362.
19. Proost P, De Wolf-Peeters C, Conings R, Opdenakker G, Billiau A, VanDamme J: Identification of a novel granulocyte chemotactic protein (GCP-2) from human tumor cells. In vitro and in vivo comparison with natural forms of GRO, IP-10, and IL-8. *J Immunol* (1993) **150**: 1000–1010.
20. Wolpe SD, Cerami A: Macrophage inflammatory proteins 1 and 2: members of a novel superfamily of cytokines. *FASEB J* (1989) **3**: 2565–2573.
21. Oquendo P, Alberta J, Wen DZ, Graycar JL, Derynck R, Stiles CD: The platelet-derived growth factor-inducible KC gene encodes a secretory protein related to platelet alpha-granule proteins. *J Biol Chem* (1989) **264**: 4133–4137.
22. Watanabe K, Kinoshita S, Nakagawa H: Purification and characterization of cytokine-induced neutrophil chemoattractant produced by epithelioid cell line of normal rat kidney (NRK-52E cell). *Biochem Biophys Res Commun* (1989) **161**: 1093–1099.
23. Farber JM: HuMig: a new human member of the chemokine family of cytokines. *Biochem Biophys Res Commun* (1993) **192**: 223–230.
24. Liao F, Rabin RL, Yannelli JR, Koniaris LG, Vanguri P, Farber JM: Human Mig chemokine: biochemical and functional characterization. *J Exp Med* (1995) **182**: 1301–1314.
25. Obaru K, Fukuda M, Maeda S, Shimada K: A cDNA clone used to study mRNA inducible in human tonsillar lymphocytes by a tumor promoter. *J Biochem* (1986) **99**: 885–894.
26. Miller MD, Krangel MS: Biology and biochemistry of the chemokines: a family of chemotactic and inflammatory cytokines. *Crit Rev Immunol* (1992) **12**: 17–46.
27. Wolpe SD, Davatelis G, Sherry B, et al.: Macrophages secrete a novel heparin-binding protein with inflammatory and neutrophil chemokinetic properties. *J Exp Med* (1988) **167**: 570–581.
28. Cochran BH, Reffel AC, Stiles CD: Molecular cloning of gene sequences regulated by platelet-derived growth factor. *Cell* (1983) **33**: 939–947.
29. Yoshimura T, Yuhki N, Moore SK, Appella E, Lerman MI, Leonard EJ: Human monocyte chemoattractant protein-1 (MCP-1). Full-length cDNA cloning, expression in mitogen-stimulated blood mononuclear leukocytes, and sequence similarity to mouse competence gene JE. *FEBS Lett* (1989) **244**: 487–493.
30. Matsushima K, Larsen CG, DuBois GC: Purification and characterisation of a novel monocyte chemotactic and activating factor produced by a human myelomonocytic cell line. *J Exp Med* (1989) **169**: 1485–1490.
31. Van Damme J, Proost P, Lenaerts J, Opdenakker G: Structural and functional identification of two human, tumor-derived monocyte chemotactic proteins (MCP-2 and MCP-3) belonging to the chemokine family. *J Exp Med* (1992) **176**: 59–64.
32. Minty A, Chalon P, Guillemot JC, et al.: Molecular cloning of the MCP-3 chemokine gene and regulation of its expression. *Eur Cytokine Network* (1993) **4**: 99–104.
33. Opdenakker G, Froyen G, Fiten P, Proost P, Van Damme J: Human monocyte chemotactic protein-3 (MCP-3): molecular cloning of the cDNA and comparison with other chemokines. *Biochem Biophys Res Commun* (1993) **191**: 535–542.
34. Uguccioni M, Loetscher P, Forssmann U, et al.: Monocyte chemotactic protein 4 (MCP-4), a novel structural and functional analogue of MCP-3 and eotaxin. *J Exp Med* (1996) **183**: 2379–2384.
35. Miller MD, Hata S, de Waal Malefyt R, Krangel MS: A novel polypeptide secreted by activated human T lymphocytes. *J Immunol* (1989) **143**: 2907–2916.

36. Schall TJ, Jongstra J, Dyer BJ, Jorgensen J, Clayberger C, Davis MM: A human T cell-specific molecule is a member of a new gene family. *J Immunol* (1988) **141**: 1018–1025.
37. Chang HC, Hsu F, Freeman GJ, Griffin JD, Reinherz EL: Cloning and expression of a gamma-interferon-inducible gene in monocytes: a new member of a cytokine gene family. *Int Immunol* (1989) **1**: 388–397.
38. Ying S, Meng Q, Taborda Barata L, *et al.*: Human eosinophils express messenger RNA encoding RANTES and store and release biologically active RANTES protein. *J Exp Med* (1996) **182**: 1169–1174.
39. Jose PJ, Griffiths-Johnson DA, Collins PD, *et al.*: Eotaxin: a potent eosinophil chemoattractant cytokine detected in a guinea pig model of allergic airways inflammation. *J Exp Med* (1994) **179**: 881–887.
40. Meurer R, Van Riper G, Feeney W, *et al.*: Formation of eosinophilic and monocytic intradermal inflammatory sites in the dog by injection of human RANTES but not human monocyte chemoattractant protein 1, human macrophage inflammatory protein 1 alpha, or human interleukin 8. *J Exp Med* (1993) **178**: 1913–1921.
41. Ponath PD, Qin S, Ringler DJ, *et al.*: Cloning of the human eosinophil chemoattractant, eotaxin: expression, receptor binding, and functional properties suggest a mechanism for the selective recruitment of eosinophils. *J Clin Invest* (1996) **97**: 604–612.
42. Kelner GS, Kennedy J, Bacon KB, *et al.*: Lymphotactin: a cytokine that represents a new class of chemokine. *Science* (1994) **266**: 1395–1399.
43. Kasahara K, Strieter RM, Chensue SW, Standiford TJ, Kunkel SL: Mononuclear cell adherence induces neutrophil chemotactic factor/interleukin-8 gene expression. *J Leukoc Biol* (1991) **50**: 287–295.
44. Metinko AP, Kunkel SL, Standiford TJ, Strieter RM: Anoxia–hyperoxia induces monocyte-derived interleukin-8. *J Clin Invest* (1992) **90**: 791–798.
45. Schroder JM, Persoon NL, Christophers E: Lipopolysaccharide-stimulated human monocytes secrete, apart from neutrophil-activating peptide 1/interleukin 8, a second neutrophil-activating protein. NH_2-terminal amino acid sequence identity with melanoma growth stimulatory activity. *J Exp Med* (1990) **171**: 1091–1100.
46. Iida N, Grotendorst GR: Cloning and sequencing of a new gro transcript from activated human monocytes: expression in leukocytes and wound tissue. *Mol Cell Biol* (1990) **10**: 5596–5599.
47. Zipfel PF, Balke J, Irving S, Kelly K, Siebenlist U: Mitogenic activation of human T cells induces two closely related genes which share structural similarities with a new family of secreted factors. *J Immunol* (1989) **142**: 1582–1590.
48. Lipes MA, Napolitano M, Jeang KT, Chang NT, Leonard WJ: Identification, cloning, and characterization of an immune activation gene. *Proc Natl Acad Sci USA* (1988) **85**: 9704–9708.
49. Berkman N, Jose P, Williams T, Barnes PJ, Chung KF: Corticosteroid inhibition of macrophage inflammatory protein-1a expression in human monocytes and alveolar macrophages. *Am J Physiol* (1995) **269**: L443–L452.
50. VanOtteren GM, Standiford TJ, Kunkel SL, Danforth JM, Burdick MD, Strieter RM: Expression and regulation of macrophage inflammatory protein-1 alpha by murine alveolar and peritoneal macrophages. *Am J Respir Cell Mol Biol* (1994) **10**: 8–15.
51. Sporn SA, Eierman DF, Johnson CE, Morris J, Martin G, Ladner M: Monocyte adherence results in selective induction of novel genes sharing homology with mediators of inflammation and tissue repair. *J Immunol* (1990) **144**: 4434–4441.
52. Schall TJ, O'Hehir RE, Goeddel DV, Lamb JR: Uncoupling of cytokine mRNA expression and protein secretion during the induction phase of T cell anergy. *J Immunol* (1992) **148**: 381–387.
53. Ziegler SF, Tough TW, Franklin TF, Armitage RJ, Alderson MR: Induction of macrophage inflammatory protein-1β gene expression in human monocytes by lipopolysaccharide and IL-7. *J Immunol* (1991) **147**: 2234–2239.
54. Strieter RM, Kasahara K, Allen RM, *et al.*: Cytokine-induced neutrophil-derived interleukin-8. *Am J Pathol* (1992) **141**: 397–407.
55. Galy AH, Spits H: IL-1, IL-4, and IFN-gamma differentially regulate cytokine production and

cell surface molecule expression in cultured human thymic epithelial cells. *J Immunol* (1991) **147**: 3823–3830.
56. Kasama T, Strieter RM, Standiford TJ, Burdick MD, Kunkel SL: Expression and regulation of human neutrophil-derived macrophage inflammatory protein 1 alpha. *J Exp Med* (1993) **178**: 63–72.
57. Braun RK, Franchini M, Erard F, et al.: Human peripheral blood eosinophils produce and release interleukin-8 on stimulation with calcium ionophore. *Eur J Immunol* (1993) **23**: 956–960.
58. Yousefi S, Hemmann S, Weber M, et al.: IL-8 is expressed by human peripheral blood eosinophils. Evidence for increased secretion in asthma. *J Immunol* (1995) **154**: 5481–5490.
59. Standiford TJ, Kunkel SL, Basha MA, et al.: Interleukin-8 gene expression by a pulmonary epithelial cell line. A model for cytokine networks in the lung. *J Clin Invest* (1990) **86**: 1945–1953.
60. Kwon O, Au BT, Collins PD, et al.: Tumour necrosis factor-induced interleukin-8 expression in pulmonary cultured human airway epithelial cells. *Am J Physiol* (1994) **267**: L398–L405.
61. Standiford TJ, Kunkel SL, Phan SH, Rollins BJ, Strieter RM: Alveolar macrophage-derived cytokines induce monocyte chemoattractant protein-1 expression from human pulmonary type II-like epithelial cells. *J Biol Chem* (1991) **266**: 9912–9918.
62. Elner SG, Strieter RM, Elner VM, Rollins BJ, Del Monte MA, Kunkel SL: Monocyte chemotactic protein gene expression by cytokine-treated human retinal pigment epithelial cells. *Lab Invest* (1991) **64**: 819–825.
63. Berkman N, Robichaud A, Krishnan VL, et al.: Expression of RANTES in human airway epithelial cells: effect of corticosteroids and interleukin-4, 10 and 13. *Immunology* (1995) **87**: 599–603.
64. Choi AMK, Jacoby DB: Influenza virus A infection induces interleukin-8 gene expression in human airway epithelial cells. *FEBS Lett* (1992) **309**: 327–329.
65. Nakamura H, Yoshimura K, McElvaney NG, Crystal RG: Neutrophil elastase in respiratory epithelial lining fluid of individuals with cystic fibrosis induces interleukin-8 gene expression in a human bronchial epithelial cell line. *J Clin Invest* (1992) **89**: 1478–1484.
66. Sousa AR, Lane SJ, Nakhosteen JA, Yoshimura T, Lee TH, Poston RN: Increased expression of the monocyte chemoattractant protein-1 in bronchial tissues from asthmatic subjects. *Am J Respir Cell Mol Biol* (1994) **10**: 142–147.
67. Marfaing-Koka A, Devergne O, Gorgone G, et al.: Regulation of the production of the RANTES chemokine by endothelial cells: synergistic induction by IFN plus TNF-a and inhibition by IL-4 and IL-13. *J Immunol* (1995) **154**: 1870–1878.
68. John M, Hirst SJ, Jose PJ, et al.: Human airway smooth muscle cells express and release RANTES in response to Th-1 cytokines: regulation by Th-2 cytokines and corticosteroids. *J Immunol* (1997) **158**: 1841–1847.
69. Mukaida N, Mahé Y, Matsushima K: Cooperative interaction of nuclear factor kB and cis-regulatory enhancer binding protein-like factor bonding elements in activating the interleukin-8 gene by pro-inflammatory cytokines. *J Biol Chem* (1990) **265**: 21 128–21 133.
70. Sica A, Matsushima K, Van Damme J, et al.: IL-1 transcriptionally activates the neutrophil chemotactic factor/IL-8 gene in endothelial cells. *Immunology* (1990) **69**: 548–553.
71. Mukaida N, Matsushima K: Regulation of IL-8 production and the characteristics of the receptor for IL-8. *Cytokine* (1992) **4**: 41–53.
72. Mukaida N, Harada A, Yasumoto K, Matsushima K: Properties of pro-inflammatory cell type-specific leukocyte chemotactic cytokines, interleukin 8 (IL-8) and monocyte chemotactic and activating factor (MCAF). *Microbiol Immunol* (1992) **36**: 773–789.
73. Matsushima K, Morishita K, Yoshimura T, et al.: Molecular cloning of a human monocyte-derived neutrophil chemotactic factor (MDNCF) and the induction of MDNCF mRNA by interleukin 1 and tumor necrosis factor. *J Exp Med* (1988) **167**: 1883–1893.
74. Shaw G, Kamen R: A conserved AU sequence from the 3' untranslated region of GM-CSF mRNA mediates selective mRNA degradation. *Cell* (1986) **46**: 659–667.
75. Seitz M, Dewald B, Gerber N, Baggiolini M: Enhanced production of neutrophil-activating peptide-1/interleukin-8 in rheumatoid arthritis. *J Clin Invest* (1991) **87**: 463–469.

76. Kwon OJ, Au BT, Collins PD, et al.: Inhibition of interleukin-8 expression by dexamethasone in human cultured airway epithelial cells. *Immunology* (1994) **81**: 389–394.
77. Standiford TJ, Strieter RM, Chensue SW, Westwick J, Kasahara K, Kunkel SL: IL-4 inhibits expression of IL-8 from stimulated human monocytes. *J Immunol* (1990) **145**: 1435–1439.
78. de Waal Malefyt R, Abrams J, Bennett B, Figdor CG, De Vries JE: Interleukin 10 (IL-10) inhibits cytokine synthesis by human monocytes: an auto regulatory role of IL-10 produced by monocytes. *J Exp Med* (1991) **179**: 1209–1220.
79. Nelson PJ, Kim HT, Manning WC, Goralski TJ, Krensky AM: Genomic organization and transcriptional regulation of the RANTES chemokine gene. *J Immunol* (1993) **151**: 2601–2612.
80. Standiford TJ, Kunkel SL, Liebler JM, Burdick MD, Gilbert AR, Strieter RM: Gene expression of macrophage inflammatory protein-1α from human blood monocytes and alveolar macrophages is inhibited by interleukin-4. *Am J Respir Cell Mol Biol* (1993) **9**: 192–198.
81. Nakao M, Nomiyama H, Shimada K: Structures of human genes coding for cytokine LD78 and their expression. *Mol Cell Biol* (1990) **10**: 3646–3658.
82. Berkman N, John M, Roesems G, Jose PJ, Barnes PJ, Chung KF: Inhibition of macrophage inflammatory protein-1α by interleukin-10: Differential sensitivities in human blood monocytes and alveolar macrophages. *J Immunol* (1995) **155**: 4412–4418.
83. Berkman N, Roesems G, Jose PJ, Barnes PJ, Chung KF: Interleukin-13 inhibits expression of macrophage-inflammatory protein-1a from human blood monocytes and alveolar macrophages. *Am J Respir Crit Care Med* (1996) **15**: 382–389.
84. Witt DP, Lander AD: Differential binding of chemokines to glycosaminoglycan subpopulations. *Curr Cell Biol* (1994) **4**: 394–400.
85. Webb LM, Ehrengruber MU, Clark-Lewis I, Baggiolini M, Rot A: Binding to heparan sulfate or heparin enhances neutrophil responses to interleukin 8. *Proc Natl Acad Sci USA* (1993) **90**: 7158–7162.
86. Tanaka Y, Adams DH, Hubscher S, Hirano H, Siebenlist U, Shaw S: T-cell adhesion induced by proteoglycan-immobilized cytokine MIP-1 beta. *Nature* (1993) **361**: 79–82.
87. Rothenberg ME, Owen WFJ, Siberstein DS: Human eosinophils have prolonged survival, enhanced functional properties and become hypodense when exposed to human interleukin 3. *J Clin Invest* (1988) **81**: 1986–1992.
88. Lopez AF, Williamson J, Gamble JR, et al.: Recombinant human granulocyte–macrophage colony-stimulating factor stimulates *in vitro* mature human neutrophil and eosinophil function, surface receptor expression, and survival. *J Clin Invest* (1986) **78**: 1220–1228.
89. Owen WF, Rothenberg ME, Silberstein DS, et al.: Regulation of human eosinophil viability, density and function by granulocyte/macrophage colony-stimulating factor in the presence of 3T3 fibroblasts. *J Exp Med* (1987) **166**: 129–141.
90. Collins PD, Griffiths-Johnson DA, Jose PJ, Williams TJ, Marleau S: Co-operation between interleukin-5 and the chemokine, eotaxin, to induce eosinophil accumulation *in vivo*. *J Exp Med* (1995) **182**: 1169–1174.
91. Baggiolini M, Wymann MP: Turning on the respiratory burst. *Trends Biochem Sci* (1990) **15**: 69–72.
92. Detmers PA, Lo SK, Olsen-Egbert E, Walz A, Baggiolini M, Cohn ZA: Neutrophil-activating protein 1/interleukin 8 stimulates the binding activity of the leukocyte adhesion receptor CD11b/CD18 on human neutrophils. *J Exp Med* (1990) **171**: 1155–1162.
93. Detmers PA, Powell DE, Walz A, Clark-Lewis I, Baggiolini M, Cohn ZA: Differential effects of neutrophil-activating peptide 1/IL-8 and its homologues on leukocyte adhesion and phagocytosis. *J Immunol* (1991) **147**: 4211–4217.
94. Schroder JM: The monocyte-derived neutrophil activating peptide (NAP/interleukin 8) stimulates human neutrophil arachidonate-5-lipoxygenase, but not the release of cellular arachidonate. *J Exp Med* (1989) **170**: 847–863.
95. Bussolino F, Sironi M, Bocchietto E, Mantovani A: Synthesis of platelet-activating factor by polymorphonuclear neutrophils stimulated with interleukin-8. *J Biol Chem* (1992) **267**: 14 598–14 603.
96. Kernen P, Wymann MP, von Tscharner V, et al.: Shape changes, exocytosis, and cytosolic free calcium changes in stimulated human eosinophils. *J Clin Invest* (1991) **87**: 2012–2017.

97. Warringa RA, Koenderman L, Kok PT, Kreukniet J, Bruijnzeel PL: Modulation and induction of eosinophil chemotaxis by granulocyte–macrophage colony-stimulating factor and interleukin-3. *Blood* (1991) **77**: 2694–2700.
98. Rot A, Krieger M, Brunner T, Bischoff SC, Schall TJ, Dahinden CA: RANTES and macrophage inflammatory protein Iα induce the migration and activation of normal human eosinophil granulocytes. *J Exp Med* (1992) **176**: 1489–1495.
99. Kameyoshi Y, Dorschner A, Mallet AI, Christophers E, Schroder J: Cytokine RANTES released by thrombin-stimulated platelets is a potent attractant for human eosinophils. *J Exp Med* (1992) **176**: 587–592.
100. Alam R, Stafford S, Forsythe P, Harrison R, et al.: RANTES is a chemotactic and activating factor for human eosinophils. *J Immunol* (1993) **150**: 3442–3447.
101. Meurer R, Van Riper G, Feeney W, et al.: Formation of eosinophilic and monocytic intradermal inflammatory sites in the dog by injection of human RANTES but not human monocyte chemoattractant protein 1, human macrophage inflammatory protein 1α, or human interleukin 8. *J Exp Med* (1993) **178**: 1913–1921.
102. Dahinden CA, Geiser T, Brunner T, et al.: Monocyte chemotactic protein 3 is a most effective basophil- and eosinophil-activating chemokine. *J Exp Med* (1994) **179**: 751–756.
103. Bacon KB, Camp RD: Interleukin (IL)-8-induced *in vitro* human lymphocyte migration is inhibited by cholera and pertussis toxins and inhibitors of protein kinase C. *Biochem Biophys Res Commun* (1990) **169**: 1099–1104.
104. Schall TJ, Bacon K, Toy KJ, Goeddel DV: Selective attraction of monocytes and T lymphocytes of the memory phenotype of cytokine RANTES. *Nature* (1990) **347**: 669–671.
105. Schall TJ, Bacon K, Camp RD, Kaspari JW, Goeddel DV: Human macrophage inflammatory protein alpha (MIP-1 alpha) and MIP-1 beta chemokines attract distinct populations of lymphocytes. *J Exp Med* (1993) **177**: 1821–1826.
106. Taub DD, Conlon K, Lloyd AR, Oppenheim JJ, Kelvin DJ: Preferential migration of activated CD4[+] and CD8[+] T cells in response to MIP-1α and MIP-1β. *Science* (1993) **260**: 355–357.
107. Carr MW, Roth SJ, Luther E, Rose SS, Springer TA: Monocyte chemoattractant protein 1 acts as a T-lymphocyte chemoattractant. *Proc Natl Acad Sci USA* (1994) **91**: 3652–3656.
108. Maghazachi AA, Al Aarkaty A, Schall TJ: C-C chemokines induce the chemotaxis of NK and IL-2 activated NK cells: role for G proteins. *J Immunol* (1994) **153**: 4969–4977.
109. Taub DD, Lloyd AR, Conlon K, et al.: Recombinant human interferon-inducible protein 10 is a chemoattractant for human monocytes and T lymphocytes and promotes T cell adhesion to endothelial cells. *J Exp Med* (1993) **177**: 1809–1814.
110. White MV, Yoshimura T, Hook W, Kaliner MA, Leonard EJ: Neutrophil attractant/activation protein-1 (NAP-1) causes human basophil histamine release. *Immunol Lett* (1989) **22**: 151–154.
111. Dahinden CA, Kurimoto Y, De Weck AL, Lindley I, Dewald B, Baggiolini M: The neutrophil-activating peptide NAF/NAP-1 induces histamine and leukotriene release by interleukin 3-primed basophils. *J Exp Med* (1989) **170**: 1787–1792.
112. Bischoff SC, Baggiolini M, De Weck AL, Dahinden CA: Interleukin 8-inhibitor and inducer of histamine and leukotriene release in human basophils. *Biochem Biophys Res Commun* (1991) **179**: 628–633.
113. Kuna P, Reddigari SR, Rucinski D, Oppenheim JJ, Kaplan AP: Monocyte chemotactic and activating factor is a potent histamine-releasing factor for human basophils. *J Exp Med* (1992) **175**: 489–493.
114. Alam R, Forsythe PA, Stafford S, Lett-Brown MA, Grant JA: Macrophage inflammatory protein-1α activates basophils and mast cells. *J Exp Med* (1992) **176**: 781–786.
115. Bischoff SC, Krieger M, Brunner T, Dahinden CA: Monocyte chemotactic protein 1 is a potent activator of human basophils. *J Exp Med* (1992) **175**: 1271–1275.
116. Bischoff SC, Krieger M, Brunner T, et al.: RANTES and related chemokines activate human basophil granulocytes through different G protein-coupled receptors. *Eur J Immunol* (1993) **23**: 761–767.
117. Kuna P, Reddigarl SR, Schall TJ, Rucinski D, Viksman MY, Kaplan AP: RANTES, a

monocyte and T lymphocyte chemotactic cytokine releases histamine from human basophils. *J Immunol* (1992) **149**: 636–642.
118. Walz A, Meloni F, Clark-Lewis I, von Tscharner V, Baggiolini M: [Ca^{2+}]$_i$ changes and respiratory burst in human neutrophils and monocytes induced by NAP-1/interleukin-8, NAP-2, and gro/MGSA. *J Leukoc Biol* (1991) **50**: 279–286.
119. Yoshimura T, Robinson EA, Tanaka S, Appella E, Leonard EJ: Purification and amino acid analysis of two human monocyte chemoattractants produced by phytohemagglutinin-stimulated human blood mononuclear leukocytes. *J Immunol* (1989) **142**: 1956–1962.
120. Yoshimura T, Robinson EA, Appella E, Matsushima K, Showalter SD, Leonard EJ: Three forms of monocyte-derived neutrophil chemotactic factor (MDNCF) distinguished by different lengths of the amino-terminal sequence. *Mol Immunol* (1989) **26**: 87–93.
121. Rollins BJ, Walz A, Baggiolini M: Recombinant human MCP-1/JE induces chemotaxis, calcium flux, and the respiratory burst in human monocytes. *Blood* (1991) **78**: 1112–1116.
122. Sozzani S, Luini W, Molino M, et al.: The signal transduction pathway involved in the migration induced by a monocyte chemotactic cytokine. *J Immunol* (1991) **147**: 2215–2221.
123. Miller MD, Krangel MS: The human cytokine I-309 is a monocyte chemoattractant. *Proc Natl Acad Sci USA* (1992) **89**: 2950–2954.
124. Zachariae CO, Anderson AO, Thompson HL, Appella E, Mantovani A, Matsushima K: Properties of monocyte chemotactic and activating factor (MCAF) purified from a human fibrosarcoma cell line. *J Exp Med* (1990) **171**: 2177–2182.
125. McColl SR, Hachicha M, Levasseur S, Neote K, Schall TJ: Uncoupling of early signal transduction events from effector function in human peripheral blood neutrophils in response to recombinant macrophage inflammatory proteins-1 alpha and -1 beta. *J Immunol* (1993) **150**: 4550–4560.
126. Jiang Y, Beller DI, Frendl G, Graves DT: Monocyte chemoattractant protein-1 regulates adhesion molecule expression and cytokine production in human monocytes. *J Immunol* (1992) **148**: 2423–2428.
127. Bacon KB, Premack BA, Gardner P, Schall TJ: Activation of dual T cell signaling pathways by the chemokine RANTES. *Science* (1995) **269**: 1727–1730.
128. Xia MH, Leppert D, Hauser SL, et al.: Stimulus specificity of matrix metalloproteinase dependence of human T cell migration through a model basement membrane. *J Immunol* (1996) **156**: 160–167.
129. Kimata H, Yoshida A, Ishioka C, Fujimoto M, Lindley I, Furusho K: RANTES and macrophage inflammatory protein 1a selectively enhance immunoglobulin E (IgE) and IgG4 production by human B cells. *J Exp Med* (1996) **183**: 2397–2402.
130. L'Heureux GP, Bourgoin S, Jean N, McColl SR, Naccache PH: Diverging signal transduction pathways activated by interleukin-8 and related chemokines in human neutrophils: interleukin-8, but not NAP-2 or GRO alpha, stimulates phospholipase D activity. *Blood* (1995) **85**: 522–531.
131. Kupper RW, Dewald B, Jakobs KH, Baggiolini M, Gierschik P: G-protein activation by interleukin 8 and related cytokines in human neutrophil plasma membranes. *Biochem J* (1992) **282**: 429–434.
132. Murphy PM: The molecular biology of leukocyte chemoattractant receptors. *Annu Rev Immunol* (1994) **12**: 593–633.
133. Neote K, Digregorio D, Mak JY, Horak R, Schall TJ: Molecular cloning, functional expression and signaling characteristics of a C-C chemokine receptor. *Cell* (1993) **72**: 415–425.
134. Gao J, Kuhns DB, Tiffany HL, et al.: Structure and functional expression of the human macrophage inflammatory protein 1α/RANTES receptor. *J Exp Med* (1993) **177**: 1421–1427.
135. Charo IF, Myers SJ, Herman A, Franci C, Connolly AJ, Coughlin SR: Molecular cloning and functional expression of two monocyte chemoattractant protein 1 receptors reveals alternative splicing of the carboxyl-terminal tails. *Proc Natl Acad Sci USA* (1994) **91**: 2752–2756.
136. Combadiere C, Ahuja SK, Murphy PM: Cloning and functional expression of a human eosinophil CC chemokine receptor. *J Biol Chem* (1995) **270**: 16491–16494.
137. Ponath PD, Qin S, Post TW, et al.: Molecular cloning and characterization of a human eotaxin receptor expressed selectively on eosinophils. *J Exp Med* (1996) **183**: 2437–2448.
138. Daugherty BL, Siciliano SJ, DeMartino JA, Malkowitz L, Sirotina A, Springer MS: Cloning,

expression and characterization of the human eosinophil eotaxin receptor. *J Exp Med* (1996) **183**: 2349–2354.
139. Ernst CA, Zhang YJ, Hancock PR, Rutledge BJ, Corless CL, Rollins BJ: Biochemical and biologic characterization of murine monocyte chemoattractant protein-1. Identification of two functional domains. *J Immunol* (1994) **152**: 3541–3549.
140. Power CA, Meyer A, Nemeth K, *et al.*: Molecular cloning and functional expression of a novel CC chemokine receptor cDNA from a human basophilic cell line. *J Biol Chem* (1995) **270**: 19 495–19 500.
141. Hoogewerf A, Black D, Proudfoot AE, Wells TN, Power CA: Molecular cloning of murine CC CKR-4 and high affinity binding of chemokines to murine and human CC CKR-4. *Biochem Biophys Res Commun* (1996) **218**: 337–343.
142. Combadiere C, Ahuja SK, Murphy PM: Cloning and functional expression of a human eosinophil CC chemokine receptor (erratum). *J Biol Chem* (1995) **270**: 30 235.
143. Raport CJ, Gosling J, Schweickart V, Gray PW, Charo IF: Molecular cloning and functional characterisation of a novel human CC chemokine receptor (CCR5) for RANTES, MIP-1β and MIP-1α. *J Biol Chem* (1996) **271**: 17 161–17 165.
144. Deng H, Lui R, Ellmeier W, *et al.*: Identification of a major co-receptor for primary isolates of HIV-1. *Nature* (1996) **381**: 661–666.
145. Cocchi F, DeVico AL, Garzino Demo A, Arya SK, Gallo RC, Lusso P: Identification of RANTES, MIP-1 alpha, and MIP-1 beta as the major HIV-suppressive factors produced by CD8$^+$ T cells. *Science* (1995) **270**: 1811–1815.
146. Neote K, Mak JY, Kolakowski LF Jr, Schall TJ: Functional and biochemical analysis of the cloned Duffy antigen: identity with the red blood cell chemokine receptor. *Blood* (1994) **84**: 44–52.
147. Neote K, Darbonne W, Ogez J, Horuk R, Schall TJ: Identification of a promiscuous inflammatory peptide receptor on the surface of red blood cells. *J Biol Chem* (1993) **268**: 12 247–12 249.
148. Horuk R, Colby TJ, Darbonne WC, Schall TJ, Neote K: The human erythrocyte inflammatory peptide (chemokine) receptor. Biochemical characterization, solubilization, and development of a binding assay for the soluble receptor. *Biochemistry* (1993) **32**: 5733–5738.
149. Ahuja SK, Murphy PM: Molecular piracy of mammalian interleukin-8 receptor type B by herpesvirus saimiri. *J Biol Chem* (1993) **268**: 20 691–20 694.
150. Marini M, Vittori E, Hollemburg J, Mattoli S: Expression of the potent inflammatory cytokines granulocyte–macrophage colony stimulating factor, interleukin-6 and interleukin-8 in bronchial epithelial cells of patients with asthma. *J Allergy Clin Immunol* (1992) **82**: 1001–1009.
151. Collins PD, Weg VB, Faccioli LH, Watson ML, Moqbel R, Williams TJ: Eosinophil accumulation induced by human interleukin-8 in the guinea-pig *in vivo*. *Immunology* (1993) **79**: 312–318.
152. Sanz MJ, Weg VB, Bolanowski MA, Nourshargh S: IL-1 is a potent inducer of eosinophil accumulation in rat skin. *J Immunol* (1995) **154**: 1364–1373.
153. Hallsworth MP, Soh CPC, Lane SJ, Arm JP, Lee TH: Selective enhancement of GM-CSF, TNF-α, IL-1β and IL-8 production by monocytes and macrophages of asthmatic subjects. *Eur Respir J* (1994) **7**: 1096–1102.
154. Richman-Eisenstat JB, Jorens PG, Hebert CA, Ueki I, Nadel JA: Interleukin-8: an important chemoattractant in sputum of patients with chronic inflammatory airway diseases. *Am J Physiol* (1993) **264**: L413–L418.
155. Alam R, York J, Boyars M, *et al.*: Increased MCP-1, RANTES and MIP-1α in bronchoalveolar lavage fluid of allergic asthmatic patients. *Am J Respir Crit Care Med* (1996) **153**: 1398–1404.
156. Car BD, Meloni F, Luisetti M, Semenzato G, Gialdroni-Grassi G, Walz A: Elevated IL-8 and MCP-1 in the bronchoalveolar lavage fluid of patients with idiopathic pulmonary fibrosis and pulmonary sarcoidosis. *Am J Respir Crit Care Med* (1994) **149**: 655–659.
157. Standiford TJ, Rolfe MW, Kunkel SL, *et al.*: Macrophage inflammatory protein-1α expression in interstitial lung disease. *J Immunol* (1993) **151**: 2852–2863.

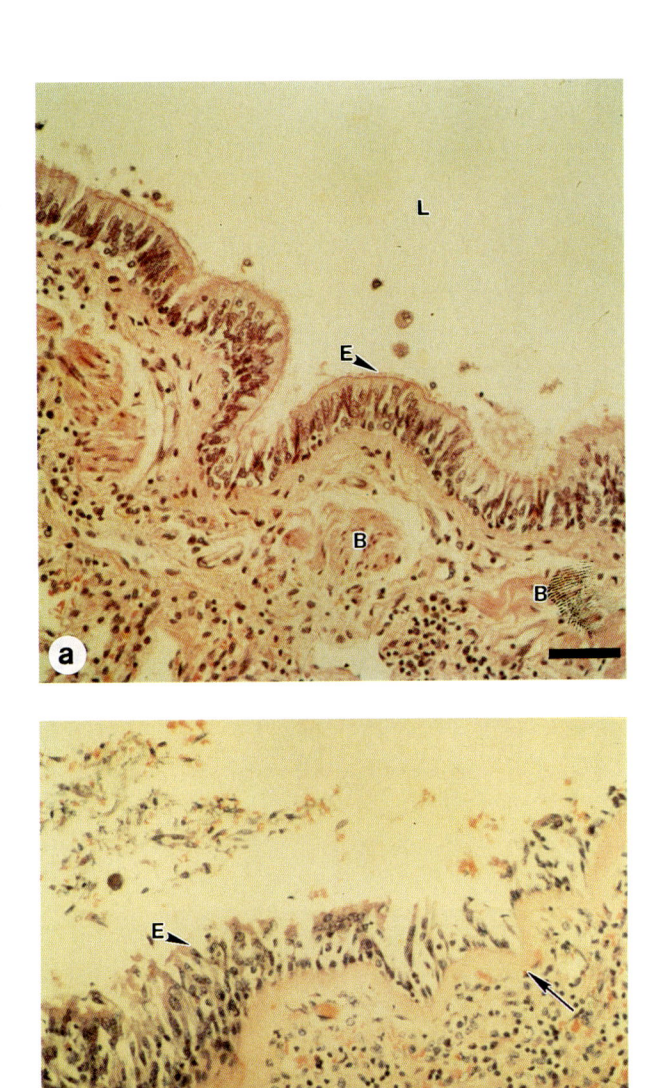

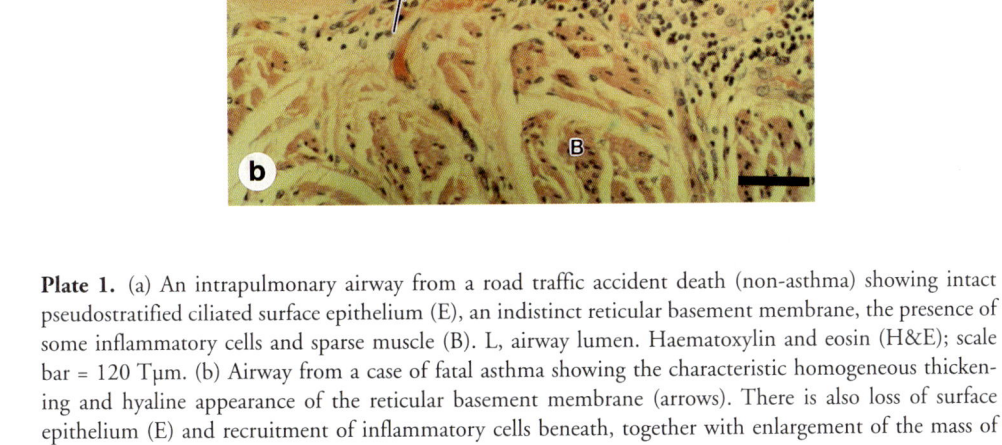

Plate 1. (a) An intrapulmonary airway from a road traffic accident death (non-asthma) showing intact pseudostratified ciliated surface epithelium (E), an indistinct reticular basement membrane, the presence of some inflammatory cells and sparse muscle (B). L, airway lumen. Haematoxylin and eosin (H&E); scale bar = 120 Tμm. (b) Airway from a case of fatal asthma showing the characteristic homogeneous thickening and hyaline appearance of the reticular basement membrane (arrows). There is also loss of surface epithelium (E) and recruitment of inflammatory cells beneath, together with enlargement of the mass of bronchial smooth muscle (B). H&E; scale bar = 120 Tμm.

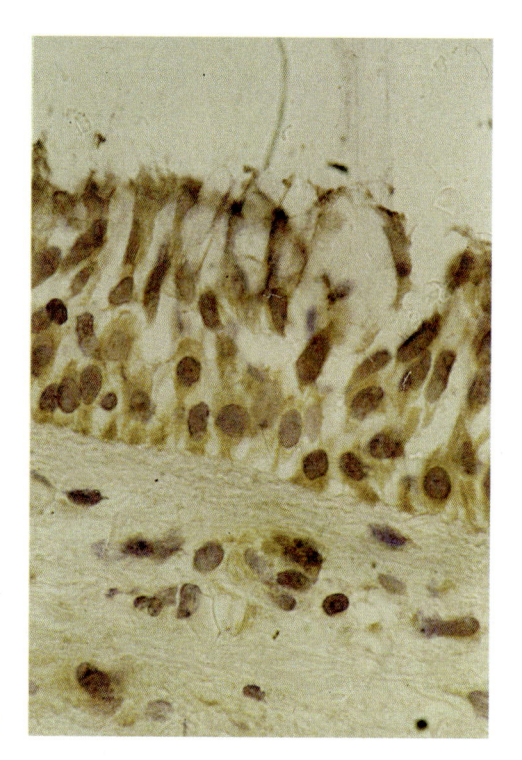

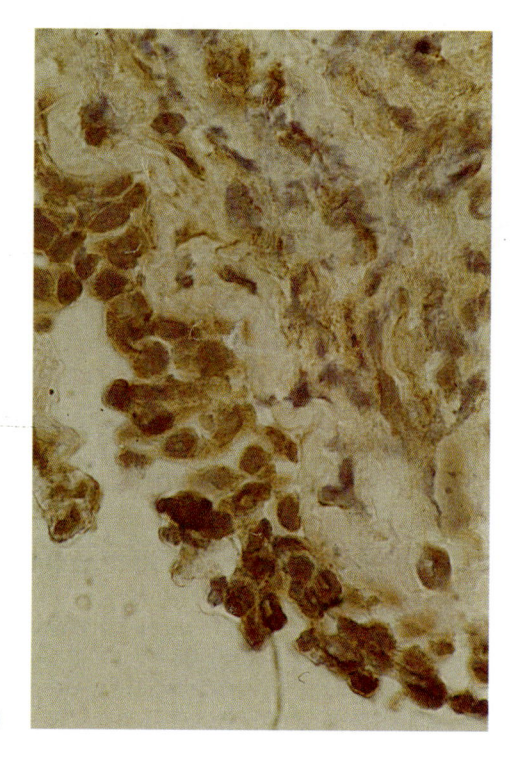

(a)

(b)

(c)

Plate 2. Immunohistochemical staining for RANTES using an anti-RANTES antibody and peroxidase–antiperoxidase technique in bronchial biopsies from a normal subject (a) and from an asthmatic subject (b, c). Positively staining airway epithelial cells and submucosal cells, probably CD4$^+$ T-cells and fibroblasts, are observed in (a) and (b). Airway smooth muscle cells are also stained in (c). (a, b, magnification × 400; c, magnification × 1000.)

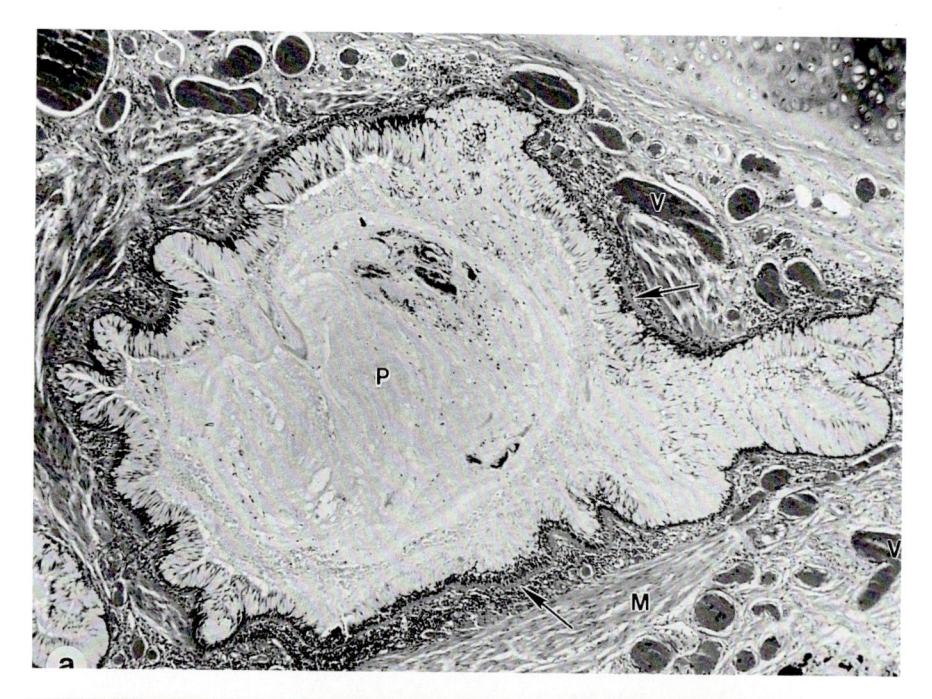

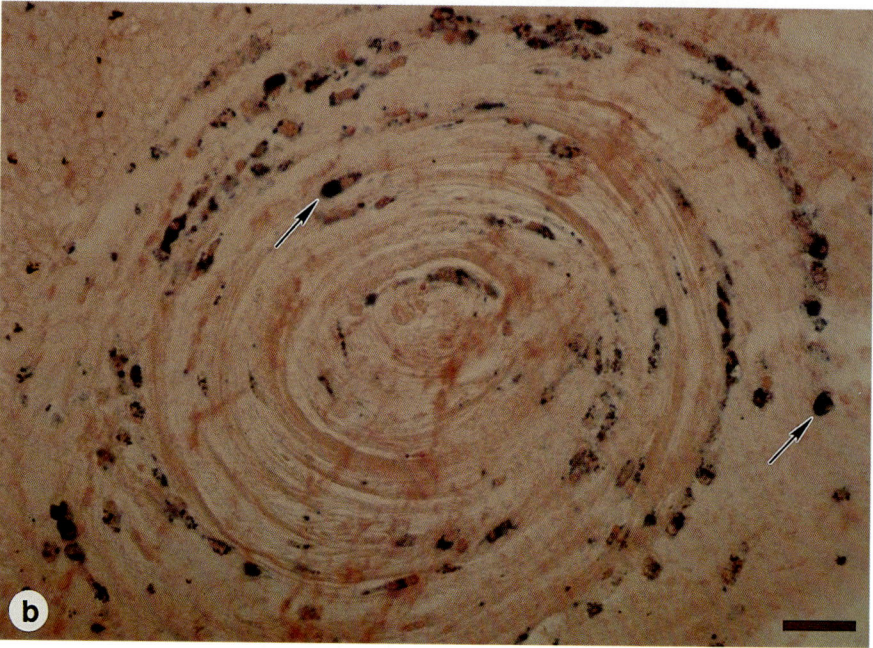

Plate 3. (a) Histological section demonstrating an airway plug (P), which consists of a mixture of secretions, inflammatory cells and, to a lesser extent, airway epithelial cells. The subepithelial reticular basement membrane is thickened (arrows) and there is an underlying zone rich in inflammatory cells, surrounded by enlarged blocks of bronchial smooth muscle (M) and dilated, congested bronchial vessels (V). Haematoxylin and eosin (H&E); scale bar = 240 Tµm. (b) Higher magnification of the plug showing the concentric lamella consisting of immunostained activated (i.e. EG2$^+$ eosinophils (arrows). Scale bar =120 Tµm.

158. Berkman N, Krishnan VL, Gilbey T, et al.: Expression of RANTES mRNA and protein in airways of patients with mild asthma. *Am J Respir Crit Care Med* (1996) **154**: 1804–1811.
159. Wang JH, Devalia JL, Xia C, Sapsford RJ, Davies RJ: Expression of RANTES by human bronchial epithelial cells *in vitro* and *in vivo* and the effect of corticosteroids. *Am J Respir Cell Mol Biol* (1996) **14**: 27–35.
160. Ying S, Taborda Barata L, Meng Q, Humbert M, Kay AB: The kinetics of allergen-induced transcription of messenger RNA for monocyte chemotactic protein-3 and RANTES in the skin of human atopic subjects: relationship to eosinophil, T cell, and macrophage recruitment. *J Exp Med* (1995) **181**: 2153–2159.
161. Cruikshank WW, Long A, Torpy RE, et al.: Early identification of IL-16 (lymphocyte chemoattractant factor) and macrophage inflammatory protein 1α (MIP-1α) in bronchoalveolar lavage fluid of antigen-challenged asthma. *Am J Respir Cell Mol Biol* (1995) **13**: 738–747.
162. Sur S, Kita H, Gleich GJ, Chenier TC, Hunt LW: Eosinophil recruitment is associated with IL-5, but not with RANTES, twenty-four hours after allergen challenge. *J Allergy Clin Immunol* (1996) **97**: 1272–1278.
163. Venge J, Lampinen M, Hakasson L, Rak S, Venge P: Identification of IL-5 and RANTES as the major eosinophil chemoattractants in the asthmatic lung. *J Allergy Clin Immunol* (1996) **97**: 1110–1115.
164. Rothenberg ME, Luster AD, Lilly CM, Drazen JM, Leder P: Constitutive and allergen-induced expression of eotaxin mRNA in the guinea pig lung. *J Exp Med* (1995) **181**: 1211–1216.

19

Lymphokines

DOUGLAS S. ROBINSON

INTRODUCTION

Lymphokines are defined as soluble factors produced by T-lymphocytes that lead to effector functions after specific immune activation. By definition lymphokines themselves are not antigen specific.[1] They are also termed cytokines, which is preferable, since most are not restricted to T-lymphocytes but are produced by a wide variety of cell types. This chapter considers the actions and relevance of T-cell-derived cytokines to airway pathology in asthma, since non-T-cell sources are covered elsewhere. The vital importance of T-lymphocytes lies in their potential for initiation of specific immune responses through interaction of antigenic peptides in the MHC groove of antigen-presenting cells with the T-cell receptor (TCR).[2]

TYPE 1 AND TYPE 2 T-CELLS

In 1986 Mosmann and Coffman[3] described two types of mouse $CD4^+$ T-helper (Th) cell clones, on the basis of the pattern of cytokines produced. Th1 clones produced interleukin (IL)-2, lymphotoxin (LT) and interferon (IFN)-γ but not IL-4 or IL-5, whereas Th2 clones produced IL-4 and IL-5 but not IL-2, LT or IFN-γ. The functional significance of this dichotomy was confirmed by studies showing that Th1 clones induced delayed-type hypersensitivity (DTH) cell-mediated reactions upon adoptive transfer, as might be predicted by the actions of IFN-γ in macrophage activation and IL-2 in T-cell activation and proliferation.[4,5] Th2 clones provided help for immunoglobulin synthesis by B-cells, but did not induce DTH.

Analysis of allergen-specific T-cell clones derived from human atopic donors showed a Th2-like cytokine profile,[6] and Parronchi et al.[7] went on to show that both Th1 and Th2 clones could be derived from humans with appropriate stimuli. Thus both human Th1 and Th2 cell responses exist, and Th2-like T-cell clones can be isolated from sites of allergic disease.[8,9] The importance of T-cell cytokines in determining human pathology was supported by findings in leprosy, where analysis of lesional skin showed cytokine mRNA for IL-2 and IFN-γ in tuberculoid leprosy skin, where DTH predominates, but a Th2 pattern of IL-4 and IL-5 in skin from subjects with lepromatous disease where antibody response predominates.[10] None the less, there are differences between human and murine T-cell cytokine patterns (Table 19.1), and many human T-cell clones produce intermediate patterns.[11,12]

T-cell clones are produced *in vitro* by expansion of T-cells, by bulk culture with either mitogens and IL-2 or antigen and IL-2, then by limiting dilution cloning to expand individual activated T-cells that respond to antigen. The repetitive proliferation and stimulation with antigen may parallel *in vivo* T-cell responses to persistent or repetitive antigen stimulation (as with ubiquitous allergens), but may not reflect short-term T-cell responses to antigen *in vivo*. Thus, although analysis of T-cell clones has given considerable insight into cytokine regulation of disease, it is important to seek *in vivo* confirmation that Th1 and Th2 cells are relevant to human immune responses.

Recently it has been shown that CD8$^+$ T-cells can also be divided on the basis of cytokine production, and the Th1 and Th2 categories have thus been broadened to type 1 and type 2 cytokine patterns.[13,14]

Table 19.1 Type 1 and type 2 cytokines.

	Type 1	Type 2	Type 1 and type 2
Murine	IFN-γ	IL-4	GM-CSF
	LT	IL-5	IL-3
	IL-2	IL-6	TNF-α
		IL-10	IL-13
Human	IFN-γ	IL-4	IL-6
	LT	IL-5	GM-CSF
	IL-2	(IL-2)	IL-3
			TNF-α
			IL-13
			IL-10

Cytokines produced from type 1 and type 2 murine and human T-cell clones. Human type 2 cells produce variable amounts of IL-2.
GM-CSF, granulocyte–macrophage colony-stimulating factor; IFN-γ, interferon γ; IL, interleukin; LT, lymphotoxin; TNF-α, tumour necrosis factor α.

ACTIONS OF TYPE 2 CYTOKINES RELEVANT TO ASTHMA

Regulation of IgE

Atopic asthma is characterized by allergen-specific IgE. Triggering high-affinity IgE receptors on mast cells and basophils produces immediate asthma symptoms through release of histamine and lipid mediators; increasingly, however, IgE has been shown to have potential in cytokine release from mast cells and basophils,[15,16] and via FcεRI on dendritic cells and monocytes can function in antigen uptake and presentation to T-lymphocytes.[17] IgE may thus play a role in chronic airway inflammation in addition to simply triggering mast cells, and epidemiological and recent immunohistochemical evidence raises the possibility that IgE may also play a role in non-atopic asthma.[18,19]

IgE production from B-cells is determined by isotype switching, with gene rearrangement and splicing to join segments determining antigen specificity (*VDJ* genes) with those determining isotype (constant or *C* genes for IgM, IgA, IgE or IgG).[20] Switching to IgE is dependent on IL-4 or IL-13.[21,22] These cytokines cause switching to a sterile mRNA transcript, and a second signal is required for definitive rearrangement to align *VDJ* and C_ε genes to produce IgE mRNA. This second signal may be via CD40 ligand (CD40L) or CD2 on the T-cell surface interacting with CD40 or CD58 on the B-cell,[23,24] or can be provided, at least *in vitro*, by soluble factors including hydrocortisone.[25]

B-cells can also present antigen complexed with MHC class II on the B-cell surface to the TCR on the T-cell: this *cognate* interaction, together with IgE switching, induces antigen-specific IgE. However, IL-4 (or IL-13) and CD40L–CD40 interaction is sufficient to switch to IgE synthesis, and the combination of these signals will activate B-cells of many different antigen specificities to produce IgE by *non-cognate* interaction, hence leading to a polyclonal increase in IgE. Since basophils, mast cells and eosinophils can produce IL-4 and have surface CD40L, these cells may contribute to amplification of IgE responses (although not in an antigen-specific manner). However, *in vitro* experiments suggested that basophils will switch to IgE but that mast cells require exogenous IL-4; eosinophils did not cause isotype switching.[26,27]

The T-cell cytokine profile is a critical determinant of IgE switching, since IL-4 or IL-13 are required for IgE synthesis and IFN-γ is inhibitory.[28] Other cytokines can influence this process: IL-6 is required for human IgE switching,[29] IgE synthesis is enhanced by IL-5, IL-10, tumour necrosis factor (TNF)-α and the chemokines macrophage inflammatory protein-1α (MIP-1α) and RANTES, whilst IgE synthesis is inhibited by IFN-γ, IL-12, transforming growth factor (TGF)-β and IL-8.[30–33] (Fig. 19.1).

Human type 2 CD4$^+$ clones will support IgE synthesis by B-cells *in vitro*.[28] The cloning of allergen-specific type 2 CD4$^+$ T-cells from tissue derived from sites of allergic disease, together with the demonstration of a type 2 cytokine profile in the airway mucosa in atopic asthma, strongly suggests that this process occurs *in vivo*.

IL-4 and IL-13

IL-4 increases expression of the vascular cell adhesion molecule (VCAM)-1 by human endothelial cells *in vitro*; since eosinophils, basophils and T-cells, but not neutrophils,

increased by
IL-5
IL-6
IL-10
TNFα
RANTES
MIP1α

IgE synthesis

IFNγ inhibits

CD40   CD40L

VDJ-Cε mRNA

DNA

IεCε mRNA

CD58   CD2

B cell   IL-4   T cell
         IL-13
         germline switching

inhibited by
IL-12
TGFβ
IL-8

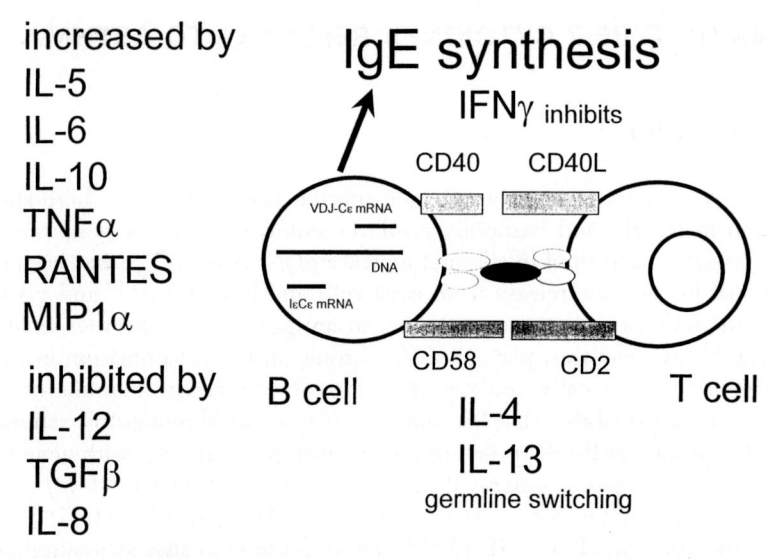

Fig. 19.1 Cytokine regulation of IgE synthesis. IL-4 or IL-13 are required for initial germline mRNA synthesis ($I\varepsilon C\varepsilon$ mRNA), but a second signal (such as CD40L or CD2) is needed for productive IgE mRNA transcription (VDJ–C_ε mRNA). IFNγ, inferferon γ; IL, interleukin; MIP-1α, macrophage inflammatory protein-1α; RANTES, regulated on activation, normal T-cell expressed, and secreted; TGFβ, transforming growth factor β; TNFα, tumour necrosis factor α.

express the counter-receptor (VLA-4), this may be relevant to accumulation of cells at sites of allergic inflammation.[34] IL-13 shares many of the actions of IL-4, via a common receptor consisting of the IL-4R α-chain and an IL-13 β-chain.[35] Both IL-4 and IL-13 are active in inducing IgE synthesis, upregulation of VCAM-1 on endothelial cells and monocyte and B-cell upregulation of both CD23 and HLA-DR. However, there is a specific IL-4 receptor present on T-lymphocytes, which do not respond to IL-13.

Eosinophils and basophils

Eosinophils and basophils differentiate from bone marrow precursors under the influence of IL-3 and granulocyte–macrophage colony-stimulating factor (GM-CSF); late maturation of eosinophil and basophil precursors is effected by IL-5.[36] These cytokines act in development of both cell types; indeed mixed eosinophil/basophil colonies are described.[37] IL-5 appears sufficient for eosinophil development, since transgenic mice expressing IL-5 with the CD2 promoter, so that all T-cells constitutively produce IL-5, have marked eosinophilia and tissue infiltration by eosinophils.[38] In guinea-pigs, intravenous IL-5 increased eosinophil influx into skin injected with eotaxin, by mobilizing bone marrow eosinophils.[39] Thus IL-5 released from the airway may act at a distant bone marrow site to increase production of and mobilize eosinophils.

IL-3, IL-5 and GM-CSF also act to prime eosinophils for chemotaxis, degranulation, cytotoxicity and synthesis of leukotriene C_4. These cytokines act to enhance survival of eosinophils by inhibition of apoptosis (see Chapter 8).

The receptors for IL-3, IL-5 and GM-CSF have been cloned and studied in some detail.[40,41] The common β-chain is important in signal transduction via the tyrosine kinase JAK2 interacting with signal transduction and activator of transcription factor (STAT) 1.[42] This may explain the similar activities of these three cytokines. They have cytokine-specific α-chains, which associate with a β-chain to form a high-affinity receptor (probably in pairs, so that an ααββ receptor complex interacts with an IL-5 dimer or two GM-CSF or IL-3 molecules[43,44]). The distribution of the α-chain determines cytokine responsiveness of various cell types: IL-5Rα is restricted to eosinophils and basophils.[45] In common with other cytokine receptors, IL-5Rα has soluble and membrane-associated isoforms, produced by alternative gene splicing to give differing mRNA.[46] The role of such soluble cytokine receptors is uncertain: they can antagonize cytokine actions, but alternatively may enhance actions by acting as carrier proteins or preventing cytokine degradation. Recent analysis of IL-5Rα isoform mRNA expression in asthma suggested that increased membrane isoform and less soluble isoform mRNA was detected in those with more severe airflow obstruction.[47]

Although rodent mast cells are responsive to IL-3, IL-4 and IL-10,[48] this is not the case for human mast cells, which appear to differentiate in response to stem cell factor (SCF) only.

T-cell derived cytokines may have other roles in asthma. TNF-α upregulates adhesion molecules on endothelial cells and may thus favour inflammatory cell recruitment. The role of IL-10 is uncertain. This cytokine is produced by both type 1 and type 2 T-cells in humans,[49,50] and inhibits proliferation and cytokine synthesis by both T-cell subtypes. It may thus act as an autoregulatory cytokine tending to turn off the immune response. Such a role is suggested in rheumatoid arthritis.[51]

EVIDENCE OF T-CELL CYTOKINE PRODUCTION IN ASTHMA

Direct evidence for allergen-specific type 2 T-cells in the airway mucosa of atopic asthmatic subjects is limited. However, Del Prete et al.[52] have derived such clones from a bronchial biopsy from a subject sensitive to pollen.

Most evidence for the involvement of type 2 cytokines in asthma has come from the study of bronchial biopsy or bronchoalveolar lavage (BAL) cell expression of cytokine mRNA. Increased numbers of IL-5 mRNA-positive cells were detected in bronchial biopsies from asthmatic subjects when compared with control subjects; this could be related to eosinophil numbers in the asthmatic subjects.[53] Examination of cytokine mRNA expression by BAL cells from atopic asthmatic and non-atopic control subjects showed increased proportions of BAL cells with mRNA transcripts for IL-4, IL-5, IL-3, GM-CSF and IL-2 in the asthmatic subjects, but no differences between the groups in numbers of cells expressing IFN-γ mRNA.[54] Furthermore, the proportion of BAL cells expressing type 2 cytokine mRNA increased at 24 h after allergen inhalation challenge when compared with diluent control inhalation in the same subjects.[55] In symptomatic asthmatic subjects treated with either prednisolone or placebo in a double-blind study, there was a reduction in both IL-4 and IL-5 mRNA-positive cells and an increase

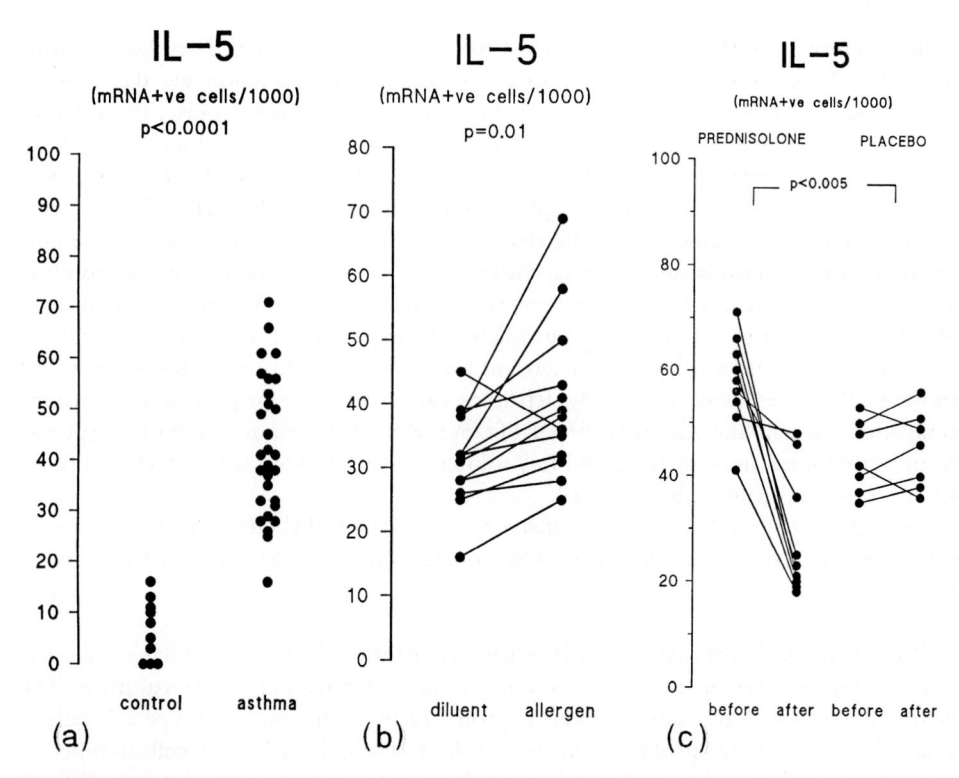

Fig. 19.2 Numbers of bronchoalveolar lavage cells from atopic asthmatic subjects that give positive *in situ* hybridization signals for IL-5 mRNA (mRNA + ve cells/1000): (a) compared with control subjects; (b) 24 h after either diluent or allergen inhalation challenge; (c) before and after a 2-week course of oral prednisolone or matched placebo.

in IFN-γ mRNA-expressing cells after steroid treatment[56] (Fig. 19.2). The relevance of IL-5 to eosinophil infiltration and the clinical manifestations of asthma was supported by relationships between the numbers of BAL cells expressing mRNA for IL-5 in baseline asthma and BAL eosinophil numbers, bronchial responsiveness to methacholine and airflow obstruction (as measured by forced expiratory volume in 1 s, FEV_1).[57] After allergen inhalation challenge, the proportion of BAL cells positive for IL-5 mRNA was related to eosinophil numbers, CD25 expression on CD4 T-cells and the fall in FEV_1 during the preceding late response.[55] Broide *et al.*[58] demonstrated T-lymphocyte expression of mRNA for GM-CSF after local allergen challenge. Elevated concentrations of IL-4 and IL-5 were detected by enzyme-linked immunosorbent assay in concentrated BAL fluid from atopic asthmatic subjects compared with control subjects, but there was no detectable IFN-γ or IL-2.[59] IL-5 protein was present in BAL fluid after local allergen challenge and also in serum of asthmatic subjects with symptomatic exacerbations.[60]

The majority of mRNA signals for IL-4 and IL-5 in BAL cells from asthmatic subjects was localized to T-cells.[54] BAL T-cell lines derived from asthmatic, but not control, subjects produced IL-3, IL-5 and GM-CSF *in vitro*.[61] Simultaneous *in situ* hybridization and immunohistochemistry again showed that the majority of mRNA signals for IL-4 and IL-5 in the asthmatic airway were localized to CD3$^+$ (T-lymphocytes), although some mRNA-positive cells were mast cells or eosinophils.[62] However, Bradding et al.[63] localized IL-4 and IL-5 immunoreactivity exclusively to mast cells in bronchial biopsies from asthmatic subjects. It is likely that this apparent discrepancy results from the difficulty in detecting immunoreactivity for cytokine protein in T-cells because these cells do not store cytokine.

IL-10 mRNA expression was detected in both T-cells and macrophages from atopic asthmatic and control subjects; IL-10 mRNA-positive cell numbers were increased in the asthmatic subjects, and further increased after allergen challenge.[64] The role of IL-10 in asthma is uncertain, although if this cytokine is acting to downregulate airway inflammation it is not entirely successful. Intervention studies blocking IL-10 would be required to substantiate such a role: anti-IL-10 antibodies did exacerbate inflammation in murine schistosomiasis.[65] IL-13 has been detected in BAL cells 24 h after allergen challenge of atopic asthmatic subjects.[66]

Although these studies suggest a predominant type 2 T-cell response in atopic asthma, increased serum concentrations of IFN-γ have been detected during asthma exacerbations[67] and an increased capacity of BAL T-cells from asthmatic subjects to produce IFN-γ to polyclonal stimuli when compared with control subjects has been demonstrated.[68] Although IFN-γ has been shown to activate eosinophil cytotoxicity,[69] the weight of evidence is in favour of a type 2 T-cell cytokine profile in asthma.

The allergen specificity of the T-cell populations expressing cytokine mRNA in biopsies or BAL from atopic asthmatic subjects remains to be established.

T-CELL CYTOKINES IN NON-ATOPIC ASTHMA

Initial measurement of cytokine concentrations in BAL fluid from asthmatic subjects with negative skin prick tests and no specific serum IgE to allergens showed increased IL-2 and IL-5 but not IL-4, in contrast to the type 2 pattern of cytokines found in BAL fluid in atopic asthmatic subjects.[59] However, these 'intrinsic' asthmatic subjects often have high total serum IgE relative to non-asthmatic subjects, and Burrows et al.[18] showed a correlation between serum IgE and asthma symptoms even in the absence of atopy. Using reverse-transcription polymerase chain reaction and *in situ* hybridization, we recently showed that both IL-4 and IL-5 mRNA expression was increased in intrinsic as well as atopic asthmatic subjects compared with control subjects; IL-4 immunoreactivity was also increased in both groups of asthmatic subjects.[70] Together with the demonstration of increased numbers of cells bearing high-affinity IgE receptors in the bronchial mucosa of both atopic and intrinsic asthmatic subjects,[19] this suggests a common immunopathology involving IgE-dependent mechanisms.

Post-mortem findings and examination of bronchial biopsies from subjects with occupational asthma due to toluene diisocyanate also show evidence of eosinophil and T-cell activation.[71] T-cell cloning from a bronchial biopsy from such an asthmatic patient isolated CD8$^+$ T-cell clones producing IL-5.[72]

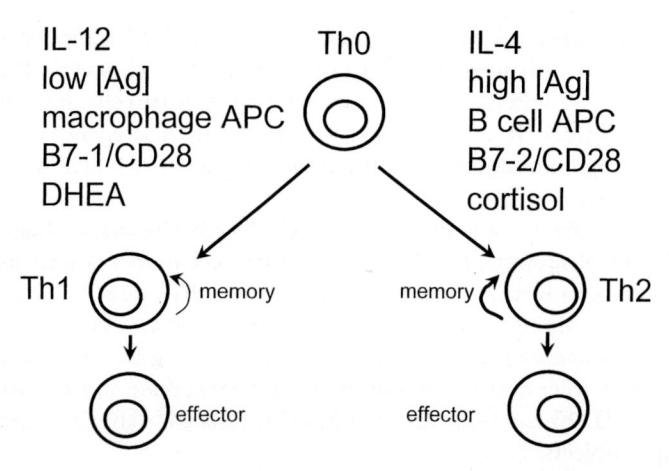

Fig. 19.3 Factors influencing development of Th1 and Th2 cell subtypes. IL-12, low antigen concentration ([Ag]), macrophages acting as antigen-presenting cell (APC), use of B7-1/CD28 costimulation or presence of dehydroepiandrosterone (DHEA) will all favour Th1 development in primary culture. These cells tend to retain their cytokine profile as memory T-cells. Cytokine production by effector cells can be modulated by other factors such as cytokine environment or nitric oxide (see text).

FACTORS DETERMINING TYPE 1 OR TYPE 2 T-CELL DEVELOPMENT

During expansion of naive CD4 T-cells from both mice and humans the best-defined factors determining development of a Th1 or Th2 population are the cytokines present during differentiation,[73–77] the antigen dose,[78] the antigen-presenting cell type and costimulating signal,[79,80] steroid hormones[81] and nitric oxide[82,83] (Fig. 19.3).

In different mouse strains there appears to be a genetically determined 'default' Th response to antigen.[77,84] Studies of the genetics of human asthma and atopy have isolated a number of polymorphisms in chromosome 5q31; these include specific changes in the IL-4 promoter that enhance IL-4 inducibility and might tend towards Th2 responses to allergen.[85,86]

ALTERATION OF ESTABLISHED TYPE 1 OR TYPE 2 CYTOKINE PROFILE

It has proved impossible to change the cytokine profile of established murine T-cell clones *in vitro*.[87] It is uncertain whether human T-cell subtypes are as fixed: induction of IFN-γ production (in addition to IL-4) in Th2 clones by IL-12 has been reported, though this may represent expansion of a minority population of naive cells.[88] It will be important to establish the kinetics of T-cell populations in the airway and the potential to alter memory responses, if future therapy is to be directed at altering the T-cell cytokine profile. It is of note that allergen immunotherapy may achieve such a change: allergen-challenged sites showed transcripts for IFN-γ and IL-2 in addition to type 2 cytokines after immunotherapy, but only IL-4 and IL-5 mRNA in placebo-treated subjects.[89]

POTENTIAL FOR INTERVENTION

A number of animal models of allergen challenge have allowed intervention to modulate cytokine production. The most compelling data are for IL-5. IL-5 gene disruption ('knock-out' mice) led to abolition of eosinophilia and bronchial hyperresponsiveness in ovalbumin-challenged mice.[90] Monoclonal antibodies to IL-5 prevented eosinophil influx in experimental allergen challenge of mice.[91] A humanized monoclonal antibody to IL-5 blocked both eosinophil influx and bronchial hyperresponsiveness to allergen challenge in a monkey model, and the effect of a single intravenous dose persisted for up to 3 months.[92] Whether such a result will be reproduced in established human asthma remains to be seen, but at least such an approach will allow testing of the eosinophil hypothesis of asthma. Similar animal experiments with IL-4 knock-out mice or blocking antibodies to IL-4 suggest that IL-4 is required for a type 2 T-cell response and airway hyperresponsiveness to airway antigen challenge.[93]

Human studies with IFN-γ in atopic disease did not show any alteration in IgE or symptoms.[94] This may reflect the difficulty of influencing established T-cell cytokine profiles. It is of note that IFN-γ does improve response rates to treatment of human leishmaniasis.[95]

Cyclosporin A improved lung function and had steroid-sparing activity in corticosteroid-dependent asthmatic subjects.[96] This may have resulted from inhibition of lymphokine production.

CONCLUSION

There is now considerable evidence for T-cell activation and a role for T-cell cytokines in asthma. T-lymphocyte activation is the only antigen-specific mechanism available for initiation of allergic sensitization, although IL-4 from non-T-cells (basophils, mast cells and eosinophils) may amplify the response and favour activation of a subsequent type 2 T-cell population. The events that determine allergen sensitization in infancy are unknown and will be an important area for further study.[97] The antigen specificity of T-cells in both atopic and non-atopic asthma is another area in need of clarification. Corticosteroids modulate the T-cell cytokine response in asthma and are the most effective anti-inflammatory drug available for asthma treatment. Whether it is possible to target specific T-cell populations or skew cytokine response remains to be seen.

REFERENCES

1. Dumonde DC, Wolstencroft RA, Panayi GS, Matthew M, Morley J, Howson WT: 'Lymphokines': non-antibody mediators of cellular immunity generated by lymphocyte activation. *Nature* (1969) **244**: 38–42.

2. Davis MM, Bjorkman PJ: T-cell antigen receptor genes and T-cell recognition. *Nature* (1988) **334**: 395–402.
3. Mosmann TR, Cherwinski H, Bond MW, Gieldin MA, Coffman RL: Two types of murine helper T cell clones. *J Immunol* (1986) **136**: 2348–2357.
4. Fong TAT, Mosmann TR: The role of IFNγ in delayed-type hypersensitivity mediated by Th1 clones. *J Immunol* (1989) **143**: 2887–2893.
5. Cher DJ, Mosmann TR: Two types of murine helper T cell clone. II. Delayed type hypersensitivity is mediated by Th1 clones. *J Immunol* (1987) **138**: 3688–3694.
6. Wierenga EA, Snoek M, de Groot C, *et al.*: Evidence for compartmentalization of functional subsets of CD4+ T lymphocytes in atopic patients. *J Immunol* (1990) **144**: 4651–4656.
7. Parronchi P, Macchia D, Piccini M-P, *et al.*: Allergen and bacterial antigen-specific T-cell clones established from atopic donors show a different profile of cytokine production. *Proc Natl Acad Sci USA* (1991) **88**: 4538–4542.
8. Romagnani S: Human Th1 and Th2: doubt no more. *Immunol Today* (1991) **12**: 256–257.
9. Maggi E, Biswas P, Del Prete G, *et al.*: Accumulation of Th2-like helper T cells in the conjunctiva of patients with vernal conjunctivitis. *J Immunol* (1991) **146**: 1169–1174.
10. Yamamura M, Uyemura K, Deans RJ, *et al.*: Defining protective responses to pathogens: cytokine profile in leprosy lesions. *Science* (1991) **254**: 277–279.
11. Umetsu DT, Jabara HH, DeKruyff RH, Abbas AK, Abrams JS, Geha RS: Functional heterogeneity among human inducer T cell clones. *J Immunol* (1988) **140**: 4211–4216.
12. Paliard X, de Waal Malefyt R, Yssel H, *et al.*: Simultaneous production of IL-2, IL-4, and IFNγ by activated human CD4+ and CD8+ T cell clones. *J Immunol* (1988) **141**: 849–855.
13. Erard F, Wild M-T, Garcia Sanz JA, Le Gros G: Switch of CD8 T cells to noncytolytic CD8−CD4− cells that make Th2 cytokines and help B cells. *Science* (1993) **260**: 1802–1805.
14. Mosmann TR, Sad S: The expanding universe of T-cell subsets: Th1, Th2 and more. *Immunol Today* (1996) **17**: 138–146.
15. Plaut M, Pierce JH, Watson CJ, Hanley-Hide J, Nordan RP, Paul WE: Mast cell lines produce lymphokines in response to cross-linkage of FcεRI or to calcium ionophores. *Nature* (1989) **339**: 64–67.
16. Okayama Y, Petit Frere C, Kassel O, *et al.*: IgE-dependent expression of mRNA for IL-4 and IL-5 in human lung mast cells. *J Immunol* (1995) **155**: 1796–1808.
17. Maurer D, Ebner C, Reininger B, *et al.*: The high affinity IgE receptor (FcεRI) mediates IgE-dependent allergen presentation. *J Immunol* (1995) **154**: 6285–6290.
18. Burrows B, Martinez FD, Halonen M, Barbee RA, Cline MG: Association of asthma with serum IgE levels and skin test reactivity to allergens. *N Engl J Med* (1989) **320**: 271–277.
19. Humbert M, Grant JA, Taborda-Barata L, *et al.*: High affinity IgE receptor bearing cells in bronchial biopsies from atopic and non-atopic asthma. *Am J Respir Crit Care Med* (1996) **153**: 1931–1937.
20. Geha RS: Regulation of IgE synthesis in humans. *J Allergy Clin Immunol* (1992) **90**: 143–150.
21. Vercelli D, Jabara HH, Arai K, Geha RS: Induction of human IgE synthesis requires interleukin 4 and T/B cell interactions involving the T cell receptor/CD3 complex and MHC class II antigens. *J Exp Med* (1989) **169**: 1295–1307.
22. Punnonen JG, Aversa G, Cocks BG, *et al.*: Interleukin 13 induces interleukin-4-independent IgG4 and IgE synthesis and CD23 expression by human B cells. *Proc Natl Acad Sci USA* (1993) **90**: 3730–3734.
23. Zhang K, Clark EA, Saxon A: CD40 stimulation provides an IFNγ-independent and IL-4-dependent differentiation signal directly to human B cells for IgE producion. *J Immunol* (1991) **146**: 1836–1842.
24. Diaz-Sanchez D, Chegini S, Zhang K, Saxon A: CD58 (LFA3) stimulation provides a signal for human isotype switching and IgE production distinct from CD40. *J Immunol* (1994) **153**: 10–19.
25. Jabara H, Vercelli D, Ahern D, Geha RS: Hydrocortisone and IL-4 induce IgE isotype switching in human B cells. *J Immunol* (1991) **147**: 1557–1560.
26. Gauchat JF, Henchoz S, Mazzei G, *et al.*: Induction of human IgE synthesis in B cells by mast cells and basophils. *Nature* (1993) **365**: 340–343.
27. Gauchat JF, Henchoz S, Fattah D, *et al.*: CD40 ligand is functionally expressed on human eosinophils. *Eur J Immunol* (1995) **25**: 863–865.

28. Del Prete G, Maggi E, Parronchi P, *et al.*: IL-4 is an essential factor for the IgE synthesis induced *in vitro* by human T cell clones and their supernatants. *J Immunol* (1988) **140**: 4193–4198.
29. Vercelli D, Jabara HH, Arai KI, Yokota T, Geha R: Endogenous interleukin 6 plays an obligatory role in interleukin 4-dependent human IgE synthesis. *Eur J Immunol* (1989) **19**: 1419–1424.
30. Pene J, Rousset F, Briere F, *et al.*: Interleukin 5 enhances interleukin 4-induced IgE production by normal human B cells. The role of soluble CD23 antigen. *Eur J Immunol* (1988) **18**: 929–935.
31. Kimata H, Yoshida A, Ishioka C, Fujimoto M, Lindley I, Furusho K: RANTES and macrophage inflammatory protein 1α selectively enhance immunoglobulin E (IgE) and IgG_4 production by human B cells. *J Exp Med* (1996) **183**: 2397–2404.
32. Kiniwa M, Gately M, Gubler U, Chizzonite R, Fargas C, Delesspesse G: Recombinant interleukin-12 supresses the synthesis of immunoglobulin E by interleukin-4 stimulated human lymphocytes. *J Clin Invest* (1992) **90**: 262–266.
33. Kimata H, Yoshida A, Ishioka C, Lindley I, Mikawa H: Interleukin 8 (IL-8) selectively inhibits immunoglobulin E production induced by IL-4 in human B cells. *J Exp Med* (1992) **176**: 1227–1231.
34. Walsh GM, Hartnell A, Mermod JJ, Kay AB, Wardlaw AJ: Human eosinophil, but not neutrophil, adherence to IL-1 stimulated HUVEC is $\alpha_4\beta_1$ (VLA-4) dependent. *J Immunol* (1991) **146**: 3419–3423.
35. Zurawski G, de Vries JE: Interleukin 13, an interleukin 4-like cytokine that acts on monocytes and B cells, but not on T cells. *Immunol Today* (1994) **15**: 19–26.
36. Clutterbuck EJ, Hirst EMA, Sanderson CJ: Human interleukin 5 (IL-5) regulates the production of eosinophils in human bone marrow cultures: comparison with IL-1, IL-3, IL-6 and GM-CSF. *Blood* (1989) **73**: 1504–1512.
37. Boyce JA, Friend D, Matsumoto R, Austen KF, Owen WF: Differentiation *in vitro* of hybrid eosinophil/basophil granulocytes: autocrine function of an eosinophil developmental intermediate. *J Exp Med* (1995) **182**: 49–55.
38. Dent LA, Strath M, Sanderson CJ: Eosinophilia in transgenic mice expressing interleukin-5. *J Exp Med* (1990) **172**: 1425–1431.
39. Collins PD, Marleau S, Griffiths-Johnson DA, Jose PJ, Williams TJ: Cooperation between interleukin 5 and eotaxin to induce eosinophil accumulation *in vivo*. *J Exp Med* (1995) **182**: 1169–1174.
40. Kitamura T, Sato N, Arai KI, Miyajima A: Expresion cloning of the human IL-3 receptor cDNA reveals a shared β subunit for the human IL-3 and GM-CSF receptors. *Cell* (1991) **66**: 1165–1174.
41. Tavernier J, Devos R, Cornelis S, *et al.*: A human high affinity interleukin 5 receptor (IL5R) is composed of an IL-5-specific α chain and a β chain shared with the receptor for GM-CSF. *Cell* (1991) **66**: 1175–1184.
42. van der Bruggen T, Caldenhoven E, Kanters D, *et al.*: Interleukin 5 signalling in human eosinophils involves JAK2 tyrosine kinase and STAT1α. *Blood* (1995) **85**: 1442–1448.
43. Muto A, Watanabe S, Miyajima A, Yokota T, Arai KI: The β subunit of human granulocyte macrophage colony stimulating factor receptor forms a homodimer and is activated via its association with the α subunit. *J Exp Med* (1996) **183**: 1911–1916.
44. Jenkins BJ, D'Andrea R, Gonda TJ: Activating point mutations in the common β subunit of the human GM-CSF, IL-3, and IL-5 receptors suggest the involvement of β subunit dimerization and cell type-specific molecules in signalling. *EMBO J* (1995) **14**: 4276–4287.
45. Lopez AF, Elliot MJ, Woodcock J, Vadas MA: GM-CSF, IL-3, and IL-5: cross competition on human haemopoietic cells. *Immunol Today* (1992) **13**: 495–500.
46. Tavernier J, Tuypens T, Plaetinck G, Verhee A, Fiers W, Devos R: Molecular basis of the membrane-anchored and two soluble isoforms of the human interleukin 5 receptor α subunit. *Proc Natl Acad Sci USA* (1992) **89**: 7041–7045.
47. Yasruel Z, Humbert M, Kotsimbos ATC, *et al.*: Expression of membrane-bound and soluble interleukin-5 alpha receptor mRNA in the bronchial mucosa of atopic and non-atopic asthmatics. *Am J Crit Care Med* (1997) **55**: 1413–1418.

48. Thompson-Snipes L, Dhar V, Bond MW, Mosmann TR, Moore KW, Rennick DM: Interleukin 10: a novel stimulatory factor for mast cells and their progenitors. *J Exp Med* (1991) **173**: 507–510.
49. Yssel H, de Waal Malefyt R, Roncarolo MG, *et al.*: IL-10 is produced by subsets of human CD4+ T cell clones and peripheral blood T cells. *J Immunol* (1992) **149**: 2378–2384.
50. Del Prete G, De Carli M, Almerigogna F, Giudizi MG, Biagiotti R, Romagnani S: Human IL-10 is produced by both type 1 helper (Th1) and type 2 helper (Th2) T cell clones and inhibits their antigen-specific proliferation and cytokine production. *J Immunol* (1993) **150**: 353–360.
51. Katsikis PD, Chu CQ, Brennan FM, Maini RN, Feldmann M: Immunoregulatory role of interleukin 10 in rheumatoid arthritis. *J Exp Med* (1994) **179**: 1517–1527.
52. Del Prete GF, de Carli M, D'Elios MM, *et al.*: Allergen exposure induces the activation of allergen-specific Th2 cells in the airway mucosa of patients with allergic respiratory disorders. *Eur J Immunol* (1993) **23**: 1445–1449.
53. Hamid Q, Azzawi M, Sun Ying, *et al.*: Expression of mRNA for interleukin 5 in mucosal bronchial biopsies from asthma. *J Clin Invest* (1991) **87**: 1541–1546.
54. Robinson DS, Hamid Q, Ying S, *et al.*: Predominant Th2-like bronchoalveolar T cell population in atopic asthma. *N Engl J Med* (1992) **326**: 295–304.
55. Robinson DS, Hamid Q, Bentley AM, Sun Ying, Kay AB, Durham SR: CD4+ T cell activation, eosinophil recruitment and interleukin 4 (IL-4), IL-5, and GM-CSF messenger RNA expression in bronchoalveolar lavage after allergen inhalation challenge in atopic asthmatics. *J Allergy Clin Immunol* (1993) **92**: 313–324.
56. Robinson DS, Hamid Q, Sun Ying, *et al.*: Prednisolone treatment in asthma is associated with modulation of bronchoalveolar lavage cell IL-4, IL-5 and IFNγ cytokine gene expression. *Am Rev Respir Dis* (1993) **148**: 401–406.
57. Robinson DS, Sun Ying, Bentley AM, *et al.*: Relationships among numbers of bronchoalveolar lavage cells expressing mRNA for cytokines, asthma symptoms, and airway methacholine responsiveness in atopic asthma. *J Allergy Clin Immunol* (1993) **92**: 397–403.
58. Broide DH, Firestein GS: Endobronchial allergen challenge in asthma: demonstration of cellular source of granulocyte macrophage colony-stimulating factor by *in situ* hybridization. *J Clin Invest* (1991) **88**: 1048–1053.
59. Walker C, Bode E, Boer L, Hansell TT, Blaser K, Virchow JC Jr: Allergic and non-allergic asthmatics have distinct patterns of T-cell activation and cytokine production in peripheral blood and bronchoalveolar lavage. *Am Rev Respir Dis* (1992) **146**: 109–115.
60. Corrigan CJ, Haczku A, Gemou-Engesaeth V, *et al.*: CD4 T-lymphocyte activation in asthma is accompanied by increased serum concentrations of interleukin-5. *Am Rev Respir Dis* (1993) **147**: 540–547.
61. Till SJ, Li B, Durham SR, *et al.*: Secretion of the eosinophil-active cytokines interleukin 5, granulocyte macrophage colony stimulating factor and interleukin 3 by bronchoalveolar lavage CD4+ and CD8+ T cell lines in atopic asthmatics and atopic and non-atopic controls. *Eur J Immunol* (1995) **25**: 2727–2731.
62. Sun Ying, Durham SR, Corrigan CJ, Hamid Q, Kay AB: Phenotype of cells expressing mRNA for Th2-type (interleukin 4 and interleukin 5) and Th1 type (interferon γ and interleukin 2) cytokines in bronchoalveolar lavage and bronchial biopsies from atopic asthmatics and normal control subjects. *Am J Respir Cell Mol Biol* (1995) **12**: 477–487.
63. Bradding P, Roberts JA, Britten KM, *et al.*: Interleukin-4, -5, and -6 and tumor necrosis factor-alpha in normal and asthmatic airways: evidence for the human mast cell as a source of these cytokines. *Am J Respir Cell Mol Biol* (1994) **10**: 471–480.
64. Robinson DS, Tsicopoulos A, Qiu Meng, Durham SR, Kay AB, Hamid Q: Increased interleukin 10 messenger RNA expression in atopic allergy and asthma. *Am J Respir Cell Mol Biol* (1996) **14**: 113–117.
65. Flores-Villanueva PO, Reiser H, Stadecker MJ: Regulation of T helper cell responses in experimental murine schistosomiasis by IL-10. *J Immunol* (1994) **153**: 5190–5199.
66. Huang SK, Xiao HQ, Kleine-Tebbe J, *et al.*: IL-13 expression at the sites of allergen challenge in patients with asthma. *J Immunol* (1995) **155**: 2688–2694.
67. Corrigan CJ, Kay AB: CD4 T-lymphocyte activation in acute severe asthma. Relationship to disease severity and atopic status. *Am Rev Respir Dis* (1990) **141**: 970–977.

68. Krug N, Madden J, Redington AE, et al.: T-cell cytokine profile evaluated at the single cell level in BAL and blood in allergic asthma. Am J Respir Cell Mol Biol (1996) 14: 319–326.
69. Valerius T, Repp R, Kalden JR, Platzer E: Effects of IFNγ on human eosinophils in comparison with other cytokines. J Immunol (1990) 145: 2950–2958.
70. Humbert M, Durham SR, Ying S, et al.: IL-4 and IL-5 mRNA and protein in bronchial biopsies from atopic and non-atopic asthma: evidence against 'intrinsic' asthma being a distinct immunopathological entity. Am J Respir Crit Care Med (1996) 154: 1497–1504.
71. Bentley AM, Maestrelli P, Saetta M, et al.: Activated T lymphocytes and eosinophils in the bronchial mucosa in isocyanate-induced asthma. J Allergy Clin Immunol (1992) 146: 170–176.
72. Maestrelli P, Del Prete GF, De Carli M, et al.: CD8 T-cell clones producing interleukin-5 and interferon-gamma in bronchial mucosa of patients with asthma induced by toluene diisocyanate. Scand J Work Environ Health (1994) 20: 376–381.
73. Swain SL, Weinberg AD, English M, Huston G: IL-4 directs the development of Th2-like helper effectors. J Immunol (1990) 145: 3796–3799.
74. Maggi E, Parronchi P, Manetti R, et al.: Reciprocal regulatory role of IFNγ and IL-4 on the development of human Th1 and Th2 clones. J Immunol (1992) 148: 2142–2147.
75. Pernis A, Gupta S, Gollob KJ, et al.: Lack of interferon γ receptor β chain and the prevention of interferon γ signalling in Th1 cells. Science (1995) 269: 245–247.
76. Szabo SJ, Jacobson NG, Dighe AS, Gubler U, Murphy KM: Developmental commitment to the Th2 lineage by extinction of IL-12 signalling. Immunity (1995) 2: 665–675.
77. Guler M, Gorham JD, Hsieh CS, et al.: Genetic susceptibility to *Leishmania*: IL-12 responsiveness in Th1 cell development. Science (1996) 271: 984–987.
78. Hosken-NA, Shibuya K, Heath AW, Murphy KM, O'Garra A: The effect of antigen dose on CD4⁺ T helper cell phenotype development in a T cell receptor-alpha beta-transgenic model. J Exp Med (1995) 182: 1579–1584.
79. Secrist-H, DeKruyff RH, Umetsu DT: Interleukin 4 production by CD4⁺ T cells from allergic individuals is modulated by antigen concentration and antigen-presenting cell type. J Exp Med (1995) 181: 1081–1089.
80. Kuchroo VK, Das MP, Brown JA, et al.: B7–1 and B7–2 costimulatory molecules activate differentially the Th1/Th2 developmental pathways: application to autoimmune disease therapy. Cell (1995) 80: 707–718.
81. Daynes RA, Araneo BA, Dowell TA, Huang K, Dudley D: Regulation of murine lymphokine production in vivo. J Exp Med (1990) 171: 979–996.
82. Liew FY, Li Y, Severn A, et al.: A possible novel pathway of regulation by murine T helper type-2 (Th2) cells of a Th1 cell activity via the modulation of the induction of nitric oxide synthase on macrophages. Eur J Immunol (1991) 21: 2489–2494.
83. Barnes PJ, Liew FY: Nitric oxide and asthmatic inflammation. Immunol Today (1995) 16: 128–130.
84. Hsieh CS, Macatonia SE, O'Garra A, Murphy KM: T cell genetic background determines default T helper phenotype development in vitro. J Exp Med (1995) 181: 713–21.
85. Marsh DG, Neely JD, Breazeale DR, et al.: Linkage analysis of IL-4 and other chromosome 5q31.1 markers and total serum immunoglobulin E concentrations. Science (1994) 264: 1152–1156.
86. Borish L, Mascali JJ, Klinnert M, Leppert M, Rosenwasser LJ: SSC polymorphisms in interleukin genes. Hum Mol Genet (1994) 3: 1710.
87. O'Garra A, Murphy K: Role of cytokines in determining T-lymphocyte function. Curr Opin Immunol (1994) 6: 458–466.
88. Yssel H, Fasler S, de Vries JE, de Waal Malefyt R: IL-12 transiently induces IFN-gamma transcription and protein synthesis in human CD4⁺ allergen-specific Th2 T cell clones. Int Immunol (1994) 6: 1091–1096.
89. Varney VA, Hamid Q, Gaga M, et al.: Influence of grass pollen immunotherapy on cellular infiltration and cytokine mRNA expression during allergen-induced late phase cutaneous responses. J Clin Invest (1993) 92: 644–651.
90. Foster PS, Hogan SP, Ramsay AJ, Matthei KI, Young IG: Interleukin 5 deficiency abolishes eosinophilia, airways hyperreactivity, and lung damage in a mouse asthma model. J Exp Med (1996) 183: 195–201.

91. Gulbenkian AR, Egan RW, Fernandez X, *et al.*: Interleukin 5 modulates eosinophil accumulation in allergic guinea pig lung. *Am Rev Respir Dis* (1992) **146**: 263–265.
92. Mauser PJ, Pitman AM, Fernandez X, *et al.*: Effects of an antibody to interleukin-5 in a monkey model of asthma. *Am J Respir Crit Care Med* (1995) **152**: 467–472.
93. Corry DB, Folkesson HG, Warnock ML, *et al.*: Interleukin 4, but not interleukin 5 or eosinophils, is required in a murine model of acute airway hyperreactivity. *J Exp Med* (1996) **183**: 109–117.
94. Li JTC, Yunginger JW, Reed CE, Jaffe HS, Nelson DR, Gleich GJ: Lack of suppression of IgE production by recombinant IFNγ: a controlled trial in patients with allergic rhinitis. *J Allergy Clin Immunol* (1990) **85**: 934–940.
95. Badaro R, Falcoff E, Badaro FS, *et al.*: Treatment of visceral leishmaniasis with pentavalent antimony and interferon gamma. *N Engl J Med* (1990) **322**: 16–21.
96. Lock SH, Kay AB, Barnes NC: Double-blind, placebo-controlled study of cyclosporin A as a corticosteroid-sparing agent in corticosteroid-dependent asthma. *Am J Respir Crit Care Med* (1996) **153**: 509–514.
97. Holt PG: Primary allergic sensitization to environmental antiens: perinatal T cell priming as a determinant of responder phenotype in adulthood. *J Exp Med* (1996) **183**: 1297–1301.

20

Other Mediators of Asthma

K. FAN CHUNG and PETER J. BARNES

INTRODUCTION

Many inflammatory mediators have been implicated in the pathogenesis of asthma.[1] Much of the evidence supporting a role for a particular mediator has rested on the evidence that the mediator can induce features that mimic some of the characteristics of asthma, such as bronchoconstriction and bronchial hyperresponsiveness. In addition, the mediator can be detected in the airway or in samples of airway fluids. However, the most compelling evidence for a role for a particular mediator rests with studies in which blocking the effects of a putative mediator, either by specific receptor blockade or by inhibition of the synthesis of the mediator, leads to objective evidence of improvement in various parameters of the asthmatic process. Receptors for most of the putative mediators in asthma have been characterized and cloned, together with the discovery of specific receptor antagonists. In this chapter, mediators not previously considered are described: histamine, platelet-activating factor (PAF), oxygen radicals, complement, serotonin, eosinophil proteins and endothelin.

HISTAMINE

Histamine has long been implicated in the pathogenesis of asthma since the discovery in 1919 that it could mimic anaphylactic bronchoconstriction in guinea-pigs.[2]

Histamine receptors

Histamine produces its effects by interacting with specific receptors on target cells. The existence of more than one receptor subtype was suggested when Ash and Schild found that the classical antihistamine pyrilamine was able to block some responses, such as contraction of guinea-pig trachea, but not others, such as gastric acid secretion.[3] The existence of a second histamine receptor subtype (H_2) was confirmed with the development of selective antagonists such as cimetidine and ranitidine. A third subtype (H_3) for which selective agonist and antagonist have been developed have also been described.[4]

Using immunohistochemical techniques to study the distribution of cyclic guanosine monophosphate, H_1 receptors have been localized to airway epithelial cells, macrophages and alveolar cells in guinea-pig lung with surprisingly little localization to airway or vascular smooth muscle.[5] However, H_1 receptors have been determined in bovine tracheal smooth muscle.[6] H_1 receptors mediate airway smooth muscle contraction, bronchial microvascular leakage and bronchial vasodilation.[7]

H_2 receptors, which stimulate adenylate cyclase, have been identified in lung using [^{3}H]tiotidine[8] but their localization is still not clear. There is some debate as to whether H_2 receptors may mediate bronchodilation. Human peripheral lung strips may show a relaxant response to histamine mediated by H_2 receptors,[9] but an H_2-selective agonist, ipromidine, has no effect on normal or asthmatic airways *in vivo*.[10] H_2 selective antagonists such as cimetidine and ranitidine have not been associated with bronchoconstriction in normal or asthmatic subjects. Histamine-induced secretion of mucous glycoproteins appears to be mediated by H_2 receptors.[11]

H_3 receptors have been differentiated using the selective agonist (R)-α-methyl-histamine and the antagonist thioperamide. They may be involved in the feedback inhibition of histamine release from mast cells or basophils. H_3 receptors may inhibit neurotransmission in parasympathetic ganglia and the release of acetylcholine from postganglionic cholinergic nerves in guinea-pig and human airways.[12] H_3 receptors have also been identified on human eosinophils,[13] but their role is uncertain.

Activation of H_1 receptors leads to the activation of phospholipase C, with subsequent secretion of inositol 1,4,5-trisphosphate and an increase in intracellular Ca^{2+} concentrations.[14] The receptor belongs to a gene family that mediates signal transduction through guanine nucleotide regulatory proteins (G proteins) and has seven putative membrane-spanning domains in an α-helical configuration. The H_1 receptor cDNA was initially cloned from bovine adrenal medulla[15] and using the bovine gene as a probe the human H_1 receptor was subsequently cloned.[16,17] Abundant expression of H_1-receptor transcripts has been demonstrated in human lung.[16]

Effects of histamine

Histamine is a potent inducer of bronchoconstriction and causes airway microvascular leakage in the bronchial microvasculature, effects that are mediated through H_1 receptors. It also induces the secretion of mucous glycoproteins from human airways, although this effect is weak compared to other secretagogues.[11] These effects are mediated by stimulation of H_2 receptors. When instilled on the human nasal mucosa, histamine increases the recovery of histamine and bradykinin, indicating an increase in both vascular and epithelial permeabilities.[18] Histamine can induce the generation of prostaglandins,[19] an effect that probably accounts for the tolerance to histamine challenge reported in mild asthmatic individuals. Histamine has a small effect as an eosinophil chemoattractant.[20]

Role of histamine in asthma

Histamine is detectable in the circulation of humans at levels of 0.2–0.4 ng/ml; levels are highest in the early hours of the morning and are elevated in stable asthmatic subjects. Modest rises have been reported in the circulation of asthmatic subjects during exercise- and allergen-induced bronchoconstriction.[21] Increased amounts of histamine can be recovered from bronchoalveolar lavage fluids of asthmatic patients at rest,[22] and after allergen challenge or after a hyperosmolar stimulus.

H_1-receptor antagonists are highly selective for H_1 receptors with little effect on H_2 and H_3 receptors. The first-generation H_1-receptor antagonists usually had some anticholinergic activity or blocked serotonin or α-adrenergic receptors, but the second-generation antagonists do not have these properties. The second-generation H_1-receptor antagonists also have low central nervous system toxicity because of their inability to cross the blood–brain barrier. It is worth mentioning that at high concentrations, some second-generation H_1-receptor antagonists such as terfenadine, astemizole and loratidine produce non-competitive inhibition at the H_1 receptor. Binding of terfenadine and astemizole is not readily reversible.[23]

One interesting property of H_1-receptor antagonists is the prevention of release of mediators of inflammation from human basophils and mast cells *in vitro*. These effects do not involve activation of the H_1 receptor.[24] *In vivo*, the second-generation H_1-receptor antagonists decrease mediator release after antigen challenge to the nasal mucosa or skin of allergic patients. There are no available data concerning the lower airways of allergic asthmatic patients. Thus, in patients with allergic rhinitis, terfenadine or loratidine reduce the amount of histamine and prostaglandin (PG)D_2 in nasal secretions.[25,26] In the allergic skin, pretreatment with cetirizine led to a reduction in the amounts of histamine, PAF and prostaglandin released following allergen challenge; in addition, there was inhibition of eosinophil and basophil influx in the skin.[27,28] However, in patients with allergic rhinitis, cetirizine is less effective in inhibiting mediator release[29] and does not appear to inhibit eosinophil numbers in nasal fluid.[30] These properties may contribute to the 'antiallergic' properties of second-generation antihistamines. The importance of these properties of antihistamines in the treatment of asthma is unknown.

H_1-receptor antagonists have been widely studied in various asthma models including allergen- and exercise-induced asthma. H_1-receptor antagonists have been shown to protect against the early bronchoconstrictor response to allergen in asthmatic subjects.[31–36] Overall, the protection afforded against allergen was not as large as that against histamine. None of these antagonists has been demonstrated to completely inhibit the early response. The time-course of the early asthmatic response is also attenuated.[37,38] Thus, a 40–80% protection of the early response for 2–15 min was observed with astemizole and terfenadine.[36,37,39,40] In one study, terfenadine afforded a 1.5–2-fold protection against allergen challenge.[32]

The effect of H$_1$-receptor antagonists on the late response to allergen has been mainly reported to be negative[31,41] with chlorpheniramine, terfenadine and loratidine. There was no consistent effect of another H$_1$-receptor antagonist, triprolidine, on the late response in wheat flour-sensitive asthmatic subjects.[42] However, azelastine was found to cause significant inhibition of the late asthmatic response, when measured as the area under the time–response curve.[43]

Against exercise-induced asthma, a variable degree of protection has been reported with various H$_1$-receptor antagonists.[44–47] However, no significant effect was observed in one study.[33] On average, a 30–50% protection has been observed in most other studies. Bronchoconstriction induced by hyperventilation, cold air hyperventilation and inhalation of non-isotonic aerosols has also been reported to be partly attenuated.[48–52] These studies indicate that histamine is partly involved in the acute responses to exercise and allergen, and that the extent to which it is involved varies from asthmatic patient to asthmatic patient.

H$_1$-receptor antagonists can produce significant bronchodilatation in asthmatic subjects when administered orally, intravenously or by inhalation.[35,39,46,53–57] Other studies, however, have failed to demonstrate such an effect.[34,47,58–60] Thus, there is a large variation in the bronchodilator response, with up to 32% increase in baseline forced expiratory volume in 1 s (FEV$_1$) after oral terfenadine in one study in children.[46] H$_1$-receptor antagonists such as chlorpheniramine, terfenadine and clemastine do not possess significant anticholinergic properties *in vivo* and their bronchodilator effect must be assumed to be secondary to H$_1$-receptor blockade. The variable effect of H$_1$-receptor antagonists on airway calibre has led to the concept of a baseline 'histamine airway tone' in asthmatic individuals.

Despite the beneficial effect, clinical studies in asthmatic subjects have not revealed startling effects. In studies involving continuous treatment over periods of 2–7 weeks with terfenadine, azelastine and cetirizine, there was no significant effect on airway responsiveness to methacholine.[61–63] Effects on symptom scores, usage of rescue medication and airway calibre measurements in atopic asthmatic subjects have not been encouraging. No difference was reported in studies comparing active to placebo treatment.[36,64,65] In more recent studies, a small improvement in symptom scores and concomitant reduction in the use of rescue medication in atopic asthmatic subjects have been shown,[61,66,67] usually obtained at doses higher than those used for allergic rhinitis. Studies in more severe and chronic asthmatic subjects have not been reported, but it is unlikely that the H$_1$-histamine receptor antagonists will prove more effective in this group.

PLATELET-ACTIVATING FACTOR

PAF is a lipid mediator of inflammation that is synthesized by a two-step procedure, with the activation of a phospholipase A$_2$ to hydrolyse components of membrane phospholipids into lyso-PAF, which is then acetylated by an acetyltransferase enzyme into PAF. PAF is rapidly inactivated by an acetylhydrolase into lyso-PAF, which is biologically inactive. A wide range of cell types produce PAF *in vitro*. In terms of the airways in asthma, the most likely sources include eosinophils, alveolar macrophages, endothelial cells and neutrophils.[68–71] Production of PAF can be stimulated by various cytokines such as granulocyte–macrophage colony-stimulating factor (GM-CSF), interleukin (IL)-1β and tumour necrosis factor (TNF)-α.[72]

PAF receptors

The existence of PAF binding sites has been studied using labelled PAF or its competitive antagonist radioligand [^{3}H]WEB2086 in membrane fractions of homogenized lung tissue.[73,74] cDNA for a PAF receptor has been cloned from guinea-pig lung[75] and from human leucocytes.[76] The PAF receptor is a member of the superfamily of receptors coupled to GTP-binding proteins. The hydropathy profile of the PAF receptor, which has seven putative membrane-spanning α-helices, is typical of other G protein-linked receptors. The inhibition of binding of PAF to its specific receptors by GTP indicates that the PAF receptor is coupled to a G protein.[77] In addition, PAF stimulates GTPase activity in cell membranes from human platelets and neutrophils.[78,79]

Effects of PAF

Interest in PAF as a potential mediator of asthma has been aroused by its potent effect in inducing bronchoconstriction, airway microvascular leakage and eosinophil chemotaxis.[80] In humans, PAF-induced bronchoconstriction is mediated through the sulphidopeptide leukotrienes.[81] In addition, there is evidence for its ability to cause a transient increase in bronchial responsiveness in normal volunteers but not in asthmatic subjects.[82,83] In normal and asthmatic subjects, PAF inhalation induces an increase in alveolar–arterial oxygen partial pressure gradient and a fall in arterial oxygen partial pressure, an effect accounted for by ventilation–perfusion mismatching and which is accompanied by neutrophil sequestration in the lung.[84–86] PAF has a wide variety of effects in activating inflammatory cells. PAF is a potent chemotactic and chemokinetic mediator for eosinophils *in vitro*[87] and promotes the adhesion of neutrophils and eosinophils to vascular endothelial cells.[88,89] PAF is potent in causing release of the granule-associated enzyme eosinophil peroxidase (EPO) from human eosinophils.[90] Activation of eosinophils by PAF induces tracheal epithelial shedding *in vitro*, associated with a slowing of ciliary beat frequency.[91] PAF induces the release of cytokines such as TNF and IL-1 from alveolar macrophages and monocytes respectively.[92,93]

PAF receptor antagonists

PAF receptor antagonists can be generally classified as either natural antagonists and their derivatives or synthetic antagonists, which may or may not be related to the structure of PAF.[94–96] The ginkgolide BN52021 was one of the first PAF antagonists to be used to demonstrate effects in antigen-induced models in the guinea-pig. It inhibited both homologous and heterologous passive anaphylaxis in guinea-pigs[97–99] and antagonizes immune bronchoconstriction when antigen is given by aerosol in passively sensitized guinea-pigs.[100] BN52021 has been shown to inhibit allergen-induced bronchial hyperresponsiveness and eosinophil influx.[101] Of the synthetic PAF antagonists, modified derivatives of the PAF molecule such as CV-3988 and Ro-19-3704 were the first to be developed,[102] followed by others unrelated to PAF structure, such as the thienotriazalo-diazepines WEB2086 and WEB2170. WEB2086 (apafant) was potent orally but with a short duration of action. More prolonged effects were observed with WEB2170 (bepafant) and particularly with WEB2347, which inhibited PAF-induced bronchoconstriction in guinea-pigs for 24 h.[103] WEB2086 inhibited allergen-induced late-phase responses of the sheep and airway eosinophilia in the guinea-pig.[104,105]

Assessment of the potency of PAF antagonists in humans has been hampered by the rapid onset of tachyphylaxis to the bronchoconstrictor response to inhaled PAF.[82] Therefore, the protective effect of PAF antagonist against PAF challenge has only been assessed against one bronchoconstrictor dose of PAF. This approach did not allow for determination of the degree of shift of the PAF dose–response curve induced by PAF receptor antagonists. For example, a newer dihydropyridine PAF antagonist, UK-74,505 (modipafant), was shown to suppress the 50% fall in specific airways conductance and the fall in peripheral neutrophil count induced by inhaled PAF after two oral doses (25 and 100 mg) at 3 h in normal volunteers.[106] At 24 h, the higher dose was still effective while the lower dose was not.

Studies in asthma

Despite the positive data found in animal studies with PAF antagonists, studies in asthma have not provided evidence for a beneficial effect. In studies of the late-phase response induced by allergen, WEB2086, UK-74,505 and MK-287 had no effect on the early- and late-phase responses after allergen challenge.[107–109] A small inhibitory effect of BN52021 on the acute bronchoconstrictor response to allergen has been found in asthmatic children.[110] Two clinical studies in asthma have now been performed, both reporting negative results.[111,112] The effect of WEB2086 at a dose of 40 mg three times daily for 6 weeks on the inhaled corticosteroid requirements of symptomatic atopic asthmatic individuals was investigated in a double-blind, randomized, placebo-controlled, parallel group study.[111] No effect was observed on inhaled corticosteroid dosage during the treatment period with WEB2086. In a preliminary study, a higher dose of WEB2086 (80 mg three times per day) provided benefit in asthmatic individuals on inhaled and oral corticosteroid therapy. However, another potent PAF antagonist, the active (+)-enantiomer of UK-74,505, UK-80,067, was not found to be clinically beneficial for symptomatic asthmatic individuals on a reducing dose of inhaled corticosteroid therapy in a multicentre, placebo-controlled, parallel group study.[112] These studies indicate that PAF antagonists are unlikely to become useful therapy for asthma.

PAF acetylhydrolase catalyses the degradation of PAF and related phospholipids. A decrease in PAF acetylhydrolase may exaggerate inflammatory and allergic responses involving PAF. Absent acetylhydrolase activity has been reported in asthmatic children[113] and this deficiency of PAF acetylhydrolase has been shown to be the result of a point mutation in exon 9 that abolishes enzymatic activity completely.[114] Whether identification of such individuals will lead to identification of individuals predisposed to asthma remains to be seen.

20 Other Mediators of Asthma

OXYGEN RADICALS

The oxygen-derived molecules include superoxide anion (O_2^-) and hydroxyl radical (OH˙) and are characterized by their reactivity towards proteins, lipids and nucleic acids. This may result in damage to membranes, receptors or enzymes, leading to alterations in cell function. These molecules are generated as part of the inflammatory response and may therefore be involved in the pathophysiology of asthma. Activation of various inflammatory cells, including macrophages, neutrophils, eosinophils and mast cells, generate O_2^- and H_2O_2, with OH^- being formed secondarily. Oxygen-radical species can act as initiators of lipid peroxidation, for which arachidonic acid is an important substrate, with the synthesis of prostaglandins and leukotrienes, Thus, thromboxane generation has been shown to be stimulated by increased lung concentrations of reactive oxygen species in the isolated perfused rabbit lung.[115] In addition, H_2O_2 can induce arachidonic acid metabolism in rat alveolar macrophages.[116] H_2O_2 may also induce the release of histamine from isolated rat peritoneal mast cells. Oxidative stress can also lead to the activation of the transcription factor, NF-κB,[117] which may lead to the increased transcription of several inducible genes that may amplify inflammation in human epithelial cells.[118] These inducible genes include nitric oxide synthase, inducible cyclooxygenase and several chemoattractant cytokines such as IL-8 and RANTES. Thus, oxygen radical may underlie the expression of various inflammatory genes.

Effects on airways

Airway smooth muscle

H_2O_2 is the oxygen radical that appears to have the major effect on airway tone and causes contraction in both bovine and guinea-pig airways.[119,120] In the guinea-pig, the contractile effect of H_2O_2 is greatly enhanced by removal of epithelium, suggesting that oxygen radicals release a relaxant factor from the epithelium. The bronchoconstriction is also reduced by indomethacin, suggesting that H_2O_2 also releases constrictor cyclo-oxygenase products.[120] Inhalation of xanthine and xanthine oxidase aerosols to generate oxygen radicals has been shown to cause bronchoconstriction in anaesthetized cats, an effect inhibited by superoxide dismutase, suggesting the involvement of superoxide anions.[121]

β-Receptor function

Oxygen-radical species may interfere with the sulphydryl groups of β-adrenergic receptors in the lung, leading to dysfunction of this receptor. Pulmonary macrophages cause a specific deterioration of β-adrenergic responsiveness in the guinea-pig trachea, an effect inhibited by catalase and thiourea, indicating that oxygen species may be involved.[122] However, direct incubation of oxygen radicals with guinea-pig airways failed to alter β-receptor function.[120]

Bronchial hyperresponsiveness

Exposure of cats to xanthine and xanthine oxidase aerosols led to increased bronchial responsiveness to acetylcholine that lasted up to 1 h.[121] Ozone, which is a potent free-radical generator, is known to induce bronchial hyperresponsiveness and neutrophil influx into the airways of dogs and of normal volunteers. Arachidonic acid metabolites are increased in lungs of rats exposed to ozone;[123] these may be metabolized by airway epithelial cells into lipoxygenase products such as leukotriene (LT)B$_4$ which has potent chemotactic activity for neutrophils.[124] In addition, ozone exposure leads to increased activation of NF-κB and expression of chemokines such as the neutrophil chemoattractants cytokine-induced neutrophil chemoattractant (CINC) and macrophage inflammatory protein (MIP)-2.[125,126] The involvement of oxygen-radical species in ozone-induced hyperresponsiveness is supported by the observation that this effect is inhibited by antioxidants.[127,128] Histamine hyperresponsiveness of guinea-pig trachea induced by LTD$_4$ *in vitro* has been shown to be inhibited by pretreatment with superoxide dismutase, indicating the involvement of superoxide anions.

Epithelial damage

Although oxygen-derived molecules such as H_2O_2 do not appear to be cytotoxic to respiratory epithelium, they markedly potentiate the cytotoxic effects of eosinophil-derived enzymes such as EPO.[129] Epithelial damage induced by eosinophils activated by PAF in guinea-pig tracheal rings *in vitro* is inhibited by catalase, supporting a role for H_2O_2.[129] Exposure of cultured epithelial cells to H_2O_2 may also increase paracellular permeability, leading to increased penetration of exogenous substances.[130]

Possible role in asthma

Several studies suggest an association between the release of reactive oxygen species and asthma. Alveolar macrophages obtained from patients with asthma release larger quantities of reactive oxygen species when compared with those from normal subjects;[131,132] the amount of reactive oxygen species released correlated with the severity of asthma.[131] In addition, significant correlations have been found between non-specific airway hyperresponsiveness to histamine aerosol and superoxide production by polymorphonuclear leucocytes in patients with chronic obstructive pulmonary disease.[133] Eosinophils obtained from venous blood of symptomatic asthmatic patients also show enhanced release of superoxide anions when stimulated by PAF or a phorbol ester.[134] Measuring oxygen-derived free radicals in the circulation is difficult because of their rapid degradation and neutralization by efficient local scavenger mechanisms. Glutathione peroxidase, which removes H_2O_2 by oxidization of reduced glutathione to oxidized glutathione, has been reported to be lower in the blood of asthmatic subjects with food and aspirin intolerance.[135] In addition, there is a reduction in whole blood glutathione peroxidase and selenium concentrations in asthmatic patients compared with control subjects,[136] selenium being required as a cofactor for glutathione peroxidase. Plasma levels of the Trolox equivalent antioxidant capacity (TEAC) was significantly reduced in patients with asthma, together with evidence for an increase in plasma lipid peroxidation, indicating oxidant–antioxidant imbalance. These abnormalities were also present in patients with chronic obstructive airways disease.[137] Increased levels of H_2O_2 have been measured in the exhaled condensates of asthmatic patients.[138]

There have been very few studies of antioxidants or free-radical scavengers in asthma. Ascorbic acid is an effective antioxidant and reduces methacholine-induced bronchoconstriction in asthmatic subjects, although this could be mediated through an alternative mechanism.[139] In one study, antioxidant supplementation with selenium and vitamins C and E did not improve bronchial hyperresponsiveness in asthmatic subjects, but the amount of supplementation may have been inadequate.[140] With the potential availability of potent antioxidants, it should be possible to determine further the role of reactive oxygen species in asthma.

COMPLEMENT

The activation sequence and generation of various components of the complement cascade are complex and the reader is referred to other reviews for a thorough description.[141,142] A series of plasma proteins is generated during the complement cascade and these may play an important role in host defence and in the pathophysiology of various disorders. In relation to asthma, we focus specifically on two components of the complement cascade system for which there is documentation of airway effects, namely C3a and C5a.

Origin and metabolism

C3a, C4a and C5a are active fragments of the complement cascade that, once formed, do not participate in the cascade itself and are known collectively as anaphylatoxins. C3a and C5a are generated by the activation of the complement pathway by both the classical and alternative pathways. C5a has 74 amino acids and contains an oligosaccharide attached at position 64, with the active site being the C-terminal pentapeptide Met-Glu-Leu-Gly-Arg. The remainder of the molecule is required for functional binding to the C5a receptor. C3a has 77 amino acids, with the active site being the C-terminal pentapeptide Leu-Gly-Leu-Ala-Arg. The remainder of the molecule is not required for functional binding to the C3a receptor.

The anaphylatoxins C3a and C5a are rapidly inactivated in plasma to the des-Arg form by the removal of the C-terminal arginine by serum carboxypeptidase N. Much of the biological activity of C3a and C5a is removed, although chemotactic activity is retained. The measurement of plasma levels of anaphylatoxins has been made possible with the development of carboxypeptidase N inhibitors.

Effects of C3a and C5a on airways

Airway smooth muscle

Intravenous injection of C5a into guinea-pigs causes bronchoconstriction by mechanisms that may involve direct effects on airway smooth muscle, as can be demonstrated *in vitro*.[143,144] However, C5a and C5a-des-Arg can induce the release of other mediators such as histamine, prostaglandins and leukotrienes from guinea-pig lung.[145,146] Both cyclo-oxygenase and lipoxygenase products appear to be important in C5a-induced contraction of airway smooth muscle preparations. C3a is a less potent inducer of airway smooth muscle contraction than C5a in the guinea-pig. This effect appears to be mediated predominantly by a cyclooxygenase product, despite the release of histamine.[147] Both C3a and C5a induce marked tachyphylaxis in airway smooth muscle preparations, although there is no cross-desensitization between them, indicating that they are likely to activate discrete receptors.[144]

Vascular effects

C5a and C3a induce vascular permeability in the skin through neutrophil activation, although the role of the neutrophil has not been fully elucidated.[148] Antagonists of PAF do not inhibit C5a-induced oedema formation in rabbit skin, despite the observation that C5a can release PAF from neutrophils.[149] In humans, C5a produces immediate wheal and flare reactions in the skin; H_1-receptor antagonists reduce the flare response but not the wheal.[150] Skin biopsies showed the presence of neutrophil infiltration, endothelial cell oedema and mast cell degranulation.

Mucus secretion

Little is known about the effects of the anaphylatoxins on airway secretion or mucociliary clearance. C3a stimulates mucous glycoprotein secretion from human airways *in vitro*, probably via a direct effect on secretory cells.[151]

Chemotaxis and cell activation

C5a and C5a-des-Arg possess chemotactic activity for neutrophils with a potency even greater than that of LTB_4.[152] C5a also has chemotactic activity for macrophages, basophils and eosinophils. By contrast, C3a is devoid of chemotactic activity. Both C5a and C5a-des-Arg also stimulate the adhesion of inflammatory cells and elicit the release of other mediators, including lysosomal enzymes, oxygen free radicals, lipoxygenase and cyclooxygenase products of arachidonic acid metabolism and PAF, from both neutrophils and eosinophils.[153-155]

Possible role in asthma

The role of anaphylatoxins in asthma is uncertain. Several clinical investigators have reported the activation of the complement cascade during asthma. Plasma C4 concentrations have been found to be elevated in childhood asthma and depressed in non-atopic adult asthmatic individuals.[156] Other investigators have not confirmed this observation. No changes in circulating complement components have been detected in allergic asthmatic subjects following either the early- or late-phase reactions after allergen provocation.[157,158] A few patients develop reduced haemolytic-component activity or C4 in arterial or venous blood following allergen provocation,[159] whereas others have reported an increase.[160] However, complement deposition has been found in the bronchial mucosa of asthmatic patients.[161] The alternative pathway can be activated in normal human serum by agents that may induce asthma, such as cotton dust, plicatic acid and house-dust mite extract.[162,163] The possibility that there may be local complement activation within the asthmatic airway cannot be excluded.

Exposure of guinea-pigs to an aerosol of C5a-des-Arg causes an increase in airway responsiveness to histamine, in association with neutrophil infiltration into the airways. The increased airway responsiveness is reduced in animals rendered neutropenic, suggesting that neutrophils contribute to the induction of bronchial hyperresponsiveness by C5a.[164]

There are few experimental data on the effect of the anaphylatoxins on human lung and one cannot extrapolate the observations in various animal species to humans. The effect of inhibitors of complement activation or of a specific antagonist of the anaphylatoxins in asthma has not yet been reported. Overall, more information is needed before the role of complement in asthma can be more clearly defined.

SEROTONIN

Serotonin (5-hydroxytryptamine, 5HT) is formed by decarboxylation of tryptophan (obtained in the diet) and is stored in secretory granules. In humans, apart from the central nervous system, serotonin is localized in neuroendocrine cells of the gastro-intestinal and respiratory tracts, in certain nerves and in secretory granules in platelets. Several types of serotonin receptors have been recognized and specific antagonists for the receptor subtypes have been developed. Of these subtypes, $5HT_2$ receptors are present on tracheal smooth muscle of the guinea-pig, mediating contraction.[165] In addition, $5HT_2$ receptors may also be located on postganglionic nerves to increase cholinergic neurotransmission in guinea-pig airways.[166] $5HT_3$ receptors are also present on nerves and stimulate neurotransmitter release from certain peripheral nerves.[167]

In several species, including guinea-pig, cat, rat, dog and monkey, serotonin induces bronchoconstriction, but there is some doubt as to its effect in human airways. Serotonin may even relax human airways *in vitro*.[168] No consistent bronchoconstrictor response has been observed in either normal or asthmatic subjects.[169,170] Serotonin may facilitate acetylcholine release from airway nerves in dogs.[171] Serotonin induces microvascular leakage in the guinea-pig.[172]

Few studies have been performed with antagonists of serotonin in asthma. Ketanserin, a $5HT_2$ antagonist, has no protective action against exercise-induced asthma.[173] The use of more specific antagonists and agonists of serotonin-receptor subtypes may help to delineate the contribution of serotonin to the pathophysiology of asthma.

EOSINOPHIL PROTEINS

The human eosinophil is characterized by the presence of eosin-staining granule contents, which are mainly made up of four proteins: eosinophil cationic protein (ECP), EPO, eosinophil-derived neurotoxin (EDN) and major basic protein (MBP).[174] These proteins have been purified and possess high isoelectric points. The cytotoxic potential of the eosinophil is mediated through the release of its granular constituents. Small amounts of MBP and ECP cause extensive damage to the epithelium, producing a histological picture similar to that seen in bronchial asthma.[175] In addition, MBP shows ciliary beat frequency. The cytotoxic effects of eosinophil proteins on airway epithelium are potentiated by the presence of halide ions and reactive oxygen species.[176] Such damage to the airway epithelium may represent a mechanism by which bronchial hyperresponsiveness is triggered. However, potentiation of the contractile response of airway tissues *in vitro* may be observed with MBP, at concentrations that cause no histological evidence of epithelial damage.[177] MBP, but not EPO or ECP, induces bronchial hyperresponsiveness *in vivo* in primates.[178]

Both ECP and MBP have been demonstrated by immunohistochemical techniques in the lung tissue of patients who have died from asthma,[179] with the extracellular deposition of ECP and MBP associated with damaged epithelium. During the late asthmatic reactions to allergen, elevated levels of ECP were measured in bronchoalveolar lavage fluid.[180] Large concentrations of MBP have been reported in sputum from asthmatic patients.[181] In addition, serum ECP levels rise significantly during the pollen season in atopic individuals with seasonal allergic symptoms, and these correlated significantly with the increase in histamine reactivity.[182] Studies in the guinea-pig indicate that administration of MBP can lead to bronchial hyperresponsiveness,[183] while neutralization of MBP inhibits antigen-induced bronchial hyperresponsiveness without effect on antigen-induced bronchial eosinophilia.[184]

The eosinophil proteins are therefore likely to be a major cause of epithelial damage in asthma. Eosinophils can also generate lipid mediators such as PAF and sulphidopeptide leukotrienes and must therefore be regarded as an important cellular source of mediators that can reproduce many of the pathophysiological features of asthma.

ENDOTHELIN

The endothelins are a family of three related peptides, each of 21 amino acids. Endothelin (ET)-1 was first discovered as a potent vasoconstrictor substance derived from cultured endothelial cells.[185] Other closely related peptides, ET-2 and ET-3, were soon discovered. All three endothelin isoforms arise from post-translational processing of preprohormones. PreproET-1 is cleaved by endopeptidases and carboxypeptidases to yield proET-1, which is further degraded by endothelin-converting enzymes to yield ET-1.

Sites of synthesis

Cells in the airways that can synthesize endothelins include bronchial epithelial cells, endothelial cells, macrophages and pulmonary neuroendocrine cells.[186–188] Cultured human bronchial epithelial cells secrete ET-1 and ET-3 in equal amounts. IgE–anti-IgE complexes can trigger ET-1 secretion by airway epithelial cells that express the low-affinity IgE receptor CD23 in some patients with asthma,[189] indicating that IgE-mediated events can stimulate ET-1 release.

Endothelin receptors

Specific high-affinity binding sites for ET-1 and ET-2 have been described in bronchial smooth muscle cells, peripheral airways, alveolar septae, endothelial cells and nerves present in airways including parasympathetic ganglia.[190] At least three receptor types for the endothelins have been described from functional responses to endothelin isopeptides and from radiolabelled endothelin binding experiments.[191–193] Two distinct receptors for endothelin (ET_A and ET_B) have been cloned, which have about 60% homology.[194–196] Both ET_A and ET_B receptors are expressed in human lung, with ET_A expression being the greatest.[196,197] The deduced structure for ET_A and ET_B receptors show that these belong to the superfamily of G protein-coupled receptors with seven hydrophobic membrane-spanning demains. ET_A receptors have high affinity for ET-1, while ET_B receptors have equal affinity for all three endothelins.

ET-1 exerts its effects through several intracellular events that include phospholipase C activation and interactions with membrane ion channels. ET-1 activates phospholipase C via a pertussis toxin-insensitive G protein following binding to its receptors.[198] This leads to a rapid increase in intracellular concentrations of inositol trisphosphate, leading to Ca^{2+} release from intracellular stores.[199,200] There is also activation of protein kinase C.[201] ET-1 also increases the transcription of various growth factor genes[202] and may interact with ATP-sensitive K^+ channels.[203,204]

Endothelin receptor antagonists

Several selective agonists and antagonists of endothelin receptors have been developed and will be valuable in defining the role of these receptors in disease, including asthma. Substitution of specific amino acids in the structure of endothelins has revealed that the ET_A receptor recognizes the N-terminal end of ET-1, whereas the ET_B receptor recognizes the C-terminal end of the molecule.[205] A selective antagonist of the ET_A receptor is BQ-123 (D-Asp-L-Pro-D-Val-L-Leu-D-Try), a cyclic pentapeptide derived from a fermentation product of *Streptomyces misakiensis*.[206] The tetrapeptide FR-139317 appears to be a selective ET_A antagonist. IRL-1038 is a selective antagonist of ET_B receptors.[207] Non-peptide antagonists have also been developed, such as the non-selective Ro-462005 and Ro-470203.[208]

Use of selective ET_B antagonists such as BQ-3020 and IRL-1620 indicates that the ET_B receptor mediates the direct constrictor effect of ET-1.[209,210] The release of prostanoids (mainly PGD_2 and PGE_2) induced by ET-1 in human airways appears to be mediated by the ET_A receptor as this is inhibited by BQ-123.[210] In the conscious sheep, however, the bronchoconstrictor response to ET-1 aerosol is blocked by a specific ET_A antagonist. This may be due to either species differences or the secondary release of constrictor mediators via ET_A receptors.[211] Proliferation of human pulmonary artery smooth muscle cells by ET-1 is inhibited by BQ-123, suggesting that the ET_A receptor mediates this effect.[212]

Effects of endothelin

Although the endothelins were originally described as vasoconstrictors, they were later found to have a wide range of other biological actions. ET-1 and ET-2 are relatively potent constrictors of human airway smooth muscle, being more potent than LTD_4.[213–215] These effects are not affected by calcium antagonists, cyclooxygenase inhibitors or leukotriene antagonists, suggesting a direct effect on airway smooth muscle mediated through ET_B receptors. Endothelins are metabolized by neutral endopeptidase (NEP) in the airways such that the potency of endothelins in human airways *in vitro* is increased in the presence of an inhibitor of NEP, phosphoramidon.[216] ET-1, but not ET-2 and ET-3, stimulates mucous glycoprotein secretion from cat airway submucosal glands via calcium ion influx.[217] ET-1 also induces the release of PGD_2 and PGE_2 from human airways, an effect mediated through ET_A receptors.[210] ET-1 constricts human bronchial arteries *in vitro*. ET-1 increases microvascular leakage in rate trachea[218] but was without effect in guinea-pig airways.[219] ET-3 enhanced neurotransmission in postganglionic cholinergic nerves in rabbits.[220] A major effect of endothelins is on pulmonary vessels, with ET-1 being a potent constrictor of human pulmonary vessels *in vitro*.[210,218,221] Endothelin binding sites in the lung parenchyma are localized to pulmonary vascular smooth muscle, particularly in arteries.[214,222,223]

ET-1 has several pro-inflammatory properties of potential relevance to asthma. ET-1 can prime neutrophils, promote neutrophil aggregation and stimulate elastase release by human neutrophils.[224] ET-1 stimulates monocytes to produce IL-6, IL-8, IL-1, TNF-α, transforming growth factor (TGF)-β, GM-CSF and PGE_2.[225] Endothelins can increase the proliferation of airway smooth muscle cells.[226,227] They can also induce the proliferation of vascular smooth muscle[228] and fibroblasts,[229] as well as increasing collagen synthesis by fibroblasts.[230] ET-1 can also stimulate the growth of airway epithelial cells.[231] ET-1 stimulates the release of arachidonic acid products from human nasal mucosa.[232] Intranasal ET-1 stimulates secretions and triggers rhinorrhoea, itching and sneezing in both allergic and non-allergic individuals.[233]

Potential role in asthma

The known properties of endothelins indicate that they may be involved in (a) bronchoconstriction, (b) airway oedema and (c) airway structural remodelling with airway smooth muscle hyperplasia and myofibroblast proliferation with subepithelial fibrosis. ET-1 has been implicated in antigen-induced bronchial hyperresponsiveness in the sheep model, since an ET_A receptor antagonist (BQ-123) reduced the late-phase airway response by 50% in allergic sheep exposed to *Ascaris suum*.

20 Other Mediators of Asthma

The evidence for the involvement of endothelins in asthma is at present indirect. Elevated concentrations of ET-1 have been detected in bronchoalveolar lavage fluid of asthmatic patients,[234–236] with decrease in the levels after treatment with inhaled steroids and β_2-adrenergic agonists; this has also been reported in patients with chronic obstructive pulmonary disease.[234] Decreased ET-1 levels in bronchoalveolar lavage fluid has been reported in patients with nocturnal asthma.[237] An increase in plasma concentration of ET-1 during exacerbations of asthma and in children with asthma has also been reported.[235,238] Elevated levels of ET-1 have been recorded during acute attacks of asthma, but the levels were similar to control patients between attacks.[239] Increased expression of ET-1 immunoreactivity in the epithelial layer of bronchial biopsies has been reported in patients with asthma.[240,241] Raised preproET-1 mRNA detected by *in situ* hybridization has also been reported.[187] Cultured bronchial epithelial cells from asthmatic individuals release more ET-1.[187]

The effect of endothelin receptor antagonists in asthma has not been studied. Because the various effects of endothelin on airways are mediated by both ET_A and ET_B receptors, the effect of a combined ET_A and ET_B receptor antagonist would appear to be more useful. The clinical studies necessary to demonstrate a beneficial effect of endothelin antagonists in preventing chronic structural remodelling would involve a long study over many months, with possibly the primary end-point being the rate of decline in lung function and in bronchial hyperresponsiveness.

REFERENCES

1. Barnes PJ, Chung KF, Page CP: Inflammatory mediators and asthma. *Physiol Rev* (1988) **40**: 49–84.
2. Dale H, Laidlaw P: The physiologic action of β-imidazolyethylamine. *J Physiol* (1911) **41**: 318–344.
3. Ash A, Schild H: Receptor mediating some actions of histamine. *Br J Pharmacol* (1966) **274**: 27–39.
4. Arrang JM, Garbarg M, Lancelot JC, *et al*.: Highly potent and selective ligands for histamine H$_3$-receptors. *Nature* (1987) **327**: 117–123.
5. Sertl K, Casale TB, Wescott SL, Kaliner MA: Immunohistochemical localization of histamine-stimulated increases in cyclic GMP in guinea pig lung. *Am Rev Respir Dis* (1987) **135**: 456–462.
6. Grandordy BM, Cuss FM, Barnes PJ: Breakdown of phosphoinositides in airway smooth muscle: lack of influence of anti-asthmatic drugs. *Life Sci* (1987) **41**: 1621–1627.
7. Evans TW, Rogers DF, Aursudkij B, Chung KF, Barnes PJ: Regional and time-dependent effects of airway inflammatory mediators on microvascular permeability in the guinea-pig. *Clin Sci* (1989) **76**: 479–485.
8. Foreman JC, Norris DB, Rising TJ, Webber SE: The binding of [^{3}H]-tiotidine to homogenates of guinea-pig lung parenchyma. *Br J Pharmacol* (1985) **86**: 475–482.
9. Vincenc K, Black J, Shaw J: Relaxation and contraction responses to histamine in the human lung parenchymal strip. *Eur J Pharmacol* (1984) **98**: 201–210.
10. White MV, Slater JE, Kaliner MA: Histamine and asthma. *Am Rev Respir Dis* (1987) **135**: 1165–1176.
11. Shelhamer J, Marom Z, Kaliner M: Immunologic and neuropharmacologic stimulation of mucous glycoprotein release from human airways *in vitro*. *J Clin Invest* (1980) **66**: 1400–1408.
12. Ichinose M, Belvisi MG, Barnes PJ: Histamine H$_3$-receptors inhibit neurogenic microvascular leakage in airways. *J Appl Physiol* (1990) **68**: 21–25.

13. Raible DG, Lenahan T, Fayvilevich Y, Kosinski R, Schulman ES: Pharmacologic characterization of a novel histamine receptor on human eosinophils. *Am J Respir Crit Care Med* (1994) **149**: 1506–1511.
14. Haaksma EE, Leurs R, Timmerman H: Histamine receptors: subclasses and specific ligands. *Pharmacol Ther* (1990) **47**: 73–104.
15. Yamashita M, Fukui H, Sugama K, *et al.*: Expression cloning of a cDNA encoding the bovine histamine H_1 receptor. *Proc Natl Acad Sci USA* (1991) **88**: 11 515–11 519.
16. Fukui H, Fujimoto K, Mizuguchi H, *et al.*: Molecular cloning of the human histamine H_1 receptor gene. *Biochem Biophys Res Commun* (1994) **201**: 894–901.
17. De Backer MD, Gommeren W, Moereels H, *et al.*: Genomic cloning, heterologous expression and pharmacological characterization of a human histamine H_1 receptor. *Biochem Biophys Res Commun* (1993) **197**: 1601–1608.
18. Baumgarten CR, Nichol RC, Naclerio RM, Lichtenstein LM, Norman PS, Proud D: Plasma kallikrein during experimentally-induced allergic rhinitis: role in kinin formation and contribution to TAME-esterase activity in nasal secretions. *J Immunol* (1986) **137**: 977–982.
19. Steel L, Platshon L, Kaliner M: Prostaglandin generation by human and guinea-pig lung tissue: comparison of parenchymal and airway responses. *J Allergy Clin Immunol* (1979) **64**: 287–293.
20. Clark RAF, Sandler JA, Gallin JI, Kaplan AP: Histamine modulation of eosinophil migration. *J Immunol* (1977) **118**: 137–145.
21. Barnes PJ, Ind PW, Brown MJ: Plasma histamine and catecholamines in stable asthmatics subjects. *Clin Sci* (1982) **62**: 661–665.
22. Flint KC, Leung KBP, Huspith BN, Brostoff J, Pearce FL, Johnson NM: Bronchoalveolar mast cells in extrinsic asthma: mechanism for the initiation of antigen specific bronchoconstriction. *Br Med J* (1985) **291**: 923–927.
23. Simons FER, Simons KJ: Antihistamines. In Middleton E, Reed CE, Ellis EF, Adkinson NF, Yuginger JW, Busse WW (eds) *Allergy: Principles and Practice*, 4th edn. St Louis, Mosby-Year Book, 1993, pp 856–892.
24. Bousquet J, Campbell A, Michel FB: Antiallergic activities of antihistamines. In Church MK, Rihoux J (eds) *Therapeutic Index of Antihistamines*. Lewinston, NY, Hogrefe & Huber, 1992, pp 57–84.
25. Naclerio RM, Kagey-Sobotka A, Lichtenstein LM, Freidhoff L, Proud D: Terfenadine, an H_1 antihistamine, inhibits histamine release *in vivo* in the human. *Am Rev Respir Dis* (1990) **142**: 167–171.
26. Bousquet J, Lebel B, Chanal I, Morel A, Michel FB: Antiallergic activity of H_1-receptor antagonists assessed by nasal challenge. *J Allergy Clin Immunol* (1988) **82**: 881–887.
27. Charlesworth EN, Kagey-Sobotka A, Norman PS, Lichtenstein LM: Effect of cetirizine on mast cell-mediator release and cellular traffic during the cutaneous late-phase reaction. *J Allergy Clin Immunol* (1989) **83**: 905–912.
28. Michel L, De Vos C, Rihoux JP, Burtin C, Benveniste J, Dubertret L: Inhibitory effect of oral cetirizine on *in vivo* antigen-induced histamine and PAF-acether release and eosinophil recruitment in human skin. *J Allergy Clin Immunol* (1988) **82**: 101–109.
29. Naclerio RM, Proud D, Kagey-Sobotka A, Freidhoff L, Norman PS, Lichtenstein LM: The effect of cetirizine on early allergic response. *Laryngoscope* (1989) **99**: 596–599.
30. Klementsson H, Andersson M, Pipkorn U: Allergen-induced increase in nonspecific nasal reactivity is blocked by antihistamines without a clear-cut relationship to eosinophil influx. *J Allergy Clin Immunol* (1990) **86**: 466–472.
31. Popa V: Effect of an H_1-blocker, chlorpheniramine, on inhalation tests with histamine and allergen in allergic asthma. *Chest* (1980) **78**: 442–445.
32. Chan TB, Shelton DM, Eiser NM: Effect of an oral H_1-receptor antagonist, terfenadine, on antigen-induced asthma. *Br J Dis Chest* (1986) **80**: 375–384.
33. Gong H, Taskin D, Dauphinee B, Djahed B, Tzu-Chin W: Effects of oral cetirizine, a selective H_1 antagonist, on allergen- and exercise-induced bronchoconstriction in subjects with asthma. *J Allergy Clin Immunol* (1990) **85**: 632–641.
34. Phillips MJ, Ollier S, Gould C, Davies RJ: Effect of antihistamines and antiallergic drugs on responses to allergen and histamine provocation tests in asthma. *Thorax* (1984) **39**: 345–351.

35. Eiser N, Mills J, Snashall P, Guz A: The role of histamine receptors in asthma. *Clin Sci* (1981) **60**: 363–370.
36. Holgate ST, Emmanuel MB, Howarth PH: Astemizole and other H_1-antihistaminic drug treatments of asthma. *J Allergy Clin Immunol* (1985) **76**: 375–380.
37. Curzen N, Rafferty P, Holgate ST: Effects of cyclo-oxygenase inhibitor, flurbiprofen and an H_1 histamine receptor antagonist, terfenadine alone and in combination on allergen induced immediate bronchoconstriction in man. *Thorax* (1987) **42**: 946–952.
38. Morgan D, Moodley I, Cundell D, Sheinman B, Smart W, Davies R: Circulating histamine and neutrophil chemotactic activity during allergen-induced asthma: the effect of inhaled antihistamines and anti-allergic compounds. *Clin Sci* (1985) **69**: 36–39.
39. Rafferty P, Beasley R, Holgate ST: The contribution of histamine to immediate bronchoconstriction provoked by inhaled allergen and adenosine 5′-monophosphate in atopic asthma. *Am Rev Respir Dis* (1987) **136**: 369–373.
40. Lai CK, Beasley R, Holgate ST: The effect of an increase in inhaled allergen dose after terfenadine on the occurrence and magnitude of the late asthmatic response. *Clin Exp Allergy* (1989) **19**: 209–216.
41. Town G, Holgate S: Comparison of the effect of loratadine on the airway and skin responses to histamine, methacholine and allergen in asthmatic subjects. *J Allergy Clin Immunol* (1990) **86**: 886–893.
42. Nakazawa T, Rakehisa T, Furukawa M, Taya T, Kobayashi S: Inhibitory effects of various drugs on dual asthmatic responses in wheat flour-sensitive subjects. *Immunology* (1976) **58**: 1–9.
43. Rafferty P, Ng WH, Phillips G, *et al.*: The inhibitory actions of azelastine hydrochloride on the early and late bronchoconstrictor responses to inhaled allergen in atopic asthma. *J Allergy Clin Immunol* (1989) **84**: 649–657.
44. Hartley JPR, Nogrady SG: Effect of an inhaled antihistamine on exercise-induced asthma. *Thorax* (1980) **35**: 675–679.
45. Finnerty JP, Holgate ST: Evidence for the roles of histamine and prostaglandins as mediators in exercise-induced asthma: the inhibitory effect of terfenadine and flurbiprofen alone and in combination. *Eur Respir J* (1990) **3**: 540–547.
46. Macfarlane PI, Heaf DP: Selective histamine blockade in childhood asthma: the effect of terfenadine on resting bronchial tone and exercise induced asthma. *Respir Med* (1989) **83**: 19–24.
47. Patel KR: Terfenadine in exercise induced asthma. *Br Med J* (1984) **288**: 1496–1497.
48. O'Hickey SP, Belcher NG, Rees PJ, Lee TH: Role of histamine release in hypertonic saline induced bronchoconstriction. *Thorax* (1989) **44**: 650–653.
49. Badier M, Beaumont D, Orehek J: Attenuation of hyperventilation-induced bronchospasm by terfenadine: a new antihistamine. *J Allergy Clin Immunol* (1988) **81**: 437–440.
50. Bewtra AK, Hopp RJ, Nair NM, Townley RG: Effect of terfenadine on cold air-induced bronchospasm. *Ann Allergy* (1989) **62**: 299–301.
51. Finnerty JP, Wilmot C, Holgate ST: Inhibition of hypertonic saline-induced bronchoconstriction by terfenadine and flurbiprofen. Evidence for the predominant role of histamine. *Am Rev Respir Dis* (1989) **140**: 593–597.
52. Finney MJ, Anderson SD, Black JL: Terfenadine modifies airway narrowing induced by the inhalation of nonisotonic aerosols in subjects with asthma. *Am Rev Respir Dis* (1990) **141**: 1151–1157.
53. Nogrady SG, Hartley JPR, Handslip PDJ, Hurst NP: Bronchodilation after inhalation of the antihistamine clemastin. *Thorax* (1978) **33**: 479–482.
54. Popa VT: Bronchodilating activity of an H_1 blocker, chlorpheniramine. *J Allergy Clin Immunol* (1977) **59**: 54–63.
55. Monie RD, White JP, Handnslip PDJ, Hartley JPR, Nogrady SG: Bronchodilator properties of inhaled clemastine. *Br J Dis Chest* (1980) **74**: 420–424.
56. Cookson WOCM: Bronchodilator action of the anti-histamine terfenadine. *Br J Clin Pharmacol* (1987) **24**: 120–121.
57. Chung KF, Morgan B, Keyes SJ, Snashall PD: Histamine dose–response relationships in normal and asthmatic subjects: importance of initial airway calibre. *Am Rev Respir Dis* (1982) **126**: 849–854.

58. Thomson N, Kerr J: Effect of inhaled H_1 and H_2 receptor antagonists in normal and asthmatic subjects. *Thorax* (1980) **35**: 428–434.
59. Natan R, Segall N, Glover G, Schocket A: The effect of H_1 and H_2 antihistamine on histamine inhalation challenges in asthmatic patients. *Am Rev Respir Dis* (1980) **120**: 1251–1258.
60. Hodges IGC, Milner AD, Stokes GM: Bronchodilator effect of two inhaled H_1-receptor antagonists, clemastine and chlorpheniramine. *Br J Dis Chest* (1983) **77**: 270–275.
61. Rafferty P, Jackson L, Smith R, Holgate ST: Terfenadine, a potent histamine H_1-receptor antagonist in the treatment of grass pollen sensitive asthma. *Br J Clin Pharmacol* (1990) **30**: 229–235.
62. Gould C, Ollier S, Aurich R, Davies R: A study of the clinical efficacy of azelastine in patients with extrinsic asthma, and its effect on airway responsiveness. *Br J Clin Pharmacol* (1988) **26**: 515–525.
63. Finnerty J, Holgate S, Rihoux J-P: The effect of 2 weeks treatment with cetirizine on bronchial reactivity to methacholine in asthma. *Br J Clin Pharmacol* (1990) **29**: 79–84.
64. Partridge MR, Saunders KB: Effect of inhaled antihistamine (clemastine) as a bronchodilator and as maintenance treatment in asthma. *Thorax* (1979) **34**: 771–776.
65. Teale C, Morrison JFJ, Pearson SB: Terfenadine in nocturnal asthma. *Thorax* (1990) **45**: 795–797.
66. Taytard A, Beaumont D, Pujet JC, Sapene M, Lewis PJ: Treatment of bronchial asthma with terfenadine: a randomized controlled trial. *Br J Clin Pharmacol* (1987) **24**: 743–746.
67. Bruttmann G, Pedrali P, Arendt C, Rihoux JP: Protective effect of cetirizine in patients suffering from pollen asthma. *Ann Allergy* (1990) **64**: 224–228.
68. Doebber TW, Wu MS: Platelet-activating factor (PAF) stimulates the PAF-synthesizing enzyme acetyl-CNA: 1-alkyl-*sn*-glycero-3-phosphocholine O2-acetyltransferase and PAF synthesis in neutrophils. *Proc Natl Acad Sci USA* (1987) **84**: 7557–7561.
69. Lee TC, Lenihan DJ, Malone B, Roddy LL, Wasserman SI: Increased biosynthesis of platelet activating factor in activated human eosinophils. *J Biol Chem* (1984) **259**: 5526–5530.
70. Arnoux B, Joseph M, Simoes AH, *et al.*: Antigenic release of PAF-acether and beta-glucuronidase from alveolar macrophages of asthmatics. *Bull Eur Physiopathol Respir* (1987) **23**: 119–124.
71. McIntyre TM, Zimmerman GA, Saton K, Prescott SM: Cultured endothelial cells synthetize both platelet-activating factor and prostacyclin in response to histamine, bradykinin and adenosine triphosphate. *J Clin Invest* (1985) **76**: 271–280.
72. Valone FH, Epstein LB: Biphasic platelet-activating factor synthesis by human monocytes stimulated with IL-1β, tumour necrosis factor, or IFNγ. *J Immunol* (1988) **141**: 3945–3950.
73. Dent G, Ukena D, Barnes PJ: PAF receptors. In Barnes PJ, Page CP, Henson PM (eds) *Platelet-activating Factor in Human Disease*. Oxford, Blackwell Scientific, 1989, pp 58–81.
74. Gomez J, Bloom JW, Yamamura HI, Halonen M: Characterization of receptors for platelet-activating factor in guinea pig lung membranes. *Am Rev Respir Cell Mol Biol* (1990) **3**: 259–264.
75. Honda Z, Nakamura M, Miki I, *et al.*: Cloning by functional expression of platelet-activating factor receptor from guinea-pig lung. *Nature* (1991) **349**: 342–346.
76. Nakamura M, Honda Z, Izumi T, *et al.*: Molecular cloning and expression of platelet-activating factor receptor from human leukocytes. *J Biol Chem* (1991) **266**: 20400–20405.
77. Hwang S, Lam M, Pong S: Ionic and GTP regulation of binding of platelet-activating factor to receptors and platelet-activating factor-induced activation of GTPase in rabbit platelet membranes. *J Biol Chem* (1986) **261**: 532–537.
78. Houslay MD, Bojanic D, Wilson A: Platelet activating factor and U44069 stimulate a GTPase activity in human platelets which is distinct from the guanine nucleotide regulatory proteins, N_s and N_i. *Biochem J* (1986) **234**: 737–740.
79. Hwang S: Identification of a second putative receptor of platelet-activating factor from human polymorphonuclear leukocytes. *J Biol Chem* (1988) **263**: 3225–3233.
80. Chung KF: Platelet-activating factor in inflammation and pulmonary disorders. *Clin Sci* (1992) **83**: 127–138.
81. Kidney J, Ridge S, Chung KF, Barnes PJ: Inhibition of PAF-induced bronchoconstriction by the oral leukotriene D4 receptor antagonist, ICI 204,219. *Am Rev Respir Dis* (1993) **147**: 215–217.

82. Cuss FM, Dixon CMS, Barnes PJ: Effects of inhaled platelet activating factor on pulmonary function and bronchial responsiveness in man. *Lancet* (1986) **ii**: 189–192.
83. Rubin AH, Smith LJ, Patterson R: The bronchoconstrictor properties of platelet-activating factor in humans. *Am Rev Respir Dis* (1987) **136**: 1145–1151.
84. Rodriguez-Roisin R, Felez MA, Chung KF, et al.: Platelet-activating factor causes ventilation–perfusion mismatch in humans. *J Clin Invest* (1994) **93**: 188–194.
85. Felez MA, Roca J, Barbera JA, Santos C, Rotger M, Chung KF: Inhaled platelet-activating factor worsens gas exchange in mild asthma. *Am J Respir Crit Care Med* (1994) **150**: 369–373.
86. Masclans JR, Barbera JA, McNee W, et al.: Salbutamol reduces pulmonary neutrophil sequestration after PAF challenge in normal subjects. *Am J Respir Crit Care Med* (1996) **154**: 529–532.
87. Wardlaw A, Chung KF, Moqbel R, et al.: Effect of inhaled platelet-activating factor in humans and bronchoalveolar lavage fluid neutrophils. *Am Rev Respir Dis* (1990) **141**: 386–392.
88. Garcia JGN, Azghani A, Callahan KS, Johnson AR: Effect of platelet-activating factor on leukocyte–endothelial cell interactions. *Thromb Res* (1988) **51**: 83–96.
89. Kimani G, Tonnesen MG, Henson PM: Stimulation of eosinophil adherence to human vascular endothelial cells *in vitro* by platelet activating factor. *J Immunol* (1988) **140**: 3161–3166.
90. Kroegel C, Yukawa T, Dent G, Chung KF, Barnes PJ: Platelet activating factor induces eosinophil peroxidase release from human eosinophils. *Immunology* (1988) **64**: 559–562.
91. Yukawa T, Read RC, Kroegel C, et al.: The effects of activated eosinophils and neutrophils on guinea pig airway epithelium *in vitro*. *Am Rev Respir Cell Mol Biol* (1990) **2**: 341–354.
92. Dubois C, Bissonette E, Rola-Pleszczynski M: Platelet-activating factor (PAF) enhances tumor necrosis factor production by alveolar macrophages. *J Immunol* (1989) **143**: 964–970.
93. Bonavida B, Mencia-Huerta JM, Braquet P: Effect of platelet activating factor on monocyte activation and production of tumor necrosis factor. *Int Arch Allergy Appl Immunol* (1989) **88**: 157–160.
94. Hosford D, Page CP, Barnes PJ, et al.: PAF-receptor antagonists. In Barnes PJ, Page CP, Henson PM (eds) *Platelet-activating Factor and Human Disease*. Oxford, Blackwell Scientific, 1990, pp 82–116.
95. Braquet P, Godfroid JJ: PAF-acether specific binding sites: 2. Design of specific antagonists. *Trends Pharmacol Sci* (1986) **7**: 397–403.
96. Chung KF, Barnes PJ: PAF antagonists: their potential therapeutic role in asthma. *Drugs* (1988) **35**: 93–103.
97. Vilain B, Lagente V, Touvay C, et al.: Pharmacological control of the *in vivo* passive anaphylactic shock by the PAF-acether antagonist compound BN52021. *Pharmacol Res Commun* (1986) **18**: 119–126.
98. Lagente V, Touvay C, Randon C, et al.: Interference of the PAF-acether antagonist BN52021 with passive anaphylaxis in the guinea-pig. *Prostaglandins* (1987) **33**: 264–274.
99. Braquet P, Guinot P, Touvay C, et al.: The role of PAF-acether on anaphylaxis demonstrated with the use of the antagonist BN52021. *Agents Actions* (1987) **21**: 97–117.
100. Cirino M, Lagente V, Lefort J, Vargaftig BB: A study with BN52021 demonstrates the involvement of PAF-acether in IgE-dependent anaphylactic bronchoconstriction. *Prostaglandins* (1986) **32**: 121–126.
101. Coyle AJ, Unwin SC, Page CP, Touvay C, Villain B, Braquet P: The effect of the selective antagonist BN 52021 on PAF and antigen-induced bronchial hyperreactivity and eosinophil accumulation. *Eur J Pharmacol* (1988) **148**: 51–58.
102. Terashita Z, Tsushima S, Yoshioka Y, Nomura H, Inada Y, Nishikawa K: CV-3988: a specific antagonist of platelet activating factor (PAF). *Life Sci* (1983) **32**: 1975–1982.
103. Heuer HO: WEB 2347: pharmacology of a new very potent and long acting hetrazepinoic PAF-antagonist and its action in repeatedly sensitized guinea-pigs. *J Lipid Mediators* (1991) **4**: 39–44.
104. Stevenson JS, Tallant M, Blinder L, Abraham WM: The effect of the PAF antagonist WEB 2086 on the early and late phase in allergic sheep. *Fed Proc* (1987) **466**: 683.
105. Lellouch-Tubiana A, Lefort J, Simon MT, Pfister A, Vargaftig BB: Eosinophil recruitment into guinea pig lungs after PAF-acether and allergen administration: modulation by

prostacyclin, platelet depletion and selective antagonists. *Am Rev Respir Dis* (1988) **137**: 948–954.
106. O'Connor BJ, Uden S, Carty TJ, Eskra JD, Barnes PJ, Chung KF: Inhibitory effect of UK,74505, a potent and specific oral platelet activating factor (PAF) receptor antagonist, on airway and systemic responses to inhaled PAF in humans. *Am J Respir Crit Care Med* (1994) **150**: 35–40.
107. Kuitert LM, Hui KP, Uthayarkumar S, *et al*.: Effect of a platelet activating factor (PAF) antagonist UK-74,505 on allergen-induced early and late response. *Am Rev Respir Dis* (1993) **147**: 82–86.
108. Freitag A, Watson RM, Matsos G, Eastwood C, O'Byrne PM: The effect of an oral platelet activating factor antagonist, WEB 2086, on allergen induced asthmatic responses. *Thorax* (1993) **48**: 594–598.
109. Bel EH, Desmet M, Rossing TH, Timmers MC, Dijkman JH, Sterk PJ: The effect of specific oral PAF-antagonist, MK-287, on antigen-induced early and late asthmatic reactions in man. *Am Rev Respir Dis* (1991) **143**: A811.
110. Hsieh KH: Effects of PAF antagonist, BN 52021, on the PAF-, methacholine- and allergen-induced bronchoconstriction in asthmatic children. *Chest* (1991) **99**: 877–882.
111. Spence DP, Johnston SL, Calverley PM, *et al*.: The effect of the orally active platelet-activating factor antagonist WEB 2086 in the treatment of asthma. *Am J Respir Crit Care Med* (1994) **149**: 1142–1148.
112. Kuitert LM, Angus RM, Barnes N, *et al*.: Effect of a novel potent platelet-activating factor antagonist, Modipafant, in clinical asthma. *Am J Respir Crit Care Med* (1995) **151**: 1331–1335.
113. Miwa M, Miyake T, Yamanaka T, *et al*.: Characterization of serum platelet-activating factor (PAF) acetylhydrolase. *J Clin Invest* (1988) **82**: 1983–1991.
114. Stafforini DM, Satoh K, Atkinson DL, *et al*.: Platelet-activating factor acetylhydrolase deficiency. A missense mutation near the active site of an anti-inflammatory phospholipase. *J Clin Invest* (1996) **97**: 2784–2791.
115. Tate RM, Morns HG, Schroeder WR, Repine JE: Oxygen metabolites stimulate thromboxane production and vasoconstriction in isolated saline-perfused rabbit lung. *J Clin Invest* (1984) **74**: 608–613.
116. Sporn PH, Peters-Golden H, Simon RH: Hydrogen-peroxide-induced arachidonic acid metabolism in the rat alveolar macrophage. *Am Rev Respir Dis* (1988) **137**: 49–56.
117. Schreck R, Rieber P, Baeuerle PA: Reactive oxygen intermediates as apparently widely used messengers in the activation of the NF-kappa B transcription factor and HIV-1. *EMBO J* (1991) **10**: 2247–2258.
118. Adcock IM, Brown CR, Kwon O, Barnes PJ: Oxidative stress induces NF kappa B DNA binding and inducible NOS mRNA in human epithelial cells. *Biochem Biophys Res Commun* (1994) **199**: 1518–1524.
119. Stewart RM, Weir EK, Montgomery MR, Niewoehner DE: Hydrogen peroxide contracts airway smooth muscle: a possible endogenous mechanism. *Respir Physiol* (1981) **45**: 333–342.
120. Rhoden KJ, Barnes PJ: Effect of hydrogen peroxide on guinea-pig tracheal smooth muscle *in vitro*: role of cyclo-oxygenase and airway epithelium. *Br J Pharmacol* (1989) **98**: 325–330.
121. Katsumata U, Miura M, Ichinose M, *et al*.: Oxygen radicals produce airway constriction and hyperresponsiveness in anesthetized cats. *Am Rev Respir Dis* (1990) **141**: 1158–1161.
122. Engels F, Oosting RS, Nijkamp F: Pulmonary macrophages induce deterioration of guinea pig tracheal β-adrenergic function through release of oxygen radicals. *Eur J Pharmacol* (1985) **111**: 143–144.
123. Shimasaki H, Takatori T, Anderson WR, Horten HL, Privett OS: Alteration of lung tepids in ozone exposed rats. *Biochem Biophys Res Commun* (1976) **68**: 1256–1262.
124. Holtzman MJ, Aizawa H, Nadel JA, Goetzl EJ: Selective generation of leukotriene B4 by tracheal epithelial cells from dogs. *Biochem Biophys Res Commun* (1983) **114**: 1071–1076.
125. Haddad E, Salmon M, Sun J, *et al*.: Dexamethasone inhibits ozone-induced gene expression of macrophage-inflammatory protein-2 in rat lung. *FEBS Lett* (1995) **363**: 285–288.
126. Haddad E, Salmon M, Koto H, Barnes PJ, Adcock I, Chung KF: Ozone induction of cytokine-induced neutrophil chemoattractant (CINC) and nuclear factor-κB in rat lung: inhibition by corticosteroids. *FEBS Lett* (1996) **379**: 265–268.

127. Matsui S, Jones GL, Woolley MJ, Lane CG, Gantovnick LS, O'Byrne PM: The effect of antioxidants on ozone-induced airway hyperresponsiveness in dogs. *Am Rev Respir Dis* (1991) **144**: 1287–1292.
128. Tsukagoshi H, Haddad A, Sun J, Barnes PJ, Chung KF: Ozone-induced airway hyperresponsiveness: role of superoxide anions, neutral endopeptidase and bradykinin receptors. *J Appl Physiol* (1995) **78**: 1015–1022.
129. Motojima S, Frigas E, Loegering DA, Gleich GJ: Toxicity of eosinophil cationic proteins for guinea pig tracheal epithelium *in vitro*. *Am Rev Respir Dis* (1989) **139**: 801–805.
130. Welsh MJ, Shasby DM, Russell MH: Oxidants increase paracellular permeability in a cultured epithelial cell line. *J Clin Invest* (1985) **76**: 1155–1168.
131. Cluzel M, Damon M, Chanez P, *et al.*: Enhanced alveolar cell luminol-dependent chemiluminescence in asthma. *J Allergy Clin Immunol* (1987) **80**: 195–201.
132. Kelly CJ, Stenton CS, Bird E, Hendrick DJ, Walters EH: Number and activity of inflammatory cells in bronchoalveolar lavage fluid in asthma and their relation to airway responsiveness. *Thorax* (1988) **43**: 684–692.
133. Postma DS, Renkema TEJ, Noordhoek JA, Faber H, Shuiter HJ, Kauffmans H: Association between nonspecific bronchial hyperreactivity and superoxide anion production by polymorphonuclear leukocytes in chronic airflow obstruction. *Am Rev Respir Dis* (1988) **137**: 57–61.
134. Chanez P, Yukawa T, Dent G, Barnes PJ, Chung KF: Generation of oxygen free radicals from blood eosinophils from asthma patients after stimulation with platelet-activating factor and pharbol ester. *Eur Respir J* (1990) **3**: 1002–1007.
135. Malmgren R, Unge G, Zetterstrom O, Theovell H, deWahl K: Lowered glutathione peroxidase activity in asthmatic patients with food and aspirin intolerance. *Allergy* (1986) **41**: 43–45.
136. Stone J, Hinks LJ, Beasley R, Holgate ST, Clayton BE: Selenium status of patients with asthma. *Clin Sci* (1989) **77**: 495–500.
137. Rahman I, Morrison D, Donaldson K, MacNee W: Systemic oxidative stress in asthma, COPD, and smokers. *Am J Respir Crit Care Med* (1996) **154**: 1055–1060.
138. Antczak A, Nowak D, Shariati B, Krol M, Piasecka G, Kumanovska Z: Increased hydrogen peroxide and lipid peroxidation products in expired breath condensates of asthmatic patients. *Eur Respir J* (1997) **10**: 1235–1241.
139. Mohsenin V, Dubois AB, Douglas JS: Effect of ascorbic acid on responses to methacholine challenge in asthmatic subjects. *Am Rev Respir Dis* (1983) **127**: 143–147.
140. Owen S, Church S, Suarez-Mendez VJ, Pearson DJ, Woodcock A: Oral antioxidant therapy and bronchial hyperreactivity. *Am Rev Respir Dis* (1990) **141**: A832.
141. Muller-Eberhard HJ: Complement. *Annu Rev Biochem* (1975) **44**: 697–724.
142. Brown EJ, Joiner KA, Frank MM: Complement. In Paul WE (ed.) *Fundamental Immunology*. New York, Raven Press (1984). pp 645–668.
143. Bodammer G, Vogt W: Actions of anaphylatoxins on circulation and respiration in the guinea-pig. *Int Arch Allergy Appl Immunol* (1967) **32**: 417–428.
144. Regal JF, Eastman AJ, Pickering RJ: C5a-induced tracheal contraction. A histamine independent mechanism. *J Immunol* (1980) **124**: 2876–2878.
145. Rocha E, Silva M, Bier O, Aronson M: Histamine release by anaphylatoxins. *Nature* (1951) **168**: 465–468.
146. Stimler NP, Bach MK, Blour CM, Hugei TE: Release of leukotrienes from guinea-pig lung stimulated by C5a des arg anaphylatoxin. *J Immunol* (1982) **128**: 2247–2252.
147. Stimler NP, Blour CM, Hugli TE: C3a-induced contraction of guinea-pig parenchyma. Role of cyclo-oxygenase metabolite. *Immunopharmacology* (1983) **5**: 251–257.
148. Wedmore CV, Williams TJ: Control of vascular permeability by polymorphonuclear leucocytes in inflammation. *Nature* (1981) **289**: 646–650.
149. Hellewell PG, Williams TJ: A specific antagonist of platelet-activating factor suppresses oedema formation in an Arthus reaction but not oedema induced by leukocyte chemoattractants in rabbit skin. *J Immunol* (1986) **137**: 302–307.
150. Yancey KB, Hammer CH, Harvalth L, Renfer L, Frank MM, Lawley TJ: Studies of human C5a as a mediator of inflammation in normal human skin. *J Clin Invest* (1985) **75**: 486–495.

151. Marom Z, Shelhamer J, Alling D, Kaliner M: The effect of corticosteroids in mucus glycoprotein secretion from human airways *in vitro*. *Am Rev Respir Dis* (1984) **129**: 62–65.
152. Movat HZ, Rettl C, Burrows CE, Johnston MG: The *in vivo* effect of leukotriene B4 on polymorphonuclear leukocytes and microcirculation. Comparison with activated complement (C5a des Arg) and enhancement of prostaglandin E2. *Am J Pathol* (1984) **115**: 233–244.
153. Clancy RM, Dahinden CA, Hugli TE: Arachidonate metabolism of human polymorphonuclear leukocytes stimulated by N-formyl-Met-Leu-Ph or complement component C5a is independent of phospholipase activation. *Proc Natl Acad Sci USA* (1983) **80**: 7200–7204.
154. Lee TC, Lenihan OJ, Malone B, Wasserman SI: Increased biosynthesis of platelet-activating factor in activated human eosinophils. *J Biol Chem* (1984) **259**: 5526–5530.
155. McCarthy K, Henson PH: Induction of lysosomal enzyme secretion by alveolar macrophages in response to the purified complement fragments C5a and C5a des arg. *J Immunol* (1979) **123**: 2511–2517.
156. Kay AB, Bacon GD, Mercer BA, Simpson H, Grafton JN: Complement components and IgE in bronchial asthma. *Lancet* (1974) **ii**: 916–920.
157. Kaufman HF, Van der Heide S, Demonchy JGR, De Vries K: Plasma histamine concentrations and complement activation during house dust mite-provoked bronchial obstructive reactions. *Clin Allergy* (1983) **13**: 219–228.
158. Durham SR, Lee TH, Cromwell O, *et al.*: Immunologic studies in allergen-induced late phase asthmatic reactions. *J Allergy Clin Immunol* (1984) **74**: 49–60.
159. Arroyave CM, Stegwewson DD, Vaughan JH, Tan FM: Plasma component changes during bronchospasm provoked in asthmatic patients. *Clin Allergy* (1977) **7**: 173–182.
160. Baur X, Dorsch W, Becker T: Levels of complement factors in human serum during immediate and late asthmatic reactions and during acute hypersensitivity pneumants. *Allergy* (1980) **35**: 383–390.
161. Callerame ML, Condemi JJ, Milton G, Bohrod MG, Vaughan JH: Immunologic reactions of bronchial tissues in asthma. *N Engl J Med* (1971) **284**: 459–464.
162. Chan-Yeung M, Gidas PC, Henson PM: Activation of complement by plicatic acid, the chemical compound responsible for asthma due to western red cedar (*Thuja plicata*). *J Allergy Clin Immunol* (1980) **65**: 333–337.
163. Srivastava N, Gupta SP, Srivastava LM: Effect of house dust mite on serum complement activation and total haemophylic activity *in vitro* in normal subjects and in patients with bronchial asthma. *Clin Allergy* (1983) **13**: 43–50.
164. Irvin CG, Berend N, Henson PM: Airways hyperreactivity and inflammation produced by aerosolisation of human C5a des arg. *Am Rev Respir Dis* (1986) **134**: 777–783.
165. Cohen ML, Schenck KW, Colbert W, Wittenauer L: Role of 5-HT$_2$ receptors in serotonin-induced contractions of non-vascular smooth muscle. *J Pharmacol Exp Ther* (1985) **232**: 770–774.
166. Macquin-Mavier I, Jarreau PH, Istin N, Harf A: 5-hydroxytryptamine-induced bronchoconstriction in the guinea-pig: effect of 5-HT$_2$ receptor activation on acetylcholine release. *Br J Pharmacol* (1991) **102**: 1003–1007.
167. Richardson BP, Engel G: The pharmacology and function of 5-HT$_3$ receptors. *Trends Neurosci* (1986) **2**: 424–428.
168. Raffestin B, Creeina J, Baullet C, Labat C, Benveniste J, Brink C: Response and sensitivity of isolated human pulmonary muscle preparations to pharmacological agents. *J Pharmacol Exp Ther* (1985) **233**: 186–194.
169. Tonnesen P: Bronchial challenge with serotonin in asthmatics. *Allergy* (1985) **40**: 136–140.
170. Cushley MJ, Wee LH, Holgate ST: The effect of inhaled 5-hydroxytryptamine (5-HT$_1$ serotonin) on airway calibre in man. *Br J Clin Pharmacol* (1986) **22**: 487–490.
171. Hahn HL, Wilson AG, Graf PD, Fischer SP, Nadel JA: Interaction between serotonin and efferent vagus nerves in dog lungs. *J Appl Physiol* (1978) **44**: 144–149.
172. Tokuyama K, Lotvall J, Barnes PJ, Chung KF: Airway narrowing after inhalation of platelet-activating factor in guinea-pig: contribution of airway microvascular leakage and edema. *Am Rev Respir Dis* (1991) **143**: 1345–1349.
173. So SY, Lam NK, Kuens S: Selective 5-HT$_2$ receptor blockade in exercise induced asthma. *Clin Allergy* (1985) **15**: 371–376.

174. Gleich GH, Adolphson CR: The eosinophilic leukocyte: structure and function. *Adv Immunol* (1986) **39**: 177–253.
175. Frigas E, Gleich GJ: The eosinophil and pathophysiology of asthma. *J Allergy Clin Immunol* (1986) **77**: 527–537.
176. Gleich GJ, Flavahan NA, Fujisaura T, Vanhoutte PM: The eosinophil as a mediator of damage of respiratory epithelium: a model for bronchial hyperreactivity. *J Allergy Clin Immunol* (1988) **81**: 776–781.
177. Flavahan NA, Slifman NR, Gleich GJ, Vanhoutte PM: Human eosinophil major basic protein causes hyperreactivity of respiratory smooth muscle: role of the epithelium. *Am Rev Respir Dis* (1988) **138**: 685–688.
178. Grundel RH, Letts LG, Gleich GJ: Human eosinophil major basic protein induces airway constriction and airway hyperresponsiveness in primates. *J Clin Invest* (1991) **87**: 1470–1473.
179. Filley WV, Holley KE, Kephart GM, Gleich GJ: Identification by immunofluoresence of eosinophil granule major basic protein in lung tissues of patients with bronchial asthma. *Lancet* (1982) **ii**: 11–16.
180. De Monchy JGR, Kauffman HK, Venge P: Bronchoalveolar eosinophilic during allergen-induced late asthmatic reactions. *Am Rev Respir Dis* (1985) **131**: 373–376.
181. Dur PJ, Ackerman SJ, Gleich GJ: Charcot–Leyden crystal protein and eosinophil major basic protein in sputum of patients with respiratory disease. *Am Rev Respir Dis* (1984) **130**: 1072–1077.
182. Rak S, Lowhagen O, Venge P: The effect of immunotherapy on bronchial hyperresponsiveness and eosinophil cationic protein in pollen allergic patients. *J Allergy Clin Immunol* (1988) **82**: 470–480.
183. Coyle AJ, Ackerman SJ, Burch R, Proud D, Irvin CG: Human eosinophil-granule major basic protein and synthetic polycations induce airway hyperresponsiveness *in vivo* dependent on bradykinin generation. *J Clin Invest* (1995) **95**: 1735–1740.
184. Lefort J, Nahori MA, Ruffie C, Vargaftig BB, Pretolani M: *In vivo* neutralization of eosinophil-derived major basic protein inhibits antigen-induced bronchial hyperreactivity in sensitised guinea-pigs. *J Clin Invest* (1996) **97**: 1117–1121.
185. Yanagisawa M, Kurihara H, Kimura S, *et al.*: A novel potent vasoconstrictor peptide produced by vascular endothelial cells. *Nature* (1988) **332**: 411–415.
186. Ehrenreich H, Anderson RW, Fox CH, *et al.*: Endothelins, peptides with potent vasoactive properties, are produced by human macrophages. *J Exp Med* (1990) **172**: 1741–1748.
187. Vittori E, Marini M, Fasoli A, De Franchis R, Mattoli S: Increased expression of endothelin in bronchial epithelial cells of asthmatic patients and effect of corticosteroids. *Am Rev Respir Dis* (1992) **146**: 1320–1325.
188. Seldeslagh KA, Lauweryns JM: Endothelin in normal lung tissue of newborn mammals: immunocytochemical distribution and co-localization with serotonin and calcitonin gene-related peptide. *J Histochem Cytochem* (1993) **41**: 1495–1502.
189. Campbell AM, Vignola AM, Chanez P, Godard P, Bousquet J: Low-affinity receptor for IgE on human bronchial epithelial cells in asthma. *Immunology* (1994) **82**: 506–508.
190. Gu XH, Casley D, Nayler W: Specific high-affinity binding sites for ^{125}I-labelled porcine endothelin in rat cardiac membranes. *Eur J Pharmacol* (1989) **167**: 281–290.
191. Martin ER, Brenner BM, Ballermann BJ: Heterogeneity of cell surface endothelin receptors. *J Biol Chem* (1990) **265**: 14 044–14 049.
192. Samson WK, Skala KD, Alexander BD, Huang FL: Pituitary site of action of endothelin: selective inhibition of prolactin release *in vitro*. *Biochem Biophys Res Commun* (1990) **169**: 737–743.
193. Harrison VJ, Randriantsoa A, Schoeffter P: Heterogeneity of endothelin–sarafotoxin receptors mediating contraction of pig coronary artery. *Br J Pharmacol* (1992) **105**: 511–513.
194. Arai H, Hori S, Aramori I, Ohkubo H, Nakanishi S: Cloning and expression of a cDNA encoding an endothelin receptor. *Nature* (1990) **348**: 730–732.
195. Nakamuta M, Takayanagi R, Sakai Y, *et al.*: Cloning and sequence analysis of a cDNA encoding human non-selective type of endothelin receptor. *Biochem Biophys Res Commun* (1991) **177**: 34–39.

196. Ogawa Y, Nakao K, Arai H, et al.: Molecular cloning of a non-isopeptide-selective human endothelin receptor. *Biochem Biophys Res Commun* (1991) **178**: 248–255.
197. Hosoda K, Nakao K, Hiroshi-Arai, et al.: Cloning and expression of human endothelin-1 receptor cDNA. *FEBS Lett* (1991) **287**: 23–26.
198. Takuwa Y, Kasuya Y, Takuwa N, et al.: Endothelin receptor is coupled to phospholipase C via a pertussis toxin-insensitive guanine nucleotide-binding regulatory protein in vascular smooth muscle cells. *J Clin Invest* (1990) **85**: 653–658.
199. Resink TJ, Scott-Burden T, Buhler FR: Endothelin stimulates phospholipase C in cultured vascular smooth muscle cells. *Biochem Biophys Res Commun* (1988) **157**: 1360–1368.
200. Kasuya Y, Takuwa Y, Yanagisawa M, Kimura S, Goto K, Masaki T: Endothelin-1 induces vasoconstriction through two functionally distinct pathways in porcine coronary artery: contribution of phosphoinositide turnover. *Biochem Biophys Res Commun* (1989) **161**: 1049–1055.
201. Griendling KK, Tsuda T, Alexander RW: Endothelin stimulates diacylglycerol accumulation and activates protein kinase C in cultured vascular smooth muscle cells. *J Biol Chem* (1989) **264**: 8237–8240.
202. Simonson MS, Wann S, Mene P, et al.: Endothelin stimulates phospholipase C, Na^+/H^+ exchange, c-fos expression, and mitogenesis in rat mesangial cells. *J Clin Invest* (1989) **83**: 708–712.
203. Kim S, Morimoto S, Koh E, Miyashita Y, Ogihara T: Comparison of effects of a potassium channel opener BRL34915, a specific potassium ionophore valinomycin and calcium channel blockers on endothelin-induced vascular contraction. *Biochem Biophys Res Commun* (1989) **164**: 1003–1008.
204. Waugh CJ, Dockrell ME, Haynes WG, Olverman HJ, Williams BC, Webb DJ: The potassium channel opener BRL 38227 inhibits binding of [^{125}I]-labelled endothelin-1 to rat cardiac membranes. *Biochem Biophys Res Commun* (1992) **185**: 630–635.
205. Sakurai T, Goto K: Endothelins. Vascular actions and clinical implications. *Drugs* (1993) **46**: 795–804.
206. Ihara M, Noguchi K, Saeki T, et al.: Biological profiles of highly potent novel endothelin antagonists selective for the ET_A receptor. *Life Sci* (1992) **50**: 247–255.
207. Urade Y, Fujitani Y, Oda K, et al.: An endothelin B receptor-selective antagonist: IRL 1038 [Cys11-Cys15]-endothelin-1(11-21). *FEBS Lett* (1994) **342**: 103.
208. Clozel M, Breu V, Gray GA, et al.: Pharmacological characterization of bosentan, a new potent orally active nonpeptide endothelin receptor antagonist. *J Pharmacol Exp Ther* (1994) **270**: 228–235.
209. Battistini B, Warner TD, Fournier A, Vane JR: Characterization of ET_B receptors mediating contractions induced by endothelin-1 or IRL 1620 in guinea-pig isolated airways: effects of BQ-123, FR139317 or PD 145065. *Br J Pharmacol* (1994) **111**: 1009–1016.
210. Hay DW, Luttmann MA, Hubbard WC, Undem BJ: Endothelin receptor subtypes in human and guinea-pig pulmonary tissues. *Br J Pharmacol* (1993) **110**: 1175–1183.
211. Abraham WM, Ahmed A, Cortes A, Spinella MJ, Malik AB, Andersen TT: A specific endothelin-1 antagonist blocks inhaled endothelin-1-induced bronchoconstriction in sheep. *J Appl Physiol* (1993) **74**: 2537–2542.
212. Zamora MA, Dempsey EC, Walchak SJ, Stelzner TJ: BQ123, an ET_A receptor antagonist, inhibits endothelin-1-mediated proliferation of human pulmonary artery smooth muscle cells. *Am J Respir Cell Mol Biol* (1993) **9**: 429–433.
213. Advenier C, Sarria B, Naline E, Puybasset L, Lagente V: Contractile activity of three endothelins (ET-1, ET-2, and ET-3) on human isolated bronchus. *Br J Pharmacol* (1990) **100**: 168–172.
214. McKay KO, Black JL, Diment LM, Armour CL: Functional and autoradiographic studies of endothelin-1 and endothelin-2 in human bronchi, pulmonary arteries, and airway parasympathetic ganglia. *J Cardiovasc Pharmacol* (1991) **17** (Suppl 7): S206–S209.
215. Henry PJ, Rigby PJ, Self JG, Preuss JM, Goldie RG: Relationship between endothelin-1 binding site densities and constrictor activities in human and airway smooth muscle. *Br J Pharmacol* (1990) **100**: 786–792.
216. Candenas ML, Naline E, Sarria B, Advenier C: Effect of epithelium removal and of enkephalin

inhibition on the bronchoconstrictor response to three endothelins of the human isolated bronchus. *Eur J Pharmacol* (1992) **210**: 291–297.
217. Shimura S, Ishihara H, Satoh M, *et al.*: Endothelin regulation of mucus glycoprotein secretion from feline tracheal submucosal glands. *Am J Physiol* (1992) **262**: L208–L213.
218. Sirois MG, Filep JG, Rousseau A, Fournier A, Plante GE, Sirois P: Endothelin-1 enhances vascular permeability in conscious rats: role of thromboxane A2. *Eur J Pharmacol* (1992) **214**: 119–125.
219. Macquin-Mavier M, Levame N, Istin, Harf A: Mechanisms of endothelin-mediated bronchoconstriction in the guinea pig. *J Pharmacol Exp Ther* (1989) **250**: 740–745.
220. McKay KO, Armour CL, Black JL: Endothelin-3 increases transmission in the rabbit pulmonary parasympathetic nervous system. *J Cardiovasc Pharmacol* (1993) **22** (Suppl 8): S181–S184.
221. McKay KO, Black JL, Armour CL: The mechanism of action of endothelin in human lung. *Br J Pharmacol* (1991) **102**: 422–428.
222. Brink C, Gillard V, Roubert P, *et al.*: Effects and specific binding sites of endothelin in human lung preparations. *Pulmon Pharmacol* (1991) **4**: 54–59.
223. Power RF, Wharton J, Zhao Y, Bloom SR, Polak JM: Autoradiographic localization of endothelin-1 binding sites in the cardiovascular and respiratory systems. *J Cardiovasc Pharmacol* (1989) **13** (Suppl 5): S50–S56.
224. Halim A, Kanayama N, el Maradny E, Maehara K, Terao T: Activated neutrophil by endothelin-1 caused tissue damage in human umbilical cord. *Thromb Res* (1995) **77**: 321–327.
225. McMillen MA, Huribal M, Kumar R, Sumpio BE: Endothelin-stimulated human monocytes produce prostaglandin E2 but not leukotriene B4. *J Surg Res* (1993) **54**: 331–335.
226. Glassberg MK, Ergul A, Wanner A, Puett D: Endothelin-1 promotes mitogenesis in airway smooth muscle cells. *Am J Respir Cell Mol Biol* (1994) **10**: 316–321.
227. Noveral JP, Rosenberg SM, Anbar RA, Pawlowski NA, Grunstein MM: Role of endothelin-1 in regulating proliferation of cultured rabbit airway smooth muscle cells. *Am J Physiol* (1992) **263**: L317–L324.
228. Hassoun PM, Thappa V, Landman MJ, Fanburg BL: Endothelin 1: mitogenic activity on pulmonary artery smooth muscle cells and release from hypoxic endothelial cells. *Proc Soc Exp Biol Med* (1992) **199**: 165–170.
229. Peacock AJ, Dawes KE, Shock A, Gray AJ, Reeves JT, Laurent GJ: Endothelin-1 and endothelin-3 induce chemotaxis and replication of pulmonary artery fibroblasts. *Am J Respir Cell Mol Biol* (1992) **7**: 492–499.
230. Kahaleh MB: Endothelin, an endothelial-dependent vasoconstrictor in scleroderma. Enhanced production and profibrotic action. *Arthritis Rheum* (1991) **34**: 978–983.
231. Murlas CG, Gulati A, Singh G, Najmabadi F: Endothelin-1 stimulates proliferation of normal airway epithelial cells. *Biochem Biophys Res Commun* (1995) **212**: 953–959.
232. Wu T, Mullol J, Rieves RD, *et al.*: Endothelin-1 stimulates eicosanoid production in cultured human nasal mucosa. *Am J Respir Cell Mol Biol* (1992) **6**: 168–174.
233. Riccio MM, Reynolds CJ, Hay DW, Proud D: Effects of intranasal administration of endothelin-1 to allergic and nonallergic individuals. *Am J Respir Crit Care Med* (1995) **152**: 1757–1764.
234. Mattoli S, Soloperto M, Marini M, Fasoli A: Levels of endothelin in the bronchoalveolar lavage fluid of patients with symptomatic asthma and reversible airflow obstruction. *J Allergy Clin Immunol* (1991) **88**: 376–384.
235. Nomura A, Uchida Y, Kamayama M, Saotome M, Oki K, Hasegawa S: Endothelin and bronchial asthma. *Lancet* (1989) **ii**: 746–747.
236. Redington AE, Springall DR, Ghatei MA, *et al.*: Endothelin in bronchoalveolar lavage fluid and its relation to airflow obstruction in asthma. *Am J Respir Crit Care Med* (1995) **151**: 1034–1039.
237. Kraft M, Beam WR, Wenzel SE, Zamora MR, O'Brien RF, Martin RJ: Blood and bronchoalveolar lavage endothelin-1 levels in nocturnal asthma. *Am J Respir Crit Care Med* (1994) **149**: 946–952.
238. Chen WY, Yu J, Wang JY: Decreased production of endothelin-1 in asthmatic children after immunotherapy. *J Asthma* (1995) **32**: 29–35.

239. Aoki T, Kojima T, Ono A, *et al.*: Circulating endothelin-1 levels in patients with bronchial asthma. *Ann Allergy* (1994) **73**: 365–369.
240. Springall DR, Howarth PH, Connihan H, Djukanovic R, Holgate ST, Polak JM: Endothelin immunoreactivity of airway epithelium in asthmatic patients. *Lancet* (1991) **337**: 697–701.
241. Ackerman V, Carpi S, Bellini A, Vassalli G, Marini M, Mattoli S: Constitutive expression of endothelin in bronchial epithelial cells of patients with symptomatic and asymptomatic asthma and modulation by histamine and interleukin-1. *J Allergy Clin Immunol* (1995) **96**: 618–627.

21

Nitric Oxide

PETER J. BARNES

INTRODUCTION

There is increasing evidence that endogenous nitric oxide (NO) plays a key role in physiological regulation of airway functions and is implicated in airway diseases, including asthma.[1-3] NO has many effects on airway function and is produced in increased amounts in asthma. Greater understanding of the role of endogenous NO will provide new insights into regulation of the airways in asthma and may provide new therapeutic approaches in the future.

GENERATION OF NO

NO is derived from the amino acid L-arginine via the enzyme NO synthase (NOS), of which at least three isoforms exist [4] (Fig. 21.1). There are two constitutive enzymes (cNOS), one first described in brain being localized to neural tissue (nNOS, NOS type I) and the other to endothelial cells (eNOS, NOS type III), although it has now become apparent that both enzymes are also expressed in other cells, such as epithelial cells. Both enzymes are activated by a rise in intracellular calcium ions (Ca^{2+}) and produce small amounts of NO that serve a local regulatory function. By contrast, the inducible form of NOS (iNOS, NOS type II) is not normally expressed but is induced by inflammatory cytokines and endotoxin. This form of the enzyme is less dependent on a rise in calcium as calmodulin is tightly bound to the enzyme, and once induced is activated and produces much larger amounts of NO than cNOS isoforms. NO derived from cNOS is involved in

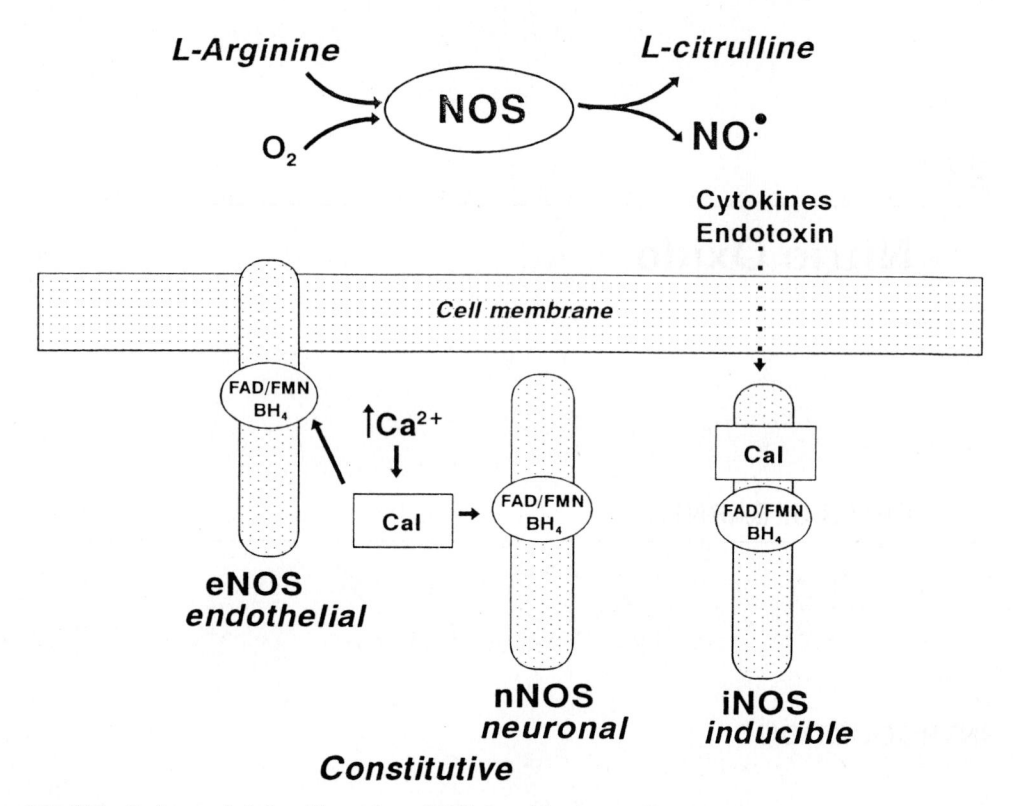

Fig. 21.1 Isoforms of nitric oxide synthase (NOS) found in airways. Two constitutively expressed enzymes, found predominantly in endothelial cells (eNOS, type III) and neurones (nNOS, type I), are activated by an increase in intracellular calcium ions (Ca^{2+}), which activates calmodulin (Cal). An inducible isoform (iNOS, type II) is largely independent of Ca^{2+} as calmodulin is integral to the enzyme.

physiological regulation of airway function, whereas NO derived from iNOS is involved in inflammatory diseases of the airways and in host defence against infection.

Expression of NOS in airways

Immunohistological studies have identified the presence of all three isoforms of NOS in human airways.[5-8] eNOS is localized to endothelial cells in the bronchial circulation, but there is also evidence for eNOS expression in epithelial cells.[9] nNOS is localized to cholinergic nerves in airways[10] but has also been reported in epithelial cells.[11]

iNOS may be expressed in several types of cell in response to cytokines, endotoxin or oxidants.[12] In asthmatic airways, there is increased immunocytochemical staining for iNOS, which is localized predominantly to airway epithelial cells;[13] there is also localization to inflammatory cells in asthmatic airways, including macrophages and eosinophils.[14]

In both a murine epithelial cell line and primary cultured human airway epithelial cells, a mixture of pro-inflammatory cytokines ('cytomix') containing tumour necrosis factor

21 Nitric Oxide

(TNF)-α, interleukin (IL)-1β and interferona (IFN)-γ increases NO production and iNOS immunoreactivity and mRNA.[11,15,16] In a human epithelial cell line (A549) and in rat type II pneumocytes, oxidants and ozone increase iNOS expression.[17,18] This is associated with activation of the transcription factor nuclear factor-κB (NF-κB), which is involved in the transcriptional expression of several inducible genes. NF-κB is of critical importance in increasing the transcription of the iNOS gene[19] and may be activated in several types of pulmonary cell by pro-inflammatory cytokines.[20,21] Glucocorticoids inhibit the induction of iNOS in epithelial cells[15,16] and this is likely to be via a direct inhibitory interaction with NF-κB.[20–22] Since NO is toxic to a wide variety of microorganisms, NO production by surface epithelial cells may therefore act as an important barrier to invasion of the respiratory tract by inhaled organisms.[23]

Inhibitors of NOS

Progress in understanding the role of NO in health and disease has been dependent on the development of specific NOS inhibitors. The first inhibitors to be developed were analogues of L-arginine, such as N^G-monomethyl-L-arginine (L-NMMA) and N^G-nitro-L-arginine methyl ester (L-NAME), which are non-selective inhibitors of NOS, and aminoguanidine, which selectively inhibits iNOS. More potent and selective inhibitors are now in development.

EFFECTS OF NO ON AIRWAY FUNCTION

NO has many effects on airway function, although the effects of endogenous NO depend on the site of production and on the amount produced (Fig. 21.2). The effects of NO also depend upon the production of peroxynitrite (ONOO$^-$) due to the presence of superoxide anions. NO combines avidly with superoxide anions to form the much more stable peroxynitrite, which generates the highly reactive hydroxyl anion (OH$^-$). Peroxynitrite induces hyperresponsiveness in animals[24] and is likely to be generated in asthma since several inflammatory cells, particularly eosinophils and neutrophils (during acute exacerbations), generate superoxide anions[25,26] (Fig. 21.3). Peroxynitrite leads to nitrosylation of tyrosine residues on proteins and nitrotyrosine may be detected immunocytochemically, providing evidence of local generation of peroxynitrite.[27–29] The presence of nitrotyrosine has recently been demonstrated in asthmatic airways, providing evidence for peroxynitrite generation within the airways; the amount of nitrotyrosine immunostaining is correlated with airway hyperresponsiveness, as measured by methacholine challenge.[14]

Vascular effects

NO is a potent vasodilator in the bronchial circulation and may play an important role in regulating airway blood flow, as in the pulmonary circulation.[30–33] Endogenous NO may increase the exudation of plasma by increasing blood flow to leaky postcapillary venules,

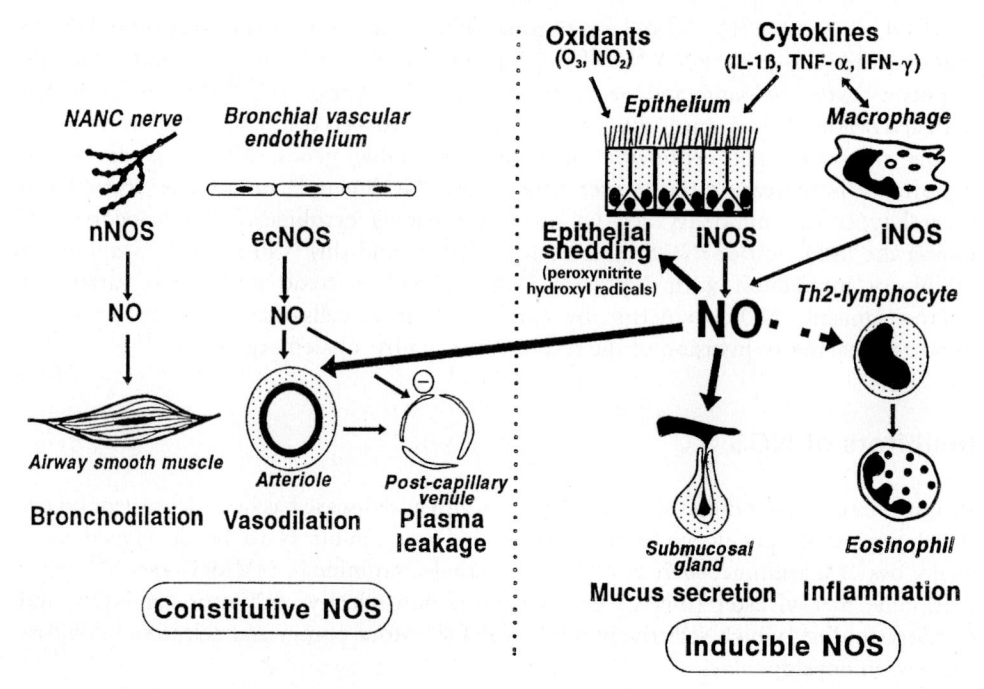

Fig. 21.2 Sources and effects of nitric oxide (NO) in the airways. IFN-γ, interferon γ; IL-1β, interleukin 1β; TNF-α, tumour necrosis factor α.

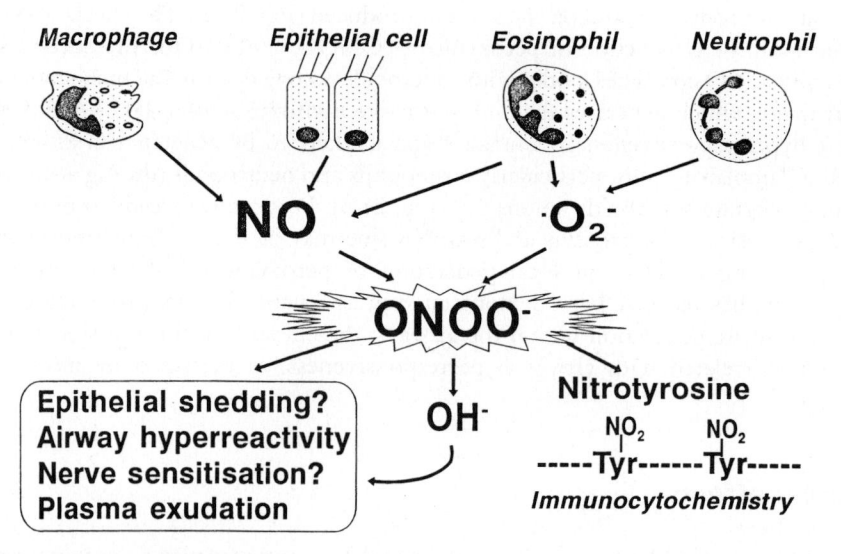

Fig. 21.3 Peroxynitrite (ONOO$^-$) generation from reaction of nitric oxide (NO) and superoxide anions ($\cdot$O$_2^-$). Peroxynitrite generates hydroxyl radicals and results in nitrosylation of tyrosine (Tyr) residues on proteins, which may be detected by immunocytochemistry.

thus increasing airway oedema.[34] However, NOS inhibitors applied to the airway surface *increase* plasma exudation, suggesting that basal release of NO has an inhibitory effect on microvascular leakage.[35] This paradox is resolved by the differing effects of NO depending on the amount produced. Thus in rat airways L-NAME increases basal leakiness whereas after endotoxin exposure, when iNOS is induced, L-NAME inhibits leakage.[36] Thus, the effect of endogenous NO on plasma exudation may depend on the amount produced and the site of production. In the context of asthma, the increased production of NO is likely to result in increased plasma exudation. Furthermore, if peroxynitrite is generated in asthma this may lead to the formation of hydroxyl radicals that also increase airway plasma exudation.[37]

NO is a potent relaxant of human pulmonary vessels.[31,38] Excessive production of NO in asthma may contribute to the ventilation–perfusion mismatch that occurs in asthmatic patients, particularly during asthma exacerbations.

Airway smooth muscle

NO and NO donor compounds also relax human airway smooth muscle *in vitro* via activation of guanylyl cyclase and an increase in cyclic GMP.[39,40] High concentrations of inhaled NO produce bronchodilatation and protect against cholinergic bronchoconstriction in guinea-pigs *in vivo*.[41] In humans, inhalation of high concentrations of NO (80 p.p.m.) has no effect on lung function in normal subjects and produces only weak and variable bronchodilatation in asthmatic patients.[42-44] This suggests that NO is more potent as a vasodilator than as a bronchodilator.

Airway secretions

L-NAME increases baseline airway mucus secretions, suggesting that NO derived from cNOS normally inhibits mucus secretion.[45] However NO donors increase mucus secretion in human airways *in vitro*.[46] In cultured guinea-pig airways after exposure to TNF-α and other inflammatory stimuli there is increased secretion of mucus, which is inhibited by L-NMMA, suggesting that large amounts of NO generated by iNOS stimulate mucus secretion.[47]

Endogenous NO may also be important in regulating mucociliary clearance, since an NOS inhibitor decreases ciliary beat frequency in bovine airway epithelial cells.[48]

Neurotransmission

NO is now established as a neurotransmitter of inhibitory non-adrenergic non-cholinergic (i-NANC) nerves or bronchodilator nerves in the airways of several species[49] (Fig. 21.4). In proximal human airways there is a prominent i-NANC bronchodilator neural mechanism, which assumes particular functional importance as it is the only endogenous bronchodilator pathway in human airways. The neurotransmitter of this i-NANC pathway in human airways is NO, since NOS inhibitors virtually abolish this neural response.[50-52] Furthermore, i-NANC stimulation of human airways

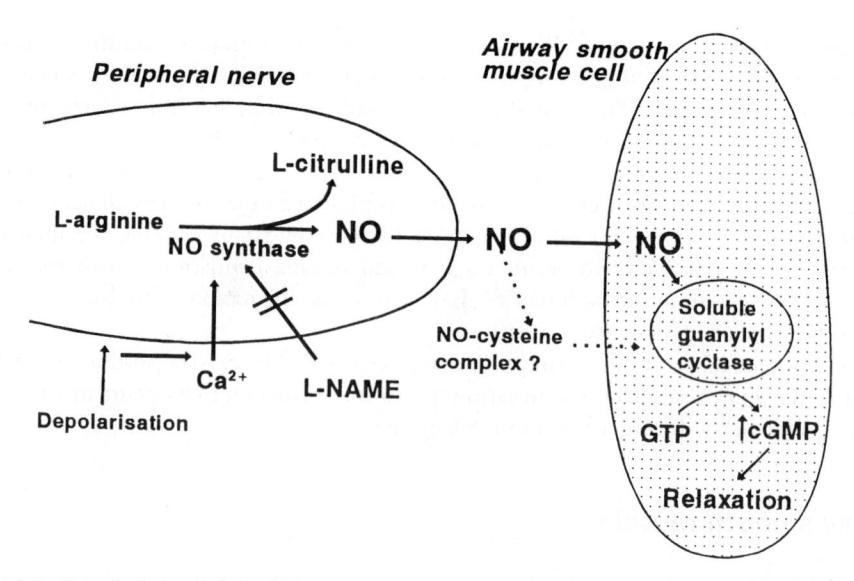

Fig. 21.4 Neuronal NO synthase (nNOS) generates NO when the nerve is depolarized. NO diffuses from the nerve ending (possibly reacting with cysteine to form a more stable intermediary compound) to activate soluble guanylyl cyclase in airway smooth muscle cells, with the formation of cyclic guanosine 3′,5′-monophosphate (cGMP) resulting in bronchodilatation.

results in an increase in cyclic GMP without any increase in cyclic AMP.[39] The density of nNOS-immunoreactive nerves is greatest in proximal airways and diminishes peripherally, which is consistent with a reduction in i-NANC responses in more peripheral airways.[6] NOS is predominantly localized to parasympathetic (cholinergic) nerves and may be colocalized with vasoactive intestinal polypeptide (VIP), although the functional role of endogenous VIP in human airways is obscure.[51] NO may be coreleased with acetylcholine from cholinergic nerves and may modulate cholinergic neural responses. NOS inhibitors increase cholinergic neural bronchoconstriction in human and guinea-pig airways.[53–55] This appears to be due to functional antagonism at the level of airway smooth muscle, rather than an effect on acetylcholine release from cholinergic nerves.[55,56]

It is possible that NO neurotransmission is impaired in inflammatory diseases of the airway, since production of superoxide anions by inflammatory cells, such as neutrophils and eosinophils, would lead to more rapid degradation of neurally released NO. This would predict enhanced cholinergic bronchoconstriction (Fig. 21.5). We have not detected any reduction in i-NANC responses in the airways of patients with mild asthma (transplant donors), but this does not preclude a defect in an inflammatory exacerbation of asthma. However, in patients with cystic fibrosis who undergo transplantation for end-stage disease, there is a profound reduction in i-NANC responses in the airways, in which there is an intense neutrophilic inflammatory response.[57] Interestingly, i-NANC responses are also markedly reduced in airways that have been extrinsically denervated by transplantation, indicating that extrinsic nerves are needed for the normal functioning of 'nitrergic' nerves in the airways.

21 Nitric Oxide

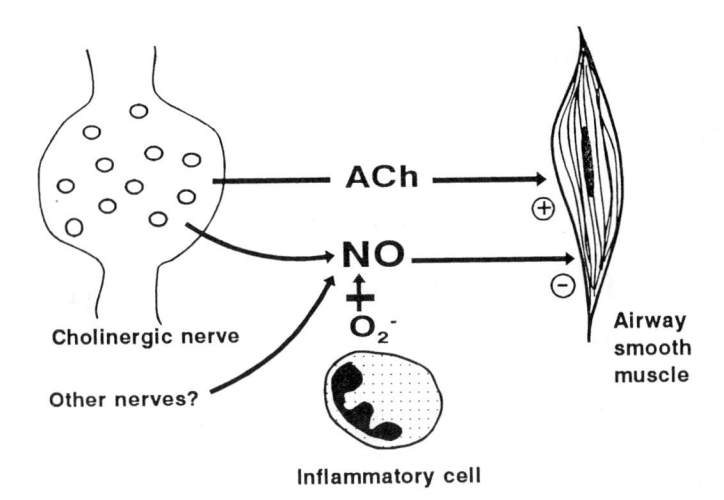

Fig. 21.5 Nitric oxide (NO) modulates cholinergic neural effects mediated via acetylcholine (ACh). In inflammation NO may be removed by superoxide anions (O_2^-) generated from inflammatory cells and this may therefore diminish the 'braking' effect of NO, resulting in exaggerated cholinergic bronchoconstriction.

Cytotoxic effects

High concentrations of NO are cytotoxic and are involved in basic defence against microorganisms. Targeted disruption ('knock-out') of the iNOS gene in mice results in a marked increase in susceptibility to infections.[58,59] It is possible that NO is toxic to epithelial cells in the airways and may contribute to epithelial shedding in asthma. These effects are likely to be mediated via the formation of peroxynitrite.

Inflammatory effects

There is increasing evidence that high concentrations of NO may have effects on the immune system and the inflammatory response. NO inhibits Th1 lymphocytes in mice and thus favours the development of a Th2 response with eosinophilia[60,61] (Fig. 21.6). There is also evidence that NO promotes the chemotaxis of eosinophils, since L-NAME blocks eosinophil recruitment in the lungs.[62]

EXHALED NO

Gustafsson and colleagues[63] first demonstrated that NO can be detected in the exhaled air of animals and normal human subjects and this has subsequently been confirmed in many studies.[64–70] Furthermore, the concentration of exhaled NO is increased in patients with inflammatory diseases of the airways such as asthma[65,66,71] and is reduced by glucocorticoid therapy.[72,73] This suggests that exhaled NO may provide a non-invasive means of

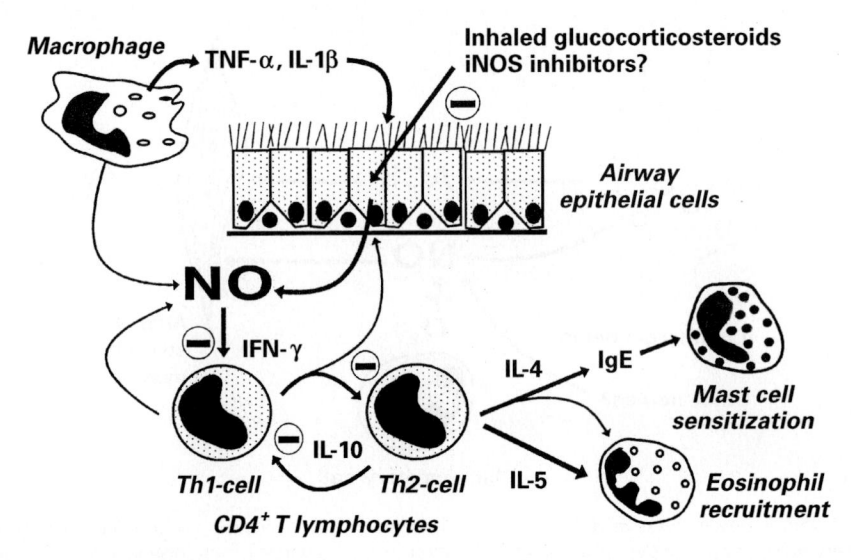

Fig. 21.6 Nitric oxide (NO) may have effects on the immune system in the airways. NO suppresses T helper (Th)1 lymphocytes, which may favour the expansion of Th2 cells that drive allergic inflammation. IFN-γ, interferon γ; IL, interleukin; TNF-α, tumour necrosis factor α.

monitoring inflammation in asthmatic airways and thus the measurement of exhaled NO has attracted increasing interest.[74]

Measurement of exhaled NO

Most studies have measured NO in exhaled air by chemiluminescence and detection depends on the photochemical reaction between NO and ozone generated in the analyser.[75] The specificity of exhaled NO measurements by chemiluminescence has recently been confirmed using gas chromatography–mass spectrometry.[69] Several NO analysers are now commercially available, but may need to be converted for on-line measurement of NO in exhaled air. Most analysers are sensitive to <1 part per billion (p.p.b.) of NO and this is adequate for studies of exhaled air. NO may be detected by direct expiration into the analyser (Fig. 21.7) or by collection into an impermeable reservoir or balloon for later analysis.

Several technical factors may affect the measurement of exhaled NO and it is important that the technique should be specified, so that comparisons between studies is possible. Breath-holding results in an increase in exhaled NO, which may reflect accumulation of NO in the upper or lower respiratory tracts.[67,76] High concentrations of NO have been detected in the upper respiratory tract and nasopharynx, with particularly high concentrations in the paranasal sinuses.[77–79] This has suggested that exhaled NO may largely reflect NO derived from the upper airways rather than the lower airways. Thus manoeuvres that block the upper respiratory tract markedly reduce exhaled NO concentrations[80] and much lower levels of NO are recorded from the lower respiratory tract of patients with tracheostomies that exclude the upper respiratory tract.[77,78]

21 Nitric Oxide

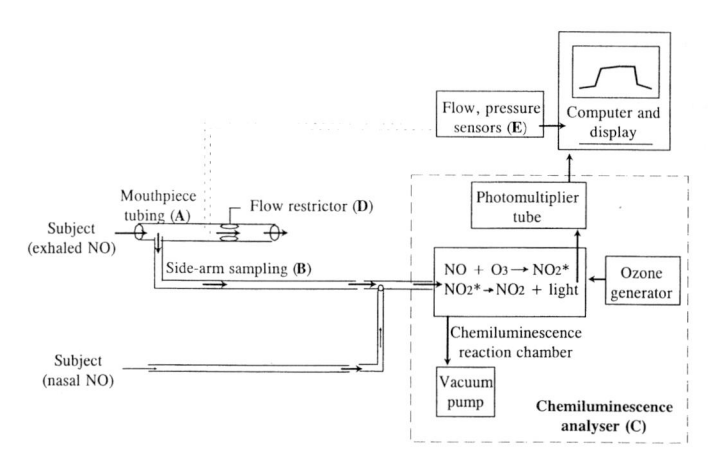

Fig. 21.7 Measurement of exhaled nitric oxide (NO) by chemiluminescence analyser using a single slow expiration.

Expiration against resistance prevents any nasal contamination, as this leads to isolation of the nasopharynx from the oropharynx by elevation of the soft palate (Fig. 21.8). This has recently been confirmed using argon as a tracer gas.[81] Thus slow expiration against resistance produces levels of exhaled NO in the expired air that are identical to those measured by direct sampling via a bronchoscope from the lower respiratory tract in both normal and asthmatic patients.[82,83] During quiet tidal breathing, however, there may be nasal contamination of the exhaled NO as there is communication between the nasopharynx and oropharynx. This means that collection of expired air in a reservoir during tidal breathing may overestimate exhaled NO levels from the lower respiratory tract due to nasal contamination.

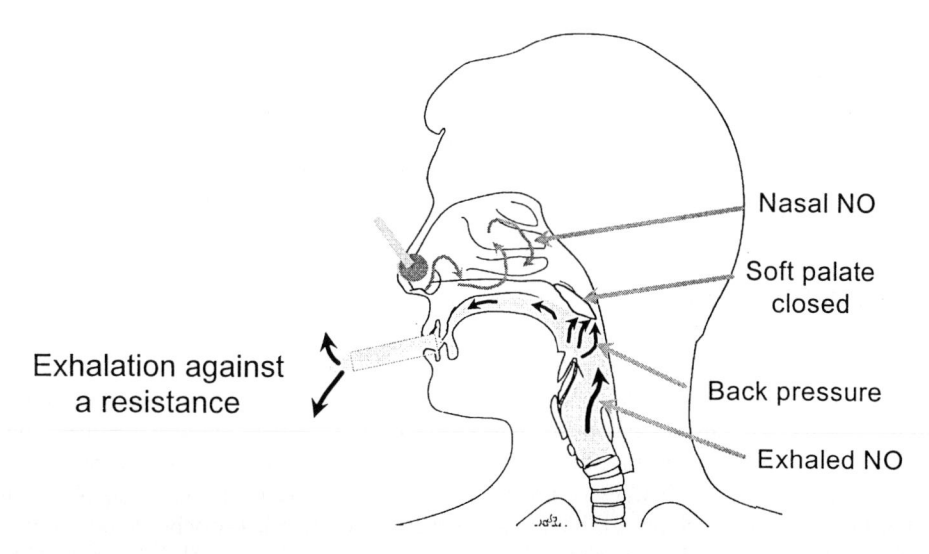

Fig. 21.8 Expiration against resistance causes closure of the soft palate and thus prevents contamination of exhaled air with the high concentration of nitric oxide (NO) within the nose.

Source of NO in exhaled air

The cellular source of NO in the lower respiratory tract is not yet certain. Studies with perfused porcine lungs suggest that exhaled NO originates at the alveolar surface rather than from the pulmonary circulation,[84] and may be derived from eNOS expressed in the alveolar walls of normal lungs.[5] Studies in ventilated perfused lungs of guinea-pigs show that exhaled NO is reduced during perfusion with calcium-free solutions, suggesting that NO is derived from cNOS, which is calcium dependent.[68] Airway epithelial cells may also express both eNOS and nNOS and may therefore contribute to NO in the lower respiratory tract.[9,11,85] In inflammatory diseases, it is likely that the increase in exhaled NO is due to induction of iNOS. Indeed increased NOS activity has been demonstrated in lung tissue of patients with asthma.[86] In asthmatic patients there is evidence for increased expression of iNOS in airway epithelial cells,[13] although even epithelial cells from normal individuals appear to express iNOS.[85] Pro-inflammatory cytokines induce the expression of iNOS in murine epithelial cells and cultured human airway epithelial cells[11,15,16] and it is likely that these same cytokines are released in asthmatic inflammation. iNOS is also expressed in other cell types, including alveolar macrophages and eosinophils.[14] Furthermore, glucocorticoids inhibit the induction of iNOS in epithelial cells *in vitro*[15,16] and *in vivo*[87] and reduce exhaled NO levels in asthmatic patients to normal.[73]

The levels of NO in the nose and nasopharynx are much higher than those recorded in expiration at the mouth, suggesting that upper airways may be the major contributor to exhaled NO, at least in normal individuals.[77,78,80,82,88] However, the lower respiratory tract is likely to contribute some of the exhaled NO, even in normal individuals. NO has been detected in the exhaled air of tracheotomized rabbits, rats, guinea-pigs and humans[63,77] and via bronchoscopy in normal individuals.[82,83] The products of NO metabolism, nitrite and nitrothiols, are also present in bronchoalveolar lavage of normal subjects.[40] Simultaneous measurement of expired CO_2 and NO demonstrate that the peak in exhaled NO precedes the peak value of CO_2 (end-tidal), suggesting that NO is derived from airways rather than alveoli.[67] Although it is likely that nasal NO contributes to the levels of exhaled NO in normal individuals, it is unlikely to contribute to the elevated levels found in inflammatory airway disease. Direct sampling via fibreoptic bronchoscopy in asthmatic patients shows a similar elevation of NO in trachea and main bronchi to that recorded at the mouth, thus indicating that the elevated levels in asthma are derived from the lower airways.[82,83]

Exhaled NO in asthma

Several studies have reported an elevation of exhaled NO in patients with asthma[65,66,71,89] (Figs 21.9 and 21.10). The increase in exhaled NO does not appear to be closely related to asthma severity or to airway responsiveness (measured by methacholine challenge)[66] but there is a relationship to eosinophil counts in the sputum[90] and with the expression of iNOS in the airways.[14] Changes in bronchial calibre have no effect on exhaled NO as neither bronchoconstriction with histamine or methacholine nor bronchodilatation with salbutamol have any effect on the measurements in asthmatic patients.[91–93] Immunocytochemical staining of bronchial biopsies has demonstrated increased expression of iNOS in

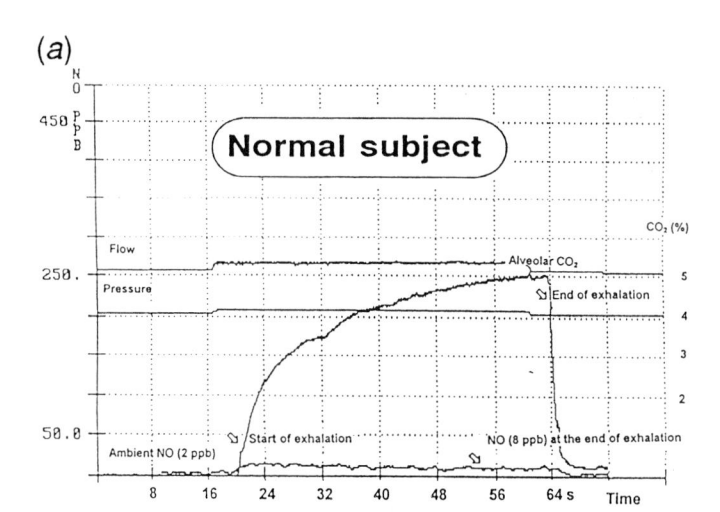

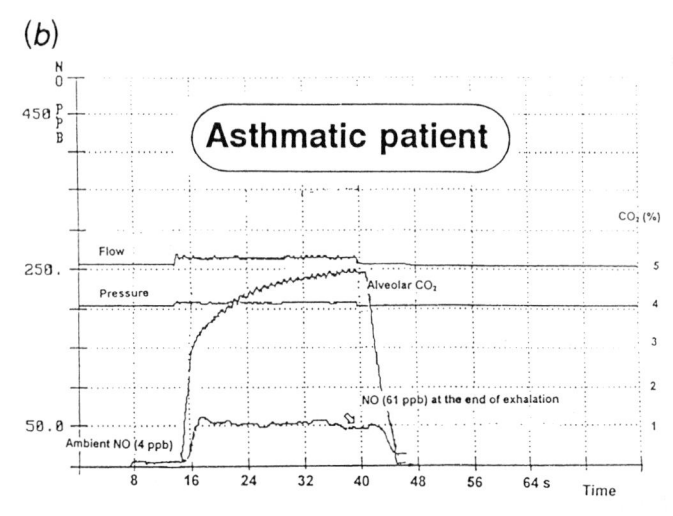

Fig. 21.9 Exhaled nitric oxide (NO) in a normal subject and a patient with asthma: (a) exhaled NO (8 p.p.b.) and carbon dioxide (CO_2) from a normal subject with constant flow and pressure; (b) exhaled NO (61 p.p.b.) and CO_2 from a patient with asthma measured under the same conditions.

epithelial cells and inflammatory cells in asthmatic compared with non-asthmatic subjects,[13,14] suggesting that pro-inflammatory cytokines present in asthmatic airways have induced its expression, resulting in increased NO production in the lower airways. After inhaled allergen challenge in asthmatic patients there is no change of exhaled NO during the early bronchoconstrictor response, but a progressive elevation during the late response.[91] In patients who have no late response to allergen (single responders), there is no change in exhaled NO throughout the study period. This suggests that increased NO is associated with the inflammatory late response and may be a reflection of iNOS expression in response to inflammatory cytokines. In sensitized guinea-pigs, allergen challenge is associated with increased NO production during the late response and this is preceded by iNOS mRNA expression.[94] Whether increased NO production is merely a

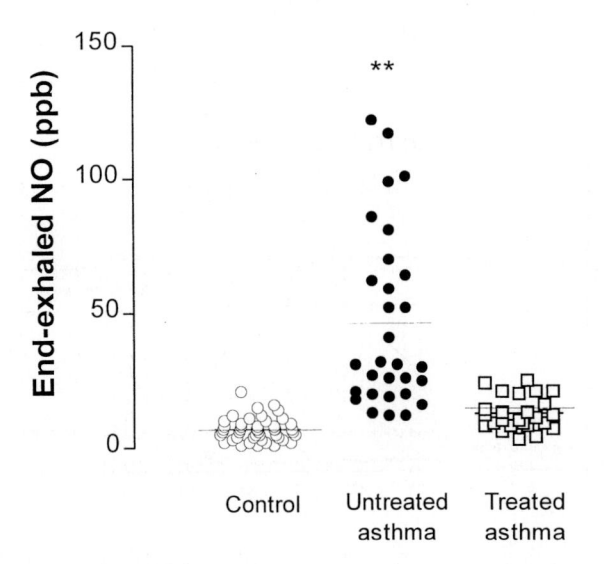

Fig. 21.10 Exhaled nitric oxide (NO) measured in parts per billion (p.p.b.) in normal controls (○), patients with untreated asthma (●) and asthmatic patients treated with inhaled steroids (□).

marker of the cytokine-mediated inflammation or contributes to the airway narrowing (secondary to vasodilatation and increased plasma exudation) during the late response is not yet certain and studies with NOS inhibitors are needed. There is also an increase in exhaled NO during exacerbations of asthma[95,96] and when the dose of inhaled glucocorticoids is reduced.[97] By contrast, there is no increase in exhaled NO after bronchoconstriction induced by histamine (direct effect on airway smooth muscle) or by adenosine (via activation of airway mast cells).[91,98] These findings suggest that exhaled NO may reflect airway inflammation in asthma and may be used as a means of monitoring inflammatory events in the lower airways.

Effects of therapy

Exhaled NO levels are significantly lower in patients with asthma who are treated with inhaled glucocorticoids, suggesting that inhaled steroids reduce exhaled NO.[66,99] An oral glucocorticoid, prednisolone (30 mg for 3 days), has no effect on exhaled NO in normal individuals, but decreases the elevated levels of exhaled NO in asthmatic patients.[72] This suggests that the exhaled NO in normal subjects is derived from cNOS (unaffected by steroids), whereas the elevated levels in asthma are derived from iNOS, which is inhibited by glucocorticoids. In asthmatic patients, a double-blind study of inhaled budesonide shows a progressive reduction in exhaled NO down to normal values after 3 weeks of therapy.[73] The reduction in exhaled NO is progressive and may reflect direct inhibitory effects of glucocorticoids on induction of iNOS, via a direct blockade of NF-κB, and an indirect effect due to reduced synthesis of the pro-inflammatory cytokines that lead to iNOS expression in airway epithelial cells. Biopsy studies have confirmed that iNOS expression in asthmatic airway epithelial cells is reduced in patients treated with inhaled steroids.[87]

21 Nitric Oxide

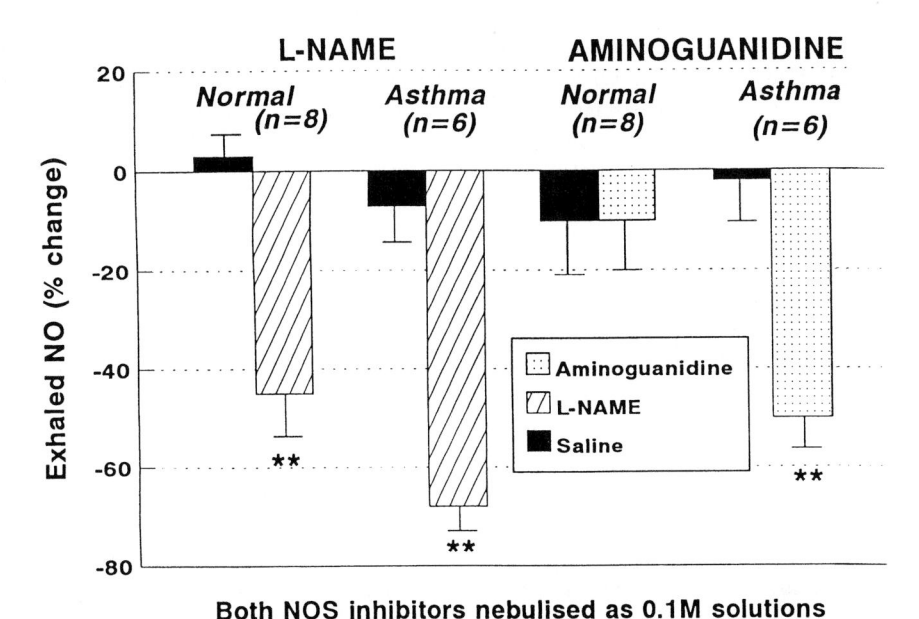

Fig. 21.11 Effect of nitric oxide (NO) synthase inhibitors L-NAME and aminoguanidine on exhaled NO in normal and asthmatic subjects. The non-selective L-NAME inhibits exhaled NO in both groups, whereas aminoguanidine, which is selective for inducible NOS, inhibits exhaled NO significantly only in patients with asthma. **, $P < 0.01$ compared with control.

Neither short-acting nor long-acting inhaled β_2-agonists reduce exhaled NO in asthmatic patients.[93] This is in keeping with other studies showing no anti-inflammatory effect of inhaled β_2-agonists in asthma and add further support to the view that exhaled NO may be useful in assessing anti-inflammatory effect of inhaled asthma treatments.

Single inhalations of L-NMMA and L-NAME (via a nebulizer) result in reduced exhaled NO in normal and asthmatic patients[66,72,100] (Fig. 21.11). Interestingly, there is no fall in forced expiratory volume in 1 s (FEV$_1$), even in asthmatic patients with highly reactive airways, suggesting that basal production of NO is not important in basal airway tone. Although infusion of L-NMMA in normal subjects causes an increase in blood pressure,[101,102] neither nebulized L-NAME nor L-NMMA have any effect on heart rate or blood pressure, suggesting that inhibition of NOS is confined to the respiratory tract. While L-NMMA and L-NAME are non-selective inhibitors of cNOS and iNOS, aminoguanidine has some selectivity for iNOS.[103,104] Inhalation of aminoguanidine has no effect on exhaled NO in normal subjects, but significantly reduces exhaled NO in patients with asthma,[105] adding further support to the view that the elevated exhaled NO in asthma is derived from iNOS.

Clinical implications

Measurement of NO in exhaled air may be a relatively simple way of monitoring inflammation in inflammatory airways disease and may also provide a means of investigating the anti-inflammatory effects of treatments, such as glucocorticoids and novel anti-inflammatory therapies. There is now persuasive evidence that levels of NO are increased in association with airway inflammation and are decreased with anti-inflammatory treatments. Correlation of exhaled NO with more direct measurements of inflammation in the airways, such as induced sputum, bronchoalveolar lavage and bronchial biopsies, is now needed. There is a correlation between exhaled NO and the number of eosinophils in induced sputum of asthmatic patients, but this is only a weak correlation and it is unlikely that expression of iNOS will reflect all of the inflammatory changes present in asthmatic airways.[90]

The great advantage of exhaled NO is that the measurement is completely non-invasive and can therefore be performed repeatedly and also in children[105] and patients with severe airflow obstruction,[95] where more invasive techniques are not possible. The measurement, however, is not specific: exhaled NO is increased in inflammation due to asthma, bronchiectasis[99] and respiratory tract infections. This means that absolute values are less important than serial measurements in individual patients. The value of this approach has been demonstrated in asthmatic patients where the dose of inhaled steroid is changed, resulting in increased levels when the dose is reduced and lower levels when the dose is increased.[97] Because exhaled NO is reduced by anti-inflammatory treatments, it may be useful for monitoring whether therapy is adequate. The technique may also have application in the monitoring of anti-inflammatory effects of new antiasthma drugs, such as selective phosphodiesterase inhibitors, leukotriene antagonists and synthesis inhibitors, and immunomodulators. Because the measurement is precise and reasonably reproducible, it may facilitate the measurement of dose–response effects with anti-inflammatory treatments, which is difficult at present. Thus, it is possible to discriminate the effects of budesonide 100 μg daily from budesonide 400 μg daily using exhaled NO, which would be difficult using other clinical parameters unless very large numbers of patients were selected.[106]

The currently available analysers for exhaled NO are expensive, but in the future it is likely that technological advances will make it possible to miniaturize these analysers so that they are portable and may even be used at home in conjunction with peak flow meters. This may lead to their application in epidemiological studies and may be a useful screening measurement for community studies.

THERAPEUTIC IMPLICATIONS

NO donor compounds relax human airways *in vitro*[39,40] and this has raised the possibility that such compounds may prove to be useful as novel bronchodilators. However, previous studies of nitrates in asthma have shown little bronchodilator or bronchoprotective effects.[1] The major problem with this class of drug is the cardiovascular side-effects, which limit the dose that can be given, as NO has a greater relaxant effect on vascular than on airway smooth muscle.

Pro-inflammatory cytokines, such as TNF-α and IL-1β, are produced by activated macrophages in asthmatic airways[107] and presumably induce iNOS expression in epithelial cells.[13,14] In addition, atmospheric pollutants such as ozone may also increase iNOS expression, via activation of NF-κB.[17,18,108] Because the actions of NO are multiple, it is not clear whether endogenous NO would be beneficial or harmful in asthma. One possibility is that endogenous NO acts as an endogenous bronchodilator system and there is evidence that NOS inhibition increases the constrictor response to histamine in guinea-pigs.[109] However, inhalation of NOS inhibitors in asthmatic patients does not increase airway obstruction or increase the bronchoconstrictor responses to histamine.[100] It is more likely that increased NO production in asthma may have an amplifying effect on airway inflammation.[61] The release of NO from epithelial cells could increase airway blood flow (hyperaemia) and plasma exudation (airway oedema), but may also have an effect on the immune response. NO has an inhibitory effect on Th1 lymphocytes, which produce IFN-γ, but no effect on Th2 cells, which produce IL-4 and IL-5.[110] If airway epithelial cells produce NO this would favour the proliferation of Th2 cells in response to allergen, since IFN-γ suppresses the proliferation and activation of Th2 cells. This would result in increased IL-4 and IL-5 secretion, leading to the characteristic eosinophilic inflammation of asthma. Glucocorticoids would inhibit this inflammatory response by preventing the induction of epithelial NO and thus increasing IFN-γ, which in turn would inhibit the production of IL-4 and IL-5. This has indeed been reported in biopsy studies of asthmatic patients after treatment with glucocorticoids.[111] Increased NO production in the respiratory tract of asthmatic patients may also contribute to the ventilation–perfusion inequalities in asthmatic lungs that may lead to hypoxia, particularly during an acute exacerbation. Formation of peroxynitrite by the reaction between NO and superoxide anions generated from eosinophils may contribute to airway hyperresponsiveness in asthma and epithelial cell shedding.

This suggests that an inhibitor of iNOS may be effective in controlling asthmatic inflammation. Such a drug could be given by inhalation thus avoiding any systemic effects. Inhibition of iNOS may be a novel therapeutic approach to the treatment of asthma and selective iNOS inhibitors are now in development. Aminoguanidine has some selectivity for iNOS compared with cNOS isoforms,[103] and a single inhalation of aminoguanidine reduces the elevated NO in patients with asthma yet has no effect in normal individuals.[100] Regular treatment with inhaled iNOS inhibitors might control asthmatic inflammation by inhibiting epithelial production of NO in the same way that inhaled steroids appear to do so. This might reduce the requirements for inhaled glucocorticoids. More selective iNOS inhibitors are currently in development.

REFERENCES

1. Barnes PJ, Belvisi MG: Nitric oxide and lung disease. *Thorax* (1993) **48**: 1034–1043.
2. Gaston B, Drazen JM, Loscalzo J, Stamler JS: The biology of nitrogen oxides in the airways. *Am J Respir Crit Care Med* (1994) **149**: 538–551.
3. Barnes PJ: Nitric oxide and airway disease. *Ann Med* (1995) **27**: 389–393.
4. Nathan C, Xie Q: Regulation of biosynthesis of nitric oxide. *J Biol Chem* (1994) **269**: 13725–13728.
5. Kobzik L, Bredt DS, Lowenstein CJ, *et al*.: Nitric oxide synthase in human and rat lung: immunocytochemical and histochemical localization. *Am J Respir Cell Mol Biol* (1993) **9**: 371–377.
6. Ward JK, Belvisi MG, Springall DR, *et al*.: Human iNANC bronchodilatation and nitric oxide-immunoreactive nerves are reduced in distal airways. *Am J Respir Cell Mol Biol* (1995) **13**: 175–184.

7. Tracey WR, Xue C, Klinghoffer V, et al.: Immunocytochemical detection of inducible NO synthase in human lung. *Am J Physiol* (1994) **266**: L722–L727.
8. Furukawa K, Harrison DG, Saleh D, Shennib H, Chagnon FP, Giaid A: Expression of nitric oxide synthase in human nasal mucosa. *Am J Respir Crit Care Med* (1996) **153**: 847–850.
9. Shaul PW, North AJ, Wu LC, et al.: Endothelial nitric oxide synthase is expressed in cultured bronchiolar epithelium. *J Clin Invest* (1994) **94**: 2231–2236.
10. Fischer A, Mundel P, Mayer B, Preissler U, Philippin B, Kummer W: Nitric oxide synthase in guinea-pig lower airway innervation. *Neurosci Lett* (1993) **149**: 157–160.
11. Asano K, Chee CBE, Gaston B, et al.: Constitutive and inducible nitric oxide synthase gene expression, regulation and activity in human lung epithelial cells. *Proc Natl Acad Sci USA* (1994) **91**: 10 089–10 093.
12. Morris S, Billiar TR: New insights into the regulation of inducible nitric oxide synthesis. *Am J Physiol* (1994) **266**: E829–E839.
13. Hamid Q, Springall DR, Riveros-Moreno V, et al.: Induction of nitric oxide synthase in asthma. *Lancet* (1993) **342**: 1510–1513.
14. Giaid A, Saleh D, Lim S, Barnes PJ, Ernst P: Formation of peroxynitrite in asthmatic airways. *Am J Respir Crit Care Med* (1997) in press.
15. Robbins RA, Springall DR, Warren JB, et al.: Inducible nitric oxide synthase is increased in murine lung epithelial cells by cytokine stimulation. *Biochem Biophys Res Commun* (1994) **198**: 1027–1033.
16. Robbins RA, Barnes PJ, Springall DR, et al.: Expression of inducible nitric oxide synthase in human bronchial epithelial cells. *Biochem Biophys Res Commun* (1994) **203**: 209–218.
17. Adcock IM, Brown CR, Kwon OJ, Barnes PJ: Oxidative stress induces NF-κB DNA binding and inducible NOS mRNA in human epithelial cells. *Biochem Biophys Res Commun* (1994) **199**: 1518–1524.
18. Punjabi CJ, Laskin JD, Pendino KJ, Goller NL, Durham SK, Laskin DL: Production of nitric oxide by rat type II pneumocytes: increased expression of inducible nitric oxide synthase following inhalation of a pulmonary irritant. *Am J Respir Cell Mol Biol* (1994) **11**: 165–172.
19. Xie Q, Kashiwarbara Y, Nathan C: Role of transcription factor NF-aB/Rel in induction of nitric oxide synthase. *J Biol Chem* (1994) **269**: 4705–4708.
20. Adcock IM, Brown CR, Gelder CM, Shirasaki H, Peters MJ, Barnes PJ: The effects of glucocorticoids on transcription factor activation in human peripheral blood monoclucear cells. *Am J Physiol* (1995) **37**: C331–C338.
21. Adcock IM, Shirasaki H, Gelder CM, Peters MJ, Brown CR, Barnes PJ: The effects of glucocorticoids on phorbol ester and cytokine stimulated transcription factor activation in human lung. *Life Sci* (1994) **55**: 1147–1153.
22. Ray A, Prefontaine KE: Physical association and functional antagonism between the p65 subunit of transcription factor NF-KB and the glucocorticoid receptor. *Proc Natl Acad Sci USA* (1994) **91**: 752–756.
23. Liew FY, Cox FF: Nonspecific resistance mechanisms: the role of nitric oxide. *Immunol Today* (1991) **12**: A17–A21.
24. Sadeghi-Hashjin G, Folkerts G, Henricks PAJ, et al.: Peroxynitrite induces airway hyper-responsiveness in guinea pigs *in vitro* and *in vivo*. *Am J Respir Crit Care Med* (1996) **153**: 1697–1701.
25. Barnes PJ: Reactive oxygen species and airway inflammation. *Free Radic Biol Med* (1990) **9**: 235–243.
26. Calhoun WJ, Reed HE, Moest OR, Stevens CA: Enhanced superoxide production by alveolar macrophages and air-space cells, airway inflammation and alveolar macrophage density after segmental antigen bronchoprovocation in allergic subjects. *Am Rev Respir Dis* (1992) **145**: 317–325.
27. Ischhiropoulos H, Zhu L, Beckman JS: Peroxynitrite formation from macrophage-derived nitric oxide. *Arch Biochem Biophys* (1992) **298**: 446–451.
28. Haddad IY, Pataki G, Hu P, Galliani C, Becjman JS, Matalon S: Quantification of nitrotyrosine levels in lung sections of patients and animals with acute lung injury. *J Clin Invest* (1994) **94**: 2407–2413.

29. Beckman JS, Koppenol WH: Nitric oxide, superoxide, and peroxynitrite: the good, the bad, and the ugly. *Am J Physiol* (1996) **271**: C1432–C1437.
30. Higenbottam TW: Lung disease and pulmonary endothelial nitric oxide. *Exp Physiol* (1995) **134**: 855–864.
31. Crawley DF, Liu SF, Evans TW, Barnes PJ: Inhibitory role of endothelium-derived nitric oxide in rat and human pulmonary arteries. *Br J Pharmacol* (1990) **101**: 166–170.
32. Liu SF, Crawley DE, Barnes PJ, Evans TW: Endothelium derived nitric oxide inhibits pulmonary vasoconstriction in isolated blood perfused rat lungs. *Am Rev Respir Dis* (1991) **143**: 32–37.
33. Martinez C, Cases E, Vila JM, *et al.*: Influence of endothelial nitric oxide on neurogenic contraction of human pumlonary arteries. *Eur Respir J* (1995) **8**: 1328–1332.
34. Kuo H, Liu S, Barnes PJ: The effect of endogenous nitric oxide on neurogenic plasma exudation in guinea pig airways. *Eur J Pharmacol* (1992) **221**: 385–388.
35. Erjefält JS, Erjefält I, Sundler F, Persson CGA: Mucosal nitric oxide may tonically suppress airway plasma exudation. *Am J Respir Crit Care Med* (1994) **150**: 227–232.
36. Bernareggi M, Mitchell JA, Barnes PJ, Belvisi MG: Dual action of nitric oxide on airway plasma leakage. *Am J Respir Crit Care Med* (1997) **155**: 869–874.
37. Lei Y-H, Barnes PJ, Rogers DF: Involvement of hydroxyl radicals in neurogenic airway plasma exudation and bronchoconstriction in guinea pigs *in vivo*. *Br J Pharmacol* (1996) **117**: 449–454.
38. Barnes PJ, Liu SF: Regulation of pulmonary vascular tone. *Phamacol Rev* (1995) **47**: 87–118.
39. Ward JK, Barnes PJ, Tadjkarimi S, Yacoub MH, Belvisi MG: Evidence for involvement of cGMP in neural bronchodilator responses in human trachea. *J Physiol* (1995) **483**: 525–536.
40. Gaston B, Reilly J, Drazen JM, *et al.*: Endogenous nitrogen oxides and bronchodilator S-nitrosolthiols in human airways. *Proc Natl Acad Sci USA* (1993) **90**: 10 957–10 961.
41. Dupuy PM, Shore SA, Drazen JM, Frostell C, Hill WA, Zapol WM: Bronchodilator action of inhaled nitric oxide in guinea pigs. *J Clin Invest* (1992) **90**: 421–428.
42. Högman M, Frostell CG, Hedenström H, Hedenstierna G: Inhalation of nitric oxide modulates adult human bronchial tone. *Am Rev Respir Dis* (1993) **148**: 1474–1478.
43. Sanna A, Kurtansky A, Veriter C, Stanescu D: Bronchodilator effect of inhaled nitric oxide in healthy men. *Am J Respir Crit Care Med* (1994) **150**: 1702–1709.
44. Kacmarek RM, Ripple R, Cockrill BA, Bloch KJ, Zapol WM, Johnson DC: Inhaled nitric oxide: a bronchodilator in mild asthmatics with methacholine-induced bronchospasm. *Am J Respir Crit Care Med* (1996) **153**: 128–135.
45. Ramnarine SI, Khawaja AM, Barnes PJ, Rogers DF: Nitric oxide inhibition of basal and neurogenic mucus secretion. *Br J Pharmacol* (1996) **118**: 998–1002.
46. Nagaki M, Shimura MN, Irokawa T, Sasaki T, Shirato K: Nitric oxide regulation of glycoconjugate secretion from feline and human airways *in vitro*. *Respir Physiol* (1995) **102**: 89–95.
47. Adler KB, Fischer BN, Li H, Choe NH, Wright DT: Hypersecretion of mucin in response to inflammatory mediators by guinea pig tracheal epithelial cells *in vitro* is blocked by inhibition of nitric oxide synthase. *Am J Respir Cell Mol Biol* (1995) **13**: 526–530.
48. Jain B, Lubinstein I, Robbins RA, Leise KL, Sisson JH: Modulation of airway epithelial cell ciliary beat frequency by nitric oxide. *Biochem Biophys Res Commun* (1993) **191**: 83–88.
49. Belvisi MG, Ward JR, Mitchell JA, Barnes PJ: Nitric oxide as a neurotransmitter in human airways. *Arch Int Pharmacodyn Ther* (1995) **329**: 111–120.
50. Belvisi MG, Stretton CD, Barnes PJ: Nitric oxide is the endogenous neurotransmitter of bronchodilator nerves in human airways. *Eur J Pharmacol* (1992) **210**: 221–222.
51. Belvisi MG, Stretton CD, Miura M, *et al.*: Inhibitory NANC nerves in human tracheal smooth muscle: a quest for the neurotransmitter. *J Appl Physiol* (1992) **73**: 2505–2510.
52. Bai TR, Bramley AM: Effect of an inhibitor of nitric oxide synthase on neural relaxation of human bronchi. *Am J Physiol* (1993) **264**: L425–L430.
53. Belvisi MG, Stretton CD, Barnes PJ: Nitric oxide as an endogenous modulator of cholinergic neurotransmission in guinea pig airways. *Eur J Pharmacol* (1991) **198**: 219–221.
54. Belvisi MG, Miura M, Stretton CD, Barnes PJ: Endogenous vasoactive intestinal peptide and

nitric oxide modulate cholinergic neurotransmission in guinea pig trachea. *Eur J Pharmacol* (1993) **231**: 97–102.
55. Ward JK, Belvisi MG, Fox AJ, et al.: Modulation of cholinergic neural bronchoconstriction by endogenous nitric oxide and vasoactive intestinal peptide in human airways *in vitro*. *J Clin Invest* (1993) **92**: 736–743.
56. Brave SR, Hobbs AJ, Gibson A, Tucker JF: The influence of L-N^G-nitro-arginine on field stimulation induced contractions and acetylcholine release in guinea pig isolated tracheal smooth muscle. *Biochem Biophys Res Commun* (1991) **179**: 1017–1022.
57. Belvisi MG, Ward JK, Tadjarimi S, Yacoub MH, Barnes PJ: Inhibitory NANC nerves in human airways: differences in disease and after extrinsic denervation. *Am Rev Respir Dis* (1993) **147**: A286
58. Wei X, Charles IG, Smith A, et al.: Altered immune responses in mice lacking inducible nitric oxide synthase. *Nature* (1995) **375**: 408–411.
59. Laubach VE, Shesely EG, Smithies O, Sherman PA: Mice lacking inducible nitric oxide synthase are not resistant to lipopolysaccharide induced death. *Proc Natl Acad Sci USA* (1995) **92**: 10688–10692.
60. Taylor-Robinson AW, Phillips RS, Severin A, Moncada S, Liew FY: The role of TH1 and TH2 cells in a rodent malaria infection. *Science* (1993) **260**: 1931–1934.
61. Barnes PJ, Liew FY: Nitric oxide and asthmatic inflammation. *Immunol Today* (1995) **16**: 128–130.
62. Ferreira HHA, Medeiros MV, Lima CSP, et al.: Inhibition of eosinophil chemotaxis by chronic blockade of nitric oxide biosynthesis. *Eur J Pharmacol* (1996) **310**: 201–207.
63. Gustaffsson LE, Leone AM, Persson M, Wiklund NP, Moncada S: Endogenous nitric oxide is present in the exhaled air of rabbits, guinea-pigs and humans. *Biochem Biophys Res Commun* (1991) **181**: 852–857.
64. Borland C, Cox Y, Higenbottam T: Measurement of exhaled nitric oxide in man. *Thorax* (1993) **48**: 1160–1162.
65. Alving K, Weitzberg E, Lundberg JM: Increased amount of nitric oxide in exhaled air of asthmatics. *Eur Respir J* (1993) **6**: 1268–1270.
66. Kharitonov SA, Yates D, Robbins RA, Logan-Sinclair R, Shinebourne E, Barnes PJ: Increased nitric oxide in exhaled air of asthmatic patients. *Lancet* (1994) **343**: 133–135.
67. Persson MG, Wiklund NP, Gustafsson LE: Endogenous nitric oxide in single exhalation, and the change during exercise. *Am Rev Respir Dis* (1993) **148**: 1210–1214.
68. Persson MG, Midtvedt T, Leone AM, Gustafsson LE: Ca^{2+}-dependent and Ca^{2+}-independent exhaled nitric oxide, presence in germ-free animals and inhibition by arginine analogues. *Eur J Pharmacol* (1994) **264**: 13–20.
69. Leone AM, Gustafsson LE, Francis PL, Persson MG, Wiklund NP, Moncada S: Nitric oxide in exhaled breath in humans: direct GC-MS confirmation. *Biochem Biophys Res Commun* (1994) **201**: 883–887.
70. Robbins RA, Floreani AA, van Essen SG, et al.: Measurment of nitric oxide by three different techniques. *Am J Respir Crit Care Med* (1996) **153**: 1631–1635.
71. Persson MG, Zetterstrom O, Argenius V, Ihre E, Gustafsson LE: Single-breath oxide measurements in asthmatic patients and smokers. *Lancet* (1994) **343**: 146–147.
72. Yates DH, Kharitonov SA, Robbins RA, Thomas PS, Barnes PJ: Effect of a nitric oxide synthase inhibitor and a glucocorticosteroid on exhaled nitric oxide. *Am J Respir Crit Care Med* (1995) **152**: 892–896.
73. Kharitonov SA, Yates DH, Barnes PJ: Regular inhaled budesonide decreases nitric oxide concentration in the exhaled air of asthmatic patients. *Am J Respir Crit Care Med* (1996) **153**: 454–457.
74. Barnes PJ, Kharitonov SA: Exhaled nitric oxide: a new lung function test. *Thorax* (1996) **51**: 218–220.
75. Archer S: Measurement of nitric oxide in biological models. *FASEB J* (1993) **7**: 349–360.
76. Kharitonov SA, Barnes PJ: Effect of pressure and flow on measurement of exhaled and nasal nitric oxide. *Am J Respir Crit Care Med* (1997) **155**: A825.
77. Gerlach H, Rossaint R, Pappert D, Knorr M, Falke KJ: Autoinhalation of nitric oxide after endogenous synthesis in nasopharynx. *Lancet* (1994) **343**: 518–519.

78. Lundberg JON, Weitzberg E, Nordvall SL, Kuylenstierna R, Lundberg JM, Alving K: Primarily nasal origin of exhaled nitric oxide and absence in Kartagener's syndrome. *Eur Respir J* (1994) **8**: 1501–1504.
79. Lundberg JON, Farkas-Szallasi T, Weitzberg E, *et al.*: High nitric oxide production in human paranasal sinuses. *Nature Med* (1995) **1**: 370–373.
80. Kimberley B, Nejadnik B, Giraud GD, Holden WE: Nasal contribution to exhaled nitric oxide at rest and during breathholding in humans. *Am J Respir Crit Care Med* (1996) **153**: 829–836.
81. Kharitonov SA, Barnes PJ: There is no contribution to exhaled nitric oxide during exhalation against resistance or during breath-holding. *Thorax* (1997) **52**: 540–544.
82. Kharitonov S, Chung KF, Evans DJ, O'Connor BJ, Barnes PJ: Increased exhaled nitric oxide in asthma is derived from the lower respiratory tract. *Am J Respir Crit Care Med* (1996) **153**: 1773–1780.
83. Massaro AF, Mehta S, Lilly CM, Kobzik L, Reilly JJ, Drazen JM: Elevated nitric oxide concentrations in isolated lower airway gas of asthmatic subjects. *Am J Respir Crit Care Med* (1996) **153**: 1510–1514.
84. Cremona G, Higenbottam T, Takao M, Hall L, Bower EA: Exhaled nitric oxide in isolated pig lungs. *J Appl Physiol* (1995) **78**: 59–63.
85. Guo FH, de Raeve HR, Rice TW, Stuehr DJ, Thunnissen FBJM, Erzurum SC: Continuous nitric oxide synthesis by inducible nitric oxide synthase in normal human airway epithelium *in vivo*. *Proc Natl Acad Sci USA* (1995) **92**: 7809–7813.
86. Belvisi MG, Barnes PJ, Larkin S, *et al.*: Nitric oxide synthase activity is elevated in inflammatory lung diseases. *Am J Respir Crit Care Med* (1995) **151**: A699.
87. Springall DR, Meng Q, Redington A, Howarth PH, Evans TJ, Polak JM: Inducible nitric oxide synthase in asthmatic airway epithelium is reduced by corticosteroid therapy. *Am J Respir Crit Care Med* (1995) **151**: A833.
88. Du Bois AB, Douglas JS, Leaderer BP, Mohsenin V: The presence of nitric oxide in the nasal cavity of normal humans. *Am J Respir Crit Care Med* (1994) **149**: A197.
89. Robbins RA, Floreani AA, von Essen SG, *et al.*: Measurement of exhaled nitric oxide by three different techniques. *Am J Respir Crit Care Med* (1996) **153**: 1631–1635.
90. Jatakanon A, Lim S, Chung KF, Barnes PJ: Correlation between exhaled nitric oxide, sputum eosinophils and methacholine responsiveness. *Am J Respir Crit Care Med* (1997) **155**: A819.
91. Kharitonov SA, O'Connor BJ, Evans DJ, Barnes PJ: Allergen-induced late asthmatic reactions are associated with elevation of exhaled nitric oxide. *Am J Respir Crit Care Med* (1995) **151**: 1894–1899.
92. Garnier P, Fajac I, Dessanges JF, Dall'Ava-Santucci J, Lockhart A, Dinh-Xuan AT: Exhaled nitric oxide during acute changes in airways calibre in asthma. *Eur Respir J* (1996) **9**: 1134–1138.
93. Yates DH, Kharitonov SA, Scott DM, Worsdell M, Barnes PJ: Short and long acting β_2-agonists do not alter exhaled nitric oxide in asthma. *Eur Respir J* (1997) **10**: 1483–1488.
94. Endo T, Uchida Y, Nomura A, Ninomiya H, Sakamoto T, Hasegawa S: Increased production of nitric oxide in the immediate and late response models of guinea pig experimental asthma. *Am J Respir Crit Care Med* (1995) **151**: A177.
95. Massaro AF, Gaston B, Kita D, Fanta C, Stamler J, Drazen JM: Expired nitric oxide levels during treatment for acute asthma. *Am J Respir Crit Care Med* (1995) **152**: 800–803.
96. Kharitonov SA, Yates D, Robbins RA, Logan-Sinclair R, Shinebourne EA, Barnes PJ: Endogenous nitric oxide is increased in the exhaled air of asthmatic patients. *Am J Respir Crit Care Med* (1994) **149**: A198.
97. Kharitonov SA, Yates DH, Chung KF, Barnes PJ: Changes in the dose of inhaled steroid affect exhaled nitric oxide levels in asthmatic patients. *Eur Respir J* (1996) **9**: 196–201.
98. Kharitonov SA, Evans DJ, Barnes PJ, O'Connor BJ: Bronchial provocation challenge with histamine or adenosine 5' monophosphate does not alter exhaled nitric oxide in asthma. *Am J Respir Crit Care Med* (1995) **151**: A125.
99. Kharitonov SA, Wells AU, O'Connor BJ, Hansell DM, Cole PJ, Barnes PJ: Elevated levels of exhaled nitric oxide in bronchiectasis. *Am J Respir Crit Care Med* (1995) **151**: 1889–1893.
100. Yates DH, Kharitonov SA, Thomas PS, Barnes PJ: Endogenous nitric oxide is decreased in

asthmatic patients by an inhibitor of inducible nitric oxide synthase. *Am J Respir Crit Care Med* (1996) **154**: 247–250.
101. Haynes WG, Noon JP, Walker BR, Webb DJ: Inhibition of nitric oxide synthesis increases blood pressure in healthy humans. *J Hypertens* (1993) **11**: 1375–1380.
102. Stammler JS, Loh E, Roddy M, Currie XE, Creager MA: Nitric oxide regulates broad systemic and pulmonary vascular resistance in normal humans. *Circulation* (1994) **89**: 2035–2040.
103. Misko TP, Moore WM, Kasten TP, *et al.*: Selective inhibition of inducible nitric oxide synthase by aminoguanidine. *Eur J Pharmacol* (1993) **233**: 119–125.
104. Hasan K, Heesen BJ, Corbett JA, *et al.*: Inhibition of nitric oxide formation by guanidines. *Eur J Pharmacol* (1993) **249**: 101–106.
105. Lundberg JON, Nordvall SL, Weitzberg E, Kollberg H, Alving K: Exhaled nitric oxide in paediatric asthma and cystic fibrosis. *Arch Dis Child* (1996) **75**: 323–326.
106. Kharitonov SA, Jatakanon A, Lim S, O'Connor BJ, Barnes PJ: Dose-dependent reduction in exhaled nitric oxide in patients with asthma regularly treated with 100mg, 400mg budesonide in double-blind placebo-controlled parallel group study. *Am J Respir Crit Care Med* (1997) **155**: A290.
107. Barnes PJ: Cytokines as mediators of chronic asthma. *Am J Respir Crit Care Med* (1994) **150**: S42–S49.
108. Liu SF, Haddad E, Adcock IM, *et al.*: Induction of NO synthase after sensitization and allergen challenge in Brown Norway rat lung. *Br J Pharmacol* (1997) **121**: 1241–1246.
109. Nijkamp FP, van der Linde HJ, Folkerts G: Nitric oxide synthesis inhalations induce airway hyperresponsiveness in guinea pig *in vivo* and *in vitro*. *Am Rev Respir Dis* (1993) **148**: 727–734.
110. Taylor-Robinson AW, Liew FY, Severn A: Regulation of the immune response by nitric oxide differentially produced by Th1 and Th2 cells. *Eur J Immunol* (1994) **24**: 980–984.
111. Robinson DS, Hamid Q, Ying S, *et al.*: Prednisolone treatment in asthma is associated with modulation of bronchoalveolar lavage cell IL-4, IL-5 and IFN-γ cytokine gene expression. *Am Rev Respir Dis* (1993) **148**: 401–406.

22

Neural Control of Airway Function in Asthma

PETER J. BARNES

INTRODUCTION

There is compelling evidence for the involvement of neural mechanisms in the pathophysiology of asthma, contributing to the symptoms and possibly to the inflammatory response.[1,2] There is a close interrelationship between inflammation and neural responses in the airways, since inflammatory mediators may influence the release of neurotransmitters via activation of sensory nerves leading to reflex effects and via stimulation of prejunctional receptors that influence the release of neurotransmitters.[3] In turn, neural mechanisms may influence the nature of the inflammatory response, either reducing inflammation or exaggerating the inflammatory response (neurogenic inflammation) (Fig. 22.1). Complex interactions between various components of the autonomic nervous system are now recognized. Inflammation may therefore affect neurotransmission in many complex ways. Adrenergic nerves may modulate cholinergic neurotransmission in the airways[4] and sensory nerves may influence neurotransmission in parasympathetic ganglia and at postganglionic nerves.[5] This means that changes in the function of one neural pathway may have effects on other neural pathways.

Autonomic nerves regulate many aspects of airway function, including airway smooth muscle tone, secretions, blood flow, microvascular permeability and the migration and release of inflammatory cells.[1,6] Neural control of human airways is complex and the contribution of neurogenic mechanisms to the pathophysiology of airway disease is still debated. Because changes in bronchomotor tone in asthma occur rapidly, it was suggested many years ago that there might be an abnormality in autonomic neural control of the airways, with an imbalance between excitatory and inhibitory pathways, resulting in excessively reactive airways. Several neural mechanisms are involved in the

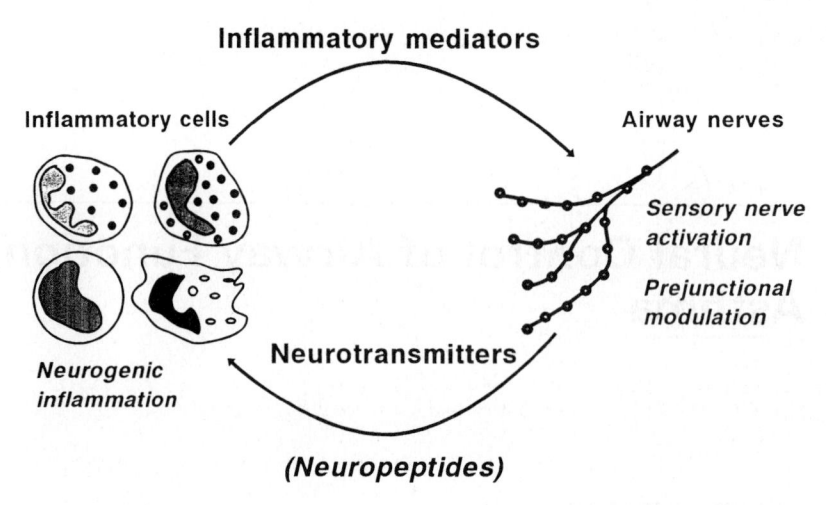

Fig. 22.1 Interaction between inflammation and neural control.

regulation of airway smooth muscle tone (Fig. 22.2) Several types of autonomic defect have been proposed in asthma, including enhanced cholinergic, α-adrenergic and non-adrenergic non-cholinergic (NANC) excitatory mechanisms, or reduced β-adrenergic and NANC bronchodilator mechanisms. Various abnormalities in airway control have been documented in asthma, and it now seems likely that these are secondary to the disease or its treatment rather than primary abnormalities.

Cotransmission

Although it was once the dogma that each nerve has its own unique transmitter, it is now apparent that almost every nerve contains multiple transmitters (Fig. 22.3). Thus airway

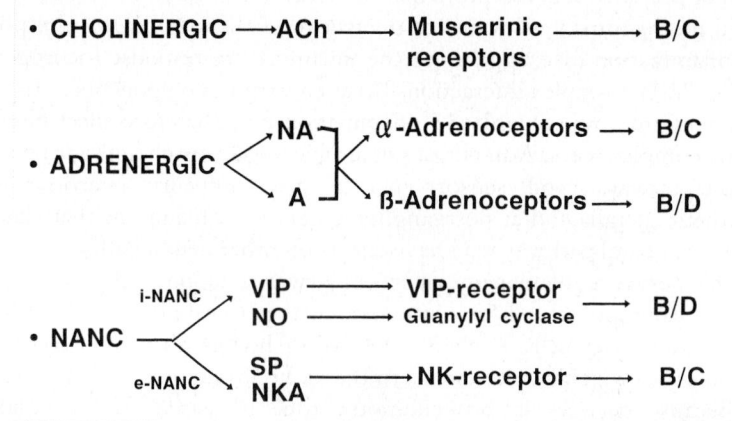

Fig. 22.2 Autonomic control of airway smooth muscle tone. There are neural mechanisms resulting in bronchoconstriction (B/C) and bronchodilatation (B/D). ACh, acetylcholine; NA, noradrenaline; A, adrenaline; VIP, vasoactive intestinal polypeptide; NO, nitric oxide; i-NANC, inhibitory non-adrenergic non-cholinergic nerves; e-NANC, excitatory non-adrenergic non-cholinergic nerves; NK, neurokinin.

22 Neural Control of Airway Function in Asthma

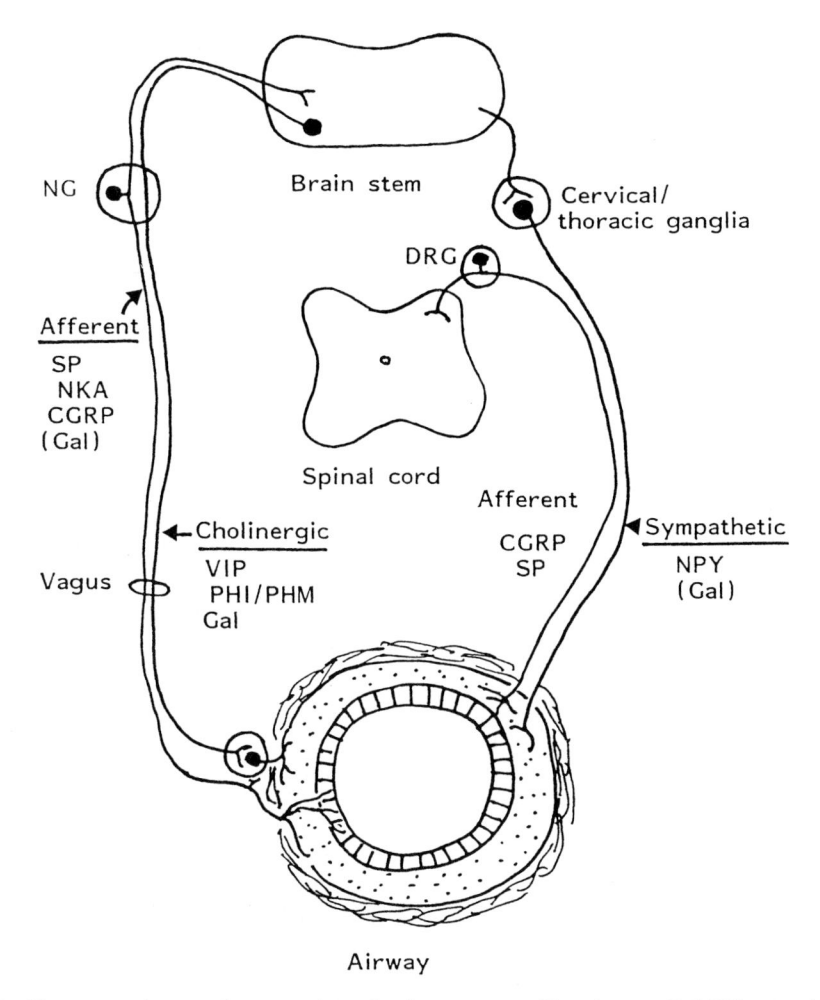

Fig. 22.3 Neurotransmitters and cotransmitters in airway nerves. SP, substance P; NKA, neurokinin A; CGRP, calcitonin gene-related peptide; Gal, galanin; VIP, vasoactive intestinal polypeptide; PHI/PHM, peptide histidine isoleucine/methionine; NPY, neuropeptide Y; DRG, dorsal root ganglion; NG, nodose ganglion.

parasympathetic nerves, in which the primary transmitter is acetylcholine (ACh) also contain the neuropeptides vasoactive intestinal polypeptide (VIP), peptide histidine isoleucine/methionine (PHI/PHM), pituitary adenylate cyclase activating peptide (PACAP), helospectins, galanin and nitric oxide (NO). These cotransmitters may have either facilitatory or antagonistic effects on target cells, or may influence the release of the primary transmitter via prejunctional receptors (Fig. 22.4). Thus VIP modulates the release of ACh from airway cholinergic nerves.[7-9] Sympathetic nerves that release noradrenaline may also release neuropeptide Y (NPY) and enkephalins, whereas afferent nerves (in which the primary transmitter may be glutamate) may contain a variety of peptides, including substance P (SP), neurokinin A (NKA), calcitonin gene-related peptide (CGRP), galanin, VIP and cholecystokinin-octapeptide (see Chapter 24).

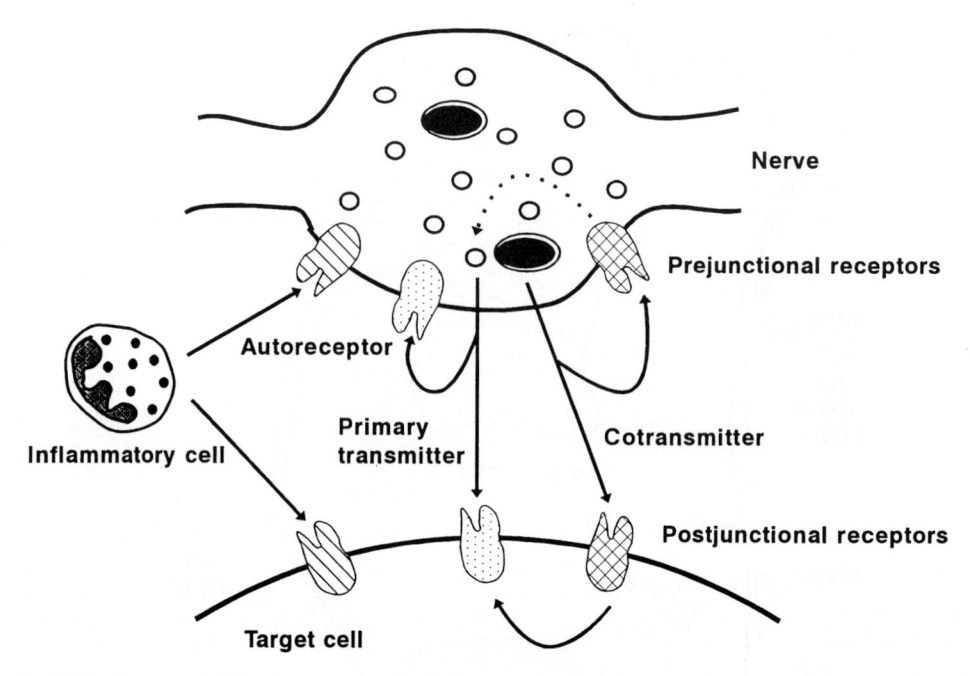

Fig. 22.4 Neuromodulation in airway nerves. Neurotransmitters and cotransmitters act on both prejunctional and postjunctional receptors. Inflammatory mediators may also act on prejunctional receptors.

AFFERENT NERVES

At least three types of afferent nerve are recognized in the lower respiratory tract.[10] Slowly adapting receptors act as stretch receptors in airway smooth muscle, whereas myelinated rapidly adapting (irritant) receptors and unmyelinated C fibres are superficially localized and may be activated by inflammatory and irritant stimuli in the airway lumen. Dolor (pain) is one of the classical signs of inflammation and sensitization and activation of sensory nerves is a common feature of acute and chronic inflammation. It is therefore likely that sensory nerves are involved in asthmatic inflammation, but in the lower respiratory tract the symptoms of cough and chest tightness may reflect sensory nerve activation rather than pain (Fig. 22.5).

Afferent nerves are likely to be activated in asthma by a number of inflammatory mediators. In addition, mediators of inflammation may sensitize airway afferents so that they are more readily triggered by endogenous or exogenous irritants. Relatively little is known about the properties of airway afferent nerves. Studies in animals *in vivo* are confounded by the effect of reflexes and anaesthetics, whereas studies in humans are difficult to conduct. An *in vitro* single-fibre model in guinea-pig proximal airways has been used in order to characterize the properties of airway afferent fibres.[11] We have demonstrated the presence of myelinated Aδ fibres, which respond to mechanical stimulation and low pH solutions (protons), and unmyelinated C fibres that respond to capsaicin, bradykinin and hypertonic saline, but not to histamine, serotonin or prostaglandins.[11-13] Both prostacyclin and platelet-activating factor appear to sensitize afferent

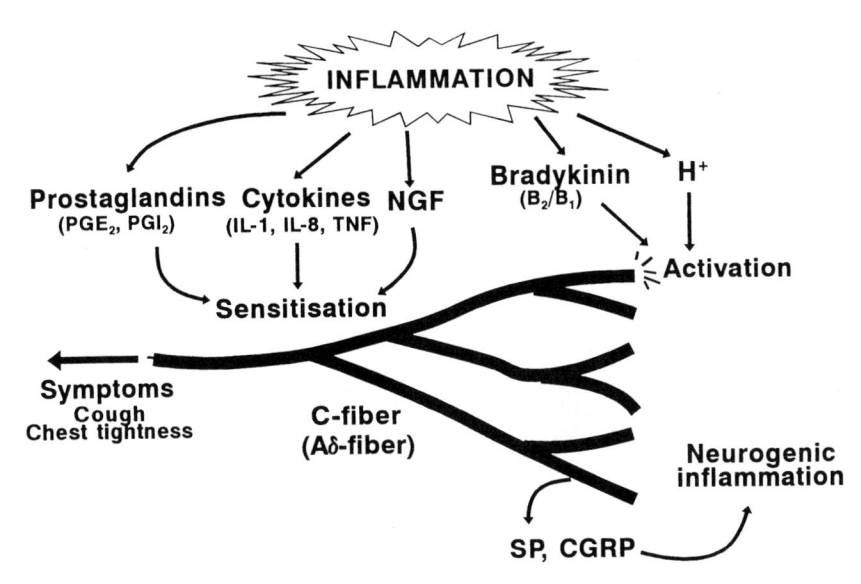

Fig. 22.5 C fibres in airways may be sensitized and activated by multiple inflammatory mediators, resulting in symptoms such as cough and chest tightness. Release of neuropeptides, such as substance P (SP) and calcitonin gene-related peptide (CGRP), may also result in neurogenic inflammation. PG, prostaglandin; IL, interleukin; TNF, tumour necrosis factor; NGF, nerve growth factor.

nerves to the activating effects of bradykinin.[13] As well as activating C fibres bradykinin itself appears to have a sensitizing effect, and this may underlie the effect of angiotensin-converting enzyme (ACE) inhibitors in inducing cough.[14] The capsaicin antagonist, capsazepine, inhibits capsaicin-induced activation of C fibres, as expected, but also inhibits activation by protons, suggesting that there is release of an endogenous activator of capsaicin-binding sites.[15]

Chronic inflammation leads to hyperaesthesia and lowered threshold of activation of sensory nerves. The molecular basis of hyperalgesia is an area of active research, although there is little information about this in airway sensory nerves.[16]

Afferent nerves may play an important role in the symptoms of asthma (cough, chest tightness), but may also be involved in the activation of cholinergic reflexes and in the release of pro-inflammatory peptide neurotransmitters.

PARASYMPATHETIC NERVES

Parasympathetic control of airways

Cholinergic nerve fibres travel down the vagus nerve and relay in parasympathetic ganglia, which are located within the airway wall.[1,6,17,18] From these ganglia short postganglionic fibres travel to airway smooth muscle and submucosal glands (Fig. 22.6). In animals, electrical stimulation of the vagus nerve causes release of ACh from cholinergic nerve terminals, with activation of muscarinic cholinergic receptors on

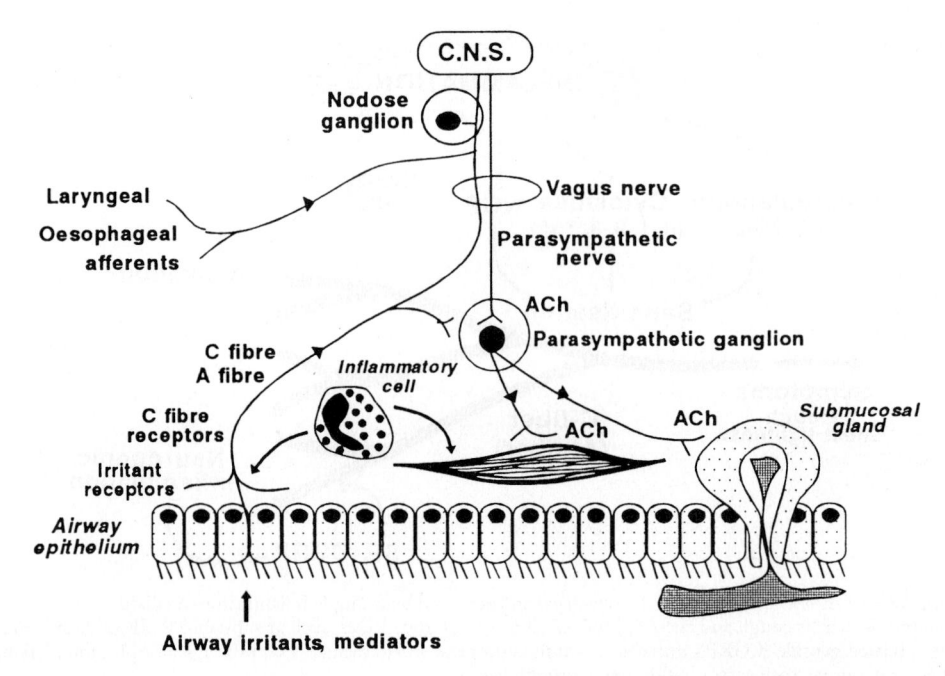

Fig. 22.6 Cholinergic control of airway smooth muscle. Preganglionic and postganglionic parasympathetic nerves release acetylcholine (ACh) and can be activated by airway and extrapulmonary afferent nerves.

smooth muscle and gland cells, which results in bronchoconstriction and mucus secretion. Prior administration of a muscarinic receptor antagonist, such as atropine, prevents vagally induced bronchoconstriction.

Cholinergic innervation is greatest in large airways and diminishes peripherally.[17,19] Studies in animals have demonstrated that cholinergic nerve effects are greatest in large airways and minimal in small airways. Receptor mapping studies have demonstrated a high density of muscarinic receptors in smooth muscle of large airways but few in peripheral airways of ferrets,[20] although in human airways muscarinic receptors are also seen in peripheral airways.[21] In humans, studies that have tried to distinguish large and small airway effects have shown that cholinergic bronchoconstriction predominantly involves larger airways, whereas β-adrenergic agonists are equally effective in large and small airways.[22] This relative diminution of cholinergic control in small airways may have important clinical implications, since anticholinergic drugs are likely to be less useful than β-agonists when bronchoconstriction involves small airways.

In animals, there is a certain degree of resting bronchomotor tone caused by tonic parasympathetic activity. This tone can be reversed by atropine and enhanced by administration of an inhibitor of acetylcholinesterase (which normally rapidly inactivates ACh released from nerve terminals). Normal human subjects also have resting bronchomotor tone, since atropine causes bronchodilatation.[23]

A wide variety of stimuli are able to elicit reflex cholinergic bronchoconstriction.[18] Sensory afferent endings, which include irritant receptors and unmyelinated nerve endings (C fibres), are found in airway epithelium, larynx and nasopharynx.[24] Sensory receptors may be triggered by many stimuli, including dust, cigarette smoke, mechanical stimulation and chemical mediators such as histamine, prostaglandins and bradykinin, which lead to reflex bronchoconstriction. Reflex bronchoconstriction may be inhibited by anticholinergic drugs, which have a variable effect on airway obstruction depending on the degree of cholinergic reflex bronchoconstriction.

Modulation of cholinergic neurotransmission

Many agonists may modulate cholinergic neurotransmission via prejunctional receptors on postganglionic nerves.[3] Some receptors increase (facilitate) whereas others inhibit the release of ACh. There are differences between species in the presence of prejunctional receptors. In guinea-pig airways prejunctional α_2-adrenergic receptors are inhibitory to neurotransmission,[25] whereas in human airways β_2-adrenergic receptors are inhibitory.[26,27] Sensory nerves may also affect cholinergic neurotransmission, since depletion of sensory neuropeptides by capsaicin pretreatment is inhibitory to cholinergic neurotransmission;[5] tachykinins facilitate neurotransmission at both ganglionic level (via NK_1 receptors) and postganglionic nerves (via NK_2 receptors) (see Chapter 24).

Inflammatory mediators may also influence cholinergic neurotransmission via prejunctional receptors. For example, thromboxane and prostaglandin (PG)D_2 facilitate ACh release from postganglionic nerves in the airways. Facilitation may also occur at parasympathetic ganglia in the airways; these structures are surrounded by inflammatory cells and have an afferent nerve input. There is evidence for facilitated transmission in sensitized animals exposed to allergen, although the mediators are not yet identified.[28] Electrophysiological recordings show a prolonged potentiation of neurotransmission in ganglia after allergen exposure in sensitized guinea-pigs.[29] This may be mimicked by PGD_2. Histamine, surprisingly, has an inhibitory effect on ganglionic neurotransmission and on postganglionic nerves in guinea-pig and human airways, mediated through H_3 receptors.[30,31] These receptors are activated by low concentrations of histamine and may represent a safety device to inhibit reflex bronchoconstriction, whereas larger concentrations of histamine may override this mechanism and cause bronchoconstriction directly by acting on H_1 receptors in airway smooth muscle.

The molecular mechanisms responsible for neuromodulation of cholinergic neurotransmission in the airways have recently been elucidated. Activation of the prejunctional receptors that modulate ACh release appear to open large-conductance calcium-activated potassium channels (maxi-K channels), since blockade of these channels by charybdotoxin and ibriotoxin abolishes the neuromodulatory effect of several agonists in guinea-pig and human airways.[32]

Muscarinic receptors

Cholinergic effects on the airways are mediated by muscarinic receptors on target cells in the airways. Five subtypes of muscarinic receptor have now been cloned and M_1, M_2, M_3 and M_4 receptors have been identified in lungs.[33] The contractile response of airway smooth muscle to ACh is mediated via M_3 receptors in most species, including humans, since the selective antagonists 4-diphenylacetoxy-N-methylpiperidine methiodide (4-

DAMP) and hexahydro-siladifenidol are selective potent inhibitors. Autoradiographic studies have demonstrated the expression of M_3 receptors in airway smooth muscle of large and small human airways.[21] This has been confirmed by *in situ* hybridization studies with M_3-specific cDNA probes.[34] Pharmacological studies indicate that the receptor which mediates the contractile response to cholinergic agonists in airway smooth muscle is an M_3 receptor coupled to phosphoinositide hydrolysis.[35] Airway smooth muscle muscarinic receptor activation results in rapid phosphoinositide hydrolysis[36,37] and the formation of inositol 1,4,5-trisphosphate, which releases calcium ions from intracellular stores (Fig. 22.7).

Binding studies of airway smooth muscle, however, indicate a preponderance of M_2 receptors, which inhibit adenylyl cyclase.[38] Recent studies indicate that these M_2 receptors may play a role in functional antagonism and counteract the bronchodilator action of β-agonists in some species,[39,40] although this has not been seen in guinea-pig and human airway smooth muscle.[41,42] The functional role of M_2 receptors in these species is therefore uncertain, but it is possible that these receptors may regulate some other function, such as proliferative responses.

Muscarinic receptors that inhibit the release of ACh from cholinergic nerves have been described in the airways of several species including humans.[3,33] These muscarinic receptors appear to be located prejunctionally on postganglionic parasympathetic nerves and have a powerful inhibitory influence on ACh release (Fig. 22.8). Muscarinic autoreceptors have been demonstrated in guinea-pig, cat, rat, dog and human airways *in vitro*.[3,43] Measurement of ACh release from airway cholinergic nerves has demonstrated that in humans the prejunctional receptor is an M_2 receptor,[44] whereas in guinea-pigs an M_4 receptor may be involved.[45] Non-selective anticholinergics, such as ipratropium

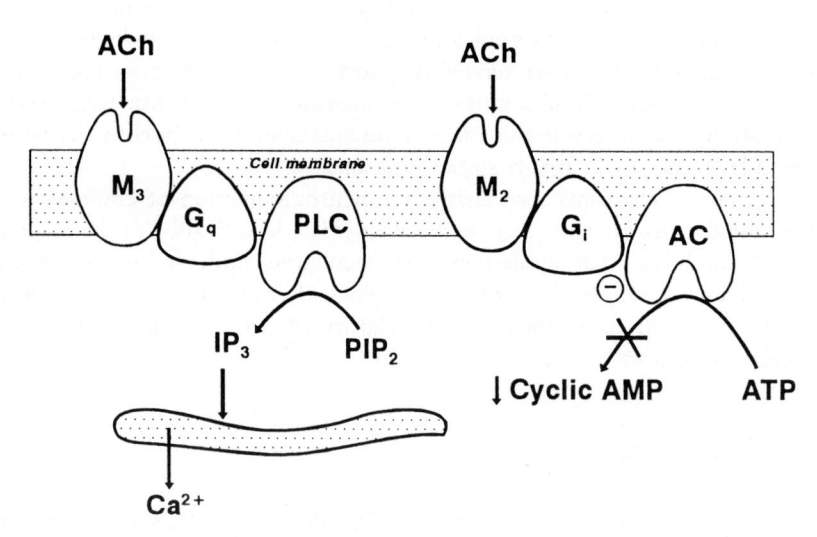

Fig. 22.7 Muscarinic receptor subtypes in airway smooth muscle. M_3 receptors are coupled via a G protein (G_q) to phospholipase C (PLC), resulting in the generation of inositol 1,4,5-trisphosphate (IP_3) and release of calcium ions (Ca^{2+}) from intracellular stores. By contrast, M_2 receptors are coupled via an inhibitory G protein (G_i) to the inhibition of adenylyl cyclase (AC), with a fall in cyclic AMP concentration. This may oppose the increase in cyclic AMP stimulated by β-agonists. PIP_2, phosphoinositide bisphosphate.

22 Neural Control of Airway Function in Asthma

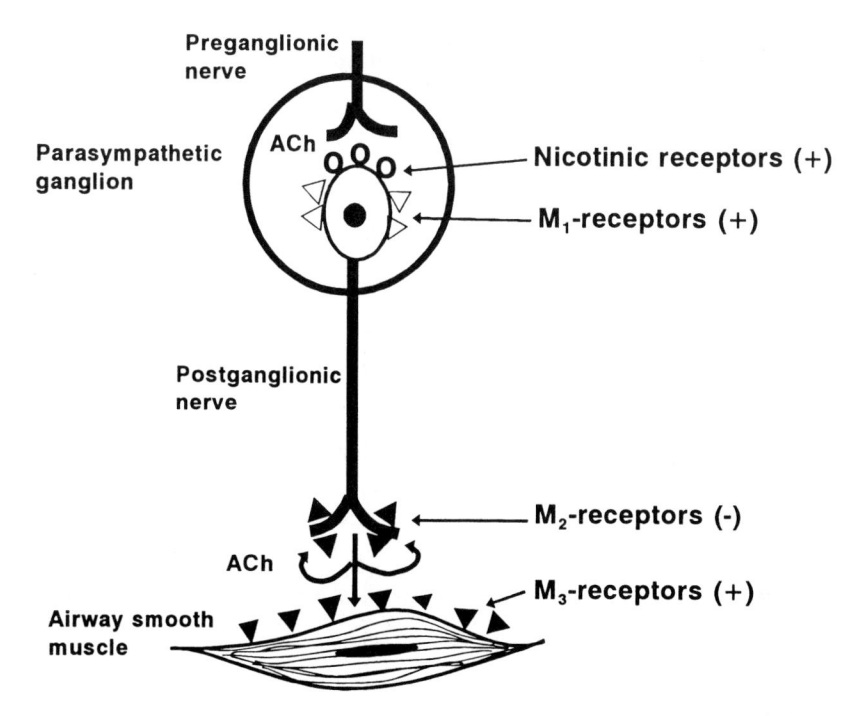

Fig. 22.8 Muscarinic receptor subtypes in airways. M_2 receptors on postganglionic cholinergic nerve terminals inhibit the release of acetylcholine (ACh), thus reducing the stimulation of postjunctional M_3 receptors, which constrict airway smooth muscle.

bromide, thus increase ACh release via an inhibitory effect on prejunctional M_2 receptors.[44] In normal human subjects pilocarpine, which selectively stimulates the prejunctional receptors, has an inhibitory effect on cholinergic reflex bronchoconstriction induced by SO_2, suggesting that these inhibitory receptors are functional *in vivo* and, presumably, serve to limit cholinergic bronchoconstriction.[46,47] In asthmatic patients pilocarpine has no such inhibitory action, indicating that there might be some dysfunction of the autoreceptor, which would result in exaggerated cholinergic reflex bronchoconstriction.[46,48] A functional defect in muscarinic autoreceptors may also explain why β-blockers produce such marked bronchoconstriction in asthmatic patients, since any increase in cholinergic tone due to blockade of inhibitory β-adrenergic receptors on cholinergic nerves would normally be switched off by M_2 receptors in the nerves and a lack of such receptors may lead to increased ACh release, resulting in exaggerated bronchoconstriction.[49] Support for this idea is provided by the protective effect of oxitropium bromide against propranolol-induced bronchoconstriction in asthmatic patients.[50,51]

The mechanism by which M_2 autoreceptors on cholinergic nerves may become dysfunctional is not certain. It is possible that chronic inflammation in airways may lead to downregulation of M_2 receptors, which may have an important functional effect if the density of prejunctional muscarinic receptors is relatively low. Experimental studies have demonstrated that influenza virus may inactivate M_2 rather than M_3 receptors.[52] This may

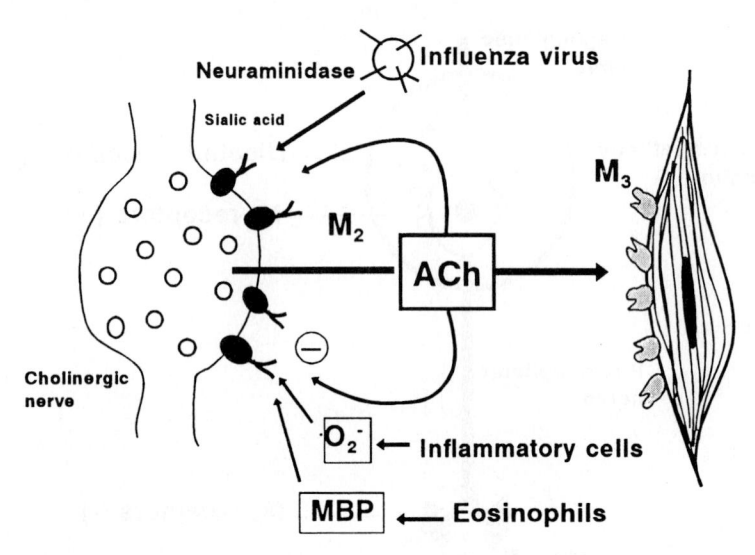

Fig. 22.9 Muscarinic autoreceptors in disease. There may be a defect in M_2-receptor function in asthma, possibly due to effects of neuraminidase released in viral infection or the effects of major basic protein (MBP) or superoxide anions (O_2^-) released by eosinophils. This would lead to increased release of acetylcholine (ACh) and therefore enhanced cholinergic neural bronchoconstriction.

be related to the action of viral neuraminidase on sialic acid residues of M_2 receptors.[53] This provides a possible explanation for increased airway reactivity after influenza infections. Inflammatory mediators such as major basic protein may also result in impaired function of prejunctional M_2 receptors, leading to increased ACh release when cholinergic reflexes are activated[54] (Fig. 22.9). Allergen challenge results in impaired prejunctional M_2-receptor function in guinea-pigs and several mechanisms may be involved.[55] Protein kinase C activation, which may occur after exposure to inflammatory mediators, and pro-inflammatory cytokines result in reduced transcription of the M_2-receptor gene.[56,57] This mechanism may contribute to cholinergic bronchoconstriction in acute exacerbations of asthma.

Role in asthma

Because many of the stimuli that produce bronchospasm in asthma activate sensory nerves and reflex bronchoconstriction in animals, it was logical to suggest that asthma may be due to exaggerated cholinergic reflex mechanisms. There is some evidence that cholinergic tone is increased in asthmatic airways.[58] There are several mechanisms by which cholinergic tone might be increased in asthma.

(1) Increased afferent receptor stimulation by inflammatory mediators, such as histamine or prostaglandins, which may be released from mast cells and other inflammatory cells in the asthmatic airway or from bradykinin formed from precursors in exuded plasma.

(2) Increased release of ACh from cholinergic nerve terminals by an action on cholinergic nerve endings themselves, or by an increase in nerve traffic through cholinergic ganglia (local airway reflex).[3]
(3) Abnormal muscarinic receptor expression, via either an increase in M_3 receptors or reduction in M_2 receptors. There is no evidence for increased M_1- or M_3-receptor expression in asthmatic lung,[59] but there is functional evidence for a defect in M_2-receptor function that may be secondary to the inflammatory process, as discussed above.
(4) Decrease in the neuromodulators (VIP, NO) that have a 'braking' effect on neurotransmission (see Chapter 24).

The effect of ACh on asthmatic airways is exaggerated, and is a manifestation of the non-specific hyperresponsiveness of the airways so characteristic of asthma. However, asthmatic airways are hyperresponsive to many spasmogens in addition to ACh; mediators such as histamine, leukotrienes and prostaglandins have a *direct* contractile effect on bronchial smooth muscle that is not blocked by anticholinergic drugs (see Fig. 22.7). Anticholinergic agents will only counteract the cholinergic reflex component of bronchoconstriction, which may be less prominent in human airways than animal studies had indicated. By contrast β_2-receptor agonists reverse bronchoconstriction irrespective of the mechanism, since they act as functional antagonists.

Although anticholinergics are less effective than β_2-receptor agonists as bronchodilators in chronic asthma, several studies suggest that they may be almost as effective in acute exacerbations,[60-62] indicating that cholinergic bronchoconstriction is the major component of airway narrowing in asthma attacks (and, by implication, in fatal asthma). In patients with chronic asthma that has been poorly controlled, there is a progressive decline in lung function over the years,[63,64] which presumably results from chronic inflammation. Vagal tone increases the airway narrowing further, and for geometric reasons will have a greater effect on airway resistance in narrowed airways. This may explain why anticholinergics are often of greater use in chronic asthmatic patients with a major element of fixed airway obstruction.

ADRENERGIC CONTROL

The airways are also under adrenergic control, which includes sympathetic nerves (that release noradrenaline), circulating catecholamines (predominantly adrenaline) and α- and β-adrenergic receptors (Fig. 22.10). The fact that β-adrenergic antagonists cause bronchoconstriction in asthmatic patients, but not in normal individuals, suggests that adrenergic control of airway smooth muscle may be abnormal in asthma.

Sympathetic innervation

Although sympathetic bronchodilator nerves have been demonstrated in several species, including cats, dogs and guinea-pigs, most evidence suggests that adrenergic nerves do not control human airway smooth muscle directly.[1] However, sympathetic nerves may

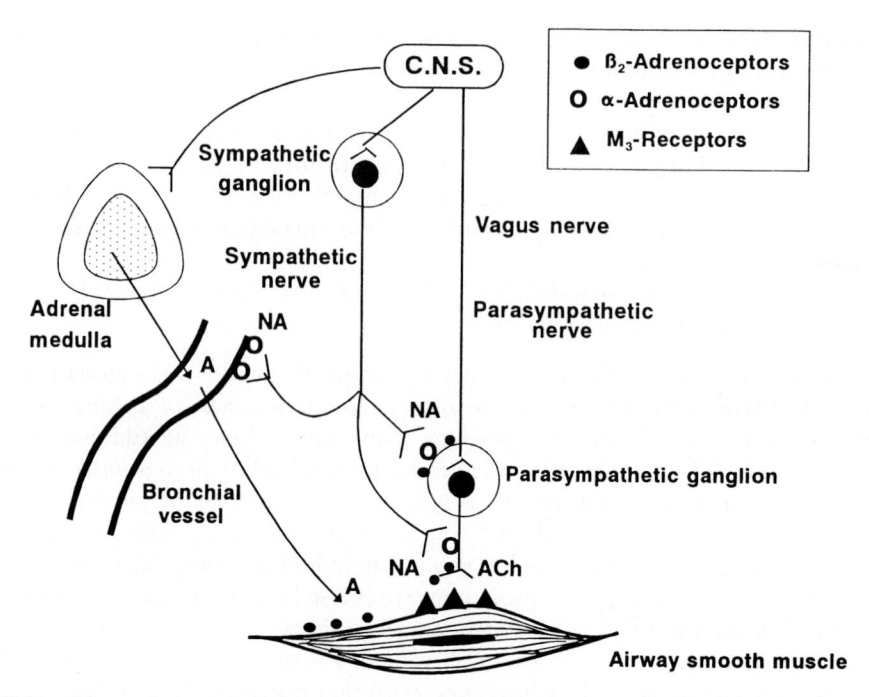

Fig. 22.10 Adrenergic control of airway smooth muscle. Sympathetic nerves release noradrenaline (NA), which may modulate cholinergic nerves at the level of the parasympathetic ganglion or postganglionic nerves, rather than directly at smooth muscle in human airways. Circulating adrenaline (A) is more likely to be important in adrenergic control of airway smooth muscle.

influence cholinergic tone of airway smooth muscle via adrenoceptors localized on parasympathetic ganglia and prejunctionally on postganglionic nerves.[3] Sympathetic nerves have been localized to the vicinity of cholinergic nerve profiles in human airways,[65] but there is no convincing evidence that endogenously released noradrenaline has effects on cholinergic control of human airways, at least *in vitro*.[27] However, sympathetic nerves may play an important role in the regulation of airway blood flow and in mucus secretion.

Circulating catecholamines

Since sympathetic nerves do not directly control airway smooth muscle, it seems probable that circulating catecholamines may play a more important role in regulation of bronchomotor tone[66] (see Chapter 23). Although the catecholamines, noradrenaline, adrenaline and dopamine, are present in the circulation, only adrenaline has physiological effects, and is secreted by the adrenal medulla. Since β-blockers cause bronchoconstriction in asthmatic patients, but not in normal subjects, this suggests that adrenergic drive to the airways is important in defending against bronchoconstriction; in the absence of adrenergic innervation this drive might be provided by circulating adrenaline. However, plasma adrenaline concentrations are not elevated in asthmatic patients, even in those who exhibit bronchoconstriction to intravenous propranolol. Even during acute exacerbations of asthma there is no elevation of plasma adrenaline, suggesting that severe bronchoconstriction is not a stimulus to adrenaline release. Furthermore, plasma adrenaline is not elevated during provoked bronchoconstriction induced by a variety of challenges.

β-Adrenergic receptors

β-Adrenergic receptors regulate many aspects of airway function, including airway smooth muscle tone.[67] Autoradiographic mapping has demonstrated that β-receptors are widely distributed in lung and are localized to many cell types, including airway smooth muscle from trachea down to terminal bronchioles.[68] In some species both β_1- and β_2-receptors have been demonstrated functionally in airway smooth muscle; the presence of β_1-receptors is related to the presence of sympathetic innervation of airway smooth muscle.[69] The lack of a functional sympathetic innervation is consistent with the autoradiographic evidence that in humans only β_2-receptors are expressed in smooth muscle at all airway levels.[68] *In situ* hybridization shows widespread expression of the β_2-receptor gene in human airways.[70] The amount of β_2-receptor mRNA in airway smooth muscle is high relative to the low receptor density; this may indicate a rapid turnover of β_2-receptors and may account for the relative resistance of airway smooth muscle to the development of tolerance. Functional studies also demonstrate that relaxation of both central and peripheral human airways is mediated solely via β_2-receptors.[71,72] β_3-Receptors have been demonstrated in canine bronchi *in vitro*,[73] but there is no evidence for these receptors in airway smooth muscle of human airways.[74]

The β-agonists act as *functional antagonists* and inhibit or reverse the contractile response, irrespective of the constricting stimulus.[75] This is a property of particular importance in asthma, since several spasmogens are likely to be involved (including leukotriene D_4, histamine, acetylcholine and bradykinin).

The intracellular mechanisms involved in mediating the relaxant effect of β-agonists in airway smooth muscle have been extensively investigated. Stimulation of β-receptors via an increase in intracellular cyclic 3′,5′-adenosine monophosphate (cAMP) concentrations activates protein kinase A, which phosphorylates several proteins that result in relaxation.[76] In airway smooth muscle protein kinase A inhibits myosin light chain phosphorylation, inhibits phosphoinositide hydrolysis and promotes Ca^{2+}/Na^+ exchange,[77] thus resulting in a fall in intracellular $[Ca^{2+}]$, and stimulates Na^+/K^+ ATPase.[78] These effects are only observed at relatively high concentrations of β-agonist when maximal relaxation responses have been exceeded. An important effect of β-agonists is the opening of large-conductance calcium-activated potassium (maxi-K) channels. Charybdotoxin and iberiotoxin inhibit the bronchodilator responses to β-agonists and to other agents that elevate cAMP.[79,80] These effects are observed at low concentrations of β-agonists in human airways *in vitro*, suggesting that this is a major mechanism of airway smooth muscle response to β-agonists.[81] As discussed above, β-receptors may activate maxi-K channels in airway smooth muscle cells directly via the α-subunit of G_s.[82] This suggests that relaxation of airway smooth muscle can occur independently of a rise in intracellular cAMP and may explain why there is a discrepancy between the low concentration of β-agonists needed to relax airway smooth muscle and the relatively high concentrations needed to elevate cAMP concentrations. Furthermore it explains why forskolin, which causes a large increase in intracellular cAMP concentration in airway smooth muscle, is a relatively poor relaxant of airway smooth muscle.[83] There is great interest in this area of research as it is possible that selective agonists of maxi-K channels may be novel bronchodilators, and several such compounds are now identified.

The β-agonists may also modulate neurotransmission in airways via prejunctional receptors on airway parasympathetic nerves.[3] In canine and feline airways, exogenous noradrenaline and endogenously released catecholamines inhibit cholinergic nerve-induced bronchoconstriction to a greater extent than an equivalent contraction induced by acetylcholine, indicating a prejunctional effect.[84] This effect in dogs is mediated via prejunctional β_1-receptors localized to postganglionic cholinergic nerves,[4] although other studies suggest that β_2-receptors are also involved.[85,86] The β-agonists may also modulate neurotransmission in parasympathetic ganglia via an effect on preganglionic nerve endings.[87] In human trachea and bronchi, β-agonists modulate cholinergic neurotransmission *in vitro* via prejunctional β_2-receptors on postganglionic cholinergic nerves.[26,27,88] Although there are close anatomical associations between adrenergic and cholinergic nerves in human airways,[65] stimulation of endogenous noradrenaline release from sympathetic nerves by tyramine has no modulatory effect.[27] It is more likely that circulating adrenaline regulates prejunctional β_2-receptors in human airways. The clinical relevance of prejunctional β_2-receptors in human airways may relate to β-blocker-induced asthma, since β-blockers may inhibit the tonic inhibitory action of circulating adrenaline, resulting in an increase in ACh release, as discussed above.[33]

Abnormalities in asthma

The possibility that β-receptors are abnormal in asthma has been extensively investigated. The suggestion that there is a primary defect in β-receptor function in asthma has not been substantiated and any defect in β-receptors is likely to be secondary to the disease, perhaps as a result of inflammation or as a consequence of adrenergic therapy. Some studies have demonstrated that airways from asthmatic patients fail to relax normally to isoprenaline, suggesting a possible defect in β-receptor function in airway smooth muscle.[89–91] Whether this is due to a reduction in β-receptors, a defect in receptor coupling, or some abnormality in the biochemical pathways leading to relaxation is not yet known, although the density of β-receptors in airway smooth muscle appears to be normal.[92,93] There is no reduction in the density of β_1- or β_2-receptors in asthmatic lung, either at the receptor or mRNA level.[59] In patients who die from an asthma attack, the density of β-receptors in airway smooth muscle may even be increased, despite the reduced bronchodilator response to β-agonists, suggesting a defect in the coupling of β-receptors.[94]

There is some evidence that pro-inflammatory cytokines may affect β_2-receptor function. IL-1β reduces the bronchodilator effect of isoprenaline *in vitro* and *in vivo* and this appears to be due to uncoupling of β_2-receptors due to increased expression of the inhibitory G protein, G_i.[95,96] However, recent studies of β_2-receptor expression in asthmatic airways obtained by biopsy have demonstrated only small defects in coupling after local allergen challenge.[97]

α-Adrenergic receptors

α-Adrenergic receptors that mediate bronchoconstriction have been demonstrated in airways of several species, but may only be demonstrated under certain experimental conditions. There is now considerable doubt about the role of α-receptors in the regulation of tone in human airways, however, since it has proved difficult to demonstrate their presence functionally or by autoradiography,[98] and α-blocking drugs do not appear to be as effective as bronchodilators. It is possible that α-receptors may play an important role in regulating airway blood flow, which may indirectly influence airway responsiveness, and there is some evidence that α-agonists may *reduce* airway narrowing in exercise-induced asthma.[99]

REFERENCES

1. Barnes PJ: Neural control of human airways in health and disease. *Am Rev Respir Dis* (1986) **134**: 1289–1314.
2. Barnes PJ: Is asthma a nervous disease? *Chest* (1995) **107**: 119S–124S.
3. Barnes PJ: Modulation of neurotransmission in airways. *Physiol Rev* (1992) **72**: 699–729.
4. Danser AHJ, van den Ende R, Lorenz RR, Flavahan NA, Vanhoutte PM: Prejunctional beta1-adrenoceptors inhibit cholinergic neurotransmission in canine bronchi. *J Appl Physiol* (1987) **62**: 785–790.
5. Stretton CD, Belvisi MG, Barnes PJ: The effect of sensory nerve depletion on cholinergic neurotransmission in guinea pig airways. *J Pharmacol Exp Ther* (1992) **260**: 1073–1080.
6. Barnes PJ: Neural control of airway function: new perspectives. *Mol Aspects Med* (1990) **11**: 351–423.
7. Ellis JL, Farmer SG: Modulation of cholinergic neurotransmission by vasoactive intestinal peptide and peptide histidine isoleucine in guinea pig tracheal smooth muscle. *Pulmon Pharmacol* (1989) **2**: 107–112.
8. Hakoda H, Ito Y: Modulation of cholinergic neurotransmission by the peptide VIP, VIP antiserum and VIP antagonists in dog and cat trachea. *J Physiol* (1990) **428**: 133–154.
9. Stretton CD, Belvisi MG, Barnes PJ: Modulation of neural bronchoconstrictor responses in the guinea pig respiratory tract by vasoactive intestinal peptide. *Neuropeptides* (1991) **18**: 149–157.
10. Coleridge HM, Coleridge JCG: Afferent nerves in the airways. In Barnes PJ (ed) *Autonomic Control of the Respiratory System*. London, Harwood, 1997, pp 39–58.
11. Fox AJ, Barnes PJ, Urban L, Dray A: An *in vitro* study of the properties of single vagal afferents innervating guinea-pig airways. *J Physiol* (1993) **469**: 21–35.
12. Fox AJ, Barnes PJ, Dray A: Stimulation of afferent fibres in the guinea pig trachea by non-isosmotic and low chloride solutions and its modulation by frusemide. *J Physiol* (1995) **482**: 179–187.
13. Fox AJ, Dray A, Barnes PJ: The activity of prostaglandins and platelet-activating factor on single airway sensory fibres of the guinea pig *in vitro*. *Am J Respir Crit Care Med* (1995) **151**: A110.
14. Fox AJ, Lalloo UG, Belvisi MG, Bernareggi M, Chung KF, Barnes PJ: Bradykinin-evoked sensitization of airway sensory nerves: a mechanism for ACE-inhibitor cough. *Nature Med* (1996) **2**: 814–817.
15. Fox AJ, Urban L, Barnes PJ, Dray A: Effect of capsazepine against capsaicin- and proton-evoked excitation of single airway C-fibres and vagus nerve from the guinea pig. *Neurosci* (1995) **67**: 741–752.
16. Dray A, Urban L, Dickenson A: Pharmacology of chronic pain. *Trends Pharmacol Science* (1994) **15**: 190–197.

17. Richardson JB: Nerve supply to the lung. *Am Rev Respir Dis* (1979) **119**: 785–802.
18. Widdicombe JG, Karlsson, J, Barnes PJ: Cholinergic mechanisms in bronchial hyperresponsiveness and asthma. In Kaliner, MA, Barnes PJ, Persson CGA (eds) *Asthma: its Pathology and Treatment*. New York, Marcel Dekker, 1991, pp 327–356.
19. Barnes PJ: Cholinergic control of airway smooth muscle. *Am Rev Respir Dis* (1987) **136**: S42–S45.
20. Barnes PJ, Basbaum CB, Nadel JA: Autoradiographic localization of autonomic receptors in airway smooth muscle: marked differences between large and small airways. *Am Rev Respir Dis* (1983) **127**: 758–762.
21. Mak JCW, Barnes PJ: Autoradiographic visualization of muscarinic receptor subtypes in human and guinea pig lung. *Am Rev Respir Dis* (1990) **141**: 1559–1568.
22. Ingram RHJ, Wellman JJ, McFadden ERJ, Mead J: Relative contribution of large and small airways to flow limitation in normal subjects before and after atropine and isoproterenol. *J Clin Invest* (1977) **59**: 696–703.
23. Vincent NJ, Knudson R, Leith DF, Macklem PT, Mead J: Factors influencing pulmonary resistance. *J Appl Physiol* (1970) **29**: 236–243.
24. Karlsson J, Sant'Ambrogio G, Widdicombe JG: Afferent neural pathways in cough and reflex bronchconstriction. *J Appl Physiol* (1988) **65**: 1007–1023.
25. Grundstrom N, Andersson RGG, Wikberg JES: Prejunctional alpha-2-adrenoceptors inhibit contraction of tracheal smooth muscle by inhibiting cholinergic neurotransmission. *Life Sci* (1981) **28**: 2981–2986.
26. Bai TR, Lam R, Prasad FYF: Effects of adrenergic agonists and adenosine on cholinergic neurotransmission in human tracheal smooth muscle. *Pulmon Pharmacol* (1989) **1**: 193–199.
27. Rhoden KJ, Meldrum LA, Barnes PJ: Inhibition of cholinergic neurotransmission in human airways by β_2-adrenoceptors. *J Appl Physiol* (1988) **65**: 700–705.
28. McCaig DJ: Comparison of autonomic responses in the trachea isolated from normal and albumin sensitive guinea-pigs. *Br J Pharmacol* (1987) **92**: 809–816.
29. Undem BJ, Riccio MM, Weinreich D, Ellis JL, Myers AC: Neurophysiology of mast cell-nerve interactions in the airways. *Int Arch Allergy Immunol* (1995) **107**: 199–201.
30. Ichinose M, Stretton CD, Schwartz J, Barnes PJ: Histamine H_3-receptors inhibit cholinergic neurotransmission in guinea-pig airways. *Br J Pharmacol* (1989) **97**: 13–15.
31. Ichinose M, Barnes PJ: Inhibitory histamine H_3-receptors on cholinergic nerves in human airways. *Eur J Pharmacol* (1989) **163**: 383–386.
32. Miura M, Belvisi MG, Stretton CD, Yacoub MH, Barnes PJ: Role of K^+ channels in the modulation of cholinergic neural responses in guinea pig and human airways. *J Physiol* (1992) **455**: 1–15.
33. Barnes PJ: Muscarinic receptor subtypes in airways. *Life Sci* (1993) **52**: 521–528.
34. Mak JCW, Baraniuk JN, Barnes PJ: Localization of muscarinic receptor subtype mRNAs in human lung. *Am J Respir Cell Mol Biol* (1992) **7**: 344–348.
35. Roffel AF, Elzinga CRS, Zaagsma J: Muscarinic M_3-receptors mediate contraction of human central and peripheral airway smooth muscle. *Pulmon Pharmacol* (1990) **3**: 47–51.
36. Chilvers ER, Challiss RAJ, Barnes PJ, Nahorski SR: Mass changes of inositol (1,4,5)trisphosphate in trachealis muscle following agonist stimulation. *Eur J Pharmacol* (1989) **164**: 587–590.
37. Chilvers ER, Barnes PJ, Nahorski SR: Characterisation of agonist-stimulated incorporation of [^{3}H]myo-inositol into inositol phospholipids and [^{3}H]inositol phosphate formation in guinea pig tracheal smooth muscle. *Biochem J* (1989) **262**: 739–746.
38. Roffel AF, Elzinga CRS, van Amsterdam RGM, de Zeeuw RA, Zaagsma J: Muscarinic M_2-receptors in bovine tracheal smooth muscle: discrepancies between binding and function. *Eur J Pharmacol* (1988) **153**: 73–82.
39. Yang CM, Chow S, Sung T: Muscarinic receptor subtypes coupled to generation of different second messengers in isolated tracheal smooth muscle cells. *Br J Pharmacol* (1991) **104**: 613–618.
40. Fernandes LB, Fryer AD, Hirschman CA: M_2 muscarinic receptors inhibit isoproterenol-induced relaxation of canine airway smooth muscle. *J Pharmacol Exp Ther* (1992) **262**: 119–126.
41. Roffel AF, Meurs M, Elzinga CRS, Zaagsma J: Muscarinic M_2 receptors do not participate in the functional antagonism between methacholine and isoprenaline in guinea pig tracheal smooth muscle. *Eur J Pharmacol* (1993) **249**: 235–238.

42. Watson N, Magnussen H, Rabe KF: Antagonism of β-adrenoceptor-mediated relaxations of human bronchial smooth muscle by carbachol. *Eur J Pharmacol* (1995) **275**: 307–310.
43. Minette PA, Barnes PJ: Prejunctional inhibitory muscarinic receptors on cholinergic nerves in human and guinea-pig airways. *J Appl Physiol* (1988) **64**: 2532–2537.
44. Patel HJ, Barnes PJ, Takahashi T, Tadjkarimi S, Yacoub MH, Belvisi MG: Characterization of prejunctional muscarinic autoreceptors in human and guinea-pig trachea *in vitro*. *Am J Respir Crit Care Med* (1995) **152**: 872–878.
45. Kilbinger H, van Bardeleben RS, Siefken H: Is the presynaptic muscarinic receptor in guinea-pig trachea an M_2-receptor? *Life Sci* (1993) **52**: 577.
46. Minette PAH, Lammers J, Dixon CMS, McCusker MT, Barnes PJ: A muscarinic agonist inhibits reflex bronchoconstriction in normal but not in asthmatic subjects. *J Appl Physiol* (1989) **67**: 2461–2465.
47. Ayala LE, Ahmed T: Is there a loss of a protective muscarinic receptor mechanism in asthma? *Chest* (1991) **96**: 1285–1291.
48. Barnes PJ: Muscarinic autoreceptors in airways: their possible role in airway disease. *Chest* (1989) **96**: 1220–1221.
49. Barnes PJ: Muscarinic receptor subtypes: implications for lung disease. *Thorax* (1989) **44**: 161–167.
50. Ind PW, Dixon CMS, Fuller RW, Barnes PJ: Anticholinergic blockade of beta-blocker induced bronchoconstriction. *Am Rev Respir Dis* (1989) **139**: 1390–1394.
51. Okayama M, Shen T, Midorikawa J, *et al*.: Effect of pilocarpine on propranolol-induced bronchoconstriction in asthma. *Am J Respir Crit Care Med* (1994) **149**: 76–80.
52. Fryer AD, Jacoby DB: Effect of inflammatory cell mediators on M_2 muscarinic receptors in the lungs. *Life Sci* (1993) **52**: 529–536.
53. Gies J, Landry Y: Sialic acid is selectively involved in the interaction of agonists M_2 muscarinic acetylcholine receptors. *Biochem Biophys Res Commun* (1988) **150**: 673–680.
54. Jacoby DB, Gleich GJ, Fryer AD: Human eosinophil major basic protein is an endogenous allosteric antagonist at the inhibitory muscarinic M_2 receptor. *J Clin Invest* (1993) **91**: 1314–1318.
55. ten Berge REJ, Santing RE, Hamstra TJ, Roffel AF, Zaagsma J: Dysfunction of muscarinic M_2-receptors after the early allergen reaction: possible contribution to bronchial hyperresponsiveness in allergic guinea pigs. *Br J Pharmacol* (1995) **114**: 881–887.
56. Rousseau E, Gagnon J, Lugnier C: Soluble and particulate cyclic nucleotide phosphodiesterases characterized from airway epithelial cells. *FASEB J* (1993) **7**: A143
57. Haddad E-B, Rousell J, Lindsay MA, Barnes PJ: Synergy between TNF-a and IL-1b in inducing down-regulation of muscarinic M_2 receptor gene expression. *J Biol Chem* (1996) **271**: 32 586–32 592.
58. Molfino NA, Slutsky AS, Julia-Serda G, *et al*.: Assessment of airway tone in asthma. *Am Rev Respir Dis* (1993) **148**: 1238–1243.
59. Haddad E-B, Mak JCW, Barnes PJ: Expression of b-adrenergic and muscarinic receptors in human lung. *Am J Physiol* (1996) **270**: L947–L953.
60. Rebuck AS, Chapman KR, Abboud R, *et al*.: Nebulized anticholinergic and sympathomimetic treatment of asthma and chronic obstructive airways disease in the emergency room. *Am J Med* (1987) **82**: 59–64.
61. O'Driscoll BR, Taylor RJ, Horsley MG, Chambers DU, Bernstein A: Nebulised salbutamol with and without ipratropium bromide in acute airflow obstruction. *Lancet* (1989) **i**: 1418–1420.
62. Ward MJ, Macfarlane JT, Davies D: A place for ipratropium bromide in the treatment of severe acute asthma. *Br J Dis Chest* (1985) **79**: 374–378.
63. Brown JP, Greville WH, Finucane KE: Asthma and irreversible airflow obstruction. *Thorax* (1984) **39**: 131–136.
64. Peat JK, Woolcock AJ, Cullen K: Rate of decline of lung function in subjects with asthma. *Eur J Respir Dis* (1987) **70**: 171–179.
65. Daniel EE, Kannan M, Davis C, Posey-Daniel V: Ultrastructural studies on the neuromuscular control of human tracheal and bronchial muscle. *Respir Physiol* (1986) **63**: 109–128.
66. Barnes PJ: Endogenous catecholamines and asthma. *J Allergy Clin Immunol* (1986) **77**: 791–795.

67. Barnes PJ: Beta-adrenergic receptors and their regulation. *Am J Respir Crit Care Med* (1995) **152**: 838–860.
68. Carstairs JR, Nimmo AJ, Barnes PJ: Autoradiographic visualization of beta-adrenoceptor subtypes in human lung. *Am Rev Respir Dis* (1985) **132**: 541–547.
69. Barnes PJ, Nadel JA, Skoogh B, Roberts JM: Characterization of beta-adrenoceptor subtypes in canine airway smooth muscle by radioligand binding and physiological responses. *J Pharmacol Exp Ther* (1983) **225**: 456–461.
70. Hamid QA, Mak JC, Sheppard MN, Corrin B, Venter JC, Barnes PJ: Localization of β_2-adrenoceptor messenger RNA in human and rat lung using *in situ* hybridization: correlation with receptor autoradiography. *Eur J Pharmacol* (1991) **206**: 133–138.
71. Goldie RG, Paterson JW, Spina D, Wale JL: Classification of β-adrenoceptors in human isolated bronchus. *Br J Pharmacol* (1984) **81**: 611–615.
72. Nials AT, Coleman RA, Johnson M, Magnussen H, Rabe RF, Vardey CJ: Effect of β-adrenoceptor agonists in human bronchial smooth muscle. *Br J Pharmacol* (1993) **110**: 1112–1126.
73. Tamoki J, Yamauchi F, Chiyotani A, Yamawaki I, Takeuchi S, Konno K: Atypical β-adrenoceptor (β_3-adrenoceptor) mediated relaxation of canine isolated bronchial smooth muscle. *J Appl Physiol* (1993) **74**: 297–302.
74. Martin CAE, Naline E, Bakdach H, Advenier C: β_3-Adrenoceptor agonists BRL 37344 and SR 58611A do not induce relaxation of human, sheep and guinea-pig airway smooth muscle *in vitro*. *Eur Respir J* (1994) **7**: 1610–1615.
75. Torphy TJ, Rinard GA, Rietola MG, Mayer SE: Functional antagonism in canine tracheal smooth muscle: inhibition by methacholine of the mechanical and biochemical responses to isoproterenol. *J Pharmacol Exp Ther* (1983) **227**: 694–699.
76. Giembycz MA, Raeburn D: Putative substrates for cyclic nucleotide-dependent protein kinases and the control of airway smooth muscle tone. *J Auton Pharmacol* (1991) **166**: 365–398.
77. Twort CAC, van Breemen C: Human airway smooth muscle in cell culture: control of the intracellular calcium store. *Pulmon Pharmacol* (1989) **2**: 45–53.
78. Gunst SJ, Stropp JQ: Effect of Na-K adenosine triphosphatase activity on relaxation of canine tracheal smooth muscle. *J Appl Physiol* (1988) **64**: 635–641.
79. Jones TR, Charette L, Garcia ML, Kaczorowski GJ: Selective inhibition of relaxation of guinea-pig trachea by charybdotoxin, a potent Ca^{++}-activated K^+ channel inhibitor. *J Pharmacol Exp Ther* (1990) **225**: 697–706.
80. Jones TR, Charette L, Garcia ML, Kaczorowski GJ: Interaction of iberiotoxin with β-adrenoceptor agonists and sodium nitroprusside on guinea pig trachea. *J Appl Physiol* (1993) **74**: 1879–1884.
81. Miura M, Belvisi MG, Stretton CD, Yacoub MH, Barnes PJ: Role of potassium channels in bronchodilator responses in human airways. *Am Rev Respir Dis* (1992) **146**: 132–136.
82. Kume H, Graziano MP, Kotlikoff MI: Stimulatory and inhibitory regulation of calcium-activated potassium channels by guanine nucleotide binding proteins. *Proc Natl Acad Sci USA* (1992) **89**: 11 051–11 055.
83. Waldeck B, Widmark E: Comparison of the effects of forskolin and isoprenaline on tracheal, cardiac and skeletal muscles from guinea-pig. *Eur J Pharmacol* (1985) **112**: 349–353.
84. Baker DG, Don U: Catecholamines abolish vagal but not acetylcholine tone in the cat trachea. *J Appl Physiol* (1987) **63**: 2490–2498.
85. Ito Y: Pre- and post-junctional actions of procaterol, a beta-adrenoceptor stimulant, on dog tracheal tissue. *Br J Pharmacol* (1988) **95**: 268–274.
86. Janssen LJ, Daniel EE: Characterization of the prejunctional β-adrenoceptors in canine bronchial smooth muscle. *J Pharmacol Exp Ther* (1991) **254**: 741–749.
87. Skoogh B, Svedmyr N: β_2-Adrenoceptor stimulation inhibits ganglionic transmission in ferret trachea. *Pulmon Pharmacol* (1989) **1**: 167–172.
88. Aizawa H, Inoue H, Miyazaki N, Ikeda T, Shigematsu N, Ito Y: Effects of procaterol, a β_2-adrenoceptor stimulant, on neuroeffector transmission in human bronchial tissue. *Respiration* (1991) **58**: 163–166.
89. Cerrina J, Ladurie ML, Labat C, Raffestin B, Bayol A, Brink C: Comparison of human bronchial muscle response to histamine *in vivo* with histamine and isoproterenol agonists *in vitro*. *Am Rev Respir Dis* (1986) **134**: 57–61.

90. Goldie RG, Spina D, Henry PJ, Lulich KM, Paterson JW: *In vitro* responsiveness of human asthmatic bronchus to carbachol, histamine, β-adrenoceptor agonists and theophylline. *Br J Clin Pharmacol* (1986) **22**: 669–676.
91. Bai TR: Abnormalities in airway smooth muscle in fatal asthma: a comparison between trachea and bronchus. *Am Rev Respir Dis* (1991) **143**: 441–443.
92. Spina D, Rigby PJ, Paterson JW, Goldie RG: Autoradiographic localization of beta-adrenoceptors in asthmatic human lung. *Am Rev Respir Dis* (1989) **140**: 1410–1415.
93. Sharma RK, Jeffery PK: Airway β-adrenoceptor number in cystic fibrosis and asthma. *Clin Sci* (1990) **78**: 409–417.
94. Bai TR, Mak JCW, Barnes PJ: A comparison of beta-adrenergic receptors and *in vitro* relaxant responses to isoproterenol in asthmatic airway smooth muscle. *Am J Respir Cell Mol Biol* (1992) **6**: 647–651.
95. Hakonarson H, Herrick DJ, Serrano PG, Grunstein MM: Mechanism of cytokine-induced modulation of β-adrenoceptor responsiveness in airway smooth muscle. *J Clin Invest* (1996) **97**: 2593–2600.
96. Koto H, Mak JCW, Haddad E-B, *et al.*: Mechanisms of impaired β-adrenergic receptor relaxation by interleukin-1β *in vivo* in rat. *J Clin Invest* (1996) **98**: 1780–1787.
97. Penn RB, Shaver JR, Zangrilli JG, *et al.*: Effects of inflammation and acute β-agonist inhalation on $β_2$-AR signaling in human airways. *Am J Physiol* (1996) **271**: L601–L608.
98. Spina D, Rigby PJ, Paterson JW, Goldie RG: α-Adrenoceptor function and autoradiographic distribution in human asthmatic lung. *Br J Pharmacol* (1989) **97**: 701–708.
99. Dinh-Xuan AT, Chaussain M, Regnard J, Lockart A: Pretreatment with an inhaled alpha 1-adrenergic agonist, methoxamine, reduces exercise-induced asthma. *Eur Respir J* (1989) **2**: 409–414

23

Humoral Control of Airway Tone

NEIL C. THOMSON

INTRODUCTION

Airway smooth muscle tone can be altered by a variety of stimuli, including hormones and vasoactive peptides reaching the lungs from the bloodstream, neurotransmitters released from nerve endings and molecules released locally from other cells within the airways (Fig. 23.1). The physical properties and the chemical and biological content of inspired air can also influence airway function. The degree of airway narrowing or relaxation produced by these stimuli depends on the amount, contractility and length–tension relationship of the smooth muscle, the loads opposing shortening produced by surrounding structures and by the thickness of the airway wall.[1] This chapter reviews the role of circulating humoral factors and oxygen tension in the regulation of airway tone in normal subjects and asthmatic patients.

VASOACTIVE PEPTIDES

Circulating catecholamines

Circulating adrenaline is released from the adrenal medulla into the circulation and may reduce bronchial smooth muscle tone directly by stimulating β_2-adrenergic receptors on airway smooth muscle or indirectly by reducing acetylcholine release from cholinergic nerves.[2] The lack of a bronchoconstrictor effect of β-adrenergic antagonists in normal subjects suggests that in this group basal concentrations of circulating adrenaline are

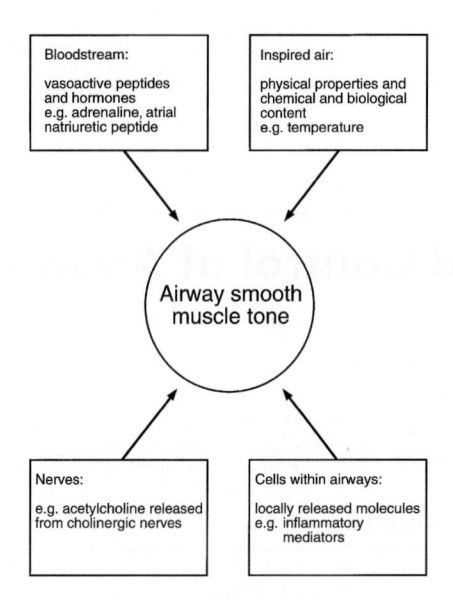

Fig. 23.1 Factors influencing airway smooth muscle tone.

probably not important in the regulation of resting bronchomotor tone. In contrast, β-adrenergic antagonists cause bronchoconstriction in some asthmatic patients, which in the absence of an important sympathetic nerve supply to airway smooth muscle suggests a role for basal concentrations of circulating adrenaline in the maintenance of airway tone in asthma, perhaps particularly in those patients in whom resting airway calibre is already reduced[1] (see Chapter 33).

Basal adrenaline concentrations and the circadian variation in adrenaline concentrations in asthmatic patients appears to be similar to those found in normal subjects.[3-5] Although Bates et al.[6] reported that plasma adrenaline levels at 10 p.m. were lower in patients with nocturnal asthma compared with a non-nocturnal asthma group, correction of the nocturnal fall in plasma adrenaline does not alter the peak flow rate values of patients with nocturnal asthma.[7] These findings, together with the report of nocturnal asthma occurring in a patient after adrenalectomy,[8] suggest that a fall in plasma adrenaline at night is not a dominant factor in nocturnal asthma.

Adrenaline is not released in response to allergen- or drug-induced bronchoconstriction *per se* and so does not appear to have an important homeostatic role in the regulation of airway calibre during bronchoconstriction to these stimuli.[9,10] Even during acute exacerbations of asthma there may be no elevation in plasma adrenaline level,[11,12] although very high adrenaline concentrations have been found in some patients with acute severe asthma.[13] The elevated adrenaline concentrations achieved after strenuous exercise[14] cause bronchodilation in both normal and asthmatic subjects[3,5,15] and may act to counteract bronchospasm induced by exercise in asthma.[16] Although a blunted catecholamine response to exercise in asthmatic patients has been reported by some investigators,[17] other studies have found no significant difference in either the peak plasma catecholamine level between normal and asthmatic subjects or the response to increasing levels of exercise.[14,18]

Noradrenaline, which has β_1- and weak β_2-adrenergic activity in addition to α-adrenergic effects, acts as a neurotransmitter in the sympathetic nervous system but overspills into the circulation. Infusion of noradrenaline, producing circulating concentrations within the physiological and pathophysiological range, has no effect on airway calibre in either normal or asthmatic subjects.[3,5] The third catecholamine present in the blood, dopamine, also has no influence on bronchomotor tone in humans.[19]

Natriuretic peptides

Natriuretic peptides are a family of hormones that have an important role in salt and water homeostasis.[20] The human natriuretic peptides include atrial natriuretic peptide (ANP), brain natriuretic peptide (BNP), C-type natriuretic peptide (CNP) and urodilatin. Most natriuretic peptides are produced primarily in the heart but are released also in other tissues including the kidneys, lungs and central nervous system. Specific ANP receptors have been localized to lung including airway smooth muscle,[21] of which some may be the ANP_C or clearance receptor subtype,[22] although the receptor subtype(s) in human airway smooth muscle is unknown. In isolated human airway tissue, ANP has a direct relaxant effect and confers protection against agonist-induced contraction.[22–25] Two principal mechanisms have been proposed for the inactivation of ANP: degradation by the enzyme neutral endopeptidase (NEP) and binding to a non-guanylyl cyclase clearance receptor (ANP_C receptor). NEP is widely distributed within the airways[26] and plays a role in modulating the effect of ANP on airway smooth muscle.[24,25]

An intravenous infusion of exogenous ANP has important actions on airway function including bronchodilation and the modification of bronchial reactivity to inhaled histamine and to fog challenge.[27–32] The rise in plasma ANP levels during exercise[33] is similar to that obtained during the lowest rates of ANP infusion and suggest that these elevations may lead to an attenuation of bronchospasm. Elevated plasma ANP levels are found in patients with cardiac failure[34] and cor pulmonale[35] and under these circumstances ANP may also play a protective role on the airways. However, circulating ANP at physiological concentrations appears unlikely to have any influence on bronchomotor tone in normal subjects.[29]

Angiotensin II

The renin–angiotensin system plays an important role in fluid and electrolyte homeostasis through the actions of the octapeptide angiotensin II. Angiotensin II is formed from angiotensinogen by the action of renin and then angiotensin-converting enzyme (ACE), 60–80% of production occurring within the pulmonary vascular endothelium.[36] An alternative ACE-independent pathway, possibly mediated by several inflammatory proteases,[37] may also cause the formation of angiotensin II.

The effect of physiological concentrations of angiotensin II on basal bronchial tone of normal individuals is not known, whereas infusion of angiotensin II in mild asthmatic patients to plasma levels found in acute asthma causes bronchoconstriction[38] (Fig. 23.2). Angiotensin II, although causing only weak contraction of isolated human and bovine bronchial rings, potentiates the effects of methacholine and endothelin (ET)-1 *in vitro*.[39,40]

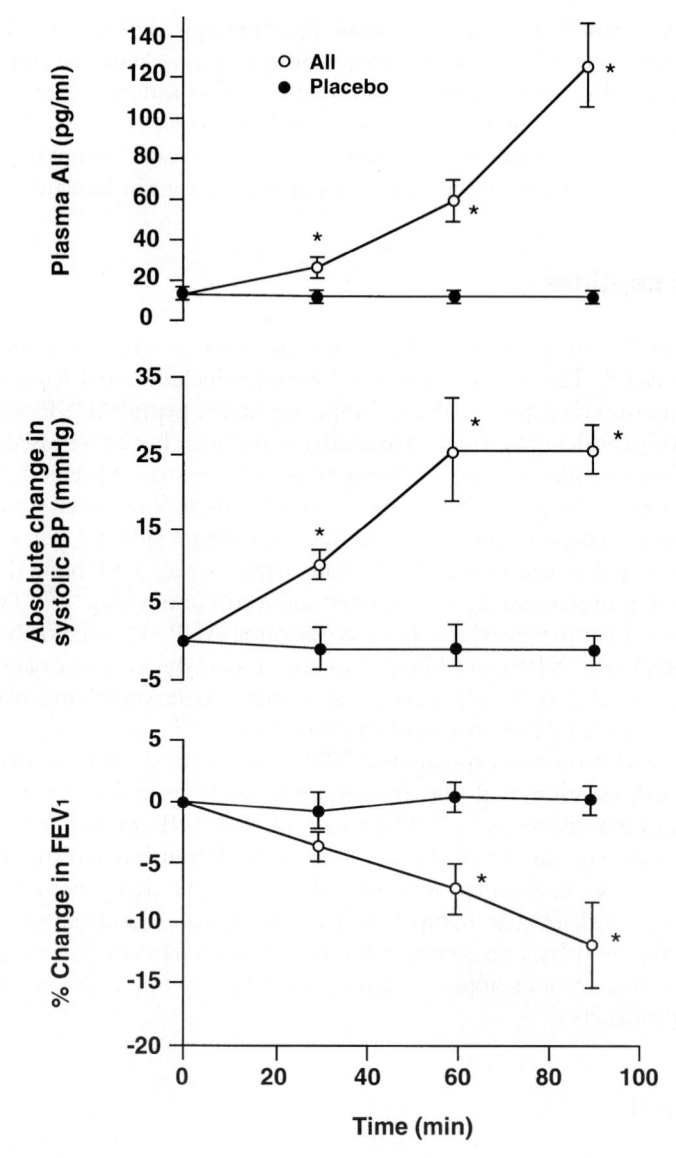

Fig. 23.2 Effect of infused angiotensin II on plasma levels of angiotensin II (AII), change in systolic blood pressure (BP) and change in forced expiratory volume in 1 s (FEV$_1$) from baseline values in asthmatic patients ($n = 8$). *$P < 0.05$ vs. placebo. Reproduced from ref. 38, with permission.

In patients with mild asthma, angiotensin II at subthreshold concentrations potentiates methacholine-induced bronchoconstriction[40], but has no effect on histamine-evoked bronchoconstriction either *in vitro* or *in vivo*.[41] These results suggest a role for angiotensin II as a putative mediator in asthma, although its effect on different spasmogens may be variable.

The renin–angiotensin system is activated in acute severe asthma but not in stable chronic asthma.[12,38] The mechanism of activation is unclear[12] but nebulized β_2-agonists cause elevation of renin and angiotensin II in normal and mild asthmatic subjects through an ACE-dependent pathway.[42,43] This may occur via stimulation of β-adrenoceptors on juxtaglomerular cells, but the levels of angiotensin II seen in acute severe asthma are higher, suggesting the existence of an alternative pathway of angiotensin II formation. Exercise activates the renin–angiotensin system,[44,45] raising the possibility that elevated angiotensin II levels during exercise could contribute to exercise-induced bronchospasm.

Endothelins

The human endothelin family comprises three structurally and pharmacologically distinct 21 amino acid peptides, termed ET-1, ET-2 and ET-3. ET-1 is present in the plasma of normal individuals and raised plasma levels have been found during attacks of acute severe asthma.[12] ET-1 is one of the most potent bronchoconstrictor peptides yet isolated, producing prolonged and potent contractions in animal airways *in vivo* by intravenous[46] and aerosol[47] administration as well as *in vitro*.[48,49] Inhaled ET-1 acts as a potent bronchochonstrictor when administered to asthmatic patients and they exhibit bronchial hyperreactivity to ET-1 compared with normal subjects.[50]

ET-1 is produced or released locally within the respiratory tract in concentrations higher than those in the plasma.[51] Immunoreactivity to endothelin can be detected in bronchial epithelium, submucosal glands and airway smooth muscle.[52,53] Endothelin receptors have been found in human airway smooth muscle and are predominantly of the ET_B subtype.[54] Endothelin reaching the airways from the bloodstream or released from airway cells could possibly contribute to bronchial tone in both normal and asthmatic subjects. The recent availability of endothelin-receptor antagonists should help establish whether endothelins have an important influence on airway smooth muscle function in humans.

HORMONES

Cortisol

Pharmacological doses of intravenous cortisol have no short-term effect on airway calibre in normal subjects.[55] Although glucocorticoids can potentiate the response to catecholamines in isolated bronchial tissue, the effect occurs only at supraphysiological concentrations.[56,57] These results suggest that endogenous cortisol is unlikely to have an important direct effect on airway tone in normal individuals. In asthma the role of physiological concentrations of circulating cortisol in airway function is uncertain. In nocturnal asthma, the nadir in the circadian variation in plasma cortisol occurs 4 h before maximal bronchoconstriction,[58,59] although the delayed action of cortisol means that it could still have an influence on airway calibre. Kallenbach *et al*.[60] found a reduced nadir of plasma cortisol in patients with nocturnal asthma compared with a group without nocturnal asthma, but this finding may have been influenced by previous corticosteroid therapy. Other studies have found no direct association between plasma cortisol concentrations and nocturnal asthma.[5] Furthermore, the infusion of physiological concentrations of hydrocortisone, eliminating the fall in plasma cortisol at night, does not prevent the nocturnal fall in peak flow rate in most asthmatic patients,[59] suggesting that the circulating cortisol level is not the only factor in determining nocturnal asthma.

Thyroid hormones

The relationship between asthma and thyroid disease provides indirect evidence of a role for thyroid hormones in maintaining airway function. The development of hyperthyroidism can be associated with a deterioration in asthma control, with subsequent improvement in symptoms following appropriate treatment.[61–63] Conversely, the occurrence of hypothyroidism has been reported to be associated with improvement in asthma control, which relapses following subsequent thyroxine replacement.[64]

Several possible mechanisms have been suggested by which thyroid hormones could influence airway smooth muscle tone and responsiveness. Firstly, β-adrenergic airway responsiveness has been reported to be inversely related to thyroxine levels both *in vitro*, in guinea-pig trachea,[65] and *in vivo*, in non-asthmatic subjects.[66] Following treatment of hyperthyroidism or hypothyroidism airway β-adrenergic responses return to euthyroid levels.[66] It is unlikely that alterations in β-adrenergic activity are due to changes in circulating catecholamine levels[67] or β-adrenergic receptor numbers[68] but it is possible that thyroxine acts at a post-receptor site within the smooth muscle. Secondly, Cockcroft *et al.*[69] reported a decrease in non-specific bronchial reactivity in an asthmatic patient after treatment of hyperthyroidism. However, studies examining the effects of different circulating thyroid hormone levels on non-specific reactivity in non-asthmatic individuals have produced conflicting results.[70–74] Thirdly, thyroxine may alter the metabolism of arachidonic acid since prostaglandin breakdown has been shown to be reduced in hyperthyroid rats.[75] The involvement of thyroid hormones in lung function, however, may be unrelated to a direct effect on the airways. Respiratory muscle weakness, which may occur in hyperthyroidism,[71] could contribute to the dyspnoea that commonly accompanies thyrotoxicosis and this action may heighten the degree of breathlessness experienced by a patient with pre-existing airway disease.

Sex hormones

Progesterone has an important role in reducing the contractility of uterine smooth muscle during pregnancy; this effect may be due to its influence on gap junction formation between smooth muscle cells. It has been suggested that progesterone might cause similar effects on bronchial smooth muscle. Progesterone could influence airway smooth muscle tone by other mechanisms: indirectly by potentiating the effect of catecholamines[57] or through its immunosuppressive properties. Progesterone levels and airway responsiveness do not show a clear relationship during either pregnancy or the menstrual cycle, although changes in the levels of other hormones may obscure an effect of progesterone on the airways.[76,77] It is of interest that intramuscular progesterone has a beneficial effect in some women with severe premenstrual asthma.[78]

Oestrogen possesses both immunostimulatory and immunosuppressive properties and causes increased acetylcholine activity in the lungs of animals,[79] which could result in an

23 Humoral Control of Airway Tone

increase or decrease in airway tone. A recent preliminary report suggested that oestrogen treatment may have steroid-sparing effects in postmenopausal asthmatic women,[80] although conversely hormone replacement therapy has been associated with an increased risk of developing asthma.[81]

Other hormones

Hormones such as adrenomedullin[82] and glucagon[83] have bronchodilator actions but whether they have a role in the control of airway smooth muscle tone in humans has not been investigated.

CIRCULATING INFLAMMATORY MEDIATORS

Inflammatory mediators including histamine, cysteinyl leukotrienes and thromboxane metabolites have been detected in the plasma and/or urine during acute asthma attacks (see Chapters 16 and 20). It remains unclear, however, whether these circulating mediators have any influence on airway tone. Although both leukotriene and H_1-receptor antagonists cause mild bronchodilation, it seems likely that this effect is due mainly to the inhibition of locally produced mediators within the lungs rather than those reaching the airways from the systemic circulation.

OXYGEN AND CARBON DIOXIDE

Airway tone and reactivity are influenced by oxygen and carbon dioxide levels. The effects of hypoxia on isolated airway smooth muscle is normally to decrease active tension,[84] whereas the effects on basal airway tone in humans remain controversial. Isocapnic hypoxia has been reported to cause an increase[83,84] or no change[87,88] in airway resistance. More recently Julia-Serda *et al.*[89] using an acoustic reflection technique to measure the cross-sectional area of the trachea and main airways, found that isocapnic hypoxia produced tracheobronchial dilation. Hyperoxia in dogs may either have no effect on resting airway tone[90] or cause mild bronchodilation.[91] In humans hyperoxia has been reported to relieve bronchoconstriction in patients with chronic obstructive pulmonary disease,[92] but was without effect on baseline airway tone in asthma.[93]

Changes in oxygen tension may alter the response of the airways to bronchoconstrictor and bronchodilator stimuli. Hypoxia *in vitro* potentiates both methacholine- and endothelin-induced bronchoconstriction in isolated bovine tissue,[94,95] whereas in isolated canine trachealis muscle[84] and guinea-pig parenchymal lung strips[96] it reduces contractile responses. In human bronchial tissue, acute hypoxia attenuates the bronchoconstrictor response to histamine.[97] *In vivo*, hypoxia potentiates the effects of histamine on the airways of dogs[90] and histamine- and carbachol-mediated bronchoconstriction in sheep.[98] This latter effect was abolished by simultaneous infusion of sodium cromoglycate, suggesting that alveolar hypoxia may stimulate increased mast cell degranulation or act via neural pathways. In asthmatic patients, methacholine-induced bronchoconstriction is potentiated by isocapnic hypoxia[99,100] (Fig. 23.3). Little is known about the effect of hypoxia on the responses evoked by bronchodilators. The ability of the bronchodilators salbutamol and ANP to reverse methacholine-induced tone in bovine[94] and rat[101] isolated bronchi is dependent upon the oxygen tension of the environment. In human bronchial tissue, acute hypoxia attenuates the bronchodilator response to salbutamol *in vitro* but has no effect on salbutamol-induced bronchodilation in patients with asthma.[102]

Hyperoxia attenuates endothelin- and methacholine-mediated constriction in isolated bovine bronchial rings.[94,95] Vidruk and Sorkness[90] demonstrated, in dogs, that histamine-induced bronchoconstriction was attenuated by hyperoxia. In asthmatic patients, methacholine challenge has been reported to be attenuated[103] or unaffected[93,104] by hyperoxia. Increases in airway tone observed during exercise in asthmatic patients are also attenuated by hyperoxia,[105] although the precise mechanism of this effect remains controversial.

Carbon dioxide airway tension has also been shown to influence airway tone. In humans, reductions in total lung resistance have been observed during hypercapnic respiration in both normal and asthmatic subjects.[106] In normal subjects hypocapnic hyperventilation causes bronchoconstriction, an effect not observed during isocapnic

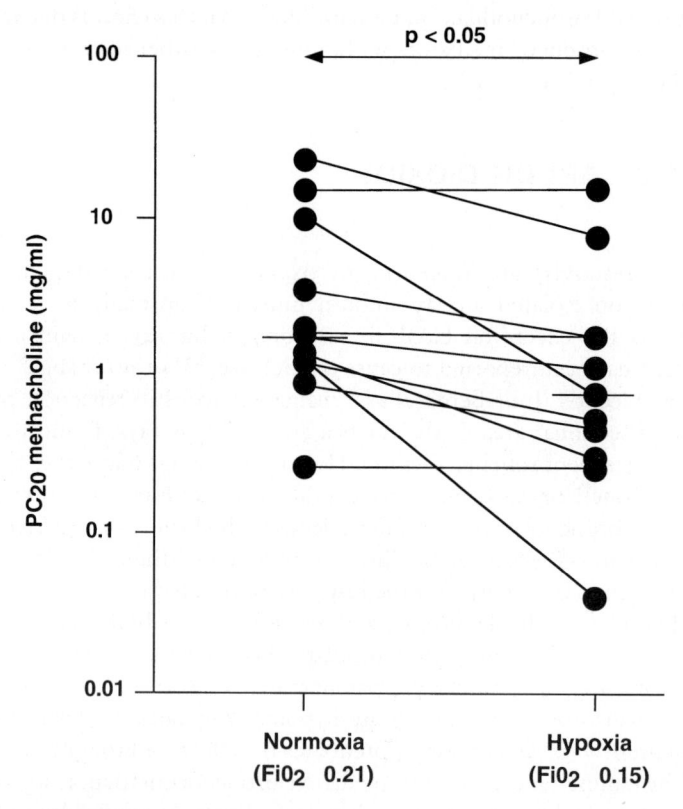

Fig. 23.3 Effect of hypoxia (inspired oxygen tension, FIO_2 0.15) on PC_{20} methacholine in asthmatic patients ($n = 11$). Reproduced from ref. 100, with permission.

hyperventilation.[107] Elshout et al.[106] suggested that asthmatic patients developed greater bronchoconstriction following hypocapnia compared with normal subjects.

CONCLUSIONS

Circulating hormones and vasoactive peptides appear to play a minor role in the physiological regulation of airway tone in normal individuals. Adrenaline is the only hormone known to influence bronchomotor tone and it is only during strenuous exercise that concentrations are elevated sufficiently to cause bronchodilation.

Circulating hormones play a more important role in the regulation of airway tone in diseased states of the airways such as asthma and possibly in other disorders such as cor pulmonale, congestive cardiac failure, respiratory failure and thyroid diseases. Circulating adrenaline has a role in the maintenance of resting airway tone in asthma, perhaps particularly in those patients in whom resting airway calibre is already reduced. The elevated adrenaline and ANP concentrations achieved after vigorous exercise may act to counteract exercise-induced asthma. It has not been established in asthma whether elevated circulating angiotensin II or ET-1 levels achieved during exercise, or more particularly in acute severe asthma, contribute to the bronchospasm. Airway tone and reactivity are influenced by oxygen levels, although the precise effects of a given oxygen tension on airway function in health and airway disease are not clearly established.

REFERENCES

1. Moreno RH, Hogg JC, Pare PD: Mechanics of airway narrowing. *Am Rev Respir Dis* (1986) **133**: 1171–1180.
2. Barnes PJ: Neural control of human airways in health and disease. *Am Rev Respir Dis* (1986) **134**: 1289–1314.
3. Berkin KE, Inglis GC, Ball SG, Thomson NC: Airway responses to low concentrations of adrenaline and noradrenaline in normal subjects. *Q J Exp Physiol* (1985) **70**: 203–209.
4. Berkin KE, Inglis GC, Ball SG, Thomson NC: Effect of low dose adrenaline and noradrenaline infusions on airway calibre in asthmatic patients. *Clin Sci* (1986) **70**: 347–352.
5. Barnes PJ, Fitzgerald G, Brown M, Dollery C: Nocturnal asthma and changes in circulating epinephrine, histamine and cortisol. *N Engl J Med* (1980) **303**: 263–267.
6. Bates ME, Clayton M, Calhoun W, et al.: Relationship of plasma epinephrine and circulating eosinophils to nocturnal asthma. *Am J Respir Crit Care Med* (1994) **149**: 667–672.
7. Morrison JFJ, Teale C, Pearson SB, et al.: Adrenaline and nocturnal asthma. *Br Med J* (1990) **301**: 473–476.
8. Morice A, Sever P, Ind PW: Adrenaline, bronchoconstriction and asthma. *Br Med J* (1986) **293**: 539–540.
9. Larsson K, Grunneberg R, Hjemdahl P: Bronchodilation and inhibition of allergen-induced bronchoconstriction by circulating epinephrine in asthmatic subjects. *J Allergy Clin Immunol* (1985) **75**: 586–593.
10. Larsson K, Carlens P, Bevegård S Hjemdahl P: Sympathoadrenal responses to bronchoconstriction in asthma: an invasive and kinetic study of plasma catecholamines. *Clin Sci* (1995) **88**: 439–446.
11. Ind PW, Causson RC, Brown MJ, Barnes PJ: Circulating catecholamines in acute asthma. *Br Med J* (1985) **290**: 267–279.

12. Ramsey SG, Dagg KD, McKay IC, Lipworth BJ, McSharry C, Thomson NC: Investigations on the activation of the renin–angiotensin system in acute severe asthma. *Eur Respir J* (1997) **10**: 2766–2771.
13. Clarke B, Ind PW, Causson R, Barnes PJ: Bronchodilation and catecholamine responses to induced hypoglycaemia in acute asthma. *Clin Sci* (1985) **69**: 35P.
14. Berkin KE, Walker G, Inglis GC, Ball SG, Thomson NC: Circulating adrenaline and noradrenaline concentrations during exercise in patients with exercise induced asthma and normal subjects. *Thorax* (1988) **43**: 295–299.
15. Warren JB, Dalton N: A comparison of the bronchodilator and vasopressor effects of exercise levels of adrenaline in man. *Clin Sci* (1983) **64**: 475–479.
16. Knox AJ, Campos-Gongora H, Wisniewski A, MacDonald IA, Tattersfield AE: Modification of bronchial reactivity by physiological concentrations of plasma epinephrine. *J Appl Physiol* (1992) **73**: 1004–1007.
17. Barnes PJ, Brown MJ, Silverman M, Dollery CT: Circulating catecholamines in exercise and hyperventilation induced asthma. *Thorax* (1981) **36**: 435–440.
18. Gilbert IA, Lennen KA, McFadden ER: Sympathoadrenal response to repetitive exercise in normal and asthmatic subjects. *J Appl Physiol* (1988) **64**: 2667–2674.
19. Thomson NC, Patel KR: Effect of dopamine on airways conductance in normals and extrinsic asthmatics. *Br J Clin Pharmacol* (1978) **5**: 421–424.
20. Ruskoaho H: Atrial natriuretic peptide: synthesis, release, and metabolism. *Pharmacol Rev* (1992) **44**: 479–602.
21. Von Schroeder HP, Nishimura E, McIntosh, CHS, Buchan AMJ, Wilson N, Laidsome JR: Autoradiographic localisation of binding sites for atrial natriuretic factor. *Can J Physiol Pharmacol* (1985) **63**: 1373–1377.
22. James S, Burnstock G: Atrial and brain natriuretic peptides sharing binding sites on cultured cells from the rat trachea. *Cell Tissue Res* (1991) **265**: 555–565.
23. Thomson NC: Atrial natriuretic peptides. In Raeburn D, Giembycz MA (eds) *Airways Smooth Muscle: Peptide Receptors, Ion Channels and Signal Transduction*. Basel, Birkhauser Verlag, 1995.
24. Angus RM, Nally JE, McCall R, Young LC, McGrath JC, Thomson NC: Modulation of the effect of atrial natriuretic peptide in human and bovine bronchi by phosphoramidon. *Clin Sci* (1994) **86**: 291–295.
25. Nally JE, Clayton RA, Thomson NC, McGrath JC: The interaction of α-human natriuretic peptide (ANP) with salbutamol, sodium nitroprusside and isosorbide dinitrate in human bronchial smooth muscle. *Br J Pharmacol* (1994) **113**: 1328–1332.
26. Nadel JA: Neutral endopeptidase modulates neurogenic inflammation. *Eur Respir J* (1991) **4**: 745–754.
27. Hulks G, Jardine A, Connell JMC, Thomson NC: Bronchodilator effect of atrial natriuretic peptide in asthma. *Br Med J* (1989) **299**: 1081–1082.
28. Chanez P, Mann C, Bousquet J, et al.: Atrial natriuretic factor (ANF) is a potent bronchodilator in asthma. *J Allergy Clin Immunol* (1990) **86**: 321–324.
29. Hulks G, Jardine A, Connell JMC, Thomson NC: Effect of atrial natriuretic factor on bronchomotor tone in the normal human airway. *Clin Sci* (1990) **79**: 51–55.
30. Hulks G, Jardine A, Connell JMC, Thomson NC: Influence of elevated plasma levels of atrial natriuretic factor on bronchial reactivity in asthma. *Am Rev Respir Dis* (1991) **143**: 778–782.
31. McAlpine LG, Hulks G, Thomson NC: Effect of atrial natriuretic peptide given by intravenous infusion on bronchoconstriction induced by ultrasonically nebulized distilled water (FOG). *Am Rev Respir Dis* (1992) **146**: 912–915.
32. Angus RM, McCallum MJA, Thomson NC: The bronchodilator, cardiovascular and cyclic guanylyl monophosphate (cGMP) response to high dose infused atrial natriuretic peptide in asthma. *Am Rev Respir Dis* (1993) **147**: 1122–1125.
33. Hulks G, Mohammed AF, Jardine AG, Connell JMC, Thomson NC: Circulating plasma levels of atrial natriuretic peptide and catecholamines in response to maximal exercise in normal and asthmatic subjects. *Thorax* (1991) **46**: 824–828.
34. Raine AEG, Erne P, Burgisser E, et al: Atrial natriuretic peptide and atrial pressure in patients with congestive cardiac failure. *N Engl J Med* (1986) **315**: 533–537.

35. Burghuber OC, Harterr E, Punzengruber C, Weissel M, Woloszczuk W: Human atrial natriuretic peptide secretions in precapillary pulmonary hypertension. *Chest* (1988) **92**: 31–37.
36. Morton JJ: Biochemical aspects of the angiotensins. In Robertson JIS, Nicholls MG (eds) *The Renin–Angiotensin System*. London, Gower Medical Publishing, 1993, pp 9.1–9.12.
37. Husain A: The chymase–angiotensin system in humans. *J Hypertens* (1993) **11**: 1155–1159.
38. Millar EA, Angus RA, Hulks G, Morton JJ, Connell JMC, Thomson NC: Activity of the renin–angiotensin system in acute severe asthma and the effect of angiotensin II on lung function. *Thorax* (1994) **49**: 492–495.
39. Nally JE, Clayton RA, Wakelam MJO, Thomson NC, McGrath JC: Angiotensin II enhances responses to endothelin-1 in bovine bronchial smooth muscle. *Pulmon Pharmacol* (1994) **7**: 409–413.
40. Miller EA, Nally JE, Thomson NC: Angiotensin II potentiates methacholine-induced bronchoconstriction in human airway both *in vitro* and *in vivo*. *Eur Respir J* (1995) **8**: 1838–1841.
41. Ramsey SG, Clayton RA, Dagg KD, Thomson LJ, Nally JE, Thomson NC: Effect of angiotensin II on histamine-induced bronchoconstriction in the human airway both *in vitro* and *in vivo*. *Respir Med* (1997) **91**: 609–615.
42. Millar EA, McInnes GT, Thomson NC: Investigation of the mechanism of β_2-agonist-induced activation of the renin–angiotensin system. *Clin Sci* (1995) **88**: 433–437.
43. Millar EA, Connell JMC, Thomson NC: The effect of nebulized albuterol on the activity of the renin–angiotensin system in asthma. *Chest* (1997) **111**: 71–74.
44. Kosunen KJ, Pakarinen AJ: Plasma renin, angiotensin II, and plasma and urinary aldosterone in running exercise. *J Appl Physiol* (1976) **41**: 26–29.
45. Milledge JS, Catley DM: Renin, aldosterone and converting enzyme during exercise and acute hypoxia in humans. *J Appl Physiol* (1982) **52**: 320–323.
46. Macquin-Mavier I, Levame M, Istin N, Harf A: Mechanisms of endothelin mediated bronchoconstriction in the guinea pig. *J Pharmacol Exp Ther* (1989) **250**: 740–745.
47. Lagente V, Chabrier PE, Mencia-Huerta JM, Braquet P: Pharmacological modulation of the bronchopulmonary action of the vasoactive peptide, endothelin, administered by aerosol in the guinea pig. *Biochem Biophys Res Commun* (1989) **158**: 625–632.
48. Advenier C, Sarrina B, Naline E, Puybasset L, Lagente V: Contractile activity of three endothelins (endothelin-1, endothelin-2 and endothelin-3) on the human isolated bronchus. *Br J Pharmacol* (1990) **100**: 168–172.
49. Nally JE, McCall R, Young LC, Wakelam MJO, Thomson NC, McGrath JC: Mechanical and biochemical responses to endothelin-1 and endothelin-3 in human bronchi. *Eur J Pharmacol* (1994) **288**: 53–60.
50. Chalmers GW, Little SA, Patel KR, Thomson NC: Endothelin-1-induced bronchoconstriction in asthma. *Am J Respir Crit Care Med* (1997) **156**: 382–388.
51. Chalmers GW, Thomson LJ, MacLeod KJ, *et al*.: Endothelin-1 levels in induced sputum samples from asthmatic and normal subjects. *Thorax* (1997) **52**: 625–627
52. Marciniak SL, Plumpton C, Barker PJ, Huskisson NS, Davenport AP: Localisation of immunoreactive endothelin and proendothelin in human lung. *Pulmon Pharmacol* (1992) **5**: 175–182.
53. McKay KO, Black JL, Diment M, Armour CL: Functional and autoradiographic studies of endothelin-1 and endothelin-2 in human bronchi, pulmonary arteries, and airway parasympathetic ganglia. *J Cardiovasc Pharmacol* (1991) **17** (Suppl 7): 206–209.
54. Knott PG, D'Aprile AC, Henry PJ, Hay DWP, Goldie RG: Receptors for endothelin-1 in asthmatic human peripheral lung. *Br J Pharmacol* (1995) **114**: 1–3.
55. Ramsdell JW, Berry CC, Clausen JL: The immediate effects of cortisol on pulmonary function in normals and asthmatics. *J Allergy Clin Immunol* (1983) **71**: 69–74.
56. Geddes BA, Jones TR, Dvorsky RJ, Lefcoe NM: Interaction of glucocorticoids and bronchodilators on isolated guinea pig tracheal and human bronchial smooth muscle. *Am Rev Respir Dis* (1974) **110**: 420–427.
57. Foster PS, Goldie RG, Paterson JW: Effect of steroids on beta-adrenoceptor mediated relaxation of pig bronchus. *Br J Pharmacacol* (1983) **78**: 441–445.

58. Reinberg A, Ghata J, Sidi E: Nocturnal asthma attacks: their relationship to the circadian adrenal cycle. *J Allergy* (1963) **34**: 323–330.
59. Soutar CA, Costello J, Ijaduola O, Turner-Warwick M: Nocturnal and morning asthma: relationship to plasma corticosteroid and response to cortisol infusion. *Thorax* (1975) **30**: 436–440.
60. Kallenbach JM, Panz VR, Joffe BI, *et al.*: Nocturnal events related to 'morning dipping' in bronchial asthma. *Chest* (1988) **93**: 751–757.
61. Elliott CA: Occurrence of asthma in patients manifesting evidence of thyroid dysfunction. *Am J Surg* (1929) **7**: 333–337.
62. Ayres J, Clark TJH: Asthma and the thyroid. *Lancet* (1981) **ii**: 1110–1111.
63. Lipworth BJ, Dhillon DP, Clark RA, Newton RW: Problems with asthma following treatment of thyrotoxicosis. *Br J Dis Chest* (1988) **82**: 310–314.
64. Bush RK, Ehrlick EN, Reed CE: Thyroid disease and asthma. *J Allergy Clin Immunol* (1977) **59**: 398–401.
65. Taylor SE: Additional evidence against universal modulation of β-adrenoceptor responses by excessive thyroxine. *Br J Pharmacol* (1983) **78**: 639–644.
66. Harrison RN, Tattersfield AE: Airway response to inhaled salbutamol in hyperthyroid and hypothyroid patients before and after treatment. *Thorax* (1984) **39**: 34–39.
67. Coulombe P, Dussault JH, Walker P: Plasma catecholamine concentrations in hyperthyroidism and hypothyroidism. *Metabolism* (1976) **25**: 973–979.
68. Scarpace PJ, Abrass IB: Thyroid hormone regulation of rat heart, lymphocyte and lung beta-adrenergic receptors. *Endocrinology* (1981) **108**: 1007–1011.
69. Cockcroft DW, Silverberg JDH, Dosman JA: Decrease in nonspecific bronchial reactivity in an asthmatic patient following treatment of hyperthyroidism. *Ann Allergy* (1978) **41**: 160–163.
70. Irwin RS, Pratter MR, Stivers DH, Braverman LE: Airway reactivity and lung function in triiodothyronine-induced thyrotoxicosis. *J Appl Physiol* (1985) **58**: 1485–1488.
71. Kendrick AH, O'Reilly JF, Laslo G: Lung function and exercise performance in hyperthyroidism before and after treatment. *Q J Med* (1988) **68**: 615–627.
72. Roberts JA, McLellan AR, Alexander WD, Thomson NC: Effect of hyperthyroidism on bronchial reactivity in non-asthmatic patients. *Thorax* (1989) **44**: 603–604.
73. Israel RH, Poe RH, Cave WT, Greenblatt DW, DePapp Z: Hyperthyroidism protects against carbachol-induced bronchospasm. *Chest* (1987) **91**: 242–245.
74. Wieshammer S, Keck FS, Shäuffelen AC, Von Beauvais H, Seibold H, Hombach V: Effects of hypothyroidism on bronchial reactivity in non-asthmatic subjects. *Thorax* (1990) **45**: 947–950.
75. Hoult JRS, Moore P: Thyroid disease, asthma, and prostaglandins. *Br Med J* (1978) **i**: 366.
76. Juniper EF, Daniel EE, Roberts RS, *et al.*: Improvement in airway responsiveness and asthma severity during pregnancy. *Am Rev Respir Dis* (1989) **140**: 924–931.
77. Juniper EF, Kline PA, Roberts RS, Hargreave FE, Daniel EE: Airway responsiveness to methacholine during the natural menstrual cycle and the effect of the oral contraceptives. *Am Rev Respir Dis* (1987) **135**: 1039–1042.
78. Beynon HLC, Garbett ND, Barnes PJ: Severe premenstrual exacerbations of asthma: effect of intramuscular progesterone. *Lancet* (1988) **ii**: 370–372.
79. Abdul-Karim RW, Marshall LD, Nesbitt REL: Influence of estradiol-17β on the acetylcholine content of the lung. *Am J Obstet Gynecol* (1970) **107**: 641–644.
80. Celedon JC, Sherman CB, Myers J, Wheeler C, Passero MA, Kern DG: Estrogens as steroid-sparing agents in postmenopausal asthmatic women. *Am J Respir Crit Care Med* (1995) **151**: A675.
81. Troisi RJ, Spiezer FE, Willet WC, Trichopoulos D, Rosner B: Menopause, postmenopausal estrogen preparations, and the risk of adult-onset asthma. *Am J Respir Crit Care Med* (1995) **152**: 1183–1184.
82. Kanazawa H, Kurihara N, Hirata K, Kudoh S, Kawaguchi T, Takeda T: Adrenomedullin, a newly discovered hypotensive peptide, is a potent bronchodilator. *Biochem Biophys Res Commun* (1994) **205**: 251–254.
83. Sherman MS, Lazar EJ, Eichacker P: A bronchodilator action of glucagon. *J Allergy Clin Immunol* (1988) **81**: 908–911.

84. Stephens NL, Chiu BS: Mechanical properties of tracheal smooth muscle and effects of O_2, CO_2, and pH. *Am J Physiol* (1970) **219**: 1001–1008.
85. Sterling GM: The mechanism of bronchoconstriction due to hypoxia in man. *Clin Sci* (1968) **35**: 105–114.
86. Saunders NA, Betts MF, Pengelly LD, Rebuck AS: Changes in lung mechanics induced by acute isocapnic hypoxia. *J Appl Physiol* (1977) **42**: 413–419.
87. Goldstein RS, Zamel N, Rebuck AS: Absence of effects of hypoxia on small airway function in humans. *J Appl Physiol* (1979) **47**: 251–256.
88. Tam EK, Geffroy BA, Myers DJ, Seltzer J, Sheppard D, Boushey HA: Effect of eucapnic hypoxia on bronchomotor tone and on the bronchomotor response to dry air in asthmatic subjects. *Am Rev Respir Dis* (1985) **132**: 690–693.
89. Julia-Serda G, Molfino NA, Furlott HG, *et al.*: Tracheobronchial dilation during isocapnic hypoxia in conscious humans. *J Appl Physiol* (1993) **75**: 1728–1733.
90. Vidruk EH, Sorkness RL: Histamine induced tracheal constriction is attenuated by hyperoxia and exaggerated by hypoxia. *Am Rev Respir Dis* (1985) **132**: 287–291.
91. Green M, Widdicombe JG: The effects of ventilation of dogs with different gas mixtures on airway calibre and lung mechanics. *J Physiol* (1966) **186**: 363–381.
92. Libby DM, Briscoe WA, King TKC: Relief of hypoxia related bronchoconstriction by breathing 30 per cent oxygen. *Am Rev Respir Dis* (1981) **123**: 171–175.
93. Dagg KD, Thomson LJ, Ramsay SG, Thomson NC: Effect of acute hyperoxia on the bronchodilator response to salbutamol in stable asthmatic patients. *Thorax* (1996) **51**: 853–854.
94. Clayton RA, Nally JE, Thomson NC, McGrath JC: Effect of oxygen tension on responses evoked by methacholine and bronchodilators in isolated bovine bronchial rings. *Pulmon Pharmacol* (1996) **9**: 123–128.
95. Nally JE, Bunton DC, Thomson NC: Potentiated endothelin-1-mediated contractions are reversed by indomethacin in bovine bronchi. *Am J Respir Crit Care Med* (1995) **151**: A227.
96. Paterson NAM, Hamilton JT, Yaghi A, Miller DS: Effect of hypoxia on responses of respiratory smooth muscle to histamine and LTD4. *J Appl Physiol* (1988) **64**: 435–440.
97. Dagg KD, Clayton RA, Thomson LJ, Chalmers G, Thomson NC: The effect of acute hypoxia on histamine-evoked bronchoconstriction in human isolated bronchi and asthmatic patients. *Am J Respir Crit Care Med* (1997) **155**: A157.
98. Ahmed T, Marchette B: Hypoxia enhances nonspecific bronchial reactivity. *Am Rev Respir Dis* (1985) **132**: 839–844.
99. Denjean A, Roux C, Herve P, *et al.*: Mild isocapnic hypoxia enhances the bronchial response to methacholine in asthmatic subjects. *Am Rev Respir Dis* (1988) **138**: 789–793.
100. Dagg KD, Thomson LJ, Clayton RA, Ramsay SG, Thomson NC: Effect of acute alterations in inspired oxygen tension on methacholine-induced bronchoconstriction in patients with asthma. *Thorax* (1997) **52**: 453–457.
101. Clayton RA, Nally JE, Thomson NC, McGrath JC: Reversal of agonist-induced contraction in isolated bronchial rings from chronically hypoxic and control rats. *Thorax* (1996) **50**: 46P.
102. Dagg KD, Clayton RA, Thomson LJ, Chalmers G, Thomson NC: The effect of acute hypoxia on albuterol-evoked bronchodilation both *in vivo* and *in vitro* in man. *Am J Respir Crit Care Med* (1997) **155**: A154.
103. Inoue H, Inoue C, Okayama M, Sekizawa K, Hida W, Takishima T: Breathing 30 per cent oxygen attenuates bronchial responsiveness to methacholine in asthmatic patients. *Eur Respir J* (1989) **2**: 506–512.
104. Wollner A, Ben-Dov I, Bar-Yishay E: Effect of hyperoxia on bronchial response to inhaled methacholine. *Allergy* (1991) **46**: 35–39.
105. Resnick AD, Chandler Deal E, Ingram RH, McFadden ER: A critical assessment of the mechanism by which hyperoxia attenuates exercise induced asthma. *J Clin Invest* (1979) **64**: 541–549.
106. Elshout FJJ, Herwaarden CLA, Folgering HThM: Effects of hypercapnia and hypocapnia on respiratory resistance in normal and asthmatic subjects. *Thorax* (1991) **46**: 28–32.
107. Sterling GM: The mechanism of bronchoconstriction due to hypocapnia in man. *Clin Sci* (1968) **34**: 277–285

24

NANC Nerves and Neuropeptides

PETER J. BARNES

INTRODUCTION

Many neuropeptides are localized to sensory, parasympathetic and sympathetic neurones in the respiratory tract[1,2] (Table 24.1). These peptides have potent effects on bronchomotor tone, airway secretions, the bronchial circulation and inflammatory and immune cells. Although the precise physiological roles of each peptide are not yet fully understood, some clues are provided by their localization and functional effects. Many of the inflammatory and functional effects of neuropeptides are relevant to asthma and there is compelling evidence for the involvement of neuropeptides in the pathophysiology and symptomatology of asthma.[3] The purpose of this chapter is to discuss effects of airway neuropeptides that are relevant to the pathophysiology of asthma and whether this might lead to new therapeutic approaches in the future.

Experimental approaches

Several approaches have been used to investigate the role of neuropeptides in asthma. The effects of exogenous neuropeptides on various target cells relevant to asthma *in vitro* and their effects on airway function *in vivo* have been widely studied in animals and humans.[3] This approach is valuable in revealing the potential effects of a particular neuropeptide, but it is not possible to know exactly what the local concentration of a particular peptide might be. Furthermore there are striking differences between species. Even data in normal human airways may not be relevant to the situation in the diseased

Table 24.1 Neuropeptides in the respiratory tract.

Peptide	Localization
Vasoactive intestinal polypeptide Peptide histidine isoleucine/methionine Peptide histidine valine 42 Helodermin Helospectins I and II Pituitary adenylate cyclase-activating peptide 27 Galanin	Parasympathetic (afferent)
Substance P Neurokinin A Neuropeptide K Calcitonin gene-related peptide Gastrin-releasing peptide	Afferent
Neuropeptide Y	Sympathetic
Somatostatin Enkephalin Cholecystokinin octapeptide	Afferent/uncertain

airway, where there might be alterations in neuropeptide receptor expression and metabolic breakdown.

A more informative approach is to investigate the action of specific blockers or enhancers or to study depletion of the relevant peptide, since this can reveal the role of the endogenous neuropeptide. Again it is possible that the disease state may alter the synthesis, release or metabolism of a particular peptide or its receptors and therefore produce changes in the effects of blocking drugs. It is only recently that potent specific neuropeptide receptor blockers have become available for clinical studies and these will prove to be important tools in the investigation of the role of neuropeptides in disease.

Several animal models of asthma have been investigated, but none of these closely mimics the chronic eosinophilic inflammation characteristic of asthma and they have been poorly predictive of drugs that will have clinical efficacy. The only certain way to evaluate the role of neuropeptides in asthma is to study the effect of specific antagonists or inhibitors in patients with the disease. Specific neuropeptide antagonists suitable for clinical use are now under development and studies are already underway in asthma. Again there may be pitfalls in this approach, as it is usual practice to select patients with mild asthma for such studies. It is possible that neuropeptides are relevant only in certain types of asthma or in more severe and intractable disease. Furthermore it may be difficult to evaluate the effects of neuropeptides on airway function in clinical studies if their main action is on mucosal inflammation, mucus secretion or on airway blood flow, since techniques to evaluate these responses are difficult in patients.

Neuropeptide interactions

In this chapter each neuropeptide is considered separately, but it is important to recognize that neuropeptides act as cotransmitters of classical autonomic nerves and that each peptide may have interactions with other nerves, resulting in complex effects on a tissue. An abnormality in one neuropeptide component may therefore have effects on the release and effects of other neuropeptides and autonomic neurotransmitters. For example, vasoactive intestinal polypeptide (VIP) from parasympathetic nerves and neuropeptide Y (NPY) from sympathetic nerves may have an inhibitory effect on the release of acetylcholine from parasympathetic nerves and neuropeptides from sensory nerves,[4] and many other such interactions are reported.[4] Depletion of neuropeptides from sensory nerves with capsaicin may markedly reduce cholinergic neurotransmission in guinea-pigs but at the same time may enhance inhibitory non-adrenergic non-cholinergic responses.[5] Another possible interaction is that neuropeptides may affect the expression of autonomic receptors, either by influencing the intracellular pathways activated by the receptor or even by regulating the gene expression of receptors.

NON-ADRENERGIC NON-CHOLINERGIC (NANC) NERVES

In addition to classical cholinergic and adrenergic innervation of airways, there are neural mechanisms that are not blocked by cholinergic or adrenergic antagonists.[6] NANC nerves were first described in the gut and therefore their existence in the respiratory tract is to be expected. NANC nerves were initially conceived as a 'third' nervous system in the lungs, but it rapidly became apparent that several distinct neural mechanisms are included and that there are no distinct NANC nerves, although NANC effects are mediated via the release of neurotransmitters from classical neural pathways (Fig. 24.1). NANC mechanisms result in both bronchodilatation and bronchoconstriction, vasodilatation and vasoconstriction, and mucus secretion, indicating that several neurotransmitters are likely to be involved.

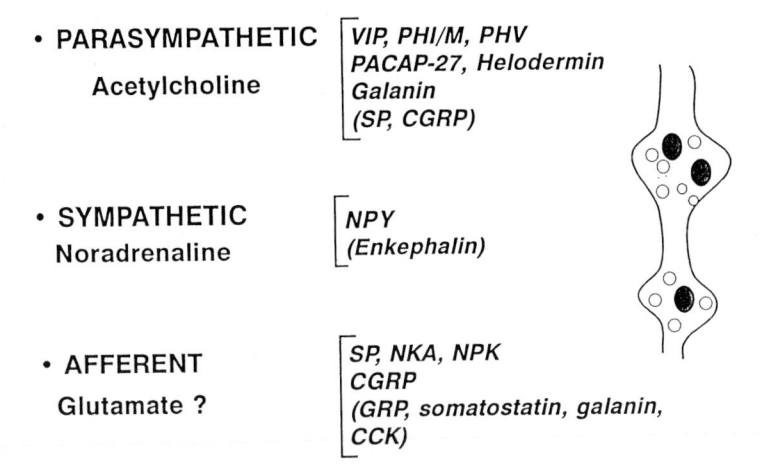

- **PARASYMPATHETIC**
  Acetylcholine

  [VIP, PHI/M, PHV
  PACAP-27, Helodermin
  Galanin
  (SP, CGRP)]

- **SYMPATHETIC**
  Noradrenaline

  [NPY
  (Enkephalin)]

- **AFFERENT**
  Glutamate ?

  [SP, NKA, NPK
  CGRP
  (GRP, somatostatin, galanin, CCK)]

Fig. 24.1 Cotransmission in airway nerves: classical autonomic nerves release multiple neuropeptides. CCK, cholecystokinin; CGRP, calcitonin gene-related peptide; GRP, gastrin-releasing peptide; NKA, neurokinin A; NPK, neuropeptide K; NPY, neuropeptide Y; PACAP-27, pituitary adenylate cyclase-activating peptide 27; PHI/M, peptide histidine isoleucine/methionine; PHV, peptide histidine valine; SP, substance P; VIP, vasoactive intestinal polypeptide.

Inhibitory NANC nerves

Inhibitory NANC (i-NANC) nerves relax airway smooth muscle and therefore serve as a neural bronchodilator mechanism. They have been demonstrated *in vitro* by electrical field stimulation after adrenergic and cholinergic blockade in several species, including humans.[7] In human airway smooth muscle the i-NANC mechanism is the only neural bronchodilator pathway, since there is no functional sympathetic innervation to airway smooth muscle, and there has been considerable research into the identity of the neurotransmitter(s). i-NANC nerves have also been demonstrated *in vivo* in some species by electrical stimulation of the vagus after adrenergic and cholinergic blockade. Stimulation of this pathway produces pronounced and long-lasting bronchodilatation, which may be inhibited by ganglion blockers. This pathway may be activated reflexly by mechanical or chemical stimulation of the larynx. In human subjects *in vivo*, mechanical stimulation of the larynx or chemical stimulation with capsaicin in the presence of adrenergic and cholinergic blockers have also demonstrated reflex reversal of induced tone.[8]

At first it was believed that purines might be the neurotransmitters in i-NANC nerves in airways, but the evidence does not support this view, since purine antagonists do not reduce the i-NANC response and reuptake inhibitors do not potentiate this response. There is evidence that neuropeptides may mediate the i-NANC response in some species; there is convincing evidence that VIP is involved in some species, such as guinea-pig and cat, but this cannot be demonstrated in human airways[9] (see below).

Recent evidence has demonstrated that nitric oxide (NO) may be the neurotransmitter of i-NANC responses in airways.[10] The neuronal isoform of NO synthase is localized to peripheral nerves in human airways[11] and inhibition of NO synthase by L-N^G-nitro-arginine methyl ester (L-NAME) markedly reduces i-NANC responses in human airways.[12–14]

Whether i-NANC responses are impaired in asthma is not yet certain. *In vitro*, the i-NANC response is not reduced compared with responses in normal airways.[15] In patients with mild asthma, no evidence for an impaired NANC bronchodilator reflex has been observed *in vivo*.[8,16] However, this does not preclude a defect in more severely affected asthmatic individuals, in whom the degree of airway inflammation may be greater. In sensitized guinea-pigs exposed to allergen, a reduction in i-NANC responses has been reported.[17] This is presumably due to the release of enzymes or reactive oxygen species from inflammatory cells in the airways. However, the contribution of VIP to i-NANC responses in human airways is not established (although it is possible that it is released with certain neural activation patterns not mimicked by electrical field stimulation *in vitro*) and increased degradation of this peptide in asthma may have a relatively minor effect on airway tone.

Excitatory NANC nerves

Electrical stimulation of guinea-pig bronchi, and occasionally trachea *in vitro*, and vagus nerve *in vivo* produces a component of bronchoconstriction that is not inhibited by atropine[18] (Fig. 24.2). This bronchoconstrictor response has been termed the excitatory NANC (e-NANC) response and there is convincing evidence that it is mediated by the

24 NANC Nerves and Neuropeptides

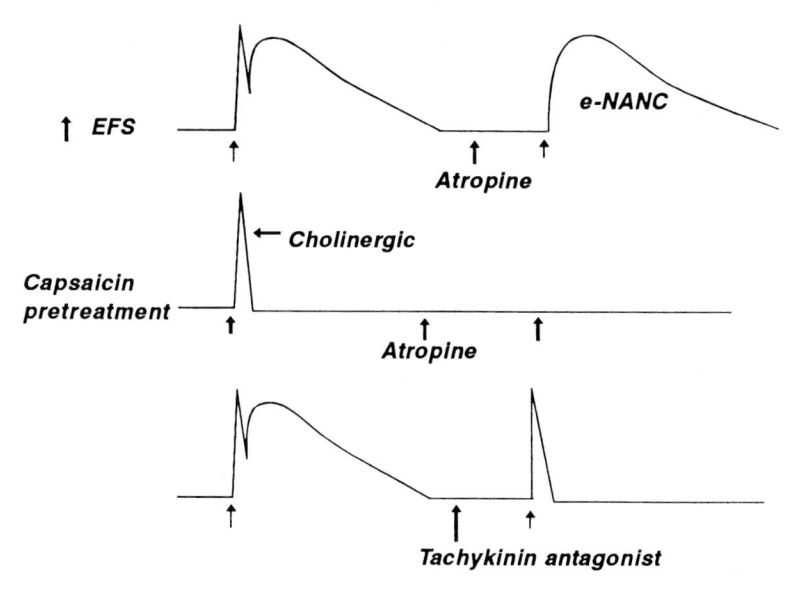

Fig. 24.2 Excitatory non-adrenergic non-cholinergic (e-NANC) response in guinea-pig bronchi *in vitro* after electrical field stimulation (EFS).

retrograde release of tachykinins from unmyelinated sensory nerves (C fibres). A similar e-NANC response has occasionally been reported in human airways *in vitro*, but this is not consistent.[19]

Other NANC responses

Other NANC responses in addition to effects on airway smooth muscle have been described in airways. NANC-mediated secretion of mucus has been demonstrated in cats *in vivo* using vagal nerve stimulation, and in ferret airways *in vitro* using electrical field stimulation.[20] NANC regulation of airway blood flow has been demonstrated in several species, with both vasodilator and vasoconstrictor effects.[21] NANC-mediated plasma extravasation has also been demonstrated in some species. These NANC secretory and vascular effects are likely to be mediated by a variety of neuropeptides, and in some instances by purines and NO.

Cotransmission

Although NANC nerves were originally envisaged as an anatomically separate nervous system, it is now more likely that NANC neural effects are mediated by the release of neurotransmitters from classical autonomic nerves (see Table 24.1). Thus the i-NANC responses in airway smooth muscle are likely to be mediated by the release of cotransmitters such as NO and VIP from cholinergic nerves. NANC vasoconstrictor responses are mediated by the release of NPY from adrenergic nerves. e-NANC bronchoconstrictor responses are mediated by the release of tachykinins from unmyelinated sensory nerves. The physiological relevance of cotransmission is likely to be related to the 'fine tuning' of classical autonomic nerves, although the role of cotransmitters may become more apparent in disease.

Coexistence of several peptides within the same nerve is commonly described in the peripheral nervous system, and multiple combinations are possible, giving rise to the concept of 'chemical coding' of nerve fibres. VIP and peptide histidine isoleucine (PHI) usually coexist since they are derived from the same precursor peptide coded by a single gene. Galanin is often present with VIP in cholinergic neurones. In sensory nerves, substance P (SP), neurokinin A (NKA) and calcitonin gene-related peptide (CGRP) often coexist, although some sensory nerves may also contain galanin and VIP.[22] Similarly, adrenergic nerves that contain NPY may also contain somatostatin, galanin, VIP and enkephalin. Thus there is a complex distribution of neuropeptides in the innervation of the airways, with the same peptides occurring in different types of nerve (see Fig. 24.1). The physiological significance of this complexity is not yet clear, but it seems likely that there may be functional interactions between the multiple neuropeptides released and the classical transmitters that allow complex integration and regulation of functions in the airway.

Neuropeptides are often released by high-frequency firing and therefore may only be coreleased with classical neurotransmitters with certain patterns of neural activation. Little is known about the optimal conditions for neuropeptide release, but it seems likely that release may be favoured by certain physiological and pathophysiological conditions. Furthermore, little is known about the effect of repeated neural activation on the synthesis and release of neuropeptides, but it is possible that in certain diseases, when chronic nerve irritation may occur, that there may be increased neuropeptide gene expression, synthesis and release.

VIP AND RELATED PEPTIDES

VIP-immunoreactive nerves are widely distributed throughout the respiratory tract in humans[2] and there is also evidence for the presence of the closely related peptides, peptide histidine methionine (PHM) and pituitary adenylate cyclase-activating peptide 27 (PACAP-27).[22,23] VIP may be localized to parasympathetic and sensory nerves.[24]

Airway effects

VIP is a potent relaxant of human bronchi *in vitro* but has little effect on peripheral airways[25] (Fig. 24.3). This is consistent with autoradiographic mapping studies which show that VIP receptors are expressed in airway smooth muscle of proximal but not distal human airways.[26] This suggests that VIP, released from parasympathetic nerves in proximal airways, may act as an endogenous bronchodilator and may counteract cholinergic bronchoconstriction. VIP acts as a functional antagonist by increasing cyclic AMP concentrations in airway smooth muscle, but also inhibits the release of acetylcholine from airway cholinergic nerves at a ganglionic and postganglionic level via

24 NANC Nerves and Neuropeptides

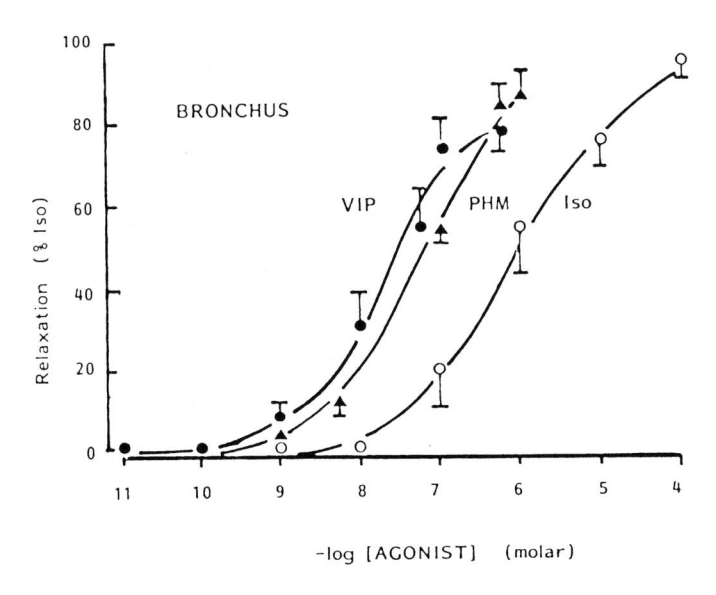

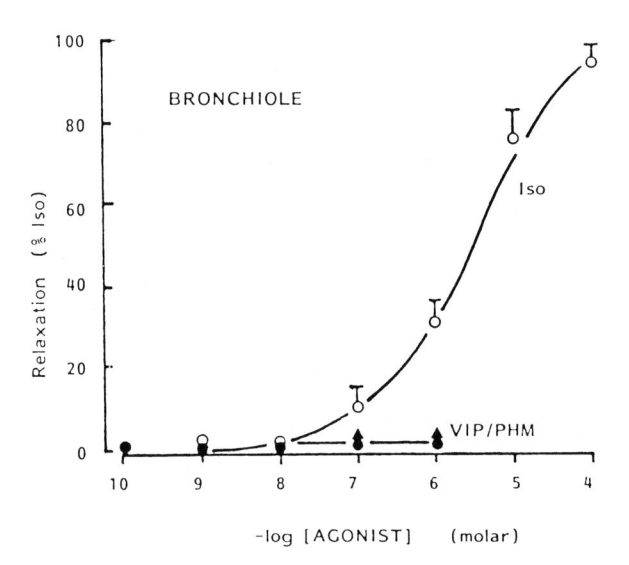

Fig. 24.3 Effect of vasoactive intestinal polypeptide (VIP), peptide histidine methionine (PHM) and isoprenaline (Iso) on human airways *in vitro*.

prejunctional receptors.[27–30] Although VIP is a potent bronchodilator after intravenous administration in cats,[31] it has no effect on airway function in normal human subjects, despite profound vascular effects which limit the dose that can be administered.[32] VIP is approximately 10-fold more potent as a vasodilator than as a bronchodilator *in vitro*[33] and this is reflected by a higher density of VIP receptors in pulmonary vascular compared with airway smooth muscle.[26] In asthmatic patients, inhaled VIP has no bronchodilator effect, although a β-adrenergic agonist in the same subjects is markedly effective.[34] Inhaled VIP has a small protective effect against the bronchoconstrictor effect of histamine[34] but no effect against exercise-induced bronchoconstriction.[35] This lack of potency of inhaled VIP may be explained by the epithelium, since this possesses proteolytic enzymes and may present a barrier to diffusion.

It is likely that VIP is more important as a regulator of airway blood flow than airway smooth muscle tone. VIP increases airways blood flow in dogs and pigs, being more potent on tracheal than bronchial vessels,[36] and may provide a mechanism for increasing blood flow to contracted smooth muscle. Thus, if VIP is released from cholinergic nerves, it may improve muscular perfusion during cholinergic contraction (Fig. 24.4).

VIP also stimulates mucus secretion, measured by ^{35}S-labelled glycoprotein secretion, in ferret airway *in vitro*[37] and there is a high density of VIP receptors in human airway submucosal glands.[26] VIP has a surprising inhibitory effect on glycoprotein secretion from human tracheal explants,[38] but the effects of VIP on mucus secretion may be complex and may depend on the drive to gland secretion. VIP is a potent stimulant of chloride ion transport and therefore water secretion in dog tracheal epithelium,[39] suggesting that VIP may be a regulator of airway water secretion and therefore mucociliary clearance. The high density of VIP receptors on epithelial cells of human airways suggests that VIP may regulate ion transport and other epithelial functions in human airways.[26]

VIP inhibits release of mediators from pulmonary mast cells[40] and may have several other anti-inflammatory actions in airways. VIP may interact with T-lymphocytes and has the potential to act as a local immunomodulator in airways.[41] VIP and a stable analogue (Ro 25-1553) inhibit the release of the cytokines interleukin (IL)-2 and IL-4 but not interferon (IFN)-γ from T-lymphocytes.[42] VIP also inhibits the proliferation of airway smooth muscle[43] and fibroblast proliferation,[44] indicating that VIP may prevent airway remodelling in asthma.

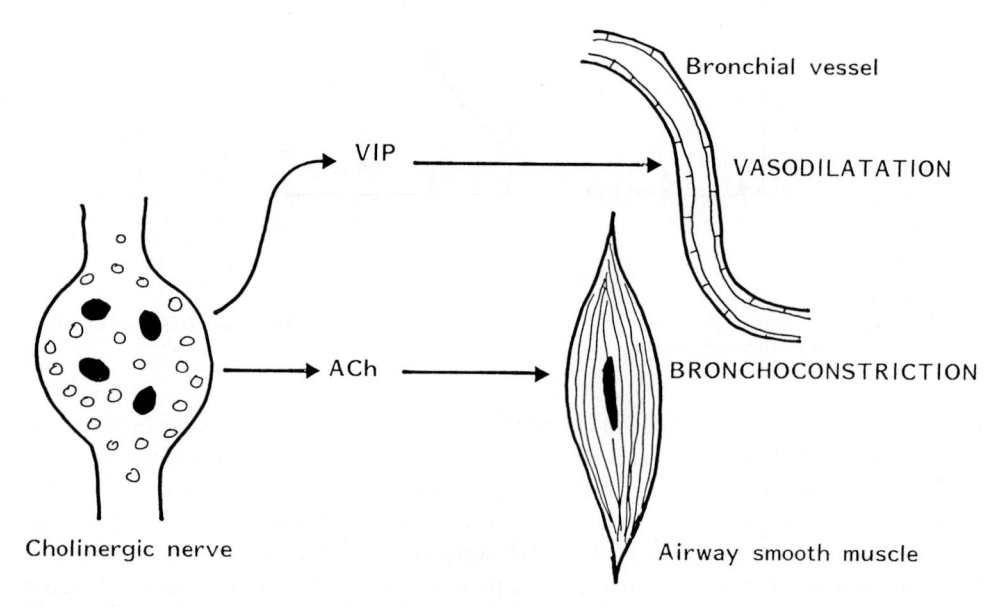

Fig. 24.4 Vasoactive intestinal polypeptide (VIP) is released from cholinergic nerves in the airways. Since VIP is a potent vasodilator it may increase blood flow to airways that are contracted with acetylcholine (ACh).

VIP as i-NANC neurotransmitter

Several lines of evidence implicate VIP as a neurotransmitter of i-NANC nerves in airways, although this is species dependent.[7] In guinea-pig trachea, α-chymotrypsin, which degrades VIP, blocks responses to exogenous VIP and results in a reduction in i-NANC response by about 50%;[45] antiserum to VIP also reduces i-NANC responses.[46] Furthermore, a cyclic AMP selective phosphodiesterase inhibitor enhances the i-NANC response in guinea-pig trachea, suggesting that the neurotransmitter increases cyclic AMP; this would be consistent with the effect of VIP.[47] In human airways *in vitro*, α-chymotrypsin, under conditions that completely block the bronchodilator response to exogenous VIP, has no effect on the pronounced i-NANC response, strongly suggesting that neither VIP nor related peptides (PACAP-27, PHM) also susceptible to degradation by α-chymotrypsin are involved in the i-NANC response in human airways[9] (Fig. 24.5). In guinea-pig airways, VIP and related peptides account for about half of the i-NANC response. Recent studies have demonstrated that NO accounts for the remaining response,[48,49] whereas in human and feline airways NO appears to account for all of the i-NANC response,[14,50] indicating that VIP is likely to play little role in regulating airway smooth muscle tone.

It is probably misleading to think of i-NANC nerves as a discrete bronchodilator pathway; it is more likely that they function as a braking mechanism to cholinergic bronchoconstriction, particularly as both transmitters may be coreleased from parasympathetic nerves in the airways. In guinea-pig trachea, α-chymotrypsin increases cholinergic nerve-induced bronchoconstriction (presumably by removing the braking action of endogenously released VIP),[51] but this is not the case in human airways.[13] However, NO acts as a modulator of cholinergic bronchoconstriction in both species.[13,52]

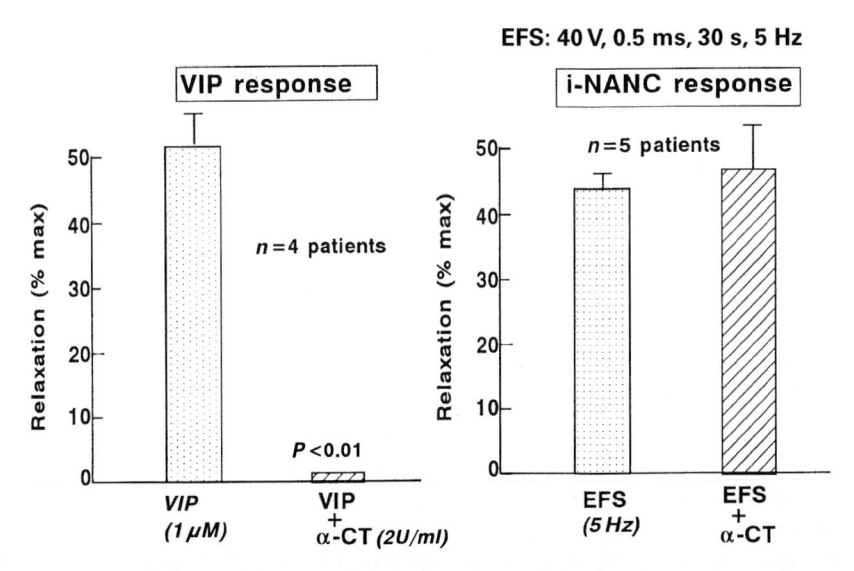

Fig. 24.5 Evidence against a role for vasoactive intestinal polypeptide (VIP) in the inhibitory non-adrenergic non-cholinergic (i-NANC) response in human airways. While the VIP response is blocked by the enzyme α-chymotrypsin (α-CT), the enzyme has no effect on the iNANC response. EFS, electrical field stimulation.

Role in asthma?

The role of VIP in the pathophysiology of asthma is far from certain in the absence of potent specific antagonists. Indeed it is not clear whether VIP has beneficial or deleterious effects, since its vasodilator and mucous secretory effects predominate over any effect on airway smooth muscle in the airways, and particularly in more peripheral airways. A striking absence of VIP-immunoreactive nerves has been described in the lungs of patients with asthma in tissues largely obtained at post-mortem.[53] The loss of VIP immunoreactivity from all tissues including pulmonary vessels is so complete that it seems unlikely to represent a fundamental absence of VIP-immunoreactive nerves in asthma. More likely is the possibility that enzymes, such as mast cell tryptase, are released from inflammatory cells in asthma and that these rapidly degrade VIP when sections are cut.[54] Biopsies taken from patients with mild asthma suggest that VIP-immunoreactive nerves appear normal in asthma[55] and the VIP content of asthmatic lungs is normal.[56] Nor is there any abnormality in the distribution of VIP receptors in the airways of asthmatic patients.[57] VIP antibodies, which would neutralize the effects of VIP, have also been described in the plasma of asthmatic patients, but as they are found just as often in non-asthmatic patients their significance is doubtful.[58] Low plasma concentrations of VIP have been described during exacerbations of asthma, although the significance of this observation is difficult to evaluate as the source of VIP in plasma is unknown.[59]

While it seems unlikely that there would be any primary abnormality in VIP innervation in the airways of patients with asthma, it is possible that a secondary abnormality may arise as a result of the inflammatory process in the airway. Mast cell tryptase degrades VIP[60] and is known to be elevated in asthmatic airways.[61] Inhibition of tryptase potentiates the bronchodilator response to VIP in human airways *in vitro*[62] and increases the *in vitro* responsiveness of canine airways.[63] Tryptase released from mast cells in the asthmatic airway may then more rapidly degrade VIP and related peptides released from airway cholinergic nerves. This would remove a 'brake' from cholinergic nerves and lead to exaggerated cholinergic reflex bronchoconstriction (Fig. 24.6).

VIP-related peptides

Several other peptides have not been identified in the mammalian nervous system that are similar in structure and effect to VIP.

PHI

PHI and its human equivalent PHM have a marked structural similarity to VIP, with 50% amino acid sequence homology. PHI and PHM are encoded by the same gene as VIP and both peptides are synthesized from the same prohormone. PHI has a similar immunocytochemical distribution in lung to VIP: PHI-immunoreactive nerves supply airway smooth muscle (especially larger airways), bronchial and pulmonary vessels, submucosal glands and airway ganglia.[2] There are some differences between VIP and PHI, since PHI is less potent as an airway vasodilator[64] and more potent as a stimulant of secretion than VIP.[65] In human bronchi *in vitro*, PHM is a potent relaxant and is equipotent to VIP.[25]

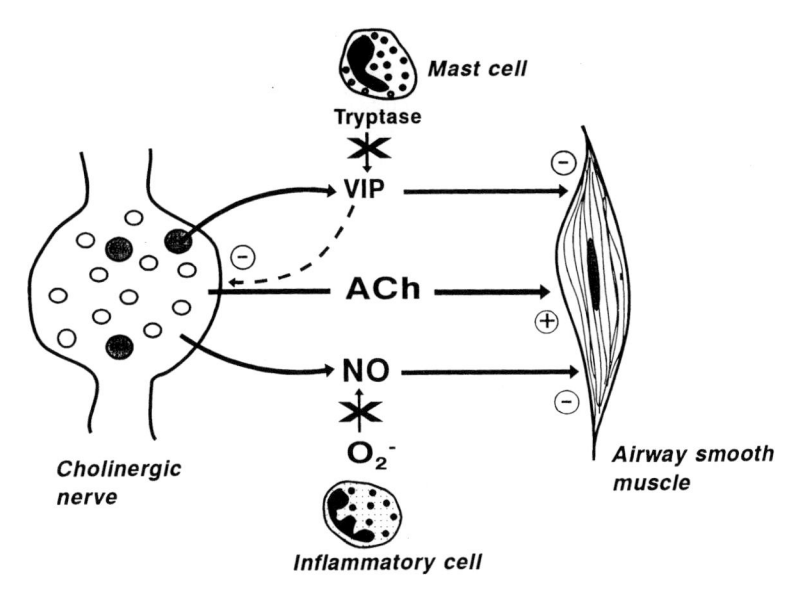

Fig. 24.6 Vasoactive intestinal polypeptide (VIP) and nitric oxide (NO) may be coreleased from cholinergic nerves and act as functional antagonists of cholinergic bronchoconstriction. In addition, they may act prejunctionally to inhibit acetylcholine (ACh) release. In asthma, enzymes such as tryptase released from airway mast cells may rapidly degrade VIP, and oxygen free radicals, such as superoxide anions (O_2^-), from inflammatory cells may inactivate NO, thus leading to exaggerated cholinergic neural bronchoconstriction.

Peptide histidine valine

Peptide histidine valine (PHV-42 is an N-terminally extended precursor of VIP. PHV is a potent bronchodilator of guinea-pig airways *in vitro*,[66] but when infused in asthmatic patients has no demonstrable bronchodilator effect.[67] It is not yet clear whether this peptide is released from airway nerves.

Helodermin and helospectins

Helodermin is a 35 amino acid peptide of similar structure to VIP that has been isolated from the salivary gland venom of the Gila monster lizard. Helodermin immunoreactivity has been localized to airway nerves and the peptide has similar effects to VIP but a longer duration of action. Helodermin is a potent relaxant of airway smooth muscle *in vitro*, and helodermin immunoreactivity has been reported in trachea.[68] Helodermin appears to activate a high-affinity form of the VIP receptor.[69] Helospectins I and II are two closely related peptides that have recently been localized to nerves within the respiratory tract;[68,70] both peptides potently relax guinea-pig airways.[68]

PACAP

PACAP, a 38 amino acid peptide isolated from sheep hypothalamus, and PACAP-27, a truncated fragment, have marked sequence homology with VIP and have been demonstrated in the peripheral nervous system.[71] PACAP immunoreactivity has a similar distribution to VIP in airways of several species and may be localized to cholinergic and also to capsaicin-sensitive afferent nerves.[70] The effects of PACAP-27 are similar to those of VIP. PACAP relaxes guinea-pig tracheal and bronchial smooth muscle *in vitro* and is approximately three times less potent than VIP.[72] PACAP, like VIP, also inhibits the release of inflammatory mediators from chopped lung tissue.[72] There appears to be a particularly high density of receptors for PACAP in lung tissue.[73] The reason for the coexistence of so many similar peptides with similar effects is not clear and until specific antagonists are developed it will be difficult to elucidate.

TACHYKININS

SP and NKA, but not neurokinin B, are localized to sensory nerves in the airways of several species. SP-immunoreactive nerves are abundant in rodent airways but sparse in human airways.[74–76] Rapid enzymatic degradation of SP in airways, and the fact that SP concentrations may decrease with age and possibly after cigarette smoking, could explain the difficulty in demonstrating this peptide in some studies. SP-immunoreactive nerves in the airway are found beneath and within the airway epithelium, around blood vessels and, to a lesser extent, within airway smooth muscle. SP-immunoreactive nerves fibres also innervate parasympathetic ganglia, suggesting a sensory input that may modulate ganglionic transmission and so result in ganglionic reflexes.

SP in the airways is localized predominantly to capsaicin-sensitive unmyelinated nerves in the airways, but chronic administration of capsaicin only partially depletes the lung of tachykinins, indicating the presence of a population of capsaicin-resistant SP-immunoreactive nerves, as in the gastrointestinal tract.[19,24] Similar capsaicin denervation studies are not possible in human airways; however, after extrinsic denervation by heart–lung transplantation there appears to be a loss of SP-immunoreactive nerves in the submucosa.[77]

Effects on airways

Airway smooth muscle

Tachykinins have many different effects on the airways that may be relevant to asthma; these effects are mediated via NK_1 receptors (preferentially activated by SP) and NK_2 receptors (activated by NKA). Tachykinins constrict smooth muscle of human airways *in vitro* via NK_2 receptors.[78,79] The contractile response to NKA is significantly greater in smaller human bronchi than in more proximal airways, indicating that tachykinins may have a more important constrictor effect on more peripheral airways,[80] whereas cholinergic constriction tends to be more pronounced in proximal airways. This is consistent with the autoradiographic distribution of tachykinin receptors, which are distributed to small and large airways. *In vivo* SP does not cause bronchoconstriction or cough, either by intravenous infusion[81,82] or by inhalation,[81,83] whereas NKA causes bronchoconstriction after both intravenous administration[82] and inhalation in asthmatic subjects.[83] Mechanical removal of airway epithelium potentiates the bronchoconstrictor response to tachykinins,[84,85] largely because the neutral endopeptidase (NEP), which is a key enzyme in the degradation of tachykinins in airways, is strongly expressed on epithelial cells.

Airway secretions

SP stimulates mucus secretion from submucosal glands in human airways *in vitro*[86] and is a potent stimulant to goblet cell secretion in guinea-pig airways.[87] Mucus secretion is mediated via NK_1 receptors.[88,89] SP is likely to mediate the increase in goblet cell discharge after vagus nerve stimulation and exposure to cigarette smoke.[90,91]

Vascular effects

Stimulation of the vagus nerve in rodents causes microvascular leakage, which is prevented by prior treatment with capsaicin or by a tachykinin antagonist, indicating that release of tachykinins from sensory nerves mediates this effect. Amongst the tachykinins, SP is most potent at causing leakage in guinea-pig airways[92] and NK_1 receptors have been localized to postcapillary venules in the airway submucosa.[93] Inhaled SP also causes microvascular leakage in guinea-pigs and its effect on the microvasculature is more marked than its effect on airway smooth muscle.[94]

Tachykinins have potent effects on airway blood flow. Indeed the effect of tachykinins on airway blood flow may be the most important physiological and pathophysiological role of tachykinins in airways. In canine and porcine trachea both SP and NKA cause a marked increase in blood flow.[95,96] Tachykinins also dilate canine bronchial vessels *in vitro*, probably via an endothelium-dependent mechanism.[97] Tachykinins also regulate bronchial blood flow in pig; stimulation of the vagus nerve causes a vasodilatation mediated by the release of sensory neuropeptides, although it is likely that CGRP coreleased from these nerves is more important.[96]

Inflammatory effects

Tachykinins may also interact with inflammatory and immune cells,[98,99] although whether this is of pathophysiological significance remains to be determined. SP degranulates certain types of mast cell, such as those in human skin, although this is not mediated via a tachykinin receptor.[100] There is no evidence that tachykinins degranulate lung mast cells.[101] SP has a degranulating effect on eosinophils;[102] again the degranulation is related to high concentrations of peptide and, as for mast cells, is not mediated via a tachykinin receptor. At lower concentrations, tachykinins have been reported to enhance eosinophil chemotaxis.[103] Tachykinins may activate alveolar macrophages[104] and monocytes to release inflammatory cytokines, such as IL-6.[105] Tachykinins and vagus nerve stimulation also cause transient vascular adhesion of neutrophils in the airway circulation.[106]

SP stimulates proliferation of blood vessels (angiogenesis)[107] and may therefore be involved in the new vessel formation found in asthmatic airways. SP and NKA also stimulate the proliferation and chemotaxis of human lung fibroblasts, suggesting that tachykinins may contribute to the fibrotic process in chronic asthma.[108]

Neural effects

In guinea-pig trachea, tachykinins also potentiate cholinergic neurotransmission at postganglionic nerve terminals; an NK_2 receptor appears to be involved.[109] There is also potentiation at ganglionic level,[110,111] which appears to be mediated via an NK_1 receptor.[111] Endogenous tachykinins may also facilitate cholinergic neurotransmission, since capsaicin pretreatment results in a significant reduction in cholinergic neural responses both *in vitro* and *in vivo*.[112,113] However, in human airways there is no evidence for a facilitatory effect on cholinergic neurotransmission,[114] although such an effect has been reported in the presence of potassium channel blockers.[115]

Metabolism

Tachykinins are subject to degradation by at least two enzymes, angiotensin-converting enzyme (ACE) and NEP.[116] ACE is predominantly localized to vascular endothelial cells and therefore degrades intravascular peptides. ACE inhibitors, such as captopril, enhance bronchoconstriction due to intravenous SP[117,118] but not inhaled SP.[119] NKA is not a good substrate for ACE, however. NEP appears to be the most important enzyme for the breakdown of tachykinins in tissues. Inhibition of NEP by phosphoramidon or thiorphan markedly potentiates bronchoconstriction *in vitro* in animal[120] and human airways[121] and after inhalation *in vivo*.[119] NEP inhibition also potentiates mucus secretion in response to tachykinins in human airways.[86] NEP inhibition enhances e-NANC and capsaicin-induced bronchoconstriction, due to the release of tachykinins from airways sensory nerves.[84,122] The activity of NEP in the airways appears to be an important factor in determining the effects of tachykinins; any factors that inhibit the enzyme or its expression may be associated with increased effects of exogenous or endogenously released tachykinins. Several of the stimuli known to induce bronchoconstrictor responses in asthmatic patients have been found to reduce the activity of airway NEP[116] (Fig. 24.7).

CGRP

CGRP-immunoreactive nerves are abundant in the respiratory tract of several species. CGRP is costored and colocalized with SP in afferent nerves.[123] CGRP has been extracted from, and is localized to, human airways.[76,124] CGRP is found in trigeminal, nodose-jugular and dorsal root ganglia[2] and has also been detected in neuroendocrine cells of the lower airways.

CGRP is a potent vasodilator that has long-lasting effects. CGRP is an effective dilator of human pulmonary vessels *in vitro* and acts directly on receptors on vascular smooth muscle.[125] It also potently dilates bronchial vessels *in vitro*[125] and produces a marked and long-lasting increase in airway blood flow in anaesthetized dogs[126] and conscious sheep *in vivo*.[127] Receptor mapping studies have demonstrated that CGRP receptors are localized predominantly to bronchial vessels rather than to smooth muscle or epithelium in human airways.[128] It is possible that CGRP may be the predominant mediator of arterial

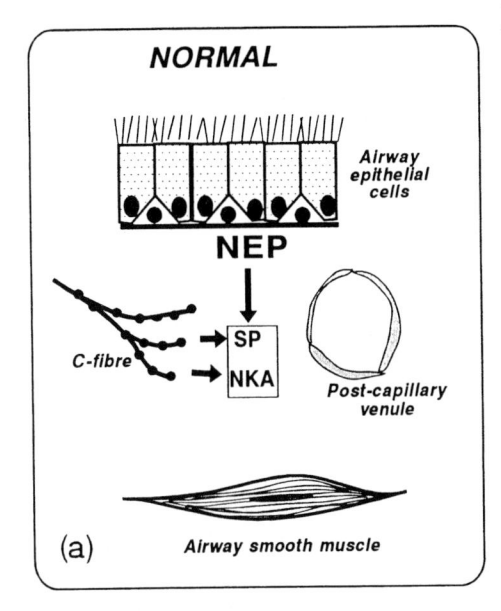

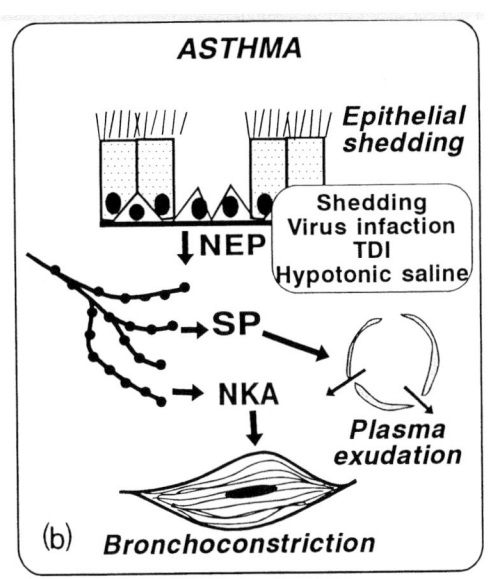

Fig. 24.7 Interaction of tachykinins with airway epithelium. When epithelium is intact, neutral endopeptidase (NEP) degrades substance P (SP) and neurokinin A (NKA) released from sensory nerves (a). In asthmatic airways, when epithelium is shed or NEP downregulated any tachykinins released will have an exaggerated effect (b). TDI, toluene diisocyanate.

vasodilatation and increased blood flow in response to sensory nerve stimulation in the bronchi.[96] CGRP may be an important mediator of airway hyperaemia in asthma.

By contrast, CGRP has no direct effect of airway microvascular leak.[92] In the skin, CGRP potentiates the leakage produced by SP, presumably by increasing the blood delivery to the sites of plasma extravasation in the postcapillary venules[129] (Fig. 24.8).

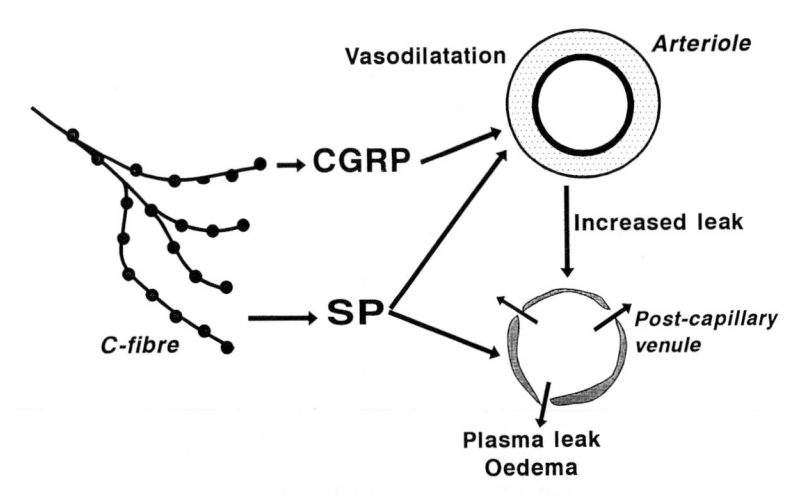

Fig. 24.8 Effect of sensory neuropeptides in airway vessels. Substance P (SP) causes vasodilatation and plasma exudation, whereas calcitonin gene-related peptide (CGRP) causes vasodilatation of arterioles, which may theoretically increase plasma extravasation by increasing blood delivery to leaky postcapillary venules.

This does not occur in guinea-pig airways when CGRP and SP are coadministered, possibly because blood flow in the airways is already high,[92] although an increased leakage response has been reported in rat airways.[130] It is possible that potentiation of leak may occur when the two peptides are released together from sensory nerves.

CGRP causes constriction of human bronchi *in vitro*,[124] although receptor mapping studies suggest few, if any, CGRP receptors on airway smooth muscle in human or guinea-pig airways;[128] this suggests that the bronchoconstrictor response reported in human airways may be mediated indirectly. In guinea-pig airways, CGRP causes constriction via the release of the potent constrictor peptide endothelin.[131]

CGRP has a weak inhibitory effect on cholinergically stimulated mucus secretion in ferret trachea[132] and on goblet cell discharge in guinea-pig airways.[87] This is probably related to the low density of CGRP receptors on mucous secretory cells, but does not preclude the possibility that CGRP might increase mucus secretion *in vivo* by increasing blood flow to submucosal glands.

CGRP injection into human skin causes a persistent flare and biopsies have revealed an infiltration of eosinophils.[133] CGRP itself does not appear to be chemotactic for eosinophils, although proteolytic fragments of the peptide are active,[134] suggesting that CGRP released into the tissues may lead to eosinophilic infiltration.

CGRP inhibits the proliferative response of T-lymphocytes to mitogens and specific receptors have been demonstrated on these cells.[135] CGRP also inhibits macrophage secretion and the capacity of macrophages to activate T-lymphocytes.[136] This suggests that CGRP has potential anti-inflammatory actions in the airways. CGRP may also play a role in epithelial repair as it stimulates airway epithelial cell proliferation.[137]

NEUROGENIC INFLAMMATION

Sensory nerves may be involved in inflammatory responses through the antidromic release of neuropeptides from nociceptive nerves or C fibres via a local (axon) reflex (Fig. 24.9). The phenomenon is well documented in several organs, including skin, eye, gastrointestinal tract and bladder.[138] There is also increasing evidence that neurogenic inflammation occurs in the respiratory tract[139,140] and it is possible that it may contribute to the inflammatory response in asthma.[141-143]

Neurogenic inflammation in animal models

There are several lines of evidence that neurogenic inflammation may be important in animal models that may have relevance to asthma. These models have usually been in rodents, where tachykinin effects are pronounced but may not be predictive of the role of tachykinins in human airways. Several experimental approaches have been used to assess the role of sensory neuropeptides in animal models of asthma; these include studies of depletion with capsaicin, enhancement with inhibitors of NEP, tachykinin receptor antagonists and inhibitors of sensory neuropeptide release.

Capsaicin pretreatment to deplete neuropeptides from C fibres results in degeneration of C fibres in neonatal animals or depletion of sensory neuropeptides after acute treatment

24 NANC Nerves and Neuropeptides

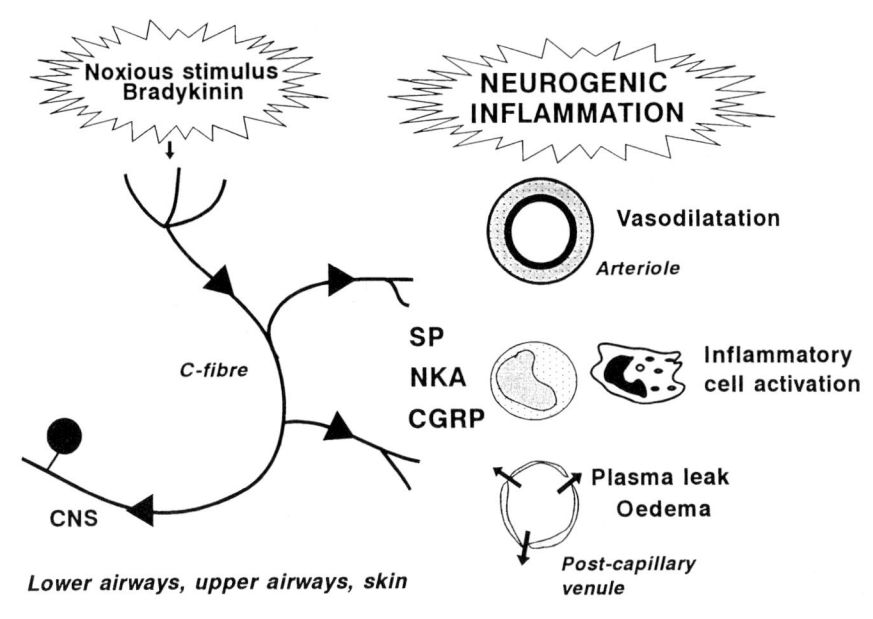

Fig. 24.9 Neurogenic inflammation involves the retrograde release of neuropeptides from C-fibre nerve endings. CNS, central nervous system; CGRP, calcitonin gene-related peptides; NKA, neurokinin A; SP, substance P.

in adult animals. In rat trachea, capsaicin pretreatment inhibits the microvascular leakage induced by irritant gases, such as cigarette smoke;[144] and inhibits goblet cell discharge and microvascular leak induced by cigarette smoke in guinea-pigs.[91] Capsaicin-sensitive nerves may also contribute to the bronchoconstriction and microvascular leak induced by isocapnic hyperventilation,[145] inhaled sodium metabisulphite,[146] nebulized hypertonic saline[147] and toluene diisocyanate[148] in rodents. In guinea-pigs, capsaicin pretreatment has little or no effect on the acute bronchoconstrictor or plasma exudation response to allergen inhalation in sensitized animals.[149] Administration of capsaicin increases airway responsiveness in guinea-pigs to cholinergic agonists; this effect is prevented by prior treatment with capsaicin, suggesting that capsaicin-sensitive nerves release products that increase airway responsiveness.[150] In pigs, capsaicin pretreatment inhibits the vasodilator response to allergen (which may be mediated by the release of CGRP).[151] In a model of chronic allergen exposure in guinea-pigs, capsaicin pretreatment results in complete inhibition of airway hyperresponsiveness, without any change in the eosinophil inflammatory response.[152] In rabbits, neonatal capsaicin treatment inhibits the airway hyperresponsiveness associated with neonatal allergen sensitization, although this does not appear to be associated with any change in content of sensory neuropeptides in lung tissue.[153] This suggests that capsaicin-sensitive nerves may play a role in chronic inflammatory responses to allergen. There has been speculation that mast cells in the airways might be influenced by capsaicin-sensitive nerves. Histological studies have demonstrated a close proximity between mast cells and sensory nerves in airways and antidromic stimulation of the vagus nerve leads to mast cell mediator release.[154]

The activity of NEP may be an important determinant of the extent of neurogenic inflammation in airways; inhibition of NEP in rodent by thiorphan or phosphoramidon has been shown to enhance neurogenic inflammation in various rodent models. NEP is not specific to tachykinins and is also involved in the metabolism of other bronchoactive peptides, including kinins and endothelins. Certain virus infections enhance e-NANC responses in guinea-pigs[155] and *Mycoplasma* infection enhances neurogenic microvascular leakage in rats,[139] an effect mediated by inhibition of NEP activity. Influenza virus infection of ferret trachea *in vitro* and of guinea-pigs *in vivo* inhibits the activity of epithelial NEP and markedly enhances the bronchoconstrictor responses to tachykinins.[156] Similarly, Sendai virus infection potentiates neurogenic inflammation in rat trachea.[157] This may explain why respiratory tract virus infections are so deleterious to patients with asthma. Hypertonic saline also impairs epithelial NEP function, leading to exaggerated tachykinin responses,[147] and cigarette smoke exposure has a similar effect, which can be explained by an oxidizing effect on the enzyme.[158] Toluene diisocyanate, albeit at rather high doses, also reduces NEP activity and this may be a mechanism contributing to the airway hyperresponsiveness that may follow exposure to this chemical.[159] Thus, many of the agents that lead to exacerbations of asthma appear to reduce the activity of NEP at the airway surface, thus leading to exaggerated responses to tachykinins (and other peptides) and so to increased airway inflammation.

Specific peptide and non-peptide tachykinin antagonists have now been developed and provide a more specific tool to investigate the role of tachykinins in animal models. The NK_1-receptor antagonist CP 96,345 blocks the plasma exudation response to vagus nerve stimulation and to cigarette smoke in guinea-pig airways[160,161] without affecting the bronchoconstrictor response, which is blocked by the NK_2-antagonist SR 48,968.[79] Similar results have been obtained with the very potent NK_1-selective antagonist FK 888.[162] CP 96,345 also blocks hyperpnoea- and bradykinin-induced plasma exudation in guinea-pigs,[163,164] but has no effect on the acute plasma exudation induced by allergen in sensitized animals.[164]

Several agonists act on prejunctional receptors on airway sensory nerves to inhibit the release of neuropeptides and neurogenic inflammation.[165] Opioids are the most effective inhibitory agonists, acting via prejunctional μ receptors, and have been shown to inhibit cigarette smoke-induced discharge from goblet cells in guinea-pig airways *in vivo*[166] and to inhibit ozone-induced hyperreactivity in guinea-pigs, which appears to be mediated via sensory nerves.[167] Several other agonists are also effective and may act by opening a common calcium-activated large-conductance potassium channel in sensory nerves.[168] Openers of other potassium channels, which achieve the same hyperpolarization of the sensory nerve, are also effective in blocking neurogenic inflammation in rodents[169] and block cigarette smoke-induced goblet cell secretion in guinea-pigs.[170]

Neurogenic inflammation in asthma?

Although it was proposed several years ago that neurogenic inflammation and peptides released from sensory nerves might be important as an amplifying mechanism in asthmatic inflammation[141] (Fig. 24.10), there is little evidence to date to support this idea, despite the extensive work in rodent models. This is partly because it has proved difficult to apply the same approaches to human volunteers.

24 NANC Nerves and Neuropeptides

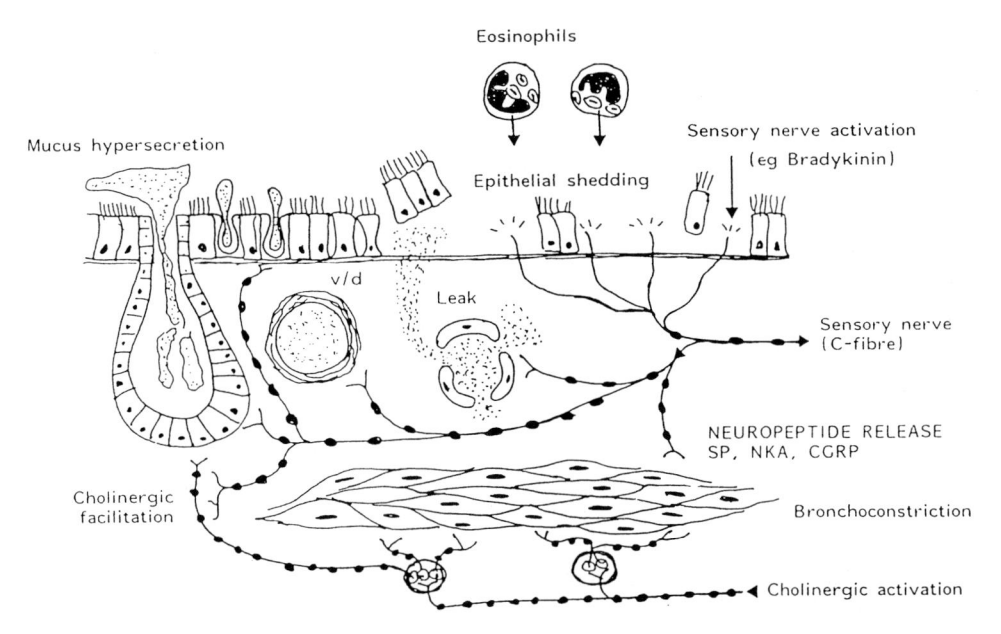

Fig. 24.10 Possible neurogenic inflammation in asthmatic airways via retrograde release of peptides from sensory nerves via an axon reflex. Substance P (SP) causes vasodilatation, plasma exudation and mucus secretion, whereas neurokinin A (NKA) causes bronchoconstriction and enhanced cholinergic reflexes and calcitonin gene-related peptide (CGRP) vasodilatation. v/d, vasodilatation.

Sensory nerves in human airways

In comparison with rodent airways, SP- and CGRP-immunoreactive nerves are very sparse in human airways. Quantitative studies indicate that SP-immunoreactive fibres constitute only 1% of the total number of intraepithelial fibres, whereas in guinea-pig they comprise 60% of the number.[171] Chronic inflammation may lead to changes in the pattern of innervation, through the release of neurotrophic factors from inflammatory cells. Thus in chronic arthritis and inflammatory bowel disease there is an increase in the density of SP-immunoreactive nerves.[172] A striking increase in SP-like immunoreactive nerves has been reported in the airway of patients with fatal asthma,[173] although this has not been seen in biopsies of asthmatic patients.[55] Elevated concentrations of SP in bronchoalveolar lavage sputum of patients with asthma have reported, with a further rise after allergen challenge.[174,175] SP also increases in the bronchoalveolar lavage of normal volunteers exposed to ozone, possible because of a reduction in NEP activity.[176]

Cultured sensory neurones are stimulated by nerve growth factor (NGF), which markedly increases the transcription of preprotachykinin A gene, the major precursor peptide for tachykinins.[177] Since NGF may be released from several types of inflammatory cell, it is possible that this could lead to increased tachykinin synthesis and increased nerve growth. Increased preprotachykinin gene has been detected in the nodose ganglion of guinea-pigs after allergen challenge.[178]

Sensory nerve activation

Sensory nerves may be activated in airway disease. In asthmatic airways the epithelium is often shed, thereby exposing sensory nerve endings. Sensory nerves in asthmatic airways may be 'hyperalgesic' as a result of exposure to inflammatory mediators such as prostaglandins and certain cytokines (such as IL-1β and tumour necrosis factor α). Hyperalgesic nerves may then be activated more readily by other mediators, such as kinins. Capsaicin induces bronchoconstriction and plasma exudation in guinea-pigs and increases airway blood flow in pigs.[151] In humans, capsaicin inhalation causes cough and a *transient* bronchoconstriction, which is inhibited by cholinergic blockade and is probably due to a laryngeal reflex.[179,180] This suggests that neuropeptide release does not occur in human airways, although it is possible that insufficient capsaicin reaches the lower respiratory tract because the dose is limited by coughing. In patients with asthma, there is no evidence that capsaicin induces a greater degree of bronchoconstriction than in normal individuals. Bradykinin is a potent bronchoconstrictor in asthmatic patients and also induces coughing and a sensation of chest tightness, which closely mimics a naturally occurring asthma attack.[181,182] Yet it is a weak constrictor of human airways *in vitro*, suggesting that its potent constrictor effect is mediated indirectly. In guinea-pigs bradykinin instilled into the airways causes bronchoconstriction that is reduced significantly by a cholinergic antagonist (as in asthmatic patients[182]) and also by capsaicin pretreatment.[183] The plasma leakage induced by inhaled bradykinin is inhibited by an NK_1-receptor antagonist.[164,184] Bradykinin activates airways C fibres and also sensitizes them to other activating stimuli.[185,186] This indicates that bradykinin activates sensory nerves in the airways and that part of the airway response is mediated by release of constrictor peptides from capsaicin-sensitive nerves. In asthmatic patients an inhaled non-selective tachykinin antagonist FK 224 has recently been shown to reduce the bronchoconstrictor response to inhaled bradykinin and also to block the cough response in those subjects that coughed in response to bradykinin.[187]

Studies with NEP inhibitors

Intravenous acetorphan, which is hydrolysed to thiorphan, potentiates the wheal and flare response to intradermal SP but there is no effect on baseline airway calibre or on bronchoconstriction induced by the 'neurogenic' trigger sodium metabisulphite.[188] The lack of effect could be due to inadequate inhibition of NEP in the airways, particularly at the level of the epithelium. Nebulized thiorphan potentiates the bronchoconstrictor response to inhaled NKA in normal and asthmatic subjects,[189,190] but there was no effect on baseline lung function in asthmatic patients, indicating that there is unlikely to be any basal release of tachykinins. NEP is strongly expressed in the human airway, but there is no evidence based on immunocytochemical staining or *in situ* hybridization that it is defective in asthmatic airways.[191] The fact that after inhaled thiorphan the bronchoconstrictor response to inhaled NKA is further enhanced in asthmatic subjects provides supportive functional data that NEP function may not be impaired, at least in mild asthma.[190] Of course, it is possible that NEP may become dysfunctional after viral infections or exposure to oxidants and thus contribute to asthma exacerbations.

Studies with tachykinins

Inhaled SP or intravenous SP infusions have no significant effect on airway function in normal or asthmatic volunteers;[81-83] this may be because changes mediated via NK_1 receptors, such as increased mucus secretion, increased airway blood flow or increased plasma exudation, cannot easily be measured in patients. However, inhalation of SP does increase responsiveness to methacholine in asthmatic patients, possibly through the induction of airway oedema.[192] In inflammatory bowel disease, there is evidence for a marked upregulation of tachykinin receptors, particularly in the vasculature, suggesting that chronic inflammation may lead to changes in tachykinin receptor expression.[193] In patients with allergic rhinitis, an increased vascular response to nasally applied SP is observed.[194] There is also evidence that NK_1-receptor gene expression may be increased in the lungs of asthmatic patients, which may be due to increased transcription in response to activation of transcription factors, such as AP-1, in human lung by pro-inflammatory cytokines.[195] Increased NK_2-receptor expression has also been described in asthmatic lungs.[196]

Modulation of neurogenic inflammation

There are several approaches to inhibit neurogenic inflammation and these have been used to study the role of neurogenic inflammation in asthma[165] (Fig. 24.11). Activation of sensory nerves may be inhibited by local anaesthetics, but it has proved to be very difficult

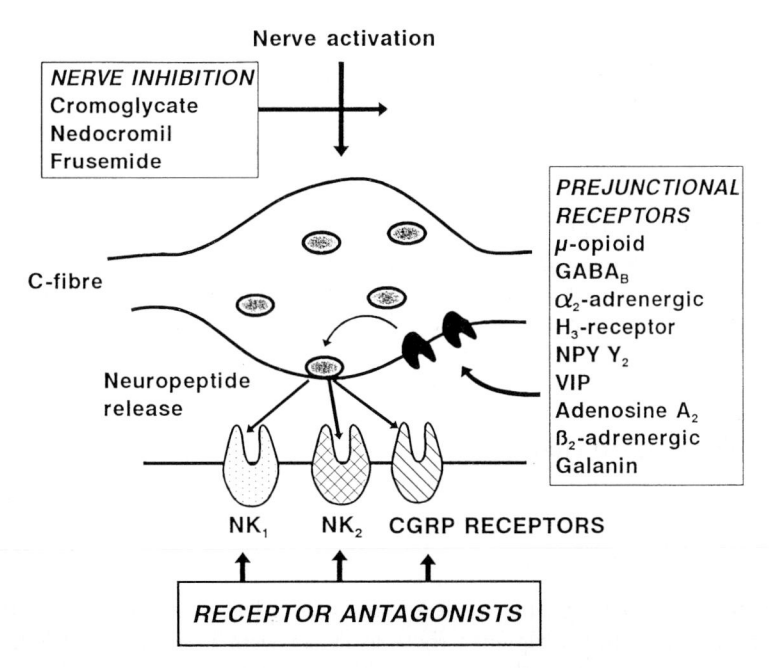

Fig. 24.11 Modulation of neurogenic inflammation in airways. CGRP, calcitonin gene-related peptide; GABA, γ-aminobutyric acid; NK, neurokinin; NPY, neuropeptide Y; VIP, vasoactive intestinal polypeptide.

to achieve adequate local anaesthesia of the respiratory tract. Inhalation of local anaesthetics, such as lignocaine, have not been found to have consistent inhibitory effects on various airway challenges and, indeed, may even promote bronchoconstriction in some patients with asthma.[197] Cromones may have direct effects on airway C fibres and this might contribute to their antiasthma effect.[198] Nedocromil sodium is effective against bradykinin-induced and sulphur dioxide-induced bronchoconstriction in asthmatic patients,[199,200] which are believed to be mediated by activation of sensory nerves in the airways. In addition, nedocromil sodium inhibits e-NANC neural bronchoconstriction due to tachykinin release from sensory nerves in guinea-pig bronchi *in vitro*, indicating an effect on release of sensory neuropeptides as well as on activation.[201] The loop diuretic frusemide, given by nebulization, behaves in a similar fashion to nedocromil sodium and inhibits metabisulphite-induced bronchoconstriction in asthmatic patients[202] and also e-NANC and cholinergic bronchoconstriction in guinea-pig airways *in vitro*.[203] In addition, nebulized frusemide also inhibits certain types of cough,[204] providing further evidence for an effect on sensory nerves.

Many drugs act on prejunctional receptors to inhibit the release of neuropeptides, as discussed above. Opioids are the most effective inhibitors, but an inhaled μ-opioid agonist, the pentapeptide BW443C, was found to be ineffective in inhibiting metabisulphite-induced bronchoconstriction, which is believed to act via neural mechanisms.[205] One problem with BW443C is that it may be degraded by NEP in the airway epithelium and therefore may not reach an adequate concentration in the vicinity of the airway sensory nerves. Another agent that has a prejunctional modulatory effect is the H_3-receptor agonist α-methyl histamine, although inhalation of α-methyl histamine had no effect on either resting tone or metabisulphite-induced bronchoconstriction in asthmatic patients.[206]

Tachykinin antagonists have now been developed for clinical studies. The non-selective tachykinin antagonist FK224 was reported to inhibit bradykinin-induced bronchoconstriction in asthma,[187] although the interpretation of this result is difficult as this drug does not appear to prevent the bronchoconstrictor effects of inhaled NKA.[207] The potent non-peptide NK_1-receptor antagonist CP-99994 had no effect on baseline lung function or on hypertonic saline-induced bronchoconstriction in asthmatic patients,[208] whereas a peptide NK_1-receptor antagonist FK-888 had a small beneficial effect on the rate of recovery from exercise-induced asthma but did not influence the peak bronchoconstrictor response.[209] It is possible that some effect might be seen in more severe asthma or in patients with virally induced exacerbations, but such studies are difficult with new drugs.

OTHER NEUROPEPTIDES

NPY

NPY is localized to adrenergic nerves, but is also colocalized with VIP in some species. In heart–lung transplantation recipients, there is an apparent increase in NPY-like immunoreactive nerves, suggesting that normally there may be some descending inhibitory influence to the expression of this peptide.[77] NPY may also be found within parasympa-

24 NANC Nerves and Neuropeptides

thetic ganglia, where it coexists with VIP since sympathectomy does not completely deplete NPY. There is a population of NPY- and VIP-immunoreactive fibres in guinea-pig trachea that are sympathetic in origin but do not contain noradrenaline.[210]

NPY is most important in the regulation of airway blood flow. NPY causes a long-lasting reduction in tracheal blood flow in anaesthetized dogs[126] but has no direct effect on canine bronchial vessels *in vitro*,[97] suggesting a preferential effect on resistance vessels in the airway. NPY may constrict resistance vessels, reducing mucosal blood flow and thus microvascular leak through reduction in the perfusion of permeable postcapillary venules;[211] this has been observed in human nasal mucosa.[212] NPY has no direct effect on airway smooth muscle of guinea-pig,[213] but may cause bronchoconstriction via release of prostaglandins.[214] NPY has a modulatory effect on cholinergic transmission of postganglionic cholinergic nerves.[213] This appears to be a direct effect on prejunctional NPY receptors, rather than secondary to any effect on α-adrenoceptors. NPY also has a modulatory effect on e-NANC bronchoconstriction both *in vitro* and *in vivo*, and this effect is surprisingly long-lasting.[213,215] NPY has no direct effect on secretion from ferret airways, although it has complex effects on stimulated secretion. NPY enhances both cholinergic and adrenergic stimulation of mucus secretion, but inhibits stimulated serous cell secretion.[65]

The role of NPY in asthma is unknown. By reducing airway blood flow (and possibly airway microvascular leak), together with its modulatory action on cholinergic and sensory nerves, it may play a beneficial role in downregulating inflammatory effects. There is no obvious defect in NPY-immunoreactive nerves in asthmatic airways.[55]

Gastrin-releasing peptide (GRP)

GRP/bombesin-immunoreactive nerves are present in the lower respiratory tract of several animal species, including humans, and is probably localized to sensory nerves.[2] GRP and bombesin-like peptides may play important roles in lung maturation and epithelial differentiation. Bombesin is a potent bronchoconstrictor in guinea-pigs *in vivo*.[216] However *in vitro* it has no effect on either proximal airways or on lung strips, indicating that it produces bronchoconstriction indirectly, although the mechanism is not yet clear. Bombesin has a constrictor effect on airway vessels.[126] GRP and bombesin are potent stimulants of airway mucus secretion in human airways *in vitro* and stimulate both serous cell lactoferrin and mucous glycoconjugate secretion. *In vivo*, topical application of bombesin results in increased secretion from mucous and serous cells of the nose.[217] Its role in asthma is unknown.

Cholecystokinin

Cholecystokinin octapeptide (CCK$_8$) has been identified in low concentration in lungs and airways of several species. CCK$_8$ is a potent constrictor of guinea-pig and human airways *in vitro*.[218] The bronchoconstrictor response is potentiated by epithelial removal and by phosphoramidon, suggesting that it is degraded by epithelial NEP. The bronchoconstrictor effect of CCK$_8$ is also potentiated in guinea-pigs sensitized and exposed to inhaled allergen, possibly because allergen exposure reduces epithelial NEP function. CCK_8 acts directly on airway smooth muscle and is potently inhibited by the specific CCK antagonist L363,851, indicating that CCK_A receptors (peripheral type) are involved. CCK_8 has no apparent effect on cholinergic neurotransmission, either at the level of parasympathetic ganglia or at postganglionic nerve terminals. While few CCK-immunoreactive nerves are present in airways, it may still have a significant effect on airway tone if these particular neural fibres are activated selectively.

Somatostatin

Somatostatin has been localized to some afferent nerves, although the concentration detectable in lung is low. Somatostatin has no direct action on airway smooth muscle *in vitro*, but appears to potentiate cholinergic neurotransmission in ferret airways.[219]

Galanin

Galanin is widely distributed in the respiratory tract innervation of several species. It is colocalized with VIP in cholinergic nerves of airways and is present in parasympathetic ganglia.[220,221] It is also colocalized with SP/CGRP in sensory nerves and dorsal root, nodose and trigeminal ganglia.[2] Galanin has no direct effect on airway tone in guinea-pigs but modulates e-NANC neurotransmission.[222,223] It has no effect on airway blood flow in dogs[126] but inhibits SP-induced mucus secretion.[224] The antagonist galantide has no effect on airway function *in vitro*, and the physiological role of galanin in airways remains a mystery.

Enkephalins

Leucine-enkephalin has been localized to neuroendocrine cells in airways[225] and [Met]enkephalin-Arg^6-Gly^7-Leu^8-immunoreactive nerves have been described in guinea-pig and rat lungs, with a similar distribution to VIP.[226] The anatomical origins and functional roles of the endogenous opioids is not clear since the opioid antagonist naloxone has no effect on neurally mediated airway effects.[227,228] However it is possible that these opioid pathways may be selectively activated from brainstem centres under certain conditions. Exogenous opioids potently modulate neuropeptide release from sensory nerves in airways via μ-opioid receptors.[227–229]

ROLE OF NEUROPEPTIDES IN ASTHMA

The presence of so many neuropeptides in the respiratory tract raises questions about their physiological role. It is now appreciated that many of these peptides are cotransmitters in classical autonomic nerves and may be regarded as modulators of autonomic effects, perhaps acting to 'fine tune' airway functions,[230] and to modulate the release of other neurotransmitters.[4] Although much of the research on neuropeptides in the airways has previously concentrated on their effects on airway smooth muscle, it is now clear that the most potent effects of many of the relevant peptides are on airway vasculature and secretions, and that neuropeptides may have an important role in regulating the mucosal surface of the airways. Another important area which is very relevant to asthma is whether neuropeptides influence the immune system and in particular the immune cells involved in asthma.[99] There is likely to be increasing research in the area of neuroimmune interaction and in some species there is already evidence for neuropeptide innervation of bronchus-associated lymphoid tissue.[231] The possibility that inflammatory cells, such as macrophages, lymphocytes and eosinophils, may themselves produce neuropeptide-like peptides under certain conditions is also an important area of future research.[99]

The lack of understanding of the physiological role of individual peptides is largely due to the lack of specific antagonists that can be given safely to humans. Several selective neuropeptide receptor antagonists are now under development, which may soon be available for clinical studies. Although there has been optimism that inhibitors or mimics of neuropeptides might have therapeutic application in asthma, it is most unlikely that such drugs would have a major advantage over existing agents. However, it is also possible that such drugs may have value in the treatment of other inflammatory airway diseases, such as chronic obstructive pulmonary disease, cystic fibrosis and bronchiectasis.

REFERENCES

1. Barnes PJ, Baraniuk J, Belvisi MG: Neuropeptides in the respiratory tract. *Am Rev Respir Dis* (1991; **144**: 1187–1198, 1391–1399.
2. Uddman R, Hakanson R, Luts A, Sundler F: Distribution of neuropeptides in airways. In Barnes PJ (ed) *Autonomic Control of the Respiratory System*. London, Harvard Academic, 1997, pp 21–37.
3. Barnes PJ: Neuropeptides and asthma. In Kaliner MA, Barnes PJ, Kunkel GHH, Baraniuk JN (eds) *Neuropeptides in Respiratory Medicine*. New York, Marcel Dekker, 1994, pp 285–311.
4. Barnes PJ: Modulation of neurotransmission in airways. *Physiol Rev* (1992) **72**: 699–729.
5. Stretton CD, Belvisi MG, Barnes PJ: Sensory nerve depletion potentiates inhibitory NANC nerves in guinea pig airways. *Eur J Pharmacol* (1990) **184**: 333–337.
6. Barnes PJ: Neural control of human airways in health and disease. *Am Rev Respir Dis* (1986) **134**: 1289–1314.
7. Lammers JWJ, Barnes PJ, Chung KF: Non-adrenergic, non-cholinergic airway inhibitory nerves. *Eur Respir J* (1992) **5**: 239–246.
8. Lammers J-WJ, Minette P, McCusker M, Chung KF, Barnes PJ: Capsaicin-induced bronchodilatation in mild asthmatic subjects: possible role of nonadrenergic inhibitory system. *J Appl Physiol* (1989) **67**: 856–861.
9. Belvisi MG, Stretton CD, Miura M, *et al.*: Inhibitory NANC nerves in human tracheal smooth muscle: a quest for the neurotransmitter. *J Appl Physiol* (1992) **73**: 2505–2510.
10. Belvisi MG, Ward JR, Mitchell JA, Barnes PJ: Nitric oxide as a neurotransmitter in human airways. *Arch Int Pharmacodyn Ther* (1995) **329**: 111–120.
11. Ward JK, Belvisi MG, Springall DR, *et al.*: Human iNANC bronchodilatation and nitric oxide-immunoreactive nerves are reduced in distal airways. *Am J Respir Cell Mol Biol* (1995) **13**: 175–184.

12. Belvisi MG, Stretton CD, Barnes PJ: Evidence that nitric oxide is the neurotransmitter of inhibitory non-adrenergic non-cholinergic nerves in human airways. *Eur J Pharmacol* (1992) **210**: 221–222.
13. Ward JK, Belvisi MG, Fox AJ, et al.: Modulation of cholinergic neural bronchoconstriction by endogenous nitric oxide and vasoactive intestinal peptide in human airways *in vitro*. *J Clin Invest* (1993) **92**: 736–743.
14. Bai TR, Bramley AM: Effect of an inhibitor of nitric oxide synthase on neural relaxation of human bronchi. *Am J Physiol* (1993) **264**: L425–L430.
15. Belvisi MG, Ward JK, Tadjarimi S, Yacoub MH, Barnes PJ: Inhibitory NANC nerves in human airways: differences in disease and after extrinsic denervation. *Am Rev Respir Dis* (1993) **147**: A286.
16. Michoud M-C, Jeanneret-Grosjean A, Cohen A, Amyot R: Reflex decrease of histamine-induced bronchoconstriction after laryngeal stimulation in asthmatic patients. *Am Rev Respir Dis* (1988) **138**: 1548–1552.
17. Miura M, Noue H, Ichinose M, Kimura K, Katsumata U, Takishima T: Effect of nonadrenergic, noncholinergic inhibitory nerve stimulation on the allergic reaction in cat airways. *Am Rev Respir Dis* (1990) **141**: 29–32.
18. Andersson RG, Grundstrom N: The excitatory noncholinergic, nonadrenergic nervous system of the guinea-pig airways. *Eur J Respir Dis* (1983) **131** (Suppl): 141–157.
19. Lundberg JM, Saria A, Lundblad L, et al.: Bioactive peptides in capsaicin-sensitive C-fiber afferents of the airways: functional and pathophysiological implications. In Kaliner M, Barnes PJ (eds) *The Airways: Neural Control in Health and Disease*. New York, Marcel Dekker, 1987, pp 417–445.
20. Webber SE, Lim JCS, Widdicombe JG: The effects of calcitonin gene-related peptide on submucosal gland secretion and epithelial albumin transport in ferret trachea *in vitro*. *Br J Pharmacol* (1991) **102**: 79–84.
21. Widdicombe JG: The NANC system and airway vasculature. *Arch Int Pharmacodyn Ther* (1990) **303**: 83–90.
22. Uddman R, Sundler F: Neuropeptides in the airways: a review. *Am Rev Respir Dis* (1987) **136**: S3–S8.
23. Lundberg JM, Fahrenkrug J, Hokfelt T, et al.: Coexistence of peptide histidine isoleucine (PHI) and VIP in nerves regulating blood flow and bronchial smooth muscle tone in various mammals including man. *Peptides* (1984) **5**: 593–606.
24. Dey RD, Altemus JB, Michalkiewicz M: Distribution of vasoactive intestinal peptide- and substance P-containing nerves originating from neurons of airway ganglia in cat bronchi. *J Comp Neurol* (1991) **304**: 330–340.
25. Palmer JBD, Cuss FMC, Barnes PJ: VIP and PHM and their role in nonadrenergic inhibitory responses in isolated human airways. *J Appl Physiol* (1986) **61**: 1322–1328.
26. Carstairs JR, Barnes PJ: Visualization of vasoactive intestinal peptide receptors in human and guinea pig lung. *J Pharmacol Exp Ther* (1986) **239**: 249–255.
27. Martin JG, Wang A, Zacour M, Biggs DF: The effects of vasoactive intestinal polypeptide on cholinergic neurotransmission in isolated innervated guinea pig tracheal preparations. *Respir Physiol* (1990) **79**: 111–122.
28. Ellis JL, Farmer SG: Modulation of cholinergic neurotransmission by vasoactive intestinal peptide and petide histidine isoleucine in guinea pig tracheal smooth muscle. *Pulmon Pharmacol* (1989) **2**: 107–112.
29. Stretton CD, Belvisi MG, Barnes PJ: Modulation of neural bronchoconstrictor responses in the guinea pig respiratory tract by vasoactive intestinal peptide. *Neuropeptides* (1991) **18**: 149–157.
30. Hakoda H, Ito Y: Modulation of cholinergic neurotransmission by the peptide VIP, VIP antiserum and VIP antagonists in dog and cat trachea. *J Physiol* (1990) **428**: 133–154.
31. Diamond L, O'Donnell M: A nonadrenergic vagal inhibitory pathway to feline airways. *Science* (1980) **208**: 185–188.
32. Palmer JBD, Cuss FMC, Warren JB, Barnes PJ: The effect of infused vasoactive intestinal peptide on airway function in normal subjects. *Thorax* (1986) **41**: 663–666.
33. Greenberg B, Rhoden K, Barnes PJ: Relaxant effects of vasoactive intestinal peptide and

peptide histidine isoleucine in human and bovine pulmonary arteries. *Blood Vessels* (1987) **24**: 45–50.
34. Barnes PJ, Brown MJ: Venous plasma histamine in exercise and hyperventilation-induced asthma in man. *Clin Sci* (1981) **61**: 159–162.
35. Bungaard A, Enehjelm SD, Aggestrop S: Pretreatment of exercise-induced asthma with inhaled vasoactive intestinal peptide. *Eur J Respir Dis* (1983) **64**: 427–429.
36. Matran R, Alving K, Martling C, Lacroix JS, Lundberg JM: Vagally mediated vasodilatation by motor and sensory nerves in the tracheal and bronchial circulation of the pig. *Acta Physiol Scand* (1989) **135**: 29–37.
37. Peatfield AC, Barnes PJ, Bratcher C, Nadel JA, Davis B: Vasoactive intestinal peptide stimulates tracheal submucosal gland secretion in ferret. *Am Rev Respir Dis* (1983) **128**: 89–93.
38. Coles SJ, Said SI, Reid LM: Inhibition by vasoactive intestinal peptide of glycoconjugate and lysozyme secretion by human airways *in vitro*. *Am Rev Respir Dis* (1981) **124**: 531–536.
39. Nathanson I, Widdicombe JH, Barnes PJ: Effect of vasoactive intestinal peptide on ion transport across dog tracheal epithelium. *J Appl Physiol* (1983) **55**: 1844–1848.
40. Undem BJ, Dick EC, Buckner CK: Inhibition by vasoactive intestinal peptide of antigen-induced histamine release from guinea pig minced lung. *Eur J Pharmacol* (1983) **88**: 247–250.
41. O'Dorisio MS, Shannaon BT, Fleshman DJ, Campolito LB: Identification of high affinity receptors for vasoactive intestinal peptide on human lymphocytes of B cell lineage. *J Immunol* (1989) **142**: 3533–3536.
42. Tang H, Welton A, Ganea D: Neuropeptide regulation of cytokine expression: effects of VIP and Ro 25-1553. *J Interferon Cytokine Res* (1995) **15**: 993–1003.
43. Maruno K, Absood A, Said SI: VIP inhibits basal and histamine-stimulated proliferation of human airway smooth muscle cells. *Am J Physiol* (1995) **12**: L1047–L1051.
44. Harrison NK, Dawes KE, Kwon OJ, Barnes PJ, Laurent GJ, Chung KF: Effects of neuropeptides in human lung fibroblast: proliferation and chemotaxis. *Am J Physiol* (1995) **12**: L278–L283.
45. Ellis JL, Framer SG: Effects of peptidases on nonadrenergic, noncholinergic inhibitory responses of tracheal smooth muscle: a comparison with effects on VIP- and PHI-induced relaxation. *Br J Pharmacol* (1989) **96**: 521–526.
46. Ellis JL, Farmer SG: The effects of vasoactive intestinal peptide (VIP) antagonists, and VIP and peptide histidine isoleucine antisera on nonadrenergic, noncholinergic relaxations of tracheal smooth muscle. *Br J Pharmacol* (1989) **96**: 513–520.
47. Rhoden KJ, Barnes PJ: Potentiation of non-adrenergic non-cholinergic relaxation in guinea pig airways by a cAMP phosphodiesterase inhibitor. *J Pharmacol Exp Ther* (1990) **282**: 396–402.
48. Li CG, Rand MJ: Evidence that part of the NANC relaxant response of guinea-pig trachea to electrical field stimulation is mediated by nitric oxide. *Br J Pharmacol* (1991) **102**: 91–94.
49. Tucker JF, Brane SR, Charalambous L, Hobbs AJ, Gibson A: L-N^G-nitro arginine inhibits non-adrenergic, non-cholinergic relaxations of guinea pig isolated tracheal smooth muscle. *Br J Pharmacol* (1990) **100**: 663–664.
50. Belvisi MG, Stretton CD, Barnes PJ: Nitric oxide is the endogenous neurotransmitter of bronchodilator nerves in human airways. *Eur J Pharmacol* (1992) **210**: 221–222.
51. Belvisi MG, Miura M, Stretton CD, Barnes PJ: Endogenous vasoactive intestinal peptide and nitric oxide modulate cholinergic neurotransmission in guinea pig trachea. *Eur J Pharmacol* (1993) **231**: 97–102.
52. Belvisi MG, Stretton CD, Barnes PJ: Nitric oxide as an endogenous modulator of cholinergic neurotransmission in guinea pig airways. *Eur J Pharmacol* (1991) **198**: 219–221.
53. Ollerenshaw S, Jarvis D, Woolcock A, Sullivan C, Scheibner T: Absence of immunoreactive vasoactive intestinal polypeptide in tissue from the lungs of patients with asthma. *N Engl J Med* (1989) **320**: 1244–1248.
54. Barnes PJ: Vasoactive intestinal peptide and asthma. *N Engl J Med* (1989) **321**: 1128–1129.
55. Howarth PH, Springall DR, Redington AE, Djukanovic R, Holgate ST, Polak JM: Neuropeptide-containing nerves in bronchial biopsies from asthmatic and non-asthmatic subjects. *Am J Respir Cell Mol Biol* (1995) **13**: 288–296.

78. Naline E, Devillier P, Drapeau G, et al.: Characterization of neurokinin effects on receptor selectivity in human isolated bronchi. *Am Rev Respir Dis* (1989) **140**: 679–686.
79. Advenier C, Naline E, Toty L, et al.: Effects on the isolated human bronchus of SR 48968, a potent and selective nonpeptide antagonist of the neurokinin A (NK$_2$) receptors. *Am Rev Respir Dis* (1992) **146**: 1177–1181.
80. Frossard N, Barnes PJ: Effect of tachykinins on small human airways. *Neuropeptides* (1991) **19**: 157–162.
81. Fuller RW, Maxwell DL, Dixon CMS, et al.: The effects of substance P on cardiovascular and respiratory function in human subjects. *J Appl Physiol* (1987) **62**: 1473–1479.
82. Evans TW, Dixon CM, Clarke B, Conradson TB, Barnes PJ: Comparison of neurokinin A and substance P on cardiovascular and airway function in man. *Br J Pharmacol* (1988) **25**: 273–275.
83. Joos G, Pauwels R, van der Straeten ME: Effect of inhaled substance P and neurokinin A in the airways of normal and asthmatic subjects. *Thorax* (1987) **42**: 779–783.
84. Frossard N, Rhoden KJ, Barnes PJ: Influence of epithelium on guinea pig airway responses to tachykinins: role of endopeptidase and cyclooxygenase. *J Pharmacol Exp Ther* (1989) **248**: 292–298.
85. Devillier P, Advenier C, Drapeau G, Marsac J, Regoli D: Comparison of the effects of epithelium removal and of an enkephalinase inhibitor on the neurokinin-induced contractions of guinea pig isolated trachea. *Br J Pharmacol* (1988) **94**: 675–684.
86. Rogers DF, Aursudkij B, Barnes PJ: Effects of tachykinins on mucus secretion on human bronchi *in vitro*. *Eur J Pharmacol* (1989) **174**: 283–286.
87. Kuo H, Rhode JAL, Tokuyama K, Barnes PJ, Rogers DF: Capsaicin and sensory neuropeptide stimulation of goblet cell secretion in guinea pig trachea. *J Physiol* (1990) **431**: 629–641.
88. Ramnarine SI, Hirayama Y, Barnes PJ, Rogers DF: 'Sensory–efferent' neural control of mucus secretion: characterization using tachykinin receptor antagonists in ferret trachea *in vitro*. *Br J Pharmacol* (1994) **113**: 1183–1190.
89. Meini S, Mak JCW, Rohde JAL, Rogers DF: Tachykinin control of ferret airways: mucus secretion, bronchoconstriction and receptor mapping. *Neuropeptides* (1993) **24**: 81–89.
90. Tokuyama K, Kuo H, Rohde JAL, Barnes PJ, Rogers DF: Neural control of goblet cell secretion in guinea pig airways. *Am J Physiol* (1990) **259**: L108–L115.
91. Kuo H, Barnes PJ, Rogers DF: Cigarette smoke-induced airway goblet cell secretion: dose dependent differential nerve activation. *Am J Physiol* (1992) **7**: L161 L167.
92. Rogers DF, Belvisi MG, Aursudkij B, Evans TW, Barnes PJ: Effects and interactions of sensory neuropeptides on airway microvascular leakage in guinea pigs. *Br J Pharmacol* (1988) **95**: 1109–1116.
93. Sertl K, Wiedermann CJ, Kowalski ML, et al.: Substance P: the relationship between receptor distribution in rat lung and the capacity of substance P to stimulate vascular permeability. *Am Rev Respir Dis* (1988) **138**: 151–159.
94. Lotvall JO, Lemen RJ, Hui KP, Barnes PJ, Chung KF: Airflow obstruction after substance P aerosol: contribution of airway and pulmonary edema. *J Appl Physiol* (1990) **69**: 1473–1478.
95. Salonen RO, Webber SE, Widdicombe JG: Effects of neuropeptides and capsaicin on the canine tracheal vasculature *in vivo*. *Br J Pharmacol* (1988) **95**: 1262–1270.
96. Matran R, Alving K, Martling CR, Lacroix JS, Lundberg JM: Effects of neuropeptides and capsaicin on tracheobronchial blood flow in the pig. *Acta Physiol Scand* (1989) **135**: 335–342.
97. McCormack DG, Salonen RO, Barnes PJ: Effect of sensory neuropeptides on canine bronchial and pulmonary vessels *in vitro*. *Life Sci* (1989) **45**: 2405–2412.
98. McGillis JP, Organist ML, Payan DG: Substance P and immunoregulation. *Fed Proc* (1987) **14**: 120–123.
99. Daniele RP, Barnes PJ, Goetzl EJ, et al.: Neuroimmune interactions in the lung. *Am Rev Respir Dis* (1992) **145**: 1230–1235.
100. Lowman MA, Benyon RC, Church MK: Characterization of neuropeptide-induced histamine release from human dispersed skin mast cells. *Br J Pharmacol* (1988) **95**: 121–130.
101. Ali H, Leung KBI, Pearce FL, Hayes NA, Foremean JC: Comparison of histamine releasing activity of substance P on mast cells and basophils from different species and tissues. *Int Arch Allergy Appl Immunol* (1986) **79**: 121–124.

102. Kroegel C, Giembycz MA, Barnes PJ: Characterization of eosinophil activation by peptides. Differential effects of substance P, mellitin, and f-met-leu-phe. *J Immunol* (1990) **145**: 2581–2587.
103. Numao T, Agrawal DK: Neuropeptides modulate human eosinophil chemotaxis. *J Immunol* (1992) **149**: 3309–3315.
104. Brunelleschi S, Vanni L, Ledda F, Giotti A, Maggi CA, Fantozzi R: Tachykinins activate guinea pig alveolar macrophages: involvement of NK2 and NK1 receptors. *Br J Pharmacol* (1990) **100**: 417–420.
105. Lotz M, Vaughn JH, Carson DM: Effect of neuropeptides on production of inflammatory cytokines by human monocytes. *Science* (1988) **241**: 1218–1221.
106. Umeno E, Nadel JA, Huang HT, McDonald DM: Inhibition of neutral endopeptidase potentiates neurogenic inflammation in the rat trachea. *J Appl Physiol* (1989) **66**: 2647–2652.
107. Fan T, Hu DE, Guard S, Gresham GA, Watling KJ: Stimulation of angiogenesis by substance P and interleukin-1 in the rat and its inhibition by NK_1 or interleukin-1 receptor antagonists. *Br J Pharmacol* (1993) **110**: 43–49.
108. Harrison NK, Dawes KE, Kwon OJ, Barnes PJ, Laurent GJ, Chung KF: Effects of neuropeptides in human lung fibroblast proliferation and chemotaxis. *Am J Physiol* (1995) **12**: L278–L283.
109. Hall AK, Barnes PJ, Meldrum LA, Maclagan J: Facilitation by tachykinins of neurotransmission in guinea-pig pulmonary parasympathetic nerves. *Br J Pharmacol* (1989) **97**: 274–280.
110. Myers AC, Undem BJ: Electrophysiological effects of tachykinins and capsaicin on guinea-pig parasympathetic ganglia. *J Physiol* (1993) **470**: 66–79.
111. Watson N, Maclagan J, Barnes PJ: Endogenous tachykinins facilitate transmission through parasympathetic ganglia in guinea-pig trachea. *Br J Pharmacol* (1993) **109**: 751–759.
112. Martling C, Saria A, Andersson P, Lundberg JM: Capsaicin pretreatment inhibits vagal cholinergic and noncholinergic control of pulmonary mechanisms in guinea pig. *Naunyn Schmiedebergs Arch Pharmacol* (1984) **325**: 343–348.
113. Stretton CD, Belvisi MG, Barnes PJ: The effect of sensory nerve depletion on cholinergic neurotransmission in guinea pig airways. *Br J Pharmacol* (1989) **98**: 782P.
114. Belvisi MG, Patacchini R, Barnes PJ, Maggi CA: Facilitatory effects of selective agonists for tachykinin receptors on cholinergic neurotransmission: evidence for species differences. *Br J Pharmacol* (1994) **111**: 103–110.
115. Black JL, Johnson PR, Alouvan L, Armour CL: Neurokinin A with K^+ channel blockade potentiates contraction to electrical stimulation in human bronchus. *Eur J Pharmacol* (1990) **180**: 311–317.
116. Nadel JA: Neutral endopeptidase modulates neurogenic inflammation. *Eur Respir J* (1991) **4**: 745–754.
117. Shore SA, Stimler-Gerard NP, Coats SR, Drazen JM: Substance P induced bronchoconstriction in guinea pig. Enhancement by inhibitors of neutral metalloendopeptidase and angiotensin converting enzyme. *Am Rev Respir Dis* (1988) **137**: 331–336.
118. Martins MA, Shore SA, Gerard NP, Gerald C, Drazen JM: Peptidase modulation of the pulmonary effects of tachykinins in tracheal superfused guinea pig lungs. *J Clin Invest* (1990) **85**: 170–176.
119. Lotvall JO, Skoogh B, Barnes PJ, Chung KF: Effects of aerosolized substance P on lung resistance in guinea pigs: a comparison between inhibition of neutral endopeptidase and angiotensin-converting enzyme. *Br J Pharmacol* (1990) **100**: 69–72.
120. Sekizawa K, Tamaoki J, Graf PD, Basbaum CB, Borson DB, Nadel JA: Enkephalinase inhibitors potentiate mammalian tachykinin-induced contraction in ferret trachea. *J Pharmacol Exp Ther* (1987) **243**: 1211–1217.
121. Black JL, Johnson PRA, Armour CL: Potentiation of the contractile effects of neuropeptides in human bronchus by an enkephalinase inhibitor. *Pulmon Pharmacol* (1988) **1**: 21–23.
122. Djokic TD, Nadel JA, Dusser DJ, Sekizawa K, Graf PD, Borson DB: Inhibitors of neutral endopeptidase potentiate electrically and capsaicin-induced non-cholinergic contraction in guinea pig bronchi. *J Pharmacol Exp Ther* (1989) **248**: 7–11.

123. Martling CR: Sensory nerves containing tachykinins and CGRP in the lower airways: functional implications for bronchoconstriction, vasodilation, and protein extavasation. *Acta Physiol Scand* (1987) Suppl **563**: 1–57.
124. Palmer JBD, Cuss FMC, Mulderry PK, et al.: Calcitonin gene-related peptide is localized to human airway nerves and potently constricts human airway smooth muscle. *Br J Pharmacol* (1987) **91**: 95–101.
125. McCormack DG, Mak JCW, Coupe MO, Barnes PJ: Calcitonin gene-related peptide vasodilation of human pulmonary vessels: receptor mapping and functional studies. *J Appl Physiol* (1989) **67**: 1265–1270.
126. Salonen RO, Webber SE, Widdicombe JG: Effects of neuropeptides and capsaicin on the canine tracheal vasculature *in vivo*. *Br J Pharmacol* (1988) **95**: 1262–1270.
127. Parsons GH, Nichol GM, Barnes PJ, Chung KF: Peptide mediator effects on bronchial blood velocity and lung resistance in conscious sheep. *J Appl Physiol* (1992) **72**: 1118–1122.
128. Mak JCW, Barnes PJ: Autoradiographic localization of calcitonin gene-related peptide binding sites in human and guinea pig lung. *Peptides* (1988) **9**: 957–964.
129. Khalil Z, Andrews PV, Helme RD: VIP modulates substance P induced plasma extravasation *in vivo*. *Eur J Pharmacol* (1988) **151**: 281–287.
130. Brockaw JJ, White GW: Calcitonin gene-related peptide potentiates substance P-induced plasma extravasation in the rat trachea. *Lung* (1992) **170**: 89–93.
131. Ninommiya H, Uchida Y, Endo T, et al.: The effects of calcitonin gene-related peptide on tracheal smooth muscle of guinea pigs *in vitro*. *Br J Pharmacol* (1996) **119**: 1341–1346.
132. Webber SG, Lim JCS, Widdicombe JG: The effects of calcitonin gene related peptide on submucosal gland secretion and epithilial albumin transport on ferret trachea *in vitro*. *Br J Pharmacol* (1991) **102**: 79–84.
133. Pietrowski W, Foreman JC: Some effects of calcitonin gene related peptide in human skin and on histamine release. *Br J Dermatol* (1986) **114**: 37–46.
134. Haynes LW, Manley C: Chemotactic response of guinea pig polymorphonucleocytes *in vivo* to rat calcitonin gene related peptide and proteolytic fragments. *J Physiol* (1988) **43**: 79P.
135. Umeda Y, Arisawa H: Characterization of the calcitonin gene related peptide receptor in mouse T lymphocytes. *Neuropeptides* (1989) **14**: 237–242.
136. Nong YH, Titus RG, Riberio JM, Remold HG: Peptides encoded by the calcitonin gene inhibit macrophage function. *J Immunol* (1989) **143**: 45–49.
137. White SR, Hershenson MB, Sigrist KS, Zimmerman A, Solway J: Proliferation of guinea pig tracheal epithelial cells induced by calcitonin gene-related peptide. *Am J Respir Cell Mol Biol* (1993) **8**: 592–596.
138. Maggi CA, Meli A: The sensory efferent function of capsaicin sensitive sensory nerves. *Gen Pharmacol* (1988) **19**: 1–43.
139. McDonald DM: Neurogenic inflammation in the respiratory tract: actions of sensory nerve mediators on blood vessels and epithelium of the airway mucosa. *Am Rev Respir Dis* (1987) **136**: S65–S72.
140. Solway J, Leff AR: Sensory neuropeptides and airway function. *J Appl Physiol* (1991) **71**: 2077–2087.
141. Barnes PJ: Asthma as an axon reflex. *Lancet* (1986) **i**: 242–245.
142. Barnes PJ: Sensory nerves, neuropeptides and asthma. *Ann NY Acad Sci* (1991) **629**: 359–370.
143. Joos GF, Germonpre PR, Kips JC, Peleman RA, Pauwels RA: Sensory neuropeptides and the human lower airways: present state and future directions. *Eur Respir J* (1994) **7**: 1161–1171.
144. Lundberg JM, Saria A: Capsaicin-induced desensitization of the airway mucosa to cigarette smoke, mechanical and chemical irritants. *Nature* (1983) **302**: 251–253.
145. Ray DW, Hernandez C, Leff AR, Drazen JM, Solway J: Tachykinins mediate bronchoconstriction elicited by isocapnic hyperpnea in guinea pigs. *J Appl Physiol* (1989) **66**: 1108–1112.
146. Sakamoto T, Elwood W, Barnes PJ, Chung KF: Pharmacological modulation of inhaled metabisulphite-induced airway microvascular leakage and bronchoconstriction in guinea pig. *Br J Pharmacol* (1992) **107**: 481–488.
147. Umeno E, McDonald DM, Nadel JA: Hypertonic saline increases vascular permeability in the rat trachea by producing neurogenic inflammation. *J Clin Invest* (1990) **85**: 1905–1908.
148. Thompson JE, Scypinski LA, Gordon T, Sheppard D: Tachykinins mediate the acute

148. (continued) increase in airway responsiveness by toluene diisocyanate in guinea-pigs. *Am Rev Respir Dis* (1987) **136**: 43–49.
149. Lötvall JO, Hui KP, Löfdahl C, Barnes PJ, Chung KF: Capsaicin pretreatment does not inhibit allergen-induced airway microvascular leakage in guinea pig. *Allergy* (1991) **46**: 105–108.
150. Hsing T, Garland A, Ray DW, Hershenson MB, Leff AR, Solway J: Endogenous sensory neuropeptide release enhances non specific airway responsiveness in guinea pigs. *Am Rev Respir Dis* (1992) **146**: 148–153.
151. Alving K, Matran R, Lacroix JS, Lundberg JM: Allergen challenge induces vasodilation in pig bronchial circulation via a capsaicin sensitive mechanism. *Acta Physiol Scand* (1988) **134**: 571–572.
152. Matsuse T, Thomson RJ, Chen X, Salari H, Schellenberg RR: Capsaicin inhibits airway hyperresponsiveness, but not airway lipoxygenase activity nor eosinophilia following repeated aerosolized antigen in guinea pigs. *Am Rev Respir Dis* (1991) **144**: 368–372.
153. Riccio MM, Manzini S, Page CP: The effect of neonatal capsaicin in the development of bronchial hyperresponsiveness in allergic rabbits. *Eur J Pharmacol* (1993) **232**: 89–97.
154. Undem BJ, Riccio MM, Weinreich D, Ellis JL, Myers AC: Neurophysiology of mast cell–nerve interactions in the airways. *Int Arch Allergy Immunol* (1995) **107**: 199–201.
155. Saban R, Dick EC, Fishlever RI, Buckner CK: Enhancement of parainfluenza 3 infection of contractile responses to substance P and capsaicin in airway smooth msucle from guinea pig. *Am Rev Respir Dis* (1987) **136**: 586–591.
156. Jacoby DB, Tamaoki J, Borson DB, Nadel JA: Influenza infection increases airway smooth muscle responsiveness to substance P in ferrets by decreasing enkephalinase. *J Appl Physiol* (1988) **64**: 2653–2658.
157. Piedimonte G, Nadel JA, Umeno E, McDonald DM: Sendai virus infection potentiates neurogenic inflammation in the rat trachea. *J Appl Physiol* (1990) **68**: 754–760.
158. Dusser DJ, Djoric TD, Borson DB, Nadel JA: Cigarette smoke induces bronchoconstrictor hyperresponsiveness to substance P and inactivates airway neutral endopeptidase in the guinea pig. *J Clin Invest* (1989) **84**: 900–906.
159. Sheppard D, Thompson JE, Scypinski L, Dusser DJ, Nadel JA, Borson DB: Toluene diisocyanate increases airway responsiveness to substance P and decreases airway neutral endopeptidase. *J Clin Invest* (1988) **81**: 1111–1115.
160. Lei Y, Barnes PJ, Rogers DF. Inhibition of neurogenic plasma exudation in guinea pig airways by CP-96,345, a new non-peptide NK$_1$-receptor antagonist. *Br J Pharmacol* (1992) **105**: 261–262.
161. Delay-Goyet P, Lundberg JM: Cigarette smoke-induced airway oedema is blocked by the NK$_1$-antagonist CP-96,345. *Eur J Pharmacol* (1991) **203**: 157–158.
162. Hirayama Y, Lei YH, Barnes PJ, Rogers DF: Effects of two novel tachykinin antagonists FK 224 and FK 888 on neurogenic plasma exudation, bronchoconstriction and systemic hypotension in guinea pigs *in vivo*. *Br J Pharmacol* (1993) **108**: 844–851.
163. Garland A, Jordan JE, Kao R, *et al*.: Neurokinin-1 receptor blockade with (I)CP-96,345 inhibits hyperpnea-induced bronchoconstriction in guinea pigs. *Am Rev Respir Dis* (1992) **145**: A45.
164. Sakamoto T, Barnes PJ, Chung KF: Effect of CP-96,345, a non-peptide NK$_1$-receptor antagonist against substance P-, bradykinin-, and allergen-induced airway microvascular leak and bronchoconstriction in the guinea pig. *Eur J Pharmacol* (1993) **231**: 31–38.
165. Barnes PJ, Belvisi MG, Rogers DF: Modulation of neurogenic inflammation: novel approaches to inflammatory diseases. *Trends Pharmacol Sci* (1990) **11**: 185–189.
166. Kuo H, P., Rohde JAL, Barnes PJ, Rogers DF: Morphine inhibition of cigarette smoke induced goblet cell secretion in guinea pig trachea *in vivo*. *Respir Med* (1990) **84**: 425.
167. Yeadon M, Wilkinson D, Darley-Usmar V, O'Leary VJ, Payne AN: Mechanisms contributing to ozone-induced bronchial hyperreactivity in guinea pigs. *Pulmon Pharmacol* (1992) **5**: 39–50.
168. Stretton CD, Miura M, Belvisi MG, Barnes PJ: Calcium-activated potassium channels mediate prejunctional inhibition of peripheral sensory nerves. *Proc Natl Acad Sci USA* (1992) **89**: 1325–1329.
169. Ichinose M, Barnes PJ: A potassium channel activator modulates both noncholinergic and

cholinergic neurotransmission in guinea pig airways. *J Pharmacol Exp Ther* (1990) **252**: 1207–1212.
170. Kuo H, Rohde JAL, Barnes PJ, Rogers DF: K^+ channel activator inhibition of neurogenic goblet cell secretion in guinea pig trachea. *Eur J Pharmacol* (1992) **221**: 385–388.
171. Bowden J, Gibbins IL: Relative density of substance P-immunoreactive nerve fibres in the tracheal epithelium of a range of species. *FASEB J* (1992) **6**: A1276.
172. Holzer P: Local effector functions of capsaicin-sensitive sensory nerve endings: involvement of tachykinins, calcitonin gene related peptide, and other neuropeptides. *Neuroscience* (1988) **24**: 739–768.
173. Ollerenshaw SL, Jarvis D, Sullivan CE, Woolcock AJ: Substance P immunoreactive nerves in airways from asthmatics and non-asthmatics. *Eur Respir J* (1991) **4**: 673–682.
174. Nieber K, Baumgarten CR, Rathsack R, Furkert J, Oehame P, Kunkel G: Substance P and b-endorphin-like immunoreactivity in lavage fluids of subjects with and without asthma. *J Allergy Clin Immunol* (1992) **90**: 646–652.
175. Tomaki M, Ichinose M, Nakajima N, *et al.*: Elevated substance P concentration in sputum after hypertonic saline inhalation in asthma and chronic bronchitis patients. *Am Rev Respir Dis* (1993) **147**: A478.
176. Hazbun ME, Hamilton R, Holian A, Eschenbacher WL: Ozone-induced increases in substance P and 8 epi-prostaglandin F_{2a} in the airways of human subjects. *Am J Respir Cell Mol Biol* (1993) **9**: 568–572.
177. Lindsay RM, Harmar AJ: Nerve growth factor regulates expression of neuropeptide genes in sensory neurons. *Nature* (1989) **337**: 362–364.
178. Fischer A, Philippin B, Saria A, McGregor G, Kummer W: Neuronal plasticity in sensitized and challenged guinea pigs: neuropeptides and neuropeptide gene expression. *Am J Respir Crit Care Med* (1994) **149**: A890.
179. Fuller RW, Dixon CMS, Barnes PJ: The bronchoconstrictor response to inhaled capsaicin in humans. *J Appl Physiol* (1985) **85**: 1080–1084.
180. Midgren B, Hansson L, Karlsson JA, Simonsson BG, Persson CGA: Capsaicin-induced cough in humans. *Am Rev Respir Dis* (1992) **146**: 347–351.
181. Barnes PJ: Bradykinin and asthma. *Thorax* (1992) **47**: 979–983.
182. Fuller RW, Dixon CMS, Cuss FMC, Barnes PJ: Bradykinin-induced bronchoconstriction in man: mode of action. *Am Rev Respir Dis* (1987) **135**: 176–180.
183. Ichinose M, Belvisi MG, Barnes PJ: Bradykinin-induced bronchoconstriction in guinea-pig *in vivo*: role of neural mechanisms. *J Pharmacol Exp Ther* (1990) **253**: 1207–1212.
184. Sakamoto T, Tsukagoshi H, Barnes PJ, Chung KF: Role played by NK_2 receptors and cyclooxygenase activation in bradykinin B_2 receptor-mediated airway effects in guinea pigs. *Agents Actions* (1993) **111**: 117.
185. Fox AJ, Barnes PJ, Urban L, Dray A: An *in vitro* study of the properties of single vagal afferents innervating guinea-pig airways. *J Physiol* (1993) **469**: 21–35.
186. Fox AJ, Lalloo UG, Belvisi MG, Bernareggi M, Chung KF, Barnes PJ: Bradykinin-evoked sensitization of airway sensory nerves: a mechanism for ACE-inhibitor cough. *Nature Med* (1996) **2**: 814–817.
187. Ichinose M, Nakajima N, Takahashi T, Yamauchi H, Inoue H, Takishima T: Protection against bradykinin-induced bronchoconstriction in asthmatic patients by a neurokinin receptor antagonist. *Lancet* (1992) **340**: 1248–1251.
188. Nichol GM, O'Connor BJ, Le Compte JM, Chung KF, Barnes PJ: Effect of neutral endopeptidase inhibitor on airway function and bronchial responsiveness in asthmatic subjects. *Eur J Clin Pharmacol* (1992) **42**: 495–498.
189. Cheung D, Bel EH, den Hartigh J, Dijkman JH, Sterk PJ: An effect of an inhaled neutral endopeptidase inhibitor, thiorphan, on airway responses to neurokinin A in normal humans *in vivo*. *Am Rev Respir Dis* (1992) **145**: 1275–1280.
190. Cheung D, Timmers MC, Zwinderman AH, den Hartigh J, Dijkman JH, Sterk PJ: Neutral endopeptidase activity and airway hyperresponsiveness to neurokinin A in asthmatic subjects *in vivo*. *Am Rev Respir Dis* (1993) **148**: 1467–1473.
191. Baraniuk JN, Ohkubo O, Kwon OJ, *et al.*: Localization of neutral endopeptidase (NEP) mRNA in human bronchi. *Eur Respir J* (1995) **8**: 1458–1464.

192. Cheung D, van der Veen H, den Hartig J, Dijkman JH, Sterk PJ: Effects of inhaled substance P on airway responsiveness to methacholine in asthmatic subjects. *J Appl Physiol* (1995) **77**: 1325–1332.
193. Mantyh CR, Gates TS, Zimmerman RP, *et al.*: Receptor binding sites for substance P but not substance K or neuromedin K are expressed in high concentrations by arterioles, venules and lymph nodes in surgical specimens obtained from patients with ulcerative colitis and Crohns disease. *Proc Natl Acad Sci USA* (1988) **85**: 3235–3259.
194. Devillier P, Dessanges JF, Rakotashanaka F, Ghaem A, Boushey HA, Lockhart A: Nasal response to substance P and methacholine with and without allergic rhinitis. *Eur Respir J* (1988) **1**: 356–361.
195. Adcock IM, Peters M, Gelder C, Shirasaki H, Brown CR, Barnes PJ: Increased tachykinin receptor gene expression in asthmatic lung and its modulation by steroids. *J Mol Endocrinol* (1993) **11**: 1–7.
196. Bai TR, Zhou D, Weir T, *et al.*: Substance P (NK_1)- and neurokinin A (NK_2)-receptor gene expression in inflammatory airway diseases. *Am J Physiol* (1995) **269**: L309–L317.
197. McAlpine LG, Thomson NC: Lidocaine-induced bronchoconstriction in asthmatic patients. Relation to histamine airway responsiveness and effect of preservative. *Chest* (1989) **96**: 1012–1015.
198. Barnes PJ: Effect of nedocromil sodium on airway sensory nerves. *J Allergy Clin Immunol* (1993) **92**: 182–186.
199. Dixon CMS, Fuller RW, Barnes PJ: The effect of nedocromil sodium on sulphur dioxide induced bronchoconstriction. *Thorax* (1987) **42**: 462–465.
200. Dixon N, Jackson DM, Richards IM: The effect of sodium cromoglycate on lung irritant receptors and left ventricular receptors in anasthetized dogs. *Br J Pharmacol* (1979) **67**: 569–574.
201. Verleden GM, Belvisi MG, Stretton CD, Barnes PJ: Nedocromil sodium modulates non-adrenergic non-cholinergic bronchoconstrictor nerves in guinea-pig airways *in vitro*. *Am Rev Respir Dis* (1991) **143**: 114–118.
202. Nichol GM, Alton EWFW, Nix A, Geddes DM, Chung KF, Barnes PJ: Effect of inhaled furosemide on metabisulfite- and methacholine induced bronchoconstriction and nasal potential difference in asthmatic subjects. *Am Rev Respir Dis* (1990) **142**: 576–580.
203. Elwood W, Lotvall JO, Barnes PJ, Chung KF: Loop diuretics inhibit cholinergic and non-cholinergic nerves in guinea pig airways. *Am Rev Respir Dis* (1991) **143**: 1340–1344.
204. Ventresca GP, Nichol GM, Barnes PJ, Chung KF: Inhaled furosemide inhibits cough induced by low chloride content solutions but not by capsaicin. *Am Rev Respir Dis* (1990) **142**: 143–146.
205. O'Connor BJ, Chen-Wordsell M, Barnes PJ, Chung KF: Effect of an inhaled opioid peptide on airway responses to sodium metabisulphite in asthma. *Thorax* (1991) **46**: 294P.
206. O'Connor BJ, Lecomte JM, Barnes PJ: Effect of an inhaled H_3-receptor agonist on airway responses to sodium metabisulphite in asthma. *Br J Clin Pharmacol* (1993) **35**: 55–57.
207. Joos GF, Van Schoor J, Kips JC, Pauwels RA: The effect of inhaled FK224, a tachykinin NK-1 and NK-2 receptor antagonist, on neurokinin A-induced bronchoconstriction in asthmatics. *Am J Respir Crit Care Med* (1996) **153**: 1781–1784.
208. Fahy J, Wong HH, Geppetti P, *et al.*: Effect of an NK_1 receptor antagonist (CP-99,994) on hypertonic saline-induced bronchoconstriction and cough in male asthmatic subjects. *Am J Respir Crit Care Med* (1995) **152**: 879–884.
209. Ichinose M, Miura M, Yamauchi H, *et al.*: A neurokinin 1-receptor antagonist improves exercise-induced airway narrowing in asthmatic patients. *Am J Respir Crit Care Med* (1996) **153**: 936–941.
210. Bowden JJ, Gibbins IL: Vasoactive intestinal peptide and neuropeptide Y coexist in non-adrenergic sympathetic neurons to guinea pig trachea. *J Auton Nerv Syst* (1992) **38**: 1–20.
211. Takahashi T, Ichinose M, Yamauchi H, *et al.*: Neuropeptide Y inhibits neurogenic inflammation in guinea pig airways. *J Appl Physiol* (1993) **75**: 103–107.
212. Baraniuk JN, Silver PB, Kaliner MA, Barnes PJ: Neuropeptide Y is a vasoconstrictor in human nasal mucosa. *J Appl Physiol* (1992) **73**: 1867–1872.
213. Stretton CD, Barnes PJ: Modulation of cholinergic neurotransmission in guinea pig trachea by neuropeptide Y. *Br J Pharmacol* (1988) **93**: 672–678.

214. Cadieux A, Benchekroun MT, St Pierre S, Fournier A: Bronchoconstrictive action of neuropeptide Y (NPY) on isolated guinea pig airways. *Neuropeptides* (1989) **13**: 215–219.
215. Matran R, Martling C-R, Lundberg JM: Inhibition of cholinergic and nonadrenergic, noncholinergic bronchoconstriction in the guinea-pig mediated by neuropeptide Y and alpha2-adrenoceptors and opiate receptors. *Eur J Pharmacol* (1989) **163**: 15–23.
216. Belvisi MG, Stretton CD, Barnes PJ: Bombesin-induced bronchoconstriction in the guinea pig: mode of action. *J Pharmacol Exp Ther* (1991) **258**: 36–41.
217. Baraniuk JN, Silver PB, Lundgren JP, Cole P, Kaliner MA, Barnes PJ: Bombesin stimulates mucous cell and serous cell secretion in human nasal provocation tests. *Am J Physiol* (1992) **262**: L48–L52.
218. Stretton CD, Barnes PJ: Cholecystokinin octapeptide constricts guinea-pig and human airways. *Br J Pharmacol* (1989) **97**: 675–682.
219. Sekizawa K, Graf PD, Nadel JA: Somatostatin potentiates cholinergic neurotransmission in ferret trachea. *J Appl Physiol* (1989) **67**: 2397–2400.
220. Dey RD, Mitchell HW, Coburn RF: Organization and development of peptide-containing neurons in the airways. *Am J Respir Cell Mol Biol* (1990) **3**: 187–188.
221. Cheung A, Polak JM, Bauer FE, *et al.*: The distribution of galanin immunoreactivity in the respiratory tract of pig, guinea pig, rat, and dog. *Thorax* (1985) **40**: 889–896.
222. Guiliani S, Amann R, Papini M, Maggi CA, Meli A: Modulatory action of galanin on responses due to antidromic activation of peripheral terminals of capsaicin sensitive sensory nerves. *Eur J Pharmacol* (1989) **163**: 91–96.
223. Takahashi T, Belvisi MG, Barnes PJ: Modulation of neurotransmission in guinea-pig airways by galanin and the effect of a new antagonist galantide. *Neuropeptides* (1994) **26**: 245–251.
224. Wagner U, Fehmann HC, Bredenbroker D, Yu F, Barth PJ, von Wichert P: Galanin and somatostatin inhibition of substance P-induced airway mucus secretion in the rat. *Neuropeptides* (1995) **28**: 59–64.
225. Cutz E: Neuroendocrine cells of the lung: an overview of morphological characteristics and development. *Exp Lung Res* (1982) **3**: 185–208.
226. Shimosegawa T, Foda HD, Said SI: [Met]enkephalin-Arg6-Gly7-Leu8-immunoreactive nerves in guinea pig and rat lungs: distribution, origin, and coexistence with vasoactive intestinal polypeptide immunoreactivity. *Neuroscience* (1990) **36**: 737–750.
227. Belvisi MG, Rogers DF, Barnes PJ: Neurogenic plasma extravasation: inhibition by morphine in guinea pig airways *in vivo*. *J Appl Physiol* (1989) **66**: 268–272.
228. Belvisi MG, Chung KF, Jackson DM, Barnes PJ: Opioid modulation of non-cholinergic neural bronchoconstriction in guinea-pig *in vivo*. *Br J Pharmacol* (1988) **95**: 413–418.
229. Frossard N, Barnes PJ: μ-Opioid receptors modulate non-cholinergic constrictor nerves in guinea-pig airways. *Eur J Pharmacol* (1987) **141**: 519–521.
230. Barnes PJ: Airway neuropeptides: roles in fine tuning and in disease. *News Physiol Sci* (1989) **4**: 116–120.
231. Nohr D, Weihe E: The neuroimmune link in the bronchus-associated lymphoid tissue (BALT) of cat and rat: peptides and neural markers. *Brain Behav Immun* (1991) **5**: 84–101.

25

Transcription Factors

ROBERT NEWTON, PETER J. BARNES AND
IAN M. ADCOCK

BASAL AND REGULATED TRANSCRIPTION

The lung consists of a diverse range of cell types possessing various cell-specific characteristics, yet containing the same genetic material. Regulation of gene expression is therefore essential to cell differentiation and maturation during lung development and in its mature state. However, cells also respond in a regulated way to cytokines, oxidative stress, viral infections and other inflammatory stimuli by the expression of appropriate response genes.[1] Such *de novo* gene expression requires synthesis of messenger RNA (mRNA) (transcription) from the DNA template followed by protein synthesis (translation).

Extracellular signals are generally communicated to the interior of the cell via receptors on the cell surface. Ligand–receptor interaction initiates intracellular signalling cascades (signal transduction) that result in activation of specific DNA-binding proteins or transcription factors.[2,3] Binding of these factors to recognition sequences in the control regions (promoters) of target genes is communicated to the basal transcription machinery, causing activation of RNA polymerase II (RNA pol II)-dependent transcription.

Initiation of transcription requires a variety of proteins that can be divided into the 'basal transcription machinery', consisting of RNA pol II and associated factors, often referred to as general transcription factors, and an array of specific transcription factors or activators that are responsible for conferring both gene and stimulus specificity to the transcriptional response.[4] Essentially, the general transcription factors are involved in recognition of core promoter elements and stabilization of RNA pol II at the transcription start site[5] (Fig. 25.1). The general transcription factor TFIID, which consists of the TATA box-binding protein (TBP) and at least 12 other TBP-associated factors (TAFs), plays an important role in promoter recognition by binding the TATA box.[5] This is the

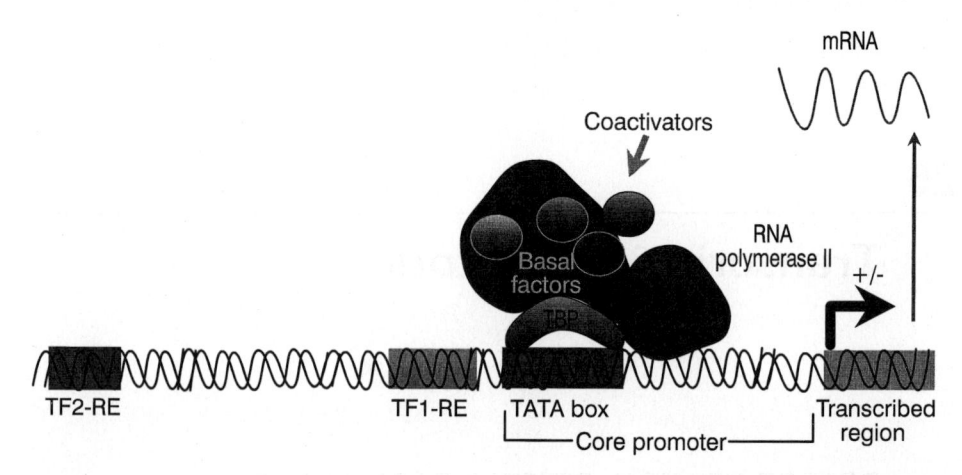

Fig. 25.1 The basal transcription machinery. Eukaryotic RNA polymerase II-dependent genes require the coordinate binding of basal or general transcription factors, including TFIIA, B, D, E, F and H to the core promoter. TATA box-binding protein (TBP), a constituent of TFIID, binds the TATA box element and together with the basal factors and RNA polymerase II makes up the basal transcription initiation complex (BTIC).

most conserved basal promoter element and is usually found 25 to 30 base pairs (bp) upstream of transcription start.[6] However, RNA pol II, along with the general transcription factors, can only direct basal transcription and additional factors are required for regulated or inducible transcription.

Regulated transcription results from interaction between *cis*-acting regulatory DNA sequences (enhancers and silencers) in the promoters of target genes and sequence-specific *trans*-acting specific transcription factors. This may cause increased (transactivation) or decreased (transrepression) transcription of the associated gene. These regulatory elements are commonly found in the immediate 5′-region upstream of transcription start. However, more distal regions and elements within the gene itself may also be required for correct transcriptional control. Transcriptional activation occurs either via direct contacts between the specific transcription factor and the basal transcription complex or indirectly through cofactors, which may not themselves contact the DNA[4,6,7] (Fig. 25.2). Thus information that is transferred from the cell surface via cytoplasm becomes integrated in the nucleus to produce regulated transcription. This review concentrates on nuclear factor-κB (NF-κB) as a model for inducible activation and briefly deals with other transcription factors thought to be important in airway inflammation.[8]

NF-κB, THE REL FAMILY OF PROTEINS AND IκB PROTEINS

NF-κB was first described as a factor that bound to the immunoglobulin κ light chain enhancer (5′-GGGACTTTCC-3′),[9] but is now recognized as an almost ubiquitous activator of immune and acute phase genes.[10] Agents that activate NF-κB include the inflammatory cytokines tumour necrosis factor (TNF)-α and interleukin (IL)-1, lipopolysaccharide (LPS) and other bacterial products, viruses, ultraviolet light and oxidative stress[10] (Fig. 25.3). Consequently, pathological conditions such as inflammation, sepsis

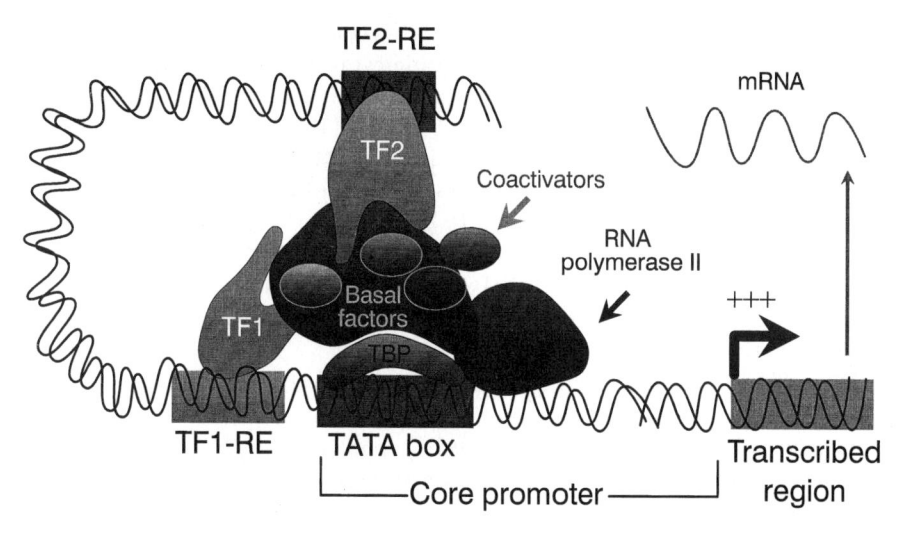

Fig. 25.2 Activated transcription. Binding of activated transcription factors to distal *cis* elements, transcription factor-1 response element (TF1-RE) and transcription factor-2 response element (TF2-RE), may cause DNA conformational changes and juxtaposition of the activated transcription factor (TF1 or TF2) to the basal transcription initiation complex (BTIC). Interaction of these factors with the BTIC may cause further conformational changes in the complex and lead to activation of RNA polymerase II.

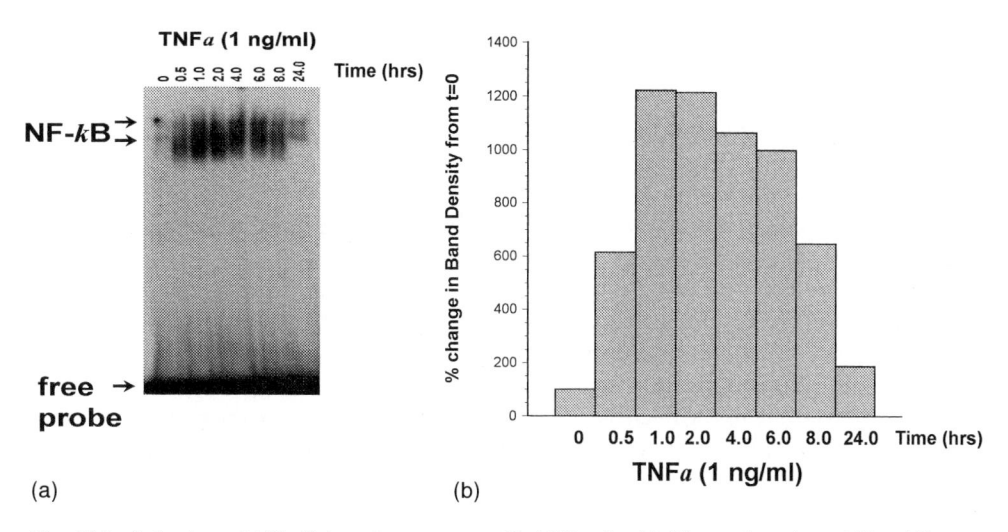

Fig. 25.3 Induction of NF-κB in pulmonary type II A549 cells. (a) Electrophoretic mobility shift assay showing the effect of tumour necrosis factor (TNF)-α over time on NF-κB DNA-binding complexes (arrowed) within the nucleus. Confluent cells were incubated for various time points up to 24 h in the presence of 1 ng/ml TNF-α. Nuclear proteins were extracted and 5 μg incubated with ^{32}P-labelled double-stranded oligonucleotides encoding the consensus NF-κB DNA-binding site. Binding of activated NF-κB to the DNA probe retards its progress through a 7.5% non-denaturing polyacrylamide gel and gives a measure of the amount of activated NF-κB present. (b) Densitometric analysis of the retarded bands in (a) indicates graphically the rapid and sustained induction of activated NF-κB within the nucleus of these cells.

and viral infection may be expected to result in NF-κB activation. Once activated, NF-κB plays an important role in a range of immunological responses. For example, IL-1 and TNF-α not only activate NF-κB but are themselves upregulated by NF-κB. Additionally, many other cytokines and chemokines, including IL-6, IL-8, monocyte chemoattractant protein (MCP)-1, macrophage inflammatory protein (MIP)-1α and RANTES are also partially regulated by NF-κB.[10] NF-κB is also important in T-cell activation, where it is involved in upregulation of both IL-2 and IL-2 receptors.[11] Indirect involvment of NF-κB in migration of inflammatory cells to inflammatory sites also occurs via expression of various adhesion molecules on endothelial cell surfaces.[10] NF-κB therefore plays a central role in inflammation and knowledge of its mode of action is necessary for a molecular understanding of inflammatory disease.

The DNA-binding activity originally described as NF-κB is now known to consist of heterodimers of Rel family transcription factors, which in mammalian cells includes the protooncogene c-Rel (Rel), p50/p105 (NF-κB1), p65 (RelA), p52/p100 (NF-κB2) and RelB.[10] These proteins share a ∼300 amino acid region known as the Rel homology domain (RHD) that shows about 35–61% identity and contains the DNA-binding region. RHD proteins are capable of dimerization via contacts in the RHD and bind DNA as heterodimers or often as homodimers. RelB, c-Rel and p65 do not bind DNA efficiently but have potent transactivation domains that are critical in transcriptional activation.[10] This contrasts with the main DNA-binding subunits p50 and p52, which are poor transactivators unless dimerized with p65, RelB or c-Rel and are synthesized as the precursor molecules p105 and p100 respectively. These precursors contain an IκB-like inhibitory region, which on proteolytic cleavage releases the DNA-binding N-terminal domain (p50 or p52). Classically, p50/p65 NF-κB heterodimers are the most abundant of the transactivating complexes. However, p50 homodimers are also commonly found constitutively in the nuclei of many cells and may play a role in silencing transcription.[10]

Primary regulation of Rel transcription factors is by sequestration of NF-κB heterodimers in the cytoplasm as inactive complexes by inhibitory molecules known as IκBs[10,12] (Fig. 25.4). Inducing agents result in kinase activation and subsequent phosphorylation and degradation of the IκB molecule.[13] This allows dissociation of the cytoplasmic NF-κB–IκB complexes and translocation of active NF-κB heterodimers to the nucleus where they bind κB elements, causing activation of transcription.[10,12]

The inhibitory IκB proteins can be divided into two distinct classes. The first consists of the precursor forms of the p50 and p52 subunits of NF-κB. The precursors, p100 and p105, are capable of binding transactivation subunits; on activation, proteolytic cleavage releases p50 or p52 respectively as active heterodimers with the transactivation subunit.[13] These may translocate to the nucleus or become bound by IκB. Since the p50/p105 gene is regulated by NF-κB, activation of NF-κB results in *de novo* synthesis of p50/p105, thereby replacing lost p105.[14]

The second class of IκBs are typified by IκBα and IκBβ. These combine with NF-κB heterodimers within the cytoplasm to prevent nuclear translocation.[10,12] IκBα regulates NF-κB through an autoregulatory feedback loop whereby IκBα lost on activation is resynthesized by an NF-κB dependent mechanism. NF-κB upregulates IκBα mRNA due to the presence of multiple NF-κB sites in the IκBα promoter.[15] The rapid *de novo* synthesis of IκBα may shut down the NF-κB response and ensure only transient activation of responsive genes.

25 Transcription Factors

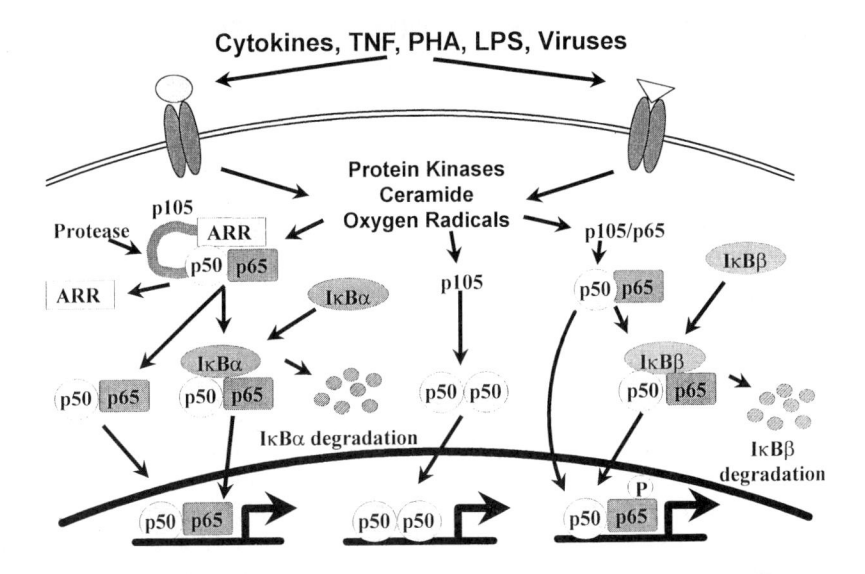

Fig. 25.4 Activation of NF-κB. Signals received by various receptors are transduced to NF-κB and IκB complexes through multiple signalling pathways, which may include protein kinases, ceramide and oxygen radicals. The inactive cytoplasmic NF-κB complexes exist in two major forms. One type of complex contains the p50 precursor, p105, whilst the other consists of NF-κB complexed with IκB molecules. These complexes may dissociate in response to different stimuli. p105 is proteolytically cleaved by a specific protease to release the IκB-like ankyrin repeat region (ARR) and p50, which undergoes homodimerization or heterodimerization with transactivation subunits. Additionally, p50 homodimers may bind to DNA but will not cause transactivation and may therefore function as endogenous silencers. The second type of cytoplasmic NF-κB complex typically consists of p50/p65 heterodimers and IκBs (IκBα or IκBβ). On activation, a specific IκB kinase phosphorylates IκB causing rapid degradation of IκB within proteosomes. The p50/p65 heterodimer translocates to the nucleus and on binding κB elements transactivates transcription. LPS, lipopolysaccharide; PHA, phytohaemagglutinin; TNF, tumour necrosis factor.

This scheme, however, fails to explain how inducers such as LPS or IL-1β cause persistent NF-κB activation. IκBβ and IκBα are both present in the lung and both bind p65- and c-Rel-containing heterodimers with similar affinity. Importantly, IκBβ is not NF-κB inducible and not therefore subject to autoregulatory feedback.[16] It has also been suggested that NF-κB, released by IκBβ, is modified by phosphorylation, thus preventing nuclear degradation or sequestration by IκBα and aiding persistent activation.[16]

Although NF-κB activation by degradation of IκB is a simple model, it can be seen that overall activation comprises multiple overlapping processes. Cell-surface events may result in differential kinase activation, causing phosphorlyation of distinct IκB/NF-κB pools. Consequently, transient activation via IκBα or persistent activation by IκBβ, as well as processing of precursors and possible mechanisms involving novel IκBs, may differentially control NF κB activity.[16–19] In addition, alternative combinations of Rel heterodimers and homodimers may also cause differential gene regulation. This paradigm of NF κB-induced expression also invites speculation as to novel targets for theraputic intervention. For instance, stimulation of IκB synthesis, or repression of p50 or p65, may be used to control NF-κB as well as direct intervention to prevent signalling events prior to activation.

56. Lilly CM, Bai TR, Shore SA, Hall AE, Drazen JM: Neuropeptide content of lungs from asthmatic and nonasthmatic patients. *Am J Respir Crit Care Med* (1995) **151**: 548–553.
57. Sharma RK, Jeffery PK: Airway VIP receptor number is reduced in cystic fibrosis but not asthma. *Am Rev Respir Dis* (1990) **141**: A726.
58. Paul S, Said SI, Thompson AB, et al.: Characterization of autoantibodies to vasoactive intestinal peptide in asthma. *J Neuroimmunol* (1989) **23**: 133–142.
59. Cardell LO, Uddman R, Edvinsson L: Low plasma concentration of VIP and elevated levels of other neuropeptides during exacerbations of asthma. *Eur Respir J* (1994) **7**: 2169–2173.
60. Caughey GH: Roles of mast cell tryptase and chymase in airway function. *Am J Physiol* (1989) **257**: L39–L46.
61. Wenzel SE, Fowler AA, Schwartz LB: Activation of pulmonary mast cells by bronchoalveolar allergen challenge. In vivo release of histamine and tryptase in atopic subjects with and without asthma. *Am Rev Respir Dis* (1988) **137**: 1002–1008.
62. Tam EK, Franconi GM, Nadel JA, Caughey GH: Protease inhibitors potentiate smooth muscle relaxation induced by vasoactive intestinal peptide in isolated human bronchi. *Am J Respir Cell Mol Biol* (1990) **2**: 449–452.
63. Sekizawa K, Caughey GH, Lazarus SC, Gold WM, Nadel JA: Mast cell tryptase causes airway smooth muscle hyperresponsiveness in dogs. *J Clin Invest* (1989) **83**: 175–179.
64. Laitinen LA, Laitinen A, Salonen RO, Widdicombe JG: Vascular actions of airway neuropeptides. *Am Rev Respir Dis* (1987) **136**: 559–564.
65. Webber SE: The effects of peptide histidine isoleucine and neuropeptide Y on mucous volume output from ferret trachea. *Br J Pharmacol* (1988) **55**: 40–54.
66. Yiangou Y, DiMarzo V, Spokes RA, Panico M, Morris HR, Bloom SR: Isolation, characterization, and pharmacological actions of peptide histidine valine 42, a novel preprovasoactive intestinal peptide derived peptide. *J Biol Chem* (1987) **262**: 14 010–14 013.
67. Chilvers ER, Dixon CMS, Yiangou Y, Bloom SR, Ind PW: Effect of peptide histidine valine on cardiovascular and respiratory funtion in normal subjects. *Thorax* (1988) **43**: 750–755.
68. Cardell LO, Sundler F, Uddman R: Helospectin/helodermin-like peptides in guinea pig lung: distribution and dilatory effects. *Regul Pept* (1993) **45**: 435–443.
69. Robberecht P, Waelbroeck M, deNeef P, Camus JC, Coy DH, Christophe J: Pharmacological characterization of VIP receptors in human lung membranes. *Peptides* (1988) **9**: 339–345.
70. Uddman R, Luts A: Pituitary adenylate cyclase activity peptide (PACAP), a new vasoactive intestinal peptide (VIP)-like peptide in the respiratory tract. *Cell Tissue Res* (1991) **265**: 197–201.
71. Miyata A, Jiang L, Dahl RD, et al.: Isolation of a neuropeptide corresponding to the N-terminal 27 residues of the pituitary adenylate cyclase activating polypeptide with 38 residues (PACAP38). *Biochem Biophys Res Commun* (1990) **170**: 643–648.
72. Conroy DM, St Pierre S, Sirois P: Relaxant effects of pituitary adenylate cyclase activating peptide (PACAP) on epithelium-intact and denuded guinea pig trachea: a comparison with vasoactive intestinal peptide. *Neuropeptides* (1995) **29**: 121–127.
73. Gottschall PE, Tatsumo I, Miyata A, Arimura A: Characterization and distribution of binding sites for the hypothalamic peptide pituitary adenylate cyclase activating polypeptide. *Endocrinology* (1990) **127**: 272–277.
74. Martling CR, Theodorsson-Norheim E, Lundberg JM: Occurrence and effects of multiple tachykinins: substance P, neurokinin A, and neuropeptide K in human lower airways. *Life Sci* (1987) **40**: 1633–1643.
75. Laitinen LA, Laitinen A, Haahtela T: A comparative study of the effects of an inhaled corticosteroid, budesonide, and of a β_2 agonist, terbutaline, on airway inflammation in newly diagnosed asthma. *J Allergy Clin Immunol* (1992) **90**: 32–42.
76. Komatsu T, Yamamoto M, Shimokata K, Nagura H: Distribution of substance-P-immunoreactive and calcitonin gene-related peptide-immunoreactive nerves in normal human lungs. *Int Arch Allergy Appl Immunol* (1991) **95**: 23–28.
77. Springall DR, Polak JM, Howard L, et al.: Persistence of intrinsic neurones and possible phenotypic changes after extrinsic denervation of human respiratory tract by heart–lung transplantation. *Am Rev Respir Dis* (1990) **141**: 1538–1546.

AP-1 AND RELATED TRANSCRIPTION FACTORS

AP-1 was identified as a factor that mediated transcription in response to protein kinase C activation by phorbol esters such as 12-O-tetradecanoyl phorbol-13-acetate (TPA) via consensus DNA sequences known as TPA response elements (TREs).[20] TREs are bound with low affinity by c-Jun homodimers but are more strongly bound by c-Jun/c-Fos heterodimers, the predominant AP-1 complex in many cell types.[21] Growth factors, cytokines, T-cell activators and other mitogens that act predominantly through mitogen-activated protein (MAP) kinase-dependent signalling pathways also activate AP-1.[20,22] This occurs by post-translational modification of pre-existing subunits and by rapid upregulation of the genes, particularly c-Fos, encoding AP-1 subunits.[22] AP-1 proteins combine as heterodimers and homodimers by virtue of a basic leucine zipper (bZIP) region, which is responsible for the dimerization required for DNA binding.[23]

Although AP-1 activity is almost ubiquitous, the tissue distribution and expression patterns of Fos and Jun family members are quite distinct and the roles played by these complexes may be extremely diverse.[24] For example, Fra-1, a c-Fos homologue, heterodimerizes with Jun proteins and shows similar DNA-binding affinity to the equivalent c-Fos/c-Jun heterodimer. However, Fra-1/Jun dimers cannot transactivate transcription due to a transactivation domain in c-Fos that is lacking in Fra-1.[25] Moreover, AP-1 proteins form functionally distinct dimeric complexes with members of the related transcription factor family ATF/CREB.[26]

Signalling events that result in activation of the MAP kinases, Jun N-terminal kinase (JNK)-1 and JNK-2, potentiate transactivation by direct phosphorylation of c-Jun at N-terminal serine residues.[27] This step requires recruitment of JNK to a specific docking site and the presence of a specificity-conferring region flanking the phosphorylation site.[28] Thus c-Jun, which has both these regions, is efficiently phosphorylated by JNK. In contrast, JunB, which has the docking region but not the specificity region, is not phosphorylated, whilst JunD, which contains the specificity region but not the docking region, is only phosphorylated when heterodimerized with a docking competent partner. Conversely, dephosphorylation of residues adjacent to the DNA-binding domain may play a role in phorbol ester activation of AP-1.[29] Furthermore, AP-1 activity can also be modulated by an inhibitory protein (IP-1), which specifically blocks DNA binding.[30]

CCAAT/ENHANCER-BINDING PROTEINS AND NF-IL-6

CCAAT/enhancer-binding proteins (C/EBP) are important in IL-1-, IL-6- and LPS-dependent signal transduction and play a major role in the induction of many immune and inflammatory response genes.[31,32] These proteins are encoded by separate genes and have recently been reclassified as C/EBPα, C/EBPβ (formally NF-IL-6), C/EBPγ and C/EBPδ.[33] C/EBP proteins also belong to the bZIP class of transcription factors and bind as heterodimers and homodimers to C/EBP sites.[33] Transcriptional activation depends on the specific binding site and may involve activator and repressor forms.[34,35] In the case of C/EBPβ, rapid *de novo* synthesis and phosphorylation of C/EBPβ is required for full

25 Transcription Factors

activation.[36] Additionally, the C/EBP proteins interact via the bZIP domain with other transcription factors to coactivate transcription.

JAK–STAT PATHWAY

Pro-inflammatory and mitogenic stimuli generally activate gene transcription via NF-κB, AP-1 and C/EBP transcription factors. However, many other cytokines use signalling pathways involving Janus kinases (JAK) and transcription factors known as signal transducers and activators of transcription (STAT).[31,37,38] The specificity and functional redundancy of these cytokines is partially explained by the fact that their cognate receptors are typically dimers of a common or public subunit and a cytokine-specific subunit.[31] For example, receptors for IL-2, IL-4, IL-7, IL-9, IL-13 and IL-15 share a common γ-subunit (γc) in addition to the cytokine-specific subunits.[31,38]

A general scheme for cytokine-induced STAT activation involves ligand-induced receptor dimerization (Fig. 25.5). This causes juxtapositioning of the receptor-associated

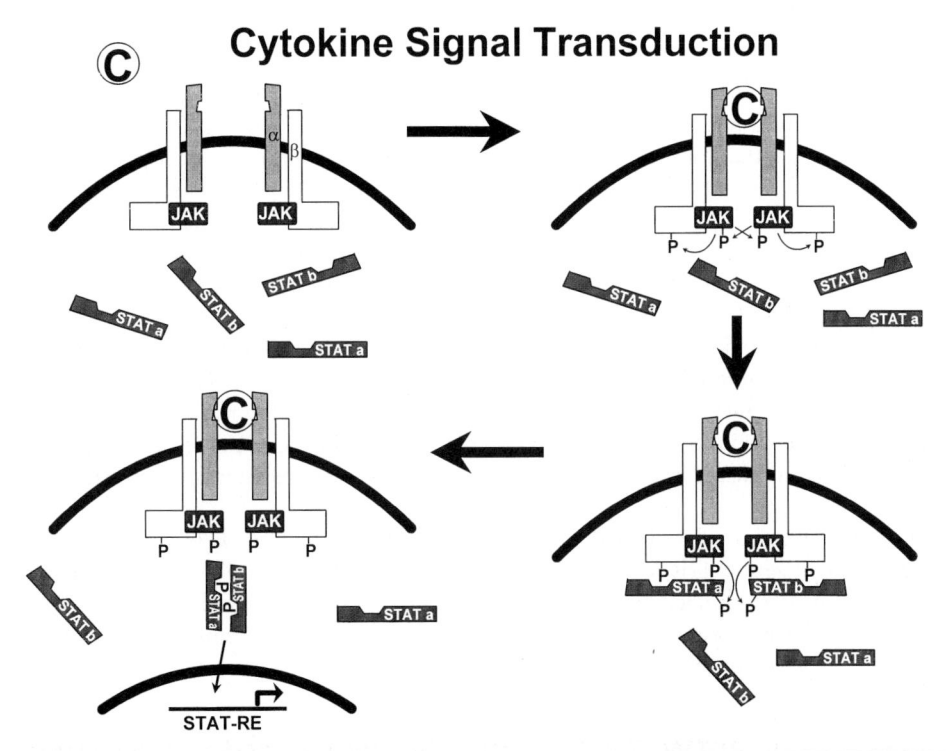

Fig. 25.5 General scheme of STAT activation. Cytokine (C) binding to the cytokine-specific receptor subunit (α) causes receptor dimerization. This brings the two JAKs, bound to the public receptor subunits (β), into close proximity, allowing reciprocal phosphorylation. The JAKs, now activated, further phosphorylate the public receptor subunits, enabling binding by specific STAT proteins. Subsequent phosphorylation of the STAT proteins by JAKs allows homodimerization or heterodimerization and activation of the STAT complex. This can then translocate to the nucleus to activate transcription at STAT-specific response elements (STAT-RE).

JAKs, allowing (reciprocal) phosphorylation of each other.[38] The two JAKs, now activated, phosphorylate tyrosines on the receptor molecules, which then act as docking sites to selectively bind particular STATs.[39] These are in turn phosphorylated by the JAKs to allow STAT homodimerization or heterodimerization and translocation to the nucleus, causing transcriptional activation.[38]

Currently, six general members of this family have been identified (STAT1–6).[38] Although the STAT, or combination of STAT proteins, activated by a specific cytokine appears to be specific, the intermediate JAKs involved may be the same for a variety of STAT proteins or receptor complexes. Thus cytokines, acting through distinct receptors, may activate the same JAKs and still produce cytokine-specific as well as common responses.[17,39]

GLUCOCORTICOID RECEPTORS

The glucocorticoid receptor (GR) and other steroid receptors are members of the nuclear receptor superfamily. Glucocorticoids are able to reduce inflammation by mimicking endogenous glucocorticoids that are important in homeostasis. Binding of GR as a dimer to DNA at glucocorticoid response elements (GREs) (5'-GGTACAnnnTGTTCT-3') is required for transactivation of transcription.[40,41]

GR is expressed in most cell types and studies in human lung suggest high levels in the airway epithelium and endothelium of bronchial vessels.[42] Inactive GR is bound to a protein complex that includes two subunits of the heat-shock protein, hsp90, and other inhibitory proteins, which act as molecular chaperones to prevent nuclear localization of GR[40] (Fig 25.6). On binding steroid, hsp90 dissociates allowing nuclear localization and DNA binding by the activated GR–steroid complex.[40] However, the steroid ligand also appears necessary for dimerization and transactivation, probably due to conformational changes that affect DNA binding.[40,43]

The position and number of GREs relative to transcription start are important determinants of the response to steroids. Thus increased numbers of GREs and proximity to the TATA box increases steroid inducibility.[44] In addition, the relative abundance and binding of other transcription factors, or coactivators, in the vicinity of the GRE may also strongly influence the inducibility and steroid responsiveness of particular cell types.[45,46]

The mechanisms of GR-mediated gene repression are less well understood. GR binding to negative GREs (nGREs) was proposed as a possible mechanism (Fig. 25.6).[47] However, glucocorticoid-repressible genes do not necessarily possess GREs or nGREs in their promoters, suggesting that other mechanisms of inhibition, for instance involving cross-talk with other pathways, are more important.[43,48]

CROSS-TALK BETWEEN TRANSCRIPTION FACTORS AND THEIR TRANSDUCTION PATHWAYS

The fact that many inflammatory genes, which are regulated by AP-1, NF-κB and C/EBP, can be downregulated by glucocorticoids indicates the possible importance of cross-talk

Classical mechanisms of steroid action

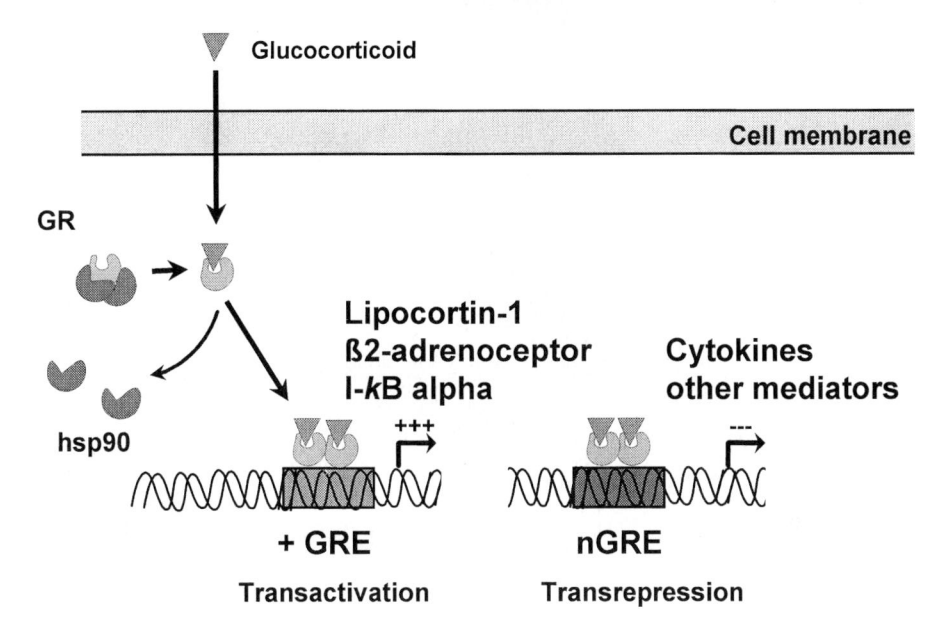

Fig. 25.6 Classical mechanisms of steroid action. Glucocorticoids, as lipophilic molecules, diffuse readily through cell membranes into the cytoplasm. Upon ligand binding, glucocorticoid receptors (GR) are activated by release of the inhibitory 90-kDa heat-shock proteins (hsp90) to reveal the nuclear localization signal. Activated GR translocates to the nucleus where it binds glucocorticoid response elements (GRE) as a dimer and upregulates steroid-responsive genes such as lipocortin-1, β_2-adrenoceptor and IκBα. Alternatively it was postulated that GR may bind repressor sequences (nGRE), causing repression of a variety of pro-inflammatory genes such as those for cytokines, chemokines and other mediators.

between these signal transduction pathways.[8,49] For example, activation of the IL-8 promoter by NF-κB and C/EBPb is inhibited by dexamethasone, primarily via the NF-κB site.[50] Conversely, activation of a GRE-dependent promoter by dexamethasone was inhibited by overexpression of p65.[51] These and other studies show direct protein–protein interactions between NF-κB and GR, which prevent NF-κB DNA binding and/or transactivation and partly account for the anti-inflammatory properties of glucocorticoids[50–52] (Fig. 25.7). Recently, a mechanism of glucocorticoid repression of NF-κB mediated transcription has been described whereby steroids rapidly induce IκBα mRNA and protein synthesis[53,54] (Fig. 25.8). Newly synthesized IκB interacts with, and binds to, NF-κB heterodimers within the cytoplasm, and probably nucleus,[55] thereby inhibiting NF-κB DNA binding and activation. However, it now appears that these mechanisms of NF-κB inhibition by glucocorticoids may be of lesser importance in endothelial[56] or airway epithelial cells.[57]

In a similar manner to repression of NF-κB, glucocorticoid repression of AP-1-dependent genes occurs by at least two mechanisms. GR either blocks Fos/Jun DNA binding or represses their ability to transactivate transcription.[58,59] This repression is independent of GR DNA binding, but requires the bZIP region of c-Jun.[60] However,

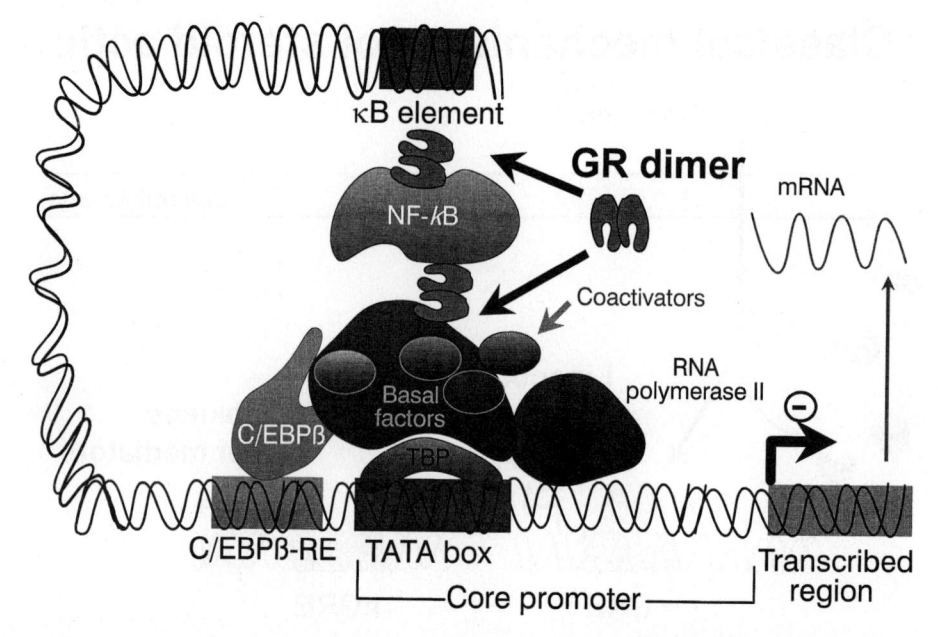

Fig. 25.7 Direct repression of NF-κB-dependent transcription by activated glucocorticoid receptor (GR). NF-κB and other transcription factors such as C/EBPβ activate the basal transcription initiation complex (BTIC) by causing conformational changes in the structure of BTIC to enhance RNA polymerase II activity. Activated GR may interfere with NF-κB-dependent transcription by directly interacting with p65 and thereby preventing NF-κB from binding to DNA or transactivating the BTIC or both.

repression of AP-1 may also occur without any apparent alteration in the DNA binding at AP-1 sites.[61]

Cross-coupling between transcription factors, such as that between NF-κB or AP-1 and GR, are now being routinely described. For instance, NF-κB interacts with the C/EBP family of transcription factors to cause either increased or decreased transactivation depending on the promoter context.[62–64] C/EBP and AP-1 can also cooperatively interact to cause transcriptional activation.[35] In addition, both C/EBP and STAT proteins may also activate transcription synergistically.[64]

One further level of cross-talk may occur via interactions with the transcriptional coactivators CREB-binding protein (CBP) and p300.[65] CBP was identified as a protein that bound phospho-CREB to activate transcription from cAMP-regulated enhancers (CREs).[66] However, it has since become clear that CBP (and p300) have multiple activating domains and can functionally interact with many transcription factors, including STATs,[67] NF-κB,[68] the AP-1 component c-Jun[69] as well as the steroid hormone receptors.[70] Thus nuclear receptors, such as GR, may cause repression of AP-1-dependent transcription via competition with CBP.[71]

Taken together these findings illustrate how stimulus-dependent transcription may require activation of, and interaction between, various diverse transcription factor families. However, the exact response may also depend upon the cell type due to the presence of endogenous transcription factors or coactivators that effect activation of the basal transcription complex.

Induction of IκBα by glucocorticoids

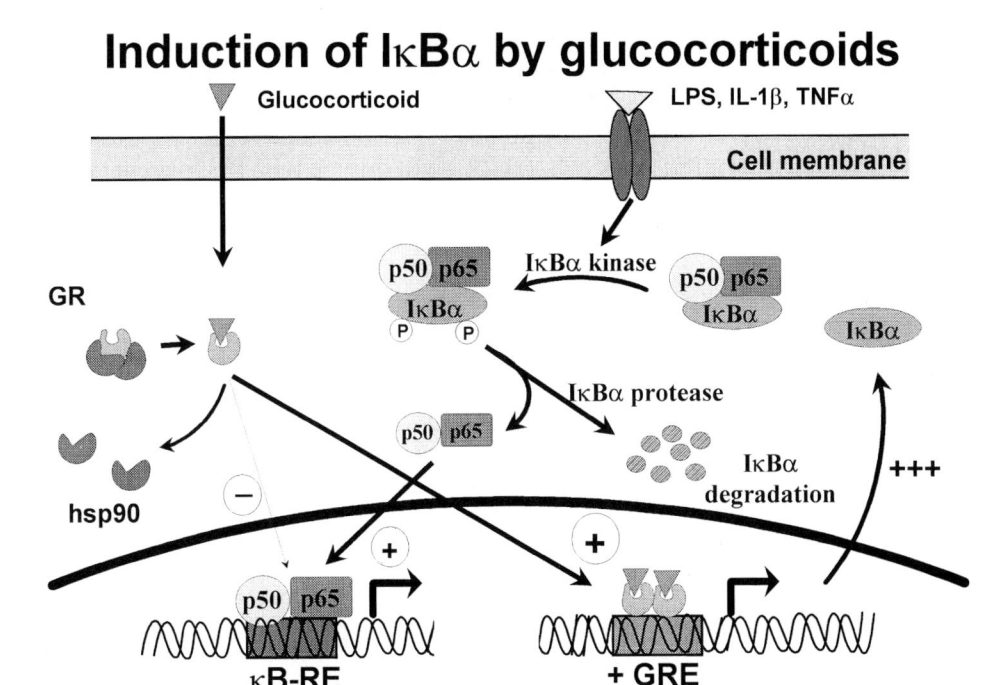

Fig. 25.8 Repression of NF-κB-dependent transcription by induction of IκBα. Recent evidence shows that in T-cells the expression of IκBα can be markedly upregulated by glucocorticoids. Although not characterized at present, this may occur via binding of activated glucocorticoid receptor (GR) to glucocorticoid response elements (GREs) or other response elements in conjunction with other factors to activate transcription of the IκBα gene. This results in rapid *de novo* synthesis of IκBα, which can then bind to active NF-κB in both the cytosol and nucleus to prevent activation of transcription. LPS, lipopolysaccharide; IL-1β, interleukin 1β; TNFα, tumour necrosis factor α.

TRANSCRIPTION FACTORS IN ASTHMA

Little information is currently available concerning the expression and activation status of transcription factors in asthma. However, low levels of activated AP 1 and NF-kB are detectable in most cells and activation of these transcription factors is rapidly increased by many factors, including multiple cytokines,[52] histamine[72] and various eicosanoids,[73] that are associated with airway inflammation in asthma. The effect of these mediators on c-Fos expression may account for the increased levels of c-Fos found in the airway epithelium of asthmatic patients.[74] In addition, elevated levels of pro-inflammatory cytokines,[75,76] such as IL-1β, TNF-α, IL-6 and IL-8, in asthmatic airways suggests that acute phase transcription factors, such as NF-κB, may also be activated. Indeed, expression of p65, as well as increased NF-κB DNA binding, is also found in biopsies and induced sputum from asthmatic subjects.[77] Similar results have been reported in other inflammatory diseases, such as rheumatoid arthritis[78] and autoimmune encephalomyelitis.[79]

During inflammation numerous other mediators, such as nitric oxide and eicosanoids, are released in addition to cytokines.[80,81] Synthesis of these mediators, along with the induction of various adhesion molecules and other receptors, are probably induced by combinations of AP-1, NF-κB and C/EBP, illustrating the pro-inflammatory actions of these transcriptions factors given prior activation by pro-inflammatory cytokines. This is further illustrated by analysis of the promoter regions of many cytokine, cytokine receptor and other inflammatory genes, which reveals numerous sites for regulation of these pro-inflammatory genes by the above transcription factors (Fig. 25.9). Thus excess activation of these transcription factors could be responsible for the prolonged inflammatory release of cytokines in inflammation and asthma and may in some individuals represent primary molecular defects.

However many other cytokines also play important roles in the chronic inflammation seen in asthma and the pattern of cytokine expression largely determines the nature and persistence of the inflammatory response. For instance, the cytokines granulocyte–macrophage colony-stimulating factor and IL-5 are predominantly modulators of eosinophil survival and function and are elevated at sites of allergic inflammation and in asthmatic airways.[75,76,82]. These cytokines primarily exert their cellular effects via the JAK–STAT pathway to activate STAT5.[38,83] In addition, important T cell effector

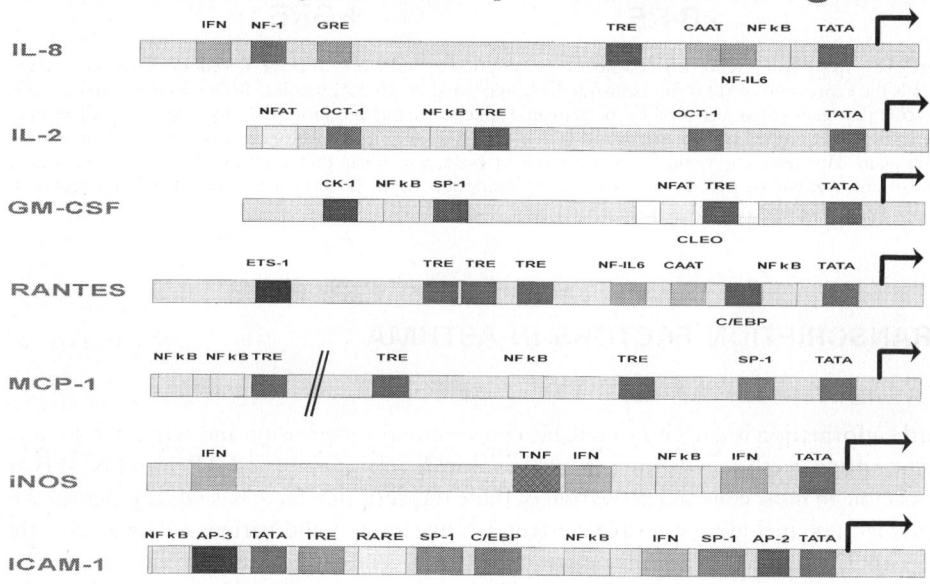

Fig. 25.9 Promoter sequences of inflammatory genes. The upstream regions of a variety of genes encoding important targets for glucocorticoid repression are represented. Many of these promoter sequences, although being repressed by glucocorticoids, do not possess either positive or negative glucocorticoid response elements (GREs). The presence of multiple binding sites for common pro-inflammatory transcription factors, such as AP-1 (TRE), NF-κB and C/EBPβ (NF-IL6), indicates the ubiquitous nature of many of these factors and illustrates the importance of other specific activators in the regulation of gene expression. GM-CSF, granulocyte–macrophage colony-stimulating factor; ICAM-1, intercellular adhesion molecule 1; IL, interleukin; iNOS, inducible nitric oxide synthase; MCP-1, monocyte chemoattractant protein-1.

molecules, such as IL 2, which also signal through the JAK–STAT pathway, are raised in asthmatic patients.[38,76]

Thus continued elucidation of signal transduction pathways and the mechanisms of action of transcription factors, as well as interactions between these pathways, will greatly enhance our understanding of inflammatory diseases such as asthma. Such analyses also have therapeutic potential in the control of lung disease. Glucocorticoids exert their anti-inflammatory effects largely by binding to transcription factors that have been activated. Other drugs that regulate the activity of specific transcription factors may also be developed in the future. The identification of novel targets, such as Jun kinases or IκB, or other proteins involved in signal transduction, may lead to the development of new, more specific drugs that are better able to control inflammation.

REFERENCES

1. He X, Rosenfeld MG: Mechanisms of complex transcriptional regulation: implications for brain development. *Neuron* (1991) **7**: 183–196.
2. Pabo CO, Sauer RT: Transcription factors: structural families and principles of DNA recognition. *Annu Rev Biochem* (1992) **61**: 1053–1095.
3. Johnson PF, McKnight SL: Eukaryotic transcriptional regulatory proteins. *Annu Rev Biochem* (1989) **58**: 799–839.
4. Roeder RG: The complexities of eukaryotic transcription initiation: regulation of preinitiation complex assembly. *Trends Genet* (1991) **16**: 402–408.
5. Buratowski S: The basics of basal transcription by RNA polymerase II. *Cell* (1994) **77**: 1–3.
6. Goodrich JA, Cutler G, Tjian R: Contracts in context: promoter specificity and macromolecular interactions in transcription. *Cell* (1996) **84**: 825–830.
7. Nordheim A: CREB takes CBP to tango. *Nature* (1994) **370**: 177–178.
8. Barnes PJ, Adcock I: Anti-inflammatory actions of steroids: molecular mechanisms. *Trends Pharmacol Sci* (1993) **14**: 436–441.
9. Sen R, Baltimore D: Multiple nuclear factors interact with the immunoglobulin enhancer sequences. *Cell* (1986) **46**: 705–716.
10. Sienbenlist U, Franzoso G, Brown K: Structure, regulation and function of NF-κB. *Annu Rev Cell Biol* (1994) **10**: 405–455.
11. Lenardo MJ, Baltimore D: NF-κB: a pleiotropic mediator of inducible and tissue-specific gene control. *Cell* (1989) **58**: 227–229.
12. Beg AA, Baldwin AS: The I kappa B proteins: multifunctional regulators of Rel/NF-kappa B transcription factors. *Genes Dev* (1993) **7**: 2064–2070.
13. Beg AA, Finco TS, Nantermet PV, Baldwin AS: Tumor necrosis factor and interleukin-1 lead to phosphorylation and loss of I kappa B alpha: a mechanism for NF-kappa B activation. *Mol Cell Biol* (1993) **13**: 3301–3310.
14. Cogswell PC, Scheinman RI, Baldwin AS: Promoter of the human NF-κB p50/p105 gene: regulation by NF κB subunits and by c-Rel. *J Immunol* (1993) **150**: 2794–2804.
15. Ito CY, Kazantsev AG, Baldwin AS: Three NF-κB sites in the IκBα promoter are required for induction of gene expression by TNFα. *Nucleic Acids Res* (1994) **22**: 3787–3792.
16. Thompson JE, Phillips RJ, Erdjument Bromage H, Tempst P, Ghosh S: IκB-β regulates the persistent response in a biphasic activation of NF-κB. *Cell* (1995) **80**: 573–582.
17. Donald R, Ballard DW, Hawiger J: Proteolytic processing of NF-κB/IκB in human monocytes. *J Biol Chem* (1995) **270**: 9–12.
18. Albertella MR, Campbell RD: Characterization of a novel gene in the human major histocompatibility complex that encodes a potential new member of the I kappa B family of proteins. *Hum Mol Genet* (1994) **3**: 793–799.

19. Ray P, Zhang DH, Elias JA, Ray A: Cloning of a differentially expressed IκB-related protein. *J Biol Chem* (1995) **270**: 10 680–10 685.
20. Angel P, Karin M: The role of Jun, Fos and the AP-1 complex in cell-proliferation and transformation. *Biochim Biophys Acta* (1991) **1072**: 129–157.
21. Ransone LJ, Verma IM: Nuclear proto-oncogenes fos and jun. *Annu Rev Cell Biol* (1990) **6**: 539–557.
22. Karin M: The regulation of AP-1 activity by mitogen-activated protein kinases. *J Biol Chem* (1995) **270**: 16 483–16 486.
23. Latchman DS: Eukaryotic transcription factors. *Biochem J* (1990) **270**: 281–289.
24. Ryseck RP, Bravo R: c-JUN, JUN B, and JUN D differ in their binding affinities to AP-1 and CRE consensus sequences: effect of FOS proteins. *Oncogene* (1991) **6**: 533–542.
25. Suzuki T, Okuno H, Yoshida T, Endo T, Nishina H, Iba H: Difference in transcriptional regulatory function between c-Fos and Fra-2. *Nucleic Acids Res* (1991) **19**: 5537–5542.
26. Hai T, Curran T: Cross-family dimerization of transcription factors Fos/Jun and ATF/CREB alters DNA binding specificity. *Proc Natl Acad Sci USA* (1991) **88**: 3720–3724.
27. Derijard B, Hibi M, Wu IH, et al.: JNK1: a protein kinase stimulated by UV light and Ha-Ras that binds and phosphorylates the c-Jun activation domain. *Cell* (1994) **76**: 1025–1037.
28. Kallunki T, Deng T, Hibi M, Karin M: c-Jun can recruit JNK to phosphorylate dimerization partners via specific docking interactions. *Cell* (1996) **87**: 929–939.
29. Boyle WJ, Smeal T, Defize LH, et al.: Activation of protein kinase C decreases phosphorylation of c-Jun at sites that negatively regulate its DNA-binding activity. *Cell* (1991) **64**: 573–584.
30. Auwerx J, Sassone Corsi P: AP-1 (Fos-Jun) regulation by IP-1: effect of signal transduction pathways and cell growth. *Oncogene* (1992) **7**: 2271–2280.
31. Kishimoto T, Taga T, Akira S: Cytokine signal transduction. *Cell* (1994) **76**: 253–262.
32. Akira S, Kishimoto T: IL-6 and NF-IL6 in acute-phase response and viral infection. *Immunol Rev* (1992) **127**: 25–50.
33. Stein B, Cogswell PC, Baldwin AS: Functional and physical associations between NF-κB and C/EBP family members: a Rel domain–bZIP interaction. *Mol Cell Biol* (1993) **13**: 3964–3974.
34. Cao Z, Umek RM, McKnight SL: Regulated expression of three C/EBP isoforms during adipose conversion of 3T3-L1 cells. *Genes Dev* (1991) **5**: 1538–1552.
35. Klamper L, Lee TH, Hsu W, Vilcek J, Chen Kiang S: NF-IL6 and AP-1 cooperatively modulate the activation of the TSG-6 gene by tumor necrosis factor alpha and interleukin-1. *Mol Cell Biol* (1994) **14**: 6561–6569.
36. Nakajima T, Kinoshita S, Sasagawa T, et al.: Phosphorylation at threonine-235 by a ras-dependent mitogen-activated protein kinase cascade is essential for transcription factor NF-IL6. *Proc Natl Acad Sci USA* (1993) **90**: 2207–2311.
37. Ihle JN, Witthuhn BA, Quelle FW, et al.: Signaling by the cytokine receptor superfamily: JAKs and STATs. *Trends Biochem Sci* (1994) **19**: 222–227.
38. Schindler C, Darnell JE: Transcriptional responses to polypeptide ligands: the JAK–STAT pathway. *Annu Rev Biochem.* (1995) **64**: 621–651.
39. Stahl N, Farruggella, TJ, Boulton TG, Zhong Z, Darnell JE, Yancopoulos GD: Choice of STATs and other substrates specified by modular tyrosine-based motifs in cytokine receptors. *Science* (1995) **267**: 1349–1353.
40. Truss M, Beato M: Steroid hormone receptors: interaction with deoxyribonucleic acid and transcription factors. *Endocr Rev* (1993) **14**: 459–479.
41. Luisi BF, Xu WX, Otwinowski Z, Freedman LP, Yamamoto KR, Sigler PB: Crystallographic analysis of the interaction of the glucocorticoid receptor with DNA. *Nature* (1991) **352**: 497–505.
42. Adcock IM, Gilbey T, Gelder CM, Chung KF, Barnes PJ: Glucocorticoid receptor localization in normal and asthmatic lung. *Am J Respir Crit Care Med* (1996) **154**: 771–782.
43. Beato M: Gene regulation by steroid hormones. *Cell* (1989) **56**: 335–344.
44. Wright APH, Gustafsson JA: Mechanism of synergistic transcriptional transactivation by the human glucocorticoid receptor. *Proc Natl Acad Sci USA* (1991) **88**: 8283–8287.
45. Strahle U, Schmid W, Schutz G: Synergistic action of the glucocorticoid receptor with transcription factors. *EMBO J* (1988) **7**: 3389–3395.
46. Eggert M, Mows CC, Tripier D, et al.: A fraction enriched in a novel glucocorticiod receptor-

25 Transcription Factors

interacting protein stimulates receptor-dependent transactivation *in vitro*. *J Biol Chem* (1995) **270**: 30 755–30 759.
47. Johnson PF, McKnight SL: Eukaryotic transcriptional regulatory proteins. *Annu Rev Biochem* (1989) **58**: 799–839.
48. Levine M, Manley JL: Transcriptional repression of eukaryotic promoters. *Cell* (1989) **59**: 405–408.
49. Ponta H, Cato AC, Herrlich P: Interference of pathway specific transcription factors. *Biochim Biophys Acta* (1992) **1129**: 255–261.
50. Mukaida N, Morita M, Ishikawa Y, *et al.*: Novel mechanism of glucocorticoid mediated gene repression. Nuclear factor-kappa B is target for glucocorticoid mediated interleukin 8 gene. *J Biol Chem* (1994) **269**: 13 289–13 295.
51. Ray A, Prefontaine KE: Physical association and functional antagonism between the p65 subunit of transcription factor NF-kappa B and the glucocorticoid receptor. *Proc Natl Acad Sci USA* (1994) **91**: 752–756.
52. Adcock IM, Shirasaki H, Gelder CM, Peters MJ, Brown CR, Barnes PJ: The effects of glucocorticoids on phorbol ester and cytokine stimulated transcription factor activation in human lung. *Life Sci* (1994) **55**: 1147–1153.
53. Scheinman RI, Cogswell PC, Lofquist AK, Baldwin AS: Role of transcriptional activation of I-κBα in mediation of immunosuppression by glucocorticoids. *Science* (1995) **270**: 283–286.
54. Auphan N, DiDonato JA, Rosette C, Helmberg A, Karin M: Immunosuppression by glucocorticoids: inhibition of NF-κB activity through induction of I-κB synthesis. *Science* (1995) **270**: 286–290.
55. Zabel U, Henkel T, Silva MDS, Baeuerle PA: Nuclear uptake control of NF-κB by MAD-3, an IκB protein present in the nucleus. *EMBO J* (1993) **12**: 201–211.
56. Brostjan C, Anrather J, Csizmadia V, *et al.*: Glucocorticoid-mediated repression of NFκB activity in endothelial cells does not involve induction of IκBα synthesis. *J Biol Chem* (1996) **271**: 19 612–19 616.
57. Adcock IM, Newton R, Barnes PJ: NF-kappaB involvement in IL-1β induction of GM-CSF and COX-2: inhibition by glucocorticoids does not require I-kappaB. *Biochem Soc Trans* (1997) **25**: S154.
58. Jonat C, Rahmsdorf HJ, Park KK, *et al.*: Antitumor promotion and antiinflammation: down-modulation of AP-1 (Fos/Jun) activity by glucocorticoid hormone. *Cell* (1990) **62**: 1189–1204.
59. Yang-Yen HF, Chambard JC, Sun YL, *et al.*: Transcriptional interference between c-Jun and the glucocorticoid receptor: mutual inhibition of DNA binding due to direct protein–protein interaction. *Cell* (1990) **62**: 1205–1215.
60. Schule R, Rangarajan P, Kliewer S, *et al.*: Functional antagonism between oncoprotein c-Jun and the glucocorticoid receptor. *Cell* (1990) **62**: 1217–1226.
61. Konig H, Ponta H, Rahmsdorf HJ, Herrlich P: Interference between pathway-specific transcription factors: glucocorticoids antagonize phorbol ester induced AP-1 activity without altering AP-1 site occupation *in vivo*. *EMBO J* (1992) **11**: 2241–2246.
62. LeClair KP, Blanar MA, Sharp PA: The p50 subunit of NF-kappa B associates with the NF-IL6 transcription factor. *Proc Natl Acad Sci USA* (1992) **89**: 8145–8149.
63. Vietor I, Oliveira IC, Vilcek J: CCAAT box enhancer binding protein α (C/EBPα) stimulates κB element-mediated transcription in transfected cells. *J Biol Chem* (1996) **271**: 5595–5602.
64. Kordula T, Travis J: The role of Stat and C/EBP transcription factors in the synergistic activation of rat serine protease inhibitor-3 gene by interleukin-6 and dexamethasone. *Biochem J* (1996) **313**: 1019–1027.
65. Arany Z, Sellers WR, Livingston DM, Eckner R: E1A-associated p300 and CREB-associated CBP belong to a conserved family of coactivators. *Cell* (1994) **77**: 799–800.
66. Kwok RPS, Lundblad JR, Chrivia JC, *et al.*: Nuclear protein CBP is a coactivator for the transcription factor CREB. *Nature* (1994) **370**: 223–226.
67. Zhang JJ, Vinkemeier U, Gu W, Chakravarti D, Horvath CM, Darnell JE: Two contact regions between Stat1 and CBP/p300 in interferon γ signaling. *Proc Natl Acad Sci USA* (1996) **93**: 15 092–15 096.
68. Perkin ND, Felzien LK, Betts JC, Leung K, Beach DH, Nabel GJ: Regulation of NF-κB by cyclin-dependent kinases associated with the p300 coactivator. *Science* (1997) **275**: 523–527.

69. Arias J, Alberts AS, Brindle P, et al.: Activation of cAMP and mitogen responsive genes relies on a common nuclear fctor. *Nature* (1994) **370**: 226–229.
70. Chakravarti D, LaMorte VJ, Nelson MC, et al.: Role of CPB/p300 in nuclear receptor signalling. *Nature* (1996) **383**: 99–103.
71. Kamei Y, Xu L, Heinzel T, et al.: A CBP integrator complex mediates transcriptional activation and AP-1 inhibition by nuclear receptors. *Cell* (1996) **85**: 403–414.
72. Panettieri RA, Yadvish PA, Kelly AM, Rubinstein NA, Kotlikoff MI: Histamine stimulates proliferation of airway smooth muscle and induces c-fos expression. *Am J Physiol* (1990) **259**: L365–L371.
73. Mazer B, Domenico J, Sawami H, Gelfand EW: Platelet-activating factor induces an increase in intracellular calcium and expression of regulatory genes in human B lymphoblastoid cells. *J Immunol* (1991) **146**: 1914–1920.
74. Demoly P, Basset Seguin N, Chanez P, et al.: c-fos proto-oncogene expression in bronchial biopsies of asthmatics. *Am J Respir Cell Mol Biol* (1992) **7**: 128–133.
75. Marini M, Vittori E, Hollemborg J, Mattoli S: Expression of the potent inflammatory cytokines, granulocyte–macrophage-colony-stimulating factor and interleukin-6 and interleukin-8 in bronchial epithelial cells of patients with asthma. *J Allergy Clin Immunol* (1992) **89**: 1001–1009.
76. Broide DH, Lotz M, Cuomo AJ, Coburn DA, Federman EC, Wasserman SI: Cytokines in symptomatic asthma airways. *J Allergy Clin Immunol* (1992) **89**: 958–967.
77. Hart L, Krishnan V, Adcock IM, Barnes PJ, Chung KF: Activation of transcription factor, nuclear factor-κB, in asthma. *Am J Respir Crit Med* (1998) in press.
78. Marok R, Winyard PG, Coumbe A, et al.: Activation of the transcription factor nuclear factor-κB in human inflamed synovial tissue. *Arthritis Rheum* (1996) **39**: 583–591.
79. Kaltschmidt, C, Kaltschmidt B, Lannes-Vieira J, et al.: Transcription factor NF-κB is activated in microglia during experimental autoimmune encephalomyelitis. *J Neuroimmunol* (1994) **55**: 99–106.
80. Barnes PJ, Liew FY: Nitric oxide and asthmatic inflammation. *Immunol Today* (1995) **16**: 128–130.
81. Lee TH: Eicosanoids in asthma. In Robinson C (ed) *Lipid Mediators in Allergic Diseases of the Respiratory Tract*. Boca Raton FL, CRC Press. (1994), pp 121–145.
82. Broide DH, Paine MM, Firestein GS: Eosinophils express interleukin 5 and granulocyte macrophage-colony-stimulating factor mRNA at sites of allergic inflammation in asthmatics. *J Clin Invest* (1992) **90**: 1414–1424.
83. Mui ALF, Wakao H, O'Farrell AM, Harada N, Miyajima A: Interleukin-3, granulocyte–macrophage colony stimulating factor and interleukin-5 transduce signals through two STAT5 homologs. *EMBO J* (1995) **14**: 1166–1175.

26

Airway Remodelling

TONY R. BAI, CLIVE R. ROBERTS AND P.D. PARÉ

INTRODUCTION

In many asthmatic subjects there are structural changes in the airways that result in measurable increases in the thickness of the airway wall, an alteration in the extracellular matrix components of the wall, and both hyperplasia and hypertrophy of resident cells such as smooth muscle cells.[1-7] This is an active process, termed 'remodelling', that involves cell growth, cell death, cell migration and production or degradation of extracellular matrix. Although asthma is characterized, indeed in part defined, by a significant reversible component, there is an element of fixed obstruction especially in those who have a long history of symptoms. In addition, the airway hyperresponsiveness to non-specific stimuli that accompanies asthma is persistent even in periods of symptomatic remission and after optimal anti-inflammatory therapy. Our thesis is that the above changes, as well as other fundamental functional changes found in asthmatic patients such as paroxysmal airway narrowing, increased maximal airway narrowing and paradoxical or deficient responses to deep inspiration, can be explained in large part by airway remodelling.

STRUCTURAL CHANGES IN THE AIRWAY WALLS IN ASTHMA

Figure 26.1 shows an intraparenchymal airway in cross-section. The wall can be divided into three compartments: the inner wall, consisting of epithelium, basement membrane, lamina propria and submucosa; the outer wall, consisting of the loose connective tissue

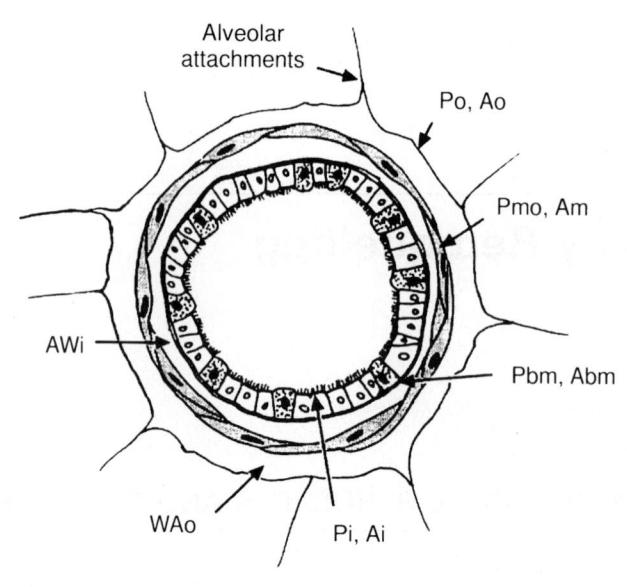

Fig. 26.1 Cross-section through a membranous airway: the three layers of the airway wall are illustrated. Pi, internal perimeter; Po, outer perimeter; Pmo, outer muscle perimeter; Pbm, basement membrane perimeter; Am, muscle area; Ai luminal area; WAo, outer wall area; Wai, inner wall area; Abm, basement membrane area; Ao, outer area.

between the muscle layer and the surrounding parenchyma (the adventitia); and the smooth muscle layer.[4] Figure 26.2 shows an overview of the potential mechanisms that can lead to persistent functional and structural changes. Sufficient allergen exposure, particularly in the first few years of life, can lead to the development of chronic allergic inflammation in the airways of genetically susceptible individuals; other risk factors for the development of the inflammatory response may include viral respiratory infection, exposure to environmental tobacco smoke and atmospheric pollution and a diet low in antioxidants and other factors.

The structural changes that occur in the airways are caused by the deposition and remodelling of connective tissue components, hypertrophy and hyperplasia of tissue cells and new vessel formation in the bronchial vasculature. These alterations combine to produce airway wall thickening. Airway wall thickening can have profound effects on airway function. Airway mechanics may be changed because of quantitative changes in airway wall compartments and/or by changes in the biochemical composition or material properties of the various constituents of the airway wall.

Figure 26.3 shows a schematic mechanical model of the forces and dimensions of the airway wall and illustrates how changes in dimensions can alter airway narrowing in response to smooth muscle stimulation. Figure 26.3A depicts the normal situation at equilibrium. The horizontal shaded bars represent the airway walls and airway narrowing is simulated by an approximation of the bars. The springs outside the bars represent lung elastic recoil, which tends to dilate the airways; the tension in these springs is balanced by the tension in the springs inside the bars, which represent the connective tissue elements in the airway wall. The cell between the bars represents the airway smooth muscle; when it is stimulated to contract it narrows the airway by approximating the bars until the

26 Airway Remodelling

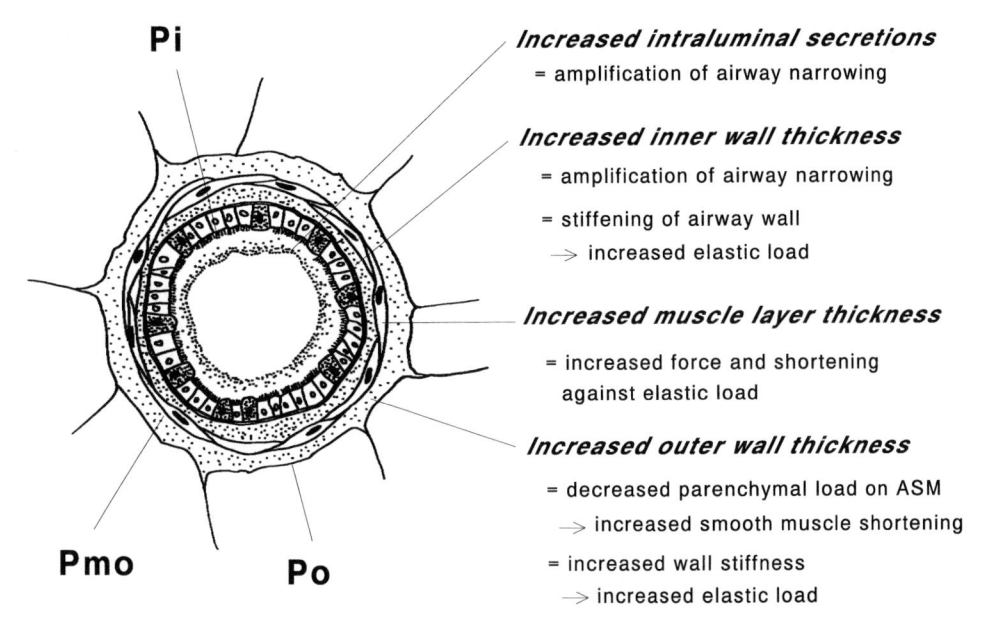

Fig. 26.2 This schema illustrates the functional consequences of airway structural changes. ASM, airway smooth muscle; Pi, internal perimeter; Pmo, outer muscle perimeter; Po, outer perimeter.

Labels in figure:
- Pi
- Pmo
- Po
- *Increased intraluminal secretions*
  = amplification of airway narrowing
- *Increased inner wall thickness*
  = amplification of airway narrowing
  = stiffening of airway wall
  → increased elastic load
- *Increased muscle layer thickness*
  = increased force and shortening against elastic load
- *Increased outer wall thickness*
  = decreased parenchymal load on ASM
  → increased smooth muscle shortening
  = increased wall stiffness
  → increased elastic load

maximal force it can generate is balanced by lung elastic recoil. Figure 26.3B shows the effect of thickening of the inner airway wall. In addition to narrowing the airway lumen, thickening of this layer will exaggerate the effect of any smooth muscle shortening. An increased volume of intraluminal secretions could also amplify the effects of smooth muscle shortening in addition to decreasing baseline airway luminal area. Figure 26.3C shows the effect of thickening of the outer wall area. Thickening of this layer causes a relaxation in the springs representing the lung parenchyma, i.e. a decrease in parenchymal

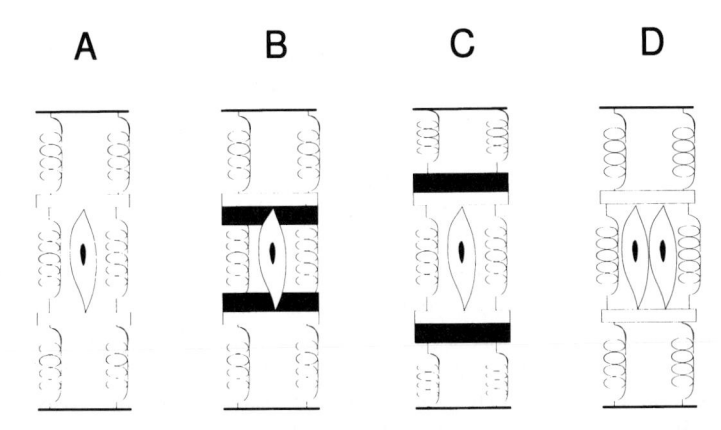

Fig. 26.3 A mechanical model of airway wall structures and forces to illustrate how the structural alterations can cause exaggerated narrowing in response to stimulation of airway smooth muscle contraction. See text for details.

tethering. When stimulated the smooth muscle in such an airway will shorten more before the elastic load provided by the parenchymal recoil prevents further narrowing. Figure 26.3D shows the effect of increased airway smooth muscle thickness. If the force-generating capacity of the muscle increases in parallel with its mass, the muscle will be able to shorten more against the elastic load provided by parenchymal recoil. Increased thickness of the airway wall internal to the smooth muscle layer can amplify the airway narrowing produced by airway smooth muscle shortening.

While it has been recognized for some time that the airway walls of asthmatic subjects are thickened,[7,8] it was not possible to perform a systematic study of the quantitative changes in airway wall dimensions because a standardized measure of airway size was not available to allow a valid comparison between control and asthmatic subjects. The demonstration by James *et al.*[9,10] that the airway basement membrane perimeter is relatively constant after smooth muscle contraction or changes in lung volume has allowed a number of investigators to examine the relationship between airway wall compartment areas and airway size.[11-16] This is most easily done by examining the relationships between airway internal perimeter (Pi) or basement membrane perimeter (Pbm) and the areas occupied by the respective tissue components. The slopes and intercepts of these relationships can be constructed and compared using valid techniques for pooling data such as random effects regression.[17] These results confirm that patients with fatal asthma show a marked increase in airway wall thickness that involves all layers of the airway wall. There are less data on patients who have had asthma but who died for other reasons or had a lobectomy. However, the available data suggest that the airway wall dimensions in these subjects are intermediate between the fatal asthmatic and the control or normal subjects.[13,18] Thus, an increase in airway wall dimensions does not simply reflect a terminal event in patients with severe asthma. Comparing all published reports to date, total wall area has been reported to be increased from 50 to 300% in fatal asthma and from 10 to 100% in non-fatal asthma.

EXTRACELLULAR MATRIX

Airway wall thickening in asthma involves increased collagen deposition. Roche and coworkers[19] have shown that the thickened subepithelial 'basement membrane' in asthma consists of a dense layer rich in fibrillar collagens under a normal subepithelial basal lamina. This distinct collagenous matrix layer is typically doubled in thickness, from 5–8 μm (normal) to 10–15 μm (asthma), and contains types I, III and V collagen and fibronectin but not basal lamina components (type IV collagen, laminin). This collagenous matrix may be synthesized by associated myofibroblasts, since myofibroblast number correlates with the magnitude of subepithelial thickening.[20,21] Similar structural changes have been observed in patients with mild asthma and in occupational asthma associated with exposure to a variety of chemicals.[22] In some individuals with toluene diisocyanate (TDI)-induced asthma, cessation of exposure to TDI leads, after 6–20 months, to decreased subepithelial collagen thickness and to decreased numbers of subepithelial fibroblasts associated with decreased numbers of mast cells and lymphocytes.[23] This suggests that these changes are potentially reversible, but the mechanism of this reversal is unknown.

The mechanical effects of changes in abundance of collagen types in the subepithelial matrix are unknown but are likely to depend on the precise architecture and chemistry of the collagens deposited.[24-27] The collagen fibrils in the subepithelial collagen layer in the airways of asthmatic patients appear to be more densely packed than normal;[19] although the significance of this is unknown, it is probable that both increased collagen fibril density and thickening of this layer would increase both the tensile stiffness and resistance to deformation of the airway wall, thus tending to oppose smooth muscle contraction and airway narrowing. Airway distensibility has been shown to be decreased in asthmatic patients[28,29] and this could be explained by excess collagen deposition in the subepithelial layer.

In addition to collagen, the adhesive glycoprotein fibronectin and the antiadhesive glycoprotein tenascin appear to be deposited in the airway wall in asthmatic patients.[30] These may be synthesized by epithelial cells in response to inflammatory mediators; fibronectin synthesis by bovine bronchial epithelial cells is stimulated by transforming growth factor (TGF)-β[31] and tenascin synthesis by transformed human bronchial epithelial cells is stimulated by tumour necrosis factor α and interferon γ.[32]

The airway walls contain proteoglycans with their characteristic polysaccharides, the glycosaminoglycans. Specific proteoglycans and glycosaminoglycans of the extracellular matrix influence tissue biomechanics, fluid balance, cellular functions and growth factor and cytokine biological activities. Changes in glycosaminoglycan metabolism occur early in a number of animal models of inflammation,[33] suggesting that changes in proteoglycan metabolism may contribute to altered extracellular matrix properties in asthma. We have used immunohistochemistry to localize hyaluronan (HA) and the proteoglycans versican and decorin in surgical and post-mortem lung samples from individuals with normal lung function and individuals in whom asthma was the cause of death.

HA and versican were localized in and around the smooth muscle bundles in the airways.[34] Decorin was found in areas rich in type I collagen. In airways from asthmatic patients, staining for all the proteoglycans was particularly prominent around smooth muscle cells and in the submucosa, i.e. between the smooth muscle and the epithelial layer. The matrix of the thickened airway walls stained particularly intensely for versican and HA, especially between and around the smooth muscle bundles, areas that appear to be 'space' following routine formalin fixation and paraffin embedding. These 'spaces' appear to be hydrated proteoglycan-rich domains in life.[34]

Though the functional correlates of proteoglycan deposition in the airway wall are unknown, hydrated proteoglycans may contribute to the increased volume of the submucosa in asthmatic patients[13] and may contribute to altered airway mechanics. HA–versican aggregates could influence the compressive stiffness of the airway wall and have an effect on airway interstitial fluid balance through their osmotic activity. The glycosaminoglycans are highly negatively charged at neutral pH.[35,36] The high concentration of HA and versican may result in an osmotic swelling pressure. The reversible redistribution of glycosaminoglycan-bound water contributes to compressive stiffness of airway walls. The deposition of a versican–HA complex in the submucosal region between the muscle and basement membrane could contribute to exaggerated airway narrowing as depicted in Fig. 26.2. Conversely, deposition of HA and versican between the smooth muscle and epithelium could increase tissue turgor and thus increase the resistance of the airway wall to deformation under loading. In addition, accumulation of a relatively incompressible matrix around smooth muscle cells in the airways might provide a parallel elastic afterload to oppose smooth muscle shortening.

The mechanisms underlying changes in extracellular matrix composition in asthma are incompletely understood but are the subject of intense investigation. A number of growth factors and cytokines released by inflammatory cells, or released by epithelial cells secondary to stimulation in allergic inflammation, have the capacity to drive altered extracellular matrix metabolism by mesenchymal cells in the airway wall. Eosinophil and mast cell numbers are increased in asthma, driven by a Th2 response. Both inflammatory cells and stimulated epithelial and mesenchymal cells (including smooth muscle cells) have the capacity to release TGF-β1, a growth factor that induces matrix deposition, and the potent fibroblast mitogens platelet-derived growth factor (PDGF) and insulin-like growth factor (IGF)-1. This combination of mitogens and growth factors is known to induce matrix synthesis in other systems and is a potentially powerful mechanism for remodelling of the architecture of the airway wall in asthma.

A recent study[37] showed that steady-state mRNA levels for TGF-β1, as well as the pattern of expression of the latent precursor and mature forms of TGF-β1, were similar in lung tissue from individuals with asthma, a group of individuals with chronic obstructive pulmonary disease (COPD) and a control group of cigarette smokers who had normal lung function. Similarly, there were no clear differences between these same groups of patients in the expression of mRNA for the collagen-associated proteoglycan decorin, a putative regulator of TGF-β1 biological activity (C. R. Roberts and A. K. Burke, unpublished results). The precursor protein for TGF-β1 was detected in epithelial cells, implying epithelial cell synthesis of this growth factor. The fact that abundant mRNA and protein for TGF-β1 was found in the 'control' group is of questionable significance since the control group were chronic smokers, albeit without airflow obstruction. TGF-β1 mRNA levels in mononuclear cells from brochoalveolar lavage from asthmatic and normal subjects have been shown to be similar.[38] Eosinophils also express TGF-β and their abundance in asthma would be expected to contribute to increased local, if not total tissue levels of this growth factor. Another potential contributor to increased matrix synthesis is PDGF. Eosinophils from asthmatic subjects express higher levels of PDGF-B mRNA than normal subjects;[39] this agent is mitogenic for mesenchymal cells including fibroblasts and smooth muscle cells. Aubert[40] examined the presence and distribution of PDGF and PDGF receptor mRNA and protein in the lungs and airways of a small group of patients with fatal asthma as well as control subjects and patients with COPD. PDGF mRNA levels tended to be greater in asthmatic than normal subjects and lower in COPD than normal subjects. The PDGF mRNA levels were significantly greater in patients with asthma compared to patients with COPD. In addition, there was a significant association between PDGF mRNA levels and PDGF receptor mRNA levels, suggesting that there is a link between the expression of this growth factor and its receptor.

Human airway epithelial cells have been shown to secrete fibroblast mitogenic activity, at least 50% of which is attributable to IGF-1.[41] This growth factor stimulates collagen production by dermal fibroblasts *in vitro*,[42] suggesting a further mechanism by which epithelial cells might stimulate collagen production and cell proliferation in the underlying matrix.

Corticosteroids have multiple effects but may prevent remodelling by both decreasing influx of inflammatory cells (perhaps without decreasing the amount of mediators such as

TGF-β1 per cell[43]) and exerting specific inhibitory effects on synthesis of matrix molecules, including collagen.[44]

The airway wall remodelling that occurs in chronic asthma must be accompanied by degradation of matrix components in addition to synthesis and deposition of new matrix. Ultrastructural evidence for elastin and cartilage degradation in some individuals with asthma has been reported.[34,45] The proteinase(s) responsible for the matrix changes are unknown and there are a number of possibilities. Neutrophil elastase, cathepsin G and lysosomal cysteine proteinases such as cathepsins B and L are able to degrade collagen, elastin and proteoglycans. Latent cathepsin B is present in the sputum of chronic bronchitic patients.[46] Latent cysteine proteinases can be activated by a number of means, including direct activation by neutrophil proteinases[46] and activation by a cartilage-specific mechanism that is not yet understood.[47] Indirect mechanisms for cartilage destruction include interleukin 1-driven resorption of airway cartilage by chondrocytes mediated by matrix metalloproteinases,[48] as has been described in cartilage destruction in inflammatory joint diseases. Proteinases released by mast cells, including tryptase, may also be responsible for degradation of a range of matrix macromolecules in asthma. Degradation of airway cartilage could contribute to airflow obstruction by decreasing airway wall stiffness, which would decrease maximal expiratory flow rates from the lung. Cartilage degradation could also decrease the force required for the smooth muscle to constrict the airways.

Degradation of elastin, as shown by Bousquet et al.,[45] and possibly degradation of other matrix molecules may have similar effects on airway wall mechanics. Matrix degradation and increased proteoglycan synthesis are associated with tissue swelling during development[49] and proteolysis in the airway wall matrix may facilitate oedema in asthma. Mast cell degranulation is associated with oedema of the airway wall and an acute decrease in interstitial pressure.[50] Proteoglycan synthesis, in concert with matrix degradation, may influence tissue swelling. The force required to deform the matrix constitutes an afterload that must be overcome by smooth muscle during shortening. Degradation of matrix elements could increase the deformability of the airway wall and thus decrease its ability to act as a load on the muscle. Consistent with this hypothesis, *in vitro* studies by Bramley and colleagues[51] suggest that mild proteolysis in the extracellular matrix associated with airway smooth muscle allows increased force generation and shortening by strips of human airway smooth muscle. Degradation of smooth muscle-associated matrix as a consequence of chronic inflammation has been postulated to exacerbate the increased smooth muscle contractility in asthma.[51,52]

Since tethering of the parenchyma to both the smooth muscle and perichondrium is believed to limit smooth muscle shortening, we suggest that degradation of collagen connected to smooth muscle bundles is a prerequisite for the uncoupling of airway smooth muscle from parenchymal tethering that has been suggested by Macklem[5] and others[53] to be an important component of asthma.

SMOOTH MUSCLE

A number of studies show that the airway smooth muscle layer is markedly thickened in patients with chronic asthma.[54,55] Part of this thickening could have been artefactual, since the airways of asthmatic subjects are often contracted and narrowed by the time of post-mortem. However, correction of airway smooth muscle area for basement membrane perimeter indicates that in patients with fatal asthma peripheral airway smooth muscle area is approximately doubled, while in patients who have asthma but die of other causes lesser degrees of airway smooth muscle thickening are observed. There is evidence that the increase in smooth muscle is due to hypertrophy of existing airway smooth muscle cells as well as hyperplasia. Ebina *et al*.[14,56] have reported two patterns of airway smooth muscle hypertrophy and hyperplasia. In their 'type 1' asthmatic subjects, airway smooth muscle mass was increased only in central bronchi where hyperplasia predominated. In 'type 2' asthmatic subjects, there was increased muscle throughout the tracheobronchial tree and the increased muscle was characterized by hyperplasia as well as hypertrophy, especially in peripheral airways. Thomson *et al*.[57] suggested that the increase in airway smooth muscle area that has been reported in asthma could have been overestimated. These investigators measured the airway smooth muscle area in the large central airways of five asthmatic subjects and showed no significant difference compared with a matched control group. They used 1.5-mm sections of plastic-embedded tissue and discriminated between smooth muscle cells and their surrounding matrix. They reasoned that the plane of section, the use of thick sections and a failure to distinguish between smooth muscle cells and their associated extracellular matrix could explain an overestimation of smooth muscle area in other studies. However, Thomson *et al*. studied only large central cartilaginous airways, and most of the increase in smooth muscle area that has been reported is in peripheral airways.

The increase in airway smooth muscle mass in asthma can have a simple geometric effect on airway narrowing, much like the effect of thickening of the submucosal region of the airway wall, and can narrow the airways as well as amplifying the effect of smooth muscle shortening. However, an increase in smooth muscle mass, if associated with a parallel and concomitant increase in force-generating ability of the muscle, will have the additional effect of allowing the airway smooth muscle to shorten excessively against the elastic loads provided by the lung parenchyma and parallel elastic elements. Unfortunately, there have been few studies in which the functional properties of airway smooth muscle from asthmatic subjects have been measured and corrected for the amount of smooth muscle in the preparation. Bramley *et al*.[52] have reported increased maximal isotonic shortening and increased isometric force generation in a single asthmatic bronchial smooth muscle specimen, despite a normal amount of smooth muscle. De Jongste *et al*.[58] have reported increased maximal force generation in a few samples of central airways of asthmatic subjects; however, they did not correct the force for the smooth muscle mass. Bai[59] found increased force generation and decreased relaxation in the airway smooth muscle obtained from patients with fatal asthma, even after correction for tissue weight. No studies have shown increased airway smooth muscle sensitivity in asthmatic subjects compared with control subjects.

Although one might expect that an increase in airway smooth muscle mass would be accompanied by an increase in force generation, this is not necessarily the case. Vascular smooth muscle proliferation induced in rabbits by hyperoxia produces an increase in smooth muscle mass, but a decrease in the maximal stress-generating ability of the vascular smooth muscle.[60] When airway smooth muscle is stimulated to proliferate *in vitro*, the muscle differentiates from a contractile to a more motile phenotype, concomitant with decreased smooth muscle α-actin content and increased γ-actin and non-muscle

26 Airway Remodelling

myosin content with increasing time in culture.[61] Similar dedifferentiation of vascular smooth muscle occurs in vascular remodelling associated with atherosclerosis.[62] It is possible that chronic stimulation by cytokines and growth factors in the inflamed airway walls of asthmatic subjects results in proliferation and dedifferentiation of the airway smooth muscle, making it less contractile.

Lambert[63,64] and Wiggs et al.[65] have studied the folding pattern of the normal bronchial mucosa and the possible effects of asthma on folding pattern. They suggest that the folding pattern may be controlled by the stiffness of the subepithelial layer. The formation of a large number of folds in the normal airway could provide a load on the airway smooth muscle and that would tend to prevent airway closure at low lung volume. They hypothesize that the increased thickness of the subepithelial layer in asthma could produce fewer folds and thus increase the tendency of the peripheral airways to narrow excessively.

SUMMARY

In summary, structural changes in the airway walls involving extracellular matrix remodelling are prominent features of asthma. These changes are likely driven by mediators released as a consequence of chronic allergic inflammation. It is clear that changes in matrix have the capacity to influence airway function in asthma. However, it is not clear how each of the many changes that occur in the airway wall contribute to altered airway function in asthma. Collagen deposition in the subepithelial matrix and HA and versican deposition around and internal to the smooth muscle would be expected to oppose the effect of smooth muscle contraction. Conversely, geometric considerations would result in exaggerated airway narrowing for a given degree of smooth muscle shortening, as the airway wall is thickened by the deposition of these molecules internal to the smooth muscle. Elastin and cartilage degradation in the airway walls would be expected to result in decreased airway wall stiffness and increased airway narrowing for a given amount of force generated by the smooth muscle. Degradation of matrix associated with the smooth muscle may both decrease the stiffness of the parallel elastic component and uncouple smooth muscle from the load provided by lung recoil, allowing exaggerated smooth muscle shortening. Increase in muscle mass may be associated with an increase, a decrease or no change in smooth muscle contractility. If an increase in muscle mass was not associated with any other phenotypic changes it would be expected to contribute to exaggerated airway narrowing.

ACKNOWLEDGEMENTS

Our work is supported by the Medical Research Council of Canada and the British Columbia Lung Association.

REFERENCES

1. Jeffery PK: Structural changes in asthma. In Page C, Black J, (eds) *Airways and Vascular Remodelling*. London, Academic Press, 1994, pp 3–21.
2. James AL, Lougheed D, Pearce-Pinto G, Ryan G, Musk B: Maximal airway narrowing in a general population. *Am Rev Respir Dis* (1992) **146**: 895–899.
3. Woolcock A, Salome CM, Yan K: The shape of the dos–response curve to histamine in asthmatic and normal subjects. *Am Rev Respir Dis* (1984) **130**: 71–75.
4. Bai A, Eidel DH, Hogg JC, et al.: Proposed nomenclature for quantifying subdivisions of the bronchial wall. *J Appl Physiol* (1994) **77**: 1011–1014.
5. Macklem PT: Theoretical basis of airway instability. Roger S. Mitchell lecture. *Chest* (1995) **107**: 87S–88S.
6. Lambert RK, Wiggs BR, Kuwano K, Hogg JC, Paré PD: Functional significance of increased airway smooth muscle in asthma and COPD. *J Appl Physiol* (1993) **74**: 2771–2781.
7. Huber HL, Koessler KK; The pathology of bronchial asthma. *Arch Intern Med* (1992) **30**: 689–760.
8. Houston JC, de Nevasquez S, Trounce JR: A clinical and pathological study of fatal cases of status asthmaticus. *Thorax* (1953) **8**: 207–213.
9. James AL, Hogg JC, Dunn LA, Paré PD: The use of internal perimeter to compare airway size and to calculate smooth muscle shortening. *Am Rev Respir Dis* (1988) **138**: 136–139.
10. James AL, Paré PD, Hogg JC: Effects of lung volume, bronchoconstriction, and cigarette smoke on morphometric airways dimensions. *J Appl Physiol* (1988) **64**: 913–919.
11. Bosken CH, Wiggs BR, Paré PD, Hogg JC: Small airway dimensions in smokers with obstruction to airflow. *Am Rev Respir Dis* (1990) **142**: 563–570.
12. Wiggs BR, Bosken C, Paré PD, James A, Hogg JC: A model of airway narrowing in asthma and in chronic obstructive pulmonary disease *Am Rev Respir Dis* (1992) **145**: 1251–1258.
13. Kuwano K, Bosken CH, Paré PD, Bai TR, Wiggs BR, Hogg JC: Small airways dimensions in asthma and in chronic obstructive pulmonary disease. *Am Rev Respir Dis* (1993) **148**: 1220–1225.
14. Ebina M, Yaegashi H, Chiba R, Takahashi T, Motomiya M, Tanemura M: Hyperreactive site in the airway tree of asthmatic patients revealed by thickening of bronchial muscles. *Am Rev Respir Dis* (1990) **141**: 1327–1332.
15. Tiddens HA, Paré PD, Hogg JC, Hop WC, Lambert R, de Jongste JC: Cartilaginous airway dimensions and airflow obstruction in human lungs. *Am J Respir Crit Care Med* (1995) **152**: 260–266.
16. Riess A, Wiggs B, Verburgt L, Wright JL, Hogg JC, Paré PD: Morphologic determinants of airway responsiveness in chronic smokers. *Am J Respir Crit Med* (1996) **154**: 1444–1449.
17. James AL, Paré PD, Hogg JC: The mechanics of airway narrowing in asthma. *Am Rev Respir Dis* (1989) **139**: 242–246.
18. Carroll N, Elliot J, Morton A, James A: The structure of large and small airways in nonfatal and fatal asthma. *Am Rev Respir Dis* (1993) **147**: 405–410.
19. Roche WR, Beasley R, Williams JH, Holgate ST: Subepithelial fibrosis in the bronchi of asthmatics. *Lancet* (1989) 520–524.
20. Brewster CEP, Howarth PH, Djukanovic R, Wilson J, Holgate ST, Roche WR: Myofibroblasts and subepithelial fibrosis in bronchial asthma. *Am J Respir Cell Mol Biol* (1990) **3**: 507–511.
21. Jeffery PK, Godfrey RWA, Adelroth E, Nelson F, Rogers A, Johansson S-A: Effects of treatment on airway inflammation and thickening of reticular collagen in asthma: a quantitative light and electron microscopic study. *Am J Respir Crit Care Med* (1992) **145**: 890–899.
22. Boulet LP, Boulet M, Laviolette M, et al.: Airway inflammation after removal from the causal agent in occupational asthma due to high and low molecular weight agents. *Eur Respir J* (1994) **7**: 1567–1575.
23. Saetta M, Maestrelli P, Turato G, et al.: Airway wall remodelling after cessation of exposure to isocyanates in sensitized asthmatic subjects. *Am J Respir Crit Care Med* (1995) **151**: 489–494.
24. Adachi E, Hayashi T: *In vitro* formation of fine fibrils with a D-periodic banding pattern from type V collagen. *Collagen Relat Res* (1985) **5**: 225–232.

25. Birk DE, Fitch JM, Babiarz JP, Doane KJ, Linsenmayer TF: Collagen fibrillogenesis *in vitro*: interaction of types I and V collagen regulate fibril diameter. *J Cell Sci* (1990) **95**: 649–657.
26. Vogel KG, Paulsson M, Heinegard D: Specific inhibition of type I and type II collagen fibrillogenesis by the small proteoglycan of tendon. *Biochem J* (1984) **223**: 587–597.
27. Scott JE: Proteoglycan–fibrillar collagen interactions. *Biochem J* (1988) **252**: 313–323.
28. Wilson JW, Li X, Pain MC: The lack of distensibility of asthmatic airways. *Am Rev Respir Dis* (1993) **148**: 806–809.
29. Colebatch HJH, Greaves IA, Ng CKY: Pulmonary mechanics in diagnosis. In de Kock MA, Nadel JA, Lewis CM (eds) *Mechanics of Airway Obstruction in Human Respiratory Disease*. Cape Town, AA Balkema, 1979, pp 25–47.
30. Laitinen LA, Laitinen A: Modulation of bronchial inflammation: corticosteroids and other therapeutic agents. *Am Rev Respir Crit Care Med* (1994) **10**: S87–S90.
31. Romberger DJ, Beckmann JD, Claasen L, Ertl RF, Rennard SI: Modulation of fibronectin production of bovine bronchial epithelial cells by transforming growth factor-beta. *Am J Respir Cell Mol Biol* (1992) **7**: 149–155.
32. Harkonen E, Virtanen I, Linnala A, Laitinen LL, Kinnula VL: Modulation of fibronectin and tenascin production in human bronchial epithelial cells by inflammatory cytokines *in vitro*. *Am J Respir Cell Mol Biol* (1995) **13**: 109–115.
33. Blackwood RA, Cantor JO, Moret J, Mandl I, Turino GM: Glycosaminoglycan synthesis in endotoxin-induced lung injury. *Proc Soc Exp Biol Med* (1983) **174**: 343–349.
34. Roberts CR: Is asthma a fibrosis disease? *Chest* (1995) **107**: 111S–117S.
35. Wight TN, Heinegard DK, Hascall VC: Proteoglycans: structure and function. In Hay ED (ed) *Cell Biology of the Extracellular Matrix*. New York, Plenum Press, 1991, pp 45–78.
36. Wight TN: Cell biology of arterial proteoglycans. *Arteriosclerosis* (1989) **9**: 1–20.
37. Aubert J-D, Dalal BI, Bai TR, Roberts CR Hayashi S, Hogg JC: Transforming growth factor-β1 gene expression in human airways. *Thorax* (1994) **49**: 225–232.
38. Deguchi Y: Spontaneous increase of transforming growth factor beta production by bronchoalveolar mononuclear cells of patients with systemic autoimmune diseases affecting the lung. *Ann Rheum Dis* (1992) **51**: 362–365.
39. Ohno I, Nitta Y, Yamaguchi K, *et al.*: Eosinophils as a potential source of platelet-derived growth factor B-chain (PDGF-β) in nasal polyps and bronchial asthma. *Am J Respir Cell Mol Biol* (1995) **13**: 639-647.
40. Aubert J-D: Platelet-derived growth factor and its receptor in lungs from patients with asthma and chronic airflow obstruction. *Am J Physiol* (1994) **266**: L655–L663.
41. Cambrey AD, Kwon OJ, Gray AJ, *et al.*: Insulin-like growth factor I is a major fibroblast mitogen produced by primary cultures of human airway epithelial cells. *Clin Sci* (1995) **89**: 611–617.
42. Ghahary A, Shen Y, Nedelec B, Scott P, Tredget E: Enhanced expression of mRNA for insulin-like growth factor 1 in post-burn hypertrophic scar tissue and its fibrogenic role in dermal fibroblasts. *Mol Cell Biochem* (1995) **148**: 25–32.
43. Khalil N, Whitman C, Zuo L, Danielpour D, Greenberg A: Regulation of alveolar macrophage transforming growth factor beta secretion by corticosteroids in bleomycin-induced pulmonary inflammation in the rat. *J Clin Invest* (1993) **92**: 1812–1818.
44. Hamalainen L, Oikarinen J, Kivirikko KI: Synthesis and degradation of type I procollagen in cultured human skin fibroblasts and the effect of cortisol. *J Biol Chem* (1985) **260**: 720–725.
45. Bousquet J, Chanez P, Lacoste JY, *et al.*: Asthma: a disease remodelling the airways. *Allergy* (1992) **47**: 3–11.
46. Buttle DJ, Abrahamson M, Burnett D, *et al.*: Human sputum cathepsin B degrades proteoglycan, is inhibited by α_2-macroglobulin and is modulated by neutrophil elastase cleavage of cathepsin B precursor and cystatin C. *Biochem J* (1991) **276**: 325–331.
47. Roberts CR, Opazo-Saez A, Burke A: Cleavage of airway cartilage proteoglycans in papain-induced airflow obstruction: a model for obstructive lung disease. Submitted.
48. Saklatvala J. Sarsfield SJ: How do interleukin 1 and tumour necrosis factor induce degradation of proteoglycan in cartilage? In Glauert AM (ed) *The Control of Tissue Damage*. New York, Elsevier, 1988, pp 97–108.

49. Toole BP: Proteoglycans and hyaluronan in morphogenesis and differentiation. In Hay ED (ed) *Cell Biology of Extracellular Matrix* 2nd edn. New York, Plenum Press, 1991, pp 305–341.
50. Koller ME, Woie K, Reed RK: Increased negativity of interstitial fluid pressure in rat trachea after mast cell degranulation. *J Appl Physiol* (1993) **74**: 2135–2139.
51. Bramley, AM, Thomson RJ, Roberts CR, Schellenberg RR: Hypothesis: excessive bronchoconstriction in asthma is due to decreased airway elastance. *Eur Respir J* (1994) **7**: 337–341.
52. Bramley AJ, Roberts CR, Schellenberg RR: Collagenase increases shortening of human bronchial smooth muscle *in vitro*. *Am J Respir Crit Care Med* (1995) **152**: 1513–1517.
53. Robinson P Okazawa M, Bai T, Paré PD: *In vivo* loads on airway smooth muscle: the role of noncontractile airway structures. *Can J Physiol Pharmacol* (1992) **70**: 602–606.
54. Dunnill MS, Massarella GR, Anderson JA: A comparison of the quantitative anatomy of the bronchi in normal subjects, in status asthmaticus, in chronic bronchitis, and in emphysema. *Thorax* (1969) **24**: 176–179.
55. Heard BE, Hossain S: Hyperplasia of bronchial muscle in asthma. *J Pathol* (1973) **110**: 319–331.
56. Ebina M, Takahashi T, Chiba T, Motomiya M: Cellular hypertrophy and hyperplasia of airway smooth muscles underlying bronchial asthma. A 3-D morphometric study. *Am Rev Respir Dis* (1993) **48**: 720–726.
57. Thomson RJ, Bramley AM, Schellenberg RR: Airway muscle stereology: implications for increased shortening in asthma. *Am J Respir Crit Care Med* (1996) **154**: 749–757.
58. de Jongste JC, Mons H, Bonata IL, Kerrebijn KF: *In vitro* responses of airways from an asthamtic patient. *Eur J Respir Dis* (1987) **71**: 23–29.
59. Bai TR: Abnormalities in airway smooth muscle in fatal asthma. *Am Rev Respir Dis* (1990) **141**: 552–557.
60. Coflesky JT, Jones RC, Reid LM, Evans JN: Mechanical properties and structure of isolated pulmonary arteries remodeled by chronic hyperoxia. *Am Rev Respir Dis* (1987) **136**: 388–394.
61. Halayko AJ, Salari H, Ma H, Stevens NL: Markers of airway smooth muscle cell phenotype. *Am J Physiol* (1996) **270**: L1040–L1051.
62. Karnovsky MJ, Edelman ER: Heparin/heparan sulphate regulation of vascular smooth muscle cell behaviour. In Page C, Black J (eds) *Airways and Vascular Remodelling*. London, Academic Press, 1994, pp 45–71.
63. Lambert RK: Role of bronchial basement membrane in airway collapse. *J Appl Physiol* (1991) **71**: 666–673.
64. Lambert RK, Codd SL, Alley MR, Pack RJ: Physical determinants of bronchial mucosal foldng. *J Appl Physiol* (1994) **77**: 1206–1216.
65. Wiggs BR, Hrousis C, Drazen J, Kampi RD: The implications of airway wall buckling in asthmatic airways. *J Appl Physiol* (1998) in press.

27

Pathophysiology of Asthma

PETER J. BARNES

INTRODUCTION

This chapter aims to provide a brief overview of asthma mechanisms that integrates some of the detailed information provided in preceding chapters into a clinical framework. These chapters highlight the complexity of asthma, with the involvement of many different inflammatory cells, multiple mediators, and with complex acute and chronic inflammatory effects on the airways. Since the last edition of this book was published there have been important advances in our understanding, particularly with the application of new molecular and cell biology techniques and with the development of new drugs to dissect the complex interacting pathways that are activated in asthma. Yet despite these considerable advances there are many fundamental questions about asthma that remain to be answered.

Our views on asthma have changed in the last decade, with the recognition that chronic inflammation underlies the clinical syndrome. In the past it was assumed that the basic defect in asthma lay in abnormal contractility of airway smooth muscle, giving rise to variable airflow obstruction and the common symptoms of intermittent wheeze and shortness of breath. However studies of airway smooth muscle from asthmatic patients have shown no consistent evidence for increased contractile responses to spasmogens such as histamine *in vitro*, indicating that asthmatic airway smooth muscle is not fundamentally abnormal and suggesting that it is the *control* of airway calibre *in vivo* that is abnormal.

ASTHMA AS AN INFLAMMATORY DISEASE

It had been recognized for many years that patients who die of asthma attacks have grossly inflamed airways. The airway lumen is occluded by a tenacious mucus plug composed of plasma proteins exuded from airway vessels and mucus glycoproteins secreted from surface epithelial cells. The airway wall is oedematous and infiltrated with inflammatory cells, which are predominantly eosinophils and lymphocytes. The airway epithelium is invariably shed in a patchy manner and clumps of epithelial cells are found in the airway lumen. Occasionally there have been opportunities to examine the airways of asthmatic patients who die accidentally and similar, though less marked, inflammatory changes have been observed.[1] More recently it has been possible to examine the airways of asthmatic patients by fibreoptic and rigid bronchoscopy, bronchial biopsy and bronchoalveolar lavage (BAL). Direct bronchoscopy reveals that the airways of asthmatic patients are often reddened and swollen, indicating acute inflammation. Lavage has revealed an increase in the numbers of lymphocytes, mast cells and eosinophils and evidence for activation of macrophages in comparison with non-asthmatic controls. Biopsies have provided evidence for increased numbers and activation of mast cells, macrophages, eosinophils and T-lymphocytes. These changes are found even in patients with mild asthma who have few symptoms, and this suggests that inflammation may be found in all asthmatic patients who are symptomatic. Indeed, inflammation may even be present in episodic asthmatic patients at a time when there are no symptoms or in atopic individuals who are not asthmatic. This suggests that the inflammation needs to reach a certain threshold to result in symptoms.

The relationship between inflammation and clinical symptoms of asthma is not clear. There is evidence that the degree of inflammation is related to airway hyperresponsive-

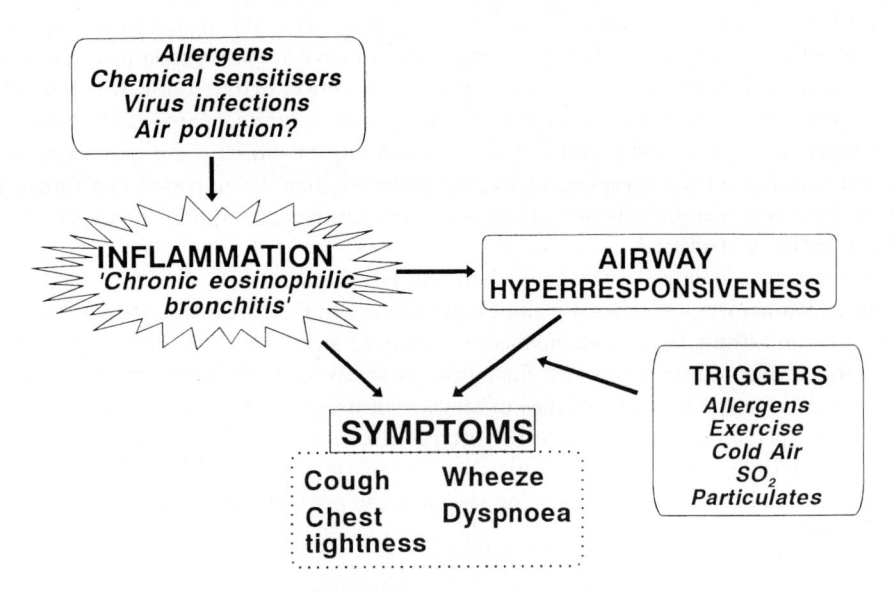

Fig. 27.1 Inflammation in the airways of asthmatic patients leads to airway hyperresponsiveness and symptoms.

27 Pathophysiology of Asthma

ness (AHR), as measured by histamine or methacholine challenge. Increased airway responsiveness is an exaggerated airway narrowing in response to many stimuli that is characteristic of asthma; the degree of AHR relates to asthma symptoms and the need for treatment. Inflammation of the airways may increase airway responsiveness, which thereby allows triggers that would not normally narrow the airways to do so (Fig. 27.1). However, inflammation may also directly lead to an increase in asthma symptoms, such as cough and chest tightness, which are the equivalent of pain in other inflammatory diseases), by sensitization and activation of airway sensory nerve endings.

Although most attention has been focused on the acute inflammatory changes seen in asthmatic airways (bronchoconstriction, plasma exudation, mucus secretion), asthma is a *chronic* inflammatory disease, with inflammation persisting over many years in most patients. Superimposed on this chronic inflammatory state are acute inflammatory episodes that correspond to exacerbations of asthma. It is clearly important to understand the mechanisms of acute and chronic inflammation in asthmatic airways and to investigate the long-term consequences of this chronic inflammation on airway function.

INFLAMMATORY CELLS

Many different inflammatory cells are involved in asthma, although the precise role of each cell type is not yet certain (Fig. 27.2). It is evident that no single inflammatory cell is able to account for the complex pathophysiology of asthma, but some cells predominate in asthmatic inflammation.

Mast cells are clearly important in initiating the acute bronchoconstrictor responses to allergen and probably to other indirect stimuli, such as exercise and hyperventilation (via osmolality or thermal changes) and fog. However, there are questions about the role of mast cells in more chronic inflammatory events, and it seems more probable that other

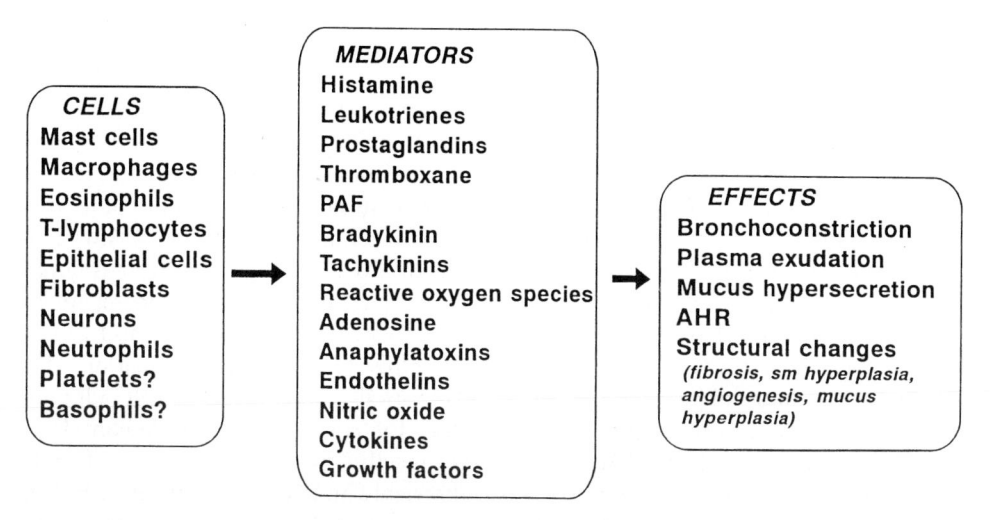

CELLS	MEDIATORS	EFFECTS
Mast cells	Histamine	Bronchoconstriction
Macrophages	Leukotrienes	Plasma exudation
Eosinophils	Prostaglandins	Mucus hypersecretion
T-lymphocytes	Thromboxane	AHR
Epithelial cells	PAF	Structural changes
Fibroblasts	Bradykinin	(fibrosis, sm hyperplasia,
Neurons	Tachykinins	angiogenesis, mucus
Neutrophils	Reactive oxygen species	hyperplasia)
Platelets?	Adenosine	
Basophils?	Anaphylatoxins	
	Endothelins	
	Nitric oxide	
	Cytokines	
	Growth factors	

Fig. 27.2 Many cells and mediators are involved in asthma and lead to several effects on the airways. AHR, airway hyperresponsiveness; PAF, platelet-activating factor; sm, smooth muscle.

cells such as macrophages, eosinophils and T-lymphocytes are more important in the chronic inflammatory process, including AHR.

Macrophages, which are derived from blood monocytes, may traffic into the airways in asthma and may be activated by allergen via low-affinity IgE receptors (FcεRII).[2] The enormous repertoire of macrophages allows these cells to produce many different products, including a large variety of cytokines that may orchestrate the inflammatory response. Macrophages have the capacity to initiate a particular type of inflammatory response via the release of a certain pattern of cytokines. Macrophages may both increase and decrease inflammation, depending on the stimulus. Alveolar macrophages normally have a *suppressive* effect on lymphocyte function, but this may be impaired in asthma after allergen exposure.[3] Macrophages may therefore play an important anti-inflammatory role, preventing the development of allergic inflammation. Macrophages may also act as antigen-presenting cells, which process allergen for presentation to T-lymphocytes, although alveolar macrophages are far less effective in this respect than macrophages from other sites, such as the peritoneum.[4] By contrast dendritic cells, which are specialized macrophage-like cells in the airway epithelium, are very effective antigen-presenting cells[4] and may therefore play a very important role in the initiation of allergen-induced responses in asthma.

Eosinophil infiltration is a characteristic feature of asthmatic airways and differentiates asthma from other inflammatory conditions of the airway. Indeed, asthma might more accurately be termed 'chronic eosinophilic bronchitis' (a term first used as early as 1916). Allergen inhalation results in a marked increase in eosinophils in BAL fluid at the time of the late reaction, and there is a close relationship between eosinophil counts in peripheral blood or BAL and AHR. Eosinophils are linked to the development of AHR through the release of basic proteins and oxygen-derived free radicals.[5]

An important area of research is now concerned with the mechanisms involved in *recruitment* of eosinophils into asthmatic airways. Eosinophils are derived from bone marrow precursors. After allergen challenge eosinophils appear in BAL fluid during the late response; this is associated with a decrease in peripheral eosinophil counts and with the appearance of eosinophil progenitors in the circulation. The signal for increased eosinophil production is presumably derived from the inflamed airway. Eosinophil recruitment initially involves adhesion of eosinophils to vascular endothelial cells in the airway circulation, their migration into the submucosa and their subsequent activation. The role of individual adhesion molecules, cytokines and mediators in orchestrating these responses has been extensively investigated. Adhesion of eosinophils involves the expression of specific glycoprotein molecules on the surface of eosinophils (integrins) and the expression of molecules such as intercellular adhesion molecule (ICAM)-1 on vascular endothelial cells. An antibody directed at ICAM-1 markedly inhibits eosinophil accumulation in the airways after allergen exposure and also blocks the accompanying hyperresponsiveness.[6] However, ICAM-1 is not selective for eosinophils and cannot account for the selective recruitment of eosinophils in allergic inflammation. The adhesion molecules VLA-4, expressed on eosinophils, and vascular cell adhesion molecule (VCAM)-1 appear to be more selective for eosinophils;[7] interleukin (IL)-4 increases the expression of VCAM-1 on endothelial cells.[8] Eosinophil migration may be due to the effects of lipid mediators, such as leukotrienes and possibly platelet-activating factor (PAF), or to the effects of cytokines such as granulocyte–macrophage colony-stimulating factor (GM-CSF) and IL-5, which may be very important for the survival of eosinophils in the airways and may 'prime' eosinophils to exhibit enhanced responsiveness. Eosinophils from asthmatic patients show exaggerated responses to PAF and phorbol esters compared with eosinophils from atopic non-asthmatic individuals;[9] this is further increased by allergen challenge,[10] suggesting that they may have been primed by exposure to cytokines in the circulation. There are several mediators involved in the migration of eosinophils from the circulation to the surface of the airway. The most potent and selective agents appear to be chemokines, such as RANTES, eotaxin and macrophage chemotactic protein (MCP)-4, which are expressed in epithelial cells.[11,12] There appears to be a cooperative interaction between IL-5 and chemokines, so that both cytokines are necessary for the eosinophilic response in airways.[13] Once recruited to the airways, eosinophils require the presence of various growth factors, of which GM-CSF and IL-5 appear to be the most important. In the absence of these growth factors eosinophils undergo programmed cell death (apoptosis).

The role of neutrophils in human asthma is less clear. Neutrophils are found in the airways of patients with chronic bronchitis and patients with bronchiectasis, who do not have the degree of AHR found in asthma; neutrophils are rarely seen in the airways of patients with chronic asthma. However, in patients who die suddenly of asthma large numbers of neutrophils are seen in the airways,[14] although this may reflect the rapid kinetics of neutrophil recruitment compared with eosinophil inflammation. Indeed, in the late response to allergen, neutrophils infiltrate the airways and then disappear, whereas eosinophils accumulate.[15]

T-lymphocytes play a very important role in coordinating the inflammatory response in asthma through the release of specific patterns of cytokines, resulting in the recruitment and survival of eosinophils and the maintenance of mast cells in the airways. T-lymphocytes are coded to express a distinctive pattern of cytokines, which may be similar to that described in the murine Th2 type of T-lymphocyte that characteristically express IL-4, IL-5 and IL-13.[16] This programming of T-lymphocytes is presumably due to antigen-presenting cells such as dendritic cells, which may migrate from the epithelium to regional lymph nodes or which interact with lymphocytes resident in the airway mucosa. There appears to be an imbalance of Th cells in asthma, with the balance tipped away from the normally predominant Th1 cells in favour of Th2 cells (Fig. 27.3). The balance between Th1 cells and Th2 cells may be determined by locally released cytokines such as IL-12, which tips the balance in favour of Th1 cells, and IL-4, which favours Th2 cells. There is some evidence that early infections might promote Th1-mediated responses to predominate and that a lack of infection in childhood may favour Th2 cell expression and thus atopic diseases.[17]

STRUCTURAL CELLS

Structural cells of the airways, including epithelial cells, fibroblasts and even airway smooth muscle cells, may also be an important source of inflammatory mediators, such as cytokines and lipid mediators, in asthma.[18–20] In addition, epithelial cells may play a key role in translating inhaled environmental signals into an airway inflammatory response and are probably a major target cell for inhaled glucocorticoids (Fig. 27.4).

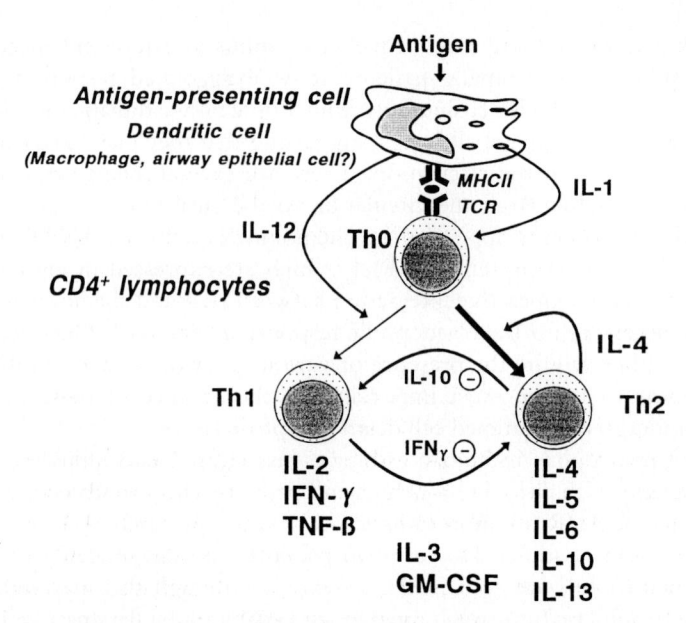

Fig. 27.3 Asthma is characterized by a preponderance of Th2 cells over Th1 cells. GM-CSF, granulocyte–macrophage colony-stimulating factor; IFN-γ, interferon γ; IL, interleukin; TCR, T-cell receptor; TNF-β, tumour necrosis factor β.

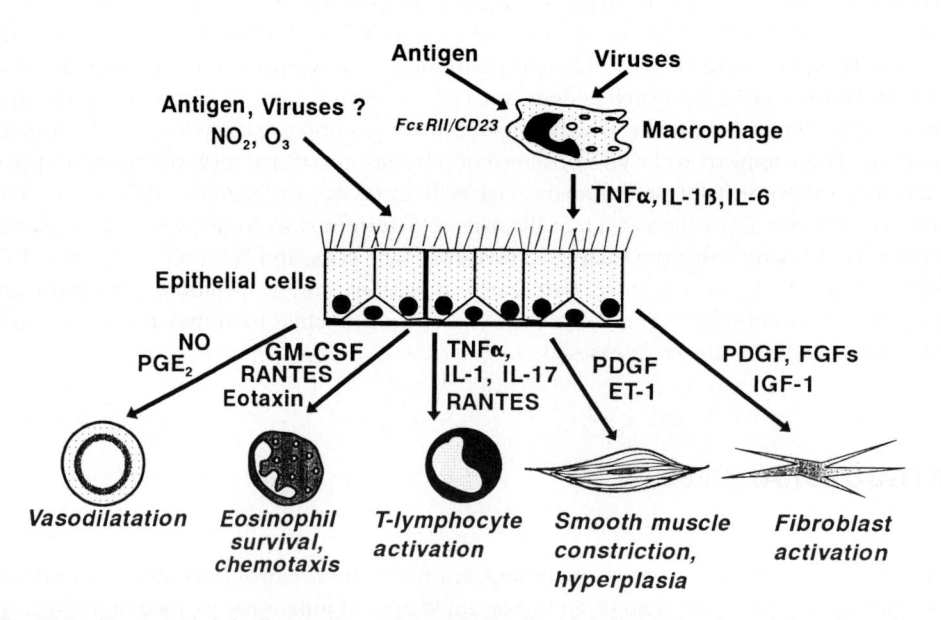

Fig. 27.4 Airway epithelial cells may play an active role in asthmatic inflammation through the release of many inflammatory mediators and cytokines. ET-1, endothelin-1; FGFs, fibroblast growth factors; GM-CSF, granulocyte–macrophage colony-stimulatory factor; IGF-1, insulin-like growth factor 1; IL, interleukin; NO, nitric oxide; PDGF, platelet-derived growth factor; TNF-α, tumour necrosis factor α.

INFLAMMATORY MEDIATORS

Many different mediators have been implicated in asthma and they may have a variety of effects on the airways that could account for the pathological features of asthma[21] ((see Fig. 27.2). Mediators such as histamine, prostaglandins and leukotrienes contract airway smooth muscle, increase microvascular leakage, increase airway mucus secretion and attract other inflammatory cells. Because each mediator has many effects, the role of individual mediators in the pathophysiology of asthma is not yet clear. Indeed the multiplicity of mediators makes it unlikely that antagonizing a single mediator will have a major impact in clinical asthma. However, recent clinical studies with leukotriene antagonists suggest that cysteinyl leukotrienes may play a predominant role.

The cysteinyl leukotrienes LTC_4, LTD_4 and LTE_4 are potent constrictors of human airways and have been reported to increase AHR and may play an important role in asthma[22] (see Chapter 16). The recent development of potent specific leukotriene antagonists has made it possible to evaluate the role of these mediators in asthma. Potent LTD_4 antagonists protect (by about 50%) against exercise- and allergen-induced bronchoconstriction,[23] suggesting that leukotrienes contribute to bronchoconstrictor responses. Chronic treatment with leukotriene antagonists improves lung function and symptoms in asthmatic patients, although the degree of improvement is not as great as seen with an inhaled glucocorticoid. It is only through the use of specific antagonists that the role of individual mediators of asthma may be defined. For example, PAF is a potent inflammatory mediator that mimics many of the features of asthma, including eosinophil recruitment and activation and induction of AHR, yet even potent PAF antagonists do not appear to control asthma symptoms, at least in chronic asthma.[24,25]

Cytokines

Cytokines are increasingly recognized to be important in chronic inflammation and play a critical role in orchestrating the type of inflammatory response (Fig. 27.5). Many inflammatory cells (macrophages, mast cells, eosinophils and lymphocytes) are capable of synthesizing and releasing these proteins, and structural cells such as epithelial cells and endothelial cells may also release a variety of cytokines and may therefore participate in the chronic inflammatory response.[26] While inflammatory mediators like histamine and leukotrienes may be important in the acute and subacute inflammatory responses and in exacerbations of asthma, it is likely that cytokines play a dominant role in chronic inflammation. Almost every cell is capable of producing cytokines under certain conditions. Research in this area is hampered by a lack of specific antagonists, although important observations have been made using specific neutralizing antibodies. The cytokines that appear to be of particular importance in asthma include the lymphokines secreted by T-lymphocytes: IL-3, which is important for the survival of mast cells in tissues; IL-4, which is critical in switching B-lymphocytes to produce IgE and for expression of VCAM-1 on endothelial cells; and IL-5, which is of critical importance in the differentiation, survival and priming of eosinophils. There is increased gene expression of IL-5 in lymphocytes in bronchial biopsies of patients with symptomatic asthma.[27] Other cytokines, such as IL-1, IL-6, tumour necrosis factor (TNF)-α and GM-

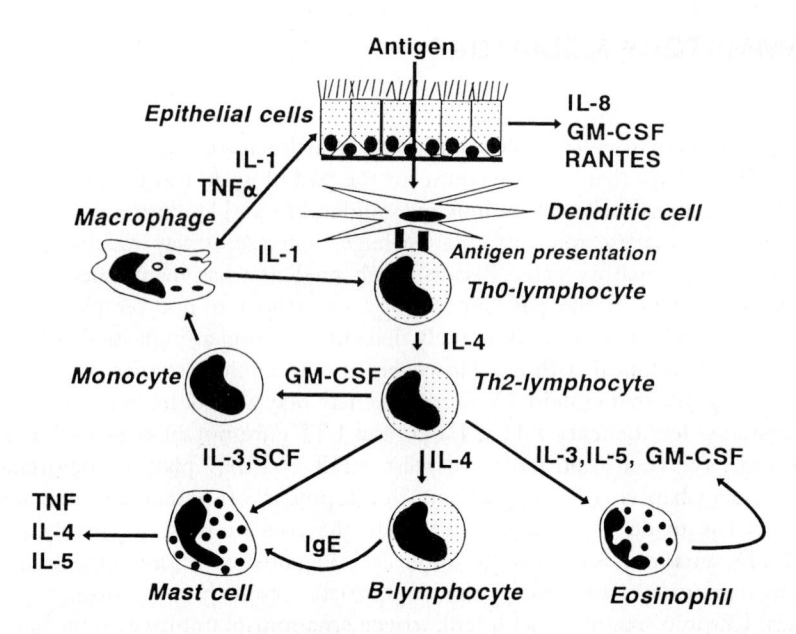

Fig. 27.5 The cytokine network in asthma. Many inflammatory cytokines are released from inflammatory and structural cells in the airway and orchestrate and perpetuate the inflammatory response. GM-CSF, granulocyte–macrophage colony-stimulating factor; IL, interleukin; SCF, stem cell factor; TNF, tumour necrosis factor.

CSF, are released from a variety of cells, including macrophages and epithelial cells, and may be important in amplifying the inflammatory response. TNF-α may be an amplifying mediator in asthma and is produced in increased amounts in asthmatic airways.[28] Inhalation of TNF-α increased airway responsiveness in normal individuals.[29]

Endothelins

Endothelins are peptide mediators that are potent vasoconstrictors and bronchoconstrictors.[30] They also induce airway smooth muscle cell proliferation and fibrosis and may therefore play a role in the chronic inflammation of asthma. There is evidence for increased expression of endothelins in asthma, particularly in airway epithelial cells.[31]

Nitric oxide

Nitric oxide (NO) is produced by several cells in the airway by NO synthases (NOS).[32] An inducible form of the enzyme (iNOS) is expressed in epithelial cells of asthmatic patients[33] and can be induced by cytokines in airway epithelial cells.[34] This may account for the increased concentration of NO in the exhaled air of asthmatic patients.[35] NO itself is a potent vasodilator and this may increase plasma exudation in the airways; it may also amplify the Th2-mediated response.[36]

EFFECTS OF INFLAMMATION

The chronic inflammatory response has several effects on the target cells of the airways, resulting in the characteristic pathophysiological changes associated with asthma (Fig. 27.6). Important advances have recently been made in understanding these changes, although their role in asthma symptoms is often not clear.

Airway epithelium

Airway epithelial shedding may be important in contributing to AHR and may explain how several different mechanisms, such as ozone exposure, certain virus infections, chemical sensitizers and allergen exposure, can lead to the development of AHR, since all these stimuli may lead to epithelial disruption. Epithelium may be shed as a consequence of inflammatory mediators, such as eosinophil basic proteins and oxygen-derived free radicals, together with various proteases released from inflammatory cells. Epithelial cells are commonly found in clumps in the BAL or sputum (Creola bodies) of asthmatic patients, suggesting that there has been a loss of attachment to the basal layer or basement membrane. Epithelial damage may contribute to AHR in a number of ways, including

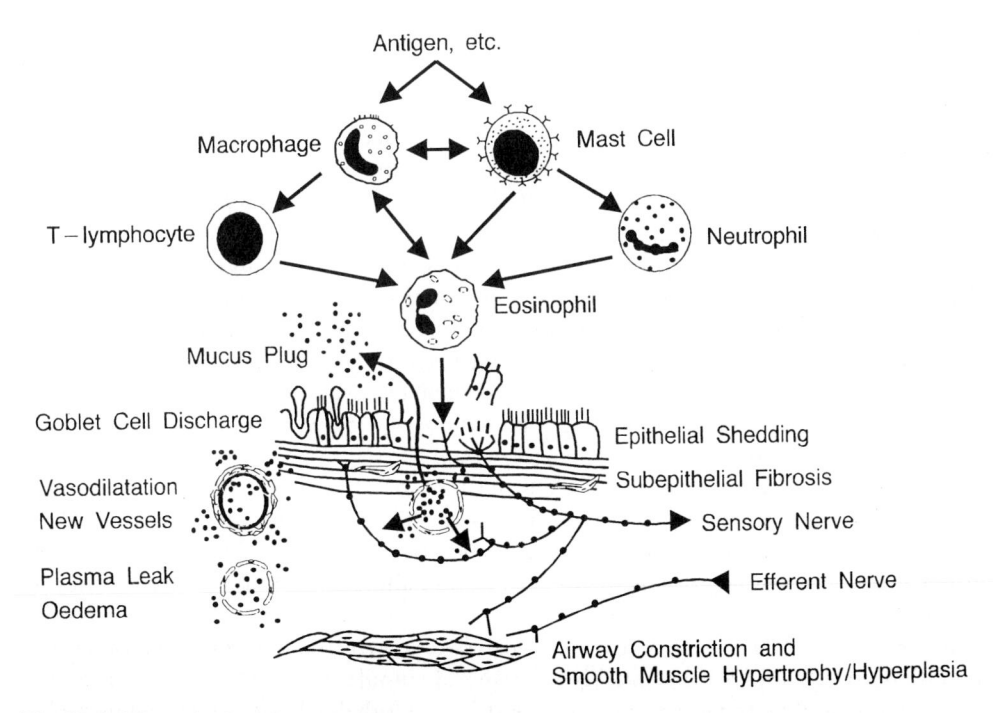

Fig. 27.6 The pathophysiology of asthma is complex, with the participation of several interacting inflammatory cells that result in acute and chronic inflammatory effects on the airway.

loss of barrier function to allow penetration of allergens, loss of enzymes (such as neutral endopeptidase) that normally degrade inflammatory mediators, loss of a relaxant factor (so-called epithelial-derived relaxant factor) and exposure of sensory nerves, which may lead to reflex neural effects on the airway.

Fibrosis

An apparent increase in the basement membrane has been described in fatal asthma, although similar changes have been described in the airways in other conditions.[1] Electron microscopy of bronchial biopsies in asthmatic patients demonstrates that this thickening is due to subepithelial fibrosis.[37] Types III and V collagen appear to be laid down and may be produced by myofibroblasts, which are situated under the epithelium. The mechanism of fibrosis is not yet clear but several cytokines, including transforming growth factor (TGF)-β and platelet-derived growth factor (PDGF), may be produced by epithelial cells or macrophages in the inflamed airway.[26] The role of fibrosis in asthma is unclear, as subepithelial fibrosis may be see even in patients with mild asthma at the onset of disease and it is not certain whether it has any functional consequences. Fibrosis may also occur deeper within the airway wall and may occur within the airway smooth muscle layer.

Airway smooth muscle

There is still debate about the role of abnormalities in airway smooth muscle in asthmatic airways. *In vitro*, airway smooth muscle from asthmatic patients usually shows no increased responsiveness to spasmogens. Reduced responsiveness to β-agonists has also been reported in post-mortem or surgically removed bronchi from asthmatic patients, although the number of β-adrenergic receptors is not reduced, suggesting that these receptors have been uncoupled.[38] These abnormalities of airway smooth muscle may be a reflection of the chronic inflammatory process. For example, the reduced β-adrenergic responses in airway smooth muscle could be due to phosphorylation of the stimulatory G protein coupling β-adrenergic receptors to adenylyl cyclase that results from the activation of protein kinase C by the stimulation of airway smooth muscle cells by inflammatory mediators.[39]

Inflammatory mediators may modulate the ion channels that serve to regulate the resting membrane potential of airway smooth muscle cells, thus altering the level of excitability of these cells. Furthermore, modulation of the activation kinetics of other ion channels by key inflammatory mediators can lead to altered contractile characteristics of smooth muscle.

In asthmatic airways there is also a characteristic *hypertrophy* and *hyperplasia* of airway smooth muscle,[40] which is presumably the result of stimulation of airway smooth muscle cells by various growth factors, such as PDGF, or endothelin-1 released from inflammatory cells.

Vascular responses

Vasodilatation occurs in inflammation, yet little is known about the role of the airway circulation in asthma, partly because of the difficulties involved in measuring airway blood flow. The bronchial circulation may play an important role in regulating airway calibre, since an increase in the vascular volume may contribute to airway narrowing. Increased airway blood flow may be important in removing inflammatory mediators from the airway, and may play a role in the development of exercise-induced asthma.[41]

Microvascular leakage is an essential component of the inflammatory response and many of the inflammatory mediators implicated in asthma produce this leakage.[42,43] There is good evidence for microvascular leakage in asthma and it may have several consequences on airway function, such as increased airway secretions, impaired mucociliary clearance, formation of new mediators from plasma precursors (such as kinins) and mucosal oedema, that may contribute to airway narrowing and increased AHR.

Mucus hypersecretion

Mucus hypersecretion is a common inflammatory response in secretory tissues. Increased mucus secretion contributes to the viscid mucus plugs that occlude asthmatic airways, particularly in fatal asthma. There is evidence for hyperplasia of submucosal glands, which are confined to large airways, and increased numbers of epithelial goblet cells. This increased secretory response may be due to inflammatory mediators acting on submucosal glands and to stimulation of neural elements. Little is understood about the control of goblet cells, which are the main source of mucus in peripheral airways, although cholinergic, adrenergic and sensory neuropeptides may be important in stimulating secretion.[44]

Neural effects

There has recently been a revival of interest in neural mechanisms in asthma.[45] Autonomic nervous control of the airways is complex: in addition to classical cholinergic and adrenergic mechanisms, non-adrenergic non-cholinergic (NANC) nerves and several neuropeptides have been identified in the respiratory tract.[46] Several studies have investigated the possibility that defects in autonomic control may contribute to AHR and asthma, and abnormalities of autonomic function, such as enhanced cholinergic and α-adrenergic responses or reduced β-adrenergic responses, have been proposed. Current thinking suggests that these abnormalities are likely to be secondary to the disease rather than primary defects.[45] It is possible that airway inflammation may interact with autonomic control by several mechanisms.

Inflammatory mediators may act on various prejunctional receptors on airway nerves to modulate the release of neurotransmitters.[47] Thus thromboxane and prostaglandin (PG)D$_2$ facilitate the release of acetylcholine from cholinergic nerves in canine airways, whereas histamine inhibits cholinergic neurotransmission at both parasympathetic ganglia and postganglionic nerves via H$_3$ receptors. Inflammatory mediators may also activate sensory nerves, resulting in reflex cholinergic bronchoconstriction or release of inflammatory neuropeptides. Inflammatory products may also sensitize sensory nerve endings in the airway epithelium, so that the nerves become hyperalgesic. Hyperalgesia and pain (dolor) are cardinal signs of inflammation, and in the asthmatic airway may mediate cough and chest tightness, which are characteristic symptoms of asthma. The precise mechanisms of hyperalgesia are not yet certain, but mediators such as prostaglandins and certain cytokines may be important.

Bronchodilator nerves that are non-adrenergic are prominent in human airways and it has been suggested that these nerves may be defective in asthma.[48] In animal airways, vasoactive intestinal polypeptide (VIP) has been shown to be a neurotransmitter of these nerves, and a striking absence of VIP-immunoreactive nerves has been reported in the lungs from patients with severe fatal asthma.[49] However, it is likely that this loss of VIP immunoreactivity is due to degradation by tryptase released from degranulating mast cells in the airways of asthmatic patients. In human airways the bronchodilator neurotransmitter appears to be NO.[50]

Airway nerves may also release neurotransmitters that have inflammatory effects. Thus neuropeptides such as substance P (SP), neurokinin A and calcitonin gene-related peptide may be released from sensitized inflammatory nerves in the airways and these peptides increase and extend the ongoing inflammatory response.[51] There is evidence for an increase in SP-immunoreactive nerves in airways of patients with severe asthma,[52] which may be due to proliferation of sensory nerves and increased synthesis of sensory neuropeptides as a result of nerve growth factors released during chronic inflammation, although this has not been confirmed in patients with mild asthma.[53] There may also be a reduction in the activity of enzymes, such as neutral endopeptidase, which degrade neuropeptides such as SP.[54] There is also evidence for increased gene expression of the receptor that mediates the inflammatory effects of SP.[55] Thus chronic asthma may be associated with increased neurogenic inflammation, which may provide a mechanism for perpetuating the inflammatory response even in the absence of initiating inflammatory stimuli.

Acute and chronic inflammation

Asthma is characterized by acute inflammatory episodes, which may occur after upper respiratory tract virus infections or exposure to a large amount of inhaled allergen, resulting in bronchoconstriction, plasma exudation and oedema and mucus secretion. However, asthma is also a chronic inflammatory process, partly driven by exposure to low-level environmental allergens, such as house-dust mite and moulds, and this may result in structural changes in the airway walls (remodelling) that lead to progressive narrowing of airways (Fig. 27.7). This may account for the accelerated decline in airway function seen in asthmatic patients over several years.[56,57] These changes include increased thickness of airway smooth muscle, fibrosis (which is predominantly subepithelial), increased mucus-secreting cells and increased numbers of blood vessels (angiogenesis). These changes may not be reversible with therapy. These changes may occur in some patients to a greater extent than others and may be increased by other factors such as concomitant cigarette smoking.

27 Pathophysiology of Asthma

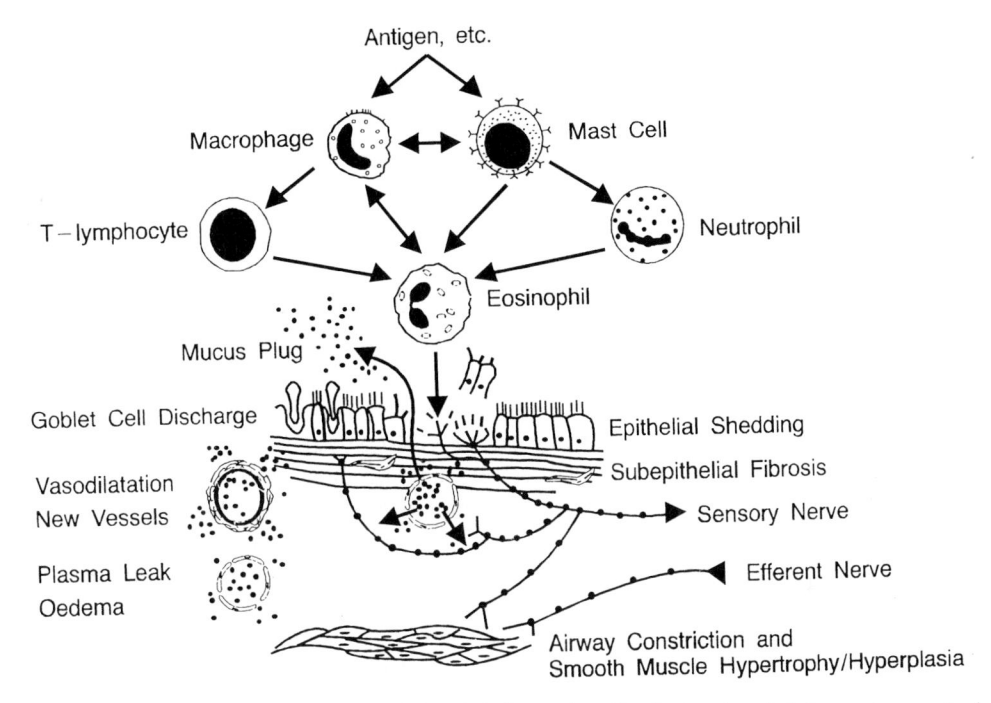

Fig. 27.7 Asthma involves both acute and chronic inflammation. Continuing chronic inflammation may lead to structural changes that may underlie irreversible narrowing of the airways in asthma.

TRANSCRIPTION FACTORS

The chronic inflammation of asthma is due to increased expression of multiple inflammatory proteins (cytokines, enzymes, receptors, adhesion molecules). In many cases these inflammatory proteins are induced by transcription factors, DNA-binding factors that increase the transcription of selected target genes.[58] One transcription factor that may play a critical role in asthma is nuclear factor-κB (NF-κB), which can be activated by multiple stimuli, including protein kinase C activators, oxidants and pro-inflammatory cytokines (such as IL-1β and TNF-α). NF-κB is the predominant transcription factor regulating the expression of iNOS, the inducible form of cyclooxygenase (COX-2), chemokines (IL-8, RANTES, macrophage inhibitory protein (MIP)-1α, eotaxin), pro-inflammatory cytokines (IL-1β, TNF-α, GM-CSF) and adhesion molecules (ICAM-1, VCAM-1).[59,60] NF-κB in epithelial cells may play a pivotal role in amplifying inflammation in diseases such as asthma.[61]

ANTI-INFLAMMATORY MECHANISMS IN ASTHMA

Although most emphasis has been placed on inflammatory mechanisms, there may be important anti-inflammatory mechanisms that may be defective in asthma, resulting in increased inflammatory responses in the airways. Endogenous cortisol may be important as a regulator of the allergic inflammatory response and nocturnal exacerbation of asthma may be related to the circadian fall in plasma cortisol. Blockade of endogenous cortisol secretion by metyrapone results in an increase in the late response to allergen in the skin.[62] Cortisol is converted to the inactive cortisone by the enzyme 11β-hydroxysteroid dehydrogenase, which is expressed in airway tissues.[63] It is possible that this enzyme functions abnormally in asthma or may determine the severity of asthma.

Various cytokines have anti-inflammatory actions. IL-1 receptor antagonist (IL-1ra) inhibits the binding of IL-1 to its receptors and therefore has a potential anti-inflammatory potential in asthma. It is reported to be effective in an animal model of asthma.[64] IL-12 and interferon (IFN)-γ enhance Th1 cells and inhibit Th2 cells and there is some evidence that IL-12 expression may be impaired in asthma.[65] IL-10, which was originally described as a cytokine synthesis inhibitory factor, inhibits the expression of multiple inflammatory cytokines (TNF-α, IL-1β, GM-CSF) and chemokines, as well as inflammatory enzymes (iNOS, COX-2).[66] It may produce these widespread anti-inflammatory actions by inhibiting NF-κB.[67] There is evidence that IL-10 secretion and gene transcription are defective in macrophages and monocytes from asthmatic patients;[68,69] this may lead to enhancement of inflammatory effects in asthma and may be a determinant of asthma severity.

Other mediators may also have anti-inflammatory and immunosuppressive effects. PGE_2 has inhibitory effects on macrophages, epithelial cells and eosinophils and exogenous PGE_2 inhibits allergen-induced airway responses; its endogenous generation may account for the refractory period after exercise challenge.[70] However, it is unlikely that endogenous PGE_2 is important in most asthmatic patients since non-selective cyclooxygenase inhibitors only worsen asthma in a minority of patients (aspirin-induced asthma). Other lipid mediators may also be anti-inflammatory, including 15-hydroxy-eicosatetraenoic acid (15-HETE) that is produced in high concentration by airway epithelial cells. 15-HETE and lipoxins may inhibit cysteinyl leukotriene effects on the airways.[71] The peptide adrenomedullin, which is expressed in high concentrations in lung, has bronchodilator activity[72] and also appears to inhibit the secretion of cytokines from macrophages.[73] Its role in asthma is currently unknown.

Airway and alveolar macrophages have a predominantly suppressive effect in asthma and inhibit T-cell proliferation.[74,75] The mechanism of macrophage-induced immunosuppression is not yet certain, but PGE_2 and IL-10 secretion may contribute. There is some evidence that the immunosuppressive effect of macrophages is reduced in asthmatic patients after allergen challenge *in vitro*, thus favouring T-cell proliferation.[3]

GENETIC INFLUENCES

There is now extensive research on the genetics of asthma, although most of this research relates to the genetics of atopy.[76] Atopy is clearly determined by genetic factors and several genes appear to be involved, although there are differences between different populations. There is evidence for linkage between markers on chromosome 11q13, which may relate to polymorphism of the gene coding for the β-chain of the high-affinity IgE receptor (FcϵRI), and 5q31, which codes for a cytokine cluster IL-3, IL-4, IL-5, IL-9,

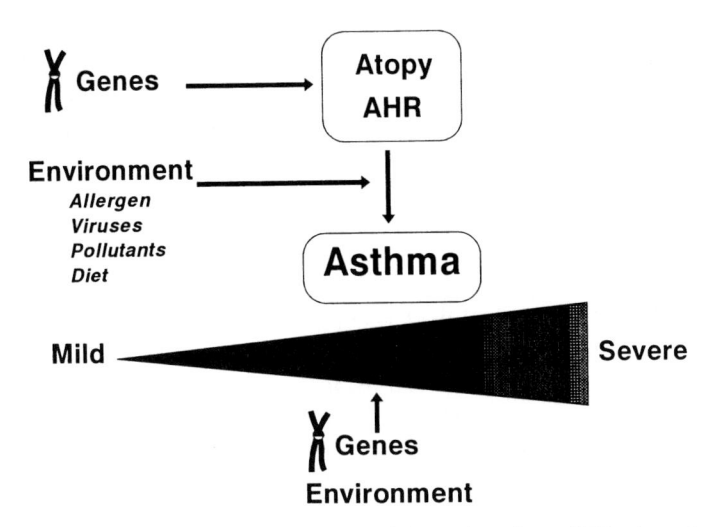

Fig. 27.8 Interaction between genetic and environmental factors in asthma. AHR, airway hyperresponsiveness.

IL-13 and GM-CSF. There are also associations between these linkages and AHR, although it is difficult to dissociate changes in airway reactivity from atopy. Atopy is the most important risk factor for the development of asthma, but while understanding the genetics of atopy will shed light on the nature of allergic inflammation, it may not be very informative in understanding asthma. It is likely that environmental factors (viral/bacterial infections, allergen exposure, diet) may be more important in determining whether an atopic individual becomes an asthmatic patient. However, once asthma is established, genetic factors may be important in determining the severity of the disease and its response to therapy (Fig. 27.8). Polymorphisms have been described in many genes involved in the inflammatory process and may occur in coding and promoter regions, resulting in increased production of inflammatory mediators, such as cytokines for example.[77] It is likely that a combination of genetic polymorphisms will determine the natural history and outcome of asthma, but so far few of these have been identified. One example that may be of clinical relevance is the polymorphisms of the β_2-adrenoceptor gene that result in the structural changes in β_2-receptor structure associated with functional changes.[78] A common polymorphism, resulting in a glycine substitution for arginine at position 16 (Arg16→Gly), is associated with increased downregulation and desensitization of the receptor in response to β_2-receptor agonists and is found with a greater frequency in patients with nocturnal asthma;[79] a Gln27→Glu polymorphism, which has less tendency to desensitize, is associated with a lower level of AHR.

UNANSWERED QUESTIONS

Although our understanding of asthma has advanced very rapidly in recent years and this has led to a fundamental change in the approach to therapy, many important questions remain unanswered.[80]

(1) Why is the prevalence of asthma increasing throughout the world as a consequence of 'Westernization'? It is not clear what environmental factors are most important for the increase in atopic diseases, but it is likely that several factors are operating together. These factors include diet (reduced intake of antioxidants, reduced unsaturated fats), lack of early childhood infections (with consequent tendency to develop Th2-driven responses), greater exposure to allergens in the home (tight housing, mattresses, central heating providing more favourable environment), cigarette smoking (pregnancy and early childhood exposure) and possibly air pollution due to road traffic.

(2) Why does asthma once established become chronic? Occupational asthma due to chemical sensitizers, such as toluene diisocyanate, is a good example where the causal agent is known. Removal from exposure to the sensitizer within 6 months of development of asthma symptoms results in complete resolution of asthma, whereas longer exposure is associated with persistent asthma even when avoidance of exposure is complete.[81] Thus once inflammation is established it may continue independently of the causal mechanism.

(3) What are the critical environmental factors involved in the early development of asthma in atopic individuals? How important are early infections, allergen exposure, indoor and outdoor air pollution?

(4) Are there different types of asthma, characterized by a common immunological and inflammatory response? Intrinsic asthma, where there is no identifiable atopy, looks very similar to allergic asthma clinically and immunocytochemically, yet there are likely to be immunological differences.

(5) How does inflammation of the airways translate into clinical symptoms of asthma? Airway thickening, as a consequence of the inflammatory response, may contribute to increased responsiveness to spasmogens.[82] However, there is no obvious relationship between the inflammatory response in airways and asthma severity; patients with mild asthma may have a similar eosinophil response to patients with severe asthma, suggesting that there are other factors that determine clinical severity. Furthermore, the fact that inhaled long-acting β_2-receptor agonists, which have no anti-inflammatory effects, provide better control of asthma symptoms than increasing the dose of inhaled steroids in asthmatic patients who are not controlled on low doses of steroids suggests that factors other than inflammation (abnormalities of airway smooth muscle?) are important.[83–85]

(6) How important are genetic factors (genetic polymorphisms) in determining the phenotype of asthma, such as disease of differing severity and varying responsiveness to steroids, β_2-receptor agonists and other therapies? By determining the profile of genetic polymorphisms in an individual patient, using novel gene-chip technology, it may be possible to predict the outcome of asthma and its therapy, thus optimising treatment.

(7) Why does asthma usually go into remission in adolescence? If the endogenous mechanisms involved in switching off asthma could be identified, this might have important implications for the development of new therapeutic approaches to asthma in the future.

REFERENCES

1. Dunnill MS: The pathology of asthma, with special reference to the changes in the bronchial mucosa. *J Clin Pathol* (1960) **13**: 27–33.
2. Lee TH, Lane SJ: The role of macrophages in the mechanisms of airway inflammation in asthma. *Am Rev Respir Dis* (1992) **145**: S27–S30.
3. Spiteri MA, Knight RA, Jeremy JY, Barnes PJ, Chung KF: Alveolar macrophage-induced suppression of peripheral blood mononuclear cell responsiveness is reversed by *in vitro* allergen exposure in bronchial asthma. *Eur Respir J* (1994) **7**: 1431–1438.
4. Holt PG, McMenamin C: Defense against allergic sensitization in the healthy lung: the role of inhalation tolerance. *Clin Exp Allergy* (1989) **19**: 255–262.
5. Gleich GJ: The eosinophil and bronchial asthma: current understanding. *J Allergy Clin Immunol* (1990) **85**: 422–436.
6. Wegner CD, Gundel L, Reilly P, Haynes N, Letts LG, Rothlein R: Intracellular adhesion molecule-1 (ICAM-1) in the pathogenesis of asthma. *Science* (1990) **247**: 456–459.
7. Pilewski JM, Albelda SM: Cell adhesion molecules in asthma: homing activation and airway remodeling. *Am J Respir Cell Mol Biol* (1995) **12**: 1–3.
8. Lamas AM, Mulroney CM, Schleimer RP: Studies of the adhesive interaction between purified human eosinophils and cultured vascular endothelial cells. *J Immunol* (1988) **140**: 1500–1510.
9. Chanez P, Dent G, Yukawa T, Barnes PJ, Chung KF: Generation of oxygen free radicals from blood eosinophils from asthma patients after stimulation with PAF or phorbol ester. *Eur Respir J* (1990) **3**: 1002–1007.
10. Evans DJ, Lindsay MA, O'Connor BJ, Barnes PJ: Priming of circulating human eosinophils following exposure to allergen challenge. *Eur Respir J* (1996) **9**: 703–708.
11. Berkman N, Krishnan VL, Gilbey T, O'Connor BJ, Barnes PJ: Expression of RANTES mRNA and protein in airways of patients with mild asthma. *Am J Respir Crit Care Med* (1996) **15**: 382–389.
12. Ponath PD, Qin S, Ringler DJ, *et al.*: Cloning of the human eosinophil chemoattractant eotaxin. Expression, receptor binding and functional properties provide a mechanism for the selective recruitment of eosinophils. *J Clin Invest* (1996) **97**: 604–612.
13. Collins PD, Marleau S, Griffiths-Johnson DA, Jose PJ, Williams TJ: Cooperation between interleukin-5 and the chemokine eotaxin to induce eosinophil accumulation *in vivo*. *J Exp Med* (1995) **182**: 1169–1174.
14. Sur S, Crotty TB, Kephart GM, *et al.*: Sudden onset fatal asthma: a distinct entity with few eosinophils and relatively more neutrophils in the airway submucosa. *Am Rev Respir Dis* (1993) **148**: 713–719.
15. Montefort S, Gratziov C, Goulding D, *et al.*: Bronchial biopsy evidence for leukocyte infiltration and upregulation of leukocyte–endothelial and adhesion molecules 6 hours after local allergen challenge of sensitized asthmatic airways. *J Clin Invest* (1994) **93**: 1411–1421.
16. Mosman TR, Sad S: The expanding universe of T-cell subsets: Th1, Th2 and more. *Immunol Today* (1996) **17**: 138–146.
17. Shirakawa T, Enomoto T, Shimazu S, Hopkin JM: The inverse association between tuberculin responses and atopic disorder. *Science* (1997) **275**: 77–79.
18. Levine SJ: Bronchial epithelial cell–cytokine interactions in airway epithelium. *J Invest Med* (1995) **43**: 241–249.
19. Devalia JL, Davies RJ: Airway epithelial cells and mediators of inflammation. *Respir Med* (1993) **6**: 405–408.
20. Saunders MA, Mitchell JA, Seldon PM, Barnes PJ, Giembycz MA, Belvisi MG: Release of granulocyte–macrophage colony-stimulating factor by human cultured airway smooth muscle cells: suppression by dexamethasone. *Br J Pharmacol* (1997) **120**: 545–546.
21. Barnes PJ, Chung KF, Page CP: Inflammatory mediators and asthma. *Pharmacol Rev* (1988) **40**: 49–84.
22. Arm JP, Lee TH: Sulphidopeptide leukotrienes in asthma. *Clin Sci* (1993) **84**: 501–510.
23. Chung KF: Leukotriene receptor antagonists and biosynthesis inhibitors: potential breakthrough in asthma therapy. *Eur Respir J* (1995) **8**: 1203–1213.

24. Spence DPS, Johnston SL, Calverley PMA, et al.: The effect of the orally active platelet-activating factor antagonist WEB 2086 in the treatment of asthma. *Am J Respir Crit Care Med* (1994) **149**: 1142–1148.
25. Kuitert LM, Angus RM, Barnes NC, et al.: The effect of a novel potent PAF antagonist, modipafant, in chronic asthma. *Am J Respir Crit Care Med* (1995) **151**: 1331–1335.
26. Barnes PJ: Cytokines as mediators of chronic asthma. *Am J Respir Crit Care Med* (1994) **150**: S42–S49.
27. Hamid Q, Azzawi M, Sun Ying, et al.: Expression of mRNA for interleukin-5 in mucosal bronchial biopsies from asthma. *J Clin Invest* (1991) **87**: 1541–1549.
28. Kips JC, Tavernier JH, Joos GF, Peleman RA, Pauwels RA: The potential role of tumor necrosis factor a in asthma. *Clin Exp Allergy* (1993) **23**: 247–250.
29. Thomas PS, Yates DH, Barnes PJ: Tumor necrosis factor-a increases airway responsiveness and sputum neutrophils in normal human subjects. *Am J Respir Crit Care Med* (1995) **152**: 76–80.
30. Barnes PJ: Endothelins and pulmonary diseases. *J Appl Physiol* (1994) **77**: 1051–1059.
31. Springall DR, Howarth PH, Counihan H, Djukanovic R, Holgate ST, Polak JM: Endothelin immunoreactivity of airway epithelium in asthmatic patients. *Lancet* (1991) **337**: 697–701.
32. Barnes PJ: Nitric oxide and airway disease. *Ann Med* (1995) **27**: 389–393.
33. Hamid Q, Springall DR, Riveros-Moreno V, et al.: Induction of nitric oxide synthase in asthma. *Lancet* (1993) **342**: 1510–1513.
34. Robbins RA, Barnes PJ, Springall DR, et al.: Expression of inducible nitric oxide synthase in human bronchial epithelial cells. *Biochem Biophys Res Commun* (1994) **203**: 209–218.
35. Kharitonov SA, Yates D, Robbins RA, Logan-Sinclair R, Shinebourne E, Barnes PJ: Increased nitric oxide in exhaled air of asthmatic patients. *Lancet* (1994) **343**: 133–135.
36. Barnes PJ, Liew FY: Nitric oxide and asthmatic inflammation. *Immunol Today* (1995) **16**: 128–130.
37. Roche WR, Beasley R, Williams JH, Holgate ST: Subepithelial fibrosis in the bronchi of asthmatics. *Lancet* (1989) **i**: 520–524.
38. Bai TR, Mak JCW, Barnes PJ: A comparison of beta-adrenergic receptors and *in vitro* relaxant responses to isoproterenol in asthmatic airway smooth muscle. *Am J Respir Cell Mol Biol* (1992) **6**: 647–651.
39. Grandordy BM, Mak JCW, Barnes PJ: Modulation of airway smooth muscle β-adrenoceptor function by a muscarinic agonist. *Life Sci* (1994) **54**: 185–191.
40. Ebina M, Yaegashi H, Chiba R, Takahashi T, Motomiya M, Tanemura M: Hyperreactive site in the airway tree of asthmatic patients recoded by thickening of bronchial muscles: a morphometric study. *Am Rev Respir Dis* (1990) **141**: 1327–1332.
41. McFadden ER: Hypothesis: exercise-induced asthma as a vascular phenomenon. *Lancet* (1990) **335**: 880–883.
42. Persson CGA: Plasma exudation and asthma. *Lung* (1988) **166**: 1–23.
43. Chung KF, Rogers DF, Barnes PJ, Evans TW: The role of increased airway microvascular permeability and plasma exudation in asthma. *Eur Respir J* (1990) **3**: 329–337.
44. Kuo H, Rhode JAL, Tokuyama K, Barnes PJ, Rogers DF: Capsaicin and sensory neuropeptide stimulation of goblet cell secretion in guinea pig trachea. *J Physiol* (1990) **431**: 629–641.
45. Barnes PJ: Is asthma a nervous disease? *Chest* (1995) **107**: 119S–124S.
46. Barnes PJ, Baraniuk J, Belvisi MG: Neuropeptides in the respiratory tract. *Am Rev Respir Dis* (1991) **144**: 1187–1198, 1391–1399.
47. Barnes PJ: Modulation of neurotransmission in airways. *Physiol Rev* (1992) **72**: 699–729.
48. Lammers JWJ, Barnes PJ, Chung KF: Non-adrenergic, non-cholinergic airway inhibitory nerves. *Eur Respir J* (1992) **5**: 239–246.
49. Ollerenshaw S, Jarvis D, Woolcock A, Sullivan C, Scheibner T: Absence of immunoreactive vasoactive intestinal polypeptide in tissue from the lungs of patients with asthma. *N Engl J Med* (1989) **320**: 1244–1248.
50. Belvisi MG, Stretton CD, Barnes PJ: Nitric oxide is the endogenous neurotransmitter of bronchodilator nerves in human airways. *Eur J Pharmacol* (1992) **210**: 221–222.
51. Barnes PJ: Sensory nerves, neuropeptides and asthma. *Ann NY Acad Sci* (1991) **629**: 359–370.

27 Pathophysiology of Asthma

52. Ollerenshaw SL, Jarvis D, Sullivan CE, Woolcock AJ: Substance P immunoreactive nerves in airways from asthmatics and non-asthmatics. *Eur Respir J* (1991) **4**: 673–682.
53. Howarth PH, Springall DR, Redington AE, Djukanovic R, Holgate ST, Polak JM: Neuropeptide-containing nerves in bronchial biopsies from asthmatic and non-asthmatic subjects. *Am J Respir Cell Mol Biol* (1995) **13**: 288–296.
54. Nadel JA: Neutral endopeptidase modulates neurogenic inflammation. *Eur Respir J* (1991) **4**: 745–754.
55. Adcock IM, Peters M, Gelder C, Shirasaki H, Brown CR, Barnes PJ: Increased tachykinin receptor gene expression in asthmatic lung and its modulation by steroids. *J Mol Endocrinol* (1993) **11**: 1–7.
56. Peat JK, Woolcock AJ, Cullen K: Rate of decline of lung function in subjects with asthma. *Eur J Respir Dis* (1987) **70**: 171–179.
57. Brown JP, Greville WH, Finucane KE: Asthma and irreversible airflow obstruction. *Thorax* (1984) **39**: 131–136.
58. Barnes PJ, Adcock IM: Transcription factors in asthma. *Clin Exp Allergy* (1995) **27** (Suppl 2): 46–49.
59. Siebenlist U, Franzuso G, Brown R: Structure, regulation and function of NF-κB. *Annu Rev Cell Biol* (1994) **10**: 405–455.
60. Barnes PJ, Karin M: Nuclear factor-κB: a pivotal transcription factor in chronic inflammatory diseases. *N Engl J Med* (1997) **336**: 1066–1071.
61. Barnes PJ, Adcock IM: NF-κB: a pivotal role in asthma and a new target for therapy. *Trends Pharmacol Sci* (1997) **18**: 46–50.
62. Herrscher RF, Kasper C, Sullivan TJ: Endogenous cortisol regulates immunoglobulin E-dependent late phase reactions. *J Clin Invest* (1992) **90**: 593–603.
63. Schleimer RP: Potential regulation of inflammation in the lung by local metabolism of hydrocortisone. *Am J Respir Cell Mol Biol* (1991) **4**: 166–173.
64. Selig W, Tocker J: Effect of interleukin-1 receptor antagonist on antigen-induced pulmonary responses in guinea-pigs. *Eur J Pharmacol* (1992) **213**: 331–336.
65. Naseer T, Minshall EM, Leung DY, *et al.*: Expression of IL-12 and IL-13 mRNA in asthma and their modulation in response to steroids. *Am J Respir Crit Care Med* (1997) **155**: 845–851.
66. Ho AS, Moore KW: Interleukin-10 and its receptor. *Ther Immunol* (1994) **1**: 173–185.
67. Wang P, Wu P, Siegel MI, Egan RW, Billah MM: Interleukin(IL)-10 inhibits nuclear factor kappa B activation in human monocytes. IL-10 and IL-4 suppress cytokine synthesis by different mechanisms. *J Biol Chem* (1995) **270**: 9558–9563.
68. Borish L, Aarons A, Rumbyrt J, Cvietusa P, Negri J, Wenzel S: Interleukin-10 regulation in normal subjects and patients with asthma. *J Allergy Clin Immunol* (1996) **97**: 1288–1296.
69. John M, Lim S, Seybold J, *et al.*: Inhaled corticosteroids increase IL-10 but reduce MIP-1a, GM-CSF and IFN-γ release from alveolar macrophages in asthma. *Am J Respir Crit Care Med* (1997) in press.
70. Pavord ID, Tattersfield AE: Bronchoprotective role for endogenous prostaglandin E2. *Lancet* (1995) **344**: 436–438.
71. Lee TH: Lipoxin A4: a novel anti-inflammatory molecule? *Thorax* (1995) **50**: 111–112.
72. Kanazawa H, Kurihara N, Hirata K, Kudo S, Kawaguchi T, Takeda T: Adrenomedullin, a newly discovered hypotensive peptide, is a potent bronchodilator. *Biochem Biophys Res Commun* (1994) **205**: 251–254.
73. Kamoi H, Kanazawa H, Hirata K, Kurihara N, Yano Y, Otani S: Adrenomedullin inhibits the secretion of cytokine-induced neutrophil chemoattractant, a member of the interleukin-8 family, from rat alveolar macrophages. *Biochem Biophys Res Commun* (1995) **211**: 1031–1035.
74. Holt PG: Regulation of antigen-presenting cell function(s) in lung and airway tissues. *Eur Respir J* (1993) **6**: 120–129.
75. Upham JW, Strickland DH, Bilyk N, Robinson BW, Holt PG: Alveolar macrophages from humans and rodents selectively inhibit T-cell proliferation but permit T-cell activation and cytokine secretion. *Immunology* (1995) **84**: 142–147.
76. Sandford A, Weir T, Pare P: The genetics of asthma. *Am J Respir Crit Care Med* (1996) **153**: 1749–1765.
77. Wilson AG, Duff GW: Genetic traits in common diseases. *Br Med J* (1995) **310**: 1482–1483.

78. Hall IP: β_2-Adrenoceptor polymorphisms: are they clinically important? *Thorax* (1996) **51**: 351–353.
79. Turki J, Pak J, Green S, Martin R, Liggett SB: Genetic polymorphism of the β_2-adrenergic receptor in nocturnal and non-nocturnal asthma: evidence that Gly 16 correlates with the nocturnal phenotype. *J Clin Invest* (1995) **95**: 1635–1641.
80. Woolcock AJ, Barnes PJ: Asthma: the important questions—part 3. *Am J Respir Crit Care Med* (1996) **153**: S1–S31.
81. Chan-Yeung M, Malo J-L: Occupational asthma. *N Engl J Med* (1995) **333**: 107–112.
82. Pare PD, Bai TR: The consequences of chronic allergic inflammation. *Thorax* (1995) **50**: 328–332.
83. Greening AP, Ind PW, Northfield M, Shaw G: Added salmeterol versus higher-dose corticosteroid in asthma patients with symptoms on existing inhaled corticosteroid. *Lancet* (1994) **344**: 219–224.
84. Woolcock A, Lundback B, Ringdal N, Jacques L: Comparison of addition of salmeterol to inhaled steroids with doubling the dose of inhaled steroids. *Am J Respir Crit Care Med* (1996) **153**: 1481–1488.
85. Pauwels RA, Lofdahl C-G, Postma DS, *et al.*: Additive effects of inhaled formoterol and budesonide in reducing asthma exacerbations: a one-year controlled study. *N Engl J Med* (1997) **337**: 1405–1411.

28

Allergens

D.W. COCKCROFT

INTRODUCTION

Asthma is currently somewhat arbitrarily defined as symptoms associated with variable airflow obstruction.[1] Inhaled allergens producing IgE-mediated responses are amongst the many triggers of airflow obstruction and symptoms in asthma.[1] The thinking regarding the importance of allergens in the pathogenesis of asthma has undergone major revision. Allergens are not only triggers of bronchospasm (like exercise, cold air, smoke, irritants, etc.) but also inducers of both inflammation[2,3] and airway hyperresponsiveness*.[4,5] Allergens are, therefore, now recognized as *inducers*, along with low molecular weight chemical sensitizers,[6–9] viral respiratory tract infections[10] and, occasionally, extremely high levels of inhaled noxious gases or fumes.[11] Unlike triggers, inducers cause true asthma exacerbations and circumstantial evidence points to them as causes of asthma. Evidence from several population studies points to inhalant atopic allergens as an important, perhaps *the* most important, cause of airway hyperresponsiveness and asthma. The shift in our understanding of the role of allergens in asthma, from that of one amongst many triggers of symptoms to an important cause of the disease itself, has important therapeutic relevance.

In this chapter, naturally occurring complete allergens that trigger IgE-mediated type I atopic hypersensitivity are discussed. Evidence implicating allergens, particularly inhaled allergens, as a cause of asthma is reviewed. IgE-mediated responses to haptens (drugs,

---

* Throughout this chapter, unless otherwise stated, the term 'airway (hyper)responsiveness' refers to the non-allergic (hyper)responsiveness to histamine, cholinergic agonists, exercise, etc., that is a characteristic feature of symptomatic asthma.

some occupational low molecular weight chemical sensitizers) and parasites are not reviewed, nor is asthma caused by other immune mechanisms, e.g. IgG4.[12]

ATOPY

Atopy is the tendency to develop IgE antibodies to commonly encountered environmental allergens by natural exposure, in which the route of entry of allergen is across intact mucosal surfaces.[13] The recognized familial nature of atopy is probably genetic; possibilities include either simple genetic inheritance with variable and low (<50%) penetrance or, more likely, multiple gene inheritance (genetic heterogeneity).[14] Environmental factors are likely relevant also. The pathophysiological basis of atopy remains unknown, although the likelihood appears to favour allergen handling perhaps at the mucosal surface rather than increased capacity to produce IgE.[15]

The prevalence of atopy in random populations, generally defined as the presence of positive(s) on skin-prick tests with a small battery of indigenous allergens, ranges from 30 to almost 50%.[16-18] The peak period of sensitization is in the third decade; thereafter the prevalence falls.[13] Our experience with a random young population would suggest that about 50% of atopic subjects will have symptoms likely referrable to atopy, which will include asthma in about 50%.[18] Thus, atopy is common, affecting about 1 in 3, with about 1 in 6 having symptomatic atopy and about 1 in 12 having atopic asthma.

INHALED ALLERGENS

Nature of allergens

Inhaled complete allergens that provoke asthma by IgE-mediated mechanisms are organic high molecular weight (20 000–40 000) protein or protein-containing molecules, which may be derived from any phylum of either the plant or animal kingdoms (including bacteria).[19,20] In clinical non-occupational settings, the important inhalant allergens fall into four groups: pollen, fungal spores, animal danders and household mite/insects.[21]

Pollen allergens that trigger asthma are predominantly from wind-pollinated plants, namely trees, grass and weeds.[21] The relevant allergens and seasonal fluctuations will vary with geography and climate, with tree pollens in spring months, grasses in the summer and weeds in late summer and autumn.[20] Although whole pollen grains may have limited access to the lower respiratory tract,[22] the relationship of pollen to clinical asthma is convincing.[21]

Atmospheric fungal spores of many groups of fungi are smaller and more respirable than pollen, and are recognized as causing atopic sensitization. Their role in triggering asthma is less certain than pollen.[20,21] Fungal spore types and seasons will also vary with geographic and climatic (temperature/humidity) conditions. Fungal spores are often associated with decaying vegetation, resulting in a late summer and autumn peak for common fungal spores, e.g. *Alternaria, Cladosporium, Aspergillus, Sporobolomyces*, etc.[20] A spring peak for atmospheric fungal spores may be seen in some areas, especially where late melting of snow cover leads to so-called snow mould.[20] Thus, atmospheric fungal spores may be responsible for autumn or spring/autumn asthma symptoms. Fungi may also be present inside living areas in moist basements, food storage areas and waste receptacles.[20] *Aspergillus* may cause a distinct clinical syndrome, allergic bronchopulmonary aspergillosis, which will be covered separately.

Household animals,[20] particularly cats and dogs but also small animals (gerbils, hamsters, rabbits, etc.) and birds, may release allergens in secretions (e.g. saliva) or excretions (e.g. urine, faeces). Large animals, particularly horses, may also provoke atopic sensitivity.

House dust, likely due to its content of mite antigens from various *Dermatophagoides* species or insect antigens such as cockroach,[23] is an important source of atopic sensitization. *Dermatophagoides* spp. in particular are likely the most important cause of atopic sensitization worldwide. Again, climatic conditions are important, since areas of low indoor relative humidity do not favour growth of house-dust mites.[20,24]

Other allergens are encountered less frequently, often in occupational settings, and include various plant parts (castor bean, cocoa bean, tobacco leaf, psyllium [laxative], vegetable gums, etc.), insect dusts, bacterial enzymes and, in the very highly sensitized, even atmospheric molecular levels of foods (e.g. cooking fish).[20]

Patterns of allergen response

Bronchial responses to inhaled allergens have been assessed primarily by the somewhat artificial inhalation tests in the laboratory with aqueous allergen extracts.[4,5,25] Nevertheless, the results of such challenges, especially the late sequelae, appear to be clinically relevant,[26] and allergen inhalation tests allow study of both the pharmacology and pathophysiology of allergen-induced asthma. Bronchial responses to allergen can be divided into early and late sequelae.

The early or immediate asthmatic response (EAR) is an episode of airflow obstruction that is maximal 10–20 min after allergen inhalation and resolves spontaneously in 1–2 h.[25,27] This response is likely predominantly bronchospastic.

The late sequelae include the late asthmatic response (LAR),[4,5,25,27–29] allergen-induced increase in airway responsiveness[4,5,26] and recurrent nocturnal asthma,[30] all of which are likely due, in whole or in part, to airway inflammation.[2,31]

The LAR is an episode of airflow obstruction that develops after spontaneous resolution of the EAR between 3 and 5 h after exposure, occasionally earlier, rarely later.[4,5,25,28,29] Resolution usually begins by 6–8 h but may require in excess of 12 h.[25,27] Modest late responses respond well to bronchodilators,[32] although unpublished observations suggest bronchodilators may be required often (e.g. up to 2-hourly). More severe late airway obstruction is not always completely reversible by bronchodilator.[27] Examples of early and late responses are shown in Fig. 28.1.

Allergen-induced increase in airway responsiveness (e.g. to histamine/methacholine) occurs following both experimental[4,5,26,31] and natural[26,33,34] allergen exposure. This is correlated with the occurrence and severity of the late response, often appearing with small, previously ignored, late responses (5–15% fall in forced expiratory volume in 1 s, FEV_1).[4,5] Airway responsiveness is not yet enhanced in most subjects 2 h after exposure,[35] appears to have developed at 3 h,[36–39] is present at 7–8 h[4,5] and may persist

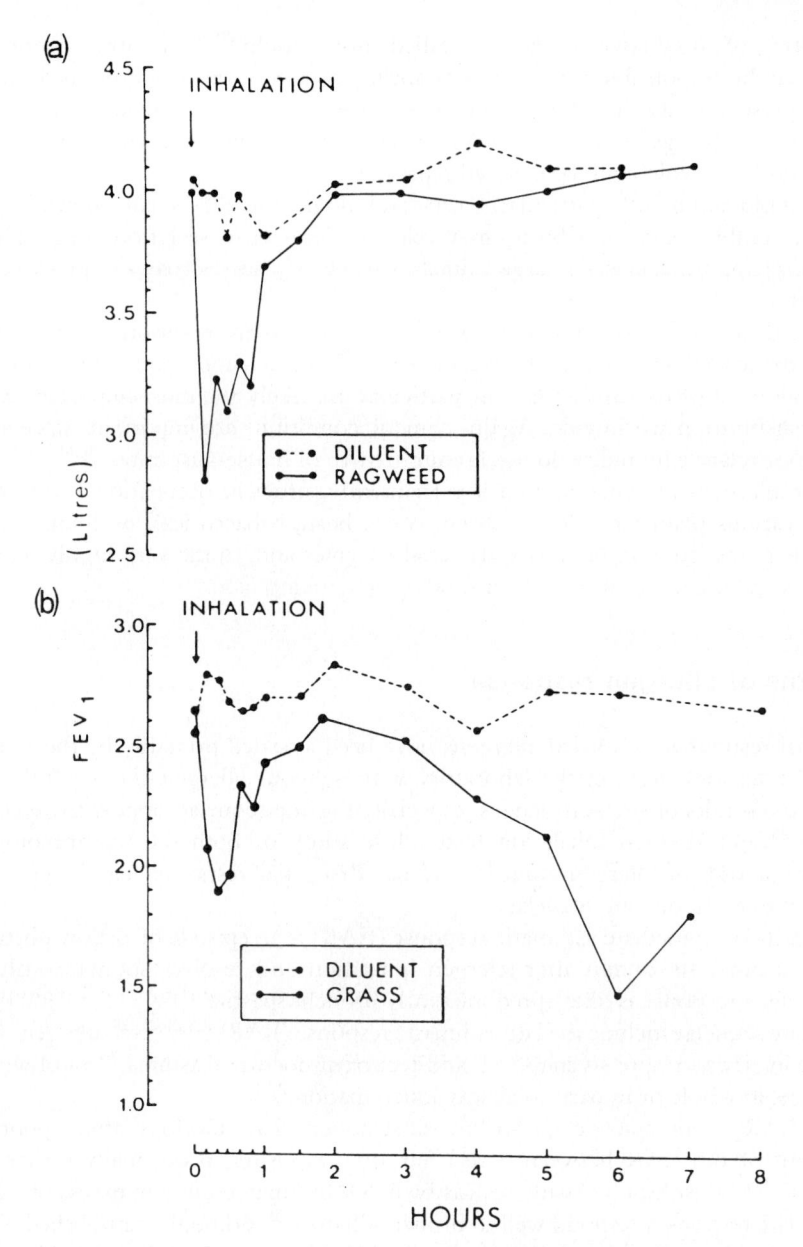

Fig. 28.1 Early and dual asthmatic responses to allergen: (a) an isolated early asthmatic response following ragweed pollen inhalation; (b) a dual asthmatic response (in another subject) following grass pollen inhalation. FEV_1, forced expiratory volume in 1 s.

for days, occasionally worsening despite return of airway calibre to baseline[5] (Fig. 28.2). As expected,[40] the increased airway responsiveness is associated with symptoms of asthma,[4,5] including recurrent nocturnal asthma.[30] Both the LAR[2,31] and the increased airway responsiveness[31] are associated with increases in airway inflammation. The

28 Allergens

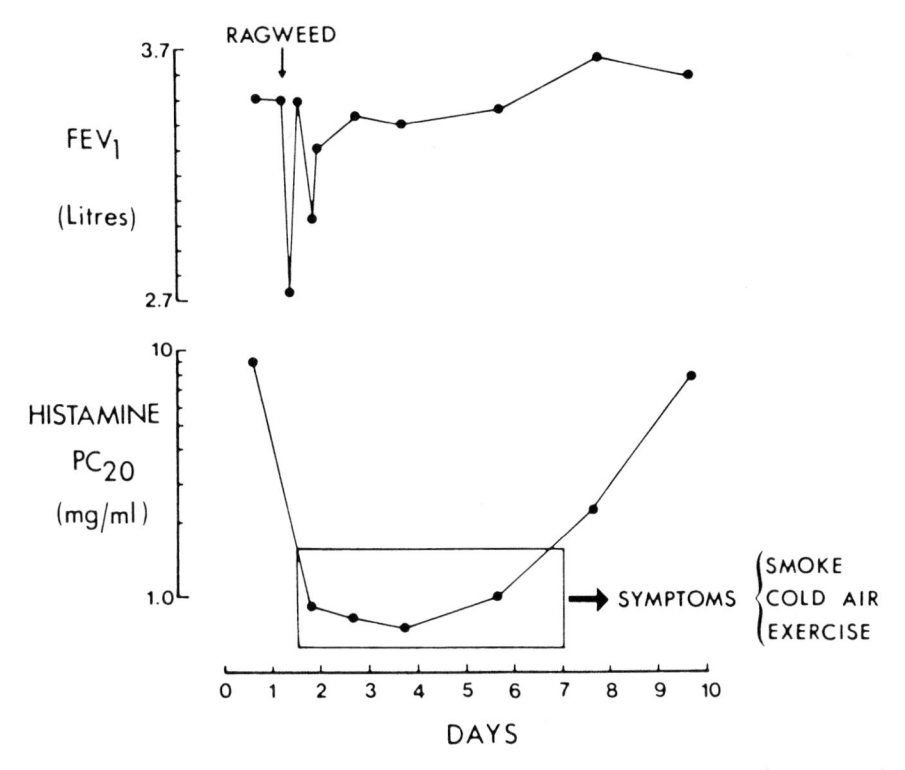

Fig. 28.2 Allergen-induced increase in non-allergic bronchial responsiveness to inhaled histamine. A dual asthmatic response, with spontaneous recovery, occurred after a single inhalation of ragweed pollen extract. Bronchial responsiveness to inhaled histamine, expressed as the provocation concentration causing a 20% fall (PC_{20}) in forced expiratory volume in 1 s (FEV_1), increased after allergen exposure and was associated with asthma symptoms on exposure to non-allergic stimuli. Reproduced from ref. 39, with permission.

occurrence of seasonal increases in airway responsiveness and airway inflammation[41] provide support for both the relevance of the bronchoprovocation model and the importance of the inflammation in the pathogenesis of the late sequelae.

Pharmacology

Pharmacological inhibition of bronchial responses to inhaled allergen has been studied for both its potential therapeutic relevance and further understanding of the pathophysiology of the responses.

Inhaled β₂-adrenergic agonists

Inhaled β₂-adrenergic agonists are the best inhibitors of the EAR[29,42–45] due to a combination of effects on smooth muscle and on mediator release.[46,47] Despite the latter, short-acting inhaled β₂-agonists (salbutamol, terbutaline, etc.) do not inhibit the LAR[29,42,43,45] or the increased airway responsiveness.[45] The long-acting inhaled β₂-

agonists salmeterol and formoterol completely inhibit or mask all aspects of the allergen-induced airway response[48,49] likely due to functional antagonism rather than anti-inflammatory effect.[49] Regular use of inhaled β_2-agonists for a week or more enhances both the EAR[50–52] and the LAR[53,54] and possibly the allergen-induced airway inflammation.[54] Administration of a larger dose of allergen after an inhaled β_2-agonist may lead to a larger LAR.[55] These features, failure to inhibit the LAR, enhanced airway responses and ability to tolerate a larger dose of allergen, may be relevant in β_2-agonist-worsened asthma control.[56]

Anticholinergic agents

Muscarinic blockers cause variable minor inhibition of the EAR[44,57–59] and likely no inhibition of the LAR.[57,59] Allergen-induced airway hyperresponsiveness appears uninfluenced by anticholinergic agents.[59] The enhanced bronchial responsiveness to histamine that develops following allergen inhalation is no more responsive to atropine than it was prior to allergen inhalation.[60]

Theophylline

Ingested theophylline offers partial protection against both the EAR and the LAR[61–64] and variable protection against the induced airway hyperresponsiveness.[62–64] It is not clear whether this is a functional antagonism or an anti-inflammatory effect.

Sodium cromoglycate (SCG)

Inhaled SCG given prior to allergen exposure inhibits both the early and late asthmatic responses,[27,29,42,45,62,65] as well as the allergen-induced increased responsiveness to both histamine[45] and methacholine.[62] Nedocromil sodium appears to have similar effects on allergen-induced asthmatic responses.[66] SCG given after the EAR will slightly delay but not inhibit the LAR.[67]

Corticosteroids

A single dose of inhaled corticosteroid, given prior to allergen, has no influence on the EAR but provides effective, often complete inhibition of the LAR.[27,29,42,45,68–71] A single dose given after the EAR will inhibit the LAR.[67] Longer treatment periods with inhaled corticosteroids will partially inhibit the EAR as well.[50,70,71] Corticosteroid-induced improvement in airway responsiveness[72,73] provides only partial explanation.[50] Reduction in mucosal mast cells[74–76] is likely more important.

Antihistamines

H_1-receptor blockers partially inhibit the early portion of the EAR.[29,42,77–79] Newer H_1-receptor blockers may also show some inhibition of the LAR;[79] further studies are necessary.

28 Allergens 513

Antiallergic drugs

Ingested antiallergic drugs such as ketotifen and repirinast have produced variable effects on allergen-induced asthma.[80–87] Most studies have failed to show any significant protection.[80,81,86,87]

Other agents

Non-steroidal anti-inflammatory agents, particularly indomethacin, appear to have no effect or perhaps enhance the EAR;[88] there is conflicting evidence regarding the LAR.[89–91] Allergen-induced increase in airway responsiveness appears to be partially inhibited by indomethacin.[91] A thromboxane synthetase inhibitor had no effect on allergen-induced early or late responses or increased airway responsiveness.[92] Interference with the leukotriene pathway with leukotriene receptor antagonist,[93] 5-lipoxygenase inhibitors[94] or 5-lipoxygenase activating protein inhibitors[95] produces some inhibition of EAR and LAR; the effect is at most modest. A platelet-activating factor (PAF) antagonist proved ineffective against allergen-induced asthma.[96] Inhaled frusemide provides inhibition of both EAR and LAR.[97] Allergen injection therapy has produced variable results in modulating the EAR[98,99] but may be particularly effective against the LAR.[100] A novel recombinant anti-IgE molecule directed against the Fc component of IgE is very effective at inhibiting both the EAR[101] and LAR.[102]

Mechanisms

The mechanisms of allergen-induced asthmatic responses have been studied out of necessity in humans by indirect means. Animal studies, *in vitro* studies on excised human tracheobronchial smooth muscle, drug-inhibition studies and, more recently, bronchoalveolar lavage have all been used to assess mechanisms.

The EAR appears to be almost purely bronchospastic in nature. Its complete inhibition and rapid reversal by β_2-agonists would support this. Histamine,[103] prostaglandin D_2[104] and likely leukotrienes[105] are released acutely from mast cells following allergen challenge. The failure of H_1-receptor blockers to completely inhibit the EAR suggests that histamine is only partially responsible. One *in vitro* study has shown that human tracheal smooth muscle contraction in response to anti-IgE can be inhibited completely by the combination of H_1-receptor blockers, cyclooxygenase inhibitors and lipoxygenase inhibitors, implying that the EAR is caused by the combined release of histamine, prostaglandins and leukotrienes.[106]

The time-course of the LAR, its steroid responsiveness and the presence of precipitins in some of the early patients who had allergic bronchopulmonary aspergillosis raised the possibility of a type III precipitin-mediated response.[27] However, lack of fever, systemic symptoms and demonstrable precipitins to most common allergens which also produced LARs[29,42] did not support a type III mechanism. The analogous late allergic cutaneous response has been shown to be IgE dependent[107,108] and anti-human IgE (anti-Fab) inhalation, producing what is believed to be a pure IgE-mediated response, has been shown to produce an LAR.[109] Thus, the LAR is part of the late sequelae of the IgE-mediated allergic reaction. This is further supported by its marked inhibition by an anti-Fc anti-IgE molecule.[102]

The failure of bronchodilators to completely reverse the LAR led early to the speculation that oedema and/or inflammation as well as bronchospasm must be involved. With the development of animal models of the LAR to allergen,[110,111] and induced airways hyperresponsiveness to both non-allergic and allergic stimuli,[112–114] it is apparent that airway inflammation with eosinophils and/or neutrophils is an important requirement for both the LAR and associated increased airway responsiveness. This has been well demonstrated in a rabbit allergic model where both the LAR and induced airway hyperresponsiveness were inhibited by polymorphonuclear leucocyte deletion.[114] Bronchoalveolar lavage and induced sputum from human subjects during allergen-induced LAR have shown that eosinophils,[2,3,31] metachromatic cells[31] and, to a lesser extent, neutrophils[3] are increased; this supports the importance of inflammation in the pathogenesis of human late responses and increased responsiveness. There are increasing data pointing to involvement of T-lymphocytes in the development of allergic inflammation.[115]

The nature of the mediators or chemotactic factors[116] responsible for the late asthmatic inflammatory sequelae is uncertain. Attention has been focused on mediators with the potential to recruit inflammatory cells, and T-cell-derived cytokines,[115] leukotrienes,[115] thromboxanes[115,117] and PAF[115,118] have all been considered potential mediators of these sequelae.

A plausible diagrammatic scheme regarding the mechanism of early and late responses is shown in Fig. 28.3. The direct immediate effects of bronchoactive mediators leads to the bronchospastic EAR. The indirect effects, likely requiring recruitment of inflammatory cells, lead to the late sequelae for which both bronchospasm and inflammation are felt to be important. The inflammation may further enhance the bronchial smooth muscle contraction by direct release of bronchoactive mediators from the recruited inflammatory cells, by indirect release of mediators from the same or other effector cells within the

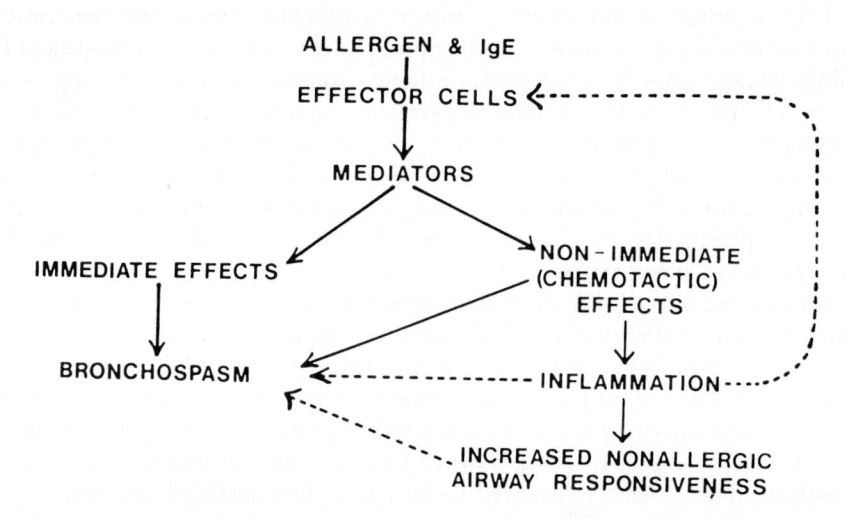

Fig. 28.3 Pathogenesis of early and late allergen-induced asthmatic sequelae. See text for explanation.

airways and, of course, by the enhanced airway responsiveness, which will increase the bronchial smooth muscle response to any stimulus.

Allergens as a cause of asthma

Increased airway responsiveness to non-allergic stimuli is a ubiquitous feature in bronchial asthma.[40] In the perennial atopic asthmatic subject, increased airway responsiveness appears to be of two types. Firstly, there is a chronic baseline level of airways hyperresponsiveness that appears to be non-reversible and non-responsive to environmental control or sodium cromoglycate and incompletely responsive to corticosteroids. Secondly, there is superimposed transient hyperresponsiveness occurring as a result of allergen exposure. This latter is responsive to environmental avoidance,[119,120] corticosteroid therapy[121] and sodium cromoglycate[34] or nedocromil.[122] This may explain the conflicting results obtained in investigations of these medications' inhibitory effects against histamine- and methacholine-induced airway responses. In subjects with only seasonal allergic asthma many,[123] if not most,[124] would appear to have only the transient seasonal increases in airway responsiveness with normal responsiveness out of season.

The clinical relevance of the allergen-induced asthmatic responses in the pathogenesis of asthma has been reviewed.[39] The late asthmatic (inflammatory) sequelae much more closely resemble, both clinically and pharmacologically, naturally occurring perennial and seasonal allergic asthma. A vicious circle hypothesis has been proposed (Fig. 28.4). Allergen exposure in sensitized individuals may lead to both early and late sequelae. The early responses may be missed either because they are mild, absent (as in occupational asthma) or mistaken for non-allergic responses, or because they are completely inhibited by regular inhaled β_2-agonists. By contrast, the more important late sequelae may be difficult to associate with the exposure because of the delayed and prolonged nature of these responses. The enhanced airway responsiveness will lead to increased responses on further exposure to the allergen, creating the vicious cycle. It will also lead to increased asthma symptoms on exposure to non-allergic triggers such as smoke, dust, exercise and cold air, these latter often being more easily identified by the patient as triggering symptoms. The potential for erroneous diagnosis as intrinsic or mixed (extrinsic/intrinsic) asthma is obvious. Therapeutic relevance is discussed below.

The role of allergen exposure as a cause of the airway hyperresponsiveness of the perennial atopic asthmatic patient has been more speculative. The relationship between atopy and asthma is well recognized.[13] Although difficult to demonstrate much relationship between atopy and airway responsiveness in clinic populations,[123,125] random population surveys in three countries (Canada, Australia, UK) have demonstrated a strong correlation between the degree of atopy and the prevalence of both airway hyperresponsiveness (Fig. 28.5) and asthma.[18,126–130] A longitudinal study has identified early-onset atopy, in particular, as a risk factor for airway hyperresponsiveness.[131] The occurrence of airway hyperresponsiveness and atopy much more often than would be expected by chance strongly suggests that allergic asthma is more than just the coincidental occurrence of the two. Although some factor, be it congenital (genetic or otherwise) or acquired (e.g. airways inflammation at critical point(s) in life), could conceivably lead to both atopy and airway hyperresponsiveness (or alternatively airway hyperresponsiveness might predispose to atopy, a concept for which there is no scientific

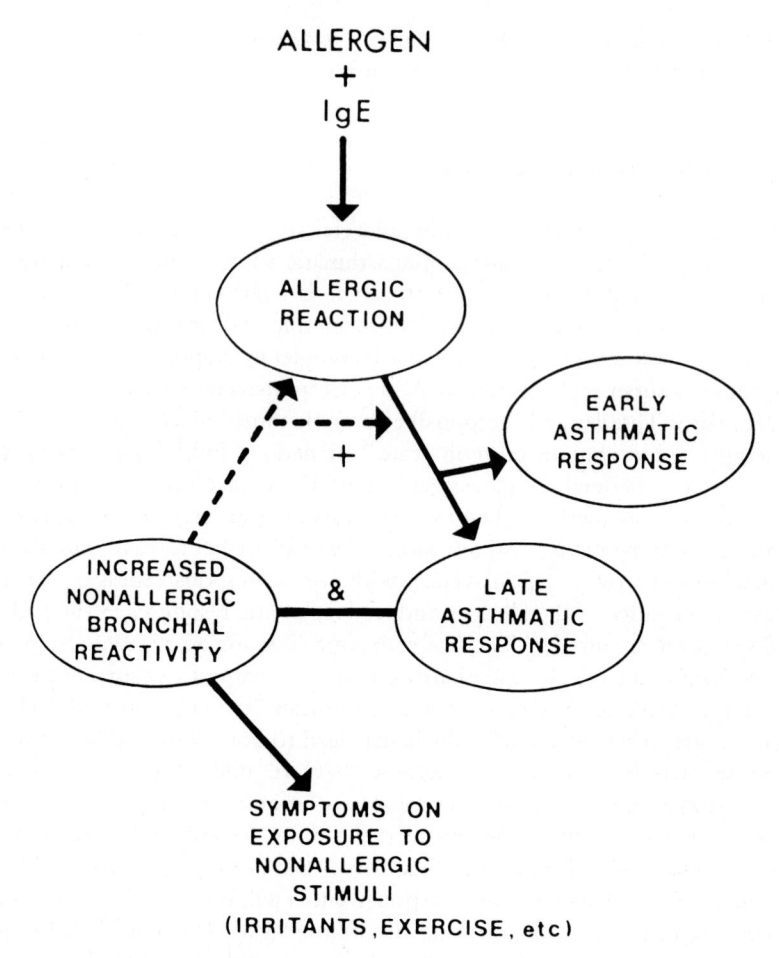

Fig. 28.4 Diagram of hypothesis explaining development and maintenance of perennial allergen-induced asthma. Reproduced from ref. 39, with permission.

validity), it is speculated that atopy (allergic airway inflammation) causes airway hyperresponsiveness. Two unique and exciting observations support this. First, a striking increase in the prevalence of both atopy and asthma amongst the natives of Papua New Guinea has been documented in longitudinal studies.[132] This appears to be secondary to civilization; with modern ways came blankets and with blankets came mites leading to atopy leading to asthma. Second, neonatal induction of IgE sensitization to ragweed in dogs followed by repeated ragweed exposure leads to the development of chronic airway hyperresponsiveness.[133]

These observations, in addition to the transient allergen-induced increases in airway responsiveness[4,5] and the apparent permanent airway hyperresponsiveness[134,135] caused by occupational exposures that also cause airway inflammation and transient increases in airway responsiveness,[6-9] make it attractive to speculate that the severity and/or duration of airways allergic reaction(s), particularly early in life, may lead to persistent hyper-

28 Allergens

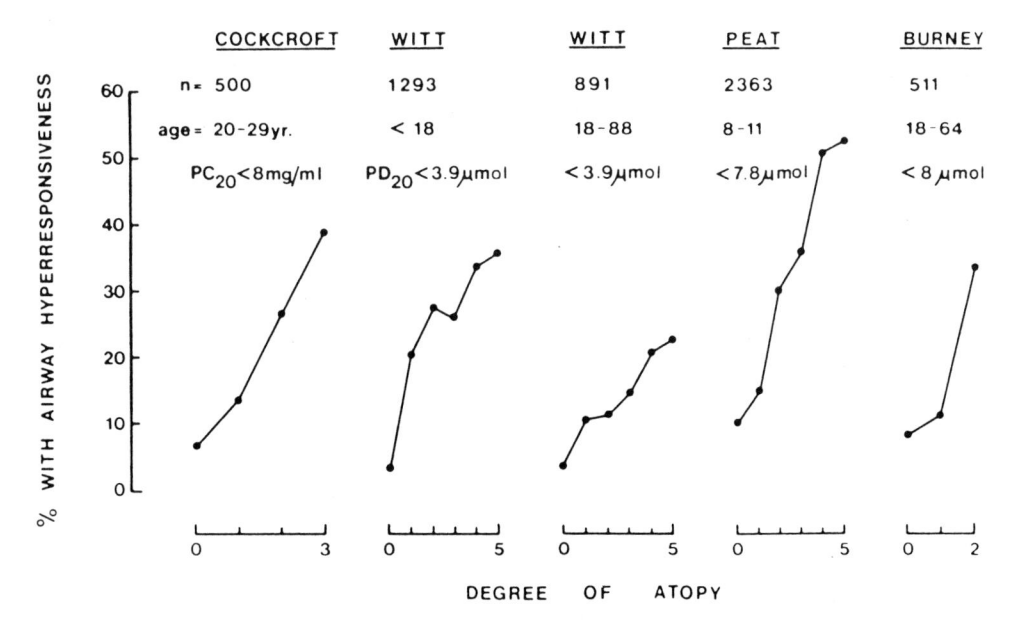

Fig. 28.5 Degree of atopy (various scales) from non-atopic (0) to highly atopic (highest number) on the horizontal axis vs. prevalence of airway hyperresponsiveness (%) on the vertical axis from four population studies.[18,127–130] Reproduced from ref. 130, with permission.

responsiveness. The 'cumulative total of airways allergic reactions' should depend both upon environmental factors and the degree of atopy (recognizing that the latter may also be dependent to some extent upon the former).

Diagnosis

The diagnosis of allergic asthma rests predominantly with the history. Historical features that must be looked for include the allergens and potential allergens to which the patient might be exposed at home, at work, at school or during recreational activities. The temporal relationship between symptoms and exposure must be examined with regard to the above-noted mechanisms (see particularly Fig. 28.4). Often low-grade sensitivity to an allergen may not be appreciated by the patient because of the lack, the mildness or the pharmacological (especially β-agonist) inhibition of early responses. In such patients, immediate hypersensitivity is often better appreciated by the relationship of eye, nose or cutaneous (following an animal scratch for example) symptoms immediately following exposure. The respiratory symptoms often lag behind and persist for several days after exposure. Seasonal variation in symptoms should give clues to sensitivity to atmospheric pollen and fungal spores; the precise seasonal variation and its relationship to various allergens will depend upon various climatic factors.

Following a good history, allergic skin testing, preferably by the prick technique, can be used to confirm suspected sensitivities.

Treatment

The treatment of allergen-induced asthma is the same as for asthma in general and is outlined in detail in the following chapters. The mechanisms outlined above, however, stress the importance of environmental control as well as the importance of the non-bronchodilator medications (sodium cromoglycate, inhaled corticosteroids) in the treatment of allergen-induced airway hyperresponsiveness, a particularly important aspect in the pathogenesis of allergic asthma. The likelihood that allergen exposure over the years can cause permanent airway hyperresponsiveness leads to the plausible speculation that early environmental control in subjects who are highly atopic or at risk for being highly atopic may have a role in the prophylaxis of asthma, in addition to its above-noted role in the treatment of asthma. Likewise, early use of anti-inflammatory therapeutic strategies might also improve the prognosis of allergic asthma by reducing either the persistence or the severity of persistent airway hyperresponsiveness.

ALLERGIC BRONCHOPULMONARY ASPERGILLOSIS

A distinctive clinical syndrome occurs when atopic individuals have organisms, against which they have IgE antibodies, growing in their airways. The prototype and by far the commonest of these 'allergic bronchopulmonary infestations' is allergic bronchopulmonary aspergillosis.[136–139] Other fungi, including *Helminthosporium* spp.,[140,141] the closely related *Curvularia* and *Drechslera* spp.,[142] *Stemphylium*,[143] *Cladosporium*,[144] *Fusarium*,[145] *Candida*[146] and possibly even bacteria such as *Pseudomonas*,[147] may cause a similar syndrome. The syndrome that these organisms can produce involves a complex immunological and mechanical pathogenesis.

Pathogenesis

The pathogenesis of allergic bronchopulmonary aspergillosis initially involves acquisition of IgE-mediated type I hypersensitivity to the fungus. Then, exposure to viable *Aspergillus* fungal spores released from decaying vegetation, particularly in the autumn and, in more temperate areas, in the winter months, will provoke a type I allergic reaction with reduced airway calibre and probably mucus hypersecretion. The spores then germinate and grow in the mucus within the lumen of the airways. The mucus will be held in place by the accompanying airway contraction, allowing the development of mucus plugs containing fungal hyphae and producing 'cast-like' outlines of the bronchial tree.

The presence in the airway of an allergen to which the subject is sensitive will lead to very high levels of both allergen-specific and total serum IgE as well as an intense peripheral and bronchial eosinophilia. The constant presence of a relatively large concentration of fungal antigen within the airways will also lead in most, but not all, individuals to the development of IgG precipitating antibodies. Likewise, other immunological responses including cell-mediated immunity may be stimulated.

The IgE–antigen reaction occurring continuously in the airways will lead to exacerbation of clinical asthma. The other immunological reactions may be responsible for bronchial and parenchymal destruction, with the development of (proximal) bronchiectasis and (upper lobe) interstitial pulmonary fibrosis. Fleeting or fixed pulmonary infiltrates may be produced either on the basis of immunological reactions within the lung or by the mechanical effect of obstruction of major bronchi by mucus plugs.

Clinical features

Allergic bronchopulmonary aspergillosis is a fairly common condition in some areas; the precise prevalence is not certain. Amongst asthmatic individuals the prevalence of type I (IgE) sensitivity to *Aspergillus* is approximately 25%;[148–150] the prevalence of type III (IgG) sensitivity to *Aspergillus* is approximately half this.[151,152] By contrast, on the dry Canadian prairies, *Aspergillus* skin sensitivity is uncommon and we have seen no new cases of allergic bronchopulmonary aspergillosis in over 20 years. Other organisms (e.g. *Helminthosporium*) are involved only rarely. The clinical picture is generally that of a subject with pre-existing atopy, and usually previous asthma, presenting with exacerbation of asthma accompanied by the expectoration of characteristic firm brown plugs.[136] Pulmonary infiltrates (with eosinophils) may be seen, occasionally but not always accompanied by febrile episodes. Chronic disease may manifest as bronchiectasis with chronic or recurrent pulmonary infection or pulmonary fibrosis with progressive dyspnoea or both.[136,137]

Diagnosis (laboratory findings)

The two features that must be present in all subjects with allergic bronchopulmonary aspergillosis are type I hypersensitivity to *Aspergillus* and the presence of *Aspergillus* in the airways. However, it is not always possible to grow the organism.[153] Specific IgG precipitating antibodies are found in about 90% of cases.[138] Other features commonly seen include intense peripheral and bronchial eosinophilia,[130] marked elevations of total serum IgE[154] and transient pulmonary infiltrates;[136–139] these all tend to correlate with activity of disease. Chronic changes in established or recurrent disease include radiographic demonstration of an unusual and essentially pathognomonic *proximal* bronchiectasis[155] and, in more severe cases, progressive upper lobe interstitial pulmonary fibrosis similar to tuberculosis and other upper lobe scarring conditions.[137]

Treatment

Allergic bronchopulmonary infestations with *Aspergillus* or other fungal organisms are not true infections; association with or progression to invasive fungal infections or mycetoma formation is rare. Treatment is thus directed against the asthma and the immunological abnormalities. This generally means intensive asthma treatment with attention paid to the administration of systemic corticosteroids in doses sufficient to suppress clinical and laboratory features of the disease.[153,156,157] Total serum IgE may be useful to predict exacerbations.[157] With such treatment, the prognosis is favourable; however, unlike other forms of allergic asthma, undertreatment can lead to substantial permanent bronchopulmonary damage.

INGESTED/INJECTED ALLERGENS

Isolated bronchial asthma induced by allergens introduced into the body via routes other than inhalation is uncommon but has been reported.[158,159] Both ingested allergens (foods)[159] and injected allergens (hyposensitization injections, insect bites and stings)[160] have the potential to produce type I-mediated hypersensitivity. Such reactions are generally manifested by one or more of the following symptoms: angioedema, urticaria, anaphylactic shock, rhinitis, conjunctivitis and bronchospasm.[158,160] Occasionally such reactions appear to be centred primarily within the lung.[159,161] It is likely that these most often represent systemic allergic reactions in asthmatic individuals who have a pre-existing high level of airway hyperresponsiveness and therefore develop disproportionately severe bronchospasm. Although significant asthmatic responses to ingested and injected allergens are much less frequent than those to inhaled allergens, when they do occur the rapidity and severity of the response may be striking. Often, a well-controlled asthmatic individual will experience sudden, at times apparently unexplained, life-threatening exacerbation of asthma that is unlikely to occur secondary to inhaled allergens. We have seen patients in whom severe sudden life-threatening episodes of asthma appear to be due to ingestion of small amounts of a food to which they were very sensitive (nuts, shellfish), and two cases in whom circumstantial evidence pointed to unrecognized insect (blackfly) bites as the cause of severe unexplained status asthmaticus.

Although uncommon, when confronted with a patient who has described sudden, otherwise unexplained, life-threatening episodes of asthma superimposed upon otherwise well-controlled asthma, severe IgE-mediated hypersensitivity to ingested foods (or drugs) or injected insect allergens (or allergen injections) must be considered. In addition, non-IgE-mediated responses to non-steroidal anti-inflammatory agents, β-adrenergic blocking drugs, cholinesterase inhibiting insecticides and certain food additives such as metabisulphites may cause sudden severe bronchospasm.

ACKNOWLEDGEMENTS

The author would like to thank Jacquie Bramley, Brenda Gore and Karen Murdock for their assistance in the preparation of this manuscript.

REFERENCES

1. Scadding JG: Definitions and clinical categories of asthma. In Clark TJH, Godfrey S (eds) *Asthma*. London, Chapman & Hall, 1983, pp 1–11.

2. de Monchy JGR, Kauffman HF, Venge P, et al.: Bronchoalveolar eosinophilia during allergen-induced late asthmatic reactions. *Am Rev Respir Dis* (1985) **131**: 373–376.
3. Metzger WJ, Richerson WB, Worden K, Monick M, Hunninghake GW: Bronchoalveolar lavage of allergic asthmatic patients following allergen bronchoprovocation. *Chest* (1986) **89**: 477–483.
4. Cockcroft DW, Ruffin RE, Dolovich J, Hargreave FE: Allergen-induced increase in nonallergic bronchial reactivity. *Clin Allergy* (1977) **7**: 503–513.
5. Cartier A, Thomson NC, Frith PA, Roberts R, Hargreave FE: Allergen-induced increase in bronchial responsiveness to histamine: relationship to the late asthmatic response and change in airway caliber. *J Allergy Clin Immunol* (1982) **70**: 170–177.
6. Lam S, Wong R, Yeung M: Nonspecific bronchial reactivity in occupational asthma. *J Allergy Clin Immunol* (1979) **63**: 28–34.
7. Lam S, LeRiche J, Phillips D, Chan-Yeung M: Cellular and protein changes in bronchial lavage fluid after late asthmatic reaction in patients with red cedar asthma. *J Allergy Clin Immunol* (1987) **80**: 44–50.
8. Mapp CE, Polato R, Maestrelli P, Hendrick DJ, Fabbri LM: Time course of the increase in airway responsiveness associated with late asthmatic reactions to toluene diisocyanate in sensitized subjects. *J Allergy Clin Immunol* (1985) **75**: 568–572.
9. Fabbri LM, Boschetto P, Zocca E, et al.: Bronchoalveolar neutrophilia during late asthmatic reactions induced by toluene diisocyanate. *Am Rev Respir Dis* (1987) **136**: 36–42.
10. Empey DW, Laitinen LA, Jacobs L, Gold WM, Nadel JA: Mechanisms of bronchial hyperreactivity in normal subjects after upper respiratory tract infection. *Am Rev Respir Dis* (1976) **113**: 131–139.
11. Brooks SM, Weiss MA, Bernstein IL: Reactive airways dysfunction syndrome (RADS). Persistent asthma syndrome after high level irritant exposures. *Chest* (1985) **88**: 376–384.
12. Gwynn CM, Smith JM, Leon GL, Stanworth DR: Role of IgG_4 subclass in childhood asthma. *Lancet* (1978) **1**: 910–911.
13. Pepys J: Atopy. In Gell PGH, Coombs RRA, Lachman PJ (eds) *Clinical Aspects of Immunology*. Oxford, Blackwell Scientific Publications, 1975, pp 877–902.
14. Sandford A, Weir T, Pare P: The genetics of asthma. *Am J Respir Crit Care Med* (1996) **153**: 1749–1765.
15. Leskowitz S, Salvaggio JE, Schwartz HJ: An hypothesis for the development of atopic allergy in man. *Clin Allergy* (1972) **2**: 237–246.
16. Woolcock AJ, Colman MH, Jones MW: Atopy and bronchial reactivity in Australian and Melanesian populations. *Clin Allergy* (1978) **8**: 155–164.
17. Brown WG, Halonen MJ, Kaltenborn WT, Barbee RA: The relationship of respiratory allergy, skin test reactivity, and serum IgE in a community population sample. *J Allergy Clin Immunol* (1979) **63**: 328–335.
18. Cockcroft DW, Murdock KY, Berscheid BA: Relationship between atopy and bronchial responsiveness to histamine in a random population. *Ann Allergy* (1984) **53**: 26–29.
19. King TP: Immunochemical properties of antigens that cause disease. In Weis EB, Stein M (eds) *Bronchial Asthma: Mechanisms and Therapies*. Boston, Little, Brown, 1993, pp 43–49.
20. Platts-Mills TAE, Solomon WR: Aerobiology and inhalant allergens. In Middleton E Jr, Reed CE, Ellis EF, Adkinson NF Jr, Yuninger JW, Busse WW (eds) *Allergy Principles and Practice*, 4th edn. St Louis, Mosby YearBook, 1993, pp 469–528.
21. Dolovich J, Zimmerman B, Hargreave FE: Allergy in asthma. In Clark TJH, Godfrey S (eds) *Asthma*. London, Chapman & Hall, 1983, pp 132–157.
22. Busse WW, Reed CE, Hoehne JH: Where is the allergic reaction in ragweed asthma? *J Allergy Clin Immunol* (1972) **50**: 289–293.
23. Pollart SM, Chapman MD, Fiocco GP, Rose G, Platts-Mills TAE: Epidemiology of acute asthma: IgE antibodies to common inhalant allergens as a risk factor for emergency room visits. *J Allergy Clin Immunol* (1989) **83**: 875–882.
24. Murray AB, Ferguson AC, Morrison B: The seasonal variation of allergic respiratory symptoms induced by house dust mites. *Ann Allergy* (1980) **45**: 347–350.
25. Robertson DG, Kerigan AT, Hargreave FE, Chalmers R, Dolovich J: Late asthmatic responses induced by ragweed pollen allergen. *J Allergy Clin Immunol* (1974) **54**: 244–254.

26. Boulet LP, Cartier A, Thomson NC, Roberts RS, Dolovich J, Hargreave FE: Asthma and increases in nonallergic bronchial responsiveness from seasonal pollen exposure. *J Allergy Clin Immunol* (1983) **71**: 399–406.
27. Pepys J: Immunopathology of allergic lung disease. *Clin Allergy* (1973) **3**: 1–22.
28. Herxheimer H: The late bronchial reaction in induced asthma. *Int Arch Allergy Appl Immunol* (1952) **3**: 323–333.
29. Booij-Noord H, deVries K, Sluiter HJ, Orie NGM: Late bronchial obstructive reaction to experimental inhalation of house dust extract. *Clin Allergy* (1972) **2**: 43–61.
30. Newman Taylor AJ, Davies RJ, Hendrick DJ, Pepys J: Recurrent nocturnal asthmatic reactions to bronchial provocation tests. *Clin Allergy* (1979) **9**: 213–219.
31. Pin I, Freitag AP, O'Byrne PM, et al.: Changes in the cellular profile of induced sputum after allergen-induced asthmatic responses. *Am Rev Respir Dis* (1992) **145**: 1265–1269.
32. Dorsch W, Baur X, Emslander HP, Fruhmann G: Zur pathogenese und therapie der allergeninduzierten verzogerten bronchialostruktion. *Prax Klin Pneumol* (1980) **34**: 461–468.
33. Altounyan REC: Changes in histamine and atropine responsiveness as a guide to diagnosis and evaluation of therapy in obstructive airways disease. In Pepys J, Franklands AW (eds) *Disodium Cromoglycate in Allergic Airways Disease*. London, Butterworth, 1970, pp 47–53.
34. Lowhagen O, Rak S: Modification of bronchial hyperreactivity after treatment with sodium cromoglycate during pollen season. *J Allergy Clin Immunol* (1985) **75**: 460–467.
35. Cockcroft DW, Murdock KY: Changes in bronchial responsiveness to histamine at intervals after allergen challenge. *Thorax* (1987) **42**: 302–308.
36. Millilo G: Discussion. In *International Conference on Bronchial Hyperreactivity*. Oxford, The Medicine Publishing Foundation, 1982, p 17.
37. Durham SR, Graneek BJ, Hawkins R, Newman Taylor AJ: The temporal relationship between increases in airway responsiveness to histamine and late asthmatic responses induced by occupational agents. *J Allergy Clin Immunol* (1987) **79**: 398–406.
38. Thorpe J, Steinberg D, Bernstein D, Bernstein IL, Murlas C: Bronchial hyperreactivity occurs soon after the immediate asthmatic response in dual responders. *Am Rev Respir Dis* (1986) **133**: A93.
39. Cockcroft DW: Mechanism of perennial allergic asthma. *Lancet* (1983) **ii**: 253–256.
40. Hargreave FE, Ryan G, Thomson NC, et al.: Bronchial responsiveness to histamine or methacholine in asthma: measurement and clinical significance. *J Allergy Clin Immunol* (1981) **68**: 347–355.
41. Djukanovic R, Feather I, Gratziou C, et al.: Effect of natural allergen exposure during the grass pollen season on airways inflammatory cells and asthma symptoms. *Thorax* (1996) **51**: 575–581.
42. Orie NGM, Van Lookeren Campagne JG, Knol K, Booij-Noord H, de Vries K: Late reactions in bronchial asthma. In Pepys J, Yamamura I (eds) *Intal in Bronchial Asthma*. Proceedings of the 8th International Congress of Allergology, Tokyo, 1974, pp 17–29.
43. Hegardt B, Pauwels R, Van Der Straeten M: Inhibitory effect of KWD 2131, terbutaline, and DSCG on the immediate and late allergen-induced bronchoconstriction. *Allergy* (1981) **36**: 115–122.
44. Ruffin RE, Cockcroft DW, Hargreave FE: A comparison of the protective effect of Sch1000 and fenoterol on allergen-induced asthma. *J Allergy Clin Immunol* (1978) **61**: 42–47.
45. Cockcroft DW, Murdock KY: Comparative effects of inhaled salbutamol, sodium cromoglycate and beclomethasone dipropionate on allergen-induced early asthmatic response, late asthmatic responses and increased bronchial responsiveness to histamine. *J Allergy Clin Immunol* (1987) **79**: 734–740.
46. Church MK, Young KD: The characteristics of inhibition of histamine release from human lung fragments by sodium cromoglycate, salbutamol, and chlorpromazine. *Br J Pharmacol* (1983) **78**: 671–679.
47. Howarth PH, Durham SR, Lee TH, Kay B, Church MK, Holgate ST: Influence on albuterol, cromolyn sodium, and ipratropium bromide on the airway and circulating mediator responses to allergen bronchial provocation in asthma. *Am Rev Respir Dis* (1985) **132**: 986–992.
48. Twentyman OP, Finnerty JP, Harris A, Palmer J, Holgate ST: Protection against allergen-induced asthma by salmeterol. *Lancet* (1990) **336**: 1338–1342.

49. Wong BJ, Dolovich J, Ramsdale EH, et al.: Formoterol compared with beclomethasone and placebo on allergen-induced asthmatic responses. *Am Rev Respir Dis* (1992) **146**: 1158–1160.
50. Cockcroft DW, McParland CP, Britto SA, Swystun VA, Rutherford BC: Regular inhaled salbutamol and airway responsiveness to allergen. *Lancet* (1993) **342**: 833–837.
51. Cockcroft DW, Swystun VA, Bhagat R: Interaction of inhaled β_2 agonist and inhaled corticosteroid on airway responsiveness to allergen and methacholine. *Am J Respir Crit Care Med* (1995) **152**: 1485–1489.
52. Bhagat R, Swystun VA, Cockcroft DW: Salbutamol-induced increased airway responsiveness to allergen and reduced protection vs. methacholine: dose–response. *J Allergy Clin Immunol* (1996) **97**: 47–52.
53. Cockcroft DW, O'Byrne PM, Swystun VA, Bhagat R: Regular use of inhaled albuterol and the allergen-induced late asthmatic response. *J Allergy Clin Immunol* (1995) **96**: 44–49.
54. Gauvreau GM, Watson RM, Jordana M, Cockcroft D, O'Byrne PM: The effect of regular inhaled salbutamol on allergen-induced airway responses and inflammatory cells in blood and induced sputum. *Am J Respir Crit Care Med* (1995) **151**: A39.
55. Lai CKW, Twentyman OP, Holgate ST: The effect of an increase in inhaled allergen dose after rimiterol hydrobromide on the occurrence and magnitude of the late asthmatic response and the associated change in nonspecific bronchial responsiveness. *Am Rev Respir Dis* (1989) **140**: 917–923.
56. van Schayck CP, Cloosterman SGM, Hofland ID, van Herwaarden CLA, van Weel C: How detrimental is chronic use of bronchodilators in asthma and chronic obstructive pulmonary disease? *Am J Respir Crit Care Med* (1995) **151**: 1317–1319.
57. Yu DYC, Galant SP, Gold WM: Inhibition of antigen-induced bronchoconstriction by atropine in asthmatic patients. *J Appl Physiol* (1972) **32**: 823–828.
58. Orehek J, Gayrard P, Grimaud Ch, Charpin J: Bronchoconstriction provoquee par inhalation d'allergene dans l'asthme: Effet antagoniste d'un anticholinergique de synthese. *Bull Eur Physiopathol Respir* (1975) **11**: 193–201.
59. Cockcroft DW, Ruffin RE, Hargreave FE: Effect of Sch 1000 in allergen-induced asthma. *Clin Allergy* (1978) **8**: 361–372.
60. Boulet LP, Latimer KM, Roberts RS, et al.: The effect of atropine on allergen-induced increases in bronchial responsiveness to histamine. *Am Rev Respir Dis* (1984) **130**: 368–372.
61. Pauwels R, van Renterghem D, Van Der Straeten M, Johannesson N, Persson GA: The effect of theophylline and enprofylline on allergen-induced bronchoconstriction. *J Allergy Clin Immunol* (1985) **76**: 583–590.
62. Cockcroft DW, Murdock KY, Gore BP, O'Byrne PM, Manning P: Theophylline does not inhibit allergen-induced increase in airway responsiveness to methacholine. *J Allergy Clin Immunol* (1989) **83**: 913–920.
63. Crescioli S, Spinazzi A, Plebani M, et al.: Theophylline inhibits early and late asthmatic reactions induced by allergens in asthmatic subjects. *Ann Allergy* (1991) **66**: 245–251.
64. Hendeles L, Harman E, Huang D, O'Brien R, Blake K, Delafuente J: Theophylline attenuation of airway responses to allergen: comparison with cromolyn metered-dose inhaler. *J Allergy Clin Immunol* (1995) **95**: 505–514.
65. Pepys J, Chan M, Hargreave FE, McCarthy DS: Inhibitory effects of disodium cromoglycate on allergen-inhalation tests. *Lancet* (1968) **ii**: 134–137.
66. Dahl R, Pedersen B: Influence of nedocromil sodium on the dual asthmatic reaction after allergen challenge: a double-blind, placebo-controlled study. *Eur J Respir Dis* (1986) **69** (Suppl 147): 263–265.
67. Cockcroft DW, McParland CP, O'Byrne PM, et al.: Beclomethasone given after the early asthmatic response inhibits the late response and the increased methacholine responsiveness and cromolyn does not. *J Allergy Clin Immunol* (1993) **91**: 1163–1168.
68. Booij-Noord H, Orie NGM, deVries K: Immediate and late bronchial obstructive reactions to inhalation of house dust and protective effects of disodium cromoglycate and prednisolone. *J Allergy Clin Immunol* (1971) **48**: 344–354.
69. Pepys J, Davies RJ, Breslin ABX, Hendricks DJ, Hutchcroft BJ: The effects of inhaled

beclomethasone dipropionate (Becotide) and sodium cromoglycate on asthmatic reactions to provocation tests. *Clin Allergy* (1974) **4**: 13–24.
70. Van Der Star JG, Berg WC, Steenhuis EJ, de Vries K: Invloed van beclometason-dipropionaat per aerosol op de obstructieve reactie in de bronchien na huisstofinhalatie. *Ned T Geneesk* (1976) **120**: 1928–1932.
71. Burge PS, Efthimiou J, Turner-Warwick M, Nelmes PTJ: Double-blind trials of inhaled beclomethasone dipropionate and fluocortin butyl ester in allergen-induced immediate and late reactions. *Clin Allergy* (1982) **12**: 523–531.
72. Du Toit JI, Salome CM, Woolcock AJ: Inhaled corticosteroids reduce the severity of bronchial hyperresponsiveness in asthma but oral theophylline does not. *Am Rev Respir Dis* (1987) **136**: 1174–1178.
73. Woolcock AJ, Yan K, Salome CM: Effect of therapy on bronchial hyperresponsiveness in the long-term management of asthma. *Clin Allergy* (1988) **18**: 165–176.
74. Laitinen LA, Laitinen A, Haahtela T: A comparative study of the effects of an inhaled corticosteroid, budesonide, and of a β_2-agonist, terbutaline, on airway inflammation in newly diagnosed asthma. *J Allergy Clin Immunol* (1992) **90**: 32–42.
75. Djukanovic R, Wilson JW, Britton YM, *et al.*: Effect of an inhaled corticosteroid on airway inflammation and symptoms of asthma. *Am Rev Respir Dis* (1992) **145**: 669–674.
76. Trigg CJ, Manolitsas ND, Wang J, *et al.*: Placebo-controlled immunopathologic study of four months of inhaled corticosteroids in asthma. *Am J Respir Crit Care Med* (1994) **150**: 17–22.
77. Holgate ST, Emanuel MB, Howarth PH: Astemizole and other H_1-antihistaminic drug treatment of asthma. *J Allergy Clin Immunol* (1985) **76**: 375–380.
78. Rafferty P, Beasley R, Holgate S: The contribution of histamine to immediate bronchoconstriction provoked by inhaled allergen and adenosine 5' monophosphate in atopic asthma. *Am Rev Respir Dis* (1987) **136**: 369–373.
79. Hamid M, Rafferty P, Holgate ST: The inhibitory effect of terfenadine and flurbiprofen on early and late-phase bronchoconstriction following allergen challenge in atopic asthma. *Clin Exp Allergy* (1990) **20**: 261–267.
80. Wells A, Taylor B: A placebo-controlled trial of ketotifen (HC20-511, Sandoz) in allergen induced asthma and comparison with disodium cromoglycate. *Clin Allergy* (1979) **9**: 237–240.
81. Pelikan Z, Pelikan M: Early and late asthmatic response to allergen challenge and their pharmacologic modulation. *Ann Allergy* (1985) **55**: 318.
82. Pauwels R, Lamont H, Van Der Straeten M: Comparison between ketotifen and DSCG in bronchial challenge. *Clin Allergy* (1978) **8**: 289–293.
83. Craps L, Greenwood C, Radielovic P: Clinical investigation of agents with prophylactic anti-allergic effects in bronchial asthma. *Clin Allergy* (1978) **8**: 373–382.
84. Klein G, Urbanek R, Matthys H: Long-term study of the protective effect of ketotifen in children with allergic bronchial asthma. The value of a provocation test in assessment of treatment. *Respiration* (1981) **41**: 128–132.
85. Adachi M, Kobayashi H, Aoki N, *et al.*: A comparison of the inhibitory effects of ketotifen and disodium cromoglycate on bronchial responses to house dust, with special reference to the late asthmatic response. *Pharmatherapeutics* (1984) **4**: 36–42.
86. Cockcroft DW, Keshmiri M, Murdock KY, Gore BP: Allergen-induced increase in airway responsiveness is not inhibited by acute treatment with ketotifen or clemastine. *Ann Allergy* (1992) **68**: 245–250.
87. Patel PC, Rutherford BC, Lux J, Cockcroft DW: The effect of repirinast on airway responsiveness to methacholine and allergen. *J Allergy Clin Immunol* (1992) **90**: 782–788.
88. Fish JE, Ankin MG, Adkinson NF Jr, Peterman VI: Indomethacin modification of immediate-type immunologic airway responses in allergic asthmatic and nonasthmatic subjects. *Am Rev Respir Dis* (1981) **123**: 609–614.
89. Nakazawa T, Toyoda T, Furukawa M, Taya T, Kobayashi S: Inhibitory effects of various drugs on dual asthmatic responses in wheat flour-sensitive subjects. *J Allergy Clin Immunol* (1976) **58**: 1–9.
90. Fairfax AJ: Inhibition of the late asthmatic response to house dust mite by non-steroidal anti-inflammatory drugs. *Prostaglandins Leukotrienes Med* (1982) **8**: 239–248.

91. Kirby JG, Hargreave FE, Cockcroft DW, O'Byrne PM: Indomethacin inhibits allergen-induced airway hyperresponsiveness but not allergen-induced asthmatic responses. *J Appl Physiol* (1989) **66**: 578–583.
92. Manning PJ, Stevens WH, Cockcroft DW, O'Byrne PM: The role of thromboxane in allergen-induced asthmatic responses. *Eur J Respir Dis* (1991) **4**: 667–672.
93. Taylor IK, O'Shaughnessy KM, Fuller RW, Dollerty CT: Effect of cysteinyl-leukotriene receptor antagonist ICI 204.219 on allergen-induced bronchoconstriction and airway hyperreactivity in atopic subjects. *Lancet* (1991) **337**: 690–694.
94. Hui KP, Taylor IK, Taylor GW, et al.: Effect of a 5-lipoxygenase inhibitor on leukotriene generation and airway responses after allergen challenge in asthmatic patients. *Thorax* (1991) **46**: 184–189.
95. Diamant Z, Timmers MC, van der Veen H, et al.: The effect of MK-0591, a novel 5-lipoxygenase activating protein inhibitor, on leukotriene biosynthesis and allergen-induced airway response in asthmatic subjects *in vivo*. *J Allergy Clin Immunol* (1995) **95**: 42–51.
96. Wilkens H, Wilkens JH, Bosse S, et al.: Effects of an inhaled PAF-antagonist (WEB 2086 BS) on allergen-induced early and late asthmatic responses and increased bronchial responsiveness to methacholine. *Am Rev Respir Dis* (1991) **143**: A812.
97. Bianco S, Pieroni MG, Refini RM, et al.: Protective effect of inhaled furosemide on allergen-induced early and late asthmatic reactions. *N Engl J Med* (1989) **321**: 1069–1073.
98. Ortolani C, Pastorello E, Moss RB, et al.: Grass pollen immunotherapy: a single year double-blind, placebo-controlled study in patients with grass pollen-induced asthma and rhinitis. *J Allergy Clin Immunol* (1984) **73**: 283–290.
99. Van Metre TE, Marsh DG, Adkinson NF, et al.: Immunotherapy for cat asthma. *J Allergy Clin Immunol* (1988) **82**: 1055–1068.
100. Warner JD, Soothill JF, Price JF, Hey EN: Controlled trial of hyposensitization to *Dermatophagoides pteronyssinus* in children with asthma. *Lancet* (1978) **ii**: 912–915.
101. Boulet LP, Chapman KR, Coté J, et al.: Inhibitory effects of an anti-IgE antibody E25 on allergen-induced early asthmatic response. *Am J Respir Crit Care Med* (1997) **155**: 1835–1840.
102. Fahy JV, Fleming HE, Wong HH, et al.: The effect of anti-IgE monoclonal antibody on the early- and late-phase responses to allergen inhalation in asthmatic subjects. *Am J Respir Crit Care Med* (1997) **155**: 1828–1834.
103. Lee TH, Brown MJ, Nagy L, Causon R, Walport HJ, Kay AB: Exercise-induced release of histamine and neutrophil chemotactic factor in atopic asthmatics. *J Allergy Clin Immunol* (1982) **70**: 73–81.
104. Murray JJ, Tonnel AB, Brash AR, et al.: Release of prostaglandin D_2 into human airways during acute antigen challenge. *N Engl J Med* (1986) **315**: 800–804.
105. Lewis RA, Austen KF: The biologically active leukotrienes: biosynthesis, metabolism, receptors, functions and pharmacology. *J Clin Invest* (1984) **73**: 889–897.
106. Schellenberg RR, Duff MJ, Foster A: Human bronchial responses to anti-IgE *in vitro*. *Clin Invest Med* (1985) **8**: A41.
107. Dolovich J, Hargreave FE, Chalmers R, Shier KJ, Gauldie J, Bienenstock J: Late cutaneous allergic responses in isolated IgE-dependent reactions. *J Allergy Clin Immunol* (1973) **52**: 38–46.
108. Solley GO, Gleich GJ, Jordon RE, Schroeter AL: The late phase of the immediate wheal and flare skin reactions. *J Clin Invest* (1976) **58**: 408–420.
109. Kirby JG, Robertson DG, Hargreave FE, Dolovich J: Asthmatic responses to inhalation of anti-human IgE. *Clin Allergy* (1986) **16**: 191–194.
110. Schampain MP, Behrens BL, Larsen GL, Henson PM: An animal model of late pulmonary responses to *Alternaria* challenge. *Am Rev Respir Dis* (1982) **126**: 493–498.
111. Abraham WM, Delehunt JC, Yerger L, Marchette B: Characterization of a late phase pulmonary response after antigen challenge in allergic sheep. *Am Rev Respir Dis* (1983) **128**: 839–844.
112. Chung KF, Becker AB, Lazarus SC, Frick OL, Nadel JA, Gold WM: Antigen-induced hyperresponsiveness and pulmonary inflammation in allergic dogs. *J Appl Physiol* (1985) **58**: 1347–1353.

113. O'Byrne PM, Walters EH, Gold ED, et al.: Neutrophil depletion inhibits airway hyperresponsiveness induced by ozone exposure. Am Rev Respir Dis (1984) **130**: 214–219.
114. Murphy KR, Wilson MC, Irvin CG, et al.: The requirement for polymorphonuclear leukocytes in the late asthmatic response and heightened airways reactivity in an animal model. Am Rev Respir Dis (1986) **134**: 62–68.
115. Beer DJ, Rocklin RE: Immunoregulation: role of macrophages and T-cells. In Weis EB, Stein M (eds) Bronchial Asthma: Mechanisms and Therapies. Boston, Little, Brown, 1993, pp 147–164.
116. Metzger WJ, Richerson HB, Wasserman SI: Generation and partial characterization of eosinophil chemotactic activity and neutrophil chemotactic activity during early and late phase asthmatic response. J Allergy Clin Immunol (1986) **78**: 282–290.
117. Chung KF, Aizawa H, Becker AB, Frick O, Gold WM, Nadel JA: Inhibition of antigen-induced airway hyperresponsiveness by a thromboxane synthetase inhibitor (OKY-046) in allergic dogs. Am Rev Respir Dis (1986) **134**: 258–261.
118. Basran GS, Page CP, Paul W, Morley J: Platelet-activating factor: a possible mediator of the dual response to allergen? Clin Allergy (1984) **14**: 75–79.
119. Murray AB, Ferguson AC, Morrison B: The seasonal variation of allergic respiratory symptoms induced by house dust mites. Ann Allergy (1980) **45**: 347–350.
120. Platts-Mills TAE, Mitchell EB, Nock P, Tovey ER, Moszoro H, Wilkins SR: Reduction of bronchial hyperreactivity during prolonged allergen avoidance. Lancet (1982) **ii**: 675–678.
121. Sotomayor H, Badier M, Vervloet D, Orehek J: Seasonal increase of carbachol airway responsiveness in patients allergic to grass pollen. Am Rev Respir Dis (1984) **130**: 56–58.
122. Dorward AJ, Roberts JA, Thomson NC: Effect of nedocromil sodium on histamine airway responsiveness in grass-pollen sensitive asthmatics during the pollen season. Clin Allergy (1986) **16**: 309–315.
123. Cockcroft DW, Killian DN, Mellon JJA, Hargreave FE: Bronchial reactivity to inhaled histamine: a method and clinical survey. Clin Allergy (1977) **7**: 235–243.
124. Cockcroft DW, Berscheid BA, Murdock KY, Gore BP: Sensitivity and specificity of histamine PC_{20} measurements in a random selection of young college students. J Allergy Clin Immunol (1992) **89**: 23–30.
125. Bryant DH, Burns MW: The relationship between bronchial histamine reactivity and atopic status. Clin Allergy (1976) **6**: 373–381.
126. Cookson WOCM, Musk AW, Ryan G: Association between asthma history, atopy, and non-specific bronchial responsiveness in young adults. Clin Allergy (1984) **16**: 425–432.
127. Witt C, Stuckey MS, Woolcock AJ, Dawkins RC: Positive allergy prick skin tests associated with bronchial histamine responsiveness in an unselected population. J Allergy Clin Immunol (1986) **77**: 698–702.
128. Peat JK, Britton WJ, Salome CM, Woolcock AJ: Bronchial hyperresponsiveness in two populations of Australian school children: III. Effect of exposure to environmental allergens. Clin Allergy (1987) **17**: 291–300.
129. Burney PFJ, Britton JR, Chinn S, et al.: Descriptive epidemiology of bronchial reactivity in an adult population: results from a community study. Thorax (1987) **42**: 38–44.
130. Cockcroft DW, Hargreave FE: Relationship between atopy and airway responsiveness. In Sluiter HJ, van der Lende R (eds) Bronchitis IV. Royal Vangorcum, Assen, The Netherlands, 1988, pp 23–32.
131. Peat JK, Salome CM, Woolcock AJ: Longitudinal changes in atopy during a 4-year period: relation to bronchial hyperresponsiveness and respiratory symptoms in a population sample of Australian schoolchildren. J Allergy Clin Immunol (1990) **85**: 65–74.
132. Dowse GK, Turner KJ, Stewart GA, Alpers MP, Woolcock AJ: The association between Dermatophagoides mites and the increasing prevalence of asthma in village communities within the Papua New Guinea highlands. J Allergy Clin Immunol (1985) **75**: 75–83.
133. Becker AB, Hershkovich J, Simons FER, Simons KJ, Lilley MK, Kepron MW: Development of chronic airway hyperresponsiveness in ragweed sensitized dogs. J Appl Physiol (1989) **66**: 2691–2697.
134. Chan-Yeung M, Lam S, Koener S: Clinical features and natural history of occupational asthma due to Western Red Cedar (Thuja plicata). Am J Med (1982) **72**: 411–415.

135. Mapp CE, Corona PC, De Marzo N, Fabbri L: Persistent asthma due to isocyanates: a follow-up study of subjects with occupational asthma due to toluene diisocyanate (TDI). *Am Rev Respir Dis* (1988) **137**: 1326–1329.
136. Malo JL, Hawkins R, Pepys J: Studies in chronic allergic bronchopulmonary aspergillosis 1: clinical and physiological findings. *Thorax* (1977) **32**: 254–261.
137. Malo JL, Papys J, Simon G: Studies in chronic allergic bronchopulmonary aspergillosis 2: radiological findings. *Thorax* (1977) **32**: 262–268.
138. Malo JL, Longbottom J, Mitchell J, Hawkins R, Pepys J: Studies in chronic allergic bronchopulmonary aspergillosis 3: immunological findings. *Thorax* (1977) **32**: 269–274.
139. Malo JL, Inouye T, Hawkins R, Simon G, Turner-Warwick M, Pepys J: Studies in chronic allergic bronchopulmonary aspergillosis 4: comparison with a group of asthmatics. *Thorax* (1977) **32**: 275–280.
140. Dolan CT, Weed LA, Dines DE: Bronchopulmonary helminthosporiosis. *Am J Clin Pathol* (1970) **53**: 235–242.
141. Matthiesson AM: Allergic bronchopulmonary disease caused by fungi other than *Aspergillus*. *Thorax* (1981) **36**: 719.
142. McAleer R, Kroenert DB, Elder JL, Froudist JH: Allergic bronchopulmonary disease caused by *Curvularia lunata* and *Drechslera hawaiiensis*. *Thorax* (1981) **36**: 338–344.
143. Benatar SR, Allan B, Hewitson RP, Don PA: Allergic bronchopulmonary stemphyliosis. *Thorax* (1980) **35**: 515–518.
144. Moreno-Ancillo A, Diaz-Pena J-M, Ferrer A, *et al.*: Allergic bronchopulmonary cladosporiosis in a child. *J Allergy Clin Immunol* (1996) **97**: 714–715.
145. Backman KS, Roberts M, Patterson R: Allergic bronchopulmonary mycosis caused by *Fusarium vasinfectum*. *Am J Respir Crit Care Med* (1995) **152**: 1379–1381.
146. Voisin C, Tonnel AB, Jacob M, Thermol P, Malin P, Lahoutte C: Infiltrats pulmonaires avec grande eosinophilie sanguine associes a une candidose bronchique. *Rev Fr Allergie Immunol Clin* (1976) **16**: 279–281.
147. Gordon DS, Hunter RG, O'Reilly RJ, Conway BP: *Pseudomonas aeruginosa* allergy and humoral antibody-mediated hypersensitivity pneumonia. *Am Rev Respir Dis* (1973) **108**: 127–131.
148. Longbottom JL, Pepys J: Pulmonary aspergillosis: diagnostic and immunologic significance of antigens and C-substance in *Aspergillus fumigatus*. *J Pathol Bacteriol* (1964) **88**: 141–151.
149. Hendrick DJ, Davies RJ, D'Souza MF, Pepys J: An analysis of prick skin test reactions in 656 asthmatic patients. *Thorax* (1975) **30**: 2–8.
150. Malo JL, Paquin R: Incidence of immediate sensitivity to *Aspergillus fumigatus* in a North American asthma population. *Clin Allergy* (1979) **9**: 377–384.
151. Hoehne JH, Reed CE, Dickie HA: Allergic bronchopulmonary aspergillosis is not rare. *Chest* (1973) **63**: 177–181.
152. Malo JL, Paquin R, Longbottom JL: Prevalence of precipitating antibodies to different extracts of *Aspergillus fumigatus* in a North American asthmatic population. *Clin Allergy* (1981) **11**: 333–341.
153. McCarthy DS, Pepys J: Allergic bronchopulmonary aspergillosis. *Clin Allergy* (1971) **1**: 261–286.
154. Patterson R, Fink JN, Pruzansky JJ, *et al.*: Serum immunoglobulin levels in pulmonary allergic aspergillosis and certain other lung disease, with special reference to immunoglobulin E. *Am J Med* (1973) **54**: 16–22.
155. Scadding JG: The bronchi in allergic bronchopulmonary aspergillosis. *Scand J Respir Dis* (1967) **48**: 372–377.
156. Safirstein BH, D'Souza MF, Simon G, Tai EHC, Pepys J: Five year follow-up of allergic bronchopulmonary aspergillosis. *Am Rev Respir Dis* (1973) **108**: 450–459.
157. Wang JLF, Patterson R, Roberts M, Ghory AC: The management of allergic bronchopulmonary aspergillosis. *Am Rev Respir Dis* (1979) **120**: 87–92.
158. Metcalfe DD: The diagnosis of food allergy: Theory and practice. In Spector SL (ed) *Provocation Challenge Procedures: Bronchial, Oral, Nasal, and Exercise*. Boca Raton, FL, CRC Press, 1983, pp 119–132.
159. Bock SA, Lee W-Y, Remigio LK, May CD. Studies of hypersensitivity reactions to foods in infants and children. *J Allergy Clin Immunol* (1978) **62**: 327–334.

160. Orange RP, Donsky GJ: Anaphylaxis. In Middleton E Jr, Reed CE, Ellis EF (eds) *Allergy Principles and Practice*. St Louis, CV Mosby, 1978, pp 563–573.
161. Gluck JC, Pacin MP: Asthma from mosquito bites: a case report. *Ann Allergy* (1986) **56**: 492–493.

29

Occupational Asthma

A.J. NEWMAN TAYLOR

INTRODUCTION: INITIATORS AND PROVOKERS OF ASTHMA

Occupational asthma is asthma induced by an agent inhaled at work. Agents encountered at work that initiate asthma should be distinguished from agents that provoke asthma. *Initiators* are able to induce asthma and cause airway inflammation and airway hyperresponsiveness. *Provokers* of asthma incite acute transient airway narrowing in individuals with hyperresponsive airways but do not initiate asthma, cause airway inflammation or increase airway responsiveness.

Initiators cause airway inflammation and induce asthma either by causing toxic damage to the airway epithelium ('irritant'-induced asthma) or as the outcome of an acquired specific hypersensitivity response ('hypersensitivity'-induced asthma). Irritant inducers include respiratory irritants such as chlorine and sulphur dioxide inhaled in toxic concentrations. Viral respiratory tract infections may initiate asthma by a similar mechanism. Hypersensitivity inducers include inhaled proteins (such as animal excreta, flour and enzymes), other complex biological molecules (such as wood resin acids) and low molecular weight chemicals (such as isocyanates and acid anhydrides) that bind covalently to body proteins to form haptens.

Provokers of acute airway narrowing may be physical, such as exercise and cold air inhalation, chemical such as sulphur dioxide or pharmacological such as histamine and methacholine. Avoiding exposure to an initiator can reduce airway responsiveness and the severity of asthma; avoiding a provoker will reduce the frequency of provoked attacks, but not the severity of asthma or airway hyperresponsiveness.

Both initiators and provokers of asthma may be encountered at work. Cold air in storage rooms and outdoors, exertion and irritant chemicals may provoke asthma in individuals with hyperresponsive airways; the effects of different provokers on the airways are probably additive. Irritant-induced asthma has also been described as 'reactive airways dysfunction syndrome' (RADS). Although less common than hypersensitivity-induced occupational asthma, several cases have been reported of patients without previous respiratory symptoms who, within hours of exposure to toxic chemicals in high concentration, develop respiratory symptoms and airway hyperresponsiveness that usually resolves spontaneously but can persist. The majority of reported cases of asthma induced by an agent inhaled at work fulfil the criteria of an acquired hypersensitivity response and the title of occupational asthma is often reserved for cases of asthma induced by sensitization.

IRRITANT-INDUCED ASTHMA

Irritant-induced asthma (RADS) is persistent asthma and airway hyperresponsiveness that develops after acute inhalation of an irritant chemical in toxic concentrations.[1] The development of respiratory symptoms and presence of airway hyperresponsiveness within hours of exposure to the cause distinguishes irritant- from hypersensitivity-induced asthma, where the disease develops only after a latent interval, usually of months or years, from initial exposure. The diagnostic criteria proposed for irritant-induced asthma are shown in Table 29.1.

The frequency of irritant-induced asthma is not known but an estimate can be made from the voluntary reporting scheme for occupational lung disease made by chest and occupational physicians in United Kingdom—Surveillance of Work and Occupational Respiratory Disease (SWORD). Of 623 cases of inhalation accidents reported in 3.5 years between January 1990 and July 1993, 11 (1.75%) had developed asthma.[2] The initiating agents in these cases were predominantly well-recognized respiratory irritants such as chlorine, oxides of nitrogen, sulphur dioxide and ammonia.

The majority of cases of irritant-induced asthma have been described in case reports and series, which are necessarily highly selected, and without objective measurement of lung function made before the inhalation accident. However, one study of hospital employees exposed to a spill of 100% acetic acid in a hospital laboratory studied all those exposed.[3] An exposure–response relationship was found between estimated intensity of exposure to the acid and the prevalence of asthmatic symptoms and airway hyperrespon-

Table 29.1 Criteria for the diagnosis of irritant-induced asthma.

1. Absence of preceding respiratory complaints is documented
2. The onset of symptoms occurred after a single specific exposure incident or accident
3. The exposure was to a gas, smoke, fume or vapour that was present in very high concentrations and had irritant qualities
4. The onset of symptoms occurred within 24 h after the exposure and persisted for at least 3 months
5. Symptoms were consistent with asthma, with cough, wheezing and dyspnoea predominating
6. Pulmonary function tests may show airflow obstruction
7. Appropriate challenge testing demonstrates increasing airway responsiveness
8. Other types of pulmonary disease are excluded

siveness: the risk of developing asthma was 10 times higher for those most highly exposed compared with those less exposed to the spill. The pathological changes observed in the airways of patients with irritant-induced asthma include bronchial epithelial cell injury with desquamation, marked subepithelial thickening with fibrosis, and infiltration with plasma cells and lymphocytes but not eosinophils.

CAUSES OF HYPERSENSITIVITY-INDUCED OCCUPATIONAL ASTHMA

Many different agents encountered at work can stimulate a hypersensitivity response and cause asthma. Some of the more important are shown in Table 29.2. Proteins and other complex molecules of biological origin may be encountered in a wide variety of circumstances. These include agriculture, the storage and transport of crops, food production, forestry and carpentry, the use of laboratory animals and the commercial exploitation of microbes as sources of food, antibiotics and enzymes. Pinewood resin (colophony), widely used in the electronics industry as a soft solder flux, fumes at the temperature of soldering.

Synthetic chemicals that cause asthma when inhaled are fewer in number but exposure to them occurs in a wide variety of occupations. Isocyanates are probably the most important. They are bifunctional and trifunctional molecules used commercially to polymerize polyhydroxyl and polyglycol compounds to form polyurethanes. Isocyanates also react with water to form carbon dioxide, a reaction used in the production of flexible polyurethane foams. The polyurethane reaction is exothermic and the heat generated is sufficient to evaporate isocyanates with high vapour pressures, such as toluene diisocyanate (TDI) and hexamethylene diisocyanate (HDI). Diphenyl methane diisocyanate (MDI) and naphthalene dissocyanate (NDI), whose vapour pressures are lower, evaporate when heat is applied. Spraying two-part polyurethane paints generates airborne isocyanates in high concentration both as vapour and droplets. Polyurethanes are widely

Table 29.2 Causes of hypersensitivity-induced occupational asthma.

	Proteins	Low-molecular-weight chemicals
Animal	Excreta of rats, mice, etc., locusts, grain mites	
Vegetable	Grain/flour Green coffee bean Ispaghula	Plicatic acid (western red cedar) Colophony (pinewood resin)
Microbial	Harvest moulds *Bacillus subtilis* enzyme	Antibiotics, e.g. penicillins, cephalosporins
'Minerals'		Acid anhydrides Isocyanates Complex platinum salts Polyamines Reactive dyes

used and exposure to isocyanates occurs in many situations, e.g. the manufacture of rigid and flexible polyurethane foams, in inks and laminating adhesives in flexible packaging and the use of two-part polyurethane varnishes and paints, particularly when sprayed (e.g. on cars and aircraft). Acid anhydrides such as phthalic anhydride (PA) are used as curing agents for epoxy and alkyd resins and in the manufacture of the plasticizer dioctyl phthalate. Complex platinum salts are essential intermediates in platinum refining. Reactive dyes are increasingly used to bind a colour (chromophore) covalently to textiles: asthma occurs among both the manufacturers and users.

IMPORTANCE OF HYPERSENSITIVITY-INDUCED OCCUPATIONAL ASTHMA

The contribution of occupational causes to the prevalence of asthma in the community is not known. Estimates in different countries have varied between 2 and 15% but their basis has generally not been secure. In a recent population-based study of adults aged between 20 and 44 years in Spain, Kojevinas et al.[4] estimated the risk of asthma attributable to occupational exposure (after adjusting for age, sex, residence and smoking) as between 5% (1 in 20) and 6.7% (1 in 15). The highest risks occurred in laboratory technicians, spray painters and bakers.

The SWORD voluntary reporting scheme in the UK, which started in January 1989, has provided valuable information about the relative importance of the different causes of occupational asthma and its incidence in different occupations. During the three years 1989–1991, 5576 new cases of occupational lung disease were reported to SWORD, of which 28% were occupational asthma (the single most common category). The causes most frequently identified were isocyanates, flour, laboratory animals, colophony and wood dust. Disease incidence rates, using denominators from the Labour Force Survey, showed a very high risk of asthma among paint sprayers, laboratory staff, plastics and metal treatment workers and in welding and electronic assembly.[5]

The rates reported by SWORD are considerably lower than those reported in Finland, one of the few countries where occupational diseases are registered. The incidence in Finland in 1981 was 71 per million compared with 22 per million in the UK in 1989. Much of this difference is probably due to lower rates of ascertainment and reporting in the UK, where the true incidence has been estimated to be some three times that reported.

OCCUPATIONAL ASTHMA AND HYPERSENSITIVITY

Occupational asthma fulfils the criteria for an acquired specific hypersensitivity response.

(1) It occurs in only a proportion, usually a minority, of those exposed to its cause.
(2) It develops only after an initial symptom-free period of exposure, which is usually weeks or months but can be years.
(3) In those who develop asthma, airway responses (both reduction in calibre and in non-specific responsiveness) are provoked by inhalation of the specific agent in concen-

29 Occupational Asthma

trations that were previously tolerable and which do not provoke similar responses in others equally exposed (i.e. in concentrations not toxic to mucosal surfaces).

These characteristics have stimulated a search for evidence of a specific immunological response to the causes of occupational asthma, both proteins and low molecular weight chemicals. Until recently, most attention has been directed towards the identification of specific IgE and IgG antibodies. In general when demonstrated, IgE and IgG4 have been found in exposed populations to be associated with disease and total specific IgG with exposure. Specific IgE was associated with asthma and IgG with exposure in those working with laboratory animals[6] and specific IgE and IgG4 with asthma and IgG with exposure to acid anhydrides.[7,8] Few studies have investigated the participation of T-lymphocytes in occupational asthma. In one study, T-lymphocytes in biopsies from nine patients with isocyanate-induced asthma, in common with patients with both extrinsic and intrinsic asthma, showed the characteristics of Th2 lymphocytes, with evidence of mRNA production of interleukin (IL)-4 and IL-5 but not interferon γ.[9]

Specific IgE antibody, inferred from an immediate skin test response to a water-soluble extract of the specific protein or hapten–protein conjugate or its identification in serum by radioallergosorbent test (RAST), has been identified in patients with occupational asthma caused by inhaled proteins of animal, vegetable or microbial origin. These include the excreta and secreta of laboratory animals, small mammals[10] and locusts,[11] wheat and rye flow[12] and proteolytic enzymes.[13] Specific IgE has also been identified in the sera of patients with asthma caused by some low molecular weight chemicals, particularly acid anhydrides[7,14,15] and reactive dyes.[16] A study to examine the determinants of allergenicity of low molecular weight chemicals compared the properties of two β-lactam antibiotics: clavulanic acid, which is not allergenic, and a carbapeneam, MM2283, which can cause asthma and stimulate IgE antibody production in humans.[17] The characteristics relevant to allergenicity were (a) reactivity with body proteins, (b) homogeneity with respect to the chemical hapten and (c) stability of the conjugate formed.

Specific IgE antibody has been identified in only some 15% of cases of isocyanate-induced asthma. This may reflect the difficulties of working with reactive chemicals in *in vitro* systems or failure to prepare the relevant *in vivo* chemical–protein conjugate for the *in vitro* test. Reactants of the isocyanate–water reaction are likely to form in the water-saturated respiratory tract and may bind to tissue proteins and form immunogens. Failure to find convincing evidence of a specific immunological response in cases of isocyanate-induced asthma has led to suggestions that it may be the outcome of a pharmacological rather than an immunological mechanism. In support of this, TDI was found to inhibit the *in vitro* stimulation of adenylyl cyclase by isoprenaline in a dose-dependent fashion,[18] possibly by covalent binding of the isocyanate group to the membrane receptor, and to provoke asthma by β-adrenoceptor inhibition in those with pre-existing airway hyperresponsiveness. However, this fails to explain the well-documented latent interval between exposure to TDI and the development of asthma, and the failure of TDI to provoke asthma in patients with asthma and airway hyperresponsiveness from other causes.[19] Furthermore, inhalation of TDI induces an increase in non-specific airway responsiveness in sensitized individuals without pre-test airway hyperresponsiveness;[20] TDI also fails to inhibit isoprenaline-induced tracheal smooth muscle relaxation.[21] The more recent findings of Th2 lymphocytes in biopsies from cases of isocyanate-induced asthma is also consistent with a specific immunological response.[9]

The development of molecular biological techniques and their application in identifying specific mRNA in T-lymphocytes has provided a powerful tool to investigate further the immunological basis of these low molecular weight chemicals where evidence of associated IgE antibody, for whatever reason, is not obtainable.

DETERMINANTS OF HYPERSENSITIVITY-INDUCED OCCUPATIONAL ASTHMA

Four separate factors have been reported to contribute to the development of occupational asthma in populations exposed to its causes: intensity of exposure, atopy, tobacco smoking and HLA phenotype.

Exposure

Although the most directly amenable to control, exposure has received the least attention. In part this has been due to the difficulty in measuring aeroallergen concentration. The development of inhibition immunoassays has now allowed measurement and several recently reported studies have found evidence for a relationship between measured intensity of exposure and the prevalence of sensitization and asthma. Cullinan et al.[22] found an exposure–response relationship between airborne rat urine protein concentration and the prevalence of both skin test reactions to rat urine protein and respiratory symptoms. Juniper et al.[23] studied a cohort of enzyme detergent workers and found the incidence of skin-prick test responses to alcalase was greatest in those most heavily exposed. Coutts et al.[24] found the prevalence of work-related nasal and lower respiratory symptoms increased with increasing frequency of exposure during the working week. The prevalence of work-related respiratory symptoms and airway hyperresponsiveness in bakery workers was greater in those who had ever worked in dustier conditions[25] and Burge et al.[26] found a gradient of work-related respiratory symptoms in relation to measured concentration of airborne colophony.

Atopy

Atopy, defined in immunological terms as those who readily produce IgE antibodies on contact with environmental allergens encountered in everyday life, is commonly identified as the presence of one or more immediate skin-prick test responses to common inhalant allergens (which in the UK would include grass pollen, *Dermatophagoides pteronyssinus* and cat fur). The prevalence in workforces of atopy, defined in this way, has been consistently reported as between one-quarter and one-third. Asthma and IgE antibody induced by several causes of occupational asthma have been reported to occur more commonly among atopic individuals. This association is best described for asthma caused by laboratory animals, *Bacillus subtilis* enzymes and complex platinum salts. Several studies have shown asthma to be some four to five times more prevalent in atopic than non-atopic laboratory animal workers.[27,28] In their cohort study of enzyme detergent workers, Juniper et al.[23] found the incidence of a skin test response to alcalase was greater among atopic individuals at each level of exposure. Similarly, Cullinan et al.[22] found atopy increased the risk of sensitization to rat urine proteins at each level of exposure. Dally et al.[29] found an increased incidence of skin-prick test responses to ammonium hexachloroplatinate in atopic individuals in a platinum refinery workforce. However, a subsequent study of the same population found smoking to be a more important risk factor.[30] However, for several causes of occupational asthma, such as isocyanates and plicatic acid, atopic individuals seem at no greater risk of developing asthma than non-atopic individuals.

Tobacco smoking

Tobacco smoking has been reported to increase the risk of developing asthma and specific IgE antibody to several different causes of occupational asthma. Specific IgE antibody or an immediate skin test response has been found some four to five times more frequently in smokers than non-smokers exposed to tetrachlorophthalic anhydride (TCPA),[31] green coffee bean and ispaghula[32] and ammonium hexachloroplatinate.[30] The risk of developing asthma is also increased, although less than for specific IgE. All seven cases of TCPA-induced asthma reported by Howe et al.[7] were cigarette smokers; the risk of asthma in platinum refinery workers and snow-crab processing workers[33] was increased some two-fold. Smoking also interacted with intensity of exposure to increase the risk of sensitization to complex platinum salts in platinum refinery workers.[34] The greatest risk occurred in smokers in high-exposure jobs; no cases occurred in non-smokers in low-exposure jobs. The risk was similar and intermediate in non-smokers in high-exposure and smokers in low-exposure jobs.

The mechanism of this 'adjuvant' effect of tobacco smoking is unknown, but may be related to injury, whatever its cause, to the respiratory mucosa concurrently with inhalation of novel antigens. Inhaled tobacco smoke potentiated the IgE response to inhaled but not subcutaneous ovalbumin.[35] Other respiratory irritants can exert a similar effect. The proportion of cynomolgus monkeys who developed asthma and a positive skin test after inhalation of complex platinum salts was increased in the animals who inhaled ozone concurrently.[36] Similarly, the frequency of IgE antibody production and airway responses provoked by inhaled ovalbumin were increased in a dose-dependent fashion in guinea-pigs that inhaled sulphur dioxide concurrently with the sensitizing dose of ovalbumin.[37]

HLA phenotype

Recently the association of specific IgE antibody and asthma caused by agents inhaled at work and HLA class II alleles has been investigated. Young et al.,[38] in a case-referent study of acid anhydride workers, found a significant excess of HLA-DR3 in cases with specific IgE to trimellitic anhydride (8 of 11 cases vs. 2 of 14 referents; odds ratio 8.14) but not in cases with specific IgE to phthalic anhydride (2 of 11 cases vs. 2 of 14 referents). Bignon et al.[39] investigated HLA class II alleles in cases of isocyanate-induced asthma. They found that the allele DQB1*0503 and the allelic combination

DQB1*0201/0301 were increased and the allele DQB1*0501 and the DQA1*0101 DQB1*0501 DR1 haplotype were significantly reduced in cases of isocyanate-induced asthma compared with unaffected isocyanate-exposed workers. These initial observations need replication, but do suggest that genetic susceptibility, at least for inhaled low molecular weight chemical haptens, is an important determinant of sensitization and asthma.

DIAGNOSIS OF HYPERSENSITIVITY-INDUCED OCCUPATIONAL ASTHMA

Accurate and early diagnosis of cases of occupational asthma is important. Remission of respiratory symptoms and restoration of normal lung function, including non-specific airway responsiveness, can follow avoidance of exposure to the specific initiating cause. Furthermore, the evidence available suggests that non-specific airway hyperresponsiveness and respiratory symptoms are more likely to persist in those who continue to be exposed to the initiating cause after asthma develops. Avoidance of exposure frequently requires a change of work that, particularly in the present economic climate, can lead to loss of employment. Accurate diagnosis is also essential if those whose asthma is not occupationally caused are to avoid being advised to change or leave their work unnecessarily.

The diagnosis of occupational asthma requires:

(1) differentiation of asthma from other causes of respiratory symptoms, in particular chronic airflow limitation and on occasions hyperventilation;
(2) differentiation of occupational from non-occupational asthma;
(3) differentiation of asthma whose primary cause is an agent inhaled at work from asthma not primarily caused by an agent inhaled at work but aggravated by a non-specific irritant factor, such as sulphur dioxide and cold air, to which exposure occurs in the workplace.

Work-related asthma is usually suggested by the history. Asthma induced by an agent inhaled at work usually occurs in an individual exposed to an agent recognized to cause occupational asthma. It develops only after an initial symptom-free period during which time the patient has been exposed without symptoms to the concentrations that now provoke asthma. Characteristically, symptoms progress in severity during the working week and improve during periods away from work and at weekends or during holidays. The patient may also be aware of others who have developed similar respiratory symptoms at his/her place of work.

Non-specific stimuli provoke asthmatic reactions that occur within minutes of exposure to an irritant and usually resolve within 1–2 h of avoidance of exposure. The onset of asthma may have preceded initial exposure to the irritant and asthma does not significantly improve when away from work. Non-specific irritants such as organic solvents, which may have a characteristic and unpleasant small, may also provoke a hyperventilation response; breathing difficulties are associated with tingling of the fingers, headaches and dizziness.

INVESTIGATION OF HYPERSENSITIVITY-INDUCED OCCUPATIONAL ASTHMA

Provocation of an asthmatic response, both airway narrowing and increased non-specific airway responsiveness, by inhalation of a specific agent provides evidence in the individual of a cause and effect relationship. However, inhalation testing in all suspected cases of occupational asthma is impracticable and unjustifiable. In the majority of cases a confident diagnosis can be made from knowledge of exposure to a recognized cause of occupational asthma and the characteristic history, supported by the results of serial peak expiratory flow (PEF) measurements, immunological tests or both. Inhalation testing should be reserved for occasions when these investigations do not provide an adequate basis for advice about future employment.

Serial PEF measurements

Asthma can be attributed with confidence to an agent inhaled at work where exposure to it in the workplace reproducibly provokes increasingly severe airway narrowing. Repeated measurements of airway calibre, most conveniently made as PEF, need to be made during a period long enough to allow observation of the consistency of any changes and their relationship to periods at work. This requires measurements to be made repeatedly during each day for a period of several weeks, most commonly 4 weeks. This can only be done by subjects making and recording their own results. Such self-recording of PEF measurements is now widely used. Patients are lent a peak flow meter and asked to record the best of three measurements of PEF made every 2 h from waking to sleeping over a period of 1 month in the first instance. To allow sufficient time for lung function to recover from exposure to an agent at work, it is helpful if the 1 month includes a longer period away from work than a weekend, preferably a 1 or 2 week holiday. Self-recording requires patient compliance and honesty. The measurements may be conveniently summarized to show the maximum, minimum and mean PEF measurements for each day (Fig. 29.1); differences between periods at work and periods away from work can be observed. This method of patient investigation has proved, in the hands of those experienced in its use, to be reliable and a relatively sensitive and specific index of occupational asthma. Graneek[40] compared the results of several PEF recordings with inhalation testing as the 'gold standard'. Patients who did not show evidence of asthma on PEF records (i.e. <20% within-day variability) did not have an asthmatic reaction provoked by inhalation tests (unless they had not been exposed to the cause of their asthma during the period of peak flow recording). Patients with evidence of work-related asthma on PEF records had an asthmatic reaction provoked by inhalation testing with the relevant agent. The major diagnostic difficulties were patients with evidence of asthma on serial PEF records without a work relationship, of whom a proportion subsequently had an asthmatic reaction provoked by an inhalation test. The most common cause for this false-negative PEF test was insufficient time away from work to allow significant improvement in PEF to occur.

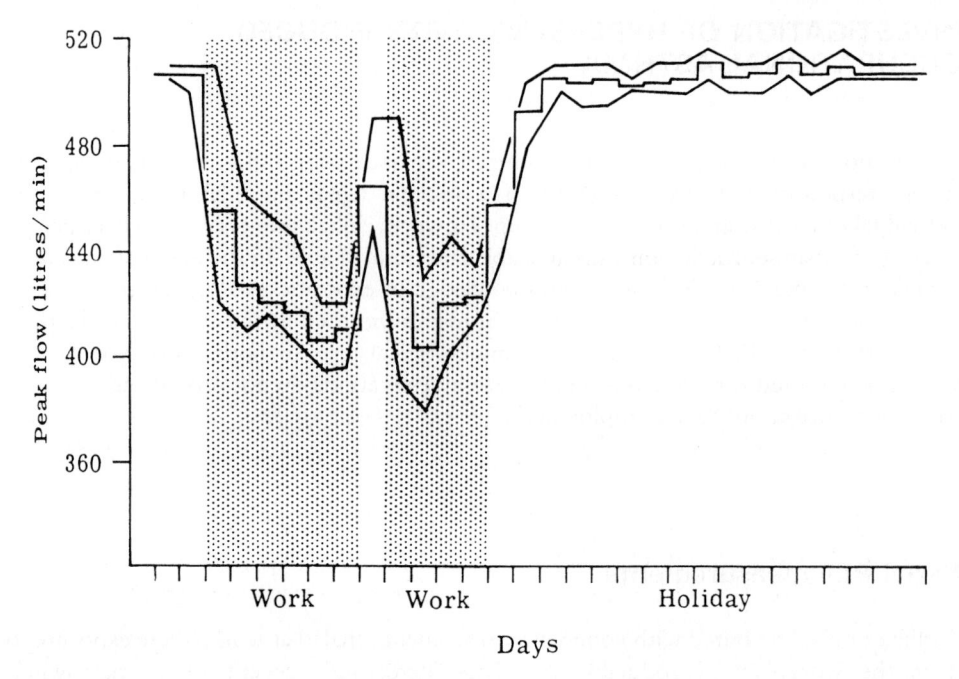

Fig. 29.1 Serial peak flow measurements showing pattern of work-related asthma. The best, worst and average peak flow results are plotted for each day. The shaded areas are periods at work, the unshaded areas periods away from work. Peak flow deteriorates at work and partially improves during 1-day absence from work. Complete improvement occurs within first 3 days of holiday and continues throughout this period.

Immunological investigations

The application of immunological tests in the investigation of occupational asthma has widened because of:

(1) identification of the nature and source of relevant allergens (e.g. the identification of urine and saliva of laboratory animals as a major source of allergenic protein), allowing the preparation of immunologically relevant test extracts;
(2) preparation of hapten–protein conjugates suitable for immunological testing (e.g. acid anhydride–human serum albumin conjugates and reactive dye–human serum albumin conjugates);
(3) development of reliable methods for identification of specific IgE antibody in serum.

Extracts of several of the causes of occupational asthma can be used to elicit skin test reactions and to identify specific IgE antibody in serum. These include the *B. subtilis* enzyme alcalase, urine and salivary proteins of laboratory animals, excreta of locusts, wheat and rye flour proteins, harvest moulds (including *Alternaria tenuis* and *Cladosporium herbarum*) and grain mites (such as *Acaris siro* and *Leptidoglyphus destructor*). In addition, hapten–protein conjugates, suitable for skin testing and identification of specific IgE antibody, have been prepared for acid anhydrides and reactive dyes. Complex platinum

29 Occupational Asthma

salts such as ammonium hexochloroplatinate can elicit immediate skin-prick test responses without the need for conjugation to human serum albumin.

The value of such tests in the diagnosis of occupational asthma depends upon their sensitivity and specificity in populations exposed to the particular cause. Extracts of urine protein obtained from rats and mice have been shown in several studies to be a sensitive and relatively specific index of asthma, but not of rhinitis, conjunctivitis or urticaria. Similarly, an immediate skin-prick test response to specific IgE antibody, identified by RAST, to human serum albumin conjugates of the acid anhydrides PA, TMA and TCPA are more sensitive than specific indices of occupational asthma caused by these agents.

Inhalation tests

There are four major indications for inhalation testing in the diagnosis of occupational asthma.

(1) Where the agent thought to be responsible for causing asthma has not previously been demonstrated to do so.
(2) Where an individual with occupational asthma is exposed at work to more than one potential cause.
(3) Where asthma is of such severity that uncontrolled exposure in the work environment is not justifiable, therefore eliminating the possibility of serial PEF measurements.
(4) Where the diagnosis of occupational asthma remains in doubt after other investigations, including serial PEF and immunological tests where appropriate, have been completed.

Inhalation tests undertaken solely for legal purposes are not justifiable.

The aim in occupational-type inhalation tests is to expose individuals under single-blind conditions to the possible cause of their asthma in circumstances that resemble as closely as possible the conditions of their exposure at work. Wherever possible, atmospheric concentrations of the inhaled agent should be based on knowledge of the concentrations experienced at work, and the physical conditions of exposure, e.g. size of dust particles, whether vapour or aerosol and the temperatures to which the materials are heated, should be similar to those at work.

The different methods used in inhalation testing depend primarily on the physical state of the test material. Soluble allergens, such as urine proteins of laboratory animals, are inhaled as nebulized extracts in solution. Volatile organic liquids such as TDI may be painted on to a flat surface in increasing concentrations on different days. The atmospheric concentration of vapour can be measured with an appropriate monitor. Exposure to dusts, such as antibiotics, complex platinum salts and acid anhydrides, is made by tipping the test material, usually diluted in dried lactose, between two trays. The atmospheric concentration achieved is surprisingly reproducible and can be measured by use of a personal dust sampler.

Estimation of airway responses provoked by inhalation tests should now include measurements of changes in both airway calibre and, ideally, non-specific airway responsiveness. Changes in airway calibre are most conveniently determined by regular measurements of forced expiratory volume in 1 s (FEV_1) and forced vital capacity

(FVC) or PEF before and for at least 24 h after the test. Changes in airway responsiveness can be made by estimating the concentration of inhaled histamine that provokes a 20% fall in FEV_1 (PC_{20}) before the test and at 3 h and 24 h after the test. The changes in airway calibre and non-specific responsiveness observed are compared with those following a control challenge test, each test being made on a separate day (Fig. 29.2).

The patterns of change in airway calibre provoked by inhalation testing are distinguished by their time of onset and duration. Immediate responses occur within minutes and resolve spontaneously within 1–2 h. These may be provoked by both allergic (e.g. grass pollen) and non-allergic (e.g. inhaled histamine or sulphur dioxide) stimuli. The response depends upon the concentration of the provoking agent and the degree of pre-existing non-specific airway responsiveness. Such immediate responses are not

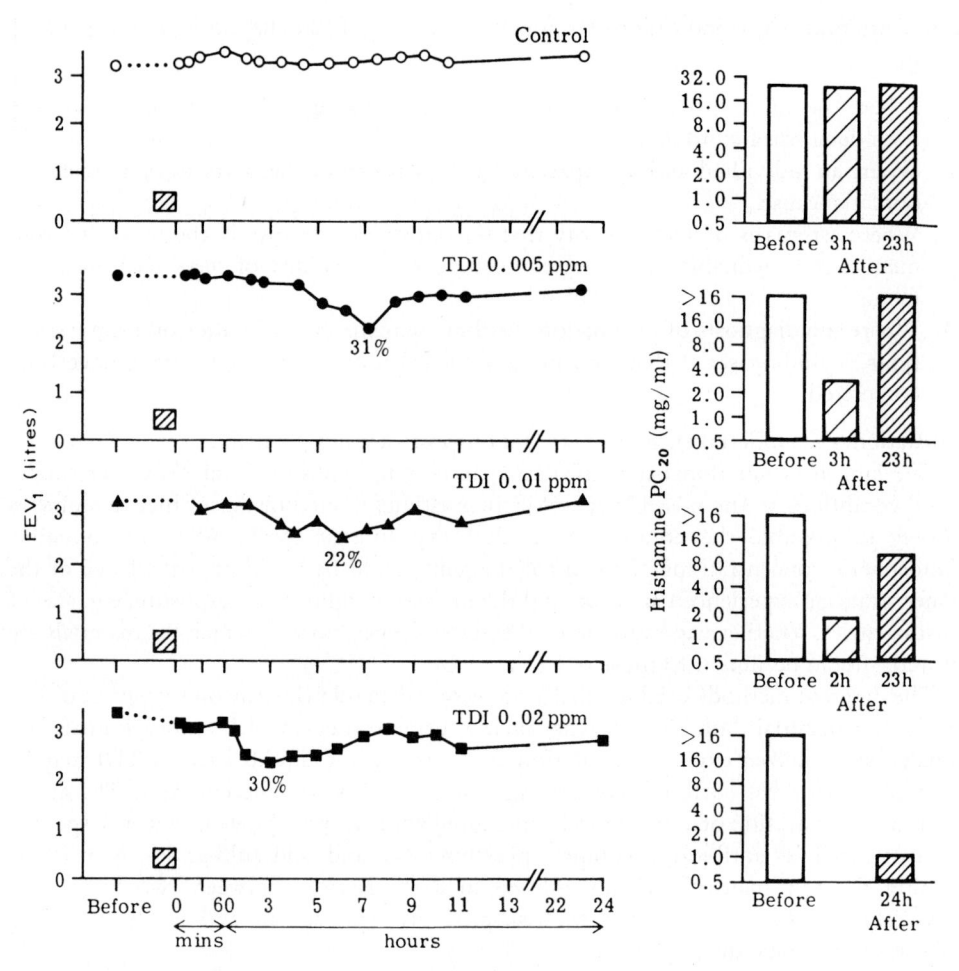

Fig. 29.2 Changes in airway calibre (forced expiratory volume in 1 s, FEV_1) and airway responsiveness (histamine PC_{20}) following inhalation test with control and increasing concentrations of toluene diisocyanate (TDI). FEV_1 and PC_{20} are stable during control day. Inhalation of TDI provokes a non-immediate asthmatic response with an increase in airway responsiveness at 3 h but not 24 h at the lowest exposure concentration and at 24 h at the higher concentrations.

associated with an increase in non-specific airway responsiveness. Late responses develop 1 h or more after the inhalation test, usually after some 3–4 h and may persist for 24–36 h. Unlike the immediate response, late responses are associated with an increase in non-specific responsiveness, which can be identified 3 h after the test prior to the onset of the late asthmatic response and, less reliably, at 24 h after the test (Fig. 29.2).

A dual response is an immediate response followed by a late response. Recurrent nocturnal responses may be provoked by a single inhalation test exposure, with asthmatic responses occurring during several successive nights with partial or complete remission during the intervening days. These responses are almost certainly the manifestation of an induced increase in non-specific airway responsiveness.[40]

The question to be answered in undertaking inhalation testing is whether the specific agent inhaled at work has induced asthma in the particular individual. The most satisfactory way to answer this question is to determine whether or not inhalation of the specific agent increases non-specific airway responsiveness, in which case the particular agent can be regarded as an inducing cause in that particular individual. As an increase in non-specific airway responsiveness is closely associated with the development of a late asthmatic response, provocation of a late asthmatic response can reasonably be used as a surrogate for an increase in non-specific airway responsiveness.[41] Non-specific irritants may provoke immediate responses in individuals with hyperresponsive airways but in subtoxic concentrations do not provoke either an increase in non-specific airway responsiveness or a late asthmatic reaction.

OUTCOME OF HYPERSENSITIVITY-INDUCED OCCUPATIONAL ASTHMA

Asthma induced by an agent inhaled at work may become chronic, persisting for several years, if not indefinitely, after avoidance of exposure to its initiating cause. This seems particularly, although not exclusively, to occur in individuals with asthma caused by low molecular weight chemicals, among whom asthma can persist in more than half. Venables et al.[42] followed up six cases of asthma caused by the acid anhydride TCPA. Despite 4 years' avoidance of exposure, five had respiratory symptoms consistent with persistent airway hyperresponsiveness and histamine PC_{20} was increased in the five in whom it was measured. The rate of decline of specific IgE to a TCPA–human serum albumin conjugate during the period of avoidance of exposure was exponential, with a $t_{\frac{1}{2}}$ of 1 year, and parallel in all six subjects, making it very improbable their continuing asthma was caused by further inadvertent exposure.

Malo et al.[43] investigated snow-crab workers with occupational asthma, diagnosed by inhalation tests, up to 5 years from their last exposure. All denied further exposure to crabmeat by either inhalation or ingestion. Respiratory symptoms persisted in all 31, of whom 26 had a measurable methacholine PC_{20}. Although FEV_1, FEV_1/FVC and PC_{20} improved during the initial period of avoidance of exposure, FEV_1 and FEV_1/FVC plateaued by 1 year and PC_{20} by 2 years.

Continuing asthma in these patients seems likely to be a manifestation of chronic airway inflammation which, although initiated by the agent inhaled at work, persists in its absence. Paggiaro et al.[44] investigated 10 patients with TDI-induced asthma with respiratory symptoms and airway hyperresponsiveness 4–40 months from their last exposure. Bronchial biopsies were obtained from eight and showed basement membrane thickening with infiltration of the mucosa by eosinophils, lymphocytes and neutrophils; in four patients in whom airway responsiveness was unchanged the proportion of eosinophils in fluid recovered at bronchoalveolar lavage was increased, whereas this was the case in only one of five whose airway responsiveness had improved.

The only important reported determinant of chronicity in occupational asthma is duration of exposure to the initiating cause after the onset of respiratory symptoms. Chan Yeung et al.[45] found that 60% of 136 cases of asthma caused by western red cedar (*Thuja plicata*) continued to have asthma when examined on average after 4 years' avoidance of exposure. In those with continuing asthma, the interval from onset of symptoms to diagnosis was on average 2.5 years longer than in those whose symptoms had resolved.

MANAGEMENT OF OCCUPATIONAL ASTHMA

Patients who develop occupational asthma in whom the specific cause is identified should be advised to avoid further exposure to the cause of their asthma. This seems particularly important where low molecular weight chemicals, such as isocyanates,

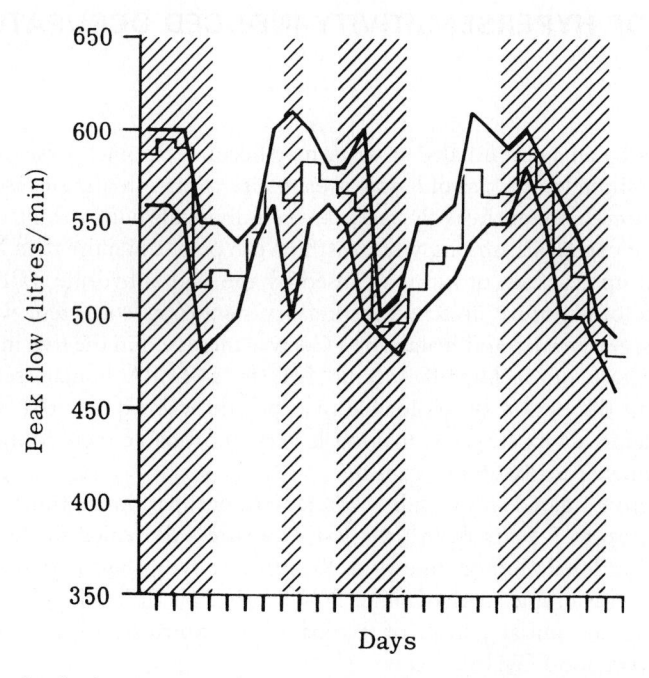

Fig. 29.3 Work-related asthma in an isocyanate worker after relocation. Serial peak flow records demonstrate that despite avoidance of exposure to isocyanate (diphenyl methane diisocyanate, MDI) at work he has consistent deteriorations in peak flow measurement while at work and is likely to have continuing inadvertent exposure to MDI.

29 Occupational Asthma

plicatic acid or acid anhydrides, are responsible as these seem particularly likely to cause continuing airway hyperresponsiveness and asthma after complete avoidance of exposure. Information from follow-up cases of asthma due to western red cedar suggest that the risk of this occurring is least for those who avoid exposure within a short time of the onset of asthma.

Such advice may require individuals to change or leave their job which, for social or financial reasons, may not be possible. This can be a particular problem for highly trained individuals, such as experimental scientists, whose livelihood depends on their knowledge and experience of working with laboratory animals. Such individuals and others sensitized to biological dusts who are unable, at least in the short term, to change their job should be advised to minimize their animal contact and to wear respiratory protection, most conveniently laminar flow equipment, when in contact with animals. In addition, background prophylaxis such as sodium cromoglycate will minimize the risk of the provocation of asthma by indirect allergen contact, as from dust on colleagues' clothing. None the less, it should be emphasized that such measures are temporary and in the long term the means should be sought to avoid exposure to the cause of asthma.

When individuals remain in employment exposed to the cause of their asthma, either directly or indirectly, the effectiveness of relocation or of respiratory protection needs to be monitored. This can be conveniently done by serial self-recordings of PEF to determine whether or not asthma persists and if so if it is work related (Fig. 29.3).

REFERENCES

1. Brookes SM, Weiss MA, Bernstein K: Reactive airways dysfunction syndrome (RADS): persistent asthma syndrome after high level irritant exposures. *Chest* (1985) **88**: 376–384.
2. Ross DJ, McDonald JC: Asthma following inhalation accidents reported to the SWORD project. *Ann Occup Hyg* (1996) **40**: 645–650.
3. Kern DG: Outbreak of the reactive airways dysfunction syndrome after a spill of glacial acetic acid. *Am Rev Respir Dis* (1991) **144**: 1058–1064.
4. Kojevinas M, Anto JM, Soriano JB, Tobias A, Burney P: The risk of asthma attributable to occupational exposures. *Am J Respir Crit Care Med* (1996) **154**: 137–143.
5. Meredith SK, McDonald JC: Work related respiratory disease in the United Kingdom 1989–1992: report of the SWORD project. *Occup Med* (1994) **44**: 183–189.
6. Platt Mills TAE, Longbottom J, Edwards J, Cockcroft A, Wilkins S: Occupational asthma and rhinitis related to laboratory animals: serum IgE and IgG antibodies to the rat urinary allergen. *J Allergy Clin Immunol* (1987) **79**: 505–515.
7. Howe W, Venables K, Topping M, *et al.*: Tetrachlorophthalic anhydride asthma: evidence for specific IgE antibody. *J Allergy Clin Immunol* (1983) **71**: 5–11.
8. Foster H, Topping M, Newman Taylor AJ: Specific IgG and IgG$_4$ antibody to tetrachloropthalic anhydride. *Allery Proc* (1988) **9**: 296.
9. Bentley AM, Maestrelli P, Fabbri LM, *et al.*: Immunohistology of the bronchial mucosa in occupational, intrinsic and extrinsic asthma. *J Allergy Clin Immunol* (1991) **87**: 246A.
10. Newman Taylor AJ, Longbottom JL, Pepys J: Respiratory allergy to urine proteins of rats and mice. *Lancet* (1997) **ii**: 847–849.
11. Tee RD, Gordon DJ, Hawkins ER, *et al.*: Occupational allergy to locusts: an investigation of the sources of the allergen. *J Allergy Clin Immunol* (1988) **81**: 517–525.
12. Bjorksten F, Backman A, Jarvinen AJ, Savilahti EK, Syvanen P, Karkkarinen T: Immunoglobulin E specific to wheat and rye flour. *Clin Allergy* (1977) **7**: 473–483.

13. Pepys J, Wells ED, D'Souza M, Greenburg M: Clinical and immunological responses to enzymes of *Bacillus subtilis* in factory workers and consumers. *Clin Allergy* (1973) **3**: 143–160.
14. Maccia CA, Bernstein IL, Emmett EA, Brooks SM: *In vitro* demonstration of specific IgE in phthalic anhydride sensitivity. *Am Rev Respir Dis* (1976) **113**: 701–704.
15. Zeiss CR, Patterson R, Pruzansky JJ, Miller MM, Rosenburg M, Levitz D: Trimellitic anhydride induced airways syndromes. *J Allergy Clin Immunol* (1977) **60**: 96–103.
16. Luczynska CM, Topping MD: Specific IgE antibodies to reactive dye–albumin conjugates. *J Immunol Methods* (1986) **95**: 177–186.
17. Edwards RG, Dewdney JM, Dobrzanski RJ, Lee D: Immunogenicity and allergenicity studies on two beta lactam structures, a clavam, clavulanic acid and a carbapeneam: structure activity relationships. *Int Arch Allergy Appl Immunol* (1988) **85**: 184–189.
18. Davies RJ, Butcher BT, O'Neil CE, Salvaggio JE: The *in vitro* effect of toluene di-isocyanate on lymphocyte cyclic adenosine monophosphate production by isopoterenol, prostaglandin and histamine. *J Allergy Clin Immunol* (1977) **60**: 223–229.
19. Lozewiz S, Assoufi BK, Hawkins R, Newman Taylor AJ: Outcome of asthma induced by isocyanates. *Br J Dis Chest* (1987) **81**: 14–22.
20. Durham SR, Grannek BJ, Hawkins R, Newman Taylor AJ: The temporal relationship between increases in airway responsiveness to histamine and late asthmatic responses induced by occupational agents. *J Allergy Clin Immunol* (1987) **79**: 398–406.
21. Mackay RT, Brooks SM: Effect of toluene di-isocyanate on beta adrenergic receptor function. *Am Rev Respir Dis* (1983) **148**: 50–53.
22. Cullinan P, Lowson D, Nieuwenhuijsen MJ, et al.: Work-related symptoms, sensitisations and estimated exposure in workers not previously exposed to laboratory rats. *Occup Environ Med* (1994) **57**: 589–592.
23. Juniper CP, How MJ, Goodwin BFJ, Kinshott AJC: *Bacillus subtilis* enzymes: a 7 year clinical epidemiological and immunological study of an industrial allergen. *J Soc Occup Med* (1977) **27**: 3–12.
24. Coutts II, Lozewitcz S, Dally MD, Newman Taylor AJ, Burge PS, Rogers JD: Respiratory symptoms related to work in a factory manufacturing cimetidine tablets. *Br Med J* (1984) **288**: 1418.
25. Musk AW, Venables KM, Crook B, et al.: Respiratory symptoms, lung function and sensitisation to flour in a British bakery. *Br J Ind Med* (1989) **46**: 636–642.
26. Burge PS, Edge G, Hawkins R, White V, Newman Taylor AJ: Occupational asthma in a factory making cored solder containing colophony. *Thorax* (1981) **36**: 828–834.
27. Slovak AMJ, Hill RN: Laboratory animal allergy: a clinical survey of an exposed population. *Br J Ind Med* (1981) **38**: 38–41.
28. Venables KM, Tee RD, Hawkins ER, et al.: Laboratory animal allergy in a pharmaceutical company. *Br J Ind Med* (1988) **45**: 660–666.
29. Dally MB, Hunter JV, Hughes EG, et al.: Hypersensitivity to platinum salts: a population study. *Am Rev Respir Dis* (1980) **4**: A120.
30. Venables KM, Dally MB, Nunn A, et al.: Smoking and occupational allergy in a platinum refinery. *Br Med J* (1989) **299**: 939–942.
31. Venables KM, Topping MD, Howe W, Luczynska CM, Hawkins R, Newman Taylor AJ: Interaction of smoking and atopy in producing specific IgE antibody against a hapten protein conjugate. *Br Med J* (1985) **290**: 201–204.
32. Zetterstrom O, Osterman K, Machado L, Johansson SGO: Another smoking hazard revised: serum IgE concentrations and increased risk of occupational allergy. *Br Med J* (1981) **283**: 1215–1217.
33. Cartier A, Malo J, Forest F, et al.: Occupational asthma in snow-crab processing workers. *J Allergy Clin Immunol* (1984) **74**: 261–269.
34. Calverley AE, Rees D, Dowdeswell RJ, Linnett PJ, Kielkowski D: Platinum salt sensitivity in refinery workers: incidence and effects of smoking and exposure. *Occup Environ Med* (1995) **52**: 661–666.
35. Zetterstrom O, Nordvall SL, Bjorksten B, Ahlstedt S, Sterlander M: Increased IgE antibody responses to rats exposed to tobacco smoke. *J Allergy Clin Immunol* (1985) **75**: 594–598.

36. Biagini RE, Moorman WJ, Lewis TR, Bernstein IL: Ozone enhancement of platinum asthma in a primate model. *Am Rev Respir Dis* (1986) **134**: 719–725.
37. Riedel F, Kramer M, Scheibenbogen C, Rieger CHC: Effects of SO_2 exposure on allergic sensitisation in the guinea pig. *J Allergy Clin Immunol* (1988) **82**: 527–534.
38. Young RP, Barker RD, Pile KD, Cookson WOCM, Newman Taylor AJ: The association of HLA-DR3 with specific IgE to inhaled acid anhydrides. *Am J Respir Crit Care Med* (1995) **151**: 219–221.
39. Bignon JS, Aron Y, Ju LT, *et al.*: HLA class II alleles in isocyanate-induced asthma. *Am J Respir Crit Care Med* (1994) **149**: 1–75.
40. Graneek BJ: Serial peak flows and bronchial challenge tests. *Thorax* (1988) **43**: 803.
41. Newman Taylor AJ, Davies RJ, Hendrik DJ, Pepys J: Recurrent nocturnal asthmatic reactions to bronchial provocation tests. *Clin Allergy* (1979) **9**: 213–219.
42. Venables KM, Topping MD, Nunn AJ, Howe W, Newman Taylor AJ: Immunologic and functional consequences of chemical (tetrachlorophthalic anhydride) induced asthma after 4 years of avoidance of exposure. *J Allergy Clin Immunol* (1987) **80**: 212–218.
43. Malo JC, Cartier A, Ghezzo H, Lafrance M, Cante M, Lehrer SB: Patterns of improvement in spirometry, bronchial hyper responsiveness and specific IgE antibody levels after cessation of exposure in occupational asthma caused by snow-crab processing. *Am Rev Respir Dis* (1988) **138**: 807–812.
44. Paggiaro P, Bacci E, Paoetto P, *et al.*: Bronchoalveolar lavage and morphology of the airways after cessation of exposure to asthmatic subjects to toluene di-isocyanate. *Chest* (1990) **98**: 536–542.
45. Chan Yeung M, McClean L, Paggiaro PL: Follow up study of 232 patients with occupational asthma caused by Western Red Cedar (*Thuja plicata*). *J Allergy Clin Immunol* (1987) **79**: 792–796.

30

Infections

WILLIAM W. BUSSE, ELLIOT C. DICK, ROBERT F.
LEMANSKE JR AND JAMES E. GERN

INTRODUCTION

Respiratory tract infections are important and frequent causes of wheezing. This is especially true in infants, where wheezing often accompanies viral respiratory infections; however, as these infants grow older, wheezing episodes with respiratory infections become less and less frequent. In addition, respiratory infections provoke episodes of wheezing in patients with existing asthma. For the patient with asthma, viral respiratory infections are perhaps the most frequent cause of asthma exacerbations. These virally provoked episodes of asthma can be severe, and the mechanisms by which various respiratory infections affect lung function is influenced by many factors, including the patient's age, the particular respiratory virus, gender, environmental factors and the presence of asthma. This chapter focuses on the various patterns by which respiratory infections cause wheezing episodes.

EPIDEMIOLOGY OF RESPIRATORY INFECTIONS AND WHEEZING

Relationship of viral respiratory infections to episodes of wheezing in infants

Wheezing with respiratory infections is extremely common in early childhood.[1] It is estimated that the prevalence of wheezing during the first 5 years of life varies from 30 to 60%.[2] In the majority of children who experience wheezing with respiratory infections, these episodes of wheezing become less frequent as the child grows older. However, a question still remains as to whether the initial episode of wheezing with a viral respiratory illness is an important factor in the eventual development of asthma. Although a significant body of information suggests an association between respiratory tract illnesses in early life and the later development of airway dysfunction, this relationship is difficult to establish and indicates the complexity of factors that surround the development of bronchial hyperresponsiveness and eventual expression of asthma. Furthermore, the source of subjects for study, i.e. follow-up of hospitalized patients vs. outpatients, contributes to the difficulty of understanding this problem.

Eisen and Bacal[3] found that children hospitalized for bronchiolitis prior to age 2 years had an increased risk for asthma. Rooney and Williams[4] also evaluated, retrospectively, the records of infants hospitalized for bronchiolitis at 18 months or younger; allergic manifestations and a family history of asthma were more frequent in children who eventually experienced one or more episodes of wheezing. Finally, McConnochie and Roghmann[5] identified 77 patients who had bronchiolitis at 25 months or younger and compared their outcome to children without a history of bronchiolitis. When these children were evaluated approximately 7 years later, only upper respiratory allergy, bronchiolitis and passive smoking exposure were found to be independent predictors of wheezing following bronchiolitis. Consequently, it is apparent that the final conclusions on the relationship between respiratory infections in infancy and later asthma must consider a host of influences, including parental smoking, underlying airway responsiveness and gender.

A similarly important issue to resolve is the relationship between respiratory infections and the pathogenesis of airway hyperresponsiveness. To evaluate the effect of bronchiolitis on airway responsiveness, Sims et al.[6] identified 8-year-old children who had respiratory syncytial virus (RSV) respiratory infections and quantitated bronchial 'lability' by exercise tests. Compared with appropriate controls, the fall in the peak flow with exercise was greater in children who had bronchiolitis; however, airway reactivity to exercise was not different between children with or without subsequent episodes of wheezing. Since other variables confounded their study, Sims et al.[6] could not prove that respiratory infections led to the later development of asthma.

Other efforts have been made to ascertain if viral lower respiratory tract infections (LRIs) in early life cause persistent pulmonary function abnormalities. Pullan and Hey[7] evaluated 130 children admitted to hospital during the first 5 years of life with RSV LRIs; 42% of the hospitalized children had future episodes of wheezing, while only 19% of control subjects experienced similar airway symptoms. However, few patients (6.2% vs. 4.5% of controls) had troublesome respiratory symptoms by 10 years of age. Furthermore, although a three-fold increase in bronchial responsiveness was found in the children with bronchiolitis, atopy was not increased. Analogous conclusions were reached by Weiss et al.[8] when they assessed the outcome of an antecedent acute respiratory illness on airway responsiveness and atopy in young adults. Airway responsiveness, evaluated by eucapnic hyperventilation to subfreezing air, was increased in children with a previous history of either croup or bronchiolitis, or greater than two acute lower respiratory illnesses.

The possibility has also been raised that a predisposition to wheezing in infancy depends more on intrinsic airway structure than atopy.[9] This position is supported by the high degree of airway responsiveness found in infancy in physiological evaluations[10–12]

30 Infections

and the incidence of wheezing with respiratory infections.[13] However, it is difficult to precisely assess airway responsiveness in young children due to limitation of lung size and other age-related factors.

To help clarify the relationship between premorbid lung function and wheezing with respiratory illnesses, Martinez et al.,[14] conducted a prospective study of respiratory illness in infancy and childhood. Lung function values were determined *prior* to any LRIs. Included in these measurements were tidal expiratory patterns, specifically the time to peak tidal expiratory flow (Tme) divided by total expiratory time (TE), or the Tme/TE ratio; Morris and Lane[15] had shown that decreasing Tme/TE ratios correlated with lower lung function in patients with progressive chronic obstructive lung disease.

Of the infants studied by Martinez et al.,[14] 36 developed an LRI and 24 wheezed with at least one of these infections. There was no difference in preinfection lung function between those infants who did not have an LRI and those with an infection but no wheezing (Table 30.1). However, infants who wheezed with the respiratory infection had diminished Tme/TE values and reduced expiratory system conductance when measured *prior* to wheezing with the infection. These data suggest that alterations in lung function are compatible with reduced airway conductance or a slow respiratory system time constant that precedes and predicts wheezing with respiratory infections in infants. Furthermore, it appears that a given child's response to infection is determined not only by the infection but also by pre-existing lung function.

Taussig et al.[16] also noted that lower levels of lung function predispose to wheezing with LRI, as opposed to the infection *per se*. The precise nature of this predisposition remains to be defined but may lie in airway geometry, airway–parenchymal interaction, or mucosal and smooth muscle response. Furthermore, this pulmonary–structural predisposition may be enhanced by an exaggerated IgE response to viral infection,[17,18] resulting in more inflammation and severe wheezing with hospitalization. Since the majority of infants who develop wheezing with LRIs do not wheeze throughout life,[19] it is likely that pulmonary function abnormalities that favour wheezing with viral infections are modified with the growth and development of the lung. Long-term outcome then seems to be more closely linked to the persistence of ongoing airway damage or bronchospasm associated with the development of atopy and true clinical asthma.

Further, and possibly definitive, insight into the relationship between wheezing with early respiratory infections and the later development of asthma has come from a unique

Table 30.1 Pulmonary function and outcome of lower respiratory tract infection (LRI).

Index	No LRI	LRI (no wheeze)	LRI (wheeze)	P value
Tme/TE	31.2 ± 9.2 (88)	31.4 ± 8.5 (12)	25.4 ± 6.9* (24)	0.01
GRS (litre/s per cmH$_2$O)	0.035 ± 0.009 (30)	0.036 ± 0.010 (6)	0.028 ± 0.006† (11)	0.04
FRC (ml)	103.2 ± 16.7 (71)	102.5 ± 15.6 (8)	97.1 ± 20.8 (15)	0.63
$\dot{V}$max (ml/s)	131.2 ± 47.9 (77)	119.1 ± 44.0 (11)	118.6 ± 51.2 (21)	0.5

Values are means $\pm$ SD. The number of subjects is shown in parentheses. All values are age or length adjusted. FRC, functional residual capacity.
* $P < 0.01$ for the comparison with the no LRI group.
† $P < 0.05$ for the comparison with the no LRI group.
Reproduced from ref. 14, with permission.

Table 30.2 Maximal expiratory flow at functional residual capacity ($\dot{V}$maxFRC) during the first year of life and at 6 years of age, according to history of wheezing.*

Age	No wheezing No.	$\dot{V}$maxFRC (ml/sec)	Transient early wheezing No.	$\dot{V}$maxFRC (ml/sec)	Late-onset wheezing No.	$\dot{V}$maxFRC (ml/sec)	Persistent wheezing No.	$\dot{V}$maxFRC (ml/sec)	F	P value
<1 year	67	123.3 (110.0–138.0)	21	70.6 (52.2–93.8)†	21	107.1 (87.5–129.6)	16	104.6 (73.6–144.5)	5.95	<0.001
6 years	260	1262.1 (1217.4–1308.1)	104	1097.7 (1034.9–1163.5)†	81	1174.9 (1111.1–1241.1)	81	1069.7 (906.9–1146.5)‡	9.60	<0.001

* A total of 125 children underwent pulmonary function testing during the first year of life and 526 were tested at 6 years of age. Values for $\dot{V}$maxFRC are geometric means (95% confidence intervals). The F-test and associated P values indicate significant differences in lung function between the four groups.
† $P < 0.01$ for comparison with the children who never wheezed and $P < 0.05$ for comparisons with the children with late-onset wheezing and persistent wheezing, by Duncan's multiple-comparison test.
‡ $P < 0.01$ for comparison with the children who never wheezed, by Duncan's multiple-comparison test.
Reproduced from ref. 20, with permission.

analysis by Martinez et al.[20] in a long-term prospective study at the University of Arizona. They identified a number of factors that affect wheezing before the age of 3 years and their relationship to wheezing at 6 years of age. The study population has previously been reported[14] and consists of newborns enrolled between 1980 and 1984 with follow-up information at 3 and 6 years of age. Key assessments in infancy included cord-serum IgE levels, pulmonary function testing before any lower respiratory tract illness had occurred, measurement of serum IgE at 9 months of age, and a questionnaire completed by the children's parents when the child was 1 year old. The children were classified into four groups: no wheezing, transient wheezing, late-onset wheezing or persistent wheezing (Table 30.2, opposite). At 6 years of age, serum IgE, pulmonary function testing and allergy skin testing were repeated and these factors were assessed in relationship to their history of wheezing. At 6 years of age, 20% of the children had at least one lower respiratory illness with wheezing during the first 3 years of life, but with no wheezing at age 6 years. These children had diminished airway function before the age of 1 year and, at 6 years of age, were more likely to have mothers who smoked but not mothers with asthma and did not have evidence of atopy, i.e. elevated serum IgE or skin test reactivity. In addition, 15% had no wheezing before the age of 3 years but had wheezing at age 6 years and 13.7% had wheezing both before 3 years of age and at 6 years of age. Those with late-onset wheezing and persistent wheezing were more likely to have mothers with a history of asthma, elevated serum IgE levels and diminished lung function at 6 years of age (Tables 30.2 and 30.3).

A number of conclusions can be drawn from this prospective longitudinal study. Most infants who wheeze early in life have a transient condition and this predilection to

Table 30.3 Total serum IgE levels and prevalence of positive skin tests for reactivity to aeroallergens in children 6 years old, according to history of wheezing.*

Category	Serum IgE No.	Mean (95% CI) (IU/ml)†	Positive skin test No.	Prevalence (%)
No wheezing	222	28.1 (22.4–35.3)	317	33.8
Transient early wheezing	95	31.0 (22.3–43.1)	125	38.4
Late-onset wheezing	68	42.1 (26.6–66.0)	97	55.7‡
Persistent wheezing	75	65.6 (45.3–94.4)§	90	51.1¶
		$F = 4.94$		$\chi^2 = 19.5$
		$P = 0.002$		$P < 0.001$

* Of the 826 children, 629 underwent skin testing for reactivity to aeroallergens and 460 had measurements of serum IgE at 6 years of age. The F-test and the χ^2 test (and the corresponding P values) indicate significant differences in serum IgE levels and the prevalence of positive skin tests, respectively, in association with the differences in patterns of wheezing.
† To convert values for IgE to micrograms per litre, multiply by 2.4. CI denotes confidence interval.
‡ $P < 0.001$ for comparison with the children who never wheezed.
§ $P < 0.01$ for comparisons with the children who never wheezed and those with transient early wheezing, by Duncan's multiple-comparison test.
¶ $P = 0.003$ for comparison with the children who never wheezed.
Reproduced from ref. 20, with permission.

wheezing with respiratory infection is associated with diminished lung function; these children are not at enhanced risk for asthma or allergic disease later in life. The groups of particular interest regarding the role of respiratory infections and development of asthma are those infants with either persistent or late-onset wheezing. Children with asthma at age 6 years appear to have a predisposition to asthma, as indicated by a maternal history of asthma and elevated serum IgE levels in the first years of life. Moreover, these children have reduced lung function by 6 years of age. The role of respiratory infections in the actual development of asthma in those with a predisposition to asthma is less clear; that is, do respiratory infections have any relationship to the subsequent development of asthma? This issue remains to be fully elucidated.

Insight to factors that may determine a host's response to viral infections and its subsequent effects on lung function has been gained by animal studies. Sorkness et al.[21] infected neonatal rats with Sendai virus. At 7 and 13–16 weeks afer infection, the rats that had been infected had lower PO_2 values and increased lung resistance when compared with controls. Furthermore, the infected animals were more sensitive to aerosolized methacholine. Lung histology in the infected animals revealed increased numbers of bronchiolar mast cells and eosinophils. In subsequent studies, these investigators have found some rat strains, i.e. Brown Norway, are more susceptible to the effects of the virus and more likely to have persistently altered lung function. Such studies in an animal model are compatible with the observations by Martinez et al.[20] and indicate that many factors determine the eventual influence of respiratory infections on airway function, i.e. wheezing and bronchial hyperresponsiveness, including a predisposition to produce IgE antibodies. However, what needs to be clarified are the mechanisms by which respiratory viruses interact with IgE-dependent factors and promote an asthma-like picture.

Viral respiratory infections provoke asthma

For decades clinicians associated asthma exacerbations with respiratory infections. With prospective studies, it became apparent that viral, not bacterial, upper respiratory infections (URIs) triggered these asthma attacks. With the use of more sensitive techniques to identify respiratory viruses, the relationship between respiratory infections, particularly viral URIs, and asthma has become even more convincing and important.

In early studies, McIntosh and coworkers[22] prospectively studied 32 young children, aged 1–5 years, with severe asthma. The aetiology of each URI was carefully evaluated and confirmed by culture and serological tests for viral antibody titres. Over 2 years, there were 102 confirmed viral respiratory infections and 139 episodes of wheezing; 58 episodes (42%) of wheezing occurred in relationship to viral respiratory infections, of which RSV was the most prevalent and likely to provoke asthma. Respiratory bacteria, *Haemophilus influenzae*, *Streptococcus pneumoniae*, β-haemolytic streptococcus and *Staphylococcus aureus*, were also cultured but their presence did not correlate with an asthma attack.

A prospective outpatient study from the University of Wisconsin evaluated 16 children, aged 3–11 years, with histories of four or more asthma attacks associated with respiratory illnesses during the previous year.[23] Detailed clinical records profiled each child's asthma severity, which was further substantiated with biweekly examinations. During an asthma exacerbation or apparent URI, additional bacteria and virus cultures were collected and asthma symptoms carefully quantitated. The 16 children experienced 61 episodes of asthma; 42 occurred in conjunction with a symptomatic respiratory infection and 24 were confirmed to be of viral aetiology by culture and/or serum haemagglutination titres. In this study, rhinovirus was the most frequently identified virus in association with wheezing. Some patients had episodes of asymptomatic viral infection, but asthma was not worsened. Only one episode of wheezing coincided with a bacterial infection.

To extend these observations and to further determine if there was a relationship between different respiratory infections and wheezing, Mertsola et al.[24] identified 54 patients, aged 1–6 years, who had recurrent attacks of wheezing. Over a 3-month observation period, these patients experienced 115 episodes of upper or lower respiratory symptoms. Of these episodes, laboratory evidence of a viral or *Mycoplasma* infection was found in 45%. More relevant to our discussion was the observation that wheezing occurred in 58% of the laboratory-confirmed viral respiratory infections. Moreover, of the children with wheezing with respiratory infections, one-third were admitted to hospital because of severe dyspnoea. Finally, of the agents isolated, coronavirus and rhinovirus were the most frequent isolates.

There is also evidence that the viruses which cause wheezing with respiratory tract infections may be age dependent; for example, infants wheeze with RSV while older children have exacerbations of asthma with rhinovirus.[25] To extend these observations, Duff and colleagues[26] examined the relationship of viral infections, passive smoke exposure and IgE antibody to inhaled allergens in infants and children who were treated for acute episodes of wheezing in an emergency department. The investigators identified 99 subjects, aged 2 months to 16 years of age, and 57 control patients, aged 6 months to 16 years of age, who were seen in a paediatric emergency room. In acutely wheezing children less than 2 years of age, viruses were detected by culture of nasal washings in 70% of the patients; the most commonly identified virus was RSV. Positive virus cultures for RSV were noted in only 20% of the appropriate age-matched controls. In contrast, nasal washes yielded positive cultures in only 31% of children over 2 years of age who presented with acute wheezing. However, the most frequently identified respiratory virus in this age group was rhinovirus.

When these investigators evaluated risk factors associated with wheezing during the viral infections, interesting and insightful correlates were noted (Table 30.4). For children less than 2 years of age, coexisting allergic diseases were not a risk factor. Rather, passive exposure to cigarette smoke, as indicated by urinary cotinine values $\geqslant 10\ \mu g/ml$, was a risk factor for wheezing. The profile in children over 2 years of age with wheezing and viral URIs was quite different. In these children, the presence of allergy, i.e. positive radioallergosorbant test (RAST) values, and not passive cigarette smoke exposure, was the dominant risk factor for wheezing with viral respiratory infections. Observations by Duff et al.[26] confirm previous information that RSV is the major viral infection causing wheezing in infants, whereas rhinovirus is the causative respiratory infection associated with wheezing in older children. Moreover, wheezing in infants is not increased by the presence of allergic disease but rather made more probable if cigarette smoke exposure is present. In the older child, allergies are a major risk factor for wheezing with rhinovirus; these data also imply that a unique relationship may exist between rhinovirus and allergy, and this interaction may influence the development of wheezing with this particular respiratory virus.

Table 30.4 Odds ratios for wheezing among children with a positive radioallergosorbent test (RAST), virus culture or elevated cotinine level.

Risk factors	Odds ratio* Age < 2 years	Odds ratio* Age > 2 years
RAST	ND†	4.5 (2.0–10.2)†
Virus	8.2 (1.3–51.0)§	3.7 (1.3–10.6)§
Cotinine $\geqslant 10$ μg/ml	4.7 (1.0–21.3)§	0.6 (0.2–2.2)
RAST and virus	ND†	10.8 (1.9–59.0)‡

*Reported odds ratios are univariate analyses followed in parentheses by 95% confidence intervals.
† Not determined because of insufficient positive patients.
‡ $P < 0.001$. The odds ratio of 10.8 was estimated assuming that one control patient was positive for both RAST and virus.
§ $P < 0.05$.
Reproduced from ref. 26, with permission.

The detection of respiratory viruses, particularly rhinovirus, by culture is difficult because of the fastidiousness of viruses in culture. Consequently, determination of the association between rhinovirus infections and provocation of asthma using virus culture may give an underestimation of the true frequency between colds and asthma. Johnston et al.[27] developed a polymerase chain reaction (PCR) assay to detect picornavirus infections. This approach has an increased sensitivity and thus greater likelihood to detect causative viruses in relationship to clinical symptoms of a cold and eventual influence on asthma.

Using PCR technology, along with culture, Johnston and colleagues[28] studied the association between upper and lower respiratory viral infections and acute exacerbations of asthma in school-age children; 108 children, 9–11 years of age, were followed over a 13-month period. Children or their parents recorded the peak expiratory flow rate twice daily and daily upper and lower respiratory tract symptoms were scored as to severity. With the onset of respiratory symptoms, or a significant fall in peak expiratory flow values, the child was visited by an investigator and nasal aspirates were obtained for virus detection.

There were a number of important observations by Johnston and colleagues in this large study. First, there was a seasonal pattern to the appearance of respiratory symptoms and changes in peak flow values of the asthma patients; the months of November–December and April appeared to be the most frequent times for colds and asthma exacerbations. Second, respiratory viruses were detected in approximately 80% of episodes with respiratory tract symptoms or falls in the peak flow (Table 30.5). Third, rhinovirus was the most frequently detected microorganism and constituted over 60% of the viruses detected (Table 30.6). Fourth, the severity of asthma symptoms in many of these subjects was severe. From these observations, and those of others, it is apparent that rhinovirus, the cause of the common cold, is the respiratory infection most likely to exacerbate wheezing in individuals with existing asthma. It is possible that this relationship simply reflects the fact that rhinovirus is the most common cause of infectious respiratory symptoms, i.e. the common cold. However, this association may also mean

Table 30.5 Number, type, severity and duration of reported and unreported episodes, from symptom and peak expiratory flow rate diaries. Values are medians (interquartile ranges) unless stated otherwise.

Type of episode	Reported	Unreported	Total
Upper respiratory tract			
No. of episodes	253	529	782
Duration (days)	7 (4–12)	4 (2–7)	5 (3–8)
Severity (symptom score)	6 (4–8)	3 (2–3)	3 (2–5)
Lower respiratory tract			
No. of episodes	200	376	576
Duration (days)	7 (5–12)	4 (2–7)	5 (3–9)
Severity (symptom score)	4 (3–7)	2 (1–3)	2 (2–4)
Peak expiratory flow			
No. of episodes	153	293	446
Duration (days)	14 (7–24)	8 (4–18)	10 (5–21)
Severity (maximum fall from median, litre/min)	81 (54–107)	43 (29–67)	52 (33–85)

Reproduced from ref. 28, with permission.

Table 30.6 Viruses detected during reported respiratory episodes.

Virus	Polymerase chain reaction	Culture	Immuno-fluorescence	Antibody rise by ELISA	Total
Rhinovirus/enterovirus	146	47			147*
Coronavirus	17	14		21	38
Influenza viruses		14	10	20	21
Parainfluenza viruses 1, 2 and 3		6	6	18	21
Respiratory syncytial virus		6	6	12	12
Other		2	1	2	3

* 84 found to be rhinovirus on further testing; remainder were unidentified picornaviruses.
ELISA, enzyme-linked immunosorbent assay.
Reproduced from ref. 28, with permission.

that a unique relationship exists between rhinovirus, allergic airway disease and exacerbations of asthma.

In contrast to observations in children, the relationship between respiratory infections and episodes of wheezing is not as striking in adults. When both children and adults were evaluated, respiratory viruses were more frequently identified during episodes of asthma in children under 10 years of age compared with adults. Eight adult subjects evaluated by Minor et al.[29] had only three documented virus infections with increased wheezing. Although an explanation for such findings is not clear, it is likely that asthma flares were more difficult to sharply delineate and the pattern of wheezing often appeared chronic rather than episodic in adults. Similarly, when Hudgel and coworkers[30] evaluated 19 adult asthma patients over a 15-month period, 76 episodes of asthma were recorded but only eight could be documented with a viral URI. Although clinically apparent viral respiratory infections precipitate asthma in adults, the relative frequency appears less than in children. In another study, Huhti et al.[31] evaluated adults admitted to hospital with asthma exacerbations. The viral detection rate was 19.0%, with RSV, parainfluenza and influenza the more frequent isolates.

Using virus culture and PCR techniques, Nicholson and coworkers[32] evaluated the relationship of symptomatic colds and asthma exacerbations in 138 adult asthmatic subjects. These investigators found symptomatic colds were reported in 80% of episodes with symptoms of wheeze, chest tightness or breathlessness. In 24% of laboratory-proved episodes of non-bacterial respiratory infections, there was a reduction in peak flow ⩾50 litre/min. Infections with rhinoviruses, coronaviruses, influenza, RSV, parainfluenza and *Chlamydia* were all associated with objective measures of airflow limitation. The findings of Nicholson et al. are further evidence of the importance of respiratory viruses, particularly rhinovirus, to asthma exacerbations, and that these episodes can be severe and occur in adults. Therefore, evidence exists to underscore the importance of viral respiratory infections to asthma exacerbations in patients of all ages.

There is little evidence that bacterial respiratory infections, other than sinusitis, provoke asthma. Berman et al.[33] performed transtracheal aspirates on 27 adult patients with asthma during an infectious exacerbation. Bacteria cultured from transtracheal aspirates were sparse and, more importantly, did not correlate with clinical illness. Moreover, transtracheal aspirates from normal subjects, without respiratory infections, yielded a similar degree of bacterial colonization. A common complication of colds is the development of sinusitis, which has been proposed as a factor in the worsening of asthma. Although the association between paranasal sinusitis and bronchial asthma has long been observed, it is conceivable that sinusitis and asthma may coexist as complications of the same respiratory infection. However, some feel that an aetiological relationship exists between sinusitis and asthma.[34,35] Although these observations and associations are intriguing and clinically important, additional studies are needed to clarify how sinusitis may influence asthma.

MECHANISMS OF VIRUS-INDUCED AIRWAY HYPERRESPONSIVENESS

A number of mechanisms have been proposed to explain how viral respiratory infections provoke asthma (Table 30.7). When evaluating these possibilities, it is important to realize that the effect respiratory viruses have on airway function will depend upon many factors, including the subject's gender, environmental factors such as cigarette smoke exposure, age, baseline lung function and atopic status, along with the particular respiratory virus. However, there is now convincing evidence that rhinoviruses are the predominant respiratory infection provoking wheezing in patients with asthma and the existence of allergic disease appears to be an important risk factor. Therefore, it is proposed that respiratory viruses, particularly rhinoviruses, may interact with those factors that contribute to or determine existing allergic airway disease and enhance existing bronchial inflammation. The end-result of the effects of the virus on underlying

30 Infections

Table 30.7 Mechanisms of virus-induced asthma.

Direct injury to airway epithelium
    Altered epithelial function
    Altered autonomic nervous system regulation of airway smooth muscle function
Enhanced allergic inflammation
    Enhanced eosinophilic recruitment
    Eosinophilic infiltration of the airway
    Generation of pro-inflammatory cytokines
    Enhanced inflammatory mediator release
Altered β-adrenergic function
Sinopulmonary reflex
Production of virus-specific IgE antibodies

---

allergic airway disease will be an enhanced likelihood for wheezing. Although other factors likely participate in this process, the following discussion examines how respiratory viruses either influence IgE-dependent events or promote the development of allergic inflammation.

Interaction of respiratory viruses with IgE-dependent events stimulate virus-specific IgE antibody production

Welliver et al.[17] tested 79 children, all less than 12 months of age and with documented RSV infection, for the presence of IgE-specific antibody to the infecting virus. Their clinical patterns of illness were divided into four groups: (a) upper respiratory tract illness; (b) pneumonia without wheezing; (c) pneumonia and wheezing; and (d) wheezing (bronchiolitis). Nasal secretions from each patient were measured for IgE-specific antibody to RSV, and RSV IgE antibody titres were compared with the patient's clinical illness. IgE titres to RSV were highest in patients with evidence of airway obstruction, e.g. pneumonia/wheezing or bronchiolitis. Parainfluenza virus-specific IgE responses were also examined in individuals with upper respiratory illness alone or bronchiolitis by Welliver et al.[18] Parainfluenza-specific IgE antibody were detected in 8 of 12 with bronchiolitis but only 1 of 10 with upper respiratory disease alone. These studies suggest that some respiratory viruses stimulate IgE-specific antibody responses and the degree to which this occurs is associated with the clinical manifestations of upper vs. lower airway disease.

    To further evaluate the significance of virus-specific IgE antibody, Welliver et al.[36] prospectively monitored 38 infants for 48 months after an initial episode of bronchiolitis. Only 20% of infants with undetectable titres of RSV IgE had subsequent episodes of documented wheezing. In contrast, 70% of those children with high RSV IgE antibody titres continued to experience wheezing when evaluated 4 years later. A portion of these same subjects was evaluated at 7–8 years of age[37] and underwent skin testing to common inhalant allergens, pulmonary function testing and methacholine provocation to determine bronchial responsiveness. For the entire study groups, RSV IgE responses were evaluated in relationship to family history of asthma, passive smoke exposure, number of recurrent wheezing episodes, number of positive skin tests and the degree of airway responsiveness to methacholine. The investigators found that an association exists between RSV IgE responses and the development of any wheezing following bronchiolitis; the nature, however, of the association between RSV IgE antibodies and any recurrent wheezing is unclear. Therefore, the consequence of an initial IgE response to the infecting virus with regard to long-term influence on airway function and the development of asthma is unresolved.

Interaction of rhinovirus on allergic inflammation

Although rhinoviruses have emerged as the most common respiratory infection causing asthma exacerbations,[28] there are a number of paradoxes to this relationship. First, rhinoviruses are infections that appear localized to the upper airway; identification of rhinovirus by culture in the lower airway is unusual[38] and pneumonic infiltrates rare. Second, even though rhinoviruses infect airway epithelium, the inflammatory response is often unimpressive. For example, Fraenkel and coworkers[39] performed nasal biopsies before and after an experimental rhinovirus infection. In both normal and atopic groups, there were no significant changes in inflammatory cells during the cold or the convalescent period compared with baseline values. Thus, immunohistochemical analysis of rhinovirus-infected upper airways does not reveal inflammation. Third, rhinovirus infections tend to be self-limited and later sequelae are rare. Therefore, the mechanisms of lower airway dysfunction are not readily explained by an overzealous inflammatory airway response to this particular virus.

Effect of rhinovirus infections on airway responsiveness and allergic inflammation

The frequent association of asthma with respiratory infections, and in particular rhinovirus illnesses, may relate to the frequency with which rhinoviruses are the cause of colds. This frequency, however, does not explain the apparently serious lower airway events that are associated with rhinovirus infections in subjects with asthma. A major goal of current research is to establish how this particular class of respiratory viruses affects lower airways so profoundly and how these effects can last for weeks beyond the acute respiratory illness.

In an initial series of experiments to evaluate the effects of respiratory infection on airway responsiveness, we infected subjects with rhinovirus experimentally and determined its effect on a variety of airway functions.[40] The patients selected for study were evaluated on three separate occasions: at baseline, during acute infection and during recovery. Of particular interest was the effect of the infection on the immediate and late-phase allergic response (LAR) to inhaled antigen. All 10 patients had a rhinovirus respiratory infection at time of study. During the acute rhinovirus respiratory infection, airway responsiveness to histamine significantly increased over baseline values. Likewise, acute airway reactivity to inhaled antigen was increased. The change in airway responsiveness to histamine and antigen was similar, suggesting that the effects of the respiratory infection on responsiveness to antigen related to alterations in non-specific bronchial reactivity.

Prior to rhinovirus inoculation, only 1 of 10 patients had an LAR to inhaled antigen. However, during the acute respiratory infection, 8 of 10 patients experienced late-phase airway obstruction to inhaled antigen challenge. Furthermore, when evaluated during the recovery period (4 weeks after rhinovirus inoculation), five of the seven patients available for testing still had LARs to inhaled antigen. These observations indicate that rhinovirus respiratory infection not only increased airway responsiveness but also changed the pattern of the airway response to inhaled antigen and promoted the development of allergic inflammation, as indicated by the development of the LAR.

The recovery pattern of airway hyperresponsiveness following the viral respiratory infection was also evaluated. Although increased airway responsiveness was still detected 4 weeks after rhinovirus inoculation, there was a trend towards recovery in both the reaction to inhaled histamine and the response to antigen. However, as already discussed, the increased frequency of LARs to antigen was still noted 4 weeks after the viral infection, suggesting that a viral respiratory infection has a greater, and possibly more lasting, effect on factors that participate in the development of LARs.

Virus-associated airway hyperresponsiveness is a multifactorial process involving a complex interplay of IgE-dependent reactions, epithelial activation or damage, autonomic nervous system dysfunction and, of particular interest and relevance to our discussion, enhanced allergic inflammation.[41] In IgE-mediated reactions, the tissue response, be it the skin, nose or airway, is influenced by IgE sensitization of mast cells and basophils, release of bronchospastic and inflammatory mediators from sensitized cells, and the response of the target organ, which in asthma is bronchial smooth muscle. To evaluate the effects of respiratory viruses on a facet of the immediate hypersensitivity response, i.e. mediator release, basophil histamine secretion was determined. In an *in vitro* model system, peripheral blood mononuclear cells were incubated with influenza A virus and the effect of this exposure upon IgE-dependent basophil histamine release and leukotriene (LT)C_4 was measured. Following incubation with virus, both histamine and LTC_4 secretion was enhanced. To define the role of T-cells in this process, T-cells were removed by magnetic head separation with anti-CD3 antibody. In the absence of T-cells, virus incubation did not enhance histamine release.[42] Furthermore, analysis of virus-treated peripheral blood mononuclear cells revealed that the virus caused the generation of interferon (IFN)-γ. These findings suggest that T-cells, and their cytokine products such as IFN-γ, may play an integral role in the processes by which respiratory viruses enhance basophil histamine release, and are likely to affect other cell functions.

The use of bronchoscopy with lavage and biopsy has provided direct assessment and analysis of airway mediators and the histology of inflammatory events in asthma. We have used fibreoptic bronchoscopy to perform segmental bronchoprovocation with antigen and bronchoalveolar lavage (BAL) to assess immediate and late responses of the airway to antigen and to study the effects of a rhinovirus illness on these allergic events.[43] To conduct these experiments, a bronchoscope is introduced into the airways, bronchial segments are identified and an allergen is administered directly on to the airway mucosa. The immediate response to antigen is characterized by a brisk rise in mast cell mediators. When lavage of the previously allergen-challenged segments is performed 48 h later, there is a highly cellular, eosinophil-rich response detected. Based on our previous observations that a rhinovirus 16 infection increases bronchial responsiveness and LARs to inhaled antigen, we used segmental antigen challenge and lavage to test our hypothesis

Table 30.8 Effect of rhinovirus infection on airway histamine release to antigen.

Before cold	Cold	After cold
1.1 ± 0.8	4.5 ± 3.3*	6.0 ± 3.8*

Values (ng/ml) are means ± SD, $n = 8$.
* $P < 0.04$ compared to pre-cold value.

that the rhinovirus upper respiratory illness may enhance the lower airways allergic inflammatory response to antigen.

Histamine is released into the airway immediately following antigen challenge and represents mast cell activation. We found histamine release into the airway immediately following local antigen challenge to be significantly potentiated by rhinovirus infection (Table 30.8). These observations indicate that an acute rhinovirus illness can enhance airway mast cell mediator release. These findings are similar to earlier reports and indicate an effect of the infection on mast cell activation.[44] Interestingly, and relevant to earlier observations that rhinovirus effects on airway function are long-lasting, enhanced histamine release to antigen was still present 6 weeks after the acute infection; at this time, the individuals were symptom-free. Our findings suggest that viral effects on allergic responses can persist for weeks beyond the acute illness and that such persistence can explain the long-lasting effects on airway function in asthma from a cold.

When lavage is performed 48 h after antigen challenge, there is a characteristic and often dramatic influx of eosinophils into the lavage fluid. We found significant enhancement of eosinophil recruitment to the airway 48 h after antigen when the subjects had an acute rhinovirus infection (Fig. 30.1). Like the observations with enhanced and persistent histamine release, not only was a greater eosinophil recruitment to antigen challenge noted during the acute infection, but this respiratory virus effect persisted into the recovery period 6 weeks later. These findings suggest a possible explanation for the increase in LARs to antigen noted during acute rhinovirus infection; if allergen exposure during rhinovirus infection is associated with enhanced eosinophil recruitment to the airway, allergic inflammation and an LAR are more likely to occur.

Fraenkel and colleagues[45] provide additional insight into the mechanisms by which rhinovirus infections promote airway inflammation and the likelihood for increased asthma during colds. Both normal and asthmatic subjects were recruited for their study. The investigators obtained bronchial mucosal biopsies before and during an acute experimental rhinovirus infection and evaluated the changes in inflammatory cell infiltrate due to the virus and airway responsiveness. An increase in the airway response to histamine during the cold was associated with increases in submucosal lymphocytes. Further, there was an increase in epithelial eosinophils during the experimental cold (Fig. 30.2). In asthmatic compared with normal subjects, the increase in mucosal eosinophils persisted into the convalescent period. Although the changes in various features of immunohistology were not dramatic, they did follow a pattern that is consistent with our findings of enhanced eosinophil recruitment to the airways and were correlated with alterations in lung physiosiveness. Collectively, there is evidence from airway physiology that rhinovirus infections enhance bronchial responsiveness and histological features of

30 Infections

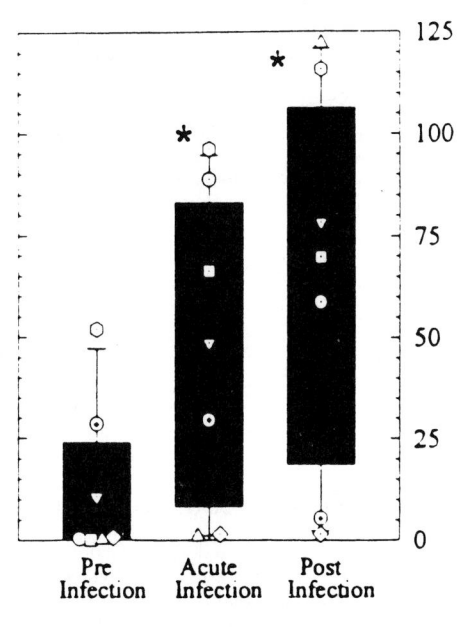

Fig. 30.1 Bronchoalveolar lavage (BAL) eosinophils 48 h after segmental antigen challenge in subjects experimentally infected with rhinovirus 16. On the ordinate is the number of eosinophils observed in BAL fluids obtained 48 h after segmental antigen challenge (late response) in seven subjects with allergic rhinitis (right panel, dotted symbols). On the abscissa are the three study periods. The wide bars represent the 25th to 75th percentiles, the whisker bars show the 5th and 95th percentiles, and the centre crossbar shows the median. Asterisks denote statistically significant differences between the postinfection and acute infection period when compared with preinfection period by post hoc testing. There was a significant potentiating effect of rhinovirus 16 infection ($P = 0.005$, ANOVA) and a significant difference in overall eosinophil recruitment related to atopy ($P < 0.001$, ANOVA), with allergic subjects exhibiting greater recruitment. In addition, there was a significant difference in the response to rhinovirus 16 infection depending on atopy ($P < 0.03$, ANOVA), with allergic subjects exhibiting heightened rhinovirus 16-related antigen-driven eosinophil recruitment. Reproduced from ref. 43, with permission.

allergic inflammation, i.e. there is a greater likelihood for LARs to occur. Furthermore, from studies that directly sample the lower airway, there is evidence of increased eosinophil recruitment and mucosal eosinophilia at the time of rhinovirus infection, and these abnormalities persist for weeks beyond the acute illness. All these findings indicate that rhinovirus infections can perpetuate features of allergic inflammation and may thus explain an increase in the likelihood of asthma.

The questions now remaining to be answered concern the mechanisms by which rhinovirus infections enhance allergic airway inflammation. Using the nasal mucosal response to rhinovirus as a surrogate of the airway response to rhinovirus, Fraenkel et al.[39] biopsied nasal tissue during an acute rhinovirus infection. Mast cells, eosinophils, lymphocytes and neutrophils were identified by appropriate monoclonal antibodies. There were no significant changes in the numbers of inflammatory cells present during the cold or the convalescent period compared with baseline values. Although these observations contrast with biopsy findings in the lower airway,[45] changes in airway

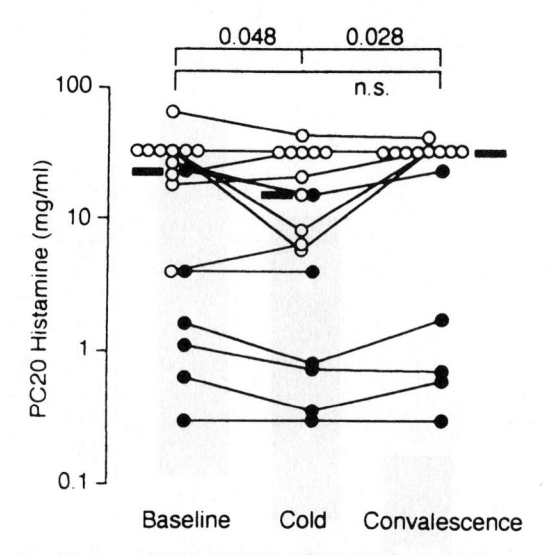

Fig. 30.2 Epithelial eosinophil counts before, during and after infection with rhinovirus 16. Normal (open circles) and asthmatic (closed circles) subjects, with medians for the whole group indicated by the horizontal bar. The eosinophil numbers increase with the cold, with P values as indicated. The figure demonstrates the persistence of elevated eosinophil numbers into convalescence in the asthmatic subjects, whereas the numbers in the normal subjects return to preinfection levels ($P = 0.043$). Reproduced from ref. 45, with permission.

histology were not dramatic. These findings suggest that other mechanisms, i.e. enhanced mediator release or, more likely, cytokine generation, are responsible for changes in airway function during a cold.

Respiratory viruses, including rhinovirus, stimulate production of pro-inflammatory cytokines

In our study with segmental antigen challenge,[43] the lavage fluid was analysed for tumour necrosis factor (TNF)-α. Lavage samples obtained 48 h after antigen challenge had a significant rise in TNF-α; these changes were noted in both normal controls and allergic subjects. Our findings raise the possibility that rhinovirus infection causes secretion of TNF-α and if enhanced secretion of cytokines were to occur in the allergic subject, features of allergic inflammation are more likely to occur. These findings also raise the possibility that rhinovirus can directly stimulate cytokine production. Linden et al.[46] measured IFN-γ, interleukin (IL)-1β, granulocyte–macrophage colony-stimulating factor (GM-CSF), IL-4 and IL-6 in nasal secretions of subjects with allergic rhinitis. During an experimental coronavirus cold, there was an increase in IFN-γ but not the other cytokines. The increase in IFN-γ is expected and represents an antiviral response to the virus. However, IFN-γ can have pro-inflammatory effects, i.e. promotion of adhesion protein expression and enhanced survival of eosinophils.[47] Moreover, these observations indicate that rhinovirus infections can stimulate cytokine production *in vivo*.

To extend these observations, live rhinovirus was incubated with human mononuclear cells and lung macrophages. Gern et al.[48] found that these cells took up the virus, but it did not replicate inside the cells. None the less, rhinovirus was able to activate monocytes and airway macrophages and cause secretion of TNF-α. These observations indicate that rhinovirus can cause synthesis and release of pro-inflammatory cytokines, which may occur independent of cell infection.

To identify the mechanisms by which rhinovirus promotes allergic inflammation, it is essential that the immune response to this virus be established so that the factors, and cells, involved in the promotion of allergic inflammation are identified. In determining the effect of rhinovirus on the immune response, it has been established that the receptor for the majority (90%) of rhinovirus strains is intercellular adhesion molecule (ICAM)-1.[49] The other 10% of rhinovirus strains use the low-density lipoprotein receptor.[50] Interaction of rhinovirus with these receptors is the first step in the host's response to this virus. With this information, we have found that a major rhinovirus, rhinovirus 16, binds to mononuclear cells via the ICAM-1 receptor, with binding occurring predominantly to monocytes but not lymphocytes. When rhinovirus 16 was incubated with a mixture of monocytes and lymphocytes, the following events occurred: there was a large increase in the number of $CD3^+$ cells that express the early activation marker CD69; when the $CD3^+/CD69^+$ cells were isolated and analysed, they were found to have increased expression of mRNA for IFN-γ. These *in vitro* observations suggest that rhinovirus activates monocytes to generate cytokines that stimulate lymphocytes. Once activated, these lymphocytes ($CD3^+/CD69^+$) generate a variety of cytokines, including IFN-γ. Although IFN-γ is normally important in antiviral activity, it also possesses a number of pro-inflammatory functions, including increased expression of endothelial and epithelial ICAM-1 (more receptors for rhinovirus) and the promotion of eosinophil survival. Therefore, in addition to activation of a small number of rhinovirus-specific lymphocytes, rhinovirus can cause a 'non-specific' activation of lymphocytes through the generation of cytokines from monocytes; these cytokines can, in turn, promote either viral elimination or inflammation. Perhaps the sequelae of this particular response are more likely if allergic inflammation already exists.

To extend this possibility, Einarsson and coworkers[51] evaluated the effect of respiratory viruses on stromal cell production of IL-11. IL-11 is an extremely cationic cytokine and has the capacity to induce airway hyperresponsiveness in the murine lung. These investigators found that rhinoviruses, along with RSV and parainfluenza, were potent stimulators of IL-11. In contrast, cytomegalovirus and adenovirus were weaker activators and are viruses not associated with asthma exacerbations. Furthermore, increased levels of IL-11 were found in nasal secretions of children with URIs and were more likely to appear in the nasal aspirates of those children with reactive airway disease. These observations are another example of the mechanism by which respiratory viruses can enhance airway inflammation through the generation of specific cytokines and thus increase the likelihood of asthma during a cold.

Using a mouse model, Coyle *et al*.[52] characterized the immune response to virus peptides and the resulting cytokine patterns. These investigators found that bystander $CD4^+$ Th2 immune responses to ovalbumin switched peptide-specific $CD8^+$ T-cells in the lung to IL-5 producers. When these IL-5-producing $CD8^+$ T-cells were challenged, via the airways, with virus peptide, a significant eosinophil infiltration was induced. Furthermore, *in vitro* studies indicated that IL-4 could switch virus-specific $CD8^+$ T-cells to IL-5 production. These results demonstrate a network by which allergic eosinophilic inflammation can occur during a respiratory viral infection and how this mechanism would impair $CD8^+$ T-cell responses (less IFN-γ production) and thus delay virus clearance. Such observations indicate not only a network to promote allergic inflammation but also a mechanism by which virus particles may persist for extended periods.

Our studies in humans, and those of Coyle et al.[52] in the mouse, suggest a possible link between eosinophils and airway changes during viral respiratory infections. When Garofalo et al.[53] sampled nasolaryngeal samples from children with RSV respiratory illnesses, the eosinophil protein, eosinophil cationic protein, was significantly increased in samples from children with bronchiolitis when compared with subjects with upper airway disease or RSV pneumonia but no wheezing. These data suggest that eosinophil activation by RSV may cause airway injury and the presence of wheezing. To extend these findings with in vitro experiments, Kimpen et al.[54] incubated isolated human eosinophils with RSV. RSV directly activated eosinophils to release superoxide and LTC_4. Furthermore, incubation of eosinophils with RSV primed the cell for further activation and generation of inflammatory mediators. Both observations indicate that an airway-active virus, in this case RSV, can interact with eosinophils and cause them to release pro-inflammatory mediators. As indicated by the authors, these effects of the virus could play a role in the pathogenesis of RSV bronchiolitis and may also be important for the development of more long-term bronchial hyperresponsiveness associated with viral infections. It is also possible that other respiratory viruses, such as rhinovirus, may have similar effects during a cold.

SUMMARY

The mechanisms involved in the development of airway hyperresponsiveness, airway obstruction and recurrent wheezing with viral respiratory infections have yet to be fully appreciated. None the less, current evidence indicates that viruses, or products of virus-infected cells, influence the inflammatory property and potential of many cells. Precisely how these effects of the virus translate into increased airway injury, responsiveness and obstruction will require further work. As the mechanisms of these interactions are established, so will improved understanding of asthma pathogenesis and treatment.

ACKNOWLEDGEMENTS

Support for this chapter has come from grants NIH HL 44098, AI-26609, KO8-01828 and General Clinical Research Grant RR-03186.

REFERENCES

1. Wilson NM: The significance of early wheezing. Clin Exp Allergy (1994) 24: 522–529.
2. Tager IB, Hanrahan JP, Torleson TD, et al.: Lung function, pre- and post-natal smoke exposure, and wheezing in the first year of life. Am Rev Respir Dis (1993) 147: 811–817.

3. Eisen AH, Bacal HL: The relationship of acute bronchiolitis to bronchial asthma: a 4-to-14 year follow-up. *Pediatrics* (1963) **31**: 859–861.
4. Rooney JC, Williams HE: The relationship between proven viral bronchiolitis and subsequent wheezing. *J Pediatr* (1971) **79**: 744–747.
5. McConnochie KM, Roghmann KJ: Bronchiolitis as a possible cause of wheezing in childhood. *Pediatrics* (1984) **74**: 1–10.
6. Sims DG, Downham MAPS, Gardner PS: Study of 8-year-old children with a history of respiratory syncytial virus bronchiolitis in infancy. *Br Med J* (1978) **1**: 11–14.
7. Pullan CR, Hey EN: Wheezing, asthma, and pulmonary dysfunction 10 years after infection with respiratory syncytial virus in infancy. *Br Med J* (1982) **284**: 1665–1669.
8. Weiss ST, Tager IB, Munoz A, Speizer FE: The relationship of respiratory infections in early childhood to the occurrence of increased levels of bronchial responsiveness and atopy. *Am Rev Respir Dis* (1985) **131**: 573–578.
9. Morgan WJ: Viral respiratory infection in infancy: provocation or propagation? *Semin Respir Med* (1990) **11**: 306–313.
10. Tepper RS: Airway reactivity in infants: a positive response to methacholine and metaproterenol. *J Appl Physiol* (1987) **62**: 1155–1159.
11. Geller DE, Morgan WJ, Cota K: Airway response to cold, dry air in normal infants. *Pediatr Pulmonol* (1988) **4**: 90–97.
12. LeSouef PN, Geelhoed GC, Turner DJ, Morgan SEG, Landau LI: Response of normal infants to inhaled histamine. *Am Rev Respir Dis* (1989) **139**: 62–66.
13. Wright AL, Taussig LM, Ray CG, *et al.*: The Tucson children's respiratory study. II. Lower respiratory tract illness in the first year of life. *Am J Epidemiol* (1989) **129**: 1232–1246.
14. Martinez FD, Morgan WJ, Wright AL, *et al.*: Diminished lung function as a predisposing factor for wheezing respiratory illnesses in infants. *N Engl J Med* (1988) **319**: 1112–1117.
15. Morris MJ, Lane DJ: Tidal expiratory flow patterns in air-flow obstruction. *Thorax* (1981) **36**: 135–142.
16. Taussig LM, Harris TR, Lebowitz MD: Lung function in infants and young children: functional residual capacity, tidal volume, and respiratory rate. *Am Rev Respir Dis* (1977) **116**: 233–239.
17. Welliver RC, Wong DT, Sun M, Middleton E Jr, Vaughan RS, Ogra PL: The development of respiratory syncytial virus specific IgE and the release of histamine in nasopharyngeal secretions after infection. *N Engl J Med* (1981) **305**: 841–846.
18. Welliver RC, Wong DT, Middleton E Jr, Sun M, McCarthy RN, Ogra PL: Role of parainfluenza virus-specific IgE in pathogenesis of croup and wheezing subsequent to infection. *J Pediatr* (1982) **101**: 889–896.
19. Voter KZ, Henry MM, Stewart PW, Henderson FW: Lower respiratory illness in early childhood and lung function and bronchial reactivity in adolescent males. *Am Rev Respir Dis* (1988) **137**: 302–307.
20. Martinez FD, Wright AL, Tanssig LM, *et al.*: Asthma and wheezing in the first six years of life. *N Engl J Med* (1995) **332**: 133–138.
21. Sorkness R, Lemanske RF, Castleman WL: Persistent airway hyperresponsiveness after neonatal viral bronchiolitis in rats. *J Appl Physiol* (1991) **70**: 375–383.
22. McIntosh K, Ellis EF, Hoffman LS, Lybass TG, Eller JJ, Fulginiti VA: The association of viral and bacterial respiratory infections with exacerbations of wheezing in young asthmatic children. *J Pediatr* (1973) **83**: 578–590.
23. Minor TE, Dick EC, DeMeo AN, Ouellette JJ, Cohen M, Reed CE: Viruses as precipitants of asthmatic attacks in children. *JAMA* (1974) **227**: 292–298.
24. Mertsola J, Ziegler T, Ruuskanen O, *et al.*: Recurrent wheezy bronchitis and viral respiratory infections. *Arch Dis Child* (1991) **66**: 124–129.
25. Pattemore PK, Johnston Sl, Bardin PG: Viruses as precipitants of asthma symptoms. I. Epidemiology. *Clin Exp Allergy* (1992) **22**: 325–336.
26. Duff AL, Pomeranz ES, Gelbre LE, *et al.*: Risk factors for acute wheezing in infants and children: viruses, passive smoke, and IgE antibodies to inhalant allergens. *Pediatrics* (1993) **92**: 535–540.
27. Johnston SL, Sanderson G, Pattemore PK, *et al.*: Use of polymerase chain reaction for

diagnosis of picornavirus infection in subjects with and without respiratory symptoms. *J Clin Microbiol* (1993) **31**: 111–117.
28. Johnston SL, Pattemore PK, Sanderson G, *et al.*: Community study of role of viral infections in exacerbations of asthma in 9–11 year old children. *Br Med J* (1995) **310**: 1225–1228.
29. Minor TE, Dick EC, Baker JW, Ouellette JJ, Cohen M, Reed CE: Rhinovirus and influenza A infections as precipitants of asthma. *Am Rev Respir Dis* (1976) **113**: 149–153.
30. Hudgel DW, Lanston E Jr, Selner JC, McIntosh K: Viral and bacterial infections in adults with chronic asthma. *Am Rev Respir Dis* (1979) **120**: 393–397.
31. Huhti C, Mokka T, Nikoskelaineu J, Haloners P: Association of viral and mycoplasma infections with exacerbations of asthma. *Ann Allergy* (1974) **33**: 145–149.
32. Nicholson KG, Kent J, Ireland DC: Respiratory viruses and exacerbations of asthma in adults. *Br Med J* (1993) **307**: 982–986.
33. Berman SZ, Mathison DA, Stevenson DD, Tan EM, Vaughan JH: Transtracheal aspiration studies in asthmatic patients in relapse with 'infective' asthma and in subjects without respiratory disease. *J Allergy Clin Immunol* (1975) **56**: 206–214.
34. Slavin RG, Cannon RF, Friedman WH, Palitang E, Sundaram M: Sinusitis and bronchial asthma. *J Allergy Clin Immunol* (1980) **66**: 250–257.
35. Rachelefsky GS, Katz RM, Siegel SC: Chronic sinus disease associated with reactive airway disease in children. *Pediatrics* (1984) **73**: 526–529.
36. Welliver RC, Sun M, Rinaldo D, Ogra PL: Predictive value of respiratory syncytial virus-specific IgE response for recurrent wheezing following bronchiolitis. *J Pediatr* (1986) **109**: 776–780.
37. Welliver RC, Duffy L: The relationship of RSV-specific immunoglobulin E antibody responses in infancy, recurrent wheezing, and pulmonary function at age 7–8 years. *Pediatr Pulmonol* (1993) **15**: 19–27.
38. Halperin SA, Eggleston PA, Beasley P, *et al.*: Exacerbations of asthma in adults during experimental rhinovirus infection. *Am Rev Respir Dis* (1985) **132**: 976–980.
39. Fraenkel DJ, Bardin PG, Sanderson G, *et al.*: Immunohistochemical analysis of nasal biopsies during rhinovirus experimental colds. *Am J Respir Crit Care Med* (1994) **150**: 1130–1136.
40. Lemanske RF Jr, Dick EC, Swenson CA, Vrtis RF, Busse WW: Rhinovirus upper respiratory infection increases airway reactivity in late asthmatic reactions. *J Clin Invest* (1989) **83**: 1–10.
41. Frick WE, Busse WW: Respiratory infections: their role in airway responsiveness and pathogenesis of asthma. *Clin Chest Med* (1988) **9**: 539–549.
42. Huftel MA, Swenson CA, Borcherding WR, *et al.*: The effect of T cell depletion on enhanced basophil histamine release after *in vitro* incubation with live influenza virus. *Am J Respir Cell Mol Biol* (1992) **7**:434–440.
43. Calhoun WJ, Dick EC, Schwartz LB, Busse WW: A common cold virus, rhinovirus 16, potentiates airway inflammation after segmental antigen bronchoprovocation in allergic subjects. *J Clin Invest* (1994) **94**: 2200–2208.
44. Calhoun WJ, Swenson CA, Dick EC, *et al.*: Experimental rhinovirus 16 infection potentiates histamine release after antigen bronchoprovocation in allergic subjects. *Am Rev Respir Dis* (1991) **144**: 1267–1273.
45. Fraenkel DJ, Bardin PG, Sanderson G, *et al.*: Lower airways inflammation during rhinovirus colds in normal and in asthmatic subjects. *Am J Respir Crit Care Med* (1995) **151**: 879–886.
46. Linden M, Greiff L, Andersson M, *et al.*: Nasal cytokines in common cold and allergic rhinitis. *Clin Exp Allergy* (1995) **25**: 166–172.
47. Wallen W, Kita H, Weiler D, Gleich GJ: Glucocorticoids inhibit cytokine-mediated eosinophil survival. *J Immunol* (1991) **147**: 3490–3495.
48. Gern JE, Dick EC, Lee WM, *et al.*: Rhinovirus enters but does not replicate inside monocytes and airway macrophages. *J Immunol* (1996) **156**: 621–627.
49. Staunton DE, Merluzzi VJ, Rothlein R, *et al.*: A cell adhesion molecule, ICAM-1, is the major surface receptor for rhinovirus. *Cell* (1989) **56**: 849–853.
50. Hofer F, Gruenberger M, Kowalski H, *et al.*: Members of the low density lipoprotein receptor family mediate cell entry of a minor-group common cold virus. *Proc Natl Acad Sci USA* (1994) **91**: 1839–1842.
51. Einarsson O, Geba GP, Zhu Z, Landry M, Elias TA: Interleukin-11: stimulation *in vivo* and *in*

vitro by respiratory viruses and induction of airways hyperresponsiveness. *J Clin Invest* (1996) **97**: 915–924.
52. Coyle AJ, Erand F, Bertrand C, *et al.*: Virus-specific CD8+ cells can switch to interleukin 5 production and induce airway eosinophilia. *J Exp Med* (1995) **181**: 1229–1233.
53. Garofalo R, Kimpen JLL, Welliver RC, Ogra PL: Eosinophil degranulation in the respiratory tract during naturally acquired respiratory syncytial virus infection. *J Pediatr* (1992) **120**: 28–32.
54. Kimpen JLL, Garofalo R, Welliver RC, Ogra PL: Activation of human eosinophils *in vitro* by respiratory syncytial virus. *Pediatr Res* (1992) **32**: 160–164.

31

Asthma Provoked by Exercise, Hyperventilation and the Inhalation of Non-isotonic Aerosols

SANDRA D. ANDERSON

INTRODUCTION

Exercise-induced asthma (EIA) or exercise-induced bronchoconstriction (EIB) are synonymous terms used to describe the transient increase in airways resistance that follows vigorous exercise in most patients with asthma. In addition to increasing airway resistance, exercise causes transient hyperinflation and arterial hypoxemia.[1] It is the increase in ventilation rate, rather than exercise itself, that is the stimulus for provoking an attack of asthma.[2] For this reason voluntary isocapnic hyperventilation is often used instead of exercise to measure bronchial responsiveness to hyperpnoea and the asthma provoked by it is called hyperventilation-induced asthma (HIA). The similarity between the changes in lung function provoked by exercise and isocapnic hyperventilation has led to the assumption that the stimulus, mechanism and pathway by which these challenges provoke an attack of asthma are probably the same. However there are some differences in relation to the effect of drugs, particularly when air of subfreezing temperature is inhaled. Further there are differences between isocapnic hyperventilation and exercise in relation to the refractory period and the cardiovascular, metabolic and hormonal changes, all of which have the potential to modify responses during and possibly even after challenge.[3]

The stimulus by which hyperpnoea with dry air induces an attack of asthma is the loss of water from the respiratory mucosa.[4] Evaporative water loss is thought to cause the airways to narrow when the increase in the concentration of ions leads to an increase in osmolarity of the airway surface liquid (ASL) of the respiratory tract,[5] often referred to as the periciliary fluid. Unfortunately, there have been no direct measurements showing an increase in osmolarity of the ASL of the lower airways after exercise, although hyperosmolar changes do occur in the nose in response to hyperpnoea,[6] and at rest when the upper airways are bypassed, as occurs in laryngectomized subjects.[7] Studies that have compared the responses to exercise and hyperventilation with the inhalation of aerosols of hyperosmolar saline have shown no inconsistencies with this hypothesis and have confirmed that hyperosmolarity is a potent stimulus for provoking airway narrowing in asthmatic subjects.[8-12] The studies on the effects of hypertonic aerosols followed the study of Allegra and Bianco[13] that showed that inhaling an aerosol of distilled water could also provoke an attack of asthma. This is often referred to as 'fog-induced' asthma. These observations have led to the use of non-isotonic aerosols to increase or decrease the osmolarity of the ASL and thus induce acute narrowing of the airways.[9-16]

It is estimated that the volume of fluid lining the large airways is small and less than 1 ml in the first 10 generations.[5,17] The osmolarity is approximately 359 mosmol and the major ions contributing to this are chloride and sodium.[18] The small volume of the ASL creates the potential for rapid change in its osmolarity. Only microlitres (equivalent to milligrams) of water need to be lost by evaporation and only microlitres of non-isotonic fluid need to be deposited on the ASL to cause a rapid and significant change in ion concentration.

Exercise, hyperventilation and the inhalation of non-isotonic aerosols, the so-called 'physical challenges', are thought to cause the release of chemical mediators from cells in the airway mucosa in response to a change in the osmotic environment. The release of chemotactic factors suggests that cell activation is also a consequence of these challenges.[19-22] The severity of the airway responses to these 'physical' challenges is thought to be an index of the severity of inflammation of the airways. Thus these challenges are now considered as useful for diagnosing asthma in the pulmonary function laboratory and standardized protocols have been developed.[15,16]

The airway narrowing provoked by these challenges is inhibited by the acute administration of β_2-adrenoceptor agonists, sodium cromoglycate, nedocromil sodium, leukotriene antagonists and the antihistamine terfenadine,[1,23-27] and reduced by the chronic administration of aerosols steroids.[28-30] For these reasons, these challenges are useful for identifying and assessing the benefits, or otherwise, of asthma treatment. They have been used widely in both adults and children.

The recognition that an attack of asthma can be provoked by a change in the osmolarity in the respiratory tract has advanced our understanding of non-immunologically mediated factors that can provoke an attack of asthma. It has also drawn to our attention the importance of maintaining a normal balance of water in the respiratory tract. This is particularly important as many events encountered in daily life have the potential to change the osmolarity of the airways. For example, exercise, accidental aspiration of non-isotonic fluid whilst swimming or diving, infection, smoke and inhalation of aerosol particles of therapeutic and non-therapeutic substances may all lead to an alteration in airway osmolarity.

RESPIRATORY WATER LOSS AND CONDITIONING OF INSPIRED AIR

Under most ambient conditions inspired air is cooler and drier than alveolar air and it must be heated and humidified as it enters the body. At rest, air is conditioned as it passes over the nasal mucosa and is almost fully conditioned by the time it reaches the pharynx. The nasal mucosa has a surface area of around 160 cm^2 and not only provides heat and water vapour to the inspired air but also, because it cools during inspiration, conserves heat and water during expiration.[31] As the rate of ventilation increases during exercise, the increase in inspiratory resistance causes a switch from nasal to mouth breathing. Therefore the burden to heat and humidify the inspired air is transferred to the intrathoracic airways.

Direct measurements of temperature flux occurring during hyperpnoea in human airways have demonstrated that the number of generations of airways required to condition the inspired air fully will vary depending upon the temperature and humidity of the inspired air and on the rate of ventilation.[32,33] Furthermore, these and other studies have demonstrated that the number of generations involved in conditioning inspired air increases with time. This implies that, during hyperpnoea, the rate of return of water to the respiratory mucosa is not sufficient to provide continuous humidification from the same generations of airways. This concept can be more easily understood when one considers the need to humidify air, even at low rates of ventilation, when the upper airways are bypassed during mechanical ventilation.[34]

The amount of water required to humidify the inspired air will depend on its water content and the rate of ventilation. When completely dry air with a temperature between 23 and 42°C is inhaled, the net loss of water during exercise[35,36] and isocapnic hyperventilation[8,37] is between 29 and 35 μg (or mg)/litre of air. A considerable proportion of this water will be provided by the intrathoracic airways, which will be cooled as a result of the latent heat of vaporization of water. Because of this cooling process, some of the water will be returned to the mucosa on expiration and the net loss of water from the airways below the pharynx is small. It has been estimated that the water lost during moderate exercise may be as little as 1–7 μl (or mg)/litre of air.[17,38] If the loss of water is accumulated over 1 min during exercise while ventilating at 80 litre/min, a loss of 80–560 μl of water loss would result. Considering even these small volumes in relation to available water (Table 31.1), it can easily be appreciated that a transient alteration in osmolarity could occur if replacement is not instantaneous.[17]

Changing the temperature of the inspired air has the potential to change the site of the osmotic stimulus. When the air is cool, and therefore dry, more generations of airways will need to be recruited in order to heat and humidify the air completely. In some asthmatic subjects there is an increase in severity of the airway response after exercise breathing cold air.[39] This may be due to a greater number of airways being recruited and made hyperosmolar by evaporative water loss when breathing cold air compared with the number recruited when breathing dry air of temperate conditions. The fact that inspiring cold air does not enhance the airway response[37,40] in all subjects may reflect a limitation in the number of airways that can become hyperosmolar. Beyond the 14th generation it is unlikely that the osmolarity of the ASL changes, as there is enough water available to humidify the air fully before it reaches the alveoli. Because of the substantial amount of water available in the small airways, there is no need for water to be replaced rapidly in the large airways. The potential for the osmolarity to increase in the large airways as a consequence is, however, of great physiological significance for the patient with asthma.

Several years ago there was some debate as to whether cooling of the airways or hyperosmolarity of the periciliary fluid was the primary stimulus to asthma provoked by hyperpnoea.[17,35,40,41] It appears that abnormal airway cooling is not a prerequisite for

Table 31.1 Surface area and volume of periciliary fluid available in the first 17 generations of human airways.

Airway generation no.	Surface area (cm²)	Periciliary fluid volume (μl) at depth of 5 μm	Periciliary fluid volume (μl) at depth of 10 μm	Cumulative surface area (cm²)	Cumulative volume (μl) of periciliary fluid at depth of 5 μm	Cumulative volume (μl) of periciliary fluid at depth of 10 μm
0	68	34	68	68	34	68
1	37	18	37	105	52	105
2	21	10	21	124	63	126
3	9	4	9	135	67	135
4	29	15	29	164	82	164
5	37	18	37	201	100	201
6	51	25	51	252	126	252
7	70	35	70	322	160	322
8	95	47	95	417	208	417
9	135	67	135	552	276	552
10	190	95	190	742	371	742
11	275	137	275	1015	507	1015
12	397	198	397	1412	706	1412
13	584	292	584	1996	998	1996
14	870	435	870	2866	1433	2866
15	1312	656	1312	4178	2089	4178
16	1905	950	1901	6183	3091	6183
17	3100	1550	3100	9283	4641	9283

The depth of the periciliary fluid (sol layer) has been taken to be uniform at 5 and 10 μm. Reproduced from ref. 17, with permission.

airways to narrow in response to hyperpnoea. Rather, airway cooling is an epiphenomenon of exercise and not essential to the development of EIA. Further, there are now reports that airway cooling is an inhibitory rather than excitatory stimulus.[9,42–44] This may account for the observation that inhaling hot dry air during exercise or hyperventilation is a potent stimulus to airway narrowing.[35,37,39] The effect of heating the inspired air may be to concentrate the osmotic stimulus in the larger airways.

Both duration and intensity of exercise are important determinants of the severity of the airway response[1] and the reason for this is likely to relate to the number of airways involved in the process of conditioning air.

ROLE OF THE BRONCHIAL CIRCULATION

Although it is now generally acknowledged that airway cooling *per se* is not the mechanism whereby hyperpnoea provokes asthma, it has been suggested that rapid rewarming of the airways may be important.[45,46] An hypothesis, put forward in 1990, stated that 'rapid expansion of the blood volume in perivascular plexi may be an important cause of the airway narrowing after exercise and isocapnic hyperventilation'.[46]

McFadden and his colleagues proposed that the bronchial microvasculature constricts in response to cooling of the airways during exercise, and that, on cessation of exercise, there is a reactive hyperaemia of the microvasculature with consequent changes in permeability leading to submucosal oedema.

Studies in humans at rest[47] and in animals report an increase in blood flow in response to inhaling dry air[48] and there is no evidence in humans to show that the bronchial microvasculature is constricted with exercise. However, the concept of vasoconstriction and a compromised bronchial circulation during exercise is of interest in that it is this circulation that provides the water to the airways. If this circulation is decreased rather than increased with exercise, it would allow a change in osmolarity to occur more rapidly and would reduce the rate of clearance of released mediators. Further, oedema may occur as a response to an increase in osmolarity of the submucosa if dehydation was severe.

The concept of rapid expansion of the bronchial microvasculature accounting for airway narrowing is difficult to justify with regard to the responses to drugs such as sodium cromoglycate and nedocromil sodium and the rapid reversal of EIA and HIA by the administration of β_2-adrenoceptor-agonists.[1] None of these drugs has profound effects on vascular smooth muscle. Further, EIA and HIA may occur when no abnormal cooling of the airways has occurred[35,49] or during exercise before rewarming takes place.[17,49–51] Further, studies in animals report only modest changes in airway calibre when the bronchial vasculature is dilated or oedema occurs independent of smooth muscle contraction.[52] The inability of the rapid rewarming hypothesis to account for many of the established facts about EIA and HIA has led to a vigorous debate on the topic.[41,53,54]

These arguments do not detract from the importance of this circulation in the bronchial hyperresponsiveness observed in asthma, particularly as oedema of the airway wall can serve to amplify the airway narrowing produced by contraction of bronchial smooth muscle.[55] Further, this circulation is important for maintaining mucociliary clearance[56] and it is likely to be the major source of ASL under conditions of acute dehydration. There are other potential sources for replenishment of ASL. For example, some water will be returned to the larger airways during expiration as warm humid air passes over the cooled mucosal surface. Other possible sources include the submucosal glands and the mucociliary escalator. Taking into account the length of the respiratory tract, considerable time would be required to replace ASL from the mucociliary escalator as it moves at about 5 mm/min. The depth and amount of ASL is most likely controlled locally by the passage of water across the epithelial cells and through paracellular channels in response to the movement of sodium and chloride ions.[57] Because the volume of fluid moving up the mucociliary escalator is large, the epithelial cells are thought to absorb rather than secrete water by the movement of ions. As sodium ions move from the lumen into the epithelial cells, chloride ions move passively and with these, water. Under conditions of drying and hyperosmotic stress, it is likely that water moves from the epithelial cells and other cells close to the airway lumen towards the airway lumen. Water will move from the submucosal space to restore cell volume. The osmotic gradient initiated by the movement of water towards the airway lumen is possibly the stimulus for the release of nitric oxide and increase in bronchial blood flow in response to airway drying and hyperosmolarity. Studies using hyperosmolar solutions in animals have demonstrated vasodilatation of the tracheal circulation and that addition of the nitric oxide synthesis inhibitor prevents the vasodilatation.[58,59] Further, the biochemical events involved in the restoration of cell volume may trigger the release of mediators from epithelial cells and mast cells.[60]

GENERATION AND DEPOSITION OF NON-ISOTONIC AEROSOLS IN THE RESPIRATORY TRACT

Distilled water and 3.6–4.5% saline have been the most common non-isotonic solutions used for bronchial provocation in both the laboratory and the field for epidemiological purposes.[11,13–15,61–64] Because of their high output (66–100 μl/litre) of aerosol, ultrasonic nebulizers rather than jet nebulizers are used to generate the aerosol particles. The mass median aerodynamic diameter (MMAD) of the particles ranges from 2 to 10 μm for most ultrasonic nebulizers. When the MMAD is 3–5 μm, between 15 and 35% of the aerosol particles inhaled are predicted to deposit in the human respiratory tract.[15] A volume of 80–480 μl (0.08–0.48 ml)/min has been predicted to deposit in the lower airways under these conditions.[15] Even at the lower rate a significant change in osmolarity could occur in ASL in less than 5 min.

COMPARISON BETWEEN CHALLENGE WITH EXERCISE AND HYPERVENTILATION AND CHALLENGE WITH NON-ISOTONIC AEROSOLS

Because the volume of ASL is so small and distributed over a large surface area, no attempt has been made to measure the osmolarity of ASL before, during and after exercise or hyperventilation. Rather, the airway responses to exercise and hyperventilation have been compared with a known hyperosmolar stimulus. Some findings are illustrated in Fig. 31.1 and demonstrate that there is excellent concordance between responses to exercise and hyperventilation and the inhalation of hyperosmolar saline.[8–10,12,65] One study,[9] which compared the responses to multiple 1-min exposures of 4.5% saline aerosol and isocapnic hyperventilation, demonstrated a difference in the time-course of the maximal response (Fig. 31.2). During aerosol challenge 86% of the maximal reduction in forced expiratory volume in 1 s (FEV_1) occurred, whereas only 54% of the response was documented during challenge with isocapnic hyperventilation. This finding is in keeping with the suggestion that airway cooling may act to inhibit responses to hyperventilation.[42–44] Cross-refractoriness has been demonstrated between exercise and hyperosmolar saline, suggesting that the mechanisms for the development of the refractoriness is the same[65] (Fig. 31.3).

Although asthmatic subjects are sensitive to inhaling both water and hypertonic aerosols, the responses in some individuals are not well correlated,[10] suggesting some differences in the development of the response or deposition of the aerosol.

There has been some attention given to comparing responses to isocapnic hyperventilation and exercise with those to distilled water.[66–69]

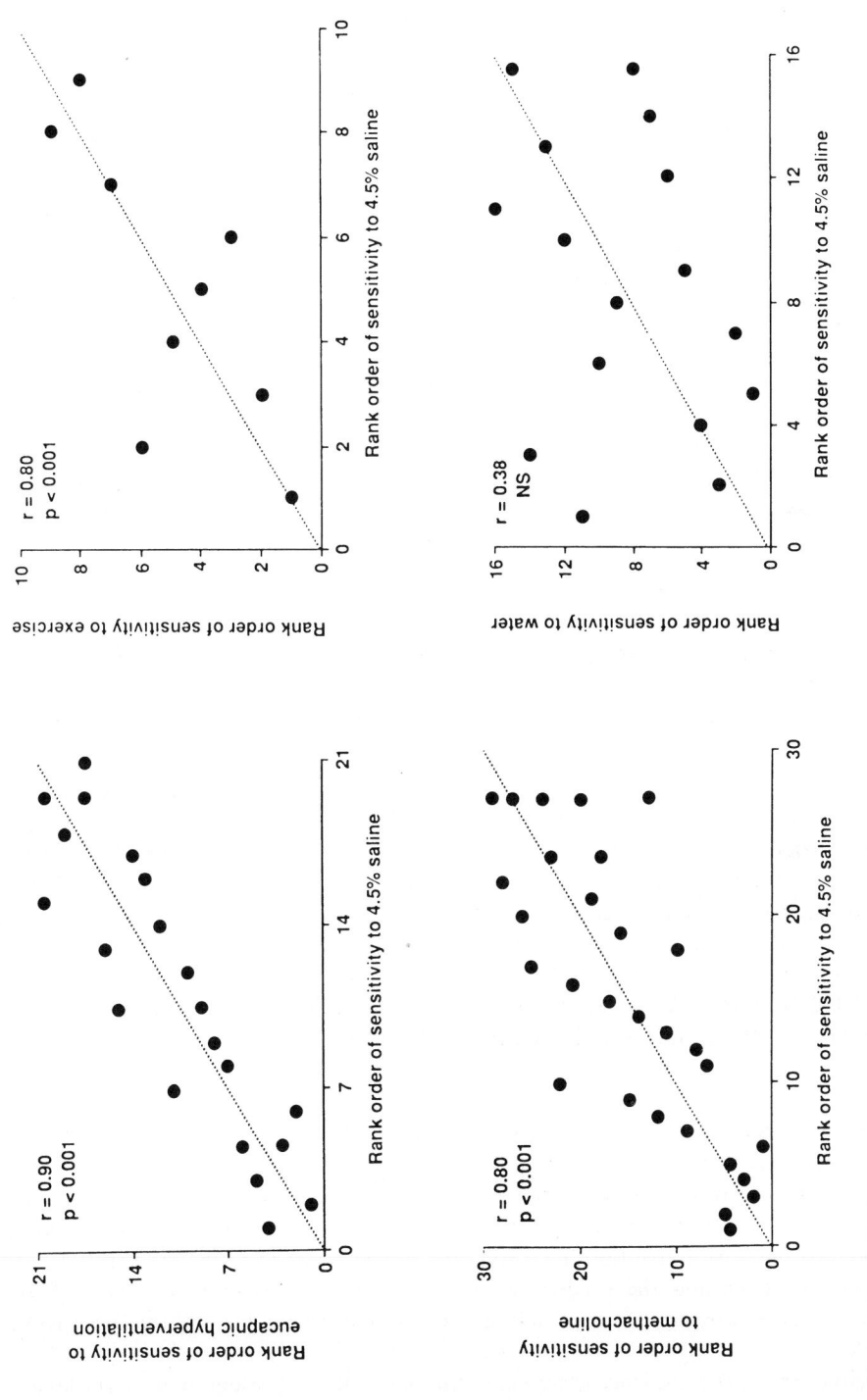

Fig. 31.1 Relationship between sensitivity to 4.5% saline inhalation and eucapnic hyperventilation, exercise, methacholine and water in asthmatic subjects, whose responses were ranked from most sensitive to least sensitive. A Spearman's rank order correlation coefficient is given. Data based on the results presented by ref 10.

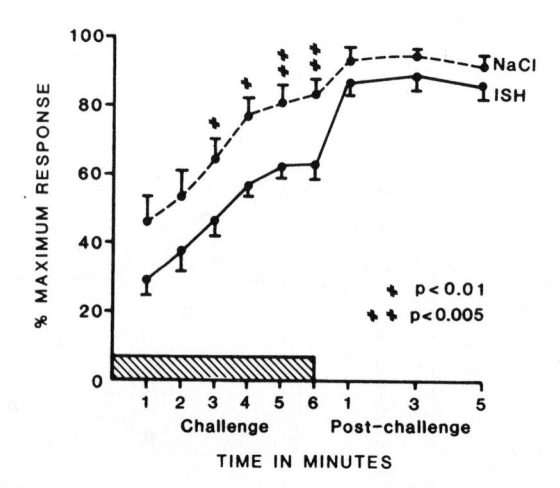

Fig. 31.2 A comparison of the percentage of the maximum response recorded for the reduction in forced expiratory volume in 1 s (FEV$_1$) during the last 6 min of challenge with 4.5% saline and isocapnic hyperventilation (ISH) in nine asthmatic subjects. Data taken from ref 9.

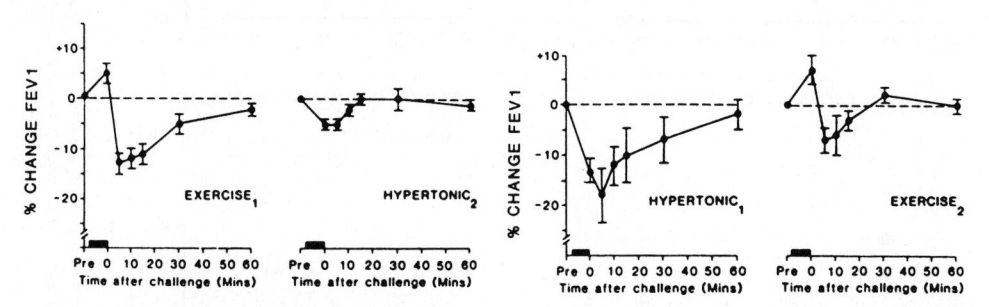

Fig. 31.3 The mean ± 1 SEM percentage change in forced expiratory volume in 1 s (FEV$_1$) from baseline following exercise (60% of maximal predicted oxygen consumption) and hypertonic challenge (3.6% saline) in four asthmatic subjects. The data demonstrate the cross-refractoriness between these two challenge tests. Reproduced from ref. 65, with permission.

MECHANISM BY WHICH A CHANGE IN OSMOLARITY AND AIRWAY DRYING INDUCE AIRWAY NARROWING

There are a number of sites in the airways where a change in osmolarity could act as a stimulus to induce airway narrowing[69] (Fig. 31.4). However it is likely that the bronchoconstrictor response to hyperpnoea and to the deposition of particles of non-isotonic aerosols is initiated by events that occur in, or close to, the airway lumen. This is based on the observation that both EIA and HIA are prevented by inhaling warm humid air during challenge and the prediction that small volumes of non-isotonic aerosols deposited in the respiratory tract are unlikely to alter osmolarity beyond the submucosal space.

There is considerable evidence in humans that mast cells are present in or on the airway epithelium and are in an ideal situation to respond to a change in the osmotic

31 Provoked by Exercise, Hyperventilation and Non-isotonic Aerosols

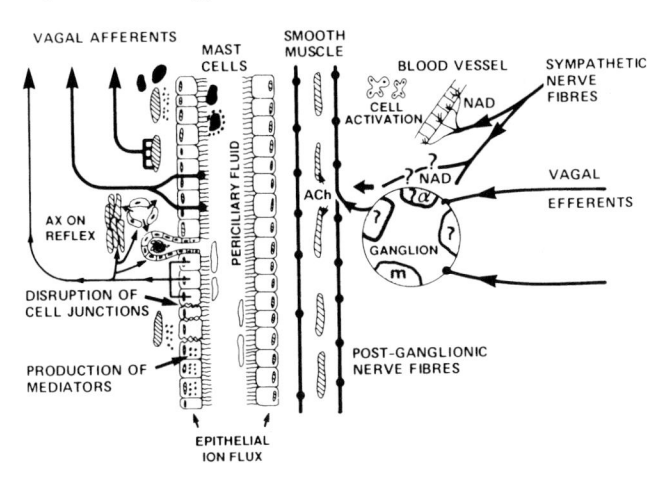

Fig. 31.4 Schematic diagram of the human airways illustrating the various sites that may be affected by an alteration of osmolarity. ACh, acetylcholine; NAD, noradrenaline. Modified from ref. 69.

environment.[70] It is also known that human lung mast cells and basophils can degranulate and release mediators in response to either an increase or a decrease in osmolarity.[71–73] Flint et al.[74] demonstrated that mast cells, recovered during bronchoalveolar lavage (BAL) in patients with asthma, release histamine in response to a hyperosmolar stimulus and that this release is inhibited by sodium cromoglycate. These observations are in keeping with the *in vitro* findings of Eggleston et al.[71] who have shown that dispersed human lung mast cells release histamine in response to hyperosmolar solutions and that this release is dissociated from release of leukotrienes and prostaglandins.[75] Studies of the effect of changing temperature on the hyperosmolar release of histamine from mast cells and basophils have shown that release is greatest at 32°C,[71] a temperature easily achieved in the large airways during exercise.[32] Furthermore, the optimum osmolarity for release of histamine was between 600 and 800 mosmol.[1] This is a value similar to that which would be induced *in vivo* by either evaporative water loss or the inhalation of aerosols of hyperosmolar saline of around 4.5%.

In vivo mediator release studies do not correlate so well with these findings. An increase in plasma histamine is not a universal finding in patients after challenge with exercise and non-isotonic aerosols.[21,22,76] Studies of isolated segment lavage[77] and BAL[78] have been disappointing in their failure to demonstrate large changes in mediator levels in response to changes in osmolarity and to exercise, although leukotrienes were found after hyperventilation with dry air.[79] The techniques are limited in their ability to sample adequately from the airways and the sampling is not simultaneous with the airway response. Mediators have been found in BAL after isocapnic hyperventilation with dry air.[79] Further, bronchoconstriction can occur in response to minute quantities of mediators, making their presence difficult to detect in lavage fluid. Recent studies using induced sputum for analysis of mediators suggests that this technique may overcome many of the problems previously encountered. Further, the airway response to hypertonic saline and the collection of inflammatory cells can be made at the same time.[80] Although histamine may not be the most potent mediator, it must contribute to airway narrowing after these challenges because antihistamines markedly reduce the changes in lung function caused by exercise, isocapnic hyperventilation and non-isotonic aerosols.[81,82] As the major source of histamine is either mast cells or basophils, it is likely that mast cells are involved and release their mediators in response to evaporative water loss and a change in osmolarity of the ASL.

There are other cells that could also participate in events leading to airway narrowing, such as epithelial cells and nerve cells. For example, epithelial cells produce leukotrienes, which have the potential to contract bronchial smooth muscle.[83] These cells may also produce prostaglandin E_2[84] in response to an increase in osmolarity. This could be an important mode of protection against bronchoconstriction and would function best in those who have an intact epithelium. A change in osmolarity in the airway lumen could stimulate the release of other substances associated with the inflammatory response, such as nitric oxide,[58,59] and this could change the permeability of the cells to water. There are contradictory findings from studies on human airways *in vitro*. Some investigators have found an osmolarity-dependent contraction of human isolated airways,[85] while others have found relaxation.[86]

The possibility that a change in osmolarity can stimulate vagal afferent nerve activity has been considered[87-89] and there is recent evidence in animals that activation of sensory nerve endings is important in the bronchoconstriction induced by changes in osmolarity.[90-93] Elevated levels of substance P have been measured in patients with asthma and chronic bronchitis.[93] However, this finding needs to be confirmed in larger numbers of subjects before any conclusion is reached as to whether this is released by the hypertonic saline used to induce the sputum.

Cells associated with inflammation, such as eosinophils and neutrophils, are not normally present in the respiratory mucosa but may be attracted to the airways by changes in airway osmolarity and subsequent release of mediators. Evidence to support a role for these cells in bronchial responsiveness to changes in osmolarity comes from the knowledge that chemotactic activity increases in serum in response to these challenges[19-22] and drugs that modify airway inflammation decrease airway responsiveness to these challenges.[29,30] However, it is unlikely that the cell activation and inflammatory events that follow allergen challenge are exactly reproduced by osmotic challenge, the reason being that there are only small, if any, changes in non-specific bronchial hyperresponsiveness after these challenges compared with large changes after allergen challenge.[94,95]

EFFECT OF PHARMACOLOGICAL AGENTS

Most drugs commonly used to treat or prevent attacks of asthma will inhibit or modify the airway responses to the 'physical' challenges.[23-30,96,97] The most effective drugs are the β_2-adrenoceptor agonists, sodium cromoglycate and nedocromil sodium. The anticholinergic agents are not as effective as the β_2-adrenoceptor agonists although, in some patients, they do inhibit the responses.[25,87-89,98,99] It should be noted, however, that most of these drugs increase the threshold at which the airways narrow in response to these challenges. Providing the stimulus persists or is made more potent the airways will still narrow in the presence of the drugs. This may have important implications from both a treatment and a mechanistic standpoint.

For example, regular treatment with inhaled steroids has been shown to increase the threshold at which the stimulus provokes airway narrowing, i.e. sensitivity, and the rate at which the airways narrow, i.e. reactivity, to these challenges.[29,30,96,100,101] However, chronic treatment with aerosol beclomethasone does not necessarily reduce the capacity of the airways to narrow excessively in response to 4.5% saline aerosol[30, 96,97] or distilled water.[101] Similar findings have been made with exercise,[23] and hyperventilation[102] where patients have severe airway responses even though they are receiving high doses of beclomethasone and have lung function within the normal range. The effect of the newer steroid fluticasone is not known, although regular treatment with budesonide can markedly reduce severity of EIA[28,29] and responses to hyperosmolar saline.[30,100] Thus the particular steroid used in the treatment of bronchial hyperresponsiveness may be as important as the dose.

There has been renewed interest in the role of antihistamines in modifying the airway responses to 'physical' challenges because of the development of non-sedating drugs that do not have anticholinergic activity. Terfenadine is an example of this and it is effective in inhibiting responses to these challenges.[28,81,82] Although antihistamines can be given as aerosols that inhibit these challenges,[103,104] they are not commercially available in the aerosol formulation.

The cyclooxygenase inhibitor flurbiprofen, when given orally in a dose of 100 mg, has been shown to reduce the bronchoconstrictor response to both exercise[80] and hypertonic saline.[105] The reduction was less with flurbiprofen compared with terfenadine, suggesting that histamine was a more important mediator in these challenges. Further, when flurbiprofen was given before continuous challenge, rather than a progressive one, it afforded no protection against hypertonic saline. Leukotriene antagonists have also been shown to inhibit responses when delivered as an aerosol[106] and by tablet before exercise.[26] Their effect on hypertonic challenge has not been reported.

There are a number of drugs not used in the treatment of asthma but which have been investigated for their effectiveness in preventing the response to these challenges. The most interesting development has come from the work of Bianco and colleagues with the diuretic frusemide. They showed that this drug, when given as an aerosol in a dose of 28 mg, was effective in preventing EIA,[107] while it was ineffective when administered in the same dose as a tablet. Inhaled frusemide prevents airway responses to challenge with water,[108] hyperventilation,[109] hypertonic saline,[110] allergen,[111] metabisulphite[112] and adenosine.[113] Inhaled frusemide reduces the frequency of cough to low chloride ion concentration but not capsaicin.[114] Frusemide has been shown to prevent the release of histamine and leukotrienes from human lung fragments *in vitro*.[115]

There have been many studies on the effect of calcium antagonists, α-adrenoceptor antagonists and the sodium nitrates on these challenges.[116–120] The precise mode of action of these drugs in reducing the response is not known but a vasodilator action on the bronchial vasculature could be important. Vasodilatation of the bronchial vasculature would permit a faster return of water and clearance of bronchoconstrictor mediators. Paradoxically, the α-adrenergic agonist methoxamine has also been shown to modify EIA,[121] which makes some of these findings difficult to interpret.

There have been a number of studies using local anaesthetics to prevent airway narrowing to these challenges. It would appear that lignocaine prevents the cough but does not necessarily prevent the bronchoconstriction induced by these challenges.[89,122] More studies are required to exclude the possibility that sensory nerves are involved because of the recent finding that hyperosmolarity is a stimulus to neuropeptide release.[90,92,93]

The study of the effects of pharmacological agents, given both as aerosol and by tablet, has given us a better understanding of the possible mechanisms whereby these challenges provoke acute airway narrowing. The relative failure of orally administered drugs such as β_2-adrenoceptor agonists, methylxanthines and anticholinergics agents to prevent the responses to these challenges at a time when bronchodilation is evident is of interest.[25,123,124] It suggests that the concentration of a drug at the smooth muscle receptor required to induce relaxation is less than that required to prevent contraction.[125] It also implies that events other than contraction of bronchial smooth muscle may contribute to airway narrowing. When drugs are administered as aerosols they reach the airways in high concentrations and have easy access to cells close to the airway lumen. For example, β_2-adrenoceptor agonists are potent inhibitors of mediator release from mast cells.[126] They also have the potential to alter ion transport across the epithelial cell to improve the delivery of water to the airway lumen,[57] and enhance the mucociliary escalator. All these modes of action may contribute to their ability to prevent the airway narrowing induced by evaporative water loss and changes in osmolarity of ASL.

It is quite clear that a drug need not have a relaxing effect on smooth muscle to prevent the responses to these challenges. Sodium cromoglycate, nedocromil sodium and frusemide are good examples of this. The finding that frusemide has a similar profile to sodium cromoglycate and nedocromil sodium, in terms of blocking many provocative stimuli, has improved our understanding of the possible mechanism whereby these drugs act to prevent airway narrowing. Recent studies have shown that nedocromil sodium and sodium cromoglycate block chloride ion channels in human tissue.[127] They are also known to block the chloride ion channel responsible for cell volume regulation.[128] We have recently proposed that their mode of action, in preventing the release of mediators, may be their ability to increase cell volume, thereby delaying the biochemical events that accompany regulatory volume increase following hyperosmotic stress.[60] Thus in the presence of sodium cromoglycate or nedocromil sodium, biochemical events that accompany regulatory volume increase may be altered. This could change degranulation of mast cells in some way and inhibit subsequent release of mediators. The potential of these same drugs to prevent the release of neuropeptides from sensory nerves is also of current interest.

The precise mechanism whereby aerosol steroids reduce the sensitivity and reactivity to these challenges is unknown but there are many possibilities. A decrease in sensitivity may reflect an improvement in the integrity of the airway epithelium, resulting in an improved mechanical barrier. Thus nerve endings may be protected from the stimulus and simply less mediator may reach the smooth muscle. Steroids have the potential to reduce airway oedema and the amplifying effect that bronchial smooth muscle contraction has on airway calibre. Steroids may also improve water transport to the airway submucosa by their effect on the Na^+ K^+ ATPase pump. By reducing the numbers of mast cells, steroids act to reduce the amount of bronchoconstricting mediators.[129] In addition, by reducing the number of other inflammatory cells, the smooth muscle may become less sensitive to contractile agents in the presence of steroids.[130]

The most effective of all the agents that block EIA is water vapour.[4,39,40] Although it was originally considered that water prevented EIA by modifying the effects of heat loss,[39] it is now appreciated that water blocks EIA by preventing respiratory water loss and the subsequent changes in osmolarity of ASL.[4] Although the inhalation of fully saturated air at body temperature (37°C, 44 mg H_2O/litre) usually prevents EIA, bronchoconstriction can still occur with these conditions.[4] As 33 mg/litre is normally lost during expiration, inhaling air with a water concentration greater than this may create a hypotonic environment and this is also a stimulus to asthma. This observation highlights the importance of maintaining a normal balance of water and ions in the respiratory tract.

If the only significant amount of water loss that occurs during exercise is from the first 14 generations of the airways, it may be that all that is required to prevent EIA is to ensure that sufficient water is given back to these airways during exercise. The small amounts lost may explain the benefits obtained from respiratory moisture exchangers, such as masks and filters, which permit a small amount of water vapour to be inhaled with each breath.[131] In the future drugs that act specifically to stimulate transport of water to these airways may be used to control EIA.

REFERENCES

1. Anderson SD, Silverman M, Godfrey S, et al.: Exercise-induced asthma: a review. *Br J Dis Chest* (1975) **69**: 1–39.
2. Deal EC, McFadden ER, Ingram RH, et al.: Hyperpnoea and heat flux: initial reaction sequence in exercise-induced asthma. *J Appl Physiol: Respir Environ Exercise Physiol* (1979) **46**: 476–483.
3. Godfrey S, Bar-Yishay E: Exercise-induced asthma revisited. *Respir Med* (1993) **87**: 331–344.
4. Anderson SD, Schoeffel RE, Follet R, et al.: Sensitivity to heat and water loss at rest and during exercise in asthmatic patients. *Eur J Respir Dis* (1982) **63**: 459–471.
5. Anderson SD: Is there a unifying hypothesis for exercise-induced asthma? *J Allergy Clin Immunol* (1984) **73**: 660–665.
6. Togias AG, Proud D, Lichtenstein LM, et al.: The osmolarity of nasal secretions increases when inflammatory mediators are released in response to inhalation of cold, dry air. *Am Rev Respir Dis* (1988) **137**: 625–629.
7. Potter JL, Matthews LW, Spector S, et al.: Studies on pulmonary secretions. II. Osmolarity of the ionic environment of pulmonary secretions from patients with cystic fibrosis, bronchiectasis and laryngectomy. *Am Rev Respir Dis* (1967) **96**: 83–87.
8. Smith CM, Anderson SD: Hyperosmolarity as the stimulus to asthma induced by hyperventilation? *J Allergy Clin Immunol* (1986) **77**: 729–736.
9. Smith CM, Anderson SD: A comparison between the airway response to isocapnic hyperventilation and hypertonic saline in subjects with asthma. *Eur Respir J* (1989) **2**: 36–43.
10. Smith CM, Anderson SD: Inhalational challenge using hypertonic saline in asthmatic subjects: a comparison with responses to hyperpnoea, methacholine and water. *Eur Respir J* (1990) **3**: 144–151.
11. Belcher NG, Lee TH, Rees PJ: Airway responses to hypertonic saline, exercise and histamine challenges in bronchial asthma. *Eur Respir J* (1989) **2**: 44–48.
12. Boulet LP, Turcotte H: Comparative effects of hyperosmolar saline inhalation and exercise in asthma. *Immunol Allergy Practice* (1989) **11**: 93–100.
13. Allegra L, Bianco S: Non-specific bronchoreactivity obtained with an ultrasonic aerosol of distilled water. *Eur J Respir Dis* (1980) **61**: 41–49.
14. Smith CM, Anderson SD: Inhalation provocation tests using non-isotonic aerosols. *J Allergy Clin Immunol* (1989) **84**: 781–790.
15. Anderson SD, Smith CM, Rodwell LT, et al.: The use of non-isotonic aerosols for evaluating

bronchial hyperresponsiveness. In Spector S (ed) *Provocation Challenge Procedures*. New York, Marcel Dekker, (1995), pp 249–278.
16. Sterk PJ, Fabbri LM, Quanjer PhH, et al.: Airway responsiveness: standardized challenge testing with pharmacological, physical and sensitizing stimuli in adults. *Eur Respir J* (1993) **6**(Suppl 16): 53–83.
17. Anderson SD, Daviskas E, Smith CM: Exercise-induced asthma: a difference in opinion regarding the stimulus. *Allergy Proc* (1989) **10**: 215–226.
18. Boat TF, Matthews L-RW: Chemical composition of human tracheo-bronchial secretions. In Dulfano MJ (ed) *Fundamentals and Clinical Pathology of Sputum*. Springfield, IL, Thomas, (1973), pp 243–273.
19. Venge P, Henriksen J, Dahl R, et al.: Exercise-induced asthma and the generation of neutrophil chemotactic activity. *J Allergy Clin Immunol* (1990) **85**: 498–504.
20. Venge P, Henriksen J, Dahl R: Eosinophils in exercise-induced asthma. *J Allergy Clin Immunol* (1991) **88**: 699–704.
21. Shaw RJ, Anderson SD, Durham SR, et al.: Mediators of hypersensitivity and 'fog'-induced asthma. *Allergy* (1985) **40**: 48–57.
22. Belcher NG, Murdoch RD, Dalton N, et al.: A comparison of mediator and catecholamine release between exercise- and hypertonic saline-induced asthma. *Am Rev Respir Dis* (1988) **137**: 1026–1032.
23. Anderson SD, Rodwell LT, Du Toit J, et al.: Duration of protection of inhaled salmeterol in exercise-induced asthma. *Chest* (1991) **100**: 1254–1260.
24. Comis A, Valletta EA, Sette L, et al.: Comparison of nedocromil sodium and sodium cromoglycate administered by pressurized aerosol, with and without a spacer device in exercise-induced asthma in children. *Eur Respir J* (1993) **6**: 523–526.
25. Boulet L-P, Turcotte H, Tennina S: Comparative efficacy of salbutamol, ipratropium and cromoglycate in the prevention of bronchospasm induced by exercise and hyperosmolar challenges. *J Allergy Clin Immunol* (1989) **83**: 882–887.
26. Finnerty JP, Wood-Baker R, Thomson H, et al.: Role of leukotrienes in exercise-induced asthma. Inhibitory effect of ICI 204219, a potent leukotriene D$_4$ receptor antagonist. *Am Rev Respir Dis* (1992) **145**: 746–749.
27. del Bufalo C, Fasano L, Patalano F, et al.: Inhibition of fog-induced bronchoconstriction by nedocromil sodium and sodium cromoglycate in intrinsic asthma: a double-blind, placebo-controlled study. *Respiration* (1989) **55**: 181–185.
28. Patel KR: Terfenadine in exercise-induced asthma. *Br Med J* (1984) **285**: 1496–1497.
29. Waalkans HJ, van Essen-Zandvliet EEM, Gerritsen J, et al.: The effect of an inhaled corticosteroid (budesonide) on exercise-induced asthma in children. *Eur Respir J* (1993) **6**: 652–656.
30. Anderson SD, du Toit JI, Rodwell LT, et al.: Acute effect of sodium cromoglycate on airway narrowing induced by 4.5 percent saline aerosol. Outcome before and during treatment with aerosol corticosteroids in patients with asthma. *Chest* (1994) **105**: 673–680.
31. Anderson SD, Togias A: Dry air and hyperosmolar challenge in asthma and rhinitis. In Busse W, Holgate S (eds) *Asthma and Rhinitis*. Blackwell Scientific Publications, Boston, 1994, pp 1178–1195.
32. McFadden ER, Pichurko BM, Bowman HF, et al.: Thermal mapping of the airways in humans. *J Appl Physiol* (1985) **58**: 564–570.
33. McFadden ER, Pichurko BM: Intraairway thermal profiles during exercise and hyperventilation in normal man. *J Clin Invest* (1985) **76**: 1007–1010.
34. Chalon J, Loew DAY, Malebranche J: Effects of dry anesthetic gases on tracheobronchial ciliated epithelium. *Anesthesiology* (1972) **37**: 338–343.
35. Anderson SD, Schoeffel RE, Black JL, et al.: Airway cooling as the stimulus to exercise-induced asthma. A re-evaluation. *Eur J Respir Dis* (1985) **67**: 20–30.
36. Tabka Z, Ben Jebria A, Guenard H: Effect of breathing dry warm air on respiratory water loss at rest and during exercise. *Respir Physiol* (1987) **67**: 115–125.
37. Eschenbacher WL, Sheppard D: Respiratory heat loss is not the sole stimulus for bronchoconstriction induced by isocapnic hyperpnoea with dry air. *Am Rev Respir Dis* (1985) **131**: 894–901.

38. Daviskas E, Gonda I, Anderson SD: Local airway heat and water vapour losses. *Respir Physiol* (1991) **84**: 115–132.
39. Deal EC, McFadden ER, Ingram RH, et al.: Role of respiratory heat exchange in production of exercise-induced asthma. *J Appl Physiol: Respir Environ Exercise Physiol* (1979) **46**: 467–475.
40. Hahn A, Anderson SD, Morton AR, et al.: A re-interpretation of the effect of temperature and water content of the inspired air in exercise-induced asthma. *Am Rev Respir Dis* (1984) **130**: 575–579.
41. Anderson SD, Daviskas L: An evaluation of the airway cooling and rewarming hypothesis as the mechanism for exercise induced asthma. In Holgate S, Kay AB, Lichtenstein L, Austen F. (eds) *Asthma: Physiology, Immunopharmacology, and Treatment*. London, Academic Press, 1993, 323–335.
42. Freed AN, Kelly LJ, Menkes HA: Airflow-induced bronchospasm: imbalance between airway cooling and airway drying? *Am Rev Respir Dis* (1987) **136**: 595–599.
43. Blackie SP, Hilliam C, Village R, et al.: The time course of bronchoconstriction in asthmatics during and after isocapnic hyperventilation. *Am Rev Respir Dis* (1990) **142**: 1133–1136.
44. Freed AN, Stream CE: Airway cooling: stimulus specific modulation of airway responsiveness in the canine lung periphery. *Eur Respir J* (1991) **4**: 568–574.
45. McFadden ER, Lenner KAM, Strohl KP: Postexertional airway rewarming and thermally induced asthma. New insights into pathophysiology and possible pathogenesis. *J Clin Invest* (1986) **78**: 18–25.
46. McFadden ER: Hypothesis: exercise-induced asthma as a vascular phenomenon. *Lancet* (1990) **335**: 880–882.
47. Agostini P, Arena V, Doria E, et al.: Inspired gas relative humidity affects systemic to pulmonary bronchial blood flow in humans. *Chest* (1990) **97**: 1377–1380.
48. Baile EM, Dahlby RW, Wiggs BR, et al.: Effect of cold and warm dry air hyperventilation on canine airway blood flow. *J Appl Physiol* (1987) **62**: 526–532.
49. Zawadski DK, Lenner KA, McFadden ER: Comparison of intraairway temperatures in normal and asthmatic subjects after hyperpnoea with hot, cold, and ambient air. *Am Rev Respir Dis* (1988) **138**: 1553–1558.
50. Smith CM, Anderson SD, Walsh S, et al.: An investigation of the effects of heat and water exchange in the recovery period after exercise in children with asthma. *Am Rev Respir Dis* (1989) **140**: 598–605.
51. Argyros GJ, Phillips YY, Rayburn DB, et al.: Water loss without heat flux in exercise-induced bronchospasm. *Am Rev Respir Dis* (1993) **147**: 1419–1424.
52. Freed AN, Omori C, Schofield BH: The effect of bronchial blood flow on hyperpnoea-induced airway obstruction and injury. *J Clin Invest* (1995) **96**: 1221–1229.
53. Anderson SD, Daviskas E: Exercise-induced asthma as a vascular phenomenon. *Lancet* (1990) **335**: 1410–1412.
54. Anderson SD, Daviskas E: The airway microvasculature and exercise-induced asthma. *Thorax* (1992) **47**: 748–752.
55. Moreno RH, Hogg JC, Pare PD: Mechanics of airway narrowing. *Am Rev Respir Dis* (1986) **133**: 1171–1180.
56. Deffebach ME, Charan NB, Lakshminarayan S, et al.: The bronchial circulation. Small, but a vital attribute of the lung. State of the Art. *Am Rev Respir Dis* (1987) **135**: 463–481.
57. Boucher RC: Human airway ion transport. *Am J Respir Crit Care Med* (1994) **150**: 271–281, 581–593.
58. Smith TL, Prazma J, Coleman CC, et al.: Control of the mucosal microcirculation in the upper respiratory tract. *Otolaryngol Head Neck Surg* (1993) **109**: 646–652.
59. Prazma J, Coleman CC, Shockley WW, et al.: Tracheal vascular response to hypertonic and hypotonic solutions. *J Appl Physiol.* (1994) **76**: 2275–2280.
60. Anderson SD, Rodwell LT, Daviskas E, et al.: The protective effect of nedocromil sodium and other drugs on airway narrowing provoked by hyperosmolar stimuli: a role for the airway epithelium. *J Allergy Clin Immunol* (1996) **98**: S124–S134.
61. Rabone S, Phoon WO, Anderson SD, et al.: Hypertonic saline bronchial challenge in adult epidemiological field studies. *Occup Med* (1996) **46**: 177–185.

62. Anderson SD, Brannan J, Trevillion L, *et al.*: Lung function and bronchial provocation tests for intending divers with a history of asthma. *SPUMS J* (1995) **25**: 233–248.
63. Riedler J, Reade T, Dalton M, *et al.*: Hypertonic saline challenge in an epidemiological survey of asthma in children. *Am J Respir Crit Care Med* (1994) **150**: 1632–1639.
64. Hopp RJ, Christy J, Bewtra AK, *et al.*: Incorporation and analysis of ultrasonically nebulized distilled water challenges in an epidemiological study of asthma and bronchial reactivity. *Ann Allergy* (1988) **60**: 129–133.
65. Belcher NG, Rees PJ, Clark TJH, *et al.*: A comparison of the refractory periods induced by hypertonic airway challenge and exercise in bronchial asthma. *Am Rev Respir Dis* (1987) **135**: 822–825.
66. Fourie PR, Joubert JR: Determination of airway hyper-reactivity in asthmatic children: a comparison among exercise, nebulized water, and histamine challenge. *Pediatr Pulmonol* (1988) **4**: 2–7.
67. Obata T, Iikura Y: Comparison of bronchial reactivity to ultrasonically nebulized distilled water, exercise and methacholine challenge test in asthmatic children. *Ann Allergy* (1994) **72**: 167–172.
68. Lemire TS, Hopp RJ, Bewtra A, *et al.*: Comparison of ultrasonically nebulized distilled water and cold-air hyperventilation challenges in asthmatic patients. *Chest* (1989) **95**: 958–961.
69. Anderson SD: Issues in exercise-induced asthma. *J Allergy Clin Immunol* (1985) **76**: 763–772.
70. Pesci A, Foresi A, Bertorelli G, *et al.*: Histochemical characteristics and degranulation of mast cells in epithelium and lamina propria of bronchial biopsies from asthmatic and normal subjects. *Am Rev Respir Dis* (1993) **147**: 684–689.
71. Eggleston PA, Kagey-Sobotka A, Lichtenstein LM: A comparison of the osmotic activation of basophils and human lung mast cells. *Am Rev Respir Dis* (1987) **135**: 1043–1048.
72. Rimmer J, Bryant DH: Effect of hypo- and hyper-osmolarity on basophil histamine release. *Clin Allergy* (1986) **16**: 221–230.
73. Silber G, Proud D, Warner J, *et al.*: *In vivo* release of inflammatory mediators by hyperosmolar solutions. *Am Rev Respir Dis* (1988) **137**: 606–612.
74. Flint KC, Leung KBP, Pearce FL, *et al.*: Human mast cells recovered by bronchoalveolar lavage: their morphology, histamine release and the effects of sodium cromoglycate. *Clin Sci* (1985) **68**: 427–432.
75. Eggleston PA, Kagey-Sobotka A, Proud D, *et al.*: Disassociation of the release of histamine and arachidonic acid metabolites from osmotically activated basophils and human lung mast cells. *Am Rev Respir Dis* (1990) **141**: 960–964.
76. Anderson SD, Bye PTP, Schoeffel RE, *et al.*: Arterial plasma histamine levels at rest, during and after exercise in patients with asthma: effects of terbutaline aerosol. *Thorax* (1981) **36**: 259–267.
77. Makker HK, Walls AF, Goulding D, *et al.*: Airway effects of local challenge with hypertonic saline in exercise-induced asthma. *Am J Respir Crit Care Med* (1994) **149**: 1012–1019.
78. Broide DH, Eisman S, Ramsdell JW, *et al.*: Airway levels of mast cell-derived mediators in exercise-induced asthma. *Am Rev Respir Dis* (1990) **141**: 563–568.
79. Pliss LB, Ingenito EP, Ingram RH, *et al.*: Assessment of bronchoalveolar cell and mediator response to isocapnic hyperpnoea in asthma. *Am Rev Respir Dis* (1990) **142**: 73–78.
80. Gibson PG, Woolley KL, Carty K, *et al.*: Safety of combined hypertonic saline inhalation challenge to induce sputum and measure airway responsiveness. *Am J Respir Crit Care Med* (1995) **151**: A401.
81. Finnerty JP, Holgate ST: Evidence for the roles of histamine and prostaglandins as mediators in exercise-induced asthma: the inhibitory effect of terfenadine and flurbiprofen alone and in combinations. *Eur Respir J* (1990) **3**: 540–547.
82. Finney MJB, Anderson SD, Black JL: Terfenadine modifies airway narrowing induced by the inhalation of non-isotonic aerosols in subjects with asthma. *Am Rev Respir Dis* (1990) **141**: 1151–1157.
83. Holtzman MJ, Aizawa H, Nadel JA, *et al.*: Selective generation of leukotriene B_4 by tracheal epithelial cells from dogs. *Biochem Biophys Res Commun* (1983) **114**: 1071–1076.
84. Assouline G, Leibson V: Stimulation of prostaglandin output from rat stomach by hypertonic solutions. *Eur J Pharmacol* (1977) **44**: 271–273.

85. Jongejan RC, de Jongste JC, Raatgeep RC, et al.: Effects of changes in osmolarity on isolated human airways. *J Appl Physiol* (1990) **68**: 1568–1575.
86. Finney MJB, Anderson SD, Black JL: Effect of a non-isotonic solution on human isolated airway smooth muscle. *Respir Physiol* (1987) **69**: 277–286.
87. Anderson SD, Schoeffel RE, Finney M: Evaluation of ultrasonically nebulised solutions as a provocation in patients with asthma. *Thorax* (1983) **38**: 284–291.
88. Sheppard D, Epstein J, Holtzman MJ, et al.: Dose-dependent inhibition of cold air-induced bronchoconstriction by atropine. *J Appl Physiol: Respir Environ Exercise Physiol* (1982) **53**: 169–174.
89. Makker HK, Holgate ST: The contribution of neurogenic reflexes to hypertonic saline-induced bronchoconstriction in asthma. *J Allergy Clin Immunol* (1993) **92**: 82–88.
90. Umeno E, McDonald OM, Nadel JA: Hypertonic saline increases vascular permeability in the rat trachea by producing neurogenic inflammation. *J Clin Invest* (1990) **851**: 1905–1908.
91. Pisarri TE, Jonson A, Coleridge HM, et al.: Vagal afferent and reflex responses to changes in surface osmolarity in lower airways in dogs. *J Appl Physiol* (1992) **73**: 2305–2313.
92. Garland A, Jordan JE, Necheles J, et al.: Hypertonicity, but not hypothermia, elicits substance P release from rat C-fiber neurons in primary culture. *J Clin Invest* (1995) **95**: 2359–2366.
93. Tomaki M, Ichinose M, Miura M, et al.: Elevated substance P content in induced sputum from patients with asthma and patients with chronic bronchitis. *Am J Respir Crit Care Med* (1995) **151**: 613–617.
94. Smith CM, Anderson SD, Black JL: Methacholine responsiveness increases after ultrasonically nebulized water, but not after ultrasonically nebulized hypertonic saline in patients with asthma. *J Allergy Clin Immunol* (1987) **79**: 85–92.
95. Malo JL, Cartier A, L'Acheveque J, et al.: Bronchoconstriction due to isocapnic cold air inhalation minimally influences bronchial hyperresponsiveness to methacholine in asthmatic subjects. *Bull Eur Physiopathol Respir* (1986) **22**: 473–477.
96. Rodwell LT, Anderson SD, Seale JP: Inhaled steroids modify bronchial responses to hyperosmolar saline. *Eur Respir J* (1992) **5**: 953–962.
97. Rodwell LT, Anderson SD, du Toit J, et al.: Nedocromil sodium inhibits the airway response to hyperosmolar challenge in patients with asthma. *Am Rev Respir Dis* (1992) **146**: 1149–1155.
98. Groot CAR, Lammers J-WJ, Festen J, et al.: The protective effects of ipratropium bromide and terbutaline on distilled water-induced bronchoconstriction. *Pulmon Pharmacol* (1994) **7**: 59–63.
99. Tranfa CME, Vatrella A, Parrella R, et al.: Effect of ipratropium bromide and/or sodium cromoglycate pretreatment on water-induced bronchoconstriction in asthma. *Eur Respir J* (1995) **8**: 600–604.
100. du Toit JI, Anderson SD, Jenkins CR, Woolcock AJ, Rodwell LT: Airway responsiveness in asthma: bronchial challenge with histamine and 4.5% sodium chloride before and after budesonide. *Allergy Asthma Proc* (1997) **18**: 7–14.
101. Groot CAR, Lammers J-WJ, Molema J, et al.: Effects of inhaled beclomethasone and nedocromil sodium on bronchial hyperresponsiveness to histamine and distilled water. *Eur Respir J* (1992) **5**: 1075–1082.
102. Smith CM, Anderson SD, Seale JP: The duration of the combination of fenoterol hydrobromide and iptratropium bromide in protecting against asthma provoked by hyperpnoea. *Chest* (1988) **94**: 709–717.
103. O'Byrne PM, Thomson NC, Morris M, et al.: The protective effect of inhaled chlorpheniramine and atropine on bronchoconstriction stimulated by airway cooling. *Am Rev Respir Dis* (1983) **128**: 611–617.
104. Rodwell LT, Anderson SD, Seale JP: Inhaled clemastine inhibits airway narrowing caused by aerosols of non-isotonic saline. *Eur Respir J* (1991) **4**: 1126–1134.
105. Finnerty JP, Wilmot C, Holgate ST: Inhibition of hypertonic saline-induced bronchoconstriction by terfenadine and flurbiprofen. Evidence for the predominant role of histamine. *Am Rev Respir Dis* (1989) **140**: 593–597.
106. Makker HK, Lau LC, Thomson HW, et al.: The protective effect of inhaled leukotriene D_4

receptor antagonist ICI 204,219 against exercise-induced asthma. *Am Rev Respir Dis* (1993) **147**: 1413–1418.
107. Bianco S, Vaghi A, Robuschi M, *et al.*: Prevention of exercise-induced bronchoconstriction by inhaled frusemide. *Lancet* (1988) **ii**: 252–255.
108. Robuschi M, Gambaro G, Spagnotto S, *et al.*: Inhaled frusemide is highly effective in preventing ultrasonically nebulised water bronchoconstriction. *Pulmon Pharmacol* (1989) **1**: 187–191.
109. Rodwell LT, Anderson SD, du Toit J, *et al.*: Different effects of inhaled amiloride and frusemide on airway responsiveness to dry air challenge in asthmatic subjects. *Eur Respir J* (1993) **6**: 855–861.
110. Rodwell LT, Anderson SD, Du Toit JI, *et al.*: The effect of inhaled frusemide on airway sensitivity to inhaled 4.5% sodium chloride aerosol in asthmatic subjects. *Thorax* (1993) **48**: 208–213.
111. Bianco S, Pieroni MG, Refini RM, *et al.*: Protective effect of inhaled furesomide on allegen-induced early and late asthmatic reactions. *N Engl J Med* (1989) **321**: 1069–1073.
112. Nichol GM, Alton EWFW, Nix A, *et al.*: Effect of inhaled frusemide on metabisulfite- and methacholine-induced bronchoconstriction and nasal potential difference in asthmatic subjects. *Am Rev Respir Dis* (1990) **142**: 576–580.
113. Polosa R, Lau LCK, Holgate ST: Inhibition of adenosine 5′-monophosphate- and methacholine-induced bronchoconstriction in asthma by inhaled frusemide. *Eur Respir J* (1990) **3**: 665–672.
114. Ventresca PG, Nichol GM, Barnes PJ, *et al.*: Inhaled frusemide inhibits cough induced by low chloride content solutions but not by capsaicin. *Am Rev Respir Dis* (1990) **142**: 143–146.
115. Anderson SD, Wei He, Temple DM: Frusemide inhibits antigen-induced release of sulfidopeptide-leukotrienes and histamine from sensitized human lung fragments. *N Engl J Med* (1991) **324**: 131.
116. Patel KR, Peers E: Felopidine, a new calcium antagonist, modifies exercise-induced asthma. *Am Rev Respir Dis* (1988) **138**: 54–56.
117. Kivity S, Ganem R, Greif J, *et al.*: The combined effect of nifedipine and sodium cromoglycate on the airway response to inhaled hypertonic saline in patients with bronchial asthma. *Eur Respir J* (1989) **2**: 513–516.
118. Patel KR, Kerr JW, MacDonald EB, *et al.*: The effect of thymoxamine and cromolyn sodium on postexercise bronchoconstriction in asthma. *J Allergy Clin Immunol* (1976) **57**: 285–292.
119. Barnes PJ, Wilson NM, Vickers H: Prazosin, an alpha$_1$-adrenoceptor antagonist, partially inhibits exercise-induced asthma. *J Allergy Clin Immunol* (1981) **68**: 411–415.
120. Tullett WM, Patel KR: Isosorbide dinitrate and isoxsuprine in exercise induced asthma. *Br Med J* (1983) **286**: 1934–1935.
121. Dinh Xuan AT, Chaussain M, Regnard J, *et al.*: Pretreatment with an inhaled alpha$_1$-adrenergic agonist, methoxamine, reduces exercise-induced asthma. *Eur Respir J* (1989) **2**: 409–414.
122. Eschenbacher WL, Boushey HA, Sheppard D: Alterations in osmolarity of inhaled aerosols cause bronchoconstriction and cough, but absence of a permeant anion causes cough alone. *Am Rev Respir Dis* (1984) **129**: 211–215.
123. Anderson SD: Bronchial challenge by ultrasonically nebulized aerosols. *Clin Rev Allergy* (1985) **3**: 427–439.
124. Anderson SD: Exercise-induced asthma. In Middleton E, Reed C, Ellis E, Adkinson NF, Yunginger JW (eds) *Allergy: Principles and Practice*, 3rd edn. St Louis, CV Mosby, 1988, pp 1156–1175.
125. Jenne JW, Tashkin DP: Bronchodilators and bronchial provocation. In Spector SL (ed) *Provocative Challenge Procedures. Background and Methodology*. Mount Kisco, NY, Futura Publishing, 1989, pp 451–517.
126. Morr H: Immunological release of histamine from human lung. 1. Studies on the beta2 sympathomimetic stimulator fenoterol. *Respiration* (1979) **38**: 163–167.
127. Alton EWFW, Kingsleigh-Smith DJ, Munkonge FM, *et al.*: Asthma prophylaxis agents alter the function of an airway epithelium chloride channel. *Am J Respir Cell Mol Biol* (1996) **14**: 380–387.

128. Paulmichl M, Norris AA, Rainey DK: Role of chloride channel modulation in the mechanism of action of nedocromil sodium. *Int Arch Allergy Clin Immunol* (1995) **107**: 416.
129. Jeffery PK, Godfrey RW, Adelroth E, *et al.*: Effects of treatment on airway inflammation and thickening of basement membrane reticular collagen in asthma. A quantitive light and electron microscopic study. *Am Rev Respir Dis* (1992) **145**: 890–899.
130. Hallahan AR, Armour CL, Black JL: Products of neutrophils and eosinophils increase the responsiveness of human isolated bronchial tissue. *Eur Respir J* (1990) **3**: 554–558.
131. Eiken O, Kaiser P, Holmer I, Baer R: Physiological effects of a mouth-borne heat exchanger during heavy exercise in a cold environment. Ergonomics (1989) **32**: 645–653.

32

Atmospheric Pollutants

RUDOLF A. JÖRRES AND HELGO MAGNUSSEN

The potential risk for human health due to air pollution has been studied extensively during the past two decades. Among the different atmospheric pollutants, ozone, nitrogen dioxide, sulphur dioxide and particles have gained major attention.

OZONE

Ozone is mainly generated from hydrocarbons and nitrogen dioxide in the presence of ultraviolet radiation. Average levels are lowest during the winter and highest in the summer. Concentrations normally peak in the afternoon. In most areas of industrialized countries, annual average levels of ozone range between 20 and 40 p.p.b. (parts per billion, v/v). Annual and daily cycles, however, are often associated with concentrations that reach or exceed 100 p.p.b.

Epidemiological studies have consistently demonstrated the occurrence of symptoms and a decline in spirometric volumes[1,2] even after short-term exposures to ozone.[3] There is increasing evidence for an association between short-term ozone levels and hospital admissions for respiratory disorders in subjects with pre-existing airway diseases.[4,5] Furthermore, symptoms of chronic respiratory disease[6] and the annual rate of decline of forced expiratory volume in 1 s (FEV_1) were found to be associated with long-term exposure to ozone.

According to experimental exposure studies, ozone causes pain on deep inspiration and a transient reduction in inspiratory capacity as reflected in FEV_1 and forced vital capacity (FVC).[7] Parameters indicative of small airways narrowing can show longer-lasting or more exaggerated effects.[8] Compared with spirometric responses, changes in airways resistance are minor. Prolonged exposures over 6.6 h with nearly continuous exercise have demonstrated that 80 p.p.b. of ozone is sufficient to elicit a response.[9] Ozone responses are a function of concentration, minute ventilation and time, with remarkably large interindividual variability.[10] On average, subjects with mild asthma showed similar or only slightly greater responses to ozone than normal subjects.[11–13] When ozone exposures are repeated, adaptation or tolerance of lung function responses can occur.[14]

Ozone causes a transient increase in methacholine responsiveness[11,13,15] that is not abolished in repeated exposures[14] and does not correlate with changes in lung function.[13] Exercise-induced asthma is not exaggerated by ozone,[16] in contrast to allergen responses. Subjects with allergic asthma exhibited increased bronchial allergen responsiveness after exposure to 120 p.p.b. ozone for 1 h at rest.[17] Similarly, a significant increase in bronchial allergen responses has been found after a 3-h exposure to 250 p.p.b. ozone including intermittent moderate exercise.[13] The same study reported a slight but statistically significant bronchial allergen response following ozone but not filtered air exposure in subjects with allergic rhinitis. Furthermore, nasal allergen responses, in terms of eosinophils and eosinophil cationic protein, were enhanced by ozone.[18]

Owing to its relatively low solubility in water, ozone can penetrate into the lung periphery causing lipid peroxidation and loss of functional groups of biomolecules. β_2-Adrenoceptor agonists[19,20] and atropine[19] only prevented the ozone-induced increase in airways resistance. Indomethacin reduced ozone-induced decrements in spirometric volumes[21] without effects on methacholine responsiveness.[22] After treatment with budesonide, the increase in pulmonary resistance and the neutrophil influx caused by ozone in dogs were decreased but the increase in airway responsiveness was not prevented.[23] There are animal data that indicate a role for antioxidants in modulating the response to ozone but it is not clear whether antioxidants affect individual susceptibility to ozone in human subjects.

Ozone induces cellular and biochemical changes detectable within bronchoalveolar lavage (BAL) fluid,[12,15,24–29] with neutrophil influx as a prominent feature. Airway inflammation occurs in the upper and lower airways[26–29] and appears to be stronger in subjects with asthma compared with healthy subjects.[12,29] Furthermore, responses are associated with elevated levels of prostaglandin (PG)E_2, PGF$_{2\alpha}$ and thromboxane (Tx)B_2 but not leukotrienes in BAL fluid[15,24,25,27] and of cytokines and chemokines such as interleukin (IL)-6, IL-8 and granulocyte–monocyte colony-stimulating factor.[12,25,26,28,29] Possibly, stimulation of non-myelinated C fibres and release of neuropeptides is involved in ozone responses. Indeed, levels of substance P[30] and bradykinin[27] in BAL fluid were elevated after ozone exposure. In addition, ozone causes cellular damage and an increase in epithelial permeability, as substantiated by lactate dehydrogenase (LDH), total protein, fibronectin, fibrinogen, IgG and albumin.[24–29] Lung function responses to ozone in terms of FEV_1 or FVC do not correlate with airway inflammation.[28] There may be, however, a relationship between fibrinogen, as a marker of vascular permeability, and indices of small airways function.[8,27]

32 Atmospheric Pollutants

NITROGEN DIOXIDE

Nitrogen dioxide (NO_2) is derived predominantly from motor vehicles, power stations and industrial processes. It also occurs in the workplace, for example in bus garages and in welding, and can contribute to indoor air pollution. Concentrations of NO_2 in urban areas are on average below 50 p.p.b. but may show peak values of 100–400 p.p.b. Indoor concentrations exceeding 500 p.p.b. have been measured in kitchens when an unventilated gas stove was used.

Outdoor and indoor NO_2 levels have been found to be correlated with the frequency[31,32] or duration[33] of respiratory illness and lung function impairment[31] in children, even at low levels.[34] Results in adults and patients with respiratory diseases are less consistent. Some investigators found an association between NO_2 levels and respiratory symptoms or peak flow rates in patients with asthma,[35] others not. Most of these studies are hampered by the strong associations between different air pollutants.

Experimental exposure studies have demonstrated that NO_2 does not affect lung function to a major extent in healthy subjects or subjects with mild asthma.[36–42] Patients with chronic obstructive pulmonary disease (COPD) showed deteriorations after inhaling 300 p.p.b. NO_2.[43] In another study, however, COPD patients did not deteriorate, on average, although disease severity and exposure conditions were similar.[44]

Subjects with asthma can show enhanced airway responsiveness to histamine[38] or methacholine[37] after inhalation of NO_2. In this respect, they are more sensitive than healthy subjects.[45] The findings in asthmatic subjects, however, have not been confirmed by all investigators.[36,41] In addition, NO_2 has been reported to enhance exercise-induced bronchoconstriction and the airway response to hyperventilation of cold air,[46] but there are also negative studies.[39,41] The bronchoconstrictor response to hyperventilation of air containing a fixed concentration of SO_2 was found to be enhanced by NO_2.[40] It is likely that this enhancement was associated with the hyperventilation response and not specific for SO_2.[47] Apparently, the effects of NO_2 in experimental studies are strongly influenced by the choice of subjects and the study protocol. NO_2 exposure can lead to an enhanced bronchial allergen response in subjects with asthma, in terms of both early- and late-phase responses.[48] Other authors have found a tendency towards effects of NO_2 on allergen responsiveness but significant effects only when NO_2 was given in combination with SO_2.[49]

The bronchoconstriction experienced by subjects with chronic bronchitis after inhalation of NO_2 could be attenuated by antihistamines, whereas atropine or β_2-adrenoceptor agonists were not effective.[50] Owing to its oxidative properties, NO_2 may lead to lipid peroxidation and generate reaction products with constituents of the airway surface. Therefore, another way to achieve protection against NO_2 could be with antioxidants. Indeed, pretreatment with vitamins C and E diminished the degree of lipid peroxidation in BAL fluid[51] and pretreatment with vitamin C inhibited the increase in methacholine responsiveness induced by NO_2.[52]

BAL studies have demonstrated that in healthy subjects exposure to 3000–4000 p.p.b. of NO_2 caused a reduction in the inhibiting capacity of the α_1-proteinase inhibitor,[53] in contrast to continuous exposure to a lower concentration or discontinuous exposure with intermittent peaks.[54] Cell numbers in BAL fluid were altered in a concentration-dependent manner; numbers of mast cells and lymphocytes were increased up to 1 day after exposure,[55,56] with slight differences in the cellular response pattern between smokers and non-smokers.[57] In subjects with mild asthma, exposure to 1000 p.p.b. NO_2 led to an increase in the concentrations of TxB_2 and PGD_2 and a decrease in the concentration of 6-keto-$PGF_{1\alpha}$,[42] without significant changes in cell counts. Healthy subjects showed only a slight increase in the level of TxB_2. Although in most studies concentrations of total protein, albumin and LDH in BAL fluid were not altered by NO_2, changes in alveolar permeability may occur with a delay of several hours.

SULPHUR DIOXIDE

Sulphur dioxide (SO_2) is mainly produced by combustion of fossil fuels. In addition to its presence in ambient air, SO_2 can be found in high concentrations in some workplaces. Currently, in most areas average annual ambient air concentrations are below 20 p.p.b. During air pollution episodes, however, concentrations of more than 100 p.p.b. can occur, with peaks exceeding 200 p.p.b. In some areas, in particular eastern Europe, much higher concentrations of SO_2 are still to be found. Levels of SO_2 are often closely linked to those of particulate matter.

With regard to acute effects of SO_2, epidemiological studies have demonstrated small and transient reductions of FEV_1 and FVC in children[58] or adults with airway obstruction.[59] Furthermore, elevated levels of SO_2 were associated with increased daily mortality[60] and respiratory morbidity due to asthma[61] or bronchitis.[62] It is often difficult to separate the acute effects of SO_2 from those of black smoke or suspended particles. Some studies have found that symptom scores were more strongly affected by total suspended particulate matter than by SO_2[63] and the Six Cities Study demonstrated that mortality was more closely related to particle levels than to SO_2.[64] Data indicate that subjects with pre-existing airway diseases are at special risk from SO_2 and particulate matter and that the risk from particles appears to be higher than that from SO_2. This is difficult to assess experimentally in human subjects. The results of the East–West comparison indicate that high levels of SO_2 and particulate matter *per se* are not associated with increased prevalences of atopy or asthma.[65,66]

Experimental exposure studies have demonstrated that SO_2 is a potent bronchoconstrictor, particularly in subjects with asthma.[67] The degree of bronchoconstriction is determined by minute ventilation and the route of inhalation.[68] When ventilation rates are increased, subjects with asthma can develop bronchoconstriction at levels of 250–600 p.p.b. or lower.[69,70] Increased flow rates reduce the percentage of the highly water-soluble SO_2 absorbed in the upper airways. In dry or dry and cold air, SO_2 responses are stronger than in humidified air.[71] Although non-specific airway hyperresponsiveness appears to be a prerequisite for SO_2 hyperresponsiveness, the degree of SO_2 responsiveness does not correlate with the degree of airway responsiveness to methacholine or histamine.[72] When SO_2 exposures are repeated within short time intervals, tolerance to the bronchoconstrictor response occurs.[73] SO_2 itself does not exert major effects on non-specific airway responsiveness[73] or allergen responsiveness.[49]

It is important to analyse the potential of drugs to protect against the effects of air pollutants in subjects with asthma. β_2-Adrenoceptor agonists,[74] disodium cromoglycate,[75] nedocromil sodium[76] and theophylline[77] either attenuated or blocked the bronchoconstrictor response to SO_2. In contrast, inhaled steroids showed only weak protection[78] and ipratropium bromide had no effect.[79] These data are compatible with the hypothesis that relaxation of the bronchial smooth muscle or blocking of C fibres and irritant receptors are ways to reduce the response to SO_2.

Some of the cellular and biochemical events induced by SO_2 have been studied by BAL in healthy subjects. Following inhalation of SO_2 in high concentrations, total cell number, total and relative numbers of mast cells and lymphocytes, and total numbers of macrophages increased.[80] Changes occurred within 4 h, with their maximum after 24 h, and had disappeared 3 days after exposure.

SUMMARY

Epidemiological and experimental exposure studies have shown that air pollutants exert detrimental effects on airway function and integrity and that subjects with pre-existing airway diseases such as asthma are at higher risk than normal subjects.

REFERENCES

1. Kinney PL, Ware JH, Spengler JD: A critical evaluation of acute ozone epidemiology results. *Arch Environ Health* (1988) **43**: 168–173.
2. Castillejos M, Gold DR, Dockery D, et al.: Effects of ambient ozone on respiratory function and symptoms in Mexico City schoolchildren. *Am Rev Respir Dis* (1992) **145**: 276–282.
3. Braun-Fahrländer C, KŸnzli N, Domenighetti G, et al.: Acute effects of ambient ozone on respiratory function of Swiss schoolchildren after a 10-minute heavy exercise. *Pediatr Pulmonol* (1994) **17**: 169–177.
4. Cody RP, Weisel CP, Birnbaum G, Lioy PJ: The effect of ozone associated with summertime photochemical smog on the frequency of asthma visits to hospital emergency departments. *Environ Res* (1992) **58**: 184–194.
5. Burnett RT, Dales RE, Raizenne ME, et al.: Effects of low ambient levels of ozone and sulfates on the frequency of respiratory admissions to Ontario hospitals. *Environ Res* (1994) **65**: 172–194.
6. Euler G, Abbey D, Hodgkin J, Magie A: Chronic obstructive pulmonary disease symptom effects of long-term cumulative exposure to ambient levels of total oxidants and nitrogen dioxide in California Seventh-Day Adventists Residents. *Arch Environ Health* (1988) **43**: 279–285.
7. Hazucha MJ, Bates DV, Bromberg PA: Mechanisms of action of ozone on the human lung. *J Appl Physiol* (1989) **67**: 1535–1541.
8. Weinmann GG, Bowes SM, Gerbase MW, et al.: Response to acute ozone exposure in healthy men. *Am J Respir Crit Care Med* (1995) **151**: 33–40.
9. Horstman DH, Folinsbee LJ, Ives PJ, et al.: Ozone concentration and pulmonary response relationships for 6.6-hour exposures with five hours of moderate exercise to 0.08, 0.10, and 0.12 ppm. *Am Rev Respir Dis* (1990) **142**: 1158–1163.
10. McDonnell WF, Smith MV: Description of acute ozone response as a function of exposure rate and total inhaled dose. *J Appl Physiol* (1994) **76**: 2776–2784.
11. Kreit JW, Gross KB, Moore TB, et al.: Ozone-induced changes in pulmonary function and bronchial responsiveness in asthmatics. *J Appl Physiol* (1989) **66**: 217–222.

12. Basha MA, Gross KB, Gwizdala CJ, et al.: Bronchoalveolar lavage neutrophilia in asthmatic and healthy volunteers after controlled exposure to ozone and filtered purified air. *Chest* (1994) **106**: 1757–1765.
13. Jörres R, Nowak D, Magnussen H: The effect of ozone exposure on allergen responsiveness in subjects with asthma or rhinitis. *Am J Respir Crit Care Med* (1996) **153**: 56–64.
14. Folinsbee LJ, Horstman DH, Kehrl HR, et al.: Respiratory responses to repeated prolonged exposure to 0.12 ppm ozone. *Am J Respir Crit Care Med* (1994) **149**: 98–105.
15. Seltzer J, Bigby BG, Stulbarg M, et al.: O_3-induced change in bronchial reactivity to methacholine and airway inflammation in humans. *J Appl Physiol* (1986) **60**: 1321–1326.
16. Weymer AR, Gong H, Lyness A, Linn WS: Pre-exposure to ozone does not enhance or produce exercise-induced asthma. *Am J Respir Crit Care Med* (1994) **149**: 1413–1419.
17. Molfino NA, Wright SC, Katz I, et al.: Effect of low concentrations of ozone on inhaled allergen responses in asthmatic subjects. *Lancet* (1991) **338**: 199–203.
18. Peden DB, Setzer RW, Devlin RB: Ozone exposure has both a priming effect on allergen-induced responses and an intrinsic inflammatory action in the nasal airways of perennially allergic asthmatics. *Am J Respir Crit Care Med* (1995) **151**: 1336–1345.
19. Beckett WS, McDonnell WF, Horstman DH, House DE: Role of the parasympathetic nervous system in acute lung response to ozone. *J Appl Physiol* (1985) **59**: 1879–1885.
20. Gong H, Bedi JF, Horvath SM: Inhaled albuterol does not protect against ozone toxicity in non-asthmatic athletes. *Arch Environ Health* (1988) **43**: 46–53.
21. Schelegle ES, Adams WC, Siefkin AD: Indomethacin pretreatment reduces ozone-induced pulmonary function decrements in human subjects. *Am Rev Respir Dis* (1987) **136**: 1350–1354.
22. Ying RL, Gross KB, Terzo TS, Eschenbacher WL: Indomethacin does not inhibit the ozone-induced increase in bronchial responsiveness in human subjects. *Am Rev Respir Dis* (1990) **142**: 817–821.
23. Stevens WHM, Ädelroth E, Wattie J, et al.: Effect of inhaled budesonide on ozone-induced airway hyperresponsiveness and bronchoalveolar lavage cells in dogs. *J Appl Physiol* (1994) **77**: 2578–2583.
24. Koren HS, Devlin RB, Graham DE, et al.: Ozone-induced inflammation in the lower airways of human subjects. *Am Rev Respir Dis* (1989) **139**: 407–415.
25. Devlin RB, McDonnell WF, Mann R, et al.: Exposure of humans to ambient levels of ozone for 6.6 hours causes cellular and biochemical changes in the lungs. *Am J Respir Cell Mol Biol* (1991) **4**: 72–81.
26. Aris RM, Christian D, Hearne PQ, et al.: Ozone-induced airway inflammation in human subjects as determined by airway lavage and biopsy. *Am Rev Respir Dis* (1993) **148**: 1363–1372.
27. Weinmann GG, Liu MC, Proud D, et al.: Ozone exposure in humans: inflammatory, small and peripheral airway responses. *Am J Respir Crit Care Med* (1995) **152**: 1175–1182.
28. Balmes JR, Chen LL, Scannell C, et al.: Ozone-induced decrements in FEV_1 and FVC do not correlate with measures of inflammation. *Am J Respir Crit Care Med* (1996) **153**: 904–909.
29. Scannell C, Chen L, Aris RM, et al.: Greater ozone-induced inflammatory responses in subjects with asthma. *Am J Respir Crit Care Med* (1996) **154**: 24–29.
30. Hazbun ME, Hamilton R, Holian A, Eschenbacher WL: Ozone-induced increases in substance P and 8-epi-prostaglandin $F_{2\alpha}$ in the airways of human subjects. *Am J Respir Cell Mol Biol* (1993) **9**: 568–572.
31. Speizer FE, Ferris Jr B, Bishop YMM, Spengler JD: Respiratory disease rates and pulmonary function in children associated with NO_2 exposure. *Am Rev Respir Dis* (1980) **121**: 3–10.
32. Neas LM, Dockery DW, Ware JH, et al.: Association of indoor nitrogen dioxide with respiratory symptoms and lung function in children. *Am J Epidemiol* (1991) **134**: 204–219.
33. Braun-Fahrländer C, Ackermann-Liebrich U, Schwartz J, et al.: Air pollution and respiratory symptoms in preschool children. *Am Rev Respir Dis* (1992) **145**: 42–47.
34. Hasselblad V, Eddy DM, Kotchmar DJ: Synthesis of environmental evidence: nitrogen dioxide epidemiology studies. *J Air Waste Manage Assoc* (1992) **42**: 662–671.
35. Lebowitz MD, Collins L, Holberg CJ: Time series analyses of respiratory responses to indoor and outdoor environmental phenomena. *Environ Res* (1987) **43**: 332–341.
36. Hazucha MJ, Ginsberg JF, McDonnell WF, et al.: Effects of 0.1 ppm nitrogen dioxide on airways of normal and asthmatic subjects. *J Appl Physiol* (1983) **54**: 730–739.

37. Mohsenin V: Airway responses to nitrogen dioxide in asthmatic subjects. *J Toxicol Environ Health* (1987) **22**: 371–380.
38. Bylin G, Hedenstierna G, Lindvall T, Sundin B: Ambient nitrogen dioxide concentrations increase bronchial responsiveness in subjects with mild asthma. *Eur Respir J* (1988) **1**: 606–612.
39. Avol EL, Linn WS, Peng RC, *et al.*: Experimental exposures of young asthmatic volunteers to 0.3 ppm nitrogen dioxide and to ambient air pollution. *Toxicol Ind Health* (1989) **5**: 1025–1034.
40. Jörres R, Magnussen H: Airways response of asthmatics after a 30 min exposure, at resting ventilation, to 0.25 ppm NO_2 or 0.5 ppm SO_2. *Eur Respir J* (1990) **3**: 132–137.
41. Jörres R, Magnussen H: Effect of 0.25 ppm nitrogen dioxide on the airway response to methacholine in asymptomatic asthmatic patients. *Lung* (1991) **169**: 77–85.
42. Jörres R, Nowak D, Grimminger F, *et al.*: The effect of 1 ppm nitrogen dioxide on bronchoalveolar lavage cells in normal and asthmatic subjects. *Eur Respir J* (1995) **8**: 416–424.
43. Morrow PE, Utell MJ, Bauer MA, *et al.*: Pulmonary performance of elderly normal subjects and subjects with chronic obstructive pulmonary disease exposed to 0.3 ppm nitrogen dioxide. *Am Rev Respir Dis* (1992) **145**: 291–300.
44. Hackney JD, Linn WS, Avol EL, *et al.*: Exposures of older adults with chronic respiratory illness to nitrogen dioxide. A combined laboratory and field study. *Am Rev Respir Dis* (1992) **146**: 1480–1486.
45. Folinsbee LJ: Does nitrogen dioxide exposure increase airways responsiveness? *Toxicol Ind Health* (1992) **8**: 273–283.
46. Bauer MA, Utell MJ, Morrow PE, *et al.*: Inhalation of 0.30 ppm nitrogen dioxide potentiates exercise-induced bronchospasm in asthmatics. *Am Rev Respir Dis* (1986) **134**: 1203–1208.
47. Rubinstein I, Bigby BG, Reiss TF, Boushey HA: Short-term exposure to 0.3 ppm nitrogen dioxide does not potentiate airway responsiveness to sulfur dioxide in asthmatic subjects. *Am Rev Respir Dis* (1990) **141**: 381–385.
48. Tunnicliffe WS, Burge PS, Ayres JG: Effect of domestic concentrations of nitrogen dioxide on airway responses to inhaled allergen in asthmatic patients. *Lancet* (1994) **344**: 1733–1736.
49. Devalia JL, Rusznak C, Herdman MJ, *et al.*: Effect of nitrogen dioxide and sulphur dioxide on airway response of mild asthmatic patients to allergen inhalation. *Lancet* (1994) **344**: 1668–1671.
50. von Nieding G, Krekeler H: Pharmakologische Beeinflussung der akuten NO_2-Wirkung auf die Lungenfunktion von Gesunden und Kranken mit einer chronischen Bronchitis. *Int Arch Arbeitsmed* (1971) **29**: 55–63.
51. Mohsenin V: Lipid peroxidation and antielastase activity in the lung under oxidant stress: role of antioxidant defenses. *J Appl Physiol* (1991) **70**: 1456–1462.
52. Mohsenin V: Effect of vitamin C on NO_2-induced airway hyperresponsiveness in normal subjects. A randomized double-blind experiment. *Am Rev Respir Dis* (1987) **136**: 1408–1411.
53. Mohsenin V, Gee JBL: Acute effect of nitrogen dioxide exposure on the functional activity of α_1-protease inhibitor in bronchoalveolar lavage fluid of normal subjects. *Am Rev Respir Dis* (1987) **136**: 646–650.
54. Johnson DA, Frampton MW, Winters RS, *et al.*: Inhalation of nitrogen dioxide fails to reduce the activity of human lung alpha-1-proteinase inhibitor. *Am Rev Respir Dis* (1990) **142**: 758–762.
55. Sandström T, Stjernberg N, Eklund A, *et al.*: Inflammatory cell response in bronchoalveolar lavage fluid after nitrogen dioxide exposure of healthy subjects: a dose–response study. *Eur Respir J* (1991) **3**: 332–339.
56. Sandström T, Andersson MC, Kolmodin-Hedman B, *et al.*: Bronchoalveolar mastocytosis and lymphocytosis after nitrogen dioxide exposure in man: a time–kinetic study. *Eur Respir J* (1990) **3**: 138–143.
57. Helleday R, Sandström T, Stjernberg N: Differences in bronchoalveolar cell response to nitrogen dioxide exposure between smokers and nonsmokers. *Eur Respir J* (1994) **7**: 1213–1220.
58. Brunekreef B, Lumens M, Hoek G, *et al.*: Pulmonary function changes associated with an air pollution episode in January 1987. *J Air Pollut Control Ass* (1989) **39**: 1444–1447.
59. Wichmann HE, Sugiri D, Islam MS, *et al.*: Pulmonary function and carboxyhaemoglobin during the smog episode in January 1987. *Zentralbl Bakteriol Mikrobiol Hyg Ser B* (1988) **187**: 31–43.

60. Derriennic F, Richardson S, Mollie A, Lellouch J: Short-term effects of sulphur dioxide pollution on mortality in two French cities. *Int J Epidemiol* (1989) **18**: 186–197.
61. Walters S, Griffiths RK, Ayres JG: Temporal association between hospital admissions for asthma in Birmingham and ambient levels of sulphur dioxide and smoke. *Thorax* (1994) **49**: 133–140.
62. Sunyer J, Saez M, Murillo C, *et al.*: Air pollution and emergency room admissions for chronic obstructive pulmonary disease: a 5-year study. *Am J Epidemiol* (1993) **137**: 701–705.
63. Euler G, Abbey D, Hodgkin J, Magie A: Chronic obstructive pulmonary disease symptom effects of long-term cumulative exposure to ambient levels of total suspended particulates and sulfur dioxide in California Seventh-Day Adventists Residents. *Arch Environ Health* (1987) **42**: 213–222.
64. Dockery DW, Pope CA, Xiping X, *et al.*: An association between air pollution and mortality in six U.S. cities. *N Engl J Med* (1993) **329**: 1753–1759.
65. von Mutius E, Martinez FD, Fritzsch C, *et al.*: Prevalence of asthma and atopy in two areas of West and East Germany. *Am J Respir Crit Care Med* (1994) **149**: 358–364.
66. Nowak D, Heinrich J, Jörres R, *et al.*: Prevalence of respiratory symptoms, bronchial hyperresponsiveness and atopy among adults: West and East Germany. *Eur Respir J* (1996) **9**: 2541–2552.
67. Sheppard D, Wong WS, Uehara CF, *et al.*: Lower threshold and greater bronchomotor responsiveness of asthmatic subjects to sulfur dioxide. *Am Rev Respir Dis* (1980) **122**: 873–878.
68. Bethel RA, Erle DJ, Epstein J, *et al.*: Effect of exercise rate and route of inhalation on sulfur-dioxide-induced bronchoconstriction in asthmatic subjects. *Am Rev Respir Dis* (1983) **128**: 592–596.
69. Linn WS, Venet TG, Shamoo DA, *et al.*: Respiratory effects of sulfur dioxide in heavily exercising asthmatics. A dose–response study. *Am Rev Respir Dis* (1983) **127**: 278–283.
70. Sheppard D, Saisho A, Nadel JA, Boushey HA: Exercise increases sulfur dioxide-induced bronchoconstriction in asthmatic subjects. *Am Rev Respir Dis* (1981) **123**: 486–491.
71. Sheppard D, Eschenbacher WL, Boushey HA, Bethel RA: Magnitude of the interaction between the bronchomotor effects of sulfur dioxide and those of dry (cold) air. *Am Rev Respir Dis* (1984) **130**: 52–55.
72. Magnussen H, Jörres R, Wagner HM, von Nieding G: Relationship between the airway response to inhaled sulfur dioxide, isocapnic hyperventilation, and histamine in asthmatic subjects. *Int Arch Occup Environ Health* (1990) **62**: 485–491.
73. Sheppard D, Epstein J, Bethel RA, *et al.*: Tolerance to sulfur dioxide-induced bronchoconstriction in subjects with asthma. *Environ Res* (1983) **30**: 412–419.
74. Koenig JQ, Marshall SG, Horike M, *et al.*: The effects of albuterol on SO_2-induced bronchoconstriction in allergic adolescents. *J Allergy Clin Immunol* (1987) **79**: 54–58.
75. Myers DJ, Bigby BG, Boushey HA: The inhibition of sulfur dioxide-induced bronchoconstriction in asthmatic subjects by cromolyn is dose-dependent. *Am Rev Respir Dis* (1986) **133**: 1150–1153.
76. Bigby B, Boushey H: Effects of nedocromil sodium on the bronchomotor response to sulfur dioxide. *J Allergy Clin Immunol* (1993) **92**: 195–197.
77. Koenig JQ, Dumler K, Rebolledo V, *et al.*: Theophylline mitigates the bronchoconstrictor effects of sulfur dioxide in subjects with asthma. *J Allergy Clin Immunol* (1992) **89**: 789–794.
78. Wiebicke W, Jörres R, Magnussen H: Comparison of the effects of inhaled corticosteroids on the airway response to histamine, methacholine, hyperventilation, and sulfur dioxide in subjects with asthma. *J Allergy Clin Immunol* (1990) **86**: 915–926.
79. McManus MS, Koenig JQ, Altman LC, Pierson WE: Pulmonary effects of sulfur dioxide exposure and ipratropium bromide pretreatment in adults with nonallergic asthma. *J Allergy Clin Immunol* (1989) **83**: 619–626.
80. Sandström T, Stjernberg N, Andersson MC, *et al.*: Cell response in bronchoalveolar lavage fluid after exposure to sulfur dioxide: a time–response study. *Am Rev Respir Dis* (1989) **140**: 1828–1831.

33

Drug-induced Asthma

PETER J. BARNES AND NEIL C. THOMSON

Relatively few drugs result in worsening of asthma. Aspirin and non-steroidal anti-inflammatory drugs are covered in Chapter 34.

β-BLOCKERS

Worsening or precipitation of asthma by β-adrenergic receptor antagonists was observed shortly after their introduction into clinical practice.[1,2] Although this property of β-blockers is well recognized, occasional reports of fatal asthma precipitated by β-blockers still occur and severe asthma may be precipitated, even in individuals with relatively mild asthma.[3] The dose of β-blocker required to precipitate bronchoconstriction may be low and there are several reports of severe asthma precipitated by eye drops of timolol, a non-selective β-blocker used to treat glaucoma.[4,5] Propafenone, a new antiarrhythmic agent with a structure similar to propranolol, has also been reported to cause bronchoconstriction in asthmatic individuals.[6]

The severity of bronchoconstrictor response to a given β-blocker is not predictable and does not appear to relate closely to the degree of airway hyperresponsiveness. The amount of bronchodilatation with a β-agonist may be an indication of sensitivity to β-blockers,[7] and patients with chronic obstructive airways disease are less likely to develop deterioration in lung function after a β-blocker.[8,9]

Non-selective β-blockers are more likely to precipitate bronchospasm than β_1-selective drugs. Thus, atenolol, acebutolol and metoprolol may give less fall in lung function than propranolol in asthmatic subjects.[10–14] Moreover, any fall in lung function found with β_1-selective antagonists is reversible by an inhaled β_2-agonist. This may be explained by the fact that there are no bronchodilator β_1-receptors in human airways.[15,16] Non-selective β-blockers with intrinsic sympathomimetic activity, such as pindolol, are also less likely to produce bronchoconstriction than those without,[10] although the bronchoconstriction with any non-selective β-blocker cannot be reversed by a β_2-agonist.

Since the severity of bronchoconstriction that will result from a β-blocker is not predictable, it is safest to avoid *all* β-blockers, even if β_1 selective, in all patients with airway obstruction. Safe alternative therapies exist for both hypertension (such as thiazides, calcium antagonists, angiotensin-converting enzyme (ACE) inhibitors, hydralazine, clonidine, prazosin and α-methyldopa) and ischaemic heart disease (calcium antagonists, nitrates).

Possible mechanisms

Despite the fact that β-blocker-induced asthma has been recognized for over 20 years, the mechanism is still not certain. Normal subjects do not develop any deterioration in lung function after β-blockers[17,18] and do not show an increase in sensitivity to bronchoconstricting agents such as histamine or methacholine.[19] This suggests that some endogenous activation of β-receptors may be important in asthmatic subjects to counteract bronchoconstrictor mechanisms (neural and inflammatory mediators).

Because most cases of bronchoconstriction were associated with propranolol, it was suggested that the mechanism was unrelated to β-blockade and could be explained by some non-specific property of propranolol (such as membrane-stabilizing activity). In rodents it was found that the inactive enantiomer D-propranolol, which has no significant β-blocking effect, was as potent as L-propranolol in causing bronchoconstriction,[20,21] and D-propranolol was as potent as L-propranolol in releasing histamine from guinea-pig lung.[22] In asthmatic subjects, however, while infused racemic DL-propranolol causes bronchoconstriction, D-propranolol does not, and the bronchoconstrictor response is related to the degree of β-blockade as assessed by isoprenaline dose–response curves.[23]

β-Blockers presumably antagonize some tonic adrenergic bronchodilator tone present in asthmatic patients but not in normal subjects. Since human airway smooth muscle has no demonstrated adrenergic innervation,[24] this might suggest that circulating catecholamines provide this drive.[25] Yet circulating catecholamines are not elevated in asthmatic subjects,[25] even in those subjects who have demonstrable bronchoconstriction after propranolol.[23] In any case, the concentrations of adrenaline in plasma (<0.3 nmol/litre) are too low to have a direct effect on human airway smooth muscle tone. This has suggested that β-blockers may inhibit the action of catecholamines on some other target cell, such as airway mast cells or cholinergic nerves. Mediator release from human lung mast cells is potently inhibited by β-agonists.[26] The effect of β-blockers may, therefore, be an increase in mediator release, which may be more marked in the 'leaky' mast cells of asthmatic individuals. This idea is supported by the observation that cromolyn sodium, a mast cell 'stabiliser', prevents the bronchoconstriction produced by inhaled propranolol.[27] However, after intravenous propranolol, no increase in plasma histamine has been detected.[23]

A more likely explanation for β-blocker-induced asthma is that there is an increase in neural bronchoconstrictor mechanisms. β_2-Adrenergic receptors on cholinergic nerves in

33 Drug-induced Asthma

human airways[28,29] may be tonically activated by adrenaline to modulate acetylcholine (ACh) release and therefore to dampen cholinergic tone. Blockade of these receptors would therefore increase the amount of ACh released tonically, but this would be compensated for by the increased stimulation of prejunctional M_2 autoreceptors, which would act homeostatically to inhibit any increase in ACh release[30] and therefore no increase in airway tone would occur, even with high doses of a β-blocker. By contrast, in asthmatic patients β-blockers inhibit prejunctional β-receptors in the same way, increasing the release of ACh;[31] however, there may be a defect in M_2-receptor function in asthmatic airways, so that the increased release in ACh cannot be compensated. Thus increased ACh reaches M_3 receptors on airway smooth muscle. In addition, bronchoconstrictor responses to ACh are exaggerated in asthma, a manifestation of airway hyperresponsiveness, and thus two interacting amplifying mechanisms may lead to marked bronchoconstriction (Fig. 33.1). Since the apparent defect in M_2 receptors may occur even in patients with mild asthma,[32,33] this explains why β-blockers may be dangerous even in patients with relatively mild asthma. Evidence to support this theory is provided by the inhibitory effect of an inhaled anticholinergic drug oxitropium bromide on β-blocker-induced asthma.[34]

In patients with more severe asthma there may be an additional neural mechanism by which β-blockers may cause bronchoconstriction. β_2-Adrenergic receptors inhibit the release of tachykinins from airway sensory nerves;[35,36] thus β-blockers may increase the release of these neuropeptides, thereby increasing bronchoconstriction and airway inflammation. While this mechanism may not be relevant in patients with mild asthma, in whom cholinergic mechanisms appear to account for the bronchoconstrictor response

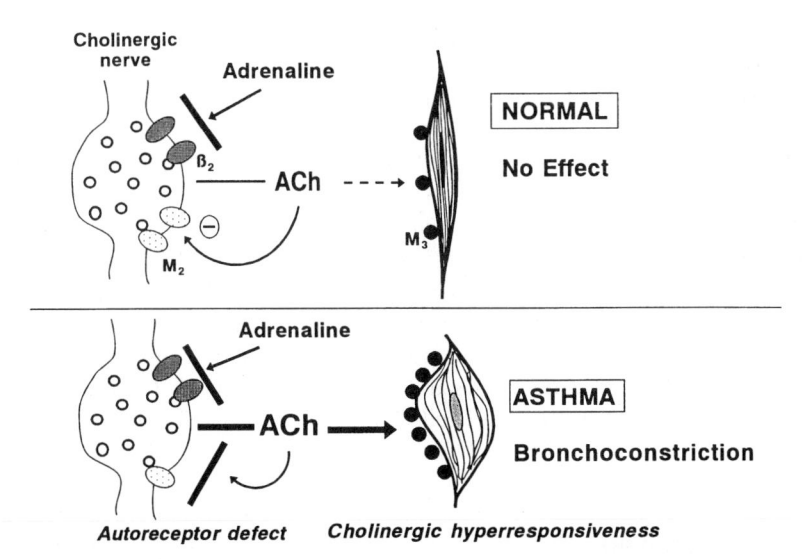

Fig. 33.1 Possible mechanism of β-blocker-induced asthma. Blockade of prejunctional β_2-receptors on cholinergic nerves in normal individuals results in increased release of acetylcholine (ACh), but this is compensated by stimulation of prejunctional muscarinic M_2 receptors to inhibit any increase in ACh. In patients with asthma, prejunctional M_2 receptors are dysfunctional, so that there is a net release of ACh; ACh also has a greater bronchoconstrictor effect on the airways due to airway hyperresponsiveness.

to β-blockers,[34] it may be relevant in more severely affected asthmatic patients in whom cholinergic mechanisms do not appear to be as important.

Another possible mechanism may be related to the recently recognized phenomenon of inverse agonism.[37] It has been found that some mutants of the β_2-receptor have constitutive activity and activate the coupling protein G_s, even in the absence of occupation by agonist.[38] In this situation β-blockers function as inverse agonists and have an inhibitory effect on baseline function. It is possible that in asthmatic patients β_2-receptors are constitutively active, so that β-blockers result in adverse effects. Different β-blockers have differing potencies as inverse agonists that are unrelated to their β-blocking potency. Thus, propranolol is a potent inverse agonist whereas pindolol is not and this may relate to the different tendency of these two agents to induce asthma.

ADDITIVES

Several chemicals used as additives in drug preparations and food have been associated with worsening of asthma and should, where possible, be avoided.

Metabisulphite

Bisulphites and metabisulphites (E220, 221, 222, 226 and 227) are antioxidants used as preservatives in several foods, including wines (especially sparkling wines), beer, fruit juices, salads and medications. Characteristically, they produce bronchoconstriction within 30 min of ingestion[39,40] and this may account for several cases of 'food allergy'. The mechanism of metabisulphite-induced asthma is probably explained by release of sulphur dioxide (SO_2) after ingestion that is then inhaled, since nebulized metabisulphite solutions generate SO_2 in sufficient quantities to provoke bronchoconstriction in asthmatic subjects.[41,42]

Tartrazine

Tartrazine (E102), a yellow dye, is used as a colouring in many foods, beverages (such as orange squash) and pharmaceutical preparations. Tartrazine sensitivity is relatively common and may affect 4% of asthmatic individuals, especially children.[43,44] Ingestion of tartrazine may result in urticarial rashes and bronchoconstriction. The mechanism may depend upon mediator release from mast cells. There is no association between tartrazine sensitivity and aspirin-induced asthma.[43]

Benzalkonium chloride

Benzalkonium chloride is a bactericidal compound added to certain nebulizer solutions, such as ipratropium bromide nebulizer solution. Paradoxical bronchoconstriction with nebulized ipratropium bromide was reported in asthma and was initially ascribed to the

33 Drug-induced Asthma

hypotonicity of the nebulizer solution.[45] When this was corrected by the use of isotonic solutions, bronchoconstriction was still occasionally reported, which might be explained by the presence of the preservatives benzalkonium chloride and ethylenediaminetetraacetic acid (EDTA) in the nebulizer solution.[46] Nebulization of ipratropium bromide with preservatives causes significant bronchoconstriction in a proportion of asthmatic patients, whereas nebulization of ipratropium bromide alone gives the expected bronchodilator response.[46,47] Nebulization of the preservative alone induces a bronchoconstrictor response in some patients. The presence of benzalkonium chloride in beclomethasone dipropionate nebulizer solution may also explain the bronchoconstriction that has been reported with this solution.[48] The mechanism of bronchoconstriction may be due to release of mediators from mast cells, perhaps due to a non-specific effect on the cell membrane.

Monosodium glutamate

Monosodium glutamate (MSG, E621) is added to food as a flavour enhancer. It is found in soy sauce, spices, stock cubes, hamburgers and in Chinese restaurant food. Some people react with sweating, flushing and numbness of the chest; in patients with asthma this may be accompanied by wheezing, which may begin several hours after ingestion ('Chinese restaurant asthma syndrome').[49,50] Precipitation of asthma symptoms by MSG is uncommon, however.

ACE INHIBITORS

There has recently been concern that drugs which inhibit ACE, such as captopril and enalapril, might lead to exacerbation of asthma, since ACE is an enzyme that degrades the bronchoconstrictor mediator bradykinin. However, exacerbation of asthma has only very rarely been reported in patients with asthma after ACE inhibitors.

Administration of a potent ACE inhibitor (ramipril) to a group of individuals with mild asthma showed no change in lung function or bronchial reactivity to inhaled histamine, nor was there any increase in bronchoconstrictor response to inhaled bradykinin.[51] As many as 20% of hypertensive patients treated with ACE inhibitors may develop an irritant cough,[52] although this is unrelated to the presence of underlying airway disease or atopic status. Perhaps this might be related to inhibition of bradykinin metabolism, with resultant stimulation of unmyelinated C fibres in the larynx. In an animal model of ACE inhibitor cough, a bradykinin antagonist blocks the cough.[53] ACE inhibitor cough is prevented by cyclooxygenase inhibitors,[54] and may be due to prostaglandin release by endogenous bradykinin, since PGE_2 and $PGF_{2\alpha}$ increase cough sensitivity.[55,56] ACE inhibitor cough is also prevented by sodium cromoglycate.[57]

There is some evidence that asthma may be precipitated by ACE inhibitors and a retrospective cohort study suggested that bronchospasm was twice as common in patients treated with ACE inhibitors compared to the reference group treated with lipid-lowering drugs.[58] There was some evidence that bronchospasm was more common in patients with cough, but these symptoms were not always associated and prior bronchospasm surprisingly was not associated with increased airway symptoms in patients started on ACE inhibitors. In a controlled trial in asthmatic and hypertensive patients (with and without cough), there was no change in lung function following administration of captopril and no increase in reactivity to histamine or bradykinin;[59] similar findings were obtained in a group of asthmatic patients given ACE inhibitors for 3 weeks, although one subject (of 21) developed increased wheezing.[60] This suggests that ACE inhibitors are unlikely to worsen asthma in the majority of patients, although there may be occasional patients in whom this occurs. When there is a possibility that asthma has been worsened or precipitated by an ACE inhibitor, the drug should be withdrawn and an alternative agent selected.

LOCAL ANAESTHETICS

Several studies have found that aerosols of the local anaesthetics bupivacaine and lignocaine (lidocaine) cause bronchoconstriction in a proportion of asthmatic patients.[61–66] The degree of bronchial reactivity to histamine does not predict the development or extent of bronchoconstriction following lignocaine inhalation.[66] The mechanism of local anaesthetic-induced bronchoconstriction is unclear. Pretreatment with anticholinergic drugs partially attenuates the bronchoconstrictor response to aerosols of local anaesthetics, suggesting that they may be acting in part via a vagal reflex pathway.[61,62,64] Inhaled local anaesthetics may selectively inhibit non-adrenergic non-cholinergic bronchodilator nerves and so allow unopposed vagal tone. Some evidence for this is provided by the demonstration that lignocaine inhalation blocks non-adrenergic non-cholinergic reflex bronchodilatation in human subjects, leading to a reflex bronchoconstrictor response.[67] It is important to be aware that some asthmatic patients may develop bronchoconstriction with topical local anaesthetics during fibreoptic bronchoscopy. All asthmatic patients should therefore receive premedication with a bronchodilator prior to bronchoscopy.

OTHER DRUGS

Many other drugs have been reported to lead to exacerbation of asthma in occasional patients. Bronchoconstriction may constitute part of an anaphylactic reaction to a drug, such as penicillin. Other drugs, such as opiates, may cause direct degranulation of mast cells.

Bronchodilator aerosols may occasionally cause a paradoxical bronchoconstriction. This is presumed to be due to the propellant (freon) or other additives (such as oleic acid, which is used as a surfactant).[68,69] The mechanism of bronchoconstriction may be via a cholinergic reflex.

REFERENCES

1. McNeill RS: Effect of β-adrenergic blocking agent, propranolol, on asthmatics. *Lancet* (1964) **ii**: 1101–1102.
2. McNeill RS, Ingram CG: Effect of propranolol on ventilatory function. *Am J Cardiol* (1966) **18**: 473–475.
3. Graft DF, Fowles J, McCoy CE, Lager RA: Detection of beta-blocker use in people with asthma. *Ann Allergy* (1992) **69**: 449–453.
4. Shoene RB: Timolol-induced bronchospasm in asthmatic bronchitis. *JAMA* (1981) **245**: 1460.
5. Dunn TL, Gerber MJ, Shen AS, Fernandez E, Iserman MD, Cherniak RM: The effect of topical ophthalmic instillation of timolol and betexolol on lung function in asthmatic subjects. *Am Rev Respir Dis* (1986) **133**: 264–268.
6. Hill MR, Gotz VP, Harman E, McLeod I, Hendeles L: Evaluation of the asthmogenicity of propafenone, a new antiarrhythmic drug. *Chest* (1986) **90**: 698–702.
7. van Herwaarden CLA: β-Adrenoceptor blockade and pulmonary function in patients suffering from chronic obstructive lung disease. *J Cardiovasc Pharmacol* (1983) **5**: S46–S50.
8. Perks W, Chatterjee S, Croxson R, Cruikshank J: Comparison of atenolol and oxprenolol in patients with angina or hypertension and coexistent chronic airways obstruction. *Br J Clin Pharmacol* (1978) **5**: 101–106.
9. Lammers JWJ, Folgering HTM, van Herwaarden CLA: Ventilatory effects of long-term treatment with pindolol and metoprolol in hypertensive patients with chronic obstructive lung disease. *Br J Clin Pharmacol* (1985) **20**: 205–210.
10. Benson MK, Berrill WT, Cruikshank JM, Sterling GS: A comparison of four β-adrenoceptor antagonists in patients with asthma. *Br J Clin Pharmacol* (1978) **5**: 415–419.
11. Greefhorst AM, van Herwaarden CLA: Comparative study of the ventilatory effects of three β_1-selective blocking drugs in asthmatic patients. *Eur J Clin Pharmacol* (1981) **20**: 417–421.
12. Ruffin RE, Frith MB, Anderton RC, Kumana CR, Newhouse MT, Hargreave FE: Selectivity of beta-adrenoceptor antagonist drugs assessed by histamine bronchial provocation. *Clin Pharmacol Ther* (1979) **25**: 536–540.
13. Lammers JWJ, Folgering HTM, van Herwaarden CLA: Ventilatory effects of $beta_1$-receptor selective blockade with bisoprolol and metoprolol in asthmatic patients. *Eur J Clin Pharmacol* (1984) **27**: 141–145.
14. Wilcox PG, Ahmad D, Darke AC, Parsons J, Carruthers SG: Respiratory and cardiac effects of metoprolol and bevantolol in patients with asthma. *Clin Pharmacol Ther* (1986) **39**: 29–34.
15. Zaagsma J, van der Heijden PJCM, van der Schaar MWG, Bank CMC: Comparison of functional β-adrenoceptor heterogeneity in central and peripheral airway smooth muscle of guinea pig and man. *J Recept Res* (1983) **3**: 89–106.
16. Carstairs JR, Nimmo AJ, Barnes PJ: Autoradiographic visualization of β-adrenoceptor subtypes in human lung. *Am Rev Respir Dis* (1985) **132**: 541–547.
17. Tattersfield AE, Leaver DG, Pride NB: Effects of β-adrenergic blockade and stimulation on normal human airways. *J Appl Physiol* (1973) **35**: 613–619.
18. Zaid G, Beall GN: Bronchial response to beta-adrenergic blockade. *N Engl J Med* (1966) **275**: 580–584.
19. Townley RG, McGeady S, Bewtra A: The effect of beta-adrenergic blockade on bronchial sensitivity to acetyl-beta-methacholine in normal and allergic rhinitis subjects. *J Allergy Clin Immunol* (1976) **57**: 358–366.
20. Maclagan J, Ney UM: Investigation of the mechanisms of propranolol induced bronchoconstriction. *Br J Pharmacol* (1979) **66**: 409–418.
21. Ney UM: Propranolol-induced airway hyperreactivity in guinea-pigs. *Br J Pharmacol* (1983) **79**: 1003–1009.
22. Terpstra GK, Raaijmakers JAM, Wassink GA: Propranolol-induced bronchoconstriction: a non-specific side-effect of β-adrenergic blocking therapy. *Eur J Pharmacol* (1981) **73**: 107–108.
23. Ind PW, Barnes PJ, Brown MJ, Dollery CT: Plasma histamine concentration during propranolol induced bronchoconstriction. *Thorax* (1985) **40**: 903–909.

24. Barnes PJ: Neural control of human airways in health and disease. *Am Rev Respir Dis* (1986) **134**: 1289–1314.
25. Barnes PJ: Endogenous catecholamines and asthma. *J Allergy Clin Immunol* (1986) **77**: 791–795.
26. Church MK, Hiroi J: Inhibition of IgE-dependent histamine release from human dispersed lung mast cells by anti-allergic drugs and salbutamol. *Br J Pharmacol* (1987) **90**: 421–429.
27. Koeter GH, Meurs H, Kauffman HF, De Vries K: The role of the adrenergic systems in allergy and bronchial hyperreactivity. *Eur J Respir Dis* (1982) **63** (Suppl 121): 72–78.
28. Rhoden KJ, Meldrum LA, Barnes PJ: Inhibition of cholinergic neurotransmission in human airways by β_2-adrenoceptors. *J Appl Physiol* (1988) **65**: 700–705.
29. Aizawa H, Inoue H, Miyazaki N, Ikeda T, Shigematsu N, Ito Y: Effects of procaterol, a beta$_2$-adrenoceptor stimulant, on neuroeffector transmission in human bronchial tissue. *Respiration* (1991) **58**: 163–166.
30. Barnes PJ: Muscarinic receptor subtypes: implications for lung disease. *Thorax* (1989) **44**: 161–167.
31. Barnes PJ: Modulation of neurotransmission in airways. *Physiol Rev* (1992) **72**: 699–729.
32. Minette PAH, Lammers J, Dixon CMS, McCusker MT, Barnes PJ: A muscarinic agonist inhibits reflex bronchoconstriction in normal but not in asthmatic subjects. *J Appl Physiol* (1989) **67**: 2461–2465.
33. Ayala LE, Ahmed T: Is there a loss of a protective muscarinic receptor mechanism in asthma? *Chest* (1991) **96**: 1285–1291.
34. Ind PW, Dixon CMS, Fuller RW, Barnes PJ: Anticholinergic blockade of beta-blocker induced bronchoconstriction. *Am Rev Respir Dis* (1989) **139**: 1390–1394.
35. Kamikawa Y, Shimo Y: Inhibitory effects of catecholamines on cholinergically and non-cholinergically mediated contractions of guinea-pig isolated bronchial muscle. *J Pharm Pharmacol* (1990) **42**: 131–134.
36. Verleden GM, Belvisi MG, Rabe KF, Miura M, Barnes PJ: β_2-Adrenoceptors inhibit NANC neural bronchoconstrictor responses *in vitro*. *J Appl Physiol* (1993) **74**: 1195–1199.
37. Leff P: Inverse agonism: theory and practice. *Trends Pharmacol Sci* (1995) **16**: 256–259.
38. Bond RA, Leff P, Jonhson TD, *et al.*: Physiological effects of inverse in transgenic mice with myocardial overexpression of the β_2-adrenoceptor. *Nature* (1995) **374**: 272–275.
39. Stevenson DD, Simon RA: Sulfites and asthma. *J Allergy Clin Immunol* (1984) **74**: 469–472.
40. Stevenson DD, Simon RA: Sensitivity to ingested metabisulfites in asthmatic subjects. *J Allergy Clin Immunol* (1981) **68**: 26–32.
41. Schwartz HJ, Chester EH: Bronchospastic responses to aerosolised metabisulfite in asthmatic subjects: potential mechanisms and clinical implications. *J Allergy Clin Immunol* (1984) **74**: 511–513.
42. Nichol GM, Nix A, Chung KF, Barnes PJ: Characterisation of bronchoconstrictor responses to sodium metabisulphite aerosol in atopic asthmatic and non-asthmatic subjects. *Thorax* (1989) **44**: 1009–1014.
43. Stevenson DD, Simon RA, Lumry WR, Mathison DA: Adverse reactions to tartrazine. *J Allergy Clin Immunol* (1986) **78**: 182–191.
44. Tarlo SM, Broder I: Tartrazine and benzoate challenge and dietary avoidance in chronic asthma. *Clin Allergy* (1982) **12**: 303–310.
45. Mann JS, Howarth PH, Holgate ST: Bronchoconstriction induced by ipratropium bromide in asthma: relation to hypotonicity. *Br Med J* (1984) **289**: 469
46. Beasley CRW, Rafferty P, Holgate ST: Bronchoconstrictor properties of preservatives in ipratropium bromide (Atrovent) nebuliser solution. *Br Med J* (1987) **294**: 1197–1198.
47. Raferty P, Beasley R, Holgate ST: Comparison of the efficacy of preservative from ipratropium bromide and Atrovent nebuliser solution. *Thorax* (1988) **43**: 446–450.
48. Clark RJ: Exacerbation of asthma after nebulised beclomethasone dipropionate. *Lancet* (1986) **1**: 574–575.
49. Allen DH, Baker GJ: Chinese restaurant asthma. *New Engl J Med* (1981) **305**: 1114–1115.
50. Allen DH, Delomery J, Baker G: Monosodium L-glutamate induced asthma. *J Allergy Clin Immunol* (1987) **80**: 530–537.
51. Dixon CMS, Fuller RW, Barnes PJ: The effect of an angiotensin converting enzyme inhibitor,

ramipril, on bronchial responses to inhaled histamine and bradykinin in asthmatic subjects. *Br J Clin Pharmacol* (1987) **23**: 91–93.
52. Fuller RW: Cough associated with angiotensin converting enzyme inhibitors. *J Hum Hypertens* (1989) **3**: 159–161.
53. Fox AJ, Lalloo UG, Belvisi MG, Bernareggi M, Chung KF, Barnes PJ: Bradykinin-evoked sensitization of airway sensory nerves: a mechanism for ACE-inhibitor cough. *Nature Med* (1996) **2**: 814–817.
54. McEwan JR, Choudry NB, Fuller RW: The effect of sulindac on the abnormal cough reflex associated with dry cough. *J Pharmacol Exp Ther* (1990) **225**: 161–164.
55. Chaudry NB, Fuller RW, Pride NB: Sensitivity of the human cough reflex: effect of inflammatory mediators prostaglandin E_2, bradykinin and histamine. *Am Rev Respir Dis* (1989) **40**: 137–141.
56. Nichol G, Nix A, Barnes PJ, Chung KF: Prostaglandin $F_{2\alpha}$ enhancement of capsaicin induced cough in man: modulation by beta$_2$-adrenergic and anticholinergic drugs. *Thorax* (1990) **45**: 694–698.
57. Hargreaves MR, Benson MK: Inhaled sodium cromoglycate in angiotensin-converting enzyme inhibitor cough. *Lancet* (1995) **345**: 13–16.
58. Wood R: Bronchospasm and cough as adverse reactions to the ACE inhibitors captopril, enalapril and lisinopril. A controlled retrospective cohort study. *Br J Clin Pharmacol* (1995) **39**: 265–270.
59. Overlack A, Muller B, Schmidt L, Scheid ML, Muller M, Stumpe KD: Airway responses and cough induced by angiotensin converting enzyme inhibition. *J Hum Hypertens* (1992) **6**: 387–392.
60. Kaufman J, Schmitt S, Barnard J, Busse W: Angiotensin-converting enzyme inhibitors in patients with bronchial responsiveness and asthma. *Chest* (1992) **101**: 922–925.
61. Thomson NC: The effect of different pharmacological agents on respiratory reflexes in normal and asthmatic subjects. *Clin Sci* (1979) **56**: 235–241.
62. Weiss EB, Patwardhan AV: The response to lidocaine in bronchial asthma. *Chest* (1977) **72**: 429–438.
63. Griffin MP, McFadden ER, Ingram RH, Pardee S: Controled analysis of the effects of inhaled lignocaine in exercise-induced asthma. *Thorax* (1982) **37**: 741–745.
64. Fish JE, Peterman VI: Effect of inhaled lidocaine on airway function in asthmatic patients. *Respiration* (1979) **37**: 201–297.
65. Miller WC, Awe R: Effect of nebulized lidocaine on reactive airways. *Am Rev Respir Dis* (1975) **111**: 739–741.
66. McAlpine LG, Thomson NC. Lidocaine-induced bronchoconstriction in asthmatic patients. Relation to histamine airway responsiveness and effect of preservative. *Chest* (1997) **96**: 1012–1015.
67. Lammers J, Minette P, McCusker M, Chung KF, Barnes PJ: Nonadrenergic bronchodilator mechanisms in normal human subjects *in vivo*. *J Appl Physiol* (1988) **64**: 1817–1822.
68. Engel T, Heinig JH, Malling H-J, Scharing B, Nikander K, Masden F: Clinical comparison of inhaled budesonide delivered either by pressurized metered dose inhaler or Turbuhaler. *Allergy* (1989) **44**: 220–225.
69. Yarbrough J, Lyndon RN, Mansfield E, Ting S: Metered dose inhaler induced bronchospasm in asthmatic patients. *Ann Allergy* (1985) **55**: 25–27.

34

Aspirin-induced Asthma

ANDRZEJ SZCZEKLIK

HISTORY AND DEFINITION

Reports of anaphylactic reactions to aspirin began to appear shortly after the introduction of aspirin into therapy, i.e. almost 100 years ago. Cases of violent, acute bronchospasm following aspirin ingestion were among the first to be reported. The association of aspirin sensitivity, asthma and nasal polyps was described by F. Widal and colleagues in 1922. This clinical entity, subsequently named aspirin triad, was popularized by the studies of Samter and Beers,[1] who in the late 1960s presented a perceptive description of the clinical picture of the syndrome. It was then realized that many other non-steroidal anti-inflammatory drugs (NSAIDs) share with aspirin the potential to trigger asthmatic reactions in the sensitive patients. In recent years aspirin-induced asthma (AIA) has attracted the attention of biochemists, pharmacologists and clinicians, since it constitutes a remarkable model for studying mechanisms operating in asthma, rhinitis and polyposis.

AIA is a clear-cut clinical syndrome with a distinct clinical picture.[2,3] Precipitation of asthma attacks by aspirin and other NSAIDs is the hallmark of the syndrome. It affects about 10% of adults with asthma, more often women than men, but is rare in asthmatic children. The majority of patients have a negative family history.

PATHOGENESIS

Allergic mechanism

Clinical reactions precipitated by aspirin in sensitive patients with asthma are reminiscent of immediate-type hypersensitivity reactions. Therefore, an underlying antigen–antibody mechanism has been suggested. However, skin tests with aspirin are negative, and numerous attempts to demonstrate specific antibodies against aspirin or its derivatives were unsuccessful. Neither differences in bioavailability of aspirin nor the formation of salicylic acid seem to contribute to aspirin-elicited reactions.[4] Furthermore, in patients with AIA asthmatic attacks can be precipitated not only by aspirin but by several other analgesics with different chemical structures, which makes immunological cross-reactivity most unlikely.

Cyclooxygenase theory

This theory[5] proposes that precipitation of asthma attacks by aspirin is not based on an antigen–antibody reaction but stems from the pharmacological action of the drug,[6] namely specific inhibition in the respiratory tract of the enzyme cyclooxygenase (COX). The original observations[7,8] that drug intolerance can be predicted on the basis of its *in vitro* inhibition of COX have been consistently reaffirmed during the ensuing years.[2]

Evidence in favour of the COX theory can be summarized as follows: (a) NSAIDs with anti-COX activity invariably precipitate bronchoconstriction in aspirin-sensitive patients; (b) NSAIDs that do not affect COX activity do not provoke bronchospasm; (c) there is a positive correlation between the potency of NSAIDs to inhibit COX *in vitro* and their potency to induce asthma attacks in sensitive patients; and (d) after aspirin desensitization, cross-desensitization to other NSAIDs that inhibit COX also occurs.

The enzyme, which appears to be central to the mechanism of aspirin intolerance, has recently become the subject of broad interest[9] since its isoforms were discovered. We now know that COX exists in at least two isoforms, COX-1 and COX-2, encoded by distinct genes. The constitutive isoform, COX-1, expressed in most tissues, has clear physiological functions, while the inducible isoform, COX-2, is induced by pro-inflammatory stimuli in a number of cells, including human pulmonary epithelial cells, fibroblasts, alveolar macrophages and blood monocytes. Cytokine induction of cytosolic phospholipase A_2 and COX-2 mRNA is suppressed by glucocorticoids in epithelial cells.[10] Immunostaining for COX-1 and COX-2 is similar in bronchial biopsies of AIA patients compared with aspirin-tolerant asthmatic patients.[11,12]

Aspirin, indomethacin and piroxicam, which at low doses precipitate asthmatic attacks in sensitive patients, are much more potent inhibitors of COX-1 than COX-2. Salicylate is practically devoid of activity on COX-1 in intact cells, but has half the potency of aspirin in inhibiting COX-2 in certain cell lines. Nimesulide, a drug known to inhibit COX-2 preferentially, was very well tolerated by AIA patients at a dose 100 mg, but at the higher dose of 400 mg it induced mild pulmonary obturation.[13] New selective COX inhibitors, which are some 1000-fold more potent against COX-2 than against COX-1,[9] might soon provide a new interesting tool for probing COX isoforms in AIA.

Involvement of leukotrienes

In AIA, inhibition of COX is associated with release of cysteinyl leukotrienes, which have emerged as important mediators of asthma and may be particularly prominent in aspirin-induced respiratory reactions. Indeed, their biological effects are consistent with most symptoms observed in AIA. Furthermore, eosinophils and mastocytes, two cells that play an essential role in AIA, have the capacity to produce large quantities of cysteinyl leukotrienes. Some patients with AIA excrete two- to ten-fold higher amounts of leukotriene (LT)E$_4$ in urine than other asthmatic patients who tolerate aspirin well. However, when baseline urinary LTE$_4$ levels in 10 AIA patients were compared with those in 31 aspirin-tolerant asthmatic patients,[14] there was a substantial overlap between the groups and no correlation was found between urinary LTE$_4$ and histamine concentration (PD$_{20}$) causing 20% fall in forced expiratory volume in 1 s (FEV$_1$). There is no doubt that aspirin challenge results in temporary, though significant, increase in urinary LTE$_4$ excretion.[15] Cysteinyl leukotrienes are also released into the nasal cavity following nasal challenge with aspirin,[16] and into bronchi following inhalation challenge with lysine-aspirin.[17] This is accompanied by inhibition of thromboxane (Tx)B$_2$ and prostaglandin (PG)E$_2$, while 15-lipoxygenase metabolites remain unaltered.

Cells expressing the principal enzymes of the 5-lipoxygenase pathway have been recently identified and quantified in bronchial mucosal biopsies of AIA patients, aspirin-tolerant asthmatic patients and normal subjects.[12] Immunostaining for 5-lipoxygenase, 5-lipoxygenase activating protein (FLAP) and LTA$_4$ hydrolase was similar. However, in AIA bronchial biopsies there was over-representation of cells expressing LTC$_4$ synthase, the essential enzyme for cysteinyl leukotriene synthesis, compared with significantly fewer cells expressing the enzyme in biopsies from aspirin-tolerant asthmatic and normal subjects. Cells positive for LTC$_4$ synthase were predominantly eosinophils (EG2$^+$) and a small number were mast cells (AA1$^+$). The gene for LTC$_4$ synthase has been localized to chromosome 5q, close to other candidate asthma genes, and a polymorphism causing constitutive overexpression of LTC$_4$ synthase in aspirin-sensitive patients is an attractive hypothesis to explain this leukotriene-dependent asthma.[12]

The proof for the critical role of cysteinyl leukotrienes has been provided with the advent of leukotriene antagonists. These compounds either (a) inhibit leukotriene synthesis by blocking 5-lipoxygenase or FLAP or (b) block specific cysteinyl leukotriene receptors. Premedication with leukotriene synthesis inhibitors or cysteinyl leukotriene receptor antagonists results in marked attenuation of aspirin-precipitated nasal and bronchial reactions,[18,19] while histamine antagonists have little effect.[20] Bronchodilatation has been also observed with leukotriene antagonists, indicating that cysteinyl leukotrienes have an effect on intrinsic airway tone in AIA.[18]

Release of mediators at the site of the reaction

Local instillation of aspirin in the nose or bronchi of sensitive patients, followed by nasal washing or bronchoalveolar lavage (BAL), allows investigation of the tissue response to aspirin and the course of the reaction. Kowalski et al.[21] reported that intranasal challenge with aspirin led to increased vascular permeability and early influx of eosinophils into nasal secretions of aspirin-intolerant patients. This was accompanied by an increase in concentrations of eosinophil cationic protein and tryptase, and the development of clinical symptoms, consisting of rhinorrhoea, sneezing and nasal congestion. No such changes were detected in asthmatic patients who tolerated aspirin well. Cysteinyl leukotrienes were not measured in this study, but their enhanced release has been demonstrated convincingly in previous reports.[16,19] On the contrary, studies of the response of histamine and PGD_2 yielded inconsistent results. Some authors[19] reported a marked rise in the levels of these two mediators in nasal washings following oral aspirin challenge, while others[22] did not confirm these findings. Studies in atopic individuals indicate that PGD_2 measurements in nasal secretion might not be a reliable marker for mast cell activation.[23]

Segmental bronchial challenge with aspirin has been recently performed in two well-matched groups of patients: those with AIA and those with asthma tolerant to aspirin.[24] At baseline the two groups did not differ with respect to BAL fluid concentrations of COX products, cysteinyl leukotrienes, histamine, tryptase, interleukin (IL)-5 or eosinophil number. At 15 min after intrabronchial instillation of 10 mg L-lysine-aspirin, there was a statistically significant rise in cysteinyl leukotrienes, IL-5 and eosinophil number in BAL fluid of AIA but not aspirin-tolerant patients. Mean histamine concentrations rose in response to aspirin, approaching the level of statistical significance. Aspirin significantly depressed PGE_2 and TxB_2 in both groups; however, mean PGD_2, $PGF_{2\alpha}$ and $9\alpha,11\beta$-PGF_2 decreased only in aspirin-tolerant patients. In individual AIA subjects, PGD_2 levels showed variability in response to aspirin, from marked increase to depression. Interestingly, Warren et al.[25] reported that segmental bronchial challenge with indomethacin led to a rise in PGD_2 and histamine in BAL fluid of three aspirin-sensitive asthmatic subjects compared with three aspirin-tolerant asthmatic subjects. Thus, bronchial aspirin challenge causes a specific eicosanoid response in aspirin-sensitive asthmatic individuals. By removing bronchodilator PGE_2 and leaving unchecked bronchoconstrictor PGD_2 and $PGF_{2\alpha}$, aspirin might further tip the eicosanoid balance towards bronchial obstruction, the balance already being disturbed by cysteinyl leukotriene overproduction.

PGE$_2$ and the switch in eicosanoid metabolism

It has been speculated that blocking of the COX pathway would result in increased bioavailability of arachidonic acid and its shunting to leukotrienes, although there is no experimental proof for this. Perhaps, in AIA aspirin promotes release of leukotrienes by removing PGE_2,[26] a dominant COX product of airway.[27] In vivo both aspirin-induced bronchoconstriction and the accompanying surge in urinary LTE_4 excretion can be completely abrogated by prior inhalation of PGE_2.[28,29] NSAIDs may therefore trigger adverse reactions by reducing PGE_2-dependent suppression of cysteinyl leukotriene synthesis in the lung. However, additional reasons are necessary to explain why NSAIDs do not trigger a similar rise in cysteinyl leukotrienes in aspirin-tolerant asthmatic individuals. The finding of enhanced LTC_4 synthase expression in AIA[12] may resolve this paradox. NSAIDs may reduce PGE_2 synthesis equieffectively in all subjects, but more cells generating LTC_4 are then liberated from the suppression in AIA lung than in other subjects, leading to detectable cysteinyl leukotriene release and bronchoconstriction only in AIA patients. Of additional importance might be the profound inhibitory effects of PGE_2 on inflammatory cells as well as inhibition of cholinergic transmission.[27]

Chronic inflammation of airway

Recent research has concentrated on aspirin-precipitated bronchial respiratory reactions, because they are a hallmark of AIA and provide a unique model for study. It has to be remembered, however, that AIA is a chronic disease that runs a protracted course even if NSAIDs are totally avoided. Bronchial biopsy studies reveal persistent inflammation of the airways with marked eosinophilia, epithelial disruption, cytokine production and upregulation of adhesion molecules.[3,11,12] Such a pathological process could result from a non-IgE-mediated reaction to an endogenous or exogenous antigen. In a relatively large sample of patients, a strong positive association between the presence of HLA-DPB1*0301 and AIA has been recently reported.[30] Interestingly, in another study of a small group of AIA patients a possible increase in the frequency of DPB1*1041 was observed.[31] Since the later allele differs from DPB1*0301 by only a single amino acid in the first hypervariable region of the cell (β1) distal domain, it is possible that the two alleles may be functionally similar. The strength of the association suggests that HLA-DPB1*0301 may itself confer susceptibility to AIA. Therefore, AIA may be in part due to an inflammatory reaction to an antigen, possibly an autoantigen or a chronic viral infection.[26] These possibilities are supported by the finding of elevated markers of autoimmunity[32,33] and enhanced IgG4 synthesis[34] in patients with AIA.

The course of AIA is reminiscent of persistent viral infection.[26] Virus could either (a) modify the genetic message for COX, making it prone to produce, in response to NSAIDs, unknown metabolites that stimulate leukotriene production[24] or (b) evoke an immunological response, perhaps dominated by specific cytotoxic lymphocytes and eosinophils, suppressed by PGE_2.[26] With the advent of molecular biology techniques it might be possible to probe the structure of COX genes and protein or look for the presence of a hypothetical virus in airway tissue of patients with AIA.

CLINICAL PRESENTATION

Although the onset of symptoms before puberty or after the age of 60 has been well documented, in most patients the first symptoms appear during the third decade. Women are affected more often than men; the family history is negative. The typical patient starts to experience intense vasomotor rhinitis characterized by intermittent and profuse watery rhinorrhoea. Over a period of months or years, chronic nasal congestion appears and physical examination reveals nasal polyps. Bronchial asthma and intolerance to aspirin develop during subsequent stages of the illness. The intolerance presents as a unique picture: within an hour after ingestion of aspirin an acute asthma attack develops, often accompanied by rhinorrhoea, conjunctival irritation and scarlet flushing of the head and neck. Aspirin is a common precipitating factor of life-threatening attacks of asthma; in a recent large survey, 25% of asthmatic patients requiring emergency mechanical ventilation were found to be aspirin intolerant.[35]

Nasal polyps are a common finding in AIA. They were diagnosed in 61 of our 100 consecutive asthmatic patients with aspirin intolerance confirmed by provocation tests. Their appearance was preceded by chronic rhinitis, which on average lasted 3.5 years, but in sporadic cases took up to 21 years. In about 60% of cases the first diagnosis of both aspirin intolerance and bronchial asthma was made prior to diagnosis of nasal polyps.

Asthma runs a protracted course, despite the avoidance of aspirin and cross-reactive drugs. Skin tests with aspirin are always negative. Atopy traits, contrary to early reports, are not rare and in fact may be more common than in the general population. Blood eosinophil count is elevated and eosinophils are present in the airways. Serum IgG4 subclass level is often elevated and half of the patients have autoantibodies against single-stranded DNA at a low titre in serum.[33]

Not only aspirin but several other NSAIDs precipitate attacks. Their chemical structures differ widely, which makes their chemical cross-reactivity most unlikely. Major offenders that precipitate bronchoconstriction include indomethacin, fenamic acids, ibuprofen, fenoprofen, ketoprofen, naproxen, diclofenac, piroxicam, tiaprofenic acid, aminopyrine, noramidopyrine, sulphinpyrazone, phenylbutazone and fenflumizole. All these drugs are contraindicated in patients with AIA. Not all of them produce adverse symptoms with the same frequency. This depends on the drug's anti-cox potency, dosage and individual sensitivity. If necessary, patients with AIA can safely take sodium salicylate, salicylamide, choline magnesium trisalicylate, dextropropoxyphene, benzydamine, guacetisal (guaiacolic ester of acetylsalicylic acid) and chloroquine. Most patients also tolerate paracetamol well at a dose not exceeding 1000 mg. Tartrazine, a yellow azo dye used for colouring drinks, foods, drugs and cosmetics, very rarely triggers adverse reactions.

DIAGNOSIS

While a patient's clinical history might raise suspicion of AIA, the diagnosis can be established with certainty only by aspirin challenge. There are no *in vitro* tests suitable for routine clinical diagnosis.

Patients are challenged when their asthma is in remission and their FEV_1 is $>65\%$ of predicted values. They continue regular medication, including corticosteroids, but stop sympathomimetics and methylxanthines for 10 h, antihistamines for 48 h, and sodium cromoglycate and ketotifen 72 h before the challenge. Glucocorticosteroids attenuate aspirin-precipitated reactions. There are three types of provocation tests, depending on the route of administration: oral, inhaled and nasal.

Oral challenge tests are most commonly performed. They consist of administration of increasing doses of aspirin (starting dose 10–20 mg) and placebo using a single-blind procedure and with careful monitoring of clinical symptoms, pulmonary function tests and parameters reflecting nasal patency for 6 h following administration of the drug. The reaction is considered positive if a decrease in $FEV_1 > 15$–20% of baseline occurs, accompanied by symptoms of bronchial obstruction and irritation of nose or eyes. In most patients the threshold dose evoking positive reactions varies between 40 and 100 mg aspirin. Adverse symptoms are relieved by inhalation of a β_2-adrenoceptor agonist; if necessary, aminophylline and steroids can be administered intravenously.

In inhalation challenge tests the increase in L-lysine-aspirin dosage is achieved every 30–45 min and the test is completed in one morning. It is therefore faster than oral challenge, which often takes 2–3 days, but the symptoms provoked are restricted only to the bronchopulmonary tract.

Nasal provocation testing is an attractive research model and can be also used as a diagnostic procedure on an outpatient basis; however, its value is limited by lower sensitivity compared with oral or inhalation challenges.

In the majority of patients, aspirin intolerance, once developed, remains for the rest of their lives. Repeated aspirin challenges are therefore positive, though some variability in intensity and spectrum of symptoms occurs. However, in an occasional patient a positive aspirin challenge may became negative after a period of a few years.

DIFFERENTIAL DIAGNOSIS

AIA should be clearly differentiated from other forms of aspirin-associated reactions.[2,3,8] Up to 40% of patients with chronic urticaria develop an obvious increase in wheals and swelling after taking aspirin. These reactions occur usually when urticaria is active; though the reason for them is not known, it appears that different mechanisms may be responsible in different patients.

Pyrazolone drugs can precipitate life-threatening reactions in both AIA and in patients allergic to pyrazolones.[3] The two groups should be clearly differentiated. In patients with pyrazolone allergy: (a) noramidopyrine and aminophenazone induce anaphylactic shock and/or urticaria; (b) skin tests with these drugs are highly positive; (c) phenylbutazone, sulphinpyrazone and several other COX inhibitors, including aspirin, can be taken with impunity; and (d) chronic bronchial asthma is present only in one-fifth of the patients.

There is a distinct subgroup of asthmatic individuals who respond favourably to aspirin and other COX inhibitors.[3] It has been suggested that pharmacological removal of a product of arachidonic acid cyclooxygenation from the respiratory tract of patients responding favourably to aspirin-like drugs helps them to overcome their airway obstruction. These patients are clinically indistinguishable from those in whom aspirin provokes bronchoconstriction. Although the syndrome is infrequent, and appears to affect less than 1% of adult asthmatic individuals, it is worth bearing in mind by the practising physician since a trial of aspirin can result in improvement.

PREVENTION AND TREATMENT

Patients with AIA should avoid aspirin, all products containing it and other analgesics that inhibit COX. If necessary, they can take safely, even for prolonged periods, certain agents that do not inhibit COX, like choline magnesium trisalicylate, dextropropoxyphene, azapropazone or benzydamine. Usually, they will also tolerate paracetamol well, at a dose not exceeding 1000 mg daily.

In most AIA patients, the state of aspirin tolerance can be introduced and maintained using desensitization. It is performed by giving incremental doses of aspirin over 2–3 days, until the well-tolerated dose of 600 mg is achieved. Aspirin is then administered regularly at a daily dose of 600–1200 mg. The patients usually experience improvement in their underlying chronic respiratory symptoms, especially in the nose, during the maintenance of the desensitized state for months or even years.[36] The state of desensitization is possible because after each dose of aspirin there is a refractory period of 2–5 days, during which aspirin and other COX inhibitors can be taken with impunity. The mechanism of desensitization is unknown; it might involve reduction of airways responsiveness to LTE_4 due to cysteinyl receptor downregulation or depression of leukotriene production.

Whether desensitized or not, most patients with AIA need regular therapy to control the symptoms of their disease. The therapy does not differ from that of other types of asthma. Long-term treatment with systemic corticosteroids is necessary in at least half of the patients.

Leukotriene antagonist drugs might soon find a place in chronic treatment of AIA. In the placebo-controlled, crossover Swedish–Polish study,[36] 40 AIA patients received 6 weeks' treatment with the 5-lipoxygenase inhibitor zileuton and placebo. Zileuton produced a significant improvement in airway function, decrease in nasal obstruction, return of smell and reduction in airway responsiveness to histamine. A recently concluded, double-blind, placebo-controlled, parallel-group, 4-week study assessed the therapeutic effects of montelukast, a cysteinyl leukotriene receptor antagonist;[37] 80 aspirin-intolerant asthmatic patients, incompletely controlled with corticosteroids, were randomized to receive montelukast or placebo once daily at bedtime. Patients on montelukast had less days with asthma exacerbations, more asthma-free days and significant improvement in parameters of asthma control, including FEV_1 and peak expiratory flow rate.

REFERENCES

1. Samter M, Beers RF: Intolerance to aspirin. Clinical studies and consideration of its pathogenesis. *Ann Intern Med* (1968) **68**: 975–983.
2. Stevenson DD: Diagnosis, prevention and treatment of adverse reactions to aspirin and nonsteroidal anti-inflammatory drugs. *J Allergy Clin Immunol* (1984) **74**: 617–622.
3. Szczeklik A: Aspirin-induced asthma. In Vane JR, Botting RM (eds) *Aspirin and Other Salicylates*. London, Chapman & Hall, 1992, pp 548–575.
4. Dahlen B, Boreus LO, Anderson P, *et al*.: Plasma acetylsalicylic acid and salicylic acid levels during aspirin provocation in aspirin-sensitive subjects. *Allergy* (1994) **49**: 43–49.
5. Szczeklik A: The cyclooxygenase theory of aspirin-induced asthma. *Eur Respir J* (1990) **3**: 588–593.
6. Vane JR: Inhibition of prostaglandin synthesis as a mechanism of action for aspirin-like drugs. *Nature* (1971) **231**: 232–234.
7. Szczeklik A, Gryglewski RJ, Czerniawska-Mysik G: Relationship of inhibition of prostaglandin biosynthesis by analgesics to asthma attacks in aspirin-sensitive patients. *Br Med J* (1975) **1**: 67–69.
8. Szczeklik A, Gryglewski RJ, Czerniawska-Mysik G: Clinical patterns of hypersensitivity to nonsteroidal antiinflammatory drugs and their pathogenesis. *J Allergy Clin Immunol* (1977) **60**: 276–284.
9. Bazan N, Botting J, Vane JR: *New Targets in Inflammation. Inhibitors of COX-2 or Adhesion Molecules*. Dordrecht, Kluwer Academic and William Harvey Press, 1996.

10. Newton R, Kuitert LM, Slater DM, *et al.*: Cytokine induction of cytosolic phospholipase A2 and cyclooxygenase-2 mRNA is suppressed by glucocorticoids in human epithelial cells. *Life Sci* (1997) **60**: 67–78.
11. Nasser SMS, Pfister R, Christie PE, *et al.*: Inflammatory cell populations in bronchial biopsies from aspirin-sensitive asthmatic subjects. *Am J Respir Crit Care Med* (1996) **153**: 90–96.
12. Sampson AP, Coburn AS, Sladek K, *et al.*: Profound overexpression of leukotriene C4 synthase in aspirin-intolerant asthmatic bronchial biopsies. *Int Arch Allergy Immunol* (1997) **113**: 355–357.
13. Bianco S, Robuschi M, Petrigni G, *et al.*: Efficacy and tolerability of nimesulide in asthmatic patients intolerant to aspirin. *Drugs* (1993) **46** (Suppl 1): 115–120.
14. Smith CM, Hawksworth RJ, Thien FC, *et al.*: Urinary leukotriene E4 in bronchial asthma. *Eur Respir J* (1992) **5**: 693–699.
15. Christie PE, Tagari P, Ford-Hutchinson AW, *et al.*: Urinary leukotriene E4 concentrations increase after aspirin challenge in aspirin-sensitive asthmatic subjects. *Am Rev Respir Dis* (1991) **143**: 1025–1029.
16. Picado C, Ramis I, Rosello J, *et al.*: Release of peptido-leukotrienes into nasal secretions after local instillation of aspirin in aspirin-sensitive asthmatic patients. *Am Rev Respir Dis* (1992) **145**: 65–69.
17. Sladek K, Dworski R, Soja J, *et al.*: Eicosanoids in bronchoalveolar lavage fluid of aspirin-intolerant patients with asthma after aspirin challenge. *Am J Respir Crit Care Med* (1994) **149**: 940–946.
18. Dahlen B, Margolskee DJ, Zetterstrom O, *et al.*: Effect of the leukotriene receptor antagonist MK-0679 on baseline pulmonary function in aspirin sensitive asthmatic subjects. *Thorax* (1993) **48**: 1205–1210.
19. Fischer AR, Rosenberg MA, Lilly CM, *et al.*: Direct evidence for a role of the mast cell in the nasal response to aspirin in aspirin-sensitive asthma. *J Allergy Clin Immunol* (1994) **94**: 1046–1056.
20. Szczeklik A, Serwonska M: Inhibition of idiosyncratic reactions to aspirin in asthmatic patients by clemastine. *Thorax* (1979) **34**: 654–657.
21. Kowalski ML, Grzegorczyk J, Wojciechowska B, *et al.*: Intranasal challenge with aspirin induces cell influx and activation of eosinophils and mast cells in nasal secretions of ASA-sensitive patients. *Clin Exp Allergy* (1996) **26**: 807–814.
22. Kowalski ML, Sliwinska-Kowalska M, Igarashi Y, *et al.*: Nasal secretions in response to acetylsalicylic acid. *J Allergy Clin Immunol* (1993) **91**: 580–598.
23. Wong D-Y, Smitz J, Clement P: Prostaglandin D2 measurement in nasal secretions is not a reliable marker for mast cell activation in atopic patients. *Clin Exp Allergy* (1995) **25**: 1228–1234.
24. Szczeklik A, Sladek K, Dworski R, *et al.*: Bronchial aspirin challenge causes specific eicosanoid response in aspirin sensitive asthmatics. *Am J Respir Crit Care Med* (1996) **154**: 1608–1614.
25. Warren MS, Sloan SJ, Westcott JY, *et al.*: LTE4 increases in bronchoalveolar lavage fluid (BALF) of aspirin-intolerant asthmatics (AIA) after instillation of indomethacin. *J Allergy Clin Immunol* (1995) **95**: 170.
26. Szczeklik A: Aspirin-induced asthma as a viral disease. *Clin Allergy* (1988) **18**: 15–20.
27. Pavord ID, Tattersfield AE: Bronchoprotective role for endogenous prostaglandin E2. *Lancet* (1995) **345**: 436–438.
28. Szczeklik A, Mastalerz L, Nizankowska E, *et al.*: Protective and bronchodilator effects of prostaglandin E and salbutamol in aspirin-induced asthma. *Am J Respir Crit Care Med* (1996) **153**: 567–571.
29. Sestini P, Armetti L, Gambaro G, *et al.*: Inhaled PGE2 prevents aspirin-induced bronchoconstriction and urinary LTE4 excretion in aspirin-sensitive asthma. *Am J Respir Crit Care Med* (1996) **153**: 572–575.
30. Dekker JW, Nizankowska E, Schmitz-Schumann M, *et al.*: Aspirin-induced asthma and HLA-DRB1 and HLA-DPB1 genotypes. *Clin Exp Allergy* (1997) **27**: 574–577.
31. Lympany PA, Welsh KI, Christie PE, *et al.*: An analysis with sequence-specific oligonucleotide probes of the association between aspirin-induced asthma and antigens of the HLA system. *J Allergy Clin Immunol* (1993) **92**: 114–123.

32. Lasalle P, Delneste Y, Gosset P, et al.: T and B cell immune response to a 55-kDa endothelial cell-derived antigen in severe asthma. *Eur J Immunol* (1993) **23**: 796–803.
33. Szczeklik A, Nizankowska E, Serafin A, et al.: Autoimmune phenomena in bronchial asthma with special reference to aspirin intolerance. *Am J Respir Crit Care Med* (1995) **152**: 1753–1756.
34. Szczeklik A, Schmitz-Schumann M, Nizankowska E, et al.: Altered distribution of IgG subclasses in aspirin-induced asthma: high IgG4, low IgG1. *Clin Exp Allergy* (1992) **22**: 283–287.
35. Marquette CH, Saulnier F, Leroy O, et al.: Long-term prognosis for near-fatal asthma. A 6-year follow-up study of 145 asthmatic patients who underwent mechanical ventilation for near-fatal attack of asthma. *Am Rev Respir Dis* (1992) **146**: 76–81.
36. Dahlen S-E, Nizankowska E, Dahlen B, et al.: The Swedish–Polish treatment study with the 5-lipoxygenase inhibitor Zileuton in aspirin-intolerant asthmatics. *Am J Respir Crit Care Med* (1995) **151**: A376.
37. Kuna P, Malmstrom K, Dahlen S-E, et al.: Montelukast (MK-0476), a cys-LT1 receptor antagonist, improves asthma control in aspirin-intolerant asthmatic patients. *Am J Respir Crit Care Med* (1997) **155**: A975.

35
Allergen Avoidance

ADNAN CUSTOVIC AND ASHLEY WOODCOCK

INTRODUCTION

The idea of the association between environmental factors and respiratory disease is not a new one. W. Storm Van Leeuwen wrote in 1927: 'In our endeavours to find the cause of the attack ... we utilised the known fact that the environment of the asthmatic patient is, as a rule, of primary importance in determining the intensity and frequency of his attacks.'[1] Italian physician Gerolamo Cardano (1501–1576) used environmental control to treat asthma in what can now be considered as the first recorded (and successful) example of allergen avoidance.[2] Cardano was invited to Scotland by John Hamilton, Archbishop of St Andrews, who was a powerful and important figure. After a prolonged consideration of the case Cardano made a recommendation that the Archbishop should get rid of his feather bedding. This was followed by a 'miraculous' remission of his intractable asthma.

The idea of a dust-free room for the study and treatment of asthma and other allergic disorders was introduced in 1925 by Simon Stein Leopold (1892–1957) and his brother Charles (1896–1960).[3] Storm van Leeuwen (1882–1933) created a 'climate' chamber in Holland in 1927, in an attempt to re-create the environment of the high-altitude sanatoria that were already known to benefit asthma sufferers, and demonstrated that the asthmatic patients improved if moved from their homes into the chamber.[4] In 1928 Dekker reported that measures aimed at reducing the amount of dust in bedrooms of asthmatic patients allergic to house dust had astonishing effect.[5]

Asthma is one of the few treatable conditions that have increased in prevalence and severity in recent years.[6–16] This increase seems to be related to changes in lifestyle and housing, and particularly to increased exposure to indoor allergens (house-dust mites, cats, dogs).[17-19] The relationship between allergen exposure and asthma can be divided into two phases: (a) exposure of a genetically predisposed individual that leads to *primary sensitization*; and (b) ongoing exposure of an already sensitized individual that contributes to the development of chronic bronchial hyperresponsiveness and asthma *symptoms*.

INDOOR ALLERGENS AS A CAUSE OF ASTHMA

The development of techniques for measuring the environmental exposure to allergens[20-31] has enabled a series of epidemiological studies of the relationship between allergens and asthma to be undertaken. These have produced strong evidence in different parts of the world that sensitization and exposure to indoor allergens is a primary cause of asthma, particularly in children and young adults (summarized in Table 35.1).[32-88] Exposure and sensitization to house-dust mite allergens is firmly established as an environmental risk factor for asthma in most parts of the world.[32-36,45-79] More recently, studies in climatic regions with both high and low mite allergen levels have provided evidence of the importance of exposure to other indoor allergens, particularly those from cats, dogs and cockroaches.[37-40,43,44,80-82]

House-dust mites

Epidemiological studies suggest a relationship between mite allergen levels, sensitization and asthma symptoms. Charpin *et al.*[32] compared subjects living in the Alps (Briancon), where exposure to house-dust mite allergens is low, with those living at sea level (Martigues), where exposure to house-dust mite allergens is high.[32] The prevalence of both sensitization to mites and asthma in adults and children was found to be significantly higher in those living in Martigues compared with Briancon.[32,33]

The results of recent studies from Australia provide striking evidence on the role of mite allergens in childhood asthma.[34-36] A large random sample of 8–11-year-old children was studied in six regions of New South Wales with differing levels of mite allergens in homes.[36] The percentage of children who were sensitized to dust mites increased with increasing exposure. Furthermore, those who were skin test positive to mites were at significant risk for asthma, and the magnitude of risk increased with increasing exposure (Fig. 35.1). After adjusting for sensitization to other allergens, the risk of dust mite-sensitized children having asthma approximately doubled for every doubling of *Der p* I levels.[36]

Cats and dogs

In recent studies in Los Alamos, New Mexico, USA, where mite allergen levels are low but pet allergens high, cat and dog allergens were found to be important risk factors for sensitization and asthma. Sporik *et al.*[37] and Ingram *et al.*[38] showed that both asthma and IgE-mediated sensitization to cat and dog allergens correlated with levels of exposure to

35 Allergen Avoidance

Table 35.1 Sensitization and exposure to indoor allergens: important risk factors for asthma.

Reference	Geographic location	Study design	Odds ratio (OR)/ significance
House-dust mites: sensitization and exposure as a risk factor for asthma			
18, 34–36	New South Wales, Australia	Exposure: risk for sensitization and asthma	2.7–20.9 (adjusted OR)
60	Sao Paulo, Brazil	Sensitization: risk for asthma	$P < 0.001$
32, 33	Martigue and Briancon, France	Exposure: risk for sensitization	$P < 0.02$
56	Germany	Exposure: risk for sensitization	5–11
61	London, UK	Exposure: risk for sensitization	$P < 0.01$
57, 58	Stockholm, Sweden	Sensitization/sensitization and exposure: risk for asthma	4.9; 25.7 (adjusted OR)
62	Tuscon, USA	Sensitization: risk for asthma	$P < 0.001$
63, 64	Dunedin, New Zealand	Skin test size and airway reactivity	$P < 0.0001$ (mite, cat), $P < 0.02$ (dog)
65	Raleigh, North Carolina, USA	Sensitization: risk for asthma	5.2 (mite), 15.5 (cat)
66, 67	Dunedin, New Zealand	Sensitization: risk for asthma	6.7 (mite), 4.2 (cat), 3.7 (dog)
69	Bergen, Norway	Sensitization: risk for bronchial responsiveness	7.36 (hazard ratio)
49	Manchester, UK	Exposure and asthma severity	$P < 0.01$
50, 51	Groningen, The Netherlands	Exposure and airway hyperresponsiveness	$P < 0.05$
52	The Netherlands	Exposure and respiratory symptoms in infants	3.8
53	France	Exposure and medication requirements	$P < 0.01$
70	Sweden	Exposure and asthma symptoms	$P < 0.05$
71, 72	Sydney, Australia	Exposure and asthma severity	$P = 0.04$, $P = 0.003$
73	The Netherlands	Exposure, PEF variability and symptoms	$P < 0.05$
Prospective studies			
74, 75	Isle of Wight, UK	Exposure: risk for sensitization and asthma	6.1; $P < 0.01$; 16.1 3.2; $P = 0.002$
76, 77	Isle of Wight, UK	Exposure: risk for sensitisation	$P < 0.01$
45	Poole, UK	Sensitization/sensitization and exposure: risk for asthma	19.7; 4.8
55, 78	South-western Germany	Sensitization/sensitization and exposure: risk for asthma	2.2–4; 2.6–3.6; 2.8
79	USA	Sensitization: risk for asthma	
47	Stockholm and N. Sweden	Exposure: risk for persistence of sensitization	30 (CI 4.8–184)

(continued)

Table 35.1 *Continued.*

Reference	Geographic location	Study design	Odds ratio (OR)/ significance
Cat and/or dog: sensitization and exposure as a risk factor for asthma			
37	Los Alamos, USA	Sensitization: risk for asthma	6.2; $P < 0.001$
38	Los Alamos, USA	Sensitization/sensitization and exposure: risk for asthma	$P < 0.001$
44	France	Early exposure: risk for sensitization	$P < 0.01$
80	UK	Early exposure: risk for sensitization	$P < 0.01$
43	Finland	Early exposure: risk for sensitization	
81	South-western Germany	Exposure: risk for sensitisation	2.7
Cockroach			
39	Baltimore, USA	Exposure: risk for sensitization	$P < 0.001$
40	Strasbourg, France	Exposure: risk for sensitization and asthma	$P < 0.01$
82	Chicago, USA	Sensitization: risk for asthma	$P < 0.001$
Sensitization and exposure as a risk factor in emergency room studies			
83	Charlottesville, USA	Sensitization: risk for asthma	4.5
84	Poole, UK	Sensitization/sensitization and exposure: risk for asthma	$P < 0.001$
85	Charlottesville, USA	Sensitization: risk for asthma	$P < 0.001$
86	Wilmington, USA	Sensitization/sensitization and exposure: risk for asthma	6.2–16.3; $P < 0.001$
87	Atlanta, USA	Sensitization and exposure: risk for severe asthma	9.5; $P < 0.001$
88	Bristol, UK	Increase in exposure: risk for emergency admission	

CI, confidence interval; PEF, peak expiratory flow.

Fel d I and *Can f* I among schoolchildren; 67% of asthmatic children were sensitized to dog and 62% to cat, and dog and cat sensitization was the strongest predictor of asthma in this locality.[38] These results strongly suggest that in areas with high levels of cat and dog allergens in homes, asthma will be associated with sensitization to cats and dogs.

Cockroaches

A recent study suggested that in inner-city areas with a high proportion of cockroach-infested houses sensitization to cockroach allergens was common, highlighting the importance of these allergens as a risk factor for asthma.[39] In Strasbourg, cockroach-

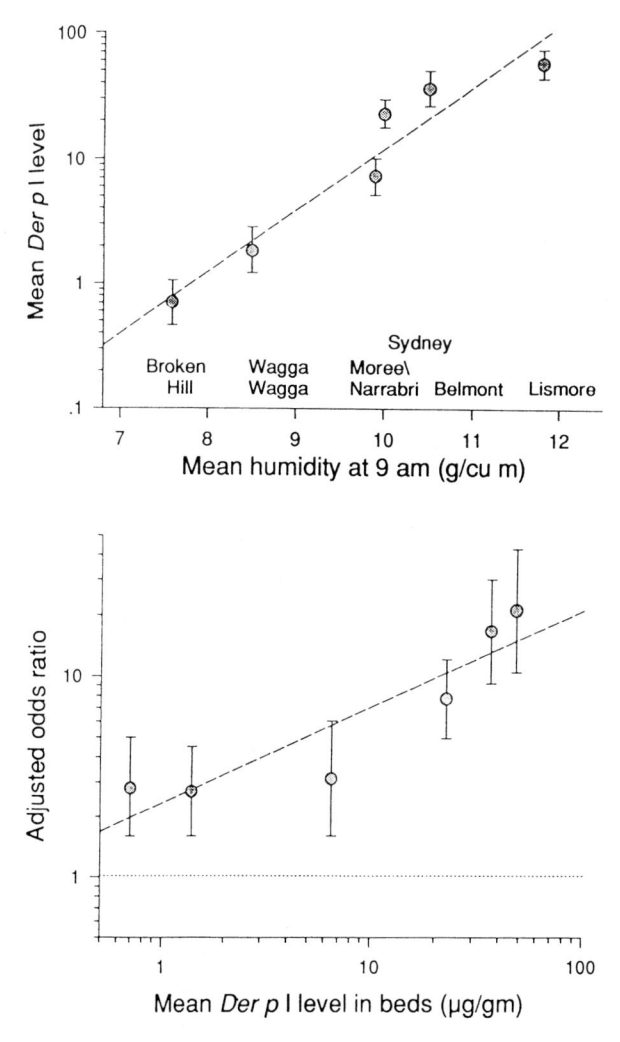

Fig. 35.1 House-dust mite allergen levels in six regions of New South Wales, Australia. Relationship of mean *Der p* I levels in each region to mean annual humidity at 9.00 a.m. and odds ratio for house dust mite-allergic children to have current asthma plotted against mean *Der p* I level in each region, shown with regression through the points. Reproduced from ref. 36, with permission.

sensitive patients with asthma were found to have high levels of cockroach allergens in their houses.[40] Cockroach sensitization is not confined to inner-city populations but occurs wherever substandard housing or apartment buildings sustain cockroach infestation.

PRIMARY SENSITIZATION

Early infancy has been identified as a critical period for primary sensitization to mite allergen. Evidence to support this view comes from studies relating atopy to month of birth. Children born just before the birch pollen season in Scandinavia have a higher risk of sensitization to birch pollen than those born after the season.[41] Similar observations have been made in relation to house-dust mite exposure received by children born in the autumn when mite numbers tend to be higher than at other times of the year.[42] Exposure to cats and dogs in early infancy are associated with specific IgE sensitization and allergic disease later in childhood.[43,44] In a cohort of 68 children prospectively followed from birth until the age of 11, Sporik et al.[45] found that exposure to a Der p I level >10 $\mu g/g$ of household dust measured in infancy was associated with a 4.8-fold relative risk of developing atopic asthma by the age of 11. Furthermore, exposure to high allergen levels at the age of 1 year was inversely related to the age of onset of asthma in mite-sensitive children, i.e. the higher the exposure to mite allergens within first year of life, the earlier the onset of asthma.

However, sensitization to an allergen can occur at any age[46] and may be related to dose and individual susceptibility. For example, in the highlands of Papua New Guinea a significant number of adults developed sensitization to mites and asthma following the introduction of blankets that were subsequently found to contain large numbers of dust mites.[47] Nevertheless, in general infants appear to be more susceptible to the development of sensitization than older individuals. The possibility of specific allergy conversion from sensitization to non-sensitization is related to allergen exposure and is mainly the result of a favourable indoor environment with low levels of mite allergens.[48]

ALLERGEN EXPOSURE AND ASTHMA SEVERITY

The severity of asthma makes an important impact on patients' lives, the healthcare system and the quantity of medical care that an individual patient requires. Several studies suggest that features in asthma may be related to allergen exposure.[49–53] We have recently shown that the clinical severity of asthma in patients sensitized to house-dust mites is related to the level of exposure to mite allergens in patients' beds[49] (Fig. 35.2).

Peat et al.[34] have compared two population samples of Australian children between the ages of 8 and 11 living in Lismore (a hot, humid, coastal region with high mite allergen levels) and Moree/Narrabri (a dry, inland region with low mite allergen levels). Although the prevalence of mite sensitivity was similar in both regions (28.6% vs. 26.4%), bronchial hyperreactivity (assessed by histamine challenge test) in children sensitized to mites was more severe in the humid coastal area with very high mite levels.[34,36]

The relationship between exposure and asthma symptoms in already sensitized individuals is complex. Some sensitized patients will react to very low doses of allergen, whilst in others the level required to cause symptoms will be considerably higher.[54] None the less, a pattern emerges in which sensitized patients exposed to high levels of allergens usually have more severe disease than those exposed to low allergen levels.[49] Avoiding exposure, therefore, is the logical way to treat asthma when the offending allergen can be identified, and should be an integral part of the overall management of the disease.

35 Allergen Avoidance

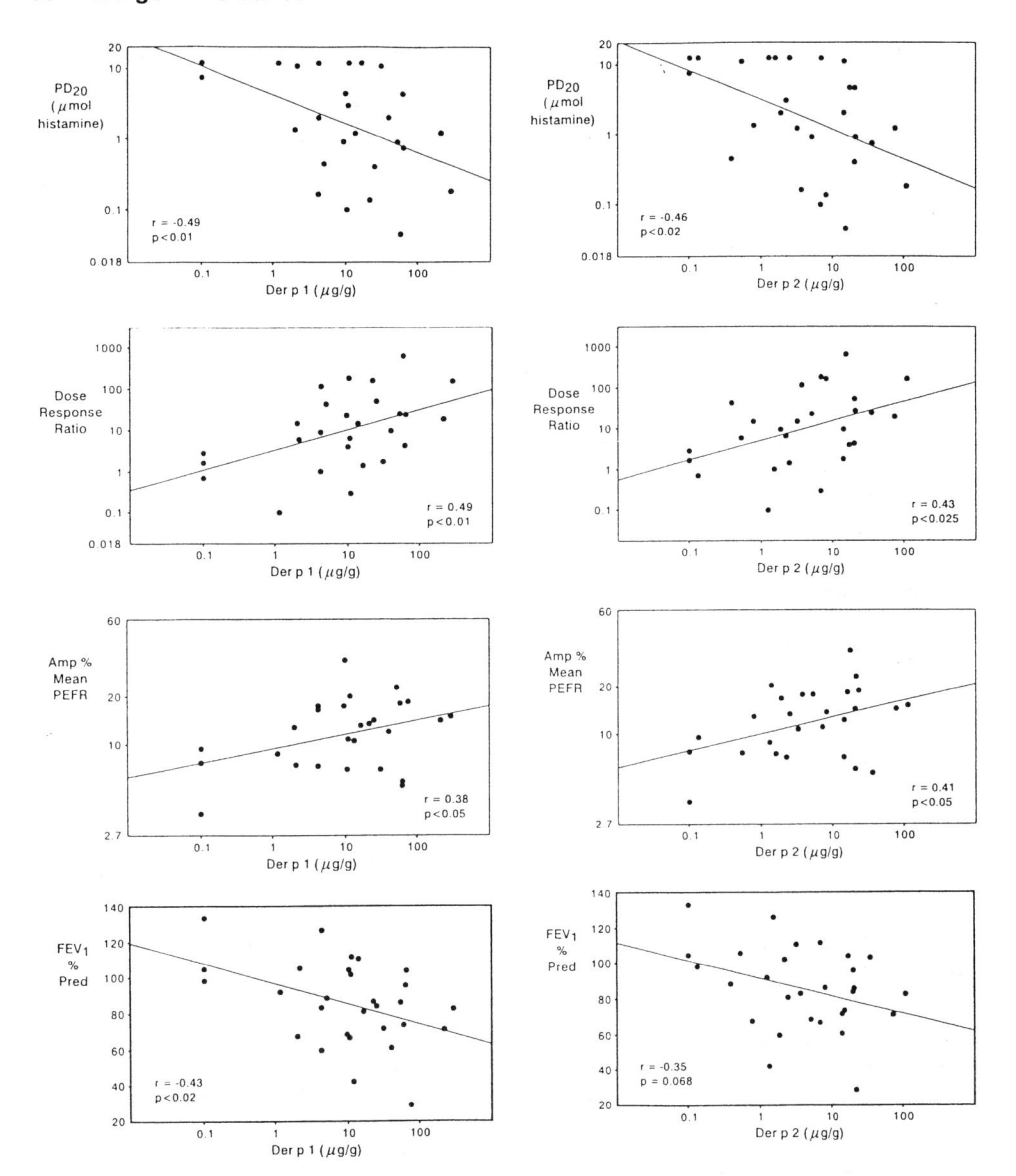

Fig. 35.2 Correlation between mite allergens (*Der p* II and *Der p* II) in bed and measures of asthma severity: bronchial hyperreactivity, expressed as PD_{20} (μmol histamine) and dose–response ratio; amplitude percent mean peak expiratory flow rate (PEFR); and percent predicted forced expiratory volume in 1 s (FEV_1). Reproduced from ref. 49, with permission.

ARE THRESHOLD VALUES USEFUL?

The First International Workshop on Mite Allergens and Asthma proposed two threshold levels for mite allergens that represent risk for sensitization and asthma: ≥ 2 μg group I mite allergen/g was regarded as a risk for the development of IgE antibody and asthma;

a level of ⩾10 μg group I mite allergen/g was regarded as a risk for an acute attack of asthma.[17] Kuehr et al.[55] reported that, in children at risk, exposure to ⩾2 μg Der p I/g represents a significant risk for mite sensitization, and suggested that this level be regarded as a minimal avoidance target for primary prevention of asthma. The threshold for sensitization (⩾2 μg Group I mite allergen/g) has been confirmed in a number of studies,[48,56–59] whilst the concept of threshold levels for exposure to indoor allergens representing a risk for asthma symptoms has been rejected. The US Institute of Medicine Report on the Health Effects of Indoor Allergens analysed the dose–response relationship between exposure and sensitization using the data from several studies, and reported a significant positive correlation between cumulative exposure to mite allergen and the risk of allergic sensitization[59] (Fig. 35.3). The provisional thresholds representing the risk for sensitization to different indoor allergens are listed in Table 35.2. The level of exposure

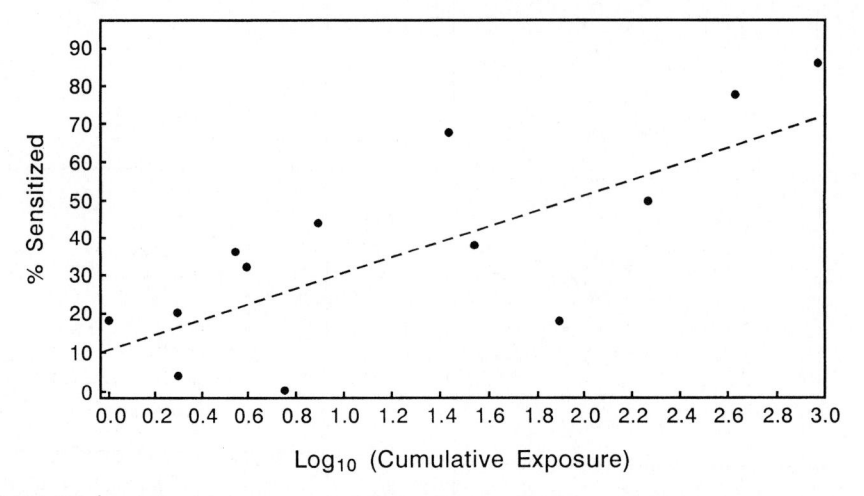

Fig. 35.3 Direct relationship between cumulative exposure to dust mite allergen (group I μg/g dust) and sensitization. Reproduced from ref. 59, with permission.

Table 35.2 Proposed threshold values for indoor allergens.

	Levels of allergens in settled dust associated with:	
	IgE sensitization	Allergic symptoms
House-dust mite:		
Group I allergen (e.g. Der p I, Der f I)	2 μg/g	10 μg/g*
Mite counts	>100 mites/g	>500 mites/g*
Cat: Fel d I	>1 μg/g†	>8 μg/g*‡
Dog: Can f I	>2 μg/g†	>10 μg/g*‡
Cockroach: Bla g II	>2 units/g†	?

* A level above which the majority of sensitized individuals who are going to develop symptoms will do so. This level increases the risk of acute asthma, but cannot be considered a threshold value.
† A level below which exposed individuals at risk are unlikely to develop sensitization. The threshold level representing a risk for sensitization has not yet been determined for cat, dog and cockroach allergens.
‡ The levels usually found in homes with cats/dogs. Not a real threshold value, but rather a level distinguishing homes with pets from those without pets.

35 Allergen Avoidance

that would provide an acceptable risk (i.e. the target for avoidance measures) has not yet been determined.

ALLERGEN AVOIDANCE

If allergen exposure is an important cause of asthma, then reducing patients' exposure should improve their asthma control. Two kinds of studies have demonstrated the effectiveness of allergen reduction in the treatment of asthma: (a) those in which patients were removed from their homes (reviewed in Table 35.3); and (b) those in which measures aimed at reduction in allergen levels were applied in patients' houses (reviewed in Table 35.4).

Table 35.3 Asthma, high altitude and removal of patients from their homes.

Reference	Location (altitude)	Study design	Clinical outcome
Epidemiological studies			
32, 33	Briancon, France (1326 m)	Comparison of asthma and allergies at altitude and sea level (adults and children)	Higher prevalence of mite sensitization and asthma at sea level (Marseille) than at altitude (Briancon)
37	Los Alamos, USA (2200 m)	Cross-sectional study investigating risk factors for childhood asthma	Association of asthma and sensitization to cats, but not to house-dust mites or cockroaches
38	Los Alamos, USA (2200 m)	Cross-sectional study investigating risk factors for childhood asthma	Association of asthma and sensitization and exposure to dogs and cats, but not to mites or cockroaches
High-altitude sanatoria			
89, 90	Davos, Switzerland (1560 m)	House-dust sensitive children (1 year stay)	Clinical improvement; reduction in BHR (histamine)
96	Davos, Switzerland (1560 m) Font-Romeu, France	212 children (Davos) 37 children (Font-Romeu)	Improvement in symptoms and reduction in medication
91	Misurina, Italy (1756 m)	14 mite-allergic children (8 month stay)	Improvement of lung function; reduction in BHR (exercise); reduction in medication
92	Misurina, Italy (1756 m)	20 allergic children (80 day stay)	Drop in antigen-induced basophil histamine release; reduction in BHR (methacholine) and IgE
97	Misurina, Italy (1756 m)	12 mite-sensitive children (6 + 3 months; 3 months at home)	Change in serum ECP and EPX and total IgE during exposure (3 months summer holidays at home)

(continued)

Table 35.3 *Continued.*

Reference	Location (altitude)	Study design	Clinical outcome
High-altitude sanatoria			
98	Davos, Switzerland (1560 m)	17 mite-sensitive children (5 week stay)	Decreased number of eosinophils and expression of T-lymphocyte activation markers; lung function improved
93	Misurina, Italy (1756 m)	Mite-allergic children (9 month stay)	Decrease in total and specific IgE; reduction in BHR (exercise, histamine and allergen challenge)
99	Misurina, Italy (1756 m)	12 mite-allergic children (3 month stay)	Decrease in PEF variability and improvement in BHR; after 3 weeks at home PEF and BHR worsened
100	Davos, Switzerland (1560 m)	16 allergic children (1 month stay)	Reduction in BHR (AMP challenge); improvement in PEF variability; reduction in eosinophils
101	Misurina, Italy (1756 m)	16 mite-sensitive children (3 month stay)	Reduction in BHR (methacholine); decrease in the percentage of sputum eosinophils
Removal of patients from their homes			
3	Philadelphia, USA: patients moved to special dust-free treatment room	2 case reports: 1 adult asthmatic patient and 1 patient with chronic sinusitis	Marked improvement in house-dust sensitive asthmatic patient, but not in patient with sinusitis
4	Lieden, The Netherlands: patients moved to allergen free air chamber	∼500 adult asthmatic patients (several days to several weeks)	75% completely or almost completely cured, 15% improved, 10% unaltered
94	London, UK: patients living in 'dust-free' hospital rooms	9 adult asthmatic patients (4 month stay)	Improvement in BHR (histamine); improvement in PEF and reduction in medication
95	Aarhus, Denmark: patients moved to the 'healthy buildings'	14 asthmatic patients (5 and 15 month stay)	Improvement in lung function, PEF, medication score, symptom score and IgE
102	Nagoya, Japan: patients hospitalized to clean room	30 mite-sensitive patients with atopic dermatitis (3–4 week stay)	Improvement in symptoms, long-term remission, decrease in eosinophils and mite-specific IgG

AMP, adenosin monophosphate; BHR, bronchial hyperresponsiveness; ECP, eosinophil cationic protein; EPX, eosinophil protein X; PEF, peak expiratory flow.

35 Allergen Avoidance

Table 35.4 Clinical studies of measures aimed at reducing allergen levels in patients' houses.

Reference and location	Study design and duration	Avoidance measures	Effect on mites/allergen	Clinical outcome
103 Leeds, UK	Ch, As, MS; $n = 14$; UC; 3–12 months	Mattress encased (plastic covers); synthetic pillows; bedding washed weekly; dusting, vacuuming	Reduction in mite counts (from 80 to 2; $P < 0.01$)	Improvement in symptom scores (9 to 1.89; $P < 0.05$)
104 Cardiff, UK	Ad, As, MS; $n = 32$; crossover PC; 6 weeks	Mattress encased (plastic covers); vacuum-cleaning of the bed; laundering of the bedding	Not monitored	No improvement in daily PEF reading or drug usage
105 Cardiff, UK	Ch, As, MS; $n = 53$; PC; 8 weeks	Mattress, carpets and upholstery vacuumed; blankets, sheets laundered; bedding washed; feather pillows, quilts replaced; soft toys removed	No difference in mite counts before and after treatment	Both active and control group improved, no difference between the groups
106 Cardiff, UK	Ch, As, MS; $n = 21$; crossover, C; 1 month + 1 month	New sleeping bags, pillows and blankets; mattress encased (plastic covers); carpets vacuumed	Colonization occurred on new bedding after second study period	PEF variability lower during the treated period, but the difference NS; majority with higher PEF during the treated period ($P < 0.01$)
107 Aarhus, Denmark	Ad, As and/or AR, HDS; $n = 23$; UC; 6 months	Mattress encased (plastic covers), $n = 3$; synthetic pillows, $n = 22$; bedroom carpet removed, $n = 7$; dusting, vacuuming	Not monitored in the study group over time	Beneficial effect reported by 15 patients, no change by 4
108 Aarhus, Denmark	Ad, Ch, As, MS; $n = 46$; C; 12 weeks run-in + 12 weeks intervention	Mattress vacuumed twice; synthetic pillows and quilts; bedding washed; bedroom carpet removed; bedroom aired + no plants	Difference between groups in BC ($P < 0.01$) but not in LC or M	Improvement in active vs. control group: PEF (NS, both improved); symptoms ($P < 0.05$); medication (NS)
109 Vancouver, Canada	Ch, As, MS and/or HDS; $n = 20$; C; 1 month	Mattress, pillows encased (vinyl covers); toys, carpets and upholstery removed (bedroom); washing, dusting, vacuuming	Not monitored	Improvement active vs. control group: symptoms ($P < 0.01$); medication ($P < 0.5$); PEF ($P < 0.05$); BHR ($P < 0.001$)

(continued)

Table 35.4 *Continued.*

Reference and location	Study design and duration	Avoidance measures	Effect on mites/allergen	Clinical outcome
110 Liverpool, UK	Ad, As; $n = 50$; C; 1 year	Mattress, pillows encased (plastic covers); synthetic duvets; bedroom carpet, upholstery removed ($n = 7$); washing, dusting, vacuuming	Significant fall in mite counts in the active ($P < 0.001$), but not in the control group	Improvement in MS. As in active group: FEV$_1$/FVC ($P < 0.02$); PEF ($P < 0.05$); BHR (PC$_{20}$)($P < 0.01$); medication ($P < 0.05$); total IgE ($P < 0.05$)
111 Leeds, UK	Ch, As; $n = 26$; C; group A, 12/52 avoidance; group B, 6/52 observation + 6/52 avoidance	Mattress, pillows encased (plastic covers); synthetic bedding; soft toys and pets excluded from bedroom; vacuuming	Mite counts: A 40 (start), 1.2 (6/52), 0.8 (12/52); B 22 (start), 10 (6/52), 2 (12/52)	Fall in total serum IgE in MS Ch ($P < 0.005$); BHR, symptoms, medication use and PEF all NS
112 Glasgow, UK	Ad, As, MS; $n = 21$; C; 8 weeks	Mattress and bedroom carpet treated with liquid nitrogen; washing, dusting, vacuuming; soft toys, plants and upholstery excluded from bedroom	Fall in number of intact mites in active group ($P < 0.01$); no change in control	Active vs. control: fall in the number of hours wheezing ($P < 0.05$); reduction in BHR ($P < 0.02$); total and specific IgE (NS)
113 London, UK	Ch, As, MS; $n = 46$; DB PC; 24 weeks	Mattress sprayed once every 2 weeks for 3 months with either natamycin or placebo; mattress vacuumed	Small, NS trend to a fall in *Der p* I in both groups	No change in BHR, symptoms and lung function
114 Derby, UK	Ad and Ch, As and/or AR and/or AD, MS; $n = 25$; UC; 12 months	Acarosan foam on mattress and bedding and moist powder on carpets and soft furniture	Reduction in *Der p* I level	As ($n = 12$): 7 better, 5 no change; AR ($n = 8$): 6 improved 2 no change; AD ($n = 5$): 2 improved
115 Berlin, Germany	Ch, As, MS; $n = 24$; DB PC; 12 months	Group A: mattress, pillow and quilt covered, carpets sprayed (3% tannic acid) 4 monthly; group B: mattress and carpet treated with benzyl benzoate; group C: placebo on mattress and carpet	Significant decrease in *Der* I in group A ($P < 0.005$); no change in groups B and C	Significant increase in BHR (PC$_{20}$) in the encasing regimen group (A): within group $P < 0.01$. No change in groups B and C: between groups $P < 0.05$

(continued)

Table 35.4 *Continued.*

Reference and location	Study design and duration	Avoidance measures	Effect on mites/allergen	Clinical outcome
116 Strasbourg, France	Ad, Ch, As, MS; n = 26; DB PC; 12 months	Benzyl benzoate foam or placebo on mattress and upholstery; benzyl benzoate powder or placebo on carpets	No significant difference in *Der* I between the groups	Active vs. placebo: clinical score, drug score, lung function, PEF all NS
117 Sydney, Australia	Ad, Ch, As; PC; 3 months run-in + 6 months treatment	Active: tannic acid/acaricide to mattress, pillow, duvet, blankets, carpets and upholstery; mattress, pillow and quilt covered. Placebo: inactive spray	At 2 weeks *Der p* I fell to 29% of baseline ($P = 0.04$ compared to placebo); 3 and 6 months, NS	Significant improvement in symptoms in both groups, but active vs. placebo NS; lung function and BHR in active vs. placebo NS
118 Washington, USA	Ad, As; n = 12; DB PC; 12 months	Benzyl benzoate powder (n = 6) or placebo (n = 6)	No change in mite allergen content in BC or LC	No difference in lung function and PEF between the groups
119 Bristol, UK	Ch, As, MS; n = 49; DB PC; 6 months	Benzyl benzoate powder or placebo on BC; benzyl benzoate foam or placebo on mattress, pillow and quilt; mattress, pillow and quilt covered (active or placebo); washing, dusting, vacuuming; soft toys excluded	M: 100% reduction in active vs. 53% reduction in placebo ($P < 0.001$); BC: active vs. placebo NS	Active vs. placebo: PEF (NS); BHR (histamine) (NS); lung function (FEV_1) ($P < 0.05$); symptoms ($P < 0.05$); medication use ($P < 0.01$)
Air cleaners and ionizers				
120 Auckland, New Zealand	Ch, As, MS; n = 10; C, crossover; 8 weeks (4 + 4)	Electrostatic precipitator in the child's bedroom	Not monitored	Control vs. active period: PEF (NS); medication use NS
121 Brisbane, Australia	Ad, Ch, As, MS; n = 9; PC, crossover; 4 weeks (2 + 2)	Active period: mattress and pillow covered; washing, dusting, vacuuming; dust retardant and anti-static spray; active electrostatic filter or HEPA filter. Placebo: inactivated air filter	Not monitored	Control vs. active period: symptom scores (NS); PEF (NS)

(continued)

Table 35.4 *Continued.*

Reference and location	Study design and duration	Avoidance measures	Effect on mites/allergen	Clinical outcome
Air cleaners and ionizers				
122 Buffalo, USA	Ad, Ch, As, AR, MS; $n = 32$; DB PC, crossover; 8 weeks (4 + 4)	Active period: HEPA filter cleaner. Placebo period: placebo filter	Not monitored	Control vs. active period: symptom and medication scores (NS); last 2 weeks of each period: nasal congestion, discharge eye irritation ($P < 0.05$); asthma symptoms (NS)
123 Ancona, Italy	Ad, Ch, As, MS; $n = 9$; PC, crossover; 16 weeks (8 + 8)	Active period: HEPA filter cleaner. Placebo period: placebo filter. Routine house cleaning	No difference in reservoir levels of mite allergens between the periods; fall within both groups ($P < 0.05$)	Control vs. active period: AR symptoms (NS); lung function (NS); PEF (NS); BHR (methacholine) (NS)
124 London, UK	Ch, As, MS; $n = 20$; DB PC crossover; 12 weeks (6 + 6)	Active period: active ionizers. Placebo period: placebo ionizers	Active vs. control period: airborne *Der p* I ($P < 0.0001$)	Active vs. control period: PEF (NS); symptom scores (NS) (trend towards increased cough during active period); medication (NS)
125 Manchester, UK	Ad, As, MS; $n = 12$; crossover (active + passive period, 30 + 24 days)	Active period: HEPA filter cleaner. Passive period: no HEPA filter cleaner	Airborne *Der p* I below detection limit in two-thirds of samples	Active vs. passive period: symptom scores (NS); lung function (NS); BHR (histamine) (NS); PEF (NS)
Allergen avoidance in allergic rhinitis				
126 Utrecht, Netherlands	Ch, Ad, AR; $n = 20$; DB PC parallel group; 12 months	Benzyl benzoate or placebo on mattress, upholstery, soft toys and carpets at 0 and 6 months; intensive cleaning	Active vs. control group: Acarex test $P < 0.05$	Active vs. control matched pairs: symptom scores ($P < 0.05$); physicians' assessment (NS); medication (NS); total IgE ($P < 0.01$)

(continued)

35 Allergen Avoidance

Table 35.4 *Continued.*

Reference and location	Study design and duration	Avoidance measures	Effect on mites/allergen	Clinical outcome
Allergen avoidance in atopic dermatitis				
127 Swansea, UK	Ch, Ad, AD, MS; $n = 18$; UC; 6 weeks	Mattress encased (plastic covers); regular vacuuming of bedding, bedroom carpets and curtains	Not monitored	15 patients improved, 3 remained unchanged
128 UK	Ch, Ad, AD, MS; $n = 37$; UC; 4–56 weeks	Mattress encased (plastic covers); regular vacuuming of mattress; carpets removed or vacuumed	Not monitored	19% complete remission, 41% almost clear, 27% better, 13% unchanged
129 Glasgow, UK	Ad, Ch, AD, MS; $n = 20$; PC; 12 weeks	NV, natamycin spray and vacuuming ($n = 6$); Nv, natamycin spray and no vacuuming ($n = 4$); nV, placebo and vacuuming ($n = 5$); nv, placebo and no vacuuming ($n = 5$)	Mite counts: 26% fall in NV ($P < 0.01$), 50% fall in nV ($P < 0.01$), 15% rise in Nv, 25% fall in nV (both NS)	Sympton scores: improvement rates 24% Nv, 20.4% nv, 8.4% NV and 0.7% nV. Fall in mite-specific IgE: NV ($P < 0.05$)
130 Liverpool, UK	Ad, Ch, AD, MS; $n = 48$; DB PC; 6 months	Active: mattress, pillow and quilt covered, carpets sprayed (benzyl benzoate + tannic acid), high filtration vacuum cleaner. Control: placebo covers and spray, standard vacuum cleaner	*Der p* I in carpets: median reduction 91% active, 89% control	Active vs. control: change in eczema severity score ($P < 0.01$); final eczema severity score ($P < 0.01$); mean final area affected ($P < 0.01$)

Ad, adults; AD, atopic dermatitis; AR, allergic rhinitis; As, asthma; BC, bedroom carpet; BHR, bronchial hyperresponsiveness; C, controlled; Ch, children; DB, double blind; Der I, *Der p* I + *Der f* I; FEV$_1$, forced expiratory volume in 1 s; FVC, forced vital capacity; HDS, house dust sensitive; HEPA, high-efficiency particulate air; LC, living room carpet; M, mattress; MS, mite sensitive; NS, not significant; P, placebo; PEF, peak expiratory flow; UC, uncontrolled.

High-altitude studies

Mite allergen levels are generally low at altitude, where humidity is too low to support mite populations. Previously mentioned studies in France suggested that living in a mite-free environment at high altitude (Briancon) reduces the risk of sensitization and development of respiratory symptoms.[32,33] Mite-sensitive children with asthma taken

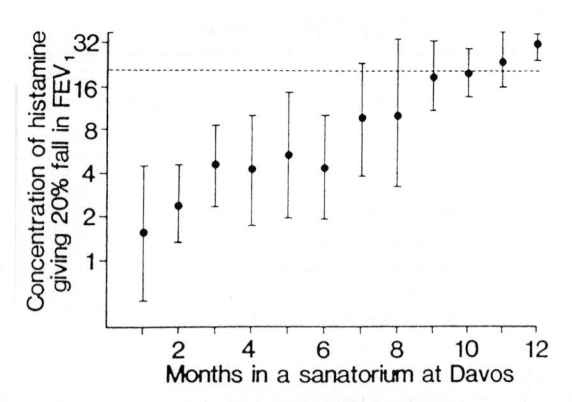

Fig. 35.4 Progressive reduction in non-specific bronchial hyperreactivity (histamine) in 10 mite-allergic children with asthma who moved from home to the mite-free environment of a sanatorium in Davos, Switzerland. FEV_1, forced expiratory volume in 1 s. Reproduced from ref. 90, with permission.

from their homes in Holland to the mite-free environment of Davos in Switzerland had a progressive reduction in non-specific bronchial hyperresponsiveness (BHR) over a period of 1 year[89,90] (Fig. 35.4). Similarly, a progressive reduction in symptoms occurred when asthmatic children were removed from their homes in northern Italy and admitted to a residential home at Misurina in the Italian Alps, 1756 m above sea level.[91] In further studies at Misurina, Piacentini et al.[92] reported a significant decrease in mite antigen-induced basophil histamine release, mite-specific serum IgE level and methacholine BHR in 20 asthmatic children, with reversal of this trend after 15 days of allergen exposure at sea level; Peroni et al.[93] demonstrated a significant reduction in total and mite-specific serum IgE after 3 and 9 months at Misurina, with significant increase 3 months after returning home. This study also reported a decrease in late allergen-induced bronchial reaction after 6 and 9 months at Misurina, with enhancement of BHR by mite allergen-specific bronchial challenge. These results suggest that mite allergen avoidance leads to a decrease of airway inflammation, with consequent improvement in non-specific BHR and symptoms, and that re-exposure results in rapid relapse.[93]

Attempts to create mite allergen-free conditions at lower altitudes by admitting adult patients to the 'allergen-free' environment of a hospital room resulted in improvement in airway reactivity and reduction in treatment requirements.[94] However, the benefits were transient, so that 3 months following discharge the bronchial reactivity measurements and medication requirements had increased. Removal of patients into new 'healthy homes' equipped with mechanical ventilation resulted in an increase in lung function of 30% compared with a control group, paralleled by a decrease in the use of medication of 60%.[95]

Allergen avoidance in patients' homes

The task of creating an allergen-free environment in patients' homes has proved to be a difficult one. There are conflicting data on the effectiveness of allergen avoidance carried out in houses, primarily because the early studies were small, poorly controlled and used measures that were not aggressive enough to reduce exposure and consequently failed to show beneficial effect. Once sufficiently aggressive measures were used and follow-up was adequate, an improvement in asthma symptoms, reduction in medication use and decrease in bronchial reactivity in both children and adults have been demonstrated.[109,110,115,119] The results of allergen avoidance trials are summarized in Table 35.4.[103–130]

ALLERGEN AVOIDANCE: PRACTICAL MEASURES

The real challenge facing practising physicians is to create a low-allergen environment in patients' homes. The integrated strategies aiming at achieving effective allergen control should be flexible to suit individual needs and at the same time not prohibitively expensive. Many different avoidance measures have been tested with sometimes widely exaggerated claims; only a few have been subjected to randomized controlled trials. It is important to make a clear distinction between those measures that have been tried only in the laboratory and those tested in clinical trials.

Distribution and aerodynamic properties of indoor allergens: relevance to avoidance

Knowledge of the sources of allergens, their airborne characteristics and particle size distribution is essential for the design of successful strategies to reduce personal exposure and asthma severity.

Allergens from mites, cats and dogs have dramatically different aerodynamic characteristics.[131–141] Airborne mite allergens can be detected only after vigorous disturbance, whilst airborne *Fel d* I and *Can f* I are readily measured in houses without artificial disturbance.[131–135] Although the majority of both cat and dog allergens are contained on relatively large particles (>10 μm diameter), there is a smaller ($\sim 20\%$) but probably clinically very relevant proportion of these allergens associated with small particles (<5 μm diameter)[131,132,135] (Figs 35.5 and 35.6). This underlies the difference in the clinical presentation between mite- and pet-sensitive asthmatic individuals. House-dust mite allergic patients are usually unaware of the relationship between mite exposure at home and asthma symptoms, and even a carefully taken allergy history cannot unequivocally implicate mites as a cause of symptoms. This is not surprising, as the pattern of exposure reflects the nature of mite allergens, i.e. low-grade, chronic exposure occurring predominantly overnight in bed. Exposure to high levels of airborne mite allergens capable of inducing symptoms only occurs immediately after vigorous disturbance (e.g. during cleaning). However, the large particles that carry mite allergen contain a large quantity of allergen and even a small number of such particles may cause a considerable inflammatory response in the airways. In contrast, patients allergic to cats or dogs often develop symptoms of acute asthma within minutes of entering a home with a cat or dog, or simply by stroking an animal, due to the inhalation of large amounts of *Fel d* I or *Can f* I on small particles that can penetrate deep into the respiratory tract[132,135] (Table 35.5).

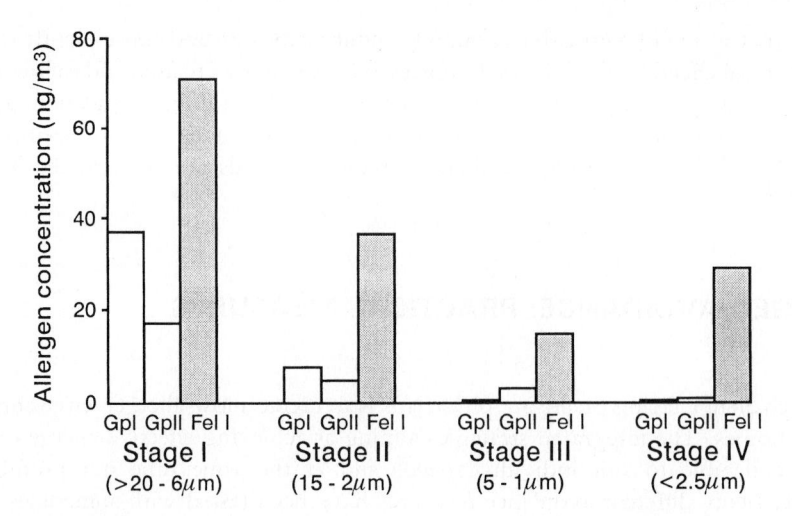

Fig. 35.5 Airborne mite and cat allergen levels on cascade-impactor stages during disturbance. Values are the mean levels for 30 min sampling (15 min of disturbance and 15-min after disturbance) in seven houses. Reproduced from ref. 131, with permission.

For the development of allergen avoidance strategies it is very important to know where patients receive most of their exposure. Since the highest levels of mite allergens within the home are found in bed and patients spend 6–8 h every night in close contact with their mattress, pillow and bedding, the bed is the most important source of exposure

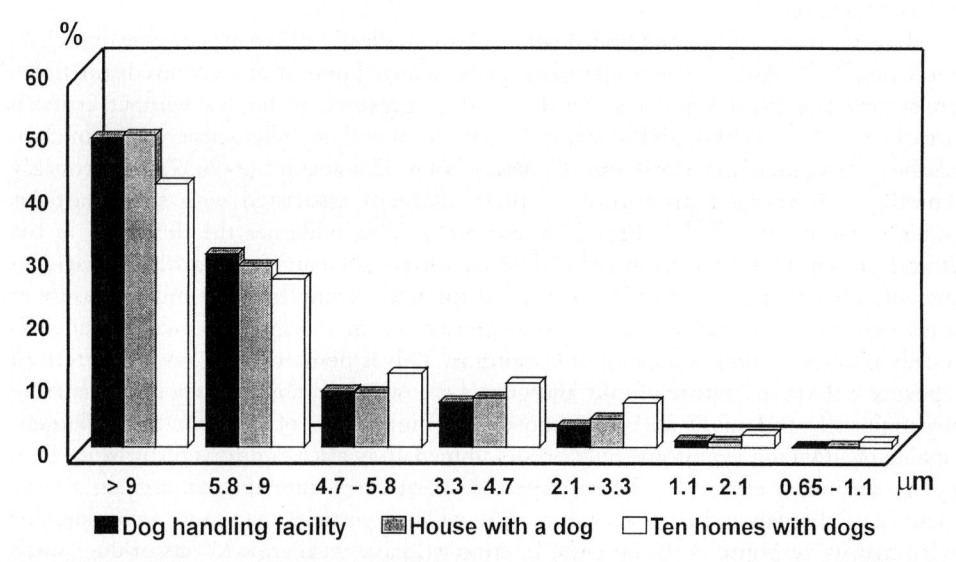

Fig. 35.6 Comparison of the particle size distribution of airborne *Can f* I (mean daily percentages) on Andersen sampler stages. Values are the mean levels for 8-h sampling in the absence of disturbance collected from three different sampling areas (dog-handling facility, 10 homes with dogs and home with a dog). Reproduced from ref. 132, with permission.

35 Allergen Avoidance

Table 35.5 Comparison of the aerodynamic properties of indoor allergens.*

Allergen	Particle size (μm)	Airborne level (ng/m³)	
		Undisturbed	*Disturbed*
Mite			
Group I	>10	<0.2	20–90
Group II	>10	<0.2	13–26
		Homes with cats	*Homes without cats*
Cat: *Fel d* I	>5 (75%)†	1.8–578	<0.2–88
	<5 (25%)†		
		Homes with dogs	*Homes without dogs*
Dog: *Can f* I	>5 (80%)†	0.3–100	<0.2–1.2
	<5 (20%)†	(undisturbed)	(undisturbed)
		Undisturbed	*Disturbed*
Cockroach			
Bla g I	>10	<0.01	9
Bla g II	>10	<0.02	2

* Mean data from studies carried out 1990–1996: see refs 131–133, 135–137, 141.
† Samples obtained with animal in the house.

and lowering exposure in the bedroom is the primary target of allergen avoidance. In contrast, the majority of exposure to allergens of domestic pets occurs in living areas other than the bedroom, and this must be taken into account when planning avoidance strategies.

Control of house-dust mites and mite allergens

Physical measures

The primary method for reducing mite allergen levels are physical measures aimed at controlling mite microhabitats.

Bed and bedding: covers and washing. The single most effective and probably most important avoidance measure is to cover the mattress, pillows and duvet with covers that are impermeable to mite allergens. These covers were initially made of plastic and uncomfortable to sleep on. However, water vapour-permeable fabrics have been developed that are both mite allergen impermeable and comfortable to sleep on. Allergen levels are dramatically reduced after the introduction of such covers.[142] Covers should be robust, easily fitted and easily cleaned, as their effectiveness is reduced if they are damaged. Mite allergen can accumulate on the covers, possibly by circulation from the carpet,[143] and it is important that covers are wiped at each change of bedding. All exposed bedding should be washed at 55°C, as McDonald and Tovey[144] have shown this to be the temperature at which all the mites in the bedding are killed. Although the cold cycle of laundry washing (30°C) dramatically reduces allergen levels, most of the mites survive it.[144] Additives for the detergents that provide a concentration of 0.03% benzyl benzoate, or dilute solutions of essential oils in normal and low temperature washing provide alternative methods of mite control.[145,146]

Asthmatic patients are often told to avoid using feather pillows and to replace them with those filled with synthetic material. This postulate has been challenged recently, first with the finding that synthetic pillows were a risk factor for severe asthma,[147] and then with the report that polyester-filled pillows contained more mite allergens than those filled with feathers.[148]

Removal of mite microhabitats. Carpets are an important microhabitat for mite colonization and a possible source of aeroallergen from which bedding can be reinfested, especially during cleaning. Even new mattresses can become a significant source of exposure after short period of time (<4 months); levels in mattresses closely correlate with those in the carpets already in the bedroom.[149] Ideally, carpets should be replaced with polished wood or vinyl flooring. Exposure of carpets to direct strong sunlight for at least 3 h creates a microenvironment lethal to mites and this simple and effective treatment may be used in loosely fitted carpets in certain climatic areas.[150] Alternatively, steam cleaning may be used as a method of killing mites and reducing allergen levels in carpets.[151,152] Curtains should be washable ($\geqslant 55°C$) and all dust-accumulating objects kept in closed cupboards. Soft toys should be kept to a minimum and be washable.

Vacuum cleaning. Intensive vacuum cleaning can remove large amounts of dust from carpets, reducing the size of the allergen reservoir.[153] However, older vacuum cleaners (with inadequate exhaust filtration) significantly increase airborne $Der\ p$ I levels.[154,155] In a study of the effect of vacuum cleaners on the concentration and particle size distribution of airborne cat allergen, Woodfolk *et al.*[156] demonstrated that some models increased total airborne allergen. These results suggest that atopic asthmatic patients should use HEPA (high efficiency particulate air)-filter vacuum cleaners with double-thickness bags, although the benefits have not been established in a clinical trial. Ducted systems offer similar advantages.

Humidity control. High levels of humidity in the microhabitats are essential for mite population growth and reducing humidity may be an effective control method. However, detailed models of the humidity profile of domestic microclimates are not available. Reducing central humidity alone may be ineffective in reducing humidity in mite microhabitats (e.g. in the middle of beds or at the base of carpets). A single portable dehumidifier placed centrally in the house is not adequate to decrease indoor humidity to the level capable of retarding mite population growth and decreasing mite allergens in the type of houses predominantly found in a mild and humid climate like the UK.[157] Mechanical ventilation heat recovery (MVHR) units have been suggested as a means of reducing mite numbers in homes by decreasing indoor humidity. These units exchange humid indoor air for dry outdoor air, recovering heat from the outgoing air in the process (i.e. the system itself does not actively dry the air). Several studies from Scandinavian countries have reported successful control of house-dust mites within domestic dwellings by the use of mechanical ventilation units.[158–160] However, MVHR units failed to reduce indoor humidity to levels capable of retarding mite population growth and decreasing mite allergens in the UK.[161] Such modifications in domestic design seem more likely to be

35 Allergen Avoidance

applicable in climates with cold, dry winters where incoming air is of sufficiently low humidity to retard mite growth.[162] Allergen avoidance measures should be allergen specific but also specific to a particular geographic area, with housing and climatic conditions being taken into account.

Chemical measures

Ever since mites were shown to be a major source of allergens in house dust, a variety of ways of killing them have been evaluated.[163] However, data from controlled clinical trials suggest lack of consistent benefit.[113,115,118]

Acaricides. A number of different chemicals that kill mites (acaricides) have been identified and shown to be effective under laboratory conditions (reviewed in ref. 164). However, data on whether these chemicals can be successfully applied to carpets, mattresses and upholstered furniture are still conflicting. Le Mao et al.[165] reported that long-term mite avoidance can be maintained by twice-yearly treatments with benzyl benzoate, although other studies could not confirm this.[118,166] Hayden et al.[167] demonstrated that the method of application of the benzyl benzoate moist powder onto carpets is very important. When carpets were treated for 4 h, only a very modest effect was observed; allowing the powder to remain on the carpet for 12–18 h with repeated brushing, followed by vigorous vacuum cleaning, reduced the concentration of mite allergens 1 month later. Allergen levels rebounded after 2 months, suggesting that repeated application every 2–3 months is necessary to control mite allergen levels.[167] Thus, the main problem of chemical treatment is not its ability to kill mites but its ability to penetrate into carpet and soft furnishing. These issues have not yet been addressed adequately. Furthermore, rigorous safety and toxicity studies on the long-term effects of acaricides need to be undertaken if they are to be used in clinical circumstances.

Liquid nitrogen. Mites can be killed by freezing with liquid nitrogen. Again, there are conflicting data on the efficacy of liquid nitrogen in the home.[166,168] Unfortunately, the technique can only be carried out by a trained operator, which further limits its use, especially since treatment will need to be repeated regularly. When used, both acaricides and liquid nitrogen should be combined with intensive vacuum cleaning following administration.

Tannic acid. The protein-denaturing properties of tannic acid are well recognized and it has been recommended for the reduction of indoor allergen levels in house dust. Woodfolk et al.[169] confirmed the profound allergen-denaturing properties of tannic acid, but also demonstrated that high levels of proteins in the dust (e.g. cat allergen in a home with a cat) blocked its effects. This suggests that $\geqslant 1\%$ tannic acid solution could reduce mite allergen levels, but only with aggressive vacuum cleaning being carried out before the treatment and in homes without pets. Solutions containing both an acaricide and tannic acid were shown to reduce skin-test reactivity of extracts prepared from patients' house dust and to have a temporary effect on mites and mite allergens.[170–172]

Air filtration and ionizers. A variety of air filtration units and ionizers have been shown to have little or no effect on symptoms and pulmonary function in mite-sensitive

Table 35.6 Proposed measures for reducing house-dust mite allergen exposure.

Encase mattress, pillow and quilt in impermeable covers
Wash all bedding in the hot cycle (55–60°C) weekly
Replace carpets with linoleum or wood flooring
If carpets cannot be removed, treat with acaricides and/or tannic acid
Minimize upholstered furniture or replace with leather furniture
Keep dust-accumulating objects in closed cupboards
Use a vacuum cleaner with integral HEPA filter and double-thickness bags
Replace curtains with blinds or easily washable (hot cycle) curtains
Hot wash/freeze soft toys

asthmatic individuals (Table 35.4).[120–125] This is not surprising, as mite allergens become airborne only following vigorous disturbance.[131,133,134] These devices do not impact on the allergen reservoirs, which represent the major source of exposure.

A large number of proprietary mite allergen control products are currently available on the market, with claims of clinical efficacy that have not been adequately tested. House-dust mites live in different sites throughout the house and it is unlikely that any single measure can solve the problem of exposure. An integrated approach (e.g. including barrier methods, dust removal, removal of mite microhabitats, etc.) is needed if a comprehensive reduction in mite allergen exposure is to be achieved (Table 35.6).

Housing design

The habitation of domestic dwellings results in the creation of house dust, including the accumulation of human skin scales. Suitable relative humidity and temperature in the microhabitat are essential for the survival of mites. However, even in the same geographic area there is still a marked difference in mite allergen levels between houses (even those in close proximity). This suggests that the design of houses has a profound effect on allergen levels and that houses with low allergen levels may be possible in any climate. (The difference in mite allergen levels between domestic dwellings and hospitals indicates that even in geographic areas where climatic conditions facilitate mite population growth, measures including encasing mattresses, regular hot washing of bedding, vigorous cleaning procedures, minimal mite microhabitats and low indoor humidity can contribute to low concentrations of mite allergens within the indoor environment.)[173] Indoor factors are crucially important in determining mite allergen levels and, when identified, may be used in the design and construction of low-allergen houses. This provides an opportunity to encourage decisions that should allow lower-allergen living.

Pet allergen avoidance

Up to 60% of asthmatic patients show IgE-mediated hypersensitivity to cat and/or dog allergen and approximately one-third of these sensitized individuals live in a home with a pet. In some parts of the world the complete avoidance of pet allergens can be extremely

35 Allergen Avoidance

difficult, as sensitized patients can be exposed to offending allergens not only in the home but also in homes without pets and in public buildings and public transport.[132,174-176] Asthma is often severe and difficult to control in pet-sensitized asthmatic individuals who continue to be exposed to the huge levels of specific allergen because they refuse to get rid of the family pet. Therefore, every effort should be made to reduce exposure to the very high levels of pet allergens seen in homes where pets may coexist with a sensitized individual.

Breed, sex and castration

It has been thought for a long time that cat allergen is spread on hair by the animal grooming or licking itself. However, the allergen is produced primarily in the sebaceous glands and in the basal squamous epithelial cells of the skin,[177,178] with very high levels reported in cat anal secretions.[179] *Fel d* I production is under hormonal control,[180] and it has been reported that castration of 1.5–2-year-old male cats resulted in a three- to five-fold reduction of *Fel d* I concentration in skin washings; testosterone treatment of the castrated cats restored the *Fel d* I levels to precastration values.[181] It has recently been suggested that *Fel d* I production is higher in male than female cats.[182] However, it is not known whether patients allergic to cats could benefit by owning a female rather than a male cat, or by castrating their male cats.

The other important question is whether one breed of cat (or dog) can produce more allergens, or different allergens, than any other. Since all cats belong to the same species, it is not likely that different breeds would exhibit breed-specific allergen molecules.[183] It is possible that there are variations in the relative concentration of allergen between different breeds (e.g. short hair and long hair).[183]

Removal of the animal from the home

Obviously, the best way to reduce exposure to cat or dog allergen in sensitized asthmatic individuals is to remove the animal from the home. However, even after permanent removal of a cat from the home, it can take many months before reservoir allergen levels return to normal.[184] Unfortunately, despite continued symptoms many patients allergic to cats and/or dogs insist on keeping their pet.

Control of airborne allergen levels with a pet in the home

Airborne pet allergen levels are about four-fold greater in rooms with a pet than without, indicating that the immediate presence of a pet contributes to airborne allergen concentrations.[132] When it is not possible to remove the animal, the pet should be kept out of the bedroom and preferably outdoors or in a well-ventilated area (e.g. kitchen).

Several studies have investigated the effect of washing the cat on *Fel d* I levels. A recent study showed no beneficial effect of washing, Allerpet-C spray and acepromazine in decreasing cat allergen shedding, although only 2 litres of water were used to wash a cat.[185] Large quantities of allergen can be removed from cats by immersion in tap water, resulting in a decreased airborne allergen concentration.[186-188] Washing dogs thoroughly in a bath using shampoo significantly reduces the levels of dog allergen in fur and dander samples.[189]

HEPA-filter air cleaners can significantly reduce airborne concentrations of cat and dog allergens in homes with pets,[189] and vacuum cleaners with built-in HEPA filters remove allergen from dust reservoirs without leaking *Fel d* I and *Can f* I.[190] As carpets may accumulate allergen levels up to 100-fold greater than polished floors, carpeting and soft furnishing should be removed.[188]

Since removal of the family pet is unfortunately rarely a viable option, we currently advise patients who are allergic to cats or dogs to follow the avoidance measures listed below.

(1) Keep the pet out of the main living areas and bedrooms.
(2) Install HEPA-filter air cleaners in the main living areas and bedrooms.
(3) Wash the pet as often as possible.
(4) Thoroughly clean upholstered furniture or replace with leather furniture.
(5) Replace carpets with linoleum or wood flooring.
(6) Fit allergen-impermeable bedding covers.
(7) Use a vacuum cleaner with integral HEPA filter and double-thickness bags.

However, the size of the clinical benefit afforded by the proposed avoidance measures has not yet been established.

ALLERGEN AVOIDANCE IN ASTHMA PREVENTION

Hide et al.[74,76,77] have investigated the effect of avoidance of mite allergens and certain foods from birth onwards on the development of atopy and asthma. The results indicated that eczema and episodic wheezing might be prevented. Even though the reduction in *Der p* I level in the active group was relatively modest (to approximately 6 μg/g), a significant reduction in sensitization to mites was observed. A number of prospective cohorts are currently being followed in order to determine whether the prevalence of asthma can be reduced by the early avoidance of house-dust mite allergens and the absence of pets in home of neonates at high risk.

CONCLUSIONS

Minimizing the impact of identified environmental risk factors (house-dust mites, cats, dogs) is a first management step to reduce the severity of asthma. Although environmental control is difficult, it must become an integral part of the overall management of sensitized asthmatic patients. We predict major changes in the domestic environment over next decade, with removal of mite microhabitats (e.g. hard flooring), reduction in indoor humidity and the covering of all beds. If the increased exposure to indoor allergens has contributed to the observed increase in asthma prevalence, the important issue of whether asthma can be prevented by allergen avoidance in infants at high risk of developing allergic disease needs to be addressed without delay.

REFERENCES

1. Storm van Leeuwen W, Einthoven W, Kremer W: The allergen proof chamber in the treatment of bronchial asthma and other respiratory diseases. *Lancet* (1927) **i**: 1287–1289.
2. Dana CL: The story of a great consultation. *Ann Med Hist* (1921) **13**: 122.
3. Leopold SS, Leopold CS: Bronchial asthma and allied allergic disorders. Preliminary report of a study under controlled conditions of environment, temperature and humidity. *JAMA* (1925) **84**: 731–735.
4. Storm van Leeuwen W: Asthma and tuberculosis in relation to 'climate allergens'. *Br Med J* (1927) **ii**: 344–347.
5. Dekker H: Asthma und milben. *Munch Med Wochenschr* (1928) **75**: 515–516.
6. Gergen PJ, Weiss KB: The increasing problem of asthma in the United States. *Am Rev Respir Dis* (1992) **146**: 823–824.
7. Burney P, Chinn S, Rona RJ: Has the prevalence of asthma increased in children? Evidence from the national study of health and growth 1973–86. *Br Med J* (1990) **300**: 1306–1310.
8. Robertson CF, Heycock E, Bishop J, Nolan T, Olinski A, Phelan P: Prevalence of asthma in Melbourne schoolchildren: changes over 26 years. *Br Med J* (1991) **302**: 1116–1118.
9. Shaw RA, Crane J, O'Donnell TV, Porteous LE, Coleman ED: Increasing asthma prevalence in a rural New Zealand adolescent population: 1975–1989. *Arch Dis Child* (1990) **65**: 1319–1323.
10. Haahtela T, Lindholm H, Bjorksten F, Koskenvuo K, Laitinen LA: Prevalence of asthma in Finnish young men. *Br Med J* (1990) **301**: 266–268.
11. Burr ML, Limb ES, Andrae S, Barry DMJ, Nagel F: Childhood asthma in four countries: a comparative survey. *Int J Epidemiol* (1994) **23**: 341–347.
12. Robertson CF, Bishop J, Sennhauser FH, Mallol J: International comparison of asthma prevalence in children: Australia, Switzerland, Chile. *Pediatr Pulmonol* (1993) **16**: 219–226.
13. Nishima S: A study of the prevalence of bronchial asthma in school children in western districts of Japan: comparison between the studies in 1982 and 1992 with the same methods and same districts. *Arerugi* (1993) **42**: 192–204.
14. Leung R, Jenkins M: Asthma, allergy and atopy in southern Chinese school students. *Clin Exp Allergy* (1994) **24**: 353–358.
15. Luyt DK, Burton PR, Simpson H: Epidemiological study of wheeze, doctor diagnosed asthma and cough in pre-school children in Leicestershire. *Br Med J* (1993) **306**: 1386–1390.
16. Peat JK, van der Berg RH, Green WF, Mellis CM, Leeder SR, Woolcock AJ: Changing prevalence of asthma in Australian children. *Br Med J* (1994) **308**: 1591–1596.
17. Platts-Mills TAE, de Weck AL: Dust mite allergens and asthma: a worldwide problem. *J Allergy Clin Immunol* (1989) **83**: 416–427.
18. Platts-Mills TAE, Thomas W, Aalberse RC, Vervloet D, Chapman MD: Dust mite allergens and asthma: report of a second international workshop. *J Allergy Clin Immunol* (1993) **89**: 1046–1057.
19. Peat JK: Prevention of asthma. *Eur Respir J* (1996) **9**: 1545–1555.
20. Chapman MD, Hayman PW, Wilkins SR, Brown MB, Platts Mills TAE: Monoclonal immunoassays for the major dust mite (*Dermatophagoides*) allergens, Der p I and Der f I and quantitative analysis of the allergen content of mite and house dust extracts. *J Allergy Clin Immunol* (1987) **80**: 184–194.
21. Luczynska CM, Arruda LK, Platts-Mills TAE, Miller JD, Lopez M, Chapman MD: A two site monoclonal antibody ELISA for the quantification of the major *Dermatophagoides* spp. allergens, Der p I and Der f I. *J Immunol Methods* (1989) **118**: 227–235.
22. Chapman MD, Aalberse RC, Brown MJ, Platts-Mills TAE: Monoclonal antibodies to the major feline allergen Fel d I. II. Single step affinity purification of Fel d I, N terminal sequence analysis and development of a sensitive two site immunoassay to assess Fel d I exposure. *J Immunol* (1988) **140**: 812–818.
23. Lombardero M, Carreira J, Duffort O: Monoclonal antibody based radioimmunoassay for the quantitation of the major cat allergen (Fel d I or Cat-1). *J Immunol Methods* (1988) **108**: 71–76.

24. DeGroot H, van Swieten P, Lind P, Aalberse RC: Monoclonal antibodies to the major feline allergen *Fel d* I. I. Biologic activity of affinity purified *Fel d* I and of *Fel d* I depleted extracts. *J Allergy Clin Immunol* (1988) **82**: 778–786.
25. Lind P: Enzyme linked immunosorbent assay for determination of major excrement allergens of house dust mite species *D. pteronyssinus*, *D. farinae* and *D. microceras*. *Allergy* (1986) **41**: 442–451.
26. Yasueda H, Mita H, Yui Y, Shida T: Measurement of allergen associated with house dust mite allergy I. Development of sensitive radioimmunoassays for the two groups of *Dermatophagoides* mite allergens, *Der p* I and *Der p* II. *Int Arch Allergy Appl Immunol* (1990) **90**: 182–189.
27. Ovsyannikova IG, Vailes L, Li Y, Hayman PW, Chapman MD: Monoclonal antibodies to Group II *Dermatophagoides* spp. allergens: murine immune response, epitope analysis, and development of a two-site ELISA. *J Allergy Clin Immunol* (1994) **94**: 537–546.
28. Pollart SM, Mullins DE, Vailes LD, Sutherland WM, Chapman MD: Identification, quantitation and purification of cockroach allergens using monoclonal antibodies. *J Allergy Clin Immunol* (1991) **87**: 511–521.
29. Pollart SM, Smith TF, Morris EC, Platts-Mills TAE, Chapman MD: Environmental exposure to cockroach allergens: analysis using a monoclonal antibody-based enzyme immunoassay. *J Allergy Clin Immunol* (1991) **87**: 505–510.
30. DeGroot H, Goei KGH, van Swieten P, Aalberse RC: Affinity purification of a major and minor allergen from dog extract: serologic activity of affinity purified *Can f* I and of *Can f* I depleted extract. *J Allergy Clin Immunol* (1991) **87**: 1056–1065.
31. Schou C, Hansen GN, Lintner T, Lowenstein H: Assay for the major dog allergen, *Can f* I: investigation of house dust samples and commercial dog extracts. *J Allergy Clin Immunol* (1991) **88**: 847–853.
32. Charpin D, Kleisbauer JP, Lanteaume A, et al.: Asthma and allergy to house dust mites in population living in high altitudes. *Chest* (1988) **93**: 758–761.
33. Charpin D, Birnbaum J, Haddi E, et al.: Altitude and allergy to house dust mites. *Am Rev Respir Dis* (1991) **143**: 983–986.
34. Peat JK, Tovey E, Mellis CM, Leeder SR, Woolcock AJ: Importance of house dust mite and *Alternaria* allergens in childhood asthma: an epidemiological study in two climatic regions of Australia. *Clin Exp Allergy* (1993) **23**: 812–820.
35. Peat JK, Tovey E, Gray EJ, Mellis CM, Woolcock AJ: Asthma severity and morbidity in a population sample of Sydney schoolchildren: Part II. Importance of house dust mite allergens. *Aust NZ J Med* (1994) **24**: 270–276.
36. Peat JK, Tovey ER, Toelle BG, et al.: House dust mite allergens: a major risk factor for childhood asthma in Australia. *Am J Respir Crit Care Med* (1996) **152**: 144–146.
37. Sporik R, Ingram MJ, Price W, Sussman JH, Honsinger RW, Platts Mills TAE: Association of asthma with serum IgE and skin test reactivity to allergens among children living at high altitude: tickling the dragon's breath. *Am J Respir Crit Care Med* (1995) **151**: 1388–1392.
38. Ingram JM, Sporik R, Rose G, Honsinger R, Chapman MD, Platts-Mills TAE: Quantitative assessment of exposure to dog (*Can f* 1) and cat (*Fel d* 1) allergens: relationship to sensitisation and asthma among children living in Los Alamos, New Mexico. *J Allergy Clin Immunol* (1995) **96**: 449–456.
39. Sarpong SB, Hamilton RG, Eggleston PA, Adkinson NF Jr: Socioeconomic status and race as risk factors for cockroach allergen exposure and sensitisation in children with asthma. *J Allergy Clin Immunol* (1996) **97**: 1393–1401.
40. De Blay F, Kassell O, Chapman MD, Ott M, Verot A, Pauli G: Mise en evidence des allergens majeurs des blattes par test ELISA dans la poussière domestique. *Presse Med* (1992) **21**: 1685.
41. Bjorksten F, Suoniemi I, Koski V: Neonatal birch pollen contact and subsequent allergy to birch pollen. *Clin Allergy* (1980) **10**: 581–591.
42. Korsgaard J, Dahl R: Sensitivity to house dust mite and grass pollen in adults. Influence of the month of birth. *Clin Allergy* (1983) **13**; 529–536.
43. Vanto T, Koivikko A: Dog hypersensitivity in asthmatic children. *Acta Paediatr Scand* (1983) **72**: 571–575.
44. Desjardins A, Benoit C, Ghezzo H, et al.: Exposure to domestic animals and risk of immunologic sensitisation in subjects with asthma. *J Allergy Clin Immunol* (1993) **91**: 979–986.

45. Sporik R, Holgate S, Platts-Mills TAE, Cogswell J: Exposure to house dust mite allergen (*Der p* I) and the development of asthma in childhood. *N Engl J Med* (1990) **323**: 502–507.
46. Sporik R, Chapman MD, Platts Mills TAE: House dust mite exposure as a cause of asthma. *Clin Exp Allergy* (1992) **22**: 897–906.
47. Wickman M, Korsgaard J: Transient sensitisation to house dust mites: a study on the influence of mite exposure and sex. *Allergy* (1996) **51**: 511–513.
48. Dowse GK, Turner KJ, Stewart GA, Alpers MP, Woolcock AJ: The association between *Dermatophagoides* mites and the increasing prevalence of asthma in village communities within the Papua New Guinea highlands. *J Allergy Clin Immunol* (1985) **75**: 75–83.
49. Custovic A, Taggart SCO, Francis HC, Chapman MD, Woodcock A: Exposure to house dust mite allergens and the clinical activity of asthma. *J Allergy Clin Immunol* (1996) **98**: 64–72.
50. Van der Heide S, de Monchy JGR, de Vries K, Bruggink TM, Kauffman HF: Seasonal variation in airway hyperresponsiveness and natural exposure to house dust mite allergens in patients with asthma. *J Allergy Clin Immunol* (1994) **93**: 470–475.
51. Van der Heide S, de Monchy JGR, de Vries K, Dubois AEJ, Kauffman HF: Seasonal differences in airway hyperresponsiveness in asthmatic patients: relationship with allergen exposure and sensitisation to house dust mites. *Clin Exp Allergy* (1997) **27**: 627–633.
52. Van Strien RT, Verhoeff AP, van Wijnen JH, Doekes G, de Meer G, Brunekreef B: Infant respiratory symptoms in relation to mite allergen exposure. *Eur Respir J* (1996) **9**: 926–931.
53. Vervloet D, Charpin D, Haddi E, *et al*.: Medication requirements and house-dust mite exposure in mite sensitive asthmatics. *Allergy* (1991) **46**: 554–558.
54. Platts-Mills TAE, Hayden ML, Woodfolk JA, Call RS, Sporik R: House dust mite avoidance regimens for the treatment of asthma. In David TJ (ed) *Recent Advances in Paediatrics*, vol. 13. Edinburgh, Churchill Livingstone, 1995, pp 45–58.
55. Kuehr J, Frischer T, Meinert R, *et al*.: Mite allergen exposure is a risk for the incidence of specific sensitization. *J Allergy Clin Immunol* (1994) **94**: 44–52.
56. Lau S, Falkenhorst G, Weber A, *et al*.: High mite-allergen exposure increases the risk of sensitization in atopic children and young adults. *J Allergy Clin Immunol* (1989) **84**: 718–725.
57. Wickman M, Nordvall SL, Pershagen G, Sundell J, Schwartz B: House dust mite sensitisation in children and residential characteristics in a temperate region. *J Allergy Clin Immunol* (1991) **88**: 89–95.
58. Wickman M, Nordvall SL, Pershagen G, Korsgaard J, Johansen N: Sensitisation to domestic mites in a cold temperate region. *Am Rev Respir Dis* (1993) **148**: 58–62.
59. Pope AM, Patterson R, Burge H (eds): *Indoor Allergens: Assessing and Controlling Adverse Health Effects*. Washington, DC, National Academy Press, 1993.
60. Arruda LK, Rizzo MC, Chapman MD, *et al*.: Exposure and sensitisation to dust mite allergens among asthmatic children in Sao Paulo, Brazil. *Clin Exp Allergy* (1991) **21**: 433–439.
61. Price JA, Pollock I, Little SA, Longbottom JL, Warner JO: Measurement of airborne mite antigen in homes of asthmatic children. *Lancet* (1990) **336**: 895–897.
62. Burrows B, Martinez FD, Halonen M, Barbec RA, Cline MG: Association of asthma with serum IgE levels and skin test reactivity to allergens. *N Engl J Med* (1989) **320**: 271–277.
63. Burrows B, Sears MR, Flannery EM, Herbison GP, Holdaway MD: Relationship of bronchial responsiveness assessed by methacholine to serum IgE, lung function, symptoms, and diagnoses in 11-year-old New Zealand children. *J Allergy Clin Immunol* (1992) **90**: 376–385.
64. Burrows B, Sears MR, Flannery EM, Herbison GP, Holdaway MD: Relationship of bronchial responsiveness to allergy skin test reactivity, lung function, respiratory symptoms, and diagnoses in thirteen-year-old New Zealand children. *J Allergy Clin Immunol* (1995) **95**: 548–556.
65. Henderson FW, Henry MM, Ivins SS, *et al*.: Correlates of recurrent wheezing in school-age children. *Am J Respir Crit Care Med* (1995) **151**: 1786–1793.
66. Sears MR, Herbison GP, Holdaway MD, Hewitt CJ, Flannery EM, Silva PA: The relative risk of sensitivity to grass pollen, house dust mite and cat dander in the development of childhood asthma. *Clin Exp Allergy* (1989) **19**: 419–424.
67. Sears MR, Burrows B, Flannery EM, Herbison GP, Holdaway MD: Atopy in childhood. I.

Gender and allergen related risks for development of hay fever and asthma. *Clin Exp Allergy* (1993) **23**: 941–948.
68. Sears MR, Burrows B, Herbison GP, Holdaway MD, Flannery EM: Atopy in childhood. II. Relationship to airway responsiveness, hay fever and asthma. *Clin Exp Allergy* (1993) **23**: 949–956.
69. Omenaas E, Bakke P, Eide GE, Elsayed S, Gulsvik A: Serum house dust mite antibodies: predictor of increased bronchial responsiveness in adults of a community. *Eur Respir J* (1996) **9**: 919–925.
70. Bjornsson E, Norback D, Janson C, et al.: Asthmatic symptoms and indoor levels of microorganisms and house dust mites. *Clin Exp Allergy* (1995) **25**: 423–431.
71. Marks GB, Tovey ER, Toelle BG, Wachinger S, Peat JK, Woolcock AJ: Mite allergen (*Der p* 1) concentration in houses and its relation to the presence and severity of asthma in a population of Sydney schoolchildren. *J Allergy Clin Immunol* (1995) **96**: 441–448.
72. Marks GB, Tovey ER, Green W, Shearer M, Salome CM, Woolcock AJ: The effect of changes in house dust mite allergen exposure on the severity of asthma. *Clin Exp Allergy* (1995) **25**: 114–118.
73. Zock JP, Brunekreef B, Hazebrook-Kampschreurn AAJM, Roosjen CV: House-dust mite allergen in bedroom floor dust and respiratory health of children with asthmatic symptoms. *Eur Respir J* (1994) **7**: 1254–1259.
74. Arshad SH, Matthews S, Gant C, Hide DW: Effect of allergen avoidance on development of allergic disorder in infancy. *Lancet* (1992) **339**: 1493–1497.
75. Arshad SH, Hide DW: Effect of environmental factors on the development of allergic disorders in infancy. *J Allergy Clin Immunol* (1992) **90**: 235–241.
76. Hide DW, Matthews S, Matthews L, et al.: Effect of allergen avoidance in infancy on allergic manifestation at age two years. *J Allergy Clin Immunol* (1994) **93**: 842–846.
77. Hide DW, Matthews S, Tariq S, Arshad SH: Allergen avoidance in infancy and allergy at 4 years of age. *Allergy* (1996) **51**: 89–93.
78. Kuehr J, Frisher T, Meinert R, et al.: Sensitisation to mite allergens is a risk factor for early and late onset of asthma and for persistence of asthmatic signs in children. *J Allergy Clin Immunol* (1995) **95**: 655–662.
79. Ohman JL, Sparrow D, MacDonald MR: New onset wheezing in an older male population: evidence of allergen sensitization in a longitudinal study. *J Allergy Clin Immunol* (1993) **91**: 752–757.
80. Warner JA, Little SA, Pollock I, Longbottom JL, Warner JO: The influence of exposure to house dust mite, cat, pollen and allergens in the homes on primary sensitisation in asthma. *Pediatr Allergy Immunol* (1991) **1**: 79–86.
81. Kuehr J, Frischer T, Karmaus W, et al.: Early childhood risk factors for sensitisation at school age. *J Allergy Clin Immunol* (1992) **90**: 358–363.
82. Kang BC, Johnson J, Veres-Thorner C: Atopic profile of inner-city asthma with a comparative analysis on the cockroach-sensitive and ragweed-sensitive subgroups. *J Allergy Clin Immunol* (1993) **92**: 802–811.
83. Duff AL, Pomeranz ES, Gelber LE, et al.: Risk factors for acute wheezing in infants and children: viruses, passive smoke, and IgE antibodies to inhalant allergens. *Pediatrics* (1993) **92**: 535–540.
84. Sporik R, Platts Mills TAE, Cogswell JJ: Exposure to house dust mite allergen of children admitted to hospital with asthma. *Clin Exp Allergy* (1993) **23**: 740–746.
85. Pollart SM, Reid MJ, Fling JA, Chapman MD, Platts Mills TAE: Epidemiology of emergency room asthma in northern California: association with IgE antibody to ryegrass pollen. *J Allergy Clin Immunol* (1988) **82**: 224–230.
86. Gelber LE, Seltzer LH, Bouzoukis JK, Pollart SM, Chapman MD, Platts-Mills TAE: Sensitisation and exposure to indoor allergens as risk factor for asthma among patients presenting to hospital. *Am Rev Respir Dis* (1993) **147**: 573–578.
87. Call RS, Smith TF, Morris E, Chapman MD, Platts Mills TAE: Risk factors for asthma in inner city children. *J Pediatr* (1992) **121**: 862–866.
88. Chavarria JF, Carswell F: House dust mite exposure and emergency hospital admission of asthmatic children. *Int Arch Allergy Immunol* (1992) **99**: 466–467.

89. Kerrebijn KF: Endogenous factors in childhood CNSLD: methodological aspects in population studies. In Orie NGM, Van der Lende R (eds) *Bronchitis III*. The Netherlands, Royal Van Gorcum Assen, 1970, pp 38–48.
90. Platts Mills TAE, Chapman MD: Dust mites: immunology, allergic disease, and environmental control. *J Allergy Clin Immunol* (1987) **80**: 755–775.
91. Boner AL, Niero E, Antolini I, Valletta EA, Gaburro D: Pulmonary function and bronchial hyperreactivity in asthmatic children with house dust mite allergy during prolonged stay in the Italian Alps (Misurina 1756 m). *Ann Allergy* (1985) **54**: 42–45.
92. Piacentini GL, Martinati L, Fornari A, *et al.*: Antigen avoidance in a mountain environment: influence on basophil releasability in children with allergic asthma. *J Allergy Clin Immunol* (1993) **92**: 644–650.
93. Peroni DG, Boner AL, Vallone G, Antolini I, Warner JO: Effective allergen avoidance at high altitude reduces allergen-induced bronchial hyperresponsiveness. *Am J Respir Crit Care Med* (1994) **149**: 1442–1446.
94. Platts Mills TAE, Tovey ER, Mitchell EB, Moszoro H, Nock P, Wilkins SR: Reduction of bronchial hyperreactivity during prolonged allergen avoidance. *Lancet* (1982) **ii**: 675–678.
95. Harving H, Korsgaard J, Dahl R: Clinical efficacy of reduction in house dust mite exposure in specially designed, mechanically ventilated 'healthy' homes. *Allergy* (1994) **49**: 866–880.
96. Morrison Smith J: The use of high altitude treatment for childhood asthma. *Practitioner* (1981) **225**: 1663–1666.
97. Boner AL, Peroni DG, Piacentini GL, Venge P: Influence of allergen avoidance at high altitude on serum markers of eosinophil activation in children with allergic asthma. *Clin Exp Allergy* (1993) **23**: 1021–1026.
98. Simon HU, Grotzer M, Nikolaizik WH, Blaser K, Schoni MH: High altitude climate therapy reduces peripheral blood T lymphocyte activation, eosinophilia, and bronchial obstruction in children with house-dust mite allergic asthma. *Pediatr Pulmonol* (1994) **17**: 304–311.
99. Valletta EA, Comis A, Del Col G, Spezia E, Boner AL: Peak expiratory flow variation and bronchial hyperresponsiveness in asthmatic children during periods of antigen avoidance and reexposure. *Allergy* (1995) **50**: 366–369.
100. van Velzen E, van den Bos JW, Benckhuijsen JAW, van Essel T, de Bruijn R, Aalbers R: Effect of allergen avoidance at high altitude on direct and indirect bronchial hyperresponsiveness and markers of inflammation in children with allergic asthma. *Thorax* (1996) **51**: 582–584.
101. Piacentini GL, Martinati L, Mingoni S, Boner AL: Influence of allergen avoidance on the eosinophil phase of airway inflammation in children with allergic asthma. *J Allergy Clin Immunol* (1996) **97**: 1079–1084.
102. Sanda T, Yasue T, Ooashi M, Yasue A: Effectiveness of house dust-mite allergen avoidance through clean room therapy in patients with atopic dermatitis. *J Allergy Clin Immunol* (1992) **89**: 653–657.
103. Sarsfield JK, Gowland G, Toy R, Norman ALE: Mite sensitive asthma of childhood. Trial of avoidance measures. *Arch Dis Child* (1974) **49**: 716–721.
104. Burr ML, Leger ASST, Neale E: Anti-mite measures in mite-sensitive adult asthma. A controlled trial. *Lancet* (1976) **i**: 333–335.
105. Burr ML, Dean BV, Merrett TG, Neale E, Leger ASST, Verrier-Jones ER: Effect of anti-mite measures on children with mite-sensitive asthma: a controlled trial. *Thorax* (1980) **35**: 506–512.
106. Burr ML, Neale E, Dean BV, Verrier-Jones ER: Effect of a change to mite-free bedding on children with mite-sensitive asthma: a controlled trial. *Thorax* (1980) **35**: 513–514.
107. Korsgaard J: Preventive measures in house-dust allergy. *Am Rev Respir Dis* (1982) **125**: 80–84.
108. Korsgaard J: Preventive measures in mite asthma. A controlled trial. *Allergy* (1983) **38**: 93–102.
109. Murray AB, Ferguson AC: Dust-free bedrooms in the treatment of asthmatic children with house dust or house dust mite allergy: a controlled trial. *Pediatrics* (1983) **71**: 418–422.
110. Walshaw MJ, Evans CC: Allergen avoidance in house dust mite sensitive adult asthma. *Q J Med* (1986) **58**: 199–215.

111. Gillies DRN, Littlewood JM, Sarsfield JK: Controlled trial of house dust mite avoidance in children with mild to moderate asthma. *Clin Allergy* (1987) **17**: 105–111.
112. Dorward AJ, Colloff MJ, MacKay NS, McSharry CM, Thomson NC: Effect of house dust mite avoidance on adult atopic asthma. *Thorax* (1988) **43**: 98–102.
113. Reiser J, Ingram D, Mitchell EB, Warner JO: House dust mite allergen levels and an anti-mite mattress spray (natamycin) in the treatment of childhood asthma. *Clin Exp Allergy* (1990) **20**: 561–567.
114. Morrow Brown H, Merrett TG: Effectiveness of an acaricide in management of house dust mite allergy. *Ann Allergy* (1991) **67**: 25–31.
115. Ehnert B, Lau-Schadendorf S, Weber A, Buettner P, Schou C, Wahn V: Reducing domestic exposure to dust mite allergen reduces bronchial hyperreactivity in sensitive children with asthma. *J Allergy Clin Immunol* (1993) **90**: 135–138.
116. Dietermann A, Bessot JC, Hoyet C, Ott M, Verot A, Pauli G: A double-blind, placebo controlled trial of solidified benzyl benzoate applied in dwellings of asthmatic patients sensitive to mites: clinical efficacy and effect on mite allergens. *J Allergy Clin Immunol* (1993) **91**: 738–746.
117. Marks GB, Tovey ER, Green W, Shearer M, Salome CM, Woolcock AJ: House dust mite allergen avoidance: a randomised controlled trial of surface chemical treatment and encasement of bedding. *Clin Exp Allergy* (1994) **24**: 1078–1083.
118. Huss RW, Huss K, Squire EN, *et al.*: Mite allergen control with acaricide fails. *J Allergy Clin Immunol* (1994) **94**: 27–31.
119. Carswell F, Birmingham K, Oliver J, Crewes A, Weeks J: The respiratory effect of reduction of mite allergen in the bedroom of asthmatic children: a double-blind, controlled trial. *Clin Exp Allergy* (1996) **26**: 386–396.
120. Mitchell EA, Elliott RB: Controlled trial of an electrostatic precipitator in childhood asthma. *Lancet* (1980) **ii**: 559–561.
121. Bowler SD, Mitchell CA, Miles J: House dust mite control and asthma: a placebo control trial of cleaning air filtration. *Ann Allergy* (1985) **55**: 498–500.
122. Reisman RE, Mauriello PM, Davis GB, Georgitis JW, DeMasi JM: A double-blind study of the effectiveness of a high-efficiency particulate air (HEPA) filter in the treatment of patients with perennial allergic rhinitis and asthma. *J Allergy Clin Immunol* (1990) **85**: 1050–1057.
123. Antonicelli L, Bilo MB, Pucci S, Schou C, Bonifazi F: Efficacy of an air cleaning device equipped with a high efficiency particulate air filter in house dust mite respiratory allergy. *Allergy* (1991) **46**: 594–600.
124. Warner JA, Marchant JL, Warner JO: Double blind trial of ionisers in children with asthma sensitive to the house dust mite. *Thorax* (1993) **48**: 330–333.
125. Warburton CJ, Niven RMcL, Pickering CAC, Hepworth J, Francis HC: Domiciliary air filtration units, symptoms and lung function in atopic asthmatics. *Respir Med* (1994) **88**: 771–776.
126. Kniest FM, Young E, Van Praag MCG, *et al.*: Clinical evaluation of a double-blind dust-mite avoidance trial with mite-allergic rhinitic patients. *Clin Exp Allergy* (1991) **21**: 39–47.
127. Roberts DLL: House dust mite allergen avoidance and atopic dermatitis. *Br J Dermatol* (1984) **110**: 735–736.
128. August PJ: House dust mite causes atopic eczema. A preliminary study. *Br J Dermatol* (1984) **111** (Suppl 26): 10–11.
129. Colloff MJ, Lever RS, McSharry C: A controlled trial of house dust mite eradication using natamycin in homes of patients with atopic dermatitis: effect on clinical status and mite populations. *Br J Dermatol* (1989) **121**: 199–208.
130. Tan BB, Weald D, Strickland I, Friedmann PS: Double blind controlled trial of effects of house dust mite allergen avoidance on atopic dermatitis. *Lancet* (1996) **347**: 15–18.
131. De Blay F, Heymann PW, Chapman MD, Platts-Mills TAE: Airborne dust mite allergens: comparison of Group II mite allergens with Group I mite allergen and cat allergen Fel d I. *J Allergy Clin Immunol* (1991) **88**: 919–926.
132. Custovic A, Green R, Fletcher A, *et al.*: Aerodynamic properties of the major dog allergen, Can f 1: distribution in homes, concentration and particle size of allergen in the air. *Am J Respir Crit Care Med* (1997) **155**: 94–98.

133. Custovic A, Taggart SCO, Niven RMcL, Woodcock A: Monitoring exposure to house dust mite allergens. *J Allergy Clin Immunol* (1995) **96**: 134–135.
134. Sporik R, Chapman M, Platts-Mills T: Airborne mite antigen (letter). *Lancet* (1990) **336**: 1507–1508. 135. Luczynska CM, Li Y, Chapman MD, Platts-Mills TAE: Airborne concentrations and particle size distribution of allergen derived from domestic cats (*Felis domesticus*): measurement using cascade impactor, liquid impinger and a two site monoclonal antibody assay for Fel d I. *Am Rev Respir Dis* (1990) **141**: 361–367.
136. Sakaguchi M, Inouye S, Yasueda H, Tatehisa I, Yoshizawa S, Shida T: Measurement of allergen associated with house dust mite allergy. II. Concentrations of airborne mite allergens (*Der* I and *Der* II) in the house. *Int Arch Allergy Appl Immunol* (1990) **90**: 190–193.
137. Sakaguchi M, Inouye S, Sasaki R, Hashimoto M, Kobayashi C, Yasueda H: Measurement of airborne mite allergen exposure in individual subjects. *J Allergy Clin Immunol* (1996) **97**: 1040–1044.
138. Van Metre TE, Marsh DG, Adkinson NF, *et al.*: Dose of cat (*Felis domesticus*) allergen 1 (*Fel d* I) that induces asthma. *J Allergy Clin Immunol* (1986) **78**: 72–75.
139. Swanson MC, Campbell AR, Klauck MJ, Reed CE: Correlation between levels of mite and cat allergens in settled and airborne dust. *J Allergy Clin Immunol* (1989) **83**: 776–783.
140. Wentz PE, Swanson MC, Reed CE: Variability of cat allergen shedding. *J Allergy Clin Immunol* (1990) **85**: 94–98.
141. Wood RA, Laheri AN, Eggleston PA: The aerodynamic characteristics of cat allergen. *Clin Exp Allergy* (1993) **23**: 733–739.
142. Owen S, Morgenstern M, Hepworth J, Woodcock A: Control of house dust mite in bedding. *Lancet* (1990) **335**: 396–397.
143. Tovey E, Marks G, Shearer M, Woolcock A: Allergens and occlusive bedding covers. *Lancet* (1993) **342**: 126.
144. McDonald LG, Tovey E: The role of water temperature and laundry procedures in reducing house dust mite populations and allergen content of bedding. *J Allergy Clin Immunol* (1992) **90**: 599–608.
145. Bischoff ERC, Fischer A, Liebenberg B, Kniest FM: Mite control with low temperature washing. 1. Elimination of living mites on carpet pieces. *Clin Exp Allergy* (1996) **26**: 945–952.
146. McDonald LG, Tovey E: The effectiveness of benzyl benzoate and some essential plant oils as laundry additives for killing house dust mites. *J Allergy Clin Immunol* (1993) **92**: 771–772.
147. Strachan DP, Carey IM: Home environment and severe asthma in adolescence: a population based case-control study. *Br Med J* (1995) **311**: 1053–1056.
148. Kemp TJ, Siebers RW, Fishwick D, O'Grady GB, Fitzharris P, Crane J: House dust mite allergen in pillows. *Br Med J* (1996) **313**: 916.
149. Custovic A, Green R, Smith A, Chapman MD, Woodcock A: New mattresses: how fast do they become significant source of exposure to house dust mite allergens? *Clin Exp Allergy* (1996) **26**: 1243–1245.
150. Tovey ER, Woolcock AJ: Direct exposure of carpets to sunlight can kill all mites. *J Allergy Clin Immunol* (1994) **93**: 1072–1074.
151. Colloff MJ, Taylor C, Merrett TG: The use of domestic steam treatment for the control of house dust mites. *Clin Exp Allergy* (1995) **25**: 1061–1066.
152. Custovic A, Taggart S, Chapman M, Fletcher A, Woodcock A: Steam cleaning and *Der p* 1 concentration in carpets. In Tovey E, Fifoot A, Sieber L (eds) *Mites, Asthma and Domestic Design II*. Sydney, University Printing Service, 1995, pp 45–47.
153. Collof MJ, Ayres J, Carswell F *et al.*: The control of allergens of dust mites and domestic pets: a position paper. *Clin Exp Allergy* (1992) **22** (Suppl 2): 1–28.
154. Kalra S, Owen SJ, Hepworth J, Woodcock A: Airborne house dust mite antigen after vacuum cleaning. *Lancet* (1990) **336**: 449.
155. Hegarty JM, Rouhbakhsh S, Warner JA, Warner JO: A comparison of the effect of conventional and filter vacuum cleaners on airborne house dust mite allergen. *Respir Med* (1995) **89**: 279–284.
156. Woodfolk JA, Luczynska CM, de Blay F, Chapman MD, Platts Mills TAE: The effect of vacuum cleaners on the concentration and particle size distribution of airborne cat allergen. *J Allergy Clin Immunol* (1993) **91**: 829–837.

157. Custovic A, Taggart SCO, Kennaugh JH, Woodcock A: Portable dehumidifiers in the control of house dust mites and mite allergens. *Clin Exp Allergy* (1995) **25**: 312–316.
158. Korsgaard J, Iversen M: Epidemiology of house dust mite allergy. *Allergy* (1991) **46** (Suppl 1): 14–18.
159. Wickman M, Emenius G, Egmar A-C, Axelsson G, Pershagen G: Reduced mite allergen levels in dwellings with mechanical exhaust and supply ventilation. *Clin Exp Allergy* (1994) **24**: 109–114.
160. Korsgaard J: Mechanical ventilation and house dust mites: a controlled investigation. In Van Moerbeke D (ed) *Dust Mite Allergens and Asthma*. Brussels, UCB Institute of Allergy, 1991, pp 87–89.
161. Fletcher A, Pickering CAC, Custovic A, Simpson J, Kennaugh J, Woodcock A: Reduction in humidity as a method of controlling mites and mite allergens: the use of mechanical ventilation in British domestic dwellings. *Clin Exp Allergy* (1996) **26**: 1051–1056.
162. Colloff MJ: Dust mite control and mechanical ventilation: when the climate is right. *Clin Exp Allergy* (1994) **24**: 94–96.
163. Heller-Haupt A, Busvine JR: Tests of acaricides against house dust mites. *J Med Entomol* (1974) **11**: 551–558.
164. Colloff MJ: House dust mites—part II. Chemical control. *Pestic Outlook* (1990) **1**: 3–8.
165. Le Mao J, Liebenberg B, Bischoff E, David B: Changes in mite allergen levels in homes using an acaricide combined with cleaning agents: a 3-year follow-up study. *Indoor Environ* (1992) **1**: 212–218.
166. Kalra S, Crank P, Hepworth J, Pickering CAC, Woodcock AA: Domestic house dust mite allergen (*Der p* I) concentrations after treatment with solidified benzyl benzoate (Acarosan) or liquid nitrogen. *Thorax* (1993) **48**: 10–14.
167. Hayden ML, Rose G, Diduch KB, *et al.*: Benzyl-benzoate moist powder: investigation of acarical activity in cultures and reduction of dust mite allergens in carpets. *J Allergy Clin Immunol* (1992) **89**: 536–545.
168. Colloff MJ: Use of liquid nitrogen in the control of house dust mite populations. *Clin Allergy* (1986) **16**: 411–417.
169. Woodfolk JA, Hayden ML, Miller JD, Rose G, Chapman MD, Platts-Mills TAE: Chemical treatment of carpets to reduce allergen: a detailed study of the effects of tannic acid on indoor allergens. *J Allergy Clin Immunol* (1994) **94**: 19–26.
170. Green WF: Abolition of allergens by tannic acid. *Lancet* (1984) **ii**: 160.
171. Warner JA, Marchant JL, Warner JO: Allergen avoidance in the homes of atopic asthmatic children: the effect of Allersearch DMS. *Clin Exp Allergy* (1993) **23**: 279–286.
172. Tovey ER, Marks GB, Matthews M, Green WF, Woolcock A: Changes in mite allergen *Der p* I in house dust following spraying with a tannic acid/acaricide solution. *Clin Exp Allergy* (1992) **22**: 67–74.
173. Babe KS, Arlian LG, Confer PD, Kim R: House dust mite (*Dermatophagoides pteronyssinus* and *Dermatophagoides farinae*) prevalence in the rooms and hallways of a tertiary care hospital. *J Allergy Clin Immunol* (1995) **95**: 801–805.
174. Custovic A, Taggart SCO, Woodcock A: House dust mite and cat allergen in different indoor environments. *Clin Exp Allergy* (1994) **24**: 1164–1168.
175. Custovic A, Green R, Taggart SCO, *et al.*: Domestic allergens in public places II: dog (*Can f* 1) and cockroach (*Bla g* 2) allergens in dust and mite, cat, dog and cockroach allergens in air in public buildings. *Clin Exp Allergy* (1996) **26**: 1246–1252.
176. Egmar L, Emenius G, Axelsson G, Pershagen G, Wickman G: Direct and indirect exposure to cat (*Fel d* I) and dog (*Can f* I) allergens in homes. *J Allergy Clin Immunol* (1993) **91**: 324.
177. Charpin C, Mata P, Charpin D, Lavant MN, Allasia C, Vervloet D: *Fel d* I allergen distribution in cat fur and skin. *J Allergy Clin Immunol* (1991) **88**: 77–82.
178. Dabrowski AJ, Van Der Brempt X, Soler M, *et al.*: Cat skin as an important source of *Fel d* I allergen. *J Allergy Clin Immunol* (1990) **86**: 462–465.
179. Dornelas de Andrede A, Birnbaum J, Magalon C, *et al.*: *Fel d* I levels in cat anal glands. *Clin Exp Allergy* (1996) **26**: 178–180.
180. Zielonka TM, Charpin D, Berbis P, Lucciani P, Casanova D, Vervloet D: Hormonal control of cat allergen (*Fel d* I) production by sebaceous glands. *J Allergy Clin Immunol* (1993) **91**: 327.

35 Allergen Avoidance

181. Zielonka TM, Charpin D, Berbis P, Luciani P, Casanova D, Vervloet D: Effects of castration and testosterone on *Fel d* 1 production by sebaceous glands of male cats: I. Immunological assessment. *Clin Exp Allergy* (1994) **24**: 1169–1173.
182. Jalil-Colome J, Dornelas de Andrade A, Birnbaum J, *et al*.: Sex difference in *Fel d* 1 allergen production. *J Allergy Clin Immunol* (1996) **98**: 165–168.
183. Schou C: Defining allergens of mammalian origin. *Clin Exp Allergy* (1993) **23**: 7–14.
184. Wood RA, Chapman MD, Adkinson NF Jr, Eggleston PA: The effect of cat removal on allergen content in the household dust samples. *J Allergy Clin Immunol* (1989) **83**: 730–734.
185. Klucka CV, Ownby DR, Green J, Zoratti E: Cat shedding is not reduced by washings, Allerpet-C spray, or acepromasine. *J Allergy Clin Immunol* (1995) **95**: 1164–1171.
186. Glinert R, Wilson P, Wedner HJ: *Fel d* I is markedly reduced following sequential washing of cats. *J Allergy Clin Immunol* (1990) **85**: 225.
187. Avner D, Woodfolk JA, Platts-Mills TAE: Washing cats: quantitation of *Fel d* 1 allergen removed by water immersion. *J Allergy Clin Immunol* (1995) **95**: 262.
188. De Blay F, Chapman MD, Platts Mills TAE: Airborne cat allergen *Fel d* I: environmental control with cat *in situ*. *Am Rev Respir Dis* (1991) **143**: 1334–1339.
189. Green R, Custovic A, Smith A, Chapman MD, Woodcock A: Avoidance of dog allergen *Can f* 1 with the dog *in situ*: washing the dog and use of a HEPA air filter. *J Allergy Clin Immunol* (1996) **97**: 302.
190. Green RM, Custovic A, Smith A, Chapman MD, Woodcock A: Testing vacuum cleaners: leakage of dust containing *Can f* 1. *Thorax* (1995) **50** (Suppl 2): A72.

36

β-Adrenoceptor Agonists

I.P. HALL AND A.E. TATTERSFIELD

INTRODUCTION

Drugs that act as selective β_2-adrenoceptor agonists are the main bronchodilator agents for the treatment of asthma. Recent advances in the field of molecular pharmacology have led to a much clearer understanding of the mechanisms whereby β-agonists cause airway smooth muscle relaxation and bronchodilatation. In addition, the development of long-acting β_2-agonists has widened the range of agents available for clinical use. The first section of this chapter deals with the molecular pharmacology of β-adrenoceptor agonists and the structure–activity relationships of these agents. The second part considers the clinical pharmacology of β_2-agonists and discusses the clinical benefits and problems associated with their use.

MOLECULAR PHARMACOLOGY

β-Adrenoceptor populations

Three different human β-adrenoceptors have been characterized by molecular cloning and molecular pharmacological techniques. β-Adrenoceptors were initially separated into β_1 and β_2 adrenoceptors by Lands et al.[1] according to differences in the tissue-selective physiological effects of a range of β-adrenoceptor agonists. More recent studies have demonstrated that the lipolytic effects of β-adrenoceptor agonists in adipocytes are mediated through different β-adrenoceptors. The discovery of selective agonists and the

Table 36.1 Major action of β-adrenergic agonists at different receptor subtypes.

Tissue	Receptor	Response
Airways	β_2	Bronchodilatation, reduction in mediator release from mast cells, increased mucus production, increased ciliary activity
Heart	β_1/β_2	Tachycardia, inotropic action
Blood vessels	β_2	Dilatation, fall in blood pressure, compensatory reflex increase in heart rate
Uterus	β_2	Relaxation
Metabolic	β_2/β_3	Increase in glucose, insulin, lactate, pyruvate, non-esterified fatty acids, glycerol and ketone bodies; decrease in potassium, phosphate, calcium and magnesium
Muscle	β_2	Tremor

subsequent molecular characterization of an atypical β-adrenoceptor from a human genomic library has led to the acceptance of the β_3-adrenoceptor as a separate entity.[2,3] Atypical β-adrenoceptors have been shown to be present in several tissues, including adipocytes,[4] stomach fundus,[5] ileum[6] and heart,[7] although inadequate information exists at present to classify all of these atypical β-adrenoceptors as being of the β_3 subtype. To date there is no evidence for the presence of the β_3 subtype in the lungs. The major actions of sympathomimetic amines at their different receptors are summarized in Table 36.1.

The distribution of β-adrenoceptors within the airways has been studied in a variety of species including humans. Smooth muscle contains almost exclusively β_2-adrenoceptors, whereas the terminal airways and alveoli contain a mixed population of β_2 and β_1 receptors.[8,9] In addition, epithelial and mucosal glands contain β_2-adrenoceptors, which may be important in regulating mucus secretion, and β_2-adrenoceptors are present on inflammatory cells within the airways including the alveolar macrophage.[10]

Molecular biology of the β_2-adrenoceptor

The majority of the intracellular effects of β-adrenoceptor stimulation are mediated through the activation of adenylate cyclase and the production of the intracellular second messenger cyclic AMP.

Both β_1 and β_2-adrenoceptors have now been cloned and their primary sequences deduced, revealing structural similarities with many other 'classical' membrane receptors that belong to a superfamily of seven transmembrane-spanning domain, G protein-coupled receptors (Fig. 36.1). Using site-directed mutagenesis and chimeric receptor techniques a number of workers have shown that amino acid residues within the hydrophobic core of the β_2-adrenoceptor are involved in ligand binding involving residues on the third, fourth and fifth transmembrane domains,[11] suggesting a binding pocket buried within the membrane bilayer. The third intracellular loop of the β-adrenoceptor contains two amphiphilic regions thought to be important for coupling to G proteins.[12] In addition, a number of phosphorylation sites exist predominantly in the third intracytoplasmic loop and in the C-terminal tail of the receptor that are important in the regulation of β-adrenoceptor function.

36 β-Adrenoceptor Agonists

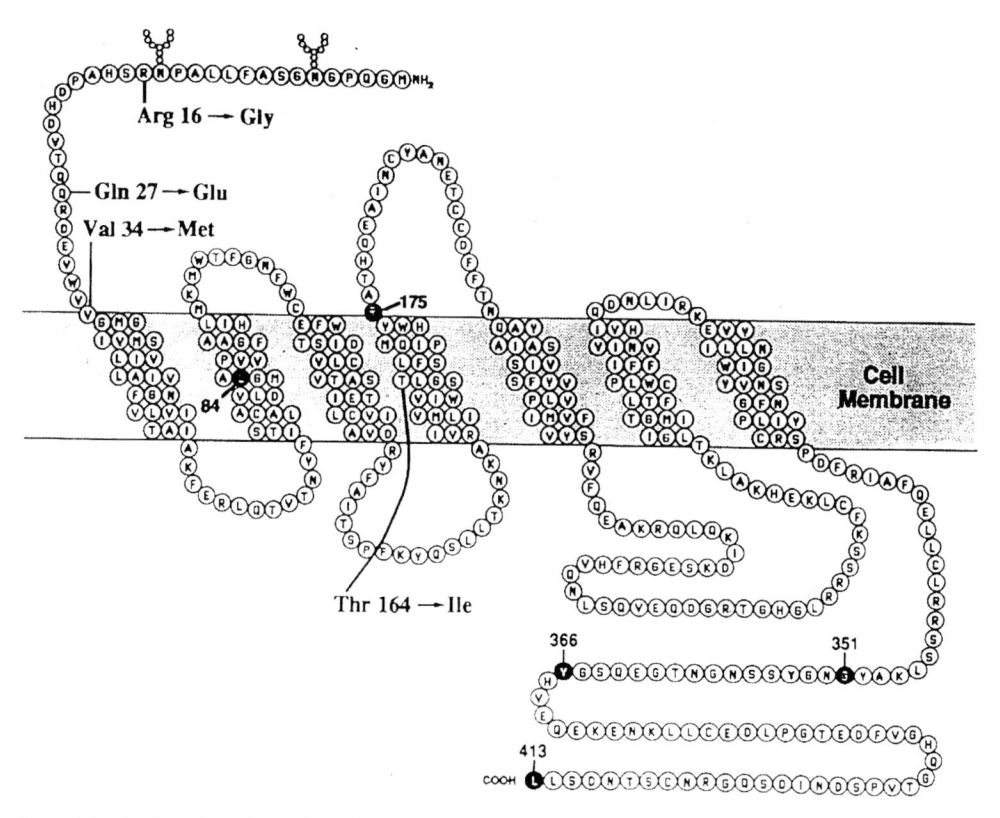

Fig. 36.1 Predicted conformation of the human β_2-adrenoceptor. Note the seven transmembrane-spanning α-helices, the extracellular N-terminal, the intracellular C-terminal and three intracellular loops. The figure also shows known mutations within the receptor. 'Silent' mutations (i.e. those where a single-base substitution does not alter the amino acid code) are shown dark; the mutations resulting in amino acid substitutions are indicated. Reproduced from ref. 27, with permission.

Linkage of β-adrenoceptors to G proteins

Following agonist binding to the β-adrenoceptor, the activation of adenylate cyclase requires activation of the intermediary G protein G_s. Adenylate cyclase can also be inhibited by a range of receptors acting via an inhibitory G protein, G_i. G proteins exist as heterotrimers composed of three subunits termed α, β and γ.[13] Following agonist interaction with the β-adrenoceptor, GDP is released from the α-subunit of G_s allowing GTP to bind; this leads to dissociation of the activated α-subunit from the β-adrenoceptor–βγ complex and it is the free α-subunit that is responsible for stimulating adenylate cyclase and hence the production of cyclic AMP. The processes involved in the synthesis of cyclic AMP are summarized in Fig. 36.2. Cyclic AMP activates protein kinase A, and it is protein kinase A that produces the majority of the cellular effects of β-adrenoceptor stimulation. However, there is some evidence that $G_{s\alpha}$ can have direct effects, for example by activating K^+ channels independent of cyclic AMP.[14]

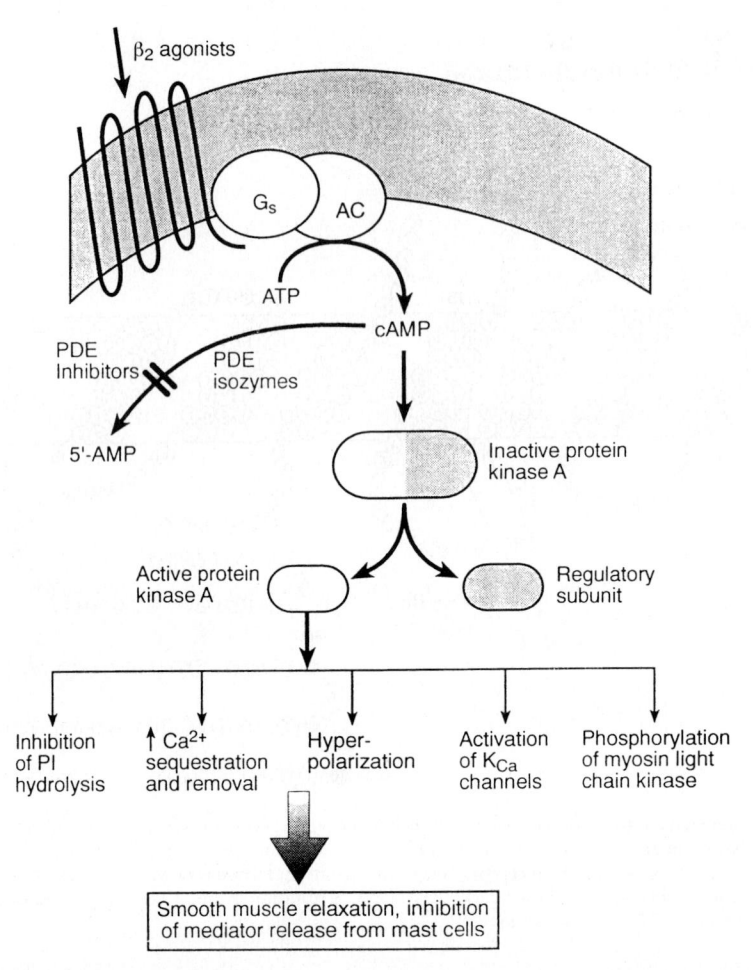

Fig. 36.2 Effects of β-adrenoceptor activation. Following binding of agonist, the receptor–G$_s$ complex releases GDP, allowing GTP to bind resulting in the dissociation of the α-subunit of G$_s$, which is responsible for activating adenylate cyclase (AC). AC catalyses the conversion of ATP to cyclic AMP, which in turn activates protein kinase A, leading to the intracellular effects of cyclic AMP. Following activation of AC, the α-subunit of G$_s$ recombines with GDP and the βγ(subunit producing the inactive form of G$_s$. Cyclic AMP is broken down by non-selective and cyclic AMP-selective phosphodiesterases (PDE) to 5′-AMP. Reproduced from ref. 143, with permission.

Intracellular actions of cyclic AMP

Much recent work has focused upon the intracellular actions of cyclic AMP in airway smooth muscle cells in an attempt to explain the relaxant effects of β$_2$-adrenoceptor agonists in this tissue. It has become increasingly clear that, rather than having one specific action resulting in smooth muscle relaxation, cyclic AMP is able to modulate a number of processes which are important in governing the contractile state of the cell (see ref. 15 for a review).

Table 36.2 Mechanisms through which cyclic AMP can potentially relax airway smooth muscle.

Reduction of intracellular calcium levels
Inhibition of inositol phospholipid hydrolysis
Sequestration of intracellular calcium
Calcium extrusion
Altered sensitivity of contractile apparatus to calcium
Activation of calcium-gated K^+ channels
Membrane hyperpolarization
Stimulation of Na^+/K^+ ATPase

Pharmacomechanical coupling in airway smooth muscle in response to spasmogens such as histamine and acetylcholine is now believed to occur as a consequence of a rise in intracellular calcium, which itself results from the binding of the intracellular second messenger inositol 1,4,5-trisphosphate to a specific receptor on the sarcoplasmic reticulum.[16,17] Elevation of cell cyclic AMP content modulates a number of key processes involved in pharmacomechanical coupling, leading to relaxation (Table 36.2).

Regulation of β-adrenoceptors

Both the number of β-adrenoceptors on a given cell and the affinity of the β-adrenoceptor for agonists are closely regulated. Traditionally it has been thought that the β-adrenoceptor can exist in either a high- or low-affinity state; both states have equal affinities for β-adrenoceptor antagonists, whereas the high-affinity state has a 50–100-fold higher affinity for agonists.[18] The proportion of receptors in the high-affinity state is increased by the addition of an agonist. However, recent studies, particularly in transformed cell systems or transgenic animals expressing β-adrenoceptors at supraphysiological levels, have provided evidence for constitutive activity of the $β_2$-adrenoceptor (i.e. receptor activity in the absence of agonist).[19] This is important because it provides an alternative framework for the action of antagonists and has led to the idea that some antagonists, rather than binding to the active site of the receptor in a competitive fashion, could alternatively stabilize the inactive form of the receptor and therefore act as an 'inverse agonist'.

Following prolonged exposure to high concentrations of agonist, the number of β-adrenoceptors present in the cell membrane falls (downregulation). The function of β-adrenoceptors can also be modulated rapidly following exposure to agonist by phosphorylation: at least two families of kinases are involved, namely cyclic AMP-independent kinases (termed βARK, β-adrenergic receptor kinase) and protein kinase A. The relative contribution of these two mechanisms remains unclear but work using either dominant negative or antisense approaches will help clarify this issue. In addition to regulation by exposure to β-agonist, expression and coupling of $β_2$-adrenoceptors can be altered by long-term exposure to glucocorticoids (which upregulate expression) and pro-inflammatory cytokines (which may be elevated following virus infection) such as interleukin 1β, which downregulates coupling.[20–22]

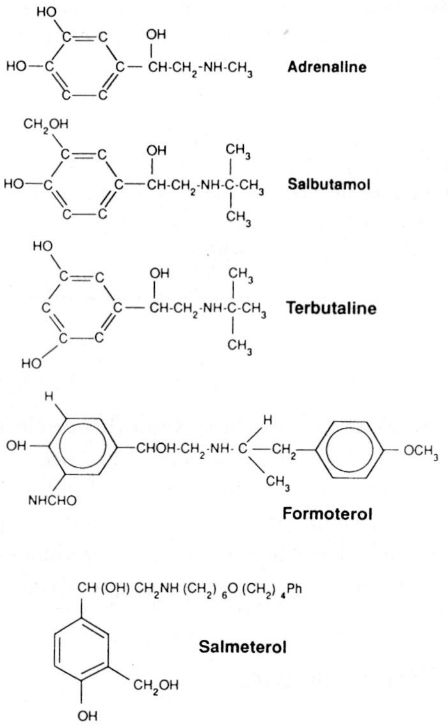

Fig. 36.3 Structure of a range of β-agonists.

Structure–activity relationships of β-agonists

The basic structure of the naturally occurring catecholamine adrenaline and a range of clinically important β_2-agonists are shown in Fig. 36.3. All these agents are composed of a benzene ring with a chain of two carbon atoms and either an amine head or a substituted amine head. When hydroxyl (OH) groups are present in positions 3 and 4 on the benzene ring, the structure is termed a catechol nucleus and the compound is a catecholamine. When these hydroxyl groups are either substituted or repositioned the resulting drug generally becomes less potent than the synthetic catecholamine isoprenaline; however, this disadvantage is outweighed in the case of agents such as salbutamol and terbutaline by the resultant resistance to metabolic degradation by catechol-O-methyltransferase (COMT) (see below). Substitutions on the α-carbon atom block oxidation by monoamine oxidase (MAO) (see below). Drugs with large substitutions on the amine head generally show greater selectivity for the β_2-adrenoceptor, although they may have reduced potency with respect to isoprenaline.

Degradation of β-agonists

The major mechanism terminating the pharmacological effects of adrenaline and noradrenaline is uptake into sympathetic nerve endings (uptake 1); uptake into other sympathetically innervated tissues such as smooth muscle (uptake 2) is partially responsible for terminating the action of catecholamines, including isoprenaline.

Metabolic degradation by COMT, the dominant enzyme present in tissues innervated by sympathetic nerves, is responsible for terminating the action of catecholamines taken up by these tissues via uptake 2. Non-catecholamines, with a substitution or repositioning of the hydroxyl groups at positions 3 and 4 on the benzene ring, are resistant to inactivation by COMT.

Oxidation by MAO is the major route for catecholamine metabolism in sympathetic nerve endings. Substitution of the terminal amine head confers resistance to inactivation by MAO. Both MAO and COMT are widely distributed in the gut, liver, lung and kidney.

Exogenously administered β-adrenoceptor agonists can be conjugated to sulphates or glucuronides in the liver, gut wall and probably in the lung. Following ingestion, the drugs are partially conjugated during first-pass metabolism; in the case of salbutamol this accounts for roughly 50% of the oral dose.[23]

Long-acting $β_2$-adrenoceptor agonists

Recent interest has centred around the development of longer-acting $β_2$-agonists such as formoterol and salmeterol. Salmeterol has a large non-polar N-substituent that is thought to interact with specific sites in the fourth transmembrane domain of the receptor, which accounts for its prolonged activity.[24] With formoterol, the longer duration of action may be due to the addition of a pyridine nucleus (see Fig. 36.3). Formoterol has a long duration of action when given by inhalation but not when given orally.[25]

The development of bambuterol, a prodrug broken down to terbutaline, represents another approach to try to increase the duration of action and selectivity of $β_2$-agonists.[26] In fact bambuterol is a prodrug of a prodrug, since it is metabolized first in the liver and gut to produce a lipophilic drug that is selectively taken up by the lung, where it is broken down to terbutaline.

$β_2$-Adrenoceptor polymorphisms

In 1993 a number of common variants of the $β_2$-adrenoceptor gene were described.[27] In all, nine single-base substitutions were identified within the coding region of the receptor on chromosome 5q, which is an intronless gene on chromosome 5q.31, but because of redundancy in the amino acid code only four of these base substitutions result in amino acid substitutions in the receptor itself (see Fig. 36.1). Whilst the polymorphism at amino acid 34 appears to have no functional consequences, the polymorphisms at amino acids 16, 27 and 164 all produce changes in the way the receptor behaves.

Most work has been performed on the polymorphisms at amino acids 16 (Arg→Gly) and 27 (Gln→Glu). Both of these are common within the general population and also in asthmatic individuals; the approximate allelic frequency in the Caucasian population being 35% and 65% for Arg16 and Gly16 and 55% and 45% for Gln27 and Glu27, respectively. Both of these polymorphisms alter the way in which the receptor downregulates following exposure to agonist. The Gly16 form of the receptor downregulates to a greater extent than the Arg16 form of the receptor in both transfected cell systems and in primary cultures of human airway smooth muscle cells. In contrast, the Glu27 form of the receptor downregulates to a lesser extent than the Gln27 form of the receptor in both of these systems.[28,29] Clinical studies have suggested that these polymorphisms may have relevance in asthmatic individuals. In particular, the Gly16 form of the receptor appears to be associated with nocturnal asthma[30] and the Glu27 form of the receptor with less reactive airways in asthmatic individuals.[31] Neither polymorphism appears to be a significant risk factor for developing asthma in adults.[32]

The polymorphism at amino acid 164 (Thr→Ile) is much rarer, with an allelic frequency of about 1% in the populations studied to date. Hence homozygous individuals with the Ile164 polymorphism have not been identified and functional studies have not been done *in vivo*. Interestingly, *in vitro* this polymorphism alters the agonist-binding characteristics of the receptor, resulting in a lower affinity for ligands containing the β-carbon hydroxyl group.[33]

CLINICAL PHARMACOLOGY

This section describes the pharmacological actions of β-agonists relevant to asthma.

Airways

β-Agonists have several actions that may affect airway function. The extent to which each contributes to the airway response is uncertain and is likely to differ in different circumstances. The main actions are as follows.

(1) Smooth muscle relaxation: β_2-agonists are functional antagonists since they antagonize the effects of a wide variety of bronchoconstrictor agents on airway smooth muscle *in vitro*[34,35] and *in vivo*.[36]

(2) Inhibition of mediator release from inflammatory cells: β-agonists inhibit the release of histamine from sensitized human lung,[37] basophils[38] and mast cells[39] and they inhibit the increase in various circulating mediators that occurs in response to exercise or antigen challenge in patients with asthma.[40,41]

(3) Inhibition of cholinergic neurotransmission: β-agonists inhibit cholinergic neurotransmission *in vitro*[42] and this may be the mechanism whereby β-blocking drugs cause bronchoconstriction.[43]

(4) Reduced vascular permeability: although β-agonists can reduce permeability and oedema in response to mediators such as histamine and leukotrienes in animals, this is not consistent[44] and the relevance of these observations to therapeutic doses of β-agonists in asthma is uncertain.

(5) Increased mucociliary clearance: this probably occurs as a result of increased ciliary beat frequency and increased production of periciliary cytosol.[45]

β_2-Agonists cause bronchodilatation in subjects with airflow obstruction. Most normal subjects show an increase in airway conductance or flow rates from a partial flow–volume curve, though not in measurements that involve a full inflation such as the forced expiratory volume in 1 s (FEV_1). This is because airway hysteresis exceeds parenchymal hysteresis in normal subjects.[46]

β-Agonists also cause marked protection against all non-specific constrictor stimuli such as histamine, methacholine, eucapnic hyperventilation and exercise in subjects with asthma.[36] A single large dose of inhaled $β_2$-agonist usually causes an acute shift in the dose–response curve to histamine or methacholine of between two and four doubling doses. When given prior to an allergen challenge all β-agonists inhibit the early response to allergen; the long-acting β-agonists[47,48] and high doses of the shorter-acting β-agonists[49] also inhibit the late response, probably due to functional antagonism related to their longer action.

The protective effect of β-agonists against constrictor stimuli is reduced following regular treatment with both the short-acting[50] and long-acting[51] $β_2$-agonists (see p. 665). The loss of protection has been documented with histamine,[50] methacholine,[52] AMP,[52] antigen[53] and exercise[54] and is likely to occur with all stimuli (Fig. 36.4). Regular treatment with a short-acting β-agonist has also been associated with a small increase in bronchial reactivity 12–59 h after cessation of treatment when the bronchodilator effect has worn off[50,55–57] (Fig. 36.4); in some studies this has been associated with a fall in FEV_1.[57] This rebound effect has not been seen following salmeterol or formoterol.

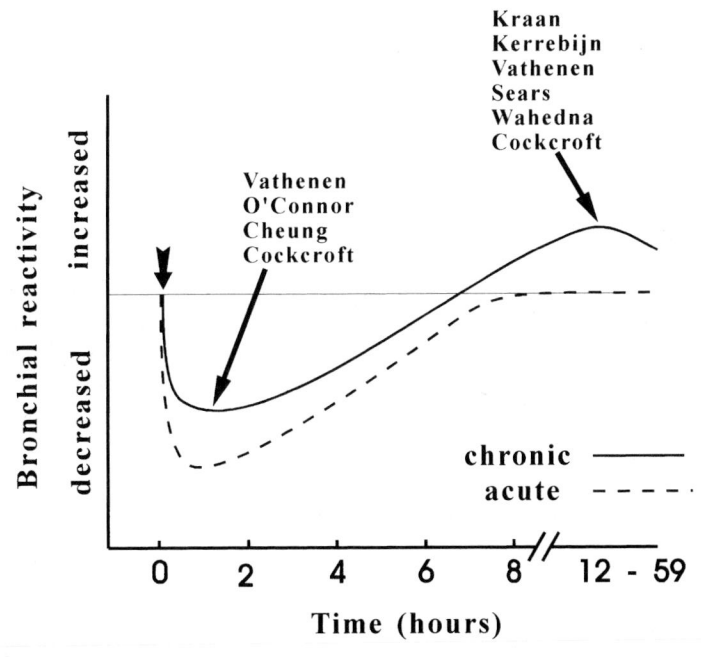

Fig. 36.4 Change in bronchial responsiveness to a β-agonist following a single dose (broken line) and after regular treatment (sold line). A single dose protects against a bronchoconstrictor challenge (reduction in bronchial responsiveness). With regular treatment, this protective effect of a β-agonist against the constrictor challenge is attenuated and, once the bronchodilator effect has worn off, there is a rebound increase in bronchial responsiveness. The names indicate the first authors of studies demonstrating these effects (see refs 50–53, 55–57, 70).

Cardiac effects

β-Agonists cause tachycardia due to stimulation of cardiac β_1 and β_2 receptors and secondary compensatory effects following β_2-mediated vasodilatation. When autonomic dysfunction prevents compensatory changes, as in quadriplegic patients, β-agonists cause a large fall in systemic vascular resistance and blood pressure despite an increase in cardiac output.[58] Tachycardia and palpitations occur less often with β_2-selective agonists, particularly when they are given by the inhaled route, although they can occur with higher doses[59] and with systemic therapy. The extent to which β-agonists may precipitate dysrhythmias is more controversial (see p. 665). β-Agonists cause a small fall in pulmonary artery pressure in patients with asthma.[60]

Arterial oxygen tension

By increasing cardiac output and hence increasing mixed venous oxygen tension, β-agonists may cause arterial oxygen tension (PaO_2) to rise; PaO_2 can fall, however, following pulmonary vasodilatation and increased perfusion of poorly ventilated areas, as shown with inert gas studies in which blood flow to areas with a low ventilation–perfusion ratio doubled following isoprenaline.[60] The fall in PaO_2 is usually small, around 0.5 kPa (3–4 mmHg), although occasional patients show larger changes.

Metabolic changes

β-agonists cause a wide range of metabolic changes (see Table 36.1) but these rarely cause problems apart from the fall in serum potassium. This has ranged up to 0.9 mmol/litre with conventional single doses of β-agonist by the inhaled or parenteral route[61] though larger changes are seen with higher doses.[59,62] The extent to which these acute changes are sustained with long-term treatment is uncertain. Ketoacidosis has been precipitated by β-agonists in diabetic patients.[63]

Other adverse effects

A fine tremor can be a nuisance with β-agonists, particularly when given orally or in high doses by nebulizer; it tends to decrease with prolonged treatment as tolerance develops. Cramp occurs relatively infrequently with β-agonists, and headaches may occur with high doses.

DIFFERENCES BETWEEN β-AGONISTS

Partial or full agonists

Some β_2-selective agonists such as salmeterol, and to a lesser extent salbutamol and terbutaline, are partial agonists *in vitro* compared with full agonists such as isoprenaline and fenoterol. This would be important if bronchodilatation is limited by smooth muscle relaxation, as the bronchodilator response would be less than that seen with a full agonist. There is, however, no evidence that the maximal airway response to any β_2-agonist is reduced relative to isoprenaline or fenoterol.[64,59] A second potential problem is that a partial agonist is also a partial antagonist, since by occupying receptors it may prevent access to drugs with fuller agonist activity. Salmeterol, for example, as a partial agonist with a long duration of action may limit the effectiveness of salbutamol, a fuller agonist, if this was given during an acute attack. Salmeterol did not inhibit the response to salbutamol in our study of patients with stable asthma[65] and did not reduce the FEV_1 achieved with salbutamol in another.[66] These data are reassuring, although an interaction in acute asthma, the situation where such an effect is most likely to occur, has not been studied.

Duration of action

The duration of action of all β-agonists increases with dose; 5 mg inhaled salbutamol, for example, has a considerably longer action than 0.5 mg. Nevertheless, β-agonists can be divided into three broad groups according to duration of action following inhalation of conventional doses: (a) the catecholamines, isoprenaline and rimiterol, which have a very short action of 1–2 h; (b) those conventionally described as short acting, such as salbutamol and terbutaline, which are active for 3–6 h, although fenoterol may be slightly shorter acting;[67] (c) the long-acting β-agonists salmeterol and formoterol, which cause bronchodilatation for at least 12 h.[68,69]

Bronchial selectivity

All the β-agonists currently used to treat asthma are selective for β_2-adrenoceptors *in vitro* compared with isoprenaline. The airway or serum potassium response to a β-agonist (both β_2 effects) can be related to the heart rate response (β_1 and β_2 effects) to provide a functional measure of bronchial selectivity *in vivo*. The response *in vivo* is affected by β_2-selectivity of the drug but it is also influenced by other factors including reflex responses, homeostatic mechanisms and pharmacokinetic factors. Such studies suggest that salbutamol, terbutaline, formoterol and salmeterol have similar airway selectivity[25,62] and that fenoterol has slightly less.[59]

EFFICACY AND SAFETY OF INHALED β-AGONISTS

There are considerable differences in efficacy and hence in the indications for the long-acting β-agonists compared with the short-acting β-agonists such as salbutamol and terbutaline and the two are therefore considered separately.

Short-acting β-agonists

Intermittent use

The selective β_2-agonists are extremely effective when used intermittently to treat attacks of asthma whether mild or severe. They work rapidly when given by inhalation and this route of administration allows higher doses to be deposited in the airways with less systemic absorption and fewer adverse effects. Severe attacks of asthma are treated with high doses of the short-acting β-agonists such as salbutamol and terbutaline (see Chapter 42).

Regular use

It was widely assumed until recently that since β-agonists were clearly effective when given for acute attacks of asthma, their use on a regular basis must also be beneficial. This view was challenged in 1990 when Sears et al.[70] reported the results of a 6-month crossover study comparing regular fenoterol (400 μg q.i.d.) with placebo in patients with asthma. Using a composite predetermined hierarchical score more patients were worse during the fenoterol period, with lower FEV_1 and morning peak flow rates, greater symptoms, increased steroid use[70] and shorter time to relapse.[71] This study led to several further studies[72–78] comparing the effect of treating asthma with regular salbutamol or placebo for periods ranging from 2 to 52 weeks (Table 36.3). Most studies showed no significant difference between regular salbutamol and placebo in terms of symptom control and morning and evening lung function. A detailed time profile of airway function on the first and last day of treatment in two of these studies, both large-parallel group studies comparing placebo, salmeterol 42 μg b.d. and salbutamol 180 μg q.i.d. for 3 months, showed a large on–off effect with salbutamol, with good bronchodilatation initially after each dose but a fall to baseline or slightly below baseline by 6 h.[74,75] Three of the eight studies[72,73,78] showed some benefit from regular salbutamol, though one was only 2 weeks long[72] and one, which assessed quality of life, concluded that the differences were clinically unimportant.[73] A further study in patients with asthma or chronic obstructive pulmonary disease (COPD) showed a more rapid fall in FEV_1 over 12 months with regular compared to intermittent β-agonist use, although there was no difference in exacerbations, symptoms or quality of life.[79] The difference in decline in FEV_1 was not seen on further follow-up, though the number of subjects remaining was small.[80]

Although the conclusions from the different studies have varied, the differences between studies have in fact been small and could easily be explained by differences in methodology, the varying duration of treatment or the much higher relative dose of

36 β-Adrenoceptor Agonists

Table 36.3 Effect of regular treatment with short-acting β-agonists.

Reference	Number studied	Drug and dose	Duration (weeks)	Outcome compared with placebo
70	64	Fenoterol 400 μg q.i.d.	24	Using a composite measure, patients worse on fenoterol
72	341	Salbutamol 200 μg q.i.d.	2	Small improvement in symptoms with salbutamol
73	140	Salbutamol 200 μg q.i.d.	4	Main outcome quality of life; marginal benefit from salbutamol considered by authors to be clinically unimportant
74	234	Salbutamol 200 μg q.i.d.	12	No difference in symptoms or in a.m. or p.m. lung function
75	257	Salbutamol 200 μg q.i.d.	12	No difference in symptoms or in a.m. or p.m. lung function
76	255	Salbutamol 200 μg q.i.d.	15	No significant differences; trends towards deterioration with salbutamol
77	17	Salbutamol 200 μg q.i.d.	16	PEF higher on regular salbutamol, no difference in symptoms or qualify of life
78	367	Salbutamol 200 μg q.i.d.	4	Better evening (but not a.m.) PEF and more symptom-free days on salbutamol
79	94	Salbutamol 400 μg q.i.d.	52	Decline in FEV_1 greater with regular β-agonist treatment

FEV_1, forced expiratory volume in 1 s; PEF, peak expiratory flow.

fenoterol in the study by Sears et al.[70] The main conclusion to be drawn is that regular treatment with β-agonists differs little from placebo and provides at best little benefit to patients.

Comparison with nedocromil and inhaled corticosteroids. Regular treatment with salbutamol for 4 weeks was less effective than nedocromil[81] and regular terbutaline was considerably less effective than inhaled budesonide in two long-term studies in children and adults with regard to symptom control, morning peak flow rate and supplemental β-agonist use.[82,83] In the adults treated with terbutaline,[82] FEV_1, peak expiratory flow (PEF) and bronchial responsiveness were similar before and after 2-years treatment but this overestimates the efficacy of terbutaline, since almost half the patients had either withdrawn during the study or had theophylline added to their treatment; a placebo limb is needed to estimate the true effect of β-agonists.

Safety

Several million puffs of $β_2$-agonist are inhaled every day in the UK and the incidence of documented side-effects is extremely low. The main problems reported by patients are tremor, headache, palpitations and flushing due to stimulation of β-receptors in organs other than the lung. Metabolic studies show various changes with β-agonists, of which hypokalaemia is the most important.

The main concern with β-agonists relates to their possible role in the epidemics of asthma deaths in the 1960s[84,85] and subsequently in New Zealand.[86] The increase in asthma deaths showed a temporal association with sales of high-dose isoprenaline in the 1960s and a geographical association with countries with the high-dose formulations. Three case-control studies in New Zealand suggested that patients dying from asthma were more likely to have received fenoterol than salbutamol,[87-89] though whether the relation was due to an adverse effect of fenoterol or to the selective use of fenoterol in patients with more severe asthma is still debated. The epidemics were associated with a non-selective (isoprenaline) and less β_2-selective (fenoterol) β-agonist and both drugs were marketed in doses that were high and which caused more systemic effects in pharmacological studies.[59,64] Further non-epidemic data from Saskatchewan[90] showed a dose–response relation between β-agonist usage and life-threatening attacks of asthma, which may well be a reflection of the fact that patients with more severe asthma are more likely to take more β-agonists and more likely to die. However, such an explanation does not explain why there were epidemics of asthma deaths.

The arguments in this debate are complex and are discussed elsewhere. Serious adverse effects are clearly very rare with the currently used shorter-acting β-agonists but they could nevertheless be important. For example, a fatal outcome in 1 in 8000 patients using a β-agonist would be undetectable in a prospective trial but, were this the case, it would account for half the asthma deaths in those aged under 55 in the UK. On balance it seems likely that the 1960s epidemic was related to the use of isoprenaline and the more recent epidemic in New Zealand was related in part at least to the use of fenoterol.

If β-agonists were responsible for the epidemics of asthma deaths and possibly for some deaths at other times, what are the possible mechanisms for such an effect? It has been argued that because β-agonists give good symptomatic relief patients might rely too heavily on them and take insufficient prophylactic treatment, corticosteroids in particular. The additional treatment available in the 1960s was probably not widely used however, consisting mainly of theophylline and oral steroids, and it is difficult to explain why New Zealand alone should have run into a second epidemic in the 1970s. Thus, although inadequate or inappropriate treatment undoubtedly underlies many asthma deaths, it does not explain why epidemics occurred over a relatively short period of time.

Development of tolerance to β-agonists. There has been considerable interest in the possibility that patients taking high doses of β-agonists may develop tolerance to their effects and some early retrospective evidence suggested that this had occurred in patients taking extremely high doses, up to an inhaler a day, of isoprenaline.[91,92] This stimulated a large number of prospective studies in patients with asthma[93] from which the following can be summarized:

(1) Most studies in patients with asthma have not shown a reduced bronchodilator response to the short-acting β-agonists following regular β-agonist treatment. This contrasts with studies in normal subjects where the bronchodilator response to a β-agonist has declined following regular treatment.[94-96]

(2) The tremor and heart rate response to a β-agonist clearly declines following regular treatment.[97,98]
(3) The protection afforded by a β₂-agonist against bronchoconstrictor challenge is reduced following regular β-agonist treatment[50,52,53,94] (see p. 659).
(4) A rebound increase in bronchial responsiveness has been seen following cessation of treatment with β-agonists[50,55–57,70,94] and in some studies this has been associated with a fall in FEV_1.[94,57]

Of these findings, the rebound increase in bronchial responsiveness is the mechanism most likely to have an adverse clinical effect. Although the increase in bronchial responsiveness is not large, of the order of one doubling dose for histamine, it may be important for an occasional patient, and could contribute very rarely in causing a severe or even fatal attack of asthma.

Dysrhythmias. Selective and non-selective β-agonists have caused myocardial necrosis, a coronary steal and death from ventricular tachydysrhythmias when given alone in high doses to animals[99,100] or, when combined with theophylline, in lower doses comparable to those given in humans.[101–103] The dysrhythmic effects of β-agonists are more pronounced in the presence of hypoxaemia.

Patients who die from asthma usually die outside hospital and the finding of widespread mucus plugging at post-mortem does not exclude a dysrhythmia as the final event. Trying to determine whether β-agonists have caused a dysrhythmia or cardiac toxicity in humans is difficult and relies on case reports and some prospective studies in small numbers of subjects (see ref. 93). Adverse cardiac effects during treatment with a β-agonist are well described. Although the patients were sometimes taking other drugs such as corticosteroids or methylxanthines, these data suggest that β-agonists, particularly when given in high doses, can occasionally contribute to the development of pulmonary oedema, myocardial ischaemia and infarction.

Several prospective studies that have looked at cardiac effects of β-agonists given alone or in combination with methylxanthines, with roughly half showing no increase in cardiac dysrhythmias and half finding a significant increase, albeit of clinically unimportant dysrhythmias.[93] All the studies contained a small number of subjects, however, and most excluded patients with cardiac problems. If dysrhythmia or myocardial ischaemia is a complication of β-agonists, it is clearly very uncommon and these studies had insufficient power to assess this. An incidence of 1 in 1000 patients on high-dose β-agonists would be clinically important but many thousands of patients would need to be studied to detect such an effect. On balance it seems likely that β-agonists, particularly if taken in high doses, can on rare occasions cause serious dysrhythmias and death in a vulnerable patient. This is more likely in patients who are hypokalaemic and hypoxaemic as a result of their asthma or treatment.

Whether β-agonists cause occasional deaths and whether, or to what extent, they contributed to the epidemics of asthma deaths in the 1960s and more recently in New Zealand will continue to be debated. However, any such effect has to be placed in perspective against the enormous symptomatic benefit they offer to patients and the extent to which they save lives, which is even more difficult to ascertain.

Long-acting β-agonists

The longer-acting β-agonists salmeterol and formoterol have a different role in the management of asthma. Both are normally given on a regular twice-daily basis and should not be used for relief of episodes of bronchoconstriction.

Comparison with shorter-acting β-agonists and placebo

A large number of clinical studies have compared regular treatment with salmeterol and, more recently, formoterol with placebo or regular short-acting β-agonists. The controlled prospective studies in adults lasting for 4 weeks or more are shown in Tables 36.4 and 36.5.[73–75,78,104–117]

Table 36.4 Controlled studies of 4 weeks' duration or more in adults in which the clinical response to salmeterol was compared with the response to placebo or a short-acting β-agonist.

Reference	No. of patients (no. on salmeterol)	Duration (weeks)	Twice-daily salmeterol dose (μg)	Control drug
104	692 (520)	4	12.5, 50, 100	Placebo
105	427 (282)	6	50	Placebo
109	119 (55)	12	100	Placebo
74	234 (78)	12	42	Placebo + Salbutamol 180 μg q.i.d.
75	257 (84)	12	42	Placebo + Salbutamol 180 μg q.i.d.
73	140 (46)	12	50	Placebo + Salbutamol 200 μg q.i.d.
106	667 (334)	12	50	Salbutamol 200 μg bd
107	388 (190)	12	50	Salbutamol 400 μg q.i.d.
108	25180 (14113)	16	50	Salbutamol 200 μg q.i.d.
78	367 (367)*	4	50	Placebo + Salbutamol 400 μg quid
114	120 (53)	6	50	Salbutamol 400 μg q.i.d.
115	190 (96)	6	100	Salbutamol 400 μg q.i.d.
116	89 (89)*	52	50	Placebo

* Crossover study.

Table 36.5 Controlled studies of 4 weeks' duration or more in adults in which formoterol was compared with a short-acting β-agonist.

Reference	No. of patients (no. on formoterol)	Duration (weeks)	Twice-daily formoterol dose (μg)	Control drug
112	16*	4	24	Salbutamol 400 μg b.d.
110	145 (73)	12	12	Salbutamol 200 μg q.i.d.
113	18 (10)	52	12	Salbutamol 200 μg b.d.
111	35 (19)	4	24†	Salbutamol 400 μg b.d.

* Crossover study.
† Could take more as necessary.

When compared with placebo salmeterol and formoterol have shown sustained bronchodilatation, a reduction in symptoms and, in the case of salmeterol, an improvement in asthma-specific quality of life.[73,114] In a recent crossover study, in which patients were able to adjust inhaled steroid treatment within limits, salmeterol 50 µg b.d. for 6 months was associated with a reduction in inhaled steroid use, symptoms and bronchodilator use and an increase in FEV_1 and PEF compared with the 6 months on placebo.[116] A large multicentre study found that 24 µg formoterol for 6 months reduced symptoms and bronchodilator use and increased PEF compared with placebo.[117] Since the long-acting β-agonists do not reduce airway inflammation as judged by the measurement of inflammatory cells in bronchoalveolar lavage,[118] there has been some concern about their effects on asthma exacerbations. Most studies have shown no difference in exacerbations, although a recent, large, year-long study showed a reduction in mild and severe exacerbations when formoterol was given to patients using both low and high doses of an inhaled steroid.[119]

When salmeterol and formoterol have been compared with regular shorter-acting β-agonists (invariably salbutamol to date) similar conclusions can be drawn. With the long-acting β-agonists, bronchodilatation is better maintained throughout the day, daytime and night-time symptoms are reduced and quality of life is improved. In the studies showing a detailed profile over 12 daytime hours, bronchodilatation with salmeterol was sustained unchanged after treatment for 3 months and associated with a reduction in both daytime and night-time symptoms.[74,75] The response appeared to be similar whether or not patients were taking an inhaled corticosteroid. There were fewer asthma exacerbations in patients randomized to salmeterol rather than salbutamol (2.91% vs. 3.79%) in a large post-marketing surveillance study of 25 180 patients[108] but there was no placebo group for comparison.

Comparison with inhaled corticosteroids

There are no published data on the relative effects of the long-acting β-agonists compared with an inhaled corticosteroid in steroid-naive patients in adults. Two large parallel-group studies in patients with symptomatic asthma have looked at the effect of adding salmeterol (50 µg b.d.) or increasing the dose of beclomethasone dipropionate from 200 to 500 µg twice daily[120] or from 500 to 1000 µg twice daily over 6 months.[121] PEF and daytime and night-time symptoms improved to a greater extent in the salmeterol group in both studies, with no difference between groups in the number of exacerbations. A third group receiving salmeterol 100 µg b.d. in the study by Woolcock et al.[121] showed similar efficacy to the salmeterol 50 µg b.d. group.

Comparison with other bronchodilators

Salmeterol 50 µg b.d. provided better symptom control and fewer side-effects than a combination of theophylline 300 mg and ketotifen 1 mg b.d.[122] and better asthma control (improved PEF and reduction in symptoms) than theophylline alone following individual dose titration of theophylline.[123]

Nocturnal and exercise-induced asthma

It was expected that the long-acting β-agonists would be helpful in nocturnal asthma and this has been seen with both drugs in most of the larger studies detailed in Tables 36.4 and 36.5. A study looking specifically at nocturnal asthma showed an improvement in evening and morning PEF and some reduction in the nocturnal fall in PEF with salmeterol 50 or 100 μg b.d. for 2 weeks compared with placebo.[124] There was no difference between the two doses of salmeterol, though sleep architecture only improved significantly with the 50-μg b.d. dose. In another study the overnight fall in FEV_1 was less following formoterol 12 μg compared with salbutamol 200 μg.[125]

When given as a single dose both salmeterol and formoterol have caused a reduction in exercise-induced bronchoconstriction over 12 h.[126–128] The protective effect was attenuated, however, following regular salmeterol 50 μg b.d. for 4 weeks.[54]

Safety

The adverse effects reported with the long-acting β-agonists are largely those expected of a β-agonist: tremor, headache, cramps and palpitations. The frequency of these side-effects with salmeterol 50 μg b.d. or formoterol 12 μg b.d. has generally been low and similar to those seen with regular salbutamol 200 μg q.i.d. Higher doses of salmeterol (100 μg b.d.) have been associated with an increased incidence of side-effects, with no further benefit over the 50-μg dose in three of four published studies.[121,104,124,129] Adverse effects, particularly tremor and palpitations, have been higher with the 100 μg dose[104,129] and included one death.[124] There is less information on the relative effect of regular treatment with different doses of formoterol, although a recent study in patients with mild to moderate asthma showed near maximum benefit from 6 μg b.d. compared with 12 and 24 μg b.d.[130]

Some specific safety issues with the long-acting β-agonists require comment.

Bronchoconstriction. Bronchoconstriction can occur following inhalation of salmeterol by metered dose inhaler and, although not common, can be severe.[131,132] It is probably due to the propellant and appears to be more likely in patients whose asthma is deteriorating or poorly controlled. It has not been described with formoterol probably because bronchodilatation occurs more rapidly and counteracts the bronchoconstriction.

Partial agonism. Salmeterol is a partial agonist compared with salbutamol and could therefore, as a partial *antagonist*, reduce the effectiveness of a fuller β-agonist such as terbutaline or salbutamol given during an acute attack of asthma. There are no good data to show that this is a clinical problem (see p. 661) and no evidence that salmeterol causes an increase in severe exacerbations, as might be expected if salmeterol was inhibiting the response to rescue β-agonist use.

Systemic activity. Salmeterol shows marked β_2-selectivity *in vitro* and when given in recommended doses cardiovascular effects are generally minimal. Dose–response studies in humans show a steep increase in heart rate and QT_c interval and a marked fall in serum potassium (a β_2-mediated effect) with higher doses,[62,65,133] indicating a relatively modest therapeutic window (Fig. 36.5). These studies also show a dose equivalence for systemic

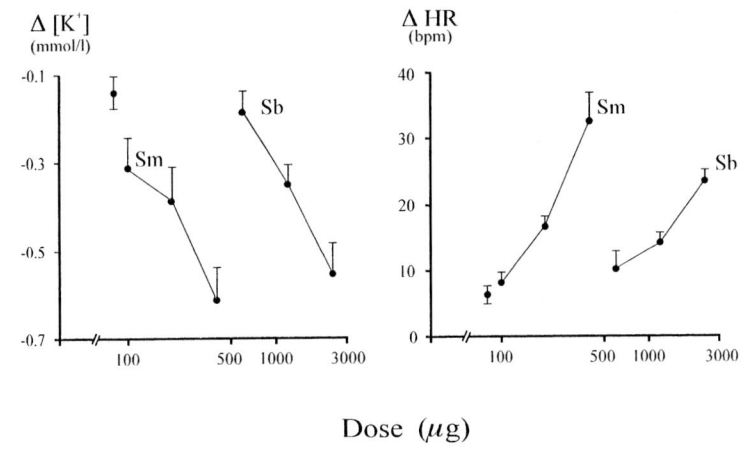

Fig. 36.5 Change in heart rate (ΔHR) and serum potassium concentration (Δ[K$^+$]) with increasing doses of salbutamol (Sb) and salmeterol (Sm) from a non-cumulative dose–response study in healthy subjects. Although there is little change in either measure with recommended doses of salmeterol or salbutamol, the study demonstrates a fairly modest therapeutic window when increasing doses are given. Reproduced from ref. 62, with permission.

effects of around eight compared with salbutamol, i.e. 50 μg salmeterol is equivalent to 400 μg salbutamol. These data suggest that salmeterol has been marketed at a relatively high dose compared with salbutamol, which may not be important as long as the drug is taken twice daily. The studies highlight the propensity of salmeterol to cause systemic effects if taken in doses above those recommended. These studies were carried out in fit subjects and larger effects may occur in an occasional patient with a vulnerable myocardium and/or other risk factors.

Tachyphylaxis or tolerance. Some studies have shown reduced bronchodilator responsiveness to salbutamol after regular treatment with salmeterol[66] and to formoterol after regular treatment with formoterol,[134–136] but the effects have been small and, in one instance, the maximum response achieved was unchanged.[66] There was some reduction in the bronchodilator effect of formoterol after the first few days of treatment in the FACET study but thereafter it was maintained throughout the year-long study. The protection afforded by salmeterol and formoterol against bronchoconstrictor challenge such as exercise and methacholine is reduced with their continued use,[51,54,136,137] as seen with all β-agonists. This means that the protective effect of the drugs against exercise is likely to be less than would be anticipated from single-dose studies. There is no evidence of a rebound bronchoconstrictor effect after regular treatment with the long-acting β-agonists,[51,135,136] which may be related to the more gradual offset of action of these drugs compared with the shorter-acting β-agonists.

ORAL β-AGONISTS

When given orally rather than by inhalation, single doses of β-agonists cause more systemic adverse effects and less bronchodilatation. This adverse ratio of benefit to side-effects is presumably maintained with regular treatment, though little has been published on the long-term effects of oral β-agonists. Oral salbutamol was associated with loss of the protective effect of salbutamol against exercise-induced bronchoconstriction.[138] Bambuterol, a terbutaline carbamate prodrug, was designed to reduce the systemic adverse effects associated with oral β-agonists.[139] It is protected against first-pass metabolism following oral absorption and has high affinity for lung tissue and a long half-life (20 h), which allows once-daily dosing. The 20 mg dose caused bronchodilatation and a reduction in nocturnal asthma.[140,141] Although its effect on symptoms was small in these studies, its effect on diurnal PEF and nocturnal wakening was larger in a recent study and comparable with the effects of salmeterol 50 μg b.d.[142]

REFERENCES

1. Lands AM, Arnold A, McAuliff JP, et al.: Differentiation of receptor systems activated by sympathomimetic amines. *Nature* (1967) **214**: 597–598.
2. Emorine LJ, Marullo S, Briend-Sutren MM, et al.: Molecular characterisation and the human β_3-adrenergic receptor. *Science* (1989) **245**: 1118–1121.
3. Zaagsma J, Nahorski SR: Is the adipocyte β-adrenoceptor a prototype for the recently cloned atypical 'β_3-adrenoceptor'? *Trends Pharmacol Sci* (1990) **11**: 3–7.
4. Engel G, Hoyer D, Berthold R, et al.: $\pm^{125}$[Iodo]-cyanopindolol, a new ligand for β-adrenoceptors: identification and quantitation of subclasses of β-adrenoceptors in guinea-pig. *Naunyn-Schmiedebergs Arch Pharmacol* (1981) **317**: 277–285.
5. Coleman RA, Denyer LH, Sheldrick KE: β-Adrenoceptors in guinea-pig gastric fundus: are they the same as the 'atypical' β-adrenoceptors in rat adipocyte? *Br J Pharmacol* (1987) **90**: 40P.
6. Bond RA, Clarke DE: Agonist and antagonist characterisation of a putative adrenoceptor with distinct pharmacological properties from the alpha and beta subtypes. *Br J Pharmacol* (1988) **95**: 723–734.
7. Kaumann AJ: Is there a third heart β-adrenoceptor? *Trends Pharmacol Sci* (1989) **10**: 316–320.
8. Carswell H, Nahorski SR: β-Adrenoceptor heterogeneity in guinea pig airways: comparison of functional and receptor labelling studies. *Br J Pharmacol* (1983) **79**: 965–971.
9. Goldie RG, Papidimitriou SM, Paterson SW, et al.: Autoradiographic localisation of β-adrenoceptors in pig lung using [^{125}I]-iodocyanopindolol. *Br J Pharmacol* (1986) **88**: 621–628.
10. Hjemdahl P, Larsson K, Johansson MC, et al.: β-Adrenoceptors in human alveolar macrophages isolated by elutriation. *Br J Clin Pharmacol* (1990; **30**: 673–682.
11. Dixon RAF, Sigal IS, Rands E, et al.: Ligand binding to the β-adrenergic receptor involves its rhodopsin-like core. *Nature* (1987) **326**: 73–77
12. Strader CD, Sigal IS, Dixon AF: Mapping the functional domains of the β-adrenergic receptor. *Am J Respir Cell Mol Biol* (1989) **1**: 81–86.
13. Harden TK: The role of guanine nucleotide regulatory proteins in receptor-selective direction of inositol lipid signalling. In Michell RH, Drummond AH, Downes CP (eds) *Inositol Lipids in Cell Signalling*. London, Academic Press, 1989, pp 113–133.
14. Kume H, Hall IP, Washabau RJ, Takagi K, Kotlikoff MI: β-Adrenergic agonists regulate K(Ca) channels in airway smooth muscle by cAMP-dependent and -independent mechanisms. *J Clin Invest* (1994) **93**: 371–379.
15. Torphy TJ, Hall IP: Cyclic AMP and the control of airways smooth muscle tone. In Raeburn D, Giembycz MA (eds) *Airways Smooth Muscle: Biochemical Control of Contraction and Relaxation*. Basel, Birkhauser Verlag, 1994, pp 215–233.
16. Hashimoto T, Hirata M, Ito Y: A role for inositol 1,4,5-trisphosphate in the initiation of agonist-induced contraction of dog tracheal smooth muscle. *Br J Pharmacol* (1985) **86**: 191–201.

17. Hall IP, Chilvers ER: Inositol phosphates and airway smooth muscle. *Pulmon Pharmacol* (1989) **2**: 113–120.
18. Lefkowitz RJ, Delean A, Hoffman BB, *et al.*: Molecular pharmacology of adenylate cyclase coupled α and β-adrenergic receptors. *Adv Cyclic Nucleotide Res* (1981) **14**: 145–161.
19. Band RA, Leff P, Johnson TD, *et al.*: Physiological effects of inverse agonists in transgenic mice with myocardial expression of the β2 adrenoceptor. *Nature* (1995) **374**: 272–276.
20. Bouvier M, Mausdorff WP, De Blasi A, *et al.*: Removal of phosphorylation sites from the $β_2$-adrenergic receptor delays onset of agonist-promoted desensitisation. *Nature* (1988) **333**: 370–373.
21. Shore SA, Laporte J, Hall IP, Hardy E, Panettieri RA: Effect of IL-1β on responses of cultured human airway smooth muscle cells to bronchodilator agonists. *Am J Respir Cell Mol Biol* (1997) **16**: 702–711.
22. Koto H, Mak JC, Haddad EB, *et al.*: Mechanism of impaired β-adrenoceptor induced airway relaxation by interleukin 1β *in vivo* in the rat. *J Clin Invest* (1996) **98**: 1780–1787.
23. Morgan DJ, Paull JD, Richmond BH, Wilson-Evered E, Ziccone SP: Pharmacokinetics of intravenous and oral salbutamol and its sulphate conjugate. *Br J Clin Pharmacol* (1986) **22**: 587–593.
24. Green SA, Spasoff AP, Coleman RA, Johnson M, Liggett SB: Sustained activation of a G protein coupled receptor via 'anchored' agonist binding. Molecular localization of the salmeterol exosite within the $β_2$-adrenergic receptor. *J Biol Chem* (1996) **39**: 24 029–24 035.
25. Lofdahl C-G, Svedmyr N: Formoterol fumarate, a new $β_2$-adrenoceptor agonist. *Allergy* (1989) **44**: 264–271.
26. Holstein-Rathlou NH, Laursen LC, Madsen F, Svendsen UG, Gnosspelius Y, Weeke B: Bambuterol: dose response study of a new terbutaline prodrug in asthma. *Eur J Clin Pharmacol* (1986) **30**: 7–11.
27. Reihsaus E, Innis M, MacIntyre N, Liggett SB: Mutations in the gene encoding for the $β_2$-adrenergic receptor in normal and asthmatic subjects. *Am J Respir Cell Mol Biol* (1993) **8**: 334–339.
28. Green SA, Turki J, Innis M, Liggett SB: Amino-terminal polymorphisms of the human $β_2$-adrenergic receptor impart distinct agonist-promoted regulatory properties. *Biochemistry* (1994) **33**: 9414–9419.
29. Green SA, Turki J, Bejarano P, Hall IP, Liggett SB: Influence of $β_2$ adrenergic receptor genotypes on signal transduction in human airway smooth muscle cells. *Am J Respir Cell Mol Biol* (1995) **13**: 25–33.
30. Turki J, Pak J, Green SA, Martin RJ, Liggett SB: Genetic polymorphisms of the $β_2$-adrenergic receptor in nocturnal and nonnocturnal asthma. Evidence that Gly16 correlates with the nocturnal phenotype. *J Clin Invest* (1995) **95**: 1635–1641.
31. Hall IP, Wheatley A, Wilding P, Liggett SB: Association of Glu 27 $β_2$-adrenoceptor polymorphism with lower airway reactivity in asthmatic subjects. *Lancet* (1995) **345**: 1213–1214.
32. Hall IP: $β_2$ Adrenoceptor polymorphisms: are they clinically important? *Thorax* (1996) **51**: 351–353.
33. Green SA, Cole G, Jacinto M, Innis M, Liggett, SB: A polymorphism of the human $β_2$-adrenergic receptor within the fourth transmembrane domain alters ligand binding and functional properties of the receptor. *J Biol Chem* (1993) **268**: 23 116–23 121.
34. Mathe AA, Astrom A, Persson N-A: Some bronchoconstricting and bronchodilating responses of human isolated bronchi: evidence for the existence of alpha-adrenoceptors. *J Pharm Pharmacol* (1971) **23**: 905–910.
35. Davis C, Conolly ME, Greenacre JK: Beta-adrenoceptors in human lung, bronchus and lymphocytes. *Br J Clin Pharmacol* (1980) **10**: 425–432.
36. Tattersfield AE: Effect of beta agonists and anticholinergic drugs on bronchial reactivity. *Am Rev Respir Dis* (1987) **136**: S64–S68.
37. Assem ESK, Schild HO: Inhibition by sympathomimetic amines of histamine release induced by antigen in passively sensitized human lung. *Nature* (1969) **224**: 1028–1029.
38. Marone G, Kagey-Sobotka A, Lichtenstein LM: Effects of arachidonic acid and its

metabolites on antigen-induced histamine release from human basophils *in vitro*. *J Immunol* (1979) **123**: 1669–1677.
39. Peters SP, Schulman ES, Schleimer RP, Macglashan DW, Newball HH, Lichtenstein LM: Dispersed human lung mast cells. Pharmacological aspects and comparison with human lung tissue fragments. *Am Rev Respir Dis* (1982) **126**: 1034–1039.
40. Howarth PH, Durham SR, Lee TH, Kay AB, Church MK, Holgate ST: Influence of albuterol, cromolyn sodium and ipratropium bromide on the airway and circulating mediator responses to allergen bronchial provocation in asthma. *Am Rev Respir Dis* (1985) **132**: 986–992.
41. Venge P, Dahl R, Peterson CGB: Eosinophil granule proteins in serum after allergen challenge of asthmatic patients and the effects of anti-asthmatic medication. *Int Arch Allergy Appl Immunol* (1988) **87**: 306–312.
42. Skoogh BE: Transmission through airway ganglia. *Eur J Respir Dis* (1983) **64** (Suppl 131): 159–170.
43. Barnes PJ: Muscarinic receptor subtypes: implications for lung disease. *Thorax* (1989) **44**: 161–167.
44. Chung KF, Rogers DF, Barnes PJ, Evans TW: The role of increased airway microvascular permeability and plasma exudation in asthma. *Eur Respir J* (1990) **3**: 329–337.
45. Wanner A, Salathé M, O'Riordan TC: Mucociliary clearance in the airways. *Am J Respir Crit Care Med* (1996) **154**: 1868–1902.
46. Pride NB, Ingram RH, Lim TK: Interaction between parenchyma and airways in chronic obstructive pulmonary disease and in asthma. *Am Rev Respir Dis* (1991) **143**: 1446–1449.
47. Twentyman OP, Finnerty JP, Harris A, Palmer J, Holgate ST: Protection against allergen-induced asthma by salmeterol. *Lancet* (1990) **336**: 1338–1342.
48. Wong BJ, Dolovich J, Ramsdale EH, *et al.*: Formoterol compared with beclomethasone and placebo on allergen-induced asthmatic responses. *Am Rev Respir Dis* (1992) **146**: 1156–1160.
49. Twentyman OP, Finnerty JP, Holgate ST: The inhibitory effect of nebulized albuterol on the early and late asthmatic reactions and increase in airway responsiveness provoked by inhaled allergen in asthma. *Am Rev Respir Dis* (1991) **144**: 782–787.
50. Vathenen AS, Knox AJ, Higgins BG, Britton JR, Tattersfield AE: Rebound increase in bronchial responsiveness after treatment with inhaled terbutaline. *Lancet* (1988) **i**: 554–558.
51. Cheung D, Timmers MK, Zwinderman AH, Bel EH, Dijkman JH, Sterk PJ: Long term effects of a long acting β_2-adrenoceptor agonist, salmeterol, on airway hyperresponsiveness in patients with mild asthma. *N Engl J Med* (1992) **327**: 1198–1203.
52. O'Connor BJ, Aikman SL, Barnes PJ: Tolerance to the nonbronchodilator effects of inhaled β_2-agonists in asthma. *N Engl J Med* (1992) **327**: 1204–1208.
53. Cockcroft DW, McParland CP, Britto SA, Swystun VA, Rutherford BC: Regular inhaled salbutamol and airway responsiveness to allergen. *Lancet* (1993) **342**: 833–837.
54. Ramage L, Lipworth BJ, Ingram CG, Cree IA, Dhillon DP: Reduced protection against exercise-induced bronchoconstriction after chronic dosing with salmeterol. *Respir Med* (1994) **88**: 363–368.
55. Kerrebijn KF, van Essen-Zandvliet EEM, Neijens HJ: Effect of long term treatment with inhaled corticosteroids and beta agonists on the bronchial responsiveness in children with asthma. *J Allergy Clin Immunol* (1987) **79**: 653–659.
56. Kraan J, Koeter G H, van der Mark Th W, Sluiter H , de Vries K: Changes in bronchial hyperreactivity induced by 4 weeks of treatment with antiasthmatic drugs in patients with allergic asthma: a comparison between budesonide and terbutaline. *J Allergy Clin Immunol* (1985) **76**: 628–636.
57. Wahedna I, Wong CS, Wisniewski AFZ, Pavord ID, Tattersfield AE: Asthma control during and after cessation of regular b$_2$-agonist treatment. *Am Rev Respir Dis* (1993) **148**: 707–712.
58. Pingleton SK, Schwartz O, Szymanski D, Epstein M: Hypotension associated with terbutaline therapy in acute quadriplegia. *Am Rev Respir Dis* (1982) **126**: 723–725.
59. Wong CS, Pavord ID, Williams J, Britton JR, Tattersfield AE: Bronchodilator, cardiovascular, and hypokalaemic effects of fenoterol, salbutamol, and terbutaline in asthma. *Lancet* (1990) **336**: 1396–1399.

60. Wagner PD, Dantzker DR, Iacovoni VE, Tomlin WC, West JB: Ventilation–perfusion inequality in asymptomatic asthma. *Am Rev Respir Dis* (1978) **118**: 511–525.
61. Scheinin M, Koulu M, Laurikainen E, Allonen H: Hypokalaemia and other non-bronchial effects of inhaled fenoterol and salbutamol: a placebo-controlled dose–response study in healthy volunteers. *Br J Clin Pharmacol* (1987) **24**: 645–653.
62. Bennett JA, Tattersfield AE: Time course and relative dose potency of systemic effects from salmeterol and salbutamol in healthy subjects. *Thorax* (1997) **52**: 458–464
63. Leslie D, Coats PM: Salbutamol-induced diabetic ketoacidosis. *Br Med J* (1977) **2**: 768.
64. Warrell DA, Robertson DG, Newton Howes J, *et al.*: Comparison of cardiorespiratory effects of isoprenaline and salbutamol in patients with bronchial asthma. *Br Med J* (1970) **i**: 65–70.
65. Smyth ET, Pavord ID, Wong CS, Wisniewski AF, Williams J, Tattersfield AE: Interaction and dose equivalence of salbutamol and salmeterol in patients with asthma. *Br Med J* (1993) **306**: 543–545.
66. Grove A, Lipworth BJ: Bronchodilator subsensitivity to salbutamol after twice daily salmeterol in asthmatic patients. *Lancet* (1995) **346**: 201–206.
67. Gray BJ, Frame MH, Costello JF: A comparative double blind study of the bronchodilator effects and side effects of inhaled fenoterol and terbutaline administered in equipotent doses. *Br J Dis Chest* (1982) **76**: 341–350.
68. Ullman A, Svedmyr N: Salmeterol, a new long acting inhaled β_2-adrenoceptor agonist: comparison with salbutamol in adult asthmatic patients. *Thorax* (1988) **43**: 674–678.
69. Wallin A, Sandström T, Rosenhall L, Melander B: Time course and duration of bronchodilatation with formoterol dry powder in patients with stable asthma. *Thorax* (1993) **48**: 611–614.
70. Sears MR, Taylor DR, Print CG, *et al.*: Regular inhaled β-agonist treatment in bronchial asthma. *Lancet* (1990) **336**: 1391–1396.
71. Taylor DR, Sears MR, Herbison GP, *et al.*: Regular inhaled β-agonist in asthma: effects on exacerbations and lung function. *Thorax* (1993) **48**: 134–138.
72. Chapman KR, Kesten S, Szalai JP: Regular vs as-needed inhaled salbutamol in asthma control. *Lancet* (1994) **343**: 1379–1382.
73. Juniper EF, Johnston PR, Borkhoff CM, Guyatt GH, Boulet L-P, Haukioja A: Quality of life in asthma clinical trials: comparison of salmeterol and salbutamol. *Am J Respir Crit Care Med* (1995) **151**: 66–70.
74. Pearlman DS, Chervinski P, LaForce C, *et al.*: A comparison of salmeterol with albuterol in the treatment of mild-to-moderate asthma. *N Engl J Med* (1992) **327**: 1420–1425.
75. D'Alonzo GE, Nathan RA, Henochowicz S, Morris RJ, Ratner P, Rennard SI: Salmeterol xinafoate as maintenance therapy compared with albuterol in patients with asthma. *JAMA* (1994) **271**: 1412–1416.
76. Drazen JM, Israel E, Boushey HA, *et al.*: Comparison of regularly scheduled with as-needed use of albuterol in mild asthma. *N Engl J Med* (1996) **335**: 841–847.
77. Apter AJ, Reisine ST, Willard A, *et al.*: The effect of inhaled albuterol in moderate to severe asthma. *J Allergy Clin Immunol* (1996) **98**: 295–301.
78. Leblanc P, Knight A, Kreisman H, Borkhoff CM, Johnston PR: A placebo-controlled, crossover comparison of salmeterol and salbutamol in patients with asthma. *Am J Respir Crit Care Med* (1996) **154**: 324–328.
79. van Schayck CP, Dompeling E, van Herwaarden CLA, *et al.*: Bronchodilator treatment in moderate asthma or chronic bronchitis: continuous or on demand? A randomised controlled study. *Br Med J* (1991) **303**: 1426–1431.
80. van Schayck CP, Dompeling E, van Herwaarden CLA: Continuous versus on demand use of bronchodilators in non-steroid asthma and chronic bronchitis: four-year follow-up randomised controlled study. *Br J Gen Pract* (1995) **45**: 239–244.
81. Wasserman SI, Furukawa CT, Henochowicz SI, *et al.*: Asthma symptoms and airway hyperresponsiveness are lower during treatment with nedocromil sodium than during treatment with regular inhaled albuterol. *J Allergy Clin Immunol* (1995) **95**: 541–547.
82. Haahtela T, Järvinen M, Kava T, *et al.*: Comparison of a β_2-agonist, terbutaline, with an inhaled corticosteroid, budesonide, in newly detected asthma. *N Engl J Med* (1991) **325**: 388–392.

83. van Essen-Zandvliet EE, Hughes MD, Waalkens HJ, et al.: Effects of 22 months of treatment with inhaled corticosteroids and/or β_2-agonists on lung function, airway responsiveness, and symptoms in children with asthma. *Am Rev Respir Dis* (1992) **146**: 547–554.
84. Inman WHW, Adelstein AM: Rise and fall of asthma mortality in England and Wales in relation to use of pressurised aerosols. *Lancet* (1969) **ii**: 279–285.
85. Stolley PD: Asthma mortality: why the United States was spared an epidemic of deaths due to asthma. *Am Rev Respir Dis* (1972) **105**: 883–890.
86. Jackson RT, Beaglehole R, Rea HH, Sutherland DC: Mortality from asthma: a new epidemic in New Zealand. *Br Med J* (1982) **285**: 771–774
87. Crane J, Pearce N, Flatt A, et al.: Prescribed fenoterol and death from asthma in New Zealand, 1981–83: case-control study. *Lancet* (1989) **i**: 917.
88. Pearce N, Grainger J, Atkinson M, et al.: Case control study of prescribed fenoterol and death from asthma in New Zealand, 1977–81. *Thorax* (1990) **45**: 170–175.
89. Granger J, Woodman K, Pearce N, et al.: Prescribed fenoterol and death from asthma in New Zealand, 1981–87: a further case-control study. *Thorax* (1991) **46**: 105.
90. Spitzer WO, Suissa S, Ernst P, et al.: The use of β-agonists and the risk of death and near death from asthma. *N Engl J Med* (1992) **326**: 501.
91. Van Metre TE: Adverse effects of inhalation of excessive amounts of nebulised isoproterenol in status asthmaticus. *J Allergy* (1969) **43**: 101–113.
92. Reisman RE: Asthma induced by adrenergic aerosols. *J Allergy* (1970) **46**: 162–177.
93. Tattersfield AE, Britton JR: Beta adrenoceptor agonists. In Barnes PJ, Rodger IW, Thomson NC (eds) *Asthma: Basic Mechanisms and Clinical Management*. London, Academic Press, 1988, pp. 527–554.
94. Tattersfield AE: Clinical studies of β-agonists in adults. In Pauwels R, O'Byrne P (eds) β-Agonists in Asthma Treatment. New York, Marcel Dekker, 1998 (in press).
95. Holgate ST, Baldwin CJ, Tattersfield AE: β-Adrenergic agonist resistance in normal human airways. *Lancet* (1977) **ii**: 375–377.
96. Harvey JE, Tattersfield AE: Airway response to salbutamol: effect of regular salbutamol inhalations in normal, atopic, and asthmatic subjects. *Thorax* (1982) **37**: 280–287.
97. Harvey JE, Baldwin CJ, Wood PJ, Alberti KGMM, Tattersfield AE: Airway and metabolic responsiveness to intravenous salbutamol in asthma: effect of regular inhaled salbutamol. *Clin Sci* (1981) **60**: 579–585.
98. Lipworth BJ, Struthers AD, McDevitt DG: Tachyphylaxis to systemic but not to airway responses during prolonged therapy with high dose inhaled salbutamol in asthmatics. *Am Rev Respir Dis* (1989) **140**: 586–592.
99. Rona G, Chappel CI, Balazs T, Gaudry R: An infarct-like myocardial lesion and other toxic manifestations produced by isoproterenol in the rat. *Arch Pathol* (1959) **67**: 443–445.
100. Todd GL, Baroldi G, Pieper GM, Clayton FC, Eliot RS: Experimental catecholamine-induced myocardial necrosis. I. Morphology, quantification and regional distribution of acute contraction band lesions. *J Mol Cell Cardiol* (1985) **17**: 317–338.
101. Joseph X, Whiteburst VE, Bloom S, Balazs T: Enhancement of cardiotoxic effects of beta-adrenergic bronchodilators by aminophylline in experimental animals. *Fundam Appl Toxicol* (1981) **1**: 443–447.
102. Nicklas RA, Whitehurst VE, Donohoe RF, Balazs T: Concomitant use of beta adrenergic agonists and methylxanthines. *J Allergy Clin Immunol* (1984) **73**: 20–24.
103. Bremner P, Burgess CD, Crane J, et al.: Cardiovascular effects of fenoterol under conditions of hypoxaemia. *Thorax* (1992) **47**: 814–817.
104. Dahl R, Earnshaw JS, Palmer JBD: Salmeterol: a four week study of a long acting β-adrenoceptor agonist for the treatment of reversible airways disease. *Eur Respir J* (1991) **4**: 1178–1184.
105. Jones KP: Salmeterol xinafoate in the treatment of mild to moderate asthma in primary care. *Thorax* (1994) **49**: 971–975.
106. Britton MG, Earnshaw JS, Palmer JBD: A twelve month comparison of salmeterol with salbutamol in asthmatic patients. *Eur Respir J* (1992) **5**: 1062–1067.
107. Lundback B, Rawlinson DW, Palmer JBD: Twelve month comparison of salmeterol and salbutamol as dry powder formulations in asthmatic patients. *Thorax* (1993) **48**: 148–153.

108. Castle W, Fuller R, Hall J, Palmer J: Serevent nationwide surveillance study: comparison of salmeterol with salbutamol in patients who require regular bronchodilator treatment. *Br Med J* (1992) **306**: 1034–1037.
109. Boyd G on behalf of a UK Study Group: Salmeterol xinafoate in asthmatic patients under consideration for maintenance oral corticosteroid therapy. *Eur Respir J* (1995) **8**: 1494–1498.
110. Kesten S, Chapman KR, Broder I, *et al.*: A three-month comparison of twice daily inhaled formoterol versus four times daily inhaled albuterol in the management of stable asthma. *Am Rev Respir Dis* (1991) **144**: 622–625.
111. Midgren B, Melander B, Persson G: Formoterol, a new long acting β_2-agonist, inhaled twice daily, in stable asthmatic subjects. *Chest* (1992) **101**: 1019–1022.
112. Wallin A, Melander B, Rosenhall L, Sandström T, Wåhlander L: Formoterol, a new long acting β_2-agonist for inhalation twice daily, compared with salbutamol in the treatment of asthma. *Thorax* (1990) **45**: 259–261.
113. Arvidsson P, Larsson S, Lofdahl C-G, Melander B, Svedmyr N, Wahlander L: Inhaled formoterol during one year in asthma: a comparison with salbutamol. *Eur Respir J* (1991) **4**: 1168–1173.
114. Rutten-van Mölken MPMH, Custers F, Vandoorslaer EKA, *et al.*: Comparison of performance of four instruments in evaluating the effects of salmeterol on asthma quality of life. *Eur Respir J* (1995) **8**: 888–898.
115. Faurschou P, Steffensen I, Jacques L: on behalf of a European Respiratory Study Group. Effect of addition of inhaled salmeterol to the treatment of moderate-to-severe asthmatics uncontrolled on high-dose inhaled steroids. *Eur Respir J* (1996) **9**: 1885–1890.
116. Wilding P, Clark M, Thompson Coon J, *et al.*: Effect of long term treatment with salmeterol on asthma control. *Br Med J* in press.
117. van der Molen T, Turner MO, Postma DS, Sears MR, for the Canadian and the Dutch D2522 investigators: An international multi-centre randomized controlled trial of formoterol in asthmatics requiring inhaled corticosteroid. *Eur Respir J* (1995) **8**: 2S.
118. Gardiner PV, Ward C, Booth H, Allison A, Hendrick DJ, Walters EH: Effect of eight weeks of treatment with salmeterol on bronchoalveolar lavage inflammatory indices in asthmatics. *Am J Respir Crit Care Med* (1994) **150**: 1006–1011.
119. Pauwels RA, Löfdahl C-G, Postma DS, *et al.*: Effects of inhaled formoterol and budesonide on exacerbations of asthma. *New Eng J Med* (1997) **337**: 1405–1411.
120. Greening AP, Ind PW, Northfield M, Shaw G: Added salmeterol versus higher-dose corticosteroid in asthma patients with symptoms on existing inhaled corticosteroid. *Lancet* (1994) **344**: 219–224.
121. Woolcock A, Lundbak B, Ringdal OLN, Jacques LA: Comparison of the effect of addition of salmeterol with doubling the inhaled steroid dose in asthmatic patients. *Am Rev Respir Dis* (1994) **149**: A280.
122. Muir JF, Bertin L, Georges D: French Multicentre Study Group. Salmeterol versus slow-release theophylline combined with ketotifen in nocturnal asthma: a multicentre trial. *Eur Respir J* (1992) **5**: 1197–1200.
123. Fjellbirkeland L, Gulsvik A, Palmer JBD: The efficacy and tolerability of inhaled salmeterol and individually dose-titrated, sustained-release theophylline in patients with reversible airways disease. *Respir Med* (1994) **88**: 599–607.
124. Fitzpatrick MF, Mackay T, Driver H, Douglas NJ: Salmeterol in nocturnal asthma: a double blind, placebo controlled trial of a long acting inhaled β_2-agonist. *Br Med J* (1990) **301**: 1365–1368.
125. Maesen FPV, Smeets JJ, Gubbelmans HLL, Zweers PGMA: Formoterol in the treatment of nocturnal asthma. *Chest* (1990) **98**: 866–870.
126. Kemp JP, Dockhorn RJ, Busse WW, Bleecker ER, Van As A: Prolonged effect of inhaled salmeterol against exercise-induced bronchospasm. *Am J Respir Crit Care Med* (1994) **150**: 1612–1615.
127. Robertson W, Simkins J, O'Hickey SP, Freeman S, Cayton RM: Does single dose salmeterol affect exercise capacity in asthmatic men? *Eur Respir J* (1994) **7**: 1978–1984.
128. Boner AL, Spezia E, Piovesan P, Chiocca E, Maiocchi G: Inhaled formoterol in the

prevention of exercise-induced bronchoconstriction in asthmatic children. *Am J Respir Crit Care Med* (1994) **149**: 935–939.
129. Palmer JBD, Stuart AM, Shepherd GL, Viskum K: Inhaled salmeterol in the treatment of patients with moderate to severe reversible obstructive airways disease – a 3-month comparison of the efficacy and safety of twice-daily salmeterol (100 mg) with salmeterol (50 μg). *Respir Med* (1992) **86**: 409–417.
130. Schreurs AJM, Sinninghe Damsté HEJ, de Graaff CS, Greefhorst APM: A dose–response study with formoterol Tubuhaler[1] as maintenance therapy in asthmatic patients. *Eur Respir J* (1996) **9**: 1678–1683.
131. Wilkinson JRW, Roberts JA, Bradding P, Holgate ST, Howarth PH: Paradoxical bronchoconstriction in asthmatic patients after salmeterol by metered dose inhaler. *Br Med J* (1992) **305**: 931–932.
132. Shaheen MZ, Ayres JG, Benincasa C: Incidence of acute decreases in peak expiratory flow following the use of metered dose inhalers in asthmatic patients. *Eur Respir J* (1994) **7**: 2160–2164.
133. Bennett JA, Smyth ET, Pavord ID, Wilding PJ, Tattersfield AE: Systemic effects of salbutamol and salmeterol in patients with asthma. *Thorax* (1994) **49**: 771–774.
134. Newnham DM, Grove A, McDevitt DG, Lipworth BJ: Subsensitivity of bronchodilator and systemic β_2-adrenoceptor responses after regular twice daily treatment with eformoterol dry powder in asthmatic patients. *Thorax* (1995) **50**: 497–504.
135. Booth H, Fishwick K, Harkawat R, Devereux G, Hendrick DJ, Walters EH: Changes in methacholine-induced bronchoconstriction with the long acting β_2-agonist salmeterol in mild to moderate asthmatic patients. *Thorax* (1993) **48**: 1121–1124.
136. Yates DH, Sussman HS, Shaw MJ, Barnes PJ, Chung KF: Regular formoterol treatment in mild asthma. *Am J Respir Crit Care Med* (1995) **152**: 1170–1174.
137. Verberne AAPH, Hop WCJ, Creyghton FBM, *et al.*: Airway responsiveness after a single dose of salmeterol and during four months of treatment in children with asthma. *J Allergy Clin Immunol* (1996) **97**: 938–946.
138. Gibson GJ, Greenacre JK, Konig P, Conolly ME, Pride NB: Use of exercise challenge to investigate possible tolerance to β-adrenoceptor stimulation in asthma. *Br J Dis Chest* (1978) **72**: 199–206.
139. Vilsvik JS, Langaker O, Persson G, *et al.*: Bambuterol: a new long acting bronchodilating prodrug. *Ann Allergy* (1991) **66**: 315–319.
140. Persson G, Baas A, Knight A, Larsen B, Olsson H: One month treatment with the once daily oral β_2-agonist bambuterol in asthmatic patients. *Eur Respir J* (1995) **8**: 34–39.
141. Petrie GR, Chookang JY, Hassan WU, *et al.*: Bambuterol: effective in nocturnal asthma. *Respir Med* (1993) **87**: 581–585.
142. Wallaert B and the French Bambuterol Study Group, Ostinelli J, Arnould B: Long acting β_2-agonists: a comparison of oral bambuterol and inhaled salmeterol in asthmatic patients with nocturnal symptoms. *Eur Respir J* (1995) **8**: 1S.
143. Hall IP, Tattersfield AE: *Asthma*, 3rd edn. Edited by Clark TJH, Godfrey S, Lee TH. (1992) London: Chapman and Hall. pp. 341–356.

37

Anticholinergic Bronchodilators

NICHOLAS J. GROSS

INTRODUCTION

Anticholinergic agents such as atropine exist in many plants and have consequently been used in herbal remedies for many centuries. They were introduced into Western medicine in the early 1800s and enjoyed enormous use as bronchodilators well into the present century. When adrenaline was discovered in the 1920s, followed soon by ephedrine, other adrenergic agents and then methylxanthines, their use declined. Natural anticholinergic agents such as atropine produced many side-effects that resulted in poor acceptability by patients. Interest in their use has returned with better understanding of the role of the parasympathetic system in controlling airway tone, and with the development of synthetic congeners of atropine that are topically active but much less prone to produce side-effects.[1]

RATIONALE FOR USE OF ANTICHOLINERGIC BRONCHODILATORS

Autonomic control of airway calibre

In human airways the bulk of efferent autonomic nerves are cholinergic.[2] Branches of the vagus nerve travel along the airways and synapse at peribronchial ganglia, from which short postganglionic nerves travel to smooth muscle cells and mucus glands, predominantly in the central airways (see Chapter 22). Acetylcholine is released from varicosities and terminals of the postganglionic nerves and activates muscarinic receptors, which results in the contraction of smooth muscle, release of mucus from mucus glands and, possibly, acceleration of ciliary beat frequency. A low level of cholinergic, vagal (bronchomotor) tone can be recorded in the resting state in experimental animals, but can be considerably augmented in response to a variety of stimuli.[1] Anticholinergic agents compete with acetylcholine at muscarinic receptors. This inhibits tonic and phasic cholinergic activity, permitting airways to dilate. They do not inhibit other mediators or mechanisms of smooth muscle contraction nor, indeed, do they affect the numerous other mechanisms of airway obstruction in abnormal states such as asthma.

Cholinergic bronchomotor activity can be reflexly augmented by a variety of stimuli through the neural pathways shown in Fig. 37.1. Afferent activity can arise from 'irritant receptors' and C fibres located anywhere in the upper and lower airways, and probably from the oesophagus and carotid bodies, and is transmitted via vagal afferents, through the vagal nuclei to vagal efferents and the larger airways that receive vagal innervation. Stimuli to which these receptors respond include mechanical irritation, a wide variety of irritant gases, aerosols, particles, cold dry air and specific mediators such as histamine and some bronchoconstricting eicosanoids.[3,4] Although vagally mediated bronchoconstriction has been clearly demonstrated in animals, and to some extent in humans also, the extent to which such mechanisms actually contribute to airflow limitation in patients with airways disease is not clear or even consistent among patients. Abolition of cholinergic activity by anticholinergic agents usually produces a degree of bronchodilatation but rarely entirely reverses airflow limitation; one can thus assume that vagal activity

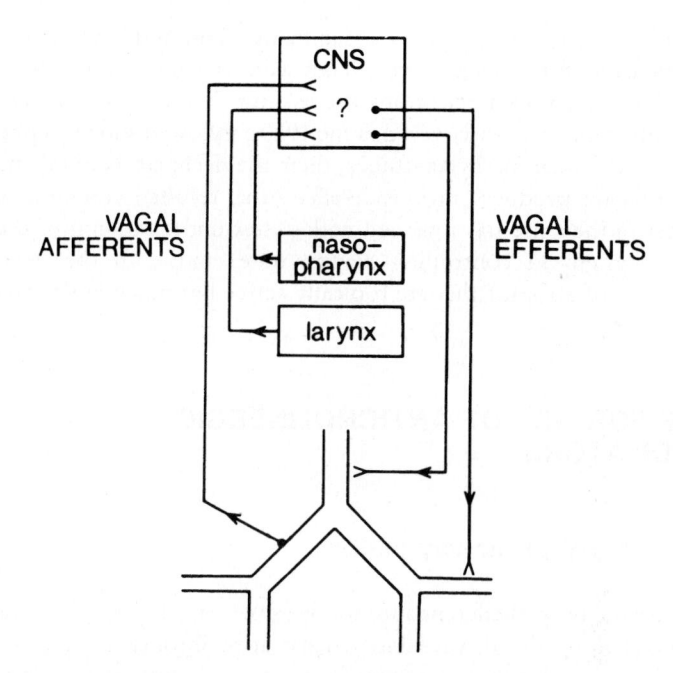

Fig. 37.1 Vagal reflex pathways from irritant receptors, through vagal afferents, central nervous system (CNS) and vagal efferents to effector cells in the airways. Reproduced from ref. 1, with permission.

accounts for only a part of the airflow obstruction in patients with asthma or chronic obstructive pulmonary disease (COPD). There is some evidence that cholinergic bronchomotor tone is increased in both asthma[5] and COPD,[6] providing further rationale for their use in these conditions.

Muscarinic receptor subtypes in airways

Molecular biology has revealed the existence of a family of muscarinic receptor subtypes, at least three of which, known as M_1, M_2 and M_3, are expressed in the lung, each of which appears to play a role in the control of airway calibre (see Chapter 22). Briefly, M_1 receptors, located in peribronchial ganglia, and M_3 receptors, located on smooth muscle cells, mediate smooth muscle contraction.[7] M_2 receptors are autoreceptors whose stimulation provides feedback inhibition of further acetylcholine release from cholinergic nerves, and thus tend to limit the bronchoconstrictor effects of parasympathetic activity. There is evidence that M_2 receptors are selectively damaged by parainfluenza virus infections as well as by some eosinophil products,[8,9] which may account at least partly for the bronchospasm associated with viral infections and asthma. Another practical aspect of this new understanding is that currently available anticholinergic bronchodilators, none of which is selective for muscarinic receptor subtypes, may be suboptimal. Attempts to develop synthetic anticholinergic agents have resulted in one, tiotropium bromide, that is selective for M_1 and M_3 receptors.[10,11] For this reason, tiotropium may prove relatively more potent as a bronchodilator than currently available agents.

PHARMACOLOGY

Two major classes of anticholinergic agents are recognized. Naturally occurring anticholinergic agents, such as atropine, scopolamine, etc., are tertiary ammonium compounds, i.e. the nitrogen atom on the tropane ring is 3-valent (Fig. 37.2). They are freely soluble in water and lipids and well absorbed from mucosal surfaces and the skin. They are thus widely distributed in the body and cross the blood–brain barrier, counteracting parasympathetic activity in almost every system and producing widespread dose-related systemic effects. Atropine, for example, in the dose that results in bronchodilatation (1.0–2.5 mg in adults) frequently produces skin flushing, mouth dryness and possibly tachycardia. In slightly higher doses it produces blurred vision, urinary retention and mental effects such as irritability, confusion and hallucinations. The therapeutic margin of atropine and its natural congeners is thus small, making these agents difficult to use.

Quaternary congeners, whose tropane nitrogen atom is 5-valent, are all synthetic, e.g. ipratropium bromide (Fig. 37.2). The charge associated with this change renders these molecules poorly absorbable from mucosal surfaces. Such agents are fully anticholinergic at the site of deposition and will, for example, dilate the pupil if delivered to the eye or dilate the bronchi if inhaled. However they are not sufficiently absorbed from these sites to produce either detectable blood levels or systemic effects even when delivered in

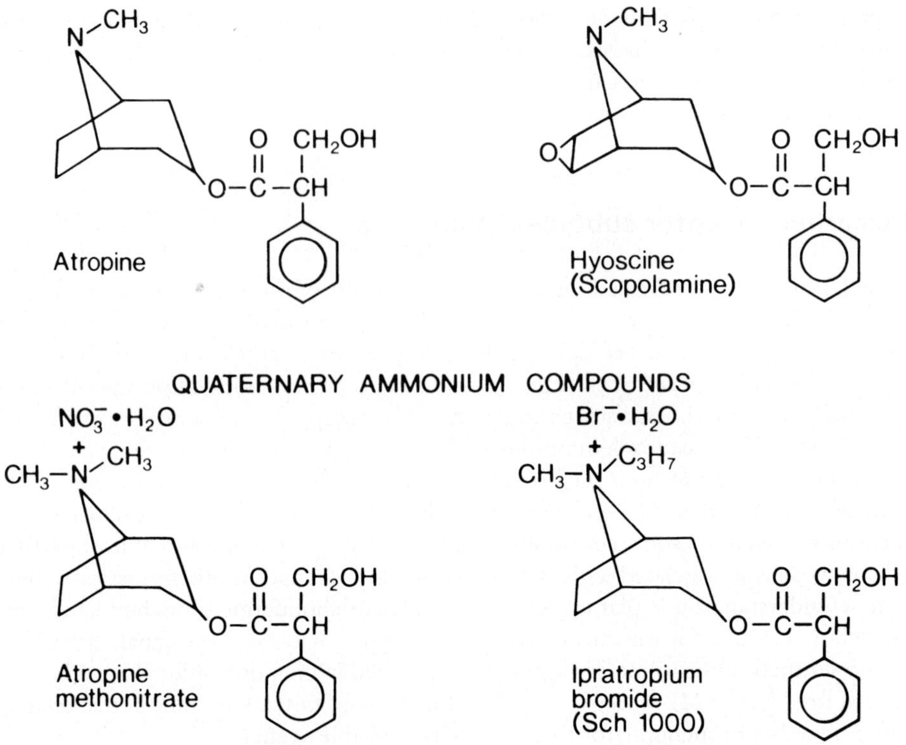

Fig. 37.2 Structures of some anticholinergic bronchodilators. Reproduced from ref. 1, with permission.

supramaximal doses.[12] These agents can thus be regarded for practical purposes as topical forms of atropine. The group includes, in addition to ipratropium, oxitropium bromide (Oxivent), atropine methonitrate, glycopyrrolate bromide (Robinul) and tiotropium bromide. The last agent, tiotropium, is of particular interest in that it is a functionally selective antagonist of the muscarinic receptor subtypes that mediate bronchoconstriction (see above) and is also extremely long acting.[10,11]

Atropine and its natural congeners exist in two optical isomeric forms only one of which is physiologically active, whereas the quaternary agents are generally synthesized in the active isomeric form, resulting in apparently greater activity of the latter. Atropine is quantitatively absorbed from the airways, reaching peak blood levels in 1 h. It has a half-life in the circulation of about 3 h in adults, but longer in children and the elderly.[1] Most of the drug is recovered unchanged in the urine, traces being found in the faeces and the breast milk of lactating women. Ipratropium pharmacokinetics have been studied in humans by radiolabelling. Administered by mouth or inhalation, the blood levels are very low, peaking at about 1–2 h and declining with a half-life of about 4 h. Its bronchodilator action is somewhat longer, probably because it is not removed from the airways by absorption. Most of an oral dose is recovered in the faeces, a small amount as inactive metabolites in the urine. It is largely excluded from the central nervous system.

CLINICAL EFFICACY

Dose–response

A complete, referenced guide to the dose–response of anticholinergic agents given by various inhalational methods is provided in a previous review.[13] For ipratropium in nebulized solution the optimal dose is 500 μg in adults and 125–250 μg in children. By metered dose inhaler (MDI) the optimal dose in younger adults with asthma is 40–80 μg, but in older patients with COPD the optimal dose is much higher, possibly 160 μg, particularly when airways obstruction is severe. Newer inhalers will employ a dry-powder form without propellants, rather than the suspension that is currently used. The optimal dose of the dry powder form may be a little lower than that for the suspension (preliminary data). For oxitropium MDI, the optimal dose is approximately 200 μg. For less commonly used agents, the optimal doses are as follows: atropine, 0.25–0.4 mg/kg; atropine methonitrate, 0.015–0.02 mg/kg; glycopyrrolate, 0.02 mg/kg.

Against specific stimuli

When given in advance of bronchospastic stimuli, anticholinergic agents provide variable degrees of protection.[1] They protect more or less completely against cholinergic agonists such as methacholine. They are also prophylactic against the bronchospasm induced in asthmatic patients by β-blocking agents and by psychogenic factors. They provide only partial protection against bronchospasm due to most other stimuli, e.g. histamine, prostaglandins, non-specific dusts and irritant aerosols, exercise and hyperventilation with cold, dry air. In most of the latter instances, adrenergic agents usually provide greater prophylaxis.

Stable asthma

A very large number of studies have compared the bronchodilator potential of anticholinergic agents with that of adrenergic agents. While many of these studies are flawed by the fact that they used recommended doses rather than optimal doses, they provide the clinician with useful information about the comparative actions of these bronchodilators. Figure 37.3, which is typical of most such studies, illustrates many of these points.[14] Anticholinergic agents are slower to reach peak effect, typically 1–2 h, compared with adrenergic agents. At their peak effect they almost invariably result in less bronchodilatation in patients with asthma. The quaternary forms may be slightly longer acting than agents such as salbutamol. Among asthmatic patients there is, however, substantial variation in responsiveness, some patients responding very little to anticholinergic agents, others responding almost as well to them as to adrenergic agents.

It has been difficult to identify subgroups of asthmatic patients who are likely to manifest the most responsiveness to anticholinergic agents. Older asthmatic individuals (over 40 years of age) may respond better than younger ones,[15] although even children aged 10–18 years have been shown to benefit[16] (see below). Individuals with intrinsic

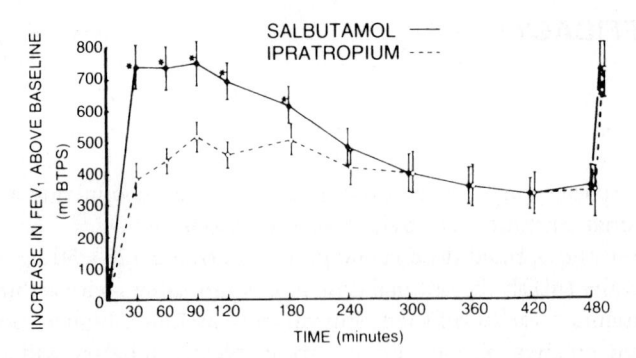

Fig. 37.3 Increase in forced expiratory volume in 1 s (FEV_1) in 25 patients with asthma after inhalation of 200 μg salbutamol by metered dose inhaler (MDI) or 40 μg ipratropium by MDI on separate days. All patients received an additional dose of salbutamol at 480 min. Asterisks denote significant differences ($P < 0.05$). Reproduced from ref. 14, with permission.

asthma and those with longer duration of asthma may also respond better than individuals with extrinsic asthma,[17] although this response also appears to be a poor predictors.[14] An individual trial remains the best way to identify responsiveness.[18]

Acute severe asthma

Most studies suggest that a β-agonist is likely to be more potent than an anticholinergic agent in acute severe asthma. The question arises whether an anticholinergic agent can add to the bronchodilatation achieved by the adrenergic agent. In the largest study on this question, Rebuck et al.[19] found that the combination of 500 μg nebulized ipratropium with 1.25 mg nebulized fenoterol resulted in significantly more bronchodilatation over the first 90 min of treatment than either agent alone. Moreover, patients with more severe airway obstruction obtained the greatest benefit from the combination. To overcome type II errors, Ward[20] performed a meta-analysis of nine similar studies involving a total of 435 patients, all but 31 of whom were adults. The pooled data clearly showed a benefit for the combination of ipratropium and a β-agonist over the β-agonist alone over the first 1–2 h of treatment. The size of the effect equated to a mean increase in peak expiratory flow rate of +44 litre/min over that obtained by a β-agonist alone, a result that was both statistically and clinically significant.

It seems appropriate to recommend that both classes of bronchodilators be given in acute severe asthma, particularly in the early hours of treatment and particularly in patients with more severe airflow obstruction.

Paediatric airways disease

Studies from Canada have compared salbutamol alone with salbutamol plus ipratropium as treatment for acute severe asthma in children. Two well-conducted studies showed that the addition of ipratropium accelerated the rate of improvement in airflow.[21,22] However, two other studies failed to show much benefit from the addition of ipratropium.[23,24] A

37 Anticholinergic Bronchodilators

large and more recent study of the same question[25] clearly showed a benefit from the addition of ipratropium that was dose related (three doses of 250 µg were better than one dose, which was better than no ipratropium). In this study, children with more severe bronchospasm at presentation clearly benefited more from the addition of ipratropium and the hospitalization rate was lower. As in adult status asthmaticus, therefore, the combination of ipratropium with an adrenergic agent is probably more effective than monotherapy, particularly in patients with more severe status.

In stable childhood asthma, the evidence for benefit from the addition of ipratropium to salbutamol is less clear. Two consensus reports reviewed the published evidence, which is not extensive, and concluded that ipratropium was safe for this purpose in the paediatric population, but that its benefit compared with an adrenergic agent alone was slight at best.[26,27] There are scattered reports of ipratropium use in other paediatric conditions such as viral bronchiolitis, cystic fibrosis, exercise-induced bronchospasm and bronchopulmonary dysplasia, but these do not provide strong and consistent evidence for the benefit of ipratropium over alternative bronchodilators.

Stable COPD

A very large number of studies have compared anticholinergic agents with other bronchodilators in patients with COPD. In general, patients with COPD do not manifest as much absolute increase in airflow to any agent or combination of agents as do patients with asthma. However, almost all are capable of some. With very few exceptions, these studies show that the anticholinergic agent provides at least as great and prolonged an increase in airflow as other agents. Most, like the largest such study,[28] show that the anticholinergic agent is a more potent bronchodilator. Even when large cumulative doses of each agent, rather than recommended doses, are given the anticholinergic agent alone achieves all the available bronchodilatation in these patients.[29,30] As this is clearly not the case in asthmatic patients, there may thus be a systematic difference between asthmatic and COPD patients with respect to their responsiveness to bronchodilators. This point is emphasized by a few studies in which patients with asthma and COPD who had similar baseline airflows have been studied side by side, e.g. Fig. 37.4.[31] This figure shows that the combination of fenoterol and theophylline resulted in more bronchodilatation than did ipratropium in the asthmatic group, but that ipratropium resulted in more bronchodilatation in the bronchitic group. Reasons for the difference between the two diagnostic groups of patients are not known but seem likely to include the fact that airflow obstruction in asthma is due to factors related to airway inflammation that are amenable at least in part to adrenergic agents but not amenable to anticholinergic agents. These factors are present to a lesser extent in patients with COPD whose major reversible component is bronchomotor tone, the latter being best reversed by anticholinergic agents.[29] Whatever the reason, COPD represents the group of patients in whom anticholinergic agents are the most useful bronchodilators.

In accord with this view, ipratropium is currently recommended as first-line treatment for stable COPD in an authoritative review[32] and in the recent official statements of the European Respiratory Society[33] and the American Thoracic Society.[34]

It should be noted, however, that the clinical utility of ipratropium (and possibly other classes of bronchodilators) is limited to their short-term relief of symptoms and that they

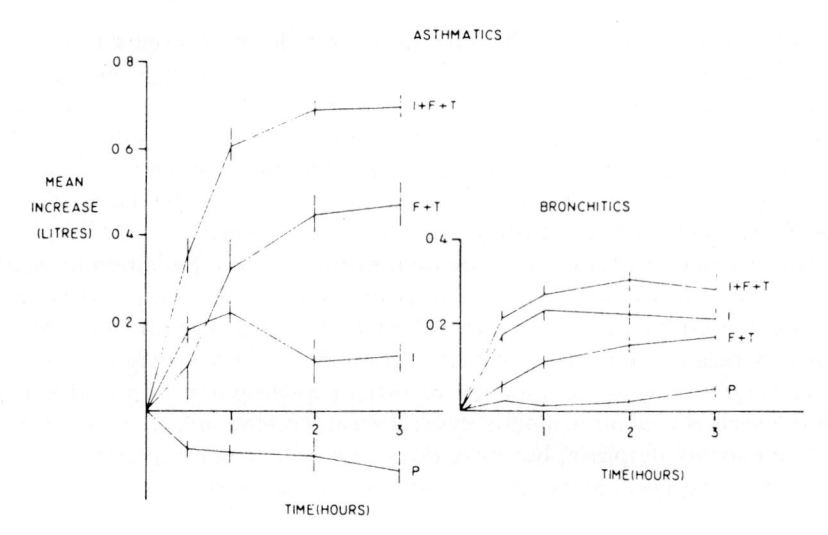

Fig. 37.4 Increase in forced expiratory volume in 1 s (FEV$_1$) of 15 patients with asthma (a) and 15 patients with chronic bronchitis (b). P, placebo metered dose inhaler (MDI); I, ipratropium 40 μg MDI; F + T, fenoterol 5 mg plus oxtriphylline 400 mg oral. Reproduced from ref. 31, with permission.

have no demonstrated long-term effect on the natural history of COPD. In the Lung Health Study, a large multicentre longitudinal study of healthy cigarette smokers, regular use of ipratropium MDI had no discernible effect on the accelerated smoking-related decline in lung function.[35]

Acute exacerbations of COPD

Three recent studies comparing the efficacy of bronchodilators in acute exacerbations of COPD found no significant differences between adrenergic and anticholinergic agents or their combination.[19,36,37]

Combinations with other bronchodilators

Combinations of different classes of bronchodilators often provide more bronchodilatation than single agents, and this effect is seen in many of the studies cited, e.g. Fig. 37.4. However, this is probably due to the fact that most clinical studies are performed with recommended rather than optimal doses of the agents. Consequently, when two or more classes of agents are given together the effects may simply be additive rather than potentiating. As anticholinergic, adrenergic and methylxanthine agents work by different mechanisms, affect different-sized airways and have different pharmacodynamic and pharmacokinetic properties, their combination is rational and is likely to result in improved bronchodilatation. No unfavourable interactions between these three classes of agents have been reported, so the greater bronchodilatation achieved by their combination is achieved without increasing the risk of side-effects. In practice, it is common to use two or even all of these agents simultaneously to manage severe airways obstruction.

The combination of more than one class of inhaled bronchodilator in a single MDI has been employed at least since the 1950s. The combination of ipratropium with the β-agonist fenoterol (Berodual and DuoVent) has been widely used since the 1970s. Such combinations provide the additional advantages that, for patients who need two agents, a single MDI containing two agents is likely to be less expensive than two MDIs, easier and more convenient for the patient to use, and therefore more likely to improve patient compliance. Because of the declining popularity of fenoterol, a new combination MDI containing ipratropium and salbutamol, both in recommended dosage, has been developed (Combivent). Clinical trials with this combination in patients with COPD[38,39,40] suggest it possesses all the advantages mentioned above. Bronchodilatation is greater during the first 4–5 h after administration, but not much prolonged over that achieved by single agents, and no increase in side-effects is incurred.

SIDE-EFFECTS

Atropine produces numerous systemic side-effects related to the inhibition of physiological functions of the parasympathetic system, as mentioned above. These effects occur in doses at or only slightly above the bronchodilator dose. Atropine is contraindicated in patients with glaucoma or prostatism. The principal advantage of quaternary anticholinergic agents is that they are so poorly absorbed from mucosae that the risk of such effects is insignificant. Even massive, inadvertent overdosage of one such agent resulted in trivial effects.[12] Ipratropium, the most widely studied quaternary anticholinergic, has been exonerated after extensive exploration for atropine-like side-effects.[41] It can, for example, be given to patients with glaucoma without affecting intraocular tension (provided it is not sprayed directly into the eye). It has been found not to affect urinary flow characteristics in older men. Nor has it been found to alter the viscosity and elasticity of respiratory mucus, or mucociliary clearance, as does atropine.[42] It has negligible effects on haemodynamics and the pulmonary circulation.[43] Consequently, quaternary anticholinergics do not carry the risk of increasing hypoxaemia, as do adrenergic agents,[44] an important consideration in exacerbations of asthma and COPD.

In normal clinical use the only side-effects that the patient might experience with ipratropium are dryness of the mouth and a brief coughing spell, which has been reported to occur in 5% of patients.[28] Rarely it can result in paradoxical bronchoconstriction. This has been variously attributed to hypotonicity of the nebulized solution, idiosyncrasy to the bromine radical, the benzalkonium preservative and a selective effect on the M_2 receptor. Paradoxical bronchoconstriction may also occur with other anticholinergic agents. Although rare, occurring in possibly 0.3% of patients, the possibility of paradoxical bronchoconstriction in a patient warrants withdrawal of the drug from that patient. Other than these two effects, very extensive investigation and the worldwide use of ipratropium for nearly two decades demonstrate a remarkably low incidence of untoward reactions.

CLINICAL RECOMMENDATIONS

The use of anticholinergic bronchodilators is best limited to the poorly absorbed quaternary forms, e.g. ipratropium, oxitropium, atropine methonitrate, glycopyrrolate, administered by inhalation. They are sometimes useful in stable asthma as adjuncts to other bronchodilator therapy, and have a demonstrated role in combination with adrenergic agents in the treatment of acute severe asthma. Their principal role is in the long-term management of stable COPD where they are probably the most efficacious bronchodilators. Because of their slow onset of action they are best used on a regular, maintenance basis, rather than p.r.n. The usual dose, two puffs of 20 μg each, is probably suboptimal for many patients with COPD and can safely be doubled or quadrupled.[45]

REFERENCES

1. Gross NJ, Skorodin MS: Anticholinergic, antimuscarinic bronchodilators. *Am Rev Respir Dis* (1984) **129**: 856–870.
2. Richardson JB: Innervation of the lung. *Eur J Respir Dis* (1982) **117** (Suppl): 13–31.
3. Widdicombe JG: The parasympathetic nervous system in airways disease. *Scand J Respir Dis* (1979) **103** (Suppl): 38–43.
4. Nadel JA: Autonomic regulation of airway smooth muscle. In Nadel JA (ed) *Physiology and Pharmacology of the Airways*. New York, Marcel Dekker, 1980, pp 217–257.
5. Shah PKD, Lakhotia M, Mehta S, Jain SK, Gupta GL: Clinical dysautonomia in patients with bronchial asthma, study with seven autonomic function tests. *Chest* (1990) **98**: 1408–1413.
6. Gross NJ, Co E, Skorodin MS: Cholinergic bronchomotor tone in COPD, estimates of its amount in comparison to normal. *Chest* (1989) **96**: 984–987.
7. Gross NJ, Barnes PJ: A short tour around the muscarinic receptor. *Am Rev Respir Dis* (1988) **138**: 765–767.
8. Fryer AD, Jacoby DB: Parainfluenza virus infection damages inhibitory M2-muscarinic receptors on pulmonary parasympathetic nerves in the guinea pig. *Br J Pharmacol* (1991) **102**: 267–271.
9. Fryer AD, Jacoby DB: Effect of inflammatory cell mediators on M2 muscarinic receptors in the lungs. *Life Sci* (1993) **52**: 529–536.
10. O'Connor BJ, Towse LJ, Barnes PJ: Prolonged effect of tiotropium bromide on methacholine-induced bronchoconstriction in asthma. *Am J Respir Crit Care Med* (1996) **154**: 876–880.
11. Maesen FP, Smeets JJ, Sledsens TJ, Wald FD, Cornelissen PJ: Tiotropium bromide, a new long-acting antimuscarinic bronchodilator: a pharmacodynamic study in patients with chronic obstructive pulmonary disease (COPD). *Eur Respir J* (1995) **8**: 1506–1513.
12. Gross NJ, Skorodin MS: Massive overdose of atropine methonitrate with only slight untoward effects. *Lancet* (1985) **ii**: 386.
13. Gross NJ, Skorodin MS: Anticholinergic agents. In Jenne JW, Murphy S (eds) *Drug Therapy*. New York, Marcel Dekker, 1987, pp 615–668.
14. Ruffin RE, Fitzgerald JD, Rebuck AS: A comparison of the bronchodilator activity of Sch 1000 and salbutamol. *J Allergy Clin Immunol* (1977) **59**: 136–141.
15. Ullah MI, Newman GB, Saunders KB: Influence of age on response to ipratropium and salbutamol in asthmia. *Thorax* (1981) **36**: 523–529.
16. Vichyanond P, Sladek WA, Syr S, Hill MR, Szefler SJ, Nelson HS: Efficacy of atropine methylnitrate alone and in combination with albuterol in children with asthma. *Chest* (1990) **98**: 637–642.
17. Jolobe OMP: Asthma versus non-specific reversible airflow obstruction, clinical features and responsiveness to anticholinergic drugs. *Respiration* (1984) **45**: 237–242.

18. Brown IG, Chan CS, Kellcy CA, Dent AG, Zimmerman PV: Assessment of the clinical usefulness of nebulised ipratropium bromide in patients with chronic airflow limitation. *Thorax* (1984) **39**: 272–276.
19. Rebuck AS, Chapman KR, Abboud R, *et al*.: Nebulized anticholinergic and sympathomimetic treatment of asthma and chronic obstructive airways disease in the emergency room. *Am J Med* (1987) **82**: 59–64.
20. Ward MJ: The role of anticholinergic drugs in acute asthma. In Gross NJ (ed) *Anticholinergic Therapy in Obstructive Airways Disease*. London, Franklin Scientific Publications, 1993, pp 155–162.
21. Beck R, Robertson C, Galdes-Sebaldt M, Levison H: Combined salbutamol and ipratropium bromide in the treatment of severe acute asthma. *J Pedriatr* (1985) **107**: 605–608.
22. Reisman J, Galdes-Sebaldt M, Kazim F, *et al*.: Frequent administration by inhalation of salbutamol and ipratropium bromide in the initial management of severe acute asthma in children. *J Allergy Clin Immunol* (1988) **82**: 1012–1018
23. Storr J, Lenney W: Nebulized ipratropium and salbutamol in asthma. *Am J Dis Child* (1986) **61**: 602–603
24. Boner AL, DeStefano G, Niero E, Vallone G, Gaburro D: Salbutamol and ipratropium bromide in the treatment of bronchospasm in asthmatic children. *Ann Allergy* (1987) **58**: 54–58.
25. Schuh H, Johnson DW, Callahan S, Canny G, Levinson H: Efficacy of frequent nebulized ipratropium bromide added to frequent high-dose albuterol therapy in severe childhood asthma. *J Pediatr* (1995) **126**: 639–645.
26. Warner JO, Getz M, Landau LI, *et al*.: Management of asthma: a consensus statement. *Arch Dis Child* (1989) **64**: 1065–1079.
27. Hargreave FE, Dolovich J, Newhouse MT: The assessment and treatment of asthma: a conference report. *J Allergy Clin Immunol* (1990) **85**: 1098–1112.
28. Tashkin DP, Ashutosh K, Bleeker E, *et al*.: Comparison of the anticholinergic ipratropium bromide with metaproterenol in chronic obstructive pulmonary disease, a 90 day multicenter study. *Am J Med* (1986) **81** (Suppl 5A): 81–86.
29. Gross NJ, Skorodin MS: Role of the parasympathetic system in airway obstruction due to emphysema. *N Engl J Med* (1984) **311**: 421–426.
30. Easton PA, Jadue C, Dhingra S, Anthonisen NR: A comparison of the bronchodilating effects of a beta-2 adrenergic agent (albuterol) and an anticholinergic agent (ipratropium bromide), given by aerosol alone or in sequence. *N Engl J Med* (1986) **315**: 735–739.
31. Lefcoe NM, Toogood JH, Blennerhassett G, Patterson NAM: The addition of an aerosol anticholinergic to an oral beta agonist plus theophylline in asthma and bronchitis. *Chest* (1982) **82**: 300–305.
32. Ferguson GT, Cherniack RM: Management of chronic obstructive pulmonary disease. *N Engl J Med* (1993) **328**: 1017–1022.
33. Siafakis NM, Vermiere P, Pride NB, *et al*.: ERS consensus statement: optimal assessment and management of chronic obstructive pulmonary disease. *Eur Respir J* (1995) **8**: 1398–1420.
34. ATS Statement: Standards for the diagnosis and care of patients with chronic obstructive pulmonary disease. *Am J Respir Crit Care Med* (1995) **152**: S77–S120.
35. Anthonisen NR, Connett JE, Kiley JP, *et al*.: Effects of smoking intervention and the use of an inhaled anticholinergic bronchodilator on the rate of decline of FEV1: the Lung Health Study. *JAMA* (1994) **272**: 1497–1505.
36. Karpel JP, Pesin J, Greenberg D, Gentry E: A comparison of the effects of ipratropium bromide and metaproterenol sulfate in acute exacerbations of COPD. *Chest* (1990) **98**: 835–839.
37. Patrick DM, Dales RE, Stark RM, Laliberte G, Dickinson G: Severe exacerbations of COPD and asthma, incremental benefit of adding irpratropium to usual therapy. *Chest* (1990) **98**: 295–297.
38. Rennard SI: Combination bronchodilator therapy in COPD. *Chest* (1995) **107** (Suppl 5): 171S–175S.
39. Petty TL: In chronic obstructive pulmonary disease, a combination of ipratropium an albuterol is more effective than either agent alone: an 85-day multicenter study. *Chest* (1994) **105**: 1411–1419.

40. Ikeda A, Nishimura K, Koyama H, Izumi T: Bronchodilating effects of combined therapy with clinical dosages of ipratropium bromide and salbutamol for stable COPD: comparison with ipratropium bromide alone. *Chest* (1995) **107**: 401–405.
41. Gross NJ: Ipratropium bromide. *N Engl J Med* (1988) **319**: 486–494.
42. Pavia D, Bateman JRM, Sheehan NF, Clarke SW: Effect of ipratropium bromide on mucociliary clearance and pulmonary function in reversible airways obstruction. *Thorax* (1979) **34**: 501–507.
43. Chapman KR, Smith DL, Rebuck AS, Leenen FHH: Hemodynamic effects of inhaled ipratropium bromide alone and in combination with an inhaled beta2-agonist. *Am Rev Respir Dis* (1983) **132**: 845–847.
44. Gross NJ, Bankwala Z: Effects of an anticholinergic bronchodilator on arterial blood gases of hypoxemic patients with COPD. *Am Rev Respir Dis* (1987) **136**: 1091–1094.
45. Leak A, O'Connor T: High dose ipratropium: is it safe? *Practitioner* (1988) **232**: 9–10.

38

Theophylline

PETER J. BARNES

INTRODUCTION

Theophylline remains the most widely prescribed antiasthma drug worldwide since it is inexpensive. In many industrialized countries, however, theophylline has become a third-line treatment only indicated in poorly controlled patients. This has been reinforced by various guidelines to therapy.[1,2] Some have even questioned whether theophylline is indicated in any patients with asthma,[3] although others have emphasized the special beneficial effects of theophylline that still give it an important place in asthma management.[4] Despite the fact that theophylline has been used in asthma therapy for over 60 years, there is still considerable uncertainty about its mode of action in asthma and its logical place in therapy. Because of problems with side-effects, there have been attempts to improve on theophylline and recently there has been increasing interest in selective phosphodiesterase (PDE) inhibitors, which may possibly improve the beneficial and reduce the adverse effects of theophylline.

HISTORICAL BACKGROUND

As long ago as 1786 William Withering recommended strong coffee as a remedy for asthma symptoms. During the last century Dr Henry Hyde Salter, himself a sufferer from asthma, also stated that strong coffee was the best treatment available for asthma. Methylxanthines such as theophylline, which are related to caffeine, have been widely used in the treatment of asthma since the 1930s. Theophylline was first identified in extracts from tea leaves by Kossel in Berlin in 1888 and was first synthesized in 1900 by the Boehringer Company. Macht and Ting from Baltimore first demonstrated the bronchodilator action of theophylline on pig airways in 1921 and in the following year Hirsch reported the beneficial effects of rectally administered theophylline in four asthmatic patients.

There was a resurgence of interest in theophylline in the 1980s when reliable slow-release preparations were developed and assays for measuring plasma theophylline concentrations became available. With the increasing use of inhaled corticosteroids, theophylline preparations have become less popular in many countries and in industrialized countries their use has declined considerably.

CHEMISTRY

Theophylline is a methylxanthine similar in structure to the common dietary xanthines caffeine and theobromine. Several substituted derivatives have been synthesized but none has any advantage over theophylline,[5] apart from the 3-propyl derivative, enprofylline, which is more potent as a bronchodilator and may have fewer toxic effects.[6] Many salts of theophylline have also been marketed, the most common being aminophylline, the ethylenediamine salt used to increase solubility at neutral pH so that intravenous administration is possible. Other salts, such as choline theophyllinate, do not have any advantage and others, such as acepifylline, are virtually inactive.[5]

MOLECULAR MECHANISMS OF ACTION

Although theophylline has been in clinical use for more than 60 years, both its mechanism of action at a molecular level and its site of action remain uncertain. Several molecular mechanisms of action have been proposed[7] (Table 38.1).

PDE inhibition

Theophylline is a weak and non-selective inhibitor of PDEs, which break down cyclic nucleotides in the cell, thereby leading to an increase in intracellular cyclic 3'5'-adenosine monophosphate (cyclic AMP) and cyclic 3',5'-guanosine monophosphate (cyclic GMP)

Table 38.1 Mechanisms of action of theophylline.

Phosphodiesterase inhibition
Adenosine receptor antagonism
Stimulation of catecholamine release
Mediator inhibition
Inhibition of intracellular calcium release

38 Theophylline

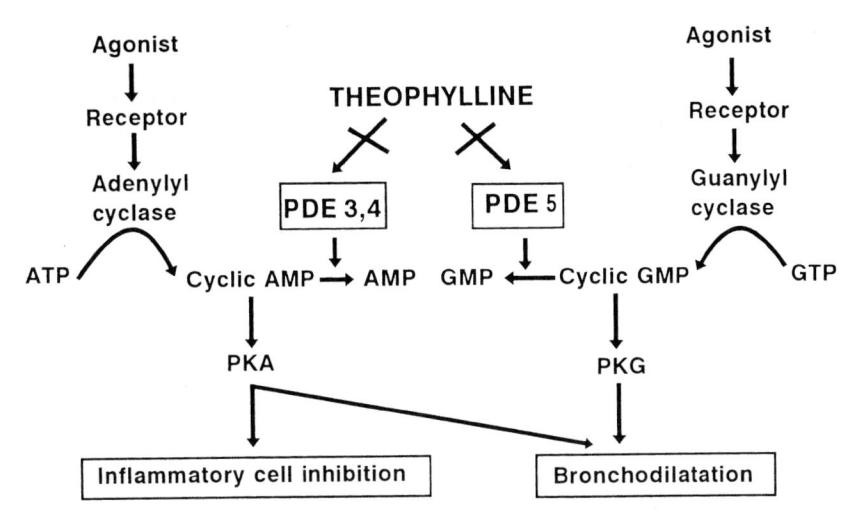

Fig. 38.1 Effect of phosphodiesterase (PDE) inhibitors on the breakdown of cyclic nucleotides in airway smooth muscle and inflammatory cells. PKA, protein kinase A; PKG, protein kinase.

concentrations (Fig. 38.1). However, the degree of inhibition is small at the concentrations of theophylline that are therapeutically relevant. Thus total PDE activity in human lung extracts is inhibited by only 5–10% by therapeutic concentrations of theophylline.[8] There is convincing evidence *in vitro* that theophylline relaxes airway smooth muscle by inhibition of PDE activity, but relatively high concentrations are needed for maximal relaxation.[9] Similarly, the inhibitory effect of theophylline on mediator release from alveolar macrophages appears to be mediated by inhibition of PDE activity in these cells.[10] There is no evidence that airway smooth muscle or inflammatory cells concentrate theophylline to achieve higher intracellular than circulating concentrations. Inhibition of PDE should lead to synergistic interaction with β-agonists, but this has not been convincingly demonstrated *in vivo*, although this might be explained because relaxation of airway smooth muscle by β-agonists may involve direct coupling of β-receptors via a stimulatory G protein to the opening of potassium channels, without the involvement of cyclic AMP.[11]

Several isoenzyme families of PDE have now been recognized and some (PDE3, PDE4, PDE5) are more important in smooth muscle relaxation.[12–14] However there is no convincing evidence that theophylline has a greater inhibitory effect on the PDE isoenzymes involved in smooth muscle relaxation. It is possible that PDE isoenzymes may have an increased expression in asthmatic airways, as a result either of the chronic inflammatory process or therapy. Elevation of cyclic AMP by β-agonists may result in increased PDE activity, thus limiting the effect of β-agonists. Indeed, alveolar macrophages from asthmatic patients appear to have increased PDE activity.[15] This would mean that theophylline might have a greater inhibitory effect on PDE in asthmatic airways compared with normal airways. Support for this is provided by the lack of bronchodilator effect of theophylline in normal subjects, compared to a bronchodilator effect in asthmatic patients.[16] Further studies on the effect of theophylline in asthmatic tissues are required.

Adenosine receptor antagonism

Theophylline is a potent inhibitor of adenosine receptors at therapeutic concentrations (both A_1 and A_2 receptors, although it is less effective against A_3 receptors), suggesting that this could be the basis for its bronchodilator effects.[17] Although adenosine has little effect on normal human airway smooth muscle *in vitro*, it constricts airways of asthmatic patients via the release of histamine and leukotrienes, suggesting that adenosine releases mediators from mast cells.[18] The receptor involved appears to be an A_3 receptor in rat mast cells,[19] although in humans there is evidence for the involvement of an A_{2b} receptor.[20] Adenosine causes bronchoconstriction in asthmatic subjects when given by inhalation.[21] The mechanism of bronchoconstriction is indirect and involves release of histamine from airway mast cells.[18,22] The bronchoconstrictor effect of adenosine is prevented by therapeutic concentrations of theophylline.[21] However, this only confirms that theophylline is capable of antagonizing the effects of adenosine at therapeutic concentrations and does not necessarily indicate that this is important for its antiasthma effect. Enprofylline, which is more potent than theophylline as a bronchodilator, has no significant inhibitory effect on adenosine receptors at therapeutic concentrations, suggesting that adenosine antagonism is an unlikely explanation for the bronchodilator effect of theophylline.[6]

However, adenosine antagonism is likely to account for some of the side-effects of theophylline, such as central nervous system stimulation, cardiac arrhythmias, gastric hypersecretion, gastro-oesophageal reflux and diuresis.

Endogenous catecholamine release

Theophylline increases the secretion of adrenaline from the adrenal medulla,[23,24] although the increase in plasma concentration is small and insufficient to account for any significant bronchodilator effect.[25]

Mediator inhibition

Theophylline antagonizes the effect of some prostaglandins on vascular smooth muscle *in vitro*,[26] but there is no evidence that these effects are seen at therapeutic concentrations or are relevant to its airway effects. Theophylline inhibits the secretion of tumour necrosis factor (TNF-α by peripheral blood monocytes[27,28] and increases the secretion of the anti-inflammatory cytokine interleukin (IL)-10.[28] Theophylline may also interfere with the action of TNF-α which may be involved in asthmatic inflammation. A related compound, pentoxifylline, prevents TNF-α-induced lung injury and enhanced hypoxic pulmonary vasoconstriction,[29,30] but its mechanism of action is not yet understood.

Calcium ion flux

There is some evidence that theophylline may interfere with calcium mobilization in airway smooth muscle. Theophylline has no effect on entry of calcium ions (Ca^{2+}) via

38 Theophylline

voltage-dependent channels, but it has been suggested that it may influence calcium entry via receptor-operated channels, release from intracellular stores or have some effect on phosphatidylinositol turnover (which is linked to release of Ca^{2+} from intracellular stores). There is no direct evidence in favour of this, other than an effect on intracellular cyclic AMP concentration due to its PDE inhibitory action. An early study suggesting that theophylline may increase Ca^{2+} uptake into intracellular stores[31] has not been followed up.

EFFECTS

Theophylline has actions on many cell types within and outside the airways (Fig. 38.2).

Airway smooth muscle

The primary effect of theophylline is assumed to be relaxation of airway smooth muscle and studies *in vitro* have shown that it is equally effective in large and small airways.[32,33] In airways obtained at lung surgery approximately 25% of preparations fail to relax with a β-agonist, but all relax with theophylline.[33] The molecular mechanism of bronchodilatation is almost certainly related to PDE inhibition, resulting in an increase in cyclic AMP.[9] The bronchodilator effect of theophylline is reduced in guinea-pig and human airways by charybdotoxin, which inhibits large-conductance Ca^{2+}-activated K^+ channels (maxi-K

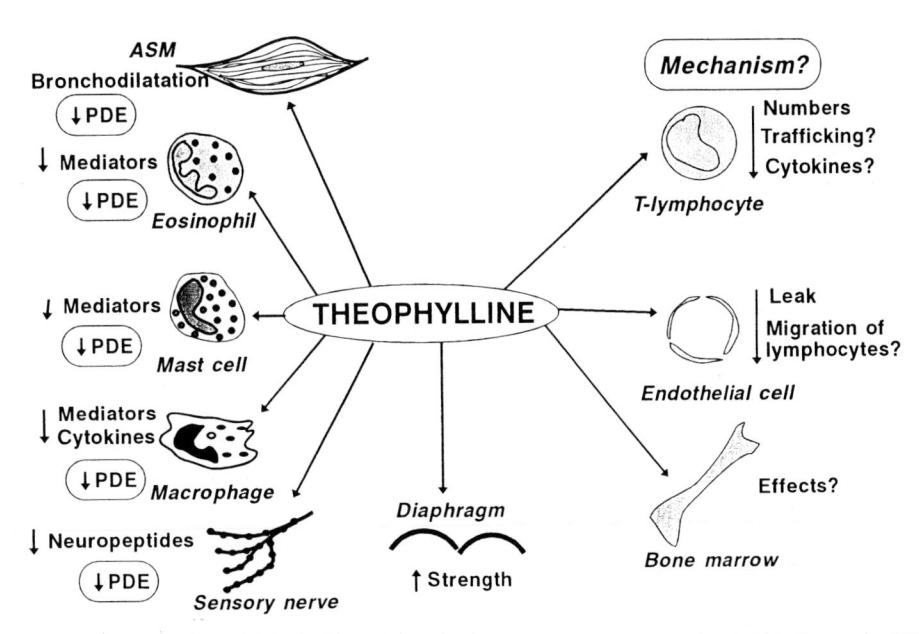

Fig. 38.2 Cellular effects of theophylline. ASM, airway smooth muscle; PDE, phosphodiesterase; PKG, protei kinase.

channels), suggesting that theophylline opens maxi-K channels via an increase in cyclic AMP.[34,35] Theophylline acts as a functional antagonist and inhibits the contractile response of several spasmogens. In airways obtained at post-mortem from patients who have died from asthma the relaxant response to β-agonists is reduced, whereas the bronchodilator response to theophylline is no different from that seen in normal airways.[36] There is now evidence that β-adrenoceptors in airway smooth muscle of patients with fatal asthma become uncoupled,[37] and theophylline may therefore have a theoretical advantage over β-agonists in severe asthma exacerbations. However, theophylline is a very weak bronchodilator at therapeutically relevant concentrations, suggesting that some other target cell may be more relevant for its antiasthma effect. In human airways the EC_{50} for theophylline is approximately 1.5×10^{-4} M, which is equivalent to 67 mg/litre assuming 60% protein binding.[33] However, as discussed above, it is important to consider the possibility that PDE activity may be increased in asthmatic airways so that theophylline may have a greater than expected effect.

In vivo, intravenous aminophylline has an acute bronchodilator effect in asthmatic patients, which is most likely due to a relaxant effect on airway smooth muscle.[38] The bronchodilator effect of theophylline in chronic asthma is small in comparison with β-agonists, however. Several studies have demonstrated a small protective effect of theophylline on histamine, methacholine or exercise challenge.[39-42] This protective effect does not correlate well with any bronchodilator effect and it is interesting that in some studies the protective effect of theophylline is observed at plasma concentrations of <10 mg/litre.[41,42] These clinical studies suggest that theophylline may have antiasthma effects unrelated to any bronchodilator action (which may only occur at very high plasma concentrations and may only be relevant in the management of acute severe asthma).

Anti-inflammatory effects

Whether theophylline has significant anti-inflammatory effects in asthma is still unresolved.[43] Theophylline inhibits histamine release from human basophils *in vitro*[44] and inhibits mediator release from chopped human lung,[45] although high concentrations are necessary and it is likely that this effect involves an increase in cyclic AMP concentration due to PDE inhibition. Theophylline also has an inhibitory effect on superoxide anion release from human neutrophils[46] and inhibits the feedback stimulatory effect of adenosine on neutrophils *in vivo*.[47] At therapeutic concentrations *in vitro* theophylline may *increase* superoxide release via an inhibitory effect on adenosine receptors, since endogenous adenosine may normally exert an inhibitory action on these cells.[48] Similar results are also seen in guinea-pig and human eosinophils.[49] At therapeutic concentrations there is an increased release of superoxide anions from eosinophils, which appears to be mediated via inhibition of adenosine A_2 receptors and is mimicked by the adenosine antagonist 8-phenyltheophylline. Inhibition of eosinophil superoxide generation occurs only at high concentrations of theophylline ($>10^{-4}$ M), which are likely to inhibit PDE. Similar results have also been obtained in human alveolar macrophages.[50] Macrophages lavaged from patients taking theophylline have been found to have a reduced oxidative burst response.[51]

In vivo, theophylline inhibits mediator-induced airway microvascular leakage in rodents when given in high doses,[52] although this is not seen at therapeutically relevant concentrations.[53] Theophylline has an inhibitory effect on plasma exudation in nasal secretions induced by allergen in patients with allergic rhinitis, although this could be secondary to inhibition of mediator release.[54]

In allergen challenge studies in asthmatic patients both intravenous theophylline and enprofylline inhibit the late response to allergen, while having relatively little effect on the early response.[55] A similar finding with allergen challenge has been reported after chronic oral treatment with theophylline.[56] This has been interpreted as an effect on the chronic inflammatory response and is supported by reduced infiltration of eosinophils into the airways after allergen challenge following low doses of theophylline.[57] In patients with nocturnal asthma, low-dose theophylline inhibits the influx of neutrophils and, to a lesser extent, eosinophils seen in the early morning.[58] Oral theophylline also inhibits the late response to toluene diisocyanate (TDI) in TDI-sensitive asthmatic patients,[59] but has no effect on the subsequent increase in methacholine responsiveness. Similarly, theophylline has no effect on the increased airway responsiveness that follows allergen challenge[60] and does not reduce airway responsiveness in asthmatic patients after chronic administration.[61] These studies indicate that theophylline may have effects on acute inflammation in the airways, but may be less effective on the chronic inflammatory process.

Immunomodulatory effects

T-lymphocytes are now believed to play a central role in coordinating the chronic inflammatory response in asthma. For many years theophylline has been shown to have several actions on T-lymphocyte function, suggesting that it might have an immunomodulatory effect in asthma. Theophylline has a stimulatory effect on suppressor ($CD8^+$) T-lymphocytes that may be relevant to the control of chronic airway inflammation[62,63] and an inhibitory effect on graft rejection.[64] *In vitro*, theophylline inhibits IL-2 synthesis in human T-lymphocytes; this effect is secondary to a rise in intracellular cyclic AMP concentration.[65,66] In allergen-induced airway inflammation in guinea-pigs, theophylline has a significant inhibitory effect on eosinophil infiltration,[67] suggesting that it may inhibit the T-cell-derived cytokines responsible for this eosinophilic response. In asthmatic patients, low-dose theophylline treatment results in an increase in activated circulating $CD4^+$ and $CD8^+$ T-cells but a decrease in these cells in the airways, suggesting that it may reduce the trafficking of activated T-cells into the airways.[68] This is supported by studies in allergen challenge, where low-dose theophylline decreases the number of activated $CD4^+$ and $CD8^+$ T-cells in bronchoalveolar lavage fluid after allergen challenge and this is mirrored by an increase in these cells in peripheral blood.[69] This is seen even in patients treated with high does of inhaled steroids, indicating that the molecular effects of theophylline are likely to be different from those of corticosteroids.

Extrapulmonary effects

For a long time it has been suggested that theophylline may exert its effects in asthma via some action outside the airways. Hyde Salter believed that one of the major effects of caffeine was its action as a cerebral stimulant. It may be relevant that theophylline is ineffective when given by inhalation until therapeutic plasma concentrations are achieved.[70] This may indicate that theophylline has effects on cells other than those in the airway. One possible target cell is the platelet and theophylline has been demonstrated to inhibit platelet activation.

An effect of theophylline that remains controversial is its action on respiratory muscles. Aminophylline was found to increase diaphragmatic contractility and to reverse diaphragm fatigue.[71] This effect has not been observed by all investigators and there are now doubts about the relevance of these observations to the clinical benefit provided by theophylline in chronic obstructive pulmonary disease (COPD).[72]

PHARMACOKINETICS

There is a close relationship between the acute improvement in airway function and serum theophylline concentration. Below 10 mg/litre therapeutic effects (at least in terms of rapid improvement in airway function) are small and above 25 mg/litre additional benefits are outweighed by side-effects, so that the therapeutic range is usually taken as 10–20 mg/litre (55–110 μM).[5] It is now apparent that non-bronchodilator effects of theophylline may be seen at plasma concentrations of <10 mg/litre and that clinical benefit may be derived from these lower concentrations. This suggests that it may be necessary to redefine the therapeutic range of theophylline based on the antiasthma effect rather than the acute bronchodilator response, which requires a higher plasma concentration. The dose of theophylline required to give therapeutic concentrations varies between subjects, largely because of differences in clearance. In addition, there may be differences in bronchodilator response to theophylline and, with acute bronchoconstriction, higher concentrations may be required to produce bronchodilatation.[73]

Theophylline is rapidly and completely absorbed, but there are large inter-individual variations in clearance due to differences in hepatic metabolism (Table 38.2). Theophylline is metabolized in the liver by the cytochrome P450/P448 microsomal enzyme system, and a large number of factors may influence hepatic metabolism. Theophylline is

Table 38.2 Factors affecting clearance of theophylline.

Increased clearance
Enzyme induction (rifampicin, phenobarbitone, ethanol)
Smoking (tobacco, marijuana)
High-protein low-carbohydrate diet
Barbecued meat
Childhood

Decreased clearance
Enzyme inhibition (cimetidine, erythromycin, ciprofloxacin, allopurinol, zileuton)
Congestive heart failure
Liver disease
Pneumonia
Viral infection and vaccination
High-carbohydrate diet
Old age

predominantly metabolized by the CYP1A2 enzyme, while at higher plasma concentrations CYP2E1 is also involved.[74]

Increased clearance

Increased clearance is seen in children (1–16 years) and in cigarette and marijuana smokers. Concurrent administration of phenytoin and phenobarbitone increases activity of P450, resulting in increased metabolic breakdown, so that higher doses may be required.

Reduced clearance

Reduced clearance is found in liver disease, pneumonia and heart failure and doses need to be halved and plasma levels monitored carefully.[75] Increased clearance is also seen with with certain drugs, including erythromycin, certain quinolone antibiotics (ciprofloxacin, but not ofloxacin), allopurinol, cimetidine (but not ranitidine), serotonin uptake inhibitors (fluvoxamine) and the 5-lipoxygenase inhibitor zileuton, which interfere with cytochrome P450 function. Thus, if a patient on maintenance theophylline requires a course of erythromycin, the dose of theophylline should be halved. Viral infections and vaccination may also reduce clearance and this may be particularly important in children. Because of these variations in clearance, individualization of theophylline dosage is required and plasma concentrations should be measured 4 h after the last dose with slow-release preparations when steady state has usually been achieved. There is no significant circadian variation in theophylline metabolism although there may be delayed absorption at night, which may relate to the supine posture.[76]

ROUTES OF ADMINISTRATION

Intravenous

Intravenous aminophylline has been used for many years in the treatment of acute severe asthma. The recommended dose is now 6 mg/kg given intravenously over 20–30 min, followed by a maintenance dose of 0.5 mg/kg per h. If the patient is already taking theophylline, or there are any factors which decrease clearance, these doses should be halved and the plasma level checked more frequently.

Oral

Plain theophylline tablets or elixir, which are rapidly absorbed, give wide fluctuations in plasma levels and are not recommended. Several effective sustained-release preparations are now available that are absorbed at a constant rate and provide steady plasma concentrations over 12–24 h.[77] Although there are differences between preparations, these are relatively minor and of no clinical significance. Both slow-release aminophylline and theophylline are available and are equally effective (although the ethylenediamine component of aminophylline has very occasionally been implicated in allergic reactions). For continuous treatment twice-daily therapy (approximately 8 mg/kg twice daily) is needed, although some preparations are designed for once-daily administration. For nocturnal asthma a single dose of slow-release theophylline at night is often effective[78,79] and is often more effective than an oral slow-release β-agonist preparation.

Once optimal doses have been determined plasma concentrations usually remain stable, providing no factors that alter clearance change.

Other theophylline salts, such as choline theophyllinate, have no advantages and some derivatives, such as acepiphylline, diprophylline and proxyphylline, are less effective.[5] Compound tablets that contain adrenergic agonists and sedatives in addition to theophylline should be avoided.

Other routes

Aminophylline may be given as a suppository, although rectal absorption is unreliable and proctitis may occur, so is best avoided. Inhalation of theophylline is irritant and ineffective. Intramuscular injections of theophylline are very painful and should never be given.

CLINICAL USE

Acute severe asthma

Intravenous aminophylline has been used in the management of acute severe asthma for over 50 years, but this use has been questioned in view of the risk of adverse effects compared with nebulized β_2-agonists. In patients with acute asthma, intravenous aminophylline is less effective than nebulized β_2-agonists[80] and should therefore be reserved for those patients who fail to respond to β-agonists. There is some evidence that the use of aminophylline in the emergency room reduces subsequent admissions to hospital with acute asthma.[81] In a meta-analysis of 13 acceptably designed clinical trials to compare nebulized β-agonists with or without intravenous aminophylline, there was no overall additional benefit from adding aminophylline.[82] This indicates that aminophylline should not be added routinely to nebulized β-agonists. Indeed addition of aminophylline may only increase side-effects.[83,84] Several deaths have been reported after intravenous aminophylline. In one study of 43 asthma deaths in southern England, there was a significantly greater frequency of toxic theophylline concentrations (21%) compared with matched controls (7%).[85] These concerns have led to the view that intravenous aminophylline should be reserved for the few patients with acute severe asthma who fail to show a satisfactory response to nebulized β-agonists. When intravenous aminophylline is used it should be given as a slow intravenous infusion with careful monitoring; plasma theophylline concentration should be measured prior to infusion.

Chronic asthma

Theophylline has little or no effect on bronchomotor tone in normal airways but reverses bronchoconstriction in asthmatic patients, although it is less effective than inhaled β-agonists and is more likely to have unwanted effects. Indeed the role of theophylline in the routine management of chronic asthma has been questioned[3] and in the various guidelines for asthma treatment theophylline is used as an additional bronchodilator if asthma remains difficult to control after high-dose inhaled steroids.[1,2,86] The introduction of long-acting inhaled β_2-agonists, such as salmeterol and formoterol, has further threatened the position of theophylline, since the side-effects of these agents may be less frequent than those associated with theophylline.

Whether theophylline has some additional benefit over and above its bronchodilator action is now an important consideration. In chronic studies, oral theophylline appears to be as effective as sodium cromoglycate in controlling young allergic asthmatic patients[87] and provides additional control of asthma symptoms even in patients talking regular inhaled steroids.[88] In one study of a group of difficult adolescent asthmatic patients who were controlled with oral and inhaled steroids, nebulized β_2-agonists, inhaled anticholinergics and cromones, in addition to regular oral theophylline, withdrawal of the oral theophylline resulted in a marked deterioration of asthma control that could not be controlled by further increase in steroids and only responded to reintroduction of theophylline.[89] This suggests that there may be a group of severe asthmatic patients who particularly benefit from theophylline. It is important to investigate these patients in more detail and to determine why theophylline, but not apparently corticosteroids, is able to benefit such patients. In a controlled trial of theophylline withdrawal in patients with severe asthma controlled only on high doses of inhaled corticosteroids, there was a significant deterioration in symptoms and lung function when placebo was substituted for the relatively low maintenance dose of theophylline.[68] There is also evidence that addition of theophylline improves asthma control to a greater extent than β_2-agonists in patients with severe asthma treated with high-dose inhaled steroids.[90] This suggests that theophylline may have a useful place in the optimal management of moderate to severe asthma and appears to provide additional control above that provided by high-dose inhaled steroids.[91]

Theophylline may be a useful treatment for nocturnal asthma and a single dose of a slow-release theophylline preparation given at night may provide effective control of nocturnal asthma symptoms.[78,92] There is evidence that slow-release theophylline preparations are more effective than slow-release oral β-agonists and inhaled β-agonists in controlling nocturnal asthma.[79,93,94] They are approximately equal in efficacy to salmeterol in controlling nocturnal asthma, but the quality of sleep is somewhat better with salmeterol compared with theophylline.[95] The mechanism of action of theophylline in nocturnal asthma may involve more than long-lasting bronchodilatation and could involve inhibition of some components of the inflammatory response, which may increase at night.[58]

Recently, the effect of addition of theophylline to inhaled steroids has been studied in patients with milder asthma. In patients with asthma not controlled on a dose of inhaled steroids (budesonide) of 800 μg daily, addition of low-dose theophylline gave better control of asthma and greater improvement in lung function than doubling the dose of inhaled steroid to 1600 μg daily.[96] Interestingly, there was a greater degree of improvement when forced vital capacity was measured than when forced expiratory volume in 1 s was measured, possibly indicating an effect on peripheral airways. Since the improvement in lung function was relatively slow, this suggests that the effect of the added theophylline may be anti-inflammatory rather than bronchodilator, particularly as the plasma concentration of theophylline in this study was rather low (mean 8.6 mg/litre) for any bronchodilator effect. This study suggests that low-dose theophylline may be preferable to increasing the dose of inhaled steroids when asthma is not controlled on moderate doses of inhaled steroids; such a therapeutic approach would be much less expensive.

The therapeutic range of theophylline is based on measurement of acute bronchodilatation in response to the acute administration of theophylline.[38] However, it is possible that the non-bronchodilator effects of theophylline, whether related to protection against bronchoconstriction or some anti-inflammatory or immunomodulatory effect, may be exerted at lower plasma concentrations, as discussed above.[97]

COPD

Theophylline may also benefit patients with COPD, increasing exercise tolerance, although without any improvement in spirometric values unless combined with an inhaled β-agonist.[98,99] However, theophylline may reduce trapped gas volume, suggesting an effect on peripheral airways, and this may explain why some patients with COPD may obtain considerable symptomatic improvement without any increase in spirometric values.[100] Although the effect of theophylline on respiratory muscle weakness was believed to be important in contributing to symptomatic improvement in patients with COPD,[101] this seems unlikely as several investigators have failed to confirm any effect on respiratory muscle function at therapeutic concentrations of theophylline.[72]

Interaction with β-agonists

If theophylline exerts its effects by PDE inhibition, then a synergistic interaction with β-agonists would be expected. Many studies have investigated this possibility, but while there is good evidence that theophylline and β-agonists have additive effects, true synergy is not seen.[102,103] This can now be understood in terms of the molecular mechanisms of action of β-agonists and theophylline. β-Agonists may cause relaxation of airway smooth muscle via several mechanisms. Classically, they increase intracellular cyclic AMP concentrations, which were believed to be an essential event in the relaxation response. It has recently become clear that β-agonists may cause bronchodilatation, at least in part, by maxi-K channels in airway smooth muscle cells that are directly linked to relaxation.[11,34,35] Maxi-K channels are opened by low concentrations of β-agonists, which are likely to be therapeutically relevant. There is now evidence that β-receptors may be coupled directly to maxi-K channels via the α-subunit of G_s[104] and therefore may induce relaxation without any increase in cyclic AMP, thus accounting for a lack of synergy. Another reason for the lack of synergy may be that cells other than airway smooth muscle may be the main target for the anti-asthma effect of theophylline.

38 Theophylline

Repeated administration of β_2-agonists may result in tolerance. While this may be explained by downregulation of β_2-receptors, an additional mechanism may involve upregulation of PDE enzymes (especially PDE4D), which then break down cyclic AMP more readily.[105] Theophylline may therefore theoretically mitigate against the development of tolerance, although this has not yet been studied clinically. Theophylline may provide useful additional bronchodilatation in patients with COPD, even when maximally effective doses of a β-agonist have been given. This means that, if adequate bronchodilatation is not achieved by a β-agonist alone, theophylline may be added to the maintenance therapy with benefit.

SIDE-EFFECTS

There is no doubt that theophylline provides clinical benefit in obstructive airway disease, but the main limitation to its use is the frequency of adverse effects.[106] Unwanted effects of theophylline are usually related to plasma concentration and tend to occur when plasma levels exceed 20 mg/litre. However, some patients develop side-effects even at low plasma concentrations. To some extent side-effects may be reduced by gradually increasing the dose until therapeutic concentrations are achieved.

The commonest side-effects are headache, nausea and vomiting, abdominal discomfort and restlessness. There may also be increased acid secretion, gastro-oesophageal reflux and diuresis. There has recently been concern that theophylline, even at therapeutic concentrations, may lead to behavioural disturbance and learning difficulties in schoolchildren,[107] although it is difficult to design adequate controls for such studies. At high concentrations, convulsions and cardiac arrhythmias may occur and there is concern that intravenous aminophylline administered in the emergency room may be a contributory factor to the deaths of some patients with severe asthma[85].

Some of the side-effects of theophylline (central stimulation, gastric secretion, diuresis and arrhythmias) may be due to adenosine receptor antagonism and may therefore be avoided by the use of drugs such as enprofylline, which has no significant adenosine antagonism at bronchodilator doses.[6] The commonest side-effects of theophylline are nausea and headaches, which are also seen with enprofylline. These side-effects may be due to inhibition of certain PDEs (e.g. PDE4 in the vomiting centre).[108]

FUTURE OF THEOPHYLLINE

Although theophylline has recently been used much less in developed countries, there are reasons for thinking that it may come back into fashion for the treatment of chronic asthma, with the recognition that it may have anti-inflammatory and immunomodulatory effects when given in low doses (plasma concentration 5–10 mg/litre).[97] At these low doses the drug is easier to use, side-effects are uncommon and the problems of drug interaction are less of a problem, thus making the clinical use of theophylline less complicated. Theophylline appears to have an effect different from that of corticosteroids and may therefore be a useful drug to combine with low-dose inhaled steroids. The molecular mechanism of the anti-inflammatory effect of theophylline is still poorly understood and is unlikely to be explained by PDE inhibition. The recent demonstration that addition of low-dose theophylline to patients not controlled on a moderate dose of inhaled steroids was better than a high dose of inhaled steroids lends some credence to this argument.[96] There is no doubt that low-dose inhaled steroids are more effective than theophylline in the initial treatment of mild asthma, but there are economic as well as clinical reasons for preferring low-dose theophylline added to low-dose inhaled steroids in preference to high-dose inhaled steroids, which have possible risks of long-term systemic side-effects in some patients. Low-dose theophylline is also less expensive than inhaled long-acting β_2-agonists and leukotriene antagonists, which might be alternative treatments for long-term control of asthma. In addition, compliance with oral therapy is likely to be greater than with inhaled therapies.[109] This suggests that low-dose theophylline may find an important place in modern asthma management in patients with moderate asthma as well as in patients with severe asthma.

REFERENCES

1. British Thoracic Society: The British guidelines on asthma management. *Thorax* (1997) **52** (Suppl 1): S1–S21.
2. Global Initiative for Asthma: *Global strategy for asthma management and prevention.* NHLBI/WHO Workshop Report. Publication 95-3659, 1995.
3. Lam A, Newhouse MT: Management of asthma and chronic airflow limitation. Are methylxanthines obsolete? *Chest* (1990) **98**: 44–52.
4. Weinberger M, Hendeles L: Theophylline in asthma. *N Engl J Med* (1996) **334**: 1380–1388.
5. Weinburger M: The pharmacology and therapeutic use of theophylline. *J Allergy Clin Immunol* (1984) **73**: 525–540.
6. Persson CGA: Development of safer xanthine drugs for the treatment of obstructive airways disease. *J Allergy Clin Immunol* (1986) **78**: 817–824.
7. Persson CGA: Overview of effects of theophylline. *J Allergy Clin Immunol* (1986) **78**: 780–787.
8. Poolson JB, Kazanowski JJ, Goldman AL, Szentivanyi A: Inhibition of human pulmonary phosphodiesterase activity by therapeutic levels of theophylline. *Clin Exp Pharmacol Physiol* (1978) **5**: 535–539.
9. Rabe KF, Magnussen H, Dent G: Theophylline and selective PDE inhibitors as bronchodilators and smooth muscle relaxants. *Eur Respir J* (1995) **8**: 637–642.
10. Dent G, Giembycz MA, Wolf B, Rabe KF, Barnes PJ, Magnussen H: Suppression of opsonized zymosan-stimulated human alveolar respiratory burst by theophylline. *Am J Respir Cell Mol Biol* (1994) **10**: 565–572.
11. Kume H, Hall IP, Washabau RJ, Takagi K, Kotlikoff MI: Adrenergic agonists regulate K_{Ca} channels in airway smooth muscle by cAMP-dependent and -independent mechanisms. *J Clin Invest* (1994) **93**: 371–379.
12. Beavo JA: Cyclic nucleotide phosphodiesterases: functional implications of multiple isoforms. *Physiol Rev* (1995) **75**: 725–748.
13. de Boer J, Philpott KJ, van Amsterdam RGM, Shahid M, Zaagsma J, Nicholson CD: Human bronchial cyclic nucleotide phosphodiesterase isoenzymes: biochemical and pharmacological analysis using selective inhibitors. *Br J Pharmacol* (1992) **106**: 1028–1034.
14. Rabe KF, Tenor H, Dent G, Webig S, Magnussen H: Phosphodiesterase isoenzymes modulating inherent tone in human airways: identification and characterization. *Am J Physiol* (1993) **264**: L458–L464.
15. Bachelet M, Vincent D, Havet N, *et al.*: Reduced responsiveness of adenylate cyclase in alveolar macrophages from patients with asthma. *J Allergy Clin Immunol* (1991) **88**: 322–328.

16. Estenne M, Yernault J, De Troyer A: Effects of parenteral aminophylline on lung mechanics in normal humans. *Am Rev Respir Dis* (1980) **121**: 967–971.
17. Pauwels RA, Joos GF: Characterization of the adenosine receptors in the airways. *Arch Int Pharmacodyn Ther* (1995) **329**: 151–156.
18. Björk T, Gustafsson LE, Dahlén S: Isolated bronchi from asthmatics are hyperresponsive to adenosine, which apparently acts indirectly by liberation of leukotrienes and histamine. *Am Rev Respir Dis* (1992) **145**: 1087–1091.
19. Fozard JR, Pfannkuche HJ, Schuurman HJ: Mast cell degranulation following adenosine A3 receptor activation in rats. *Eur J Pharmacol* (1996) **298**: 293–297.
20. Feoktiskov I, Bioggioni I: Adenosine 2b receptors evoke interleukin-8 secretion in human mast cells. An enprofylline-sensitive mechanism with implications for asthma. *J Clin Invest* (1995) **96**: 1979–1986.
21. Cushley MJ, Tattersfield AE, Holgate ST: Adenosine-induced bronchoconstriction in asthma: antagonism by inhaled theophylline. *Am Rev Respir Dis* (1984) **129**: 380–384.
22. Cushley MJ, Holgate ST: Adenosine induced bronchoconstriction in asthma: role of mast cell mediator release. *J Allergy Clin Immunol* (1985) **75**: 272–278.
23. Higbee MD, Kumar M, Galant SP: Stimulation of endogenous catecholamine release by theophylline: a proposed additional mechanism of theophylline effects. *J Allergy Clin Immunol* (1982) **70**: 377–382.
24. Ishizaki T, Minegishi A, Morishita A, *et al.*: Plasma catecholamine concentrations during a 72 hour aminophylline infusion in children with acute asthma. *J Allergy Clin Immunol* (1988) **92**: 146–154.
25. Barnes PJ: Endogenous catecholamines and asthma. *J Allergy Clin Immunol* (1986) **77**: 791–795.
26. Horrobin DF, Manku MS, Franks DJ, Hamet P: Methylxanthine phosphodiesterase inhibitors behave as prostaglandin antagonists in a perfused rat mesenteric artery preparation. *Prostaglandins* (1977) **13**: 33–40.
27. Spatafora M, Chiappara G, Merendino AM, D'Amico D, Bellia V, Bonsignore G: Theophylline suppresses the release of tumour necrosis factor-alpha by blood monocytes and alveolar macrophages. *Eur Respir J* (1994) **7**: 223–228.
28. Mascali JJ, Cvietusa P, Negri J, Borish L: Anti-inflammatory effects of theophylline: modulation of cytokine production. *Ann Allergy Asthma Immunol* (1996) **77**: 34–38.
29. Lilly CM, Sandhu JS, Ishizaka A, *et al.*: Pentoxifylline prevents tumor necrosis factor-induced lung injury. *Am Rev Respir Dis* (1989) **139**: 1361–1368.
30. Liu S, Dewar A, Crawley DE, Barnes PJ, Evans TE: Effect of tumor necrosis factor on hypoxic pulmonary vasoconstriction. *J Appl Physiol* (1992) **72**: 1044–1049.
31. Kolbeck RC, Speir WA, Carrier GO, Bransome ED: Apparent irrelevance of cyclic nucleotides to the relaxation of tracheal smooth muscle induced by theophylline. *Lung* (1979) **156**: 173–183.
32. Finney MJB, Karlson JA, Persson CGA: Effects of bronchoconstriction and bronchodilation on a novel human small airway preparation. *Br J Pharmacol* (1985) **85**: 29–36.
33. Guillot C, Fornaris M, Badger M, Orehek J: Spontaneous and provoked resistance to isoproterenol in isolated human bronchi. *J Allergy Clin Immunol* (1984) **74**: 713–718.
34. Jones TR, Charette L, Garcia ML, Kaczorowski GJ: Selective inhibition of relaxation of guinea-pig trachea by charybodotoxin, a potent Ca^{++}-activated K^+ channel inhibitor. *J Pharmacol Exp Ther* (1990) **225**: 697–706.
35. Miura M, Belvisi MG, Stretton CD, Yacoub MH, Barnes PJ: Role of potassium channels in bronchodilator responses in human airways. *Am Rev Respir Dis* (1992) **146**: 132–136.
36. Goldie RG, Spina D, Henry PJ, Lulich KM, Paterson JW: *In vitro* responsiveness of human asthmatic bronchus to carbachol, histamine, β-adrenoceptor agonists and theophylline. *Br J Clin Pharmacol* (1986) **22**: 669–676.
37. Bai TR, Mak JCW, Barnes PJ: A comparison of beta-adrenergic receptors and *in vitro* relaxant responses to isoproterenol in asthmatic airway smooth muscle. *Am J Respir Cell Mol Biol* (1992) **6**: 647–651.
38. Mitenko PA, Ogilvie RI: Rational intravenous doses of theophylline. *N Engl J Med* (1973) **289**: 600–603.

39. McWilliams BC, Menendez R, Kelly WH, Howick J: Effects of theophylline on inhaled methacholine and histamine in asthmatic children. *Am Rev Respir Dis* (1984) **130**: 193–197.
40. Cartier A, Lemire I, L'Archeveque J: Theophylline partially inhibits bronchoconstriction caused by inhaled histamine in subjects with asthma. *J Allergy Clin Immunol* (1986) **77**: 570–575.
41. Magnusson H, Reuss G, Jorres R: Theophylline has a dose-related effect on the airway response to inhaled histamine and methacholine in asthmatics. *Am Rev Respir Dis* (1987) **136**: 1163–1167.
42. Magnussen H, Reuss G, Jörres R: Methylxanthines inhibit exercise-induced bronchoconstriction at low serum theophylline concentrations and in a dose-dependent fashion. *J Allergy Clin Immunol* (1988) **81**: 531–537.
43. Persson CGA: Xanthines as airway anti inflammatory drugs. *J Allergy Clin Immunol* (1988) **81**: 615–617.
44. Lichtenstein LM, Margolis S: Histamine release *in vitro*: inhibition by catecholamines and methylxanthines. *Science* (1968) **161**: 902–903.
45. Orange RP, Kaliner MA, Laraia PJ, Austen KF: Immunological release of histamine and slow reacting substance of anaphylaxis from human lung. II. Influence of cellular levels of cyclic AMP. *Fed Proc* (1971) **30**: 1725–1729.
46. Nielson CP, Crawley JJ, Morgan ME, Vestal RE: Polymorphonuclear leukocyte inhibition by therapeutic concentrations of theophylline is mediated by cyclic 3',5' adenosine aminophosphate. *Am Rev Respir Dis* (1988) **137**: 25–30.
47. Kraft M, Pak J, Borish L, Martin RJ: Theophylline's effect on neutrophil function and the late asthmatic response. *J Allergy Clin Immunol* (1996) **98**: 251–257.
48. Schrier DJ, Imre KM: The effects of adenosine agonists on human neutrophil function. *J Immunol* (1986) **137**: 3284–3289.
49. Yukawa T, Kroegel C, Dent G, Chanez P, Ukena D, Barnes PJ: Effect of theophylline and adenosine on eosinophil function. *Am Rev Respir Dis* (1989) **140**: 327–333.
50. Dent G, Giembycz MA, Rabe KF, Barnes PJ, Magnussen H: Inhibition of human alveolar macrophage respiratory burst by theophylline: association with phosphodiesterase inhibition. *Br J Pharmacol* (1992) **105**: 86.
51. O'Neill SJ, Sitar DS, Kilass DJ: The pulmonary disposition of theophylline and its influences on human alveolar macrophage bactericidal function. *Am Rev Respir Dis* (1988) **134**: 1225–1228.
52. Erjefalt I, Persson CGA: Pharmacologic control of plasma exudation into tracheobronchial airways. *Am Rev Respir Dis* (1991) **143**: 1008–1014.
53. Boschetto P, Roberts NM, Rogers DF, Barnes PJ: The effect of antiasthma drugs on microvascular leak in guinea pig airways. *Am Rev Respir Dis* (1989) **139**: 416–421.
54. Naclerio RM, Bartenfelder D, Proud D, *et al.*: Theophylline reduces histamine release during pollen-induced rhinitis. *J Allergy Clin Immunol* (1986) **78**: 874–876.
55. Pauwels R, van Revterghem D, van der Straeten M, Johanesson N, Persson CGA: The effect of theophylline and enprophylline on allergen-induced bronchoconstriction. *J Allergy Clin Immunol* (1985) **76**: 583–590.
56. Ward AJM, McKenniff M, Evans JM, Page CP, Costello JF: Theophylline: an immunomodulatory role in asthma? *Am Rev Respir Dis* (1993) **147**: 518–523.
57. Sullivan P, Bekir S, Jaffar Z, Page C, Jeffery P, Costello J: Anti-inflammatory effects of low-dose oral theophylline in atopic asthma. *Lancet* (1994) **343**: 1006–1008.
58. Kraft M, Torvik JA, Trudeau JB, Wenzel SE, Martin RJ: Theophylline: potential antiinflammatory effects in nocturnal asthma. *J Allergy Clin Immunol* (1996) **97**: 1242–1246.
59. Mapp C, Boschetto P, Dal Vecchio L, *et al.*: Protective effect of antiasthma drugs on late asthmatic reactions and increased airway responsiveness induced by toluene diisocyanate in sensitized subjects. *Am Rev Respir Dis* (1987) **136**: 1403–1407.
60. Cockroft DW, Murdock KY, Gore BP, O'Byrne PM, Manning P: Theophylline does not inhibit allergen-induced increase in airway responsiveness to methacholine. *J Allergy Clin Immunol* (1991) **83**: 913–920.
61. Dutoit JI, Salome CM, Woolcock AJ: Inhaled corticosteroids reduce the severity of bronchial hyperresponsiveness in asthma, but oral theophylline does not. *Am Rev Respir Dis* (1987) **136**: 1174–1178.

62. Shohat B, Volovitz B, Varsano I: Induction of suppressor T cells in asthmatic children by theophylline treatment. *Clin Allergy* (1983) **13**: 487–493.
63. Fink G, Mittelman M, Shohat B, Spitzer SA: Theophylline-induced alterations in cellular immunity in asthmatic patients. *Clin Allergy* (1987) **17**: 313–316.
64. Guillou PJ, Ramsden C, Kerr M, Davison AM, Giles GR: A prospective controlled clinical trial of aminophylline as an adjunct immunosuppressive agent. *Transplant Proc* (1984) **16**: 1218–1220.
65. Didier M, Aussel C, Ferrua B, Fehlman M: Regulation of interleukin 2 synthesis by cAMP in human T cells. *J Immunol* (1987) **139**: 1179–1184.
66. Iwaz J, Kovassi E., Lafont S, Revillarl JP: Elevation of cyclic adenosine monophosphate levels independently down regulates IL-1, IL-2 receptor syntheses. *Int J Immunopharmacol* (1990) **12**: 631–637.
67. Sanjar S, Aoki S, Kristersson A, Smith D, Morley J: Antigen challenge induces pulmonary eosinophil accumulation and airway hyperreactivity in sensitized guinea pigs: the effect of anti-asthma drugs. *Br J Pharmacol* (1990) **99**: 679–686.
68. Kidney J, Dominguez M, Taylor PM, Rose M, Chung KF, Barnes PJ: Immunomodulation by theophylline in asthma: demonstration by withdrawal of therapy. *Am J Respir Crit Care Med* (1995) **151**: 1907–1914.
69. Jaffar ZH, Sullivan P, Page C, Costello J: Low-dose theophylline modulates T-lymphocyte activation in allergen-challenged asthmatics. *Eur Respir J* (1996) **9**: 456–462.
70. Cushley MJ, Holgate ST: Bronchodilator actions of xanthine derivatives administered by inhalation in asthma. *Thorax* (1985) **40**: 176–179.
71. Aubier M, De Troyer A, Sampson M, Macklem PT, Roussos C: Aminophylline improves diaphragmatic contractility. *N Engl J Med* (1981) **305**: 249–252.
72. Moxham J: Aminophylline and the respiratory muscles: an alternative view. *Clin Chest Med* (1988) **2**: 325–340.
73. Vozeh S, Kewitz G, Perruchoud A, *et al.*: Theophylline serum concentration and therapeutic effect in severe acute bronchial obstruction: the optimal use of intravenously administered aminophylline. *Am Rev Respir Dis* (1982) **125**: 181–184.
74. Zhang ZY, Kaminsky LS: Characterization of human cytochromes P450 involved in theophylline 8-hydroxylation. *Biochem Pharmacol* (1995) **50**: 205–211.
75. Jusko WJ, Gardner MJ, Mangiore A, Schentag JJ, Kopp JR, Vance JW: Factors affecting aminophylline clearance: age, tobacco, marijuana, cirrhosis, congestive heart failure, obesity, oral contraceptives, benzodiazepines, barbiturates and ethanol. *J Pharm Sci* (1979) **68**: 1358–1366.
76. Warren JB, Cuss F, Barnes PJ: Posture and theophylline kinetics. *Br J Clin Pharmacol* (1985) **19**: 707–709.
77. Weinberger M, Hendeles L: Slow-release theophylline: rationale and basis for product selection. *N Engl J Med* (1983) **308**: 760–763.
78. Barnes PJ, Greening AP, Neville L, Timmers J, Poole GW: Single dose slow-release aminophylline at night prevents nocturnal asthma. *Lancet* (1982) **i**: 299–301.
79. Heins M, Kurtin L, Oellerich M, Maes R, Sybrecht GW: Nocturnal asthma: slow-release terbutaline versus slow-release theophylline therapy. *Eur Respir J* (1988) **1**: 306–310.
80. Bowler SD, Mitchell CA, Armstrong JG: Nebulised fenoterol and i.v. aminophylline in acute severe asthma. *Eur Respir J* (1987) **70**: 280–283.
81. Wrenn K, Slovis CM, Murphy F, Greenberg RS: Aminophylline therapy for acute bronchospastic disease in the emergency room. *Ann Intern Med* (1991) **115**: 241–247.
82. Littenberg B: Aminophylline treatment in severe acute asthma: a metaanalysis. *JAMA* (1988) **259**: 1678–1689.
83. Fanta CH, Rossing TH, McFadden ER: Treatment of acute asthma: is combination therapy with sympathomimetics and methylxanthines indicated? *Am J Med* (1986) **80**: 5–10.
84. Siegel D, Sheppard D, Gelb A, Weinberg PF: Aminophylline increases the toxicity but not the efficacy of an inhaled β-agonist in the treatment of acute exacerbation of asthma. *Am Rev Respir Dis* (1985) **132**: 283–286.
85. Eason J, Makowe HLJ: Aminophylline toxicity: how many hospital asthma deaths does it cause? *Respir Med* (1989) **83**: 219–226.

86. Sheffer AL: National Heart Lung and Blood Institute National Asthma Education Programme Expert Panel Report: guidelines for the diagnosis and management of asthma. *J Allergy Clin Immunol* (1991) **88**: 425–534.
87. Furukawa CT, Shapiro SG, Bierman CW: A double-blind study comparing the effectiveness of cromolyn sodium and sustained release theophylline in childhood asthma. *Pediatrics* (1984) **74**: 435–439.
88. Nassif EG, Weinburger M, Thompson R, Huntley W: The value of maintenance theophylline in steroid-dependent asthma. *N Engl J Med* (1981) **304**: 71–75.
89. Brenner MR, Berkowitz R, Marshall N, Strunk RC: Need for theophylline in severe steroid-requiring asthmatics. *Clin Allergy* (1988) **18**: 143–150.
90. Rivington RN, Boulet LP, Cote J, et al.: Efficacy of slow-release theophylline, inhaled salbutamol and their combination in asthmatic patients on high-dose inhaled steroids. *Am J Respir Crit Care Med* (1995) **151**: 325–332.
91. Barnes PJ: The role of theophylline in severe asthma. *Eur Respir Rev* (1996) **6**: 154S–159S.
92. Martin RJ, Pak J: Overnight theophylline concentrations and effects on sleep and lung function in chronic obstructive pulmonary disease. *Am Rev Respir Dis* (1992) **145**: 540–544.
93. Zwilli CW, Neagey SR, Cicutto L, White DP, Martin RJ: Nocturnal asthma therapy: inhaled bitolterol versus sustained-release theophylline. *Am Rev Respir Dis* (1989) **139**: 470–474.
94. Fairfax AJ, Clarke R, Chatterjee SS, et al.: Controlled release theophylline in the treatment of nocturnal asthma. *J Int Med Res* (1990) **18**: 273–281.
95. Selby C, Engleman HM, Fitzpatrick MF, Sime PM, Mackay TW, Douglas NJ: Inhaled salmeterol or oral theophylline in nocturnal asthma? *Am J Respir Crit Care Med* (1997) **155**: 104–108.
96. Evans DJ, Taylor DA, Zetterstron V, et al.: A comparison of low dose budesonide plus theophylline for moderate asthma. *N Engl J Med* (1997) **337**: 1412–1418.
97. Barnes PJ, Pauwels RA: Theophylline in asthma: time for reappraisal? *Eur Respir J* (1994) **7**: 579–591.
98. Taylor DR, Buick B, Kinney C, Lowry RC, McDevitt DG: The efficacy of orally administered theophylline, inhaled salbutamol, and a combination of the two as chronic therapy in the management of chronic bronchitis with reversible airflow obstruction. *Am Rev Respir Dis* (1985) **131**: 747–751.
99. Murciano D, Avclair M, Parievte R, Aubier M: A randomized controlled trial of theophylline in patients with severe chronic obstructive pulmonary disease. *N Engl J Med* (1989) **320**: 1521–1525.
100. Chrystyn H, Mulley BA, Peake MD: Dose response relation to oral theophylline in severe chronic obstructive airway disease. *Br Med J* (1988) **297**: 1506–1510.
101. Aubus P, Cosso B, Godard P, Miche FB, Clot J: Decreased suppressor cell activity of alveolar macrophages in bronchial asthma. *Am Rev Respir Dis* (1984) **130**: 875–878.
102. Handslip PDJ, Dart AM, Davies BTI: Intravenous salbutamol and aminophylline in asthma: a search for synergy. *Thorax* (1981) **36**: 741–744.
103. Jenne JW: Theophylline as a bronchodilator in COPD and its combination with inhaled β-adrenergic drugs. *Chest* (1987) **92**: 7S–14S.
104. Kume H, Graziano MP, Kotlikoff MI: Stimulatory and inhibitory regulation of calcium-activated potassium channels by guanine nucleotide binding proteins. *Proc Natl Acad Sci USA* (1992) **89**: 11 051–11 055.
105. Giembycz MA: Phosphodiesterase 4 and tolerance to beta 2-adrenoceptor agonists in asthma. *Trends Pharmacol Sci* (1996) **17**: 331–336.
106. Williamson BH, Milligan C, Griffiths K, Sparta S, Tribe AC, Thompson PJ: An assessment of major and minor side effects of theophylline. *Aust NZ J Med* (1988) **19**: 539.
107. Rachelefsky WOJ, Adelson J, Mickey MR, et al.: Behaviour abnormalities and poor school performance due to oral theophylline use. *Pediatrics* (1986) **78**: 1113–1138.
108. Nicholson CD, Challiss RAJ, Shahid M: Differential modulation of tissue function and therapeutic potential of selective inhibitors of cyclic nucleotide phosphodiesterase isoenzymes. *Trends Pharmacol Sci* (1991) **12**: 19–27.
109. Kelloway JS, Wyatt RA, Adlis SA: Comparison of patients' compliance with prescribed oral and inhaled asthma medications. *Arch Int Med* (1994) **154**: 1349–1352.

39

Cromones

ANOOP J. CHAUHAN, NICHOLAS J. WITHERS,
THRIUMALA M. KRISHNA AND STEPHEN T. HOLGATE

INTRODUCTION

There have been major changes over the last decade in the way that asthma is perceived as a disease of airway inflammation rather than purely in terms of disordered smooth muscle function. The recognition of the cellular basis of airway inflammation has in turn allowed a greater understanding of the mechanisms of action of a variety of drugs in asthma including the *cromones*, sodium cromoglycate and nedocromil sodium. This chapter compares and contrasts the pharmacology and clinical efficacy of both drugs.

Sodium cromoglycate (SCG) was first discovered in 1965, as a result of earlier investigations with the naturally occurring antispasmodic khellin.[1] It is a derivative of cromone-2-carboxylic acid and comprises two cromone rings with two carboxylic acid groups joined by a flexible linking chain (Fig. 39.1). SCG is a white, odourless powder that is soluble in water and totally ionized at physiological pH.

Nedocromil sodium (NS) is a novel chemical entity, the sodium salt of pyraquinoline dicarboxylic acid (Fig. 39.2). Nedocromil was selected as a potential improvement on the established anti-inflammatory agent SCG because of its activity in models of immediate hypersensitivity; subsequent experimental data have shown it to have anti-inflammatory

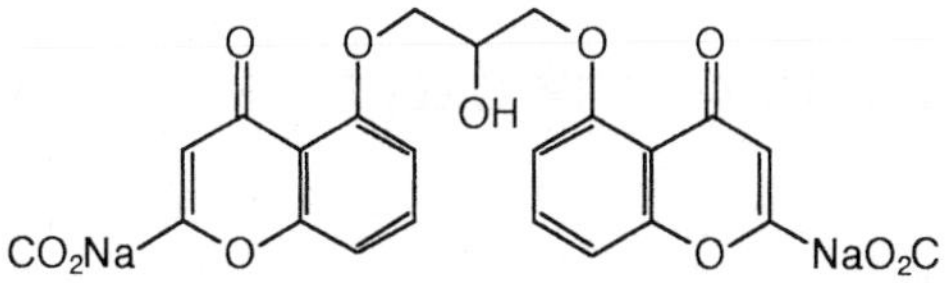

Fig. 39.1 Chemical structure of sodium cromoglycate.

ASTHMA: BASIC MECHANISMS AND CLINICAL MANAGEMENT (3rd Edn)
ISBN 0-12-079027-9

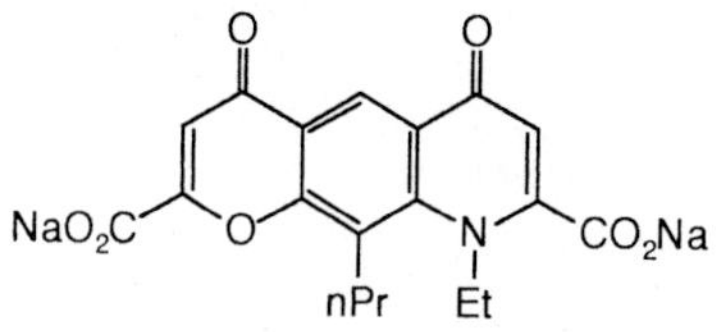

Fig. 39.2 Chemical structure of nedocromil sodium.

properties *in vitro*, in animal models and in patients with asthma.[2] Clinical trials have demonstrated its efficacy and possible superiority to SCG in the management of chronic asthma.

PHARMACOKINETICS

SCG is poorly absorbed from the gastrointestinal tract and thus its main routes of administration are by inhalation or applied topically to the eyes and nose. After inhalation approximately 10% of the dose enters the lungs and is absorbed systemically. Clearance from the lungs is rapid, 98% of the dose being removed at 24 h. After inhalation of a single 20 mg dose of SCG the maximum systemic concentration is about 9.2 ng/ml after 15 min,[3] with a mean plasma half-life of 80 min.

NS is also ideally suited for topical delivery to the airways with rapid, complete absorption through the lungs with negligible absorption in the gastrointestinal tract. Thus, following inhalation of a 4-mg dose, 10% of the dose enters the lung and 90% enters the gastrointestinal tract. It undergoes minimal absorption and is excreted for the most part in the faeces.[4] Systemic levels of NS are relatively low, with a bioavailability of about 6–9% after inhalation of 4 mg[5] and peak plasma levels of approximately 2.8 ng/ml occurring at 15 min.

SCG does not undergo systemic or presystemic metabolism in humans and unchanged drug is excreted in the urine (50% after intravenous administration and 10% after inhalation) and the bile (50% and 87% respectively).[3] The increased amount in the bile following inhalation is due to the large proportion of inhaled dose that is swallowed. The amount of SCG absorbed after inhalation is also dependent both on the delivery system and the dose of drug.[6] There is also no detectable metabolism of NS following intravenous or inhaled administration.

Both drugs are weakly bound to plasma proteins, ensuring that they do not displace other protein-bound drugs from albumin and thus the incidence of drug interactions remains low. SCG is not excreted in human breast milk and does not cross the blood–brain barrier.

EFFECTS ON INFLAMMATORY CELLS AND NERVES

Both of the cromones have been shown to have many different actions on the principal cells involved in the underlying airway inflammation in asthma (Table 39.1).

39 Cromones

Table 39.1 Comparison of the *in vitro* effects of nedocromil sodium and sodium cromoglycate on inflammatory cells isolated from human blood. Adapted with permission of Dr A.A. Norris, Fisons plc.

Cell type	Response inhibited	Nedocromil sodium	Sodium cromoglycate	Reference
Alveolar macrophage	IgE-dependent enzyme release	+++	+/−	29
Eosinophil/neutrophil	fMLP-induced activation	+++	++	13
Neutrophil	Chemotaxis induced by ZAS/PAF/fMLP/LTB$_4$	+++	+++	16, 20
Eosinophil	Chemotaxis induced by ZAS	−	+++	16
	PAF/LTB$_4$	+++	−	

+, slightly active; ++, moderately active; +++, very active; −, no activity.
See text for explanation of other abbreviations.

Mast cells

Both SCG and NS inhibit the release of a wide range of inflammatory mediators from animal and human lung mast cells following IgE-mediated antigenic stimulation. Early work *in vitro* in animal models demonstrated the inhibitory effects of SCG on histamine release from rat peritoneal mast cells following challenge with anti-IgE.[7] Subsequent studies have confirmed similar findings following exposure to non-immunologic stimuli, including phospholipase A, calcium ionophore[8] and, more recently, tumour necrosis factor (TNF)-α from anti-IgE-sensitized rat peritoneal mast cells.[9] Similarly, the release of histamine,[10] heparin,[11] protease,[12] TNF-α,[9] prostaglandin (PG)D$_2$[13] and leukotriene (LT) C$_4$ are also inhibited by NS.

Both cromones have little effect on the release of these mediators from skin mast cells, but inhibit histamine release from intestinal mast cells with greater efficacy than on any other mast cell type.[14] Furthermore, NS shows a greater inhibitory effect than SCG on histamine release from mast cells in lavage fluid both from normal volunteers and asthmatic patients.[10,15]

Eosinophils

Eosinophil chemotaxis is a key process in the airway inflammation of asthma. Many different pro-inflammatory mediators, importantly platelet-activating factor (PAF), zymosan-activated serum (ZAS), N-formyl-methionyl-leucyl-phenylalanine (fMLP), interleukin (IL)-5 and IL-8, act as eosinophil chemoattractants. NS has been shown to inhibit the chemotactic response of eosinophils to LTB$_4$, PAF[16] and IL-5[17] but not ZAS,[16] whilst the first of these studies demonstrated that SCG did not inhibit eosinophil chemotaxis via LTB$_4$ or PAF but did inhibit ZAS-induced migration, indicating different modes of action of the cromones. The action of IL-5 and IL-8 on eosinophils is potentiated in the presence of granulocyte–macrophage colony-stimulating factor (GM-CSF) and IL-3, and recently NS-induced inhibition of IL-8 chemotaxis of IL-3- and GM-CSF-primed eosinophils *in vitro* has also been demonstrated.[18]

The inhibitory effects of NS on LTC_4 release by eosinophils following stimulation with calcium ionophore and with zymosan has been supported by some studies,[19,20] whilst others have failed to confirm this observation.[21] Activated eosinophils and their products have been shown to impair the ciliary beat activity of cultured airway epithelial cells, an effect that can be inhibited by incubating eosinophils with NS.[22] Furthermore, SCG has been shown to suppress the release of cytotoxic mediators from eosinophils and neutrophils coated with complement[23] or antibody.[24]

Neutrophils

NS has been shown to inhibit neutrophil chemotaxis by PAF, LTB_4 and ZAS in a similar manner to SCG,[16,25] although other studies have shown conflicting results. One study reported no inhibitory effect of NS or chemoattractant-stimulated neutrophil migration through cellular and non-cellular barriers,[26] whilst in another study the release of LTB_4 from neutrophils was inhibited by NS only in cells taken from asthmatic patients.[27] A further study showed no inhibitory effect on LTB_4 release from neutrophils.[20]

Other inflammatory cells

Anti-IgE or antigen activates blood monocytes and airway macrophages to release preformed and freshly synthesized mediators that contribute to airway inflammation.[28] NS inhibits the release of lysosomal enzyme and IL-6 from human alveolar macrophages and oxygen radicals from peripheral monocytes,[29] following stimulation by anti-IgE. Stimulation with IL-4 promotes release of IgE antibody from tonsillar mixed mononuclear cells of non-atopic subjects, which is inhibited by SCG in a dose-dependent manner, also causing an elevation of IgG_4 release.[30]

The bronchial epithelium is now recognized to be an alternative source of many of the cytokines and other inflammatory mediators involved in the underlying inflammatory mechanisms in asthma. The bronchial epithelium of asthmatic patients shows increased synthesis and release of GM-CSF.[31] Pretreatment of cultured epithelial cells with NS inhibits the production of GM-CSF and IL-8, which is seen in untreated cells following stimulation with IL-1.[32]

Neural mechanisms

Both cromones are also recognized to exert their effects on the neurogenic mechanisms involved in asthma, with preliminary evidence from various animal models. SCG inhibits both the activity of afferent C fibres in guinea-pig[33] and dog trachea following stimulation with capsaicin[34] and the nurogenic inflammation in rat trachea after vagal stimulation.[35] These observations may explain the ability of SCG to inhibit reflex bronchoconstriction induced by cigarette smoke, metabisulphite and cold air; this may occur by the direct action of the drug on vascular endothelial cells to reduce permeability.[36]

Inhalation of adenosine 5′-monophosphate (AMP), substance P and bradykinin lead to significant airway obstruction and cough, all effects that can be inhibited by SCG. Most recent evidence has shown that SCG possesses tachykinin antagonist properties and can cause dose-dependent inhibition of substance P- and neurokinin-induced oedema in human skin without affecting the flare response.[37] NS also inhibits substance P-induced potentiation of the cholinergic response in the isolated innervated rabbit trachea[38] and substance P-induced release of histamine from human lung mast cells.[39] NS has also been shown to inhibit a non-adrenergic non-cholinergic (NANC) bronchoconstricting response in guinea-pig bronchi.[40] This is achieved by inhibition of neuropeptide release rather than end-organ antagonism and is a property not demonstrated by SCG.

EVIDENCE FOR ANTI-INFLAMMATORY ACTIONS

Many studies incorporating animal models of asthma have shown protective effects against the early- and late-phase responses following antigen challenge,[41–43] partly as a result of reducing eosinophil and neutrophil chemotaxis. More convincing evidence comes from studies conducted in humans; single doses of both NS and SCG can attenuate the early[44,45] and late-phase[46,47] responses, and the bronchial hyperresponsiveness[48] that follows antigen challenge in allergic asthmatic subjects. In contrast to animal models, NS does not inhibit the late response when administered after the early reaction.[46] Several studies have demonstrated that short-term treatment (<6 weeks) with SCG can provide protection against increases in bronchial reactivity seen in the pollen season,[49,50] whilst longer-term use can reduce bronchial reactivity in asthmatic patients.[51,52]

In asthmatic individuals, both cromones can also protect against the bronchoconstrictor response following a wide variety of stimuli, including cold air,[53] fog,[54] sulphur dioxide[55] and sodium metabisulphite.[56] NS also reduces bronchoconstriction following neurokinin A,[57] bradykinin[58] and AMP.[59] NS is also more effective than placebo in protecting against exercise-induced asthma both in children[60,61] and adults.[62–64] A dose of 4 mg of NS had a similar effect on exercise-induced bronchoconstriction to a 20-mg dose of SCG,[61,63] although SCG appears to have a longer duration of action.[63]

MECHANISMS OF ACTION

The wide spectrum of activity of the cromones indicates that they may be affecting pathways common to many cell types, including mast cells, epithelial cells and sensory nerves.

Neither of the cromones are thought to act directly on calcium channels but are known to reduce calcium influx into cells. The antigen-induced release of mast cells, in keeping with many other physiological secretory systems, is dependent on the synergistic action of an increase in intracellular calcium and an inward flow of intracellular chloride ions through specific chloride channels. Electrophysiological studies in rat peritoneal mast cells have shown that degranulation of mast cells is dependent on a sustained elevation of intracellular calcium following a transient rise in calcium from intracellular stores. The sustained phase of free calcium is a result of an opening of an intermediate-conductance chloride channel, which in turn activates a membrane calcium channel allowing calcium to enter the cell from outside.[65] Both cromones have been shown to block these chloride channels in cultured mucosal mast cells.[66] Chloride channels may play a functional role in mast cell mediator release by providing the membrane hyperpolarization necessary to support calcium influx in non-excitable cells. This view is supported by studies showing that removal of extracellular chloride ions reduces the antigen-induced release of histamine from rat mast cells.[67]

It has also been shown that SCG and NS can block the activity of a chloride channel in airway epithelial cells[68] and that SCGT can reversibly block (in a dose-dependent manner) a functionally similar channel in intestinal epithelial cells.[69] Chloride channels in sensory nerves may also be blocked by cromones from evidence *in vitro*.[40,70] The functional role of chloride channels has not been fully evaluated, though it is likely to be related to cell volume regulation.

CLINICAL STUDIES IN ASTHMA (Table 39.2)

SC

There are three therapeutic forms of SCG: pressurized aerosol (metered dose inhaler, MDI), nebulizer solution and powdered formulation (Spinhaler). A study with SCG delivered through a Spinhaler showed that the percentage of dose delivered to the lungs (of an absolute 20 mg dose) ranged between 1.1 and 3.42 mg at an inhalation rate of 120 litre/min.[71] Using similar techniques, the same group showed that with a 5 mg aerosol dose via a standard MDI, the dose delivered to the lungs was 0.88 and 1.13 mg when a spacer was used at an inhalation rate of 29 litre/min.[72] In view of these relatively small differences, it is not surprising that a 2 mg dose of SCG by MDI four times a day is comparable to a 20-mg dose by Spinhaler four times a day in managing clinical asthma.[73]

Challenge studies

The dose–response relationship of inhaled SCG has been demonstrated in a series of studies using exercise challenge and fog.[61,74–77] These studies show that increased doses of 5 mg SCG via MDI provides protection against challenge, a longer duration of protection and protects more subjects. The short-term and long-term effects of SCG on bronchial hyperresponsiveness in asthma have been investigated in at least 29 studies and have been reviewed by Hoag and McFadden.[78] They concluded that short-term use (<6 weeks) of SCG in 17 studies and long-term use (>12 weeks) in 12 studies demonstrated a reduction in airway responsiveness, though this was only significant in seven and six short-term and long-term studies respectively. Most of the studies used inhaled histamine as a provocation stimulus.

Evidence of significant protection against antigen challenge reducing both the early and late asthmatic responses are provided by studies from Cockcroft et al.[79,80] These studies compared SCG, beclomethasone dipropionate (BDP), salbutamol, ketotifen and theophylline and showed that SCG and BDP protected against the late asthmatic reaction following antigen challenge and the bronchial hyperractivity that follows.

Comparison with placebo

The clinical efficacy of 5 mg SCG by the MDI formulation has been established in placebo-controlled trials in both children and adults.[81-84] Carrasco et al.[82] showed a mean improvement of 15% in morning peak expiratory flow (PEF) in all patients and 30% in the most severe asthmatic patients (i.e. those with a diurnal variation in PEF > 15%). Bodman et al.[81] demonstrated a 35% reduction in daytime and nocturnal asthma symptoms in those treated with SCG compared with 7% in the placebo group; there was also improvement in all measures of lung function. Nebulized SCG in children has also been studied: Hiller et al.[85] showed that 11 of 17 children on SCG treatment showed significant improvement in respiratory scores when compared with placebo; in children aged 3 months to 2 years Geller-Bernstein and Levin[86] recorded improvements in respiratory symptom scores in 24 children aged 1 year or over but not in 20 children aged less than 1 year.

Comparison with corticosteroids

Early clinical trials with SCG reported that treatment with this compound could lead to reduction of corticosteroid use in some patients.[87,88] Other studies have not shown any beclomethasone-sparing affect or any other significant clinical advantage.[89-91] More recent studies in adults, however, have demonstrated that 10 mg (four times a day) of inhaled SCG is clinically equivalent to 100 μg (four times a day) of BDP[92] and also to 4 mg (four times a day) of NS.[93] In studies of childhood asthma, 10 mg SCG given four times a day was superior to both placebo and oral ketotifen[94] and the same dose given three times a day through a spacer could substitute for 200 μg of inhaled BDP without lose of asthma control.[95]

NS

Comparison with placebo

The clinical efficacy of NS in allergic asthma has been evaluated in many studies. Several of these have examined the effect of addition of this drug or placebo to the treatment regimen of patients with moderate asthma already maintained on either inhaled or oral bronchodilators.[96-100] More recently, further double-blind, placebo-controlled studies have been conducted in adults, with the age ranges varying from 12 to 74 years. All these studies show clinically and statistically significant effects of 4 mg NS on daily asthma symptom scores, pulmonary function (increases in PEF) and reduction in β_2-agonist use.[101-106] A meta-analysis of all double-blind, placebo-controlled clinical trials supplied and analysed by Fisons (irrespective of outcome) has been performed.[107] Data from 4723

Table 39.2 Recent trials comparing the efficacy of cromones with other standard treatments in asthma using a variety of study designs.

Year	Authors	Reference	Comparison	Dose per day	Duration	No of subjects	Symptoms	PEF	Use of β_2 agonist
1989	Svendsen et al.	113	NS BDP	8 mg 400 µg	6 weeks crossover	39 39	↓↓ →	↑ ↑↑	→ ↑
1990	Bergmann et al.	111	NS BDP Placebo	16 mg 400 µg	6 weeks	69 68 65	→ → —	— — —	— → →
1991	Svendsen et al.	123	NS Placebo (added to >1000 µg corticosteroid therapy)	16 mg	8 weeks	17 18	↓↓ →	↑ —	— —
1992	Groot et al.	115	NS BDP	16 mg 800 µg	8 weeks crossover	23 23	Not assessed Not assessed	— →	— ↑
1992	Wells et al.	101	NS Placebo (added to corticosteroid therapy)	8 or 16 mg	4 weeks crossover	42 12	↓↓↓ →	↑ —	↓↓ →
1992	Orefice et al.	52	NS SC BDP Placebo	16 mg 40 mg 1500 µg	12 weeks open study	42 39 40 44	↓↓ ↓↓↓ ↓↓↓ —	Not assessed Not assessed Not assessed Not assessed	↓↓ ↓↓↓ ↓↓↓ —

Year	Study	Ref	Drug	Dose	Duration	n			
1994	Faurschou et al.	92	SCG BDP	40 mg 400 µg	8 weeks	19 18	→ →	← ↑↑↑	→ ↓y
1994	de Jong et al.	104	NS Salbutamol	16 mg 800 µg	6 weeks	15 14	→ ←	← →	→ ←
1994	Clancy et al.	105	NS Placebo (added to β₂-agonist therapy in nocturnal asthma)	16 mg	8 weeks	25 25	→ —	← —	→ —
1994	Fink et al.	103	NS Placebo (added to β₂-agonist therapy)	16 mg	10 weeks	54 56	→ —	← —	→ —
1995	Creticos et al.	106	NS Placebo (added to β₂-agonist therapy)	8 mg	8 weeks	56 56	→ —	— —	→ —
1995	Wasserman et al.	109	NS Salbutamol	16 mg 720 µg	12 weeks	117 118	↓↓ ←	← —	— —
1995	Croce et al.	94	SCG Ketotifen Placebo (childhood asthma)	40 mg 2 mg	14 weeks	39 39 36	→ — —	← ← ←	→ — —

BDP, beclomethasone dipropionate; NS, nedocromil sodium; PEF, peak expiratory flow; SCG, sodium cromoglycate.

patients (2385 NS; 2338 placebo) were analysed and presented as the difference between groups for symptoms, lung function and bronchodilator use. Overall, the data supported a highly significant effect of NS for all efficacy variables, though the clinical improvements were mostly evident in mild to moderate asthmatic patients only on bronchodilator therapy.

Comparison with bronchodilators and theophyllines

Several clinical trials have compared the efficacy of NS when using it as a substitute either for inhaled bronchodilators[104] or oral theophyllines.[99,100] In the replacement of inhaled bronchodilators NS caused a significant decrease in symptom scores and diurnal PEF variation. Patients treated with salbutamol deteriorated significantly from their baseline values and showed a significant increase in plasma eosinophilia.[104] Two recent studies have compared the effects of NS with continuous β_2-agonist treatment in mild to moderately severe allergic asthmatic patients.[108,109] Manolitsas et al.[108] demonstrated that the number of activated eosinophils in the bronchial biopsy increased with treatment with salbutamol and decreased with NS. Wasserman et al.[109] and de Jong et al.[104] demonstrated a greater reduction in PEF variability and symptom scores during treatment with NS compared with salbutamol.

In one large study, a daily dose of 16 mg NS was significantly more effective than placebo in controlling asthma symptoms following withdrawal of theophyllines.[99] When NS was added to maintenance therapy and then substituted for theophylline, symptom severity and twice-daily PEF rates improved significantly within 2 weeks. A direct comparison of NS and theophyllines in a double-blind trial in 73 asthmatic patients showed no difference in patient symptom scores, inhaled bronchodilator use or spirometry.[110] In another study of moderately theophylline-dependent patients, the attention of NS gave significant improvements in asthma sympton scores and bronchodilator use.[100]

Comparison with corticosteroids

Because of the laboratory evidence of the anti-inflammatory properties of NS, numerous studies have examined its efficacy both as an alternative to inhaled corticosteroids and in reducing corticosteroid dosages in dependent patients. Two double-blind placebo-controlled studies have demonstrated that both NS and BDP significantly improved symptom scores compared with placebo.[111,112] Improvements were also seen in spirometry, but these were more significant in the corticosteroid-treated group. Several crossover studies have also shown improvements from baseline in most parameters (symptom scores and lung function) with both NS and corticosteroid treatment, but once again BDP proved to be significantly more effective than NS in several areas including response to histamine provocation.[113–115] The observation that BDP was superior to NS in two of the studies may be explained by the lower dose of NS (8 mg/day instead of 16 mg/day) used in the study by Svendsen et al.[113] and the higher dose of BDP (800 μg/day instead of 400 μg/day) used by Groot et al.[115]

The corticosteroid-sparing effects of NS have also been demonstrated in some studies in asthmatic patients, whilst others have not demonstrated this effect. In a 12-week, double-blind, placebo-controlled study of asthmatic patients taking inhaled corticosteroids plus a mean oral dose of 8 mg, Boulet et al.[116] were able to reduce a significantly greater percentage of the oral steroid dosage during treatment with NS.[116] Patients showed no changes in symptom score or spirometry but did show a significant increase in use of β_2-agonists. In a further study, asthmatic patients were rendered unstable by a 50% reduction in inhaled corticosteroid dosage; subsequent introduction of NS significantly improved symptom score and PEF compared with placebo.[117] Similar results were seen following reduction of inhaled corticosteroids by Bone et al.[118]

In patients maintained on high doses of inhaled corticosteroids (1000–2000 μg/day BDP, 4 mg NS four times daily only enabled a 31% (625 μg/day) reduction in dose compared with placebo.[102] In a group of steroid-dependent asthmatic patients taking at least 8 mg/day of prednisolone, NS showed no difference to placebo in reduction of corticosteroid dosage.[119]

The benefit of the addition of NS treatment in patients already receiving corticosteroids has been examined in several studies. In general, NS is superior to placebo in improving PEF, symptom scores and reducing β_2-agonist use[101,120–123] though the effects on reducing corticosteroid doses have not been consistent.

NS in children

Several placebo-controlled studies of NS administered by MDIs have been performed. They have shown that 4 mg NS or placebo four times daily added to predominantly bronchodilator therapy in asthmatic children led to significant improvements in asthma symptom scores and bronchodilator use.[124–129] The ages ranged from 3 to 19 years.

NS vs. SCG

Several human studies have suggested the superiority of NS over SCG in the inhibition of several bronchoconstrictor responses.[56,59,130–132] Both cromones, when administered prior to an allergen challenge, inhibit the early and late asthmatic responses. When both drugs are administered after the early response, NS (and *not* SCG) inhibits the late response. The efficacy of both cromones have also been compared in clinical trials of asthma. In one such trial NS proved to be superior to SCG in control of nocturnal and daytime symptoms in asthmatic patients already controlled on β_2-agonists and inhaled corticosteroids.[133] A second study, albeit unblinded, confirms this difference,[52] whilst Boldy et al.[93] could find no difference in efficacy between therapeutic doses of the two compounds in 69 asthmatic patients aged over 50. In more recent studies, NS (16 mg/day) was compared with BDP (1500 μg/day) and SCG (40 mg/day). NS was only more effective than SCG for symptom scores.[101] Schwartz et al.[134] demonstrated in a multicentre study that the clinical efficacy of both drugs was comparable and that both were more effective than placebo in mild to moderate asthma. The therapeutic efficacy of both cromones has recently been reviewed.[135–137]

REFERENCES

1. Proctor RK: Cromolyn sodium (drug evaluation data). *Drug Intell Clin Pharm* (1974) **8**: 20–24.
2. Abraham R, Kauffman HF, Groen H, Koeter GH, de Monchy JGR: The effect of nedocromil sodium on the early and late reaction and allergen-induced bronchial hyperresponsiveness. *J Allergy Clin Immunol* (1991) **87**: 993–1001.
3. Walker SR, Evans ME, Richards AJ, Paterson JW: The fate of ^{14}C disodium cromoglycate in man. *J Pharm Pharmacol* (1972) **24**: 525–531.
4. Auty RM, Clarke AJ: Kinetics and disposition of nedocromil sodium in man: a preliminary report. *Eur J Respir Dis* (1986) **69**(Suppl 147): 246–247.
5. Neale MG, Brown K, Foulds RA, Lal S, Morris DA, Thomas D: The pharmacokinetics of nedocromil sodium, a new drug for the treatment of reversible obstructive airways disease, in human volunteers and patients with reversible obstructive airways disease. *Br J Clin Pharmacol* (1987) **24**: 493–501.
6. Fuller FW, Collier JG: The pharmacokinetic assessment of sodium cromoglycate. *J Pharm Pharmacol* (1983) **35**: 289–292.
7. Goose J, Blair AMJN: Passive cutaneous anaphylaxis in the rat, induced with two homologous regin-like antibodies sera and its specific inhibition with disodium cromoglycate. *Immunology* (1969) **16**: 749–751.
8. Johnson HG, Bach NK: Prevention of calcium ionophore-induced release of histamine in rat mast cells by disodium cromoglycate. *J Immunol* (1975) **114**: 514–516.
9. Bissonnette EY, Enciso JA, Befus AD: Inhibition of tumour necrosis factor-alpha (TNF-alpha) release from mast cells by the anti-inflammatory drugs, sodium cromoglycate and nedocromil sodium. *Clin Exp Immunol* (1995) **102**: 78–84.
10. Leung KB, Flint KC, Brostoff J, et al.: Effects of sodium cromoglycate and nedocromil sodium on histamine secretion from human lung mast cells. *Thorax* (1988) **43**: 756–761.
11. Enerback K, Bergstrom S: Effect of nedocromil sodium on the compound exocytosis of mast cells. *Drugs* (1989) **37**(Suppl 1): 44–50.
12. Wilsoncroft P, Gaffen Z, Reynia S, Brain SD: The modulation by nedocromil sodium of proteases released from rat peritoneal mast cells capable of degrading vasoactive intestinal peptide and calcitonin gene-related peptide. *Immunopharm* (1993) **25**: 97–204.
13. Lebel B, Bousquet J, Chanez P, et al.: Spontaneous and non-specific release of histamine and PGD2 by bronchoalveolar lavage cells from asthmatic and normal subjects: effect of nedocromil sodium. *Clin Allergy* (1988) **18**: 605–613.
14. Okayama Y, Benyon RC, Rees PH, Lowman MA, Hillier K, Church MK: Inhibition profiles of sodium cromoglycate and nedocromil sodium on mediator release from mast cells of human skin, lung, tonsil, adenoid and intestine. *Clin Exp Allergy* (1992) **22**: 401–409.
15. Okayama Y, Church MK: Comparison of modulatory effect of ketotifen, sodium cromoglycate, procaterol and salbutamol in human skin, lung and tonsil mast cells. *Int Arch Allergy Immunol* (1992) **97**: 216–225.
16. Bruijnzeel PLB, Warring RAJ, Kok PTM, Kreukniet J: Inhibition of neutrophil and eosinophil induced chemotaxis by nedocromil sodium and sodium cromogycate. *Br J Pharmacol* (1990) **99**: 798–802.
17. Resler B, Sedgwick JB, Busse WW: Inhibition of interleukin-5 effects on human eosinophils by nedocromil sodium. *J Allergy Clin Immunol* (1992) **89**: 235.
18. Warringa RAJ, Mengelers HJJ, Maikoe T, Bruijnzeel PLB, Koenderman L: Inhibition of cytokine-primed eosinophil chemotaxis by nedocromil sodium. *J Allergy Clin Immunol* (1993) **91**: 802–809.
19. Sedgwick JB, Bjornsdottir U, Geiger KM, Busse WW: Inhibition of eosinophil density change and leukotriene C$_4$ generation by nedocromil sodium. *J Allergy Clin Immunol* (1992) **90**: 202–209.
20. Bruijnzeel PLB, Warringa RAJ, Kok PTM, Kreukniet J: Nedocromil sodium inhibits the

A23187 and opsonised zymosan-induced leukotriene formation by human eosinophils but not by human neutrophils. *Br J Pharmacol* (1989) **96**: 631–636.
21. Burke LA, Crea AE, Wilkinson JR, Arm JP, Spur BW, Lee TH: Comparison of the generation of platelet-activating factor and leukotriene C4 in human eosinophils stimulated by unopsonized zymosan and by the calcium ionophore A23187: the effects of nedocromil sodium. *J Allergy Clin Immunol* (1990) **85**: 26–35.
22. Devalia JL, Sapsford RJ, Rusznak C, Davies RJ: The effect of human eosinophils on cultured human nasal epithelial cell activity and the influence of nedocromil sodium. *Am J Respir Cell Mol Biol* (1992) **7**: 270–277.
23. Kay AB, Walsh GM, Moqbel R, et al.: Disodium cromoglycate inhibits activation of human inflammatory cells in vitro. *J Allergy Clin Immunol* (1987) **80**: 1–8.
24. Rand TH, Lopez AF, Gamble JR, Vadas MA: Nedocromil sodium and cromolyn (sodium cromoglycate) selectively inhibit antibody-dependent granulocyte-mediated cytotoxicity. *Int Arch Allerby Appl Immunol* (1988) **87**: 151–158.
25. Bruijnzeel PLB, Warringa RAJ, Kok PTM: Inhibition of platelet activating factor- and zymosan-activated serum-induced chemotaxis of human neutrophils by nedocromil sodium, BN 52021 and sodium cromoglycate. *Br J Pharmacol* (1989) **97**: 1251–1257.
26. Carolan EJ, Casale TB: Effects of nedocromil sodium and WEB 2086 on chemoattractant-stimulated neutrophil migration through cellular and noncellular barriers. *Ann Allergy* (1992) **69**: 323–328.
27. Radeau T, Chavis C, Godard PH, Michel FB, Crastes de Paulet A, Damon M: Arachidonate 5-lipoxygenase metabolism in human neutrophils from patients with asthma: in vitro affect of nedocromil sodium. *Int Arch Allergy Immunol* (1992) **97**: 209–215.
28. Capron A, Dessaint JP, Capron M, Joseph M, Amiesen JC, Tonnel AB: From parasites to allergy: the second receptor for IgE (F_cERII). *Immunol Today* (1992) **7**: 15–18.
29. Thorel T, Joseph M, Tsicopoulos A, Tonnel AB, Capron A: Inhibition by nedocromil sodium of IgE-mediated activation of human mononuclear phagocytes and platelets in allergy. *Int Arch Allergy Appl Immunol* (1988) **85**: 232–237.
30. Kimata H, Yoshida A, Ishioka C, Milawa H: Disodium cromoglycate (DSCG) selectively inhibits IgE production and enhances IgG4 production by human B cells in vitro. *Clin Exp Immunol* (1991) **84**: 395–399.
31. Mattoli S, Mattoso VL, Soloperto M, Allegra L, Fasoli A: Cellular and biochemical characteristics of bronchoalveolar lavage fluid in symptomatic nonallergic asthma. *J Allergy Clin Immunol* (1991) **87**: 794–802.
32. Marini M, Soloperto M, Zheng Y, Mezzetti M, Mattoli S: Protective effect of nedocromil sodium on the IL-1-induced release of GM-CSF from cultured human epithelial cells. *Pulmon Pharmacol* (1992) **5**: 61–65.
33. Erjefalt I, Persson CGA: Anti-asthma drugs attentuate inflammatory leakage of plasma into airway lumen. *Acta Physiol Scand* (1986) **128**: 653–654.
34. Dixon M, Jackson DM, Richards IM: The action of sodium cromoglycate on 'C' fibre endings in the dog lung. *Br J Pharmacol* (1980) **70**: 11–13.
35. Norris AA, Leeson ML, Jackson DM, Holroyde MC: Modulation of neurogenic inflammation in rat trachea. *Pulmon Pharmacol* (1990) **3**: 180–184.
36. Persson CGA: Cromoglycate, plasma exudation and asthma. *Trends Pharmacol Sci* (1987) **8**: 202–203.
37. Crossman DC, Dashwood MR, Taylor GW, Wellings R, Fuller RW: Sodium cromoglycate: evidence of tachykinin antagonist activity in the human skin. *J Appl Physiol* (1993) **75**: 167–172.
38. Armour CL, Johnson PRA, Black JL: Nedocromil sodium inhibits substance-P-induced potentiation of cholinergic neural responses in the isolated innervated rabbit trachea. *J Auton Pharmacol* (1991) **11**: 167–172.
39. Louis RE, Radermecker MF: Substance P-induced histamine release from human basophils, skin and lung fragments: effect of nedocromil sodium and theophylline. *Int Arch Allergy Appl Immunol* (1990) **892**: 329–333.
40. Verleden GM, Belvis MG, Stretton CD, Barnes PJ: Nedocromil sodium modulates

nonadrenergic, noncholinergic bronchoconstrictor nerves in guinea-pig airways *in vitro*. *Am Rev Respir Dis* (1991) **143**: 114–118.
41. Abraham WM, Stevenson JS, Eldridge M, Garrido R, Nieves L: Nedocromil sodium in allergen-induced bronchial responses and airway hyperresponsiveness in allergic sheep. *J Apply Physiol* (1988) **65**: 1062–1068.
42. Hutson PA, Holgate ST, Church MK: Inhibition by nedocromil sodium of early and late phase bronchoconstriction and airway cellular infiltration provoked by ovalbumin inhalation in conscious sensitized guinea-pigs. *Br J Pharmacol* (1989) **94**: 6–8.
43. Schellenberg RR, Ishida K, Thomson RJ: Nedocromil sodium inhibits airway hyperresponsiveness and eosinophilic infiltration induced by repeated antigen challenge in guinea-pigs. *Br J Pharmacol* (1991) **103**: 1842–1846.
44. Pepys J, Hargreave FE, Chan M, McCarthy DS: Inhibitory effects of disodium cromoglycate on allergen-inhalation tests. *Lancet* (1986) **ii**: 134–137.
45. Svendsen UG, Nielsen NH, Frolund L, Madsen F, Weeke B: Effects of nedocromil sodium and placebo delivered by pressurised aerosol in bronchial antigen challenge. *Allergy* (1986) **41**: 468–470.
46. Crimi E, Brusasco V, Crimi P: Effect of nedocromil sodium on the late asthmatic reaction to bronchial antigen challenge. *J Allergy Clin Immunol* (1989) **83**: 985–990.
47. Herdman M, Davies RJ: Effect of inhaled nedocromil sodium delivered by metered dose inhaler and the breath actuated Autohaler on allergen induced bronchospasm. *Thorax* (1991) **46**: 294P.
48. Twentyman OP, Varley JG, Holgate ST: Effect of beclomethasone and cromoglycate on late phase bronchoconstriction and increased hyperresponsiveness. *J Allergy Clin Immunol* (1989) **83**: 245.
49. Bleecker ER, Britt EJ, Mason PL: The effect of cromolyn on antigen and methacholine airways reactivity during ragweed season in allergic asthma. *Chest* (1982) **82**: 227.
50. Lowhagen O, Rak S: Modification of bronchial hyperreactivity after treatment with sodium cromoglycate during pollen season. *J Allergy Clin Immunol* (1985) **75**: 460–467.
51. Chabra SK, Gaur SN: Effect of long-term treatment with sodium cromoglycate on non-specific bronchial hyperresponsiveness in asthma. *Chest* (1989) **95**: 1235–1238.
52. Orefice U, Struzzo P, Dorigo R, Peratoner A: Long-term treatment with sodium cromoglycate, nedocromil sodium and beclomethasone dipropionate reduces bronchial hyperresponsiveness in asthmatic subjects. *Respiration* (1992) **59**: 97–101.
53. Juniper EF, Latimer KM, Morris MM, Roberts RS, Hargreave FE: Airway responses to hyperventilation of cold dry air: duration of protection by cromoly sodium. *J Allergy Clin Immunol* (1986) **78**: 387–391.
54. Black JL, Smith CM, Anderson SD: Cromolyn sodium inhibits the increased responsiveness to methacholine that follows ultrasonically nebulised water challenge in patients with asthma. *J Allergy Clin Immunol* (1987) **80**: 39–44.
55. Sheppard D, Nadel JA, Boushey HA: Inhibition of sulfur dioxide-induced bronchoconstriction by disodium cromoglycate in asthmatic patients. *Am Rev Respir Dis* (1981) **124**: 257–259.
56. Dixon CMS, Ind PW: Inhaled sodium metabisulphate induced bronchoconstriction: inhibition by nedocromil sodium and sodium cromoglycate. *Br J Clin Pharmacol* (1990) **30**: 371–376.
57. Crimi N, Palermo F, Oliveri R, Palermo B, Polosa R, Mistretta A: Protection of nedocromil sodium on bronchoconstriction induced by inhaled neurokinin A (NKA) in asthmatic patients. *Clin Exp Allergy* (1992) **22**: 75–81.
58. Dixon CM, Barnes PJ: Bradykinin-induced bronchoconstriction: inhibition by nedocromil sodium and sodium cromoglycate. *Br J Clin Pharmacol* (1989) **27**: 831–836.
59. Phillips GD, Scott VL, Richards R, Holgate ST: Effect of nedocromil sodium and sodium cromoglycate against bronchoconstriction induced by inhaled adenosine 5'-monophosphate. *Eur Respir J* (1989) **2**: 210–217.
60. Boner AL, Vallone G, Bennati D: Nedocromil sodium in exercise-induced bronchostriction in children. *Ann Allergy* (1989) **62**: 38–41.
61. Cavagni G, Caffarelli C, Bertolini P, Giordano S: Comparison of sodium cromoglycate (SCG)

with nedocromil (N) in exercise-induced bronchospasm (EIB) in children. *Ann Allergy* (1992) **68**: 95.
62. Albazzaz MK, Neale MG, Patel KR: Dose–response study of nebulised nedocromil sodium in exercise induced asthma. *Thorax* (1989) **44**: 816–819.
63. Konig P, Hordvik NL, Kreutz C: The preventive effect and duration of action of nedocromil sodium and cromolyn sodium on exercise-induced asthma (EIA) in adults. *J Allergy Clin Immunol* (1987) **79**: 64–68.
64. Morton AR, Ogle SL, Fitch KD: Effects of nedocromil sodium, cromolyn sodium and a placebo in exercise-induced asthma. *Ann Allergy* (1992) **68**: 143–148.
65. Penner R, Matthews G, Neher E: Regulation of calcium influx by secondary messengers in rat mast cells. *Nature* (1988) **334**: 499–504.
66. Romanin C, Reinsprecht M, Pecht I, Schindler H: Immunologically activated chloride channels involved in degranulation of rat mucosal mast cells. *EMBO J* (1991) **10**: 3603–3608.
67. Friis U, Johansen T, Hayes N, Foreman J: IgE-receptor activated chloride uptake in relation to histamine secretion from mast cells. *Br J Pharmacol* (1994) **111**: 1179–1183.
68. Alton EW, Kingsleigh-Smith DJ, Munkonge FM, *et al.*: Asthma prophylaxis agents alter the function of an airway epithelial chloride channel. *Am J Respir Cell Mol Biol* (1996) **14**: 380–387.
69. Reinsprecht M, Pecht I, Schindler H, Romanin C: Potent block of Cl$^-$ channels by antiallergic drugs. *Biochem Biophys Res Commun* (1992) **188**: 957–963.
70. Jackson DM, Pollard CE, Roberts SM: The effect of nedocromil sodium on the isolated rabbit vagus nerve. *Eur J Pharmacol* (1992) **221**: 175–177.
71. Newman SP, Hollingworth A, Clark AR: Effect of different modes of inhalation on drug delivery from a dry power inhaler. *Int J Pharm* (1994) **102**: 127–132.
72. Newman SP, Clark AR, Talalee N, Clarke SW: Lung deposition of 5 mg of Intal from a pressurised metered dose inhaler assessed by radiotracer technique. *Int J Ther* (1991) **74**: 203–208.
73. Lal S, Malhotra SM, Gribben MD: Comparison of sodium cromoglycate pressurised aerosol and powder in the treatment of asthma. *Clin Allergy* (1982) **12**: 197–201.
74. Del Bufalo D, Fasano L, Patalano F, Ruggieri F, Gunella G: Prevention of non-specific bronchial hyperreactivity. Dose-dependent effect of sodium cromoglycate metered dose aerosol. *Allergol Immunopathol* (1988) **16**: 77–80.
75. Tullett WM, Tan KM, Wall RT, Patel KR: Dose response effect of sodium cromoglyacte pressurised aerosol in exercise induced asthma. *Thorax* (1985) **40**: 41–44.
76. Patel KR, Wall RT: Dose–duration effect of sodium cromoglycate in exercise-induced asthma. *Eur J Respir Dis* (1986) **69**: 256–260.
77. Cavallo A, Cassaniti C, Glogger A, Magrini H: Action of nedocromil sodium in exercise-induced asthma in adolescents. *J Invest Allergol Clin Immunol* (1995) **5**): 286–288.
78. Hoag JE, McFadden ER: Long term effect of cromolyn sodium on non specific bronchial hyperresponsiveness: a review. *Ann Allergy* (1991) **66**: 53–63.
79. Cockcroft DW, Murdock KY: Comparative effects of inhaled salbutamol, sodium cromoglycate and beclomethasone dipropionate on allergen-induced early asthmatic responses, late asthmatic responses and increased bronchial responsiveness to histamine. *J Allergy Clin Immunol* (1987) **79**: 734–740.
80. Cockcroft DW, Keshmiri M, Murdock KY, Gore BC: Allergen-induced increase in airway responsiveness is not inhibited by acute treatment with ketotifen or clemastine. *Ann Allergy* (1992) **68**: 245–250.
81. Bodman S, Gross G, Lockley R, Mansfield L, Prenner B: Sodium cromoglycate 10 mg four times daily is an effective treatment for adult asthmatics. *Eur Respir J* (1994) **7**: 151S.
82. Carrasco E, Sepuldeva R: Comparison of 1 mg and 5 mg sodium cromoglycate metered dose inhalers in the treatment of asthma: a 12-week double blind, parallel group trial. *Curr Med Res Opin* (1989) **11**: 341–353.
83. Acheson HWK, Drury VWM, Harvard-Davies R, Smail SA: Clinical trials in general practice: lessons learned from a study testing asthma drugs. *Rev Esp Allergol* (1989) **2**: 268.

84. Kuzemko J, Fleet H, Wood CBS: The effect of inhaled sodium cromoglycate when delivered via a nebuhaler in the treatment of childhood asthma. *Thorax* (1991) **46**: 769P–770P.
85. Hiller EJ, Milner AD, Lennery W: Nebulised sodium cromoglycate in young asthmatic children. *Arch Dis Child* (1977) **52**: 875–876.
86. Geller-Bernstein C, Levin S: Nebulised sodium cromoglycate in the treatment of wheezy bronchitis in infants and young children. *Respiration* (1982) **43**: 294–298.
87. Smith JM, Devey CF: Clinical trial of disodium cromoglycate in the treatment of asthma in children. *Br Med J* (1968) **2**: 340–344.
88. Moran F, Bankier JDH, Boyd G: Disodium cromoglycate in the treatment of allergic bronchial asthma. *Lancet* (1968) **ii**: 137–139.
89. Toogood JH, Jennings B, Lefcoi NM: A clinical trial of combined cromolyn/beclomethasone treatment for asthma. *J Pediatr* (1981) **67**: 317–324.
90. Hiller EJ, Milner AD: Betamethasone 17 valerate aerosol and disodium cromoglycate in the treatment of asthma. *Br J Dis Chest* (1975) **69**: 103–106.
91. Dawood AG, Hendry AT, Walker SR: The combined use of betamethasone valerate and sodium cromoglycate in the treatment of asthma. *Clin Allergy* (1977) **7**: 161–165.
92. Faurschou P, Bing J, Edman G, Engel A: Comparison between sodium cromoglycate (MDI: metered-dose inhaler) and beclomethasone dipropionate (MDI) in treatment of adult patients with mild to moderate bronchial asthma. A double-blind, double-dummy randomized, parallel-group study. *Allergy* (1994) **49**: 659–636.
93. Boldy DA, Ayres JG: Nedocromil sodium and sodium cromoglycate in patients aged over 50 years with asthma. *Respir Med* (1993) **87**: 517–523.
94. Croce J, Negreiros EB, Mazzei JA, Isturiz G: A double-blind, placebo-controlled comparison of sodium cromoglycate and ketotifen in the treatment of childhood asthma. *Allergy* (1995) **50**: 524–752.
95. Petersen W, Daugbjerg P, Fog E, *et al.*: Sodium cromoglycate (10 mg TDS) together with terbutaline (0.5 mg TDS) can replace inhaled steroids (200 µg) in childhood asthma. *Eur Respir J* (1993) **6**: 356S.
96. Williams HN: Multi-centre clinical trial of nedocromil sodium in reversible airways disease in adults: a general practitioner collaborative study. *Curr Med Res Opin* (1989) **11**: 417–426.
97. Greif J, Fink G, Smorzik Y, Topilsky M, Bruderman I, Spitzer SA: Nedocromil sodium and placebo in the treatment of bronchial asthma. A multicenter, double-blind, parallel-group comparison. *Chest* **96**: 583–858.
98. Bianco S, Del Bono N. Grassi V, Orefice U: Effectiveness of nedocromil sodium versus placebo as additions to routine asthma therapy: a multicentre double-blind, group comparative trial. *Respiration* (1989) **56**: 204–211.
99. Cherniack, Wasserman SI, Ramsdell JW *et al.*: A double-blind multicenter group comparative study of the efficacy and safety of nedocromil sodium in the management of asthma. *Chest* (1990) **97**: 1299–1306.
100. Callaghan B, Teo NC, Clancy L: Effects of the addition of nedocromil sodium to maintenance bronchodilator therapy in the management of chronic asthma. *Chest* (1992) **101**: 787–792.
101. Wells A, Drennan C, Holst P, Jones D, Rea H, Thornley P: Comparison of nedocromil sodium at two dosage frequencies with placebo in the management of chronic asthma. *Respir Med* (1992) **86**: 311–316.
102. Wong CS, Cooper S, Britton JR, Tattersfield AE: Steroid sparing effect of nedocromil sodium in asthmatic patients on high doses of inhaled steroids. *Clin Exp Allergy* (1993) **23**: 370–376.
103. Fink JN, Forman S, Silvers WS, Soifer MM, Tashkin DP, Wilson AF: A double-blind study of the efficacy of nedocromil sodium in the management of asthma in patients using high doses of bronchodilators. *J Allergy Clin Immunol* (1994) **94**: 473–481.
104. de Jong JW, Teengs JP, Postma DS, van der Mark TW, Koeter GH, de Monchy JG: Nedocromil sodium versus albuterol in the management of allergic asthma. *Am J Respir Crit Care Med* (1994) **149**: 91–97.
105. Clancy L, Keogan S: Treatment of nocturnal asthma with nedocromil sodium. *Thorax* (1994) **49**: 1225–1227.

106. Creticos P, Burk J, Smith L, Comp R, Norman P, Findlay S: The use of twice daily nedocromil sodium in the treatment of asthma. *J Allergy Clin Immunol* (1995) **95**: 829–836.
107. Edwards AM, Stevens MT: The clinical efficacy of inhaled nedocromil sodium (Tilade) in the treatment of asthma. *Eur Respir J* (1993) **6**: 35–41.
108. Manolitsas ND, Wang J, Devalia JL, Trigg CJ, McAulay AE, Davies RJ: Regular albuterol, nedocromil sodium, and bronchial inflammation in asthma. *Am J Respir Crit Care Med* (1995) **151**: 1925–1930.
109. Wasserman SI, Furukawa CT, Henochowicz SI, et al.: Asthma symptoms and airway hyperresponsiveness are lower during treatment with nedocromil sodium than during treatment with regular inhaled albuterol. *J Allergy Clin Immunol* (1995) **95**: 541–547.
110. Crimi E, De Benedetto F, Grassi V, Orefice U, Ruggieri F: Nedocromil sodium versus theophylline in the treatment of bronchial asthma. *Allergy Clin Immunol News* (1991) **4**: 328.
111. Bergmann KC, Bauer CP, Overlack A: A placebo-controlled blinded comparison of nedocromil sodium and beclomethasone dipropionate in bronchial asthma. *Lung* (1990) **168**(Suppl 1): 230–239.
112. Bel EH, Timmers C, Hermans J, Dijkman JH, Sterk PJ: The long-term effects of nedocromil sodium and beclomethasone diproprionate on bronchial responsiveness to methacholine in nonatopic asthmatics. *Am Rev Respir Dis* (1990) **141**: 21–28.
113. Svendsen UG, Frolund L, Madsen F, Nielsen NH: A comparison of the effects of nedocromil sodium and beclomethasone dipropionate on pulmonary function, symptoms, and bronchial responsiveness in patients with asthma. *J Allergy Clin Immunol* (1989) **874**: 224–231.
114. Harper GD, Neill P, Vathenen AS, Cookson JB, Ebden P: A comparison of inhaled beclomethasone dipropionate and nedocromil sodium as additional therapy in asthma. *Respir Med* (1990) **84**: 463–469.
115. Groot CA, Lammers JW, Molema J, Festen J, van Herwaarden CL: Effect of inhaled beclomethasone and nedocromil sodium on bronchial hyperresponsiveness to histamine and distilled water. *Eur Respir J* (1992) **5**: 1075–1082.
116. Boulet LP, Cartier A, Crockcroft DW, et al.: Tolerance to reduction of oral steroid dosage in severely asthmatic patients receiving nedocromil sodium. *Respir Med* (1990) **84**: 317–323.
117. Ruffin R, Alpers JH, Kroemer DK, et al.: A 4-week Australian multicentre study of nedocromil sodium in asthmatic patients. *Eur J Respir Dis* (1986) **147**: 336–339.
118. Bone MF, Kubik MM, Keaney NP, et al.: Nedocromil sodium in adults with asthma dependent on inhaled corticosteroids: a double blind, placebo controlled study. *Thorax* (1989) **44**: 654–659.
119. Goldin JG, Bateman ED: Does nedocromil sodium have a steroid sparing effect in adult asthmatic patients requiring maintenance oral corticosteroids? *Thorax* (1988) **43**: 982–986.
120. Lal S, Malhotra S, Gribben D, Hodder D: Nedocromil sodium: a new drug for the management of bronchial asthma. *Thorax* (1984) **39**: 809–812.
121. Williams AJ, Stableforth D: The addition of nedocromil sodium to maintenance therapy in the management of patients with bronchial asthma. *Eur J Respir Dis* (1986) **69**: 340–343.
122. Fyans PG, Chatterjee PC, Chatterjee SS: Effects of adding nedocromil sodium (Tilade) to the routine therapy of patients with bronchial asthma. *Clin Exp Allergy* (1989) **19**: 521–528.
123. Svendsen UG, Jorgensen H: Inhaled nedocromil sodium as additional treatment to high dose inhaled corticosteroids in the management of bronchial asthma. *Eur Respir J* (1991) **4**: 992–999.
124. Armenio L, Bladini G, Bardare M, et al.: Double blind, placebo controlled study of nedocromil sodium in asthma. *Arch Dis Child* (1993) **68**: 193–197.
125. Leng Foo A, Lanteri CJ, Burton PR, Sly PD: The effect of nedocromil sodium on histamine responsiveness in clinically stable asthmatic children. *J Asthma* (1993) **30**: 381–390.
126. Harris J, Bukstein D, Chervinsky P, et al.: Nedocromil sodium 4 mg QID is effective in reducing asthma symptoms in ragweed-allergic asthmatic children. *Am J Respir Crit Care Med* (1994) **149**: A351.
127. Bronsky E, Ellis M, Fries S, et al.: More symptom free days for asthmatic children treated with nedocromil sodium. *Eur Respir J* (1994) **7**: 140S.

128. Harnden AR, Woods S, Haider S: Efficacy of nedocromil sodium used with a spacer device in the younger asthmatic. *Eur Respir J* (1994) **7**: 139S.
129. Hakim EA, Hide D, Kuzemko J: Can nedocromil sodium substitute for sodium cromoglycate in children with asthma? *Eur Respir J* (1994) **7**: 139S.
130. Wright W, Zhang YG, Salome CM, Woolcock AJ: Effect of inhaled preservatives on asthmatic patients. 1. Sodium metabisulphite. *Am Rev Respir Dis* (1990) **141**: 1400–1404.
131. Crimi N, Palermo F, Oliveri R, *et al.*: Effect of nedocromil on bronchospasm induced by inhalation of substance P in asthmatic subjects. *Clin Allergy* (1988) **18**: 375–382.
132. Richards R, Phillips GD, Holgate ST: Nedocromil sodium is more potent than sodium cromoglycate against AMP-induced bronchoconstriction in atopic asthmatic subjects. *Clin Exp Allergy* (1989) **19**: 285–291.
133. Lal S, Dorow PD, Benho KK, Chatterjee SS: Nedocromil sodium is more effective than cromolyn sodium for the treatment of chronic reversible obstructive airway disease. *Chest* (1993) **104**: 438–447.
134. Schwartz HJ, Blumenthal M, Brady R, *et al.*: A comparative study of the clinical efficacy of nedocromil sodium and placebo. How does cromolyn sodium compare as an active control treatment? *Chest* (1996) **109**: 945–952.
135. Holgate ST: A rationale for the use of nedocromil sodium in the treatment of asthma. *J Allergy Clin Immunol* (1996) **98**: S157–S160.
136. Holgate ST: The efficacy and therapeutic position of nedocromil sodium. *Respir Med* (1996) **90**: 391–394.
137. Holgate ST: Inhaled sodium cromoglycate. *Respir Med* (1996) **90**: 387–390.

40

Glucocorticosteroids

PETER J BARNES

INTRODUCTION

Steroids are the most effective therapy currently available for asthma and improvement with steroids is one of the hallmarks of asthma. Inhaled glucocorticoids have revolutionized asthma treatment and have now become the mainstay of therapy for patients with chronic disease.[1] There has recently been an enormous increase in our understanding of the molecular mechanisms whereby glucocorticoids suppress inflammation in asthma and this has led to changes in the way steroids are used and may point the way to the development of more specific therapies in the future.[2,3]

MOLECULAR MECHANISMS

Glucocorticoids are very effective anti-inflammatory therapy in asthma and the molecular mechanisms involved in suppression of inflammation in asthma have recently been clarified.[3,4] It is evident that steroids are highly effective because they block many of the inflammatory pathways activated in asthma.

Glucocorticoid receptors

Glucocorticoids exert their effects by binding to glucocorticoid receptors (GRs), which are localized to the cytoplasm of target cells. The affinity of cortisol binding to GRs is

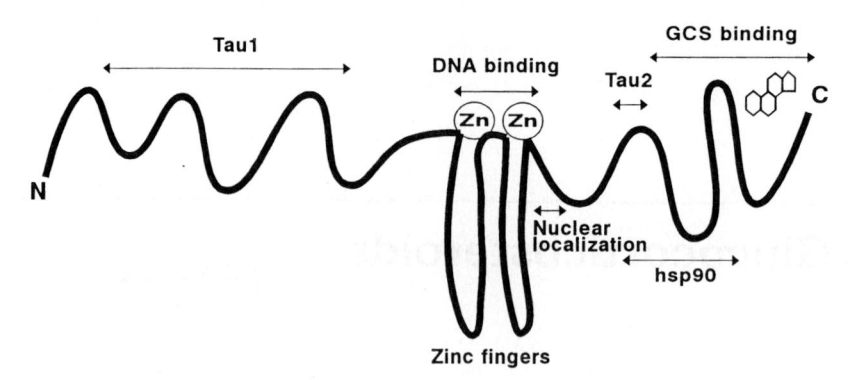

Fig. 40.1 Domains of the glucocorticoid (GCS) receptor.

approximately 30 nM, which falls within the normal range for plasma concentrations of free hormone. There is a single class of GR that binds glucocorticoids, with no evidence for subtypes of differing affinity in different tissues. Recently a splice variant of GR, termed GR-β, has been identified that does not bind glucocorticoids but binds to DNA and may therefore interfere with the action of steroids.[5] The structure of GR has been elucidated using site-directed mutagenesis, which has revealed distinct domains.[6,7] The glucocorticoid-binding domain is at the C-terminal end of the molecule; in the middle of the molecule are two finger-like projections that interact with DNA. Each of these 'zinc fingers' is formed by a zinc molecule bound to four cysteine residues (Fig. 40.1). An N-terminal domain (τ_1) is involved in transcriptional *trans*-activation of genes once binding to DNA has occurred and this region may also be involved in binding to other transcription factors.[8] Another *trans*-activating domain (τ_2) is adjacent to the steroid-binding domain and is also important for the nuclear translocation of the receptor. The inactivated GR is bound to a protein complex (300 kDa) that includes two molecules of 90-kDa heat-shock protein (hsp 90) and various other inhibitory proteins. The hsp 90 molecules act as a 'molecular chaperone' preventing the unoccupied GR localizing to the nuclear compartment. Once the glucocorticoid binds to GR, hsp 90 dissociates, thus exposing two nuclear localization signals and allowing the nuclear localization of the activated GR–steroid complex and its binding to DNA (Fig. 40. 2).

Effects on gene transcription

Glucocorticoids produce their effect on responsive cells by activating GR to regulate directly or indirectly the transcription of certain target genes.[9,10] The number of genes per cell regulated *directly* by steroids is estimated to be between 10 and 100, but many genes are regulated indirectly through an interaction with other transcription factors, as discussed below. Upon activation GR forms a dimer that binds to DNA at consensus sites termed glucocorticoid response elements (GREs) in the 5′ upstream promoter region of steroid-responsive genes. This interaction changes the rate of transcription, resulting in either induction or repression of the gene. The consensus sequence for GRE binding is the palindromic 15-base pair sequence GGTACAnnnTGTTCT (where n is any nucleo-

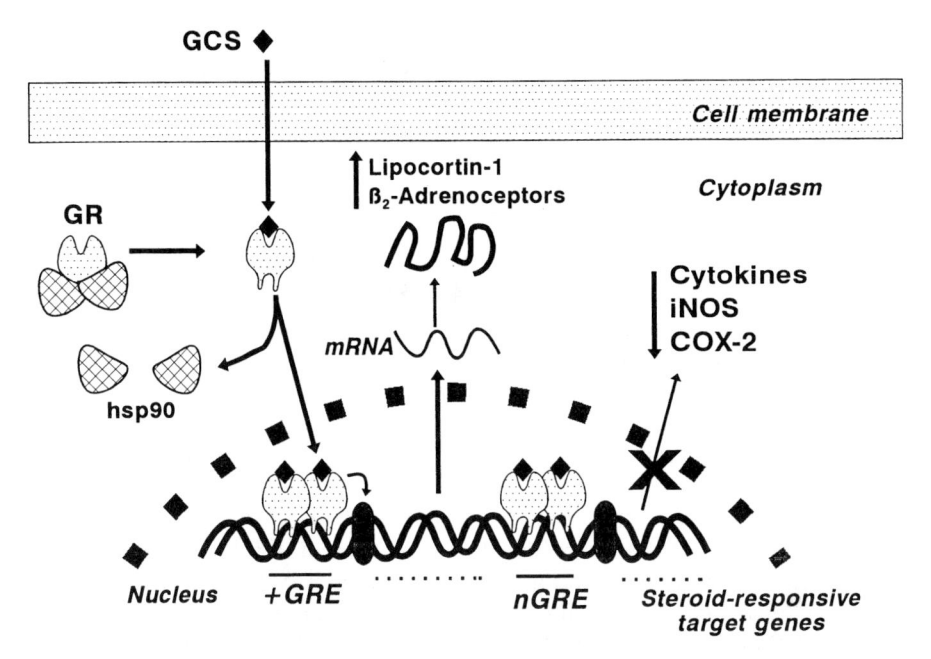

Fig. 40.2 Classical model of glucocorticoid action. The glucocorticoid (GCS) enters the cell and binds to a cytoplasmic glucocorticoid receptor (GR) that is complexed with two molecules of a 90-kDa heat-shock protein (hsp 90). GR translocates to the nucleus where, as a dimer, it binds to a glucocorticoid response element (GRE) in the 5′-upstream promoter sequence of steroid-responsive genes. GREs may increase transcription and nGREs may decrease transcription, resulting in increased or decreased mRNA and protein synthesis. COX-2, inducible cyclooxygenase; NOS, inducible nitric oxide synthase.

tide), although for repression of transcription the putative negative GRE (nGRE) has a more variable sequence (ATYACnnTnTGATCn). Crystallographic studies indicate that the zinc finger binding to DNA occurs within the major groove of DNA, with each finger interacting with one-half of the palindrome. Interaction with other transcription factors may also be important in determining differential steroid responsiveness in different cell types. Other transcription factors binding in the vicinity of GRE may influence steroid inducibility and the relative abundance of different transcription factors may contribute to the steroid responsiveness of a particular cell type. GR may also inhibit protein synthesis by reducing the stability of mRNA via enhanced transcription of specific ribonucleases that break down mRNA containing constitutive AU-rich sequences in the untranslated 3′ region, thus shortening the turnover time of mRNA.

Interaction with transcription factors

Activated GR may bind directly with other activated transcription factors as a protein–protein interaction. This could be an important determinant of steroid responsiveness and is a key mechanism whereby glucocorticoids exert their anti-inflammatory actions.[4] This interaction was first demonstrated for the collagenase gene, which is induced by the transcription factor activator protein (AP)-1, which is a heterodimer of Fos and Jun oncoproteins. AP-1, activated by phorbol esters or tumour necrosis factor (TNF)-α, forms a protein–protein complex with activated GR, and this prevents GR interacting with DNA and thereby reduces steroid responsiveness.[11] In human lung, TNF-α and phorbol esters increase AP-1 binding to DNA and this is inhibited by glucocorticoids.[12,13] GR also interacts with other transcription factors that are activated by inflammatory signals, including nuclear factor (NF)-κB in a similar manne[12–16] (Fig. 40.3). There is also evidence that β_2-agonists, via cyclic AMP formation and activation of protein kinase A, result in the activation of the transcription factor CREB that binds to a cyclic AMP-responsive element (CRE) on genes. A direct interaction between CREB and GR has been demonstrated.[17] β-Agonists increase CRE binding in human lung and epithelial cells *in vitro* and at the same time reduce GRE binding, suggesting that there may be a protein–protein interaction between CREB and GR within the nucleus.[18,19] These interactions between activated GR and transcription factors occur within the nucleus, but recent observations suggest that these protein–protein interactions may also occur in the cytoplasm.[20]

Recent evidence suggests that several transcription factors, including GR, interact with large coactivator molecules, such as CREB-binding protein (CBP) and the related p300, which bind to the basal transcription factor apparatus.[21] Since binding sites on this molecule may be limited, this may result in competition and several transcription factors, including CREB itself, NF-κB and AP-1 may compete with GR for binding, so that there is an indirect rather than a direct protein–protein interaction.

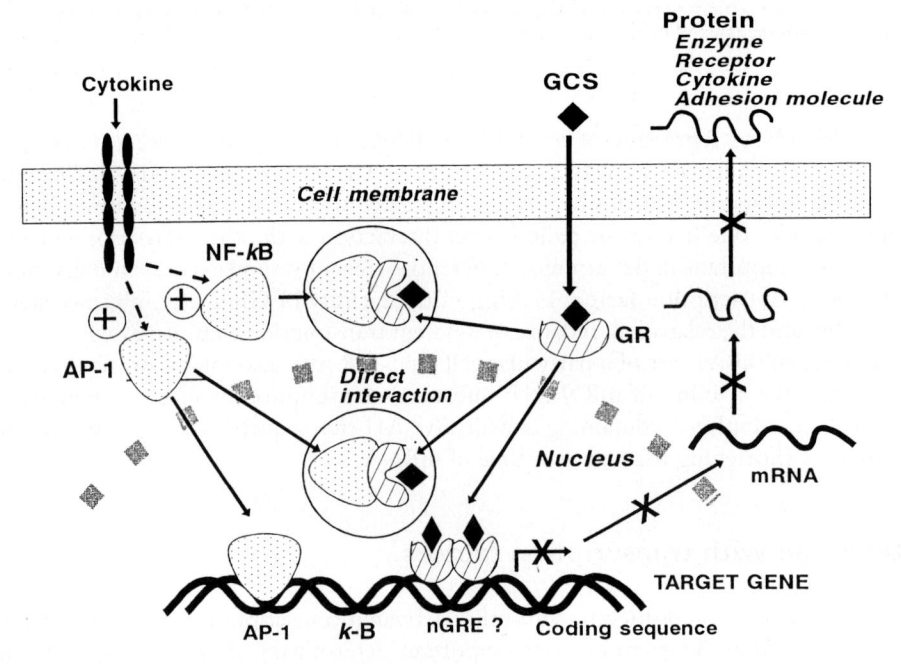

Fig. 40.3 Direct interaction between the transcription factors activator protein (AP)-1 and nuclear factor (NF)-κB and the glucocorticoid receptor (GR) may result in mutual repression. In this way steroids may counteract the chronic inflammatory effects of cytokines that activate these transcription factors. GRE, glucocorticoid response element; GCS, glucocorticoid.

Target genes in inflammation control

Glucocorticoids may control inflammation by inhibiting many aspects of the inflammatory process, increasing the transcription of anti-inflammatory genes and decreasing the transcription of inflammatory genes[2,4] (Table 40.1).

Anti-inflammatory proteins

Glucocorticoids may suppress inflammation by increasing the synthesis of anti-inflammatory proteins. Steroids increase the synthesis of lipocortin 1, a 37-kDa protein that has an inhibitory effect on phospholipase A_2 (PLA_2) and which therefore may inhibit the production of lipid mediators. Steroids induce the formation of lipocortin 1 in several cells and recombinant lipocortin 1 has acute anti-inflammatory properties.[22] However, glucocorticoids do not induce lipocortin 1 expression in all cells and this may be only one of many genes regulated by glucocorticoids. Glucocorticoids also increase the synthesis of secretory leucocyte protease inhibitor (SLPI) in human airway epithelial cells by increasing gene transcription.[23] SLPI is the predominant antiprotease in conducting airways and may be important in reducing airway inflammation by counteracting inflammatory enzymes, such as tryptase.

Interleukin (IL)-10 is an anti-inflammatory cytokine secreted predominantly by macrophages in the lung that inhibits the transcription of many pro-inflammatory cytokines and chemokines; this appears to be mediated via an inhibitory effect on NF-κB.[24] IL-10 secretion by alveolar macrophages may be impaired in asthmatic patients, resulting in increased macrophage cytokine secretion.[25,26] Glucocorticoid treatment in asthmatic patients increases IL-10 secretion by these cells, although this appears to be an indirect effect, since treatment of alveolar macrophages *in vitro* with glucocorticoids tends to decrease IL-10 secretion.[26]

Table 40.1 Effect of glucocorticoids on gene transcription.

Increased transcription
Lipocortin 1
$β_2$-Adrenoceptor
Secretory leucocyte inhibitory protein
IκB-α

Decreased transcription
Cytokines (IL-1, IL-2, IL-3, IL-4, IL-5, IL-6, IL-8, IL-11, IL-13, TNF-α, GM-CSF, RANTES, MIP-1α, eotaxin, SCF)
Inducible nitric oxide synthase (iNOS)
Inducible cyclooxygenase (COX-2)
Inducible phospholipase A_2 (cPLA$_2$)
Endothelin 1
NK$_1$ receptors
Adhesion molecules (ICAM-1, VCAM-1)

For explanation of abbreviations, see text.

β_2-Adrenoceptors

Steroids increase the expression of β_2-adrenoceptors by increasing the rate of transcription and the human β_2-adrenoceptor gene has three potential GREs.[27] Steroids double the rate of β_2-adrenoceptor gene transcription in human lung *in vitro*, resulting in increased expression of β_2-adrenoceptors.[28] Using autoradiographic mapping and *in situ* hybridization in animals to localize the increase in β_2-adrenoceptor expression, there appears to be an increase in all cell types, including airway epithelial cells and airway smooth muscle, after chronic glucocorticoid treatment.[29] This may be relevant in asthma as it may prevent downregulation in response to prolonged treatment with β_2-agonists. In rats glucocorticoids prevent the downregulation and reduced transcription of β_2-adrenoceptors in response to chronic β-agonist exposure.[29]

Cytokines

Although it is not yet possible to be certain of the most critical aspects of steroid action in asthma, it is likely that their inhibitory effects on cytokine synthesis are of particular relevance. Steroids inhibit the transcription of several cytokines that are relevant in asthma, including IL-1β, TNF-α, granulocyte–macrophage colony-stimulating factor (GM-CSF), IL-2, IL-3, IL-4, IL-5, IL-6, IL-11 and the chemokines IL-8, RANTES, macrophage chemotactic protein (MCP)-1, MCP-3, macrophage inflammatory protein (MIP)-1α and eotaxin. These inhibitory effects were at one time thought to be mediated directly via interaction of GR with an nGRE in the upstream promoter sequence of the cytokine gene, resulting in reduced gene transcription. Surprisingly, there is no apparent nGRE consensus sequence in the upstream promoter region of these cytokines, suggesting that glucocorticoids inhibit transcription indirectly. Thus, the 5' promoter sequence of the human IL-2 gene has no GRE consensus sequences, yet glucocorticoids are potent inhibitors of IL-2 gene transcription in T-lymphocytes. Transcription of the IL-2 gene is predominantly regulated by a cell-specific transcription factor, nuclear factor of activated T-cells (NF-AT), which is activated in the cytoplasm on T-cell receptor stimulation via calcineurin. A nuclear factor is also necessary for increased activation and this factor appears to be AP-1, which binds directly to NF-AT to form a transcriptional complex.[30] Glucocorticoids therefore inhibit IL-2 gene transcription indirectly by binding to AP-1, thus preventing increased transcription due to NF-AT.[31] Other examples of cytokine genes negatively regulated by glucocorticoids that do not have a GRE in their promoter region include IL-8, which is regulated predominantly via NF-κB, and RANTES, which is regulated by NF-κB and AP-1.[32] There may be marked differences in the response of different cells and of different cytokines to the inhibitory action of glucocorticoids and this may be dependent on the relative abundance of transcription factors. Thus in alveolar macrophages and peripheral blood monocytes GM-CSF secretion is more potently inhibited by glucocorticoids than IL-1β or IL-6 secretion.[33]

Inflammatory enzymes

Nitric oxide synthase (NOS) may be induced by pro-inflammatory cytokines, resulting in increased nitric oxide (NO) production (see Chapter 21). NO may amplify asthmatic inflammation and contribute to epithelial shedding and airway hyperresponsiveness through the formation of peroxynitrite. The induction of the inducible form of NOS (iNOS) is potently inhibited by glucocorticoids. In cultured human pulmonary epithelial cells, pro-inflammatory cytokines result in increased expression of iNOS and increased NO formation, due to increased transcription of the iNOS gene, and this is inhibited by glucocorticoids.[34] There is no nGRE in the promoter sequence of the iNOS gene, but NF-κB appears to be the most important transcription factor in regulating iNOS gene transcription.[35] Since TNF-β, IL-1β and oxidants activate NF-κB in airway epithelial cells, this accounts for their activation of iNOS expression.[36] Glucocorticoids may therefore prevent induction of iNOS by inactivating NF-κB, thereby inhibiting transcription.

Glucocorticoids inhibit the synthesis of several inflammatory mediators implicated in asthma through an inhibitory effect on enzyme induction. Glucocorticoids inhibit the induction of the gene coding for inducible cyclooxygenase (COX-2) in monocytes and epithelial cells and this also appears to be via NF-κB activation.[37–39] Glucocorticoids also inhibit gene transcription of a form of PLA$_2$ induced by cytokines.[39]

Steroids also inhibit the synthesis of endothelin 1 in lung and airway epithelial cells and this effect may also be via inhibition of transcription factors that regulate its expression.[40]

Inflammatory receptors

Glucocorticoids also decrease the transcription of genes coding for certain receptors. Thus the NK$_1$ receptor, which mediates the inflammatory effects of substance P in the airways, may show increased gene expression in asthma.[41] This may be inhibited by steroids through an interaction with AP-1 as the NK$_1$-receptor gene promoter region has no GRE but has an AP-1 response element.[42]

IL-1 acts on two types of surface receptor designated IL-1RI and IL-1RII. The inflammatory effects of IL-1β are mediated exclusively via IL-1RI, whereas IL-1RII has no signalling activity but binds IL-1 and therefore acts as a molecular trap that interferes with the actions of IL-1. Glucocorticoids are potent inducers of this decoy IL-1 receptor and result in release of a soluble form of the receptor, thus reducing the functional activity of IL-1.[43]

Apoptosis

Steroids markedly reduce the survival of certain inflammatory cells, such as eosinophils. Eosinophil survival is dependent on the presence of certain cytokines, such as IL-5 and GM-CSF. Exposure to steroids blocks the effects of these cytokines and leads to programmed cell death or apoptosis.[44]

Adhesion molecules

Adhesion molecules play a key role in the trafficking of inflammatory cells to sites of inflammation. The expression of many adhesion molecules on endothelial cells is induced by cytokines, and steroids may lead indirectly to reduced expression via their inhibitory effects on cytokines such as IL-1Ã and TNF-β. Steroids may also have a direct inhibitory effect on the expression of adhesion molecules, such as intercellular adhesion molecule

(ICAM)-1 and E-selectin, at the level of gene transcription.[45] ICAM-1 expression in bronchial epithelial cell lines and monocytes is inhibited by glucocorticoids.[46]

Other effects

Steroids exert a number of other anti-inflammatory effects that are not yet understood at a molecular level. Steroids have a direct inhibitory effect on plasma exudation from postcapillary venules at inflammatory sites. The onset of effect is delayed, suggesting that gene transcription and protein synthesis are involved. The mechanism for this antipermeability effect has not been fully elucidated, although there is evidence that synthesis of a 100-kDa protein distinct from lipocortin 1, termed vasocortin, may be involved.[47]

EFFECTS ON CELL FUNCTION

Steroids may have direct inhibitory actions on several inflammatory cells implicated in pulmonary and airway diseases (Fig. 40.4).

Macrophages

Steroids inhibit the release of inflammatory mediators and cytokines from alveolar macrophages *in vitro*,[33] although their effect after inhalation *in vivo* is modest.[48] Steroids

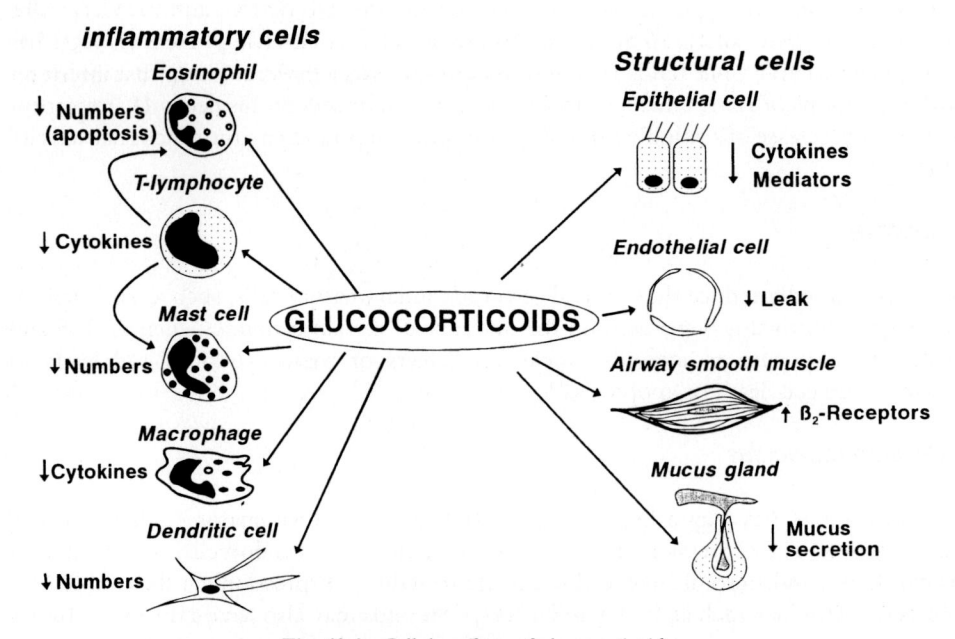

Fig. 40.4 Cellular effects of glucocorticoids.

40 Glucocorticosteroids

may be more effective in inhibiting cytokine release from alveolar macrophages than in inhibition of lipid mediators and reactive oxygen species *in vitro*.[49] Inhaled steroids reduce the secretion of chemokines and pro-inflammatory cytokines from alveolar macrophages from asthmatic patients, whereas the secretion of IL-10 is increased.[26] Oral prednisone inhibits the increased gene expression of IL-1β in alveolar macrophages obtained by bronchoalveolar lavage from asthmatic patients.[50]

Eosinophils

Steroids have a direct inhibitory effect on mediator release from eosinophils, although they are only weakly effective in inhibiting secretion of reactive oxygen species and eosinophil basic proteins.[51] Steroids inhibit the permissive action of cytokines such as GM-CSF and IL-5 on eosinophil survival[52,53] and this contributes to the reduction in airway eosinophils seen with steroid therapy. One of the best-described actions of steroids in asthma is a reduction in circulating eosinophils, which may reflect an action on eosinophil production in the bone marrow. Inhaled steroids inhibit the increase in circulating eosinophil count at night in patients with nocturnal asthma and also reduce plasma concentrations of eosinophil cationic protein.[54] After inhaled steroids (budesonide 800 μg b.d.) there is a marked reduction in the number of low-density eosinophils, presumably reflecting inhibition of cytokine production in the airways.[55]

T-lymphocytes

An important target cell in asthma may be the T-lymphocyte, since steroids are very effective in inhibition of activation of these cells and in blocking the release of cytokines likely to play an important role in the recruitment and survival of inflammatory cells involved in asthmatic inflammation. Thus glucocorticoids potently inhibit the secretion of IL-5 from T-lymphocytes.[56]

Mast cells

While steroids do not appear to have a direct inhibitory effect on mediator release from lung mast cells,[57] chronic steroid treatment is associated with a marked reduction in mucosal mast cell number.[58,59] This may be linked to a reduction in IL-3 and stem cell factor (SCF) production, which are necessary for mast cell expression at mucosal surfaces. Mast cells also secrete various cytokines (TNF-β, IL-4, IL-5, IL-6, IL-8), but whether these are inhibited by steroids has not yet been reported.

Dendritic cells

Dendritic cells in the epithelium of the respiratory tract appear to play a critical role in antigen presentation in the lung, as they have the capacity to take up allergen, process it into peptides and present it via MHC molecules on the cell surface to uncommitted T-lymphocytes.[60] In experimental animals the number of dendritic cells is markedly reduced by systemic and inhaled steroids, thus dampening the immune response in the airways.[61] Topical steroids markedly reduce the number of dendritic cells in the nasal mucosa[62] and it is likely that a similar effect would be seen in airways.

Neutrophils

Neutrophils, which are not prominent in the biopsies of asthmatic patients, are not very sensitive to the effects of steroids. Indeed systemic steroids increase peripheral neutrophil counts, which may reflect an increased survival time due to an inhibitory action on neutrophil apoptosis (in complete contrast to the increased apoptosis seen in eosinophils).[63]

Endothelial cells

GR gene expression in the airways is most prominent in endothelial cells of the bronchial circulation and airway epithelial cells. Steroids do not appear to inhibit the expression of adhesion molecules directly, although they may inhibit cell adhesion indirectly by suppression of cytokines involved in the regulation of adhesion molecule expression. Steroids may have an inhibitory action on airway microvascular leak induced by inflammatory mediators.[64,65] This appears to be a direct effect on postcapillary venular epithelial cells and and there is evidence that synthesis of a 100-kDa protein distinct from lipocortin 1, termed vasocortin, may be involved.[47] Although there have been no direct measurements of the effects of steroids on airway microvascular leakage in asthmatic airways, regular treatment with inhaled steroids decreases the elevated plasma proteins found in bronchoalveolar lavage fluid of patients with stable asthma.[66]

Epithelial cells

Epithelial cells may be an important source of inflammatory mediators in asthmatic airways and may drive and amplify the inflammatory response in the airways.[67,68] Airway epithelium may be one of the most important targets for inhaled glucocorticoids in asthma.[3,69] Steroids inhibit the increased transcription of the IL-8 gene induced by TNF-β in cultured human airway epithelial cells *in vitro*[70,71] and the transcription of the RANTES gene in an epithelial cell line.[72] Inhaled steroids inhibit the increased expression of GM-CSF and RANTES in the epithelium of asthmatic patients.[67,73,74] There is increased expression of iNOS in the airway epithelium of patients with asthma[75] and this may account for the increase in NO in the exhaled air of patients with asthma compared with normal subjects.[76] Asthmatic patients who are taking regular inhaled steroid therapy, however, do not show such an increase in exhaled NO,[76] suggesting that glucocorticoids have suppressed epithelial iNOS expression. Furthermore double-blind randomized studies show that oral and inhaled glucocorticoids reduce the elevated exhaled NO in asthmatic patients to normal values.[77,78] Glucocorticoids also decrease the transcription of other inflammatory proteins in airway epithelial cells, including COX-2,

40 Glucocorticosteroids

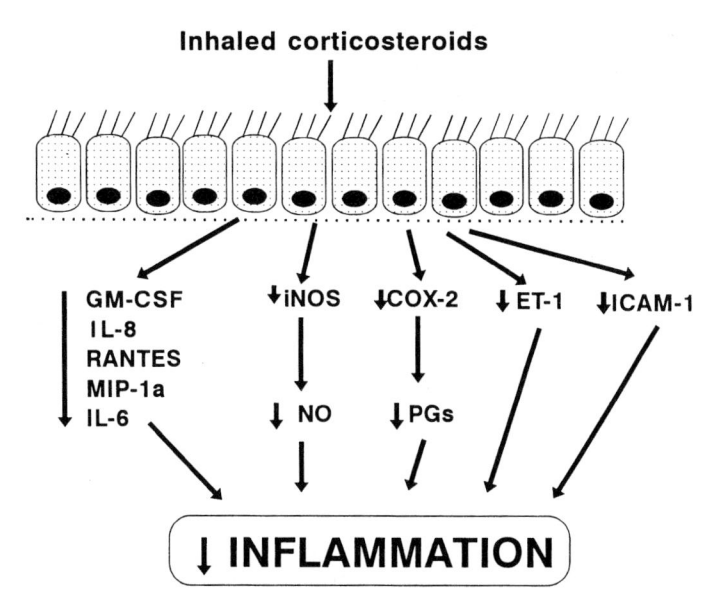

Fig. 40.5 Inhaled steroids may inhibit the transcription of several 'inflammatory' genes in airway epithelial cells and thus reduce inflammation in the airway wall. COX-2, inducible cyclooxygenase; ET-1, endothelin 1; GM-CSF, granulocyte–macrophage colony-stimulating factor; ICAM-1, intercellular adhesion molecule 1; IL, interleukin; iNOS, inducible nitric oxide synthase; MIP-1α, macrophage inflammatory protein-1α; NO, nitric oxide; PGs, prostaglandins

cPLA$_2$ and endothelin 1.[37,39,40] Airway epithelial cells may be the key cellular target of inhaled steroids; by inhibiting the transcription of several inflammatory genes inhaled steroids may reduce inflammation in the airway wall (Fig. 40.5).

Mucus secretion

Steroids inhibit mucus secretion in airways and this may be a direct action of steroids on submucosal gland cells.[79] Recent studies suggest that steroids may also inhibit the expression of mucin genes, such as MUC2 and MUC5a.[80] In addition, there are indirect inhibitory effects due to the reduction in inflammatory mediators that stimulate increased mucus secretion.

EFFECTS ON ASTHMATIC INFLAMMATION

Glucocorticoids are remarkably effective in controlling the inflammation in asthmatic airways and it is likely that they have multiple cellular effects. Biopsy studies in patients with asthma have now confirmed that inhaled steroids reduce the number and activation of inflammatory cells in the airway.[58,59,74,81,82] Similar results have been reported in bronchoalveolar lavage of asthmatic patients, with a reduction in both eosinophil number and eosinophil cationic protein concentrations, a marker of eosinophil degranulation, after inhaled budesonide.[83] These effects may be due to inhibition of cytokine synthesis in inflammatory and structural cells. There is also a reduction in activated $CD4^+$ T-cells ($CD4^+/CD25^+$) in bronchoalveolar lavage fluid after inhaled glucocorticoids.[84] The disrupted epithelium is restored and the ciliated to goblet cell ratio is normalized after 3 months of therapy with inhaled steroids.[58] There is also some evidence for a reduction in the thickness of the basement membrane,[74] although in asthmatic patients taking inhaled steroids for over 10 years the characteristic thickening of the basement membrane was still present.[85]

Effects on airway hyperresponsiveness

By reducing airway inflammation inhaled steroids consistently reduce airway hyperresponsiveness in asthmatic adults and children.[86] Chronic treatment with inhaled steroids reduces responsiveness to histamine, cholinergic agonists, allergen (early and late responses), exercise, fog, cold air, bradykinin, adenosine and irritants (such as sulphur dioxide and metabisulphite). The reduction in airway hyperresponsiveness takes place over several weeks and may not be maximal until after several months of therapy. The magnitude of reduction is variable between patients and is in the order of one to two doubling dilutions for most challenges but often fails to return to the normal range. This may reflect suppression of the inflammation but persistence of structural changes that cannot be reversed by steroids. Inhaled steroids not only make the airways less sensitive to spasmogens but they also limit the maximal airway narrowing in response to spasmogens.[87]

CLINICAL EFFICACY OF INHALED STEROIDS

Inhaled steroids are very effective in controlling asthma symptoms in asthmatic patients of all ages and severity.[1,88]

Studies in adults

Inhaled steroids were first introduced to reduce the requirement for oral steroids in patients with severe asthma and many studies have confirmed that the majority of patients can be weaned off oral steroids.[89] As experience has been gained with inhaled steroids they have been introduced in patients with milder asthma, with the recognition that inflammation is present even in patients with mild asthma.[90] Inhaled anti-inflammatory drugs have now become first-line therapy in any patient who needs to use a β_2-agonist inhaler more than once a day and this is reflected in national and international guidelines for the management of chronic asthma.[91–94] In patients with newly diagnosed asthma inhaled steroids (budesonide 600 µg twice daily) reduced symptoms and β_2-agonist inhaler usage and improved peak expiratory flow (PEF). These effects persisted over the 2 years of the study, whereas in a parallel group treated with inhaled β_2-agonists alone there was no significant change in symptoms or lung function.[95] In another study, patients with mild asthma treated with a low dose of inhaled steroid (budesonide 200 μg twice daily) showed less symptoms and a progressive improvement in lung function over several months and many patients became completely asymptomatic.[96] Similarly, inhaled beclomethasone dipropionate (BDP, 400 μg twice daily) improved asthma symptoms and lung function and this was maintained over the 2.5 years of the study.[97] There was also a significant reduction in the number of exacerbations. Although the effects of inhaled steroids on airway hyperresponsiveness may take several months to reach a plateau, the reduction in asthma symptoms occurs more rapidly.[98]

High-dose inhaled steroids have now been introduced in many countries for the control of more severe asthma. This markedly reduces the need for maintenance oral steroids and has revolutionized the management of more severe and unstable asthma.[99-101] Inhaled steroids are the treatment of choice in nocturnal asthma, which is a manifestation of inflamed airways, reducing night-time awakening and reducing the diurnal variation in airway function.[102,103]

Inhaled steroids effectively control asthmatic inflammation but must be taken regularly. When inhaled steroids are discontinued there is usually a gradual increase in symptoms and airway responsiveness back to pretreatment values,[98] although in patients with mild asthma who have been treated with inhaled steroids for a long time symptoms may not recur in some patients.[104]

Studies in children

Inhaled steroids are equally effective in children. In an extensive study of children aged 7–17 years there was a significant improvement in symptoms, peak flow variability and lung function compared with a regular inhaled β_2-agonist that was maintained over the 22 months of the study;[105] however, asthma deteriorated when the inhaled steroids were withdrawn.[106] There was a high proportion of drop-outs (45%) in the group treated with inhaled β_2-agonist alone. Inhaled steroids are also effective in younger children. Nebulized budesonide reduced the need for oral steroids and also improved lung function in children under the age of 3 years.[107] Inhaled steroids given via a large-volume spacer improved asthma symptoms and reduced the number of exacerbations in preschool children and in infants.[108,109]

Prevention of irreversible changes

Some patients with asthma develop an element of irreversible airflow obstruction, the pathophysiological basis of which is not yet understood. It is likely that they are the result of chronic airway inflammation and that they may be prevented by treatment with inhaled steroids. There is some evidence that the annual decline in lung function may be slowed by the introduction of inhaled steroids.[110] Recent evidence also suggests that delay in starting inhaled steroids may result in less overall improvement in lung function in both adults and children.[111-113]

Reduction in mortality

Whether inhaled steroids reduce the mortality from asthma is not yet established, as prospective studies are almost impossible to conduct. In a retrospective review of the risk of mortality and prescribed antiasthma medication, there was a significant apparent protection provided by regular inhaled BDP therapy (adjusted odds ratio of 0.1), although numbers were small.[114]

Comparison between inhaled steroids

Several inhaled steroids are currently prescribable in asthma, although their availability varies between countries. There have been relatively few studies comparing efficacy of the different inhaled steroids, and it is important to take into account the delivery system and the type of patient under investigation when such comparisons are made. In the UK, BDP, budesonide and fluticasone propionate (FP) are available; in the USA, BDP, flunisolide, triamcinolone and FP are available. There are few studies comparing different doses of inhaled steroids in asthmatic patients. Budesonide has been compared with BDP and in adults and children appears to have comparable antiasthma effects at equal doses.[115,116] However, there do appear to be some differences between inhaled steroids in terms of their systemic effects at comparable antiasthma doses.

PHARMACOKINETICS

The pharmacokinetics of inhaled steroids is important in determining the concentration of drug reaching target cells in the airways and in the fraction of drug reaching the systemic circulation and therefore causing side-effects.[1,117,118] Beneficial properties in an inhaled steroid are a high topical potency, a low systemic bioavailability of the swallowed portion of the dose and rapid metabolic clearance of any steroid reaching the systemic circulation. After inhalation a large proportion of the inhaled dose (80–90%) is deposited on the oropharynx and is then swallowed and therefore available for absorption via the liver into the systemic circulation (Fig. 40.6). This fraction is markedly reduced by using a large-volume spacer device with a metered dose inhaler (MDI) or by mouth-washing and discarding the washing when using dry powder inhalers. Between 10 and 20% of inhaled drug enters the respiratory tract, where it is deposited in the airways and this fraction is available for absorption into the systemic circulation. Most of the early studies on the distribution of inhaled steroids were conducted in healthy volunteers, and it is not certain what effect inflammatory disease, airway obstruction, age of the patient or concomitant medication may have on the disposition of the inhaled dose. There may be important differences in the metabolism of different inhaled steroids. BDP is metabolized to its more active metabolite beclomethasone monopropionate in many tissues including lung,[119] but there is no information about absorption or metabolism of this metabolite in humans. Flunisolide and budesonide are subject to extensive first-pass metabolism in the liver so that less reaches the systemic circulation.[120,121] Little is known about the distribution of

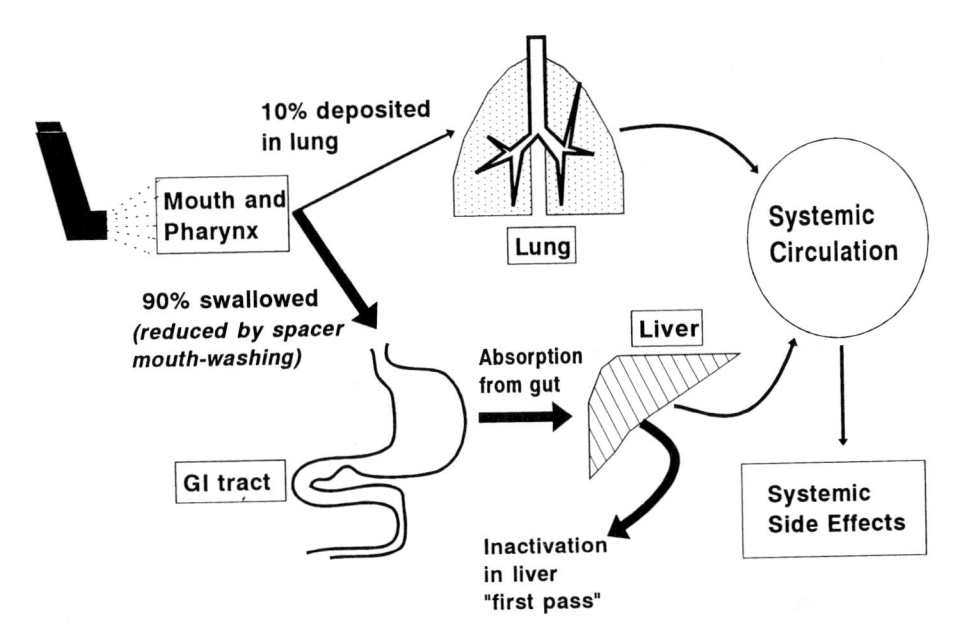

Fig. 40.6 Pharmacokinetics of inhaled steroids.

triamcinolone.[122] FP is almost completely metabolized by first-pass metabolism, which reduces systemic effects.[123]

Frequency of administration

When inhaled steroids were first introduced it was recommended that they should be given four times daily, but several studies have now demonstrated that twice-daily administration gives comparable control,[124,125] although four times daily administration may be preferable in patients with more severe asthma.[126] However, patients may find it difficult to comply with such frequent administration unless they have troublesome symptoms. For patients with mild asthma who require $\leqslant 400$ μg daily, once-daily therapy may be sufficient.[127]

SIDE-EFFECTS OF INHALED STEROIDS

The efficacy of inhaled steroids is now established in short-term and long-term studies in adults and children, although there are still concerns about side-effects, particularly in children and when high inhaled doses are needed. Several side-effects have been recognized (Table 40.2).

Table 40.2 Side-effects of inhaled steroids.

Local side-effects
Dysphonia
Oropharyngeal candidiasis
Cough

Systemic side-effects
Adrenal suppression
Growth suppression
Bruising
Osteoporosis
Cataracts
Glaucoma
Metabolic abnormalities (glucose, insulin, triglycerides)
Psychiatric disturbances

Local side-effects

Side-effects due to the local deposition of the steroid in the oropharynx may occur with inhaled steroids, but the frequency of complaints depends on the dose and frequency of administration and on the delivery system used.

Dysphonia

The commonest complaint is of hoarseness of the voice (dysphonia) and may occur in over 50% of patients using an MDI.[128,129] Dysphonia is not appreciably reduced by using spacers, but may be less with dry powder devices.[130] Dysphonia may be due to myopathy of laryngeal muscles and is reversible when treatment is withdrawn.[129] For most patients it is not troublesome but may be disabling in singers and lecturers.

Oropharyngeal candidiasis

Oropharyngeal candidiasis (thrush) may be a problem in some patients, particularly in the elderly, with concomitant oral steroids and more than twice-daily administration.[128] Large-volume spacer devices protect against this local side-effect by reducing the dose of inhaled steroid that deposits in the oropharynx.

Other local complications

There is no evidence that inhaled steroid, even in high doses, increases the frequency of infections, including tuberculosis, in the lower respiratory tract.[131,132] There is no evidence for atrophy of the airway epithelium and even after 10 years of treatment with inhaled steroids there is no evidence for any structural changes in the epithelium.[85] Cough and throat irritation, sometimes accompanied by reflex bronchoconstriction, may occur when inhaled steroids are given via an MDI. These symptoms are likely to be due to surfactants in pressurized aerosols, as they disappear after switching to a dry powder steroid inhaler device.[133]

Systemic side-effects

The efficacy of inhaled steroids in the control of asthma is undisputed, although there are concerns about systemic effects of inhaled steroids, particularly as they are likely to be used over long periods and in children of all ages.[88,134] The safety of inhaled steroids has been extensively investigated since their introduction 30 years ago. One of the major problems is to decide whether a measurable systemic effect has any significant clinical consequence and this necessitates careful long-term follow-up studies. As biochemical markers of systemic steroid effects become more sensitive, systemic effects may be seen more often, but this does not mean that these effects are clinically relevant. There are several case reports of adverse systemic effects of inhaled steroids; these are often idiosyncratic reactions, which may be due to abnormal pharmacokinetic handling of the inhaled steroid. The systemic effect of an inhaled steroid will depend on several factors, including the dose delivered to the patient, the site of delivery (gastrointestinal tract and lung), the delivery system used and individual differences in the patient's response to the steroid.

Effect of delivery systems

The systemic effect of an inhaled steroid is dependent on the amount of drug absorbed into the systemic circulation. As noted above, approximately 90% of the inhaled dose from an MDI deposits in the oropharynx and is swallowed and subsequently absorbed from the gastrointestinal tract (see Fig. 40.6). Use of a large-volume spacer device markedly reduces oropharyngeal deposition and therefore the systemic effects of inhaled steroids.[135–137] For dry powder inhalers similar reductions in systemic effects may be achieved with mouth-washing and discarding the fluid.[137] All patients using a daily dose of ≥ 800 μg of an inhaled steroid should therefore use either a spacer or mouth-washing to reduce systemic absorption. Approximately 10% of inhaled dose from an MDI enters the lung and this fraction (which presumably exerts the therapeutic effect) may be absorbed into the systemic circulation. As the fraction of inhaled steroid deposited in the oropharynx is reduced, the proportion of inhaled dose entering the lungs is increased. More efficient delivery to the lungs is therefore accompanied by increased systemic absorption, but this is offset by a reduction in the dose needed for optimal control of airway inflammation. For example, a multiple dry powder delivery system, the Turbuhaler, delivers approximately twice as much steroid to the lungs as other devices and therefore has increased systemic effects. However this is compensated for by the fact that only half the dose is required.[138]

Hypothalamic–pituitary–adrenal axis

Glucocorticoids may cause hypothalamic–pituitary–adrenal (HPA) axis suppression by reducing corticotrophin (ACTH) production, which reduces cortisol secretion by the adrenal gland. The degree of HPA suppression is dependent on dose, duration, frequency and timing of steroid administration. The clinical significance of HPA axis suppression is two-fold. Firstly, prolonged adrenal suppression may lead to reduced adrenal response to stress. There is no evidence that cortisol responses to the stress of an asthma exacerbation or insulin-induced hypoglycaemia are impaired, even with high doses of inhaled steroids.[139] Secondly, measurement of HPA axis function provides evidence for systemic effects of an inhaled steroid. Basal adrenal cortisol secretion may be measured by a morning plasma cortisol, 24-h urinary cortisol or by plasma cortisol profile over 24-h.[140] Other tests measure the HPA response following stimulation with tetracosactrin (which measures adrenal reserve) or stimulation with metyrapone and insulin (which measures the response to stress).

There are many studies of HPA axis function in asthmatic patients with inhaled steroids, but the results are inconsistent as they have often been uncontrolled and patients have also been taking courses of oral steroids (which may affect the HPA axis for weeks).[141] BDP, budesonide and FP at high doses by conventional MDI (<1600 μg daily) give a dose-related decrease in morning serum cortisol levels and 24-h urinary cortisol, although values still lie well within the normal range.[142–144] However, when a large-volume spacer is used doses of 2000 μg daily of BDP or budesonide have little effect on 24-h urinary cortisol excretion.[145] Studies with inhaled flunisolide and triamcinolone in children show no effect on 24-h cortisol excretion at doses of up to 1000 μg daily.[146,147] Stimulation tests of HPA axis function similarly show no consistent effects of doses of 1500 μg or less of inhaled steroid. At high doses (<1500 μg daily) budesonide and FP have less effect than BDP on HPA axis function.[143,148] In children no suppression of urinary cortisol is seen with doses of BDP of 800 μg or less.[149–151] In studies where plasma cortisol has been measured at frequent intervals, there was a significant reduction in cortisol peaks with doses of inhaled BDP as low as 400 μg daily,[152] although this does not appear to be dose related in the range 400–1000 μg.[153,154] The clinical significance of these effects is not certain, however.

Overall, the studies that are not confounded by concomitant treatment with oral steroids have consistently shown that there are no significant suppressive effects on HPA axis function at doses of $\leqslant$1500 μg in adults and $\leqslant$400 μg in children.

Effects on bone metabolism

Steroids lead to a reduction in bone mass by direct effects on bone formation and resorption and indirectly by suppression of the pituitary–gonadal and HPA axes, effects on intestinal calcium absorption, renal tubular calcium reabsorption and secondary hyperparathyroidism.[155] The effects of oral steroids on osteoporosis and increased risk of vertebral and rib fractures are well known, but there are no reports suggesting that long-term treatment with inhaled steroids is associated with an increased risk of fractures. Bone densitometry has been used to assess the effect of inhaled steroids on bone mass. Although there is evidence that bone density is less in patients taking high-dose inhaled steroids, interpretation is confounded by the fact that these patients are also taking intermittent courses of oral steroids.[156]

Changes in bone mass occur very slowly and several biochemical indices have been used to assess the short-term effects of inhaled steroids on bone metabolism. Bone formation has been assessed by measurement of plasma concentrations of bone-specific alkaline phosphatase, serum osteocalcin, a non-collagenous 49 amino acid peptide secreted by osteoblasts, or by procollagen peptides. Bone resorption may be assessed by urinary hydroxyproline after a 12-h fast, urinary calcium excretion and pyridinium cross-link excretion. It is important to consider the age, diet, time of day and physical activity of the patient in interpreting any abnormalities. It is also necessary to choose appropriate control groups as asthma itself may have an effect on some of the measurements, such as osteocalcin.[157] Inhaled steroids, even at doses up to 2000 µg daily, have no significant effect on calcium excretion, but acute and reversible dose-related suppression of serum osteocalcin has been reported with BDP and budesonide when given by conventional MDI in several studies.[141] Budesonide consistently has less effect than BDP at equivalent doses and only BDP increases urinary hydroxyproline at high doses.[158] With a large-volume spacer even doses of 2000 µg daily of either BDP or budesonide are without effect on plasma osteocalcin concentrations.[145] Urinary pyridinium and deoxypyridinoline cross-links, which are a more accurate and stable measurement of bone and collagen degradation, are not increased with inhaled steroids (BDP <1000 µg daily), even with intermittent courses of oral steroids.[156] It is important to monitor changes in markers of bone formation as well as bone degradation, as the net effect on bone turnover is important.

There has been particular concern about the effect of inhaled steroids on bone metabolism in growing children. A very low dose of oral steroids (prednisolone 2.5 mg) causes significant changes in serum osteocalcin and urinary hydroxyproline excretion, whereas BDP and budesonide at doses up to 800 µg daily have no effect.[157,159] It is important to recognize that the changes in biochemical indices of bone metabolism are less than those seen with low doses of oral steroids. This suggests that even high doses of inhaled steroids, particularly when used with a spacer device, are unlikely to have any long-term effect on bone structure. Careful long-term follow-up studies in patients with asthma are needed.

There is no evidence that inhaled steroids increase the frequency of fractures. Long-term treatment with high-dose inhaled steroids has not been associated with any consistent change in bone density.[160,161] Indeed, in elderly patients there may be an increase in bone density due to increased mobility.[161]

Effects on connective tissue

Oral and topical steroids cause thinning of the skin, telangiectasiae and easy bruising, probably as a result of loss of extracellular ground substance within the dermis, due to an inhibitory effect on dermal fibroblasts. There are reports of increased skin bruising and purpura in patients using high doses of inhaled BDP, although the amount of intermittent oral steroids used by these patients is not known.[162,163] Easy bruising in association with inhaled steroids is more frequent in elderly patients[164] and there are no reports of this problem in children. Long-term prospective studies with objective measurements of skin thickness are needed with different inhaled steroids.

Ocular effects

Long-term treatment with oral steroids increases the risk of posterior subcapsular cataracts and there are several case reports describing cataracts in individual patients taking inhaled steroids.[141] In a study of 48 patients who were exposed to oral and/or high-dose inhaled steroids, the prevalence of posterior subcapsular cataracts (27%) correlated with the daily dose and duration of oral steroids but not with the dose and duration of inhaled steroids.[165] In a recent cross-sectional study in patients aged 5–25 years taking either inhaled BDP or budesonide, no cataracts were found on slit-lamp examination, even in patients taking 2000 µg daily for over 10 years.[166] Recently, there has been a report of a slight increase in the risk of glaucoma in patients taking very high doses of inhaled steroids.[167]

Growth

There has been particular concern that inhaled steroids may cause stunting of growth and several studies have addressed this issue. Asthma itself (as with other chronic diseases) may have an effect on the growth pattern and has been associated with delayed onset of puberty and deceleration of growth velocity that is more pronounced with more severe disease.[168] However, asthmatic children appear to grow for longer, so that their final height is normal. The effect of asthma on growth makes it difficult to assess the effects of inhaled steroids on growth in cross-sectional studies, particularly as courses of oral steroids are a confounding factor. Longitudinal studies have demonstrated that there is no significant effect of inhaled steroids on statural growth in doses of up to 800 µg daily and for up to 5 years of treatment.[105,141,169,170] A prospective study of inhaled BDP (400 µg daily) vs. theophylline in children with mild to moderate asthma showed no effect on height, although there was some reduction in growth velocity compared with children treated with theophylline.[171] However, it is not possible to relate changes in growth velocity to final height, as other studies have demonstrated that there is a 'catch-up' period. In a longitudinal study in children aged 2–7 years with severe asthma, budesonide 200 µg daily had no effect on growth over 3–5 years.[151] In children with virally induced wheezing, BDP 400 µg daily has been reported to reduce growth compared with a placebo.[172] A meta-analysis of 21 studies, including over 800 children, showed no effect of inhaled BDP on statural height, even with high doses and long durations of therapy.[173]

Short-term growth measurements (knemometry) have demonstrated that even a low dose of an oral steroid (prednisolone 2.5 mg) is sufficient to give complete suppression of lower leg growth. However, inhaled budesonide up to 400 µg is without effect, although some suppression is seen with 800 µg and with 400 µg BDP.[174,175] The relationship between knemometry measurements and final height are uncertain, since low doses of oral steroids that have no effect on final height cause profound suppression.

Metabolic effects

Several metabolic effects have been reported after inhaled steroids, although there is no evidence that these are clinically relevant at therapeutic doses. Fasting glucose and insulin are unchanged in adults with doses of BDP up to 2000 µg daily[115] and in children with inhaled budesonide up to 800 µg daily.[176] In normal individuals, high-dose inhaled BDP may slightly increase resistance to insulin.[177] However, in patients with poorly controlled asthma high doses of BDP and budesonide paradoxically decrease insulin resistance and improve glucose tolerance, suggesting that the disease itself may lead to abnormalities in carbohydrate metabolism.[178] Neither BDP 2000 µg daily in adults nor budesonide 800 µg daily in children have any effect on plasma cholesterol or triglycerides.[115,176]

40 Glucocorticosteroids

Haematological effects

Inhaled steroids may reduce the numbers of circulating eosinophils in asthmatic patients,[55] possibly due to an effect on local cytokine generation in the airways. Inhaled steroids may cause a small increase in circulating neutrophil counts.[145,179]

Central nervous system effects

There are various reports of psychiatric disturbance, including emotional lability, euphoria, depression, aggressiveness and insomnia, after inhaled steroids. Only eight such patients have so far been reported, suggesting that this is very infrequent and a causal link with inhaled steroids has usually not been established.[141]

Safety in pregnancy

Based on extensive clinical experience inhaled steroids appear to be safe in pregnancy, although no controlled studies have been performed. There is no evidence for any adverse effects of inhaled steroids on the pregnancy, the delivery or the fetus.[141] It is important to recognize that poorly controlled asthma may increase the incidence of perinatal mortality and retard intrauterine growth; more effective control of asthma with inhaled steroids may reduce these problems.

CLINICAL USE OF INHALED STEROIDS

Inhaled steroids are now recommended as first-line therapy for all but the mildest of asthmatic patients.[1] Inhaled steroids should be started in any patient who needs to use a β-agonist inhaler for symptom control more than once daily (or possibly three times weekly). It is conventional to start with a low dose of inhaled steroid and to increase the dose until asthma control is achieved. However, this may take time and a preferable approach is to start with a dose of steroids in the middle of the dose range (400 μg twice daily) to establish control of asthma more rapidly.[180] Once control is achieved (defined as normal or best possible lung function and infrequent need to use an inhaled β_2-agonist) the dose of inhaled steroid should be reduced in a step-wise manner to the lowest dose needed for optimal control. It may take as long as 3 months to reach a plateau in response and any changes in dose should be made at intervals of 3 months or more. This strategy ('start high, go low') is emphasized in the revised BTS Guidelines for Asthma Management.[94] When doses of $\geqslant 800$ μg daily are needed, a large-volume spacer device should be used with an MDI and mouth-washing with a dry powder inhaler in order to reduce local and systemic side-effects. Inhaled steroids are usually given as a twice-daily dose in order to increase compliance. When asthma is more unstable four times daily dosage is preferable.[126] For patients who require $\leqslant 400$ μg daily once-daily dosing appears to be as effective as twice-daily dosing, at least for budesonide.[127]

The dose of inhaled steroid should be increased to 2000 μg daily if necessary, although higher doses may result in systemic effects and it may be preferable to add a low dose of oral steroid, since higher doses of inhaled steroids are expensive and have a high incidence of local side-effects. Nebulized budesonide has been advocated in order to give an increased dose of inhaled steroid and to reduce the requirement for oral steroids,[181] but this treatment is expensive and may achieve its effects largely via systemic absorption.

Most of the guidelines for asthma treatment suggest that additional bronchodilators (slow-release theophylline preparations, inhaled and oral long-acting β_2-agonists and inhaled anticholinergics) should be introduced after increasing the dose of inhaled steroid to 1600–2000 µg daily. However, an alternative approach is to introduce these treatments when patients are taking 400–800 µg inhaled steroid daily. Addition of the long-acting inhaled β_2-agonist salmeterol provides better control of asthma symptoms than doubling the dose of inhaled steroids.[182,183] Similarly, addition of low-dose oral theophylline gives better control than doubling the dose of inhaled steroid in patients not controlled on budesonide 800 µg daily.[184]

Inhaled steroids may be the most cost-effective way of controlling asthma, since reducing the frequency of asthma attacks will save on total costs.[185,186] Inhaled steroids improve the quality of life of patients with asthma and allow many patients a normal lifestyle.[187]

SYSTEMIC STEROIDS

Oral or intravenous steroids may be indicated in several situations. Prednisolone, rather than prednisone, is the preferred oral steroid as prednisone has to be converted in the liver to the active prednisolone. Prednisone may be preferable in pregnant patients as it is not converted to prednisolone in the fetal liver, thus diminishing the exposure of the fetus to glucocorticoids. Enteric-coated preparations of prednisolone are used to reduce side-effects (particularly gastric side-effects) and give delayed and reduced peak plasma concentrations, although the bioavailability and therapeutic efficacy of these preparations is similar to uncoated tablets. Prednisolone and prednisone are preferable to dexamethasone, betamethasone or triamcinolone, which have longer plasma half-lives and therefore an increased frequency of adverse effects.

Short courses of oral steroids (30–40 mg prednisolone daily for 1–2 weeks or until PEF values return to best attainable) are indicated for exacerbations of asthma; the dose may be tailed off over 1 week once the exacerbation is resolved. The tail-off period is not strictly necessary,[188] but some patients find it reassuring.

Maintenance oral steroids are only needed in a small proportion of asthmatic patients with the most severe asthma that cannot be controlled with maximal doses of inhaled steroids (2000 µg daily) and additional bronchodilators. The minimal dose of oral steroid needed for control should be used and reductions in the dose should be made slowly in patients who have been on oral steroids for long periods (e.g. by 2.5 mg per month for doses down to 10 mg daily and thereafter by 1 mg per month). Oral steroids are usually given as a single morning dose, as this reduces the risk of adverse effects since it coincides with the peak diurnal concentrations. There is some evidence that administration in the afternoon may be optimal for some patients who have severe nocturnal asthma.[189] Alternate-day administration may also reduce adverse effects, although in some patients control of asthma may not be as good on the day when the oral dose is omitted.

Intramuscular triamcinolone acetonide (80 mg monthly) has been advocated in patients with severe asthma as an alternative to oral steroids.[190,191] This may be considered in patients in whom compliance is a particular problem, but the major concern is the high frequency of proximal myopathy associated with this fluorinated steroid. Some patients who do not respond well to prednisolone are reported to respond to oral betamethasone, presumably because of pharmacokinetic handling problems with prednisolone.[192]

Steroid-sparing therapy

In patients who have serious side-effects with maintenance steroid therapy there are several treatments that have been shown to reduce the requirement for oral steroids.[193] These treatments are commonly termed steroid sparing, although this is a misleading description that could be applied to any additional asthma therapy (including bronchodilators). The amount of steroid sparing with these therapies is not impressive.

Several immunosuppressive agents have been shown to have steroid effects, including methotrexate,[194,195] oral gold[196] and cyclosporin A.[197,198] These therapies all have side-effects that may be more troublesome than those of oral steroids and are therefore only indicated as an additional therapy to reduce the requirement of oral steroids. None of these treatments is very effective, but there are occasional patients who appear to show a particularly good response. Because of side-effects these treatments cannot be considered a way to reduce the requirement for inhaled steroids. Side-effects are a problem with these immunosuppressive drugs and include nausea, vomiting, hepatic dysfunction, hepatic fibrosis, pulmonary fibrosis and increased infections with methotrexate and renal dysfunction with cyclosporin and oral gold. Several other therapies, including azathioprine, dapsone and hydroxychloroquine, have not been found to be beneficial. The macrolide antibiotic troleandomycin is also reported to have steroid-sparing effects, but this is only seen with methylprednisolone and is due to reduced metabolism of this steroid, so that there is little therapeutic gain.[199]

Acute severe asthma

Intravenous hydrocortisone is given in acute severe asthma. The recommended dose is 200 mg i.v.[93] While the value of corticosteroids in acute severe asthma has been questioned, others have found that they speed the resolution of attacks.[200] There is no apparent advantage in giving very high doses of intravenous steroids (such as methylprednisolone 1 g). Indeed, intravenous steroids have occasionally been associated with an acute severe myopathy.[201] In a recent study no difference in recovery from acute severe asthma was seen whether doses of hydrocortisone of 50, 200 or 500 mg i.v. 6-hourly were used;[202] another placebo-controlled study showed no beneficial effect of intravenous steroids.[203] Intravenous steroids are indicated in patients with acute asthma if lung function is >30% predicted and in whom there is no significant improvement with nebulized β_2-agonist. Intravenous therapy is usually given until a satisfactory response is obtained and then oral prednisolone may be substituted. Oral prednisolone (40–60 mg) has a similar effect to intravenous hydrocortisone and is easier to administer.[200,204] Oral prednisolone is the preferred treatment for acute severe asthma, providing there are no contraindications to oral therapy.[94]

GLUCOCORTICOID RESISTANCE IN ASTHMA

Although glucocorticoids are highly effective in the control of asthma and other chronic inflammatory or immune diseases, a small proportion of patients with asthma fail to respond even to high doses of oral glucocorticoids.[205–207] Resistance to the therapeutic effects of glucocorticoids is also recognized in other inflammatory and immune diseases, including rheumatoid arthritis and inflammatory bowel disease. Steroid-resistant patients, although uncommon, present considerable management problems. Recently, new insights into the mechanisms whereby glucocorticoids suppress chronic inflammation has shed new light on the molecular basis of glucocorticoid resistance in asthma.[206]

Clinical features

Glucocorticoid resistance in asthma was first described by Schwartz et al.[208] in 1968 in six asthmatic patients who did not respond clinically to high doses of systemic steroids and in whom there was also a reduced eosinopenic response.[208] Carmichael and colleagues[209] reported a larger group of patients with chronic asthma who were steroid resistant. These patients failed to improve their mean PEF by <15% after taking prednisolone 20 mg daily for at least 7 days. They differed clinically from steroid-sensitive patients only in having a longer duration of symptoms, lower morning PEF values and a more frequent family history of asthma. These patients are not Addisonian and they do not suffer from the abnormalities in sex hormones described in familial glucocorticoid resistance. Plasma cortisol and adrenal suppression in response to exogenous cortisol is normal in these patients.[210]

Complete steroid resistance in asthma is rare, but there are no population studies giving an estimate of the proportion of patients who are resistant. It is likely that most specialists would only have a few such patients in their clinic and the prevalence is probably >1 in 1000 asthmatic patients. Much more common is a reduced responsiveness to steroids, so that large inhaled or oral doses are needed to control asthma adequately. It is important to establish that the patient has asthma rather than chronic obstructive pulmonary disease (COPD), 'pseudoasthma' (a hysterical conversion syndrome involving vocal cord dysfunction), left ventricular failure or cystic fibrosis that do not respond to steroids.[211] Asthmatic patients are characterized by a variability in PEF and, in particular, a diurnal variability of <15% and episodic symptoms. It is also important to identify provoking factors (allergens, drugs, psychological problems) that may increase the severity of asthma and its resistance to therapy.

Distinction between steroid-sensitive and steroid-resistant asthmatic patients depends on the response to a high dose of oral steroids given for a reasonable period. In research studies prednisolone is usually given in a dose of 40 mg daily for 2 weeks with twice-daily monitoring of PEF. In steroid-resistant asthma, patients fail to improve the morning PEF or forced expiratory volume in 1 s (FEV_1) by <15%. Patients with steroid-resistant

40 Glucocorticosteroids

asthma show the typical diurnal variability in PEF and bronchodilate in response to inhaled β_2-agonists. Bronchial biopsy shows the typical inflammatory infiltrate of eosinophils in steroid-resistant patients.[212] It is clearly important to establish that the patient is taking the oral steroid by measurement of plasma cortisol, which is suppressed after high-dose oral steroids in both steroid-sensitive and steroid-resistant patients,[210] or by measurement of plasma prednisolone concentrations. Patients with COPD fail to improve their lung function after a course of oral steroids, but are distinguished from steriod-resistant asthmatic patients by their lack of acute bronchodilator response and absence of diurnal variability in PEF.

Another group of patients with asthma is responsive to steroids, but only in relatively high oral doses. These patients are best described as steroid dependent (i.e. dependent on oral steroids as opposed to inhaled steroids). These patients deteriorate when the dose of oral steroids is reduced. Rarely a maintenance dose of <40 mg prednisolone daily may be required and such patients may mistakenly be classified as steroid resistant. Steroid-dependent asthmatic patients usually have severe disease and are presumed to have a high level of inflammation in their airways.

Mechanisms of steroid resistance

There may be several mechanisms for resistance to the effects of glucocorticoids. Although a family history of asthma is more common in patients with steriod-resistant than steroid-sensitive asthma, little is known of the inheritance of steroid-resistant asthma. Resistance to the inflammatory and immune effects of glucocorticoids should be distinguished from the very rare familial glucocorticoid resistance, where there is an abnormality of glucocorticoid binding to GR.

Familial glucocorticoid resistance

The rare inherited syndrome familial glucocorticoid resistance is characterized by high circulating levels of cortisol without signs of symptoms of Cushing's syndrome.[213] Clinical manifestations, which may be absent, are due to an excess of non-glucocorticoid adrenal steroids, stimulated by high ACTH levels, resulting in hypertension with hypokalaemia and/or signs of androgen excess (usually hirsutism and menstrual abnormalities in females). Inheritance appears to be dominant with variable expression, but only about 12 cases have so far been reported. Several abnormalities in GR function have been described in peripheral blood leucocytes or fibroblasts from these patients. These include a decreased affinity of GR for cortisol, a reduced number of GRs, GR thermolability and an abnormality in the binding of the GR complex to DNA. The molecular basis of the disease in four patients with a reduction in GRs appears to be a point mutation in the steroid-binding domain of GR.

Resistance to anti-inflammatory actions of steroids

Resistance to the anti-inflammatory and immunomodulatory effects of glucocorticoids differs from the familial glucocorticoid resistance described above, as it is not associated with high circulating concentrations of cortisol or ACTH, and is not accompanied by hypertension, hypokalaemia or androgen excess. Furthermore, these patients are not Addisonian and show normal adrenal suppression. This suggests that any abnormality is unlikely to be due to the same abnormalities in the steroid-binding domain of GR, as described in familial glucocorticoid resistance. Indeed, recent chemical mutational analysis of GR has failed to demonstrate any major abnormality in predicted structure in steroid-resistant compared with steroid-sensitive asthma.[214] Steroid resistance may be *primary* (inherited or acquired of unknown cause) or *secondary* to some factor known to reduce glucocorticoid responsiveness (glucocorticoids themselves, cytokines, β-adrenergic agonists). There are several possible sites where abnormalities in the anti-inflammatory response to glucocorticoids in asthma may arise.

Pharmacokinetic abnormalities. The initial suggestion of Schwartz *et al.*[208] was that defective responses to steroids were due to increased clearance of the glucocorticoid, resulting in reduced clinical and eosinopenic response. There is no evidence for altered bioavailability or plasma clearance of prednisolone or methylprednisolone in patients with steroid-resistant asthma.[215–217] Metabolism of glucocorticoids may be increased by induction of P450 enzymes in response to certain drugs (e.g. rifampicin, carbamazepine), which may thus lead to a secondary steroid resistance.[117]

Antibodies to lipocortin 1. Some anti-inflammatory effects of glucocorticoids may be due to induction of lipocortin 1.[22] In some patients with steroid-resistant rheumatoid arthritis, autoantibodies to lipocortin 1 have been described.[218] However, two independent studies have failed to demonstrate the presence of IgG or IgM lipocortin 1 antibodies in either steroid-resistant or steroid-dependent asthma.[219,220]

Cellular abnormalities. Glucocorticoid resistance has been documented *in vitro* in various cells from steroid-resistant asthmatic patients. The enhanced expression of activation antigens (CR-1, CR-3 and class II HLA-DR molecules) in peripheral blood mononuclear cells (PBMC) and the growth of colonies stimulated by phytohaemagluttinin (PHA) is not inhibited by hydrocortisone in steroid-resistant asthmatic patients, in contrast to complete suppression with low concentrations of hydrocortisone (10^{-8}–10^{-9} M) in steroid-sensitive asthmatic and normal individuals.[221,222] PBMC from asthmatic patients generate a neutrophil-priming activity (a 3-kDa cytokine not yet identified) that is inhibited by glucocorticoids in steroid-sensitive but not in steroid-resistant asthmatic patients.[223] There is also evidence for defective T-lymphocyte responsiveness to glucocorticoids in steroid-resistant asthma. Dexamethasone significantly inhibits PHA-induced proliferation and IL-2 and interferon (IFN)-γ generation in peripheral T-cells from steroid-sensitive but not steroid-resistant patients.[217,224] There is no difference in the proportion of CD4$^+$ and CD8$^+$ T-cells in steroid-resistant patients, although there is increased expression of CD25 (IL-2 receptor) in steroid-resistant compared with steroid-sensitive patients, indicating a greater degree of immune activation.[224] These studies in circulating leucocytes suggest that the defect in glucocorticoid responsiveness extends outside the respiratory tract and is therefore unlikely to be secondary to inflammatory changes in the airways. In patients with steroid-resistant asthma the reduced blanching response to topical glucocorticoids applied to the skin further indicates that there is a generalized abnormality that is unlikely to be secondary to local cytokine production.[225]

Abnormalities in GR function. In familial glucocorticoid resistance there appears to be an abnormality in GR structure that results in reduced glucocorticoid binding affinity. GR binding in monocytes and T-lymphocytes of patients with steroid-resistant asthma shows either no difference in GR affinity and receptor density or a relative reduction in GR affinity.[215,224,226,227] Corrigan et al.[215] found some reduction of GR affinity in T-cells from steroid-resistant asthmatic patients but this could not account for the resistance to PHA-induced proliferative responses in cells from the same patients. Sher et al.[227] described two types of glucocorticoid resistance: a reduced affinity of GR binding confined to T-lymphocytes that reverted to normal after 48 h in culture, and a much less common reduction in GR density (in only 2/17 steroid-resistant patients) that did not normalize with prolonged incubation.[227] This suggests that there may be different types of steroid resistance in asthma. The small reduction in GR affinity is unlikely to be of functional significance and is not associated with elevated plasma cortisol concentrations, as observed in patients with familial glucocorticoid resistance. The small reduction in GR affinity may be secondary to cytokine exposure, since the normalization of GR affinity *in vitro* is prevented by a combination of IL-2 and IL-4[227] and this combination of cytokines reduces the binding affinity in nuclear GR in T-lymphocytes, although either cytokine alone has no effect.[228] This suggests that steroid resistance may occur in the airways of patients with asthma as a secondary phenomenon due to the local production of cytokines. In steroid-resistant asthmatic patients there is a significant increase in the numbers of bronchoalveolar lavage cells expressing IL-2 and IL-4 mRNA compared with steroid-sensitive asthmatic patients, but no difference in IFN-γ mRNA-positive cells. After oral prednisone for 1 week there is a reduction in IL-4-expressing cells and a rise in IFN-γ-positive cells in steroid-sensitive asthma, whereas in steroid-resistant asthma there was no fall in IL-4-positive cells and a fall in IFN-γ-positive cells.[212] This may indicate that there are different patterns of cytokine release that may contribute to steroid resistance. Although this may account for the increased requirement for glucocorticoids in more severe asthma, it is unlikely to account for the reduced steroid response seen in circulating mononuclear cells and in the skin of patients with no response to oral glucocorticoids.

There is, however, a marked reduction in GR–GRE binding in PBMC of patients with steroid-resistant asthma and Scatchard analysis has demonstrated a marked reduction in GR available for DNA binding compared with cells from patients with steroid-sensitive asthma.[229]

Interaction between GR and transcription factors. In the PBMC of patients with steroid-sensitive asthma and normal control subjects the phorbol ester PMA, which activates AP-1, results in reduced GRE binding. This inhibitory effect is significantly abrogated in the PBMC of patients with steroid-resistant asthma, indicating a likely abnormality in the interaction between GR and AP-1.[230] This defect does not appear to apply to the other transcription factors, NF-κB and CREB, that also interact with GR.[231] The abnormality in the interaction between GR and AP-1 is unlikely to be due to a defect in GR, since the protein sequence of GR in patients with steroid-resistant asthma is normal.[214] It is more likely to be due to a defect in AP-1 or its activation. Indeed, activation of c-Fos by phorbol esters is potentiated in the cells of patients with steroid-resistant compared with steroid-sensitive asthma and one of the key enzymes involved in activation of AP-1, namely Jun N-terminal (JNK) kinase, is abnormally activated in these

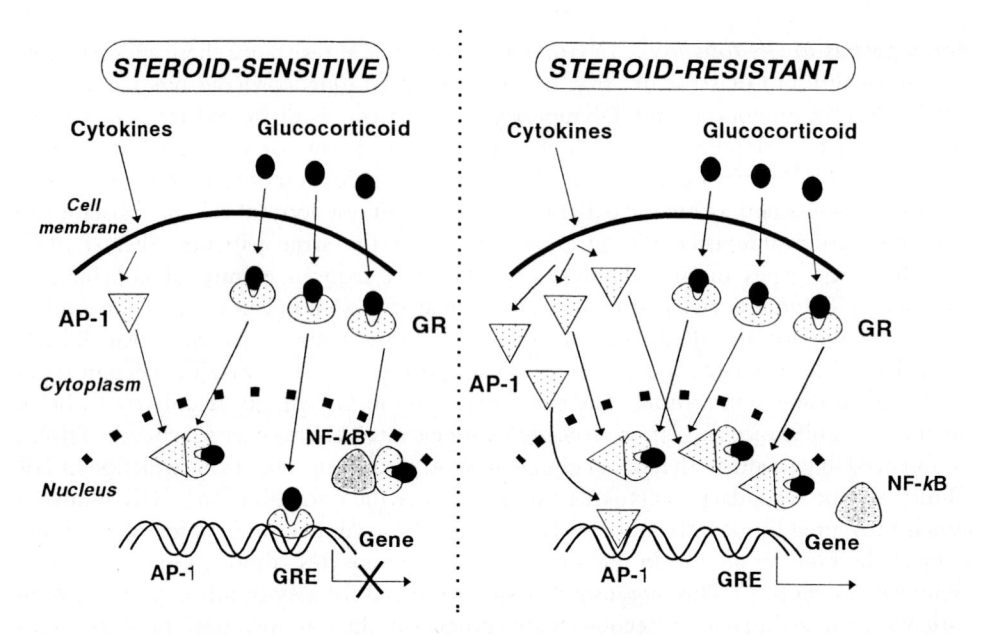

Fig. 40.7 Proposed mechanism of primary steroid resistance in asthma. Increased activation of activator protein (AP)-1 results in the consumption of glucocorticoid receptors (GR), thus preventing the anti-inflammatory action of steroids, either through binding to glucocorticoid response elements (GREs) or through inhibition of nuclear factor (NF)-κB.

patients.[232] The increased basal and cytokine-induced AP-1 activity may lead to consumption of GR, so that steroids are not able to suppress the inflammatory response, either through interaction with GRE or with other transcription factors such as NF-κB (Fig. 40.7).

An abnormality in AP-1 may also account for the selective resistance to the effects of steroid in steroid-resistant asthma, since AP-1 is more likely to be important in the regulation of some genes than in others. It would also explain why resistance is seen to the anti-inflammatory effects of steroids, since such resistance can only arise when AP-1 is activated at the inflammatory site, whereas the hormonal effects of steroids at uninflamed sites will not be impaired. Furthermore, there may also be differences in the steroid resistance of different target cells, depending upon the relative balance of transcription factors.

Secondary steroid resistance

Although complete steroid resistance is uncommon, there may be a spectrum of steroid responsiveness in asthma (Fig. 40.8). This may reflect several mechanisms that are secondary either to disease activity itself or to the effects of therapy.

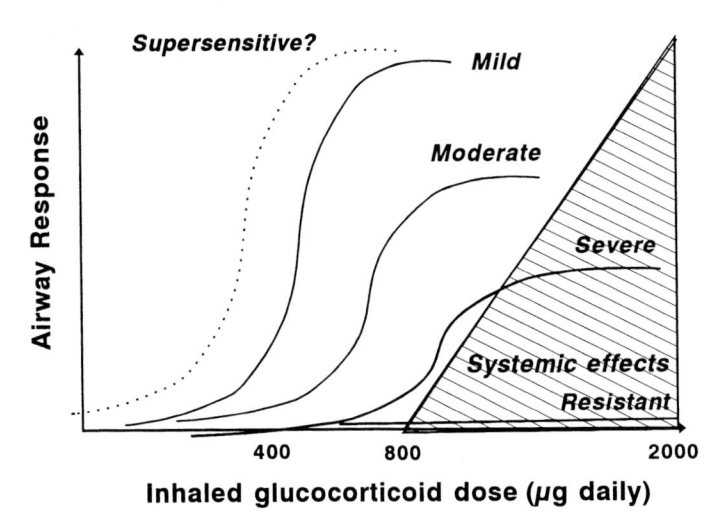

Fig. 40.8 Steroid responsiveness may vary between patients.

Downregulation of GR

Downregulation of GR in circulating lymphocytes after oral prednisolone has been demonstrated in normal individuals.[233] Whether high local concentrations of inhaled glucocorticoids reduce GR expression in surface cells of the airway, such as epithelial cells, is not yet certain. It is possible that certain individuals may be more susceptible to the effects of downregulation. If effective GR density is reduced by direct interaction with other transcription factors, such as AP-1 and NF-κB, then the downregulating effect of glucocorticoids on GR would be expected to have a greater functional consequence.

Effects of cytokines

Several pro-inflammatory cytokines, including IL-1β, IL-6 and TNF-α, activate AP-1 and NF-κB in human lung.[12,234] As all these cytokines are known to be secreted in asthmatic inflammation, this suggests that these transcription factors will be activated in the cells of asthmatic airways. These activated transcription factors may then form protein–protein complexes with activated GR, both in the cytoplasm and within the nucleus, thus reducing the number of effective GRs and thereby decreasing steroid responsiveness[4] (Fig. 40.9). In a model *in vitro* system, increased expression of c-Fos or c-Jun oncoproteins prevents the activation of mouse mammary tumour virus promoter by GR, thus creating a model of steroid resistance.[11] Addition of recombinant c-Jun or c-Fos proteins to partially purified GR results in inhibition of DNA binding.[11] Phorbol esters, which activate AP-1, result in attenuation of glucocorticoid-mediated gene activation.[235] Any reduction in glucocorticoid responsiveness would be greater as the intensity of asthmatic inflammation increased and may contribute, for example, to the failure of oral or intravenous glucocorticoids to control acute exacerbations of asthma. Once the inflammation is brought under control with large doses of oral glucocorticoids, steroid responsiveness increases again so that lower doses of inhaled or oral glucocorticoids are

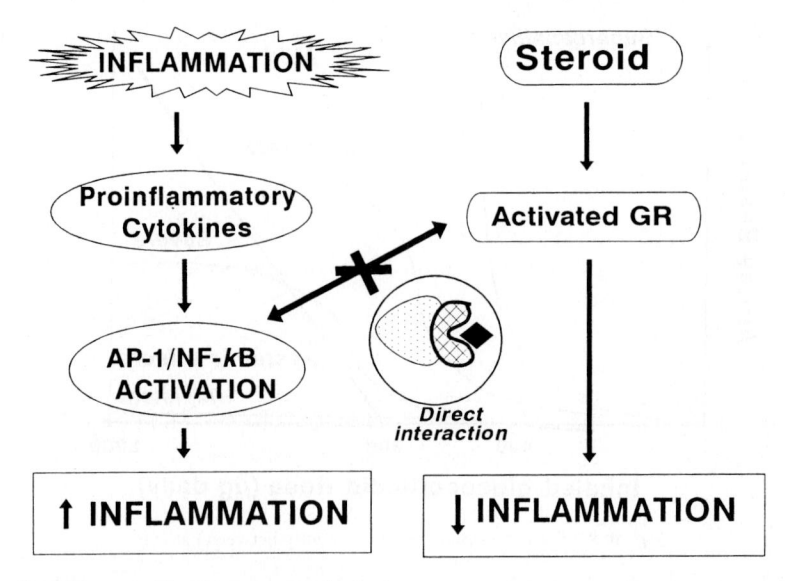

Fig. 40.9 Secondary steroid resistance may arise in the presence of cytokine-mediated inflammation through an interaction between cytokine-activated transcription factors, such as activator protein (AP)-1 and nuclear factor (NF)-κB, and the glucocorticoid receptor (GR), resulting in a reduced availability of GR for control of the inflammatory response. This can only be overcome by increasing the dose of glucocorticoid administered.

needed to control asthmatic inflammation. Increased resistance may also be due to the effects of cytokines on GR function, since high concentrations of IL-2 and IL-4 have been shown to reduce GR affinity in T-lymphocytes *in vitro*.[228] This effect would only be seen in mucosal T-cells of patients with severe asthma and it is therefore difficulty to obtain evidence to support this possibility.

Effect of β_2-agonists

High concentrations of β_2-agonists activate CREB in rat and human lung and in inflammatory cells via an increase in cyclic AMP concentration.[13,19,236] This results in reduced GRE binding due to the formation of GR–CREB complexes.[18] This predicts that high concentrations of β_2-agonists would induce steroid resistance. In asthmatic patients, while 3 weeks of treatment with an inhaled steroid blocked the airway response to inhaled allergen, concomitant treatment with inhaled steroid and a relatively large dose of inhaled β-agonist appeared to provide no significant protection against allergen challenge.[237] This suggests that high doses of inhaled β_2-agonists might interfere with the antiasthma effect of inhaled glucocorticoids. It is possible that some patients who use very high doses of inhaled β_2-agonists (over two canisters per month of MDI or regular nebulized doses) may develop a degree of steroid resistance that is overcome by increasing the dose of inhaled or oral glucocorticoid. This may account for some of the deleterious effects of high-dose β-agonists on asthma mortality and morbidity.[238–240] The use of high doses of nebulized β_2-agonists in the treatment of acute exacerbations of asthma may result in resistance to the effects of high-dose intravenous glucocorticoids in the treatment of these exacerbations. Steroid responsiveness might be restored by reducing the dose of inhaled β_2-agonists. In an uncontrolled study in steroid-dependent patients with severe asthma, gradual withdrawal of nebulized β_2-agonists resulted in a reduced requirement for oral prednisolone.[241]

Premenstrual asthma

In some women there is an increase in asthma symptoms and increased PEF variability premenstrually, which recovers at the start of menstruation.[242] This premenstrual exacerbation of asthma may be very severe in some women, necessitating ventilation. The increase in asthma premenstrually does not appear to respond well to glucocorticoids, even in high doses, yet responds well to high doses of progesterone.[242] This suggests that there is a form of steroid resistance that is regulated by the levels of endogenous female sex hormones. The mechanisms whereby a fall in progesterone and a rise in oestrogen induce reduced glucocorticoid responsiveness in some women with asthma is unknown, but may involve some sort of competition for GRE binding sites, since oestrogen and progesterone receptors have close structural similarities with GR. Recently an interaction between the progesterone receptor and NF-κB has been described.[243] This suggests that a fall in progesterone may increase NF-κB activation, but why only a small proportion of women are affected is not yet certain. It is possible that premenstrual asthma may only occur in women who already have a degree of glucocorticoid resistance.

Viral infection

It is possible that steroid resistance may evolve as a result of viral infection, since many viruses are capable of activating transcription factors that could interfere with glucocorticoid action. In children with severe steroid-dependent asthma there is evidence for persistent adenovirus infection in the airways.[244] Viruses may activate transcription factors, resulting in increased glucocorticoid resistance. Thus the EIA protein expressed by adenoviruses binds to the antioncogene retinoblastoma protein, thus increasing oncogene expression, which may increase glucocorticoid resistance.[245]

REFERENCES

1. Barnes PJ: Inhaled glucocorticoids for asthma. *N Engl J Med* (1995) **332**: 868–875.
2. Barnes PJ: Molecular mechanisms of steroid action in asthma. *J Allergy Clin Immunol* (1996) **97**: 159–168.
3. Barnes PJ: Mechanism of action of glucocorticoids in asthma. *Am J Respir Crit Care Med* (1996) **154**: S21–S27.
4. Barnes PJ, Adcock IM: Anti-inflammatory actions of steroids: molecular mechanisms. *Trends Pharmacol Sci* (1993) **14**: 436–441.
5. Bamberger CM, Bamberger AM, de Castr M, Chrousos GP: Glucocorticoid receptor β, a potential endogenous inhibitor of glucocorticoid action in humans. *J Clin Invest* (1995) **95**: 2435–2441.
6. Muller M, Renkawitz R: The glucocorticoid receptor. *Biochim Biophys Acta* (1991) **1088**: 171–182.

7. Encio PJ, Detgra-Wadleigh SD: The genomic structure of the human glucocorticoid receptor. *J Biol Chem* (1991) **266**: 7182–7188.
8. Heck S, Kullmann M, Grast A, et al.: A distinct modulating domain in glucocorticoid receptor monomers in the repression of activity of the transcription factor AP-1. *EMBO J* (1995) **13**: 4087–4095.
9. Gronemeyer H: Control of transcription activation by steroid hormone receptors. *FASEB J* (1992) **6**: 2524–2529.
10. Beato M, Herrlich P, Schutz G: Steroid hormone receptors: many actors in search of a plot. *Cell* (1995) **83**: 851–857.
11. Yang-Yen H, Chambard J, Sun Y, et al.: Transcriptional interference between c-Jun and the glucocorticoid receptor: mutual inhibition of DNA binding due to direct protein–protein interaction. *Cell* (1990) **62**: 1205–1215.
12. Adcock IM, Shirasaki H, Gelder CM, Peters MJ, Brown CR, Barnes PJ: The effects of glucocorticoids on phorbol ester and cytokine stimulated transcription factor activation in human lung. *Life Sci* (1994) **55**: 1147–1153.
13. Adcock IM, Brown CR, Gelder CM, Shirasaki H, Peters MJ, Barnes PJ: The effects of glucocorticoids on transcription factor activation in human peripheral blood mononuclear cells. *Am J Physiol* (1995) **37**: C331–C338.
14. Ray A, Prefontaine KE: Physical association and functional antagonism between the p65 subunit of transcription factor NF-κB and the glucocorticoid receptor. *Proc Natl Acad Sci USA* (1994) **91**: 752–756.
15. Scheinman RI, Gualberto A, Jewell CM, Cidlowski JA, Baldwin AS: Characterization of the mechanisms involved in transrepression of NF-κB by activated glucocorticoid receptors. *Mol Cell Biol* (1996) **15**: 943–953.
16. Caldenhoven E, Liden J, Wissink S, et al.: Negative cross-talk between RelA and the glucocorticoid receptor: a possible mechanism for the antiinflammatory action of glucocorticoids. *Mol Endocrinol* (1995) **9**: 401–412.
17. Imai F, Minger JN, Mitchell JA, Yamamoto KR, Granner DK: Glucocorticoid receptor–cAMP response element-binding protein interaction and the response of the phosphoenolpyruvate carboxykinase gene to glucocorticoids. *J Biol Chem* (1993) **268**: 5353–5356.
18. Peters MJ, Adcock IM, Brown CR, Barnes PJ: β-Agonist inhibition of steroid-receptor DNA binding activity in human lung. *Am Rev Respir Dis* (1993) **147**: A772.
19. Stevens DA, Barnes PJ, Adcock IM: β-Agonists inhibit DNA binding of glucocorticoid receptors in human pulmonary and bronchial epithelial cells. *Am J Respir Crit Care Med* (1995) **151**: A195.
20. Adcock IM, Barnes PJ: Tumour necrosis factor alpha causes retention of activated glucocorticoid receptor within the cytoplasm of A549 cells. *Biochem Biophys Res Commun* (1996) **225**: 1127–1132.
21. Janknecht R, Hunter T: A growing coactivator network. *Nature* (1996) **383**: 22–23.
22. Flower RJ, Rothwell NJ: Lipocortin-1: cellular mechanisms and clinical relevance. *Trends Pharmacol Sci* (1994) **15**: 71–76.
23. Abbinante-Nissen JM, Simpson LG, Leikauf GD: Corticosteroids increase secretory leukocyte protease inhibitor transcript levels in airway epithelial cells. *Am J Physiol* (1995) **12**: L601–L606.
24. Wang P, Wu P, Siegel MI, Egan RW, Billah MM: Interleukin (IL)-10 inhibits nuclear factor kappa B activation in human monocytes. IL-10 and IL-4 suppress cytokine synthesis by different mechanisms. *J Biol Chem* (1995) **270**: 9558–9563.
25. Borish L, Aarons A, Rumbyrt J, Cvietusa P, Negri J, Wenzel S: Interleukin-10 regulation in normal subjects and patients with asthma. *J Allergy Clin Immunol* (1996) **97**: 1288–1296.
26. John M, Lim S, Seybold J, et al.: Inhaled corticosteroids increase IL-10 but reduce MIP-1a, GM-CSF and IFN-γ release from alveolar macrophages in asthma. *Am J Respir Crit Care Med* (1997) in press.
27. Collins S, Caron MG, Lefkowitz RJ: β-Adrenergic receptors in hamster smooth muscle cells are transcriptionally regulated by glucocorticoids. *J Biol Chem* (1988) **263**: 9067–9070.
28. Mak JCW, Nishikawa M, Barnes PJ: Glucocorticosteroids increase β_2-adrenergic receptor transcription in human lung. *Am J Physiol* (1995) **12**: L41–L46.

29. Mak JCW, Nishikawa M, Shirasaki H, Miyayasu K, Barnes PJ: Protective effects of a glucocorticoid on down-regulation of pulmonary β_2-adrenergic receptors *in vivo*. *J Clin Invest* (1995) **96**: 99–106.
30. Northrop JP, Ullman KS, Crabtree GR: Characterization of the nuclear and cytoplasmic components of the lymphoid-specific nuclear factor of activated T cells (NF-AT). *J Biol Chem* (1993) **268**: 2917–2923.
31. Paliogianni F, Raptis A, Ahuja SS, Najjar SM, Boumpas DT: Negative transcriptional regulation of human interleukin 2 (IL-2) gene by glucocorticoids through interference with nuclear transcription factors AP-1 and NF-AT. *J Clin Invest* (1993) **91**: 1481–1489.
32. Nelson PJ, Kim HT, Manning WC, Goralski TJ, Krensky AM: Genomic organisation and transcriptional regulation of the RANTES chemokine gene. *J Immunol* (1993) **151**: 2601–2612.
33. Linden M, Brattsand R: Effects of a corticosteroid, budesonide, on alveolar macrophages and blood monocyte secretion of cytokines: differential sensitivity of GM-CSF, IL-1β and IL-6. *Pulmon Pharmacol* (1994) **7**: 43–47.
34. Robbins RA, Barnes PJ, Springall DR, *et al.*: Expression of inducible nitric oxide synthase in human bronchial epithelial cells. *Biochem Biophys Res Commun* (1994) **203**: 209–218.
35. Xie Q, Kashiwarbara Y, Nathan C: Role of transcription factor NF-κB/Rel in induction of nitric oxide synthase. *J Biol Chem* (1994) **269**: 4705–4708.
36. Adcock IM, Brown CR, Kwon OJ, Barnes PJ: Oxidative stress induces NF-κB DNA binding and inducible NOS mRNA in human epithelial cells. *Biochem Biophys Res Commun* (1994) **199**: 1518–1524.
37. Mitchell JA, Belvisi MG, Akarasereemom P, *et al.*: Induction of cyclo-oxygenase-2 by cytokines in human pulmonary epithelial cells: regulation by dexamethasone. *Br J Pharmacol* (1994) **113**: 1008–1014.
38. Yamamoto K, Arakawa T, Ueda N, Yamamoto S: Transcriptional roles of nuclear factor κB and nuclear factor-interleukin 6 in the tumor necrosis-α-dependent induction of cyclooxygenase-2 in MC3T3-E1 cells. *J Biol Chem* (1995) **270**: 31 315–31 320.
39. Newton R, Kuitert LM, Slater DM, Adcock IM, Barnes PJ: Induction of cPLA$_2$ and COX-2 mRNA by proinflammatory cytokines is suppressed by dexamethasone in human airway epithelial cells. *Life Sci* (1997) **60**: 67–78.
40. Vittori E, Marini M, Fasoli A, de Franchis R, Mattoli S: Increased expression of endothelin in bronchial epithelial cells of asthmatic patients and effect of corticosteroids. *Am Rev Respir Dis* (1992) **146**: 1320–1325.
41. Adcock IM, Peters M, Gelder C, Shirasaki H, Brown CR, Barnes PJ: Increased tachykinin receptor gene expression in asthmatic lung and its modulation by steroids. *J Mol Endocrinol* (1993) **11**: 1–7.
42. Ihara H, Nakanishi S: Selective inhibition of expression of the substance P receptor mRNA in pancreatic acinar AR42J cells by glucocorticoids. *J Biol Chem* (1990) **36**: 22 441–22 445.
43. Colotta F, Re F, Muzio M, *et al.*: Interleukin-1 type II receptor: a decoy target for IL-1 that is regulated by IL-4. *Science* (1993) **261**: 472–475.
44. Owens GP, Hahn WE, Cohen JJ: Identification of mRNAs associated with programmed cell death in immature thymocytes. *Mol Cell Biol* (1991) **11**: 4177–4188.
45. Cronstein BN, Kimmel SC, Levin RI, Martiniuk F, Weissmann G: A mechanism for the antiinflammatory effects of corticosteroids: the glucocorticoid receptor regulates leukocyte adhesion to endothelial cells and expression of endothelial–leukocyte adhesion molecule 1 and intercellular adhesion molecule 1. *Proc Natl Acad Sci USA* (1992) **89**: 9991–9995.
46. Van De Stolpe A, Caldenhoven E, Raaijmakers JAM, Van Der Saag PT, Koendorman L: Glucocorticoid-mediated repression of intercellular adhesion molecule-1 expression in human monocytic and bronchial epithelial cell lines. *Am J Respir Cell Mol Biol* (1993) **8**: 340–347.
47. Carnuccio R, Di Rosa M, Guerrasio B, Iuvone T, Satebin L: Vasocortin: a novel glucocorticoid-induced anti-inflammatory protein. *Br J Pharmacol* (1987) **90**: 443–445.
48. Bergstrand H, Bjrnson A, Blaschuke E, *et al.*: Effects of an inhaled corticosteroid, budesonide, on alveolar macrophage function in smokers. *Thorax* (1990) **45**: 362–368.
49. Standiford TJ, Kunkel SL, Rolfe MW, Evanoff HL, Allen RM, Srieter RW: Regulation of human alveolar macrophage and blood monocyte-derived interleukin-8 by prostaglandin E$_2$ and dexamethasone. *Am J Respir Cell Mol Biol* (1992) **6**: 75–81.

50. Borish L, Mascali JJ, Dishuck J, Beam WR, Martin RJ, Rosenwasser LJ: Detection of alveolar macrophage-derived IL-1β in asthma. Inhibition with corticosteroids. *J Immunol* (1992) **149**: 3078–3082.
51. Kita H, Abu-Ghazaleh R, Sanderson CJ, Gleich GJ: Effect of steroids on immunoglobulin-induced eosinophil degranulation. *J Allergy Clin Immunol* (1991) **87**: 70–77.
52. Lamas AM, Leon OG, Schleimer RP: Glucocorticoids inhibit eosinophil responses to granulocyte–macrophage colony-stimulating factor. *J Immunol* (1991) **147**: 254–259.
53. Wallen N, Kita H, Weiller D, Gleich GJ: Glucocorticoids inhibit cytokine-mediated eosinophil survival. *J Immunol* (1991) **147**: 3490–3495.
54. Wempe JB, Tammeling EP, Keter GH, Haransson L, Venge P, Postma DS: Blood eosinophil numbers and activity during 24 hours: effects of treatment with budesonide and bambuterol. *J Allergy Clin Immunol* (1992) **90**: 757–765.
55. Evans PM, O'Connor BJ, Fuller RW, Barnes PJ, Chung KF: Effect of inhaled corticosteroids on peripheral eosinophil counts and density profiles in asthma. *J Allergy Clin Immunol* (1993) **91**: 643–649.
56. Rolfe FG, Hughes JM, Armour CL, Sewell WA: Inhibition of interleukin-5 gene expression by dexamethasone. *Immunology* (1992) **77**: 494–499.
57. Cohan VL, Undem BJ, Fox CC, Adkinson NF, Lichtenstein LM, Schleimer RP: Dexamethasone does not inhibit the release of mediators from human lung mast cells residing in airway, intestine or skin. *Am Rev Respir Dis* (1989) **140**: 951–954.
58. Laitinen LA, Laitinen A, Haahtela T: A comparative study of the effects of an inhaled corticosteroid, budesonide, and of a $\beta_2$agonist, terbutaline, on airway inflammation in newly diagnosed asthma. *J Allergy Clin Immunol* (1992) **90**: 32–42.
59. Djukanovic R, Wilson JW, Britten YM, et al.: Effect of an inhaled corticosteroid on airway inflammation and symptoms of asthma. *Am Rev Respir Dis* (1992) **145**: 699–674.
60. Holt PG: Regulation of antigen-presenting cell function(s) in lung and airway tissues. *Eur Respir J* (1993) **6**: 120–129.
61. Nelson DJ, McWilliam AS, Haining S, Holt PG: Modulation of airway intraepithelial dendritic cells following exposure to steroids. *Am J Respir Crit Care Med* (1995) **151**: 475–481.
62. Holm AF, Fokkens WJ, Godthelp T, Mulder PG, Vroom TM, Rinjntjes E: Effect of 3 months' nasal steroid therapy on nasal T cells and langerhans cells in patients suffering from allergic rhinitis. *Allergy* (1995) **50**: 204–209.
63. Cox G: Glucocorticoid treatment inhibits apoptosis in human neutrophils. *J Immunol* (1995) **193**: 4719–4725.
64. Boschetto P, Rogers DF, Fabbri LM, Barnes PJ: Corticosteroid inhibition of airway microvascular leakage. *Am Rev Respir Dis* (1991) **143**: 605–609.
65. Erjefalt I, Persson CGA: Pharmacologic control of plasma exudation into tracheobronchial airways. *Am Rev Respir Dis* (1991) **143**: 1008–1014.
66. Van de Graaf EA, Out TA, Loos CM, Jansen HM: Respiratory membrane permeability and bronchial hyperreactivity in patients with stable asthma. *Am Rev Respir Dis* (1991) **143**: 362–368.
67. Devalia JL, Wang JH, Sapsford RJ, Davies RJ: Expression of RANTES in human bronchial epithelial cells and the effect of beclomethasone dipropionate (BDP). *Eur Respir J* (1994) **7** (Suppl 18): 98S.
68. Levine SJ: Bronchial epithelial cell–cytokine interactions in airway epithelium. *J Invest Med* (1995) **43**: 241–249.
69. Schweibert LM, Stellato C, Schleimer RP: The epithelium as a target for glucocorticoid action in the treatment of asthma. *Am J Respir Crit Care Med* (1996) **154**: S16–S20.
70. Kwon OJ, Au BT, Collins PD, et al.: Inhibition of interleukin-8 expression in human cultured airway epithelial cells. *Immunology* (1994) **81**: 389–394.
71. Kwon OJ, Au BT, Collins PD, et al.: Tumor necrosis factor-induced interleukin 8 expression in cultured human epithelial cells. *Am J Physiol* (1994) **11**: L398-L405.
72. Kwon OJ, Jose PJ, Robbins RA, Schall TJ, Williams TJ, Barnes PJ: Glucocorticoid inhibition of RANTES expression in human lung epithelial cells. *Am J Respir Cell Mol Biol* (1995) **12**: 488–496.
73. Sousa AR, Poston RN, Lane SJ, Narhosteen JA, Lee TH: Detection of GM-CSF in asthmatic

bronchial epithelium and decrease by inhaled corticosteroids. *Am Rev Respir Dis* (1993) **147**: 1557–1561.
74. Trigg CJ, Manolistas ND, Wang J, et al.: Placebo-controlled immunopathological study of four months inhaled corticosteroids in asthma. *Am J Respir Crit Care Med* (1994) **150**: 17–22.
75. Hamid Q, Springall DR, Riveros-Moreno V, et al.: Induction of nitric oxide synthase in asthma. *Lancet* (1993) **342**: 1510–1513.
76. Kharitonov SA, Yates D, Robbins RA, Logan-Sinclair R, Shinebourne E, Barnes PJ: Increased nitric oxide in exhaled air of asthmatic patients. *Lancet* (1994) **343**: 133–135.
77. Yates DH, Kharitonov SA, Robbins RA, Thomas PS, Barnes PJ: Effect of a nitric oxide synthase inhibitor and a glucocorticosteroid on exhaled nitric oxide. *Am J Respir Crit Care Med* (1995) **152: 892–896.**
78. Kharitonov SA, Yates DH, Barnes PJ: Regular inhaled budesonide decreases nitric oxide concentration in the exhaled air of asthmatic patients. *Am J Respir Crit Care Med* (1996) **153**: 454–457.
79. Shimura S, Sasaki T, Ikeda K, Yamauchi K, Sasaki H, Takishima T: Direct inhibitory action of glucocorticoid on glycoconjugate secretion from airway submucosal glands. *Am Rev Respir Dis* (1990) **141**: 1044-1099.
80. Kai H, Yoshitake K, Hisatsune A, et al.: Dexamethasone suppresses mucus production and MUC-2 and MUC-5AC gene expression by NCI-H292 cells. *Am J Physiol* (1996) **271**: L484–L488.
81. Jeffery PK, Godfrey RW, Ådelroth E, Nelson F, Rogers A, Johansson S: Effects of treatment on airway inflammation and thickening of basement membrane reticular collagen in asthma. *Am Rev Respir Dis* (1992) **145**: 890–899.
82. Burke C, Power CK, Norris A, Condez A, Schmekel B, Poulter LW: Lung function and immunopathological changes after inhaled corticosteroid therapy in asthma. *Eur Respir J* (1992) **5**: 73–79.
83. Ådelroth E, Rosenhall L, Johansson S, Linden M, Venge P: Inflammatory cells and eosinophilic activity in asthmatics investigated by bronchoalveolar lavage. *Am Rev Respir Dis* (1990) **142**: 91–99.
84. Wilson JW, Djukanovic R, Howarth PH, Holgate ST: Inhaled beclomethasone dipropionate down-regulates airway lymphocyte activation in atopic asthma. *Am J Respir Crit Care Med* (1994) **149**: 86–90.
85. Lungren R, Soderberg M, Horstedt P, Stenling R: Morphological studies on bronchial mucosal biopsies from asthmatics before and after ten years treatment with inhaled steroids. *Eur Respir J* (1988) **1**: 883–889.
86. Barnes PJ: Effect of corticosteroids on airway hyperresponsiveness. *Am Rev Respir Dis* (1990) **141**: S70–S76.
87. Bel EH, Timers MC, Zwinderman AH, Dijkman JH, Sterk PJ: The effect of inhaled corticosteroids on the maximal degree of airway narrowing to methacholine. *Am Rev Respir Dis* (1991) **143**: 109–113.
88. Kamada AK, Szefler SJ, Martin RJ, et al.: Issues in the use of inhaled steroids. *Am J Respir Crit Care Med* (1996) **153**: 1739–1748.
89. Reed CE: Aerosol glucocorticoid treatment of asthma: adults. *Am Rev Respir Dis* (1990) **140**: S82–S88.
90. Barnes PJ: A new approach to asthma therapy. *N Engl J Med* (1989) **321**: 1517–1527.
91. Sheffer AL: National Heart Lung and Blood Institute National Asthma Education Programme Expert Panel Report: guidelines for the diagnosis and management of asthma. *J Allergy Clin Immunol* (1991) **88**: 425–534.
92. Sheffer AL: International Consensus Report on Diagnosis and Management of Asthma. *Clin Exp Allergy* (1992) **22** (Suppl 1): 1–72.
93. British Thorac Society: Guidelines on the management of asthma. *Thorax* (1993) **48** (Suppl): S1–S24.
94. British Thoracic Society: The British guidelines on asthma management. *Thorax* (1997) **52** (Suppl 1): S1–S21.
95. Haahtela T, Jarvinen M, Kava T, et al.: Comparison of a β_2-agonist terbutaline with an inhaled steroid in newly detected asthma. *N Engl J Med* (1991) **325**: 388–392.

96. Juniper EF, Kline PA, Vanzieleghem MA, Ramsdale EH, O'Byrne PM, Hargreave FE: Effect of long-term treatment with an inhaled corticosteroid (budesonide) on airway hyperresponsiveness and clinical asthma in nonsteroid-dependent asthmatics. *Am Rev Respir Dis* (1990) **142**: 832–836.
97. Kerrebijn KF, Von Essen-Zandvliet EEM, Neijens HJ: Effect of long-term treatment with inhaled corticosteroids and beta-agonists on bronchial responsiveness in asthmatic children. *J Allergy Clin Immunol* (1987) **79**: 653–659.
98. Vathenen AS, Knox AJ, Wisniewski A, Tattersfield AE: Time course of change in bronchial reactivity with an inhaled corticosteroid in asthma. *Am Rev Respir Dis* (1991) **143**: 1317–1321.
99. Toogood JH: High dose inhaled steroid therapy for asthma. *J Allergy Clin Immunol* (1989) **83**: 528–536.
100. Salmeron S, Guerin J, Godard P, *et al.*: High doses of inhaled corticosteroids in unstable chronic asthma. *Am Rev Respir Dis* (1989) **140**: 167–171.
101. Lacronique J, Renon D, Georges D, Henry-Amar M, Marsac J: High-dose beclomethasone: oral steroid-sparing effect in severe asthmatic patients. *Eur Respir J* (1991) **4**: 807–812.
102. Dahl R, Pedersen B, Hgglf B: Nocturnal asthma: effect of treatment with oral sustained-release terbutaline, inhaled budesonide and the two in combination. *J Allergy Clin Immunol* (1989) **83**: 811–815.
103. Wempe JB, Tammeling EP, Postma DS, Auffarth B, Teengs JP, Kœter GH: Effects of budesonide and bambuterol on circadian variation of airway responsiveness and nocturnal asthma symptoms of asthma. *J Allergy Clin Immunol* (1992) **90**: 349–357.
104. Juniper EF, Kline PA, Vanzielegmem MA, Hargreave FE: Reduction of budesonide after a year of increased use: a randomized controlled trial to evaluate whether improvements in airway responsiveness and clinical asthma are maintained. *J Allergy Clin Immunol* (1991) **87**: 483–489.
105. van Essen-Zandvliet EE, Hughes MD, Waalkens HJ, Duiverman EJ, Pocock SJ, Kerrebijn KF: Effects of 22 months of treatment with inhaled corticosteroids and/or β_2-agonists on lung function, airway responsiveness and symptoms in children with asthma. *Am Rev Respir Dis* (1992) **146**: 547–554.
106. Waalkens HJ, van Essen-Zandvliet EE, Hughes MD, *et al.*: Cessation of long-term treatment with inhaled corticosteroids (budesonide) in children with asthma results in deterioration. *Am Rev Respir Dis* (1993) **148**: 1252–1257.
107. Ilangovan P, Pedersen S, Godfrey S, Nikander K, Novisky N, Warner JO: Nebulised budesonide suspension in severe steroid-dependent preschool asthma. *Arch Dis Child* (1993) **68**: 356–359.
108. Gleeson JGA, Price JF: Controlled trial of budesonide given by Nebuhaler in preschool children with asthma. *Br Med J* (1988) **297**: 163–166.
109. Bisgard H, Munck SL, Nielsen JP, Peterson W, Ohlsson SV: Inhaled budesonide for treatment of recurrent wheezing in early childhood. *Lancet* (1990) **336**: 649–651.
110. Dompeling E, Van Schayck CP, Molema J, Folgering H, van Grusven PM, van Weel C: Inhaled beclomethasone improves the course of asthma and COPD. *Eur Respir J* (1992) **5**: 945–952.
111. Haahtela T, Jrvinsen M, Kava T, *et al.*: Effects of reducing or discontinuing inhaled budesonide in patients with mild asthma. *N Engl J Med* (1994) **331**: 700–705.
112. Agertoft L, Pedersen S: Effects of long-term treatment with an inhaled corticosteroid on growth and pulmonary function in asthmatic children. *Respir Med* (1994) **5**: 369–372.
113. Selroos O, Backman R, Forsen K, *et al.*: When to start treatment of asthma with inhaled steroids? *Eur Respir J* (1994) **7** (Suppl 18): 151S.
114. Ernst P, Spitzer WD, Suissa S, *et al.*: Risk of fatal and near fatal asthma in relation to inhaled corticosteroid use. *JAMA* (1992) **268**: 3462–3464.
115. Ebden P, Jenkins A, Houston G, Davies BH: Comparison of two high-dose corticosteroid aerosol treatments, beclomethasone dipropionate (1500 mcg/day) and budesonide (1600 mcg/day) for chronic asthma. *Thorax* (1986) **41**: 869–874.
116. Baran D: A comparison of inhaled budesonide and beclomethasone dipropionate in childhood asthma. *Br J Dis Chest* (1987) **81**: 170–175.

40 Glucocorticosteroids

117. Szefler S: Glucocorticoid therapy for asthma: clinical pharmacology. *J Allergy Clin Immunol* (1991) **88**: 147–165.
118. Johnson M: Pharmacodynamics and pharmacokinetics of inhaled glucocorticoids. *J Allergy Clin Immunol* (1996) **97**: 169–176.
119. Wurthwein G, Rohdewald P: Activation of beclomethasone dipropionate by hydrolysis to beclomethasone-17-monophosphate. *Biopharm Drug Dispos* (1990) **11**: 381–394.
120. Chaplin MD, Rooks W, Svenson EW, Couper WC, Nerenberg C, Chu NI: Flunisotide metabolism and dynamics of a metabolite. *Clin Pharmacol Ther* (1980) **27**: 402–413.
121. Ryrfeldt A, Andersson P, Edsbacker S, Tonnesson M, Davies D, Pauwels R: Pharmacokinetics and metabolism of budesonide, a selective glucocorticoid. *Eur J Respir Dis* (1982) **63** (Suppl 122): 86–95.
122. Mollman H, Rohdewald P, Schmidt EW, Salomon V, Derendorf H: Pharmacokinetics of triamcinolone acetonide and its phosphate ester. *Eur J Clin Pharmacol* (1985) **29**: 85–89.
123. Harding SM: The human pharmacology of fluticasone dipropionate. *Respir Med* (1990) **84** (Suppl A): 25–29.
124. Toogood JH, Baskerville JC, Jennings B, Lefcoe NM, Johansson SA: Influence of dosing frequency and schedule on the response of chronic asthmatics to the aerosol steroid budesonide. *J Allergy Clin Immunol* (1982) **70**: 288–298.
125. Meltzer EO, Kemp JP, Welch MJ, Orgel HA: Effect of dosing schedule on efficacy of beclomethasone dipropionate aerosol in chronic asthma. *Am Rev Respir Dis* (1985) **131**: 732–736.
126. Malo J, Cartier A, Merland N, *et al.*: Four-times-a-day dosing frequency is better than twice-a-day regimen in subjects requiring a high-dose inhaled steroid, budesonide, to control moderate to severe asthma. *Am Rev Respir Dis* (1989) **140**: 624–628.
127. Jones AH, Langdon CG, Lee PS, *et al.*: Pulmicort Turbohaler once daily as initial prophylactic therapy for asthma. *Respir Med* (1994) **88**: 293–299.
128. Toogood JA, Jennings B, Greenway RW, Chung L: Candidiasis and dysphonia complicating beclomethasone treatment of asthma. *J Allergy Clin Immunol* (1980) **65**: 145–153.
129. Williamson IJ, Matusiewicz SP, Brown PH, Greening AP, Crompton GK: Frequency of voice problems and cough in patients using pressurised aerosol inhaled steroid preparations. *Eur Respir J* (1995) **8**: 590–592.
130. Selroos O, Backman R, Forsen KO, *et al.*: Local side effects during 4-year treatment with inhaled corticosteroids: a comparison between pressurized metered dose inhalers and Turbuhaler. *Allergy* (1994) **49**: 888–890.
131. Brogden RN, Heel RC, Speight TM, Avery GS: Beclomethasone dipropionate. A reappraisal of its pharmacodynamic properties and therapeutic efficacy after a decade of use in asthma and rhinitis. *Drugs* (1984) **28**: 99–126.
132. Brogden RN, McTavish D: Budesonide. An updated review of its pharmacological properties and therapeutic efficacy in asthma and rhinitis. *Drugs* (1992) **44**: 375–407.
133. Engel T, Heinig JH, Malling H-J, Scharing B, Nikander K, Masden F: Clinical comparison of inhaled budesonide delivered either by pressurized metered dose inhaler or Turbuhaler. *Allergy* (1989) **44**: 220–225.
134. Geddes DM: Inhaled corticosteroids: benefits and risks. *Thorax* (1992) **47**: 404–407.
135. Brown PH, Blundell G, Greening AP, Crompton GK: Do large volume spacer devices reduce the systemic effects of high dose inhaled corticosteroids? *Thorax* (1990) **95**: 736–739.
136. Brown PH, Greening AP, Crompton GK: Large volume spacer devices and the influence of high dose beclomethasone dipropionate on hypothalamo-pituitary–adrenal axis function. *Thorax* (1993) **48**: 233–238.
137. Selroos O, Halme M: Effect of a Volumatic spacer and mouth rinsing on systemic absorption of inhaled corticosteroids from a metered-dose inhaler and dry powder inhaler. *Thorax* (1991) **46**: 891–894.
138. Thorsson L, Edsbcker S, Conradson T: Lung deposition of budesonide from Turbohaler R is twice that from a pressurized metered-dose inhaler P-MDI. *Eur Respir J* (1994) **7**: 1839–1844.
139. Brown PH, Blundell G, Greening AP, Crompton GK: High dose inhaled corticosteroids and the cortisol induced response to acute severe asthma. *Respir Med* (1992) **86**: 495–497.

140. Holt PR, Lowndes DW, Smithies E, Dixon GT: The effect of an inhaled steroid on the hypothalamic–pituitary–adrenal axis: which tests should be used? *Clin Exp Allergy* (1990) **20**: 145–149.
141. Barnes PJ, Pedersen S: Efficacy and safety of inhaled steroids in asthma. *Am Rev Respir Dis* (1993) **148**: S1–S26.
142. Lfdahl CG, Mellstrand T, Svedmyr N: Glucocorticosteroids and asthma: studies of resistance and systemic effects of glucocorticosteroids. *Eur J Respir Dis* (1989) **65** (Suppl 130): 69–79.
143. Pedersen S, Fuglsang G: Urine cortisol excretion in children treated with high doses of inhaled corticosteroids; a comparison of budesonide and beclomethasone. *Eur Respir J* (1988) **1**: 433–435.
144. Clark DJ, Lipworth BJ: Adrenal suppression with chronic dosing of fluticasone propionate compared with budesonide in adult asthmatic patients. *Thorax* (1997) **52**: 55–58.
145. Brown PH, Matusiewicz SP, Shearing C, Tibi L, Greening AP, Crompton GK: Systemic effects of high dose inhaled steroids: comparison of beclomethasone dipropionate and budesonide in healthy subjects. *Thorax* (1993) **48**: 967–973.
146. Sly RM, Imseis M, Frazer M, *et al.*: Treatment of asthma in children with triamcinolone acetonide aerosol. *J Allergy Clin Immunol* (1978) **62**: 76–82.
147. Placcentini G, Sette L, Peroni DG, Bonizatto C, Bonetti S, Boner AL: Double blind evaluation of effectiveness and safety of flunisolide aerosol for treatment of bronchial asthma in children. *Allergy* (1990) **45**: 612–616.
148. Fabbri L, Burge PS, Croonenburgh L, *et al.*: Comparison of fluticasone propionate with beclomethasone dipropionate in moderate to severe asthma treated for one year. *Thorax* (1993) **48**: 817–823.
149. Prahl P: Adrenocortical suppression following treatment with beclomethasone and budesonide. *Clin Exp Allergy* (1991) **21**: 145–146.
150. Bisgaard H, Damkjaer Nielsen M, Andersen B, *et al.*: Adrenal function in children with bronchial asthma treated with beclomethasone dipropionate or budesonide. *J Allergy Clin Immunol* (1988) **81**: 1088–1095.
151. Volovitz B, Amir J, Malik H, Kauschansky A, Varsano I: Growth and pituitary–adrenal function in children with severe asthma treated with inhaled budesonide. *N Engl J Med* (1993) **329**: 1703–1708.
152. Law CM, Honour JW, Marchant JL, Preece MA, Warner JO: Nocturnal adrenal suppression in asthmatic children taking inhaled beclomethasone dipropionate. *Lancet* (1986) **i**: 942–944.
153. Tabacknik E, Zadik Z: Diurnal cortisol secretion during therapy with inhaled beclomethasone dipropionate in children with asthma. *J Pediatr* (1991) **118**: 294–297.
154. Philip M, Aviram M, Lieberman E, *et al.*: Integrated plasma cortisol concentration in children with asthma receiving long-term inhaled corticosteroids. *Pediatr Pulmonol* (1992) **12**: 84–89.
155. Hosking DJ: Effect of corticosteroids on bone turnover. *Respir Med* (1993) **87** (Suppl A): 15–21.
156. Packe GE, Douglas JG, MacDonald AF, Robins SP, Reid DM: Bone density in asthmatic patients taking high dose inhaled beclomethasone dipropionate and intermittent systemic steroids. *Thorax* (1992) **47**: 414–417.
157. Knig P, Hillman L, Cervantes CI: Bone metabolism in children with asthma treated with inhaled beclomethasone dipropionate. *J Pediatr* (1993) **122**: 219–226.
158. Ali NJ, Capewell S, Ward MJ: Bone turnover during high dose inhaled corticosteroid treatment. *Thorax* (1991) **46**: 160–164.
159. Wolthers OD, Pedersen S: Bone turnover in asthmatic children treated with oral prednisolone or inhaled budesonide. *Pediatr Pulmonol* (1993) **16**: 341–346.
160. Herrala J, Puolijoki H, Impivaara O, Liippo K, Tala E, Nieminen MM: Bone mineral density in asthmatic women on high dose inhaled beclomethasone dipropionate. *Bone* (1994) **15**: 621–623.
161. Toogood JH, Baskerville JC, Markov AE, *et al.*: Bone mineral density and the risk of fracture in patients receiving long-term inhaled steroid therapy for asthma. *J Allergy Clin Immunol* (1995) **96**: 157–166.
162. Capewell S, Reynolds S, Shuttleworth D, Edwards C, Finlay AY: Purpura and dermal thinning associated with high dose inhaled corticosteroids. *Br Med J* (1990) **300**: 1548–1551.

163. Mak VHF, Melchor R, Spiro S: Easy bruising as a side-effect of inhaled corticosteroids. *Eur Respir J* (1992) **5**: 1068-1074.
164. Roy A, Leblanc C, Paquette L, *et al.*: Skin bruising in asthmatic subjects treated with high doses of inhaled steroids: frequency and association with adrenal function. *Eur Respir J* (1996) **9**: 226–231.
165. Toogood JH, Markov AE, Baskerville J, Dyson C: Association of ocular cataracts with inhaled and oral steroid therapy during long term treatment for asthma. *J Allergy Clin Immunol* (1993) **91**: 571–579.
166. Simons FER, Persaud MP, Gillespie CA, Cheang M, Shuckett EP: Absence of posterior subcapsular cataracts in young patients treated with inhaled glucocorticoids. *Lancet* (1993) **342**: 736–738.
167. Garbe E, LeLorier J, Boivin J, Suissa S: Inhaled and nasal glucocorticoids and the risks of ocular hypertension or open-angle glaucoma. *JAMA* (1997) **227**: 722–727.
168. Russell G: Asthma and growth. *Arch Dis Child* (1993) **69**: 695–698.
169. Balfour-Lynn L: Growth and childhood asthma. *Arch Dis Child* (1986) **61**: 1049–1055.
170. Ninan T, Russell G: Asthma, inhaled corticosteroid treatment and growth. *Arch Dis Child* (1992) **67**: 703–705.
171. Tinkelman DG, Reed CE, Nelson HS, Offord KP: Aerosol beclomethasone dipropionate compared with theophylline as primary treatment of chronic, mild to moderately severe asthma in children. *Pediatrics* (1993) **92**: 64–77.
172. Doull IJ, Freezer NJ, Holgate ST: Growth of prepubertal children with mild asthma treated with inhaled beclomethasone dipropionate. *Am J Respir Crit Care Med* (1995) **151**: 1715–1719.
173. Allen DB, Mullen M, Mullen B: A meta-analysis of the effects of oral and inhaled corticosteroids on growth. *J Allergy Clin Immunol* (1994) **93**: 967–976.
174. Wolthers OD, Pedersen S: Growth in asthmatic children treated with budesonide. *Pediatrics* (1993) **90**: 517–518.
175. Wolthers O, Pedersen S: Short term growth during treatment with inhaled fluticasone dipropionate and beclomethasone dipropionate. *Arch Dis Child* (1993) **68**: 673–676.
176. Turpeinen M, Sorva R, Juntungen-Backman K: Changes in carbohydrate and lipid metabolism in children with asthma inhaling budesonide. *J Allergy Clin Immunol* (1991) **88**: 384–389.
177. Kruszynska YT, Greenstone M, Home PD: Effect of high dose inhaled beclomethasone dipropionate on carbohydrate and lipid metabolism in normal subjects. *Thorax* (1987) **42**: 881–884.
178. Kiviranta K, Turpeinen M: Effect of eight months of inhaled beclomethasone dipropionate and budesonide on carbohydrate metabolism in adults with asthma. *Thorax* (1993) **48**: 974–978.
179. Toogood JH, Baskerville J, Jennings B: Use of spacers to facilitate inhaled corticosteroid treatment of asthma. *Am Rev Respir Dis* (1984) **129**: 723–729.
180. Barnes PJ: Inhaled glucocorticoids: new developments relevant to updating the Asthma Management Guidelines. *Respir Med* (1996) **90**: 379–384.
181. Otulana BA, Varma N, Bullock A, Higenbottam T: High dose nebulized steroid in the treatment of chronic steroid-dependent asthma. *Respir Med* (1992) **86**: 105–108.
182. Greening AP, Ind PW, Northfield M, Shaw G: Added salmeterol versus higher-dose corticosteroid in asthma patients with symptoms on existing inhaled corticosteroid. *Lancet* (1994) **344**: 219–224.
183. Woolcock AJ, Barnes PJ: Asthma: the important questions—part 3. *Am J Respir Crit Care Med* (1996) **153**: S1–S31.
184. Evans DJ, Taylor DA, Steinjans V, *et al.*: Low dose inhaled steroid plus theophylline vs high dose inhaled steroid in the control of asthma. *N Engl J Med* (1997) **337**: 1412–1418.
185. Rutten-van Molken MPMH, van Doorslaer EKA, Jansen MCC, Kerstjens HAM, Rutten FFH: Costs and effects of inhaled coricosteroids and bronchodilators in asthma and chronic obstructive pulmonary disease. *Am J Respir Crit Care Med* (1995) **151**: 975–982.
186. Barnes PJ, Jonsson B, Klim J: The costs of asthma. *Eur Respir J* (1996) **9**: 636–642.
187. van Schayk CP, Dompeling E, Rutten MP, Folgering H, van den Boom G, van Weel C: The influence of an inhaled steroid on quality of life in patients with asthma or COPD. *Chest* (1995) **107**: 1199–1205.

188. O'Driscoll BR, Kalra S, Wilson M, Pickering CAC, Caroll KB, Woodcock AA: Double blind trial of steroid tapering in acute asthma. *Lancet* (1993) **341**: 324–327.
189. Beam WR, Ballard RD, Martin RJ: Spectrum of corticosteroid sensitivity in nocturnal asthma. *Am Rev Respir Dis* (1992) **145**: 1082–1086.
190. McLeod DT, Capewell SJ, Law J, MacLaren W, Seaton A: Intramuscular triamcinolone acetamide in chronic severe asthma. *Thorax* (1985) **40**: 840–845.
191. Ogirala RG, Aldrich TK, Prezant DJ, Sinnett MJ, Enden JB, Williams MH: High dose intramuscular triamcinolone in severe life-threatening asthma. *N Engl J Med* (1991) **329**: 585–589.
192. Grandordy B, Beilmatoug N, Morelle A, De Lauture D, Marac J: Effect of betamethasone on airway obstruction and bronchial response to salbutamol in prednisolone resistant asthma. *Thorax* (1987) **42**: 65–71.
193. Hill SJ, Tattersfield AE: Corticosteroid sparing agents in asthma. *Thorax* (1995) **50**: 577–582.
194. Mullarkey MF, Lammert JK, Blumenstein BA: Long-term methotrexate treatment in corticosteroid-dependent asthma. *Ann Intern Med* (1990) **112**: 577–581.
195. Shiner RJ, Nunn AJ, Chung KF, Geddes DM: Randomized, double-blind, placebo-controlled trial of methotrexate in steroid-dependent asthma. *Lancet* (1990) **336**: 137–140.
196. Nierop G, Gijzel WP, Bel EH, Zwinderman AH, Dijkman JH: Auranofin in the treatment of steroid dependent asthma: a double blind study. *Thorax* (1992) **47**: 349–354.
197. Alexander AG, Barnes NC, Kay AB: Trial of cyclosporin in corticosteroid-dependent chronic severe asthma. *Lancet* (1992) **339**: 324–328.
198. Szczeklik A, Nizankoska E, Dworski R, Domagala B, Pinis G: Cyclosporin for steroid-dependent asthma. *Allergy* (1991) **46**: 312–315.
199. Nelson HS, Hamilos DL, Corsello PR, Levesque NV, Buchameier AD, Bucher BL: A double-blind study of troleandamycin and methylprednisolone in asthmatic patients who require daily corticosteroids. *Am Rev Respir Dis* (1993) **147**: 398–404.
200. Engel T, Heinig JH: Glucocorticoid therapy in acute severe asthma: a critical review. *Eur Respir J* (1991) **4**: 881–889.
201. DeCramer M, Lacquet LM, Fagard R, Rogiers P: Corticosteroids contribute to muscle weakness in chronic airflow obstruction. *Am J Respir Crit Care Med* (1995) **150**: 11–16.
202. Bowler SD, Mitchell CA, Armstrong JG: Corticosteroids in acute severe asthma: effectiveness of low doses. *Thorax* (1992) **47**: 584–587.
203. Morell F, Orkiols R, de Gracia J, Curul V, Pujol A: Controlled trial of intravenous corticosteroids in severe acute asthma. *Thorax* (1992) **47**: 588–591.
204. Harrison BDN, Stokes TC, Hart GJ, Vaughan DA, Ali NJ, Robinson AA: Need for intravenous hydrocortisone in addition to oral prednisolone in patients admitted to hospital with severe asthma without ventilatory failure. *Lancet* (1986) **i**: 181–184.
205. Cypcar D, Busse WW: Steroid-resistant asthma. *J Allergy Clin Immunol* (1993) **92**: 362–372.
206. Barnes PJ, Adcock IM: Steroid-resistant asthma. *Q J Med* (1995) **88**: 455–468.
207. Barnes PJ, Greening AP, Crompton GK: Glucocorticoid resistance in asthma. *Am J Respir Crit Care Med* (1995) **152**: 125S–140S.
208. Schwartz, Lowell FC, Melby JC: Steroid resistance in bronchial asthma. *Am J Int Med* (1968) **69**: 493–499.
209. Carmichael J, Paterson IC, Diaz P, Crompton GK, Kay AB, Grant IWB: Corticosteroid resistance in chronic asthma. *Br Med J* (1981) **282**: 1419–1422.
210. Lane SJ, Atkinson BA, Swimanathan R, Lee TH: Hypothalamic–pituitary axis in corticosteroid-resistant asthma. *Am J Respir Crit Care Med* (1996) **153**: 1510–1514.
211. Woolcock AJ: Steroid resistant asthma: what is the clinical definition? *Eur Respir J* (1993) **6**: 743–747.
212. Leung DYM, Martin RJ, Szefler SJ, *et al.*: Dysregulation of interleukin 4, interleukin 5, and interferon γ gene expression in steroid-resistant asthma. *J Exp Med* (1995) **181**: 33–40.
213. Lamberts SWJ, Kioper JW, de Jong FH: Familial and iatrogenic cortisol receptor resistance. *J Steroid Biochem Molec Biol* (1992) **43**: 385–388.
214. Lane SJ, Arm JP, Staynov DZ, Lee TH: Chemical mutational analysis of the human glucocortiocoid receptor cDNA in glucocorticoid-resistant bronchial asthma. *Am J Respir Cell Mol Biol* (1994) **11**: 42–48.

215. Corrigan C, Brown PH, Barnes NC, et al.: Glucocorticoid resistance in chronic asthma. *Am Rev Respir Dis* (1991) **144**: 1016–1025.
216. Lane SJ, Palmer JBD, Skidmore IF, Lee TH: Corticosteroid pharmacokinetics in asthma. *Lancet* (1990) **336**: 126S.
217. Alvarez J, Surs W, Leung DY, Ikle D, Gelfand EW, Szefler SJ: Steroid-resistant asthma: immunologic and pharmacologic features. *J Allergy Clin Immunol* (1992) **89**: 714–721.
218. Goulding NJ, Podgorski MR, Hall ND, et al.: Autoantibodies to recombinant lipocortin-1 in rheumatoid arthritis and systemic lupus erythematosus. *Ann Rheum Dis* (1989) **48**: 843–850.
219. Wilkinson JR, Podgorski MR, Godolphin JL, Goulding NJ, Lee TH: Bronchial asthma is not associated with auto-antibodies to lipocortin-1. *Clin Exp Allergy* (1990) **20**: 189–192.
220. Chung KF, Podgorski MR, Goulding NJ, et al.: Circulating autoantibodies to recombinant lipocortin-1 in asthma. *Respir Med* (1991) **95**: 121–124.
221. Wilkinson JRW, Lane SJ, Lee TH: The effects of corticosteroids on cytokine generation and expression of activation antigens by monocytes in bronchial asthma. *Int Arch Allergy Clin Immunol* (1991) **94**: 220–221.
222. Poznansky MC, Gordon ACH, Douglas JG, Krajewski AS, Wyllie AH, Grant IWB: Resistance to methylprednisolone in cultures of blood mononuclear cells from glucocorticoid-resistant asthmatic patients. *Clin Sci* (1984) **67**: 639–645.
223. Wilkinson JRW, Crea AEG, Clark TJH, Lee TH: Identification and characterization of a monocyte-derived neutrophil-activating factor in corticosteroid-resistant bronchial asthma. *J Clin Invest* (1989) **84**: 1930–1941.
224. Corrigan CJ, Brown PH, Barnes NC, Tsai J, Frew AJ, Kay AB: Peripheral blood T lymphocyte activation and comparison of the T lymphocyte inhibitory effects of glucocorticoids and cyclosporin A. *Am Rev Respir Dis* (1991) **144**: 1026–1032.
225. Brown PH, Teelucksingh S, Matusiewicz SP, Greening AP, Crompton GK, Edwards CRW: Cutaneous vasoconstrictor responses to glucocorticoids in asthma. *Lancet* (1991) **337**: 576–580.
226. Lane SJ, Lee TH: Glucocorticoid receptor characteristics in monocytes of patients with corticosteroid-resistant bronchial asthma. *Am Rev Respir Dis* (1991) **143**: 1020–1024.
227. Sher ER, Leung YM, Surs W, et al.: Steroid-resistant asthma. Cellular mechanisms contributing to inadequate response to glucocorticoid therapy. *J Clin Invest* (1994) **93**: 33–39.
228. Kam JC, Szefler SJ, Surs W, Sher FR, Leung DYM: Combination IL-2 and IL-4 reduces glucocorticoid-receptor binding affinity and T cell response to glucocorticoids. *J Immunol* (1993) **151**: 3460-3466.
229. Adcock IM, Lane SJ, Brown CA, Peters MJ, Lee TH, Barnes PJ: Differences in binding of glucocorticoid receptor to DNA in steroid-resistant asthma. *J Immunol* (1995) **154**: 3000–3005.
230. Adcock IM, Lane SJ, Brown CA, Lee TH, Barnes PJ: Abnormal glucocorticoid receptor/AP-1 interaction in steroid resistant asthma. *J Exp Med* (1995) **182**: 1951–1958.
231. Adcock IM, Lane SJ, Barnes PJ, Lee TH: Enhanced phorbol ester induced c-Fos transcription and translation in steroid-resistant asthma. *Am J Respir Crit Care Med* (1996) **153**: A862.
232. Adcock IM, Brady H, Lim S, Karin M, Barnes PJ: Increased JUN kinase activity in peripheral blood monocytes from steroid-resistant asthmatic subjects. *Am J Respir Crit Care Med* (1997) **155**: A288.
233. Rosewicz S, McDonald AR, Maddux BA, Godfine ID, Miesfeld RL, Logsden CD: Mechanism of glucocorticoid receptor down-regulation by glucocorticoids. *J Biol Chem* (1988) **263**: 2581–2584.
234. Adcock IM, Brown CR, Shirasaki H, Barnes PJ: Effects of dexamethasone on cytokine and phorbol ester stimulated c-Fos and c-Jun DNA binding and gene expression in human lung. *Eur Respir J* (1994) **7**: 2117–2123.
235. Vacca A, Screpanati I, Maroder M, Petrangeli E, Frati L, Guline A: Tumor promoting phorbol ester and raw oncogene expression inhibit the glucocorticoid-dependent transcription from the mouse mammary tumor virus long terminal repeat. *Mol Endocrinol* (1989) **3**: 1659–1665.
236. Peters MJ, Adcock IM, Brown CR, Barnes PJ: β-Adrenoceptor agonists interfere with glucocorticoid receptor DNA binding in rat lung. *Eur J Pharmacol* (1995) **289**: 275–281.

237. Wong CS, Wahedna I, Pavord ID, Tattersfield AE: Effect of regular terbutaline and budesonide on bronchial reactivity to allergen challenge. *Am J Respir Crit Care Med* (1994) **150**: 1268–1278.
238. Barnes PJ, Chung KF: Questions about inhaled β_2-agonists in asthma. *Trends Pharmacol Sci* (1992) **13**: 20–23.
239. Taylor DR, Sears MR, Herbison GP, *et al.*: Regular inhaled β-agonist in asthma: effects on exacerbation and lung function. *Thorax* (1993) **48**: 134–138.
240. Suissa S, Ernst P, Boivin J, *et al.*: A cohort analysis of excess mortality in asthma and the use of inhaled β-agonists. *Am J Respir Crit Care Med* (1994) **149**: 604-610.
241 Peters MJ, Yates DH, Chung KF, Barnes PJ: β_2-Agonist dose reduction: strategy and early results. *Thorax* (1993) **48**: 1066.
242. Beynon HLC, Garbett ND, Barnes PJ: Severe premenstrual exacerbations of asthma: effect of intramuscular progesterone. *Lancet* (1988) **ii**: 370–372.
243. Kalkhoveng E, Wissink S, Van Der Saag PT, van der Burg B: Negative interaction between the Rel A (p65) subunit of NF-κB and the progesterone receptor. *J Biol Chem* (1996) **271**: 6217–6224.
244. Macek V, Soru J, Korpriva S, Marin J: Persistent adenoviral infection and chronic airway obstruction in children. *Am J Respir Crit Care Med* (1994) **150**: 7–10.
245. Whyte P, Buchovich RG, Horovitz JM: Association between an oncogene and an anti-oncogene: the adenovirus EIA protein binds to the retinoblastoma gene product. *Nature* (1988) **334**: 124–129.

41

Mediator Antagonists

PETER J. BARNES AND K. FAN CHUNG

INTRODUCTION

Many different inflammatory mediators have now been implicated in asthma,[1,2] and several specific receptor antagonists and synthesis inhibitors have been developed that will prove invaluable in working out the contribution of each mediator (Table 41.1). As many mediators probably contribute to the pathological features of asthma, it seems unlikely that a single antagonist will have a major clinical effect, compared with non-specific agents such as β-agonists and corticosteroids. However, until such drugs have been evaluated in careful clinical studies, it is not possible to predict their value. Many different mediator inhibitors are currently under evaluation in asthma as they provide the possibility of more specific therapy with reduced side-effects. The widespread use of antihistamines in rhinitis is proof of the efficacy of this strategy.

ANTIHISTAMINES

Conventional doses of non-sedating antihistamines, such as terfenadine, astemizole and loratadine, have little effect in clinical asthma.[3] Improved control of allergic rhinitis with more effective antihistamines may result in improved asthma control (and this has been clearly demonstrated with nasal steroid preparations). Several new non-sedating antihistamines have been developed, including ebastine, epinastine, mizolastine and noberastine, but it is difficult to see that they have any major advantage over existing drugs.

Table 41.1 Inflammatory mediators and inhibitors.

Mediator	Inhibitor
Histamine	Terfenadine, loratadine, cetirizine
Leukotriene D$_4$	Zafirlukast, montelukast
	Zileuton, Bay-x1005, ZD2138 (5-lipoxygenase inhibitors)
Leukotriene B$_4$	LY-293111
Prostaglandins	L745,337, NS-398 (COX-2 inhibitors)
Thromboxane	Ozagrel, vapiprost
Platelet-activating factor	Apafant, modipafant, bepafant
Bradykinin	Icatibant, WIN 64338
Adenosine	Theophylline
Reactive oxygen species	N-Acetylcysteine, ascorbic acid, nitrones
Nitric oxide	Aminoguanidine, 1400W
Endothelin	Bosentan, SB209670
Mast cell tryptase	APC-366
Eosinophil basic proteins	Heparin
Interleukin (IL)-1β	Recombinant IL-1 receptor antagonist
Tumour necrosis factor (TNF)-α	TNF antibody, TNF soluble receptors
IL-4	IL-4 antibody
IL-5	IL-5 antibody

Some new antihistamines under development appear to have 'antiallergic' properties that may make them more useful in the treatment of asthma and other allergic diseases. Cetirizine is a non-sedating antihistamine that appears to have additional properties that may be unrelated to H$_1$-receptor antagonism. Several *in vitro* studies have demonstrated that cetirizine has an inhibitory effect on eosinophil chemotaxis and adherence to endothelial cells, and *in vivo* inhibits eosinophil recruitment into asthmatic airways after allergen challenge.[4] In clinical studies, cetirizine has been shown to improve symptom scores and to reduce the use of rescue drugs in both seasonal and perennial asthma,[5,6] although it has no effect on allergen-induced asthma.[7] Astemizole has a similar profile of *in vitro* and *in vivo* effects to cetirizine. In clinical studies, it has a significant effect in reducing asthma symptoms after chronic treatment.[8] The beneficial effect of these new antihistamines may not be related to H$_1$-receptor antagonism but to some other ill-defined property of the drugs that deserves further investigation. The effects of these drugs is relatively weak, however, and they would only be useful clinically in mild asthma. Nevertheless, they may have an important place in patients with seasonal asthma who also have troublesome symptoms of rhinitis.

LEUKOTRIENE ANTAGONISTS

Elevated levels of cysteinyl leukotrienes are detected in bronchoalveolar lavage (BAL) fluid and elevated leukotriene (LT)E$_4$ levels in the urine of asthmatic patientss. Cysteinyl leukotrienes (LTC$_4$, LTD$_4$, LTE$_4$) are potent constrictors of human airways *in vitro* and *in vivo*, cause airway microvascular leakage in animals and stimulate airway mucus secretion,[9] as discussed in Chapter 16 (Fig. 41.1). These effects are all mediated in

41 Mediator Antagonists

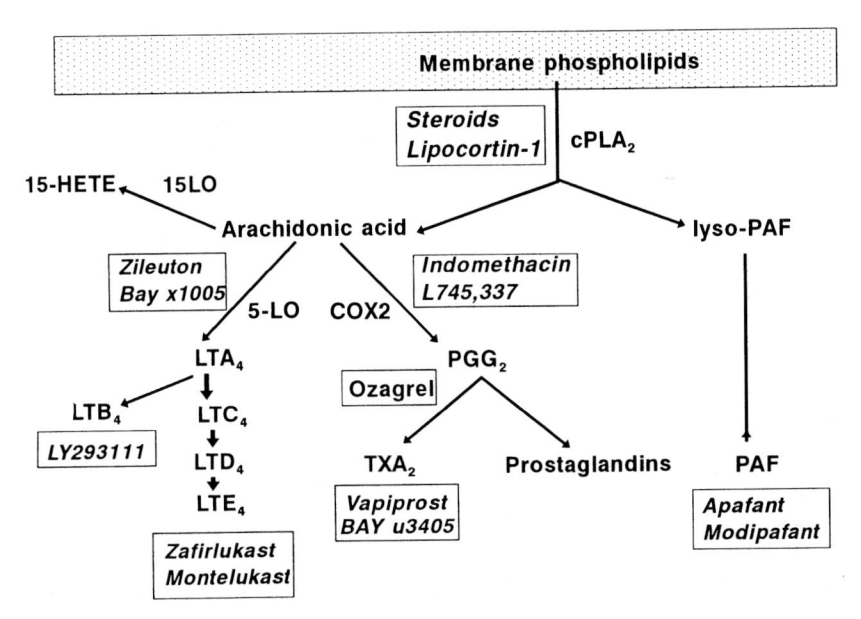

Fig. 41.1 Leukotrienes have many effects on airways mediated by cys-LT$_1$ receptors. Leukotrienes may be inhibited by inhibitors of 5-lipoxygenase (5-LO) or by blockade of cys-LT$_1$ receptors. ASA, aspirin-sensitive asthmatics; PAF, platelet-activating factor.

human airways via cys-LT$_1$ receptors. The role of cys-LT$_2$ receptors (found on pulmonary veins) is uncertain. Potent cys-LT$_1$ antagonists have been developed and several are now entering the clinic.[9–11]

Clinical studies

Zafirlukast is a relatively potent antagonist now licensed in several countries. A single 40-mg dose produces a 100-fold shift of the LTD$_4$ dose–response curve and significant antagonism is maintained for 24 h.[12] Oral administration blocks the early and late response to allergen and exercise-induced asthma.[13,14] In a 14-day study zafirlukast (twice daily) significantly improved lung function; in a 6-week study 10 mg and 20 mg twice daily improved lung function and symptom scores in patients with mild to moderate asthma.[15] In another study, addition of zafirlukast (20 mg twice daily for 3 months) improved asthma symptoms and lung function in patients with mild asthma and also reduced asthma exacerbations.[16] Orally administered pranlukast was the first leukotriene antagonist introduced and, like zafirlukast, improves lung function and reduces symptoms in patients with a range of severity of asthma.[17] It has been available in Japan for over 1 year. Montelukast is effective as a once-daily administration[18] and improves asthma control.[19] Many other LTD$_4$ antagonists have recently been developed and are in phase I and phase II testing for asthma.

Class-related side-effects do not seem to be a problem at present and this may be because leukotrienes are only produced under pathological circumstances.

Clinical use

The clinical results to date with potent LTD_4 antagonists are encouraging, particularly since these drugs are active orally. Their place in asthma is difficult to assess until their long-term anti-inflammatory potential is known. The present results suggest that they have a bronchodilator effect: they prevent induced bronchoconstriction (exercise, allergen, cold air) and in some patients there is a small bronchodilator effect,[18,20] suggesting that there is leukotriene 'tone' in asthma. They are completely effective in preventing aspirin challenge in aspirin-sensitive asthmatic patients.[21,22]

Animal studies indicate that LTD_4 causes microvascular leakage, an acute inflammatory event. Inhalation of the cysteinyl leukotriene LTE_4 results in eosinophilic inflammation of the airways,[23] and in animal studies cysteinyl leukotrienes have been shown to induce eosinophilia via the release of interleukin (IL)-5.[24] A modest inhibition of the number of eosinophils in BAL fluid has been observed after segmental allergen challenge with zafirlukast.[25] However, careful biopsy and sputum studies need to address this issue in the future.

Leukotriene antagonists may be indicated in the control of mild asthma and may be useful in reducing the requirements for inhaled steroids in more severe asthma. It is of interest that even high does of inhaled or oral steroids do not reduce leukotriene production in asthma, as measured by urinary LTE_4 excretion.[26,27]

It is unlikely that leukotriene antagonists will replace steroids; the clinical data suggests that they have a relatively modest effect on symptom control compared with steroids, but more dose-ranging studies are needed to determine their maximal potential. Leukotriene antagonists may also be useful in treating concomitant rhinitis.[28] The major advantage of leukotriene antagonists is that they are active by mouth and this may improve patient adherence with chronic therapy. So far they do not appear to have any class-related side-effects and are well tolerated. They may also have some anti-inflammatory effect. It is unlikely that they will prove to be as effective as inhaled steroids, but may be a useful therapy to add on to improve control rather than increasing the dose of inhaled steroids. Studies comparing the effects of leukotriene antagonists with inhaled steroids are now needed.

One of the features of early studies of leukotriene antagonists has been the heterogeneity of response, with some patients (approximately one- third) showing a very good response and others apparently unresponsive. This presumably reflects the varying contribution of leukotrienes in different patients and might be a reflection of polymorphism of the 5-lipoxygenase (5-LO) gene. Aspirin-sensitive asthmatic patients respond very well and it will be important to recognize other responders clinically or by some predictive test.

5-LO inhibitors

5-LO is the critical enzyme involved in the generation of leukotrienes. Several drugs have been developed that inhibit 5-LO and therefore have a similar profile of effect to leukotriene antagonists. The currently available inhibitors may be classified into direct inhibitors that act directly on the enzyme and indirect inhibitors that interfere with a membrane docking protein, 5-LO activating protein (FLAP), necessary for enzyme activation.[29,30]

Several hydroxamates and N-hydroxyureas have been described as 5-LO inhibitors and act by interfering with the redox state of the active binding site. The most extensively investigated is zileuton, but this compound is relatively weak and has a short duration of action. The effect of zileuton in challenge studies is similar to that of leukotriene antagonists, with partial inhibition of allergen- and exercise-induced asthma and complete inhibition of aspirin-induced asthma.[31,32] In addition, zileuton has been shown to reduce airway hyperresponsiveness and to decrease inflammatory cells in the airways in nocturnal exacerbations of asthma.[33,34] Beneficial effects have been observed in clinical studies (800 mg b.i.d. and 600 mg q.i.d. for 1 month) with results similar to those seen with zafirlukast.[35] Long-term studies (13 weeks) also show significant improvement in asthma control compared with placebo and a reduction in the frequency of exacerbations.[36] In a long-term study over 6 months, zileuton was shown to improve asthma control and lung function and to reduce the use of rescue β_2-agonists in patients with mild to moderate asthma.[37]

Interestingly, blood eosinophil counts are significantly reduced by long-term treatment with zileuton compared with placebo treatment.[37] Furthermore, in a study of segmental allergen challenge, pretreatment with zileuton for 8 days significantly reduced the eosinophil influx.[38] This supports the view that leukotriene antagonists may have some anti-inflammatory action in asthma.

Zileuton is now licensed in the USA, but its short duration of action means that it has to be given four times daily. Abnormal liver enzymes are relatively common and it is therefore important that liver function tests are performed in patients stating on this therapy. A more potent derivative of zileuton, ABT-761, has been developed that has a longer duration of action in normal volunteers and clinical trials in asthma are in progress.[39] Several other redox 5-LO inhibitors, including CGS-21595, DuP-654, docebenone and SC-45662, are also in clinical development.

A novel series of non-redox 5-LO inhibitors have recently been developed, the most potent of which is ZD2138, which is reported to potently inhibit 5-LO activity for 24 h after once-daily dosing (400 mg orally) in normal volunteers. Clinical studies in asthma show inhibition of aspirin-induced asthma as expected, but no inhibition of allergen-induced responses.[40,41]

MK-886 and MK-591 inhibit 5-LO activation but do not directly inhibit the enzyme; they apparently bind to FLAP in the cell membrane and prevent translocation of 5-LO from the cytosol.[30] Another FLAP antagonist in clinical studies is Bay-x1005, which is effective in allergen-induced asthma.[42] Whether FLAP inhibitors have advantages over direct inhibitors in terms of side-effects remains to be determined.

There is a theoretical advantage to the use of 5-LO inhibitors compared with leukotriene antagonists since the formation of LTB_4 and other 5-LO products, as well as cysteinyl leukotrienes, will also be inhibited. This may make the drugs more applicable to other airway diseases where LTB_4 may be involved, as well as having a wider application to other inflammatory conditions (rheumatoid arthritis, inflammatory bowel disease, psoriasis, etc.). What remains to be determined is how much enzyme inhibition is necessary at the inflammatory site for a good clinical response.

LTB₄ antagonists

LTB₄ is a potent chemotactic agent for neutrophils and has some eosinophil-activating effects. Inhaled LTB₄ has little effect on asthmatic airways[43] and it is unlikely to be an important mediator of asthma. However, it may be involved in airway diseases where neutrophil infiltration is prominent, including cystic fibrosis, bronchiectasis, chronic obstructive pulmonary disease and fibrosing alveolitis. An LTB₄ antagonist, LY-293111, had no effects on the late response or airway hyperresponsiveness following allergen challenge, but significantly reduced the neutrophil increase in BAL.[44] This suggests that LTB₄ antagonists are unlikely to be useful in asthma therapy.

PROSTAGLANDIN INHIBITORS

Thromboxane inhibitors

Several other inhibitors of lipid mediators have been developed (Fig. 41.2). Thromboxane (Tx)A₂ and stable thromboxane analogues, such as U46619, are potent bronchoconstrictors in animal and human airways.[45] Inhaled U46619 is a potent bronchoconstrictor in asthmatic patients.[46] Furthermore this thromboxane analogue also causes plasma exudation in guinea-pig airways and enhances acetylcholine release from canine airway cholinergic nerves.[47,48] The more stable metabolite of TxA₂, TxB₂, has been detected in

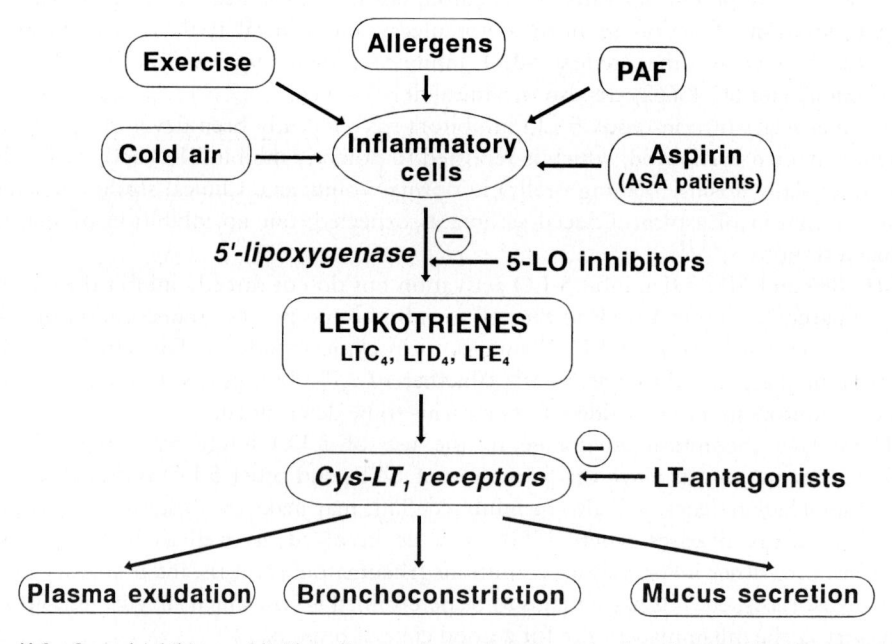

Fig. 41.2 Several inhibitors of lipid mediator pathways have been developed. COX 2, cyclooxygenase 2; cPLA₂, cytosolic phospholipase A₂; 15-HETE, 15, hydroxyeicosatetraenoic acid; 5-LO, 5-lipoxygenase; 15LO, 15-lipoxygenase; PAF, platelet-activating factor.

BAL fluid of patients with asthma and there is some evidence for increased urinary excretion of the dinor metabolite of TxB_2 in asthmatic patients.[26]

Thromboxane and prostaglandin $(PG)D_2$ cause bronchoconstriction via activation of thromboxane (TP) receptors and several TP-receptor antagonists have been developed. GR 32191 (vapiprost), Bay u3405, KW-4099 and ICI 192,605 have shown no beneficial effects in either allergen or exercise challenge in asthmatic patients.

Cyclooxygenase (COX) inhibitors, including aspirin and flurbiprofen, have little or no beneficial effect in asthma challenge studies or in clinical asthma, but this may be because they block production of both bronchoconstrictor (PGD_2, $PGF_{2\alpha}$, TxA_2) and bronchodilator (PGE_2, PGI_2) mediators. For this reason specific TxA_2 synthase inhibitors might be more useful than COX inhibitors, although they would not block the synthesis of PGD_2, which is the major prostaglandin produced by the mast cell in response to allergen challenge in human airways (as measured by BAL). Several thromboxane synthase inhibitors have been developed for use in asthma. Ozagrel (OKY 046) is a moderately potent orally active TxA_2 inhibitor now licensed for use in asthma in Japan. However there is little objective evidence that it is effective in asthma, although some subjective measurements improved in multicentre trials. Another more potent thromboxane synthase inhibitor, pirmagrel (CGS 13080), was able to completely prevent the increase in serum TxB_2 following allergen challenge in asthmatic patients; while it caused a very small reduction in the early response to allergen, there was no effect on the late response or on airway hyperresponsiveness.[49] Overall, thromboxane inhibitors have been very disappointing in asthma and there is insufficient evidence to continue their development as new antiasthma therapies.

COX-2 inhibitors

A constitutive form of COX (COX-1) is basally expressed in cells; a second form (COX-2) is expressed in normal and asthmatic airways.[50] It is possible that selective inhibition of COX-2 might be beneficial in asthma and several selective COX-2 inhibitors, such as L745,337 and NS-398, have now been developed.[51,52] These selective inhibitors may also prove to be safe in patients with aspirin-sensitive asthma, since it is possible that bronchoconstriction in these patients may be due to inhibition of PGE_2 synthesis by COX-1.

PLATELET-ACTIVATING FACTOR (PAF) ANTAGONISTS

PAF has several properties which suggest that it may play an important role in asthma; in animal studies PAF receptor antagonists have been effective in blocking allergen-induced eosinophilic inflammation and airway hyperresponsiveness.[53] Several potent PAF-receptor antagonists have now been developed and asthma has been one of the primary indications for these compounds.

Apafant (WEB 2086) was the first potent specific PAF antagonist developed and has proved to be a useful prototype drug. It is well absorbed by the oral route and very effectively blocks the effects of intradermal and inhaled PAF in normal volunteers.[54]

However, neither oral nor inhaled WEB 2086 had any effect in allergen challenge.[55,56] A potent PAF antagonist, UK 74,505, after a single oral dose inhibits the airway effects of inhaled PAF for up to 24 h,[57] yet is ineffective in allergen challenge (early or late response or subsequent airway hyperresponsiveness).[58] In multicentre clinical trials, orally administered apafant and modipafant (active enantiomer of UK 74,506) were found to have no clinical benefit in patients with asthma who were withdrawn from inhaled steroids.[58,59] However it could still be argued that the dose used in this study was too low to block endogenously released PAF. More recently, even more potent PAF antagonists have been developed and these have shown some, albeit modest, effects in asthmatic patients. More potent PAF antagonists include Y-24180, which has been shown to reduce airway hyperresponsiveness in asthmatic patients,[60] and foropafant (SR27417A), which significantly reduces the late response after inhaled allergen challenge.[61] Both antagonists have a potency greater than that of PAF for its receptor, and it is possible that even more potent antagonists might have greater effects. It is also possible that PAF antagonists may be useful in some patients with asthma or in certain circumstances, such as virus-induced exacerbations of asthma.

The results with PAF antagonists are disappointing, particularly in view of the encouraging results obtained with animal models, the fact that PAF appears to be produced in asthma and that PAF is a potent activator of human eosinophils. There may be several explanations for this. It is possible that the concentrations of antagonist are not high enough at the sites of inflammation in the airway, since PAF functions as a 'paracrine' mediator that has effects only on neighbouring cells and may therefore act at high local concentrations that are difficult to antagonize. PAF may not be important in chronic, as opposed to acute, inflammation and there is evidence in the chronically exposed allergen model in primates that this is the case. Whether this is because of tachyphylaxis of PAF receptors or PAF production is switched off in chronic inflammation is not certain. In any case, the preliminary clinical studies make it unlikely that PAF will prove to be a useful treatment for asthma in the future, although it is possible that certain types of asthma may respond.

PHOSPHOLIPASE INHIBITORS

Phospholipase (PL)A$_2$ appears to be of critical importance in the generation of all lipid mediators and would therefore appear to be a suitable target for inhibitory drugs. Non-steroidal drugs that inhibit PLA$_2$, such as mepacrine, might be expected to share the beneficial effects of steroids, although mepacrine is weak and non-specific. More potent PLA$_2$ inhibitors, such as manoalide, derived from a sponge, are also non-selective.[63] It is now clear that there may be many forms of PLA$_2$ and a high molecular weight (85-kDa) cytosolic PLA$_2$ (cPLA$_2$) that is arachidonic acid selective has been characterized.[62] This may make it more feasible to select inhibitors by random screening, but whether PLA$_2$ inhibitors would offer any advantages over 5-LO inhibitors is uncertain in the light of the fact that neither COX inhibitors nor PAF- receptor antagonists are effective in asthma.

Inhibition of PLC (phosphoinositidase C), which is the enzyme leading to phosphoinositide breakdown, could also be useful. Neomycin appears to inhibit this enzyme and is useful experimentally, but is too non-specific for clinical development. It is now recognized that there are many subtypes of PLC and in the future it might be possible to develop isoform-specific inhibitors that have a more selective effect.

PLD appears to play a critical role in the priming of inflammatory cells and therefore may be an important target in asthma.[64] Although poor inhibitors such as wortmannin are available, more selective PLD inhibitors are under development. These drugs might have particular applicability in asthma, since priming of key inflammatory cells might be an important step in the amplification of the inflammatory response.

BRADYKININ ANTAGONISTS

Bradykinin has also attracted attention as a potential mediator of asthma symptoms, since it appears to be the inflammatory mediator produced in asthma that is most likely to activate sensory nerves in the airway and mediate cough.[65] All the effects of bradykinin on the airways appear to be mediated via B_2 receptors. The first B_2-receptor antagonists developed were very weak and unstable and therefore had no clinical potential. Thus NPC567 was found to have no inhibitory effect on bradykinin responses in the nose.[66] A potent and stable bradykinin B_2-receptor antagonist, icatibant (HOE 140), a peptide derivative, has been developed and in animal studies provides long-lasting antagonism to infused or inhaled bradykinin.[67] Icatibant inhibits allergen-induced nasal blockage in the nose, indicating potential for this type of drug in treating rhinitis.[68] Studies in asthma have recently been performed with nebulized icatibant (900 μg, 3000 μg t.i.d.) in patients with severe asthma. The drug was well tolerated and there was a small improvement in lung function (approximately 10% increase in forced expiratory volume in 1 s), but no change in asthma symptoms.[69]

There are disadvantages to a peptide drug in terms of expense of synthesis, stability and delivery, so there is now a search for non-peptide B_2-receptor antagonists, equivalent to the recently developed tachykinin antagonists. Non-peptide antagonists of B_2 receptors, such as WIN 64338, have now been developed that may prove to be more useful.[70]

B_1 receptors appear to be involved in inflammatory responses and may be induced by cytokines. There is some evidence for upregulation of B_1 receptors in animal lungs,[71] indicating a potential for B_1-receptor antagonists. However, B_1 receptors have not been demonstrated in human airway smooth muscle, as judged by the failure of the B_1-selective agonists Lys-bradykinin and [desArg^9] bradykinin to induce bronchoconstriction in asthma.[72] However, it is possible that B_2 receptors may regulate airway functions other than bronchoconstriction.

ANTIOXIDANTS

Activated inflammatory cells, such as mast cells, macrophages and eosinophils, produce a variety of reactive oxygen species (ROS) that may have inflammatory and tissue-damaging effects.[73] Furthermore, oxidants in polluted air may be an important factor in worsening asthma. There is therefore a rationale in exploring the use of antioxidants in asthma. There is increasing evidence that ROS may contribute to chronic inflammatory diseases. Superoxide anions are potent activators of NF-κB and combine with nitric oxide (NO) to form toxic peroxynitrite ions. Currently available antioxidants, such as vitamins C and E, are not very effective and more potent antioxidants are currently under development for other diseases. Long-acting superoxide dismutase (PEG-SOD) is in development as an anti-inflammatory agent. More potent antioxidants have now been developed, including the spin-trap antioxidants such as nitrones,[74] although these have not yet reached the stage of clinical development.

ADENOSINE ANTAGONISTS

Adenosine is a potent constrictor of asthmatic airways and is likely to be produced by a number of inflammatory cells in the airways. Its effects on the airways are probably due to activation of mast cells via an A_2 receptor.[75] Theophylline at therapeutic concentrations blocks adenosine receptors, although it is unlikely that this contributes to its antiasthma effect since enprofylline, which has little adenosine antagonistic activity at therapeutic concentrations, is even more potent as an antiasthma agent. Whether a more selective A_2-receptor antagonist might have therapeutic potential is not yet known as such agents have not yet been developed for clinical use.

There is some uncertainty about which adenosine receptors are involved in mast cell activation. In animal studies there is some evidence that mast cells have adenosine A_{2B} receptors,[76] but other studies suggest that A_3 receptors are involved.[77] There are no selective adenosine-receptor antagonists available for clinical use. Recently an inhaled DNA antisense oligonucleotide against adenosine A_1 receptors has been shown to inhibit airway hyperresponsiveness in an animal model of asthma.[78]

NO SYNTHASE INHIBITORS

NO, produced in large amounts from the inducible isoform of NO synthase (iNOS) that is overexpressed in asthmatic airways, may have pro-inflammatory effects in asthma by increasing plasma exudation in airways and amplifying eosinophilic inflammation,[79] as discussed in Chapter 21. There is evidence in experimental animals that endogenous NO may increase plasma exudation in airways, may be involved in recruitment of eosinophils and may play a role in eosinophil survival. It follows that inhibition of iNOS may be beneficial in asthma. It is important to avoid inhibition of constitutive NOS, as this may cause hypertension and bronchoconstriction. Aminoguanidine has some selectivity for iNOS and reduces exhaled NO in asthmatic patients to a greater extent than in normal subjects.[80] Whether regular treatment with aminoguanidine is beneficial in asthma has not yet been determined. More potent and selective iNOS inhibitors are now in clinical development.[81]

ENDOTHELIN ANTAGONISTS

Endothelins are potent bronchoconstrictors and are abnormally expressed in asthmatic airways.[82,83] Endothelin (ET)-1 causes proliferation of airway smooth muscle and stimulates fibrosis, suggesting that it may play a role in the remodelling of asthmatic airways. These effects are mediated via stimulation of both ET_A receptors (fibrosis) and ET_B receptors (bronchoconstriction). Because the various effects of endothelin on the airways are mediated by both receptor subtypes, the effects of combined antagonism for ET_A and ET_B receptors would appear to be more useful. Several potent non-peptide endothelin antagonists, such as bosentan and SB209670, have now been developed.[84] However, it may be difficult to test these compounds in asthma, as long-term studies would be needed to determine whether these drugs prevent the irreversible airway narrowing of chronic asthma.

BASIC PROTEIN INHIBITORS AND HEPARIN

Eosinophil basic proteins may play an important role in the epithelial shedding and other inflammatory effects seen in asthma, suggesting that inhibition of their synthesis, release or effects may be beneficial. Heparin is strongly anionic and counteracts the effects of cationic eosinophil proteins, such as major basic protein, eosinophil cationic protein and eosinophil-derived neurotoxin. Recently, heparin has been found to have an inhibitory effect in allergen-induced airway responses in experimental animals (sheep and rabbits) and a recent clinical study suggests that inhaled heparin is able to block the early and late response to allergen[85,86] and exercise.[87] Heparin also inhibits the proliferation of airway smooth muscle and therefore may have potential as a drug to inhibit airway remodelling in asthma.[88] Several non-anticoagulant fragments of heparin have now been developed and it may be worth investigating their anti-asthma potential.

INFLAMMATORY ENZYME INHIBITORS

Mast cell tryptase has potent inflammatory effects in the airways and may increase airway responsiveness. Mast cell tryptase inhibitors may therefore be of potential value.[89] One such inhibitor, APC-366, has been shown to block late responses in allergic sheep.[90] Clinical trials with this drug are now underway.

Mast cell chymase has potent effects on mucus secretion[91] and a chymase inhibitor may therefore be useful. Heparin may function as a tryptase/chymase inhibitor since it binds to these enzymes and may therefore inactivate their effects.

Neutrophil elastase may also have potent effects on mucus secretion and several potent non-peptide neutrophil elastase inhibitors, such as ICI 200,355, have now been developed.[92] Whether these will have therapeutic efficacy in asthma remains to be determined.

COMBINED INHIBITORS

It seems very unlikely that antagonizing a single mediator of asthma will ever be as useful as therapies with a broad spectrum of activity, although such therapies may have the advantage of fewer side-effects and the possibility of oral administration. Certain types of asthmatic patient may respond much better to these more specific therapies. For example, it seems likely that aspirin-sensitive asthmatic patients may particularly benefit from LTD_4 antagonists.

There are some antagonists that appear to act as dual or non-specific mediator inhibitors, such as Sch-37370, which has both antihistamine and PAF antagonist activity, and BN 50548, which has antioxidant and PAF inhibitory effects. A combined histamine and leukotriene antagonist might be useful. These drugs may have some advantage in blocking more than one mediator, since many mediators are involved in asthma.

REFERENCES

1. Barnes PJ, Chung KF, Page CP: Inflammatory mediators and asthma. *Pharmacol Rev* (1988) **40**: 49–84.
2. Chung KF, Barnes PJ: Role of inflammatory mediators in asthma. *Br Med Bull* (1992) **48**: 135–148.
3. Howarth PH: Histamine and asthma: an appraisal based on specific H_1-receptor antagonism. *Clin Exp Allergy* (1990) **20** (Suppl 2): 31–41.
4. Walsh GM: The anti-inflammatory effects of cetirizine. *Clin Exp Allergy* (1994) **24**: 81–85.
5. Spector SL, Nicodemus CF, Corren J, *et al.*: Comparison of the bronchodilatory effects of cetirizine, albuterol, and both together versus placebo in patients with mild-to-moderate asthma. *J Allergy Clin Immunol* (1995) **96**: 174–181.
6. Grant JA, Nicodemus CF, Findlay SR, *et al.*: Cetirizine in patients with seasonal rhinitis and concomitant asthma: prospective, randomized, placebo-controlled trial. *J Allergy Clin Immunol* (1995) **95**: 923–932.
7. de Bruin Weller MS, Rijssenbeek Nouwens LH, de Monchy JG: Lack of effect of cetirizine on early and late asthmatic response after allergen challenge. *J Allergy Clin Immunol* (1994) **94**: 231–239.
8. Gould CAL, Ollier S, Aurich R, Davies RJ: A study of the clinical efficacy of azelastine in patients with extrinsic asthma and its effect on airway responsiveness. *Br J Clin Pharmacol* (1988) **26**: 515–525.
9. Henderson WR: The role of leukotrienes in inflammation. *Ann Intern Med* (1994) **121**: 684–697.
10. Chung KF: Leukotriene receptor, antagonists and biosynthesis inhibitors: potential breakthrough in asthma therapy. *Eur Respir J* (1995) **8**: 1203–1213.
11. Smith LJ: Leukotrienes in asthma. The potential therapeutic role of antileukotriene agents. *Arch Intern Med* (1996) **156**: 2181–2189.
12. Smith LJ, Glass M, Minkwitz MC: Inhibition of leukotriene D4-induced bronchoconstriction in subjects with asthma: a concentration–effect study of ICI 204,219. *Clin Pharmacol Ther* (1993) **54**: 430–436.
13. Finnerty JP, Wood-Baker R, Thomson M, Holgate ST: Role of leukotrienes in exercise-induced asthma: inhibitory effect of ICI 204,219, a potent leukotriene D_4 receptor antagonist. *Am Rev Respir Dis* (1992) **145**: 746–749.
14. Taylor IK, O'Shaughnessy KM, Fuller RW, Dollery CT: Effect of cysteinyl-leukotriene receptor antagonist ICI 204,219 on allergen-induced bronchoconstriction and airway hyperreactivity in atopic subjects. *Lancet* (1991) **337**: 690–694.

15. Spector SL, Smith LJ, Glass M: Effects of 6 weeks of therapy with oral doses of ICI 204,219, a leukotriene D_4 receptor antagonist in subjects with bronchial asthma. *Am J Respir Crit Care Med* (1994) **150**: 618–623.
16. Suissa S, Dennis R, Ernst P, Sheehy O, Wood Dauphinee S: Effectiveness of the leukotriene receptor antagonist zafirlukast for mild-to-moderate asthma. A randomized, double-blind, placebo-controlled trial. *Ann Intern Med* (1997) **126**: 177–183.
17. Barnes NC, de Jong B, Miyamoto T: Worldwide clinical experience with the first marketed leukotriene receptor antagonist. *Chest* (1997) **111**: 52S–60S.
18. Reiss TF, Sorkness C, Stricker W, *et al*.: Effects of montelukast (MK-0476), a potent cysteinyl leukotriene receptor antagonist, on bronchodilatation in asthmatic subjects ttreated with and without inhaled steroids. *Thorax* (1997) **52**: 45–48.
19. Reiss TF, Altman LC, Chervinsky P, *et al*.: Effects of montelukast (MK-0476), a new potent cysteinyl leukotriene (LTD4) receptor antagonist, in patients with chronic asthma. *J Allergy Clin Immunol* (1996) **98**: 528–534.
20. Hui KP, Barnes NC: Lung function improvement in asthma with a cysteinyl-leukotriene receptor antagonist. *Lancet* (1991) **337**: 1062–1063.
21. Christie PE, Smith CM, Lee TH: The potent and selective sulfidopeptide leukotriene antagonist SK&F 104353 inhibits aspirin-induced asthma. *Am Rev Respir Dis* (1991) **144**: 957–958.
22. Yamamoto H, Nagata M, Kuramitsu K, *et al*.: Inhibition of analgesic-induced asthma by leukotriene receptor antagonist ONO-1078. *Am J Respir Crit Care Med* (1994) **150**: 254–257.
23. Laitinen LA, Laitinen A, Haahtela T, Vilkaa V, Spur BW, Lee TH: Leukotriene E_4 and granulocyte infiltration into asthmatic airways. *Lancet* (1993) **341**: 989–990.
24. Underwood DC, Osborn RR, Newsholme SJ, Torphy TJ, Hay DWP: Persistent airway eosinophilia after leukotriene (LT) D_4 administration in the guinea pig. *Am J Respir Crit Care Med* (1996) **154**: 850–857.
25. Calhoun WJ, Williams KL, Simonson SG. Lavins BJ: Effect of zafirlukast (Accolate) on airway inflammation after segmental allergen challenge in patients with mild asthma. *Allergy* (1997) **37** (Suppl): 90.
26. Dworski R, Fitzgerald GA, Oates JA, Sheller JR: Effect of oral prednisone on airway inflammatory mediators in atopic asthma. *Am J Respir Crit Care Med* (1994) **149**: 953–959.
27. O'Shaughnessy KM, Wellings R, Gillies B, Fuller RW: Differential effects of fluticasone propionate on allergen-induced bronchoconstriction and increased urinary leukotriene E4 excretion. *Am Rev Respir Dis* (1993) **147**: 1472–1476.
28. Donnelly AL, Glass M, Minkwitz MC, Casale TB: The leukotriene D4-receptor antagonist, ICI 204,219, relieves symptoms of acute seasonal allergic rhinitis. *Am J Respir Crit Care Med* (1995) **151**: 1734–1739.
29. McMillan RM, Walker ERH: Designing therapeutically effective 5-lipoxygenase inhibitors. *Trends Pharmacol Sci* (1992) **13**: 323–330.
30. Ford-Hutchinson AW: FLAP: a novel drug target for inhibiting the synthesis of leukotrienes. *Trends Pharmacol Sci* (1991) **11**: 68–70.
31. Wenzel SE, Kamada AK: Zileuton: the first 5-lipoxygenase inhibitor for the treatment of asthma. *Ann Pharmacother* (1996) **30**: 858–864.
32. McGillis JP, Organist ML, Payan DG: Substance P and immunoregulation. *Fed Proc* (1987) **14**: 120–123.
33. Fischer AR, McFadden CA, Frantz R, *et al*.: Effect of chronic 5-lipoxygenase inhibition on airway hyperresponsiveness in asthmatic subjects. *Am J Respir Crit Care Med* (1995) **152**: 1203–1207.
34. Wenzel SE, Trudeau JB, Kaminsky DA, Cohn J, Martin RJ, Wescott JY: Effect of 5-lipoxygenase inhibition on bronchoconstriction and airway inflammation in nocturnal asthma. *Am J Respir Crit Care Med* (1995) **152**: 857–905.
35. Israel E, Rubin P, Kemp JP, *et al*.: The effect of inhibition of 5-lipoxygenase by zileuton in mild-to-moderate asthma. *Ann Intern Med* (1993) **119**: 1059–1066.
36. Israel E, Cohn J, Dube L, Drazen JM: Effect of treatment with zileuton, a 5-lipoxygenase inhibitor, in patients with asthma. A randomized controlled trial. Zileuton Clinical Trial Group. *JAMA* (1996) **275**: 931–936.

37. Liu MC, Dube LM, Lancaster J: Acute and chronic effects of a 5-lipoxygenase inhibitor in asthma: a 6-month randomized multicenter trial. Zileuton Study Group. *J Allergy Clin Immunol* (1996) **98**: 859–871.
38. Kane GC, Pollice M, Kim CJ, *et al.*: A controlled trial of the effect of the 5-lipoxygenase inhibitor, zileuton, on lung inflammation produced by segmental antigen challenge in human beings. *J Allergy Clin Immunol* (1996) **97**: 646–654.
39. Van Schoor J, Joos G, Kips JC, Drajesk JF, Carpentier PJ, Pauwels RA: The effect of ABT-761, a novel 5-lipoxygenase inhibuitor, on exercise- and adenosine-induced bronchoconstriction in asthmatic subjects. *Am J Respir Crit Care Med* (1997) **155**: 857–880.
40. Nasser SMS, Bell GS, Hawksworth RJ, *et al.*: Effect of the 5-lipoxygenase inhibitor ZD2138 on allergen-induced early and late responses. *Thorax* (1994) **49**: 743–748.
41. Nasser SMS, Bell GS, Foster S, *et al.*: Effect of the 5-lipoxygenase inhibitor ZD2138 on aspirin-induced asthma. *Thorax* (1994) **49**: 749–756.
42. Dahlen B, Kumlin M, Ihre E, Zetterstrm O, Dahlen SE: Inhibition of allergen-induced airway obstruction and leukotriene generation in atopic asthmatic subjects by the leukotriene biosynthesis inhibitor BAYx1005. *Thorax* (1997) **52**: 342–347.
43. Black PN, Fuller RW, Taylor GW, Barnes PJ, Dollery CT: Effect of inhaled leukotriene B4 alone and in combination with prostaglandin D2 on bronchial responsiveness to histamine in normal subjects. *Thorax* (1989) **44**: 491–495.
44. Evans DJ, Barnes PJ, Coulby LJ, *et al.*: The effect of a leukotriene B_4 antagonist LY293111 on allergen-induced responses in asthma. *Thorax* (1996) **51**: 1178–1184.
45. O'Byrne PM, Fuller RW: The role of thromboxane A_2 in the pathogenesis of airway hyperresponsiveness. *Eur Respir J* (1989) **2**: 782–786.
46. Saroea HG, Inman MD, O'Byrne PM: U46619-induced bronchoconstriction in asthmatic subjects is mediated by acetylcholine release. *Am J Respir Crit Care Med* (1995) **151**: 321–324.
47. Lötvall JO, Elwood W, Tokuyama K, Sakamoto T, Barnes PJ, Chung KF: Effect of thromboxane A_2 mimetic U 46619 on airway microvascular leakage in the guinea pig. *J Appl Physiol* (1991) **72**: 2415–2419.
48. Chung KF, Evans TW, Graf PD, Nadel JA: Modulation of cholinergic neurotransmission in canine airways by thromboxane mimetic U46619. *Eur J Pharmacol* (1985) **117**: 373–375.
49. Manning PJ, Stevens WH, Cockroft DW, O'Byrne P: The role of thromboxane in allergen-induced asthmatic responses. *Eur Respir J* (1991) **4**: 667–672.
50. Demoly P, Jaffuel D, Lequeux N, *et al.*: Prostaglandin H synthase 1 and 2 immunoreactivities in the bronchial mucosa of asthmatics. *Am J Respir Crit Care Med* (1997) **155**: 670–675.
51. Futaki N, Takahashi S, Yokoyama M, Arai I, Higuchi S, Otomo S: NS-398 a new anti-inflammatory agent, selectively inhibits prostaglandin synthase cyclooxygenase (COX-2) activity *in vitro*. *Prostaglandins* (1994) **47**: 55–59.
52. Chan CC, Boyce S, Brideau C, *et al.*: Pharmacology of a selective cyclooxygenase-2 inhibitor, L-745, 337: a novel nonsteroidal anti-inflammatory agent with an ulcerogenic sparing effect in rat and nonhuman primate stomach. *J Pharmacol Exp Ther* (1995) **274**: 1531–1537.
53. Barnes PJ, Chung KF, Page CP: Platelet-activating factor as a mediator of allergic disease. *J Allergy Clin Immunol* (1988) **81**: 919–934.
54. Hayes J, Ridge SM, Griffiths S, Barnes PJ, Chung KF: Inhibition of cutaneous and platelet responses to platelet activating factor by oral WEB 2086 in man. *J Allergy Clin Immunol* (1991) **88**: 83–88.
55. Wilkens M, Wilkens JH, Busse S, *et al.*: Effects of an inhaled PAF antagonist (WEB 2086BS) on allergen-induced early and late asthmatic responses and increased bronchial responsiveness to methacholine. *Am Rev Respir Dis* (1991) **143**: A812.
56. Freitag A, Watson RM, Mabos G, Eastwood C, O'Byrne PM: Effect of a platelet activating factor antagonist, WEB 2086, on allergen induced asthmatic responses. *Thorax* (1993) **48**: 594–598.
57. O'Connor BJ, Ridge SM, Chen-Wordsell YM, Uden S, Barnes PJ, Chung KF: Inhibition of airway and neutrophil responses to inhaled platelet-activating factor by an oral PAF antagonist UK 7450S. *Am Rev Respir Dis* (1991) **150**: 35–40.
58. Kuitert LM, Hui KP, Uthayarkumar S, *et al.*: Effect of the platelet activating factor antagonist UK 74,505 on the early and late response to allergen. *Am Rev Respir Dis* (1993) **147**: 82–86.

59. Spence DPS, Johnston SL, Calverley PMA, et al.: The effect of the orally active platelet-activating factor antagonist WEB 2086 in the treatment of asthma. *Am J Respir Crit Care Med* (1994) **149**: 1142–1148.
60. Hozawa S, Haruta Y, Ishioka S, Yamakido M: Effects of a platelet-activating factor antagonist Y 24180 on bronchial hyperresponsiveness in patients with asthma. *Am J Respir Crit Care Med* (1995) **152**: 1198–1202.
61. Evans DJ, Barnes PJ, Cluzel M, O'Connor BJ: Effects of a potent platelet activating factor antagonist, SR27417A, on allergen-induced asthmatic responses. *Am J Respir Crit Care Med* (1997) **156**: 11–16.
62. Lombardo D, Dennis EA: Cobra venom phospholipase A2 inhibition by manoalide: a novel type of phospholipase inhibitor. *J Biol Chem* (1985) **260**: 7234–7240.
63. Dennis EA: Diversity of group types, regulation and function of phospholipase A2. *J Biol Chem* (1994) **269**: 13 057–13 060.
64. Thompson NT, Bonser RW, Garland LG: Receptor-coupled phospholipase D and its inhibition. *Trends Pharmacol Sci* (1991) **12**: 404–407.
65. Barnes PJ: Bradykinin and asthma. *Thorax* (1992) **47**: 979–983.
66. Pongracic JA, Naclerio RM, Reynolds CJ, Proud D: A competitive kinin receptor antagonist, [DArg^0, Hyp^3, DPhe^7]-bradykinin, does not affect the response to nasal provocation with bradykinin. *Br J Clin Pharmacol* (1991) **31**: 287–294.
67. Hock FJ, Wirth K, Albus U, et al.: HOE 140 a new potent and long acting bradykinin-antagonist: *in vitro* studies. *Br J Pharmacol* (1991) **102**: 769–773.
68. Austin CE, Foreman JC, Scadding SK: Reduction by Hoe 140, the B2 kinin receptor antagonist, of antigen-induced nasal blockage. *Br J Pharmacol* (1994) **111**: 969–971.
69. Akbary AM, Wirth KJ, Scholkens BA: Efficacy and tolerability of Icatibant (Hoe 140) in patients with moderately severe chronic bronchial asthma. *Immunopharmacology* (1996) **33**: 238–242.
70. Scherrer D, Daeffler L, Trifilieff A, Gies J-P: Effects of WIN 64338, a non peptide bradykinin B$_2$-receptor antagonist, on guinea-pig trachea. *Br J Pharmacol* (1995) **115**: 1127–1128.
71. Tsukagoshi H, Haddad E-B, Barnes PJ, Chung KF: Bradykinin receptor subtypes in rat lung: effect of interleukin-1β. *J Pharmacol Exp Ther* (1995) **273**: 1257–1263.
72. Polosa R, Holgate ST: Comparative airway responses to inhaled bradykinin, kallidin and [desArg^9]-bradykinin in normal and asthmatic airways. *Am Rev Respir Dis* (1990) **142**: 1367–1371.
73. Barnes PJ: Reactive oxygen species and airway inflammation. *Free Radic Biol Med* (1990) **9**: 235–243.
74. Thomas CE, Ohlweiler DF, Carr AA, et al.: Characterization of the radical trapping activity of a novel series of cyclic nitrone spin traps. *J Biol Chem* (1996) **271**: 3097–3104.
75. Cushley MJ, Holgate ST: Adenosine induced bronchoconstriction in asthma: role of mast cell mediator release. *J Allergy Clin Immunol* (1985) **75**: 272–278.
76. Feoktiskov I, Bioggioni I: Adenosine 2b receptors evoke interleukin-8 secretion in human mast cells. An enprofylline-sensitive mechanism with implications for asthma. *J Clin Invest* (1995) **96**: 1979–1986.
77. Linden J: Cloned adenosine A3 receptors: pharmacological properties, species differences and receptor functions. *Trends Pharmacol Sci* (1994) **15**: 298–306.
78. Nyce JW, Metzger WJ: DNA antisense therapy for asthma in an animal model. *Nature* (1997) **385**: 721–725.
79. Barnes PJ, Liew FY: Nitric oxide and asthmatic inflammation. *Immunol Today* (1995) **16**: 128–130.
80. Yates DH, Kharitonov SA, Thomas PS, Barnes PJ: Endogenous nitric oxide is decreased in asthmatic patients by an inhibitor of inducible nitric oxide synthase. *Am J Respir Crit Care Med* (1996) **154**: 247–250.
81. Garvey EP, Oplinger JA, Furfine ES, et al.: 1400W is a slow tight-binding and highly selkective inhibitor of inducible nitric oxide synthase *in vitro* and *in vivo*. *J Biol Chem* (1997) **272**: 4559–4563.
82. Hay DWP, Henry PJ, Goldie RG: Is endothelin-1 a mediator in asthma? *Am J Respir Crit Care Med* (1996) **155**: 1994–1997.

83. Barnes PJ: Endothelins and pulmonary diseases. *J Appl Physiol* (1994) **77**: 1051–1059.
84. Douglas SA, Meek TD, Ohlstein EH: Novel receptor antagonists welcome a new era in endothelin biology. *Trends Pharmacol Sci* (1994) **15**: 313–316.
85. Bowler SD, Smith SM, Lavercombe PS: Heparin inhibits immediate response to antigen in the skin and lungs of allergic subjects. *Am Rev Respir Dis* (1993) **147**: 160–163.
86. Diamant Z, Timmers MC, van der Veen H, Page CP, van der Meer FJ, Sterk PJ: Effect of inhaled heparin on allergen-induced early and late asthmatic responses in patients with atopic asthma. *Am J Respir Crit Care Med* (1996) **153**: 1790–1795.
87. Garrigo J, Danta I, Ahmed T: Time course of the protective effect of inhaled heparin on exercise-induced asthma. *Am J Respir Crit Care Med* (1996) **153**: 1702–1707.
88. Johnson PR, Armour CL, Carey D, Black JL: Heparin and PGE2 inhibit DNA synthesis in human airway smooth muscle cells in culture. *Am J Physiol* (1995) **269**: L514-L519.
89. Tanaka RD, Clark JM, Warne RL, Abraham WM, Moore WR: Mast cell tryptase: a new target for therapeutic intervention in asthma. *Int Arch Allergy Immunol* (1995) **107**: 408–409.
90. Clark JM, Abraham WM, Fishman CE, *et al.*: Tryptase inhibitors block allergen-induced airway and inflammatory responses in allergic sheep. *Am J Respir Crit Care Med* (1995) **152**: 2076-2083.
91. Summers R, Sigler R, Shelhamer JH, Kaliner M: Effects of infused histamine on asthmatic and normal subjects: comparison of skin test responses. *J Allergy Clin Immunol* (1981) **67**: 456-464.
92. Sommerhoff CP, Krell RD, Williams JL, Gomes BC, Strimpler AM, Nadel JA: inhibition of human neutrophil elastase by ICI 200,355. *Eur J Pharmacol* (1991) **193**: 153–158.

42

Immunomodulators

C.J. CORRIGAN

CLINICAL NEED

Inhaled glucocorticoids form the mainstay of asthma therapy for the majority of patients. Topical glucocorticoids are relatively free of unwanted effects and offer a very favourable benefit–risk ratio. However, there remains a group of patients with chronic severe asthma whose disease is inadequately controlled by inhaled glucocorticoid therapy, even when optimal delivery has been assured, compliance has been verified and the effects of other exacerbating factors minimized.[1] In these patients oral glucocorticoids are often required to control symptoms. Patients taking oral glucocorticoids are at risk from developing the unwanted effects of these drugs, particularly with long-term administration, and may remain symptomatic despite this therapy. In addition, an important minority of patients appear to be resistant to the clinical antiasthma effects of glucocorticoids, which may be related, at least in part, to a relative resistance of their T-cells to glucocorticoid inhibition.[2] For all these patients, alternative modalities of therapy are urgently required. It is also worthy of note that, despite the development of modern medications, there are still 2000 deaths yearly caused by asthma in the UK. While some of these deaths may be related to inadequate therapy, poor perception by patients of the severity of their disease and delay in seeking help, it is certainly possible that some of them are attributable to an inadequate response to properly administered therapy.

IMMUNOSUPPRESSIVE THERAPY IN ASTHMA

Background

It is increasingly recognized that asthma is associated with chronic, cell-mediated inflammation of the bronchial mucosa in which eosinophil-active cytokine products of activated T-cells play a prominent role. Evidence suggests that glucocorticoids ameliorate asthma at least partly through inhibition of T-cells and their asthma-relevant cytokine products, particularly interleukin (IL)-5.[3-5] For this reason, there has been interest in the investigation of other 'anti-inflammatory' or 'immunosuppressive' agents for their possible therapeutic effects in asthma. Since many of these agents have potentially serious unwanted effects, attention has generally been focused on those asthmatic patients who continue to have severe disease despite properly administered, maximal topical glucocorticoid therapy and additional continuous systemic therapy. With such patients it is perceived that the benefits of amelioration of the disease and reduction or abolition of systemic glucocorticoid therapy, with its associated hazards, might outweigh the risks.

Azathioprine, methotrexate and gold salts

Evaluation of these drugs in asthma has been based on empirical observation of their 'anti-inflammatory' effects in diseases such as rheumatoid arthritis, rather than cogent hypotheses regarding their possible mechanisms of action.

A double-blind placebo-controlled study of azathioprine therapy in patients with chronic severe asthma failed to demonstrate any significant clinical benefit and was also associated with serious unwanted effects.[6] The mechanism of anti-inflammatory action of azathioprine is unclear. Although it appears to inhibit killer cell activity and antibody production, there is no evidence that it inhibits the production of asthma-relevant cytokines or the effects of these cytokines on eosinophils.

Both parenteral and oral gold salt preparations have been evaluated for their possible therapeutic benefits in chronic asthma. Parenteral gold therapy is associated with potentially serious unwanted effects such as proteinuria. A blinded crossover study of parenteral gold therapy in patients with severe asthma revealed some marginal benefits of active therapy compared with placebo, but the authors concluded that the unwanted effects were likely to preclude its routine use.[7] Oral gold salts have been evaluated for therapy of severe asthma in two double-blind, placebo-controlled, parallel-group studies. In the first, 28 patients with severe asthma taking small dosages of oral glucocorticoids were treated with oral gold salts (3 mg twice daily) or placebo for 26 weeks and progressive oral glucocorticoid dose reduction attempted.[8] Those patients treated with gold salts achieved a significantly higher reduction in oral glucocorticoid dosage (mean 4 mg daily prednisone or equivalent) compared with those taking placebo (mean 0.3 mg daily) with little change in lung function, although the frequency of disease exacerbations requiring increased oral glucocorticoids was somewhat lower in those patients receiving active therapy. Similarly, in a second study, 279 patients requiring at least 10 mg prednisone (or equivalent) daily were treated for 6 months with oral gold salts (3 mg twice daily) or placebo; 60% of patients receiving active therapy were able to reduce their daily dosages of oral glucocorticoids by at least 50% compared with 32% of those receiving placebo.[9] Lung function was not significantly altered. Unwanted effects associated with gold salt therapy included gastrointestinal upset and exacerbation of pre-existing eczema, which was so severe in a minority of cases as to necessitate withdrawal from the trial. Mechanisms of action of gold salts that may be relevant to a glucocorticoid-sparing effect in asthma are ill-defined, although these drugs do inhibit T-lymphocyte proliferation *in vitro* and have recently been shown to inhibit IL-5-mediated prolongation of eosinophil survival, albeit at high concentrations.[10]

The antimetabolite methotrexate is a folic acid antagonist. One of its actions is to inhibit synthesis of nucleotides and, ultimately, cellular DNA replication. Following an initial open study[11] in which 25 patients with chronic severe asthma apparently dependent on very large dosages of oral prednisone were able to reduce these considerably (mean dosages 27 mg to 6 mg daily) in association with methotrexate therapy (15–50 mg weekly for at least 18 months) without significant alteration of lung function, several placebo-controlled studies were subsequently performed. In a randomized double-blind study, 69 patients with oral glucocorticoid-dependent asthma (mean daily prednisolone dosage 14.2 mg) were treated with methotrexate (15 mg weekly) or placebo for 24 weeks, and daily prednisolone dosages were reduced by 2.5 mg every 4 weeks if symptoms or lung function were unchanged or improved; 50% of those patients treated with methotrexate, compared with 14% of those taking placebo, were able to reduce their oral prednisolone, although this reduction was not maintained when the study medication was discontinued.[12] Of the 38 patients receiving methotrexate, five developed substantial abnormalities in liver function tests, an unwanted effect of methotrexate that is well recognized. In another double-blind trial, 10 patients with severe asthma whose dosages of oral prednisone had been reduced to the minimum necessary for disease control were given methotrexate (15 mg weekly) or identical placebo in random order for 3 months with an intervening 1-month washout period.[13] The mean dosage of prednisone required by these patients was statistically, if clinically questionably, smaller when receiving methotrexate (12.0 mg daily) compared with placebo (8.4 mg daily). Disease severity was not significantly different in both arms of the study. Unwanted effects were considered 'mild' and included anorexia, alopecia and stomatitis. Similar findings were reported in a second double-blind crossover study.[14]

In contrast to these studies, at least three double-blind placebo-controlled studies of relatively short duration failed to demonstrate any benefit of methotrexate therapy in oral glucocorticoid-dependent asthmatic patients in terms of both reduction of oral glucocorticoid dosages and improvement in lung function.[15–17] Furthermore there have been isolated reports of deaths caused by opportunistic infection[17–19] as well as cases of pneumonitis[20] in oral glucocorticoid-dependent asthmatic patients treated with methotrexate, and one suggestion that methotrexate may actually induce asthma.[21]

The mechanism by which methotrexate could exert a possible glucocorticoid-sparing effect in asthma is completely undefined. It is theoretically possible that this could at least partly reflect inhibition of T-cell proliferation, although one study on human T-cell lines suggested that methotrexate is ineffective in inhibiting cytokine production by T-cells.[22]

Cyclosporin A

Cyclosporin A (CsA) is a lipophilic, cyclic undecapeptide derived from the fungus *Tolypocladium inflatum*. Like glucocorticoids, CsA is thought to exert its immunosuppressive effects predominantly through inhibition of T-cell proliferation and cytokine secretion,[23] although it does have potential anti-inflammatory effects on other cells (e.g. it inhibits degranulation of mast cells and basophils).[24] CsA is not myelotoxic and is now widely used for the suppression of allograft rejection. It has also been evaluated in many 'autoimmune' diseases such as rheumatoid arthritis and cirrhosis.

CsA inhibits early T-cell signalling mechanisms following ligation of the T-cell antigen receptor.[23] It is therefore effective only when present at an early stage in this activation process. It inhibits calcium-dependent cell signalling pathways, the end-result being the inhibition of T-cell proliferation and cytokine mRNA transcription. CsA binds to an ubiquitous cytoplasmic protein called cyclophilin, which has peptidyl-prolyl *cis-trans* isomerase (PPIase) activity that is inhibited on binding to CsA. However, this inhibition does not relate directly to the immunosuppressive properties of CsA, since it has been shown that certain CsA analogues that bind to and inhibit this PPIase do not inhibit T-cell activity. This effect of CsA may theoretically account for at least part of its activity in inhibiting cellular degranulation. The CsA–cyclophilin complex binds to another cytoplasmic protein called calcineurin, which is a calcium- and calmodulin-dependent serine/threonine phosphatase. Evidence suggests that the cytoplasmic subunit of a transcriptional activating factor called NF-AT requires dephosphorylation by calcineurin before it can translocate to the cell nucleus, where it combines with its nuclear subunit to become an active transcriptional activator. Thus, CsA–cyclophilin complexes prevent formation of active NF-AT in the nucleus, reducing the binding of NF-AT to cytokine gene promoter regions and hence the subsequent transcription of these genes. The fact that T-cells are particularly dependent on NF-AT for activation of transcription of IL-2, the autocrine T-cell growth factor, may explain why CsA is particularly inhibitory for T-cells rather than proliferating cells in general.

In view of the increasing evidence for a role for activated T-cells in the pathogenesis of asthma, and the predominant inhibitory effects of CsA on T-cell proliferation and cytokine production, the appraisal of CsA for a possible antiasthma effect would appear to be soundly based in theory. This has recently been undertaken in three double-blind placebo-controlled trials. In the first trial, 33 patients with chronic severe asthma who had taken oral glucocorticoids for at least the previous 3 months at a dosage of at least 5 mg of prednisolone (or equivalent) daily were treated for 12 weeks with CsA (initial daily dosage 5 mg/kg, subsequently adjusted according to whole-blood trough concentrations) and identical placebo in random order, with a 2-week washout period.[25] All subjects were non-smokers with no contraindications to CsA therapy whose forced expiratory volume in 1 s (FEV_1) and/or peak expiratory flow rate (PEFR) measurements were <75% of the predicted values. All patients showed >20% reversibility in airways obstruction in response to inhaled β_2-agonist, confirming that they did not have fixed obstructive airways disease despite the severity of their symptoms. In this trial, oral glucocorticoid therapy was maintained constant throughout the entire trial period unless it was required to be increased because of disease exacerbation. CsA therapy, as compared with placebo, was associated with moderate and statistically significant increases in morning PEFR and FEV_1 in these patients (12% and 17.6% respectively). The frequency of disease

42 Immunomodulators

exacerbations requiring increased dosages of prednisolone was also significantly reduced. CsA therapy was generally well tolerated by these patients, although one withdrew because of hypertrichosis. A second trial addressed the possible glucocorticoid-sparing effects of concomitant CsA therapy in a similar group of chronic, glucocorticoid-dependent asthmatic patients.[26] In this parallel-group study, 39 oral glucocorticoid-dependent asthmatic patients were established on maintenance dosages of glucocorticoid at the minimum level required to maintain their disease stable. After a run-in period they were randomized to receive CsA (5 mg/kg daily) or identical placebo for 36 weeks. Attempts were made to reduce prednisolone dosages at 14-day intervals if the patients' asthma had been stable or improved as judged by diary card symptom scores and measurement of lung function. Concomitant therapy with CsA allowed a reduction in the median dosage of prednisolone of 62% (10 mg to 3.5 mg) compared with only 25% (10 mg to 7.5 mg) in the patients treated with placebo. Despite this reduction in oral prednisolone dosage, mean morning PEFR values significantly increased in the patients treated with CsA by a mean of 9.4% compared with no significant change in the group taking placebo. Anticipated unwanted effects of CsA, particularly mild nephrotoxicity and hypertension, were observed but were controllable with appropriate therapy or careful monitoring of whole-blood trough concentrations. The incidences of hypertrichosis and paraesthesia were increased in the actively treated group, while eight patients treated with CsA and one treated with placebo required therapy for hypertension. However, a third study of very similar design and duration to this second study failed to show any significant effect of CsA therapy in allowing reduction of oral glucocorticoid dosages or improving lung function in a group of 34 oral glucocorticoid-dependent asthmatic patients.[27]

Overall these studies suggest that CsA therapy is effective in only a proportion of chronic, severe, oral glucocorticoid-sensitive asthmatic patients, where it may be effective in reducing disease severity and/or enabling reductions in dosages of maintenance oral glucocorticoids. It may be particularly beneficial in glucocorticoid-resistant asthmatic patients,[2] since T-cells from these patients have been shown to be relatively resistant to the inhibitory effects of glucocorticoids at therapeutic concentrations *in vitro*, whereas CsA inhibits T-cells from both glucocorticoid-sensitive and -resistant asthmatic patients with equivalent potency at therapeutic concentrations.[28,29] Nevertheless, CsA is not an ideal therapy since there are worries about its unwanted effects, particularly in the long term, on the kidney.

There is some evidence to support the hypothesis that CsA exerts a beneficial effect in asthma through inhibition of T-cell activation and cytokine production. For example, CsA has been shown to inhibit IL-5 production by activated T-cells from asthmatic patients *in vitro*,[30] while therapy of a group of severe, glucocorticoid-dependent asthmatic patients with CsA was associated with reductions in their serum concentrations of soluble CD25 (the soluble form of the T-cell IL-2 receptor released by activated T-cells).[31] Nevertheless, it remains possible that some of the additional activities of CsA, for example its ability to inhibit degranulation of granulocytes such as basophils, may also be relevant to an antiasthma effect.

Intravenous immunoglobulin therapy

The need for a pooled intravenous immunoglobulin (IVIG) preparation capable of passive transference of broad-based humoral immunity was first recognized following the description of congenital agammaglobulinaemia by Bruton. It was subsequently recognized that this therapy, originally intended to restore immune deficiency, actually appeared to have therapeutic effects in diseases involving immune effector mechanisms, most notably 'autoimmune' diseases such as immune thrombocytopenia.[32] IVIG therapy is currently being evaluated for a possible therapeutic benefit in many other such diseases, including rheumatoid arthritis, systemic lupus erythematosus, systemic vasculitis, myasthenia gravis and others.

The possible efficacy of IVIG therapy for asthma has been investigated in two open-label uncontrolled studies on children. In the first, eight children who required continuous systemic glucocorticoid therapy for asthma control were treated with IVIG 1 g/kg as a 6% solution on two consecutive days monthly for a total of 6 months (seven infusions).[33] This regimen was based on regimens used for IVIG therapy of autoimmune diseases. IVIG therapy allowed mean alternate-day prednisone dosages, which were claimed to be the lowest possible for stable maintenance, to be reduced from 32.5 mg to 11.5 mg. Mean FEV_1 (which was near normal anyway prior to therapy in most of the children) and PC_{20} methacholine did not change significantly, although diary symptom scores and β_2-agonist usage were significantly reduced. Mean serum total IgE concentrations were slightly but significantly reduced, while mean serum IgG concentrations were increased from 5.85 g/litre prior to treatment to 12.4 g/litre 4 months after completion of the infusions. Unwanted effects of therapy were apparently too trivial to warrant specific documentation. The authors concluded that IVIG therapy was associated with 'clinical improvement, sparing of oral glucocorticoids and absence of drug-induced morbidity'.

Less favourable conclusions were reached in a second study involving 14 adolescents requiring inhaled glucocorticoids (400–2000 μg daily) to keep them 'almost free from symptoms' (i.e. requiring no more than two dosages of inhaled β_2-agonist daily).[34] Owing to ethical constraints, the children were offered IVIG therapy or entry to an untreated 'reference' group in an open fashion. Nine patients choosing therapy were commenced on IVIG at an intended dosage of 1 g/kg on a single day monthly for 5 months. Treatment with the high dosages used in the previous study[33] was considered 'too expensive'. In the event, this dosage had to be further decreased to 0.5 g/kg after the first infusion because it caused severe headache, and was then progressively increased (mean dosage 0.8 g/kg). Even at these lower dosages, infusions were frequently accompanied by fever and rigors. During the infusion period, six patients were able to reduce dosages of inhaled glucocorticoid (mean 720 μg to 400 μg daily), whereas three were not. Histamine bronchial reactivity was also reduced (mean PC_{20} 0.33 mg/ml to 1.23 mg/ml), as were total symptom scores. These differences, by comparison with the 'reference' group, were not maintained after 10 additional months of follow-up. Mean serum IgG concentrations before (11.6 g/litre) and 4 months after (11.4 g/litre) the infusions were almost identical, as were serum total IgE concentrations. In addition, serum concentrations of IgG anti-IgE antibodies and eosinophil cationic protein were unchanged. These authors concluded that the effects of IVIG therapy were 'small and temporary, and the treatment complicated and expensive'. They also hypothesized that some of the improvement observed in the previous study[33] might have reflected the effects of immunoglobulin replacement, since they considered the mean pretherapy serum IgG concentration in these children (5.85 g/litre) to be abnormally low, and a beneficial effect of IVIG therapy had previously been demonstrated in asthmatic children with hypogammaglobulinaemia.[35]

How might IVIG exert an antiasthma effect? It has been postulated[33] that IVIG therapy might represent a form of passive immunotherapy, interfering with IgE-mediated reactions.[36] In support of this, IVIG therapy was associated with reduced immediate skin-prick test reactivity to a variety of allergens, although this was not accompanied by radioallergosorbent test inhibition *in vitro*[33] or evidence of elevated binding of circulating IgE to IgG autoantibodies.[34] The mechanism of this phenomenon, as well as its possible relevance to amelioration of asthma, remains obscure. IVIG has been shown to abrogate activation of both T-cells and B-cells *in vitro*.[37,38] Furthermore, soluble CD4, CD8 and HLA molecules have been identified in IVIG preparations,[39] which may act to inhibit presentation of antigen to T-cells. A further interesting possibility is that IVIG preparations might contain cytokine autoantibodies, which might inhibit the activities of asthma-relevant cytokines on inflammatory effector cells such as eosinophils.

Finally, although it is self-evident that uncontrolled open-label studies such as these are impossible to interpret, since they do not allow for the power of the placebo effect and take no account of spontaneous variability in asthma severity, they do arguably justify the need for larger, placebo-controlled trials of the possible benefits of IVIG therapy in glucocorticoid-dependent asthma.

Summary

Although immunosuppressive therapy appears to be of some benefit to a subgroup of chronic, severe, glucocorticoid-dependent asthmatic patients, many reservations remain about the use of currently available drugs in this clinical situation. The unwanted effects of specific drugs, whilst potentially acceptable in certain patients, are certainly not trivial. In view of these additional unwanted effects, it is not yet clear whether or not treatment of patients with any of these drugs will afford them a better overall benefit–risk ratio for therapy of their asthma in the long term, even if they are able to reduce their dosages of oral glucocorticoids.

Another problem is related to the fact that only a proportion of asthmatic patients appear to show a clinically favourable response to additional immunosuppressive therapy. This inevitably means that patients must be given a trial of therapy of sufficient dosage and duration clearly to confirm or exclude a beneficial clinical effect. So far, it has not been possible to identify responsive patients a priori, in the case of CsA neither by clinical and demographic data[25] nor by laboratory tests.[40] In patients who do not respond to immunosuppressive therapy, the benefit–risk ratio associated with adding it must inevitably be poorer.

Further to these problems, there is a clear risk of opportunistic infection in asthmatic patients treated with oral glucocorticoids and additional immunosuppressive agents. There remains also a small but finite increased risk of neoplasia. Patients who are pregnant or at risk of pregnancy may not be treated, and there are specific contra-indications to therapy according to the particular immunosuppressive drug employed.

In view of all these observations, it is clear that any further investigation of immunosuppressive therapy for asthma should be performed within the confines of a controlled trial. There is an urgent need to produce a global definition of precisely which patients are suitable for such trials and what constitutes an appropriate trial of therapy. Finally, nothing is known about the long-term efficacy of immunosuppressive therapy in asthma. Alterations in such efficacy could change the benefit–risk ratio for individual patients with time.

The possibility of topical immunosuppressive therapy for asthma remains an intriguing one. For example, it would appear highly desirable if CsA therapy could be administered topically, and indeed preliminary studies investigating this possibility in patients with lung transplants have appeared in the literature. There are considerable technical problems associated with this technology, since CsA is highly insoluble in aqueous media.

NEWER IMMUNOSUPPRESSIVE AGENTS

FK506 (tacrolimus) is a macrolide derived from the soil organism *Streptomyces tsukudaiensis* which, despite having a different structure to CsA, has similar immunosuppressive properties, although it is more potent. Like CsA, FK506 binds to an ubiquitous cytoplasmic protein called FK-binding protein (FKBP), which has PPIase activity.[23] The FK506–FKBP complex also binds to calcineurin, inhibiting its serine/threonine phosphatase activity and thus NF-AT translocation. FK506 is being evaluated for prevention of rejection of allografts, but its increased potency may be outweighed by similarly increased toxicity.

Rapamycin is also a macrolide derived from *Streptomyces hygroscopicus*. It competes with FK506 for binding to FKBP. Despite this, rapamycin has a fundamentally different mechanism of action from CsA and FK506.[41] It inhibits progression from G_1 to S phase of the cell cycle during T-cell proliferation and, unlike CsA, is effective even when added late after cellular activation. Rapamycin inhibits calcium-independent cell signalling pathways that result from binding of cytokine receptors on the surface of T-cells to their specific ligands. In particular, rapamycin inhibits the T-cell signal transduction pathway following binding of IL-2 to its surface receptor. It probably acts by inhibiting phosphorylation/activation of the kinase p70 S6 ($p70^{S6k}$) and by inhibiting the enzymatic activity of the cyclin-dependent kinase cdk2–cyclin E complex. Thus, rapamycin impedes progression through the G_1/S transition of the T-cell proliferation cycle, resulting in a mid-to-late G_1 arrest.[41]

Other new immunosuppressive drugs whose antiproliferative actions are relatively specific for T-cells continue to appear.[42] Brequinar sodium and mycophenolic acid are inhibitors of *de novo* synthesis of pyrimidines and purines respectively, and their actions are particularly marked on T-cells. Mycophenolic acid has already undergone promising assessment as a therapy for allograft rejection. Like CsA, pharmacological concentrations of both rapamycin and mycophenolic acid have been shown to inhibit the proliferation of T-cells from glucocorticoid-resistant asthmatic patients whose T-cells are resistant to inhibition by therapeutic concentrations of glucocorticoids *in vitro*,[29,43] although it remains to be seen whether or not these drugs will inhibit the local production of asthma-relevant cytokines.

CONCLUSIONS

In view of the evidence that activated T-cells and their cytokine products play a fundamental role in asthma pathogenesis, it is likely that T-cell inhibition will continue to form one fundamental basis of antiasthma therapy. Glucocorticoids appear to act at least partly through this mechanism, and are effective therapy for a majority of asthmatic patients with a favourable benefit–risk ratio. New approaches to therapy are still urgently required for those patients who continue to have severe disease despite optimized delivery of inhaled and superadded systemic glucocorticoid and the minimization of exacerbating factors. In these patients, the potential hazards of continuous oral glucocorticoid therapy remain a cause for grave concern. Thus, the continued search for alternative drugs that act through inhibition of T-cells and/or the actions of their cytokine products as possible therapeutic agents in asthma would seem to be justified.

Unfortunately, none of the immunosuppressive drugs so far assessed for therapy of this category of patients has yet emerged as clearly beneficial in terms of improving the long-term benefit–risk ratio of therapy. Systematized approaches to the assessment of the possible value of these drugs are required. Other, possibly more promising approaches for the future include topical delivery of immunosuppressive drugs, the use of humanized monoclonal antibodies directed against T-cells and key cytokines or their receptors, and small molecule antagonists of cytokines or other key steps in the asthma inflammatory process.

REFERENCES

1. Woolcock AJ: Steroid resistant asthma: what is the clinical definition? *Eur Respir J* (1993) **6**: 743–747.
2. Corrigan CJ, Brown P, Barnes NC, *et al*.: Glucocorticoid resistance in chronic asthma. Glucocorticoid pharmacokinetics, receptor characteristics, and inhibition of peripheral blood T cell proliferation by glucocorticoids *in vitro*. *Am Rev Respir Dis* (1991) **144**: 1016–1025.
3. Corrigan CJ, Haczku A, Gemou-Engesaeth V, *et al*.: CD4 T lymphocyte activation in asthma is accompanied by increased serum concentrations of interleukin-5: effect of glucocorticoid therapy. *Am Rev Respir Dis* (1993) **147**: 540–547.
4. Doi S, Gemou-Engesaeth V, Kay AB, Corrigan CJ: Polymerase chain reaction quantification of cytokine messenger RNA expression in peripheral blood mononuclear cells of patients with severe asthma: effect of glucocorticoid therapy. *Clin Exp Allergy* (1994) **24**: 854–867.
5. Corrigan CJ, Hamid Q, North J, *et al*.: Peripheral blood CD4, but not CD8 T lymphocytes in patients with exacerbation of asthma transcribe and translate messenger RNA encoding cytokines which prolong eosinophil survival in the context of a Th2-type pattern: effect of glucocorticoid therapy. *Am J Respir Cell Mol Biol* (1995) **12**: 567–578.
6. Hodges NG, Brewis RAL, Howell JBL: An evaluation of azathioprine in severe chronic asthma. *Thorax* (1971) **26**: 734–739.
7. Klaustermeyer WB, Noritake DT, Kwong FK: Chrysotherapy in the treatment of corticosteroid-dependent asthma. *J Allergy Clin Immunol* (1987) **79**: 720–725.
8. Nierop G, Gijzel WP, Bel EH, Zwinderman AH, Dijkman JH: Auranofin in the treatment of steroid dependent asthma: a double blind study. *Thorax* (1992) **47**: 349-354.
9. Bernstein IL, Bernstein DI, Dubb JW, Faiferman I, Wallin B: A placebo-controlled multicenter study of auranofin in the treatment of patients with corticosteroid-dependent asthma. Auranofin Multicenter Drug Trial. *J Allergy Clin Immunol* (1996) **98**: 317-324.

10. Suzuki S, Okubo M, Kaise S, Ohara M, Kasukawa R: Gold sodium thiomalate selectively inhibits interleukin-5 mediated eosinophil survival. *J Allergy Clin Immunol* (1995) **96**: 251-256.
11. Mullarkey MF, Lammert JK, Blumenstein BA: Long-term methotrexate treatment in corticosteroid-dependent asthma. *Ann Intern Med* (1990) **112**: 577–581.
12. Shiner RJ, Nunn AJ, Chung KF, Geddes DM: Randomised, double-blind, placebo-controlled trial of methotrexate in steroid-dependent asthma. *Lancet* (1990) **336**: 137–140.
13. Dyer PD, Vaughan TR, Weber RW: Methotrexate in the treatment of steroid-dependent asthma. *J Allergy Clin Immunol* (1991) **88**: 208–212.
14. Stewart GE, Diaz JD, Lockey RF, Seleznick MJ, Trudeau WL, Ledford DK: Comparison of oral pulse methotrexate with placebo in the treatment of severe glucocorticosteroid-dependent asthma. *J Allergy Clin Immunol* (1994) **94**: 482–489.
15. Kanzow G, Nowak D, Magnussen H: Short term effect of methotrexate in severe steroid-dependent asthma. *Lung* (1995) **173**: 223–231.
16. Coffey MJ, Sanders G, Eschenbacher WL, *et al.*: The role of methotrexate in the management of steroid-dependent asthma. *Chest* (1994) **105**: 117–121.
17. Erzurum SC, Leff JA, Cochran JE, *et al.*: Lack of benefit of methotrexate in severe, steroid-dependent asthma. A double-blind, placebo-controlled study. *Ann Intern Med* (1991) **114**: 353–360.
18. Morice AH, Lai WK: Fatal varicella zoster infection in a severe steroid dependent asthmatic patient receiving methotrexate. *Thorax* (1995) **50**: 1221–1222.
19. Gatnash AA, Connolly CK: Fatal chickenpox pnuemonia in an asthmatic patient on oral steroids and methotrexate. *Thorax* (1995) **50**: 422–423.
20. Tsai JJ, Shin JF, Chen CH, Wang SR: Methotrexate pneumonitis in bronchial asthma. *Int Arch Allergy Immunol* (1993) **100**: 287–290.
21. Jones G, Mierins E, Karsh J: Methotrexate-induced asthma. *Am Rev Respir Dis* (1991) **143**: 179–181.
22. Schmidt J, Fleissner S, Heimann-Weitschat I, *et al.*: Effect of corticosteroids, cyclosporin A, and methotrexate on cytokine release from monocytes and T cell subsets. *Immunopharmacology* (1994) **27**: 173–179.
23. Schreiber SL, Crabtree GR: The mechanism of action of cyclosporin A and FK506. *Immunol Today* (1992) **13**: 136–142.
24. Hultsch T, Rodriguez JL, Kaliner MA, Hohman RJ: Cyclosporine A inhibits degranulation of rat basophilic leukemia cells and human basophils. *J Immunol* (1990) **144**: 2659–2664.
25. Alexander AG, Barnes NC, Kay AB: Trial of cyclosporin in corticosteroid-dependent chronic severe asthma. *Lancet* (1992) **339**: 324–328.
26. Lock SH, Kay AB, Barnes NC: Double-blind, placebo-controlled study of cyclosporin A as a corticosteroid-sparing agent in corticosteroid-dependent asthma. *Am J Respir Crit Care Med* (1996) **153**: 509–514.
27. Nizankowska E, Soja J, Pinis G, *et al.*: Treatment of steroid-dependent bronchial asthma with cyclosporin. *Eur Respir J* (1995) **8**: 1091–1099.
28. Corrigan CJ, Brown PH, Barnes NC, Tsai JJ, Frew AJ, Kay AB: Glucocorticoid resistance in chronic asthma. Peripheral blood T lymphocyte activation and comparison of the T lymphocyte inhibitory effects of glucocorticoids and cyclosporin A. *Am Rev Respir Dis* (1991) **144**: 1026–1032.
29. Haczku A, Alexander A, Brown P, *et al.*: The effect of dexamethasone, cyclosporine and rapamycin on T lymphocyte proliferation *in vitro*: comparison of cells from patients with glucocorticoid-sensitive and glucocorticoid-resistant chronic asthma. *J Allergy Clin Immunol* (1994) **93**: 510–519.
30. Mori A, Suko M, Nishizaki Y, *et al.*: IL-5 production by CD4[+] T cells of asthmatic patients is suppressed by glucocorticoids and the immunosuppressants FK506 and cyclosporin A. *Int Immunol* (1995) **7**: 449–457.
31. Alexander AG, Barnes NC, Kay AB, Corrigan CJ: Clinical response to cyclosporin in chronic severe asthma is associated with reduction in serum soluble interleukin-2 receptor concentrations. *Eur Respir J* (1995) **8**: 574–578.
32. Bussel JB, Szatrowski TP: Uses of intravenous gammaglobulin in immune haematologic disease. *Immunol Invest* (1995) **24**: 451–456.

33. Mazer BD, Gelfand EW: An open-label study of high-dose intravenous immunoglobulin in severe childhood asthma. *J Allergy Clin Immunol* (1991) **87**: 976–983.
34. Jakobsson T, Croner S, Kjellman N-IM, Pettersson A, Vassella C, Björkstén B: Slight steroid-sparing effect of intravenous immunoglobulin in children and adolescents with moderately severe bronchial asthma. *Allergy* (1994) **49**: 413–420.
35. Page R, Friday G, Stillwagon P, Skoner D, Caliguiri L, Fireman P: Asthma and selective immunoglobulin subclass deficiency: improvement of asthma after immunoglobulin replacement therapy. *J Pediatr* (1988) **112**: 127–131.
36. Blaser K, de Weck AL: Regulation of the IgE antibody response by idiotype–anti-idiotype network. *Prog Allergy* (1982) **32**: 203–264.
37. Stohl W: Cellular mechanisms in the *in vitro* inhibition of pokeweed mitogen-induced B cell differentiation by immunoglobulin for intravenous use. *J Immunol* (1986) **126**: 4407–4413.
38. Kawada K, Terasaki PI: Evidence for immunosuppression by high-dose gamma globulin. *Exp Hematol* (1987) **15**: 133–136.
39. Blasczyk R, Westhoff U, Grosse-Wilde M: Soluble CD4, CD8 and HLA molecules in commercial immunoglobulin preparations. *Lancet* (1993) **341**: 789–790.
40. Alexander AG, Barnes NC, Kay AB, Corrigan CJ: Can clinical response to cyclosporin in chronic severe asthma be predicted by an *in vitro* T-lymphocyte proliferation assay? *Eur Respir J* (1996) **9**: 1421–1426.
41. Dumont FJ, Su Q: Mechanism of action of the immunosuppressant rapamycin. *Life Sci* (1995) **58**: 373–395.
42. Thomson AW, Starzl TE: New immunosuppressive drugs: mechanistic insights and potential therapeutic advances. *Immunol Rev* (1993) **136**: 17–98.
43. Corrigan CJ, Bungre JK, Assoufi B, Cooper AE, Seddon H, Kay AB: Intracellular steroid metabolism and relative sensitivity to glucocorticoids and immunosuppressive agents of T lymphocytes from glucocorticoid sensitive and resistant asthmatics. *Eur Respir J* (1996) **9**: 2077–2086.

43

Future Therapies for Asthma

PETER J. BARNES, IAN W. RODGER AND
NEIL C. THOMSON

INTRODUCTION

Currently available therapy for asthma is highly effective and, if used appropriately, usually has no problems in terms of adverse effects. However, some patients (5–10% of asthmatic patients) remain poorly controlled, despite what appears to be optimal therapy. There are some concerns about the safety of asthma therapy, particularly the treatment of childhood asthma, as this treatment has to be given over a very long period. Compliance with inhaled therapy, particularly controller therapy, is very poor and might be improved with oral therapy (once-daily calendar pack). Yet oral therapy presents a problem of side-effects, since the drug exerts effects throughout the body, whereas asthma is localized to the airways. This will necessitate the development of drugs that are *specific* for asthma and do not have effects on other systems or on normal physiological mechanisms (unlike β-agonists and glucocorticoids). None of the currently available therapy is curative nor has so far been shown to alter the natural history of the disease. Perhaps it is difficult to seek a cure for asthma until more is known about the molecular causes of the disease.

Despite considerable efforts by the pharmaceutical industry, it has been very difficult to develop new classes of therapeutic agent. Asthma is the most rapidly growing therapeutic market in the world, reflecting the worldwide increase in prevalence of asthma and the increasing recognition that chronic treatment is needed for many patients.

It is clearly important to understand more about the underlying mechanisms of asthma and also about how the currently used drugs work before rational improvements in therapy can be expected. There are several opportunities for new drug development in asthma, but whether these will revolutionize asthma treatment is unknown.

ASTHMA: BASIC MECHANISMS AND CLINICAL MANAGEMENT (3rd Edn)
ISBN 0-12-079027-9

Copyright © 1998 Academic Press Limited
All rights of reproduction in any form reserved

There are three major approaches to the development of new antiasthma treatments:

(1) improvement in existing classes of effective drug, e.g. long-acting inhaled β_2-agonists (salmeterol and formoterol) or long-acting anticholinergics (tiotropium bromide);
(2) development of novel compounds, based on rational developments and improved understanding of asthma, e.g. cysteinyl leukotriene antagonists and 5-lipoxygenase inhibitors, or interleukin (IL)-5 antibodies;
(3) development of novel compounds based on serendipity, e.g. frusemide.

NEW BRONCHODILATORS

Bronchodilators are presumed to act by reversing contraction of airway smooth muscle, although some may have additional effects on mucosal oedema or inflammatory cells. The biochemical basis of airway smooth muscle relaxation has been studied extensively, yet no new types of bronchodilator have had any clinical impact. The molecular basis of bronchodilatation involves an increase in intracellular cyclic adenosine 3′,5′ monophosphate (cAMP) and a reduction in cytosolic calcium ion concentration ([Ca^{2+}]) (Fig. 43.1). Recent studies suggest that the rise in cAMP is linked to the opening of Ca^{2+}-activated K^+ channels (maxi-K channels) in animal and human airway smooth muscle. However β-agonists may open maxi-K channels via direct G protein coupling to the channel, and this may occur at low concentrations of β-agonist that do not involve any increase in cAMP concentration.[1] The molecular mechanisms underlying bronchodilata-

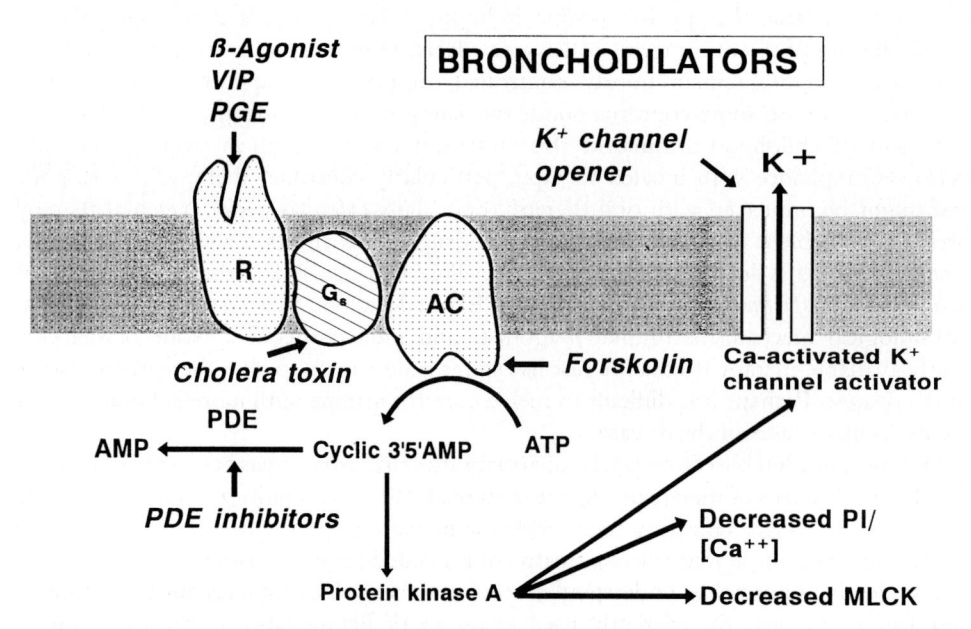

Fig. 43.1 Bronchodilator mechanisms. AC, adenylyl cyclase; MLCK, myosin light chain kinase; PDE, phosphodiesterase; PGE, prostaglandin E; PI, phosphoinositide; R, receptor; VIP, vasoactive intestinal polypeptide.

43 Future Therapies for Asthma

Table 43.1 New bronchodilators.

Long-acting inhaled β_2-agonists (e.g. salmeterol, formoterol)
Novel xanthine derivatives
Selective muscarinic antagonists (M_1/M_3 antagonists under development, e.g. tiotropium)
Potassium channel activators (e.g. levcromakalim)
Selective phosphodiesterase inhibitors
Nitrovasodilators
Vasoactive intestinal polypeptide and analogues
Atrial natriuretic peptide and analogues

---

tion may be exploited in the development of new bronchodilators, several of which are under development (Table 43.1).

β_2-Agonists

Many selective β_2-agonists are now available and there has been a search for β-agonists with even greater selectivity for β_2-adrenoceptors. However it is unlikely that any greater selectivity would be an advantage clinically, since when the drugs are given by inhalation a high degree of functional β_2-adrenoceptor selectivity is obtained. Furthermore, many of the side-effects of β-agonists (tremor, tachycardia, hypokalaemia) are mediated via β_2-adrenoceptors. There has been recent concern that inhaled β_2-agonists may be associated with increased asthma morbidity and mortality, but this is controversial and there is little evidence that the normally used doses of inhaled β_2-agonists are a problem, particularly when these drugs are used as required for symptom relief rather than on a regular basis.[2,3]

Long-acting β_2-agonists

The most important recent advance in bronchodilator therapy has been the introduction of inhaled β_2-agonists with a long duration of action, such as salmeterol and formoterol, that give bronchodilatation and protection against bronchoconstriction for over 12 h. These drugs have proved to be very useful clinically and provide additional control of asthma when added to inhaled glucocorticoids (see Chapter 36).

Drugs that increase cAMP

Understanding the molecular mechanism of β-agonists has prompted a search for other drugs that increase intracellular cAMP concentrations in airway smooth muscle cells. Several other receptors on airway smooth muscle, other than β-adrenoceptors, may activate adenylyl cyclase via a stimulatory G protein (G_s).

Vasoactive intestinal polypeptide (VIP) is a potent bronchodilator of human airways *in vitro*[4] but is ineffective in asthmatic patients *in vivo*.[5] This may reflect degradation of the peptide by airway epithelial cells. A more stable cyclic analogue of VIP (Ro-25-1553) has a more prolonged effect *in vitro* and *in vivo*,[6] although it is unlikely that VIP could be more effective than a β_2-agonist as it would have a substantially greater vasodilator effect and VIP receptors, unlike β_2-adrenoceptors, may not be expressed in peripheral airways.

Prostaglandin (PG)E$_2$ is a potent bronchodilator *in vitro* and inhibits the activation of inflammatory cells, such as eosinophils and T-lymphocytes. However, *in vivo* PGE$_2$ causes coughing and bronchoconstriction via activation of sensory nerve endings and therefore is unlikely to be useful as a therapy. The sensory nerve effects may be mediated via different PGE (EP) receptor subtypes and it may therefore be possible to develop selective EP agonists in the future.

Selective phosphodiesterase inhibitors

By inhibiting the breakdown of cAMP by phosphodiesterase (PDE), it should be possible to increase intracellular concentrations and thereby relax airway smooth muscle and also potentiate the bronchodilator effect of β-agonists. At least five isoenzyme families have now been distinguished based on substrate specificity, and the development of selective inhibitors[7] and molecular cloning studies have revealed that more than seven families may exist (Table 43.2). Furthermore, several splice variants of each gene product are known to exist, so that there may be more than 25 distinct PDE enzymes[8] (see Chapter 38).

Some PDEs are more important in smooth muscle relaxation, with the predominance of PDE3 in human airway smooth muscle. However, PDE3 inhibitors have been associated with cardiovascular abnormalities, particularly arrhythmias, and it is unlikely that they would have a safety profile that would allow their clinical development.

Methylxanthines

Theophylline has remained an important treatment in asthma for over 50 years, yet its mode of action is still unknown. It now seems unlikely that bronchodilatation plays an important role in the antiasthma effect of theophylline and increasingly likely that some

Table 43.2 Phosphodiesterase isoenzymes and inhibitors.

Family	Isoenzyme characteristics	Inhibitors
1	Ca^{2+}/calmodulin stimulated	KS 505a, vinpocetine
2	cGMP stimulated	MEP-1
3	cGMP inhibited	Milrinone, cilostamide
		SKF 94120
		SKF 94836 (siguazodan)
		Motipazone, Org 9935
4	cAMP specific	Rolipram, Ro20-1742
		Denbufylline
		CDP 840, SB 207499
5	cGMP	Zaprinast, SKF 95654
6	Photoreceptor family	(Zaprinast)
7	Rolipram insensitive, cAMP specific	?

anti-inflammatory or immunomodulatory effect is important.[9] While some effects of theophylline appear to be mediated via PDE inhibition (including common side-effects such as nausea and headache), this cannot account for all the antiasthma effects of theophylline. Recent studies suggest that theophylline may have immunomodulatory effects in the airways and may therefore have an antiinflammatory effect in asthma. These effects of theophylline are seen at relatively low plasma concentrations (>10 mg/litre). Withdrawal of theophylline in patients with severe asthma treated with high-dose inhaled glucocorticoids results in worsening of asthma symptoms and airway obstruction and a fall in peripheral blood counts of activated $CD4^+$ and $CD8^+$ T-lymphocytes, with corresponding increases in the airways.[10] This suggests that theophylline might inhibit the trafficking of activated T-cells into the airway. It is unlikely that these immunomodulatory effects of theophylline are due to PDE inhibition in view of the low plasma concentrations of theophylline and suggests that some other molecular mechanism, as yet unidentified, is responsible.

Several adverse effects of theophylline are due to antagonism of adenosine receptors. The development of enprofylline (3-propylxanthine), which retains the bronchodilator and PDE inhibitory effect but is not an adenosine antagonist, was an important advance;[11] unfortunately enprofylline was not developed due to unexpected toxicity problems.

Drugs that increase cyclic GMP

Atrial natriuretic peptide (ANP) when given by intravenous infusion produces a significant bronchodilator response and protects against bronchoconstrictor challenges when given intravenously or by inhalation.[12] It is likely that the effects of ANP on airways are mediated by stimulation of particulate guanylyl cyclase and subsequent generation of cyclic guanosine 3′,5′-monophosphate (cGMP). More stable analogues of ANP may be useful, but it is unlikely that these drugs would have any advantage over β_2-agonists as bronchodilators and the costs involved in development of such peptides would be prohibitive.

Nitrovasodilator compounds, such as isosorbide dinitrate, glyceryl trinitrate (GTN) and sodium nitroprusside, are thought to activate soluble guanylyl cyclase. A dose-dependent relaxant effect of various nitrovasodilators has been demonstrated on airway smooth muscle in a number of animal studies, and this effect appears to be mediated via stimulation of soluble guanylyl cyclase and subsequent generation of cGMP. However, these drugs have not proved to be very effective *in vivo* and, once again, the major disadvantage of such therapies is their potent vasodilator effects.

PDE5 is the major enzyme degrading cGMP and PDE5 inhibitors, such as SKF 95654, have some bronchodilator effect. However, a major disadvantage of these compounds is a propensity to cause vasodilatation.

Selective anticholinergics

There are several distinct subtypes of muscarinic receptor, which have differing physiological roles in the airway,[13] and there is some rationale for the development of selective antimuscarinics that block M_3 (with or without M_1) receptors but avoid blockade of prejunctional M_2 receptors that would lead to an increase in the release of acetylcholine. A very promising new anticholinergic is tiotropium bromide with a duration of action of <24 h and kinetic selectivity for M_1 and M_3 receptors.[14] Tiotropium bromide causes prolonged inhibition of cholinergic nerve-induced contraction of human airways *in vitro*[15] and provides prolonged bronchodilatation and protection against methacholine challenge (<36 h) in asthmatic patients.[16] While it is unlikely that a long-acting anticholinergic would have a major role in asthma therapy, this is likely to be a very valuable treatment in chronic obstructive pulmonary disease (COPD), where cholinergic tone is the major reversible element. Studies of inhaled tiotropium bromide have shown prolonged bronchodilatation in COPD after single doses, making once-daily administration possible.[17]

K+ channel openers

K^+ channels play an important role in the recovery of excitable cells after activation and in maintaining cell stability. Opening of K^+ channels therefore results in relaxation of smooth muscle and inhibition of secretion. Many different types of K^+ channel have now been recognized electrophysiologically and several selective toxins and drugs are available. Drugs that selectively activate an ATP-dependent K^+ channel in smooth muscle, such as BRL 3491 (cromakalim), have been developed for the treatment of hypertension. These drugs inhibit spontaneous and induced tone in airway smooth muscle *in vitro* and might, therefore, have a role in normalizing 'hyperreactive' airway smooth muscle. K^+ channel activators have been investigated as potential antiasthma compounds.[18] The active enantiomer of cromakalim, BRL 38227 (levcromakalim), and a more potent drug, HOE234, are relatively effective relaxants of human bronchi *in vitro* and appear equally active against several spasmogens.[19,20] In asthmatic patients levcromakalim had no bronchodilator action or protective effect against bronchoconstrictor challenge at maximally tolerated oral doses.[21] Side-effects include headache, flushing and postural hypotension, due to vasodilatation. It will therefore be necessary to develop these drugs for inhalational use in order to avoid these effects, although it may be possible to develop K^+ channel openers more selective for airway than vascular smooth muscle, in view of the diversity of K^+ channels. One such airway-selective K^+ channel opener (BRL 55834) has already been described,[22] although the degree of selectivity between airway and vascular smooth muscle is unlikely to be sufficient to avoid vasodilator side-effects.

The future success of these compounds in asthma will probably depend on whether they have any additional effects not shared with β-agonists. K^+ channel activators inhibit the release of neuropeptides from sensory nerves and modulate neurotransmission in the airways, but whether they have effects on inflammatory cells is not certain. Many different types of K^+ channel have now been characterized; cromakalim and related drugs appear to open an ATP-dependent channel (which opens in response to a fall in intracellular ATP concentrations). However, relaxation of airway smooth muscle in response to β-agonists and theophylline and modulation of neurotransmitter release in airway nerves appears to involve maxi-K channels, which are selectively blocked by charybdotoxin and iberiotoxin.[23-25] Several drugs that activate maxi-K channels have now been discovered

43 Future Therapies for Asthma

and may be developed as novel bronchodilators.[26] One such drug, NS-1619, has been shown to inhibit the activation of airway sensory nerves *in vitro* and to inhibit cough in guinea-pigs.[27]

MEDIATOR ANTAGONISTS

Many different inflammatory mediators have now been implicated in asthma, and several specific receptor antagonists and synthesis inhibitors have been developed that will prove invaluable in elucidating the contribution of each mediator (Table 43.3). As many mediators probably contribute to the pathological features of asthma, it seems unlikely that a single antagonist will have a major clinical effect, compared with non-specific agents such as β-agonists and glucocorticoids. However, until such drugs have been evaluated in careful clinical studies, it is not possible to predict their value. These drugs are also discussed in Chapter 41.

New antihistamines

Antihistamines that block H_1 receptors have no benefit in the long-term control of asthma, despite the fact that histamine is produced in asthma and mediates several effects on the airways, including activation of inflammatory cells and increased expression of adhesion molecules. New antihistamines, such as cetirizine, ebastine and astemizole, have been claimed to have additional antiasthma effects that are not mediated via H_1-receptor blockade, including an inhibitory effect on eosinophil chemotaxis and adherence to

Table 43.3 Inflammatory mediators and inhibitors.

Mediator	Inhibitor
Histamine	Terfenadine, loratadine, cetirizine
Leukotriene D_4	Zafirlukast, montelukast
	Zileuton, Bay-x1005, ZD2138
Leukotriene B_4	LY-293111
Platelet-activating factor	Apafant, modipafant, bepafant
Thromboxane	Ozagrel
Bradykinin	Icatibant, WIN 64338
Adenosine	Theophylline
Reactive oxygen species	N-Acetylcysteine, ascorbic acid, nitrones
Nitric oxide	Aminoguanidine, 1400W
Endothelin	Bosentan, SB209670
Interleukin (IL)-1β	Recombinant IL-1 receptor antagonist
Tumour necrosis factor (TNF)-α	TNF antibody, TNF soluble receptors
IL-4	IL-4 antibody
IL-5	IL-5 antibody
Mast cell tryptase	APC366
Eosinophil basic proteins	Heparin

endothelial cells and inhibition of eosinophil recruitment into asthmatic airways after allergen challenge.[28] In clinical studies, cetirizine has been shown to improve symptom scores and to reduce the use of rescue drugs in both seasonal and perennial asthma,[29,30] although it has no effect on allergen-induced asthma.[31] The effects of these drugs are relatively weak, however, and they would only be useful clinically in mild asthma. Nevertheless, they may have an important place in patients with seasonal asthma who also have troublesome symptoms of rhinitis.

Leukotriene antagonists

Potent leukotriene-receptor (cys-LT_1 receptor) antagonists (zafirlukast, montelukast, pranlukast) and 5-lipoxygenase (5-LO) inhibitors (zileuton, Bay-x1005) have now been developed and several of these drugs have recently become available in several countries.[32,33] The place of these drugs in asthma management has not yet been established. Clinical trials indicate that they have a modest but useful clinical effect in patients with mild asthma, although this is likely to be less than that provided by inhaled steroids. It is possible that they may be more effective in more severe asthma and that they may be additive to the effects of inhaled steroids, so that they will be used best as an add-on therapy in moderate to severe asthma. The major advantage of these drugs is that they are orally active and do not appear to have any class-specific side-effects. Some patients, including those with aspirin-sensitive asthma, appear to particularly benefit from leukotriene antagonist therapy, and it is likely that there will be a group of patients who do well with these drugs. It will be important to predict which patients are likely to benefit; at the moment this is best done by a 4-week trial of therapy. An acute bronchodilator response is seen with leukotriene antagonists in some patients and this might predict a good clinical response.

There do not appear to be any differences in efficacy between leukotriene antagonists and 5-LO inhibitors, suggesting that the blockade of other 5-LO products (5-hydroxyeicosatetraenoic acid, leukotriene (LT)B_4) is not important in the therapeutic response. Whether leukotriene antagonists have unequivocal anti-inflammatory effects in asthma has not yet been established. However inhalation of the cysteinyl leukotriene LTE_4 results in eosinophilic inflammation of the airways[34] and in animal studies leukotrienes have been shown to induce eosinophilia via the release of IL-5.[35] In one study there was a reduction in bronchoalveolar lavage (BAL) eosinophils after treatment with the 5-LO inhibitor zileuton[36] and in another study treatment with montelukast resulted in a fall in circulating and sputum eosinophils.[37] Further studies of the effects of leukotriene antagonists on airway inflammation in asthma are needed and in particular a comparison with the effects of inhaled steroids.

LTB_4 is a potent chemoattractant of neutrophils and it is therefore unlikely that it will play a major role in chronic asthma since neutrophils are rarely seen in the airways, although they may be prominent in more acute asthma. An LTB_4 antagonist, LY-293111, had no effects on the late response or airway hyperresponsiveness following allergen challenge, despite significantly reducing the neutrophil increase in BAL.[38] Thus, it is unlikely that LTB_4 antagonists will be useful in asthma. It is possible, however, that they may have a role in COPD, where neutrophil inflammation is prominent.

Platelet-activating factor antagonists

Platelet-activating factor (PAF) mimics many of the pathophysiological features of asthma, including induction of eosinophilic inflammation and airway hyperresponsiveness.[39] Furthermore, PAF antagonists are effective in several animal models of asthma. It was believed that PAF antagonists would therefore prove to be useful in the long-term management of asthma. However, even potent PAF antagonists, such as apafant (WEB2086) and modipafant (UK-74,505), proved to be of no value in clinical trials of chronic asthma,[40,41] despite the fact that they inhibit the effects of inhaled PAF for a prolonged period.[42] It is possible that, while effective against endogenous PAF, these drugs are still not sufficiently potent to inhibit the effects of endogenously released PAF, since PAF appears to act very locally, almost as a 'paracrine' mediator. More recently, even more potent PAF antagonists have been developed and these have shown some, albeit modest, effects in asthmatic patients. Y-24180 has been shown to reduce airway hyperresponsiveness in asthmatic patients,[43] whereas SR27417A significantly reduces the late response after inhaled allergen challenge.[44] Both antagonists have a potency greater than that of PAF for its receptor, and it is possible that even more potent antagonists might have greater effects. It is also possible that PAF antagonists may be useful in some patients with asthma or in certain circumstances, such as virally induced exacerbations of asthma.

Prostanoid inhibitors

Cyclooxygenase inhibitors, such as aspirin and indomethacin, do not seem to have any beneficial effects in asthma, despite the fact that prostanoids are produced in asthmatic inflammation and exert bronchoconstrictor and pro-inflammatory effects in the airways. One explanation is that there are prostanoids, such as PGE_2 and PGI_2, which may have anti-inflammatory effects, thus counteracting the bronchoconstrictor and inflammatory effects of thromboxane $(Tx)A_2$, PGD_2 and $PGF_{2\alpha}$. This led to the development of thromboxane receptor (TP) antagonists and thromboxane synthase inhibitors. However, these drugs have proved to be disappointing in clinical trials.[45] Cyclooxygenase exists in two forms, a constitutive form (COX-1) and a form that is inducible by endotoxin and pro-inflammatory cytokines (COX-2).[46] COX-2 is expressed in normal and asthmatic airways.[47] It is possible that selective inhibition of COX-2 might be beneficial in asthma and several selective COX-2 inhibitors, such as L745,337 and NS-398, have now been developed.[48,49] These selective inhibitors may also prove to be safe in patients with aspirin-sensitive asthma, since it is possible that bronchoconstriction in these patients may be due to inhibition of COX-1 and consequent reduction in PGE_2 synthesis.

Bradykinin antagonists

Bradykinin has also attracted attention as a potential mediator of asthma symptoms, since it appears to be the inflammatory mediator produced in asthma that is most likely to activate and sensitize sensory nerves in the airway, resulting in the troublesome symptoms of cough and chest tightness.[50] A potent and stable bradykinin B_2-receptor antagonist, icatibant, has recently been developed and may have therapeutic potential as a modulator of asthma symptoms. Icatibant inhibits some effects of allergen challenge in the nose.[51] Non-peptide antagonists of B_2 receptors, such as WIN 64338, have now been developed that may prove to be more useful.[52]

Adenosine antagonists

Adenosine is released in asthma and causes bronchoconstriction by activating mast cells, suggesting that adenosine antagonists may be useful in asthma therapy. Theophylline is a non-selective adenosine-receptor antagonist, but there is no evidence that its antiasthma effects are mediated via adenosine antagonism. There is uncertainty about which adenosine receptors are involved in mast cell activation. In animal studies there is some evidence that mast cells have adenosine A_{2B} receptors,[53] although other studies suggest that A_3 receptors are involved.[54] There are no selective adenosine- receptor antagonists available for clinical use. Recently, an inhaled DNA antisense oligonucleotide against adenosine A_1 receptors has been shown to inhibit airway hyperresponsiveness in an animal model of asthma.[55]

Antioxidants

The role of reactive oxygen species (ROS) in asthma is uncertain. ROS are generated from inflammatory cells, such as eosinophils and macrophages, and they may combine with nitric oxide (NO) to form the more stable oxidant peroxynitrite ($ONOO^-$), which in turn generates hydroxyl anions (OH^-). ROS activate the transcription factor NF-κB, which regulates the expression of many of the inflammatory proteins overexpressed in asthmatic airways. Antioxidants, such as N-acetylcysteine and ascorbic acid, have been tested in asthma but have not been shown to have any marked beneficial effects. However, these agents are relatively weak antioxidants. More potent antioxidants have now been developed, including the spin-trap antioxidants such as nitrones,[56] although these have not yet reached the stage of clinical development.

NO synthase inhibitors

NO, produced in large amounts from the inducible isoform of NO synthase (iNOS) overexpressed in asthmatic airways, may have pro-inflammatory effects in asthma by increasing plasma exudation in airways and amplifying eosinophilic inflammation[57] (see Chapter 21). It follows that inhibition of iNOS may be beneficial in asthma. It is important to avoid inhibition of constitutive NOS, as this may cause hypertension and bronchoconstriction. Aminoguanidine has some selectivity for iNOS and reduces exhaled NO in asthmatic patients to a greater extent than in normal subjects.[58] Whether regular treatment with aminoguanidine is beneficial in asthma has not yet been determined. More potent and selective iNOS inhibitors are now in clinical development.[59]

Endothelin antagonists

Endothelins are potent bronchoconstrictors and are abnormally expressed in asthmatic airways.[60–62] Endothelin (ET)-1 causes proliferation of airway smooth muscle and stimulates fibrosis, suggesting that it may play a role in the remodelling of asthmatic airways. These effects are mediated via stimulation of ET_A and ET_B receptors. Several potent non-peptide endothelin antagonists, such as bosentan and SB209670, have now been developed.[63] However, it may be difficult to test these compounds in asthma, as long-term studies would be needed to determine whether these drugs prevent the irreversible airway narrowing of chronic asthma.

Multiple mediator antagonists?

It seems very unlikely that antagonizing a single mediator of asthma will ever be as useful as therapies with a broad spectrum of activity, but such therapies may have the advantage of fewer side-effects and the possibility of oral administration. Certain types of asthmatic patient may respond much better to these more specific therapies. For example, it seems likely that aspirin-sensitive asthmatic patients may particularly benefit from LTD_4 antagonists.

CYTOKINES AND CYTOKINE INHIBITORS

A complex cytokine network is responsible for maintaining the chronic inflammation in asthma.[64] Amongst the many cytokines involved in asthma, some may be more important in determining the nature of the inflammatory response or in amplifying the inflammatory state. There are several possible approaches to inhibiting specific cytokines. These include drugs that inhibit cytokine synthesis (glucocorticoids, cyclosporin A, tacrolimus), blocking antibodies to cytokines or their receptors, soluble receptors to mop up secreted cytokines, and receptor antagonists or drugs that block the signal transduction pathways activated by cytokines (Fig. 43.2). Other cytokines appear to have an anti-inflammatory effect and may therefore be regarded as potentially therapeutic.

Anti-IL-5

IL-5 plays a key role in orchestrating the eosinophilic inflammation of asthma.[65] Blocking antibodies to IL-5 have been shown to inhibit eosinophilic inflammation and airway hyperresponsiveness in animal models of asthma, including primates.[66,67] This blocking effect may last for up to 3 months after a single injection. This makes treatment of chronic asthma with such a therapy a feasible proposition. Humanized antibodies have now been developed and are in clinical trial in asthma. There is also a search for IL-5 receptor antagonists.

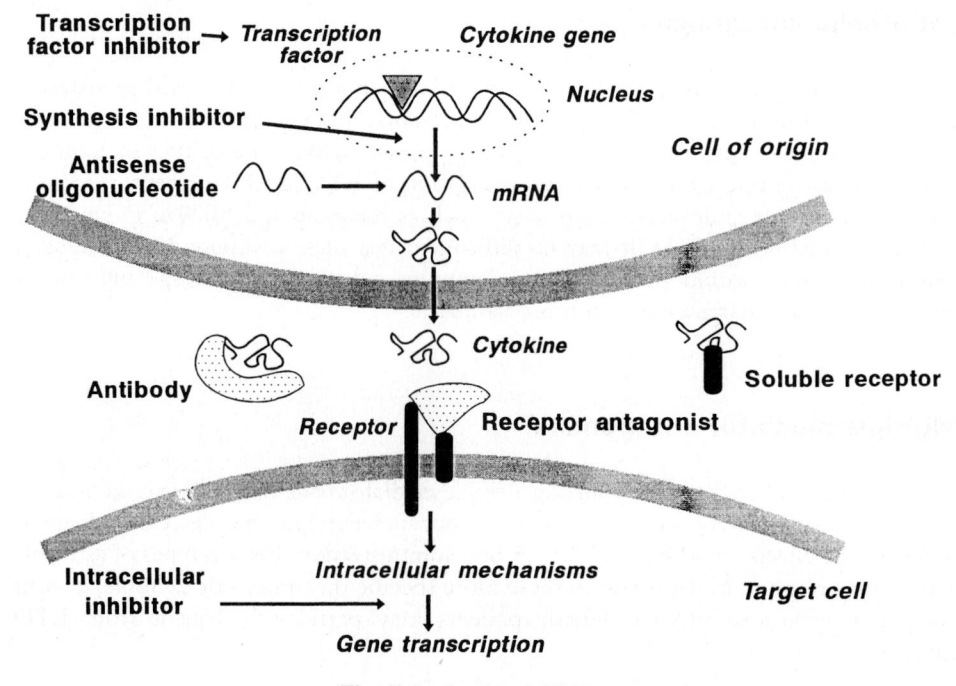

Fig. 43.2 Cytokine inhibition.

Anti-IL-4

IL-4 is critical for the synthesis of IgE by B-lymphocytes and is also involved in eosinophil recruitment to the airways. Inhibition of IL-4 may therefore be effective in inhibiting allergic diseases, and anti-IL-4 receptors are now in clinical development as a strategy to inhibit IL-4. Cell-specific IL-4 transcription appears to be activated by the protooncogene c-maf, which may provide another target for inhibition.[68]

Anti-tumour necrosis factor

TNF-α is expressed in asthmatic airways and may play a key role in amplifying asthmatic inflammation, through the activation of NF-kB and other transcription factors. In rheumatoid arthritis blocking antibodies to TNF-α have produced remarkable clinical benefits, even in patients who were relatively unresponsive to steroids.[69] Such antibodies or soluble TNF receptors are a logical approach to asthma therapy, particularly in patients with severe disease. One problem encountered in this therapy, however, is the development of antibodies that may limit the therapeutic effects after repeated administration.

Chemokine inhibitors

C-C chemokines, such as RANTES, macrophage chemotactic protein (MCP)-3, MCP-4 and eotaxin, may play a critical role in the recruitment of eosinophils into the airways of asthmatic patients (see Chapter 18). Fortunately all these chemokines act on a common receptor, the CCR-3 receptor, which is expressed predominantly on eosinophils.[70] Chemokine receptors are G protein-coupled receptors with the typical seven transmembrane-spanning segments and therefore have a simpler structure than the receptors for most cytokines. Antibodies to CCR-3 receptors have now been developed and it is possible that non-peptide inhibitors will be discovered by random screening of chemical libraries.

Anti-inflammatory cytokines

Some cytokines appear to have anti-inflammatory effects in asthmatic inflammation and therefore may be considered potentially therapeutic. While it may not be feasible to administer these proteins as long-term therapy, it may be possible to develop drugs that activate the same receptors or specific signal transduction pathways activated by these receptors.

IL-1 receptor antagonist (IL-1ra) is a cytokine that binds to IL-1 receptors and blocks the action of IL-1β. In experimental animals it reduces airway hyperresponsiveness,[71] although clinical studies of recombinant human IL-1ra in asthma have been disappointing.

Interferon (IFN)-γ inhibits Th2 cells and should therefore theoretically reduce asthmatic inflammation. Administration of IFN-γ by nebulization to asthmatic patients has not been found to be effective, however, possibly due to the difficulty in obtaining a high enough concentration locally in the airways.[72]

IL-12 is the endogenous regulator of Th1 cells and determines the balance between Th1 and Th2 cells.[73] IL-12 administration to rats inhibits allergen-induced inflammation[74] and also inhibits sensitization to allergens. IL-12 produces some of its effects by releasing endogenous IFN-γ, but it also has additional effects. Recombinant human IL-12 has now been administered to humans and appears to be safe when administered in low doses. This is therefore a potential treatment for asthma that may reset a fundamental immunological switch.

IL-10 inhibits the synthesis of many inflammatory cytokines (TNF-α, granulocyte–macrophage colony-stimulating factor (GM-CSF), chemokines) that are overexpressed in asthma. Its effects are partly mediated via inhibition of NF-κ.[75] Indeed, there may be a defect in IL-10 secretion in asthma.[76,77] Recombinant human IL-10 has proved to be remarkably effective in controlling inflammatory bowel disease, where similar cytokines are expressed, and may be given as a weekly injection.[78]

ANTI-INFLAMMATORY DRUGS

The recognition that asthma is a chronic inflammatory disease has prompted the earlier introduction of anti-inflammatory treatments. There has been an intensive search for anti-

Table 43.4 New anti-inflammatory drugs for asthma.

New glucocorticoids (mometasone, cyclesonide, RP 106541)
Immunomodulators (inhaled oxeclosporin, tacrolimus, rapamycin, mycophenolate mofetil)
PDE4 inhibitors (CDP 840, RP 73401, SB 207499)
Adhesion molecule blockers (VLA-4 antibody)
Cytokine inhibitors (anti-IL-4, anti-IL-5, anti-TNF antibodies)
Anti-inflammatory cytokines (IL-1ra, IFN-γ, IL-10, IL-12)
Anti-IgE antibody
Peptides for immunotherapy

IFN-γ, interferon γ; IL, interleukin; IL-1ra, interleukin 1 receptor antagonist; TNF, tumour necrosis factor.

inflammatory treatments as effective as glucocorticoids but with fewer side-effects. Several anti-inflammatory drugs for asthma are now in clinical development (Table 43.4).

Glucocorticoids

Glucocorticoids are the most efficacious treatment currently available for the long-term management of asthma.[79] Steroids of high topical potency, such as beclomethasone dipropionate and budesonide, are highly effective when given by inhalation. Budesonide is efficiently metabolized by hepatic first-pass metabolism and therefore has less systemic effects at higher doses than beclomethasone. Fluticasone propionate has even greater first-pass metabolism and essentially no oral bioavailability. However, all these steroids are absorbed from the lung and therefore have some systemic absorption. It order to reduce systemic effects further it may be necessary to increase the metabolism of any absorbed steroid in the circulation, such as by erythrocyte-derived enzymes. It is important to prevent rapid metabolism in lung tissue as this contributes to a reduced anti-inflammatory action, as illustrated by the poor efficacy of the 'soft' steroids butixocort and tipredane.

Because steroids are so effective in the control of asthma, an important goal of research is to identify the particular cellular and molecular mechanisms of critical importance in controlling asthmatic inflammation. It is now recognized that anti-inflammatory effects of glucocorticoids are likely to be via direct inhibition of the transcription factors, such as activator protein (AP)-1 and NF-κB, activated by inflammatory signals.[80]

PDE4 inhibitors

PDE4 is the predominant PDE in inflammatory cells, including mast cells, eosinophils, T-lymphocytes, macrophages, sensory nerves and epithelial cells (Table 43.5). This has suggested that PDE4 inhibitors would be useful in asthma therapy as an anti-inflammatory treatment. In animal models of asthma, several such drugs have been shown to induce eosinophil apoptosis and reduce eosinophil infiltration after allergen or reduce airway hyperresponsiveness, validating their potential in asthma.[81,82] One PDE4 inhibitor, CDP840, has been shown to have a weak inhibitory effect on the late response to allergen in humans, but is not being further developed.[83] However, most of the PDE4

Table 43.5 Distribution of phosphodiesterases.

	Predominant phosphodiesterase isoenzymes
Smooth muscle	
Airway smooth muscle	3, 4, 5
Vascular smooth muscle	3, 5
Inflammatory cells	
Mast cell	4
Macrophage	3, 4
Monocyte	4
Eosinophil	4
Neutrophil	4
Platelet	3, 5
T-lymphocyte	4
Endothelial cell	3, 4
Sensory nerves	4
Airway epithelial cells	4

inhibitors so far tested clinically have had unacceptable side-effects, particularly nausea, vomiting and headaches. These are the very side-effects that have been such a problem with theophylline, a non-selective PDE inhibitor.

Several steps may be possible to overcome these problems. It is conceivable that vomiting is due to inhibition of a particular subtype of PDE4. At least four human PDE4 genes have been identified and each has several splice variants.[8] This raises the possibility that subtype-selective inhibitors may be developed that might retain the anti-inflammatory effect but not the side-effects. Such subtype-selective inhibitors are now in development. Another possibility is that vomiting is due to binding of inhibitors to a non-catalytic site on the enzyme known as the high-affinity rolipram-binding site. PDE inhibitors that cause vomiting, such as rolipram, bind to this site, whereas other PDE4 inhibitors, such as SB 207,499, bind with much lower affinity and therefore are less likely to be emetogenic.[84] It is also possible that particular PDE4 isoenzymes (or even novel subtypes) may have an increased expression in asthmatic airways as a result of the chronic inflammatory process. Selective inhibition of disease-specific isoenzymes is an attractive possibility, as this might then avoid adverse effects. In patients with atopic dermatitis, there is evidence for increased expression of PDE4 in monocytes and lymphocytes in the peripheral blood, suggesting that atopic diseases may be associated with abnormal expression of PDEs in inflammatory cells.[85]

Transcription factor blockers

Transcription factors such as NF-κB and AP-1 play an important role in the orchestration of asthmatic inflammation[86] and this has prompted a search for specific blockers of these transcription factors (see Chapter 25). NF-κB is naturally inhibited by the inhibitory protein IκB, which is degraded after activation by specific kinases. Inhibitors of IκB kinases or the proteasome, the multifunctional enzyme that degrades IκB, would thus inhibit NF-κB and there is a search for such inhibitors. There are some naturally occurring inhibitors of NF-κB, such as the fungal product gliotoxin,[87] although this compound is toxic. There are concerns that inhibition of NF-κB may cause side-effects such as increased susceptibility to infections, which has been observed in gene disruption studies when components of NF-κB or IκB have been deleted.[86]

Many of the key cytokines involved in asthma transmit their signal through activation of the JAK–STAT (Janus kinase–signal transduction activated transcription) pathway.[88] Thus IL-5 activates STAT-1 and STAT-3 and IL-4 specifically activates STAT-6, making these attractive targets for inhibition. Small peptide inhibitors of STATs may be synthesized, resulting in blockade of these specific transduction pathways.

Cyclosporin A and tacrolimus inhibit T-lymphocyte function by blocking the transcription factor NF-AT (nuclear factor of activated T-cells) via a blocking action on calcineurin (see below).

Antiallergic drugs

Cromones may be effective in controlling mild asthma in some patients. They appear to have a specific action on allergic inflammation, yet the molecular mechanism of action remains uncertain. Although it was believed that the primary mode of action involved inhibiting mast cell mediator release, it has now been demonstrated that it has effects on several other inflammatory cells and on sensory nerves.[89] There is now increasing evidence that cromones may act on certain types of chloride channels expressed in mast cells and sensory nerves.[90] Both cromoglycate and nedocromil sodium must be given by inhalation and all attempts to develop orally active drugs of this type have been unsuccessful, possibly because topical administration is critical to their efficacy.

The loop diuretic frusemide (furosemide) shares many of the actions of cromones, inhibiting indirect bronchoconstrictor challenges (allergen, exercise, cold air, adenosine, metabisulphite) but not direct bronchoconstrictor challenges (histamine, methacholine) when given by inhalation.[91] The mechanism of action of frusemide is not shared by the more potent loop diuretic bumetanide, suggesting that some other mechanism than the inhibition of the $Na^+/K^+/Cl^-$ cotransporter must be involved. This is most likely to involve inhibition of the same chloride channel inhibited by cromones. Frusemide itself does not appear to be very effective when given regularly by metered dose inhaler in asthma,[92] although it is possible that more potent and long-lasting chloride channel blockers might be developed in the future.

Eosinophil inhibitors

Asthma is characterized by eosinophilic inflammation; therefore selective blockade of eosinophils is a logical strategy. Indeed, there are unlikely to be any major side-effects for such a therapeutic approach. Eosinophil infiltration into the airways and their activation may be blocked in several ways (Fig. 43.3). Eosinophil recruitment from the circulation may be blocked by antibodies to the adhesion molecules vascular cell adhesion molecule (VCAM)-1 (expressed on endothelial cells) and very late activation antigen (VLA)-4 (expressed on eosinophils). Humanized VLA antibodies are not in clinical trial in asthma and small molecules that may be suitable for oral absorption are also in development. IL-5

43 Future Therapies for Asthma 811

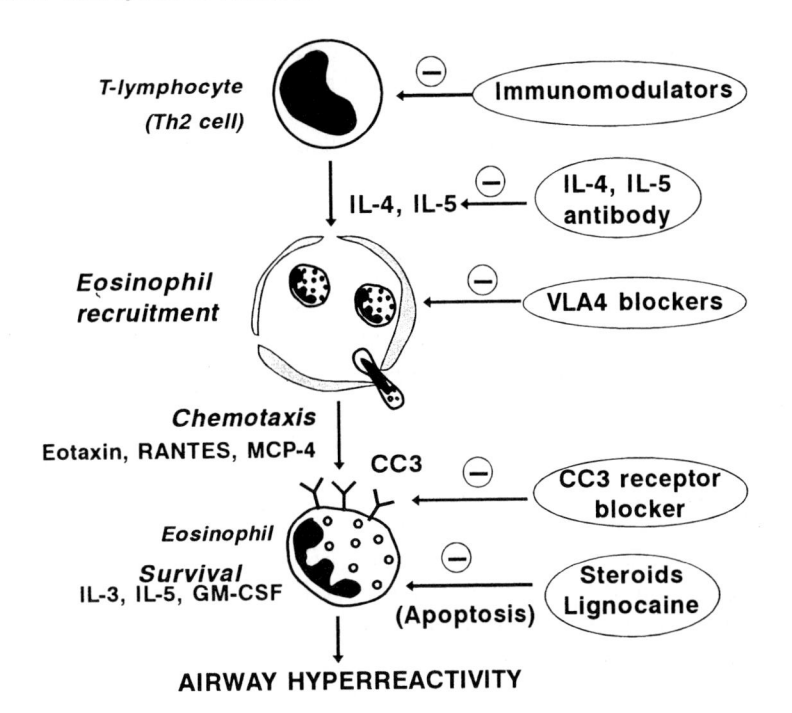

Fig. 43.3 Eosinophil inhibition. GM-CSF, granulocyte–macrophage colony-stimulating factor; IL, interleukin; MCP-4, XX; Rantes, xx; VLA4, very late activation antigen 4.

plays an important role in eosinophil recruitment and selective blockade of IL-5 may therefore be a valuable approach. Humanized IL-5 antibodies are now in clinical trial in asthma (as discussed above), but other approaches for blocking IL-5 such as transcription blockade and inhibitors of IL-5 receptors are also under investigation. Chemokines play a critical role in selectively attracting eosinophils into the airways. As discussed above, although several chemokines (RANTES, MCP-3, MCP-4, eotaxin) are selective for eosinophils, they all work through a common receptor on the eosinophil, CCR-3.[93] Since this is a typical G protein-coupled receptor with seven transmembrane-spanning domains, it may be possible to screen for CCR-3 inhibitors. Once recruited into the airways, eosinophils would normally undergo apoptotic death, but survive due to the effects of various growth factors, such as IL-3, IL-5 and GM-CSF. Glucocorticoids induce eosinophil apoptosis and other drugs may also have such an effect; the complex biochemical pathways involved in apoptosis may provide opportunities for selective eosinophil deletion.[94] It has recently been observed that the local anaesthetic lignocaine (lidocaine) increases eosinophil apoptosis;[95] in an uncontrolled trial in steroid-dependent asthmatic patients nebulized lignocaine appeared to have a steroid-sparing effect and improved asthma control.[96]

Immunomodulators

T-lymphocytes may play a critical role in initiating and maintaining the inflammatory process in asthma via the release of cytokines that result in eosinophilic inflammation, suggesting that T-cell inhibitors may be useful in controlling asthmatic inflammation. Glucocorticoids suppress inflammation in asthma partly through an inhibitory action on T-cell cytokine production. PDE4 inhibitors also have an inhibitory action on T-cell function[97] and inhibit the secretion of IL-5 from allergen-driven T-cells.[98]

The non-specific immunomodulator cyclosporin A has a steroid-sparing effect in steroid-dependent asthmatic patients,[99,100] although its efficacy is limited and side-effects, particularly nephrotoxicity, limit its widespread use.[101] The possibility of using inhaled cyclosporin A is now being explored, since in animal studies the inhaled drug is effective in inhibiting the inflammatory response in experimental asthma.[102] Immunomodulators, such as tacrolimus (FK506) and rapamycin appear to be more potent but are also toxic and would offer no real advantage. Novel immunomodulators such as mycophenolate mofetil may be less toxic and therefore of greater potential value in asthma therapy.[103]

One problem with non-specific immunomodulators, such as cyclosporin A, is that they inhibit both Th1 and Th2 cells and therefore do not reset the imbalance between these types of T-cell. They also inhibit suppressor T-cells that may modulate the inflammatory response. What is required is selective inhibition of Th2 cells and there is now a search for such drugs. The molecular steps involved in selection of Th2 cells by allergens is under intensive investigation; the costimulatory molecule B7.2, expressed on antigen-presenting cells that interacts with CD28 on T-cells, may be a target for attack.[104]

Cell adhesion blockers

It is now recognized that the infiltration of inflammatory cells into tissues is dependent on adhesion of blood-borne inflammatory cells to endothelial cells prior to migration to the inflammatory site.[105] This depends upon specific glycoprotein adhesion molecules on both leucocytes and endothelial cells that may be upregulated and show increased binding affinity in response to various stimuli such as cytokines or mediators such as PAF or leukotrienes. Monoclonal antibodies that inhibit these adhesion molecules may prevent inflammatory cell infiltration. Thus a monoclonal antibody to intercellular adhesion molecule (ICAM)-1 on endothelial cells prevents the eosinophil infiltration into airways and the increase in bronchial reactivity after allergen exposure in sensitized primates.[106] Although the interaction between VLA-4 and VCAM-1 is important for eosinophil inflammation, humanized antibodies to VLA-4 have not been developed.[107] While blocking adhesion molecules is an attractive new approach to the treatment of inflammatory disease, there may be potential dangers in inhibiting immune responses, leading to increased infections and increased risks of neoplasia.

IgE inhibition

Since release of mediators in asthma may be IgE dependent, an alternative approach is to block the activation of IgE using blocking antibodies that do not result in cell activation. Anti-IgE antibodies have been developed that inhibit allergen-induced mast cell degranulation.[108] Clinical studies in asthma show that these antibodies have some inhibitory effect in allergen-induced responses.[109] While infusions of antibody may not be feasible for the long-term treatment of mild asthma, this could be a realistic therapy for

43 Future Therapies for Asthma

patients with more severe forms of asthma. In the future it may be possible to develop smaller molecules that inhibit IgE. IL-4 inhibitors should have a similar effect, since IgE is dependent on IL-4. This is an attractive possibility, as such therapies would be effective in other inflammatory diseases, such as rhinitis and atopic dermatitis.

Immunotherapy

Although immunotherapy as currently practised has been disappointing in the therapy of asthma,[110] it is likely that more effective vaccines will be developed in the future.[111] As the complex mechanisms of antigen presentation and the interaction between antigen-presenting cells and T-lymphocyte receptors are elucidated, this may lead to the development of peptides that will block allergen-induced immune reactions.[112] Such peptides are now in clinical trials in allergic diseases.

Vaccination

The vast majority of asthma is allergic and allergy appears to be related to an imbalance between Th1 and Th2 cells. The development of allergic disease may be determined early in life by factors that affect this balance. There is a strong inverse association between a positive tuberculin test (indicating a Th1-mediated response) and atopy.[113] This suggests that it might be possible to immunize children against the risk of developing allergic diseases by stimulating local Th1-mediated immunity in the respiratory tract before sensitization occurs.[114]

GENE THERAPY

Many genes are involved in asthma.[115] Several genes determine atopic status but, more importantly, genes may also determine the severity and pattern of asthma. Genetic polymorphisms of cytokine and other genes may determine the severity of asthmatic inflammation and the response to treatment, so it may be possible to predict the outcome of asthma by screening for such polymorphisms in the future. The diversity of genes involved in asthma make gene therapy for asthma an unlikely prospect. It is possible, however, that transfer of anti-inflammatory genes may provide anti-inflammatory or inhibitory proteins in a convenient manner. Such gene transfer has been shown to be feasible in animals using viral vectors.[116] Possible anti-inflammatory proteins relevant to asthma include IL-10, IL-12 and IκB. Antisense oligonucleotides may switch off specific genes, although there are considerable problems in getting these molecules into cells. An inhaled antisense oligonucleotide directed against the adenosine A_1 receptor has been shown to reduce airway hyperresponsiveness in a rabbit model of asthma, demonstrating the feasibility of this approach in treating asthma.[55] Considering all the practical problems encountered by gene therapy makes this approach unlikely in the foreseeable future, other than for proof-of-concept studies.

CONCLUSIONS

Many different therapeutic approaches to the treatment of asthma may be possible, yet there have been few new drugs that have reached the clinic. β_2-Agonists are by far the most effective bronchodilator drugs and lead to rapid symptomatic relief. Now that inhaled β_2-agonists with a long duration of action have been developed it is difficult to imagine that more effective bronchodilators could be discovered. Similarly, inhaled glucocorticoids are extremely effective as chronic treatment in asthma and suppress the underlying inflammatory process. There is increasing evidence that earlier use of inhaled glucocorticoids may not only control asthma effectively but also prevent irreversible changes in airway function. For most patients a short-acting β_2-agonist on demand and regular inhaled steroids are sufficient to give excellent control of asthma.[79] For some patients a fixed combination β_2-agonist and steroid inhaler may be a useful development, since they will improve the compliance of inhaled steroids (which is poor because of the lack of immediate bronchodilator effect).[117]

The *ideal* drug for asthma would probably be a tablet that can be administered once daily to improve compliance. It should have no side-effects and this means that it should be specific for the abnormality of asthma (or allergy).

Future developments in asthma therapy should be directed towards the inflammatory mechanisms and perhaps more specific therapy may one day be developed. The possibility of developing a 'cure' for asthma seems remote, but when more is known about the genetic abnormalities of asthma it may be possible to search for such a therapy. Advances in molecular biology may aid the development of drugs that can specifically switch off relevant genes, but more must be discovered about the basic mechanisms of asthma before such advances are possible.

REFERENCES

1. Kume H, Hall IP, Washabau RJ, Takagi K, Kotlikoff MI: β-Adrenergic agonists regulate K_{Ca} channels in airway smooth muscle by cAMP-dependent and -independent mechanisms. *J Clin Invest* (1994) **93**: 371–379.
2. Barnes PJ, Chung KF: Questions about inhaled β_2-agonists in asthma. *Trends Pharmacol Sci* (1992) **13**: 20–23.
3. Drazen JM, Israel E, Boushey HA, *et al.*: Comparison of regularly scheduled with as needed use of albuterol in mild asthma. *N Engl J Med* (1996) **335**: 841–847.
4. Palmer JBD, Cuss FMC, Barnes PJ: VIP and PHM and their role in nonadrenergic inhibitory responses in isolated human airways. *J Appl Physiol* (1986) **61**: 1322–1328.
5. Barnes PJ, Dixon CMS: The effect of inhaled vasoactive intestinal peptide on bronchial hyperreactivity in man. *Am Rev Respir Dis* (1984) **130**: 162–166.
6. O'Donnel M, Garippa RJ, Rinaldi N, *et al.*: Ro25-1553: a novel long-acting vasoactive intestinal peptide agonist. Part 1: *in vitro* and *in vivo* bronchodilator studies. *J Pharmacol Exp Ther* (1994) **270**: 1282–1288.
7. Beavo JA: Cyclic nucleotide phosphodiesterases: functional implications of multiple isoforms. *Physiol Rev* (1995) **75**: 725–748.
8. Muller T, Engels P, Fozard J: Subtypes of the type 4 cAMP phosdphodiesterase: structure, regulation and selective inhibition. *Trend Pharmacol Sci* (1996) **17**: 294–298.

9. Barnes PJ, Pauwels RA: Theophylline in asthma: time for reappraisal? *Eur Respir J* (1994) **7**: 579–591.
10. Kidney J, Dominguez M, Taylor PM, Rose M, Chung KF, Barnes PJ: Immunomodulation by theophylline in asthma: demonstration by withdrawal of therapy. *Am J Respir Crit Care Med* (1995) **151**: 1907–1914.
11. Persson CGA: Development of safer xanthine drugs for the treatment of obstructive airways disease. *J Allergy Clin Immunol* (1986) **78**: 817–824.
12. Angus RM, Millar EA, Chalmers GW, Thomson NC: Effect of inhaled atrial natriuretic peptide and a neutral endopeptidase inhibitor on histamine-induced bronchoconstriction. *Am J Respir Crit Care Med* (1995) **151**: 2003–2005.
13. Barnes PJ: Muscarinic receptor subtyes in airways. *Life Sci* (1993) **52**: 521–528.
14. Barnes PJ, Belvisi MG, Mak JCW, Haddad E, O'Connor B: Tiotropium bromide (Ba 679 BR), a novel long-acting muscarinic antagonist for the treatment of obstructive airways disease. *Life Sci* (1995) **56**: 853–859.
15. Takahashi T, Belvisi MG, Patel H, *et al.*: Effect of Ba 679 BR, a novel long-acting anticholinergic agent, on cholinergic neurotransmission in guinea-pig and human airways. *Am J Respir Crit Care Med* (1994) **150**: 1640–1645.
16. O'Connor BJ, Towse LJ, Barnes PJ: Prolonged effect of tiotropium bromide on methacholine-induced bronchoconstriction in asthma. *Am J Respir Crit Care Med* (1996) **154**: 876–880.
17. Maesen FPV, Smeets JJ, Sledsens TJM, Wald FDM, Cornelissen JPG: Tiotropium bromide, a new long-acting antimuscarinic bronchodilator: a pharmacodynamic study in patients with chronic obstructive pulmonary disease (COPD). *Eur Respir J* (1995) **8**: 1506–1513.
18. Black JL, Barnes PJ: Potassium channels and airway function: new therapeutic approaches. *Thorax* (1990) **45**: 213–218.
19. Black JL, Armour CL, Johnson PRA, Alouan LA, Barnes PJ: The action of a potassium channel activator BRL 38227 (lemakalim) on human airway smooth muscle. *Am Rev Respir Dis* (1990) **142**: 1384–1389.
20. Miura M, Belvisi MG, Ward JK, Tadjkarini M, Yacoub MH, Barnes PJ: Bronchodilatory effects of the novel potassium channel opener HOE 234 in human airways *in vitro*. *Br J Clin Pharmacol* (1993) **35**: 318–320.
21. Kidney JC, Fuller RW, Worsdell Y, Lavender EA, Chung KF, Barnes PJ: Effect of an oral potassium channel activator BRL 38227 on airway function and responsiveness in asthmatic patients: comparison with oral salbutamol. *Thorax* (1993) **48**: 130–134.
22. Arch JR, Bowring NE, Buckle DR: Evaluation of the novel potassium channel activator BRL 55834 as an inhaled bronchodilator in guinea-pigs and rats: comparison with levcromakalim and salbutamol. *Pulmon Pharmacol* (1994) **7**: 121–128.
23. Jones TR, Charette L, Garcia ML, Kaczorowski GJ: Interaction of iberiotoxin with β-adrenoceptor agonists and sodium nitroprusside on guinea pig trachea. *J Appl Physiol* (1993) **74**: 1879–1884.
24. Miura M, Belvisi MG, Stretton CD, Yacoub MH, Barnes PJ: Role of potassium channels in bronchodilator responses in human airways. *Am Rev Respir Dis* (1992) **146**: 132–136.
25. Stretton CD, Miura M, Belvisi MG, Barnes PJ: Calcium-activated potassium channels mediate prejunctional inhibition of peripheral sensory nerves. *Proc Natl Acad Sci USA* (1992) **89**: 1325–1329.
26. Olesen SP, Munch E, Moldt P, Orejer J: Selective activation of Ca^{2+}-dependent K^+ channels by novel benzimidazolone. *Eur J Pharmacol* (1994) **251**: 53–59.
27. Fox AJ, Barnes PJ, Venkatesan P, Belvisi MG: Activation of large conductance potassium channels inhibits the afferent and efferent function of airway sensory nerves. *J Clin Invest* (1997) **99**: 513–519.
28. Walsh GM: The anti-inflammatory effects of cetirizine. *Clin Exp Allergy* (1994) **24**: 81–85.
29. Spector SL, Nicodemus CF, Corren J, *et al.*: Comparison of the bronchodilatory effects of cetirizine, albuterol, and both together versus placebo in patients with mild-to-moderate asthma. *J Allergy Clin Immunol* (1995) **96**: 174–181.
30. Grant JA, Nicodemus CF, Findlay SR, *et al.*: Cetirizine in patients with seasonal rhinitis and concomitant asthma: prospective, randomized, placebo-controlled trial. *J Allergy Clin Immunol* (1995) **95**: 923–932.

31. de Bruin Weller MS, Rijssenbeek Nouwens LH, de Monchy JG: Lack of effect of cetirizine on early and late asthmatic response after allergen challenge. *J Allergy Clin Immunol* (1994) **94**: 231–239.
32. Chung KF: Leukotriene receptor antagonists and biosynthesis inhibitors: potential breakthrough in asthma therapy. *Eur Respir J* (1995) **8**: 1203–1213.
33. Horwitz RJ, McGill KA, Busse WW: The role of leukotriene modifiers in the treatment of asthma. *N Engl J Med* (1997)
34. Laitinen LA, Laitinen A, Haahtela T, Vilkaa V, Spur BW, Lee TH: Leukotriene E$_4$ and granulocyte infiltration into asthmatic airways. *Lancet* (1993) **341**: 989–990.
35. Underwood DC, Osborn RR, Newsholme SJ, Torphy TJ, Hay DWP: Persistent airway eosinophilia after leukotriene (LT) D$_4$ administration in the guinea pig. *Am J Respir Crit Care Med* (1996) **154**: 850–857.
36. Wenzel SE, Trudeau JB, Kaminsky DA, Cohn J, Martin RJ, Wescott JY: Effect of 5-lipoxygenase inhibition on bronchoconstriction and airway inflammation in nocturnal asthma. *Am J Respir Crit Care Med* (1995) **152**: 857–905.
37. Leff JA, Pizzichini E, Efthimiadis *et al.*: Effect of montelukast (MK-0476) on airway eosinophilic inflammation in mildly uncontrolled asthma: a randomized placebo-controlled trial. *Am J Respir Crit Care Med* (1997) **55**: A977.
38. Evans DJ, Barnes PJ, Coulby LJ, *et al.*: The effect of a leukotriene B$_4$ antagonist LY293111 on allergen-induced responses in asthma. *Thorax* (1996) **51**: 1178–1184.
39. Barnes PJ, Chung KF, Page CP: Platelet-activating factor as a mediator of allergic disease. *J Allergy Clin Immunol* (1988) **81**: 919–934.
40. Spence DPS, Johnston SL, Calverley PMA, *et al.*: The effect of the orally active platelet-activating factor antagonist WEB 2086 in the treatment of asthma. *Am J Respir Crit Care Med* (1994) **149**: 1142–1148.
41. Kuitert LM, Hui KP, Uthayarkumar S, *et al.*: Effect of the platelet activating factor antagonist UK 74,505 on the early and late response to allergen. *Am Rev Respir Dis* (1993) *147*: 82–86.
42. O'Connor BJ, Uden S, Carty TJ, Eskra D, Barnes PJ, Chung KF: Effect of a potent and specific platelet activating factor (PAF) receptor antagonist on airway and systemic responses to PAF in man. *Am J Respir Crit Care Med* (1994) **150**: 35–40.
43. Hozawa S, Haruta Y, Ishioka S, Yamakido M: Effects of a platelet-activating factor antagonist Y 24180 on bronchial hyperresponsiveness in patients with asthma. *Am J Respir Crit Care Med* (1995) **152**: 1198–1202.
44. Evans DJ, Barnes PJ, Cluzel M, O'Connor BJ: Effects of a potent platelet activating factor antagonist, SR27417A, on allergen-induced asthmatic responses. *Am J Respir Crit Care Med* (1997) **156**: 11–16.
45. O'Byrne PM, Fuller RW: The role of thromboxane A$_2$ in the pathogenesis of airway hyperresponsiveness. *Eur Respir J* (1989) **2**: 782–786.
46. Mitchell JA, Larkin S, Williams TJ: Cyclooxygenase-2: regulation and relevance in inflammation. *Biochem Pharmacol* (1995) **50**: 1535–1542.
47. Demoly P, Jaffuel D, Lequeux N, *et al.*: Prostaglandin H synthase I and 2 immunoreactivities in the bronchial mucosa of asthmatics. *Am J Respir Crit Care Med* (1997) **155**: 670–675.
48. Chan CC, Boyce S, Brideau C, *et al.*: Pharmacology of a selective cyclooxygenase-2 inhibitor, L-745,337: a novel nonsteroidal anti-inflammatory agent with an ulcerogenic sparing effect in rat and nonhuman primate stomach. *J Pharmacol Exp Ther* (1995) **274**: 1531–1537.
49. Futaki N, Takahashi S, Yokoyama M, Arai I, Higuchi S, Otomo S: NS-398 a new anti-inflammatory agent, selectively inhibits prostaglandin synthase cyclooxygenase (COX-2) activity *in vitro*. *Prostaglandins* (1994) **47**: 55–59.
50. Barnes PJ: Bradykinin and asthma. *Thorax* (1992) **47**: 979–983.
51. Austin CE, Foreman JC, Scadding SK: Reduction by Hoe 140, the B$_2$ kinin receptor antagonist, of antigen-induced nasal blockage. *Br J Pharmacol* (1994) **111**: 969–971.
52. Scherrer D, Daeffler L, Trifilieff A, Gies J-P: Effects of WIN 64338, a non peptide bradykinin B$_2$-receptor antagonist, on guinea-pig trachea. *Br J Pharmacol* (1995) **115**: 1127–1128.
53. Feoktiskov I, Bioggioni I: Adenosine 2b receptors evoke interleukin-8 secretion in human mast cells. An enprofylline-sensitive mechanism with implications for asthma. *J Clin Invest* (1995) **96**: 1979–1986.

54. Linden J: Cloned adenosine A_3 receptors: pharmacological properties, species differences and receptor functions. *Trends Pharmacol Sci* (1994) **15**: 298–306.
55. Nyce JW, Metzger WJ: DNA antisense therapy for asthma in an animal model. *Nature* (1997) **385**: 721–725.
56. Thomas CE, Ohlweiler DF, Carr AA, *et al.*: Characterization of the radical trapping activity of a novel series of cyclic nitrone spin traps. *J Biol Chem* (1996) **271**: 3097–3104.
57. Barnes PJ, Liew FY: Nitric oxide and asthmatic inflammation. *Immunol Today* (1995) **16**: 128–130.
58. Yates DH, Kharitonov SA, Thomas PS, Barnes PJ: Endogenous nitric oxide is decreased in asthmatic patients by an inhibitor of inducible nitric oxide synthase. *Am J Respir Crit Care Med* (1996) **154**: 247–250.
59. Garvey EP, Oplinger JA, Furfine ES, *et al.*: 1400W is a slow, tight binding and highly selective inhibitor of inducible nitric-oxide synthase *in vitro* and *in vivo*. *J Biol Chem* (1997) **272**: 4959–4963.
60. Hay DWP, Henry PJ, Goldie RG: Is endothelin-1 a mediator in asthma? *Am J Respir Crit Care Med* (1996) **155**: 1994–1997.
61. Barnes PJ: Endothelins and pulmonary diseases. *J Appl Physiol* (1994) **77**: 1051–1059.
62. Chalmers GW, Little SA, Patel KR, Thomson NC: Endothelin-1-induced bronchoconstriction in asthma. *Am J Respir Crit Care Med* (1997) **156**: 382–388.
63. Douglas SA, Meek TD, Ohlstein EH: Novel receptor antagonists welcome a new era in endothelin biology. *Trends Pharmacol Sci* (1994) **15**: 313–316.
64. Barnes PJ: Cytokines as mediators of chronic asthma. *Am J Respir Crit Care Med* (1994) **150**: S42–S49.
65. Egan RW, Umland SP, Cuss FM, Chapman RW: Biology of interleukin-5 and its relevance to allergic disease. *Allergy* (1996) **51**: 71–81.
66. Mauser PJ, Pitman A, Witt A, *et al.*: Inhibitory effect of the TRFK-5 anti IL-5 antibody in a guinea pig model of asthma. *Am Rev Respir Dis* (1993) **148**: 1623–1627.
67. Mauser PJ, Pitman AM, Fernandez X, *et al.*: Effects of an antibody to interleukin-5 in a monkey model of asthma. *Am J Respir Crit Care Med* (1995) **152**: 467–472.
68. Ho IC, Hodge MR, Rooney JW, Glimcher LH: The proto-oncogene c-maf is responsible for tissue-specific expression of interleukin-4. *Cell* (1996) **85**: 973–983.
69. Elliott MJ, Maini RN, Feldmann M, *et al.*: Randomised double-blind comparison of diuretic monoclonal antibody to tumour necrosis factor α (cA2) versus placebo in rheumatoid arthritis. *Lancet* (1994) **344**: 1105–1110.
70. Adams DH, Lloyd AR: Chemokines: leucocyte recruitment and activation cytokines. *Lancet* (1997) **349**: 490–495.
71. Selig W, Tocker J: Effect of interleukin-1 receptor antagonist on antigen-induced pulmonary responses in guinea-pigs. *Eur J Pharmacol* (1992) **213**: 331–336.
72. Boguniewicz M, Martin RJ, Martin D, Gibson U, Celniker A: The effects of nebulized recombinant interferon-γ in asthmatic airways. *J Allergy Clin Immunol* (1995) **95**: 133–135.
73. Trinchieri G: Interleukin 12: a proinflammatory cytokine with immunoregulatory functions that bridge innate resistance and antigen-specific adaptive immunity. *Annu Rev Immunol* (1995) **13**: 252–276.
74. Gavett SH, O'Hearn DJ, Li X, Huang SK, Finkelman FD, Wills-Karp M: Interleukin 12 inhibits antigen-induced airway hyperresponsivness, inflammation and Th2 cytokine expression in mice. *J Exp Med* (1995) **182**: 1527–1536.
75. Wang P, Wu P, Siegel MI, Egan RW, Billah MM: Interleukin (IL)-10 inhibits nuclear factor kappa B activation in human monocytes. IL-10 and IL-4 suppress cytokine synthesis by different mechanisms. *J Biol Chem* (1995) **270**: 9558–9563.
76. Borish L, Aarons A, Rumbyrt J, Cvietusa P, Negri J, Wenzel S: Interleukin-10 regulation in normal subjects and patients with asthma. *J Allergy Clin Immunol* (1996) **97**: 1288–1296.
77. John M, Lim S, Seybold J, *et al.*: Inhaled corticosteroids increase IL-10 but reduce MIP-1α, GM-CSF and IFN-γ release from alveolar macrophages in asthma. *Am J Respir Crit Care Med* (1997) in press.

78. Schreiber S, Heinig T, Thiele HG, Raedler A: Immunoregulatory role of interleukin 10 in patients with inflammatory bowel disease. *Gastroenterology* (1995) **108**: 1434–1444.
79. Barnes PJ: Inhaled glucocorticoids for asthma. *N Engl J Med* (1995) **332**: 868–875.
80. Barnes PJ: Molecular mechanisms of steroid action in asthma. *J Allergy Clin Immunol* (1996) **97**: 159–168.
81. Sanjar S, Aoki S, Kristersson A, Smith D, Morley J: Antigen challenge induces pulmonary eosinophil accumulation and airway hyperreactivity in sensitized guinea pigs: the effect of anti-asthma drugs. *Br J Pharmacol* (1990) **99**: 679–686.
82. Turner CR, Andresen CJ, Smith WB, Watson JW: Effects of rolipram on responses to acute and chronic antigen exposure in monkeys. *Am J Respir Crit Care Med* (1994) **149**: 1153–1159.
83. Harbinson PL, Macleod D, Hawksworth R, et al.: The effect of a novel orally active selective PDE4 isoenzyme inhibitor (CDP840) on allergen-induced responses in asthmatic subject. *Eur Respir J* (1997) **10**: 1008–1014.
84. Barnette MS, Bartus JO, Burman M, et al.: Association of the anti-inflammatory activity of phosphodiesterase 4 (PDE4) inhibitors with either inhibition of PDE4 catalytic activity or competition for [^{3}H]rolipram binding. *Biochem Pharmacol* (1996) **51**: 949–956.
85. Chan SC, Reifsouyder D, Beavo JA, Hanifin JM: Immunochemical characterization of the distinct monocyte cyclic AMP-phosphodiesterases from patients with atopic dermatitis. *J Allergy Clin Immunol* (1993) **91**: 1179–1188.
86. Barnes PJ, Karin M: Nuclear factor-±B: a pivotal transcription factor in chronic inflammatory diseases. *N Engl J Med* (1997) **336**: 1066–1071.
87. Pahl HL, Krauss B, Schultze-Osthoff K, et al.: The immunosuppressive fungal metabolite gliotoxin specifically inhibits transcription factor NF-±B. *J Exp Med* (1996) **183**: 1829–1840.
88. Ihle JN, Witthuhn BA, Quelle FW, et al.: Signalling by the cytokine receptor superfamily: JAKs and STATs. *Trends Biochem Sci* (1994) **19**: 222–225.
89. Barnes PJ, Holgate ST, Laitinen LA, Pauwels R: Asthma mechanisms, determinants of severity and treatment: the role of nedocromil sodium. *Clin Exp Allergy* (1995) **25**: 771–787.
90. Heinke S, Szucs G, Norris A, Droogmans G, Nilius B: Inhibition of volume-activated chloride currents in endothelial cells by chromones. *Br J Pharmacol* (1995) **115**: 1393–1398.
91. Barnes PJ: Diuretics and asthma. *Thorax* (1993) **48**: 195–197.
92. Yates DH, O'Connor BJ, Yilmaz G, et al.: Effect of acute and chronic inhaled furosemide on bronchial hyperresponsiveness in mild asthma. *Am J Respir Crit Care Med* (1995) **152**: 892–896.
93. Ponath PD, Qin S, Post TW, et al.: Molecular cloning and characterization of a human eotaxin receptor expressed selectively on eosinophils. *J Exp Med* (1996) **183**: 2437–2448.
94. Anderson GP: Resolution of chronic inflammation by therapeutic induction of apoptosis. *Trends Pharmacol Sci* (1996) **17**: 438–442.
95. Ohnishi T, Kita H, Mayeno AN, et al.: Lidocaine in bronchoalveolar lavage fluid (BALF) is an inhibitor of eosinophil-active cytokines. *Clin Exp Immunol* (1996) **104**: 325–331.
96. Hunt L, Swelund H, Frigas E, Gleich GH: Nebulized lidocaine can be used successfully as glucocorticoid sparing/replacing therapy in severe glucocorticoid-dependent asthma. *Am J Respir Crit Care Med* (1996) **153**: A534.
97. Giembycz MA, Corrigan CJ, Seybold J, Newton R, Kay AB, Barnes PJ: Identification of cyclic AMP phosphodiesterases 3, 4 and 7 in human CD4$^+$ and CD8$^+$ T-lymphocytes. *Br J Pharmacol* (1996) **118**: 1945–1958.
98. Essayan DM, Huang S-K, Kagey-Sabotka A, Lichtenstein LM: Effects of nonselective and isoenzyme selective cyclic nucleotide phosphodiesterase inhibitors on antigen-induced cytokine gene expression in peripheral blood mononuclear cells. *Am J Respir Cell Mol Biol* (1995) **13**: 692–702.
99. Alexander AG, Barnes NC, Kay AB: Trial of cyclosporin in corticosteroid-dependent chronic severe asthma. *Lancet* (1992) **339**: 324–328.
100. Nizankowska E, Soja J, Pinis G, et al.: Treatment of steroid-depenent bronchial asthma with cyclosporin. *Eur Respir J* (1995) **8**: 1091–1099.
101. Barnes PJ: Immunomodulators in asthma: where do we stand? *Eur Respir J* (1996) **9**: 154S–159S.

43 Future Therapies for Asthma

102. Morley J: Cyclosporin A in asthma therapy: a pharmacological rationale. *J Autoimmun* (1992) **5** (Suppl A): 265–269.
103. Thompson AG, Starzl TC: New immunosuppressive drugs: mechanistic insights and potential therapeutic advances. *Immunol Rev* (1993) **136**: 71–98.
104. Kuchroo VK, Das MP, Brown JA, et al.: B7-1 and B7-2 costimulatory molecules activate differentially the Th1/Th2 developmental pathways: application to autoimmune disease therapy. *Cell* (1995) **80**: 707–718.
105. Pilewski JM, Albelda SM: Cell adhesion molecules in asthma: homing activation and airway remodelling. *Am J Respir Cell Mol Biol* (1995) **12**: 1–3.
106. Weg VB, Williams TJ, Lobb RR, Noorshargh S: A monoclonal antibody recognizing very late activation antigen-4 inhibits eosinophil accumulation *in vivo*. *J Exp Med* (1993) **177**: 561–566.
107. Yuan Q, Strauch KL, Lobb RR, Hemler ME: Intracellular single-chain antibody inhibits integrin VLA-4 maturation and function. *Biochem J* (1996) **318**: 591–596.
108. Saban R, Haak-Frendscho M, Zine M, et al.: Human FcεR1-IgG and humanized anti-IgE monoclonal antibody MaE11 block passive sensitization of human and rhesus monkey lung. *J Allergy Clin Immunol* (1994) **94**: 836–843.
109. Fahy JV, Fleming HE, Wong HH, et al.: The effect of an anti-IgE monoclonal antibody on the early and late phase responses to allergen inhallation in asthmatic subjects. *Am J Respir Crit Care Med* (1997) **155**: 1826–1834.
110. Barnes PJ: Immunotherapy for asthma: is it worth it? *N Engl J Med* (1996) **334**: 531–532.
111. Hoyne G-F, Lamb J-R: Peptide-mediated regulation of the allergic immune response. *Immunol Cell Biol* (1996) **74**: 180–186.
112. Yssel H, Fasler S, Lamb J, de Vries JE: Induction of non-responsiveness in human allergen specific type 2 helper cells. *Curr Opin Immunol* (1994) **6**: 847–852.
113. Shirakawa T, Enomoto T, Shimazu S, Hopkin JM: The inverse association between tuberculin responses and atopic disorder. *Science* (1997) **275**: 77–79.
114. Holt PG: A potential vaccine strategy for asthma and allied atopic diseases during infancy. *Lancet* (1994) **344**: 456–458.
115. Sandford A, Weir T, Pare P: The genetics of asthma. *Am J Respir Crit Care Med* (1996) **153**: 1749–1765.
116. Xing Z, Ohkawara Y, Jordana M, Grahern FL, Gauldie J: Transfer of granulocyte–macrophage colony-stinulating factor gene to rat induces eosinophilia, monocytosis and fibrotic lesions. *J Clin Invest* (1996) **97**: 1102-1110.
117. Barnes PJ, O'Connor BJ: Use of a fixed combination β_2-agonist and steroid dry powder inhaler in asthma. *Am J Respir Crit Care Med* (1995) **151**: 1053–1057.

44

Management of Severe Asthma

GRAHAM K. CROMPTON

INTRODUCTION

Severe acute asthma is a potentially life-threatening episode of asthma that can develop gradually over a number of days or rapidly within a few minutes. The old term 'status asthmaticus' should no longer be used. Clinically, patients are distressed by dyspnoea, chest tightness and are unable to speak full sentences. A minority of asthmatic patients are able to tolerate severe airflow limitation without showing much respiratory distress, and the severity of disease in these patients can only be assessed accurately by objective measurements, in particular arterial blood gas analysis. Most patients with severe acute asthma sit or stand with shoulder muscles braced in an attempt to assist their breathing. Ill patients are usually pale and sweaty. Obvious cyanosis indicates severe hypoxaemia. Confusion and drowsiness only accompanies gross hypoxaemia and hypercapnia. Respiratory acidosis should be regarded as evidence of impending death. Children, however, tend to develop acidaemia more readily than adults and this is often due to a combination of respiratory and metabolic acidosis.

Subcutaneous emphysema of the neck and face occurs in some patients with severe asthma and is a reflection of very high intrathoracic pressures but rarely is associated with pneumothorax.

GENERAL ASSESSMENT AND MANAGEMENT

A rapid clinical appraisal of the features mentioned above usually gives a fairly accurate assessment of disease severity. Auscultation of the chest is of limited value since patients with severe disease are unable to shift enough air to produce rhonchi, the so-called 'bilateral silent chest'. In the absence of a chest radiograph auscultation is essential, mainly to ensure that breath sounds are present over both lungs. If auscultation reveals a difference between the breath sounds and intensity of rhonchi, a pneumothorax must be suspected on the 'quiet' side. Whenever possible a chest X-ray should be performed, especially if mechanical ventilation is necessary, in order to exclude a pneumothorax, which is a rare but potentially fatal complication of severe asthma. Measurement of heart rate and the degree of pulsus paradoxus[1] are invaluable in initial assessment and subsequent monitoring of response to treatment. However, pulsus paradoxus is not always present even in severe disease and therefore can be misleading, and in distressed patients it is not always easy to measure. For these reasons, the use of pulsus paradoxus to assess disease severity and response to therapy is now not thought to be necessary in all patients.[2] The heart rate is usually rapid and almost invariably over 100–120/min. However, it must be appreciated that severe hypoxaemia can cause progressive bradycardia, which can culminate in hypoxaemic cardiac asystole unless hypoxaemia is reversed.

Ventilatory function tests such as the peak expiratory flow (PEF) and the forced expiratory volume in 1 s (FEV_1) are of limited value in the assessment of very seriously ill patients, since anxiety and distress make it difficult for them to perform forced expiratory manoeuvres and values recorded may be inaccurate. However, PEF should be recorded in all patients able to cooperate in order to provide an objective baseline, which allows subsequent rapid assessment of response to treatment. A PEF of <33% of predicted normal or of the best obtainable result if known (<200 litre/min if the best obtainable result is not known) should be regarded as evidence of a potentially life-threatening attack of asthma.[3]

Arterial blood gas analysis is essential in the assessment of disease severity and repeated measurements are necessary to evaluate response to treatment. Normal blood gas tensions with the patient breathing air exclude life-threatening disease. All ill patients are hypoxaemic. Hypoxaemia plus hypercapnia is only found in severely ill patients. Although arterial blood gas analysis is essential in the management of the very sick asthmatic patient, it is important to avoid unnecessary arterial punctures in the less ill patient. Pulse oximetry can be used to monitor response in those whose initial arterial blood gas tensions were normal, or have returned to normal with treatment, providing there is no clinical concern about the patient's progress. It has been shown in the initial investigation of patients presenting with acute asthma that an oxygen saturation of 92% or higher is unlikely to be associated with respiratory failure, defined as $PaO_2 < 8.0$ kPa or $PaCO_2 > 6$ kPa.[4] The judicious use of pulse oximetry can avoid the discomfort of arterial punctures in selected patients, but in patients in whom there is any doubt about disease severity or response to treatment arterial blood gas analysis must be performed.

General measures

Severe acute asthma causes great respiratory distress and most ill patients are terrified because they feel that they are going to die. It would seem logical, therefore, to ease distress and anxiety with a sedative or anxiolytic drug. However, although reports of deaths directly attributable to sedation are few,[4,5] sedation must be avoided since it is illogical. Patients with severe acute asthma breathe with the maximum efficient use of their respiratory muscles and in spite of this are unable to maintain normal arterial blood gas tensions. Therefore, to suppress ventilation with any form of sedation is likely to lead to deterioration rather than improvement of the basic disease state. Occasionally patients with mild or moderately severe asthma become excessively agitated and this leads to hyperventilation out of proportion to the severity of asthma. These patients usually have near-normal arterial oxygen tensions and respiratory alkalosis. The use of a benzodiazepine in a small dose may be appropriate in such circumstances. However, as a general rule, sedation in severe asthma must be avoided and should never be given without immediate access to facilities for arterial blood gas monitoring and assisted ventilation.[2]

Hydration

Intravenous fluids are rarely necessary for rehydration unless the attack has been of many hours, duration. Prolonged severe asthma does lead to fluid deprivation, since breathless patients avoid drinking as they 'cannot afford the time to swallow'. Hyperventilation associated with severe asthma also leads to an increase in obligatory fluid loss. Although fluid replacement is necessary in only a few patients, this should always be kept in mind and it is wise to have an intravenous access in all ill patients, particularly in those who do not rapidly respond to initial nebulized bronchodilator therapy. The intravenous line can be used for fluid replacement when necessary, but its main purpose is to provide immediate intravenous access for the administration of drugs should this be necessary. All ill patients are capable of developing sudden deterioration and if respiratory arrest occurs an *in situ* intravenous line is invaluable. Electrolyte imbalance can occur in some patients. Hypokalaemia can be induced by β-agonist and corticosteroid therapy, and can be made worse by over-zealous intravenous transfusion of potassium-free solutions.

SPECIFIC TREATMENT FOR SEVERE ACUTE ASTHMA

There is no fixed order for the treatments used for severe asthma since in very ill patients almost all drugs and oxygen will be given at the same time. In less critically ill patients, bronchodilator drugs alone may be given and other treatments may not be necessary. In the majority of patients with severe asthma, however, a combination of treatments is usually necessary. Response to treatment must be assessed by measurements of PEF and arterial blood gas analysis, together with careful clinical observations.

Oxygen therapy

All ill patients are hypoxaemic and require oxygen. At one time it was suggested that low inspired oxygen concentrations should be given unless blood gas monitoring was immediately available[6] in order to avoid the precipitation, or worsening, of hypercapnia. However, it is now accepted that there is no risk of causing carbon dioxide retention in an asthmatic patient who does not also have coexisting chronic obstructive bronchitis.

Oxygen should therefore be given by face mask in the highest concentration possible. In patients over the age of 50 in whom the diagnosis of asthma is in doubt, high concentrations of oxygen therapy should be used with caution in case the predominant clinical problem is chronic obstructive pulmonary disease (COPD) rather than asthma. In patients with severe COPD high-concentration oxygen therapy may worsen respiratory acidosis, but it is a greater clinical sin to deny an asthmatic patient high-concentration oxygen therapy than to cause an increase in carbon dioxide retention in a patient with COPD. Masks delivering 24 or 28% oxygen are not appropriate.[2] The most comfortable masks for distressed patients are high-concentration Venturi oxygen systems employing the Bernoulli principle. However, the recommended oxygen flow rate for these masks may have to be exceeded to maintain the desired high inspired oxygen tension in patients with rapid breathing frequencies and high tidal volumes.[7]

Hypoxaemia should be treated with oxygen in concentrations as high as necessary to maintain an arterial oxygen tension (PaO_2) of at least 9 kPa (80 mmHg) or an oxygen saturation (SaO_2) in excess of 92%. When facilities for arterial blood gas analysis are not available, and it is known that the patient suffers from asthma and not COPD, high concentrations of oxygen should be given (35–60%). Even when patients are found to have hypercapnia and hypoxaemia high concentrations of oxygen must be administered (35% or more), since such abnormalities of blood gas tensions reflect severe airflow obstruction, not a central defect of carbon dioxide responsiveness, that requires oxygen, optimal bronchodilator therapy (see p. 825) and corticosteroids (see p. 826). To deny adequate oxygenation of these patients is irrational, since deliberate perpetuation of hypoxaemia is dangerous. When hypoxaemia persists and hypercapnia worsens assisted ventilation is required (see p. 827) and this cannot be avoided by giving low-concentration oxygen therapy. In the treatment of patients with uncomplicated bronchial asthma, high concentrations of oxygen carry no dangers[8,9] and a high inspired oxygen concentration should be used in all patients. Most distressed patients dislike closely fitting face masks and should be treated with masks employing the Venturi principle, nasal prongs or both. Oxygen tents may have to be used for young children.

The administration of bronchodilator drugs, particularly when given by the intravenous route, can cause worsening of hypoxaemia. It is therefore prudent to treat patients with oxygen before, during and after such treatments. Patients with severe asthma should be given oxygen during the transfer from home to hospital.[10] The general practitioner, or family physician, should give oxygen in the home whenever possible and ensure that oxygen therapy is administered by the ambulance crew when admission to hospital is necessary.

β_2-Adrenoreceptor agonist therapy

Treatment with selective β_2-adrenoreceptor agonists is now first-line therapy in most hospitals in the UK[11] and other countries. A large dose of one of these drugs nebulized in oxygen is to be preferred to their intravenous administration,[12] since although both routes of treatment are effective[13–15] high-dose inhaled salbutamol has been shown to be more effective than conventional doses of the same drug given intravenously.[16] The dose of intravenous salbutamol necessary to achieve a better response than nebulized treatment causes unacceptable cardiovascular effects.[17] A combination of aerosol and intravenous

β_2-adrenoreceptor agonists should be avoided. Salbutamol (2.5–5.0 mg) or terbutaline (5–10 mg) should be nebulized in oxygen whenever possible. Nebulized therapy is unlikely to cause worsening of hypoxaemia,[18] but oxygen therapy should not be interrupted during treatment of hypoxaemic patients. Ultrasonic nebulizers and air compressors to drive jet nebulizers should, therefore, not be used in hospital. Aerosol therapy delivered by intermittent positive-pressure breathing (IPPB) has no advantages over simple jet nebulizer treatment.[19,20] The response to a large dose of nebulized salbutamol is rapid and continued improvement of ventilatory function should not be excepted for more than 12–20 min.[18,20] Hence, if a patient remains unwell more than 20 min after a large dose of a β_2-adrenoreceptor agonist, treatment should be repeated or additional therapy with ipratropium bromide (see below) and in some patients a xanthine derivative (see p. 826) should be given. In the unresponsive patient a β-agonist nebulized in oxygen should be given as often as necessary. This is, in effect, continuous nebulized therapy, which has been shown to be beneficial in uncontrolled studies in children with severe acute asthma.[21–23] Recently, continuously administered nebulized salbutamol has been found to be superior to the same drug given intermittently every hour for 2 h in adults with severe acute asthma.[24]

Large-volume spacer attachments to the conventional metered dose inhaler (MDI) provide an alternative to nebulizers as a method of administering high doses of β_2-adrenoreceptor agonists in the treatment of severe asthma.[25] Such devices can be used outside hospital when facilities for nebulization in oxygen are not available. Face masks specially designed for use with large-volume holding chambers make administration of drugs to children much easier than via a mouthpiece. Salbutamol (4 μg/kg body weight) or terbutaline (0.25–0.5 mg) can be given by slow intravenous injection and also by continuous intravenous infusion. Compared with aerosol therapy, intravenous administration is associated with more unwanted effects and is also more likely to cause an increase in the degree of arterial hypoxaemia. Administration of β_2-adrenoreceptor agonists by intramuscular or subcutaneous injection should be avoided whenever possible, since it is difficult to assess the time of maximum response when these drugs are given by these routes.

Anticholinergics

The quaternary ammonium compound ipratropium bromide has little value in the treatment of chronic asthma but has an important role in the management of the severe acute episode. The onset of bronchodilator action of this drug is considerably slower than that of an inhaled sympathomimetic bronchodilator, but is as effective as salbutamol.[26,27] Also, there is evidence that a combination of ipratropium bromide and an inhaled sympathomimetic is more effective than either drug given alone,[26–28] although this has been questioned.[29] Ipratropium bromide should be nebulized in a dose of 0.25–0.5 mg, but should never be given alone as primary treatment since it has a much slower onset of action than salbutamol and terbutaline and has been reported to cause bronchoconstriction.[30,31] Most bronchoconstrictor responses were probably caused by the tonicity of, and the preservative in, the original respirator solution;[32,33] however, since the first report of an adverse reaction was to ipratropium bromide inhaled from a pressurized aerosol[31] it is wise to avoid this drug as primary treatment. However, the combination of ipratropium bromide and a sympathomimetic bronchodilator should be given when there has not been a satisfactory response to initial treatment with a nebulized β_2-adrenoreceptor agonist alone, and this combination may become routine first-line initial treatment for all severely ill patients. The optimal duration of this combined therapy has not been determined, but it is unlikely to be necessary for more than 24 h in patients who have responded to initial therapy.

Xanthine derivatives

Xanthine derivatives (aminophylline, theophylline) by intravenous injection have been used for many years in the treatment of severe asthma. However, intravenous salbutamol has been shown to be at least as effective as intravenous aminophylline.[34,35] Since high-dose aerosol β_2-agonist therapy is as effective as, and has fewer side-effects than, intravenous treatment (see p. 824), nebulized β_2-adrenoreceptor agonists have replaced intravenous theophyllines as first-line treatment of severe asthma. Aminophylline is recommended as a slow intravenous injection over 20 min (5 mg/kg body weight) followed, if necessary, by a continuous infusion (0.5 mg/kg body weight).[36] An alternative dose recommendation is 0.5–0.9 mg/kg per hour without a loading dose. If the weight of the patient is unknown, doses can be estimated depending upon the patient's size (small patients, 600–1000 mg/24 h; medium-size patients, 900–1500 mg/24 h; large patients, 1100–1900 mg/24 h).[2] Unwanted effects such as nausea are common and potentially fatal central nervous system toxic effects can occur if serum levels exceed 30 mg/litre. It is therefore unwise to give an intravenous loading dose to patients already taking an oral methylxanthine preparation, unless the serum theophylline level is known. Infusions of theophylline should be adjusted to maintain a blood level within the 'therapeutic range' of 10–20 mg/litre[37] unless this is associated with unacceptable side-effects. Aminophylline (theophylline) should be used in patients who do not respond quickly to a nebulized β_2-adrenoreceptor agonist combined with ipratropium bromide. In moribund patients, however, it should be given at once together with nebulized therapy.

Corticosteroids

The value of corticosteroid therapy in the management of severe asthma was first reported in 1949[38] and confirmed by clinical trial in 1956.[39] During the last three decades corticosteroids have been used routinely by most physicians in the treatment of patients with severe asthma, although some paediatricians tend to reserve their use for the most critically ill patients. The value of these drugs in severe asthma has only rarely been questioned[40] and their use should be routine in all patients who do not respond rapidly and substantially to bronchodilator therapy.[41,42] Large doses of intravenous hydrocortisone or methylprednisolone are usually given empirically, since short-course high-dose corticosteroid treatment is free from serious unwanted effects providing the possibilities of hypokalaemia and fluid retention are born in mind. Very high doses of hydrocortisone or methylprednisolone are probably not necessary[43–45] and might cause an acute myopathy.[46] Doses of hydrocortisone producing blood levels that exceed stress-induced physiological levels have been recommended,[47] i.e. 3–4 mg/kg loading dose followed by

44 Management of Severe Asthma

the same dose by intravenous infusion 6-hourly. However, standard empirical doses of 200 mg 4–6-hourly during the first 24–48 h of treatment are often used. Intravenous corticosteroids are then replaced by prednisolone in doses of 30–60 mg daily. Oral therapy is given initially to patients who do not have life-threatening disease. However, it has been suggested that intravenous corticosteroid therapy is not necessary if large doses of oral prednisolone, together with a nebulized sympathomimetic drug plus intravenous aminophyline, are used.[48] It has been suggested that corticosteroids are only necessary in patients with severe acute asthma if there has been no response to bronchodilator and oxygen therapy after 2 h or if the patients are already taking oral corticosteroids,[49] although the concensus view remains that all patients with severe acute asthma should be treated with corticosteroids.[50]

ASSISTED VENTILATION

Assisted ventilation is rarely necessary, but has to be instituted in a few patients as a life-saving procedure. In adults a cuffed endotracheal tube is essential and a powerful volume-cycled ventilator must be used.

The indications for assisted ventilation in asthma are difficult to define, since some patients have to be electively ventilated because of lack of response to treatment together with the assumption that a crisis is impending. However, it is generally accepted that assisted ventilation is essential in patients who, in spite of full medical treatment have:

(1) a Pa_{CO_2} of >50 mmHg (6.6 kPa) and rising;
(2) a Pa_{O_2} of <50 mmHg (6.6 kPa) and falling;
(3) a pH of 7.3 or less and falling;
(4) intolerable respiratory distress;
(5) respiratory arrest;
(6) cardiorespiratory arrest.

Ventilation should also be considered in all patients who have had prolonged severe asthma and who are becoming tired and exhausted. Physical exhaustion associated with systemic hypotension is a dangerous combination. When there is any doubt about whether a patient should or should not be ventilated the safest course of action is to ventilate and not to procrastinate.

It has been suggested that the complications of ventilation of patients with asthma are a consequence of barotrauma caused by high intrapulmonary pressures.[51–53] For this reason it has been recommended that inflation pressures of >50 cmH$_2$O should be avoided by planned assisted hypoventilation, irrespective of the inevitable perpetuation of respiratory acidosis in some patients.[51] However, this approach to the ventilation of patients with severe asthma has been questioned and the restoration of normal blood gas tensions as quickly as possible, irrespective of the inflation pressures necessary to achieve this objective, has been recommended.[54] The debate about the advantages and disadvantages of adopting policies of planned hypoventilation or the more aggressive approach of trying to normalize arterial blood gas tensions as quickly as possible will continue. However, there appears to be no doubt that the combination of planned assisted hypoventilation and therapeutic bronchial lavage is associated with unacceptable pulmonary infection and death.[55] Patients treated with planned assisted hypoventilation almost invariably have to be given muscle relaxants to allow them to synchronize with the ventilator, whereas in patients in whom respiratory acidosis is normalized as quickly as possible sedation with opiates is usually sufficient. In a small minority of patients it is not possible to mechanically ventilate because of extreme high inflation pressures that exceed the pressures most volume-cycled ventilators are capable of generating. In these patients it is always worthwhile trying the effects of the anaesthetic diethyl ether, which has long been known to have a relaxant effect on airways.[56,57] The most effective modern anaesthetic agent with properties similar to diethyl ether is isoflurane,[58] which should be given if diethyl ether is not available or cannot be given via modern anaesthetic apparatus. Treatment with a mixture of helium (60–70%) and oxygen (30–40%) has been reported to improve ventilation in patients with severe acute asthma when given via a ventilator or face mask.[59]

MANAGEMENT OF CATASTROPHIC ASTHMA

Some patients suddenly develop catastrophic attacks of asthma, perhaps in spite of receiving treatment that controls the symptoms of most other patients. These individuals are, of course, at greatest risk of dying from asthma. Little can be done for the unfortunate patient whose first episode of asthma is a devastating attack, except hope that it is not fatal so that appropriate plans for the management of the next anticipated severe attack can be made. All patients who have had one life-threatening episode of asthma must be regarded as high-risk candidates for a recurrence. Not only is the asthmatic patient who has had one bad attack more likely to have a recurrence than patients who have never experienced a severe episode, but the pattern of future attacks is likely to be similar also. Once recognized, therefore, the catastrophic asthmatic patient provides a difficult therapeutic challenge since severe attacks of sudden onset can be anticipated in the future and many patients with this type of asthma can die within a short time of the attack starting. There are a number of important issues concerned in the management of patients with catastrophic asthma, which can be discussed under separate headings but are obviously intimately interrelated in clinical practice.

Identification of possible trigger factors

Hypersensitivity reactions are usually easy to recognize even in retrospect. Drug reactions are sometimes more difficult, particularly those caused by aspirin, non-steroidal anti-inflammatory drugs (NSAIDs) or combinations containing such drugs. It is unlikely that ingestion of a β-blocker would be responsible, since this should have been discovered by the general practitioner or the team responsible for resuscitation. Even if no drug cause is obvious, on general principles aspirin, NSAIDs and all β-blockers should be avoided. A card on which this information is stated should be carried by the patient. The ingestion of peanuts can cause severe bronchoconstriction associated with an anaphylactic-type reaction.

Regular treatment to be taken in order to try to prevent recurrence of the severe episode

It would appear to be prudent for all patients who have experienced a life-threatening episode of asthma to be treated with an inhaled corticosteroid in a dose of at least 800–1000 µg daily. Immediately after a devasting attack long-term prednisolone may be given in an initial dose of 10 mg daily with reductions in dose thereafter by 1 mg decrements every month until treatment is withdrawn or symptoms recur.

Recognition of the onset of a potentially life-threatening attack in the future

Most patients have no difficulty in recognizing the onset of a severe episode but sometimes deterioration can take place over hours or even days before the patient is aware of symptomatic deterioration. Patients with catastrophic asthma should therefore record their PEF regularly in order to pick up any deterioration, which should allow therapeutic intervention at the earliest possible stage. At the onset of a severe attack, large doses of a bronchodilator, e.g. 20–50 doses from a conventional inhaler, should be taken and the patient should seek medical advice immediately.

Self-treatment

As well as being advised to inhale a large dose of a β_2-agonist, perhaps via a large-volume spacer or nebulizer if possible, selected patients should be provided with preloaded syringes containing either terbutaline or adrenaline for subcutaneous or intramuscular injection. A number of such syringes should be provided so that they can be kept in strategic places such as home, briefcase, schoolbag, glove compartment of car, office drawer, etc. Oral prednisolone should also be taken in a single dose of 50 mg as soon as the onset of a severe attack has been recognized.

Admission to hospital as soon as possible

Some hospitals run formal self-admission services that have been shown to save lives.[60] If possible, arrangements for self-admission should be made for all patients who have experienced a rapid onset episode of life-threatening asthma. Preferably, admission should be by ambulance since this will allow nebulized salbutamol or terbutaline in oxygen to be given by the ambulance crew during the journey to hospital.[61,62] The ambulance crew should alert the receiving hospital of the imminent arrival of a patient with severe asthma in order that the hospital can be prepared for resuscitation and ventilation if necessary.

OTHER MEASURES

Antibiotics

Antibiotics have somehow become almost routine in the treatment of severe asthma even though there is no evidence that they are of any value.[63,64] Patients with severe acute asthma should only be given antibiotic therapy when there is an absolute indication for this treatment, such as pneumonia or overt bronchial infection.

Mucolytics

There is no evidence that the mucolytic drugs presently available have any beneficial effects in any of the diseases for which they are recommended. They are of no value in the treatment of asthma.

Bronchial lavage

The mechanical removal of mucus plugs by lavaging the bronchi with saline via an endotracheal tube or bronchoscope has been recommended in patients who require assisted ventilation. This technique has been reported to be associated with serious bronchopulmonary infection.[55] Therefore, the routine use of bronchial lavage in patients requiring assisted ventilation can no longer be advocated.

Physiotherapy

Patients with severe acute asthma cannot cough efficiently, because of severe generalized airflow obstruction. The physiotherapist has no role in the management of the severely ill patient in terms of assisted coughing and expectoration.[2]

Magnesium sulphate

Magnesium sulphate was first reported to have a bronchodilating action in 1936[65] and recently there has been a resurgence of interest in its possible role in the treatment of severe acute asthma.[66,67] These reports suggest that intravenous magnesium sulphate in doses of 1.2–3.0 g has a beneficial effect in patients with severe asthma and that it is free from serious adverse effects, although this has not been confirmed[68] and it is unlikely that this drug will have a major role in the management of acute severe asthma.

SUMMARY

All patients with severe acute asthma should survive if they reach hospital alive. Unfortunately, many patients die outside hospital or in ambulances travelling to hospital. Every attempt should be made to speed up the admission process for all patients and the organization of more formal self-admission services should be encouraged.[60] Family physicians should be persuaded to treat patients in their own homes with bronchodilators and corticosteroids, and ensure that oxygen in high concentration is administered by the ambulance crew if admission to hospital is necessary. The facility now exists in many regions of the UK for a β_2-adrenoreceptor agonist to be given in ambulances and full use of this service must be encouraged. Ideally, all ambulances should be equipped with nebulizers and all ambulance personnel trained to administer a nebulized β_2-adrenoreceptor agonist in oxygen. Whenever possible PEF should be measured before nebulized therapy is administered to provide hospital staff with objective evidence of the response to this treatment.

Hospitals accepting patients with severe acute asthma should be fully equipped for respiratory resuscitation and assisted ventilation.

REFERENCES

1. Knowles GK, Clark TJH: Pulsus paradoxus as a valuable sign indicating severity of asthma. *Lancet* (1973) **ii**: 1356–1359.
2. British Thoracic Society, Research Unit of the Royal College of Physicians of London, King's Fund Centre, National Asthma Campaign: Guidelines for management of asthma. 2. Acute severe asthma in adults and children. *Thorax* (1993) **48**(Suppl): S1–S24.
3. Carruthers DM, Harrison BDW: Arterial blood gas analysis or oxygen saturation in the assessment of acute asthma? *Thorax* (1995) **50**: 186–188.
4. Benastar SR: Fatal asthma. *N Engl J Med* (1986) **314**: 423–429.
5. Eason J, Markowe HLJ: Controlled investigation of deaths from asthma in hospitals in the North East Thames Region. *Br Med J* (1987) **294**: 1255–1258.
6. Rebuck AS, Read J: Assessment and management of severe asthma. *Am J Med* (1971) **51**: 788–798.
7. Goldstein RS, Young J, Rebuck AS: Effect of breathing pattern on oxygen concentration received from standard face masks. *Lancet* (1982) **ii**: 1188–1190.
8. Flenley DC: In *Respiratory Medicine*. London, Baillière Tindall, 1990.
9. Harrison BDW: In Brewis RAL, Corrin B, Geddes DM, Gibson GJ (eds) *Respiratory Medicine*. London, WB Saunders, 1995, 1211–1238.
10. Cochrane GM: Acute severe asthma: oxygen and high dose β agonist during transfer for all? *Thorax* (1995) **50**: 1–2.
11. O'Driscoll BR, Cochrane GM: Emergency use of nebulised bronchodilator drugs in British Hospitals. *Thorax* (1987) **42**: 491–493.
12. Crompton GK: Nebulised or intravenous beta$_2$ adrenoreceptor agonist therapy in acute asthma? *Eur Respir* (1990) **3**: 125–126.
13. Fitchett DH, McNicol MW, Riordan JF: Intravenous salbutamol in the management of status asthmaticus. *Br Med J* (1975) **1**: 53–55.
14. Streeton JA, Morgan BE: Salbutamol in status asthmaticus and severe chronic obstructive bronchitis. *Postgrad Med J* (1971) **47**(Suppl 47): 125–128.

15. Bloomfield P, Carmichael J, Petrie GR, *et al.*: Comparison of salbutamol given intravenously and by intermittent positive pressure breathing in life-threatening asthma. *Br Med J* (1979) **1**: 848–850.
16. Swedish Society of Chest Medicine: High dose inhaled versus intravenous salbutamol combined with theophylline in severe acute asthma. A multicentre study of 176 patients. *Eur Respir J* (1990) **3**: 163–170.
17. Cheong B, Reynolds SR, Rajan G, Ward MJ: Intravenous beta agonist in severe acute asthma. *Br Med J* (1988) **297**: 448–450.
18. Douglas JG, Rafferty P, Fergusson RJ, *et al.*: Nebulised salbutamol without oxygen in severe acute asthma: how effective and how safe? *Thorax* (1985) **40**: 180–183.
19. Campbell IA, Hill A, Middleton H, *et al.*: Intermittent positive-pressure breathing. *Br Med J* (1978) **1**: 1186.
20. Fergusson RJ, Carmichael J, Rafferty P. *et al.*: Nebulised salbutamol in life-threatening asthma: is IPPB necessary? *Br J Dis Chest* (1983) **77**: 255–261.
21. Kelly HW, McWilliams B, Katz R, Crowley M, Murphy S: Safety of frequent high dose nebulized terbutaline in children with acute asthma. *Ann Allergy* (1990) **64**: 229–233.
22. Katz RW, Kelly HW, Crowley M: Safety of continuously nebulized albuterol for bronchospasm in infants and children. *Pediatrics* (1993) **92**: 666–669.
23. Moler FW, Hurwitz ME, Custer JR: Improvement in clinical asthma score and $PaCO_2$ in children with severe asthma treated with continuously nebulized terbutaline. *J Allergy Clin Immunol* (1988) **81**: 1101–1109.
24. Shrestha M, Bidadi K, Gourlay S, Hayes J: Continuous vs intermittent albuterol, at high and low doses, in the treatment of severe acute asthma in adults. *Chest* (1996) **110**: 42–47.
25. Morgan MDL, Singh BV, Frame MH, Williams SJ: Terbutaline aerosol given through pear spacer in acute severe asthma. *Br Med J* (1982) **285**: 849–850.
26. Ward MJ, Fentem PH, Roderick Smith WH, Davies D: Ipratropium bromide in acute asthma. *Br Med J* (1981) **1**: 598–600.
27. Leahy BC, Gomm SA, Allen SC: Comparison of nebulised salbutamol with nebulised ipratropium bromide in acute asthma. *Br J Dis Chest* (1983) **77**: 159–163.
28. Rebuck AS, Chapman KR, Abboud R, *et al.*: Nebulised anticholinergic and sympathomimetic treatment of obstructive airways disease in the emergency room. *Am J Med* (1987) **82**: 59–64.
29. Summers QA, Tarala RA: Nebulised ipratropium in the treatment of acute asthma. *Chest* (1990) **97**: 430–434.
30. Patel KR, Tullet WM: Bronchoconstriction in response to ipratropium bromide. *Br Med J* (1983) **286**: 1318.
31. Connolly CK: Adverse reaction to ipratropium bromide. *Br Med J* (1982) **285**: 934–935.
32. Mann JS, Howarth PH, Holgate ST: Bronchoconstriction induced by ipratropium bromide in asthma: relation to hypotonicity. *Br Med J* (1984) **289**: 469.
33. Beasley CRW, Rafferty P, Holgate ST: Bronchoconstrictor properties of preservatives in ipratropium bromide (Atrovent) nebuliser solution. *Br Med J* (1987) **294**: 1197–1198.
34. Williams SJ, Parrish RW, Seaton A: Comparison of intravenous aminophylline and salbutamol in severe asthma. *Br Med J* (1975) **4**: 685.
35. Femi-Pearse D, George WO, Illechukwu ST, *et al.*: Comparison of intravenous aminophylline and salbutamol in severe asthma. *Br Med J* (1977) **1**: 491.
36. *British National Formulary*, No. 30. London, The Pharmaceutical Press, 1995.
37. Weinberger M, Hendeles L: Use of theophylline for asthma. In Clark TJH, Godfrey S (eds) *Asthma*, 2nd edn. London, Chapman & Hall, 1983, pp 336–357.
38. Bordley JE, Carey RA, Harvey AM, *et al.*: Preliminary observations on the effect of adrenocorticotropic hormone (ACTH) in allergic diseases. *Bull Johns Hopkins Hosp* (1949) **85**: 396–398.
39. Medical Research Council: Controlled trial of effects of cortisone acetate in status asthmaticus. *Lancet* (1956) **ii**: 803–806.
40. Luksza AR: Acute severe asthma treated without steroids. *Br J Dis Chest* (1982) **76**: 15–19.
41. Anonymous: Acute asthma. *Lancet* (1986) **i**: 131–133.
42. Fanta CH, Rossing TH, McFadden ER: Glucocorticoids in acute asthma: a critical controlled trial. *Am J Med* (1983) **74**: 845–851.

43. Britton MG, Collins JV, Brown D, *et al.*: High-dose corticosteroids in severe acute asthma. *Br Med J* (1976) **2**: 73–74.
44. Tanaka RM, Santiago SM, Kuhn GJ, *et al.*: Intravenous methylprednisolone in adults in status asthmaticus. *Chest* (1982) **4**: 438–440.
45. Bowler SD, Mitchell CA, Armstrong JG: Corticosteroids in acute severe asthma: effectiveness of low doses. *Thorax* (1992) **47**: 584–587.
46. Shee CD: Risk factors for hydrocortisone myopathy in acute severe asthma. *Respir Med* (1990) **84**: 229–233.
47. Collins JV, Clark TJH, Brown D, Townsend J: Intravenous corticosteroids in treament of acute bronchial asthma. *Lancet* (1970) **ii**: 1047–1050.
48. Harrison BDW, Stokes TC, Hart GJ, *et al.*: Need for intravenous hydrocortisone in addition to oral prednisolone in patients admitted to hospital with severe asthma without ventilatory failure. *Lancet* (1986) **i**: 181–184.
49. Morell F, Orriols R, de Gracia J, Curull V, Pujol A: Controlled trial of intravenous corticosteroids in severe acute asthma. *Thorax* (1992) **47**: 588–591.
50. Barnes NC: Effects of corticosteroids in acute severe asthma. *Thorax* (1992) **47**: 582–583.
51. Darioli R, Perret C: Mechanical controlled hypoventilation in status asthmaticus. *Am Rev Respir Dis* (1984) **129**: 385–387.
52. Karetzky MS: Asthma mortality: an analysis of one year's experience, review of the literature and assessment of current modes of therapy. *Medicine (Baltimore)* (1975) **54**: 481–484.
53. Branthwaite MA: The management of severe asthma. In Baderman H (ed) *Management of Medical Emergencies*. Tunbridge Wells, Pitman Medical, 1978, pp 48–56.
54. Higgins B, Greening AP, Crompton GK: Assisted ventilation in severe acute asthma. *Thorax* (1986) **41**: 464–467.
55. Luksza AR, Smith P, Coakley J, *et al.*: Acute severe asthma treated by mechanical ventilation: 10 years' experience from a district general hospital. *Thorax* (1986) **41**: 459–463.
56. Adriana J, Rovenstine EA: The effect of anaesthetic drugs upon bronchi and bronchioles of excised lung tissue. *Anesthesiology* (1943) **4**: 253–262.
57. Robertson CE, Steedman D, Sinclair CJ, *et al.*: Use of ether in life-threatening acute severe asthma. *Lancet* (1985) **i**: 187–188.
58. Maltais F, Sovilh M, Golberg P, Gottfried SB: Respiratory mechanics in status asthmaticus. Effects of inhalational anesthesia. *Chest* (1994) **106**: 1401–1406.
59. Kass JE, Castiotta RJ: Heliox therapy in acute severe asthma. *Chest* (1995) **107**: 757–760.
60. Crompton GK, Grant IWB, Bloomfield P: Edinburgh Emergency Asthma Admission Service: report on 10 years' experience. *Br Med J* (1979) **2**: 1199–1201.
61. Crompton G: The catastrophic asthmatic. *Br J Dis Chest* (1987) **81**: 321–325.
62. Fergusson RJ, Stewart CM, Wathen CG, Moffat R. Crompton GK: Effectiveness of nebulised salbutamol administered in ambulances to patients with severe acute asthma. *Thorax* (1995) **50**: 81–82.
63. Shapiro GC, Eggleston PA, Pierson WE, *et al.*: Double-blind study of the effectiveness of a broad spectrum antibiotic in status asthmaticus. *Pediatrics* (1974) **53**: 867–872.
64. Graham VAL, Milton AF, Knowles GK, Davies RJ: Routine antibiotics in hospital management of acute asthma. *Lancet* (1982) **i**: 418–420.
65. Rosello JC, Pla JC: Sulfato de magnesio en la crisis de asma. *Prensa Med Argent* (1936) **23**: 1677–1680.
66. Skobeloff EM, Spivey WH, McNamara RM, Greenspon L: Intravenous magnesium sulphate for the treatment of acute asthma in the Emergency Department. *JAMA* (1989) **262**: 1210–1213.
67. Noppen M, Vanmaele L, Impens N, Schandevyl W: Bronchodilator effect of intravenous magnesium sulphate in acute severe bronchial asthma. *Chest* (1990) **97**: 373–376.
68. Matusiewicz SP, Cusak S, Greening AP, Crompton GK: A double blind placebo controlled parallel group study of intravenous magnesium sulphate in acute severe asthma. *Eur Respir J* (1994) **7**: 14s.

45

Management of Asthma in Adults

ANN J. WOOLCOCK

INTRODUCTION

This chapter discusses the long-term management of asthma, with reference to the management of acute attacks where appropriate. Poor understanding of the causes and natural history of asthma and lack of data from long-term controlled trials of different forms of treatment means that there are no universally accepted methods of management. This has led to a variety of treatment practices and patients often receive confusing information from doctors, nurses, pharmacists and asthma educators. Conflicting information about drugs and their side-effects exists in the literature. Furthermore, drugs widely used in one country appear to be not prescribed as often in another.[1] To try to address this confusion, asthma management plans were written in several countries.[2–5] These were followed by international and global plans.[6–8] The purpose of these plans is to provide a basis for a unified approach to management and, eventually, to allow self-management by the patient. Hopefully studies will be undertaken in different countries to determine if management plans are effective in improving the long-term outcome.

In the absence of long-term data about outcomes, these management plans have been written as 'consensus' documents. Most of them are extremely detailed and not easily used by busy doctors. The answers to specific questions frequently asked by doctors, such as when to use which drug, the appropriate doses and criteria for altering the doses, are only partially addressed. Furthermore, the term 'asthma' is sometimes used to mean both the disease and the episodes of airway narrowing, which leads to confusion. Some of these problems are addressed in the six-point management plan shown in Table 45.1. It is based on the National Asthma Campaign in Australia[9] and stresses reducing the severity of the disease using both pharmacological and non-pharmacological measures. The emphasis is mainly on the management of patients with severe persistent asthma, but the principles are the same for all patients. Details of the ways in which factors that trigger and aggravate the disease can be controlled are outlined and the drugs commonly used are described. The management of patients with a poor response to conventional treatment and the likely changes in treatment that will occur in coming years are discussed at the end of the chapter.

CLASSIFICATION OF ASTHMA FOR PURPOSES OF MANAGEMENT

Persistent asthma

The airways narrow too much and too easily in response to a wide variety of provoking stimuli. It varies from mild to life-threatening in severity. Bronchial hyperresponsiveness (BHR)[10] and airway inflammation are present.[11,12]

Episodic asthma (often seasonal)

The airways narrow too much and too easily in response to specific stimuli, such as pollen allergens. Between attacks, airway function and bronchial responsiveness are normal. Episodic asthma is more common in children than adults, but occurs in some patients who are allergic to pollens and grain dusts during the season of exposure. Some patients, particularly children, have episodes of airway narrowing only during viral respiratory infections. The histological changes present during and between episodes of symptoms in patients with episodic asthma have not been reported.

Occupational asthma

The airways narrow in response to a specific substance to which the patient becomes 'sensitized' at the workplace. Usually it is episodic at first, but then becomes persistent. Occupational asthma is uncommon, but it is important to recognize patients with this disease at an early stage because it is potentially reversible. The pathological changes appear to be the same as those seen in other forms of asthma.[13]

Asthma in remission

Routine lung function tests sometimes reveal adults without symptoms, but with well-documented childhood asthma, who have had a small decrease in spirometric function, a mild degree of BHR or an increased response to bronchodilator. There are no data on which to base a decision to treat such individuals, but it would seem important to monitor their airway function over time.

45 Management of Asthma in Adults

AIMS OF MANAGEMENT

(1) To diagnose and classify the disease.
(2) To prevent symptoms.
(3) To prevent the long-term risks, including persistent airflow limitation.
(4) To prevent the side-effects of drugs.
(5) To prevent death.

Although these aims are logical and based on common sense, there have been no long-term trials to determine the best ways of achieving them. It should be recognized that the disease can be completely controlled in almost all patients, and the earlier the diagnosis and introduction of an orientated management plan, the better the outcome achieved. It is generally agreed that the aims of asthma management are largely related to prevention, yet prevention of the disease and of attacks is hardly ever mentioned in articles and book chapters relating to management, even though it is now well established that exposure to allergens is the most important risk factor for developing asthma.[14,15]

ASTHMA MANAGEMENT PLAN

Table 45.1 shows a six-step management plan. The plan stresses the assessment of the severity of asthma (when the patient is not having an attack) and indicates ways to reduce the severity of the disease.

Step 1: Find best lung function

The best achievable values for peak expiratory flow (PEF) and spirometric function, forced vital capacity (FVC) and forced expiratory volume in 1 s (FEV$_1$), are needed as a guide to the long-term control of asthma. The best PEF value obtained is used to determine a 'target' range for PEF monitoring. The PEF that is 90% of the best is calculated and this becomes the 'target' value for the patient to aim to achieve on waking. At the same time the 'best' spirometric value should be recorded at each clinic visit, while the patient is on prednisone. These values serve as reference values for long-term

Table 45.1 Summary of asthma management.

1. Find best lung function
2. Assess severity and type of asthma
3. Maintain 'target' lung function by:
   (a) Optimal medication
   (b) Avoiding causes/triggers/aggravators
4. Write an 'action plan' for exacerbations
5. Educate the patient and family
6. Review regularly

INITIAL ASSESSMENT OF SEVERITY

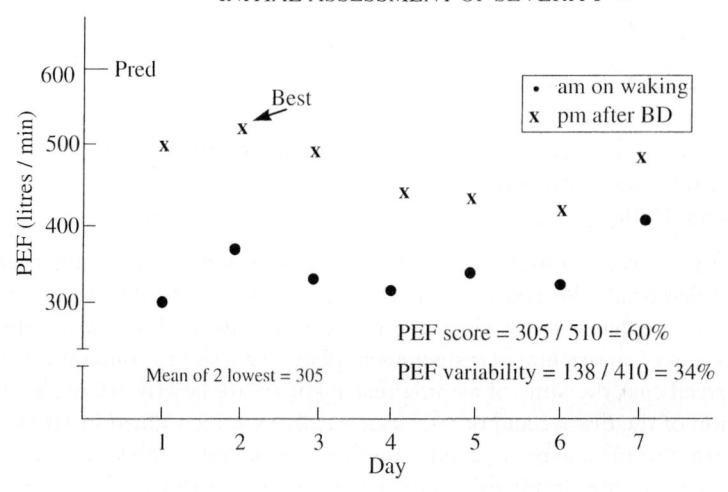

Fig. 45.1 Peak expiratory flow (PEF) readings over 7 days recorded on waking and after bronchodilators in the evening by a 25-year-old Caucasian male, height 176 cm, with severe uncontrolled asthma. The PEF 'score', the mean of the two lowest values expressed as a percentage of the 'best' recorded value, is 60% in this example. The PEF variability, the range divided by the mean for each day and averaged over 7 days, is 34%. Pred, predicted; BD, bronchodilator.

management. If the patient has not achieved a PEF value close to the predicted value (obtained from tables for age, sex, height and race) in a week of monitoring the morning PEF values before bronchodilator and the evening PEF values after using a bronchodilator, as shown in Fig. 45.1, it is possible that the best lung function has not been reached. In such a patient, a trial of oral steroids with PEF monitoring, as shown in Fig. 45.2, should be undertaken. Usually 5 days of prednisone or prednisolone 0.8 mg/kg daily is enough, but if improvement is still occurring after 5 days the medication can be continued for up to 10 days. The prednisone is then stopped and inhaled corticosteroids are continued. It is important to know if residual airflow limitation exists after maximal therapy and the 'target' for the morning PEF value.

Step 2: Assess severity and type of asthma

Severity is assessed from a careful history to exclude occupational or domestic provoking stimuli, from bronchodilator use and morning PEF expressed as a percentage of the recent best as described above.

All other aspects of treatment depend on this step. The process includes confirmation of the diagnosis and classification of the nature and severity of the disease. At present there is no 'gold standard' against which to assess severity. Table 45.2 shows a scoring system that has proved useful. The total score ranges from 1 to 12, with 0–4 for symptoms, 0–4 for bronchodilator use over a 24-h period and 0–4 for variability of PEF rates. The symptoms are wheeze, chest tightness, breathlessness and cough, alone or in combination. The frequency of symptoms is important and, in particular, waking at night regularly with wheezing or coughing is a symptom of severe disease.[16] Documenting

45 Management of Asthma in Adults

FINDING BEST PEAK FLOW RATE

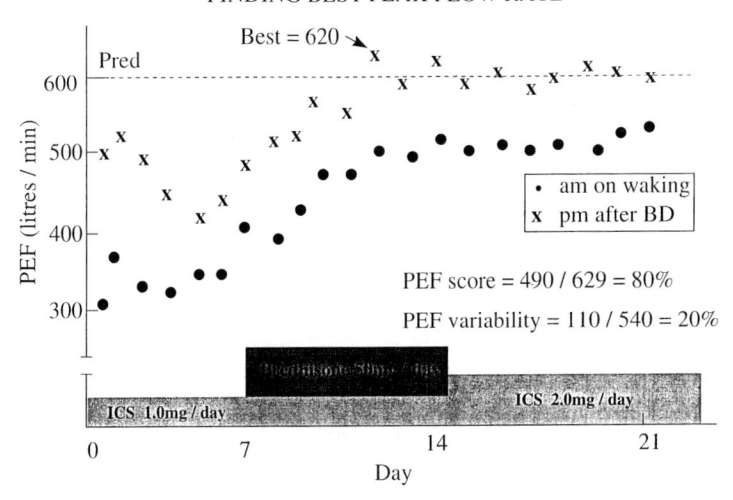

Fig. 45.2 Peak expiratory flow (PEF) recordings, from the same patient as in Fig. 45.1, over a period of 3 weeks including 7 days of oral steroids administered in order to find the 'best' possible PEF in this patient. The 'best' increased from 510 to 620. The PEF score is 80% and the variability is 20% at week 3. ICS, inhaled corticosteroids; Pred, predicted; BD, bronchodilator.

Table 45.2 Asthma severity score (not during an exacerbation).

	Score
Symptoms	
None for >6 months	0
<once/week or only with exercise	1
>daily, <weekly	2
Daily symptoms, occasionally at night	3
Waking at night with symptoms >once/week	4
Bronchodilator required	
None in last year	0
<1 week	1
<daily	2
1–3 times/24 h	3
>4 times/24 h	4
Morning PEF (as percentage of recent best PEF value	
>93	0
>85 <93	1
>78 <85	2
>70 <78	3
<70	4

Maximum possible score = 12
1–3, episodic; 4–5, mild; 6–8, moderate; 9–12, severe
Score >10 suggests that severity may be life-threatening

bronchodilator use helps in the assessment of severity, when the patient has been advised to use them only for symptoms. The presence of daily symptoms, in spite of frequent bronchodilator use can be an indicator of severe disease.

Measurement of morning PEF is important for determining the severity of asthma and for continuing management. The minimum value for morning PEF over a 7-day period, expressed as a percentage of the recent best (or even predicted) value, is a good indicator of the severity of the airway abnormality in most patients.[17,18] Values above 93% are normal, while values below 75% indicate severe disease as shown in Table 45.2. The PEF 'score', illustrated in Figs 45.1–45.3, is compared with the standard PEF 'variability' calculation, which is more cumbersome. Most patients who have moderate or severe disease need to monitor morning PEF while taking inhaled corticosteroids, especially at times when they are decreasing their doses or when they are experiencing more symptoms than usual. A week of readings, recorded both on waking and after bronchodilator use in the afternoons (Fig. 45.1), is enough to provide an initial assessment and accurate reflection of the situation, unless the patient is having, or has just had, a severe exacerbation.

A total score can be calculated and those with a score of more than 4 usually have persistent asthma. When the morning PEF is close to predicted and >93% of the recent best and symptoms are episodic, the patient can be regarded as having episodic asthma. This can be confirmed by a provocation test with histamine or methacholine. Provocation with exercise or non-isotonic aerosols[19–21] can also be used but a negative test dose not exclude the presence of BHR to histamine or methacholine. If the test is normal, the patient can be regarded as having episodic disease and be treated with β-agonists on demand. If BHR is present, the patient should be regarded as having persistent disease. This can be mild, moderate or severe, depending on the score.

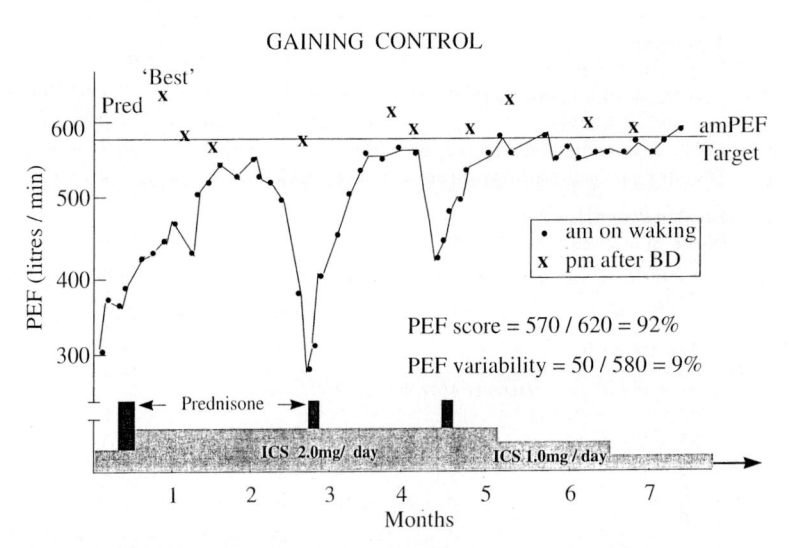

Fig. 45.3 Peak expiratory flow (PEF) values for the patient shown in Figs 45.1 and 45.2 for a period of 7 months. By month 7 the PEF score has reached 92%, variability has decreased to 9% and the dose of inhaled corticosteroids (ICS) has been reduced. There were two exacerbations that required short courses of prednisone. Pred, predicted; BD, bronchodilator.

Step 3a: Maintain 'target' lung function by optimal medication

In patients with persistent disease, drug therapy is prescribed in order to keep the patient free from symptoms and the morning PEF >93% of the recent best. A suggested scheme for drugs, using the severity score as a guideline, is outlined in Table 45.3. The aim is to reduce the severity of the disease by treating the airway inflammation with inhaled corticosteroids, reducing responses to triggers, treating aggravating factors, removing causal factors and addressing lifestyle problems.

Inhaled corticosteroids (ICS)

Mechanism of action. These drugs are effective because they have a number of actions.[22] They are topically active and known to cause vasoconstriction in the skin.[23] This vasocontriction, together with interference with the local production of inflammatory cytokines, are probably key elements in their effectiveness. Biopsy studies show that they restore the bronchial epithelium, which is commonly friable and easily shed.[24] Their effects on cells and structures deeper in the airway wall are unknown, although it appears that they have little effect on the collagen deposited beneath the basement membrane and on the release of leukotrienes from eosinophils and mast cells.

Clinical effects. When ICS are given acutely, they do not inhibit the early or late responses to allergen challenges. However, when given for several days they have some inhibitory effect on these responses. In most patients, particularly those who have not taken the drugs previously, there is a dose-related effect in improving the severity of the disease as measured by symptoms, baseline lung function and morning PEF. Improvement may take several weeks to become apparent to the patient and may continue for many months.[22] In some patients ICS appear to have little effect on BHR.[25,26]

The response to ICS appears to be dose related and the drugs can be given twice daily. Although these drugs have now been in use for many years, there is little published about their long-term effectiveness in individual patients. Clinical experience suggests that relapse occurs in many patients when they are stopped.[27] Finding the correct dose for treatment requires trial and error and the continuing use of diary cards to record symptoms and morning PEF values.

Side-effects. Some adult patients taking ICS experience dysphonia, the cause of which is unclear, and a smaller number develop thrush.[28] These problems can be prevented by the use of a spacer device,[29,30] by reducing the number of inhalations (using high-strength aerosols) and by gargling after use. In some patients antifungal agents are needed to control the local symptoms.

The systemic side-effects of ICS depend on the dose: they are rarely seen with doses below 1.0 mg and are much less than those observed with the doses of oral steroids needed to maintain the same degree of control. Biochemical evidence of adrenal suppression occurs rarely on doses of less than 1.5 mg daily; when it is present, it is probably not medically important but indicates that enough drug is being absorbed to have a systemic effect. More important clinical effects are bruising, osteoporosis and development of cataracts. All these are likely to develop when ICS are used in high doses

Table 45.3 Asthma management: medication (adults).

Drug	Severe	Moderate	Mild	Episodic	Comment
Long-term prevention					
Oral Steroid (prednisone)	1.0 mg/kg daily				Dose for exacerbation
Inhaled corticosteroid	0.8–2.0 mg/day (or more)	0.8–2.0 mg/day	0.2–0.5 mg/day	None	Dose depends on drug, form and device used
Cromoglycate, nedocromil	None	None	Variable doses	None	
LA β_2-agonist (inhaled)	Salmeterol 0.05 mg b.d.	Salmeterol 0.05 mg b.d.	? if night symptoms	None	
(or) Theophylline (slow release)	Dose variable	Dose variable	Dose variable	None	
(or) LA β_2-agonist tablets	Dose variable	Dose variable	Dose variable		
Quick relief					
MA β_2-agonist	p.r.n.	p.r.n.	p.r.n.	p.r.n.	

LA, long acting (12 h); MA, medium acting (4–8 h).
Reproduced from ref. 7, with permission.

for long periods, although there is no evidence that osteoporosis and cataracts occur in the absence of courses of oral steroids. The well-documented side-effects of steroids have caused patients, doctors and pharmacists to maintain an element of steroid phobia that is probably unjustified. Nevertheless, the long-term effects of these drugs has not been documented and it is prudent to continue to reduce the daily dose to the minimum needed to maintain control.

Dose and administration. It is not possible to suggest doses of ICS for varying severity of disease because the drugs vary in potency and the formulation (aerosol or powder) and delivery device used, all of which affect the dose prescribed. In general it makes sense to start with very high doses and to reduce the dose as soon as the severity improves. Nebulized forms (available as budesonide) may be needed initially in those with poor lung function, but the role of this form of the drug in adults is not established. Attention should be paid to teaching each patient how ICS work, how to use them and what to expect. This takes time but is one of the most crucial elements of asthma management.

Refractoriness. Steroid resistance rarely occurs and this is discussed below. To date, refractoriness to ICS has not been described in the way that refractoriness to oral steroids appears to occur, but clinical experience suggests that some patients who initially have good control become more difficult to manage, even when these drugs are used in high doses. It is possible that regular, large doses of β-agonist aerosols lead to worsening control of asthma and thus an apparent refractoriness to ICS. However, other factors, such as leukotriene production, that are unresponsive to steroids may be responsible and this is sometimes seen in aspirin-sensitive asthmatic individuals.

Systemic steroids

Mechanism of action. Corticosteroids, when administered systemically, have similar actions to those described for topical steroids, although their effect on small vessels in the airways is unknown. They take 4–6 h to have an effect and will act on all cells with steroid receptors.

Clinical effects. In the management plan, oral steroids are used as a trial to find the best lung function in patients whose PEF remains lower than the predicted value, and to treat severe exacerbations. In children, it has been shown that a single dose (30 or 60 mg) of prednisone, as well as nebulized bronchodilator, reduces the need for hospital admission;[31] this also happens in adults, although data showing this rapid effect have not been published.

Side-effects. These are well documented and include bruising, osteoporosis, cataracts, hypertension, diabetes and Cushingoid features. Some of these effects are potentially reduced by changing the patient to ICS. It takes many months to change patients to an inhaled form and care must be taken to implement all the other steps in the management plan at the same time.

Dose and administration. For finding the best lung function, it is usual to use 0.8–1.0 mg/kg body weight per day in divided doses for 5–10 days (see Fig. 45.2). Trial and error are needed to determine the symptoms and the morning PEF values that herald an exacerbation and the dose of oral steroid needed to abort a severe attack. Usually 25–50mg in divided doses for 1–2 days is sufficient. It is not necessary to reduce the dose slowly when it has been used for less than 1 week, unless experience shows that sudden withdrawal is associated with worsening asthma. ICS should not be stopped while the patient is on oral steroids. Some patients, usually those who have been on oral steroids before the advent of the inhaled forms, require daily use of oral steroids to control symptoms. However, over a period of many months, sometimes years, it is possible to withdraw all oral steroids and replace them with the newer potent inhaled forms. Some physicians use alternate-day steroids but this often results in a higher 'overall' steroid dose than the use of 'pulses' of high-dose therapy, even if these are necessary at monthly intervals.

β-Agonists

These drugs can be divided into short (1–2 h), medium (4–8 h) and long (12 h) acting. The short-acting are rarely used, the medium-acting are extensively used and the long-acting are increasingly used. β-Agonists are most effective when inhaled, but oral and intravenous forms of the medium-acting class are available. For reasons that are not understood, only the inhaled forms of β-agonists protect against provoked attacks such as exercise. Long-acting drugs are given in lower doses than medium-acting drugs. Long-acting drugs appear to be more effective in controlling asthma than in relieving acute episodes of airway narrowing.

Mechanism of action. By stimulating the β_2 receptor, these drugs increase cyclic AMP and this increases cellular calcium, leading to cell actions described elsewhere in this book. The main action of β-agonists in the airways is to relax (or prevent contraction) of smooth muscle cells. They also stabilize mast cells, preventing release of mediators after mast cells are stimulated. In this respect they are more potent than sodium cromoglycate.[32,33]

Clinical effects. Isoprenaline and ephedrine act for 1–3 h and are largely drugs of the past. They are rarely used in countries where longer-acting β-agonists are available. The first longer-acting β-agonist was orciprenaline, which is no longer used today because of its actions on the β_1 receptor. Fenoterol was then developed. It, too, is less widely used today because it was associated with the increased deaths from asthma in New Zealand in the early 1980s.[34] It is not clear if fenoterol was to blame for the deaths, which began to decrease before its market share fell, but it does seem likely that fenoterol has effects different from the other medium-acting drugs salbutamol and terbutaline. Salbutamol and terbutaline act within minutes, last for about 6 h and appear to be extremely safe. They are used by large numbers of asthmatic patients without major side-effects. As aerosols they protect against provoking stimuli for about 2 h. No tachyphylaxis to their bronchodilating effects has been found in asthmatic patients.[35]

The question as to whether salbutamol and terbutaline affect the severity of asthma when they are used regularly has not been resolved completely. In children, BHR was shown to increase when terbutaline was used alone in contrast to inhaled steroids which improved the severity.[36,37] A year-long study in New Zealand showed that fenoterol four times a day was associated with worsening control of asthma in 40 of 64 subjects who completed a trial.[38] However, the effects demonstrated by fenoterol may not apply to all β-agonists. However, it is clear that when used alone, β-agonists do not improve the overall severity of asthma. Nevertheless, many people with episodic or mild disease use them to control symptoms and the severity of their asthma dose not increase. Sales of salbutamol have been increasing worldwide in the last 20 years and there is little objective evidence that this drug has caused any problem.

Side-effects. Tremor and slight tachycardia occur acutely and are well known. These effects usually decrease with time and are rarely a problem unless the drugs are used to excess.[39] Attempts to demonstrate tachyphylaxis to the bronchodilating effects of salbutamol in asthmatic airways have been unsuccessful, although it may occur in non-asthmatic patients.[35]

Dose and administration. These drugs are usually given in the inhaled form with doses varying from 100–200 µg. Metered-dose inhalers, nebulizing solutions and dry powder forms are available in addition to tablets and syrups. β-Agonists are used for making the diagnosis (if the FEV_1 increases by 15% within 10 min of an aerosol bronchodilator this is diagnostic of asthma), for finding the 'best' lung function, for severe symptoms lasting more than 10 min and for reversing airway obstruction (e.g. when the PEF is <60–70% of the target PEF) (Table 45.4). Medium-acting β-agonists are not needed on a regular basis except in those patients whose symptoms cannot be controlled by a combination of other drugs, including ICS and long-acting β-agonists.

Long-acting β-agonists

These act for 12 h[40,41] and, in the case of salmeterol, appear to have actions other than bronchodilatation.[42] Their overall place in the management of patients with asthma is continuing to be defined. They have a stabilizing effect that allows the dose of ICS to be reduced. Salmeterol is more widely used than formoterol at present and is best used in low doses (50 µg twice daily); higher doses are rarely more effective. The drug needs to be used with care and its actions carefully explained. It is not a 'reliever' and does not alleviate asthma symptoms in the short term. The greatest advantage of these drugs is in preventing night-time symptoms and improving the quality of sleep. Side-effects do not appear to be a problem. Tachyphylaxis to bronchodilatation has not been demonstrated but there is tachyphylaxis to the protecting effects of salmeterol against stimuli such as exercise after the first few days of treatment.[43]

Table 45.4 Indications for use of β-agonist aerosols.

Acute severe attacks: these drugs may be life saving
For diagnosis: does lung function improve within minutes?
For assessment of severity: amount required plus PEF variability
For symptoms causing distress or anxiety
Before exercise to prevent airway narrowing

Sodium cromoglycate (SCG) and nedocromil sodium (NS)

SCG has been available for nearly 30 years. It prevents airway narrowing induced by allergens, exercise, SO_2 and other irritants. It may have a small effect on airway inflammation, perhaps by preventing inflammation from worsening during the allergen season. NS is newer and more potent and has been developed for adults. It is effective in reducing cough and has a small effect on BHR.

Mechanisms of action. In spite of their widespread use and a large body of literature, little is known about how these drugs work. They stabilize isolated mucosal mast cells and reduce the release of mediators,[32] but the extent to which they do this *in vivo* is not known. Recently NS has been shown to affect the chloride channel and it seems likely that their ability to antagonize the effects of osmotic changes in the airways may be due to this effect.[20,21] It is likely that they act on other inflammatory cells and afferent nerve endings, as well as on mast cells. *In vitro* NS appears to be more potent in preventing mast cell degranulation than SCG, and it may be slightly more potent in preventing attacks.

Clinical effects. In the short term, these drugs prevent exercise-induced attacks in many adults and most children. The degree to which they inhibit attacks is dose related and short-lived, because the drugs are rapidly removed from the airways during exercise. Taken before an allergen challenge they prevent both early and late reactions.[44] They also inhibit the effects of other 'indirect' provoking stimuli such as SO_2. In this respect, NS is more effective than SCG.

In the long term, in adults and children with episodic asthma these drugs prevent seasonal 'spontaneous' exacerbations so that the episodes are inhibited or decreased. SCG has the reputation of being more effective in children, probably because episodic asthma is common in children. In some studies a decrease in BHR has been observed while patients are taking SCG regularly.[45] It appears that the drug must be given in adequate doses for at least 12 weeks. In patients already taking inhaled corticosteroids, the response to SCG appears not to be as good as that to inhaled steroids,[45] although there is some evidence that NS improves the overall control of the disease in such patients.[46]

Side-effects. These drugs have virtually no side-effects. The dry powder form of SCG sometimes causes minor irritation and some patients complain that NS has an unpleasant taste.

Doses and administration. SCG is usually administered as 5 mg per puff and NS as 4 mg per puff from metered-dose aerosols; two puffs are inhaled three or four time per day. SCG is also available as 20-mg Spincaps and 20-mg nebulizer solution. These drugs can be used immediately prior to exposure to known triggers to prevent attacks.

For long-term use, they need to be given in an adequate dose and at regular intervals, initially four times a day. Once control is improved they are used three times a day. In children with moderate disease, e.g. a score of 6, the disease can be usually controlled with SCG alone. In those with a score of 7 or more, experience shows that ICS are usually needed to obtain control. The question then arises, should SCG or NS be used as well? In long-term management, a trial (adequate doses essential over a period of weeks) is indicated to determine if the overall score is improved while the patient stays on a fixed dose of ICS. In adults, once the severity is reduced (score of 8 or less), SCG or NS can be introduced as a trial to try to reduce the dose of ICS.

Anti-cholinergics

Ipratropium bromide and oxitropium are used to block cholinergic receptors.[47] They are effective bronchodilators and are useful in patients with chronic obstructive pulmonary disease. They have a slower onset than the β-agonists. They are used in conjunction with β-agonists in the treatment of acute attacks, particularly in children. At present they have little role in long-term management of asthma, although they sometimes have a place in patients in whom constant use of medium-acting β-agonists causes worsening control of the disease and who need a regular bronchodilator.

Theophylline

Mechanism of action. This drug has many actions.[48,49] The mechanism by which it bronchodilates is unknown, although it appears to increase intracellular cyclic AMP by a mechanism different from the β-agonists. It may also have some 'anti-inflammatory' effects,[50] although the mechanisms are largely unknown. There is no place for theophylline as a second drug when a β-agonist has failed to control the disease. If the patient needs more than occasional β-agonist aerosols then either ICS or NS, not theophylline, are indicated.

Clinical effects. Theophylline gives symptomatic relief to those with severe airway narrowing. There is a well-recognized group of subjects, usually dependent on long-term or frequent short-term courses of oral steroids, who need theophylline to control their symptoms. Many of these patients need doses above the usual to keep serum levels in the 'therapeutic' range. These patients apparently metabolize the drug differently. Its action in relieving symptoms is unknown. There are some patients who get immediate symptomatic relief from the drug and use it either intermittently or as a single dose at night.

Side-effects. These are well known and include nausea, headache and hyperactivity in children. The drug can cause fatal neurological and cardiovascular events if given in too high a dose. The increasing awareness of side-effects of this drug has led to its decreasing use in some countries. It has been reported to cause learning difficulties in children, perhaps because it interferes with sleep. However, side-effects can be avoided by using the drug in lower doses as an adjunct to other therapy, rather than as a primary bronchodilator. It is likely that the exact place of theophylline (which is cheap) will be defined in coming years and the side-effects largely avoided.

Dose and administration. When used as a bronchodilator, particularly in treating severe attacks, doses are adjusted to keep the serum levels within the 'therapeutic' range, therefore avoiding toxicity. The slow-release forms are most effective in reducing large fluctuations in serum levels. In patients with severe disease, especially those requiring low doses of oral prednisone in addition to ICS, they appear to have a steroid-sparing effect and can be used in lower doses. The effort and expense of monitoring serum levels is not required for low-dose treatment.

Other treatments

Other drugs are rarely needed in the long-term management of asthma. However, a wide variety of drugs are available.

Antihistamines. In general these have no place in the treatment of asthma. However, the non-sedating antihistamines have few side-effects and are useful in subjects with other allergic problems that often accompany asthma, for example urticaria, allergic reactions to foods and rhinitis. A number of patients with seasonal exacerbations of asthma find that regular use of antihistamines during 'the season' controls their symptoms.[51]

Leukotrienes and 5-lipoxygenase inhibitors. These drugs have been shown to improve symptoms and lung function during 4–6 weeks of therapy.[52–54] The place of these drugs in treatment is not yet clear. Because ICS do not affect the production of leukotrienes,[55] these drugs may have an added effect when used with ICS, but trials of this are not yet available. As with antihistamines, they are given orally and can be used once or twice daily. They have the potential to make a big advance in the control of asthma, especially in some forms of disease such as aspirin-sensitive asthma.

Ketotifen. This drug is marketed as an antihistamine and is widely used.[1,56] There are no published trials that show that it is effective in reducing the severity of the disease in those with moderate and severe persistent asthma; however, it has some 'anti-inflammatory' properties *in vitro* and in patients with mild disease can reduce symptoms to the point where other drugs are not necessary.

Gold salts. These drugs, sometimes used for the treatment of arthritis, are used in some countries for the treatment of patients with severe asthma.[57] There has been no trial published of the effects on patients already enrolled in a management plan that maximizes the use of inhaled steroids and other measures aimed at reducing severity.

Methotrexate. This drug is used commonly in the USA in patients who need high doses of oral steroids to control their asthma.[58–62] It has anti-inflammatory actions and in some subjects can be used to reduce the dose of prednisone without causing too many side-effects. Discontinuing methrotrexate usually leads to a relapse. As with gold, there has been no trial in patients who have first been treated with a strict management plan that includes high doses of ICS, as described above. Overall, its inconsistent effectiveness and side-effects lead to the conclusion that it probably has little place outside the occasional patient who responds well and does not develop side-effects.

Cyclosporin A. There is one controlled trial of cyclosporin (5 mg/kg daily) in steroid-dependent asthmatic patients.[63] Overall, the small improvements achieved seem to be outweighed by the side-effects and cost. Anecdotal use has not shown it to be a big advance.

45 Management of Asthma in Adults

Other drugs. There are many reports of attempts to use other anti-inflammatory agents, including colchicine,[64] macrolide antibiotics[65] and chloroquine.[66] They are of research interest and not needed where the drugs described in the management plan are available and used effectively.

Non-pharmacological treatments

Exercise. Exercise should be encouraged in all patients; swimming appears to be particularly beneficial. In patients who have severe disease with panic attacks and hyperventilation, swimming helps to teach them to control their breathing rate. There are numerous studies that show the effectiveness of exercise in improving overall well-being.[67,68] The acute symptoms of exercise-induced asthma can be prevented with SCG, NS or a bronchodilator aerosol prior to the exercise.[69,70]

Other measures. Acupuncture, hypnosis, meditation, Chinese herbal medicine, garlic tablets and many other remedies are used frequently by patients with asthma. These measures are almost never successful in reducing the basic severity of the disease, but often the patient feels better. Patients should not be discouraged from exploring these treatments, provided that the other measures outlined in the management plan are continued.

Step 3b: Maintain 'target' lung function by avoiding causes/triggers/aggravators

Causes

These are rarely known except for occupational sensitizers and some allergens. Once the relevant allergen/sensitizer is known, complete avoidance measures should be taken, particularly before the disease becomes persistent.

Triggers

Physical factors. These include exercise, strong smells, cold, changes in the weather, etc. They are usually easily identified by the patient. The effect of most can be minimized by appropriate use of SCG, NS or β-agonist aerosols before a known exposure, for example before exercise.

Allergens. There is no doubt that aeroallergens are important triggers of attacks in most allergic asthmatic individuals. There is also increasing evidence that asthma is caused by exposure to allergens[71] and that avoidance improves the severity of the disease.[72-74] Allergen exposure in the first years of life may be particularly important in causing severe disease in children. This, together with the fact that asthma rarely completely reverses once it becomes persistent, means that the most rational approach to asthma management is prevention. Allergen avoidance is important for families with a history of allergic disease. The allergens that appear to be important throughout the world are dust, mites, moulds, pollens and animal proteins. Large amounts of these allergens can be found in house dust, particularly in carpets. House-dust mites, both alive and dead, present the biggest problem. Present evidence suggests that most mattresses and pillows harbour mites, apart from areas with very low humidity. It seems likely that allergens are constantly inhaled and lead to continuing inflammation of the airways. The most sensible solution for patients is to live in an allergen-free environment as much as possible.

Recommendations for reducing allergen levels in houses.
(1) The humidity in houses should be minimized by good ventilation. This is probably the most important measure.
(2) All bedding (mattress, pillow and doonas (duvets)) that cannot be washed regularly in hot water should be encased in allergen-proof covers; bedding that can be washed regularly should be washed at 60°C. Sunlight is excellent and, whenever practical, bedding should be put in the sun to kill mites. Wall-to-wall carpet should be removed from bedrooms where possible. It is almost impossible to make a house in a humid environment allergen-free while carpets remain. It should be explained to the patient that vacuum cleaning the carpet removes no more than a small percentage of mites (which have legs designed to stick to carpet fibres). Furthermore, antimite sprays must be used frequently and in a quantity that penetrates the carpet completely.
(3) Sheepskins should be discarded or washed frequently with water at a temperature of at least 60°C.
(4) Babies should not be put directly on to a carpet, but rather on a cover that can be kept free from mites and allergen by hot washing.
(5) Cats and dogs should be removed completely. Cats kept outside the house remain an important source of allergen.
(6) Clothes should be kept in cupboards and washed or dry cleaned frequently.

Aggravators

It should be routine for the doctor to ask the patient about symptoms of rhinitis (nasal obstruction or sneezing), snoring and interrupted sleep, and gastric reflux (heartburn). These problems, when treated, help the overall well-being of patients and though they may have only a small effect on the underlying severity of the asthma, they cannot be ignored. β-Blockers and aspirin are also important aggravators that must be constantly asked about.

Step 4: Write an action plan for exacerbations

A written action plan is necessary for all patients, even if they have infrequent exacerbations. The general aim is to prevent a severe attack by increasing the dose of inhaled steroids or starting high-dose oral steroids. There is no place for low doses of oral steroids in preventing attacks. The plan for the patient whose PEF values are shown in Figs 45.1–45.3 is shown in Table 45.5. The plan should be written down and the patient should be able to recognize an impending attack, know which drugs to take and how to call for help.

45 Management of Asthma in Adults

Table 45.5. An action plan for exacerbations for the patient shown in Fig. 45.2 based on the peak expiratory flow (PEF) values shown in there. Suggested changes to treatment if the morning PEF falls below the target value are written as a guide to the patient.

PEF reading 'Best' = 620 Morning target = 580		How you feel (symptoms)	What to do
>550		Fine	Continue regular treatment
<550	or	'Tight' on waking	Double doses of corticosteroids
<500	or	Waking at night with asthma symptoms	Start prednisone 25 mg b.d.
			Take bronchodilator puffer
<400	or	Severe attack Take lots of bronchodilator	Take prednisone 50 mg
			Seek help

Treatment of acute attacks

If, in spite of the written plan or because the plan was not used, the patient has a severe attack, the guidelines outlined by the British Thoracic Society[75] or the Canadian Medical Association[76] should be followed. The patient should be monitored with an ear oximeter and PEF measurements and oxygen given. The drugs needed are oral or intravenous steroids in adequate doses, nebulized bronchodilating drugs and, if the response is poor after several hours, parenteral bronchodilators, usually intravenous salbutamol. Intravenous aminophylline is rarely used now, but many patients respond to it well.

Step 5: Education

Controlled trials rarely show that educating patients with asthma leads to decreased severity or improved control.[77-83] However, unless the patient, the family and, in the case of children, the teacher most involved with the child understand the nature of asthma, the aims of treatment and the management plan, full control of the disease will not be achieved. This is especially important in those with a score of more than 6. Studies of the role of education when a strict management plan is used are awaited.

Education takes time, which may not be available to all doctors. This means that an education programme should be developed in association with hospitals where asthma is treated. Each doctor needs a kit that includes pictures of airways, drugs and a number of peak flow meters. The latter can be lent until it is decided whether the patient needs one permanently (score of 6 or more). Figure 45.4 shows a way of illustrating the nature of the problem that can be understood by most people.

Step 6: Review regularly

Asthma is a chronic disease that carries a number of long-term risks, including continuing symptoms, altered lifestyle, the development of permanently abnormal lung function, the

ASTHMA

Airway - cross-section

Normal   **Asthma**   **Attack**

Anti-inflammatories   Bronchodilators

- Allergen avoidance
- Inhaled steroids

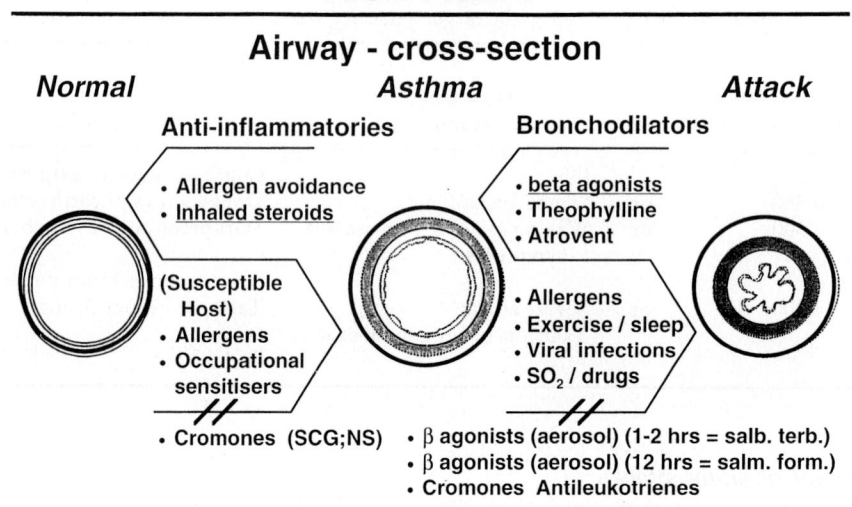

- beta agonists
- Theophylline
- Atrovent

(Susceptible Host)
- Allergens
- Occupational sensitisers

- Allergens
- Exercise / sleep
- Viral infections
- SO_2 / drugs

- Cromones (SCG;NS)
- β agonists (aerosol) (1-2 hrs = salb. terb.)
- β agonists (aerosol) (12 hrs = salm. form.)
- Cromones Antileukotrienes

Fig. 45.4 Mechanisms thought to be present in causing persistent asthma and attacks of airway narrowing. The role of different classes of drugs in treating the disease and in reversing and preventing attacks is shown. This diagram is useful for patient education. The actions of the drugs are discussed in the text.

side-effects of drugs and premature death. In those with moderate or severe persistent asthma (a score of 6 or more), it almost never remits and this means continuing care. Regular visits to the doctor, at which the diaries are reviewed, are essential. At these visits a new score (hopefully lower) is calculated, treatment changed, the action plan updated, drugs prescribed and education continued. Figure 45.3 shows the progress of the patient illustrated in Figs 45.1 and 45.2, who was treated using the management plan for 6 months. Some patients improve more rapidly than this and others take much longer. In time, patients can take more responsibility for their disease, but in the first year of treatment regular appointments, regardless of symptoms, must be made. Every patient with persistent asthma should have one doctor who is committed to her/his care for life.

TREATMENT OF THE PATIENT WITH SEVERE PERSISTENT ASTHMA

In some patients, the plan fails to reduce the severity of asthma. The reasons for this include severity and chronicity of the disease, lifestyle, lack of compliance or a combination of these. In this group of patients, long-term commitment is needed and each step in the management plan should be implemented carefully. In addition, it is helpful to ask the following questions.

(1) Does the patient have asthma? Sometimes a diagnosis is assumed in a patient who is hyperventilating, has similar symptoms or has bronchiectasis or cystic fibrosis. A biopsy is usually diagnostic and can give helpful information.

45 Management of Asthma in Adults

(2) Is the patient complying with drugs? Some patients deliberately avoid using the drugs, others forget or take them incorrectly. This may be difficult to determine, but often explains the lack of improvement. Objective studies of compliance show that the majority of patients fail to take their medications as prescribed.[84]

(3) Have the relevant allergens and aggravators been removed/treated? In particular, does the patient snore or have severe reflux? Treating the rhinitis usually allows the patient to sleep better and treatment of severe snoring is very helpful.[85]

(4) Does the patient respond to high doses of prednisone using objective measures? If so, then adequate doses of inhaled plus oral pulse therapy are needed. If not, the diagnosis should be questioned. The patient may have chronic fixed airway obstruction and be incapable of improvement.

(5) Is the environment a causal factor? If the patient knows that in a different house, region or country, asthma control is improved then a trial of living in a new environment is indicated.

(6) Would a trial of gold salts or methotrexate help? A trial with one of these drugs is indicated if the management plan has been implemented and drugs complied with for at least a year. In addition, it should be demonstrated that steroid therapy is effective but needed in amounts that cause excessive side-effects before these potentially toxic agonists are used.

(7) Does the patient have true steroid-resistant asthma? This condition is extremely rare and can only be recognized after addressing the previous points. Such patients should be investigated and steroids kept at low doses while other treatments are used.[86]

(8) Does the patient have abnormal illness behaviour? Some patients need to be sick and will only respond to asthma management if they develop another illness. A variety of psychological problems occur in patients with severe persistent asthma and the help of a psychiatrist can be of great benefit.

LIKELY FUTURE CHANGES TO MANAGEMENT

Asthma continues to be managed without the benefit of a systematic approach and without the results of long-term trials of treatment, in which the effect of specific management plans on the outcome of the disease are defined. Until such trials are done, it is unlikely that dramatic progress will be made. The following list outlines some of the likely changes to the overall approach to management of asthma, based on results of recent studies and on the increasing use of management plans.

(1) More emphasis will be placed on investigation of infants and children at the time of the first wheeze with a view to prevention.

(2) Less emphasis will be placed on the use of medium-acting β-agonists except in episodic disease.

(3) There will be increasing use of long-acting β-agonists. These drugs may replace the presently used β-agonists in patients with severe disease, especially those who wake at night. Long-acting β-agonists used in conjunction with ICS may have a role in 'aborting' the disease if used soon after symptoms begin.

(4) ICS will be used more precisely. These drugs are expensive and it is increasingly

realized that they have side-effects and may induce a state of partial refractoriness to higher doses. At present the doses used are largely arbitrary. It is possible to write protocols for determining the dose that is most likely to be effective in the individual patient for both initial control and long-term use. Short courses of high doses may be used for several weeks followed by low doses, avoiding the side-effects of continued high doses of ICS.

(5) Alternatives to metered-dose aerosols that need coordination will be used. Dry powder dispensers and mini nebulizers will become more sophisticated.
(6) New non-steroidal anti-inflammatory medication will be developed. Although not yet available, it makes sense to develop drugs that help to control the allergic reaction in the airways so that less cytokines are released and inflammation prevented.
(7) Self-management by each patient. This is already happening and will continue as management plans become widely introduced.
(8) More emphasis on non-pharmacological treatments. The role of controlling the patient's home environment, of swimming and exercise programmes and of relaxation will be much more widely accepted. It is possible that not treating some patients with mild, episodic asthma may be beneficial in the long term.
(9) Large population studies of management plans to determine those treatments that lead to the best long-term outcome will be undertaken.
(10) As the cost of drugs increases, patients, doctors and governments will need to be assured that the drugs being used have definite effects to prevent or treat symptoms or to reduce the severity of the disease. Thus the introduction of protocols that will allow doctors to determine the exact number and doses of drugs needed to obtain predetermined outcomes seems inevitable.

REFERENCES

1. Kurosawa M: Anti-allergic drug use in Japan: the rationale and the clinical outcome. *Clin Exp Allergy* (1994) **24**: 299–306.
2. Woolcock AJ, Rubinfeld AR, Seale JP, *et al.*: Asthma management plan, 1989. *Med J Aust* (1989) **151**: 650–653.
3. Hargreave FE, Dolovich J, Newhouse MT: The assessment and treatment of asthma: a conference report. *J Allergy Clin Immunol* (1990) **85**: 1098–1111.
4. British Thoracic Society: Guidelines for the treatment of asthma. 1. Chronic persistent asthma. *Br Med J* (1990) **301**: 651–653.
5. National Institutes of Health: Executive summary: guidelines for the diagnosis and management of asthma. National Asthma Education Program Expert Panel Report. US Department of Health and Human Services, 1991.
6. National Institutes of Health: Guidelines for the diagnosis and management of asthma. US Department of Health and Human Services, 1991.
7. National Heart Lung and Blood Institute: Global strategy for asthma management and prevention NHLBI/WHO Workshop Report. National Institutes of Health, National Heart Lung and Blood Institute, January 1995.
8. Ait-Khaled N, Enarson D. Management of asthma in adults: a guide for low income countries 1996. In *International Union Against Tuberculosis and Lung Disease*. Frankfurt am Main: pmi Verlagsgnippe GmbH, 1996.

9. National Asthma Campaign: *Asthma Management Handbook 1996*. Melbourne, National Asthma Campaign Ltd, 1996.
10. Woolcock AJ, Jenkins CR: Assessment of bronchial responsiveness as a guide to prognosis and therapy in asthma. *Med Clin North Am* (1990) **74**: 753–765.
11. Laitinen LA, Heino M, Laitinen A, Kava T, Haahtela T: Damage of the airway epithelium and bronchial reactivity in patients with asthma. *Am Rev Respir Dis* (1985) **131**: 599–606.
12. Beasley R, Roche WR, Roberts JA, Holgate ST: Cellular events in the bronchi in mild asthma and after bronchial provocation. *Am Rev Respir Dis* (1989) **139**: 806–817.
13. Saetta M, Stefano AD, Rosina C, Thiene G, Fabbri LM: Quantative structural analysis of peripheral airways and arteries in sudden fatal asthma. *Am Rev Respir Dis* (1991) **143**: 138–143.
14. Peat JK, Woolcock AJ: Sensitivity to common allergens: relation to respiratory symptoms and bronchial hyperresponsiveness in children from three different climatic areas of Australia. *Clin Exp Allergy* (1991) **21**: 573–581.
15. Sears MR, Herbison GP, Holdaway MD, Hewitt CJ, Flannery EM, Silva PA: The relative risks of sensitivity to grass pollen, house dust mite and cat dander in the development of childhood asthma. *Clin Exp Allergy* (1989) **19**: 419–424.
16. Martin RJ, Cicutto LC, Ballard RD: Factors related to the nocturnal worsening of asthma. *Am Rev Respir Dis* (1990) **141**: 33–38.
17. Reddel HK, Salome CM, Peat JK, Woolcock AJ: Which index of peak expiratory flow is most useful in the management of stable asthma. *Am J Respir Crit Care Med* (1995) **151**(5): 1320–1325.
18. Siersted HC, Hansen HS, Hansen N-CG, Hyldebrandt N, Mostgaard G, Oxho H: Evaluation of peak expiratory flow variability in an adolescent population sample. The Odense schoolchild study. *Am J Respir Crit Care Med* (1994) **149**: 598–603.
19. Du Toit JI, Anderson SD, Jenkins CR, Woolcock AJ, Rodwell LT: Airway responsiveness in asthma: bronchial challenge with histamine and 4.5% sodium chloride before and after budesonide. *Allergy Asthma Proc* (1997) **18**: 7–14.
20. Anderson SD, Rodwell LT, Daviskas E, Spring JF, du Toit J: The protective effect of nedocromil sodium and other drugs on airway narrowing provoked by hyperosmolar stimuli: a role for the airway epithelium? *J Allergy Clin Immunol* (1996) **98** (Suppl 5): S124–S134.
21. Anderson SD, du Toi tJI, Rodwell LT, Jenkins CR: Acute effect of sodium cromoglycate on airway narrowing induced by 4.5% saline aerosol. Outcome before and during treatment with aerosol corticosteroids in patients with asthma. *Chest* (1994) **105**: 673–680.
22. Woolcock AJ, Jenkins CR: Clinical responses to corticosteroids. In: Kaliner MA, Barnes PJ, Persson CGA (eds) *Asthma, its Pathology and Treatment*. New York: Marcel Dekker, 1991, pp 633–665.
23. Place VA, Velazquez JG, Burdick KH: Precise evaluation of topically applied corticosteroid potency. Modification of the Stoughton-McKenzie assay. *Arch Dermatol* (1970) **101**: 531–537.
24. Ollerenshaw S, Woolcock AJ: Characteristics of the inflammation in biopsies from large airways of subjects with asthma and subjects with chronic airflow limitation. *Am Rev Respir Dis* (1992) **145**: 922–927.
25. Ryan G, Latimer KM, Juniper EF, Roberts RS, Hargreave FE: Effect of beclomethasone dipropionate on bronchial responsiveness to histamine in controlled nonsteroid-dependent asthma. *J Allergy Clin Immunol* (1985) **75**: 25–30.
26. Frankel D, Latimer K, Ruffin R: Does bronchial hyperresponsiveness improve following hospitalisation for acute asthma? *Aust NZ J Med* (1990) **20** (Suppl): 522.
27. Gibson PG, Wong BJO, Hepperle MJE, et al.: A research method to induce and examine a mild exacerbation of asthma by withdrawl of inhaled corticosteroid. *Clin Exp Allergy* (1992) **22**: 525–532.
28. Toogood JH, Jennings B, Greenway RW, Chuang L: Candidiasis and dysphonia complicating beclomethasone treatment of asthma. *J Allergy Clin Immunol* (1980) **65**: 145–153.
29. Toogood JH, Jennings B, Baskerville J, Newhouse M: Assessment of a device for reducing oropharyngeal complications during beclomethasone treatment of asthma. *Am Rev Respir Dis* (1981) **123**: 113.
30. Toogood JH, Baskerville J, Jennings B, Lefcoe NM, Johansson SA: Use of spacers to facilitate inhaled corticosteroid treatment of asthma. *Am Rev Respir Dis* (1984) **129**: 723–729.

31. Storr J, Barrell E, Barry W, Lenney W, Hatcher G: Effect of a single oral dose of prednisolone in acute childhood asthma. *Lancet* (1987) **i**: 879–882.
32. Church MK, Young KD: The characteristics of inhibition of histamine release from human lung fragments by sodium cromoglycate, salbutamol and chlorpromazine. *Br J Pharmacol* (1983) **78**: 671–679.
33. Howarth PH, Durham SR, Lee TH, Kay AB, Church MK, Holgate ST: Influence of albuterol, cromolyn sodium and ipratropium bromide on the airway and circulating mediator responses to allergen bronchial provocation in asthma. *Am Rev Res Dis* (1985) *132*: 986–992.
34. Pearce N, Crane J, Burgess C, Grainger J, Beasley R: Fenoterol and asthma mortality in New Zealand. *N Z Med J* (1990) **103**: 73–75.
35. Harvey JE, Tattersfield AE: Airway response to salbutamol: effect of regular salbutamol inhalations in normal, atopic, and asthmatic subjects. *Thorax* (1982) **37**: 280–287.
36. Kerrebijn KF, van Essen-Zandvliet EEM, Neijens HJ: Effect of long-term treatment with inhaled glucocorticosteroids and beta-agonists on bronchial responsiveness in asthmatic children. *J Allergy Clin Immunol* (1987) **79**: 653–659.
37. Kraan J, Koeter GH, Van der Mark TW, Sluiter HJ, De Vries K: Changes in bronchial hyperreactivity induced by 4 weeks of treatment with antiasthmatic drugs in patients with allergic asthma: a comparison between budesonide and terbutaline. *J Allergy Clin Immunol* (1985) **76**: 628–636.
38. Sears MR, Taylor DR, Print CG, *et al.*: Regular inhaled beta-agonist treatment in bronchial asthma. *Lancet* (1990) **336**: 1391–1396.
39. Wong CS, Pavord ID, Williams J, Britton JR, Tattersfield AE: Bronchodilator, cardiovascular, and hypokalaemic effects of fenoterol, salbutamol, and terbutaline in asthma. *Lancet* (1990) **336**: 1396–1399.
40. Arvidsson P, Larsson S, Lofdahl C-G, Melander B, Wahlander L, Svedmyr N: Formoterol, a new long-acting bronchodilator for inhalation. *Eur Respir J* (1989) **2**: 325–330.
41. Ullman A, Hedner J, Svedmyr N: Inhaled salmeterol and salbutamol in asthmatic patients. An evaluation of asthma symptoms and the possible development of tachyphylaxis. *Am Rev Respir Dis* (1990) **142**: 571–575.
42. Twentyman OP, Finnerty JP, Harris A, Palmer J, Holgate ST: Protection against allergen-induced asthma by salmeterol. *Lancet* (1990) **336**: 1338–1342.
43. Cheung D, Timmers MC, Zwinderman AH, Bel EH, Dijkman JH, Sterk PJ: Long-term effects of a long-acting β_2-adrenoceptor agonist, salmeterol, on airway hyperresponsiveness in patients with mild asthma. *N Engl J Med* (1992) **327**: 1198–1203.
44. Booij-Noord H, Orie NGM, De Vries K: Immediate and late bronchial obstructive reactions to inhalation of house dust and protective effects of disodium cromoglycate and prednisolone. *J Allergy Clin Immunol* (1971) **48**: 344–354.
45. Hoag JE, McFadden ER: Long-term effect of cromolyn sodium on nonspecific bronchial hyperresponsiveness: a review. *Ann Allergy* (1991) **66**: 53–63.
46. Boulet LP, Cartier A, Cockcroft DW, *et al.*: Tolerance to reduction of oral steroid dosage in severely asthmatic patients receiving nedocromil sodium. *Respir Med* (1990) **84(4)**: 317–323.
47. Chapman KR: The role of anticholinergic bronchodilators in adult asthma and chronic obstructive pulmonary disease. *Lung* (1990) **168** (Suppl): 295–303.
48. Pauwels RA: New aspects of the therapeutic potential of theophylline in asthma. *J Allergy Clin Immunol* (1989) **83**: 548–553.
49. Hendeles L, Weinberger M: Theophylline: a 'state of the art' review. *Pharmacotherapy* (1983) **3**: 2–44.
50. Billing B, Dahlqvist R, Hornblad Y, Leidman T, Skareke L, Ripe E: Theophylline in maintenance treatment of chronic asthma: concentration-dependent additional effect of beta 2 agonist therapy. *Eur J Respir Dis* (1987) **70**: 35–43.
51. Howarth PH. Histamine and asthma: an appraisal based on specific H1-receptor antagonism. *Clin Exp Allergy* (1990) **20** (Suppl 2): 31–41.
52. Cloud ML, Enas GC, Kemp J, *et al.*: A specific LTD4/LTE4-receptor antagonist improves pulmonary function in patients with mild, chronic asthma. *Am Rev Respir Dis* (1989) **140**: 1336–1339.

53. Israel E, Rubin P, Kemp JP, et al.: The effect of inhibition of 5-lipoxygenase by zileuton in mild-to-moderate asthma. *Ann Intern Med* (1993) **119**: 1059–1066.
54. Spector SL, Smith LJ, Glass M: Effects of 6 weeks of therapy with oral doses of ICI 204, 219, a leukotriene D4 receptor antagonist, in subjects with bronchial asthma. ACCOLATE Asthma Trialists Group. *Am J Respir Crit Care Med* (1994) **150**: 618–623.
55. O'Shaughnessy KM, Wellings R, Gillies B, Fuller RW: Differential effects of fluticasone propionate on allergen-evoked bronchoconstriction and increased urinary leukotriene E4 excretion. *Am Rev Respir Dis* (1993) **147**: 1472–1476.
56. Grant SM, Goa KL, Fitton A, Sorkin EM: Ketotifen: a review of its pharmacodynamics and pharmacokinetic properties, and therapeutic use in asthma and allergic disorders. *Drugs* (1990) **40**: 412–448.
57. Klaustermeyer WB, Noritake DT, Kwong FK: Chrysotherapy in the treatment of corticosteroid-dependent asthma. *J Allergy Clin Immunol* (1987) **79**: 720–725.
58. Shiner RJ, Nunn AJ, Fan Chung K, Geddes DM: Randomised, double-blind, placebo-controlled trial of methotrexate in steroid-dependent asthma. *Lancet* (1990) **336**: 137–140.
59. Mullarkey MF, Lammert JK, Blumenstein BA: Long-term methotrexate treatment in corticosteroid-dependent asthma. *Ann Intern Med* (1990) **112**: 577–581.
60. Mullarkey MF, Blumenstein BA, Andrade WP, Bailey GA, Olason I, Wetzel CE: Methotrexate in the treatment of corticosteroid-dependent asthma. A double-blind crossover study. *N Engl J Med* (1988) **318**: 603–607.
61. Erzurum SC, Leff JA, Cochran JE, et al.: Lack of benefit of methotrexate in severe, steroid-dependent asthma. A double-blind, placebo-controlled study. *Ann Intern Med* (1991) **114**: 353–360.
62. Stewart GE, Diaz JD, Lockey RF, Seleznick MJ, Trudeau WL, Ledford DK: Comparison of oral pulse methotrexate with placebo in the treatment of severe glucocorticosteroid-dependent asthma. *J Allergy Clin Immunol* (1994) **94**: 482–489.
63. Alexander AG, Barnes NC, Kay AB: Trial of cyclosporin in corticosteroid-dependent chronic severe asthma. *Lancet* (1992) **339**: 324–328.
64. Schwartz YA, Kivity S, Ilfeld DN, et al.: A clinical and immunologic study of colchicine in asthma. *J Allergy Clin Immunol* (1990) **85**: 578–582.
65. Itkin IH, Menzel ML: The use of macrolide antibiotic substances in the treatment of asthma. *J Allergy* (1970) **45**: 146–149.
66. Charous BL: Open study of hydroxy chloroquine in the treatment of severe symptomatic or corticosteroid-dependent asthma. *Ann Allergy* (1990) **65**: 53–58.
67. Svenonius E, Kautto R, Abborelius MJ: Improvement after training of children with exercise-induced asthma. *Acta Paediatr Scand* (1983) **72**: 23–30.
68. Anderson SD: Exercise-induced asthma: stimulus, mechanism and management. In Barnes PJ, Rodgers IW, Thomson NC (eds) *Asthma: Basic Mechanisms and Clinical Management*. London, Academic Press, 1988, pp 503–522.
69. Anderson SD, Rodwell LT, Du Toit J, Young IH: Duration of protection by inhaled salmeterol in exercise-induced asthma. *Chest* (1991) **100**: 1254–1260.
70. Woolley M, Anderson SD, Quigley BM: Duration of protective effect of terbutaline sulfate and cromolyn sodium alone and in combination on exercise-induced asthma. *Chest* (1990) **97**: 39–45.
71. Sporik R, Holgate ST, Platts-Mills TAE, Cogswell JJ: Exposure to house-dust mite allergen (Der p I) and the development of asthma in childhood. A prospective study. *N Engl J Med* (1990) **323**: 502–507.
72. Platts-Mills TAE, Tovey ER, Mitchell EB, Moszoro H, Nock P, Wilkins SR: Reduction of bronchial hyperreactivity during prolonged allergen avoidance. *Lancet* (1982) **2**: 675–678.
73. Murray AM, Ferguson AC: Reduction of bronchial hyperreactivity during prolonged allergen avoidance. *Lancet* (1982) **ii**: 1212.
74. Dorward AJ, Colloff MJ, MacKay NS, McSharry C, Thomson NC: Effect of house dust mite avoidance measures on adult atopic asthma. *Thorax* (1988) **43**: 98–102.
75. British Thoracic Society: Guidelines for management of asthma in adults. 2. Acute severe asthma. *Br Med J* (1990) **301**: 797–800.

76. Beveridge RC, Grunfeld AF, Hodder RV, Verbeek PR: Guidelines for the emergency management of asthma in adults. *Can Med Assoc J* (1996) **155**: 25–37.
77. Ringsberg KC, Wiklund I, Wihelmsen L: Education of adult patients at an 'asthma school;: effects on quality of life, knowledge and need for nursing. *Eur Respir J* (1990) **3**: 33–37.
78. Bolton MB, Tilley BC, Kuder J, Reeves T, Schultz LR: The cost and effectiveness of an education program for adults who have asthma. *J Gen Intern Med* (1991) **6**: 401–407.
79. Muhlhauser I, Richter B, Kraut D, Weske G, Worth H, Berger M: Evaluation of a structured treatment and teaching programme on asthma. *J Intern Med* (1991) **230**: 157–164.
80. Mayo PH, Richman J, Harris HW: Results of a program to reduce admissions for adult asthma. *Annals Intern Med* (1990) **112**: 864–871.
81. Kotses H, Bernstein IL, Bernstein DI, *et al.*: A self-management program for adult asthma. Part 1: Development and evaluation. *J Allergy Clin Immunol* (1995) **95**: 529–540.
82. Wilson SR, Scamagas P, German DF, *et al.*: A controlled trial of two forms of self-management education for adults with asthma. *Am J Med* (1993) **94**: 564–576.
83. Bailey WC, Richards JM, Brooks CM, Soong S-J, Windsor RA, Manzella BA: A randomized trial to improve self-management practices of adults with asthma. *Arch Intern Med* (1990) **150**: 1664–1668.
84. Horn CR, Clark TJH, Cochrane GM: Compliance with inhaled therapy and morbidity from asthma. *Respir Med* (1990) **84**: 67–70.
85. Chan CS, Woolcock AJ, Sullivan CE: Nocturnal asthma: role of snoring and obstructive sleep apnea. *Am Rev Respir Dis* (1988) **137**: 1502–1504.
86. Corticosteroid action and resistance in asthma. Proceeding of a meeting. Buckinghamshire, UK. *Am J Respir Crit Care Med* (1996) **154** (Suppl 2): S1–S78.

46

Asthma in Children

SØREN PEDERSEN

INTRODUCTION

The reported prevalence rates of childhood asthma in western Europe and the USA vary from 5 to 10%, with an increase in occurrence during the last decades in many countries.[1-10] This makes asthma the most common chronic disease in children in the western world. Furthermore, asthma ranks as one of the most important causes of ill health[11-14] and creates not only a myriad of physical, emotional and social problems for the child and the family but also a financial burden on the family in most countries. Finally, there seems to be a clear increase in hospital admissions of children due to acute wheeze[15,16] and the number of children dying from asthma is still unacceptable.[7,14,17]

Although childhood and adult asthma share the same underlying pathophysiological mechanisms, there are some important anatomical, physiological, social, emotional and developmental age-related differences. Therefore, it is necessary to consider the management of this age group in its own right and not merely extrapolate from experience with adults. The present chapter endeavours to highlight and discuss some of the features and problems that are distinctive in childhood asthma and important in the day-to-day management of the disease. Within the childhood population it is convenient to consider children younger than 3 years separately, since this age group differs in many aspects from older children.

ANATOMICAL AND PHYSIOLOGICAL FACTORS

There are many reasons why infants and young children are at increased risk for symptomatic airway obstruction and, at times, do not respond well to bronchodilators or other inhaled drugs. Compared with adults, young children have disproportionally smaller peripheral airways and smaller cross-sectional airway areas,[18,19] rendering them more easily obstructed by oedema, secretions, cellular debris and smooth muscle contractions. The chest wall of an infant is more compliant than in later life and there is a relative lack of elastic recoil, which predisposes to early airway closure even during tidal breathing.[20,21] This, in combination with bronchial walls that are less rigid, produces ventilation–perfusion mismatching and lower oxygen tensions compared to adults.[22]

The collateral channels between the alveoli (pores of Kohn) and the bronchoalveolar communications (Lambert's canals) are decreased in number and size.[23] Therefore, collateral ventilation is less developed and airway obstruction is more prone to cause segmental collapse and atelectasis than in older subjects. Finally, the angle of insertion of the diaphragm is more horizontal than in adults. This means that during inspiration the infant's diaphragm tends to cause retraction of the compliant rib cage rather than elevation and increased diameter as seen in adults. This makes the diaphragm less efficient and, due to few fatigue-resistant muscle fibres,[24] less suitable to cope with prolonged increased workloads. For these reasons, young children may readily develop respiratory failure during acute episodes of asthma.

The airway smooth muscle of infants extends all the way out to the peripheral airways;[25] it is able to contract under provocation[25] and to bronchodilate under the influence of drugs.

WHEEZING ILLNESSES

Wheezing-associated lower respiratory tract illnesses in young children are extremely common, with estimates of cumulative prevalence ranging between 30 and 50%. The most common triggering factor is a viral infection, often respiratory syncytial virus (RSV), parainfluenza and/or rhinovirus. Research results of the past few years have increased our understanding of wheezing disorders in early childhood and it is now recognized that there are several distinct disorders masquerading as asthma in these age groups, each with its own cause, natural history and prognosis. There seem to be at least two major groups of recurrent wheezers in early life.[26,27]

In one group (about 35%) wheezing is simply an early manifestation of asthma. In addition to acute episodic wheeze and cough with viral infection, the pattern of illness also includes interval symptoms between the acute episodes, for instance brought on by crying, laughing, exercise or night-time. In such a child, symptoms may vary from day to day and acute severe episodes may complicate viral respiratory infections. The least frequent pattern in the community (although common in hospital outpatient practice) is the infant with persistent wheeze who even at his or her best appears to be functioning suboptimally. If well adapted, fat and content, the term 'happy wheezer' is sometimes applied. This group of children normally continues with persistent symptoms later in childhood and they tend to have an increased prevalence of family history of atopy, allergic markers and bronchial hyperresponsiveness. Without any treatment, lung functions grow significantly less than expected during the first 6 years of life.

The other group, transient infant wheezers (about 65%), simply have episodic attacks of wheeze and cough associated with evidence of viral upper respiratory infection, punctuated by symptom-free intervals of variable duration. At one extreme, some children have a single episode, either a mild attack lasting 2–3 days (often with their first RSV infection) or a severe but isolated attack of acute bronchiolitis; at the other extreme, monthly admissions to hospital may be precipitated by each passing cold. Symptoms between the acute attacks of wheeze are rare. Wheezing in this group seems to be associated with lower airway calibre that is present from birth, but which in contrast to the group with asthma does not decline further despite recurrent symptoms. These children normally have a non-atopic background, do not show bronchial hyperresponsiveness and tend to have a good prognosis; the vast majority outgrow their symptoms before the age of 3. Symptoms between the acute attacks of wheeze are rare.

These two groups just described account for most recurrent wheeze in young children. The groups share several predisposing factors, such as exposure to environmental tobacco smoke and virus and an increased risk in males. From a clinical viewpoint, it is impossible to distinguish between the two groups during the first few years of life. In addition, asthma in young children shares many symptoms, such as wheezing, recurrent fever, rhonchi, tachypnoea, chest retractions and cough, with other rare diseases including those shown in Table 46.1.

Asthma in young children is still a clinical diagnosis. No single tests are truly helpful. Recurrent episodes of wheeze in a predisposed child, in a child with other atopic symptoms or in a child with symptoms between the acute episodes is likely to be asthma. As in older children, a favourable response to a therapeutic trial with inhaled corticosteroids may be helpful in confirming/establishing the diagnosis.

NATURAL HISTORY

Accurate information about the natural history of childhood asthma is difficult to obtain because it requires that a large, entirely unselected cohort of children is assessed clinically and then followed prospectively for many years with regular clinical check-ups and lung function measurements. At present there are no truly prospective studies of asthma incidence throughout childhood reported in the literature. In most studies, parents have been interviewed at one point in time and asked to recall past episodes of wheezing in their children. When this technique has been validated by comparison with previous interviews or general practitioner records, it is apparent that the cumulative incidence of wheezing episodes is underestimated, perhaps by as much as one-third. In one study in which parents were interviewed repeatedly throughout childhood, the cumulative incidence of 'ever had asthma and/or wheezy bronchitis' was 21% by age 5, 28% by age 10 and 30% by age 16.[10]

It is often anticipated that childhood asthma is a self-limiting disorder that will improve spontaneously during adolescence. This seems to be an oversimplification that mainly applies to children with mild disease. The majority of children with moderate and

Table 46.1 Rare diseases that share symptoms with childhood asthma.

Developmental anomalies
Tracheo-oesophageal fistula and related disorders
Bronchomalacia (localized or generalized)
Stove-pipe trachea
Bronchial compression syndromes
   Vascular ring
   Anomalous origin of right subclavian artery
   Bronchial or pericardial cyst
Congenital heart disease (left–right shunting)
Granulomata or polyps
Pulmonary sequester
Bronchial stenosis

Host defence defects
Cystic fibrosis
Ciliary dyskinesia
Defects of immunity
Severe combined immune deficiency
Combined IgA and IgG_2 deficiency

Postviral syndromes
Obliterative bronchiolitis
Airway stricture or granuloma or lymphadenitis

Recurrent aspiration
Gastro-oesophageal reflux
Disorders of swallowing
Neuromuscular disease
Mechanical disorders

Perinatal disorders
Chronic lung disease of prematurity
Congenital infection

Other
Tuberculosis
Mediastinal mass
Foreign body
Rhinitis (cough)

---

severe persistent symptoms in childhood will continue to wheeze and have reduced lung function values into adult life, though many tend to improve somewhat.[28–39] Furthermore, a large proportion of children with apparent remissions in early adolescence will have recurrence of their symptoms when they grow older.[28] Finally, the term 'outgrowing asthma' may be somewhat misleading since recent studies indicate that many patients who consider themselves symptom-free still have reduced lung functions[30,31,33,39–42] and increased bronchial reactivity to specific and non-specific agents.[34,35,37]

In the clinical situation, it is difficult to reliably differentiate beforehand whether or not an individual patient will eventually outgrow the disease. The presence of atopy, low lung functions and smoking in adolescence usually indicate a less favourable prognosis in terms of outgrowing the disease.[33,39] Controversy still exists as to the importance of age of onset as a prognostic factor. Some suggest that an early debut indicates a more favourable prognosis, others that it is associated with a poorer prognosis and still others claim that age of onset has no influence on the ultimate prognosis.[31,36] At present, there is no indication that therapy or other measures can alter the natural history of asthma once it is present. However, controlled long-term studies in this area are still lacking.

RISK FACTORS AND PREVENTION

A risk factor is a characteristic present more commonly among persons who have, or later develop, a disease than among those who do not. Few risk factors are causal, in the sense that modification of a cause results in changes in the rate of disease occurrence, although all causes are risk factors. Thus, the identification of risk factors is the first step in discovering causes of a disease. Risk factors can also be used to identify high-risk groups for special study or to target healthcare.

It is well known that atopic children whose parents have asthma are at high risk for developing the disease themselves.[43–46] Crowded living conditions, frequent viral infections, high mite and dander allergen exposure and gas-cooking fumes are all risk factors for frequent wheeze during early childhood,[44,47–54] and these have been shown to negatively affect the expression of the disease once it has developed. There is no evidence, however, that environmental intervention in these areas will influence the development of reactive airway disease, but the strong association between the risk factors and the occurrence of wheeze calls for further studies.

Naturally, it is the dream of many paediatricians to find measures to prevent the development of asthma. In this respect, strict breast-feeding and delayed introduction of solid foods have been most thoroughly studied. There have been methodological flaws in many of the studies and the data have been conflicting.[47] The study by Høst et al.[55] clearly shows how difficult it may be just to control exposure correctly. None the less, it appears that at best such measures will only produce a small reduction in asthma frequency or some delay in the development in high-risk children.[44,47–54] A well-conducted study showed that in atopic breast-fed infants wheezing was reduced in atopic children for the first 2 years of life, while this protection lasted up to 7 years in the non-atopic infants. This suggests that breast-feeding protects against the viral respiratory infections that are the predominant causes of wheezing in the first 2 years of life.

Another important area is parental smoking. Many studies have shown that children of smoking parents, particularly when it is the mother who smokes, have a higher frequency of respiratory illnesses.[56] This effect is also apparent with antenatal smoking, where the influence is not on the airways directly but may affect both immunological maturation and lung growth.[57] Maternal smoking during pregnancy increases cord blood IgE in the offspring.[58] Such exposure also has an effect on infant lung function as measured by forced expiratory flows.[59] Thus, two of the main prerequisites for the development of asthma, namely allergy and airway narrowing, can both be affected by prenatal smoking. Such evidence should be enough to recommend avoidance of smoking during pregnancy and exposure to environmental tobacco smoke during childhood.

GROWTH

Many children with moderate and severe chronic asthma have a growth pattern different from normal children. They show delayed onset of puberty and preadolescent deceleration of growth velocity resembling growth retardation.[60–67] This deviant growth pattern is accompanied by a concomitant retardation in bone age so that bone age corresponds to height. This difference in growth pattern seems to be unrelated to the use of inhaled corticosteroids but is more pronounced in children with the most severe asthma. Generally, the children with severe asthma tend to grow for a longer period of time than their peers, so that before the age of 20 they will achieve normal, final predicted height.[60,61] In contrast, growth retardation caused by daily and alternate-day administration of large doses of systemic corticosteroids for extended periods of time may be permanent.[68]

The deviant growth pattern of asthmatic children complicates the interpretation of results from studies comparing the heights of asthmatic children treated with inhaled corticosteroids with the heights of normal children. Furthermore, it means that case reports of apparently reduced growth in association with asthma treatment should not lead to conclusions about cause-and-effect relationships. Only controlled longitudinal studies using carefully selected control groups of asthmatic children can be used to assess the influence of exogenous factors upon growth. Such studies are difficult to conduct, although several attempts have been made.[69] There have been flaws in the designs of the studies, but the general conclusion has been that long-term treatment of asthmatic children with doses of inhaled corticosteroids tailored to the severity of the disease does not adversely affect growth.[69] A few studies of children with mild asthma using fixed doses of beclomethasone, which were markedly higher than the dose required to control the disease,[69–71] have found significant growth retardation. Other prospective controlled studies of one or more years' duration with budesonide and fluticasone propionate have not found any adverse effect on growth[72–75] even when given in doses markedly more effective than sodium cromoglycate. Finally, a meta-analysis of 21 studies concluded that inhaled corticosteroids did not adversely affect final height.[68]

Recently, knemometry has made it possible to measure changes in short-term lower leg growth within weeks.[76] This method allows for controlled study designs and has been shown to be very good at defining doses of inhaled corticosteroids not associated with retardation of statural growth.[76] On the other hand, it has also become clear that steroid-induced changes in short-term lower leg growth is a poor predictor of long-term statural growth or final height,[77–81] which is the most relevant clinical parameter to study when evaluating the effect of exogenous steroids on growth.

PHARMACOKINETICS AND PHARMACODYNAMICS

Only the pharmacokinetics of theophylline have been studied to a satisfactory extent in all age groups of children. It appears, however, that children metabolize most drugs more rapidly than adults and that the clearance of drug varies from age group to age group and from patient to patient.[82–85] Thus, children often require quite high, individually adjusted doses of oral drugs to achieve a satisfactory effect. This is not always appreciated and consequently many drugs are used in suboptimal or ineffective doses in children. Furthermore, the tablet sizes available often do not allow the dose to be related accurately to the size of the child and this may also contribute to suboptimal therapy. The rapid clearance of drugs means that children must take oral drugs at short intervals or slow-release preparations twice daily. The latter is preferable due to improved compliance. However, the advantages of slow-release products may sometimes be outweighed by the inconsistent absorption characteristics of these products when taken in combination with food.[86] Since the various products are influenced in different and unpredictable ways, children should only be treated with slow-release preparations for which absorption characteristics have been found reliable even in the presence of food. As a rule, the bioavailability of slow-release β_2-agonist preparations is lower than the bioavailability of plain tablets and syrup. Therefore, the dose should normally be increased by 30% or more when switching from plain oral β_2-agonist therapy to slow-release therapy.[87,77] Food also reduces bioavailability; thus if the product is normally taken with meals, the dose should be increased by an additional 30%.[87,88]

In the day-to-day management of asthma, sparse knowledge about the pharmacokinetics and pharmacodynamics of the various drugs may to some extent be compensated for by careful monitoring of effect and by repeated adjustment of dose. This is time-consuming but mandatory for successful therapy.

ASSESSMENT OF THE CLINICAL CONDITION

Most physicians would agree that an optimally controlled child with asthma lives a normal life without any restrictions in physical activity and without any asthma symptoms or exacerbations. Lung functions are 'normal' and increase over time, as in healthy non-asthmatic children. Moreover, the child is knowledgeable about the disease, is confident with self-management and experiences no side-effects from the treatment. This level of control probably requires that bronchial reactivity is within the normal range and that the inflammatory changes in the airways are minimal.

Although these objectives seem fairly straightforward and are achievable in the majority of children, clinical experience and the literature suggest that they are not fulfilled in many children.[89,90] The main reason for this seems to be variations in the perception and definition of the terms 'symptom-free', 'normal lung function' and 'accurate assessment of the daily physical activity of the child'. Several studies have shown that asthma severity and the impact of the disease upon the daily life of the child are often markedly underestimated, probably because it is difficult to correctly assess symptoms and the extent to which the child has adapted her or his lifestyle to avoid symptoms. As a consequence, many children are undertreated, optimal asthma control is not achieved and effective treatment is either not prescribed or prescribed after the asthmatic condition has developed for many years.

Since reliable and accurate assessment of the clinical condition presents particular problems in children, this is discussed before an effective treatment plan is suggested.

History-taking

History-taking forms the basis of the assessment of the child's condition. For many reasons, most of the information about the child is usually obtained by careful questioning of the parents. This complicates the assessment to a great extent: many parents are unaware of important details. For example, in spite of the fact that the majority of children consider parental smoking a problem for their asthma, less than 10% of parents recognize this. Furthermore, many parents consciously or unconsciously interpret their observations and present an 'edited' version to the physician. Thus, a question about whether or not the child participates in sport activities is rarely answered by a simple 'yes' or 'no' but with a more plausible explanation such as 'he/she is not interested in sports', emphasizing that the problem is not the child's disease. Many parents do not want their child to be taking regular medication because they are afraid of side-effects and because it constantly reminds them that they have a sick child. In some countries, the cost of medicine also plays an important role in this respect. These factors may also influence the parent's description of the child's symptoms. Finally, most children with chronic asthma have had their symptoms since early childhood. Consequently, they have adjusted their lifestyle to the asthmatic condition and their parents have come to accept a state of chronic invalidism in their child as 'normal'. It requires time, skill and experience from the physician to accurately assess to what extent a child is restricted in his/her normal daily activities. Symptom control is often achieved in a child at the expense of some restrictions in their daily life.

Direct communication with the child is often more useful but not without problems either. Children have a limited vocabulary, do not readily open themselves to the physician, may not be used to long conversations with adults, are often influenced by the parents' attitudes about tobacco smoke, fear of medication and level of acceptance of symptoms, and may not want to disappoint or bother the doctor. Finally, children require some time to formulate their answers and therefore are frequently interrupted by their parents. Inaccurate and unspecific questions such as 'how is your asthma' are useless, since the answer will likely be 'fine' or 'OK' in 9 out of 10 children in whom subsequent specific and careful questioning would otherwise reveal important problems interfering with an unrestricted lifestyle.

Standardized questionnaires about asthma symptoms, such as episodic wheeze, cough, breathlessness, restriction of activity during the day, and disturbed sleep and awakenings during the night, are useful but diminish the problems with accurate history-taking only to some extent. This was clearly demonstrated in a recent study in which marked differences were found between physicians' and children's assessment of asthma control.[91] The physicians assessed that excellent asthma control had been achieved by 90% of the children, whereas only 60% of the children thought so. In addition, several studies have demonstrated that children whose asthma is believed to be optimally controlled improve markedly when they are treated with inhaled steroids. So, it seems that the term 'symptom-free' may cover a large scale of asthma control!

Lung functions

Since history-taking is difficult and often inaccurate, it is important to objectively measure lung functions in all children who can cooperate. Most children older than

3 years will be able to perform reproducible peak expiratory flow (PEF) measurements and should do this at home from time to time. Although the value of home PEF measurements has sometimes been questioned,[92] there is no doubt that such measurements may be important in individual patients. However, PEF and forced expiratory volume in 1 s (FEV_1) may be normal in the presence of quite marked small airways obstruction.[30,31,42,93] Even though the detection of small airways obstruction or hyperinflation present during an apparently symptom-free period is not of proven clinical importance, it may be of some help to select children who may benefit clinically from more aggressive treatment despite an apparently stable condition. A standardized exercise test is also very useful in this respect. A fall in lung function >15% after an exercise test performed without any inhaled premedication strongly suggests the presence of bronchial hyperreactivity that is incompatible with a normal unrestricted lifestyle in most children.

Measured lung functions are normally related to some reference value and expressed as a percentage of predicted normal. This expression may be deceptive and the percentage of predicted normal should always be evaluated in light of the substantial interindividual variations seen in the lung functions of growing children. Since we regularly see children whose best lung functions are around 110% of the predicted normal, a lung function of 90% of predicted normal may be quite good in one child whereas it may be unacceptable in another. It is useful to use the child's personal best value established during aggressive treatment as a reference value and to aim subsequently at lung functions around that level.

Impairment of lifestyle

Many children only achieve acceptable symptom control at the expense of marked restrictions in their daily life. In the clinic it may be difficult to assess to what extent a child has adjusted his/her lifestyle to minimize asthma symptoms. As for symptoms, specific and accurate questions are important in order to avoid missing important information and making incorrect conclusions. Participation in sports at school and outside school (hours per day), which role the child has in the sport activity (goalkeeper or field player), hours spent outside playing, hobbies and areas of interest, or how the child copes with long bike rides and walks should always be assessed. Often the parents are not aware of the child's level of physical activity; teachers or day-care personnel normally are, and it may be useful to get their opinion together with the child's. When in doubt, a standardized exercise test will reveal some information about the child's physical condition (how big a workload is required to generate a certain pulse rate).

It is also important to ask specific questions about which activities that the child/family avoids or participates in with some limitations. This includes restrictions in travel because the family is afraid of being too far away from their doctor, avoiding visits to friends or certain rooms/buildings because the environment (smoke, pets, dust) inadvertently provokes symptoms. Some children are not allowed to participate in sports or scout camps because their parents are afraid that they will not be able to cope with correct medication or may neglect early signs of a deterioration in asthma control.

Psychosocial behaviour

There do not seem to be any inherited characteristic personality features amongst asthmatic children, yet the incidence of emotional disorder is about twice as common as in the general population.[94] Anxiety states, school attendance problems, underachievement, dependency problems, peer group ridicule and social isolation all seem to contribute to this. Additionally, all clinics sometimes see parents who fail to give the child an opportunity to socialize, learn and separate. The result is family dysfunction with overprotection of, and overinvolvement with, the child or a negative attitude toward the child.[95,96] Once developed, such dysfunctions are very difficult to treat and they may adversely affect the child's behaviour for many years.

The majority of the problems mentioned above stem from preventable conditions such as lack of information, misunderstanding, ignorance or undertreatment of the disease.[97] Therefore, the preconditions for preventing these inexpedient disorders seems to be early diagnosis, information and effective treatment, which in turn requires an open, active attitude from the physician and effective management from early childhood. It is time-consuming but likely to pay off in the long run, since these dysfunctions are easier to prevent than to treat.

IRREVERSIBLE AIRWAY OBSTRUCTION

It is often anticipated that chronic childhood asthma does not lead to chronic irreversible airway obstruction. However, some children may develop this to a clinically important extent.[30,33,40–42,98–100] Furthermore, chest deformities due to persistent hyperinflation are not uncommon. It is difficult beforehand to reliably differentiate those individuals who are destined to develop these complications and it is not known whether they are preventable. A number of studies have found that growth of lung function was compromised in asthmatic children and that they continued to have lower levels of lung function into adult life, regardless of whether they had lost their asthmatic symptoms or not, suggesting that asthma can adversely influence lung growth. Furthermore, improvement in lung function was found to be significantly greater in patients who started inhaled corticosteroids early after the onset of asthma compared with patients who did not start the treatment until some years after onset of asthma symptoms.[72,101,102] In addition, children who started early achieved better lung function at a lower accumulated dose of budesonide during the first 4.5 years of treatment than children in whom inhaled steroids were not initiated until after more than 5 years of symptoms.[72] This indirectly suggests that airway remodelling takes place in the airways of patients with asthma and that early treatment with inhaled steroids may prevent airway remodelling and the development of irreversible structural changes.

There are elements of responsiveness amenable to treatment with inhaled steroids and others that are not. Does this resistant component represent scarring that could have been prevented by either rigorous early treatment or a different therapeutic approach altogether? Two recent trials suggest that late use of inhaled corticosteroids is associated with a smaller effect on bronchial hyperreactivity compared with early use.[101–103] The question clearly has great implications when considering the management of wheezing in children and calls for further studies.

TREATMENT

Asthma in children is rewarding to treat since the life of the child can usually be radically altered. We still have no cure for asthma so pharmacotherapy is likely to remain the cornerstone of asthma management in children for the next decade. As in adults, the various drugs can be given either systemically or by inhalation. Much speaks in favour of the inhaled route. The medication is deposited directly at the receptors in the airway, allowing a rapid onset of action. The administered therapeutic dose is small compared with other routes of administration and hence the incidence of side-effects for a given clinical effect is lower.[104,105] Exercise-induced bronchoconstriction, which is a troublesome, socially invalidating problem in the majority of asthmatic children, can only be effectively blocked by inhaled therapy. Finally, the drugs available for safe, effective, long-term treatment of the inflammatory component of the disease (inhaled corticosteroids and sodium cromoglycate) can only be given by the inhaled route. For these reasons, inhaled therapy should be the mainstay of treatment of childhood asthma.

Inhalation therapy in children is not without problems and pitfalls, however.[106] Within the childhood population, drug delivery to the patient varies with the age of the child for most (all?) inhalers,[107] and many inhalers cannot be used at all by young children. This, of course, makes it very difficult to give a general, simple answer about best inhaler choice for a child. In the following section, some of the advantages, disadvantages and limitations of the most widely used inhalers are discussed in detail in order to facilitate rational prescription of inhalers to various age groups of children.

The most important questions to consider when prescribing an inhaler for a child are listed below.

(1) Which inhaler has the best clinical effect for a given systemic effect (therapeutic ratio) in the day-to-day treatment? The systemic effect of a drug depends upon the amount of drug deposited in the intrapulmonary airways and the amount absorbed from the gastrointestinal tract. The clinical effect depends only upon the intrapulmonary deposition. Therefore, a clinically very effective inhaler will also be expected to have a potentially higher systemic effect than a clinically less effective inhaler. In contrast, the contribution of the orally deposited drug to the systemic effect is higher for an inhaler with a low intrapulmonary and high oral drug deposition, this being more pronounced for drugs with high oral bioavailability.
(2) Which inhaler reproducibly delivers the highest fraction of the delivered dose to the intrapulmonary airways in different age groups when using the inhaler optimally?
(3) Which inhaler is the simplest and easiest to use optimally for various age groups of patients?
(4) Which inhaler is preferred by the patient?

Over the years, many investigators have reported high frequencies of improper inhaler use as a direct and very important cause of treatment failure.[106] Therefore, accurate knowledge among physicians about the nature and magnitude of the problems patients experience with inhalation therapy, and about which age groups can normally use the various inhalation devices correctly, is important for correct inhaler prescription. For the average child, a simple inhalation technique, easy handling, and a smart, convenient design is probably more important than a 25% higher drug delivery to the intrapulmonary airways.

Pressurized metered dose inhalers

Ease of use

Metered dose inhalers (MDIs) are convenient, portable inhalers that are very difficult to use correctly, mainly because of the high velocity of aerosol particles (100 km/h) leaving the mouthpiece, which causes problems with correct coordination of actuation and inhalation, termination of inhalation when the cold aerosol particles reach the soft palate (cold freon effect), actuation of the aerosol into the mouth followed by inhalation through the nose, and rapid inhalation.[106,108–110] All these mistakes are associated with a reduced clinical effect. As a consequence, more than 50% of children receiving inhalation therapy with an MDI can be expected to obtain reduced or no clinical benefit from the prescribed medication.[108] Therefore, all prescriptions for an MDI in children should be accompanied by repeated, thorough tuition of correct inhaler use followed by the child's demonstration of inhalation technique. Conventional MDIs are normally not the best choice for children if alternative devices are available. Children less then 6 years old cannot be expected to learn efficient use of a conventional MDI.

Use of a breath-actuated MDI (Autohaler) will reduce tuition time and abolish coordination difficulties[111] and hence improve the dose to the intrapulmonary airways.[112] However the cold freon effect and the problem of nasal inhalation are unaffected; therefore, in children the Autohaler should mainly be reserved for patients older than 6–7 years since they can be taught a correct inhalation technique within 2–3 min and can also use it during episodes of acute wheeze.[113,114]

Inhalation technique

Actuation of the aerosol during the first part of a very slow, deep inhalation (about 30 litre/min) followed by a breath-holding pause of 10 s before exhalation has been found to enhance lung deposition and clinical effect in adults compared with fast inhalations (about 90 litre/min) with and without a breath-holding pause.[106,115–119] This technique is not easy for a child (particularly the correct coordination of inhalation and the very slow inhalation), so an optimal MDI inhalation technique is difficult.

Therapeutic ratio

After correct use of an MDI, about 80% of the dose lodges in the oropharynx, 10% is retained in the inhaler and 10% is deposited in the intrapulmonary airways.[106,118–121] This means that the therapeutic ratio is low for drugs with a high gastrointestinal bioavailability.

Spacers

Ease of use

Various holding chambers (spacers) may be attached to the mouthpiece of a conventional pressurized aerosol. These devices ensure that the aerosol particles have a slower velocity and a smaller particle size when they reach the patient. Most spacers have a valve that opens during inspiration and closes during expiration. As a consequence, spacers reduce the risk of the cold freon effect and the occurrence of coordination problems. Thus, spacers are easier to use than an MDI alone.[106] Virtually all schoolchildren can learn the use of these devices and can also use them effectively during attacks of acute bronchoconstriction when they are as effective as nebulizers.[293,123] Furthermore, spacers are the inhaler of choice in preschool children,[124–130] the majority of whom can also learn to use a spacer with a valve system, particularly if it has a face mask. During episodes of acute wheeze, however, some young children may not be able to open or close the valve system of some spacers and therefore may not gain optimal benefit.[107] The main problem with spacers is that they are often bulky and difficult to carry around. They are more suitable for prophylactic treatment given at home in the morning and evening.

Inhalation technique

Slow inhalations (around 30 litre/min) improve the effect when a straight extension tube spacer is used in children, whereas breath-holding, tilting of the head during inhalation or inhalation from functional residual capacity instead of residual volume do not influence the effect.[106,131] This is in agreement with the finding that slow, quiet tidal breathing results in an optimal effect when a Nebuhaler large-volume spacer is used.[132] Thus, optimal inhalation from a spacer is easier than for an MDI. Crying during the inhalation seems to reduce intrabronchial deposition markedly. This may be a problem in young children who are not always very cooperative.

Drug delivery to preschool children can be markedly improved by attaching a face mask to the mouthpiece of a Nebuhaler, presumably because the face mask reduces the occurrence of air leakage between the mouthpiece and the lips during the inhalation. It still remains to be shown whether the increased drug delivery to the patient using a face mask is associated with an increased clinical effect or if it is negated by reduced drug delivery to the intrapulmonary airways caused by nose breathing[133–136] when the face mask is used.

Only one dose should be fired into the spacer at a time. When two or more doses are fired in a row, the inhaled dose is reduced.[137–139]

Therapeutic ratio

All spacers reduce the oropharyngeal deposition of drug substantially.[115,118,140] The amount of drug retained in the inhaler is increased markedly by all spacers (most by the low-volume spacers) and hence the dose to the patient is reduced.[141,142] None the less, the dose delivered to the intrapulmonary airways is often the same or higher than that from an MDI.[115,118,120,140,143,144] Therefore, spacers have a favourable clinical effect/systemic effect ratio, though this has only been studied in schoolchildren not using a face mask.

Other studies

Static electricity. The half-life of respirable aerosol particles (<5 μm mass median aerodynamic diameter) depends upon the volume of the spacer[141] and the electrostatic attraction between the spacer wall and the charged aerosol.[141,142] Often the output of respirable particles from a spacer is markedly improved by use of a antistatic lining in the spacer.[138] Washing the spacer in soapy water may reduce the antistatic lining and hence reduce drug output.[141] Priming a new spacer or a newly washed spacer by firing some doses into it will improve output. It normally requires up to 15 doses to achieve optimal priming. The optimal frequency and mode of cleaning a spacer is not known. Thus, maintenance of a spacer is normally more difficult than for an MDI. A new metal spacer is now available in some countries. It totally eliminates the problems of static electricity.[141,145] As a consequence, cleaning and washing do not influence output and priming becomes redundant.

Spacer volume. The optimal volume of a spacer is not known. It is often anticipated that a low volume is advantageous in preschool children because they have a lower minute ventilation (tidal volume around 10 ml/kg) than older children.[107] However, some studies have indicated that spacer volume may not be as important as expected in these age groups.[142,146] The reason for this seems to be that young children hyperventilate markedly and hence increase their minute ventilation when a tightly fitting face mask is placed around their mouth and nose.[146] This *in vivo* finding is in agreement with an *in vitro* study which found that only at low tidal volumes (around 25 ml) did a low-volume spacer like Aerochamber enhance drug delivery,[147] while at tidal volumes of 150 ml drug delivery was greatest from Nebuhaler. As mentioned earlier, a high volume also results in a longer half-life of the respirable particles.[142]

Due to their many advantages, a variety of new spacer systems is launched every year. Though deceptively similar in appearance, it is clear that there may be marked differences in the amount of drug retained in them and in dose delivered to the intrapulmonary airways. Thus, delivery of respirable particles may vary by a factor of three between different devices.[141,142,148] Uncritical use of any new spacer device is not recommended until its value has been thoroughly documented in laboratory studies and controlled trials.

Dry powder inhalers

Ease of use

Dry powder inhalers are breath-actuated and therefore reduce or eliminate the coordination problems of actuation and inhalation seen with the MDI.[106] For many years, dry powder inhalers were single-dose inhalers and therefore less convenient but easier to use then the MDI. Some children have difficulties with correct loading and splitting of the capsules when using single-dose inhalers, particularly during episodes of acute wheeze.[106,149] In accordance with this, several recent studies have found that the new multiple-dose dry powder inhalers (Turbuhaler, Diskhaler, Diskus) are easier to use and more convenient, so these inhalers are preferred to single-dose inhalers and MDIs in schoolchildren.

The main problem with dry powder inhalers is to train the patient not to exhale through the inhaler before the inhalation, since this will blow the dose out of the inhaler.

Inhalation technique

Fast inhalations enhance the effect of all dry powder inhalers in children, whereas breath-holding, tilting of the head during inhalation or inhalation from functional residual capacity instead of residual volume do not influence the effect.[106,150,151] Thus, the inhalation technique is simple. The number of respirable particles and the effect decrease with decreasing inspiratory flow rates. The inhalation effort and the inhalation flow rate needed to generate a therapeutic aerosol vary between different dry powder inhalers and for these reasons results obtained with one inhaler cannot be used to characterize another. Until further studies are available, dry powder inhalers should preferably not be used in children younger than 5 years. However, recent studies suggest that the majority of children ⩾4 years can generate a sufficient flow rate to gain optimal benefit from Turbuhaler and Diskus.[152,153]

Therapeutic ratio

Normally, dry powder inhalers deposit a rather large proportion of drug in the oropharynx. Therefore, the therapeutic ratio is low when used to deliver drugs with a high gastrointestinal bioavailability. However, Turbuhaler results in twice as high intrabronchial deposition as treatment with an MDI with a Nebuhaler attached;[121,154] hence the therapeutic ratio for budesonide Turbuhaler is approximately the same as for an MDI with a spacer.

Nebulizers

Ease of use and convenience

Little coordination is required from the patient if continuous nebulization and a face mask with holes are used. Thus, nebulizers are rather simple to use. However, compared with other devices, nebulizers are expensive, bulky, inconvenient, time-consuming and inefficient delivery systems that are difficult to maintain and require electricity. With our present knowledge, their use for daily treatment of children with asthma should be limited to patients who cannot be taught correct use of another device or for drugs that cannot be delivered by any other inhaler system. In clinical practice, this mainly means some children younger than 3–4 years and some developmentally delayed older patients.

In spite of all the problems with nebulized therapy, nebulizers remain the delivery system of choice in the treatment of acute severe asthma in all age groups, even though the same results can often be obtained with other inhalation systems.[123,155,156,293] In the acute situation, it is advantageous that oxygen can be administered through the nebulizer at the same time as the β_2-agonist.

Inhalation technique

No controlled studies have been performed in children regarding the optimal inhalation technique. However, quiet tidal breathing is often recommended because it produces optimal results in adults.[157] Normally, this is also the only technique that can be achieved in the groups of children using nebulizers. The inhalation treatment should be given when the child is quiet, since crying at the time of treatment will reduce dose delivery to the intrapulmonary airways. Suction through the jet-nebulizer, which may occur when there are no holes in the face mask, may draw relatively large droplets that would otherwise fall back into the jet-nebulizer reservoir into the inhaled airstream.[158] This will increase the dose to the patient but not the clinical effect and therefore reduces the therapeutic ratio.

Some children object to a tightly fitting face mask and so many parents administer the inhalation treatment through a face mask 2–3 cm from the face of the child. This reduces drug delivery to the child by approximately 50%, with a corresponding increase in release of aerosol to the environment. Inhalation technique with nebulizers is rather simple, although it is not without problems nor is it as foolproof as often believed.

Output characteristics and particle size distribution are not fixed parameters. They change during ageing and with cleaning of the nebulizer. Hence, regular checks are needed. The optimal method and frequency of cleaning a nebulizer is not known. Normally it is recommended that the nebulizer (plus face mask or mouthpiece) be rinsed in hot water after each nebulization. In addition, the nebulizer and tubing should be washed in hot soapy water twice weekly and rinsed and disinfected once a week in a mixture of one cup of hot water plus two tablespoons of vinegar. This should be followed by drying by blowing air through the system. Some nebulizers are supplied with needles to unblock the feeding tubes, but generally these should not be used because they may damage the nozzles. Finally, the air inlet filter should be changed at regular intervals, depending upon the environment in which the nebulizer is being used. The maintenance of a nebulizer is difficult.

After 6 months of age, the quantity of aerosol inspired from a jet-nebulizer becomes largely independent of age due to entrainment occurring when inspiratory flow exceeds jet-nebulized flow.[135,159] However, infants and young children appear to receive less than half the dose to the intrapulmonary airways than the dose received by older children breathing through the same device, probably because the younger age groups inhale through the nose, which reduces the dose deposited in the intrapulmonary airways.[135]

Therapeutic ratio

We have no knowledge about the clinical therapeutic ratio for the various nebulizer systems. Nose breathing, which is common in young children, would be expected to filter off the large particles,[135,136] which for some drugs are likely to be absorbed and cause systemic effects without providing additional clinical effect.

Other studies

Simply varying the choice of compressor, jet-nebulizer and volume-fill has been shown to vary the mass of drug in respirable particles over a 10-fold range.[160] Due to this enormous variation, it is not meaningful to discuss comparisons with other inhalers in general. However, nebulizers are usually far less effective per milligram of drug than other inhaler systems, so higher doses are required to achieve the same clinical effect.[122,161] This difference in delivery seems to be more pronounced for steroids[162] than for β_2-agonists because fewer respirable particles are generated from a steroid suspension.

The inhaler strategy found most useful at present is based upon the considerations discussed above and is summarized in Table 46.2. The age group 0–4 years mainly uses a spacer with a valve system, except when they are severely obstructed in which case many use a nebulizer. Older children by routine are prescribed a multiple-dose dry powder inhaler. An MDI with a spacer is used for the administration of inhaled corticosteroids with high gastrointestinal bioavailability. Nebulizers are mainly reserved for severe acute attacks of bronchoconstriction. With this approach children can be taught effective inhaler use with a minimum of instructional time. It must always be remembered to consider the child's wishes, since prescription of an inhaler that the physician but not the child likes is likely to reduce compliance.

Drug strategy

The correct way to treat a child requiring daily medication is often discussed among paediatricians and important differences in strategy exist between various countries. Generally, there are two different approaches.

The traditional *step-up strategy*, which uses inhaled corticosteroids late and reserves this treatment for children with more severe disease (Fig. 46.1), does not try to establish the child's personal best lung function nor optimal control. Typically, the treatment is

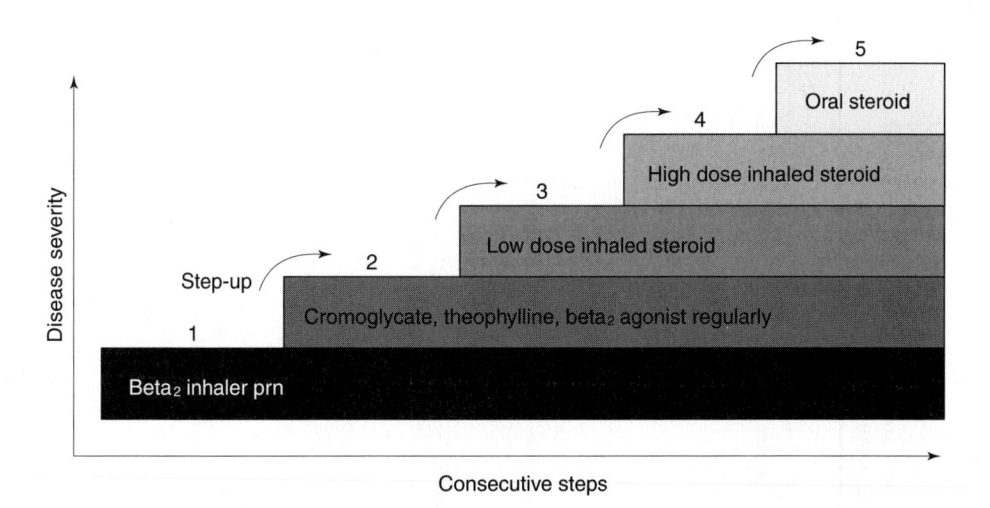

Fig. 46.1 With the step-up strategy, the first treatment level is chosen based on the initial assessment of asthma severity. If the clinical result of the treatment is assessed as being satisfactory, the child continues on the first treatment level without having his or her personal best control and lung function established. If control is not satisfactory, the treatment is gradually built up until control is assessed as being satisfactory. (Mygind N, Dahl R, Pedersen S, Pedersen CT: *Essential Allergy*, 2nd edn. (1996) Oxford: Blackwell Science, fig. 10.5.1).

Table 46.2 Age groups in which the various inhalers can be used. In addition the most common problems and optimal inhalation technique is mentioned.

	Age group	Drugs	Optimal technique	Problems
Metered dose inhaler	Children >7 years	All drugs except steroids with a high oral bioavailability	Actuation early during a slow (30 litre/min) deep inhalation followed by 10-s breath-holding	Coordination of actuation and inhalation. Cold freon effect (termination of inhalation at actuation) slow inspiration is difficult
Breath-actuated metered dose inhaler	Children >7 years	All drugs except steroids with a high oral bioavailability	Slow (30 litre/min) deep inhalation followed by 10-s breath-holding	Cold freon effect. Slow inhalation is difficult
Dry powder inhalers	Children ≥5 years	All drugs except steroids with a high oral bioavailability	Fast inhalation (minimal effective flow varies from one type of inhaler to another)	Dose lost if child exhales through the inhaler
Spacer with valve system	Children ≤5 years (use face mask)	All drugs	Slow deep inhalation (30 litre/min) or slow tidal breathing starting immediately after actuation	Bulky
	Children >5 years (no face mask)	Steroids with high oral bioavailability	Actuation of only one dose per inhalation	Static electricity reduces output (output reduced after cleaning)
Nebulizers	Children <2 years who cannot use other inhalers	All drugs (high flows required for steroids)	Slow tidal breathing. Tightly fitting face mask	Expensive. Time-consuming. Bulky. Difficult maintenance. Output changes over time
	All age groups with acute severe wheeze	Ultrasonic nebulizers cannot nebulize steroids	Mouthpiece preferable if the child can use it	

46 Asthma in Children

gradually stepped up, starting with inhaled β_2-agonists and then adding theophylline or oral β_2-agonists or sodium cromoglycate (SCG). If these drugs are assessed to be insufficient, inhaled corticosteroids are added to or replace the previous treatment. Alternatively, the clinician assesses asthma severity at the first visit and then decides which treatment should be given (typically inhaled β_2-agonists p.r.n. for mild persistent asthma; SCG, continuous theophylline or oral β_2-agonists plus inhaled β_2-agonists p.r.n. for moderate asthma; inhaled corticosteroids plus inhaled β_2-agonists p.r.n. for severe asthma, the latter being supplemented with long-acting inhaled β_2-agonists or theophylline as required). With this approach a minority (10–20%) of children with chronic asthma are assessed to require inhaled corticosteroids.

Due to the problems with severity assessment and estimation of optimal control in children previously described, this strategy carries with it a substantial risk of ending up at a treatment level which is suboptimal. One of the reasons for this is that, with this approach, the clinical condition after initiation of treatment is always compared with the situation when no or less treatment was given and *not* with personal best control. Because the family will normally be pleased with the relative improvement achieved by the initial treatment, they may not be prepared to try a more aggressive treatment or, in other words, may want to 'leave well enough alone'.

The *step-down strategy* (Fig. 46.2) uses inhaled corticosteroids early for all children with chronic symptoms. The philosophy is to establish the child's personal optimal control and best lung function before deciding the chronic treatment. With this approach the majority of children end up on continuous treatment with inhaled corticosteroids. Instead of stepping up the treatment or deciding the treatment level based on a clinical assessment of severity at the first visit, this approach tries to establish optimal control and personal best lung function in each patient, typically by a period of aggressive treatment with high-dose inhaled corticosteroids (400–800 μg/day) in combination with inhaled β_2-agonists for 6–8 weeks. After this period, the lowest amount of medication needed to maintain optimal control is determined by gradual reduction of the dose of inhaled corticosteroid every 6–8 weeks until unacceptable symptoms or decrease in lung function appear. Then the dose of inhaled corticosteroid is increased a little or additional treatment added, whichever is more acceptable. This approach allows the asthmatic child, her/his family and the physician to see how the child's life may be improved during optimal treatment in an attempt to provide a better basis for deciding the final therapy.

In the following, some studies and considerations of importance concerning treatment strategy of asthma in children are briefly discussed and summarized in order to provide the reader with information for therapeutic decision-making. In addition, the author makes his own conclusion based upon his interpretation of the literature and his personal experience obtained in a large number of controlled clinical therapeutic trials in children with asthma.

Several issues must be considered when positioning a drug in asthma management:

(1) the aims and outcome measures of the treatment;
(2) how the drug affects these aims and outcomes;
(3) the dose of drug required to achieve the aims;
(4) comparison of the clinical efficacy with the efficacy of other treatments or no treatment at all;
(5) the risk of clinically important side-effects at doses required to control the disease.

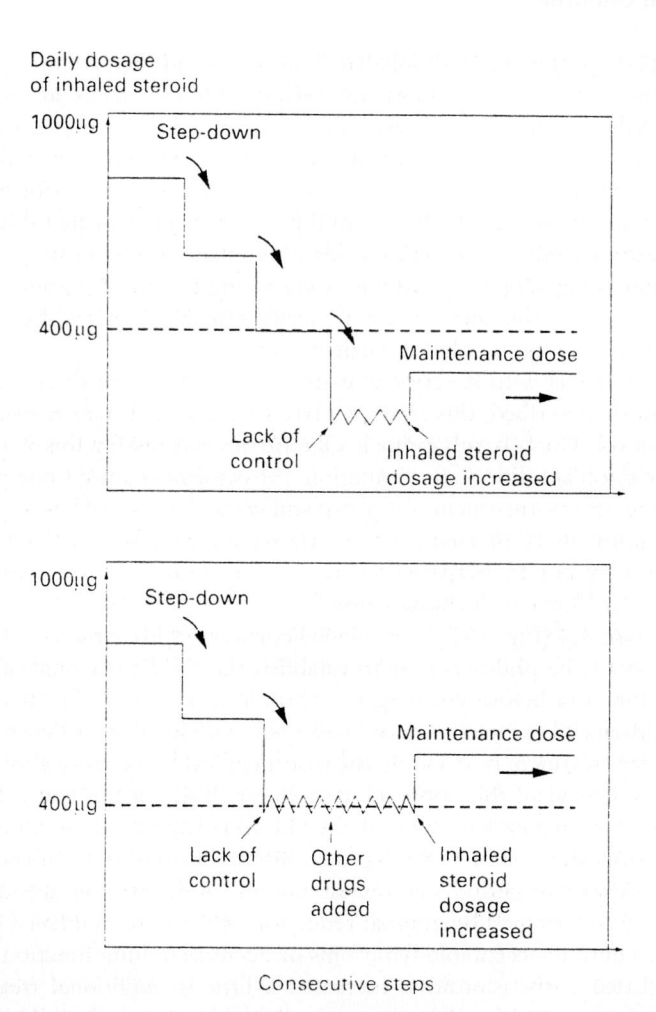

Fig. 46.2 With the step-down strategy, treatment is started with inhaled steroid in a high dose (800–1000 μg/day) in order to achieve optimal control and define personal best lung function. The dose is gradually reduced at 6–8-week intervals until optimal control is no longer maintained. The steroid dose is then increased to the previous level, and the treatment is continued on that dose if it is ⩽400 μg/day. If the required maintenance dose is >400 μg/day, treatment with other drugs is added. Only if that is not sufficient is the dose of inhaled steroid increased. (Mygind N, Dahl R, Pedersen S, Pedersen CT: *Essential Allergy*, 2nd edn. (1996) Oxford: Blackwell Science, fig. 10.5.2).

Aims of treatment

Historically, the most extensively studied outcome parameters for antiasthma drugs in children have been (a) symptoms and (b) lung function (PEF rate). During recent years, other outcome parameters have been recognized to be of similar or even greater importance than improvement in PEF rate and symptoms. These include (c) reduction in frequency and severity of acute exacerbations, (d) reduction in mortality and morbidity, (e) cost-effectiveness, (f) control of airway hyperresponsiveness, (g) normalization of the chronic inflammatory changes in the airways, (h) prevention of airway remodelling and (i) normal growth of lung function. Furthermore, the question of

46 Asthma in Children

whether early intervention with treatment can change the natural course or even cure the underlying disease are important issues to consider when the response to treatment is evaluated.

It has become clear that a certain drug or dose may control one outcome without having any significant effects upon other outcomes;[163] i.e. acceptable symptom control may be achieved without any, or little, effect upon some of the other outcomes such as exacerbations, bronchial hyperreactivity or control of inflammation in the airways. Therefore, it is important that first-line treatment influences as many outcome parameters as possible.

Effect on the clinical outcome measures

Inhaled steroids are effective in patients with asthma regardless of disease severity. This has been documented in controlled trials in both children and adults. Treatment has a significant and often marked effect upon outcome parameters (a)–(i) without any *clinically important* systemic side-effects when given in low doses. Continuous use improves daytime and night-time symptoms and lung function, reduces morbidity and mortality, the frequency of acute exacerbations and number of hospital admissions, improves bronchial hyperreactivity, reduces the chronic inflammatory changes in the airways, reduces airway remodelling and ensures normal growth of lung function.[71,72,75,91,163–202] The treatment has also been shown to be cost-effective (savings in healthcare costs are greater than spendings on medication) compared with other treatments.[72,180,199,203–205] Furthermore, some studies have suggested that inhaled steroids may also have an effect upon the natural course of the disease.[72,101,102] These effects are probably a result of the reduction in the underlying airway inflammation, although the component of increased airway responsiveness caused by structural changes in the airway may not respond to inhaled steroids. In contrast to other drugs, steroids not only shift the dose–response curve to spasmogens to the right, but also limit the maximum narrowing in response to spasmogens.[206]

Chronic treatment with all other antiasthma drugs has been shown to have a significant and reproducible effect mainly upon outcome parameters (a) and (b) in controlled trials.

Inhaled steroid dose required to achieve treatment aims

The dose of inhaled steroid required to achieve the aims of treatment seems to depend upon:

(1) outcome measure studied;
(2) duration of administration of the inhaled steroid;
(3) severity of the disease;
(4) drug/inhaler combination used for the administration;
(5) age of the patient; and
(6) duration of asthma when treatment is initiated.

As a consequence, each child may have her/his own individual dose–response curve. However, some general aspects are important:

Marked and rapid clinical improvement and changes in symptoms and lung function are seen at very low doses (around 100 μg) of inhaled steroids in most children with moderate and severe asthma.[192,193,207] These improvements in lung function and symptoms precede and reach a plateau before a reduction in airway responsiveness is seen.[208] Additional improvement in these parameters with increasing doses is rather small, often taking an additional four-fold increase in dose to produce further significant effect. The dose–response curve for the other outcome parameters is less well studied. It may be less steep in children with moderate and severe asthma,[207] although this is probably not the case in children with mild asthma, who seem to reach the top of the dose–response curve with low daily doses of inhaled steroids.[91,209] In children with moderate and severe asthma, 4-weeks treatment with a daily dose of 400 μg budesonide from a spacer (equivalent to 200 μg from Turbuhaler or 200 μg fluticasone propionate) produces about 85% of the maximum achievable protection against exercise-induced asthma.[207] This means that the vast majority of schoolchildren will achieve optimal control of symptoms and effect on PEF rates and a marked and clinically significant effect on other outcome parameters at daily doses of <400 μg/day of inhaled steroids.[91,182,189,190,192,193,207,208,210–212] The marked efficacy of low doses of inhaled steroids has also been confirmed in clinical trials in adults.[213–215]

Clinical efficacy of inhaled corticosteroids and comparisons with other antiasthma drugs

The beneficial effects of inhaled steroids in children are more pronounced than for any other antiasthma drugs, as shown in a number of studies (no direct comparisons with leukotriene antagonists have been published to date).[71,72,75,91,163,180,181,191,196,197,200] In two recent studies, including one from general practice, children with mild and moderate asthma achieved markedly better symptom control, significantly higher morning and evening PEF rates and clinic lung function, and reported fewer adverse events during treatment with 50 μg fluticasone propionate twice daily compared with children treated with SCG 20 mg four times daily.[75,91] Similar results were also reported with budesonide 400 μg/day from a spacer.[181,216] In three other studies, 200 and 400 μg/day beclomethasone propionate was significantly more effective than continuous theophylline treatment administered in optimal doses in combination with inhaled bronchodilators.[71,196,197] Furthermore, inhaled steroids were better than nedocromil,[191] combinations of all other asthma drugs,[72] salmeterol[163] and SCG.[200] In all these studies, a fixed dose of inhaled corticosteroid was used (dose titration was not performed); in light of the dose–response relationships described earlier, it is likely that similar results could have been achieved at lower doses of inhaled corticosteroid in many of the studies.

So far, no controlled clinical studies have found other drugs to be more effective than inhaled steroids in children or adults with asthma.

Early vs. late use of inhaled corticosteroids

A recent long-term study provided interesting information about the beneficial clinical effects associated with long-term continuous use of inhaled steroids.[72] The improvement in lung function was significantly greater in children who started budesonide treatment early (within 2 years) after the onset of asthma compared with children who did not start the treatment until some years after onset of asthma symptoms. In addition, children who started early had better lung function at a lower accumulated dose of budesonide during

46 Asthma in Children

the first 4.5 years of treatment than children in whom inhaled steroids were not initiated until after more than 5 years of symptoms. This suggests that early treatment with inhaled steroids may prevent airway remodelling and the development of irreversible structural changes in the airways. Recently, similar findings have been reported in adult studies, i.e. early introduction of inhaled steroids produces a better clinical effect compared with late introduction.[101,102] The findings of these four studies speak strongly for introducing inhaled steroid treatment early rather than late in the treatment schedule. Similar findings have not been reported with other drugs.

Side-effects

While the supporters of the step-up strategy normally do not question the benefits of inhaled steroids, they are still concerned about the risks of systemic side-effects of these drugs. Though such side-effects have not been demonstrated during 5 years of prospective continuous treatment,[72] it cannot be excluded that they may perhaps become a problem with treatment over much longer treatment periods. However, 20 years of clinical experience and retrospective studies do not support this assumption, although the power of such observations is not as strong as the power of controlled, prospective long-term studies. When this issue is discussed, it must be remembered that the safety of no other drug has been evaluated prospectively over such time periods.

It is clear from the efficacy studies in paediatric patients that 100–200 μg inhaled steroid per day is more effective than any other treatment in the majority of patients. Therefore, it is only relevant to assess the risk of clinically important systemic side-effects of such doses of inhaled steroids when comparing the clinical efficacy/side-effect ratio with other drugs. When reviewing the literature about the safety of inhaled steroids, it becomes clear that no controlled studies have reported any clinically relevant systemic side-effects or detectable systemic activity with such doses of inhaled corticosteroid in children older than 3 years with persistent asthma.[69,217]

Another argument against using inhaled corticosteroids in children assessed to have mild disease is that the long-term prognosis of these children is normally good. As discussed earlier, choosing other treatments implies a risk of undertreatment even if the condition is considered mild. Thus, even if these children have a good long-term prognosis, it should not keep us from treating them optimally with the most effective treatment in the mean time. Furthermore, if the majority only require continuous treatment and low doses around 100–200 μg/day for a limited number of years, safety concerns should not be an issue. Naturally, treatment should only be given if there is a good clinical response, which will sometimes not be the case in children with mild, infrequent, episodic asthma.

Conclusion

Inhaled steroids have been used for the treatment of asthma in children for more than 20 years. During this time, a substantial number of studies have been performed evaluating the safety and efficacy of this therapy. Generally, the results have been reassuring. In patients with mild and moderate asthma, low daily doses of around 100–

200 µg of inhaled steroid produce a clinical effect which, in most trials, is better than the effect of any other treatment to which it has been compared. No clinically important side-effects have been associated with treatment in this dose range. Inhaled steroids improve more outcome parameters than any other antiasthma drug. Furthermore, early intervention with inhaled steroids facilitates (may be a precondition for) long-term optimal asthma control. The marked efficacy and many beneficial effects of low doses of inhaled steroids, combined with the lack of clinically important systemic side-effects, give this treatment a favourable benefit–risk ratio compared with other treatments. This supports placing this treatment as a first-line therapy in children with asthma requiring continuous prophylactic treatment. Since the occurrence of measurable systemic effects increases with dose, the lowest dose that controls the disease should always be used. Furthermore, inhaler–steroid combinations with a high clinical efficacy/systemic effect ratio are preferred.

Finally, even if inhaled steroids should form the basis of all chronic asthma treatment in children, it is important also to realize the limitations of such treatment. Often the dose–response curve for a certain outcome parameter such as symptoms is rather flat, so that only a small increase in effect is seen with increasing doses after a marked effect of the initial dose. If this is the case, it might be better to add another drug, such as a long-acting inhaled β_2-agonist or perhaps theophylline, rather then to increase the steroid dose further (see Fig. 46.2). More studies are needed to confirm this.

Since we have adopted the step-down strategy, we have realized that the level of asthma control in many of our apparently optimally treated children could improve markedly. After a period of high-dose inhaled corticosteroids most children and their parents were not prepared to accept the previous control level with which both they and we were previously pleased. In addition to improvements in quality of life and reduced morbidity, the effect of the change in management plan is also reflected in the number of acute admissions, which has been reduced by 85%.[72]

No matter which strategy is chosen, it is important that the child and the family are involved in the therapeutic decisions, otherwise compliance is likely to be poor. Scrutiny and strict control will not improve compliance, only collaboration with the child and his/her parents will do so. This means that the family has to be educated in preventing and controlling symptoms. The child must be able to detect initial symptoms and know the appropriate action to take. This is best achieved by using self-management programmes.[218] When the treatment level has been decided, each child should be supplied with a written personal management plan for early treatment of exacerbations. This plan should include criteria to introduce or increase a treatment, criteria to reduce the increased treatment to the normal level and criteria that indicate when the patient should contact their physician and/or emergency department. Since no two children are alike, the criteria and the action to take need to be tailored individually to each patient. Addition of an oral bronchodilator, increase in the dose of inhaled corticosteroids, addition of oral steroids or a combination of these are valuable alternatives. It is important that the instructions are simple and in writing, preferably with illustrations, otherwise the risk of misunderstandings is too high.[219] Finally, patients should be taught which allergens and agents they should avoid and they should know how to adjust treatment to prevent exacerbations precipitated by exposure to such agents.

SPECIAL AGE GROUPS

Puberty and adolescence

Puberty and adolescence pose special problems. Teenagers often want to show their independence and to oppose their parents. If the parents have always been responsible for the management of the asthma, then the teenager may use their asthma in disagreements with their parents, believing that they are punishing the parents rather than themselves by not taking their medication. As a consequence, many teenagers stop their medication or take it irregularly. Some even start smoking. As a result, hospitalizations and deaths from asthma are often more common in adolescence than in childhood and adulthood.[17] This calls for special attention in this group of patients, but many countries do not have clinics with a special knowledge about these age groups. As a preventive measure, it is important that the parents and the paediatric clinics try to make the child responsible for his/her own treatment in cooperation with the healthcare system and the family in good time before puberty. The child should know that it is his/her responsibility and the other persons are there to help if necessary. The parents (especially mothers) should be warned not to be overprotective.

Young children

The best treatment strategy in children younger than 3 years is not known. There are fewer double-blind placebo-controlled studies in this age group than in schoolchildren and the well-known problems with accurate lung function measurements and challenge tests make it difficult to objectively demonstrate the beneficial effects of a treatment. Furthermore, it is difficult to distinguish children with asthma from 'transient wheezers' with virus induced wheeze. However, our present level of knowledge indicates that young children with frequent or persistent problems are candidates for a trial of continuous treatment.

Oral bronchodilators are widely used, but their effectiveness is often disappointing. For oral β_2-agonists, it is not known if this is due to insufficient doses. Controlled clinical trials are lacking. It is the author's experience that oral terbutaline in doses of 0.5–0.7 mg/kg daily may sometimes produce some effect. Though young children are normally said to be poorly responsive to inhaled β_2-agonists, several studies indicate that inhaled and oral treatment produces the same bronchodilatory, protective and clinical effects as in schoolchildren.[130,220–234] The long-acting β_2-agonists have not yet been studied in preschool children.

The majority of studies with theophylline in preschool children have evaluated only its pharmacokinetics and not its clinical effects. There are some indications that theophylline treatment offers some beneficial clinical effects and also some bronchodilation in these age groups.[235,236] However, further double-blind studies are needed to assess the optimal dose and place of theophylline relative to other treatments in these age groups, particularly in the light of its anti-inflammatory actions at low doses.

Nebulized ipratropium bromide has been shown to produce some bronchodilation in young children with wheeze.[237–240] However, results of clinical trials evaluating the value of ipratropium bromide in the day-to-day management have been disappointing.[211,241]

The clinical documentation of SCG in preschool children is sparse and the results inconsistent. Some controlled trials have not been able to demonstrate any effect, while others have indicated a significant effect,[242–246] probably of the same magnitude as theophylline.[244,247] The relative lack of controlled clinical trials and conflicting results call for further studies in these age groups to assess where in the treatment strategy SCG fits and at which doses it should be used. So far, nedocromil has not been studied in preschool children.

In contrast to other therapies, inhaled corticosteroids have been found very effective in young children with severe persistent or recurrent wheeze or symptoms.[69,124–127,248–251] However, the results with nebulized beclomethasone dipropionate have been disappointing.[252–254] This treatment plays a limited role in the routine management. In contrast, high-dose nebulized budesonide or inhaled corticosteroids given through spacers seem very effective in reducing morbidity and number of acute admissions.[127,255,256] However, the safety of long-term treatment and the optimal doses of inhaled corticosteroids still remain to be determined in these age groups. As in older children, the dose of inhaled corticosteroid should always be tailored to the severity of the disease.

SEVERE ACUTE ASTHMA/STATUS ASTHMATICUS

There is more than one way to manage an acute asthma exacerbation in a child. No two patients or situations are alike and many reviews have suggested various management plans.[257–265] The following is the author's brief suggestion of a protocol for treating acute asthma in various age groups of children. The background for the recommendations has been presented in more detail in a review article.[265] The suggestions are based upon controlled studies, clinical experience and the fact that the primary factors leading to obstruction are bronchoconstriction, inflammation, mucus plugging and oedema.

Generally, principles and modalities of emergency treatment of children are the same as in adults. The condition denotes a prolonged episode of severe bronchial obstruction that is temporarily refractory to the patient's usual therapy. Since it is a life-threatening condition that requires immediate and correct action, all physicians dealing with childhood asthma should be familiar with the paediatric doses of medications used in these situations (Table 46.3). The doses of the various drugs required depend upon response and the correct strategy is to administer enough drug for each individual patient while carefully monitoring for adverse effects and measuring clinical response. Dose recommendations should be considered as suggestions for an average patient; doses should be individualized upon the basis of the response and, for theophylline, on serum level monitoring. It is important to ensure that the correct treatment has been applied and that the treatment is improving the condition satisfactorily. Careful monitoring is therefore mandatory.

Asthma severity in children is often underestimated by physicians. The child may appear deceptively well and yet be suffering from quite marked airway obstruction (PEF around 50–60% predicted normal or personal best value). Even in the presence of

46 Asthma in Children

Table 46.3 Recommended average doses of drugs used to treat acute severe asthma in children. After the initial treatment the various doses should be adjusted individually according to the clinical effect/side-effects. Increase of theophylline dose requires measurement of serum theophylline concentration. A recent meta-analysis of 13 double-blind controlled studies on the treatment of acute asthma in children and adults concluded that there is no convincing evidence of any clinical benefit of adding theophylline to treatment with steroids and sympathomimetics.[297] Systemic dexamethasone or oral prednisolone were of little benefit in infants with acute wheeze in two studies[306,307] but of significant benefit in another.[308] Therefore, further studies are needed in this age group.

β₂-Agonists	
Nebulizer	0.2 mg/kg salbutamol or 0.4 mg/kg terbutaline
	Volume-fill = 4 ml
	Mouthpiece or tightly fitting face mask
	May be repeated at frequent intervals in doses of 0.3 mg salbutamol or 0.6 mg terbutaline every hour
Spacer device or other inhalers	One puff every minute until satisfactory response
	Maximum dose: 50 µg/kg salbutamol or 100 µg/kg terbutaline
Subcutaneous or intramuscular	10 µg/kg salbutamol or terbutaline
Intravenous (salbutamol/terbutaline)	Loading dose: 2–5 µg/kg over 5 min
	Continuous: 5 µg/kg per hour
Ipratropium bromide (only in combination with inhaled β₂-agonists)	
Nebulizer	250 µg in a volume-fill of 4 ml to all age groups
	May be repeated 4–6-hourly
Corticosteroids	
Prednisolone	Loading dose: 1–2 mg/kg (maximum 60 mg)
	Continuous: 2 mg/kg daily divided into two doses
Intravenous methylprednisolone	Loading dose: 1–2 mg/kg
	Continuous: 1 mg/kg 6-hourly
Intravenous hydrocortisone	Loading dose: 10 mg/kg
	Continuous: 5 mg/kg 6-hourly
Theophylline (for patients not receiving theophylline prior to treatment)	
Intravenous	Loading dose: 6 mg/kg lean body weight over 10min
Oral or rectal administration	Loading dose: 8–9 mg/kg lean body weight
Continuous treatment (oral or intravenous). Measurement of serum levels required	<1 year: (0.3) × (age in weeks) + 8 mg/kg in 24 h
	1–9 years: 24 mg/kg in 24 h
	9–12 years: 20 mg/kg in 24 h
	12–16 years: 18 mg/kg in 24 h
	Over 16 years: 14 mg/kg in 24 h

wheeze, many children still want to participate in other children's activities. Objective measures are therefore important and necessary for a correct assessment (Table 46.4). Even trained doctors may not be good at predicting a patient's lung function.[266] When in doubt about whether the condition should be categorized as moderate of severe, it is normally severe. Often inappropriate investigations and treatments are initiated. This is unfortunate since the severity cannot be assessed accurately by clinical examination alone.[267] Wheeziness correlates poorly with airway obstruction and the physician should always be aware of the danger of the 'silent chest'. Objective monitoring of the clinical

Table 46.4 Assessment of severity of asthma exacerbations in children. Patients in the moderate category should be considered for admission.

	Mild	Moderate	Severe
Treatment place	Home/outpatient	Home/outpatient	Hospital
Wheeze	Only end-expiratory	Loud	Loud or absent
Breathless			
Older child	Playing	Walking	Talks in single words
Infant	Crying	Difficult feeding	Stops feeding
Accessory muscles, retractions	Usually not	Moderate	Marked
Respiratory rate			
<3 months	<60/min	60–70/min	>70/min
3–12 months	<50/min	50–60/min	>60/min
1–6 years	<40/min	40–50/min	>50/min
>6 years	<30/min	30–40/min	>40/min
Pulse rate			
<1 year	<150/min	150–170/min	>170/min
1–2 years	<120/min	120–140/min	>140/min
>2 years	<110/min	110–130/min	>130/min
Pretreatment PEF	>70%	50–70%	<50%
Response to β_2-agonist	>3 h	2–3 h	<2 h
$PaCO_2$	<35 mmHg (4.7 kPa)	<40 mmHg (5.3 kPa)	>40 mmHg (5.3 kPa)
SaO_2 (on air)	>94%	92–94%	<92%

Life-threatening features
PEF <40% of best
Cyanosis
Bradycardia
Fatigue/exhaustion/reduced consciousness
Silent chest
Paradoxical thoraco-abdominal movement
Disappearance of retractions without concomitant clinical improvement

High-risk patient:[309] low admission threshold
Recent withdrawal of oral steroids
Hospitalization for asthma in past year
Earlier catastrophic attacks
Psychiatric disease/psychosocial problems
Poor compliance
Young children (develop respiratory failure more readily and are difficult to assess)

PEF, peak expiratory flow

condition of the child on arrival in the emergency room and at regular intervals thereafter is crucial for effective treatment. The key parameters to monitor are:

(1) respiratory rate,
(2) degree of hyperinflation,
(3) pulse rate,
(4) PEF rate,
(5) use of auxiliary muscles and retractions,

(6) colour,
(7) duration of effect of nebulized β_2-agonist,
(8) blood gases (especially $P\text{CO}_2$) and oxygen saturation,
(9) general clinical condition, including occurrence of side-effects.

The frequency of assessment depends upon severity, but generally more frequent monitoring (every 30 min) is required at the beginning of treatment and in the more severe cases. Standardized forms for the recording are strongly recommended. $P\text{CO}_2$ is particularly important[268] and can be measured transcutaneously in young children. If such equipment is not available, a capillary blood sample is reliable. Difficulty with arterial puncture is not an excuse for not measuring $P\text{CO}_2$. As in adults, $P\text{CO}_2$ is normally low. A normal or high $P\text{CO}_2$ in a wheezy child indicates that the condition is so severe that it requires continuous monitoring.

The drugs and doses used in the treatment of acute severe asthma in children is summarized in Table 46.2. In the following, some background results from studies in children are briefly discussed.

Nebulized β_2-agonists and systemic steroids are the cornerstone of emergency treatment. When normal therapy fails it is often due to marked inflammation and mucus plugging in the airways. Hence, it is better to initiate systemic steroids early rather than to 'wait and see how the condition develops'.[269] If the severity is milder than anticipated on arrival, the steroid treatment can always be stopped. There are no studies in children that have evaluated the doses or preparations to be used in true status asthmaticus. The author prefers methylprednisolone because of its favourable penetration into the lungs[270,271] and its less mineralocorticoid effect of fluid and electrolytes. The duration of steroid treatment depends upon the rate of recovery. Sometimes a few days' treatment is sufficient, but when the child responds slowly steroids should be administered for 10–15 days and then tapered according to the clinical condition as dictated by PEF measurements and clinical assessment.

High doses of nebulized salbutamol (0.30 mg/kg) have been found to be better than low doses (0.15 mg/kg) when given at 3-hourly intervals.[272] Furthermore, continuous or frequent low-dose nebulization produced better results than the same dose nebulized intermittently at 2–3 h intervals.[273–276]

Ipratropium bromide normally results in less bronchodilation than inhaled β_2-agonists[277] and, administered alone, these drugs have no role in the management of acute severe asthma in schoolchildren.[278] However, controlled studies have found that the combination of a β_2-agonist and an anticholinergic agent produces somewhat better results than with either drug used alone[240,278–282] and without an increase in side-effects. Though statistically highly significant, the advantages of combination therapy were rather small in most studies. This may be the reason why other studies have failed to find any benefit of such combined therapy.[283–286]

The recommended doses of intravenous β_2-agonists are partly empirical and partly derived from other studies,[287–290] only one of which was a dose–response study.[290] Undoubtedly, due to systemic absorption of drug from the lungs and gastrointestinal tract, nebulized therapy alone will be able to produce therapeutic plasma levels of β_2-agonists if sufficiently high doses and frequent administrations are used. In agreement with this, some studies have suggested that nebulized therapy can totally replace the intravenous route.[291–293]

Normally plasma levels of β_2-agonists are not measured. The dose must be adjusted based on the clinical response. It seems, however, that optimal clinical effect is rarely achieved until the child experiences some side-effects, such as tremor, palpitations or headache.[290] A small drop in blood pressure and a compensatory increase in pulse rate is often seen after systemic use or administration of high-doses of inhaled drug.[290] Normally this is not clinically important but should be kept in mind during treatment.

Though significant bronchodilating effects have been demonstrated in children, the role of theophylline in the acute management of asthma has been questioned based on the findings in a recent study.[294] No additional benefit of theophylline in children treated with steroids and frequent inhalations of β_2-agonists was reported, probably because theophylline is a weaker bronchodilator than inhaled β_2-agonists.[295,296] In accordance with this, a recent meta-analysis of 13 double-blind controlled studies on the treatment of acute asthma in children and adults did not find convincing evidence of any clinical benefit of adding theophylline to treatment with steroids and sympathomimetics.[297]

The criteria for admission to the intensive care unit and for mechanical ventilation are similar to those in adults.

Inhaled oxygen is an important part of emergency treatment and the inhaled β_2-agonists should be nebulized with oxygen instead of compressed air. The doses of inhaled β_2-agonists given in Table 46.3 are the ones used in the author's department. Many other regimens are recommended in the literature.[87,298,299] Though large-volume spacers are valuable alternatives to nebulizers outside hospital,[298] the large doses of freon and lubricant (irritants) that go with them,[293] their inferiority in oxygen delivery, and problems with opening and closing of the valve systems make them less suitable than nebulizers in hospital settings.

As soon as the child does not have to stay in bed, can be assumed the absorption of oral drugs to be reliable and intravenous therapy can be switched to oral therapy. Fortunately, full recovery after status asthmaticus in children is the rule. However, it is important to realize that the return to a completely normal state is very slow. After 1 or 2 days' treatment, children may appear deceptively well clinically. This should not lead to a premature cessation of treatment. Normally it takes at least 1 month until bronchial hyperreactivity has returned to normal. Therefore, it is important that these children monitor PEF and clinical symptoms in a diary for some weeks after discharge and that they are all treated with high-dose inhaled corticosteroids during the recovery phase, otherwise the risk of relapse or of going for many weeks with uncontrolled disease is too high.[268]

IMMUNOTHERAPY

This issue is discussed in Chapter 43. Therefore only some considerations particular for childhood asthma are mentioned here. When considering whether immunotherapy has any place in the treatment of childhood asthma, one must balance the interesting modulation of the immune response against the clinical efficacy and the risk of inducing anaphylatic reactions.

A subcommittee of the European Academy of Allergology and Clinical Immunology recently recommended that hyposensitization should not be used routinely in children under 5 years of age.[300,301] It should only be initiated when a clear, important and unavoidable allergic factor has been identified by bronchial challenge, since a positive skin-prick test and/or radioallergosorbent test do not accurately define the importance of the allergy for the asthma disease. Furthermore, avoidance of sensitizing allergens should not be possible.

Immunotherapy seems most beneficial in children with mild asthma, a condition that is normally safely and effectively treated with low doses of inhaled corticosteroids (100–200 μg/day).[300,301] Furthermore, this group of children seems to have the best prognosis for outgrowing their asthma. Once initiated, immunotherapy must be continued for as long as any other form of prophylactic treatment. For these reasons, its role in the treatment of childhood asthma still remains controversial and differences in opinion exist between different consensus reports.

GENERAL MEASURES

The various aims of asthma management are best achieved by a combination of drug treatment and general supportive measures. The most important general measures are briefly discussed.

Environmental control

Removal of important allergens and irritants such as cigarette smoke is an important requisite for good management of childhood asthma. However, when such measures are considered in children, the potential benefits must be balanced against the psychological consequences of the recommendations given. The asthmatic child is an integrated part of a family and the various recommendations may have an important impact upon family relations, since the restrictions often involve subjects who are not sick. Removal of the father's sporting dog, banning cigarette smoking or moving the asthmatic child to a sibling's room because it is bigger, easier to clean or seems to have a better indoor climate may create negative tensions and adverse effects that counterbalance the beneficial effects upon the asthma. Therefore, the physician should supply the family with knowledge and discuss the problems rather than just ordering various measures. It is up to the family to make the decisions. What is best for one family may not be good for another. Such a policy is more likely to preserve a good physician–patient relationship as well.

Young children in day-care institutions tend to get viral upper airway infections much more frequently than their peers.[51] This increases the likelihood of recurrent wheeze and it may be tempting to advise these children to stay away from the crowded day-care centres. Though likely to be beneficial, such an intervention has a heavy impact not only upon the child's social development but also upon the life of the parents, who may not be able to work full time if the child has to be isolated from contact with other children. Such advice should be reserved for the very severe cases who cannot be controlled on regular therapy with bronchodilators and inhaled corticosteroids.

Exercise

Participation in sport and play is important for normal growth and psychosocial development in children. Therefore, exercise-induced asthma is an even bigger problem in children than in adults. Children with asthma should be encouraged to play and participate in physical activities. Much effort should be put into stimulating enjoyable, balanced physical activities to build up fitness and self-confidence in the child's own environment. Many asthmatic children are physically unfit[302] and therefore may require a short physical training programme to break the vicious circle of exercise-induced asthma and unfitness. Such a programme does not have any direct effect upon the asthma, but it is likely to improve physical fitness, exercise tolerance, the child's ability to cope with the asthma, neuromuscular coordination and self-confidence,[181,302,303] all of which are important for normal integration of the child with his/her peers. It is important that this aspect of asthma management is not forgotten. Often a 6–8 weeks' training course is sufficient.

Prophylactic medication with inhaled β_2-agonists prior to exercise is the most widely used treatment. However, often children may not know in advance that they are going to be physically active. Furthermore, many children are reluctant or forget to take their medication prior to exercise. Instead, they choose not to participate wholeheartedly in the physical activity. It is better to treat this socially invalidating symptom with drugs that do not require premedication immediately prior to exercise to be effective. In this respect, continuous treatment with inhaled corticosteroids[181,211,212] and/or inhalation of a long-acting β_2-agonist such as formoterol or salmeterol in the morning is very effective.[304,305]

REFERENCES

1. Memon I, Loftus BG: Prevalence of asthma in Galway city school children. *Ir Med J* (1993) **86**: 136–137.
2. Sennhauser FH, Kuehni CE: Prevalence of childhood asthma: facts, tendencies and interpretations. *Agents Actions* (1993) **40** (Suppl): 87–99.
3. Ayres JG, Pansari S, Weller PH, et al.: A high incidence of asthma and respiratory symptoms in 4–11 year old children. *Respir Med* (1992) **86**: 403–407.
4. Robertson CF, Bishop J, Dalton M, et al.: Prevalence of asthma in regional Victorian schoolchildren. *Med J Aust* (1992) **156**: 831–833.
5. Robertson CF, Heycock E, Bishop J, Nolan T, Olinsky A, Phelan PD: Prevalence of asthma in Melbourne schoolchildren: changes over 26 years. *Br Med J* (1991) **302**: 1116–1118.
6. Clifford RD, Radford M, Howell JB, Holgate ST: Prevalence of respiratory symptoms among 7 and 11 year old schoolchildren and association with asthma. *Arch Dis Child* (1989) **64**: 1118–1125.
7. Mitchell EA, Anderson HR, Freeling P, White PT: Why are hospital admission and mortality rates for childhood asthma higher in New Zealand then in the United Kingdom? *Thorax* (1990) **45**: 176–182.
8. Asher MI: Isaac phase one: worldwide variations in the prevalence of wheezing and asthma in children. *Eur Respir J* (1996) **9** (Suppl 23): 410s.
9. Anderson HR, Butland BK, Strachan DP: Trends in prevalence and severity of childhood asthma. *Br Med J* (1994) **308**: 1600–1604.
10. Anderson HR, Pottier AC, Strachan DP: Asthma from birth to age 23: incidence and relation to prior and concurrent atopic disease. *Thorax* (1992) **47**: 537–542.

11. Hill RA, Standen PJ, Tattersfield AE: Asthma, wheezing, and school absence in primary schools. *Arch Dis Child* (1989) **64**: 246–251.
12. Braback L, Kalvesten L: Asthma in schoolchildren. Factors influencing morbidity in a Swedish survey. *Acta Paediatr Scand* (1988) **77**: 826–830.
13. Weiss KB, Gergen PJ, Hodgson TA: An economic evaluation of asthma in the United States. *N Engl J Med* (1992) **326**: 862–866.
14. Wever-Hess J, Wever AM, Yntema JL: Mortality and morbidity from respiratory diseases in childhood in the Netherlands, 1980–1987. *Eur Respir J* (1991) **4**: 429–433.
15. Hyndman SJ, Williams DR, Merrill SL, Lipscombe JM, Palmer CR: Rates of admission to hospital for asthma. *Br Med J* (1994) **308**: 1596–1600.
16. Mitchell EA, Dawson KP: Why are hospital admissions of children with acute asthma increasing? *Eur Respir J* (1989) **2**: 470–472.
17. Boman G, Foucard T, Bergström SE, Formgren H, Hedlin G: Report cases of death from asthma between 1 and 43 years of age. Läkartidningen (1993) **91**: 22.
18. Hislop A, Muir DCF, Jacobsen M, Simon G, Reid L: Postnatal growth and function of the pre-acinar airways. *Thorax* (1972) **27**: 265–274.
19. Hogg J, Williams J, Richardson J, *et al*.: Age as a factor in the distribution of lower airway conductance and in the pathologic anatomy of obstructive lung disease. *N Engl J Med* (1970) **232**: 1283–1287.
20. Bryan AC, Mansell AL, Levison H: In Hodson WA (ed) *Development of the Lung*. New York, Marcel Dekker, 1977, pp 445–460.
21. Helms P: Chest wall mechanics. In Warner JO, Metha MH (eds) *Scoliosis Prevention*. New York, Praeger, 1985, pp 184–189.
22. Newth CJ: Respiratory disease and respiratory failure: implications for the young and the old. *Br J Dis Chest* (1986) **80**: 209–217.
23. Boyden EA: Notes on the development of the lung in infancy and early childhood. *Am J Anat* (1967) **121**: 749–762.
24. Keens TG, Bryan AC, Levison H, Ianazzo CD: Development pattern of muscle fibre types in human ventilatory muscles. *J Appl Physiol* (1978) **44**: 909–913.
25. Prendiville A, Green S, Silverman S: Bronchial responsiveness to histamine in wheezy infants. *Thorax* (1987) **42**: 92–99.
26. Silverman M, Taussig LM, Martinez F, *et al*.: Early childhood asthma: what are the questions? *Am J Respir Crit Care Med* (1995) **151**: S1–42.
27. Martinez FD, Wright AL, Taussig LM, Holberg CJ, Halonen M, Morgan WJ: Asthma and wheezing in the first six years of life. *N Engl J Med* (1995) **332**: 133–138.
28. McNicol KN, Williams HE: Spectrum of asthma in children. Clinical and physiological components. *Br Med J* (1973) **4**: 7–11.
29. Blair H: Natural history of childhood asthma. 20 years follow-up. *Arch Dis Child* (1977) **52**: 613–619.
30. Friberg S, Bevegord S, Graff-Lonnevig V: Asthma from childhood to adult age. A prospective study of twenty subjects with special reference to the clinical course and pulmonary function. *Acta Paediatr Scand* (1988) **77**: 424–431.
31. Gerritsen J, Koeter GH, Postma DS, Schouten JP, Knol K: Prognosis of asthma from childhood to adulthood. *Am Rev Respir Dis* (1989) **140**: 1325–1330.
32. Balfour-Lynn L: Childhood asthma and puberty. *Arch Dis Child* (1985) **60**: 231–235.
33. Kelly WJ, Hudson I, Raven J, Phelan PD, Pain MC, Olinsky A: Childhood asthma and adult lung function. *Am Rev Respir Dis* (1988) **138**: 26–30.
34. Davé NK, Hopp RJ, Biven RE, *et al*.: Persistence of increased nonspecific bronchial reactivity in allergic children and adolescents. *J Allergy Clin Immunol* (1990) **86**: 147–153.
35. Foucard T, Sjöberg O: A prospective 12-year follow-up study of children with wheezy bronchitis. *Acta Paediatr Scand* (1984) **73**: 577–583.
36. Martin AJ, Landau LI, Phelan PD: Predicting the course of asthma in children. *Aust Paediatr J* (1982) **18**: 84–87.
37. Gerritsen J, Koëter GH, Monchy JGR, Champagne JGL, Knol K: change in airway responsiveness to inhaled house dust from childhood to adulthood. *J Allergy Clin Immunol* (1990) **85**: 1083–1089.

38. Martin AJ, McLennan LA, Landau LI, Phelan PD: The natural history of childhood asthma to adult life. *Br Med J* (1980) **280**: 1397–1400.
39. Kelly WJ, Hudson I, Phelan PD, Pain MC, Olinsky A: Childhood asthma in adult life: a further study at 28 years of age. *Br Med J* (1987) **294**: 1059–1062.
40. Martin AJ, Landau LI, Phelan PD: Lung function in young adults who had asthma in childhood. *Am Rev Respir Dis* (1980) **122**: 609–617.
41. Blackhall M: Ventilating function in subjects with childhood asthma who have become symptom free. *Arch Dis Child* (1970) **45**: 363–366.
42. Akhter J, Gaspar MM, Newcomb RW: Persistent peripheral airway obstruction in children with severe asthma. *Ann Allergy* (1989) **63**: 53–58.
43. Kjellman N: Atopic disease in seven-year-old children: incidence in relation to family history. *Acta Paediatr Scand* (1977) **66**: 465–471.
44. Horwood LJ, Fergusson DM, Hons BA, Shannon FT: Social and family factors in the development of early childhood asthma. *Pediatrics* (1985) **75**: 859–868.
45. Lubs ME: Empiric risks for genetic counseling in families with allergy. *J Pediatr* (1972) **80**: 26–31.
46. Luoma R, Koivikko A, Viander M: Development of asthma, allergic rhinitis and atopic dermatitis by the age of five years: a prospective study of 543 newborns. *Allergy* (1983) **38**: 339–346.
47. Kramer MS: Does breast feeding help protect against atopic disease? Biology, methodology and a golden jubilee of controversy. *J Pediatr* (1988) **112**: 181–190.
48. Wright AL, Holberg CJ, Martinez FD, Morgan WJ, Taussig LM: Breast feeding and lower respiratory tract illness in the first year of life. *Br Med J* (1989) **299**: 946–949.
49. Welliver JC, Sun M, Rinaldo D, Ogra PL: Predictive value of respiratory syncytial virus-specific IgE responses for recurrent wheezing following bronchiolitis. *J Pediatr* (1986) **109**: 776–780.
50. Weiss ST, Tager IB, Munzo A, Speizer FE: The relationships of respiratory infections in early childhood to the occurrence of increased levels of bronchial responsiveness and atopy. *Am Rev Respir Dis* (1985) **131**: 573–578.
51. Bisgaard H, Dalgaard P, Nyboe J: Risk factors for wheezing during infancy. *Acta Paediatr Scand* (1987) **76**: 719–726.
52. Lau S, Falkenhorst G, Weber A, *et al.*: High mite-allergen exposure increases the risk of sensitization in atopic children and young adults. *J Allergy Clin Immunol* (1989) **84**: 718–725.
53. Martinez FD, Antognoni G, Macri F, *et al.*: Parental smoking enhances bronchial responsiveness in nine-year-old children. *Am Rev Respir Dis* (1988) **138**: 518–523.
54. Sporik R, Holgate ST, Platts-Mills TAE, Cogswell JJ: Exposure to house-dust mite allergen (*Der p* I) and the development of asthma in children. *N Engl J Med* (1990) **323**: 502–507.
55. Høst A, Husby S, Østerballe O: A prospective study of cow's mild allergy in exclusively breast-fed infants. *Acta Paediatr Scand* (1988) **77**: 663–670.
56. Kershaw CR: Passive smoking, potential atopy and asthma in the first five years. *J Roy Soc Med* (1987) **80**: 683–688.
57. Taylor B, Wadsworth J: Maternal smoking during pregnancy and lower respiratory tract illness in early life. *Arch Dis Child* (1987) **62**: 76–79.
58. Magnusson CGM: Maternal smoking influences cord serum IgE and IgD levels and increases the risk of subsequent infant allergy. *J Allergy Clin Immunol* (1986) **78**: 898–904.
59. Hanrahan JP, Tager IB, Segal MR, *et al.*: The effect of maternal smoking during pregnancy on early infant lung function. *Am Rev Respir Dis* (1992) **145**: 1129–1135.
60. Balfour Lynn L: Effect of asthma on growth and puberty. *Pediatrician* (1987) **14**: 237–241.
61. Balfour Lynn L: Growth and childhood asthma. *Arch Dis Child* (1986) **61**: 1049–1055.
62. Armenio L, Baldini G, Bardare M, *et al.*: Double blind, placebo controlled study of nedocromil sodium in asthma. *Arch Dis Child* (1993) **68**: 193–197.
63. Fergusson AC, Murray AB, Tze WJ: Short stature and delayed skeletal maturation in children with allergic disease. *J Allergy Clin Immunol* (1982) **69**: 461–465.
64. Sprock A: Growth pattern in 200 children with asthma. *Ann Allergy* (1965) **23**: 608–611.
65. Hauspie R, Susanne C, Alexander F: A mixed longitudinal study of the growth in height and weight in asthmatic children. *Hum Biol* (1976) **48**: 271–276.

66. Hauspie R, Susanne C, Alexander F: Maturational delay and temporal growth retardation in asthmatic boys. *J Allergy Clin Immunol* (1977) **59**: 200–206.
67. Martin AJ, Landau LI, Phelan PD: The effect on growth of childhood asthma. *Acta Paediatr Scand* (1981) **70**: 683–688.
68. Allen DB, Mullen ML, Mullen B: A meta-analysis of the effect of oral and inhaled corticosteroids on growth. *J Allergy Clin Immunol* (1994) **93**: 967–976.
69. Pedersen S: Importance of early intervention in children: efficacy and safety. In Schleimer R, Busse W, O'Byrne P (eds) *Topical Glucocorticoids in Asthma: Mechanisms and Clinical Actions*. New York, Marcel Dekker, 1996, pp 551–560.
70. Doull IJ, Freezer NJ, Holgate ST: Growth of prepubertal children with mild asthma treated with inhaled beclomethasone dipropionate. *Am J Respir Crit Care Med* (1995) **151**: 1715–1719.
71. Tinkelman DG, Reed CE, Nelson HS, Offord KP: Aerosol beclomethasone dipropionate compared with theophylline as primary treatment of chronic, mild to moderately severe asthma in children. *Pediatrics* (1993) **92**: 64–77.
72. Agertoft L, Pedersen S: Effects of long term treatment with an inhaled corticosteroid on growth and pulmonary function in asthmatic children. *Respir Med* (1994) **88**: 373–381.
73. Merkus PJ, van Essen Zandvliet EE, Duiverman EJ, van Houwelingen HC, Kerrebijn KF, Quanjer PH: Long-term effect of inhaled corticosteroids on growth rate in adolescents with asthma. *Pediatrics* (1993) **91**: 1121–1126.
74. König P, Ford L, Galant S, et al.: A 1-year comparison of the effects of inhaled fluticasone propionate (FP) and placebo on growth in prepubescent children with asthma. *Eur Respir J* (1996) **9** (Suppl 23): 2945.
75. Price JF, Russell G, Hindmarsh PC, Weller P, Heaf DP, William J: Growth during one year of treatment with fluticasone propionate or sodium cromoglycate in children with asthma. *Pediatr Pulmonol* (1997) **24**: 178–186.
76. Wolthers OD: *Knemometry in the assessment of exogenous glucocorticosteroids in children with asthma and rhinitis*. MD Thesis, Copenhagen-Århus-Odense, 1996.
77. Marshall W: Evaluation of growth rate in height over periods of less than one year. *Arch Dis Child* (1971) **46**: 414–420.
78. Butler GE, McKie M, Ratcliffe SG: The cyclical nature of prepubertal growth. *Ann Hum Biol* (1990) **17**: 177–198.
79. Voss LD, Wilkin TJ, Balley BJR, Betts PR: The reliability of height and height velocity in the assessment of growth (the Wessex Growth Study). *Arch Dis Child* (1991) **66**: 833–837.
80. Karlberg J, Gelander L, Albertsson-Wikland K: Distinctions between short- and long-term human growth studies. *Acta Paediatr Scand* (1993) **82**: 631–634.
81. Karlberg J, Low L, Yeung CY: On the dynamics of the growth process. *Acta Paediatr Scand* (1994) **83**: 777–778.
82. Hendeles L, Weinberger M: Drugs in perspective. Theophylline. A 'State of the art' review. *Pharmacotherapy* (1983) **3**: 2–44.
83. Hendeles L, Iafrate R, Weinberger M: A clinical and pharmacokinetic basis for the selection and use of slow release theophylline products. *Clin Pharmacokinet* (1984) **9**: 95–135.
84. Pedersen S, Steffensen G, Ekman I, Tönneson M, Borgå O: Pharmacokinetics of budesonide in children after oral deposition and inhalation from two different inhalers. *Eur J Clin Pharmacol* (1987) **31**: 579–582.
85. Hultquist C, Lindberg C, Nyberg L, Kjellman B, Wettrell G: Kinetics of terbutaline in asthmatic children. *Eur J Respir Dis* (1984) **65** (Suppl 134): 195–203.
86. Pedersen S: Effects of food on the absorption of theophylline in children. *J Allergy Clin Immunol* (1986) **78**: 704–709.
87. Nyberg L, Kennedy BM: Pharmacokinetics of terbutaline given in slow-release tablets. *Eur J Respir Dis* (1984) **65** (Suppl 134): 119–139.
88. Davies DS: The fate of inhaled terbutaline. *Eur J Respir Dis* (1984) **65** (Suppl 134): 141–147.
89. Speight AN, Lee DA, Hey EN: Underdiagnosis and undertreatment of asthma in childhood. *Br Med J* (1983) **286**: 1253–1256.
90. Anderson HR, Bailey PA, Cooper JS, et al.: Morbidity and school absence caused by asthma and wheezing illness. *Arch Dis Child* (1983) **58**: 777–784.

91. Price JF, Weller PH: Comparison of fluticasone propionate and sodium cromoglycate for the treatment of childhood asthma. *Respir Med* (1995) **89**: 363–368.
92. GRASSIC. Effectiveness of routine self monitoring of peak flow in patients with asthma. *Br Med J* (1994) **308**: 564–567.
93. Cooper DM, Cutz E, Levison H: Occult pulmonary abnormalities in asymptomatic asthmatic children. *Chest* (1977) **71**: 361–365.
94. Mattson A: Psychologic aspects of childhood asthma. *Pediatr Clin North Am* (1975) **2**: 77–78.
95. Pinkerton P: Childhood asthma. *Br J Hosp Med* (1971) **9**: 331–338.
96. Liebman R, Minuchin S, Baker L: The use of structural family therapy in the treatment of intracable asthma. *Am J Psychiatry* (1974) **131**: 535–540.
97. Reddihough D, Landau L, Jones H, Richards W: Family anxieties in childhood asthma. *Aust Paediatr* (1977) **13**: 295–298.
98. Burrows B, Knudson RJ, Lebowitz MD: The relationship of childhood respiratory illness to adult obstructive airway disease. *Am Rev Respir Dis* (1977) **115**: 751.
99. Loren ML, Leung PK, Cooley RL, *et al*.: *Chest* (1978) **74**: 126.
100. Brown PJ, Greville HW, Finucane KE: Asthma and irreversible airflow obstruction. *Thorax* (1984) **39**: 131–136.
101. Haahtela T, Järvinen M, Kava T, *et al*.: Effects of reducing or discontinuing inhaled budesonide in patients with mild asthma. *N Engl J Med* (1994) **331**: 700–705.
102. Selroos O, Backman R, Forsen KO, *et al*.: The effect of inhaled corticosteroids in asthma is related to the duration of pretreatment symptoms. *Am J Respir Crit Care Med* (1994) **149**: A211.
103. Overbeek SE, Kerstjens HA, Bogaard JM, Mulder PG, Postma DS: Is delayed introduction of inhaled corticosteroids harmful in patients with obstructive airways disease (asthma and COPD)? *Chest* (1996) **1**: 35–41.
104. Thiringer G, Svedmyr N: Comparison of infused and inhaled terbutaline in patients with asthma. *Scand J Respir Dis* (1976) **57**: 17–24.
105. Williams SJ, Winner SJ, Clark TJH: Comparison of inhaled and intravenous terbutaline in acute severe asthma. *Thorax* (1981) **36**: 629–631.
106. Pedersen S: Inhaler use in children with asthma. *Dan Med Bull* (1987) **34**: 234–249.
107. Bisgaard H: Aerosol treatment of young children. *Eur Respir Rev* (1994) **4**: 15–20.
108. Pedersen S, Frost L, Arnfred T: Errors in inhalation technique and efficacy of inhaler use in asthmatic children. *Allergy* (1986) **41**: 118–124.
109. Pedersen S: Aerosol treatment of bronchoconstriction in children, with or without a tube spacer. *N Engl J Med* (1983) **308**: 1328–1330.
110. Pedersen S, Østergaard PA: Nasal inhalation as a cause of inefficient pulmonary aerosol inhalation technique in children. *Allergy* (1983) **38**: 191–194.
111. Pedersen S, Mortensen S: Use of different inhalation devices in children. *Lung* (1990) **168** (Suppl): 653–657.
112. Newman SP, Weisz AW, Talaee N, Clarke SW: Improvement of drug delivery with a breath actuated pressurized aerosol for patients with poor inhaler technique. *Thorax* (1991) **46**: 712–716.
113. Ruggins NR, Milner AD, Swarbrick A: An assessment of a new breath actuated inhaler device in acutely wheezy children. *Arch Dis Child* (1993) **68**: 477–480.
114. Pedersen S: Treatment of bronchoconstriction in children with a breath-actuated and a conventional metered dose inhaler. *J Allergy Clin Immunol* (1992) **89**: 154.
115. Dolovich M, Ruffino RE, Roberts R, Newhouse MT: Optimal delivery of aerosols from metered dose inhalers. *Chest* (1981) **80** (Suppl 65): 911–915.
116. Newman SP, Pavia D, Clarke SW: Improving the bronchial deposition of pressurized aerosols. *Chest* (1981) **80** (Suppl 65): 909–911.
117. Newman SP, Pavia D, Garland N, Clarke SW: Effect of various inhalation modes on the deposition of radioactive pressurized aerosols. *Eur J Respir Dis* (1982) **63** (Suppl 119): 57–65.
118. Newman SP: *Deposition and effects of inhalation aerosols*. Thesis, Lund, Rahms Tryckeri, 1983.
119. Borgström L: *Methodological Studies on Lung Deposition. Evaluation of Inhalation Devices and Absorption Mechanisms*. Uppsala, Acta Universitatis Uppsaliensis, 1993.

120. Newman SP, Pavia D, Morén F, Sheahan NF, Clarke SW: Deposition of pressurized aerosols in the human respiratory tracts. *Thorax* (1981) **36**: 52–55.
121. Pedersen S, Steffensen G, Borgaa O: Pharmacokinetics of budesonide after two different modes of inhalation and oral deposition in children with asthma. *Eur J Clin Pharmacol* (1995) in press.
122. Fuglsang G, Pedersen S: Comparison of Nebuhaler and Nebulizer treatment of acute severe asthma in children. *Eur J Respir Dis* (1986) **69**: 109–113.
123. Pendergast J, Hopkins J, Timms B, Van Asperen PP: Comparative efficacy of terbutaline administered by Nebuhaler and by nebulizer in young children with acute asthma. *Med J Aust* (1989) **151**: 406–408.
124. Noble V, Ruggins NR, Everad ML, Milner AD: Inhaled budesonide via a modified Nebuhaler for chronic wheezing in infants. *Arch Dis Child* (1992) **67**: 285–288.
125. Connett GJ, Warde C, Wooler E, Lenney W: Use of budesonide in severe asthmatics aged 1–3 years. *Arch Dis Child* (1993) **69**: 351–355.
126. Greenough A, Pool J, Gleeson JG, Price JF: Effect of budesonide on pulmonary hyperinflation in young asthmatic children. *Thorax* (1988) **43**: 937–938.
127. Bisgaard H, Munck SL, Nielsen JP, Petersen W, Ohlsson SV: Inhaled budesonide for treatment of recurrent wheezing in early childhood. *Lancet* (1990) **336**: 649–651.
128. Bisgaard H, Ohlsson S: PEP-spacer: an adaption for administration of MDI to infants. *Allergy* (1989) **44**: 363–364.
129. Gleeson JG, Price JF: Controlled trial of budesonide given by the nebuhaler in preschool children with asthma. *Br Med J* (1988) **297**: 163–166.
130. Pool JB, Greenough A, Gleeson JG, Price JF: Inhaled bronchodilator treatment via the nebuhaler in young asthmatic patients. *Arch Dis Child* (1988) **63**: 288–291.
131. Pedersen S: Optimal use of tube spacer aerosols in asthmatic children. *Clin Allergy* (1985) **15**: 473–478.
132. Gleeson JG, Price JF: Nebuhaler technique. *Br J Dis Chest* (1988) **82**: 172–174.
133. Bisgaard H, Mygind N: Nasal allergy. In Lessof MH, Lee TH, Keremy DM (eds) *Allergy. An International Textbook*. London, John Wiley & Sons, 1987, pp 531–552.
134. Köhler D, Fleicher W: Established facts of inhalation therapy: a review of aerosol therapy and commonly used drugs. *Lung Respiratory* (1989) **6**: 1–16.
135. Chua HL, Collis GG, Newbury AM, *et al.*: The influence of age on aerosol deposition in children with cystic fibrosis. *Eur Respir J* (1994) **7**: 2185–2191.
136. Everard ML, Hardy JG, Milner AD: Comparison of nebulised aerosol deposition in the lungs of healthy adults following oral and nasal inhalation. *Thorax* (1993) **48**: 1045–1046.
137. Newman SP, Millar AB, Lennard-Jones TR, Moren F, Clarke SW: Improvement of pressurised aerosol deposition with nebuhaler spacer device. *Thorax* (1984) **39**: 935–941.
138. O'Callaghan C, Lynch J, Cant M, Robertson C: Improvement in sodium cromoglycate delivery from a spacer device by use of an antistatic lining, immediate inhalation, and avoiding multiple actuations of drug. *Thorax* (1993) **48**: 603–606.
139. Clark AR, Rachelefsky G, Mason PL, Goldenhersh MJ, Hollingworth A: The use of reservoir devices for the simultaneous delivery of two metered-dose aerosols. *J Allergy Clin Immunol* (1990) **85**: 75–79.
140. Newman SP, Morén F, Pavia D, Little F, Clarke SW: Deposition of pressurized suspension aerosols inhaled through extension devices. *Am Rev Respir Dis* (1981) **124**: 317–320.
141. Bisgaard H, Anhoj J, Klug B, Berg E: Non-electrostatic-spacer for aerosol delivery. *Arch Dis Child* (1995) in press.
142. Ahrens R, Lux C, Bahl T, Han S: Choosing the metered-dose inhaler spacer or holding chamber that matches the patient's need: evidence that the specific drug being delivered is an important consideration. *J Allergy Clin Immunol* (1995) **96**: 288–294.
143. Morén F: Drug deposition of pressurized inhalation aerosols. I. Influence of actuator tube design. *Int J Pharmacol* (1978) **1**: 205–212.
144. Berg E: *In vitro* properties of pressurized metered dose inhalers with and without spacer devices. *J Aerosol Med* (1995) **8** (Suppl 3): 3–11.
145. Bisgaard H: A metal aerosol holding chamber devised for young children with asthma. *Eur Respir J* (1995) **8**: 856–860.

146. Agertoft L, Pedersen S: Influence of spacer device on drug delivery to young children with asthma. *Arch Dis Child* (1994) **71**: 217–220.
147. Everard ML, Clark AR, Milner AD: Drug delivery from holding chambers with attached facemask. *Arch Dis Child* (1992) **67**: 580–585.
148. Lee H, Evans HE: Evaluation of inhalation aids of metered dose inhalers in asthmatic children. *Chest* (1987) **91**: 366–369.
149. Pedersen S: Treatment of acute bronchoconstriction in children with use of a tube spacer aerosol and a dry powder inhaler. *Allergy* (1985) **40**: 300–304.
150. Richards R, Dickson CR, Renwick AG, Lewis RA, Holgate ST: Absorption and disposition kinetics of cromolyn sodium and the influence of inhalation technique. *J Pharmacol Exp Ther* (1987) **241**: 1028–1032.
151. Pedersen S, Hansen OR, Fuglsang G: Influence of inspiratory flow rate upon the effect of a Turbuhaler. *Arch Dis Child* (1990) **65**: 308–310.
152. Agertoft L, Ekelund J, Holtås E, Nikander K, Pedersen S: Impact of training on the peak inspiratory flow (PIF) in children using dry powder inhalers (DPI). *Eur Respir J* (1995) **8**: 14s.
153. Bisgaard H, Ifversen M, Klug B, Skamstrup K, Sumby B: Inspiratory flow rate through the Diskus/Accuhaler inhaler and Turbuhaler inhaler in children with asthma. *J Aerosol Med* (1995) **8**: 100 (P126).
154. Pedersen S, Steffensen G, Ohlsson SV: The influence of orally-deposited budesonide on the systemic availability of budesonide after inhalation from a Turbuhaler. *Eur J Clin Pharmacol* (1993) **36**: 211–214.
155. Lowenthal D, Kattan M: Facemasks versus mouthpieces for aerosol treatment of asthmatic children. *Pediatr Pulmonol* (1992) **14**: 192–196.
156. Scalabrin DM, Naspitz CK: Efficacy and side effects of salbutamol in acute asthma in children: comparison of oral route and two different nebulizer systems. *J Asthma* (1993) **30**: 51–59.
157. Ryan G, Dolovich MB, Obminski G, *et al.*: Standardisation of inhalation provocation tests: influence of nebulizer output, particle size and method of inhalation. *J Allergy Clin Immunol* (1981) **67**: 156–161.
158. Mercer TT, Goddard RF, Flores RL: Effect of auxiliary air flow on the output characteristics of compressed-air nebulisers. *Ann Allergy* (1969) **27**: 211–217.
159. Collis GG, Cole CH, Le Souëf PN: Dilution of nebulised aerosols by air entrainment in children. *Lancet* (1990) **336**: 341–343.
160. Newman SP, Pellow PGD, Clay MM, Clarke SW: Evaluation of jet nebulizers for use with gentamycin solution. *Thorax* (1985) **40**: 671–676.
161. Blackhall MI, O'Donnell SR: A dose–response study of inhaled terbutaline administered via nebuhaler or nebuliser to asthmatic children. *Eur J Respir Dis* (1987) **71**: 96–101.
162. O'Callaghan C: Particle size of beclomethasone dipropionate produced by 2 nebulisers and spacers. *Thorax* (1990) **45**: 109–111.
163. Verberne AA, Frost C, Roor da RJ, van der Laag H, Kerrebijn KFSO: One year treatment with salmeterol, compared with beclomethasone in children with asthma. *Am J Respir Crit Care Med* (1996) **156**(3): 688–695.
164. Laitinen LA, Laitinen A, Haahtela T: A comparative study of the effects of an inhaled corticosteroid, budesonide, and of a beta-2-agonist, terbutaline, on airway inflammation in newly diagnosed asthma. *J Allergy Clin Immunol* (1992) **90**: 32–42.
165. Djukanovic R, Wilson JW, Britten YM, *et al.*: Effect of an inhaled corticosteroid on airway inflammation and symptoms of asthma. *Am Rev Respir Dis* (1992) **145**: 699–674.
166. Jeffery PK, Godfrey RW, Adelroth E, Nelson F, Rogers A, Johansson S: Effect of treatment on airway inflammation and thickening of basement membrane reticular collagen in asthma. *Am Rev Respir Dis* (1992) **145**: 890–899.
167. Burke C, Power CK, Norris A, Condez A, Schmekel B, Poulter LW: Lung function and immunopathological changes after inhaled corticosteroid therapy in asthma. *Eur Respir J* (1992) **5**: 73–79.
168. Barnes PJ: Effect of corticosteroids on airway hyperresponsiveness. *Am Res Respir Dis* (1990) **141**: 70–76.
169. Ernst P, Habbick B, Suissa S, *et al.*: Is the association between inhaled beta-agonist use and

life-threatening asthma because of confounding by severity? *Am Rev Respir Dis* (1993) **148**: 75–79.
170. Boner AL, Piacentini GL, Bonizzato C, Dattoli V, Sette L: Effect of inhaled beclomethasone dipropionate on bronchial hyperreactivity in asthmatic children during maximal allergen exposure. *Pediatr Pulmonol* (1991) **10**: 2–5.
171. Cockroft DW, Murdoch KY: Comparative effects of inhaled salbutamol, sodium cromoglycate and BDP on allergen-induced early asthmatic responses, late asthmatic responses and increased bronchial responsiveness to histamine. *J Allergy Clin Immunol* (1987) **79**: 734–740.
172. Burge PS: The effects of corticosteroids on the immediate asthmatic reaction. *Eur J Respir Dis* (1982) **63** (Suppl 122): 163–166.
173. Dahl R, Johansson S: Importance of duration of treatment with inhaled budesonide on the immediate and late bronchial reaction. *Eur J Respir Dis* (1982) **62** (Suppl 122): 5167–5175.
174. De Baets FM, Goetyn M, Kerrebijn KF: The effect of two months of treatment with inhaled budesonide on bronchial responsiveness to histamine and house-dust mite antigen in asthmatic children. *Am Rev Respir Dis* (1990) **142**: 581–586.
175. Molema J, van Herwaarden CLA, Folgering HTM: Effect of long-term treatment with inhaled cromoglycate and budesonide on bronchial hyperresponsiveness in patients with allergic asthma. *Eur Respir J* (1989) **2**: 308–316.
176. Kraan J, Koeter GH, Van der Mark TW, Sluiter HJ, De Vries K: Changes in bronchial hyperreactivity induced by 4 weeks of treatment with antiasthmatic drugs in patients with allergic asthma: a comparison between budesonide and terbutaline. *J Allergy Clin Immunol* (1985) **76**: 628–636.
177. Kerrebijn KF, van Essen-Zandvliet EEM, Neijens HL: Effect of long-term treatment with inhaled corticosteroids and beta-agonists on bronchial responsiveness in asthmatic children. *J Allergy Clin Immunol* (1987) **79**: 653–659.
178. Dutoit JI, Salome CM, Woolcock AJ: Inhaled corticosteroids reduce the severity of bronchial hyperresponsiveness in asthma, but oral theophylline does not. *Am Rev Respir Dis* (1987) **136**: 1174–1178.
179. Bel EH, Timmers MC, Hermans JO, Dijkman JH, Sterk PJ: The longer term effects of nedocromil sodium and beclomethasone dipropionate on bronchial responsiveness to methacholine in non-atopic asthmatic subjects. *Am Rev Respir Dis* (1990) **141**: 21–28.
180. Van Essen-Zandvliet EE, Hughes MD, Waalkens HJ, Duiverman EJ, Pocock SJ, Kerrebijn KF: Effects of 22 months of treatment with inhaled corticosteroids and/or beta-2-agonists on lung function, airway responsiveness and symptoms in children with asthma. *Am Rev Respir Dis* (1992) **146**: 547–554.
181. Østergaard P, Pedersen S: The effect of inhaled disodium cromoglycate and budesonide on bronchial responsiveness to histamine and exercise in asthmatic children: a clinical comparison. In Godfrey S (ed) *Glucocorticosteroids in Childhood Asthma*. Amsterdam, Excerpta Medica, 1987, pp 69–76.
182. Benoist MR, Brouard JJ, Rufin P, *et al*.: Dissociation of symptom scores with bronchial hyperreactivity: study in asthmatic children on long-term treatment with inhaled beclomethasone dipropionate. *Pediatr Pulmonol* (1991) **13**: 71–77.
183. Smolensky MH, McGovern JP, Scott PH, Reinberg A: Chronobiology and asthma. II. Body-time-dependent differences in the kinetics and effects of bronchodilator medications. *J Asthma* (1987) **24**: 91–134.
184. White MP, MacDonald TH, Garg RA: Ketotifen in the young asthmatic: a double blind placebo controlled trial. *J Int Med Res* (1988) **16**: 107–113.
185. Bennati D, Piacentini GL, Peroni DG, Sette L, Testi R, Boner AL: Changes in bronchial reactivity in asthmatic children after treatment with beclomethasone alone or in association with salbutamol. *J Asthma* (1989) **26**: 359–364.
186. Kraemer R, Sennhauser F, Reinhardt M: Effects of regular inhalation of beclomethasone dipropionate and sodium cromoglycate on bronchial hyperreactivity in asthmatic children. *Acta Paediatr Scand* (1987) **76**: 119–123.
187. Resnick A, Greenberger PA: A corticosteroid program for prevention of hospitalization for status asthmaticus in children. *Allergy Proc* (1987) **8**: 104–107.

188. Eseverri JL, Botey J, Marin AM: Budesonide: treatment of bronchial asthma during childhood. *Allerg Immunol* (1995) **27**: 129–135.
189. Ribeiro LB: Budesonide: safety and efficacy aspects of its long-term use in children. *Pediatr Allergy Immunol* (1993) **4**: 73–78.
190. Boner AL, Comis A, Schiassi M, Venge P, Piacentini GL: Bronchial reactivity in asthmatic children at high and low altitude: effect of budesonide. *Am J Respir Crit Care Med* (1995) **151**: 1194–1200.
191. Gonzalez Perez-Yarza E, Garmendia Iglesias A, Mintegui Aramburu J, Callen Blecua M, Albisu Andrade Y, Rubio Calvo E: Prolonged treatment of mild asthma with inhaled antiinflammatory therapy. *An Esp Pediatr* (1994) **41**: 102–106.
192. Larsen JS, De Boisblanc BP, Schaberg A, et al.: Magnitude of improvement in FEV1 with fluticasone propionate. *Am J Respir Crit Care Med* (1994) **149**: A214.
193. MacKenzie CA, Weinberg EG, Tabachnik E, Taylor M, Havnen J, Crescenzi K: A placebo controlled trial of fluticasone propionate in asthmatic children. *Eur J Pediatr* (1993) **152**: 856–860.
194. Perera BJ: Efficacy and cost effectiveness of inhaled steroid in asthma in a developing country. *Arch Dis Child* (1995) **72**: 312–316.
195. Svedmyr J, Nyberg E, Øsbrink-Nilsson E, Hedlin G: Intermittent treatment with inhaled steroids for deterioration of asthma due to upper respiratory tract infections. *Acta Paediatr Int J Paediatr* (1995) **84**: 884–888.
196. Meltzer EO, Orgel HA, Ellis EF, Eigen HN, Hemstreet MPB: Long-term comparison of three combinations of albuterol, theophylline, and beclomethasone in children with chronic asthma. *J Allergy Clin Immunol* (1992) **90**: 2–11.
197. Youngchaiyud P, Permpikul C, Suthamsmai T, Wong E: A double-blind comparison of inhaled budesonide, a long-acting theophylline and their combination in the treatment of nocturnal asthma. *Allergy* (1995) **50**: 28–33.
198. Pedersen S, Agertoft L: Effect of long-term budesonide treatment on growth, weight and lung function in children with asthma. *Am Rev Respir Dis* (1993) **147**: A265.
199. Connett GJ, Lenney W, McConchie SM: The cost effectiveness of budesonide in severe asthmatics aged one to three years. *Br J Med Econ* (1993) **6**: 127–134.
200. Edmunds AT, Goldberg RS, Duper B, Devichand P, Follows RM: A comparison of budesonide 800 micrograms and 400 micrograms via Turbohaler with disodium cromoglycate via Spinhaler for asthma prophylaxis in children. *Br J Clin Res* (1994) **5**: 11–23.
201. Stromberg L: Decreasing admissions for childhood asthma to Swedish country hospital. *Acta Paediatr Scand* (1996) **85**: 173–176.
202. Wennergren G, Kristjansson S, Strannegard IL: Decrease in hospitalization for treatment of childhood asthma with increased use of antiinflammatory treatment, despite an increase in the prevalence of asthma. *J Allergy Clin Immunol* (1996) **97**: 742–748.
203. Lenney W, Wells NE, O'Neill BA: The burden of paediatric asthma. *Eur Respir Rev* (1994) **4**: 49–62.
204. Gerdtham UG, Hertzman P, Jönsson B, Boman GSO: Impact of inhaled corticosteroids on acute asthma hospitalization in Sweden 1978 to 1991. *Med-Care* (1996) **34**(12): 1188–1198.
205. Fuglsang G, Vikre-Jørgensen J, Agertoft L, et al.: Influence of salmeterol treatment upon nitric oxide level in exhaled air and bronchodilator response to terbutaline in children with mild asthma. *Pediatr Pulmonol* (1998) in press.
206. Bel EH, Timers MC, Zwinderman AH, Dijkman JH, Sterk PJ: The effect of inhaled corticosteroids on the maximal degree of airway narrowing to methacholine. *Am Rev Respir Dis* (1991) **143**: 109–113.
207. Pedersen S, Hansen OR: Budesonide treatment of moderate and severe asthma in children. A dose response study. *J Allergy Clin Immunol* (1995) **1**: 29–33.
208. Shapiro GG: Childhood asthma: update. *Pediatr Rev* (1992) **13**: 403–412.
209. Katz Y, Lebas FX, Medley HV: Double-blind placebo controlled parallel group study to compare the efficacy and safety of fluticasone propionate at two doses delivered via a Diskhaler inhaler in children with asthma. *Am J Respir Crit Care Med* (1996) **153**: A75.
210. Waalkens HJ, van Essen Zandvliet EE, Gerritsen J, Duiverman EJ, Kerrebijn KF, Knol K:

The effect of an inhaled corticosteroid (budesonide) on exercise-induced asthma in children. Dutch CNSLD Study Group. *Eur Respir J* (1993) **6**: 652–656.
211. Henriksen JM, Dahl R: Effects of inhaled budesonide alone and in combination with low-dose terbutaline in children with exercise-induced asthma. *Am Rev Respir* (1983) **128**: 993–997.
212. Henriksen JM: Effect of inhalation of corticosteroids on exercise-induced asthma: randomised double blind cross-over study of budesonide in asthmatic children. *Br Med J* (1985) **291**: 248–249.
213. Hummel S, Lehtonen L: Comparison of oral-steroid sparing by high-dose and low-dose inhaled steroid in maintenance treatment of severe asthma. *Lancet* (1992) **340**: 1483–1487.
214. Boe J, Rosenhall L, Alton M, *et al.*: Comparison of dose response effects of inhaled beclomethasone dipropionate and budesonide in the management of asthma. *Allergy* (1989) **44**: 349–355.
215. Dahl R, Lundback B, Malo J, *et al.*: A dose ranging study of fluticasone propionate in adult patients with moderate asthma. *Chest* (1993) **5**: 1352–1358.
216. Østergaard P, Pedersen S: Bronchial hyperactivity in children with perennial extrinsic asthma. In Oseid S, Edwards AM (eds) *The Asthmatic Child in Play and Sport*. London, Pitman, 1982, pp 326–331.
217. Barnes PJ, Pedersen S: Efficacy and safety of inhaled corticosteroids in asthma. *Am Rev Respir Dis* (1993) **148**: 1–26.
218. Lewis CE, Rachelefsky G, de la Sota A, *et al.*: A randomized trial of asthma care training for kids. *Pediatrics* (1984) **74**: 478–486.
219. Pedersen S: Ensuring compliance in children. *Eur Respir J* (1992) **5**: 143–145.
220. Nussbaum E, Eyzaguirre M, Galant SP: Dose–response relationship of inhaled metaproterenol sulfate in preschool children with mild asthma. *Pediatrics* (1990) **85**: 1072–1075.
221. Ahlstrom H, Svenonius E, Svensson M: Treatment of asthma in pre-school children with inhalation of terbutaline in Turbuhaler compared with Nebuhaler. *Allergy* (1989) **44**: 515–518.
222. Yuksel B, Greenough A, Maconochie I: Effective bronchodilator therapy by a simple spacer device for wheezy premature infants in the first two years of life. *Arch Dis Child* (1990) **65**: 782–785.
223. Conner WT, Dolovich MB, Frame RA, Newhouse MT: Reliable salbutamol administration in 6 to 36 month old children by means of a metered dose inhaler and aerochamber with mask. *Pediatr Pulmonol* (1989) **6**: 263–267.
224. Prendiville A, Rose A, Maxwell DL, Silvermann M: Hypoxaemia in wheezy infants after bronchodilator treatment. *Arch Dis Child* (1987) **62**: 997–1000.
225. Lodrup KC, Carlsen KH: The effect of inhaled nebulised racemic adrenaline upon lung function in infants with bronchiolitis. European Respiratory Society Meeting 1990.
226. Prendiville A, Green S, Silverman M: Airway responsiveness in wheezy infants: evidence for functional beta adrenergic receptors. *Thorax* (1987) **42**: 100–104.
227. Yuksel B, Greenough A: Effect of nebulized salbutamol in preterm infants during the first year of life. *Eur Respir J* (1991) **4**: 1088–1092.
228. Wilkie RA, Bryan MH: Effect of bronchodilator on airway resistance in ventilator-dependent neonates with chronic lung disease. *J Pediatr* (1987) **111**: 278–282.
229. Sosulski R, Abbasi S, Bhutani V, Fox W: Physiological effects of terbutaline on pulmonary function of infants with bronchopulmonary dysplasia. *Pediatr Pulmonol* (1986) **2**: 269–273.
230. Kao LC, Durand DJ, Nickerson GB: Effects of inhaled metaproterenol and atropine on the pulmonary mechanics of infants with bronchopulmonary dysplasia. *Pediatr Pulmonol* (1989) **7**: 74–80.
231. Cabal LA, Lanazabal C, Ramanathan R, *et al.*: Effects of metaproterenol on pulmonary mechanics, oxygenation and ventilation in infants with chronic lung disease. *J Pediatr* (1987) **110**: 116–119.
232. Kraemer R, Frey U, Sommer CW, Russi E: Short term effect of albuterol, delivered via a new auxiliary device, in wheezy infants. *Am Rev Respir Dis* (1991) **144**: 347–351.
233. O'Callaghan C, Milner AD, Swarbrick A: Nebulised salbutamol does have a pertective effect on airways in children under one year old. *Arch Dis Child* (1988) **63**: 479–483.
234. Ho L, Collis G, Landau LI, Le Souef PN: Effect of salbutamol on oxygen saturation in bronchiolitis. *Arch Dis Child* (1981) **66**: 1061–1064.

235. Stratton D, Carswell F, Hughes AO, Fysh WJ, Robinson P: Double-blind comparisons of slow release theophylline, ketotifen and placebo for prophylaxis of asthma in young children. *Br J Dis Chest* (1984) **78**: 163–167.
236. Groggins RC, Lenney W, Milner AD, Stokes GM: Efficacy of orally administered salbutamol and theophylline in pre-schoolchildren with asthma. *Arch Dis Child* (1980) **55**: 204–206.
237. Hodges IGC, Groggins RC, Milner AD, Stokes GM: Bronchodilator effect of inhaled ipratropium bromide in wheezy toddlers. *Arch Dis Child* (1981) **56**: 729–732.
238. O'Callaghan C, Milner AD, Swarbrick A: Spacer device with face mask attachment for giving bronchodilators to infants with asthma. *Br Med J* (1989) **298**: 160–161.
239. Groggins RC, Milner AD, Stokes GM: Bronchodilator effects of clemastine, ipratropium bromide, and salbutamol in preschool children with asthma. *Arch Dis Child* (1981) **56**: 342–344.
240. Stokes GM, Milner AD, Hodges IGC, Elphick MC, Henry RI: Nebulised therapy in acute severe bronchitis in infancy. *Arch Dis Child* (1983) **58**: 279–282.
241. Henry RL, Hiller EJ, Milner AD, Hodges IGC, Stokes GM: Nebulised ipratropium bromide and sodium cromoglycate in the first two years of life. *Arch Dis Child* (1984) **59**: 54–57.
242. Bertelsen A, Andersen JB, Busch P, et al.: Nebulised sodium cromoglycate in the treatment of wheezy bronchitis. *Allergy* (1986) **41**: 266–270.
243. Cogswell JJ, Simpkiss MJ: Nebulised sodium cromoglycate in recurrently wheezy pre-school children. *Arch Dis Child* (1985) **60**: 736–738.
244. Glass J, Archer LN, Adams W, Simpson H: Nebulised cromoglycate, theophylline, and placebo in preschool asthmatic children. *Arch Dis Child* (1981) **56**: 648–651.
245. Geller-Bernstein C, Levin S: Nebulised sodium cromoglycate in the treatment of wheezy bronchitis in infants and young children. *Respiration* (1982) **43**: 294–298.
246. Miraglia del Giudice M, Capristo A, Maiello N, Apuzzu G: Nebulized sodium cromoglycate for the treatment of asthma in children under five years of age. *Med Probl Paediatr* (1982) **121**: 122–127.
247. Furukawa CT, Shapiro GG, Bierman CW, Kraemer MJ, Wad DJ, Pierson WE: A double blind study comparing the effectiveness of cromolyn sodium and sustained release theophyllin in childhood asthma. *Pediatrics* (1984) **74**: 453–459.
248. Pedersen S: Clinical pharmacology and therapeutics. In Silverman M (ed) *Childhood Asthma and Other Wheezing Disorders*. London, Chapman & Hall, 1995, p 261–312.
249. Godfrey S, Avital A, Rosler A, Mandelberg A, Uwyyed K: Nebulised budesonide in severe infantile asthma. *Lancet* (1987) **ii**: 851–852.
250. de Jongste JC, Duiverman EJ: Nebulised budesonide in severe childhood asthma. *Lancet* (1989) **i**: 1388.
251. Volovitz B, Amir J, Malik H, Kauschansky A, Varsano I: Growth and pituitary–adrenal function in children with severe asthma treated with inhaled budesonide. *N Engl J Med* (1993) **329**: 1703–1733.
252. Webb MSC, Milner AD, Hiller EJ, Henry RI: Nebulised beclomethasone dipropionate suspension. *Arch Dis Child* (1986) **61**: 1108–1110.
253. Maayan C, Itzhaki T, Bar-Yishay E, Gross S, Tal A, Godfrey S: The functional response of infants with persistent wheezing to nebulized beclomethasone dipropionate. *Pediatr Pulmonol* (1986) **2**: 9–14.
254. Storr J, Lenney CA, Lenney W: Nebulised beclomethasone dipropionate in preschool asthma. *Arch Dis Child* (1986) **61**: 270–273.
255. Pedersen S: Studies with nebulised budesonide. In Godfrey S (ed) *Budesonide. Nebulising Suspension*. Oxford, Henry Ling, Dorset Press, 1989, pp 25–29.
256. Greenough A, Pool J, Gleeson JG, Price JF: Effect of budesonide on pulmonary hyperinflation in young asthmatic children. *Thorax* (1988) **43**: 937–938.
257. Rachelefsky GS, Warner JO: International consensus on the management of pediatric asthma: a summary statement. *Pediatr Pulmonol* (1993) **15**: 125–127.
258. Warner JO, Gotz M, Landau LI, et al.: Management of asthma: a consensus statement. *Arch Dis Child* (1989) **64**: 1065–1079.
259. Nelson DR, Sachs MI, O'Connell EJ: Approaches to acute asthma and status asthmaticus in children. *Mayo Clin Proc* (1989) **64**: 1392–1402.

260. McWilliams B, Kelly HW, Murphy S: Management of acute severe asthma. *Pediatr Ann* (1989) **18**: 774–775, 779.
261. Murphy S, Kelly HW: Management of acute asthma. *Pediatrician* (1991) **18**: 287–300.
262. Press S, Lipkind RS: A treatment protocol of the acute asthma patient in a pediatric emergency department. *Clin Pediatr* (1991) **30**: 573–577.
263. Henry RL, Robertson CF, Asher I, et al.: Management of acute asthma. Respiratory paediatricians of Australia and New Zealand. *J Paediatr Child Health* (1993) **29**: 101–103.
264. Niggemann B, Wahn U: Die Therapie des Status asthmaticus im Kindesalter. *Monatsschr Kinderheilkd* (1991) **139**: 323–329.
265. Pedersen S: Guidlines for management of acute asthma in children. In O'Byrne P, Thomson N (eds) *Manual of Asthma Management.* London, WB Saunders, 1995, pp 511–543.
266. Connett GJ, Lenney W: Use of pulse oximetry in the hospital management of acute asthma in childhood. *Pediatr Pulmonol* (1993) **15**: 345–349.
267. Canny GJ, Levison H: Pulmonary function abnormalities during apparent clinical remission in childhood asthma. *J Allergy Clin Immunol* (1988) **82**: 1–4.
268. Williams AJ, Santiago S, Weiss EB, Stein M: Status asthmaticus. *Acute Care* (1988) **14–15**: 208–228.
269. Storr J, Barry W, Barrell E, Lenney W, Hatcher G: Effect of a single dose of prednisolone in acute childhood asthma. *Lancet* (1987) **i**: 879–882.
270. Vichyanond P, Irvin CG, Larsen GL, Szefler SJ, Hill MR: Penetration of corticosteroids into the lung: evidence for a difference between methylprednisolone and prednisolone. *J Allergy Clin Immunol* (1989) **84**: 867–873.
271. Harfi H, Hanissian AS, Crawford LV: Treatment of status asthmaticus in children with high doses and conventional doses of methylprednisolone. *Pediatrics* (1978) **61**: 829–831.
272. Schuh S, Reider MJ, Canny G, et al.: Nebulized albuterol in acute childhood asthma: comparison of two doses. *Pediatrics* (1990) **86**: 509–513.
273. Papo MC, Frank J, Thompson AE: A prospective, randomized study of continuous versus intermittent nebulized albuterol for severe status asthmaticus in children. *Crit Care Med* (1993) **21**: 1479–1486.
274. Robertson CF, Smith F, Beck R, Levison H: Response to frequent low doses of nebulized salbutamol in acute asthma. *J Pediatr* (1985) **106**: 672–674.
275. Portnoy J, Nadel G, Amado M, Willsie-Ediger S: Continuous nebulization for status asthmaticus. *Ann Allergy* (1992) **69**: 71–79.
276. Singh M, Kumar L: Continuous nebulized salbutamol and oral once a day prednisolone in status asthmaticus. *Arch Dis Child* (1993) **69**: 416–419.
277. Svenonius E, Arborelius M, Wiberg R, Ekberg P: Prevention of exercise-induced asthma by drugs inhaled from metered aerosols. *Allergy* (1988) **43**: 252–257.
278. Watson WTA, Becker AB, Simons FER: Comparison of ipratropium solution, fenoterol solution and their combination administered by nebuliser and face mask to children with acute asthma. *J Allergy Clin Immunol* (1988) **82**: 1012–1018.
279. Reisman J, Galdes-Sebalt M, Kazim F, Canny G, Levison H: Frequent administration by inhalation of salbutamol and ipratropium bromide in the initial management of severe acuta asthma in children. *J Allergy Clin Immunol* (1988) **81**: 10–20.
280. Beck R, Robertson C, Galdès-Sebaldt M, Levison H: Combined salbutamol and ipratropium bromide by inhalation in the treatment of severe acute asthma. *J Pediatr* (1985) **107**: 605–608.
281. Phanichyakarn P, Kraisarin C, Sasisakulporn C: Comparison of inhaled terbutaline and inhaled terbutaline plus ipratropium bromide in acute asthmatic children. *Asian Pac J Allergy Immunol* (1990) **8**: 45–58.
282. Reisman J, Galdes-Sebalt M, Kazim F, Canny G, Levison H: Frequent administration by inhalation of salbutamol and ipratropium bromide in the initial management of severe acute asthma in children. *J Allergy Clin Immunol* (1988) **81**: 16–20.
283. Storr J, Lenney W: Nebulised ipratropium and salbutamol in asthma. *Arch Dis Child* (1986) **61**: 602–603.
284. Lenney W, Evans AP: Nebulized salbutamol and ipratropium bromide in asthmatic children. *Br J Dis Chest* (1986) **80**: 59–65.

285. Summers OA, Tarala RA: Bronchodilator efficacy of nebulized ipratropium sequentially and in combination in acute asthma. *Thorax* (1987) **42**: 731.
286. Rayner RJ, Cartlidge PHT, Upton CJ: Salbutamol and ipratropium in acute asthma. *Arch Dis Child* (1987) **62**: 840–841.
287. Edmunds AT, Godfrey S: Cardiovascular response during severe acute asthma and its treatment in children. *Thorax* (1981) **36**: 534–540.
288. Hambleton G, Stone MJ: Comparison of iv salbutamol with iv aminophylline in the treatment of severe, acute asthma in childhood. *Arch Dis Child* (1979) **54**: 391–402.
289. Bohn D, Kalloghlian A, Jenkins J, Edmunds J, Barker G: Intravenous salbutamol in the treatment of status asthmaticus in children. *Crit Care Med* (1984) **12**: 892–896.
290. Fuglsang G, Pedersen S, Borgström L: Dose–response relationships of intravenously administered terbutaline in children with asthma. *J Pediatr* (1989) **114**: 315–320.
291. Janson C, Herala M: Plasma terbutaline levels in nebulisation treatment of acute asthma. *Pulmon Pharmacol* (1991) **4**: 135–139.
292. Pedersen S: Treatment strategies for acute asthma in infants and children. *Res Clin Forums* (1993) **15**: 55–61.
293. Fuglsang G, Pedersen S: Comparison of nebuhaler and nebulizer treatment of acute severe asthma in children. *Eur J Respir Dis* (1986) **69**: 109–113.
294. Carter E, Cruz M, Chesrown S, Shieh G, Reilly K, Hendeles L: Efficacy of intravenously administered theophylline in children hospitalized with severe asthma. *J Pediatr* (1993) **122**: 470–476.
295. Barclay J, Whiting P, Mickey M, Addis G: Theophylline–salbutamol interaction: bronchodilator response to salbutamol at maximally effective plasma theophylline concentrations. *Br J Clin Pharmacol* (1981) **11**: 203–208.
296. Fanta C, Rossing T, McFadden E: Treatment of acute asthma: is combination therapy with sympathomimetics and methylxanthines indicated? *Am J Med* (1986) **80**: 5–10.
297. Littenberg B: Aminophyllin treatment in severe, acute asthma. *JAMA* (1988) **259**: 1678–1684.
298. Rubin BK, Marcusbamer S, Priel I, App EM: Emergency management of the child with asthma. *Pediatr Pulmonol* (1990) **8**: 45–57.
299. Schuh S, Parkin P, Rajan A, et al.: High versus low-dose, frequently administered, nebulized albuterol in children with severe, acute asthma. *Pediatrics* (1989) **83**: 513–518.
300. Reed CE (ed): *Immunotherapy: Yesterday's Treatment.* CV Mosby, 1986.
301. Warner JO, Kerr JW: Hypersensitisation. *Br Med J* (1987) 1179.
302. Henriksen JM, Nielsen TT: Effect of physical training on exercise-induced bronchoconstriction. *Acta Paediatr Scand* (1983) **72**: 31–36.
303. Oseid S, Edwards AM (eds): *The Asthmatic Child in Play and Sport.* London, Pitman, 1983.
304. Henriksen JM, Agertoft L, Pedersen S: Protective effect and duration of action of inhaled formoterol and salbutamol on exercise-induced asthma in children. *J Allergy Clin Immunol* (1992) **89**: 1176–1182.
305. Green CP, Price JF: Prevention of exercise induced asthma by inhaled salmeterol xinafoate. *Arch Dis Child* (1992) **67**: 1014–1017.
306. Tal A, Bavilska C, Yohai D, Bearman JE, Gorodisher R, Moses SW: Dexamethasone and salbutamol in the treatment of acute wheezing in infants. *Pediatrics* (1993) **71**: 13–18.
307. Webb MSC, Henry RL, Milner AD: Oral corticosteroids for wheezing attacks under 18 months. *Arch Dis Child* (1986) **61**: 15–19.
308. Daugbjerg P, Brenøe E, Forchammer HE: A comparison between nebulized terbutaline, nebulized corticosteroid and systemic corticosteroid for acute wheezing in children up to 18 months of age. *Acta Paediatr Scand* (1993) **82**: 547–551.
309. Strunk RC: Identification of the fatality-prone subject with asthma. *J Allergy Clin Immunol* (1989) **83**: 477–485.

47

Pharmacoeconomics of Asthma Treatments

SEAN D. SULLIVAN AND KEVIN B. WEISS

INTRODUCTION

Increasing use and costs of medical care services, particularly the substantial burden of managing chronic diseases have made medical decision-makers, public and private payers, and society acutely aware of the problem of scarce resources. Scarcity of medical care resources has focused discussion on the need to make choices among medical treatments. The choice of treatments must balance the tension between providing all medical services that are technically feasible or that patients desire and financing these services with limited resources.

Resource constraints directly and indirectly affect medical treatment decisions. Yet, little data exist to inform clinicians about the impact of alternative treatment choices on resource use and overall costs of care. Pharmacoeconomic studies can be used to improve the quality of medical and financial decision-making by providing evidence on the interrelationships between treatment choices, health outcomes and the overall cost of medical care.

For the management of asthma, there are a variety of medical treatment alternatives from which to choose: pharmaceuticals, specialty care, desensitization regimens and education programmes to name but a few. In the past, decisions about the use of treatments were made based in large part on clinical and patient factors. However, in the face of mounting budget constraints physicians and other healthcare professionals are looking toward outcomes research and pharmacoeconomic studies to assist in the selection and prescribing of asthma management strategies. Within managed healthcare, the acceptance of medical innovation is increasingly determined by technology adoption committees that make use of drug and device formularies, practice guidelines, prior

ASTHMA: BASIC MECHANISMS AND CLINICAL MANAGEMENT (3rd Edn)
ISBN 0-12-079027-9

Copyright © 1998 Academic Press Limited
All rights of reproduction in any form reserved

approval and case and disease management programs.[1] These committees are now seeking information on patient outcomes and the cost consequences of introduction of new technology.[2] Pharmacoeconomic and outcomes studies can support rational decision-making about the use of treatments for patients with asthma.

All this presumes that there are sufficient data on the costs and benefits of treatment options for asthma and that clinicians and resource decision-makers understand and use pharmacoeconomic and outcomes studies as part of the treatment selection process. The purpose of this chapter is to briefly define the discipline and uses of pharmacoeconomic studies and to review the most recent data on the costs and benefits of asthma pharmacotherapy.

PRINCIPLES AND APPLICATIONS OF PHARMACOECONOMICS

Pharmacoeconomics can be defined as a set of research methods to assess and quantify the costs and clinical consequences of medical care treatments in order to estimate the 'economic value' of the treatment in relation to alternative treatments.[3] A pharmacoeconomic evaluation of a medical treatment should incorporate evidence on the clinical consequences (efficacy and safety) and the costs and relative cost-effectiveness of treatment alternatives. Unfortunately, despite the obvious need for such information, little pharmacoeconomic evidence is available on alternative asthma management strategies.

Integrating costs and outcomes into one analysis is the primary goal of pharmacoeconomic research studies. The possible outcomes of a pharmacoeconomic study are illustrated in Fig 47.1.[4] Quadrant B illustrates a treatment that may be less efficacious or more harmful and cost more than the current treatment. Quadrant D shows a treatment that imparts lower health outcome and is less expensive. Quadrant A shows the cost–outcome relationship of most new medical technology. Health benefits are improved at a quantifiable and incremental expense to the healthcare system. Given this information,

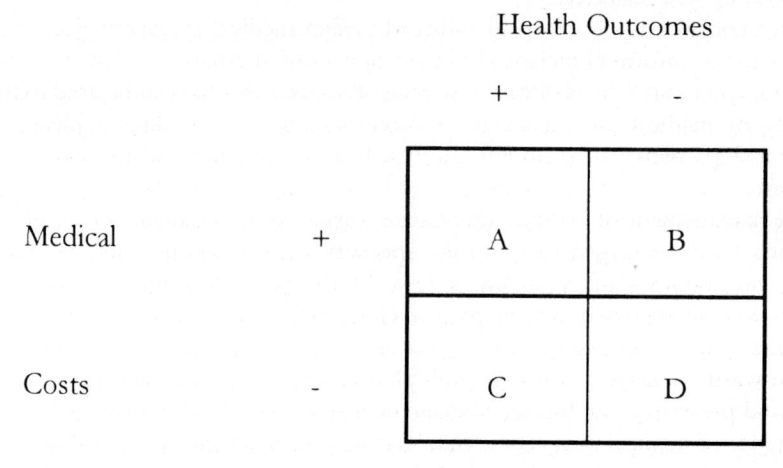

Fig. 47.1 Possible outcomes of pharmacoeconomic studies.

clinicians, patients and payers must decide whether the improvement in health outcome is worth the additional costs of care. In a healthcare system with a fixed budget, additional expenditure on new treatments deprives patients with other medical ailments of these same resources.

Finally, quadrant C depicts a dominant technology, one that improves health outcome and achieves cost savings. Quadrant C technologies are the most desirable for any healthcare system. The unique challenge of pharmacoeconomic studies is to quantify the health outcomes of medical treatments, in terms useful to decision-makers.

Decisions about which medical treatment or procedures to employ often are based upon evidence from controlled clinical trials regarding efficacy and safety. However, efficacy is not synonymous with effectiveness. Efficacy is measured under tightly controlled research conditions, often on a highly selected patient population. Effectiveness refers to the impact of the intervention or technology under routine clinical conditions when administered to a more generalized patient population. Further, clinical trials are not optimally designed for pharmacoeconomic assessment.[5] Clinical trials frequently involve blinded, placebo control groups or alternatives not widely used in clinical practice. More importantly, clinical trials are not statistically powered for evaluation of economic end-points. Trials often focus on physiological end-points rather than symptom-based, functional or health-related quality-of-life outcomes. Trial data may show that a therapy produces a small improvement in clinical outcome and that the improvement is achieved at a very high cost relative to standard care. Thus, it is often difficult to determine which therapy or combination of therapies is most efficient at achieving a desired cost-related outcome.

Cost-effectiveness and cost–benefit analyses are pharmacoeconomic methods that compare the costs and consequences of alternative healthcare interventions to provide information on how to choose among alternative treatments in order to most efficiently allocate scarce resources. This chapter focuses on cost-effectiveness analysis, partly because of its increased use in pharmacoeconomic studies but primarily because of its potential for improving decision making for the treatment of asthma.

COST–BENEFIT ANALYSIS

Cost–benefit analysis allows for the identification and comparison of the costs associated with the implementation or use of a medical programme or technology and the benefits derived from its application.[6] Both costs and benefits are defined in monetary terms and adjusted to net present values. The results of these studies are usually reflective of a wider societal point of view. Thus the analysis considers both private and social costs and benefits. The ratio of monetary benefits to overall costs provides a way to determine whether the value produced by the technology is worth the cost: the intervention is said to be cost beneficial if the benefits exceed the costs. However, many technical, social and ethical problems are associated with expressing health outcomes in monetary terms. Difficulty may arise when the benefits of the intervention are not amenable to economic valuation, such as years of life saved or improvements in psychosocial outcomes. It is often difficult to value health improvements in monetary terms. Economists rely on revealed preferences in actual markets or on hypothetical estimations derived from willingness-to-pay studies.[7] As a consequence, cost–benefit analyses tend to be used less widely.

COST-EFFECTIVENESS ANALYSIS

The most common economic evaluation technique is cost-effectiveness analysis. This analytical technique simultaneously considers the relative costs and outcomes of two or more alternative medical technologies when used to treat a similar condition.[8] Like cost–benefit analysis, the cost-effectiveness technique makes explicit the positive and negative costs and consequences of various medical technologies. However, cost-effectiveness analysis differs from cost–benefit analysis in that the health outcomes of treatments are expressed in 'natural' units such as symptom-free days or quality-adjusted years of life saved and not in monetary terms. A cost-effectiveness evaluation requires estimation of two inputs: (a) a direct measure of absolute and comparative health outcome or effectiveness; and (b) an estimate of total and incremental medical costs.

In cost-effectiveness analysis, as in other economic evaluation techniques, costs are comprehensively evaluated and not limited to the assessment of the cost of therapy. For example, if only the costs of medications are assessed in an evaluation of drug therapy for asthma, a number of important economic parameters will be disregarded. These may include the direct costs associated with the use of medical resources to treat significant adverse reactions to the drug or the savings that result from averted hospitalization and emergency department visits due to improved clinical outcome. Furthermore, important non-economic factors, such as improvements in functional status days missed from work or school and changes in quality of life, also will be ignored.

Cost-effectiveness analysis is grounded in the clinical effectiveness of healthcare interventions. Thus, the clinical outcomes of an intervention and its alternatives must be clearly understood before cost-effectiveness hypotheses can be generated and tested. Consequently, fully informed resource allocation decisions require information about the clinical effectiveness of the medical care treatment, its impact on the patient's health and functional status and the full economic implications of its use. Inevitably, pharmacoeconomic evaluations contain imperfect information, and the level and extent of uncertainty within the study necessitates further exploration; therefore, there is a need for simulations and sensitivity analyses that evaluate the impact of varying assumptions on the results of the study.

Cost–utility analysis is a special form of the cost-effectiveness model in which health outcomes are expressed in quality-adjusted life years (QALY) gained. The quality adjustment is derived from preference weights or health utilities.[3] The advantages of a cost–utility study are: (a) a QALY captures simultaneously changes in mortality and morbidity and is applicable to all disease states and treatments; (b) a QALY considers patients' preferences for health outcomes; and (c) important to analysts, cost–utility analyses conform to normative theory of decision-making under uncertainty.[10]

These studies have a conceptual appeal to researchers and are the most ideal for use in decision-making because of the features described above. The province of Ontario, Canada strongly encourages cost–utility analyses of pharmaceuticals to support formulary listing. However, technical limitations in measurement of preferences and a lack of long-term data concerning asthma interventions severely limit the use of this approach.

ASTHMA OUTCOMES FOR PHARMACOECONOMIC EVALUATION

A variety of outcome measures can be used for cost-effectiveness analyses. The choice of outcome measure for pharmacoeconomic analysis depends upon the intervention itself and the specific informational needs of the medical care decision-makers. If the results of asthma cost-effectiveness analysis studies are to be useful, the outcome variable must be relevant to the healthcare system and the providers. In addition, there must be some degree of standardization of outcome measures across studies so that different interventions can be compared when resource allocation decisions are considered.

A recent report of the National Heart, Lung and Blood Institute Workshop on Asthma Outcome Measures for Research Studies provides a useful review of the many end-points available to researchers who study asthma.[10] In addition, the National Asthma Education and Prevention Program Task Force on the Cost-effectiveness, Quality and Financing of Asthma Care has recommended the use of symptom-free time as the standard outcome measure for cost-effectiveness evaluation of asthma treatments.[11]

PHARMACOECONOMICS OF ASTHMA PHARMACOTHERAPY

Each year an increasing number of cost-effectiveness studies are published in the health and medical literature. The evaluation methods vary from simple cost identification studies to sophisticated economic models. The strongest growth in this literature is in cost-effectiveness studies of pharmaceuticals. Many of these studies are funded by the pharmaceutical industry and, as such, have generated substantial discussion about appropriate study design and bias.[12] Consequently, expert panels have been convened to develop standards in the conduct[13] and reporting[14] of cost-effectiveness studies.

Pharmacoeconomic analyses of asthma pharmacotherapy have been growing at an equal pace. Tables 47.1 and 47.2 summarize the important retrospective and prospective pharmacoeconomic evaluations of asthma pharmacotherapy.

Inhaled corticosteroids

The therapeutic management guidelines of the US National Asthma Education and Prevention Expert Panel Reports I and II (NAEPP), the International Consensus Report (ICR), the Global Initiative for Asthma Report (GINA) and the British Thoracic Society recommend as initial treatment such combination therapy for persons with moderate-to-severe asthma. However, adding inhaled corticosteroid medications to an existing regimen of inhaled or oral bronchodilator therapy contributes significantly to the overall cost of treating asthma in these patients. An important, and as yet not fully explored, research question is: Are inhaled corticosteroids in combination with bronch-

Table 47.1 Summary of non-randomized pharmacoeconomic studies of asthma pharmacotherapy.

Reference	Study method used	Sample size	Perspective	Treatments studied	Length of study	Costs measured	Health outcomes measured	Economic outcomes
15	Retrospective; pre/post quasi-experimental design	†36 adults	Societal	Budesonide	5 years; 2 years before, 3 years after	Direct	Reduction in need for oral steroid after introduction of inhaled budesonide	Estimated 55% reduction in direct costs
30	Retrospective; pre/post quasi-experimental design	53	Health system	Two groups: cromolyn users and non-users	3.2–3.8 years*	Direct	None	Estimated 92–96% reduction in health services use
16	Retrospective; econometric model	†	Societal	All inhaled corticosteroids	11 years	Direct	Reduction in hospital-bed-days and discharges for asthma	Estimated benefit–cost ratio of between 1.5:1 and 2.8:1
21	Prospective pre/post quasi-experimental design	86 children	Societal	Two groups: various doses of beclomethasone and budesonide	4 years	Direct	Reduction in acute severe attacks, hospital admissions, breakthrough wheezing, missed school days and treatment satisfaction	Estimated 83% reduction in total costs of care; $0.04 per unit increase in patient satisfaction with treatment

* Cromolyn users contributed 3.2 years of data and non-users of Cromolyn contributed 3.8 years of data.
† Unit of analysis is counties and not persons. The study represents a total of 71% of the Swedish population.

Table 47.2 Summary of randomized pharmacoeconomic studies of asthma pharmacotherapy.

Reference	Study method used	Sample size	Perspective	Treatments studied	Length of study	Costs measured	Health outcomes measured	Economic outcomes
29	Randomized controlled trial	145 adults	Health system	Two groups: salmeterol and placebo	12 weeks	Direct	Episode-free days	No statistically significant difference in clinical effectiveness, thus a cost–outcome was not calculated
22	Randomized controlled trial	556 adults	Health system	Two groups: budesonide 400 µg compared with budesonide 800 µg	12 weeks	Direct	Lung function (FEV_1) and symptoms	Not cost-effective to increase dose of budesonide from 400 µg to 800 µg in mild to moderate patients
25	Randomized controlled trial	40 children	Societal	Two groups: budesonide compared with placebo	26 weeks	Direct and indirect	Lung function (FEV_1), symptoms, symptom-free days	Budesonide is dominant therapy; saved $9.43 for each symptom-free day gained
26	Randomized controlled trial	116 children	Societal	Two groups: budesonide and salbutamol, salbutamol alone	3 years*	Direct and indirect	Lung function (FEV_1), symptom-free days, school absences	Budesonide is cost-effective; $83 per 10% improvement in FEV_1, $4.75 per symptom-free day gained
27	Randomized controlled trial	274 adults	Societal	Three groups: beclomethasone and terbutaline, ipratropium and terbutaline, terbutaline alone	2.5 years	Direct and indirect	Lung function (FEV_1, PC_{20}), symptom-free days	Beclomethasone is cost-effective; $201 per 10% improvement in FEV_1, $5 per symptom-free day gained. Ipratropium is not cost-effective
24	Randomized controlled trial	57 adults	Societal	Three groups: budesonide 400 µg or budesonide 800 µg and bronchodilator compared with bronchodilator alone	16 weeks	Direct	Lung function (PEFR), symptom scores, exacerbations, emergency room visits and willingness to pay	Budesonide is cost-beneficial at 400 µg/day but not at 800 µg/day compared with bronchodilator alone
28	Randomized controlled trial	225 children	Societal	Two groups: sodium cromoglycate 20 mg q.i.d. compared with fluticasone 50 µg b.i.d.	8 weeks	Direct	Lung function (PEFR), symptom scores and the probability of successful treatment	Fluticasone is cost-effective compared with sodium cromoglycate. Cost-effectiveness ratios vary according to outcome measure selected

* The study had a planned 3-year follow-up but only 39 patients reached a follow-up period of 22 months.
† FEV_1, forced expiratory volume in 1 s; PEFR, peak expiratory flow rate.

odilators cost-effective compared with bronchodilator alone when used to treat persons with either mild-to-moderate or moderate-to-severe asthma?

The first evidence of inhaled budesonide on health services outcomes was reported in a letter by Adelroth and Thompson.[15] The authors attempted to show the relationship between use of high-dose inhaled budesonide (800 µg/day) and asthma-related inpatient hospital days in 36 oral steroid-dependent patients with asthma over a 5-year period. The analysis employed a pre-post, quasi-experimental study design where patients served as their own controls. A dramatic reduction in inpatient admissions, days and costs was observed in patients on budesonide compared with the previous 2 years on oral steroid therapy. Cost per patient declined by over 55% per year for up to 3 years after the initiation of inhaled budesonide.

Gerdtham and colleagues[16] built on Adelroth and Thompson's initial work in Sweden by constructing a pooled, time-series economic model to determine the association between greater use of inhaled corticosteroids and asthma-related hospital days in 14 counties over an 11-year period, again using a non-experimental methodology.[16] More than 80% of inhaled corticosteroid use during this time was with budesonide. Although not a true cost–benefit analysis, the study did indicate a strong negative association between use of inhaled corticosteroids and hospital-bed-days for asthma. An approximate cost–benefit ratio was developed from the multivariate models suggestive of positive economic benefits in excess of costs on the order of between 1.5:1.0 and 2.8:1.0, depending on the analytical model.

The lack of experimental design in the study by Gerdtham and colleagues and the very small sample size of the Adelroth and Thompson study restrict the internal validity and conclusions of these two studies. However, these studies make use of an alternative evaluation strategy wherein the authors attempt to measure population effectiveness of the inhaled product in the absence of the constraints of a clinical study design.[5] The strength of the conclusions by Gerdtham and colleagues lies in the longitudinal and generalizable nature of the data.

The results from these two studies suggest a favourable economic impact of using inhaled corticosteroids. Similar reductions in inpatient care have been associated with use of other anti-inflammatory therapy in persons with chronic asthma.[7–21] However, these clinical and epidemiological studies lacked pharmacoeconomic valuation.

Several recent studies have employed experimental research designs to investigate the cost-effectiveness of inhaled corticosteroids. Campbell and colleagues[22] reported on the cost-effectiveness of increasing the daily dose of inhaled budesonide from 400 to 800 µg/day after 6 weeks in persons with mild-to-moderate asthma. Data from a 12-week randomized trial of 556 patients aged 14–84 years were used in the analysis. The health outcomes of increasing the dose of inhaled budesonide in these patients was reported elsewhere and showed that 800 µg/day of budesonide failed to improve lung function or reduce symptoms when compared with 400 µg/day.[23] The total cost of treatment (medication only) was estimated to be £3108 (about $US4660) in the 400 µg/day group (12 weeks) compared with £4662 (about $US6993) in both the 400 µg/day (6 weeks) and 800 µg/day (6 weeks) groups. The authors concluded that increasing the dose of budesonide from 400 to 800 µg/day was not a cost-effective strategy.

Similar findings were observed by O'Byrne et al.[24] in a randomized trial of budesonide 400 µg/day, 800 µg/day and placebo in 57 adult asthmatics with mild disease. Low-dose budesonide demonstrated better morning and nocturnal asthma symptom control, improved peak expiratory flow rate (PEFR) and was judged to be cost-beneficial compared with placebo. High-dose budesonide did not improve lung function or symptom scores relative to low-dose budesonide enough to justify the added cost of therapy.

In a somewhat longer study, Connett and colleagues[25] studied the cost-effectiveness of inhaled budesonide compared with placebo in a 6-month randomized trial of 40 children with persistent asthma aged 1–3 years.[25] The results indicated that budesonide produced a favourable clinical response, increasing symptom-free days when compared with placebo (195 vs. 117 days). Direct medical costs (including the cost of budesonide) and indirect costs were tabulated for the numerator of the cost-effectiveness ratio. The results suggested that budesonide is a dominant therapy, i.e. compared with placebo, budesonide increased overall effectiveness and reduced overall costs by £6.33 (about \$US9.45) per symptom-free day gained.

Rutten-van Mölken and associates[26] reported on the cost-effectiveness of adding inhaled corticosteroid to as-needed bronchodilator compared with as-needed bronchodilator alone in a 12-month randomized trial of 116 children with asthma aged 7–16 years. The investigators evaluated forced expiratory volume in 1 s (FEV_1) as the primary outcome. Frequency of symptom-free days and the number of school absences were included as secondary outcome measures. Patients randomized to inhaled corticosteroid plus as-needed bronchodilator experienced significantly increased lung function (FEV_1) and symptom-free days and reduced days missed from school relative to as-needed bronchodilator alone. Computation of the cost-effectiveness ratio indicated that, when compared with bronchodilator alone, bronchodilator plus inhaled corticosteroid increased FEV_1 by 10% at an additional total cost of about \$US83. Alternatively, the additional cost of bronchodilator plus inhaled corticosteroid was about \$US4.75 per symptom-free day gained. In this study, addition of inhaled corticosteroid to a treatment regimen of inhaled bronchodilator was more effective than bronchodilator alone but at an additional cost, the value of which depended on whether the outcome was improved lung function (FEV_1) or better symptom control.

In the largest and most comprehensive study to date, Rutten-van Mölken and associates[27] analysed data from a randomized trial of 274 adult participants (age 18–60 years) in an effort to investigate the costs and effects of adding inhaled anti-inflammatory therapy to existing inhaled β_2-agonist. Patients were selected for inclusion if they met the age criteria and had been diagnosed with moderately severe obstructive airway disease defined by pulmonary function criteria. The patients were of mixed diagnosis and could be enrolled if they had either asthma or chronic obstructive pulmonary disease. Patients were randomized to fixed-dose inhaled terbutaline plus inhaled placebo, inhaled terbutaline plus 800 µg/day of inhaled beclomethasone or inhaled terbutaline plus inhaled ipratropium bromide 160 µg/day. Patients were followed for up to 2.5 years or until premature withdrawal.

The economic objective of this study was to determine if additional treatment costs of the combination therapies were outweighed or justified by additional clinical benefits and reduced utilization of other healthcare services. The clinical results suggested that addition of the inhaled corticosteroid to fixed-dose terbutaline led to a significant improvement in pulmonary function (FEV_1 and PC_{20}) and symptom-free days, whereas addition of the inhaled ipratropium bromide to fixed-dose terbutaline produced no significant clinical benefits over placebo. The average annual monetary savings associated with the use of inhaled corticosteroid were not offset by the increase in costs from the average annual price of the inhaled product. The incremental cost-effectiveness ratio for inhaled corticosteroid was $US201 per 10% improvement in FEV_1 and $US5 per symptom-free day gained. It was not appropriate to evaluate the incremental cost-effectiveness of ipratropium bromide because of the lack of clinical benefit relative to placebo. In many ways, this study represents a model for pharmacoeconomic analysis in asthma. The resource and cost estimates are clear and precise, the study period is sufficiently long and the analytical techniques are appropriate. However, the mixed population of asthma and chronic obstructive pulmonary disease limits the utility of these data for decision-making for asthma treatment.

The cost-effectiveness of fluticasone was studied in a group of 4–12-year-old children who required inhaled prophylactic treatment for asthma.[28] Over an 8-week study period, 115 patients received sodium cromoglycate 20 mg four times daily and 110 patients received fluticasone propionate 50 µg twice daily. The effectiveness of both treatments was determined by morning and evening PEFR, daily symptom control, safety, proportion of successfully treated patients and incidence of adverse consequences. The authors concluded that over an 8-week period, fluticasone was cost-effective compared with sodium cromoglycate for prophylactic treatment using treatment success rates as the primary cost-effectiveness outcome measure.

Controlled clinical trials are necessary to investigate the efficacy and safety of pharmacotherapy. It is not clear whether such rigorous study designs are necessary for economic evaluation. Three of the economic studies just described showed that inhaled corticosteroids in low doses (400 µg/day) reduced asthma-related morbidity and that the economic benefits either offset or add to overall treatment costs. The studies by Campbell and coworkers[22] and Booth et al.[28] were brief in duration, measured only medication costs and focused on clinical measures of pulmonary function or physician-rated treatment success, whereas most primary-care clinicians and health plans are interested in symptoms. The studies by Connett and colleagues and Rutten-van Mölken and associates were somewhat longer in duration and evaluated symptom-free days as the primary outcome measure.[25–27] All three studies determined that inhaled corticosteroids improved symptom-free days compared with the bronchodilator alone, but each arrived at a different economic conclusion. The study by Connett and colleagues[25] included an estimate of indirect costs, which increased the estimate of overall economic benefit. Both studies by Rutten-van Mölken and associates[26,27] valued only medical care costs and showed that adding inhaled corticosteroid to a regimen of inhaled bronchodilator improved clinical outcomes and increased the overall cost of care.

These studies highlight the need for standardization of study design, particularly time horizon, selection of comparator therapy and standardization and valuation of economic and outcome measures. Several of the papers used clinical and economic data of 8–12 weeks in duration. Asthma is not an 8–12-week disease. Without standardization, decision-makers are likely to be confused by short-term studies and conflicting results.

Long-acting β_2 agonists

Long-acting bronchodilators such as formoterol and salmeterol represent a relatively new approach to prophylactic and symptomatic treatment for asthma. Only one published study has simultaneously evaluated the impact of a long-acting agent on clinical and economic outcomes for patients with asthma.[29] In this paper, the authors reported on a retrospective cost-effectiveness analysis of a clinical trial of 145 patients diagnosed with asthma and randomized to receive 12 weeks of maintenance therapy with either long-acting formoterol or short-acting albuterol. The primary clinical outcome measure was cumulative symptom-free days over the 12-week period. The authors concluded that there were no statistically significant differences in symptom-free days between the two treatment groups. Because of these results, no incremental cost-effectiveness ratio was calculated. For illustrative purposes, the authors simulated a range of possible clinical benefits and cost-effectiveness ratios by respecifying the symptom-free composite score to include or not include adverse events.

Inhaled cromolyn sodium

Ross and coworkers[30] made use of patient and health services records in one large group practice to estimate the economic consequences of including cromolyn sodium in the treatment regimen of asthma patients. A total of 53 patients were retrospectively identified from medical records and categorized into two groups: those who received cromolyn sodium for at least 1 year ($n = 27$) and those who received no cromolyn sodium as part of the treatment regimen ($n = 26$). Patients receiving cromolyn sodium provided an average of 3.2 years of health service utilization data and those in the comparison group provided 3.8 years of data. Medication costs for patients on cromolyn sodium were slightly higher ($US27.90 per month) than for the control group ($US25.20 per month). However, emergency department and hospital costs declined significantly for patients receiving cromolyn sodium; after the change in medication, they experienced a 96% reduction in the rate of emergency department visits and a 92% reduction in the rate of hospital admissions. The authors made no direct measurement of outcomes of therapy and did not control for symptom severity or other baseline confounding that might partly explain differences in the results. Thus this study is not a true cost-effectiveness analysis; rather, it is a cost comparison of two retrospective cohorts of asthma patients.

Other pharmacotherapy

The remaining economic studies of asthma medications identified in this review were not full cost-effectiveness evaluations. Tierce and colleagues[31] performed a retrospective cost-identification study comparing use of inhaled albuterol with inhaled metaproterenol (orciprenatine) in 1463 Michigan Medicaid patients. Asthma-related medications, physician and emergency department visits, and hospital care were assessed and valued using Medicaid prices. The authors concluded that the overall cost of care was significantly lower in the albuterol group compared with the metaproterenol group. Because this study was not randomized, questions remain about baseline comparability of the two groups.

A modest number of papers exist on the impact of other β_2-agonists on health services utilization,[32] methylprednisolone use in the emergency department[33] or aerosolized vs. metered-dose inhaler delivery of β_2-agonists.[36,37] Three of the studies did not attempt to value the intervention benefits in monetary terms; rather, these studies expressed outcomes in terms such as number of visits.[32,33,35] Further, one study was not considered because of a mixed inception sample that included patients without asthma among the evaluable patients.[34]

CONCLUSIONS

Healthcare decision-makers are interested in employing rational approaches to allocating resources among patients with chronic disease. Some policy-makers may view cost-effectiveness analyses as a means to justify rationing, but others are beginning to embrace economic evaluation methods for improving decision-making about, for example, limits on insurance coverage for specific interventions, formulary development, quality improvement programmes and, importantly, appropriate utilization of services. However, before sound decisions can be made on appropriate selection and use of interventions, more comparative data on the economic value of the treatments are needed.

Too few studies as yet adequately characterize the economic impact of the large number of interventions currently being used to manage asthma. There is no standard approach for evaluating the economic costs and benefits of medical treatments used to treat asthma or for comparing the clinical and economic benefits of alternative treatments. Researchers conduct studies with varying lengths of follow-up, use different outcome measures, include different costs in the calculation of total cost and evaluate different mixes of patients. These inconsistencies hinder efforts by decision-makers to compare clinical and economic benefits.

Despite the many shortcomings, pharmacoeconomic studies of asthma treatments need to be encouraged and nurtured. Substantial improvements and standardization are needed in the study design, study duration, sample size determination based on economic and health status end-points, selection of appropriate comparison therapies, and selection and evaluation of costs and outcomes.

REFERENCES

1. Grimes DA: Technology follies: the uncritical acceptance of medical innovation. *JAMA* (1993) **269**: 3030–3033.
2. Luce BR, Lyles CA, Rentz AM: The view from managed care pharmacy. *Health Affairs* (1996) **15**: 168–176.
3. Drummond MF, Stoddart GL, Torrance GW: *Methods for the Economic Evaluation of Health Care Programmes.* New York, Oxford University Press, 1987.

4. Banta HD, Luce BR: *Health Care Technology and its Assessment.* New York, Oxford University Press, 1993.
5. Drummond MF, Davies L: Economic analysis alongside clinical trials. Revisiting the methodological issues. *Int J Technol Assess Health Care* (1991) **7**: 561–573.
6. Warner KE, Luce BR: *Cost–Benefit and Cost-effectiveness Analysis in Health Care: Principles, Practice, and Potential.* Ann Arbor, MI, Health Administration Press, 1982.
7. Johannesson M, Weinstein MC: Designing and conducting cost–benefit analyses. In Spilker B (ed) *Quality of Life and Pharmacoeconomics in Clinical Trials*, 2nd edn. Philadelphia, Lippincott-Raven, 1996. pp. 1085–1092.
8. Weinstein MC, Stason WB: Foundations of cost-effectiveness analysis for health and medical practices. *N Engl J Med* (1977) **296**: 716–721.
9. von Neumann J, Morgenstern O: *Theory of Games and Economic Behaviour.* Princeton, Princeton University Press, 1994.
10. National Heart, Lung, and Blood Institute: Asthma outcome measures. *Am J Respir Crit Care Med* (1994) **149**: S1–S90.
11. Sullivan SD, Elixhauser A, Buist AS, Luce BR, Eisenberg J, Weiss KB: National Asthma Education and Prevention Program working group report on the cost effectiveness of asthma care. *Am J Respir Crit Care Med* (1996) **154**: S84–S95.
12. Hillman AL, Eisenberg JM, Pauly MV, *et al.*: Avoiding bias in the conduct and reporting of cost-effectiveness research sponsored by pharmaceutical companies. *N Engl J Med* (1991) **324**: 1362–1365.
13. Gold MR, Siegel JE, Russell LB, Weinstein MC: *Cost-effectiveness in Health and Medicine.* New York, Oxford University Press, 1996.
14. Drummond MF, Jefferson TO: Guidelines for authors and peer reviewers of economic submissions to the BMJ. The BMJ Economic Evaluation Working Party. *Br Med J* (1996) **313**: 275–283.
15. Adelroth E, Thompson S: Advantages of high-dose inhaled budesonide. *Lancet* (1988) **i**: 476.
16. Gerdtham UG, Hertzman P, Boman G, Jonsson B: Impact of inhaled corticosteroids on asthma hospitalization in Sweden. *Appl Econ* (1996) **28**: 1591–1599.
17. Karalus NC, Harrison AC: Inhaled high-dose beclomethasone in chronic asthma. *N Z Med J* (1987) **100**: 306–308.
18. Wennergren G, Kristjasson S, Strannegard I: Decrease in hospitalization for treatment of childhood asthma with increased use of antiinflammatory treatment, despite an increase in prevalence. *J Allergy Clin Immunol* (1996) **97**: 742–748.
19. Donahue JG, Weiss ST, Livingston JM, Goetsch MA, Greineder DK, Platt R: Inhaled steroids and the risk of hospitalization for asthma. *JAMA* (1997) **277**: 887–891.
20. Suissa S, Dennis R, Ernst P, Sheehy O, Wood-Dauphinee S: Effectiveness of the leukotriene receptor antagonist zafirlukast for mild-to-moderate asthma. A randomized, double-blind, placebo-controlled trial. *Ann Intern Med* (1997) **126**: 177–183.
21. Perera BJC: Efficacy and cost effectiveness of inhaled steroids in asthma in a developing country. *Arch Dis Child* (1995) **72**: 312–316.
22. Campbell LM, Simpson RJ, Turbitt ML, *et al.*: A comparison of the cost-effectiveness of budesonide 400 μg/day and 800 μg/day in the management of mild-to-moderate asthma in general practice. *Br J Med Econ* (1993) **6**: 67–74.
23. Rees TP, Lennox B, Timney AP, *et al.*: Comparison of increasing the dose of budesonide to 800 μg/day with a maintained dose of 400 μg/day in mild to moderate asthmatic patients. *Eur J Clin Res* (1993) **4**: 67–77.
24. O'Byrne P, Cuddy L, Taylor DW, Birch S, Morris J, Syrotuik J: Efficacy and cost benefit of inhaled corticosteroids in patients considered to have mild asthma in primary care. *Can Respir J* (1996) **3**: 169–175.
25. Connett GJ, Lenney W, McConchie SM: The cost-effectiveness of budesonide in severe asthmatics aged one to three years. *Br J Med Econ* (1993) **6**: 127–134.
26. Rutten-van Mölken MP, Van Doorslaer EK, Jansen MC, *et al.*: Cost-effectiveness of inhaled corticosteroid plus bronchodilator therapy versus bronchodilator monotherapy in children with asthma. *PharmacoEconomics* (1993) **4**: 257–270.

27. Rutten-van Mölken MP, Van Doorslaer EK, Jansen MC, Kerstjens HA, Rutten FF: Costs and effects of inhaled corticosteroids and bronchodilators in asthma and chronic obstructive pulmonary disease. *Am J Respir Crit Care Med* (1995) **151**: 975–982.
28. Booth PC, Wells NEJ, Morrison AK: A comparison of the cost effectiveness of alternative prophylactic therapies in childhood asthma. *PharmacoEconomics* (1996) **10**: 262–268.
29. Sculpher M, Buxton M: Episode-free days as endpoints in economic evaluations of asthma therapy. *PharmacoEconomics* (1993) **4**: 345–352.
30. Ross RN, Morris M, Sakowitz SR, Berman BA: Cost-effectiveness of including cromolyn sodium in the treatment program for asthma: a retrospective, record-based study. *Clin Ther* (1988) **10**: 188–203.
31. Tierce JC, Meller W, Berlow B, Gerth WC: Assessing the cost of albuterol inhalers in the Michigan and California Medicaid programs: a total cost-of-care approach. *Clin Ther* (1989) **11**: 53–61.
32. Emerman CL, Cydulka RK, Effron D, Lukens TW, Gershman H, Boehm SP: A randomized, controlled comparison of isoetharine and albuterol in the treatment of acute asthma. *Ann Emerg Med* (1991) **20**: 1090–1093.
33. Littenburg B, Gluck EH: A controlled trial of methylprednisolone in the emergency treatment of acute asthma. *N Engl J Med* (1986) **314**: 150–152.
34. Jasper AC, Mohsenifar Z, Kahan S, Goldberg HS, Koerner SK: Cost–benefit comparison of aerosol bronchodilator delivery methods in hospitalized patients. *Chest* (1987) **91**: 614–618.
35. Summer W, Elston R, Tharpe L, Nelson S, Haponik EF: Aerosol bronchodilator delivery methods: relative impact on pulmonary function and cost of respiratory care. *Arch Intern Med* (1989) **149**: 618–623.

48

Education and Self-management

MARTYN R. PARTRIDGE

INTRODUCTION

International and national guidelines on asthma management now lay out clear advice to health professionals as to how to approach their patients with asthma.[1-3] Excellent treatments of proven efficacy are available, and yet around the world suffering from the condition continues; in part, this may reflect the increasing prevalence of the condition. The primary prevention of asthma remains the major challenge of all with an interest in the condition. However we need to ascertain why suffering continues when good treatment is available. Possible explanations include:

(1) misdiagnosis or delayed diagnosis;
(2) underestimation of severity by either patient or health professional, leading to either overuse of bronchodilators or inadequate use/underuse of anti-inflammatory agents;
(3) Health professionals being out of date;
(4) Treatments being too expensive for either individual or state; or
(5) Patients being prescribed the correct treatment but for one reason or another not taking them.

HEALTH PROFESSIONAL EDUCATION AND GUIDELINES

Issues (1)–(3) above are largely concerned with health professionals' education and here the use of guidelines may be helpful. These should be regarded as a simple tool to assist the busy clinician by providing him/her with a summary of research. They are also useful for teaching medical students, nurses and other health professionals. In addition, guidelines serve as a way of setting standards and as a basis for audit. An evaluation of the effect of guidelines for a variety of conditions has shown an overwhelmingly positive effect of the introduction of guidelines on both the process of care and outcomes.[4] It seems likely that successful guidelines are those associated with active education, feedback to individual practitioners as to how their performance compares with that outlined,[5] and where advice is taken down to a district,[6] department[7] or individual consultation level.[8] This is often most easily done by the setting up of local asthma task forces, as recommended in the Global Strategy on Asthma Management.[2] These task forces, which consist of health planners and health professionals from both primary and secondary care, can then tackle local issues and problems, adopting and adapting national guidelines to make them locally relevant and induce a sense of ownership. Of course, even when all of this is undertaken, the patient may not benefit if the treatment is administered in a way which makes it likely that the patient does not take it.

COMPLIANCE

The problem of non-compliance with both treatment and other aspects of management in asthma is likely to be large and may be underestimated.[9-11] Non-compliance with treatment is either inferred from the presence of poorly controlled disease or confirmed by monitoring drug taking, which may involve measurement of drug levels in urine, plasma or saliva, or by prescription monitoring. More modern methods involve the fitting of microprocessors to the lids of bottles[12] or inhaler devices.[13,14] Such methods are inappropriate at a clinical level and it is preferable to accept that non-compliance is common and instead make efforts to consider the possible factors involved and work with the patient to tackle the underlying causes. Possibilities are shown in Table 48.1. Clearly, therapeutic regimens should, wherever possible, only involve once- or twice-daily dosing of the minimum number of medications and adequate instruction should be given about the correct use of inhalers. However, it is the non-drug factors listed in Table 48.1 that may be of greatest importance.

Very often patients fail to benefit from treatment because they simply misunderstand, forget or are not told which medicines to take and when. In one survey, only 9% of patients felt that they had been given enough information about their condition and only 27% had been given written instructions about their medication regimen.[15] Other studies have suggested that those who exhibit denial about either the diagnosis or the severity of their asthma,[16] or who are depressed,[17] may be less likely to adhere to their prescribed medication. In other cases, non-compliance may result from too little attention being paid to cultural or ethnic factors. Perceptions vary between those from differing backgrounds: if the patient's traditional upbringing and beliefs are that good things enter the body by the stomach and bad things via the lungs, they require more advice and more time to understand why the inhaled route is preferred for the delivery of asthma medications. The major issue is thus one of communication. Many years ago Korsch and Negreve[18] showed a direct relationship between patient satisfaction with communication and compliance; of

48 Education and Self-management

Table 48.1 Factors involved in non-compliance.

Drug factors
Difficulties with inhaler devices
Awkward regimes (e.g. four times daily or multiple drugs)
Side-effects
Cost of medication
Distant pharmacies

Non-drug factors
Misunderstanding or lack of instruction
Fears about side-effects
Dissatisfaction with healthcare professionals
Unexpressed/undiscussed fears or concerns
Inappropriate expectations
Poor supervision, training or follow-up
Anger about condition or its treatment
Underestimation of severity
Cultural issues
Stigmatization
Forgetfulness or complacency
Attitudes toward ill health
Religious issues

those patients who were satisfied with the communication aspects of the consultation 54% complied totally, whereas in the dissatisfied group the equivalent figure was 16%.

HOW DO WE IMPROVE COMMUNICATION?

This involves a consideration of three issues.

(1) How do we offer information to our patients?
(2) What information do we give?
(3) How do we ascertain the individual patient's needs?

How do we offer information to our patients?

The individual consultation remains the cornerstone of good communication, and it is estimated that the average doctor undertakes between 200 000 and 250 000 consultations during his/her professional lifetime. During such consultations, the doctor assesses and responds to the information needs of individual patients. Trained educators and nursing colleagues also offer information. However, spoken information, whether given by doctors, nurses or educators, is often forgotten, with half of what has been said during a consultation being forgotten within 5 min of it ending. It is essential that the spoken word is therefore reinforced by other means if important messages are not to be lost. No single reinforcing method is better than another, but some are more practical. Wherever possible all methods should be available so that the patients themselves may access the source of information with which they feel most comfortable.

Written information is the easiest to give. This may be a printed booklet containing essential basic or 'core' information; or it may be a printed booklet covering a specific topic such as asthma and exercise, asthma and pregnancy or the use of steroids in asthma. These booklets are best given in response to spoken questions, the 'core' information booklet being most likely to help if given in a way that enhances the doctor–patient or nurse–patient partnership. The doctor may thus say 'I would like you to read this booklet and mark any section which you wish to discuss further with me when we next meet'. Giving printed booklets alone may enhance knowledge but may not alter behaviour or improve compliance. However, they are probably a sensible starting point for enhancing patient satisfaction; in one survey (Table 48.2) only 9% of patients reported that they had been given enough information about asthma. More helpful is personalization of advice; whilst this may involve detailed self-management plans (see p. 923), all patients should receive written advice about their personal medication regimens and yet only 27% had done so in one survey.[15] Three published studies[19–21] suggest that such written advice improves recall and thus would seem essential if the patient is to comply with their prescribed medications. In one study amongst children with asthma, it was only the parents who had received written as well as spoken information who undertook the correct sequence of therapeutic changes.[19] In another study, 95% of those given a written treatment plan could recall the details of how to take their medicines when asked 3 months later, compared with 58% of those who had not been given a booklet containing this information on leaving hospital.[21] Giving patients simple written information about their treatment takes only a few seconds even in a busy consultation, especially if use is made of preprinted cards on to which the additional information is written (Fig. 48.1).

Whilst written information is the most practical and basic way of reinforcing the spoken word, other ways of giving information and educating patients need to be available and have advantages in some circumstances. In one study, patients who had experienced a large number of information sources expressed a preference for receiving information firstly from the doctor and secondly by watching a videotape.[22] Such videos are an excellent medium for those with poor literacy skills, and also good for teaching practical techniques such as how to use an inhaler or a peak flow monitor.[23] However, patient preference does not necessarily equate with effectiveness; in another study, patients were reported to prefer a book to the use of a tape recording, although they

Table 48.2 Feelings of patients at the time of diagnosis.*

	Total (%)
Lacked understanding of condition	44
Wanted more information	55
Felt they had been given plenty of information	9
Felt they had had a good discussion with their doctor or nurse	22

*A random sample of 2500 people on the UK National Asthma Campaign membership database was sent a questionnaire and 65% responded in November and December 1993.
Reproduced from ref. 15, with permission.

48 Education and Self-management

Your name

Family doctor

Doctor's telephone

Asthma is a common condition. It can be easily controlled with modern treatments. However, in order to look after your asthma properly you need to know a little about your medicines, and you need to know the signs that might suggest your asthma is worsening.

Preventer inhalers
(usually brown, red or orange)

These are the most important medicines. They help prevent the symptoms of asthma from appearing – coughing, wheezing, tight chest and shortness of breath. Preventers will not work unless they are taken every day, as advised by your doctor or nurse.

Your preventer is

which you should take in a daily dose of

with a (specific device)

Reliever inhalers
(usually blue)

These rescue you from asthma symptoms as they happen. You should take your reliever, as advised by your doctor or nurse.

Your reliever is

which you should take with a (specific device)

Other treatments are

Fig. 48.1 An example of part of a simple treatment card produced jointly by the UK Department of Health and the National Asthma Campaign.

actually learnt more from the tape.[24] The benefits of individual vs. group education also merit attention. Wilson and colleagues[25] compared outcomes in terms of 'bother from asthma' in a group given a workbook, a group undergoing individual education and a group having class educational sessions. All three did better than a control group, although the benefits were only significant for the actively educated group; those receiving the same education in a group did slightly better than those receiving individual education. Such benefits of group education probably reflect the benefits of support to be found in groups, which may work by reducing the stigma so commonly experienced by those with asthma.[26] Other studies have suggested a high drop-out or non-attendance rate for group sessions,[27] emphasizing that what is needed is a range of methods of teaching our patients how to help themselves, so that there is a suitable method available for everyone.

What information do we give?

Most of the published guidelines contain advice about the information that doctors perceive patients with asthma to need. This is likely to include information regarding:

(1) the diagnosis;
(2) treatment, i.e. preventers and relievers;
(3) monitoring;
(4) advice about signs that suggest worsening asthma and what to do under those circumstances;
(5) Details of local support groups.

The difficulty, of course, is that the information that the doctor thinks the patient needs to know may not be the same as what the patient actually wants to know. Each of the 160 million people worldwide with asthma have different personalities and different life experiences, and we need to check individual patient's or parent's concerns and their requirements for information.

How do we ascertain the individual patient's needs?

Different people have different expectations of both asthma and its treatment and such feelings need to be elicited. Open-ended questions such as 'What can I do for you?' or 'What do you want of me?' can be illuminating. For example, if a patient only wanted to hear that he/she would soon grow out of their asthma and the doctor fails to elicit this and instead concentrates on extolling the virtues of steroids, then the patient is unlikely to be satisfied and further educational efforts may not be successful. It is often necessary to elicit unspoken fears about treatment; the Global Initiative on Asthma Management[2] recommends the use of further open-ended questions such as 'How do you feel about this treatment?' and 'Some people worry about steroids: what do you think?' If concerns are expressed, then further information about the advantages and disadvantages of certain treatments can be discussed and again, wherever possible, reinforced with the written word. Goals for management can be similarly explored and these may need to be negotiated; it is important wherever possible to enlist the help of family and friends in the consultation. Finally, it should be remembered that not all that the patient needs to know can be given within one consultation, nor can all fears and concerns be elicited or discussed. The process should be regarded as a continuing one that proceeds over several consultations with reinforcement and revision of the messages on each occasion.

SELF-MANAGEMENT

Fear of attacks and not feeling in control are unpleasant emotions and most guidelines recommend the concept of patient 'self-management'. This entails the patient or parent being taught to adjust treatment in a variety of circumstances according to advice given to them in advance by the doctor or nurse. All patients self-manage to some extent by using their bronchodilators in response to symptoms, and self-management may only involve the patient being advised to seek medical attention if they have an increased need for relief medication or if they awake at night. For others, self-management may involve a more detailed plan such as that shown in Fig. 48.2.

Early uncontrolled trials suggested benefit from the use of such plans and there have been several controlled trials subsequently of the efficacy of administration of such plans. Such trials need to be viewed carefully before we extrapolate results; we need to define the population studied, the intervention suggested and whether there was adequate opportunity for the intervention to alter outcome. Clearly, in a group of overtreated or adequately treated subjects, exacerbations may be so infrequent that administration of a self-management plan will not alter outcome because the opportunities for it to do so are too few. Two good controlled trials of those with moderate asthma have been published[28,29] and the results may be extrapolated to common clinical situations. In the first, 35 patients were offered self-management advice similar to that shown in Fig. 48.2; 6 months later it was shown that they suffered less exacerbations and needed to use less health-service resources than a control group.[28] In the second study, 56 patients with a mean duration of asthma of 8.2 years, a mean forced expiratory volume in 1 s of 82.4% predicted and a mean daily dose of inhaled steroids of 979 μg were given a detailed self-management plan that involved doubling their inhaled steroid if their peak expiratory

48 Education and Self-management

Zone 1
Your asthma is under control if:

- Your peak flow readings are above ....
  and
  it does not disturb your sleep
  and
- it does not restrict your activities

Action

Continue your normal medicines

Zone 2
(Your doctor or nurse may decide not to use this zone)
Your asthma is getting worse if:

- Your peak flow readings have fallen to between ......
  and ......
- You are needing to use your ....
  (reliever inhaler) more than usual

- You are waking at night with asthma symptoms

Action

- Increase your
  ....
  (Preventer inhaler) to ....
- Continue to take your
  ....
  (reliever inhaler to relieve your asthma symptoms)

Zone 3
Your asthma is severe if:

- Your peak flow readings have fallen to between
  .... and ....

- You are getting increasingly breathless
- You are needing to use your ....
  (reliever inhaler) every .... hours or more often

Action
Ring your doctor or nurse

- Take ....
  Prednisolone (steroid) tablets (...... mg each) and then
  ....
  ....
  ....
  Discuss with your doctor how to stop taking the tablets
- Continue to take your
  .... (reliever and preventer inhalers) as required

Zone 4
Medical alert/emergency if:

- Your peak flow readings have fallen to below ....

- You continue to get worse

Do not be afraid of causing a fuss. Your doctor will want to see you urgently

Action
Get help immediately

- Ring your doctor immediately
- Telephone ....
  or call an ambulance
- Take ....
  Prednisolone (steroid) tablets (...... mg) each immediately
- Continue to take your
  ....
  (Reliever inhaler) as needed

Fig. 48.2 An example of a four step self-management plan.

flow fell below 85% and starting a course of steroid tablets if it fell below 70%. The patients were followed up for 12 months and results showed significantly improved outcomes, with a marked reduction in overall asthma incidents in those taught to self-manage their asthma compared with those treated in a conventional way.[29]

Despite these positive trial results many questions remain. These involve decisions about who needs such plans, whether they should be based upon objective peak flow recordings or on symptoms or both, and whether there should be three or four steps to such plans. These questions cannot be answered on the basis of current studies, nor do we have a clear idea as to who should have detailed plans as opposed to the simpler written advice mentioned earlier. Similarly, the same may not apply in children as it does in adults. It has been suggested that, in children's symptoms (or spirometry) may be a better predictor of deterioration than peak flow monitoring,[30,31] although the latter has been reported as aiding mothers in their assessment of the severity of their children's asthma.[32]

In adults, there is evidence that ability to perceive severity of asthma may vary from one patient to another; this may apply to up to 60% of patients.[33] Inability to detect deterioration in airway calibre may lead to patients seeking medical attention at a late stage of an attack; a Japanese study has suggested that these patients may figure prominently amongst those requiring mechanical ventilation for severe asthma.[34] If identified, such patients should be given self-management plans and taught objective monitoring. In the absence of clear evidence-based advice, the British Guidelines on Asthma Managemensuggest that others requiring self-management plans may be those who have been recently hospitalized, those who have had sudden severe (brittle) attacks of asthma and those who regularly require high-dose inhaled steroid treatment (or more) to control their asthma.[35] It may also be worth targeting those who attend accident and emergency departments with uncontrollable asthma, especially because recent studies have shown that one-third of these patients have recently been in hospital and one-quarter have attended the emergency department in the previous 3 months.[36]

CONCLUSION

It is no good having excellent treatments available if they are not used. Successful management of asthma involves more than writing the correct prescription. Treatment has to be offered in a way that makes it likely that it is taken and this involves good communication, reinforcement of the spoken message and acquisition by the patient of skills in self-management.

REFERENCES

1. National Heart, Lung and Blood Institute: The international consensus report on diagnosis and treatment of asthma. Publication No. 92–3091. Bethesda, MD, National Institutes of Health, 1992.
2 .National, Heart, Lung and Blood Institute: The global strategy for asthma management and prevention. Publication No. 95–3659. Bethesda, MD, National Institutes of Health, 1995.

3. British Thoracic Society, British Paediatric Association, Royal College of Physicians of London, The Kings Fund Centre, The National Asthma Campaign, et al.: Guidelines on the management of asthma. *Thorax* (1993) **48**: S1–S24. 4. Grimshaw JM, Russell IT: Effect of clinical guidelines on medical practice: a systematic review of rigorous evaluation. *Lancet* (1993) **342**: 1317–1322.
5. Neville RG, Clark RA, Hoskins G, Smith B: First national audit of acute asthma attacks in general practice 1991–1992. *Br Med J* (1993) **306**: 559–562.
6. Lim KL, Harrison BDW: A criterion based audit of in patient asthma care. *J Roy Coll Physicians Lond* (1992) **26**: 71–75.
7. Town L, Kwong T, Holst P, Beasley R: Use of management plan for treating asthma in an emergency department. *Thorax* (1990) **45**: 702–706.
8. Feder G, Griffiths C, Highton C, Eldridge S, Spence M, Southgate L: Do clinical guidelines introduced with practice based education improve care of asthmatic and diabetic patients? A randomised controlled trial in general practices in East London. *Br Med J* (1995) **311**: 1473–1478.
9. Horn CR: Compliance by asthmatic patients: how much of a problem? *Res Clin Forums* (1986) **8**: 47–53.
10. Horn CR, Clark TJH, Cochrane GM: Compliance with inhaled therapy and morbidity from asthma. *Respir Med* (1990) **84**: 67–70.
11. Evans L, Spelman N: Problems of non compliance with drug therapy. *Drugs* (1983) **25**: 63–76.
12. Cranmer JA, Mattson RH, Prevey MC, Scheyer RD, Ovellette VL: How often is medication taken as prescribed? *JAMA* (1989) **261**: 3273–3277.
13. Rand CS, Wise RA, Nide S, et al.: Metered dose inhaler adhesive in a clinical trial. *Am Rev Respir Dis* (1992) **146**: 1559–1564.
14. Cochrane GM: Compliance in asthma: a European perspective. *Eur Respir Rev* (1995) **5**: 116–119.
15. Partridge MR: Asthma: lessons from patient education. *Patient Education and Counselling* (1995) **26**: 81–86.
16. Campbell DA, Yellowlees PM, McLennan G, et al.: Psychiatric and medical features of mean near fatal asthma. *Thorax* (1995) **50**: 254–259.
17. Bosley CM, Corder ZM, Cochrane GM: Psychosocial factors and asthma. *Respir Med* (1996) **90**: 453–457.
18. Korsch BM, Negreve VF: Doctor patient communication. *Sci Am* (1972) **227**: 66–72.
19. Pedersen S: Ensuring compliance in children. *Eur Respir J* (1992) **5**: 143–145.
20. Raynor DK, Booth TG, Blenkinsopp A: Effect of computer generated reminder charts on patients compliance with drug regimens. *Br Med J* (1993) **306**: 1158–1161.
21. Sandler DA, Heaton C, Garner ST, Mitchell JRA: Giving an information booklet to patients leaving hospital. *Br Med J* (1989) **298**: 870–874.
22. Partridge MR: Asthma education: more reading or more viewing? *J Roy Soc Med* (1986) **79**: 326–328.
23. Mulloy EMT, Albalark MK, Warley ARH, Harvey JE: Video education for patients who use inhalers. *Thorax* (1987) **42**: 719–720.
24. Jenkinson D, Davison J, Jones K, Hawtin P: Comparison of effects of a self management booklet and audiocassette for patients with asthma. *Br Med J* (1988) **297**: 267–270.
25. Wilson SR, Scamagas P, German DF, et al.: A controlled trial of two forms of self management education for adults with asthma. *Am J Med* (1993) **94**: 564–576.
26. Sibbald B: Patient self care in acute asthma. *Thorax* (1989) **44**: 97–101.
27. Yoon R, McKenzie DK, Miles DA, Bauman A: Characteristics of attenders and non attenders at an asthma education programme. *Thorax* (1991) **46**: 886–890.
28. Ignacio-Garcia J, Gonzalez-Santos P: Asthma self management education programme by home monitoring of peak expiratory monitoring. *Am J Respir Crit Care Med* (1995) **151**: 353–359.
29. Lahdensuo A, Haahtela T, Herrala J, et al.: Randomised comparison of guided self management and traditional treatment of asthma over one year. *Br Med J* (1996) **312**: 748–752.
30. Sly PD, Cahill P, Willet K, Burton P: Accuracy of mini peak flow meters in indicating changes in lung function in children with asthma. *Br Med J* (1994) **308**: 572–574.

31. Uwyyed K, Springer C, Avital A, Bar-Yishay E, Godfrey S: Home recording of PEF in young asthmatics: does it contribute to management? *Eur Respir J* (1996) **9**: 872–879.
32. Lloyd BW, Ali MH: How useful do patients find home peak flow monitoring for children with asthma. *Br Med J* (1992) **305**: 1128–1129.
33. Kendrick AH, Higgs CMB, Whitfield MJ, Laszlo G: Accuracy of perception of severity of asthma: patients treated in general practice. *Br Med J* (1993) **307**: 422–424.
34. Kikuchi Y, Okabe S, Tamura G, *et al.*: Chemosensitivity and perception of dyspnoea in patients with a history of near fatal asthma. *N Engl J Med* (1994) **330**: 1329–1334.
35. British Thoracic Society, The National Asthma Campaign, The Royal College of Physicians of London, *et al.*: The British Guidelines on Asthma Management, 1995. Review and position statement. *Thorax* (1997) **52** (Suppl I): 1–20.
36. Partridge MR, Latouche D, Trako E, Thurston JGB: A national census of those attending U.K. accident and emergency departments with asthma. *J Accid Emerg Med* (1997) **14**: 16–20.

Index

acaricides 637
Acaris siro 538
ACE *see* angiotensin-converting enzyme
acebutolol 597
acetylcholine (ACh) 103, 214, 230, 258, 391, 396, 425, 599
N-acetylcysteine 220
acid anhydrides 543
ACTH 741
activation-induced cell death (AICD) 176
activator protein (AP)-1 752
additives 600–1
adenosine 5′-monophosphate (AMP) 711
adenosine antagonists 776, 804
adenosine receptor antagonism 692
adhesion molecules 239–51, 731–2
 and airway epithelium 194–5
 soluble 243–4
adhesion receptor antagonists in allergic inflammation 246
adhesion receptors 239
 endothelial 242–3
 expression in allergic disease 242–4
 leucocyte 240–2
 role in leucocyte migration in allergic disease 244–6
adolescence 883
adrenaline 409–10
α-adrenergic agonists 579
β-adrenergic agonists 429
 major action 652
β-adrenergic antagonists 409
adrenergic control 399–403
α-adrenergic receptors 399, 403

β-adrenergic receptor antagonists 597
β-adrenergic receptors 59, 399, 401
 abnormalities in asthma 402
β-adrenoceptor agonists 122, 651–76
α-adrenoceptor antagonists 579
β-adrenoceptor populations 651–2
β-adrenoceptors
 linkage to G proteins 653
 regulation 655
β_2-adrenergic agonists 42
 inhaled 511–12
β_2-adrenergic receptors 42, 43, 409, 598
β_2-adrenoceptor agonists 578, 580
 long-acting 657
 molecular pharmacology 651–8
β_2-adrenoceptor polymorphisms 657–8
β_2-adrenoceptors 730
 molecular biology 652
β_2-adrenoreceptor agonists 824–5
adrenomellin 415
adult asthma 12–16
 AHR in 14–15
 diagnosis 12–16
 education 851
 incidence of 14
 management 835–58
  future 853–4
 prevalence of 13–15
 risk factors for 15–16
afferent nerves 392–3, 425
aggravators 850
aggregations of platelets 49
agonist-mediated excitation 103–4
agonist-mediated inhibition 104

β-agonists 16–17, 19–21, 402, 682, 844–5
  actions affecting airway function 658–9
  adverse effects 660
  as functional antagonists 401
  bronchial selectivity 661
  cardiac effects 660
  clinical pharmacology 658–60
  degradation 656–7
  development of tolerance to 664–5
  differences between 661
  duration of action 661
  dysrhythmias due to 665
  effect on arterial oxygen tension 660
  efficacy and safety of inhaled 662–9
  interaction with theophylline 700–1
  long-acting 666–9, 845
  metabolic changes 660
  oral 669–70
  partial or full agonists 661
  short-acting 662–5, 666
  side-effects 845
  structure–activity relationships 656
β2-agonists 20, 151, 797, 877
  and secondary steroid resistance 754–5
  inhaled 883, 888, 890
  intravenous 887
  long-acting 797, 883, 913
  nebulized 887
  oral 883
AHR *see* airway hyperresponsiveness (AHR)
AIA *see* aspirin-induced asthma (AIA)
air filtration 637–8
air pollution factor in childhood asthma 10
airflow obstruction 190
airflow resistance 73, 215
airway calibre
  autonomic control 677–9
  patterns of change 540
airway compliance, factors reducing 69
airway conductance 9
airway cross-sectional area 214
airway drying 576–8
airway fluids 48
airway function
  effects of deep inflation 70–3
  effects of posture and sleep 76–7
  neural control of 389–407
airway hyperreactivity 122, 811
airway hyperresponsiveness (AHR) 2, 8, 9, 12, 48, 49, 56, 128, 214, 215, 488–9, 495, 520, 736
  adult asthma 14–15
  childhood asthma 11
  development and resolution 69
  virus-induced 556–64
airway microvasculature 232, 253
airway mucus *see* mucus
airway narrowing 232, 576–8
  and airway drying 576–8
  and osmolarity 576–8
  in episodic asthma 69
  in vivo 65–9
  intrapulmonary 75

  intrathoracic 69–74
  mechanisms restricting 66–8
  response to increasing severity 77–82
  sites 69–76
airway nerves 498
airway obstruction 214
  irreversible 868–9
  perception 80
  severe 82
airway pathology 47–64
airway reactivity 214–15
airway remodelling 475–86
airway resistance 71
airway responsiveness, rhinovirus infections effect on 558–62
airway smooth muscle (ASM) 65–6, 349, 352, 373
  adrenergic control 399, 400
  autonomic control of tone 390
  cell–cell junctions 90–2
  cell morphology 90
  cell structure and function 89–112
  contractility 69
  contraction 67, 69, 72, 89
    luminal effects of 68
  effect of theophylline 693–4
  factors influencing tone 410
  humoral control of tone 409–21
  hyperplasia 482, 496
  hypertrophy 482, 496
  increased mass 69
  innervation 93–6
  major features 95
  mass increase 482
  physiological functioning 101
  role of abnormalities 496
  species differences 93
  spontaneous mechanical and electrical activity 102–3
  thickening 481–2
  see also ion channels
airway surface liquid (ASL) 569–71, 573, 580
airway walls 215
  innervation 58–9
  remodelling 56
  structural changes 475–8
  thickening 56, 478
airways disease in childhood 682–3
albumin 210
allele-sharing methods 36
allergen-induced asthma, treatment 518
allergens 507–28
  as cause of asthma 515–17
  avoidance 617–49, 849–50
  avoidance in patients' homes 632–3
  chemical measures for reducing 637–8
  childhood asthma 889
  exposure and asthma severity 622
  exposure control 625
  high-altitude studies 631–2
  indoor 618–21
    distribution and aerodynamic properties 633–5
  ingested/injected 520

Index

inhaled 508–18
  mechanisms 513–15
  nature of 508–9
  pets 638–40
  pharmacology 511–13
  physical measures for reducing 636–7
  practical avoidance measures 625–40
  primary sensitization 621–2
  response patterns 509–11
  threshold values 623–5
allergic airway disease 555
allergic asthma, diagnosis 517
allergic bronchopulmonary aspergillosis 518–20
  clinical features 519
  diagnosis 519
  pathogenesis 518–19
  treatment 519
allergic disease
  expression of adhesion receptors 242–4
  leucocyte migration in 244–6
allergic inflammation 119
  adhesion receptor antagonists in 246
  process 148
  rhinovirus infections effect on 558–62
  role of basophils 121–2
allergy and occupational asthma 532–4
Alternaria alternata 40
Alternaria tenuis 538
alveolar macrophages 134
ambroxol 220
amino acid peptides 191
aminophylline 826
  intravenous 697, 698
  routes of administration 698
ammonium hexochloroplatinate 539
anaesthetics 602
anaphylatoxins 115, 351
  role in asthma 352–3
angiotensin II 411
angiotensin-converting enzyme (ACE) 299, 411, 436
angiotensin-converting enzyme (ACE) inhibitors 393, 601–2
antagonists, mediator 767–82, 801
anti-allergic drugs 513, 810
anti-asthmatic drugs 120–1
anti-IgE 118
anti-IgE antibodies 812
anti-IL-4 806
anti-IL-5 805
anti-inflammatory cytokines 807
anti-inflammatory drugs 807–13
anti-inflammatory mechanisms 499–500
anti-inflammatory medication 20
anti-inflammatory proteins 729
anti-tumour necrosis factor 806
antibiotics 830
anticholinergic agents 221, 512, 580, 602, 799–800, 825–6, 847
anticholinergic bronchodilators 677–88
  against specific stimuli 681
  clinical efficacy 681–5
  clinical recommendations 686

combinations 684–5
  dose–response 681
  pharmacology 679–80
  rationale for use 677–9
  side-effects 685
antigen
  processing and presentation 132–4
  responses 167
antigen-presenting cells (APCs) 132–4, 163, 165–8, 195
antihistamines 512, 579, 767–8, 848
  new agents 801
antioxidants 775–6, 804
AP-1 135, 464, 468, 752, 753, 809–10
apafant 773, 803
APC-366 777
APCs *see* antigen-presenting cells (APCs)
apoptosis 731
  in chronic-airway inflammation 177
arachidonic acid 102, 281, 613
  cleavage 282
  metabolism 270–1
  pathway metabolites 131–2
arterial oxygen tension 660
Ascaris suum 356
ASM *see* airway smooth muscle (ASM)
Aspergillus 509, 518–19
aspirin, anaphylactic reactions 607
aspirin-induced asthma (AIA) 607
  allergic mechanism 608
  clinical presentation 611–12
  cyclooxygenase theory 608
  diagnosis 612–13
  differential diagnosis 613
  history and definition 607
  leukotrienes in 609
  pathogenesis 608–12
  prevention 613–14
  release of mediators at site of reaction 609–10
  treatment 613–14
assisted ventilation 827–8
association studies 36
  bronchial hyperresponsiveness (BHR) 42
astemizole 122, 768, 801
asthma
  classification 836
  definition 1–2
  evidence for increasing severity 16
  future therapies 795–819
  in adults *see* adult asthma
  in childhood *see* childhood asthma
  in remission 836
  initiators of 529
  inward perviousness 262
  life-threatening 829
  pathophysiology 205–7, 487–506
  pharmacotherapy, pharmacoeconomics 907–14
  provokers of 529
  severity and progression 167
  stable 681–2
  unanswered questions 501–2
atenolol 597
ATF/CREB 464

atmospheric pollutants 589–96
  see also specific pollutants
atopy 500, 508
  genetic studies 38–41
  in childhood asthma 7, 10
  in occupational asthma 534–5
  inheritance mode 37
atrial natriuretic peptide (ANP) 799
atropine 96–7, 677, 679, 680
  side-effects 685
atropine methonitrate 680, 686
autonomic nerves 389
azathioprine 747, 784–5

β-blockers 597–600, 828
  mechanisms 598
B-cells 159
  antibody production 173–4
B-lymphocytes 161
β-receptor function 349
Bacillus subtilis 534, 538
BAL *see* bronchoalveolar lavage (BAL)
basic protein inhibitors 777
basophils 57, 113–40, 245, 315, 332–3
  human 115
  role in allergy and asthma 121–2
Bay-x1005 802
BDP *see* beclomethasone dipropionate (BDP)
beclomethasone dipropionate (BDP) 713, 716, 717, 737, 738, 742–4
bed and bedding 636–7
benzalkonium chloride 600–1
betamethasone 746
BHR *see* bronchial hyperresponsiveness (BHR)
bisulphites 600
BN50548 778
BN52021 347–8
bone metabolism 742–3
Bothrops jararaca 297
bradykinin 100, 230, 298–302, 304, 393, 601
  antagonists 775, 803–4
breathing, control 80
brequinar sodium 790
BRL 3491 800
BRL 38227 800
bromhexine 220
bronchial biopsy 333, 488
bronchial circulation 572–4
bronchial dimensions 73
bronchial epithelium 189
bronchial gland ectasia 51
bronchial goblet cell hyperplasia 51
bronchial hyperresponsiveness (BHR) 37, 190, 350, 840
  association studies 42
  genetic studies 41–3
  linkage studies 41
  mechanisms responsible for 38
bronchial lavage 830
bronchial microvasculature 573
bronchial mucosal swelling 233

bronchial smooth muscle mass enlargement 53–6
bronchial vasculature 56
bronchial wall occupied by bronchial smooth muscle 53
bronchiectasis 211
bronchiolitis 548
bronchoalveolar lavage (BAL) 48, 49, 116, 127, 260, 261, 311, 318, 333, 334, 488, 559, 561, 768
  immunologically induced release of mediators 118
  in extrinsic asthma 117–18
bronchoconstriction 89, 122, 215, 415, 668
bronchoconstrictors 76, 272
bronchodilation 415
bronchodilator nerves 373, 498
bronchodilators 16, 602, 667, 716
  new agents 796–801
  oral 883
bronchoscopy 488
budesonide 737, 738, 742–4
bumetanide 810
bupivacaine 602
BW443C 444

C chemokine 312
C-C chemokine 311, 315, 317
C-C chemokine receptors 316
C-X-C chemokine receptors 316
C-X-C chemokines 310–11, 315
Ca^{2+} channels
  voltage-dependent 98
  voltage-independent 100
Ca^{2+}-dependent Cl^- channels 97
Ca^{2+}-dependent inositol-specific phospholipase C (PLC) 100
calcitonin gene-related peptide (CGRP) 94, 391, 428, 436–8
calcium antagonists 579
calcium ion flux 692–3
calcium ionophores 131
calmodulin 103
Candida albicans 162
capsaicin 438, 439
captopril 601
carbon dioxide
  effect on airway tone 415–17
  retention 80
carbon monoxide transfer coefficient 76
carboxypeptidase N 299
cat allergens 618, 638–40
catastrophic asthma, management 828
catechol-O-methyltransferase (COMT) 656–7
catecholamines 400–1, 409–11, 692
cation channels, non-selective 97
CCAAT/enhancer-binding proteins (C/EBP) 464–5
CC-CKR3 receptor 319
CCR-3 receptors 807
cell adhesion blockers 812
cell adhesion molecules 196

Index

cells
  inflammatory 489–91
  structural 491
central nervous system, side-effects 745
cetirizine 122, 151, 768, 801
CGRP *see* calcitonin gene-related peptide (CGRP)
Charcot–Leyden crystal protein 143
Charcot–Leyden crystals 48
charybdotoxin 96, 104
chemokine inhibitors 807
chemokine receptors 316–17
chemokine superfamily 310
chemokines 164, 192, 309–27, 811
  as chemoattractants and cell activators 313–15
  cell sources 312
  discovery 310–12
  expression and release in asthma 317–18
  regulation 312–13
  structure 310–12
chemotactic factors 192–3
childhood
  airways disease in 682–3
  inhaled steroids 737
  wheezing in 547–52, 860–1
childhood asthma 2–5
  air pollution factor in 10
  allergens 889
  anatomical and physiological factors 860
  and AHR 11
  assessment of clinical condition 865–8
  atopy in 7, 10
  clinical outcome measures 879
  diet in 9–10
  diseases sharing symptoms 862
  drug strategy 875–82
  emotional disorder 868
  environment factor in 6–8
  environmental control 889
  ethnic factors in 6
  exercise 890
  factors influencing development of 5–10
  family size factor in 10
  genetic factors in 6
  growth pattern 864
  history-taking 866
  immunotherapy 888–9
  inhalation therapy 869
  irreversible airway obstruction 868–9
  lifestyle impairment 867
  lung function in 9, 866–7
  management 859–902
  natural history 861–3
  passive smoking in 8–9
  pharmacodynamics 864–5
  pharmacokinetics 864–5
  prevalence 3, 4
  prevention 863
  prognosis 11–12
  psychosocial behaviour 868
  risk factors 863
  severe acute asthma/status asthmaticus 884–8
  sex factor 6
  socioeconomic status in 10
  special age groups 883
  step-down strategy 877
  step-up strategy 875
  treatment 869–82
Chlamydia 556
cholecystokinin octapeptide (CCK$_8$) 391, 445–6
cholinergic agonists 100, 102
cholinergic nerve fibres 393–4
cholinergic neurotransmission, modulation 395
chronic bronchitis 51, 211
chronic obstructive pulmonary disease (COPD) 48, 51, 56, 69, 78, 82, 480, 662, 679, 686, 800, 802, 824
  acute exacerbations 684
  stable 683–4
  theophylline in 696, 700
ciliated cells 207
Cladosporium herbarum 538
Clara cells 207
cluster of differentiation (CD) designations 161
coagulation time 145
cockroach allergens 620–1
coinfection, T-cell responses 175
collagen
  deposition 478–9
  type III 51
  type IV 51
  type V 51
colloidal gold 257
colony-stimulating factors 193
communication improvement 919–22
complement 351–3
  effects of C3a and C5a on airways 352
  metabolism 351
  origin 351
  role in asthma 352–3
  vascular effects 352
complement receptors CR1 and CR3 129
complementarity-determining region (CDR) 163
compliance/non-compliance 918–19
compound 48/80 115, 119
congestion 56
connective tissue 743
connective tissue mast cells (CTMC) 114–15
connexins 91
COPD *see* chronic obstructive pulmonary disease (COPD)
coronaviruses 556
corticosteroids 16, 21, 82, 120, 150, 512, 518, 519, 663, 667, 713, 716–17, 826–7, 829
  inhaled 841, 875, 880–1, 884, 907–14
  side-effects 841–2
cortisol 413–14
cost-benefit analyses 905–6
cost-effectiveness analyses 905–7
  outcome measures 907
cost-utility analysis 906
cotransmission 427–8
cotransmitters 391
COX *see* cyclooxygenase (COX)
CREB 728, 751–2, 754
CREB-binding protein (CBP) 468, 728

Creola bodies 48, 49
cromakalim 800
cromoglycates 151, 810
cromolyn sodium, inhaled 913
cromones 707–24, 810
　efficacy of 714–15
CTLA-4 170, 176
Curschmann's spirals 48
cutaneous lymphocyte antigen (CLA) 240
cyclic 3′-5′-adenosine monophosphate see cyclic-AMP (cAMP)
cyclic 3′,5′-guanosine monophosphate see cyclic GMP
cyclic-AMP (cAMP) 92, 102, 151, 272, 401–2, 653, 655, 690, 693, 694, 797–8
　intracellular actions 654–5
　regulated enhancers (CREs) 468
　responsive element (CRE) 728
cyclic-GMP 104, 373, 374, 690, 799
cyclooxygenase (COX) 191, 271, 608, 610, 611, 613, 614, 773
　COX-2 773, 803
　inhibitors 303, 803
　role in asthma 272–3
cyclosporin A (CsA) 168, 747, 786–7, 812, 848
cysteinyl leukotrienes see leukotrienes
cytokine-induced neutrophil chemoattractant (CINC) 311, 350
cytokines 57, 115, 123, 129–33, 143, 151, 164, 171–2, 192–4, 257, 329–30, 493–4, 730, 805
　alteration of established type 1 or type 2 336
　anti-inflammatory 807
　in non-atopic asthma 335
　in secondary steroid resistance 753–4
　inhibitors 805
　potential for intervention 337
　pro-inflammatory 562–4
　production in asthma 333–5
　receptors 162
　signal transduction 465
　type 1 330, 336
　type 2 330, 331, 336
　see also specific cytokines
cytosolic PLA$_2$ (cPLA$_2$) 281

dapsone 747
deep inflation (DI), effects on airway function 70–3
delayed-type hypersensitivity (DTH) 329–30
dendritic cells (DCs) 127–40, 733–4
Dermatophagoides 509
Dermatophagoides farinae 169
Dermatophagoides pteronyssinus 40–1, 119, 169, 534
dexamethasone 136, 746, 750
diagnosis 1–2
　adult asthma 12–16
　allergic asthma 517
　allergic bronchopulmonary aspergillosis 519
　aspirin-induced asthma 612–13
　feelings of patients on 920
　irritant-induced asthma 530
　occupational asthma 536

diet
　in adult asthma 15
　in childhood asthma 9–10
differential diagnosis, aspirin-induced asthma 613
1,4-dihydropyridine (DHP) 98, 102
diphenyl methane diisocyanate (MDI) 531
4-diphenylacetoxy-N-methylpiperidine methiodide (4-DAMP) 395–6
DNA 210, 213, 726–8
DNA binding 466
DNA sequence 35
dog allergens 618, 638–40
drug-induced asthma 597–605
drug strategy, childhood asthma 875–82
dry powder inhalers 872–3
Duffy blood group antigen 317
dysphonia 740, 841

E-selectin 193, 240, 242–4, 246
E-selectin ligand (ESL)-1 241
early or immediate asthmatic response (EAR) 509, 512–14
ebastine 801
ECP see eosinophil cationic protein (ECP)
eczema, prevalence 7
education 917–26
　adult asthma 851
EIA see exercise-induced asthma (EIA)
eicosanoids 270, 272
　metabolism 610–11
ELAM-1 130
electrophysiological studies 99
emotional disorder, childhood asthma 868
emphysema 75
enalapril 601
end-expired lung volume (FRC) 76
endobronchial pressure measurements 73
endothelial adhesion receptors 242–3
endothelial cells 734
endothelin antagonists 777, 805
endothelin family 191
endothelin receptor antagonists 355–7
endothelin receptors 355
endothelins 100, 354–7, 411, 413, 415, 494
　effects 356
　role in asthma 356–7
　sites of synthesis 355
enkephalins 446
environment factor in childhood asthma 6–8
environmental tobacco smoke (ETS) exposure 8
eosinophil cationic protein (ECP) 142–3, 210, 354
　biological activities 144–5
　levels 150, 151
　measurement 150
　molecular characteristics 144–5
eosinophil-derived neurotoxin (EDN) 354
eosinophil function 132
eosinophil granule 55
eosinophil inhibitors 810–11
eosinophil peroxidase (EPO) 142, 354
　biological activities 146
　molecular characteristics 146

Index

eosinophil protein X/eosinophil-derived neurotoxin *see* EPX/EDN
eosinophil proteins 354
eosinophilia 57
eosinophils 117, 128, 141–57, 244, 314, 332–3, 709–10, 733
  accumulation mechanisms 148–9
  biochemistry 142–7
  biological responses 147
  degranulation 147–8
  epithelial damage induced by 350
  function 142–7
  history 142
  monitoring accumulation and activity in asthma 150
  pharmacological control 150–1
  production 141
  receptors involved secretory response 147–8
  recruitment 490
  role in asthma 149–50
  transendothelial and transepithelial migration 196
eotaxin 149, 311
epidemiology 1–33
  changing 16
  respiratory infections 547–56
  wheezing 547–56
epidermal growth factor (EGF) 240
episodic asthma 836
  airway narrowing in 69
epithelial basement membrane thickening 51
epithelial cell-derived mediators 190–4
epithelial cells 50, 187–204, 492, 734–5
  culture *in vitro* 196–7
  types in airway epithelium 188
epithelial damage induced by eosinophils 350
epithelial eosinophil counts 562
epithelial phospholipase oxygenation system 191
epithelial repair 254
epithelial restitution 263–5
epithelium 49–50, 495
  and adhesion molecules 194–5
  and hyperresponsiveness 190
  morphology 187–90
  putative role in aetiology of asthma 199
EPX/EDN 143
  biological activities 145
  molecular characteristics 145
erythromycin 220
ethnic factors in childhood asthma 6
ethylenediaminetetraacetic acid (EDTA) 601
eucapnic hyperventilation 575
European Community Respiratory Health Study (ECRHS) 13–14
excitatory junction potentials (EJPs) 94
excitatory NANC (e-NANC) nerves 426–7
exercise 72, 849
  in childhood asthma 890
  refractoriness 276–7
exercise-induced asthma (EIA) 232, 569, 572, 573, 579, 580, 668
  challenge with 574

exercise-induced bronchoconstriction (EIB) 276–7, 569
expiratory flow limitation 77
exposure in occupational asthma 534
extracellular matrix 478–81
extrathoracic airway 74–5
extravasated plasma 256
extrinsic asthma, BAL in 117–18

factor XII 145
family history 35
family size factor in childhood asthma 10
FcεRII 129
FEV_1 2, 12, 15, 56, 66, 69, 72–4, 117–18, 120, 289, 539, 574, 576, 837, 911, 912
fibrillary material 49
fibrinogen 210
fibrosis 496
FK506 168, 790, 812
FK-binding protein (FKBP) 790
FLAP antagonists 288, 609, 771
flexible fibreoptic bronchoscopy 57
flunisolide 738
flurbiprofen 579
fluticasone propionate (FP) 738, 742
forced expiratory volume in 1 s *see* FEV_1
forced vital capacity (FVC) 117, 539–40, 837
foropafant 774
FPL 55712 102
frusemide 579, 580, 810
functional residual capacity (FRC) 68, 71, 77, 80, 82
fungal infections 519
fungal spores 508, 518
furosemide 810
Fyn 168

G proteins 653, 811
galanin 391, 428, 446
gap junction 90, 91, 93, 189
  conductances 93
  density 93
gas exchange inefficiency 80
gastric reflux 850
gastrin-releasing peptide (GRP) 445
GATA-3 171
gene therapy 813
genes in asthma 35–6
genetic factors 500–1
  in childhood asthma 6
genetic markers 35
genetic mutations 37
genetic studies 37
  BHR 41–3
genetics 35–46
glacagon 415
glottis 74
glucocorticoid response elements (GREs) 466, 726–7, 752
glucocorticoids 258, 380–1, 467, 469, 471, 491, 808, 812

glucocorticoids, *continued*
  cellular effects 732
  classical model action 727
  effects on asthmatic inflammation 735–6
  effects on cell function 732–5
  effects on gene transcription 726–7
  effects on monocyte/macrophage function 134–5
  inhaled 783
  interaction with transcription factors 727–8
  molecular mechanisms 725
  receptors (GRs) 466, 725–6, 749–50
    abnormalities in function 751
    downregulation 753
    interaction with transcription factors 751–2
  resistance 748–55
    cellular abnormalities 750
    clinical features 748–9
    familial 749
  target genes in inflammation control 729–32
  topical 783
glucocorticosteroids 219, 725–66
  see also glucocorticoids
glutamate 425
glyceryl trinitrate (GTN) 799
glycopyrrolate 686
glycopyrrolate bromide 680
glycosaminoglycans 51
goblet cells 187, 207–8, 213
  exocytosis 213
  hyperplasia 210, 211, 219
gold salts 747, 784–5, 848
granulocyte chemotactic protein 2 (GCP-2) 311
granulocyte-colony stimulating factor (G-CSF) 193
granulocyte-macrophage colony-stimulating factor (GM-CSF) 50, 123, 129–34, 136, 141, 143, 147–9, 192, 193, 332, 562, 807
growth factors 194
  sources in chronic asthma 176
growth pattern, childhood asthma 864
growth stunting 744

haematological side-effects 745
β-haemolytic streptococcus 552
Haemophilus influenzae 552
HC-14 311
health professionals, education and guidelines 917–18
heliumn–oxygen mixture, response to breathing 70
helodermin 433
helospectins 391, 433
hemi-desmosomes 189
heparin 777
hereditary component 35
HETE 131, 217
5-HETE 132, 219
12-HETE 192
15-HETE 192, 271
hexamethylene diisocyanate (HDI) 531
high endothelial venules (HEV) 241

high molecular weight (HMW) kininogen 298
high performance liquid chromatography (HPLC) analysis 301
histamine 65, 72, 100, 103, 117, 118, 257, 258, 272, 301, 343–6, 395, 560
  as bronchoconstrictor mediator 122
  effects 344–5
  H_1-receptor antagonists 122
  receptors 344
  release 120–3, 132
  role in asthma 345–6
  tachyphylaxis 275–6
history-taking 2
  childhood asthma 866
HLA molecules 40
HLA phenotype 535–6
HLA polymorphism 169–70
HLA-DR antigens 195
HMW kininogen 299
Hoe140 304
HOE234 800
hormones 413–15
hospital admission 5, 16, 21, 829
house dust 638
house-dust mites 618, 636–8
housing design and mite allergen levels 638
5-HPETE 282
Human Genome Project 44
human lung mast cells 116
human umbilical vein endothelial cells (HUVEC) 240, 241, 243
humidity control 636–7
humoral control of airway smooth muscle tone 409–21
hyaluronan (HA) 479
hydration 823
hydrocortisone 414, 826
  intravenous 747
hydrogen peroxide 132
hydroxychloroquine 747
hydroxyeicosatetraenoic acids *see* HETE
hydroxyl radicals 132
hyperinflation 77, 80–2
  advantages and disadvantages 81
hyperoxia 415, 416
hypersensitivity reactions 828
hyperventilation 72, 417
hyperventilation-induced asthma (HIA) 569, 573
  challenge with 574
hypocapnia 80, 417
hypothalamic–pituitary–adrenal axis 741–2
hypoxaemia 80

I-309 311
iberiotoxin 96, 104
ICAM-1 130, 165, 175, 193–5, 242, 243, 246, 563, 812
ICAM-2 165
ICAM-3 165, 242
IgA 147, 164
IgE 119, 129, 143, 147, 164
  antibodies 508, 557–8

Index

inhibition 812–13
levels 39–40
regulation 331–3
sensitivity 519
sensitization 559
synthesis 332
IgE-dependent events 557–8
IgE-mediated mechanisms 508
IgG 164
sensitivity 519
IkB proteins 462–3, 813
IkBα proteins 462–3, 469
IkBβ proteins 462–3
immune complexes 131
immune responses 167
defects in termination 176
immunofluorescence 50
immunohistochemical analysis 558
immunological tests, occupational asthma 538–9
immunologically induced mediator release 118
immunomodulators 783–93, 811–12
immunoregulation 195–6
immunosuppressive therapy 784–90
newer agents 790
immunotherapy 813
childhood asthma 888–9
incidence of adult asthma 14
indomethacin 102, 276, 513
inducible isoform nitric oxide synthase (iNOS) 776
infections 547–67
inflammation 190
acute 498
chronic 498, 611
inflammatory cells 128, 489–91
recruitment of 57
inflammatory disease 488–9
inflammatory effects 495–8
inflammatory enzymes 730–1
inhibitors 777
inflammatory genes, promoter sequences 470
inflammatory inhibitors 767, 801
inflammatory mechanisms 128
inflammatory mediators 233, 415, 493–4, 767, 801
inflammatory products 498
inflammatory receptors 731
inflammatory responses 389, 438, 495
influenza 556
information
for patients 919–22
guidelines 921
inhalation bronchograms 73
inhalation testing, occupational asthma 539–41
inheritance modes 37–8
inhibitory non-adrenergic non-cholinergic (i-NANC) nerves 373, 426, 431
initiators of asthma 529
inositol 1,4,5-trisphosphate (IP$_3$) 100, 103
inspired air conditioning 570–2
insulin-like growth factor (IGF-1) 144–5, 480
integrins 240
and their receptors 242
intercellular adhesion molecule see ICAM-1, -2, -3

interferon
IFN 309
IFN-α 129
IFN-τ 123, 129, 132, 136, 163, 171, 175, 311, 337, 559, 562, 563, 807
interleukins 39, 309
IL-1 123, 130, 131, 164, 168, 240
IL-1 receptor antagonist (IL-1ra) 807
IL-1α 168
IL-1β 130, 168, 217, 562
IL-2 143
IL-3 115, 123, 141
IL-4 57, 123, 129, 131, 132, 163, 174, 175, 331–2, 562, 806
IL-4 inhibitors 813
IL-5 57, 123, 141, 148, 149, 151, 175, 563, 805, 810–11
IL-6 50, 123, 129, 130, 143, 217, 257, 562
IL-8 50, 130, 143, 149, 193, 312, 315, 317
IL-10 57, 131, 132, 692, 729, 807, 813
IL-12 163, 172–3, 807, 813
IL-13 123, 331–2
IL-16 163
intermittent positive-pressure breathing (IPPB) 825
International Classification of Diseases (ICD) 17
International Study of Asthma and Allergies in Childhood (ISAAC) 4, 10
interstitial cystitis 119
intrabronchial measurements 73
intraepithelial nerves 50
intravenous immunoglobulin (IVIG) therapy 788–9
intrinsic positive end-expiratory pressure (intrinsic PEEP) 80
ion channels 96–100
ionic conductance changes 103
ionizers 637–8
ipratropium 682–4, 686
side-effects 685
ipratropium bromide 679, 825, 847, 884, 887
irritant-induced asthma 530
diagnosis 530
isocapnic hyperventilation (ISH) 576
isocapnic hypoxia 416
isocyanates 542
isosorbide dinitrate 799
isovolumic maximum expiratory flow 72
ITAMs 166, 168

JAK-STAT 810
Janus kinases (JAK) 465–6, 470–1
JE 123
Jun N-terminal kinase (JNK) 751
Jun N-terminal kinase (JNK-1 and JNK-2) 464

K$^+$ channel blockers 102
K$^+$ channel openers 800
K$^+$ channels 96–7
kallikrein 298
kallikrein-kinin system 297

ketotifen 713, 848
kininase 1 299
kininase 2 299
kininogenases 298
kininogens 298, 299
kinins 297–307
  effects on airways 301–2
  enzymes releasing 298
  formation 298–9, 301
  hydrolysis 299
  mechanisms of action 302–3
  metabolism 298–9
  pharmacological properties 297, 300
  receptors 300
  structure 298–9

L-NAME 230, 371, 373, 375, 381, 426
L-NMMA 371, 373, 381
L-selectin 163, 168, 195, 241
L-type calcium channel 98, 99, 103
lactoferrin 210
lamina propria 255
laminin 51
larynx 74
late-phase asthmatic response (LAR) 47, 49, 51, 56, 122–3, 509, 512–14, 558–60
Lck 168
LD78 311
Leptidoglyphus destructor 538
leucocyte adhesion deficiency (LAD) 241
leucocyte adhesion receptors 240–2
leucocyte function-associated antigen 1 (LFA-1) 195, 246
leucocyte migration in allergic disease 244–6
leukotrienes 102, 131, 217, 271, 281–95, 493, 768–9, 848
  antagonists 579, 802
  biological effects 283–6
  formation 281–3
  in aspirin-induced asthma 609
  inhibition in chronic stable asthma 289
  LTB$_4$ 129, 131, 132, 149, 771, 772, 802
  LTC$_4$ 72, 100, 115, 118, 129, 132, 143, 559, 564
  LTD$_4$ 100, 132, 277, 288, 769, 770
  LTE$_4$ 132, 609, 614, 768, 770
  metabolic pathway for production and degradation 284
  metabolism 281–3
  receptor blockade and synthesis inhibition 288
  recovery in asthma 286–7
  role in asthma 283–9
levcromakalim 800
life-threatening, asthma 829
lifestyle impairment, childhood asthma 867
lignocaine 602
lipid mediators 191–2
lipocortin 1 antibodies 750
lipoxins 281
5-lipoxygenase (5-LO) 513
  inhibitors 770–1, 802, 848
5-lipoxygenase (5-LO) activating protein (FLAP) 288, 609, 771

liquid nitrogen 637
LOD score 36, 39–40
low molecular weight (LMW) kininogen 298, 299
lower respiratory tract infections (LRIs) 549
LPS 131
luminal effects of ASM contraction 68
luminal mucus 210, 215
lung function in childhood asthma 9, 866–7
lung parenchyma 75–6
lung tissue resistance 76
LY171883 289
LY293111 802
lymphocytes 159–86
  as sources of growth factors 176
  implication in pathogenesis of asthma 161–3
  number and precentage in lung and blood 160
  responses to inhaled foreign antigens 163–76
  subpopulations 159–61
  surface markers 159–61
lymphokines 329–42
lymphotoxin 171
lysylbradykinin 298, 299, 300

MAC-1 *see* macrophage antigen 1 (MAC-1)
macrophage antigen 1 (MAC-1) 195, 246
macrophage colony-stimulating factor (M-CSF) 129, 133
macrophage inflammatory protein (MIP)
  gene family 123
  MIP-1α 123, 311–13
  MIP-1β 123, 311–13
  MIP-2 311, 350
macrophage products modulating inflammation 129–32
macrophages 57, 127–40, 217, 490, 732–3
  effect of glucocorticoids 134–5
  functions in asthma 128–34
magnesium sulphate 830
major basic protein (MBP) 143, 354
  biological activities 146–7
  molecular characteristics 146–7
major histocompatibility complex *see* MHC class II
management
  adult asthma 835–58
    future 853–4
    planning 837–52
  catastrophic asthma 828
  childhood asthma 859–902
  severe acute asthma 821–33
    general assessment 821–3
mass median aerodynamic diameter (MMAD) 574
mast cell tryptase 777
mast cells 113–40, 128, 245, 489, 709, 733
  heterogeneity 114–15
  human 115
  human lung 116
  role in early asthmatic reactions 122
  role in late asthmatic reactions and cytokine production 122–3
M-CSF *see* macrophage colony-stimulating factor (M-CSF)
MDI *see* metered dose inhaler (MDI)

Index

mechanical ventilation heat recovery (MVHR) units 636
membrane products 55
membrane receptors 148
memory cells, responsiveness to chemoattractant signals 168–9
messenger RNA (mRNA) 92, 99, 123, 130, 143, 221, 313, 335, 459, 534
metabisulphite 600
metabolic side-effects 744
metered dose inhaler (MDI) 712, 738, 740–2, 745, 825, 870
　inhalation technique 870
　spacers 871–2
methacholine 65, 67, 72, 272, 411, 415, 416, 575
methotrexate 747, 784–5, 848
methoxamine 579
methylprednisolone 826
methylxanthines 580, 798–9
metoprolol 597
MG-CSF 133
MHC class II 133, 164, 165, 174
MHC class II antigens 132, 136
MHC class II molecules 168
microcirculation 232, 253
microvascular-epithelial exudation
　acute challenge-induced 257–9
　plasma 253–67
microvascular leakage 497
MIP see macrophage inflammatory protein (MIP)
mite allergens 636–8
mite microhabitats, removal 636
MK-591 771
MK-886 771
modipafant 803
monoamine oxidase (MAO) 656–7
monoclonal antibodies 247
monocyte chemoattractant protein 1 (MCP-1) 311
monocytes 127–40, 217, 315, 490
　effect of glucocorticoids 134–5
　functions in asthma 128–34
mononuclear phagocytes 128
　activation 129
　soluble mediators synthesized by 130
monosodium glutamate (MSG) 601
morphine 115
mortality, inhaled steroids 738
mortality rates in asthma 17–21
mucins 206–7
　abnormality 211–12
mucolytic agents 220, 830
mucosal exudation of plasma 213, 253, 259–60
mucosal inflammation 59
mucosal mast cells (MMC) 114–15
mucosal oedema 56
mucus 205–7
　abnormalities in asthma 208–13
　constituents 206
　humoral inducers of secretion 217
　hypersecretion 208–11, 213–15, 497
　　pharmacological treatment 218–21

　　therapeutic prospects 220
　hyperviscosity 213–15
　interactions with other components 212–13
　neural mechanisms of secretion 218
　plugs 208, 212
　secretion 51, 205–27, 352, 373, 735
　secretion control 216
　secretion inducers 216–18
multifunctional cytokines 193–4
multiple inert gas elimination technique (MIGET) 78
multiple mediator antagonists 805
muscarinic autoreceptors 398
muscarinic receptors 395–8, 799
　subtypes 679
mycophenolic acid 790
Mycoplasma infection 553
Mycoplasma pulmonis 162
myocardial necrosis 665

Na^+ channels, voltage-activated 97
NANC nerves 425–8, 711
　excitatory (e-NANC) 426–7
　inhibitory (i-NANC) 373, 426, 431
NANC neural control 218
NANC responses 427
naphthalene diisocyanate (NDI) 531
nasal obstruction 75
National Health and Nutrition Examination Surveys 4
natriuretic peptides 411
natural killer (NK) cells 161
nebulizers 873–5
nedocromil 663
nedocromil sodium (NS) 121, 122, 221, 303, 578, 580, 707, 810, 846
　anti-inflammatory actions 711
　chemical structure 708
　clinical studies 713–17
　in children 717
　mechanisms of action 711–12
　neural mechanisms 710–11
　vs. SCG 717
neural control of airway function 389–407
neural effects 497–8
neural mechanisms in asthma 497–8
neurogenic inflammation 438–44
　in animal models 438–40
　in asthma 440–4
　modulation 443–4
neurokinin A (NKA) 100, 391, 428, 434, 442
neuronal NO synthase (nNOS) 374
neuropeptide Y (NPY) 391, 444–5
neuropeptides 423–57, 498
　interactions 424–5
　role in asthma 446–7
neurotransmitters 391
neutral endopeptidase (NEP) 435, 436, 438, 440, 442
　inhibitors 442
neutrophil elastase 777
neutrophils 314, 491, 710, 734

NF-IL-6 464–5
NF-κB 217, 371, 460–3, 468, 751–3, 755, 809–10
nitric oxide (NO) 10, 94, 190–1, 230, 369–88, 391, 494, 804
   clinical implications 381–2
   cytotoxic effects 375
   effects of therapy 380–1
   effects on airway function 371–5
   exhaled 375–6, 378–80
   generation 369
   inflammatory effects 375
   measurement of exhaled 376–7
   neurotransmission 373–4
   source in exhaled air 378
   therapeutic implications 382–3
   vascular effects 371–2
nitric oxide synthase (NOS) 190–1, 303, 369, 494, 730–1
   expression in airways 370–1
   inducible isoform (iNOS) 776
   inhibitors 371, 381, 776, 804
   neuronal (nNOS) 374
nitrogen dioxide (NO$_2$) 591–2
nitrovasodilator compounds 799
nocturnal asthma 668
non-adrenergic non-cholinergic (NANC) nerves see NANC nerves
non-isotonic aerosols
   challenge with 574
     generation and deposition in respiratory tract 574
     inhalation 569–87
non-selective cation channels 97
non-steroidal anti-inflammatory drugs (NSAIDs) 513, 607, 610–12, 828
noradrenaline 411, 425
NPC567 303
NPY see neuropeptide Y (NPY)
NS-1619 801
nuclear factor κB see NF-κB

occupational asthma 15, 529, 836
   and allergy 532–4
   atopy in 534–5
   causes 531–2
   determinants 534–6
   diagnosis 536
   exposure in 534
   hypersensitivity-induced 537–41, 541–2
   immunological tests 538–9
   importance of 532
   inhalation testing 539–41
   initiators and provokers of 529–30
   management 542–3
ocular side-effects 743–4
oestrogen 414
open lung biopsy 57
oropharyngeal candidiasis 740
osmolarity and airway narrowing 576–8
osteocalcin 743
oxitropium 686, 847
oxitropium bromide 680

oxygen
   effect on airway tone 415–17
   inhaled 888
oxygen radicals 348–51
   effects on airways 349–50
   role in asthma 350–1
oxygen therapy 823–4
oxygenation 78–80
ozone 589–90

p300 468
P-selectin 242, 243, 244
P-selectin glycoprotein ligand (PSGL)-1 241
PAF see platelet-activating factor (PAF)
paradoxical bronchoconstriction 685
parainfluenza 556, 860
parasympathetic nerves 393–9, 425
   role in asthma 398–9
parenchymal cells 120
partial agonism 668
passive smoking in childhood asthma 8–9
pathophysiology of asthma 487–506
patient needs 922
PCF$_{2\alpha}$ 273–4
PDE see phosphodiesterase (PDE)
peak expiratory flow (PEF) 2, 537, 540, 837, 839, 840
peak expiratory flow rate (PEFR) 911, 912
peak flow variability 2
PEEP see positive end-expiratory pressure (PEEP), intrinsic
peptide histidine isoleucine (PHI) 428, 432
peptide histidine isoleucine/methionine (PHI/PHM) 391
peptide histidine methionine (PHM) 428, 432
peptide histidine valine (PHV-42) 433
peptides 298
peripheral airways 482
   functional abnormality 74
peripheral resistance 73
peroxynitrite 371
peroxynitrite (ONOO$^-$) generation 372
persistent asthma 836
pet allergen avoidance 638–40
PGP 9.5 50
pharmacoeconomics 903–16
   asthma pharmacotherapy 907–14
   principles and applications 904–5
   research studies 904
pharmacological agents 578–81
   see also specific agents
pharynx 74
phenotypes, changes in 128–34
phenylalkylamines 98
phosphatidylinositides 100
phosphodiesterase (PDE) 798
   distribution 809
   inhibitors 689, 690, 808–9
phospholipase, inhibitors 774–5
phospholipase A$_2$ (PLA$_2$) 281, 729
phospholipase C (PLC) 103
physiology 65–87

Index

physiotherapy 830
pituitary adenylate cyclase-activating peptide (PACAP) 391, 433–4
pituitary adenylate cyclase-activating peptide 27 (PACAP-27) 428, 433–4
plasma 261
    microvascular-epithelial exudation of 253–67
plasma-derived adhesive proteins 265
plasma-derived gel 263–5
plasma exudation 213, 253, 259–60
    in inflammatory airway diseases 259–60
    pathways 254–7
    roles of 261–2
plasma membranes 92
plasma proteins 210
platelet-activating factor (PAF) 131, 132, 147, 149, 217, 312, 346–8, 513
    antagonists 773–4, 803
    effects 347
    receptor antagonists 347–8
    receptors 347
    studies in asthma 348
platelet-derived growth factor (PDGF) 130–2, 480
platelet-endothelial cell adhesion molecule (PECAM) 242
platelet factor 4 (PF-4) 310
plicatic acid 543
pollen allergens 508
polymerase chain reaction (PCR) 554, 556
polymorphonuclear leucocytes (PMNs) 282
polyps 75
positive end-expiratory pressure (PEEP), intrinsic 80
post-mortem examination 48
post-mortem morphometric study 213
posture, effect on airway function 76–7
prednisolone 120, 746, 747, 827, 829
prednisone 746
pregnancy, inhaled steroids 745
prekallikrein 299
premenstrual asthma 755
prevalence of adult asthma 13–15
prevalence of childhood asthma 3, 4
primagrel 773
procaine 104
progesterone 414
prognosis of childhood asthma 11–12
propafenone 597
propranolol 597, 598
prostaglandins 217, 269–80, 772–3
    inhibitory 275
    PGD_2 115, 118, 122, 132, 230, 271, 273, 278, 395, 610
    PGE_1 230
    PGE_2 92, 129, 131–3, 143, 191, 192, 271, 610–11
    $PGF_{2\alpha}$ 100, 230, 271, 610
    PGG_2 271
    PGH_2 271
    PGI_2 271
    stimulatory 273
prostanoid inhibitors 803
prostanoid receptors 271–2

protein kinase A (PKA) 104
protein kinase C (PKC) 92, 103
provokers of asthma 529
psychosocial behaviour, childhood asthma 868
puberty 883
pulmonary function in childhood 9
pulmonary gas exchange 78–80
pulmonary mast cells 120–3
pyrazolones 613

quality-adjusted life years (QALY) 906

radioallergosorbent test (RAST) 533, 553, 554
RANTES 149, 193, 311–15, 317, 318, 730, 807
rapamycin 168, 790
rat mast cell protease (RMCP) 114
reactive airways dysfunction syndrome (RADS) *see* irritant-induced asthma
reactive nitrogen intermediates 132
reactive oxygen metabolites 132–4
reactive oxygen species (ROS) 775, 804
reflex bronchoconstriction 73
Rel family transcription factors 462
Rel homology domain (RHD) 462
renin-angiotensin system 411, 412
residual volume (RV) 71
respiratory bacteria 552
respiratory infections
    epidemiology 547–56
    viral 552–6
respiratory muscle function 80–2
respiratory muscle strength 82
respiratory syncytial virus (RSV) 9, 548, 556, 860
    bronchiolitis 564
respiratory tract infections in early childhood 9
respiratory viruses 557–8
    stimulation of pro-inflammatory cytokines 562–4
respiratory water loss 570–2
resting membrane potential 101
rhinitis 75, 850
    prevalence 7
rhinovirus infections 555, 556, 860
    effect on airway responsiveness 558–62
    effect on allergic inflammation 558–62
    interaction on allergic inflammation 558
risk factors
    adult asthma 15–16
    asthma mortality 19
    childhood asthma 863
RNA polymerase II (RNA pol II) 459
RSV *see* respiratory syncytial virus (RSV)

salbutamol 122, 669, 682, 683, 713, 887
saline inhalation sensitivity 575
salmeterol 668–9, 880
sarcoplasmic reticulum 101–2
Sch-37370 778
scopolamine 679
secretory leucocyte protease inhibitor (SLPI) 729

segregation analysis 35–6
selectins 163, 168, 193, 195, 240–4, 242, 243, 244, 246
  and counter-receptors 240–1
self-management 922–4
self-treatment 829
sensory nerves
  activation 442
  in human airways 441
serial PEF measurements 537
serotonin (5-hydroxytryptamine, 5HT) 100, 353
serous cells 187, 207
severe acute asthma 682, 747
  childhood 884–8
  general assessment, management 821–3
  management 821–33
  specific treatment 823–7
severe persistent asthma, treatment 852–3
sex factor in childhood asthma 6
sex hormones 414
SH2 168
shedding–repair processes 264
shedding–restitution-evoked processes 265
sleep, effect on airway function 76–7
small transunit outward currents (STOCs) 101
smoking and smoking-related diseases 8, 12, 535
socioeconomic status in childhood asthma 10
sodium cromoglycate (SCG) 121, 122, 221, 303, 512, 518, 578, 580, 707, 846, 877, 880
  anti-inflammatory actions 711
  challenge studies 712
  chemical structure 707
  clinical studies 712–13
  effects on inflammatory cells and nerves 708–11
  mechanisms of action 711–12
  neural mechanisms 710–11
  pharmacokinetics 708
  therapeutic forms 712
  vs. NS 717
sodium nitrates 579
sodium nitroprusside 799
soluble guanylyl cyclase 799
somatostatin 446
spirometric function 837
sputum 48, 213
  production 210
  viscosity 211, 220
SR27417A 803
Staphylococcus aureus 552
STAT 465–6, 470–1
STAT-1 810
STAT-3 810
STAT-6 174–5, 810
status asthmaticus 205
  childhood 884–8
stem cell factor (SCF) 115
steriod action, classical mechanisms 467
steroid resistance 843
  anti-inflammatory actions 749–52
  mechanisms 749–52
  secondary 752–5
steroid-sparing effects 290

steroid-sparing therapy 747
steroids 579, 580, 725
  inhaled 736–8, 879
    clinical use 745–6
    frequency of administration 739
    pharmacokinetics 738–9
    side-effects 739–45
    systemic side-effects 741
  side-effects 843, 881
  systemic 746–8, 843
STICs 101
STOCs 101
Streptococcus pneumoniae 552
Streptomyces hygroscopicus 790
Streptomyces misakiensis 355
structural cells 491
subepithelial fibroblasts 51
subepithelial oedema 56
submucosal glands 208
  enlargement 51
substance P (SP) 94, 100, 115, 119, 391, 428, 434, 435, 437
sulphur dioxide (SO_2) 10, 592–3, 600
superoxide anions 129, 132
Surveillance of Work and Occupational Respiratory Disease (SWORD) 530, 532
sympathetic innervation 399–400, 425
symptom histories 2

T-cell activation, costimulation 170–1
T-cell activator (TCA)-3 123
T-cell anergy 168
T-cell receptor (TCR) 40, 159
  CDR3/MHC-peptide 165
  complex component 171
  TCR-α complex 40
T-cell responses in viral exacerbation and coinfection 175
T-cell subsets 168
T-cells 159, 163, 563
  markers found exclusively or predominantly on 162
  mechanisms in asthma 161–3
  primary activation 165–8
  type 1 and 2 329–30, 336
T-helper cells 57
T-lymphocytes 57, 115, 128, 129, 132, 133, 136, 149, 159, 195, 246, 314–15, 329, 491, 695, 733, 811
tachykininergic tone 221
tachykinins 434–6, 443
  antagonists 444
  effects on airways 434–6
  inflammatory effects 435
  metabolism 436
  neural effects 436
  vascular effects 435
tachyphylaxis 669
tacrolimus 790, 812
tannic acid 637
tartrazine 600
TATA box-binding protein (TBP) 459

Index

TBP-associated factors (TAFs) 459
TCR see T-cell receptor (TCR)
terfenadine 122, 579
tetrachlorophthalic anhydride (TCPA) 535, 541
tetraethylammonium (TEA) 96, 101, 103, 104
TGF see transforming growth factor (TGF)
Th1 cells 491, 492, 807, 812, 813
Th1 phenotypes 171–5
Th2 cells 491, 492, 807, 812, 813
Th2 phenotypes 171–5
theophylline 512, 689–706, 713, 716, 798–9, 826, 847, 877
    anti-inflammatory effects 694–5
    cellular effects 693
    chemistry 690
    clinical use 698–700
    effect on airway smooth muscle 693–4
    extrapulmonary effects 695–6
    factors affecting clearance 696
    future 701–2
    historical background 689–90
    immunomodulatory effects 695
    in acute severe asthma 698
    in chronic asthma 699–700
    in chronic obstructive pulmonary disease (COPD) 696, 700
    increased clearance 697
    interaction with β-agonists 700–1
    mediator inhibition 692
    molecular mechanisms of action 690–3
    oral administration 697–8
    pharmacokinetics 696–7
    reduced clearance 697
    routes of administration 697–8
    side-effects 701, 847
thromboxanes 102, 269–80, 395, 772
    receptors 271–2
    TxA_2 271, 274, 278
    TxB_2 772–3
Thuja plicata 542
thyroid hormones 414
tidal intrathoracic pressures 78
tiotropium bromide 680, 800
tobacco smoking see smoking
toluene diisocyanate (TDI) 259, 478, 531, 553, 695
Tolypocladium inflatum 786
total airways resistance 74
total lung capacity (TLC) 71, 75, 77
total pulmonary resistance 73
total respiratory resistance 75
trachea 74
    structures 94
tracheobronchial circulation 229–37
    drug and mediator distribution 231–2
    heat and water exchange 232
    organization and control 229–30
    role in asthma 231–4
tracheobronchial dilation 415
tracheobronchial epithelium 187
transcription factors 135, 171, 459–74, 499, 727–8, 751–2
    and their transduction pathways 466–8

blockers 809–10
    role in asthma 469–71
    see also specific transcription factors
transforming growth factor (TGF)-α 143
transforming growth factor (TGF)-β 57, 133, 143, 194, 479
transforming growth factor (TGF)-β1 131, 480
triamcinolone 738, 746
triamcinolone acetonide 747
trigger factors 849
    identification 828
tryptase 122
tumour necrosis factor (TNF) 129, 131, 132, 309
tumour necrosis factor (TNF)-α 123, 130, 217, 240, 562, 563, 692, 806
type I-mediated hypersensitivity 520
type I sensitivity see IgE
type III sensitivity see IgG
tyracheobronchial vasculature 233

U46619 272, 772
UK-74,505 803
upper respiratory infections (URIs) 552, 556

vaccination 813
vacuum cleaning 636
vagal reflex pathways 678
vascular cellular adhesion molecule 1 see VCAM-1
vascular disease 56
vascular responses 497
vasoactive intestinal peptide (VIP) 58–9, 94, 218, 391, 409–13, 428–34
    airway effects 428–30
    as i-NANC neurotransmitter 431
    related peptides 432–4
    role in asthma 432
vasodilatation 56, 497
VCAM-1 148, 193, 194–5, 242, 243, 246, 313, 331, 810, 812
ventilation 827
ventilation–perfusion (V/Q) ratio 78–80
ventilatory pattern 78
versican 479
very late antigen (VLA)-4 148, 164, 195, 246, 247, 810, 812
VIP see vasoactive intestinal peptide (VIP)
viral exacerbation, T-cell responses 175
viral infection and steroid resistance 755
viral lower respiratory tract infections (LRIs) 548
viral respiratory infections 552–6
virus culture 556
virus-induced airway hyperresponsiveness 556–64
virus-induced asthma, mechanisms 557
voltage-activated Na^+ channels 97
voltage-dependent Ca^{2+} channels 98
voltage-independent Ca^{2+} channels 100

WEB2086 348, 774, 803
WEB2170 348
wedged-catheter techniques 73
wheezing
   epidemiology 547–56
   in childhood 547–52, 860–1
   prevalence in adults 13–14
   prevalence in childhood 3, 4
   prevalence trends of current and cumulative 5
WIN64338 775, 804
window current 100

xanthine derivatives 826

Y-24180 803

zafirlukast 289, 769
ZAP-70 168
ZD2138 771
zileuton 289, 802